Handbook of Near-Infrared Analysis

PRACTICAL SPECTROSCOPY

PUBLISHED TITLES

Handbook of Near-Infrared Analysis: Second Edition, Revised and Expanded
Edited by Donald A. Burns and Emil W. Ciurczak

Handbook of Raman Spectroscopy: From the Research Laboratory to the Process Line
Edited by Ian R. Lewis and Howell G. M. Edwards

Handbook of X-Ray Spectrometry: Second Edition, Revised and Expanded
Edited by René E. Van Grieken and Andrzej A. Markowicz

Ultraviolet Spectroscopy and UV Lasers
Edited by Prabhakar Misra and Mark A. Dubinskii

Pharmaceutical and Medical Applications of Near-Infrared Spectroscopy, Second Edition
Emil W. Ciurczak and Benoît Igne

Applied Electrospray Mass Spectrometry
Edited by Birendra N. Pramanik, A. K. Ganguly, and Michael L. Gross

Practical Guide to ICP-MS
Edited by Robert Thomas

NMR Spectroscopy of Biological Solids
Edited by A. Ramamoorthy

Handbook of Near Infrared Analysis, Third Edition
Edited by Donald A. Burns and Emil W. Ciurczak

Coherent Vibrational Dynamics
Edited by Guglielmo Lanzani, Giulio Cerullo, and Sandro De Silvestri

Practical Guide to ICP-MS: A Tutorial for Beginners, Second Edition
Robert Thomas

Practical Guide to ICP-MS: A Tutorial for Beginners, Third Edition
Robert Thomas

Pharmaceutical and Medical Applications of Near-Infrared Spectroscopy
Emil W. Ciurczak and James K. Drennen III

Measuring Elemental Impurities in Pharmaceuticals: A Practical Guide
Robert Thomas

Handbook of Near-Infrared Analysis, Fourth Edition
Edited by Emil W. Ciurczak, Benoît Igne, Jerome Workman, Jr., Donald A. Burns

For more information about this series, please visit: https://www.routledge.com/Practical-Spectroscopy/book-series/CRCPRASPECTR

Handbook of Near-Infrared Analysis

Fourth Edition

Edited by
Emil W. Ciurczak, Benoît Igne, Jerome Workman, Jr., and
Donald A. Burns

CRC Press
Taylor & Francis Group
Boca Raton London New York

CRC Press is an imprint of the
Taylor & Francis Group, an **Informa** business

Fourth edition published 2021
by CRC Press
6000 Broken Sound Parkway NW, Suite 300, Boca Raton, FL 33487-2742

and by CRC Press
2 Park Square, Milton Park, Abingdon, Oxon, OX14 4RN

First edition published by Marcel Dekker 1992
Third edition published by CRC Press 2007

CRC Press is an imprint of Taylor & Francis Group, LLC

Library of Congress Cataloging-in-Publication Data
Names: Ciurczak, Emil W., 1945- editor. | Igne, Benoït, editor. |
Workman, Jerome, editor. | Burns, Donald A., 1929- editor.
Title: Handbook of near-infrared analysis / edited by Emil W. Ciurczak,
Benoît Igne, Jerome Workman, Jr., Donald A. Burns.
Description: Fourth edition. | Boca Raton : Taylor and Francis, 2021. |
Series: Practical spectroscopy, 0148-9054 | Includes bibliographical
references and index.
Identifiers: LCCN 2020046266 (print) | LCCN 2020046267 (ebook) |
ISBN 9781138576483 (hardback) | ISBN 9781351269889 (ebook)
Subjects: LCSH: Near infrared spectroscopy.
Classification: LCC QD96.I5 H36 2021 (print) | LCC QD96.I5 (ebook) |
DDC 543/.57—dc23
LC record available at https://lccn.loc.gov/2020046266
LC ebook record available at https://lccn.loc.gov/2020046267

ISBN: 9781138576483 (hbk)
ISBN: 9780367684532 (pbk)
ISBN: 9781351269889 (ebk)

Typeset in Times
by codeMantra

Dedication to Donald Burns

This fourth edition would not have been possible, had not Don Burns decided to craft the first. His academic credentials (AB in Chemistry from Syracuse University, an MS in Organic Chemistry from University of Puget Sound, and a PhD in Biochemistry from Purdue University) didn't prepare him for a career in near-infrared. But as a trained scientist, he recognized the potential of NIRS and, as all the pioneers in the "art of NIRS," his training consisted of on-the-job-training.

He quickly recognized the dearth of general technical articles and texts, aside from agricultural or food-based publications. The idea for this book originated at an ACS meeting in 1979 when Dr. Maurits Dekker invited Don to write or edit a book on automated pesticide analysis. Discussions turned the subject into near-infrared analysis, and it seemed like an easy assignment since there were a number of experts in this field to tap for chapters. [And asked E.W. Ciurczak to be co-editor.]

He enlisted many "experts" in diverse fields to contribute chapters about their *work in* their *fields. Now, in one volume, a new, intermediate, or experienced worker in the field of NIRS could find references, examples, hints, and spectra. Since the technique was mainly used in agriculture and food (the late, great Karl Norris co-edited an excellent text, along with Phil Williams, another pioneer, on agricultural applications containing a lot of breakthrough basic science), these were the most comprehensive chapters in that first endeavor. Other chapters on polymers, pharmaceuticals, and so on were added to make the text a stand-alone reference source… decades before the Internet became ubiquitous.*

Unlike many of the authors included in the handbook, Don's forte was not publishing a plethora of articles, but also working on the "inside," making it easier for the more the visible practitioners to be successful. At Technicon, he helped design the training, understanding that no instrument could produce results unless the technician understood the complexities of the science. His paradigm was for a customer/client to come to Technicon's facility for a 1-week training on their *instrument, which was then shipped to* their *location. [This was important* circa *1980, since each instrument had its own foibles and users could become familiar with their own unit and hit the lab running.] One example of a solid contribution to the art was* indicator variables. *These helped researchers obviate unwanted variance, long before we had the tools of principal component analysis (PCA) and partial least squares (PLS) (both methods "dropped" commercially in 1987).*

He spent much of his professional career on education of new and current users of NIRS, encouraging coworkers to present, where to present, and reminding them as to when deadlines for each conference were. He made many major contributions to the spread (and acceptance) of NIRS, both by giving birth to the handbook and by crafting the NIR short course for the ACS (in which E.W. Ciurczak was honored to be included). In 1990, he received the EAS award for Accomplishments in NIRS. For years, he was on the ACS Speakers Circuit with a talk entitled "Catch 'em with Near-IR," *which dealt with counterfeit currency and fake materials like turquoise, cigars, and ivory.*

Today, Don lives in Los Alamos, NM, with his wife of many years, has three children and six grandchildren.

Contents

Applications – Biopharmaceutical

Applications – fNIR

Applications – Forestry

Applications – Petrochemicals

Applications – Pharmaceuticals

Applications – Remote Sensing and Soil Analysis

Applications – Textiles

Preface to Fourth Edition

Near-infrared spectroscopy is now being applied in every industry on Earth. It is rapid (some measurements are in the millisecond range), nondestructive, needs no sample preparation, and is usually simple to use. "Why," you might ask, "if the technique is so widespread, do we need another handbook?" For several reasons, it turns out.

While the technique is simple to use, it is complex to produce the necessary equations. Most spectroscopic methods are applied to "pure" samples. That is, a sample is ground up, dissolved in a solvent, placed into a cuvette, and measured. In this case, the Beer–Bouguer–Lambert law, taught in every undergraduate instrumentation course, is sufficient (even the basis of HPLC). When we use NIRS for solid, often complex samples, the law falls to pieces.

The pathlength, known in solutions, is not known in NIRS; the matrix, not seen in Beer's law work, often makes up the lion's share of a NIR spectrum; the matrix, designed to be noninteractive for Beer's law work, often has large interactions with the analyte(s), mostly through hydrogen bonding or scattering/absorption phenomena. As a result, the simple math employed in most spectroscopy is not helpful in NIRS. Complex algorithms, based on matrix algebra, are often needed to deconvolve the complex spectra of NIR samples.

Add to that the need to understand light scattering (diffuse reflection), polymorphic effects, particle size effects, and which type of spectrometer to use. Type of instrument? In UV/Vis, the most common unit is a diode-array unit. For mid-range infrared, most modern units are interferometer-based Fourier transform units. But NIR units can be filter-based, grating types, interferometer-based, diode arrays, linear variable filters, acousto-optic tunable filters, and several more arcane types. Clearly, choosing a proper instrument is not simple.

As a consequence of this complexity (and the fact that few professors are familiar with all the "moving parts" of NIR spectroscopy), it is rarely, if ever, covered in undergraduate curricula. In fact, it is mainly seen in graduate schools, other than research groups that apply it for their research.

For these reasons, we felt it imperative to not just list/demonstrate applications of NIRS, but to include chapters on theory, math applied to the work, instrument theory/design/application, and, of course, the work being performed by category, from food through polymer to pharmaceuticals. While the editors of this text are experienced (80+ years combined experience), we do not assume to speak on all the covered topics. We are fortunate to have gotten the support of a large number of "experts" in their chosen fields who have shared their experiences. This Handbook is, in itself, a good reference, but also lists a plethora of sources for further reading and reference. Enjoy!

Editors

Emil W. Ciurczak has a BA in Chemistry from Rutgers and advanced degrees in Chemistry from Rutgers and Seton Hall, and has worked in the pharmaceutical industry since 1970 (Ciba-Geigy, Sandoz, Berlex, Merck, and Purdue Pharma), performing method development on most types of analytical equipment. In 1983, he introduced NIR spectroscopy to pharmaceutical applications. He also consulted for several NIRS equipment manufacturers. Most research was on pharmaceutical applications of NIR where he has published over four dozen articles in refereed journals and over 250 magazine columns, and presented nearly 200 technical papers.

Since 2005, Emil has been Contributing Editor for *Pharmaceutical Manufacturing* magazine (also *Spectroscopy*; 1987–2007) and, since 2014, Contributing Editor for *Contract Pharma*. He has written and edited several texts and chapters: "Handbook of NIR Analysis" (first, second, and third editions), "Pharmaceutical and Medical Applications of NIRS" (second edition in progress), "Molecular Spectroscopy Workbench," and chapters on NIR applications to life sciences. Emil sits on several magazine editorial boards, is active in SAS, is a founder of the Council for NIRS, and was the 2002 chair for IDRC (Chambersburg Conference).

He has been teaching (college adjunct) since 1979 as well as presenting short courses (in NIRS, Instrument Troubleshooting, and PAT/QbD) for the ACS, CfPA, SPIE, ASSA, and other organizations in the United States, Europe, Asia, and South America.

Emil is currently a consultant in the field of NIRS and holds a dozen patents for NIR-based devices and software. He consults with various pharmaceutical companies, instrument manufacturers, and the FDA. He was a member of the PAT subcommittee (Validation) for the FDA and a member of the PAT Expert Committee for the USP, and was the 2004 recipient of the EAS Achievements in NIR Award. Emil agreed to be co-editor.

Benoît Igne has a master's degree from Ecole d'Ingénieurs de PURPAN, Toulouse, France, and a PhD in Agricultural Engineering from Iowa State University, Ames, Iowa, USA. He began working as a postdoctoral fellow and a program coordinator at the Duquesne University School of Pharmacy, Pittsburgh, Pennsylvania, USA. In 2014, he joined GlaxoSmithKline outside of Philadelphia, Pennsylvania, USA. There, he led the development of process analytical technologies for the monitoring, development, and control of solid, liquid, and biopharmaceutical products. His main responsibility was to integrate near-infrared spectroscopic tools onto a drug product continuous manufacturing line. Benoît received the 2015 Eastern Analytical Symposium Award for Outstanding Achievements in Near-Infrared Spectroscopy. He has published over 40 peer-reviewed articles and several books and book chapters. He has been involved with the Council for Near-Infrared Spectroscopy since 2010 and was elected President from 2016 to 2020. Benoît has played various roles in the Society of Applied Spectroscopy, including the Membership committee and the Tour Speaker program. Finally, he is the subcommittee Chair for ASTM 55.01, developing standards for the use of process analytics.

Jerome Workman, Jr., is Senior Technical Editor for LCGC and Spectroscopy; Certified Core Adjunct Professor, at National University, CA; and Principal at Biotechnology Business Associates. He was formerly Executive Vice President of Research & Engineering for Unity Scientific and Process Sensors Corp.; Vice President of Technology for Masimo Corp.; Senior Research Fellow at Kimberly-Clark Analytical Science & Technology; and Principal Scientist at PerkinElmer. Dr. Workman has played a major role in defining and developing over 25 scientific instrument advancements with novel software improvements for spectroscopic instrumentation. He has over 75 U.S. and international patent applications since 1998 with 27 issued U.S. and international patents and multiple trade secrets. He has over 500 technical publications – including 22 reference books on a broad range of spectroscopy, chemometrics, and data processing techniques – and has received awards from the Eastern Analytical Symposium, ASTM International, Coblentz Society, Society for Applied Spectroscopy, National Institute of Standards and Technology (NIST), Frost and Sullivan, and others. He has taught annual courses in spectroscopy, chemometrics, and statistics for the Association of Official Analytical Chemists (AOAC), the American Chemical Society (ACS), the Instrument Society of America (ISA), and the Federation of Analytical Chemists and Spectroscopy Societies (FACSS), and at several universities and corporations. He is a Fellow of the American Institute of Chemists (FAIC), the American Society for Testing and Materials (FASTM), and the Royal Society of Chemistry, UK (FRSC, CChem, CSci). Workman holds a BA degree *cum laude* in natural sciences, an MA in biological sciences and genetics from Saint Mary's University of Minnesota, and a PhD degree in biological chemistry from Columbia Pacific University. Upon graduation, he immediately started work as a scientist with Technicon Instruments in Tarrytown, NY. He is an alumnus of Columbia University Business School having completed the Columbia Senior Executive Program and also holds graduate Certificates in Executive Development (CIED) and in Business Excellence (CIBE) from Columbia. He also holds a Certificate in Strategy and Innovation from the M.I.T. Sloan School.

The idea for this book originated at an ACS meeting in 1979 when Dr Maurits Dekker invited Don to write or edit a book on automated pesticide analysis. Discussions turned the subject into near-infrared analysis, and it seemed like an easy assignment since there were many experts in this field.

Donald A. Burns had been promoting computerization of instruments and automation of analytical procedures for many years at *Nat'l Cash Register Co*, *Dover Medical Research Center* (Dover DE), *Abbott Laboratories*, *Technicon Instruments Corp*, and *Los Alamos Nat'l Lab* (from which he is now retired).

For more than 20 years, Don has been giving courses on near-infrared spectroscopy at national and international meetings, often sharing the podium with colleagues in the field, and occasionally including brief exhibits by instrument makers. In 1990, he received the EAS award in NIR. For several years, he was on the ACS Speakers Circuit with a talk entitled "*Catch 'em with Near-IR,*" which dealt with counterfeit currency and fake materials like turquoise, cigars, and ivory.

Don has an AB in Chemistry from Syracuse University, an MS in Organic Chemistry from the University of Puget Sound (Tacoma WA), and a PhD in Biochemistry from Purdue University. He's married, has three children and six grandchildren, and lives in Los Alamos, NM.

Contributors

Vincent Baeten
Knowledge and Valorization of Agricultural Products Department
Walloon Agricultural Research Centre
Gembloux, Belgium

Nicolas Bernard
Manufacturing and Projects Division
Total Refining & Chemicals
Harfleur, France

Jean-Claude Boulet
INRA
UMR1083, Montpellier SupAgro
University of Montpellier
Montpellier, France

Steven D. Brown
Department of Chemistry and Biochemistry
University of Delaware
Newark, Delaware

Sean M. Christian
Neolitics
Sparrows Point, Maryland, USA

Emil W. Ciurczak
Doramaxx Consulting
Golden's Bridge, New York

Donald J. Dahm
Department of Chemical Engineering
Rowan University
Glassboro, New Jersey

Kevin D. Dahm
Department of Chemical Engineering
Rowan University
Glassboro, New Jersey

Virginie Decruyenaere
Productions in Agriculture Department
Walloon Agricultural Research Centre
Gembloux, Belgium

Stephen R. Delwiche
Food Quality Laboratory
Beltsville Agricultural Research Center
USDA Agricultural Research Service
Beltsville, Maryland

José A. M. Demattê
Department of Soil Science
Luiz de Queiroz College of Agriculture
University of São Paulo
São Paulo, Brazil

Eyal Ben Dor
Department of Geography
School of Earth Science
Tel Aviv University
Tel Aviv, Israel

Kim H. Esbensen
Copenhagen, Denmark

Jess V. Ford
Manager, Sustaining Engineering
Wireline Research & Development
Weatherford, Houston, TX

Rob Frankhuizen
State Institute for Quality Control of Agricultural Products (RIKILT-DLO)
Wageningen, The Netherlands

Subhas Ghosh
College of Technology
Eastern Michigan University
Ypsilanti, Michigan

Bruno Godin
Knowledge and Valorization of Agricultural Products Department
Walloon Agricultural Research Centre
Gembloux, Belgium

Sébastien Gourvénec
Manufacturing and Projects Division
Total Refining & Chemicals
Harfleur, France

Aoife A. Gowen
Spectral Imaging Research Group
UCD School of Biosystems and Food Engineering, College of Engineering and Architecture
University College Dublin
Dublin, Ireland

Jean Guilment
ARKEMA
Laboratoire d'Etude des Matériaux (LEM)
Serquigny, France

Peter H. Hindle
Peverel Design Ltd.
Essex, United Kingdom

David W. Hopkins
NIR Consultant
Battle Creek, Michigan

Christian Huck
Department of Radiochemistry
Center for Chemistry and Biomedicine
Leopold-Franzens University
Innsbruck, Austria

W. Jeffrey Hurst
The Hershey Company
Hershey, Pennsylvania

Benoît Igne
GlaxoSmithKline
Collegeville, Pennsylvania

Mika Ishigaki
School of Science and Technology
Kwansei Gakuin University
Hyogo, Japan

Daitaro Ishikawa
Graduate School of Agricultural Science
Tohoku University
Sendai, Japan

Tom M. Jones
Sanofi-Aventis
Malvern, Pennsylvania

Gabor J. Kemeny
Middleton Spectral Vision
Middleton, Wisconsin

Lana R. Kington
Bristol-Myers Squibb
Evansville, Indiana

Zoltan Kovacs
Department of Physics and Control
Faculty of Food Science
Szent István University
Budapest, Hungary

Francis Kwofie
Department of Chemistry
Oklahoma State University
Stillwater, Oklahoma

Barry K. Lavine
Department of Chemistry
Oklahoma State University
Stillwater, Oklahoma

Robert A. Lodder
Department of Pharmaceutical Sciences
University of Kentucky
Lexington, Kentucky

Howard Mark
Mark Electronics
Suffern, New York

Gary McGeorge
Drug Product Development
Bristol-Myers Squibb
New Brunswick, New Jersey

Charles E. Miller
AspenTech Global Solution Consulting
Bedford, Massachusetts

Olivier Minet
Knowledge and Valorization of Agricultural Products Department
Walloon Agricultural Research Centre
Gembloux, Belgium

Jelena Munćan
Department of Biomedical Engineering
Faculty of Mechanical Engineering
University of Belgrade
Belgrade, Serbia

Nghia Nguyen-Do-Trong
Department of Biosystems – MeBioS Division
Katholieke Universiteit Leuven
Leuven, Belgium

Bart Nicolaï
Department of Biosystems – MeBioS Division
Katholieke Universiteit Leuven
Leuven, Belgium

Brian G. Osborne
Retired
Cowan, New South Wales, Australia

Yukihiro Ozaki
School of Science and Technology
Kwansei Gakuin University
Hyogo, Japan

Dolores Pérez-Marín
Non-Destructive Spectral Sensors Unit
Faculty of Agriculture and Forestry Engineering
University of Córdoba Campus de Rabanales
Córdoba, Spain

Gary E. Ritchie
GER Compliance
Fenton, Missouri

Rodnei Rizzo
Center for Nuclear Energy in Agriculture
University of São Paulo
São Paulo, Brazil

James Rodgers
Solutia Inc.
Gonzalez, Florida

Jean-Michel Roger
ITAP
IRSTEA, Montpellier SupAgro
University of Montpellier
Montpellier, France

Rodolfo J. Romañach
Department of Chemistry
University of Puerto Rico
Mayaguez, Puerto Rico

Wouter Saeys
Department of Biosystems – MeBioS Division
Katholieke Universiteit Leuven
Leuven, Belgium

Emiliano de Pedro Sanz
Non-Destructive Spectral Sensors Unit
Faculty of Agriculture and Forestry Engineering
University of Córdoba Campus de Rabanales
Córdoba, Spain

L. R. Schimleck
Wood Science and Engineering
Oregon State University
Corvallis, Oregon

Fabrício da Silva Terra
Center for Technological Development
Federal University of Pelotas
Pelotas, Brazil

Roumiana Tsenkova
Graduate School of Agricultural Science
Kobe University
Kobe, Japan

S. Tsuchikawa
Graduate School of Bioagricultural Sciences
Nagoya University
Nagoya, Japan

Ana Garrido-Varo
Non-Destructive Spectral Sensors Unit
Faculty of Agriculture and Forestry Engineering
University of Córdoba Campus de Rabanales
Córdoba, Spain

Phil Williams
PDK Projects Inc.
Nanaimo, Canada

Jerome Workman, Jr.
LCGC and Spectroscopy
MJH Life Sciences
Iselin, New Jersey

1 History of NIRS Development

Peter H. Hindle[*]
Peverel Design Ltd.

Updated by Emil W. Ciurczak
Doramaxx Consulting

CONTENTS

1.1 THE DISCOVERY OF NEAR-INFRARED RADIATION

Light and vision were linked since the earliest times, but the relationship was not understood. The ancient Greeks postulated that our eyes emitted rays that allowed things to be "seen." In A.D.130, Ptolemaeus measured the refraction of light for a number of transparent materials, and later, in 1305, Von Freiburg generated the structure of the rainbow by using water-filled glass spheres.

By the mid-18th century, new laws of reflection and refraction had been formulated, based on the work of scientists such as Snell, Huygens, Newton, Bradley, and Priestly. The wave and corpuscular natures of light had been proposed as well as measuring its velocity, and that it follows the inverse square law. For any student of the infrared, Herschel's discovery of near-infrared (NIR) radiation is probably of the greatest significance.

Sir William Herschel, a noted musician, became one of the finest observational astronomers of all time. Concerned with the damage to astronomers' eyes (while studying sunspots), he sought to discover which color should be filtered for safety. In 1800, his two publications [1] described a study on the heating effect within the spectrum of solar radiation. He used a large glass prism to disperse sunlight onto three thermometers with carbon-blackened bulbs. Near the "red end" of the spectrum, the heating effect became apparent, but, just beyond the red, while there was no visible light, the temperature appeared at its greatest. It was "beyond the red" (or, in Latin, *infrared*) light; but, since glass does not transmit much mid-range infrared radiation, what was seen was actually *near-infrared*. (A year later, in 1801, Johann Ritter announced his discovery of ultraviolet light.)

Herschel named this phenomenon "radiant heat" and referred to the "thermometrical spectrum." He first considered this form of energy to be different from "real" or visible light. His conclusions may appear naïve to us, but there was no one who was aware of the electromagnetic (EM) spectrum and that visible light was only a very small part of it. In 1835, Ampere, employing the newly invented thermocouple, demonstrated that NIR had the same optical characteristics as visible light and concluded that they were the same phenomenon. Ampere's contribution is important because it introduces the concept of the *extended spectrum*. This was later connected with Ritter's light (UV) since the spectrum was beginning to make scientific sense.

* emil@ciurczak.com

Herschel [2] clearly attached great importance to the analysis of light. In a letter to Professor Patrick Wilson, he concluded:

"... And we cannot too minutely enter into an analysis of light, which is the subtlest of all active principles that are concerned with the mechanism of the operation of nature."

By the beginning of the 20th century, the nature of the EM spectrum was much better understood. James Clerk Maxwell formulated his four equations determining the propagation of light; the works of Kirchhoff, Stefan, and Wien were neatly capped by Max Planck's radiation law in 1900.

Observationally, little progress had been made. Fraunhofer used a newly produced diffraction grating to resolve the sodium "D" lines in a Bunsen burner flame as early as 1823, and Kirchhoff had visually recorded the atomic spectra of many elements by the 1860s. Unfortunately, a lack of suitable detection equipment impeded progress outside the visible portion of the EM spectrum.

1.2 THE FIRST INFRARED SPECTRA

An important step was taken in the early 1880s. It was discovered that the photographic plate, invented in 1829 by Niepce and Daguerre, had some NIR sensitivity. Abney and Festing [3] were able to record the NIR spectra of organic liquids in the range of 1–1.2 µm in 1881. This work was significant; not only did it represent the first serious NIR measurements, but also represented the first real interpretations. Abney and Festing recognized both atomic groupings and the importance of the hydrogen bond in the NIR spectrum.

Stimulated by the work of Abney and Festing, Coblentz [4] constructed a spectrometer using a rock-salt prism and a sensitive thermopile connected to a mirror galvanometer. Because of its primitive construction, the instrument was highly susceptible to both vibration and thermal disturbances. After each step in the rotation of the prism, corresponding to each spectral element to be measured, Coblentz had to retire to another room in order to allow the instrument to equilibrate. Each pair of readings (with and without the sample in the beam) was made with the aid of a telescope to observe the galvanometer deflection. It took Coblentz an entire day to generate a single spectrum. Around 1905, he published a series of papers and ultimately recorded the spectra of several hundred compounds in the 1.0–15-µm (1,000–15,000 nm) wavelength region.

Coblentz discovered that no two compounds had the exact same spectrum, even when they had the same complement of elements (e.g., the isomers propan-1-ol and propan-2-ol). Each compound had a unique "fingerprint." Coblentz also noticed certain patterns in the spectra; for example, all compounds with OH groups, whether alcohols or phenols, absorb in the 2.7 µm region of the spectrum. In this way, numerous molecular groups were characterized. He also speculated the existence of harmonically related series. In short, Coblentz gave chemists a new tool, for example, spectroscopy, where he could obtain structural information about organic compounds.

Contemporaries of Coblentz were also working on exciting, new instrumental designs, which, years later, were to become the mainstay of the present-day spectrometry. Rowland developed large, ruled diffraction gratings and, notably, concave gratings, in the 1880s. In 1891, A. A. Michelson [5] published a paper describing the two-beam interferometer.

1.3 A STEADY EVOLUTION

Through the first half of the 20th century, workers extended the spectral database of organic compounds and assigned spectral features to functional groups. While infrared spectroscopy had moved away from being a scientific curiosity, it still had little practical use. Basically, suitable spectrometers did not exist, and few chemists had access to what instruments there were available. Over 50 years passed between Coblentz's original work and the routine use of spectroscopy as a tool. In fact, two-thirds of a century would pass before routine NIR measurement made its debut.

The first published quantitative NIRS measurement was the determination of atmospheric moisture at the Mount Wilson observatory by F. E. Fowle in 1912 [6] followed in 1938 by Ellis and

Bath [7] who determined the amount of water in gelatin. In the early 1940s, Barchewitz [8] analyzed fuels, while Barr and Harp [9] published the spectra of some vegetable oils. In the late 1940s, Harry Willis, working at Imperial Chemical Industries (ICI), used a prewar spectrometer to characterize polymers and later employed NIRS for the measurement of the thickness of polymer films.

The state of the art by the mid-1950s was summarized by Wilbur Kaye [10,11], in 1960 by R. E. Goddu [12], and in 1968 by K. Whetsel [13]. Until 1970, only about 50 papers had been written on work involving NIRS, despite having been discovered 150 years prior.

In the 1930s, lead sulfide (PbS) was being studied as a compound semiconductor. World War II stimulated its development as an infrared detector for heat-sensing purposes. In the 1950s, PbS became available for commercial applications as a very sensitive detector for the 1.0–2.5-µm (1000–2500 nm) wavelength region. The NIR region, at last, had a good, inexpensive, rugged detector, spurring instrument building.

Research into NIR spectra (1000–2500 nm) as opposed to the mid-infrared (MIR) (roughly 2.–15 µm) had a slow start. Many spectroscopists considered the region too confusing with many weak and overlapping peaks of numerous overtone and combination bands (making assignments difficult). Compared with the MIR region, NIR absorption features are very weak (by two or three orders of magnitude), and because of the overall complexity, baselines are hard to define. However, two strong aspects of NIR technology had been overlooked. First, the PbS detector was very sensitive and tungsten filament lamps (particularly quartz halogen) were a good source of NIR radiation for making diffuse reflection measurement possible. Second, relatively low-cost instruments could be manufactured because easily produced detectors, stable light sources, and optics made from inexpensive glass were readily available.

1.4 THE DIGITAL REVOLUTION

Modern NIR technology relies heavily on the computer (and the microprocessor in particular), not only for its ability to control and acquire data from the instrument, but also for calibration and data analysis. The fundamentals of data analysis have been known since the 1930s. Work on the diffuse scattering of light in both transmission and reflection (by Kubelka and Munk [14] in 1931) opened the door to NIR measurements on solids. In 1933, Hotelling [15] wrote a classic paper on principal components analysis (PCA), and Mahalanobis formulated a mathematical approach for representing data clustering and separation in multidimensional space.

In 1938, Alan Turing created the first programmable computer, employing vacuum tubes and relays. By the 1950s, the first commercial computer, UNIVAC, was available; by the mid-1950s, FORTRAN, the first structured, scientific language, had been developed by Backus at International Business Machines (IBM). The first personal computer (PC) was probably the Altair in 1975, followed in 1977 by the Commodore Personal Electronic Transactor (PET); the same year Bill Gates and Paul Allen found the Microsoft Corporation. IBM began marketing their first PC in 1981 and their designs set the format for compatibility. PC sales of around 300,000 in 1981 rocketed to 3 million in the next year. The PC soon became the driving force behind NIR instrumentation.

Starting in the early 1950s, there was a growing demand for fast, quantitative determinations of moisture, protein, and oil for grain exports. Karl Norris, of the United States Department of Agriculture (USDA), was charged with solving the problem for wheat. Previously, he was asked to work on another problem: how to speed the "candling" of eggs for occult blood [16]. Fortunately, he owned a Beckman DU (indicating Model version D with Ultraviolet–visible capabilities) spectrophotometer, which he used to measure the transmission properties of the shell, yolk, and white of the egg in the visible region. We made a minor modification of the spectrophotometer to make it possible to measure an intact egg, and this led to the development of a method for detecting blood in white shell eggs. Later, he used this instrument as a model for his own NIR unit (hand-made).

He took the bold step of choosing NIRS for grain analysis, working with primitive means by today's standards. In 1968, Ben-Gera and Norris published their initial work on applying multiple linear regression (MLR) to the problem of calibration relating to agricultural products (he even built his own calibration models). The early 1970s saw the birth of what was to become the laboratory instrument sector of NIR with the emergence of names like Dickey–John, Technicon, and Neotec all in the United States.

At the same time, and quite separately, online process instruments emerged. In Germany, Pier Instrument produced a two-filter sensor with tube-based electronics. In 1970, Infrared Engineering (in the United Kingdom) and Anacon (in the United States), both employing integrated circuit (IC)-based electronics, entered the marketplace. A few years later, Moisture Systems Corporation (in the United States) entered the field. Online instrumentation is now used for both continuous measurement and process control over a wide range of applications, including chemicals, pharmaceuticals, tobacco, food, drinks, and web-based products [16,17].

One unfortunate consequence of NIRS being introduced to the scientific community, at large, was that it was not seen as *rigorous* science. That is, virtually all methodologies evolved from theory (mostly in universities, later from Bell Labs) to instruments – then to general usage. The USDA empirically discovered how to utilize this wavelength region, based on available sources, gratings, and detector materials (the NIR region of the EM can be argued to extend from 3500 nm (3.5 μ) and well into the visible region (400 nm)). Thus, we defined the NIR region as 1100–2500 nm – the operational range of lead sulfide (PbS).

In addition, USDA scientists, when giving papers at meetings, routinely pointed to the spectrum at the "protein peak" or the "starch peak." Fortunately (for me, at least), when I attended a meeting at Technicon (1983), there was a chemist from DuPont who pointed to the peak (which USDA chemists called the "protein peak") and stated it was the amide-II peak or a combination of the carbonyl (C=O) stretch and N-H bending motions. I saw the same peak as protein = amide = the same one in many drug substances. This epiphany led me down the "rabbit hole" and began a 35-year journey with NIR.

During the 1980s, the microprocessor was integrated into the designs of most instruments. Much more sophisticated data acquisition and manipulation was now possible. The scope of data treatment display and interpretation was enhanced to include MLR, partial least squares (PLS), PCA, and cluster analysis. Third-party software suppliers have emerged offering a wide choice of data treatments, freeing the user from the constraints of instrument suppliers. Subsequent chapters of this book are devoted in detail to these issues.

NIR technology has evolved rapidly since 1970 and has now gained wide acceptance. In many sectors, it is now the measurement of choice. Its history is far richer than can be presented within the scope of this chapter, and so some references for further reading have been included [18–35].

REFERENCES

1. W. Herschel, *Philos. Trans. R. Soc.*, *90*: 255–283 (1800).
2. W. Herschel, 1933. *William Herschel Chronicles*, vol. 1, William Herschel Museum, Bath, UK.
3. W. Abney and E. R. Festing, *Philos. Trans. R. Soc.*, *172*: 887–918 (1881).
4. W. W. Coblentz, *Investigations of Infrared Spectra Part 1*. Publication No. 35, Carnegie Institute of Washington (1905).
5. A. Michelson, *Philos. Mag.*, *31*: 256 (1891).
6. F. E. Fowle, *Astophys. J.*, *35*: 149–162 (1912).
7. J. Ellis and J. Bath, *J. Chem. Phys.*, *6*: 723 (1938).
8. P. Barchewitz, *J. Chem. Phys.*, *45*: 40 (1943).
9. I. Barr and W. Harp, *Phys. Rev.*, *63*: 457 (1943).
10. W. Kaye, *Spectrochim. Acta*, *6*: 257 (1954).
11. W. Kaye, *Spectrochim. Acta*, *7*: 181 (1955).
12. R. E. Goddu, *Adv. Anal. Chem. Inst.*, *1*: 347 (1960).

13. B. Whetsel, et al., *Anal. Chem.*, *30*: 1598 (1958).
14. P. Kubelka and E. Munk, *Zeits. Tech. Physik.*, *12*: 593 (1931).
15. H. Hotelling, *J. Ed. Psych.*, *24*: 417–441, 489–520 (1933).
16. E. E. Finney and K. H. Norris, *Trans ASAE*, *21*: 534 (1977).
17. I. Ben-Gera and K. H. Norris, *J. Feed Sci.*, *64*: 33 (1968).
18. I. B. Benson, *Meas. Control*, *22*: 45–99 (1989).
19. P. H. Hindle, *Paper Technol. Ind.* (Sept. 1984).
20. L. A. Butler, *Cereal Foods World*, *28*: 238 (1983).
21. M. S. Day and R. B. Feam, *Lab. Practice*, *31*: 328 (1982).
22. G. Birth, *J. Food Sci.*, *44*: 949 (1979).
23. S. Borman, *Anal. Chem.*, *56*: 934 (1984).
24. D. E. Honigs, *Anal. Instrum.*, *14*: 1 (1985).
25. H. Norris, *Cereal Foods World*, *31*: 578 (1986).
26. D. L. Wetzel, *Anal. Chem.*, *55*: 1165 (1983).
27. J. B. Whetsel, *Appl. Spectrosc. Rev.*, *2*: 1 (1968).
28. K. Boer, *Trends Anal. Chem.*, *3*: 9 (1984).
29. A. Polessello and R. Giangiacomo, *CRC Crit. Rev.*, *18*: 203 (1983).
30. C. Stauffer, *Cereal Foods World*, *29*: 577 (1984).
31. K. G. Weyer, *Appl. Spectrosc. Rev.*, *21*: 1 (1985).
32. R. Moens, *Cereal Foods World*, *25*: 518 (1980).
33. E. Stark, et al., *Appl. Spectrosc. Rev.*, *22*: 335 (1986).
34. J. Workman, *J. Near Infrared Spectrosc.*, *1*: 221–245 (1993).
35. K. H. Norris, *NIR News*, *3*(1): 12 (1992).

Part I

Physics

2 Introduction to the Physics of Light Scattering

Donald J. Dahm and Kevin D. Dahm[*]
Rowan University

CONTENTS

Summary: Over the last few years, there have been many articles written on the theory and models of near-infrared (NIR) scatter. One of these was in the last edition of this handbook, coauthored by Peter Griffiths and Donald Dahm, entitled "Continuum and Discontinuum Theories of Diffuse Reflection." It was long and a bit tedious. To the reader, it may have seemed that the authors didn't want to leave anything out. Writing the article required reconciling two different points of view, which were termed "continuum" and "discontinuum" theories. It remains, we believe, a fine compilation of the status of the theory at that time. Much of what was discussed then is now "settled science," even if nature doesn't always conform to the models described. Consequently, in this edition, with Peter's encouragement, we are trying for a more "concise" treatment.

* dahm@rowan.edu

2.1 THE LAW OF BOUGUER

The basic law of spectroscopy is that of Bouguer. It is often called the "Bouguer–Lambert law." It says that when a beam of light is moving in a straight line through a uniform medium, it is attenuated according to the equation:

$$I = I_0\exp(-\mathbf{kd}), \tag{2.1}$$

where **k** is an attenuation (or extinction) coefficient and **d** is the distance traveled through the medium. The attenuation is caused by absorption and/or scattering by particles in the light path. When light travels through a medium, it may encounter a number of particles on its path. When we say a "uniform medium," this implies that when light travels a specific distance, it will encounter the same number of identical particles along this distance, regardless of the specific location and direction of the path within the medium.

If we recognize that attenuation has two causes, absorption and scatter, we can divide the {**k**} into two coefficients, {Ab+Sc}, and write:

$$I = I_0 \exp\left[-(Ab + Sc)\mathbf{d}\right]. \tag{2.2}$$

Equation 2.1 is often applied to travel of light within a solution, where scatter is considered negligible (and thus it is assumed $\mathbf{k}=A_b$). If we rearrange Equation 2.1, we can obtain:

$$\left\{\ln(I_0 / I) = \mathbf{kd}\right\} \text{ or } \left\{2.303\log_{10}(I_0 / I) = \mathbf{kd}\right\} \text{ or } \left\{\log_{10}(I_0 / I) = \mathbf{kd} / 2.303\right\}. \tag{2.3}$$

The quantity {log (I_0/I)} is known as the "decadic absorbance" of a sample, and thus, it is recognized that **Absorbance is given by some constant of proportionality (k) times the distance traveled through the sample**.

August Beer (and others) recognized that this constant of proportionality could be related to the concentration of an absorber. The constant of proportionality is best related to a "true concentration," which is an "amount per volume."[1] Taking this into account, we often write what is short-handed as "Beer's law":

$$\text{Absorbance} = \mathbf{A} = \log(I_0 / I) = \varepsilon c\mathbf{d}, \tag{2.4}$$

where ε is often called[2] the "molar extinction coefficient" and the concentration {c} is expressed in moles per liter.

In practice, when applying Beer's law in spectroscopy, we often do not worry about determining the numerical value of the absorption coefficient. Rather, we use data obtained from the standard samples to fit an empirical equation that relates absorbance to concentration, and then determine the unknown concentrations of samples from the equation, with the sample's absorbance as the input parameter. Even if we don't use Equation 2.4 directly, it informs the idea of an equation in which absorbance is proportional to concentration.

The Law of Bouguer has great practical appeal because of its simplicity, but has the limitation that it does not meaningfully distinguish between light that has been absorbed and light that has been diverted from its path: anything that does not continue traveling along the original straight-line

---

1 Beer used a spectrometer with a 10 cm pathlength, and defined the standard concentration of a material as 10 wt %. Thus, Beer's original formulation did not use a "true concentration"; it used the ratio of the actual concentration to a standard concentration. Practitioners of NIR spectroscopy are often haphazard in how they express concentration (Mark et al. 2010).

2 Some elements of the absorption spectroscopy community eschew the term "extinction coefficient" in favor of "absorption coefficient."

path is lumped into the "attenuation." This approach is insufficient for many practical analytical problems, and there are spectroscopic methods like near-infrared spectroscopy (NIRS) that explicitly rely on quantifying scattered light and distinguishing it from absorption. The rest of this chapter explores the physics underlying absorption and scatter, and how these physical principles inform the models that we use.

2.2 BASIC GEOMETRIC OPTICS (LAWS OF REFLECTION AND REFRACTION)

In geometric optics, we are concerned with how light interacts with an individual surface. The outcomes illustrated in Figure 2.1 are called "reflection" and transmission that has undergone "refraction." The term "scatter," introduced in Section 2.1, is not used in discussing geometric optics and is associated with the interactions between the light and a particle; these phenomena are discussed in the next two sections. Sometimes people confuse or conflate the terms "reflection," "refraction," and "scatter." Since geometric optics is often the first theory to which we are exposed, we may tend to continue to use the language of geometric optics, even when we have a better vocabulary available.

In discussing geometric optics, we talk about light "rays" and sometimes construct ray diagrams, as shown in Figure 2.1. A ray moves through a medium in a straight line, and the more "optically dense" the medium, the slower the speed of light in it. When the ray encounters a new medium with a different density, it speeds up or slows down. The "refractive index" for a medium can be defined as the ratio of the speed of light in a vacuum to the speed of light in the medium.[3] Thus, for our purposes, the refractive index of a material is always greater than one. For example, water has a refractive index of 1.333 (no units).

You may recall that the traveling wave equation is expressed in words as:

$$\text{velocity of wave front} = \text{frequency} \times \text{wavelength}. \tag{2.5}$$

When the light changes velocity at an interface between two media, the frequency is the same on either side of the interface, and consequently, the wavelength decreases in proportion to the velocity.

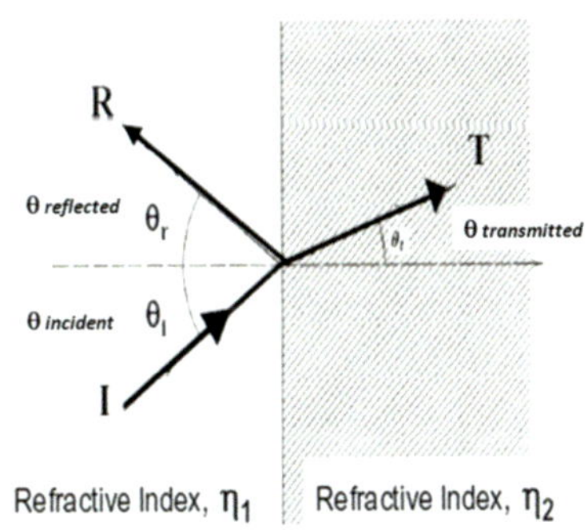

FIGURE 2.1 Schematic representation of refraction and reflections of a plane electromagnetic wave at a boundary, with symbolism related to reflection and refraction at a surface.

---

[3] Refractive index is more correctly defined by the "phase velocity" in the medium. Consequently, it is possible for refractive indices to be less than unity. The above treatment assumes that the refractive index is unaffected by wavelength of the light and absorption of the medium. In nonabsorbing regions of a spectrum, far from regions of strong absorption, the real refractive index of a material decreases slowly as the wavelength increases. This behavior is termed "normal dispersion of the refractive index." In contrast, near an absorption maximum, a material's real refractive index decreases markedly with increasing wavelength on the low-wavelength side of an absorption band, then increases rapidly through the band center, and again decreases markedly on the high-wavelength side of the band; and resumes normal dispersion at wavelengths far from the band. This behavior in the vicinity of an absorption band is called "anomalous dispersion" of the refractive index. See Griffiths and De Haseth (2007b), and Dahm and Dahm (2007).

Angle of incidence is a fundamental concept in geometric optics. "Directed illumination" describes the situation in which the directed beam is perpendicular (normal) to the surface of the sample. Directed illumination is considered to have an "angle of incidence" of zero degrees, and other angles of incidence are measured in comparison with this "normal," as illustrated in Figure 2.1.

If light strikes a surface (interface) where there is a change in refractive index, as depicted in Figure 2.1, a portion of the light is reflected and a portion is transmitted through the interface. For the reflected light, as is well known, the angle of incidence is equal to the angle of reflection, or in the nomenclature of Figure 2.1:

$$\theta_{\text{incident}} = \theta_{\text{reflected}}. \tag{2.6}$$

The "transmitted" light, as shown in Figure 2.1, has penetrated the surface but has been diverted from its original path, and the resulting "angle of refraction" is governed by Snell's law:

$$\eta_2 \sin(\theta_{\text{transmitted}}) = \eta_1 \sin(\theta_{\text{incident}}) \tag{2.7}$$

$$\frac{V_1}{V_2} = \frac{\eta_1}{\eta_2}, \tag{2.8}$$

where V is the velocity of light in the medium, η is the refractive index of the medium, θ_{incident} is the angle between the path of the incident light and the normal. $\theta_{\text{transmitted}}$, which is also known as the "angle of refraction," is the angle between the path of the transmitted light and the normal.

2.2.1 Intensity Calculations

We can also predict the intensity of reflected and transmitted light relative to the incident beam. These calculations are more involved, and if absorption is significant, they require expressing the refractive index as a complex number.

For the special case where the incident radiation is normal and the absorption is negligible, the intensity of the reflected and transmitted beams can be calculated from the refractive indices $\{\eta_1$ and $\eta_2\}$ of the two media:

$$r = \frac{(\eta_2 - \eta_1)^2}{(\eta_2 + \eta_1)^2} \tag{2.9}$$

$$t = \frac{4\,\eta_1\,\eta_2}{(\eta_2 + \eta_1)^2}, \tag{2.10}$$

where $\{r\}$ is the fraction of the incident light reflected, and $\{t\}$ is the fraction of the transmitted light.

Next, we consider the case where light is striking a surface at normal incidence, but absorption does occur. This requires us to express the index of refraction of each medium as a complex number. The real component of the index of refraction, again, allows us to quantify the change in velocity at the surface, and the resulting scatter. The imaginary component quantifies the absorption of light in the medium. Absorption was not something we needed to account for in Equations 2.6–2.8, when we were simply calculating the angles of reflection and refraction, but is certainly relevant for intensity calculations.

For the incident normal radiation, the expression for the fraction of incident light that is reflected is:

$$r = \frac{(\eta - 1)^2 + \eta^2\kappa^2}{(\eta + 1)^2 + \eta^2\kappa^2}, \tag{2.11}$$

where η is the relative refractive index of the two media, and the symbol $\{\kappa\}$ represents the absorption index. Notice that in Equation 2.11, the relative refractive index is squared. Since $\left\{i=\sqrt{-1}\right\}$, when the imaginary component of the complex number is squared, the result is that the absorption term is multiplied by $\{-1\}$, thus reducing the intensity. (The above is a rationalization, not a proof. The rigorous mathematics is more complicated than this.)

The absorption index (κ) is not the same as the absorption coefficient $\{k\}$ described in Section 2.1, but they are related to each other by:

$$k\lambda_0 = 4\pi\eta\kappa \tag{2.12}$$

The symbol $\{\lambda_0\}$ represents the wavelength of light in a vacuum.

Again, Equations 2.9–2.11 assume the incident light is normal to the surface. The mathematics for non-normal incidence is more complicated (see section H-2 of Dahm and Dahm, 2007) and will not be addressed here, but Figure 2.2 allows us to explore the outcome qualitatively. Figure 2.2 presents as an example the reflectance as a function of angle of incidence for an interface where the relative refractive index is 1.5. For nonpolarized light, the reflectivity is at a minimum at normal incidence and becomes equal to 1 at 90°. As illustrated in Figure 2.2, the case of polarized light is more complicated. At normal incidence, there is no polarization effect.

2.2.2 Direct Illumination of a Sample

The foregoing discussion of geometric optics presents what happens "when light encounters a surface." It is worthwhile to consider what, precisely, this might mean in the context of a spectroscopic experiment. An illustrative example of a sampling setup is shown in Figure 2.3. This sample is

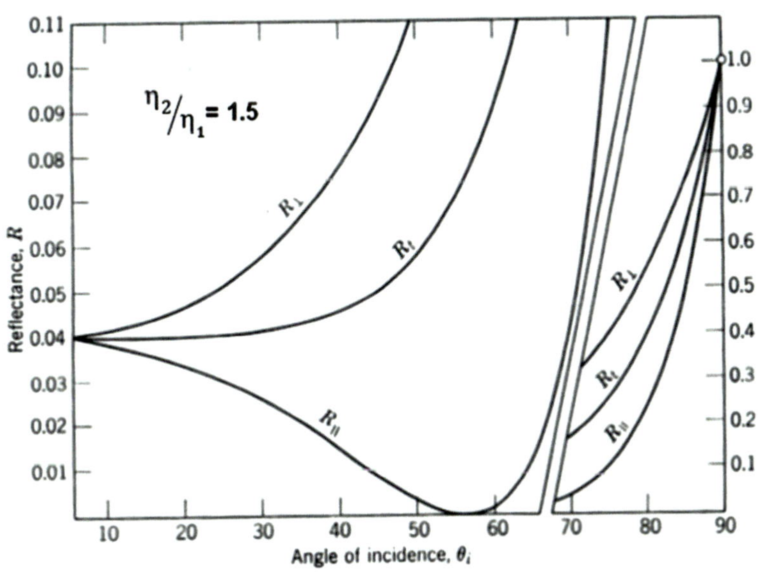

FIGURE 2.2 Fresnel reflectivity as a function of the angle of incidence for an interface where the refractive index of the object is 1.5 times that of the medium. This is characteristic of an air–glass interface. The plot shows the reflectivity for perpendicular and parallel polarization (Polarized radiation where the plane of polarization is parallel to the plane of incidence is said to be p-polarized. Polarized radiation where the plane of polarization is perpendicular to the plane of incidence is said to be s-polarized. In the absence of an external reference, the direction of polarization is undefined. The plane of incidence is the plane that contains both the incident beam and the surface normal.) as well as the average of the two.

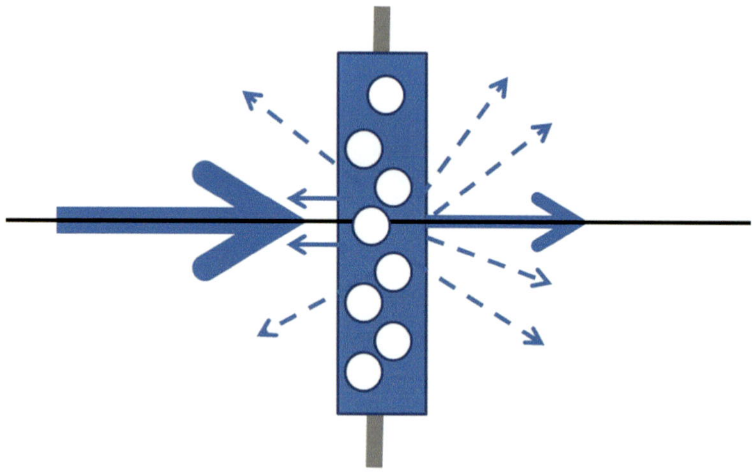

FIGURE 2.3 The direction of illumination defines a reference line. The radiation that exits the sample still traveling in the path defined by this reference line is called the "directly transmitted" light. The specular reflection goes in the opposite direction. The diffusely transmitted light and diffusely reflected light (represented as dashed arrows) occur at other angles.

"directly illuminated." That means two things: the light is "collimated" (all moving in the same direction) and it is normal (perpendicular) to the surface. In our example, the beam encounters a slab, perhaps made of plastic with a perfectly flat, smooth surface. A portion of the light is reflected at the first surface. This may be called "specular reflectance" or "regular reflection." The remaining light enters the slab. There are air pockets present, which provide surfaces that scatter light. In the figure, the particles that make up the slab are spherical. This means that the light beam will encounter various surfaces at different angles. While the theory of geometric optics may be sufficient to describe any single encounter, the net effect of all the encounters and the various angles is to produce *diffuse* reflectance; light will be emitted from the front surface of the slab at a variety of angles.

We frequently associate the word "reflection" with the formation of a mirror image. Of course, the rougher the surface of the mirror, the more distortions there are in the image. If the surface is made up of small particles, the distortions are so severe that we see no reflected image at all. But this does not mean that reflection has not occurred: the surface is not one mirror, but many, with the particle surfaces placed at various angles to the light. For each tiny mirror surface, the angle of incidence equals the angle of reflection. Since there are many such angles, the reflections send light back in many different directions. The resulting pattern of reflection is dispersed over a wide range of angles, just as is the scatter described in the previous paragraph. The difference between the two is that in regular reflection, there is no effect of sample absorption, and the spectral character of the reflected light is the same as the spectral character of the incident light. In diffuse reflection, the spectral character is that of the incident light modified by the absorption of the sample. The sum of the regular and diffuse reflection is called the "remission."

2.2.3 Diffuse Illumination

The alternate to directed radiation that is sometimes used in theoretical discussions is "diffuse irradiation" or "diffuse illumination." We might think of a diffuse beam as having equal intensity at every angle of incidence. However, the most commonly used idealized model for diffuse illumination is "Lambertian," which describes a situation in which the intensity is distributed across the angles of incidence in proportion to the cosine of the angle of incidence.

2.2.4 Terms Used

As we move from considering what happens when light encounters single surfaces to what happens in a sample as a whole, it is useful to define some terms. Light that is emitted from the "front" surface of a sample is called "remission." The regular reflection and the diffuse reflection from a sample, taken together, make up the *remitted* light. The fraction of the incident intensity that is remitted is called the "remission fraction" or "remittance" of a sample.

Similarly, all of the light that penetrates the sample and is emitted from the "back" surface of the sample is called the "transmitted" light. If light penetrates the sample without being diverted from its original incident path, then this is called "direct transmission." Any light that has been diverted from its original angle of incidence but still penetrates the sample is called "diffuse transmission." The direct transmission and the diffuse transmission taken together make up the transmitted light. The fraction of the incident intensity that is transmitted is called the "transmission fraction" or "transmittance" for a sample.

The fraction of the incident intensity that is absorbed is called the "absorption fraction" or "absorptance" of a sample. In treatments related to absorption spectroscopy, the three fractions (remission, transmission, and absorption) are assumed to sum to one. The remitted light and diffusely transmitted light are said to have been "scattered." The fate of the incident light then can also be described as the sum of that absorbed, scattered, and directly transmitted. As this section illustrates, there are actually multiple distinct physical phenomena that can contribute to what is collectively called "scatter." The following section explores absorption and scatter further, discussing phenomena associated with interactions between light and single *particles*.

2.3 ABSORPTION AND SCATTER FROM A SINGLE PARTICLE

In order for a molecule to absorb radiation, there must be a mechanism for the light energy to be turned into some other forms of energy. At the wavelengths associated with NIRS, absorptions are caused by the bond vibrations.[4] When the incident light has the same fundamental frequency (or a harmonic thereof) as a bond (often pictured as a spring), some of the light energy will be used to make the atoms involved in the bond vibrate faster. In the visible and UV range, the predominant mechanism for absorption is the promotion of electrons to higher energy levels in the atom.

The origin of the scatter is the charged particles in an atom. They vibrate in time with the electric vector of the incident light (electromagnetic radiation). Because they are lighter, the electrons vibrate far more vigorously than the nuclei and are the main source of scattered light. The scatter occurs because the accelerating (vibrating) charges emit radiation of the same[5] (as the incident beam) wavelength in all directions. The emissions from the various charged particles interfere with each other, and give rise to a scattering pattern that is dependent upon the relative placement of the vibrating charges.

As a particle becomes larger and more complex, the mathematics necessary to calculate the scattering (even approximately) becomes more complex. Below, we discuss three cases of increasing complexity. In each case, the particle is assumed to be bathed in the incident beam, and the scatter changes with angle (θ) relative to the incident beam. One common element is that in each of the three models, there is a term for the "inverse square falloff" of light intensity at a distance $\{\boldsymbol{R}\}$ from the source.

[4] The model most often used in calculating energy levels for bond vibrations is that of a "simple harmonic oscillator." A bond vibration is not strictly harmonic because of a strong repulsion between atoms when they approach each other closely.

[5] Scatter that has the same wavelength as the incident light is called "elastic scatter." If the wavelength changes, the scatter is called "inelastic." A well-known case of inelastic scatter in vibrational spectroscopy is Raman scattering.

2.3.1 Scatter from a Point Particle Having a Charge

The (Joseph J) Thompson model states that scatter from a single charged point having charge {$\boldsymbol{q}$} and mass {$\boldsymbol{m}$} at a distance {$\boldsymbol{R}$} is proportional to:

$$\left(\frac{q^4}{m^2c^4}\right)\frac{1+\cos^2(2\theta)}{2R^2}. \tag{2.13}$$

Notice there is neither a size nor a wavelength parameter in the formula, though there is an implied assumption that the wavelength is much larger than the point particle. Because electrons are orders of magnitude lighter in mass than protons and m appears in the denominator, according to this equation the scatter from an electron is over 3,000,000 times that of a proton. Consequently, the observed scatter from a particle is generally thought of as the collective scatter from all the electrons in the particle, with the need to take into account the interference of the waves originating from each.

2.3.2 Scatter from Particles with Size Smaller than the Wavelength

The Raleigh approximation says that the intensity of light scattered by a small particle, having a relative refractive index* {η}, is proportional to:

$$\left(\frac{1+\cos^2\theta}{2R^2}\right)\left(\frac{2\pi}{\lambda}\right)^4\left(\frac{\eta^2-1}{\eta^2+2}\right)^2\left(\frac{d}{2}\right)^6. \tag{2.14}$$

The scatter drops off by the fourth power of the wavelength (λ), is proportional to the 6th power of the particle diameter {$\boldsymbol{d}$}, and is a function of the scattering angle θ. Notice that the number of waves that fit into the scattering cross section of a particle of a given size is inversely proportional to the square of the wavelength, then squaring that to get intensity; we arrive at the "inverse fourth power of the wavelength." Similarly, the number of waves that fit into the particle is given by the volume, which is proportional to d^3, which when squared, gives us the d^6 term. If there is no refractive index difference between the particle and the medium, the relative[6] refractive index η is 1, and the scattered intensity becomes zero. This is only reasonable if the particle is so small compared to the wavelength that the scatter from each electron in the particle is essentially coming from the same place in the wave phase.

2.3.3 Scatter from Spherical Particles of Any Size

Mie theory is an analytical solution of Maxwell's equations that uses infinite series. The Mie equation for the coherent scatter from a uniform sphere of any size, bathed in the incident beam, is:

$$\frac{\lambda^2}{8\pi^2R^2}\left[S_1(\theta)+S_1(\theta)\right]. \tag{2.15}$$

While this equation looks innocent enough, the {**S**} terms, called "angular scattering functions," are the complicated Legendre polynomials as detailed in Equations 2.16 and 2.17. The parameters {a} and {b} are called "Mie coefficients," and are not constants, but complicated (Bessel) functions

[6] The relative refractive index is given by the refractive index of the scattering particle divided by the refractive index of the medium.

involving the size of the particle (in terms of wavelength of the incident light) and the complex refractive index.[7]

$$S_1(\theta)=\left\{\sum_{n=1}^{\infty}\frac{2n+1}{n(n+1)}\left[a_n\pi_n(\cos\theta)+b_n\tau_n(\cos\theta)\right]\right\}^2 \tag{2.16}$$

$$S_2(\theta)=\left\{\sum_{n=1}^{\infty}\frac{2n+1}{n(n+1)}\left[a_n\tau_n(\cos\theta)+b_n\pi_n(\cos\theta)\right]\right\}^2. \tag{2.17}$$

In this mathematical treatment, the sphere is treated as an assembly of points, each of which scatters light waves isotropically. The refractive index used in the calculations is a complex number of the form <scat + *i* abs> – the same usage as was described for doing intensity calculations from geometric optics in Section 2.2.

2.3.4 Relationships between Particle Size, Wavelength, and Scatter

An individual larger particle will both absorb and scatter more light than an individual smaller one. To a first approximation, the amount of absorption will be proportional to the volume of the particle, and the amount of scatter will be proportional to the surface area. We note that, using a sphere as an example, the surface area is proportional to radius squared and the volume is proportional to radius cubed; thus, the surface area-to-volume ratio decreases as radius increases. Consequently, when we compare two particulate samples of uniform composition but different particle sizes, we observe the following: for a given amount (mass) of sample, there will be more scatter from the sample composed of (many) smaller particles than from the sample composed of (fewer) larger particles. This is what is meant by the catchphrase "small particles scatter more than large ones."

There is a rough guide as to when the various treatments for scatter from a single particle discussed above are reasonable approximations. The guide is based on a "size parameter" (*Sp*), which, for a spherical particle, is the ratio of the circumference to the wavelength of the illuminating light. Often, a nonspherical particle is treated as a sphere with an average diameter, *d*:

$$Sp = \text{circumference/wavelength} = \pi d/\lambda. \tag{2.18}$$

The applicable size range for a particle with no absorption using various mathematical treatments is listed in Table 2.1 and schematically illustrated in Figure 2.4. The scatter drops off exponentially

TABLE 2.1
Regions within Figure 2.4 and Theories That Explain the Behavior Illustrated in Each

Region	Size Parameter	Theoretical Treatment	Particle Shape
3	$Sp > 2000$	Geometric optics	Plane parallel
2	0.02–2000	Mie	Spherical
1	0.002–0.02	Raleigh	Unspecified
0	$Sp < 0.002$	Essentially no scatter	Unspecified

[7] A set of Mie functions has been programmed in MATLAB® to compute the Mie coefficients, efficiency of extinction, scattering, backscattering and absorption, an asymmetry parameter, and the two angular scattering functions (Mätzler, 2002).

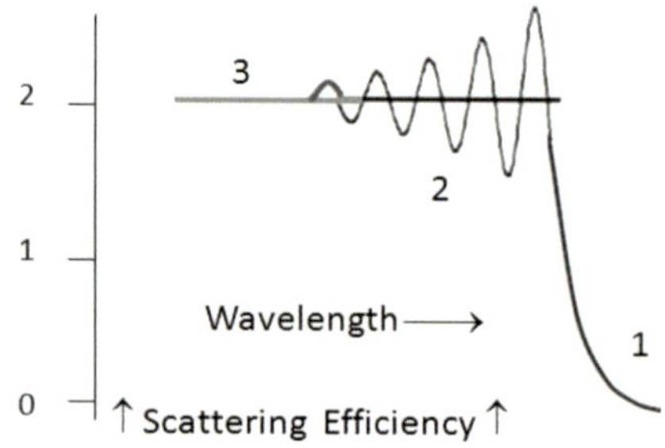

FIGURE 2.4 Scatter from a particle of a fixed size as a function of changing wavelength. A scattering efficiency of 2 implies that the particle scatters twice as much light as is incident upon the physical cross section of the particle. The maximum scattering efficiency occurs when the circumference of the sphere is equal to the wavelength. There are then maxima of decreasing magnitudes for the circumference being 2, 3, 4... times the wavelength.

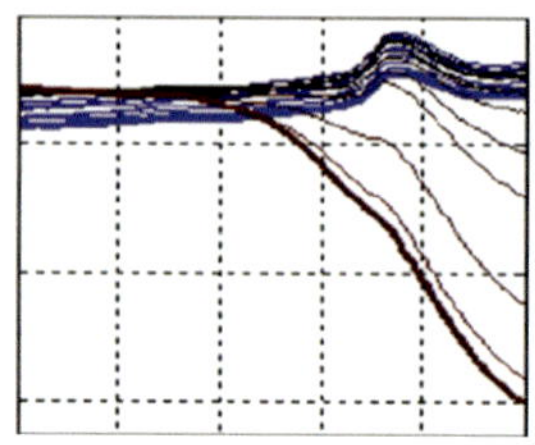

FIGURE 2.5 Log(1/T) versus wavelength from sequential spectra of coalescing particles in milk (Bogomolov et al., 2013). The bold line represents the smallest particle size, and its "exponential" falloff begins at the lowest wavelength and is most dramatic.

at the right of the figure. "Region 0" on the figure is the name we give to the extremely high wavelengths that lead to small size parameters and essentially no scatter.

Figure 2.5 shows the actual data consistent with this picture for a fixed wavelength range and changing particle size. While the exponential dropoff predicted by the Raleigh model is evident, the various minima and maxima predicted by the Mie model are washed out, because in reality the particles in a sample are not exactly the same size as each other.

The pattern of the scatter from a particle is also a function of particle size, as shown in Figure 2.6. When particles present a large (compared to the wavelength of the incident light) flat surface to the incident beam, the fraction of light scattered is relatively independent of wavelength. (For example, we observe the color of our clothes to appear the same in a mirror as they do when we look directly at them.) On the other hand, for particles much smaller than the wavelength, there is little scatter at all. When the particles are about the same size as the wavelength, there is a rapid falloff of scatter with wavelength. The exact rate of this falloff will depend on the particle size (and therefore, the wavelength range where the falloff occurs). For tiny particles, the falloff is described by the Raleigh equation.

The preceding assumes a typical case in which the particles reside in a matrix that has no absorption, such as particles in an air matrix. For the case of absorbing particles in a matrix that is also absorbing (e.g., oil droplets floating in water), there will be a different response to particle size in the two phases. We illustrate that for the case of coalescing oil drops in Figure 2.7. A continuous liquid phase (in this case, water) will absorb light as it is being transmitted. A continuous liquid phase will not scatter light except at the front surface, and this front surface of the continuous phase will be ignored in this discussion. The scatter comes from the presence of the noncontinuous phase (in this case, the oil droplets) and the change in refractive index that light experiences when it encounters a

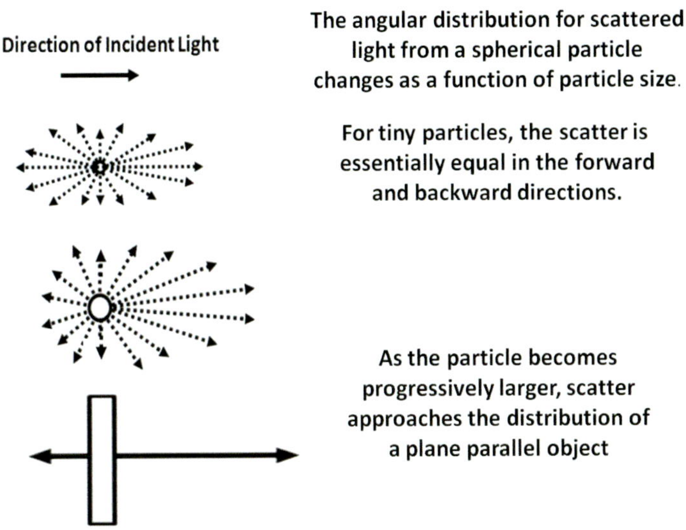

FIGURE 2.6 Examples of scattering patterns from different sizes of particle.

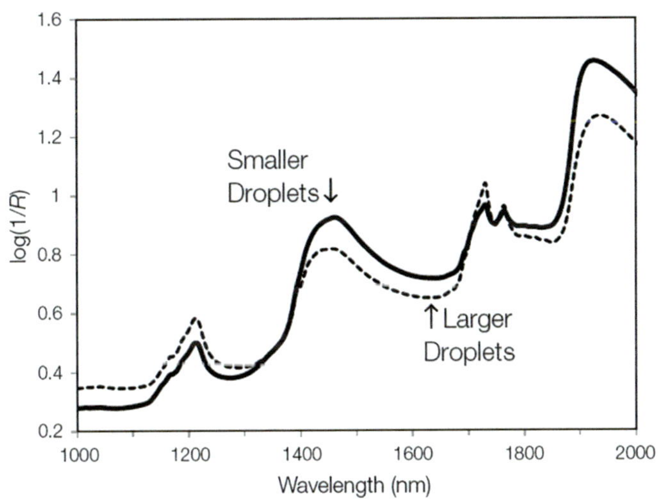

FIGURE 2.7 Remission spectra of an oil-in-water emulsion taken at two different times during which the oil droplets have coalesced.

boundary between the two phases. For a given amount (mass) of oil in the sample, scatter decreases as the drops become larger. Consequently, the two spectra in Figure 2.7 are measurably different from each other even though they are (in terms of composition) exactly the same sample; the only difference between them is the passage of time during which the oil coalesced into fewer, larger droplets.

2.4 COHERENT AND INCOHERENT SCATTER

The previous sections examined interactions between light and a single surface, and then light and a single particle. When we consider multiple scattering sites that are in proximity to each other, we need to consider how these sites can influence each other. Briefly, "coherent scatter" refers to an integrated combination of waves, while "incoherent scatter" refers to an independent summation of intensity emanating from each site.

2.4.1 Coherent Scatter

Imagine a particle filled with atoms that are scattering centers. In any direction from one atom, there are many more atoms that are located within the distance defined by the wavelength of the incident radiation. When an incident light beam enters such a particle and sequentially encounters atoms, all of the scatter is in phase with the incident beam. The scattered waves that are emitted in the "straightforward" (along the original path) direction all reinforce each other. The radiation scattered in this direction will be indistinguishable from the direct beam. There is no interference that causes a loss of intensity in that direction.

However, in any other direction, there will be waves coming together from various atoms that are at all different phases of the wave cycle. This will result in a complete annihilation of the wave within the particle. Consequently, the only atoms that will give rise to scatter from the particle are those near a surface where the atom is not surrounded by atoms to generate the distractive scatter. This is our model of scatter from a uniform particle. (A corollary of this argument is that the scatter will be proportional to the surface area of the particle, as the surface is the source of the measurable scatter.) Because the scatter is considered as interacting waves, we refer to this as "coherent scatter."

Theoretical calculations are often carried out for "semi-infinite plane parallel" objects. These are often called "sheets" or "slabs." These objects have two flat surfaces that are perfectly parallel to each other and separated by a finite "width." The object extends indefinitely in the other two directions as far as is necessary to ensure that the scatter at the "edges" of the slab is negligible. The fraction of the incident beam that is absorbed by this idealized sheet can be calculated from the thickness of the sheet, the angle of incidence, the index of refraction, and the absorption coefficient of the material making up the sheet. The calculation is straightforward, though it involves a small number of convergent series. The calculation is even simpler if the assumption that the reflectivity of a surface is the same whether the light is entering or leaving the sheet is a reasonable one.

If we have a large sheet (pane) of clear glass that is free of defects, it is an excellent example of coherent scattering from a semi-infinite plane parallel object. The only light scattered from this idealized object will follow the rules of geometric optics, with no diffusely scattered light. (In fact, a good definition of diffusely scattered light is that which does not follow the rules of geometric optics.) If an incident beam is normal (perpendicular) to the sheet, all the light will be either reflected or transmitted. All light will move in exactly the forward direction or 180° from it. For any other incident angle, there will be a bend in the light beam due to refraction. Note, too, that a mirror (or any "plane parallel layer") has both a front surface and a rear surface. Light can be reflected by either the front or the rear surface, but light reflected from the rear surface will ultimately exit the front surface at a position different from the portion originally reflected by the front surface. Furthermore, a portion of the light that is reflected from the rear surface will then be reflected by the inner side of the front surface and return to the rear surface, resulting in a portion of the "transmitted" beam being displaced as well. The continuation of this process yields a series of fringes in the observed transmitted and reflected beams. This is usually termed "beam broadening."

Within the interior of a slab (everything between the front and rear surfaces), we can model a single flux of light intensity as being attenuated (or extinguished) by absorption alone, as if there is no scatter at all. The mathematics describing this case is that of the Bouguer–Lambert law.

2.4.2 Incoherent Scatter

The previous section alludes to the fact that when we model scatter from two or more distinct particles, we could try to model the interactions of the waves from one particle with those coming from other particles. We would be integrating the scatter from all the particles, treating the scatter as "coherent." In this case, we see the medium as acting as one entity with individual wave amplitudes adding coherently. In terms of possible theoretical approaches, the opposite extreme is the case where every point in the uniform sheet (in this case often called "a slab") is a scattering center, which scatters incoherently: that is, the slab is imagined to everywhere scatter light intensity rather than light waves. In this case, we are not adding waves, but instead adding the intensity that is emerging from each particle. If the wave has a unit amplitude varying sinusoidally, it goes from −1 to 1, with a mean amplitude of 0. However, the intensity is not given by the amplitude, but by the amplitude squared. There are thus no "negative intensity" regions in the waveform. While this approach is usually termed an assumption of "incoherent scatter," we think the phrase "independent summation of intensity" is more descriptive.

There is not a physical example that perfectly represents the "incoherent scatter" case, because a uniform, continuous material will scatter coherently. However, a white ceramic material may be used to illustrate the case. There are internal scattering centers that give rise to diffuse scatter, no matter how much we polish the surface. If these scattering centers are separated by distances of several wavelengths (of the incident radiation), we begin to approximate the ideal case of incoherent scatter. Another case often modeled as if it were an ideal incoherent scattering material is that of tissue, such as human flesh. This idealized case is the one for which most of the mathematical modeling relating to absorption spectroscopy has been done.

2.5 OUR MODELS AND THEIR LIMITATIONS

The previous sections discuss the physics of primarily localized phenomena: how light interacts with surfaces and particles. Practically speaking, we can neither observe nor model literally every individual interaction occurring in a sample. While our models are informed by our understanding of the physics, we have to devise and use models that relate to what *is* measurable: namely, the light that is either remitted from or transmitted through a sample as a whole. Unfortunately, there is no single model for diffusely scattering samples that can be regarded as uniformly applicable.

2.5.1 The "Two-Flux" Modeling Approach

In discussing models and their limitations, it will again be helpful to picture our sample as one or more plane parallel layers, or slabs. Consider a flux of light intensity moving as an incident beam in a single direction through the slab. If at every point along the path (or alternatively, every time the light encounters a new layer), a portion of the light reverses direction, we will have a second flux (in the reverse direction of the incident light) being built up. This flux in turn will be diminished by light reversing direction and rejoining the first (incident) flux. We refer to this as the "two-flux"[8] case. A common arrangement in both experiment and theory is one in which two fluxes are quantified: the remission and the transmission.

As introduced in Section 2.2, "transmission" describes all light that penetrates the sample. Light can emerge from the sample travelling at many different angles relative to the initial path. In the two-flux approach, however, we don't distinguish between these various angles: all of the light that penetrates the sample, regardless of the specific path, is lumped together into a single flux called "transmission." Similarly, the "remission" is all of the light that changes direction and emerges from the "front" surface, regardless of the specific angle at which it emerges.[9]

[8] The full phrase you might see is "the two-flux solution to the equation of radiative transfer (ERT)." One may see references to the ERT in relation to absorption spectroscopy. While there is certainly an overlap between what we discuss in this chapter and the subject of radiative transfer, we tend to reserve the term "radiative transfer" for studies related to stellar atmospheric phenomena.

[9] Experimentally, we might use an integrating sphere to measure transmission and remission. If we have hemispherical detectors both "in front of" and "behind" the sample, then to an excellent approximation, all light leaving the sample will be counted as either "remission" or "transmission."

Let us picture as our sample a single, ideal slab, with incident illumination normal to the surface of the slab. If there is no scatter, then the only source of "remission" is specular reflection from the front surface. Here, the two-flux model is an "exact" mathematical description – the only fluxes are "transmitted" light that penetrates the sample along its original normal path, and "remitted" light that is reflected at 180°. Next, picture a more complex example in which the sample is a coherently scattering slab, and the incident light is not necessarily normal to the slab and is not necessarily collimated. In this case, scattered light will leave the front surface at various angles and will therefore penetrate the sample along different paths of different lengths, and therefore have different degrees of opportunity to be attenuated by absorption. Therefore, in this situation, a "two-flux" model that treats all light as simply travelling "forward" or "backward" is inherently an approximation. The higher the degree of absorption, the worse an approximation it will become. Mathematically, this approximation is particularly evident when we picture a sample composed of more than one scattering slab. Even if the light incident upon the front surface of the sample is collimated, that will only be true for the "first layer" of the sample; the light that is transmitted through the first layer and reaches the second layer will be a combination of direct and diffuse transmission. A "two-flux" model will not distinguish these components of transmission from each other.

2.5.2 Kubelka and Munk

Perhaps the best-known work on the two-flux case has been done by Kubelka. One of the kinds of samples he considered was that of an "infinitely thick" sample: a sample thick enough that it has essentially no transmission (direct or diffuse). Kubelka and Munk (1931) envisioned the sample as made up of a series of plane parallel layers, each of infinitesimal thickness, and sought to relate the behavior of the sample as a whole to the properties of the infinitesimal layer. They showed that the remission from such a sample is given by:

$$R_\infty = 1 + \frac{a_0}{r_0} - \sqrt{\frac{a_0^2}{r_0^2} + 2\frac{a_0}{r_0}}, \tag{2.19}$$

where a_0 is the fraction of light absorbed by a layer of infinitesimal thickness, and r_0 refers to the fraction of light remitted (the portion of light that reverses direction) by the layer of infinitesimal thickness. Note that the absorption and remission fractions appear in Equation 2.19 only as a ratio to each other. If one resolves the equation for $\{a_0 / r_0\}$, one obtains:

$$\frac{a_0}{r_0} = \frac{(1 - R_\infty)^2}{2R_\infty}. \tag{2.20}$$

Thus, according to Equations 2.19 and 2.20, the remission from an infinitely thick sample depends upon the ratio of absorption fraction to remission fraction, not upon their absolute magnitudes. There is, however, an implied assumption in the mathematics that a_0 and r_0 will be extremely small numbers: insignificant compared to unity. Realistically, many samples cannot plausibly be modeled as a series of many "thin" layers that each have "infinitesimal" absorption and remission fractions: what if the absorption and/or remission fraction from a single particle is significant? Subsequently, a more general case was developed (Dahm and Dahm, 1999). The sample is again modeled as a series of plane parallel layers, all of which are identical to each other. However, each slab is finite in thickness and nothing in particular is assumed about the magnitudes of the absorption and remission fractions from a single layer, beyond that the absorption, remission, and transmission fractions sum to 1. The resulting equation is:

$$\frac{(1 - R)^2 - T^2}{R} = (2 - a - 2r)\left(\frac{a}{r}\right), \tag{2.21}$$

in which the lowercase letters a and r refer to the absorption and remission fractions for a single slab, and the uppercase letters R and T refer to the remission and transmission fractions for the sample as a whole. Note that the number of "layers" that make up the sample does not appear in the equation. Consequently, while the specific values of R and T individually will depend upon the number of layers, the function defined by the left-hand side of the equation will be independent of the number of layers and therefore independent of the thickness of the sample. This function may be called the "absorption–remission function" or the "ART function," and is defined as:

$$A(R,T) = \frac{(1-R)^2 - T^2}{R}. \tag{2.22}$$

One of the limitations of this model is that its use requires a, r, and t to be "known" for a single "layer." Dahm and Dahm (1999) have proposed the "representative layer," described in Chapter 3 of this handbook, as a strategy for relating the properties of a "layer" to the properties of the individual particles that make up the sample. This still requires absorption and scattering coefficients of the individual particles to be "known." What we will emphasize here is that, as illustrated throughout this chapter, absorption and scattering are complex and interrelated phenomena. They depend not only upon the specific materials making up the sample, but also upon factors like particle size and particle geometry that are not necessarily repeatable from one sample to another. Consequently, while it is possible to quantify and use absorption and scattering coefficients, care must be taken before assuming the values apply *beyond the specific experimental context in which they were obtained*.

In addition, Equations 2.19–2.22 all employ the approximation of the "two-flux" approach, which is known to have systematic errors that become significant at high levels of absorption. The following sections explore alternatives.

2.5.3 Beyond the Two-Flux Approach

As mentioned in the previous section, the "two-flux" approach is inherently an approximation. An example of a "three-flux" approach might be as follows: a light beam enters the sample moving in the incident direction, identical to the "forward" flux in the two-flux approach. The beam undergoes scattering at points along its path. At each scattering point, a fraction of the intensity reverses direction (i.e., is scattered at 180°), substantially identical to the "backward" flux in the two-flux approach. But in addition, at each scattering point, a fraction of the intensity is scattered in each of the directions at 60° to the incident beam. A portion of each of these "third" fluxes will still ultimately leave the sample and contribute to transmission and remission, but they take a longer path from the scattering site to the edge of the sample, and will be attenuated more by absorption than were the "straight-ahead" and "straight-back" fluxes that emerged from the same scattering point.

The mathematics of the three-flux case has been worked out by Burger et al. (1997). The solutions obtained are far more complex than for the two-flux case and must be evaluated numerically. By comparing results from their three-flux case to the "infinite flux" case described next, it seems there is little advantage to increasing the number of fluxes above three.

The "exact solution"[10] for a continuous array of intensity scattering points has been developed by Ron Giovanelli (1955). At each point within the sample, a fraction $\{\alpha\}$ of the incident light is absorbed, and a fraction $\{\sigma\}$ is scattered isotropically (scattered with equal frequency in all directions). We refer to such an array as an "ideal diffuser." The solutions are very complicated. Solutions to various special cases have been published. The main use of these solutions in absorption spectroscopy is to see to what extent a particular approximation (e.g., a Monte Carlo algorithm) duplicates the results of the solution.

[10] A "mathematically exact" solution does not imply that the model will describe the real samples perfectly.

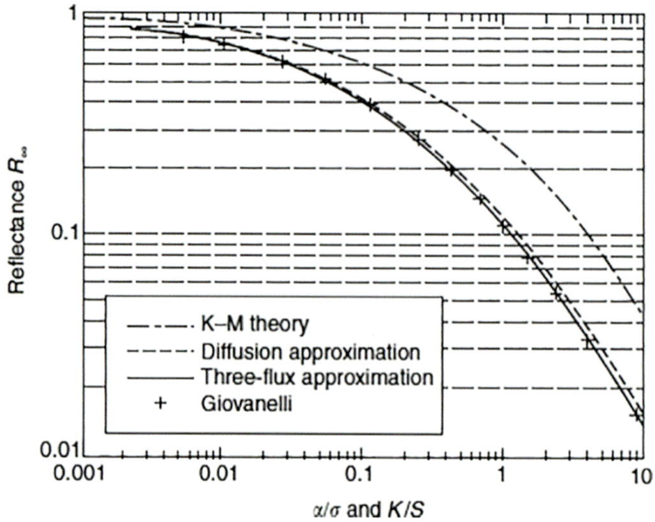

FIGURE 2.8 Diffuse reflectance R_∞ of an isotropic, optically thick sample according to the three-flux approximation, the diffusion approximation, Giovanelli and the K–M theory.

Figure 2.8 was also included in the second and third editions of this handbook. It shows several things worth noting.

1. The three-flux approximation gives results that are very similar to those of the exact solution.
2. The diffusion approximation (not discussed here) gives better agreement at lower levels of absorption than higher. This observation is consistent with the understanding that the validity of a diffusion approximation requires a photon to be, on the average, scattered many times before it is absorbed.
3. The derived Kubelka–Munk coefficients have a ratio (*K/S*) of 8/3 that of the Mie coefficients (α/σ).

It is this third point that we wish to examine further. The symbolism {*K/S*} for the ratio of the Kubelka–Munk coefficients was made standard by Kortüm (1969). However, {*S*} was not a coefficient of scatter, but of remission – only half of isotropic scatter goes "backward" and contributes to remission (Kortüm, 1969). This means that if we are going to compare K/S to Mie absorption and *scattering* coefficients, we should be looking at {*K/S*} versus {$2\ \alpha/\sigma$}.

Mie theory is for spheres of finite size, but the Kubelka–Munk coefficients are for a point of infinitesimal size. For a finite object, the total absorption is related to the volume around this infinitesimal point, and the scatter is proportional to the cross-sectional area of the object. Including these factors in the above comparison, we obtain the two-flux description of absorption and scatter by a finite sphere:

$$\frac{K}{S} = \frac{2\alpha\left(\frac{4}{3}\pi r^3\right)}{\sigma\left(\pi r^2\right)} = \frac{8\pi r}{3\sigma}. \tag{2.23}$$

So the factor of 8/3 can be understood using geometry, but isn't uniformly applicable. There is a radius term {*r*} that is left in the numerator after the cancellation. In reality, what Equation 2.23 illustrates is that as the particle becomes larger, the ratio of absorption to remission becomes larger.

An important point of this section is that the (implied) assumptions of the various theories are not compatible. Mie theory is for a finite-sized sphere that is scattering waves at every point. Kubelka–Munk theory relies on dividing a sample into a series of planes of infinitesimal thickness reflecting intensity, and is therefore best applied to samples composed of very small particles, such that the sample can be fairly modeled as a homogeneous continuum. While comparing Mie and Kubelka–Munk coefficients to each other may be an instructive thought exercise, it is possible to overextend or misuse authoritative-sounding published findings that "Kubelka–Munk coefficients and Mie coefficients are related to each other by a factor of 8/3." In reality, neither Mie coefficients nor Kubelka–Munk coefficients can be regarded as uniformly applicable properties of a material; their values are not necessarily "portable" from one physical system to another.

2.5.4 Monte_Carlo Methods

If one needs to go beyond the two-flux case, as is often necessary in studying tissue (Prahl et al., 1989), the current method of choice is the "Monte Carlo method." The Monte Carlo method is a computational algorithm that relies on a repeated random "sampling" to obtain numerical results. Monte Carlo methods are the most useful when it is difficult or impossible to use other mathematical methods. Notice that there may be a wide array of possible algorithms to model the same physical process. The phrase "we used Monte Carlo method" should not be regarded as self-explanatory by either a reader or a writer: understanding the algorithm behind the calculations and how well the algorithm matches the physical situation is key in evaluating the significance of the results.

For purposes of illustration, let us consider a very simple model, the Monte Carlo equivalent to the two-flux model.[11] Suppose the computer sends out 1000 photons into a sample of thickness "d." Each photon takes an initial step of "x" (where x is smaller than d) in the direction of the incident beam. For each of the photons, there is a probability assigned to it being absorbed, reversing direction, or continuing in the same direction. A random number generator is used to decide, consistent with the assigned probabilities, which of the fates each photon encountered. If the program records a photon being *absorbed*, it is dropped from the calculations. The remaining photons are sent on another step in their assigned directions, and the process is repeated until all photons either are absorbed or reach the front or the rear surface. When a particular photon reaches the rear surface of the sample, it is recorded as having been *transmitted*. When it reaches the front surface, it is recorded as "*remitted*." The number of photons that experienced each outcome is taken as a fraction of 1000, and our calculation is complete.

An added advantage of the Monte Carlo method is that a number of steps taken and the step size are combined to generate a distance traveled by a photon. From this distance and the speed of light, the time of travel of a particular photon can be calculated. Displaying the number of remitted or transmitted photons as a function of time generates a "time-resolved spectrum."

2.5.5 Time-Resolved Spectroscopy

Perhaps the most promising way of getting spectrum that is truly linear with absorption is through time-resolved spectroscopy. You will recall that Beer's law is:

$$\text{Absorbance} = \text{absorptivity} \times \text{concentration} \times \text{pathlength}.$$

For absorption spectroscopy in solutions, absorbance is often taken as: $\log(I_0 / I_{\text{sample}})$.

[11] When we did the numerical modeling that led to the development of the Dahm equation, we chose to keep track of results by accounting for the fraction of light that reached a specific position in the sample, rather than by tracking individual photons as described here. We originally did not claim to have used "Monte Carlo" techniques.

As photons are scattered throughout a sample, some will take a longer path and therefore a longer time. For the photons on a particular pathlength through a sample, absorbance is given by:

$$\log\left(I_{\text{path without absorption}} / I_{\text{path with absorption}}\right).$$

Consequently, if we know what the intensity would be in the absence of absorption, we can determine what the absorbance would be in the absence of scatter. Sometimes it is possible to find a place in a spectrum where there is zero absorption, and then, the determination can be made by using the intensity at that location as the $\{I_0\}$ for calculations of absorbance at wavelengths where there is absorption.

While the method has the potential for great utility, we note some challenges in quantifying the scattering. On any one "path" through the sample, there will be attenuation because of scatter and absorption. There will be a loss due to scatter every time light enters a particle. The more particles are encountered on a particular path, the more scatter loss there is. The challenge is that on some paths, there are more scattering events than others, even if the pathlength (and therefore the time of flight) is the same. Consequently, there is no one scattering coefficient (defined as fraction of intensity lost due to scatter/unit pathlength) that can uniquely describe the sample. As a further complication, frequency of scattering is itself wavelength dependent. Practitioners who attempt to use a single scattering coefficient to describe a sample (for all wavelengths and all pathlengths) are actually using some sort of averaged value, which may dilute the utility of the time-of-flight technique.

REFERENCES

Bogomolov, A., Melenteva, A. and Dahm, D. J., 2013. Fat globule size effect on visible and shortwave near infrared spectra of milk, *J. Near Infrared Spectrosc.* 21, 435–440.

Burger, T., Ploss, H. J., Kuhn, J., Ebel, S. and Fricke, J., 1997. Diffuse reflectance and transmittance spectroscopy for the quantitative determination of scattering and absorption coefficients in quantitative powder analysis, *Appl. Spectrosc.* 51(9), 1323–1329.

Dahm, D. J. and Dahm, K. D., 1999. Representative layer theory for diffuse reflectance, *Appl. Spectrosc.* 53, 647–654.

Dahm, D. J. and Dahm, K. D., 2007. *Interpreting Diffuse Reflectance and Transmittance.* NIR Publications, Chichester, UK, p. 26.

Giovanelli, R. G., 1955. Reflection by semi-infinite diffusers, *Opt. Acta*, 2: 153.

Griffiths, P. R. and Dahm, D. J., 2007a. Continuum and discontinuum theories of diffuse reflection. Chapter 3 in *Handbook of Near-Infrared Analysis*, Third Edition; D. A. Burns and E. W. Ciurczak eds., CRC Press, Boca Raton, FL, pp. 21–64.

Griffiths, P. R. and De Haseth, J. A., 2007b. *Fourier Transform Infrared Spectroscopy*, Second Edition; John Wiley & Sons, Hoboken, NJ, pp. 14–15.

Kortum, G., 1969. *Reflectance Spectroscopy*, Springer, Berlin.

Kubelka, P. and Munk, F., 1931. Ein Beitrag zur Optik der Farbanstriche, *Z.Tech. Phys. (Leipzig)*, 12, 593–601.

Mark, H., Rubinovitz, R., Heaps, D., Gemperline, P., Dahm, D. and Dahm, K., 2010. Comparison of the use of volume fractions with other measures of concentration for quantitative spectroscopic calibration using the classical least squares method. *Appl. Spectrosc.*, 64(9), 995–1006.

Mätzler, C., August 2002. MATLAB Functions for Mie Scattering and Absorption, Version 2; Research Report No. 2002–11; Institute of Applied Physics, University of Bern.

Prahl, S. A., Keijzer, M., Jacques, S. L. and Welch, A. J., 1989. A Monte Carlo model of light propagation in tissue. *SPIE Proceedings of Dosimetry of Laser Radiation in Medicine and Biology.* Institutes for Advanced Optical Technologies, 1988, Berlin, Germany.

3 Principles of Diffuse Reflectance Spectroscopy

Kevin D. Dahm[*] *and Donald J. Dahm*
Rowan University

CONTENTS

In Chapter 2, the authors provided an overview of physical principles that underlie the observed absorption and scattering that occur in spectroscopy. This chapter focuses on the construction of mathematical models. Unfortunately, there is no single, uniformly applicable mathematical treatment that "works" for all applications (whether one is thinking of spectroscopy in general or near-infrared (NIR) spectroscopy in particular). Consequently, the intent of this chapter is to place a particular emphasis on the assumptions made in the development of each model. The usefulness of a model depends upon how well the assumptions made in formulating the model align with the physical realities of the case at hand.

3.1 SPECTROSCOPY OF ABSORBING MATERIALS IN SOLUTION

If absorption of light is the only physical process occurring, then the transmission of light through a homogeneous sample is most commonly modeled by Beer's law[1]:

$$\boldsymbol{A}_{10} = -\log_{10}\frac{I}{I_0} = \kappa ct. \tag{3.1}$$

$\boldsymbol{A}_{10}$ can be called the "absorbance," or the "decadic absorbance," in order to emphasize the use of $\log_{10}$, since the use of the Napierian logarithm in Beer's law is also possible $\left(\boldsymbol{A}_e = -\ln\frac{I}{I_0}\right)$. I_0 is the intensity of light entering the sample and I is the intensity of light transmitted through the sample (illustrated in

* dahm@rowan.edu

1 "Beer's law" has been given many names, such as the "Bouguer–Lambert law," the "Beer–Lambert law," or other various combinations of the names Beer, Bouguer, and Lambert. Names not containing the name Beer, such as "Law of Bouguer" or the Bouguer–Lambert law, are usually associated with formulas that do not contain a concentration term.

FIGURE 3.1 Measurement of transmission (I/I_0) through a nonscattering sample.

Figure 3.1). I/I_0 is thus the fraction of the incident light that is transmitted through the sample, is called the "transmittance" or "transmission fraction," and is commonly given the symbol T.

Thus, Beer's law can also be written as:

$$A_{10} = -\log_{10} T = \kappa ct \text{ or } T = 10^{-\kappa ct}. \tag{3.2}$$

In either Equation 3.1 or 3.2, c is the concentration of the analyte[2] and t is the thickness of the sample.

The symbol κ in Beer's law represents the wavelength-dependent "absorption coefficient." The value of the coefficient κ depends upon the units used to express concentration and sample thickness. In essence, κ represents the inherent ability of the analyte to absorb light at the particular wavelength under consideration. Absorption is understood as occurring at the molecular level, and the ability of an individual molecule to absorb light isn't dependent on macroscopic properties (e.g., size and geometry) of the particle or sample that contains that molecule. Consequently, when we use Beer's law, we often regard the value of κ at a particular wavelength as dependent only upon the identity of the molecule itself and the medium in which it is found. Rigorously, this model is only valid for a dilute solution – that is, a solution in which the analyte (which is the species responsible for the absorption) is present in low concentrations. As shown in Figure 3.2a, in a dilute solution, the analyte can be assumed to interact only with the medium. Figure 3.2b illustrates a more concentrated solution in which some of the analyte particles are close enough together for their intermolecular interactions (e.g., induced dipoles) to be significant. Such variations in molecular environment can result in variations in the ability of the material to absorb light. Consequently, deviations from Beer's law can occur, even for nonscattering samples, at high concentrations of analyte.

In analytical spectroscopy, Beer's law is an equation with great practical appeal. It constitutes a simple linear relationship between a readily measured quantity, A_{10}, and the analyte concentration c. Beer's law is also a very intuitive equation in that we understand absorption as a molecular-level phenomenon, so it makes sense that the fraction of light absorbed is proportional to the number of molecules the light has the opportunity to interact with (which is proportional to both t and c) and is also a direct

[2] The "Beer's law" presented here is for the simple case in which there is only one analyte. In other words, we assume there is only one chemical species present that contributes to absorption and the concentration of that one species is what we are trying to model. If there is more than one analyte that contributes to the absorption, then the absorbance is a linear combination of the absorbance of each: $A_{10} = -\log_{10} T = \kappa_i c_i \, t$.

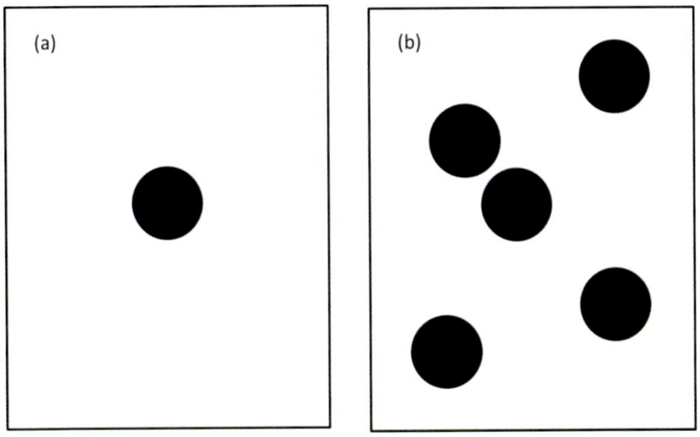

FIGURE 3.2 (a) Dilute solution of analyte and (b) concentrated solution of analyte.

function of how well those molecules absorb the light (measured by κ). Finally, Beer's law illustrates a theme in spectroscopy: A_{10} is called the "absorbance," but it is not an explicit measurement of the amount of light absorbed. Light that is absorbed no longer exists and cannot be detected by spectroscopic instruments. Rather, we infer the amount of absorption from the transmittance T (I/I_0). As absorption increases, the fraction of transmitted T decreases, and consequently, the absorbance, defined as $-\log_{10}(T)$, increases.

The preceding discussion assumes that light simply travels in a straight line as shown in Figure 3.1. Some fraction of it (I/I_0) penetrates the sample and reaches the detector, and the rest is assumed to have been absorbed. In diffuse reflectance spectroscopy, we must also account for processes that cause the light to change direction.

3.2 SCATTERING SAMPLES

Figure 3.3 illustrates a transmission experiment in which a fraction of light is absorbed by the sample, and a fraction is scattered. The detector pictured in Figure 3.3 will detect neither absorbed nor scattered light; it will only detect the "direct transmission," which is the light that penetrates the sample in a straight line. We call this the "direct transmission" because in discussing scattering samples, we will use the term "transmission" to describe all of the light that penetrates the sample, regardless of the direction of travel.

In principle, we could apply Beer's law:

$$A_{10} = -\log_{10}\frac{I}{I_0} = \varepsilon ct \tag{3.3}$$

To a scattering sample, if I is measured using a small area detector as displayed in Figure 3.3. Here, the attenuation of light represented in the quotient (I/I_0) is caused by both scattering and absorption. Consequently, ε is termed the "extinction coefficient"[3] or "attenuation coefficient." It is analogous to the absorption coefficient κ, but generalized names like "attenuation coefficient" and "extinction coefficient" are used to remind us that absorption is not the only process occurring that leads to attenuation of the beam. In principle, we could also subdivide the extinction coefficient:

$$\varepsilon = \mu_a + \mu_s, \tag{3.4}$$

with μ_a representing the contribution of absorption to attenuation and μ_s representing the contribution of scattering. Note that we could perform a single experiment measuring direct transmission

[3] The term "molar extinction coefficient" is also commonly used when the concentration c is expressed on a molar basis.

through a sample of known concentration c to obtain a value of ε for the sample, but that single experiment cannot distinguish between the amount of attenuation caused by absorption and the amount of attenuation caused by scatter.

At first glance, Beer's law might appear just as useful and practical here as was Beer's law for nonscattering samples: the sample thickness t would typically be known, so if ε is known, then once again we have a simple linear relationship between the absorbance A_{10} and the concentration c. In practice, however, there are limitations to applying Beer's law to scattering samples in this way. It was noted in Section 3.1 that the absorption coefficient κ can be regarded as a characteristic of the material, because absorption is a molecular-level phenomenon. Scattering, by contrast, is understood as a *surface* phenomenon – it occurs at the surfaces of particles. Consequently, μ_s does not only depend upon the specific materials that make up the sample. It also depends upon factors like the size and shape of particles that are not necessarily repeatable from one sample to another, and may not even be consistent within a single sample. Liu et al. (2007) published a study that is instructive because it illustrates both the utility and the limitations of Beer's law for scattering samples. The authors calculated the extinction coefficients at 506 nm for gold nanoparticles of different sizes. Plots of A_{10} vs. c were essentially linear for each *particular* nanoparticle core size. The fact that the plots were linear demonstrated that Beer's law was an excellent model, and the authors used the slopes of these plots to determine the values of ε. However, the slope of each plot was different, meaning that there was no unique "extinction coefficient for gold at 506 nm"; the observed ε (and by extension μ_s) was a function of the size of the nanoparticles.

In sum, mathematically speaking, Beer's law is readily solved for c if ε is known, but practically speaking, "knowing" the value of ε can be a more complicated concept for a scattering sample than "knowing" the value of κ for a nonscattering sample.

3.3 DIFFUSE TRANSMITTANCE AND DIFFUSE REFLECTANCE

In practice, there are many samples in which either the scattering or the absorption (or both) is so strong that there is essentially no "direct transmission"; all of the incident light is either absorbed or scattered at least once. In such a case, the small area detector pictured in Figure 3.3 would produce no significant reading, but some diffuse, scattered light may emerge from either or both of the

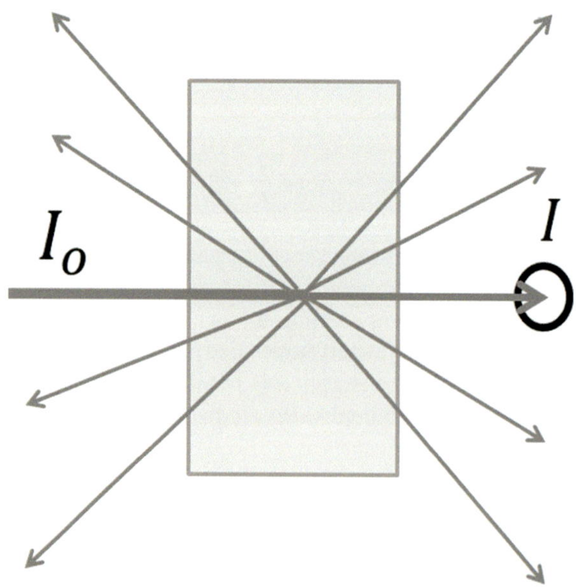

FIGURE 3.3 Measurement of direct transmission through a scattering sample.

"front" and "back" of the sample. In such a case, we can use hemispherical detectors, as illustrated in Figure 3.4, to measure the transmission T (defined as the fraction of the entering light that penetrates the sample) and the remission R (defined as the fraction of the entering light that changes direction and re-emerges from the "front" of the sample). In an ideal experimental arrangement, essentially all light exiting the sample would be detected by either of the two hemispherical detectors, and it could be assumed that the balance of the entering light was absorbed by the sample. Thus, A is defined as the fraction of entering light absorbed by the sample:

$$A = 1 - R - T. \tag{3.5}$$

An important special case is the "infinitely thick sample." This refers to a sample that no measurable amount of light penetrates (either directly or diffusely). Thus, in a transmission experiment, the sample would be indistinguishable from an infinitely thick layer. Mathematically, for an infinitely thick sample:

$$T = 0 \text{ and } A = 1 - R. \tag{3.6}$$

Various models have been proposed for diffuse reflectance processes of the type illustrated in Figure 3.4, and will be summarized throughout the remainder of this chapter. Unfortunately, there is no single modeling approach that is as uniformly applicable or as straightforward to apply as is Beer's law for nonscattering samples.

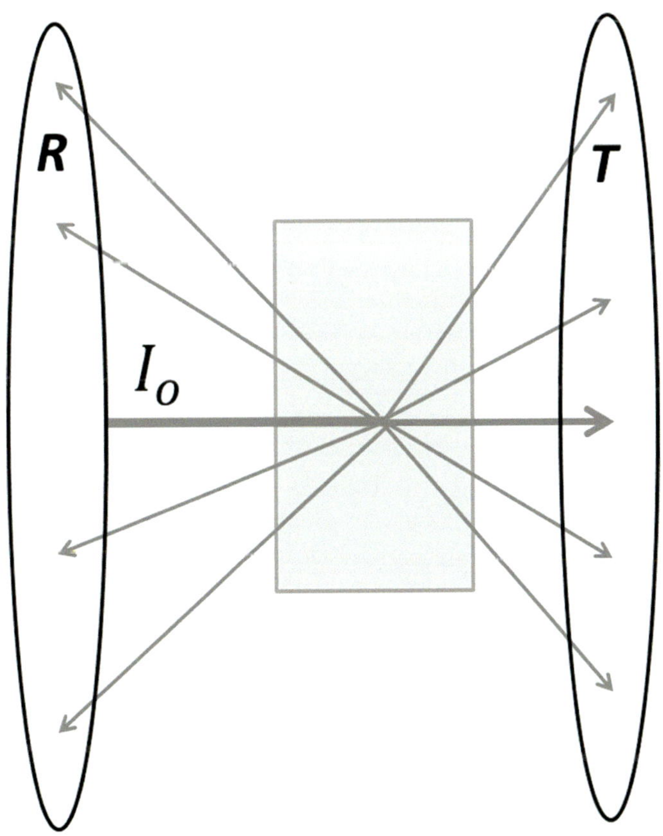

FIGURE 3.4 Measurement of remission from and transmission through a scattering sample.

3.4 SCHUSTER'S MODEL

Schuster published a classic paper entitled "Radiation Through a Foggy Atmosphere," in which he proposed a model for absorption and scattering of light in a mass of gas that "holds fine particles of matter in suspension" (Schuster, 1905). His modeling approach employed several assumptions and approximations that are instructive:

- *Division of the sample into plane parallel "layers."* Schuster subdivided the gaseous medium into distinct identical parallel layers, each of which had thickness "*dx*." Each layer has a planar front surface that is perpendicular to the incident beam, as illustrated in Figure 3.5.
- *The two-flux model.* Schuster used the approximation that light travels in two directions: forward (meaning in the direction of the incident beam) and backward. He did not distinguish between the various angles at which light could strike a layer's front surface. All of the light represented by the various solid arrows as shown in Figure 3.5 would simply be lumped into a flux that exits the first layer in a "forward" direction and reaches the front of the second layer. Similarly, all of the light represented by the dashed arrows is light that is "scattered backward" by the front layer and would contribute to the measured remission from the sample. Schuster himself acknowledged the physical

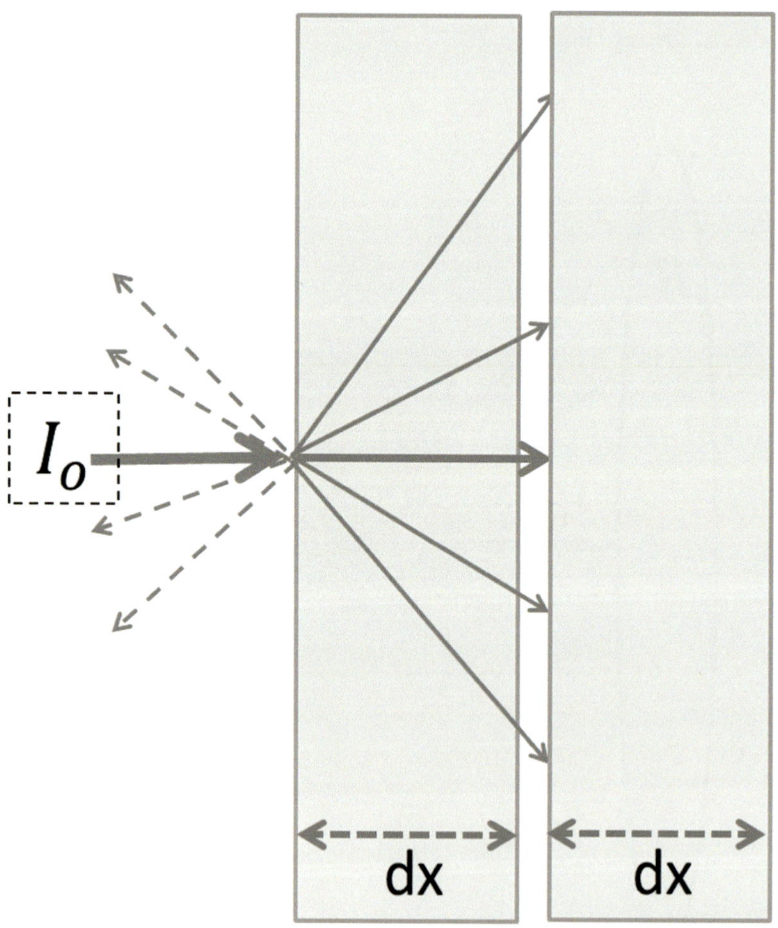

FIGURE 3.5 Scattering of light by the first plane parallel layer in a sample. Light scattered "forward" reaches the second layer, and light scattered "backward" is remitted from the sample.

limitation of this approximation, as in reality, light striking a layer at different angles will penetrate the layer along different path lengths, and therefore have different degrees of opportunity to be absorbed.[4]

- *Continuous mathematics.* Schuster defined k as the "coefficient of absorption," s as the "coefficient of scatter," and S as the "total intensity of radiation incident on a layer of small thickness dx."[5] Consequently, the fraction of the entering light that is absorbed by an individual layer is $kSdx$ and the fraction of the entering light scattered by the layer is $sSdx$. The solution strategy then involves integrating over the sample thickness to solve for the "forward" and "backward" light fluxes as functions of x. However, the use of operations like differentiation and integration inherently relies upon the premise that the quantities being modeled are continuous functions. The implications of the use of continuous mathematics are explored in more detail in Section 3.6.
- *Isotropic scatter.* "Isotropic" means that scattered light is scattered with equal intensity at all angles. In the context of a two-flux model, this simply means that of the light scattered by a layer, $sSdx$, "one-half is sent forward and one-half returned backward" (Schuster, 1905).

Schuster produced a generalized equation describing the extent of absorption and scattering for a sample, and also explored a number of simplified special cases. One special case resulting from the Schuster treatment that is of interest for diffuse reflectance spectroscopy is the "infinitely thick" sample, in which (Dahm and Griffiths, 2007):

$$\frac{(1-R_\infty)^2}{2R_\infty} = \frac{k}{s}. \tag{3.7}$$

R_∞ can be understood as the fraction of incident light that is remitted by the infinitely thick sample. The left-hand side is a function of R_∞ that is often given the symbol $f(R_\infty)$ and has been termed the "remission function" or the "Kubelka–Munk function." The contributions of Kubelka and Munk are discussed in the next section.

3.5 KUBELKA–MUNK THEORY

Kubelka and Munk (1931) conducted a theoretical investigation of the optical behavior of paint layers. While this was a very different physical system than that examined by Schuster, they too applied a two-flux model and continuous mathematics, and obtained a very similar result. In 1948, Kubelka published a paper that sought to generalize the mathematical approach. Instead of simply using

[4] The exact quote from Schuster's paper is as follows: "The equations of the last paragraph have been deduced under the assumption that the radiation throughout the absorbing mass is uniformly distributed in such a way that it does not depend on the angle between any direction considered and the normal drawn toward the same side. This supposition is obviously incorrect, for it appears that, even if it were to hold at any surface, for example, the first surface of the layer dx, absorption in that layer would destroy the uniformity owing to the greater absorption which the oblique rays suffer. To some extent the effect of scattering would act in the sense of partly restoring the equality of distribution…" (Schuster, 1905)

[5] While Schuster used the terms "coefficient of absorption" and "coefficient of scatter," Schuster's k is not functionally identical to the absorption coefficient κ as it was defined in the section on Beer's law:

$$A_{10} = -\log_{10} T = \kappa cl$$

A_{10} is inherently a dimensionless quantity. Consequently, κ in Beer's law has dimensions of both length^{-1} and concentration^{-1}, with the specific units depending upon whether c is expressed on a molar or mass basis. By contrast, concentration is not explicitly represented in Schuster's treatment. His expression for absorption by a layer is $kSdx$, with dx having the dimension of length and k having the dimension of length^{-1}. In effect, an absorption coefficient defined as in Beer's law and the concentration of the absorber are lumped together into what Schuster called "k." Because k and s appear only in the ratio k/s, the particular units used cancel and aren't important in this specific equation. But we will see other expressions in which the individual magnitudes of the absorption and scattering coefficients are important, as well as their ratio, and in this case, care must be taken with how a "coefficient" is defined.

two fluxes ("toward" and "away from" the illuminated surface of the sample), Kubelka assumed the radiation was perfectly diffuse: that radiation entering a layer (in either direction) would enter with equal frequency at all possible angles of incidence. In the original Kubelka and Munk's (1931) paper, the following relationship was reported for the remission from an infinitely thick sample:

$$f(R_\infty)=\frac{(1-R_\infty)^2}{2R_\infty}=\frac{s}{r}, \tag{3.8}$$

where s represents the "absorption constant" and r represents the "scattering constant."[6] In Kubelka's (1948) more general paper, he stated that this equation constitutes an *exact* solution of the differential equations in only two circumstances: perfectly diffuse radiation and radiation entering the layer at an angle of 60° to the surface of the layer (because a beam travelling at this angle experiences a path length that is representative of the average of all path lengths.)

Like Schuster, Kubelka and Munk arrived at an equation that states the remission function $f(R_\infty)$ from an infinitely thick sample is proportional to the ratio of the absorption coefficient to the scattering coefficient, but not dependent upon their absolute values.

3.6 CONTINUOUS AND DISCONTINUOUS MODELS

Beer's law, the Schuster model, and the Kubelka–Munk model are all "continuum" models; they were all derived using continuous mathematics. A physical implication of the continuum approach can be seen by examining Schuster's expressions for absorption and scattering within a layer of thickness dx: absorption = $kSdx$ and scatter = $sSdx$. The expressions assume that the quantities of absorption and scattering are both directly proportional to the intensity S of light that arrives at the surface of the layer. Thus, embedded in the math are the fundamental assumptions that (1) all of the light represented by S has the opportunity to be either absorbed or scattered and (2) the quantities of absorption and scatter by this individual layer do not depend upon each other. In effect, the sample is treated as a homogeneous continuum, in which either absorption or scattering can occur at any location.

There are practical physical situations for which a continuum constitutes a very reasonable model. However, there are also theoretical limitations to such a model. As a physical illustration, consider a suspension of sand particles in water. Such a sample is not homogeneous: there are locations where the sample is composed of water, there are locations that can be described as the interior of a sand particle, and there are interfaces between water and sand where the properties of the sample change, not continuously, but abruptly. It is in fact the discontinuities at the surfaces that give rise to the phenomenon of scattering, as discussed at length in Chapter 2. As a mathematical illustration, consider the layer as shown in Figure 3.6. A fraction of the entering light R is remitted, and a fraction of T is transmitted. But if remission is caused by scattering that is assumed to occur at the surface of the layer, and absorption is assumed to occur at the molecular level as the sample penetrates the layer, then the fraction R has *no opportunity to be absorbed by that layer.* The fraction of absorption by the layer thus depends upon the fraction R, a fact that the Schuster expression $kSdx$ does not capture.

Further, the use of continuous mathematics inherently relies on the premise that the mathematical relationships between variables are continuous. If I represents the forward (or backward) flux of light through a sample, a continuum model treats I as a continuous function of position x. But if a

[6] The equation Kubelka and Munk actually published in 1931 was $H_\infty = 1+\frac{s}{r}-\sqrt{\frac{s^2}{r^2}+2\frac{s}{r}}$. If this equation is rearranged so that it is explicit in s/r, the result is Equation 3.8, and the left-hand side later became known as the "Kubelka–Munk function." Note that the original equation used the symbol H, representing the German word "helligkeit." The symbol R, for remission, has long been used when the Kubelka–Munk equation is presented in an English language context, including by Kubelka himself in his 1948 paper.

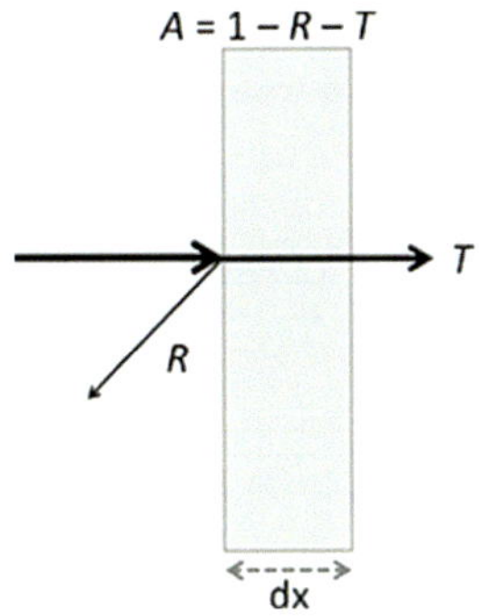

FIGURE 3.6 Interaction of light with a plane parallel layer of thickness dx; a fraction of the incident radiation is remitted (R), a fraction transmitted (T), and a fraction absorbed (A).

significant fraction of incident light is scattered at a single surface (e.g., if $R = 0.25$ in Figure 3.6), then I cannot plausibly be modeled as continuous; it undergoes a step change at each surface.

Consequently, we will now explore alternative model-building approaches that rely upon discontinuous mathematics.

3.7 THE EQUATIONS OF BENFORD

A publication by Benford (1946) presented a modeling framework that is applicable to diffuse reflectance. Here again, a sample is envisioned as divided into plane parallel layers, and a two-flux approximation is used, but discontinuous mathematics is employed. For each layer, a fraction of light T is assumed to be transmitted through the layer, a fraction A is assumed to be absorbed by the layer, and a fraction R is assumed to be the reverse direction. Benford assumed that A, R, and T are known for each individual layer, and examined the outcome when multiple layers are combined into a sample. For example, if a sample is composed of two layers x and y (see Figure 3.7) that have transmission fractions T_x and T_y, then the fraction of the incident beam that is simply transmitted through both layers is given by T_xT_y. This, however, is not the complete observed transmission for the sample since another fraction of light will be:

- Transmitted through layer x;
- Remitted by layer y, causing it to return "backward" to layer x;

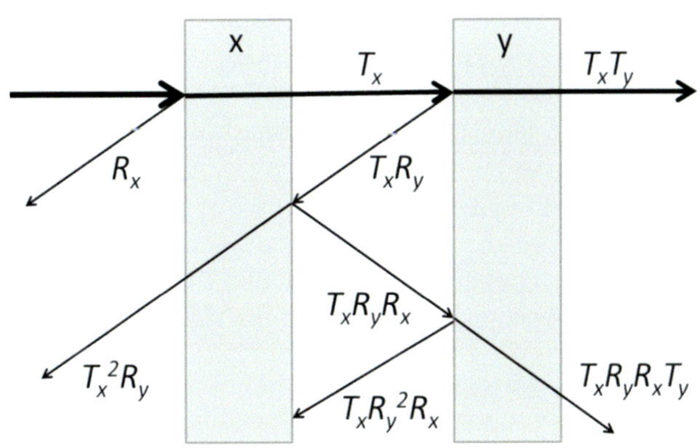

FIGURE 3.7 Benford's two-flux model, applied to the layers x and y.

- Remitted by layer x, causing it to return "forward" to layer y;
- Transmitted through layer y, exiting the sample as "transmission."

The fraction of light that undergoes this specific sequence is $T_x R_y R_x T_y$, but it is also possible for the remission back and forth between layers x and y to continue indefinitely. Benford used infinite series summations to produce the following expressions:

$$T_{x+y} = \frac{T_x T_y}{1 - R_x R_y} \tag{3.9}$$

$$R_{x+y} = R_x + \frac{T_x^2 R_y}{1 - R_x R_y} \tag{3.10}$$

$$A_{x+y} = 1 - T_{x+y} - R_{x+y}, \tag{3.11}$$

with A_{x+y}, R_{x+y}, and T_{x+y} representing the absorption, remission, and transmission observed for the two-layer sample as a whole. Note that Benford's treatment assumes that light is effected the same way by a layer regardless of the direction of travel: A_x, R_x, and T_x are assumed to be the same whether the light is moving "forward" or "backward" when it encounters layer x.

The equations above do not require that layers x and y be identical to each other. However, if a sample is composed of a uniformly distributed material, then it can be modelled as a series of identical layers. For this case, Benford derived the following expressions. If A_d, R_d, and T_d are known for a uniformly distributed sample of thickness d, then for a sample of the same uniformly distributed material of thickness $2d$, the absorption, remission, and transmission are given by:

$$T_{2d} = \frac{T_d^2}{1 - R_d^2} \tag{3.12}$$

$$R_{2d} = R_d \left(1 + T_{2d}\right) \tag{3.13}$$

$$A_{2d} = 1 - T_{2d} - R_{2d}. \tag{3.14}$$

And for a sample of the same uniformly distributed material of thickness $d/2$:

$$R_{d/2} = \frac{R_d}{1 + T_d} \tag{3.15}$$

$$T_{d/2} = \left[T_d \left(1 - R_{d/2}^2\right)\right]^{0.5} \tag{3.16}$$

$$A_{d/2} = 1 - T_{d/2} - R_{d/2}. \tag{3.17}$$

The above formulas can be used to predict A, R, and T for any thickness of a material from any other thickness. For example, if one knows A, R, and T for a sample of thickness d, then by iterative application of the "$2d$" formulas, one can calculate A, R, and T for a sample of thickness $2d$, $4d$, $8d$ etc. Table 3.1 illustrates the results of this calculation for two cases: $R_d = 0.001$ and $A_d = 0.0005$, and $R_d = 0.002$ and $A_d = 0.001$.

The results shown illustrate a few points:

- In both cases, the light transmitted through a sample of thickness $8192d$ is negligible, so these can be regarded as models for "infinitely thick" samples.

TABLE 3.1
R, T, and A for Hypothetical Samples Composed of Different Numbers of Identical Layers

# Layers	R	T	A	A(R,T)	# Layers	R	T	A	A(R,T)
1	0.0010	0.9985	0.0005	0.9987	1	0.0020	0.9970	0.0010	0.9975
2	0.0020	0.9970	0.0010	0.9987	2	0.0040	0.9940	0.0020	0.9975
4	0.0040	0.9940	0.0020	0.9987	4	0.0079	0.9881	0.0040	0.9975
8	0.0079	0.9881	0.0040	0.9987	8	0.0157	0.9764	0.0080	0.9975
16	0.0156	0.9764	0.0080	0.9987	16	0.0306	0.9535	0.0159	0.9975
32	0.0306	0.9536	0.0159	0.9987	32	0.0585	0.9100	0.0315	0.9975
64	0.0584	0.9101	0.0315	0.9987	64	0.1071	0.8310	0.0619	0.9975
128	0.1069	0.8312	0.0619	0.9987	128	0.1818	0.6986	0.1196	0.9975
256	0.1816	0.6988	0.1196	0.9987	256	0.2736	0.5047	0.2217	0.9975
512	0.2733	0.5050	0.2217	0.9987	512	0.3489	0.2753	0.3757	0.9975
1024	0.3487	0.2756	0.3757	0.9987	1024	0.3791	0.0863	0.5346	0.9975
2048	0.3788	0.0865	0.5347	0.9987	2048	0.3824	0.0087	0.6089	0.9975
4096	0.3821	0.0087	0.6091	0.9987	4096	0.3824	0.0001	0.6175	0.9975
8192	0.3822	8.93×10^{-5}	0.6177	0.9987	8192	0.3824	9.205×10^{-9}	0.6176	0.9975

- In both cases, $R > A$ for a single layer of thickness d. But for thicker samples (>1000d), the absorption fraction is actually larger than the remission fraction for the sample as a whole. This observation is a result of multiple interactions: light that is transmitted through several layers prior to reversing direction will be attenuated by absorption both while it is traveling "forward" and again while it is traveling "backward" through these layers.
- In both cases, the ratio of R_d to A_d is the same, that is, 2:1. The remission fraction R for an "infinitely thick" sample is the same (0.382) to three significant figures in both of the cases as shown in Table 3.1. This result is reminiscent of the outcome cited previously for the continuous models: the remission R_∞, and therefore the Kubelka–Munk function $f(R_\infty)$, depends upon the ratio of the absorption and scattering coefficients, but not upon their absolute values. When the probabilities of absorption and scattering *for a single interaction* are very small ($T > 0.99$), modeling the system as a homogeneous continuum makes good physical sense and the continuous and discontinuous approaches provide compatible results. However, Table 3.2 illustrates a case in which $R_d = 0.2$ and $A_d = 0.1$. Here, the ratio of R_d to A_d is again 2:1, but in this case, R_∞ for the "infinitely thick" sample is measurably

TABLE 3.2
R, T, and A for Hypothetical Samples Composed of Different Numbers of Identical Layers, Each Layer Having a Remission Fraction of 20% and an Absorption Fraction of 10%

# Layers	R	T	A	A(R,T)
1	0.2	0.7	0.1	0.750
2	0.3021	0.5104	0.1875	0.750
4	0.3887	0.2867	0.3246	0.750
8	0.4263	0.0968	0.4769	0.750
16	0.4312	0.0115	0.5573	0.750
32	0.4313	0.0002	0.5686	0.750
64	0.4313	3.192×10^{-8}	0.5687	0.750

> different from its value for the cases in Table 3.1. When there is a significant probability of absorption or scattering from a single interaction, the continuum model does not well apply and the absolute values of R and A do influence R_∞.

Also shown in Tables 3.1 and 3.2 is the function *A(R, T)*, which we term the "absorption/remission function" (or ART function) and which is defined as:

$$A(R,T)=\frac{(1-R)^2-T^2}{R}. \tag{3.18}$$

As Tables 3.1 and 3.2 demonstrate, the value of this function is independent of thickness for each sample. *A(R, T)* depends upon the values of the absorption, remission, and transmission (A, R, and T) fractions for a single layer, but is independent of the number of layers in the sample. This observation is explored further in the next section.

3.8 THE DAHM EQUATION

Tables 3.1 and 3.2 show that the absorption/remission function *A(R, T)* is constant for any thickness of a uniformly distributed material. The Dahm equation states specifically that this function is equal to:

$$A(R,T)=\frac{(1-R)^2-T^2}{R}=\frac{A(1+T-R)}{R}, \tag{3.19}$$

with A, R, and T representing the fractions of incident light that undergo absorption, remission, and transmission. Note that for the special case of an "infinitely thick" sample, T~0, the absorption/remission function simplifies to:

$$A(R,T)=\frac{(1-R)^2}{R}. \tag{3.20}$$

Thus for the special case of an infinitely thick sample, the absorption/remission function is proportional to the Kubelka–Munk function $f(R_\infty)$: they are identical aside from a factor of 2 that is typically included in the denominator of the Kubelka–Munk function.

Note, further, that for a thin layer of a material[7] in which the probabilities of absorption and remission are both very small, T~1 and R~0. In this circumstance, the right-hand side of the Dahm equation simplifies to:

$$\frac{A(1+T-R)}{R}\sim\frac{A(1+1-0)}{R}\sim\frac{2A}{R}. \tag{3.21}$$

This is reminiscent of the results from the Schuster and Kubelka–Munk treatments, in which the Kubelka–Munk function was proportional to the *ratio* of the absorption and remission coefficients, but not dependent upon their absolute values. Here again, we see that a discontinuous approach gives results that are compatible with the continuous approaches for the special case in which T~1 for a single interaction.

---

[7] It might seem odd that in discussing the left-hand side of the Dahm equation, we make reference to an infinitely thick sample, while in discussing the right-hand side we make reference to a "thin" layer that has T~1. However, this is justifiable because the absorption/remission function is constant with respect to thickness, and it is the simplest way to illustrate the physical situation in which the Dahm equation produces results that are compatible with the Schuster and Kubelka–Munk results.

The Dahm equation was derived empirically using the results from Benford's treatment, such as the example cases illustrated in Tables 3.1 and 3.2. Thus, Benford's equations represent a practical framework and strategy for modeling diffuse reflectance with discontinuous mathematics. However, the use of Benford's equations relies on A, R, and T for a "layer" being known. Most practical samples are not literally divided into discrete, identical layers. Consequently, the concept of a "representative layer" was devised to facilitate the application of a discontinuous approach like Benford's to particulate samples.

3.9 THE REPRESENTATIVE LAYER

Figure 3.8a illustrates a particulate sample in which there are two distinct materials (represented as white and black spheres) and two different particle sizes for each material. Figure 3.8b illustrates three "representative layers" of the sample. The idea behind the representative layer theory is that a sample can, for spectroscopic purposes, be modeled as if it were comprised of a series of distinct layers, each of which is representative of the sample as a whole.

To be representative, the layer must meet the following criteria (Dahm and Dahm, 2007):

- The volume fraction of each type of particle is the same in the representative layer as in the sample as a whole.
- The surface area fraction of each type of particle is the same in the representative layer as in the sample as a whole.
- The void fraction of the representative layer is the same as in the sample.
- The representative layer is nowhere more than one particle thick. Note this means the "thickness" of the representative layer is not uniform, as illustrated in Figure 3.8b. This criterion is imposed so that we can assume that a given photon of light has only one interaction with the layer. It might be transmitted, remitted, or absorbed as a result of this interaction, but it is assumed not to interact with a second particle within the same layer.

In the above discussion, when we talk about a "type" of particle, we must clearly distinguish between particles of different composition. In addition, however, we must distinguish between particles of different sizes. Recall that scattering is envisioned as a surface phenomenon and absorption is envisioned as occurring at the molecular level throughout the particle. Consequently, our expectation is that the contribution of a "type" of particle to absorption will be proportional to the volume

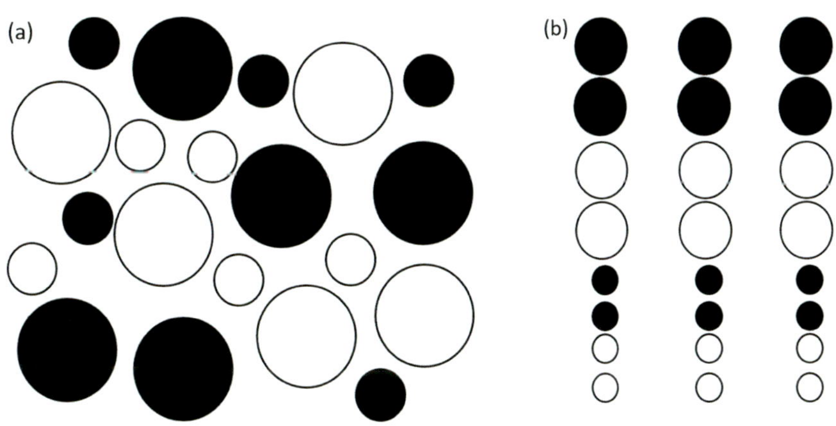

FIGURE 3.8 (a) A particulate sample that contains particles of two different compositions, each in two different sizes. (b) A series of layers that are regarded as "representative layers" for this sample.

fraction of that particle in the sample, and the contribution of a "type" of particle to scattering will be proportional to the surface area fraction of that particle in the sample. This is why our "representative layer" criteria above incorporate both volume fraction and surface area fraction. Since small particles have larger surface area-to-volume ratios than large particles, it is necessary to distinguish between them. Thus, the sample in Figure 3.8 has in total four "types" of particles: large and small particles with two different compositions.

Under these criteria, we can propose a model for the fractions of incident light that are absorbed (A_l), remitted (R_l), and transmitted (T_l) by one representative layer.

$$A_l = \sum S_j\left(1 - e^{-K_j d_j}\right) \tag{3.22}$$

$$R_l = \sum S_j b_j d_j \tag{3.23}$$

$$T_l = 1 - A_l - R_l, \tag{3.24}$$

in which:

- S_j is the fraction of cross-sectional surface area that is occupied by particles of type j.
- K_j is the effective absorption coefficient for particles of type j.
- b_j is the remission coefficient for particles of type j.
- d_j is the thickness of a particle of type j in the direction of the incident beam.
- The summation is carried out over all of the distinct "types" of particle.

In effect, S_i represents the fraction of light that will interact with a particle of type j, and K_j and b_j quantify the likelihood of that interaction resulting in absorption and remission, respectively. Surface area fractions and volume fractions for each type of particle can be defined as follows:

$$v_i = \frac{\dfrac{w_i}{\rho_i}}{\sum \dfrac{w_j}{\rho_j}} \tag{3.25}$$

$$s_i = \frac{\dfrac{w_i}{\rho_i d_i}}{\sum \dfrac{w_j}{\rho_j d_j}} \tag{3.26}$$

$$V_i = (1 - v_0)v_i \tag{3.27}$$

$$S_i = (1 - v_0)s_i, \tag{3.28}$$

in which:

- w_i is the mass fraction of particles of type i in the sample.
- v_i is the fraction of *occupied* volume composed of particles of type i.
- s_i is the fraction of particle surface area that is comprised of particles of type i.
- V_i is the fraction of *total* volume composed of particles of type i.
- S_i is the fraction of cross-sectional surface area that is comprised of particles of type i.
- ρ_i is the density of particles of type i.
- v_0 is the void fraction of the sample.

This section presents a logical way of relating the spectroscopic behavior of a "representative layer" to the properties of the individual particles that make up the layer. The values of the absorption and remission coefficients represent a challenge in this modeling approach. Absorption is calculated from the fraction of light striking each type of particle and a "Beer's law"-type calculation of the absorption by each type of particle, so the values of K_i used should ideally model the ability of the particle to absorb light, independent of other processes (scattering, remission) that also occur. However, in the context of a multilayered sample, absorption and remission do interact and influence each other, as Tables 3.1 and 3.2 illustrate. The next section describes a strategy (Dahm 2013) for separating the effects of absorption and scatter.

3.10 CONSTRUCTING A HYPOTHETICAL "INFINITESIMAL" SAMPLE

In Section 3.1, covering nonscattering samples and Beer's law, we defined the absorbance as:

$$A_{10} = -\log_{10} T.$$

In this section, we will define "absorbance" this way:

$$A_{10} = -\log_{10}(R+T) = -\log_{10}(1-A). \tag{3.29}$$

The rationale for using $R+T$ is that here we will be trying to quantify absorption specifically, and neither remitted light nor transmitted light has been absorbed. (If we define the "absorbance" for a scattering sample as $-\log_{10} T$, then what we are really quantifying is extinction, not absorption, as discussed in Section 3.2.)

Consider as an example a sample that is 1 cm thick, and an experiment in which 60% of the incident light is absorbed, 37% is remitted, and 3% is transmitted. The absorbance of this sample, calculated directly, is $A_{10} = -\log_{10}(1-0.6) = 0.3979$. Since the thickness t is 1 cm, we could then apply Beer's law as follows:

$$A_{10} = \kappa c t$$

$$0.3979 = \kappa c (1\ \text{cm})$$

$$\kappa c = 0.3979\ \text{cm}^{-1},$$

with the specific value of the absorption coefficient κ depending upon the units used to express concentration c. However, this κ is not necessarily a good representation of the inherent ability of the material to absorb light at this wavelength. The observed absorption fraction of the sample (0.6) is also influenced by scattering, which in turn affects the path lengths traveled by the light, the fraction of light that changes direction (possibly more than once) at different locations inside the sample, etc. But as a sample becomes thinner and thinner, the interdependence between absorption and scatter becomes less significant. In the extreme limit, if a sample was only one particle thick (i.e., the "representative layer" is the entire sample), then the observed absorption and remission would be the result of single interactions with single particles. While it isn't realistic to fabricate an experimental sample that is exactly one particle thick, the equations of Benford give us a mathematical way to take data from a real sample and infer the behavior of an extremely thin sample made of the same material.

Recall that Benford's equations include formulas for calculating A, R, and T for a sample of thickness $d/2$ from the values for a sample of thickness d. Repetitive application of these formulas allows one to calculate A, R, and T for layers of infinitesimal thickness. Table 3.3 shows the results when this process is applied to our example sample with d = 1 cm, $A_d = 0.6$, $R_d = 0.37$, and $T_d = 0.03$.

TABLE 3.3
Use of Benford's Equations to Calculate Absorbance for Different Thicknesses of a Sample

Thickness (cm)	A	R	T	$-\log_{10}(1-A)$
1	0.6	0.37	0.03	0.39794
1/2	0.479133	0.359223	0.161644	0.283273
1/4	0.30842	0.309237	0.382343	0.160158
1/8	0.173627	0.223705	0.602668	0.082824
1/16	0.0917	0.139583	0.768717	0.041771
1/32	0.047052	0.078918	0.874031	0.020931
1/64	0.023822	0.042111	0.934067	0.010471
1/128	0.011984	0.021773	0.966242	0.005236
1/256	0.00601	0.011074	0.982916	0.002618
1/512	0.00301	0.005584	0.991406	0.001309
1/1024	0.001506	0.002804	0.99569	0.000655

Note that Beer's law states that absorbance will be directly proportional to the thickness of the sample. Inspection of Table 3.3 shows that this is not at all true for the sample as a whole. However, for the bottom five rows (d = 1/64 cm to d = 1/1024 cm), absorbance is to an excellent approximation a linear function of thickness. Applying Beer's law to the sample with thickness 1/256 cm gives:

$$A_{10} = \kappa c t$$

$$0.002618 = \kappa c\left(\frac{1}{256}\text{cm}\right)$$

$$\kappa c = 0.6702\ \text{cm}^{-1}.$$

This is a very different value of absorption coefficient than that calculated for the sample as a whole, and one that is expected to represent more nearly the "absorbing power" of the material. Essentially, the same value is obtained from any of the thicknesses in the bottom five rows of Table 3.3; in effect, it appears that any layer with a thickness of 1/64 cm or smaller is thin enough for the effects of absorption and scattering to be considered separated and independent. Note, further, that if a hypothetical sample had $\kappa c = 0.6702\ \text{cm}^{-1}$ and a thickness of 1 cm but no scattering (and therefore no remission), we would expect according to Beer's law that:

$$A_{10} = \left(0.6702\ \text{cm}^{-1}\right)\left(1\ \text{cm}\right) = 0.6702 = -\log_{10}(1-A)$$

$$A = 0.7863.$$

Thus, a hypothetical sample that had the same thickness and the same "absorbing power" as the actual sample, but no scattering/remission, would absorb 78.63% of the incident light, rather than 60%.

There are at least two limitations to this method for separating the effects of absorption and scatter:

- It is only expected to be applicable to samples that are composed of a material that is "uniformly distributed," since that is an inherent assumption in Benford's equations.

- It wouldn't work well for samples in which single particles have a significant absorption and/or remission fraction. Suppose the sample represented by Table 3.3 is made up of individual particles that are approximately 0.5 cm thick. In such a case, our mathematical exercise is of dubious value: it is not realistic to claim that a layer that is 1/64 cm thick could be "representative" of a sample that has particles of 0.5 cm thickness. Further, according to the table, 36% of the incident light is remitted by a layer that is 0.5 cm thick. If a single 0.5 cm particle remits 36% of the incident light, then a large fraction of the incident light never has the opportunity to be absorbed, and truly separating the effects of absorption and scatter is perhaps an unattainable goal.

REFERENCES

F. Benford, "Radiation in a diffusing medium," *Journal of the Optical Society of America*, 36, 524–554 (1946).

D. Dahm and K. Dahm, *Interpreting Diffuse Reflectance and Transmittance*, NIR Publications, Chichester, UK (2007).

D. Dahm and P. Griffiths, "Continuum and discontinuum theories of diffuse reflection," In D.A. Burns, E.W. Ciurczak (eds.) *Handbook of Near Infrared Analysis*, 3rd ed., CRC Press, Boca Raton, FL (2007).

K. Dahm and D. Dahm, "Separating the effect of scatter and absorption using the representative layer," *Journal of Near Infrared Spectroscopy*, 21, 5 (2013).

P. Kubelka and F. Munk, "Ein Beitrag zur Optik der Farbanstriche," *Z.Tech. Phys. (Leipzig)*, 12 (1931).

P. Kubelka, "New contributions to the optics of intensely light-scattering materials, Part I," *Journal of the Optical Society of America*, 38, 5 (1948).

X. Liu, M. Atwater, J. Wang, and Q. Huo, "Extinction coefficient of gold nanoparticles with different sizes and different capping ligands," *Colloids and Surfaces B: Biointerfaces*, 58, 3–7 (2007).

A. Schuster, "Radiation through a foggy atmosphere," *The Astrophysical Journal*, 21, 1–22 (1905).

4 Spectral Interpretation

*Jerome Workman, Jr.**
MJH Life Sciences

CONTENTS

4.1 INTRODUCTION

Near-infrared (NIR) energy is an electromagnetic energy at a frequency of intramolecular and intermolecular vibrations. The energy band for the NIR is defined as 12,821 to 4000 cm^{-1} (780–2500 nm). The classical mid-infrared spectral range is 4000 to 400 cm^{-1} (2500–25,000 nm); the millimeter, far-infrared, or terahertz region is 400 to 10 cm^{-1} (25,000–1,000,000 nm or 0.025–1.0 mm). The portion of the visible spectrum for which the human eye is sensitive is defined as 360–780 nm. The information within Table 4.1 characterizes the spectral regions just described. The table shows the molecular interactions associated with the energy frequencies (or corresponding wavelengths) of the various regions.

Molecular spectra result from the vibrational modes (or motions) of atomic nuclei within their respective molecule type. These nuclei move together or apart in different geometric configurations, such as along a straight-line vector, rotation, and vibration. The atomic bonds vibrate by stretching,

* jworkman2002@yahoo.com

TABLE 4.1
Visible and Infrared Spectroscopic Regions

Spectral Region	Wavenumbers/ (Wavelength)	Molecular Characteristic Measured
Visible	27,778 to 12,821 cm^{-1} (360–780 nm)	Electronic transitions: color measurements
Near-infrared (NIR)	14,493 to 3333 cm^{-1} (690–3000 nm) or 12,821 to 4000 cm^{-1} (780–2500 nm)[a]	Overtone and combination bands of fundamental molecular vibrations, especially stretching and bending, as well as some deformation
Infrared (IR)	4000 to 400 cm^{-1} (2500–25,000 nm)	Fundamental molecular vibrations: stretching, bending, wagging, scissoring
Far-infrared (FIR or terahertz)	400 to 10 cm^{-1} (2.5×10^4 to 10^6 nm)	Molecular rotation, penetrates fabric, polymer sheets, and paper

[a] Official ASTM International Definition.

bending, and wagging relative to their centers of gravity. These vibrations within and between molecules exhibit a measurable spectroscopic activity that may be recorded using a spectrophotometer. The measured spectra from these instruments are highly structured and complex.

The energy level in a molecule is described as the sum of the atomic and molecular motions due to translational, rotational, vibrational, and electronic energies. Frequency of any vibration is not dependent on the amplitude of the molecular motion. The primary vibrational frequencies for any molecular bond are called the "fundamental frequencies of the molecule." The symmetry of the molecule is the most important factor in determining both the frequency and the amplitude of a molecular vibration.

4.2 SPECTRAL RESPONSE AND CONCENTRATION OF MOLECULES

The type of molecular vibration determines the frequency at which the vibration absorbs NIR energy. The amplitude of the absorption at any particular wavelength or wavenumber is determined by the molecular vibration absorptivity and the number of molecules (or concentration) of them within the spectrophotometer light beam. An assumption is made that a change in the spectrophotometer response is directly related to the concentration of molecules. This response relationship is defined using the Bouguer, Lambert, and Beer relationship most often termed "Beer's law." This Beer's law relationship is defined as absorbance (A, Au, or signal strength) of an analyte being measured using a spectrophotometer is equivalent to the product of the absorptivity (ε) of a specific type of molecular vibration, the concentration (c) of the molecules in the measurement beam, and the pathlength (l) of the sample holder within the measurement beam. This relationship between the measured spectral signal and concentration of a molecule may be expressed as:

$$A = \varepsilon c l, \tag{4.1}$$

where ε is the absorptivity (referred to as "molar extinction coefficient" by earlier physicists) in units of liter $mole^{-1} cm^{-1}$, c is the concentration of molecules in the spectrometer beam in units of mole $liter^{-1}$, and pathlength is the thickness in units of cm of the measured sample at a specific concentration. The absorptivity for any specific molecule type is calculated by careful measurements of the absorbance of a compound, generally diluted in a suitable organic solvent, and by applying the relationship:

$$\varepsilon = \frac{A}{cl}. \tag{4.2}$$

Note that for transmittance spectroscopy, where $T = 0.0–1.0$, and transmission spectroscopy, where $T = 0\%–100.0\%$, a more complete description of the various terms in Beer's law is expressed as:

$$T = \frac{I}{I_0} = 10^{-\varepsilon cl} \Rightarrow \text{Abs.} = A = -\log_{10}\left(\frac{I}{I_0}\right) = -\log_{10} T = \varepsilon cl. \tag{4.3}$$

In the expression found in (4.3), the symbols I and I_0 represent the attenuated energy detected after sample interaction and the initial energy incident to the sample, respectively. For reflectance spectroscopy, where $R = 0.0–1.0$, and reflection spectroscopy, where $R = 0.0–100.0\%$, the various relationships may be demonstrated with the following:

$$R = \frac{I}{I_0} = 10^{-\varepsilon cl} \Rightarrow \text{Abs.} = A = -\log_{10}\left(\frac{I}{I_0}\right) = -\log_{10} R = \varepsilon cl. \tag{4.4}$$

Noting the relationship exists where the change in intensity (I) of the transmitted or reflected light from a sample is a function of the change in pathlength (l) of the sample as expressed by the absorptivity (ε) of a specific analyte (or molecular type) and its concentration (c) by:

$$-\frac{\delta I}{\delta l} = \varepsilon cI \cdot (\ln 10) \tag{4.5}$$

Modern spectrophotometers with their specific software utilize these assumptions for making spectroscopic measurement and generally display spectroscopic data as transmission (T), reflection (R), and absorbance A (y-axis or ordinate axis) versus wavelength (nm, microns) or wavenumber (cm^{-1}) (as x-axis or abscissa axis).

4.3 THE NOMENCLATURE OF MOLECULAR BAND VIBRATIONS

There are multiple types of molecular vibrations that absorb at unique wavelengths or frequencies of the NIR energy depending upon the bond or vibration type. The various types of molecular vibrations that are active within the NIR region are illustrated using the representative figures. Each type of molecular vibration has a unique frequency or set of frequencies where NIR energy absorption occurs. The location of these frequencies and the associated molecular structures (spectra–structure correlations) are given in Tables 4.3, 4.6, and 4.7 and Figure 4.19. For Figures 4.1–4.18, the (A) and (B) figures show different illustrations for the same vibrational type.

4.3.1 Stretching Molecular Vibrations

Standard nomenclature for stretching vibrations indicates that this type of molecular vibration is defined as a variation of the *length* of the bond, and is indicated with the symbol, ν (Gr: nu), with a subscript that indicates the molecular oscillating group, such as:

$$\text{Example of symbols: stretching of C=O} \rightarrow \nu_{C=O}.$$

If the group consists of three atoms, instead of two, it can have two types of stretching: asymmetric stretching and symmetric stretching. In order to distinguish the two types of stretching, a superscript is introduced such as "*s*" (for the symmetrical stretching) and "*a*" (for the asymmetrical stretching).

4.3.1.1 Methyl Symmetrical Stretching of $CH_3 \rightarrow \nu^s_{CH_3}$

Hydrogen and other attached atoms are moving toward or away from the carbon atom in a similar external direction as $\mathbf{CH_3} \rightarrow \nu_s$, as shown in Figures 4.1 and 4.2.

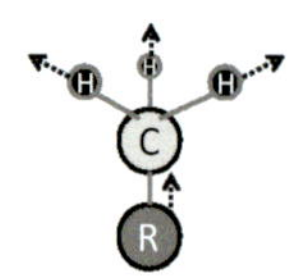

FIGURE 4.1 Methyl symmetrical stretching (A).

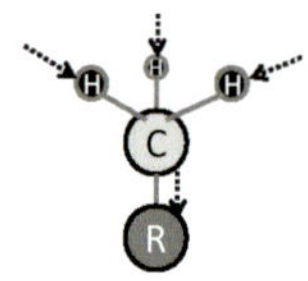

FIGURE 4.2 Methyl symmetrical stretching (B).

4.3.1.2 Methyl Asymmetric Stretching of the $CH_3 \rightarrow \nu^{a}_{CH_3}$

Hydrogen and other attached atoms are moving in opposite directions with respect to the carbon atom as $\mathbf{CH_3} \rightarrow \nu_a$, as shown in Figures 4.3 and 4.4.

4.3.1.3 Methylene Symmetrical Stretching of $CH_2 \rightarrow \nu^{s}_{CH_2}$

In this movement, the two hydrogen atoms move away from the carbon atom at the same time and direction in the outward movement, and approach the carbon atom at the same time and direction in the return movement as $\mathbf{CH_2} \rightarrow \nu_s$, as shown in Figures 4.5 and 4.6.

4.3.1.4 Methylene Asymmetric Stretching of the $CH_2 \rightarrow \nu^{a}_{CH_2}$

In this molecular motion, while one hydrogen atom is moving away from the carbon atom, the other is approaching as $\mathbf{CH_2} \rightarrow \nu_a$, as shown in Figures 4.7 and 4.8.

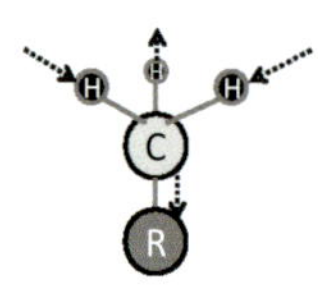

FIGURE 4.3 Methyl asymmetrical stretching (A).

FIGURE 4.4 Methyl asymmetrical stretching (B).

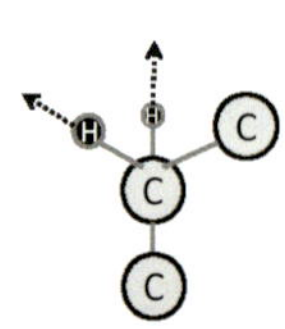

FIGURE 4.5 Methylene symmetrical stretching (A).

FIGURE 4.6 Methylene symmetrical stretching (B).

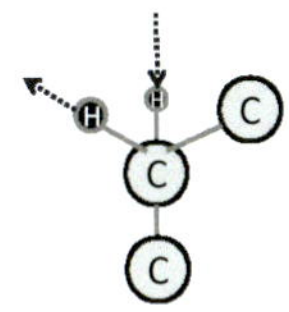

FIGURE 4.7 Methylene asymmetrical stretching (A).

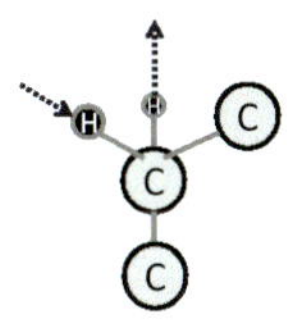

FIGURE 4.8 Methylene asymmetrical stretching (B).

4.3.1.5 Symmetrical Stretching of O-C-H → $\nu^{s}_{\text{O-C-H}}$

In this molecular motion, the two hydrogen atoms move away from the carbon atom at the same time and direction in the outward motion, and approach at the same time and direction in the return motion as **O-C-H** → ν_s. The oxygen atom is moving in a likewise external direction as the hydrogen atoms with respect to the carbon atom. This motion is illustrated in Figures 4.9 and 4.10.

4.3.1.6 Asymmetric Stretching of O-C-H → $\nu^{a}_{\text{O-C-H}}$

In this motion, while one hydrogen moves away from the carbon atom, the other two hydrogen atoms approach the carbon atom. In the second type of asymmetrical motion, the two hydrogen atoms move away from the carbon atom, while the other hydrogen atom moves toward the carbon atom. This is illustrated in Figures 4.11 and 4.12.

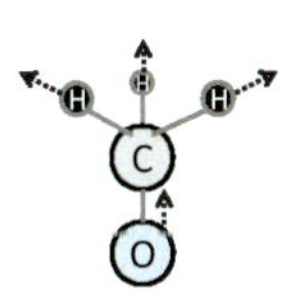

FIGURE 4.9 Symmetrical stretching of O-C-H (A).

FIGURE 4.10 Symmetrical stretching of O-C-H (B).

FIGURE 4.11 Asymmetrical stretching of O-C-H (A).

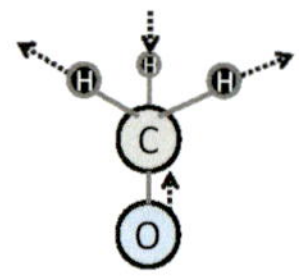

FIGURE 4.12 Asymmetrical stretching of O-C-H (B).

4.3.2 Bending Molecular Vibrations

Bending molecular motion is a variation or change in the angle of a bond occurring on the plane of the bond as indicated with the symbol, **δ**, with a subscript that indicates the oscillating group type. Examples are given in the following text.

4.3.2.1 Methylene Bending of $CH_2 \rightarrow \delta_{CH_2}$

The two methylene hydrogen atoms move toward each other or away from each other simultaneously in a bending motion as $\mathbf{CH_2} \rightarrow \boldsymbol{\delta}$. This is illustrated in Figures 4.13 and 4.14.

Note that if the molecular group consists of four atoms (e.g., one carbon and three hydrogen atoms), instead of three atoms (e.g., one carbon and two hydrogen atoms), two types of bending motions can occur: symmetrical bending and asymmetrical bending. As the standard nomenclature, symmetrical bending is indicated by adding the superscript, *s*; and asymmetric, the superscript, *a*.

4.3.2.2 Methyl Symmetrical Bending of $CH_3 \rightarrow \delta^s{}_{CH_3}$

In symmetric bending, all three angles of the C-H bonds are increased or decreased simultaneously, in an inward or outward motion as $\mathbf{CH_3} \rightarrow \delta_s$. This motion is illustrated in Figures 4.15 and 4.16.

4.3.2.3 Methyl Asymmetric Bending of $CH_3 \rightarrow \delta^a{}_{CH_3}$

In asymmetric bending, one of the three angles is lessened (in the outward movement), and the other two are increased, and vice versa, in the return movement as $\mathbf{CH_3} \rightarrow \delta_a$. Such motion is illustrated in Figures 4.17 and 4.18.

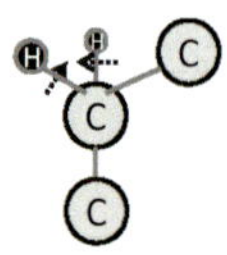

FIGURE 4.13 Methylene bending of CH_2 (A).

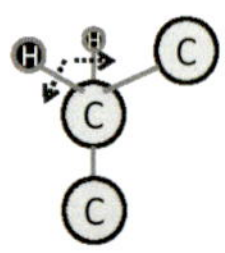

FIGURE 4.14 Methylene bending of CH_2 (B).

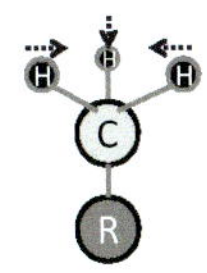

FIGURE 4.15 Methyl symmetrical bending of CH_3 (A).

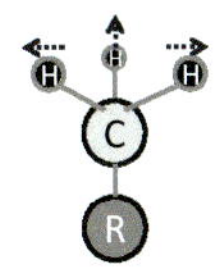

FIGURE 4.16 Methyl symmetrical bending of CH_3 (B).

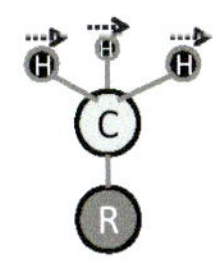

FIGURE 4.17 Methyl asymmetrical bending of CH_3 (A).

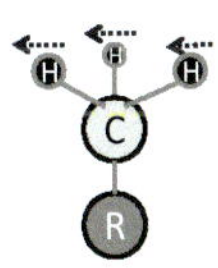

FIGURE 4.18 Methyl asymmetrical bending of CH_3 (B).

4.4 RELATIVE INTENSITIES OF DIFFERENT MOLECULAR BAND TYPES

4.4.1 Relative Band Intensities

Very high absorptivity (i.e., high absorbance or optically dense) materials with absorbance above 3–6 Au are difficult to measure accurately without the use of high-precision photometric instruments, such as a double monochromator system having stray light specifications below 0.0001% T. Measurements can be made with extremely slow scanning speeds and by dynamically opening the slits during measurement based on the transmission energy profile while scanning. These measurements should be avoided by the novice unless high-performance instrumentation and technical support are available. Table 4.2 shows the relative absorption band intensities for each particular band as well as general spectral windows for fundamental through fourth overtone C-H stretching regions of the NIR spectral region.

4.4.2 Molecular Vibrations Most Active in the NIR

Specific molecular vibrations most active in the NIR spectral region are listed here, with X-H bonds being the most active and intense. The most typical absorption bands with their respective band locations are given in Table 4.3.

The subsequent values for wavelength, wavenumber, and frequency for both the visible and the extended NIR regions are shown in Table 4.4.

TABLE 4.2
Relative Band Intensities: MIR vs. NIR for C-H Stretch

Band	Wavenumber (Wavelength) Region	Relative Intensity	Recommended Sample Cell Pathlength (for Liquid Hydrocarbons)
Fundamental (ν)	2959 to 2849 cm^{-1} (3380–3510 nm)	1	0.1–4 mm
First overtone (2ν)	5917 to 5698 cm^{-1} (1690–1755 nm)	0 .01	0.1–2 cm
Second overtone (3ν)	8873 to 8547 cm^{-1} (1127–1170 nm)	0.001	0.5–5 cm
Third overtone (4ν)	11834 to 11390 cm^{-1} (845–878 nm)	0.0001	5–10 cm
Fourth overtone (5ν)	14493 to 12987 cm^{-1} (690–770 nm)	0.00005	10–20 cm

TABLE 4.3
Relative Peak Positions for Most Common NIR Absorption Bands

NIR Overtone Band Positions (Group)	Peak (nm)	Peak (cm^{-1})
C=O ($3\nu_S$) and .C=ONH from amides	1920	5208
C=O + N-H from polyamides	1598	6258
C=O and C=OOH from carboxylic acids	1900	5263
C=O carbonyl from aliphatic hydrocarbons	1160	8621
C=O carbonyl from ketones & aldehydes	1450	6897
C=O from acids and esters	1950	5128
C-H aryl	2188–2540	4570 to 3937
C-H from aliphatic hydrocarbons	1410	7092
C-H from aliphatic hydrocarbons	1631–1762	6131 to 5675
C-H from aliphatic hydrocarbons	2308–2318	4333 to 4314
C-H from alkenes	1620	6173
C-H from alkenes	2170	4608
C-H from amides	2300	4348
C-H from amines	1735	5764
C-H from aromatic hydrocarbons	1142	8757
C-H from aromatic hydrocarbons	1680–1770	5952 to 5650
C-H from aromatic hydrocarbons	2477	4037
C-H from aromatic hydrocarbons, combination	1360–1446	7353 to 6916
C-H from cellulose	1780	5618
C-H from ethers	1701–1733	5879 to 5770
C-H from halogenated (CH_3X)	1650–1711	6061 to 5845
C-H from hydrocarbons, methyl group	1705	5865
C-H from hydrocarbons, methylene	1765	5666
C-H from ketones	1678–1695	5959 to 5900
C-H from lipids	2310–2380	4329 to 4202
C-H from nitro (CH_3) as (CH_3NO_2)	1654–1702	6046 to 5875
C-H from polysaccharides	2280–2352	4386 to 4252
C-H from silicone (dimethyl siloxane)	1748	5721
C-H from silicone (dimethyl siloxane)	2295	4357

(Continued)

TABLE 4.3 (*Continued*)
Relative Peak Positions for Most Common NIR Absorption Bands

NIR Overtone Band Positions (Group)	Peak (nm)	Peak (cm^{-1})
C-H from vinyl and vinylidene as isoprene	1613–1637	6200 to 6109
C-H methyl from aliphatic hydrocarbons	1194	8375
C-H methylene from aliphatic hydrocarbons	1211	8258
C-H/C=O combination from carbohydrates	2200	4545
C-H/C=O from lipids	2140	4673
C-H/C-C combination from cellulose	2488	4019
C-H/C-C/C-O-C combination from polysaccharides	2500	4000
C-H/O-H from alcohols, diols	1688–1732	5924 to 5774
C-N-C from proteins	2470	4049
C-N-C from amide	2530	3953
$CONH_2$ from amide A and amide II, proteins (polypeptides)	2060	4854
$CONH_2$ from proteins	2167–2174	4615 to 4600
N-H combination band from primary amides	1463–1570	1463
N-H from amides/proteins	1980–2060	5051 to 4854
N-H from polyamide	1480–1618	6757 to 6180
N-H from primary amides	1430–1490	6993 to 6711
N-H from primary amides	1960	5102
N-H from primary amides	2010–2030	4975 to 4926
N-H from primary aromatic amine	1963–1977	5094 to 5058
N-H from proteins/amino acids	2180	4587
N-H from urea	1500	6667
N-H/C=O from polyamides	2012–2127	4970 to 4701
O-H from alcohols as R-C-O-H	1580	6329
O-H from aliphatic hydrocarbons	1410	7092
O-H from alkyl alcohols (diols)	1460	6849
O-H from alkyl alcohols with single O-H group	1389	7199
O-H from cellulose	1820	5495
O-H from liquid water – near boiling	1938	5160
O-H from liquid water – near freezing	1938	5160
O-H from liquid water – room temperature	1940	5155
O-H from methanol (non-hydrogen-bonded)	1470	6803
O-H from water	1790	5587
O-H from water and polyvinyl alcohol OH	1942	5149
O-H water – near boiling	1425	7018
O-H water – near freezing	1453	6882
O-H Water – room temperature	1450	6897
S-H from thiols	1740	5747
Si-O from silicone (dimethyl siloxane)	1452	6887
Si-O-H stretch + Si-O-Si from silicone (dimethyl siloxane)	1933	5173

4.5 INTERPRETIVE NEAR-INFRARED SPECTROSCOPY

The energy absorbed by a matrix consisting of organic compounds having a variety of molecular vibrations depends upon the species (or type) of molecule present, the concentration of these individual molecular species, and the interactions between the molecules in the matrix. In order for NIR (or any other vibrational spectroscopic measurement technique) to be valid, one must

TABLE 4.4
Equivalent Wavelength, Wavenumber, and Frequency Values for the Visible and Near-Infrared Spectral Regions

Spectral Region	Centimeters (cm)	Microns (μm)	Nanometers (nm)	Wavenumber (cm^{-1})	Frequency (Hertz or Hz)
Visible	3.5×10^{-5}–7.8×10^{-5}	0.35–0.78	350–780	28,571 to 12,821	8.563×10^{14} to 3.842×10^{14}
Extended near-infrared	7.8×10^{-5} to 3.0×10^{-4}	0.78–3.0	780–3000	12,821 to 3333	3.842×10^{14} to 9.989×10^{13}

be absolutely assured that different types of molecules absorb at unique frequencies. Due to the broadband nature of NIR spectra, consisting of overlapping combination and overtone bands, the individual species are not well resolved as they are in the mid-infrared region or by using Raman spectroscopy. Many compounds absorb energy throughout the entire NIR spectral region making it difficult, if not impossible, to clearly resolve a useful reference baseline for simple peak height or peak area quantitative methods. This brings one to the same conclusion drawn by early investigators that in the NIR region, multivariate calibration techniques, such as multiple linear regression (MLR), principal components analysis (PCA), and partial least squares (PLS), were required to "interpret" the poorly resolved overlapping bands and to "compensate" for background interferences.

4.5.1 Group Frequencies

Frequencies that are characteristics of groups of atoms (termed "functional groups") are given common group names and can be assigned general NIR locations (in nanometers) for major groups associated with common vibrational bands. The locations for the fundamental frequencies, overtones, and combination bands can be determined both theoretically and empirically (Workman 1996, Goddu and Delker 1960, Mayo 1994). Tables 4.3 and 4.5 provide a source of information as to names and locations of the major functional groups active in the NIR region. The following text outlines more details regarding the characteristics of the NIR spectra.

4.5.2 Fingerprint Frequencies

Fingerprint frequencies in the infrared spectral region are due to molecular vibrations of the entire molecule rather than vibrations specific to a particular functional group. Since an infrared spectrum is unique to an individual molecule, it is termed "the fingerprint" or "fingerprint region" of that molecule – much as the fingerprint of a person is indicative of the specific identity of that person (Herschel 1880). However, in identifying the unique "fingerprint vibrations" of a molecule, information is given as to the approximate molecular formula and composition of each specific molecule.

4.5.3 Dividing the Near-Infrared Spectrum

The group frequencies are given as those bands occurring below 6667 nm, and the fingerprint frequencies are given as those above 6667 nm. Thus, for the NIR region (below 2500 nm), the fingerprint frequencies do not occur. The group frequencies below 6667 nm usually demonstrate bands of reasonable intensity, with somewhat reliable interpretation based upon frequency or wavelength spectral position.

TABLE 4.5
Example Functional Group Names and General Formulas in Simplified Molecular Input Line Entry Specification or SMILES Format

Hydrocarbons, etc.
C-C, paraffinics (or alkanes)
CH_3-$(CH_2)_N$-CH_3, normal paraffins
R-CH_3, methyl C-H
R-CH_2R', methylene C-H
R-CHR'R'', methine C-H
C1CCCCC1, cyclohexane
>C=C<, olefinic group (or alkenes)
-CH=CH_2, vinyls
R'R"C=CH_2, vinylidenes
>C=C=CH_2, allenes
-C≡C-, acetylenes (or acetylinics)
C_NH_N, aromatics
C_NH_{2N}, naphthalenes (or cycloalkanes)
Carbonyl compounds
R-C(=O) -H, aldehydes
R-C(=O) -R', ketones
>C=C<HC(=O) -R', ketenes
R-C(=O)-OR', esters
S-C(=O) -O-R', thiol esters
R-C(=O) –C(=O) -R, anhydrides
-C-O-O-C-, peroxides
-O-C(=O) –C(=O) -O-, oxalates
-O-C(=O) –, carboxy-
-C(=O) -, carbonyl group
-C(=O) –NH_2, primary amides
-C(=O) -NHR, secondary amides
-C(=O) -NR'R", tertiary amides
Ethers
-OCH_3, methoxy (or ether group)
C-O-C, ethers
Hydroxyl
-OH, hydroxyl- [O-H]
R-OH, alcohols [O-H]
-CH_2-OH, primary alcohols
R'R"CH-OH, secondary alcohols
R'R"R'"C-OH, tertiary alcohols
Ar-OH, phenolics (or phenols) [O-H]
Ar=c1ccccc1.
Nitrogen compounds
-C≡N, or C#N, nitriles
R-C≡N-, or R-C#N-, nitriles
S- C≡N^-, or S- C#N^-, thiocyanate
C≡N^-, or C#N, cyanide
C-NO, nitroso group
R'R"NNO, nitrosamines
R-NO_2^-, nitro group (or nitrite)
R-NH_2, primary amines
R'R"NH, secondary amines
R-NH_3^+, R'R"NH_2^+, amine salts
N=N, azo group
-N=N^+=N^-, -N$^-$$N^+$≡N; azides
NNO, azoxy group
R-O-N-N=O, organic nitrites
-NO_3, nitrates
ON=NO, nitroso group
H_3N^+-CH-C(=O)O^-, amino acids

4.5.4 Coupling of Vibrations

Coupling vibrations indicate that the oscillators or molecular vibrations of two or more molecules are interactive, so that the original vibrational energy states (if the vibrations could occur independently of one another) result in split energy states due to the interaction of the vibrations. Coupling is divided into two basic orders, first and second order (i.e., Fermi resonance). First-order coupling can be involved in several important infrared group frequencies. For example, CO_2 has the two separate (uncoupled oscillators) of C=O, each occurring at approximately 5405 nm (ν at ~1850 cm^{-1}), 2756.6–2972.8 nm (2ν), 1855.7–2071.9 nm (3ν), and 1405.3–1621.5 nm (4ν). The interactive (coupled oscillators) energy states occur at 4257 nm (ν at 2349 cm^{-1}, asymmetric stretch), 2171.1–2341.4 nm (2ν), 1461.6–1631.9 nm (3ν), and 1106.8–1277. 1 nm (4ν), and 7463 nm (ν at 1340 cm^{-1}, symmetric stretch), 3806.1–4104.7 nm (2ν), 2562.3–2860.8 nm (3ν), and 1940.4–2238. 9 nm (4ν). First-order coupling is involved for the multiple infrared group frequencies, including the following (Mayo 1994):

The stretches for all cumulated double bonds, X=Y=Z, e.g., C=C=N.
The stretches in XY_2, including -CH_2- and H_2O.
The stretches in XY_3 groups, including -CH_3.

The deformations of XY_3 groups, including -CH_3 and CCl_3.
The N-H in-plane bend of secondary amides, e.g., R-CO-NH-R′.

4.5.5 Fermi Resonance (or Second-Order Coupling)

Fermi resonance (Fermi 1931) is the interaction or coupling of two vibrational energy states with the resultant separation of the states where one of the states results in an overtone or a sum tone vibration. An overtone vibration occurs as the integer multiple of a fundamental vibration in frequency space. Note that absorption and resultant absorbance intensity fall off rapidly with the higher multiples or higher overtones:

1. The first overtone of a fundamental vibration (ν) is equal to $2 \times \nu$.
2. The second overtone of a fundamental vibration is equal to $3 \times \nu$.
3. The third overtone is equal to $4 \times \nu$.

Note: Overtones are a special case of sum tones where the frequencies are identical. A sum tone is the general case of an overtone where the frequencies are not equal and where a variety of vibrational energy states can occur:

1. A binary sum tone is equal to the sum of two fundamentals, e.g., $\nu_i + \nu_k$.
2. A ternary sum tone is equal to the sum of three fundamental vibrations, e.g., $\nu_i + \nu_k + \nu_m$.
3. Other sum tones can occur such as the sum of an overtone and a fundamental vibration, e.g., $2\nu_i + \nu_k$.

Three requirements are stated for Fermi resonance (Mayo 1994):

1. The zero-order frequencies must be close together (typically within $30\,cm^{-1}$).
2. The fundamental and the overtone or sum tone must have the same symmetry.
3. There must be a mechanism for the interaction of the vibrations.

Note: (1) The vibrations cannot be separate (or localized) in distinctly different parts of the molecule. (2) The vibrations must be mechanically interactive in order that the interaction of one vibration affects another.

The results of Fermi resonance are important for infrared and NIR spectroscopic techniques. Fermi resonance causes the following effects on spectral bands:

1. The resultant bands are moved in position from their expected frequencies.
2. Overtone bands are more intense than expected.
3. There may exist doublet bands where only singlets were expected.
4. Solvent changes can bring about slight shifts in frequency location of a band, and intensities can be greatly changed.

4.6 TABLES AND FIGURES FOR SPECTRAL COMPARISONS

Tables 4.6 and 4.7 and Figure 4.19 are given to provide a reference for spectra–structure correlation present in the NIR spectra. Table 4.6 displays the information in alphabetical order based on functional groupings, and Table 4.7 and Figure 4.19 demonstrate the same information in ascending wavelength order (nm).

TABLE 4.6
NIR Overtone Band Positions – By Functional Group Alphabetical Order (Workman 2016, 2000, Workman and Weyer 2012)

NIR Overtone Band Positions (by Functional Group)	Start (nm)	End (nm)
C=O ($3\nu_S$) and .C=ONH from amides	1920	1930
C=O + N-H from polyamides	1598	1608
C=O and C=OOH from carboxylic acids	1900	1910
C=O carbonyl from aliphatic hydrocarbons	1160	1170
C=O carbonyl from ketones & aldehydes	1450	1460
C=O from acids and esters	1950	1960
C=O from urea	2030	2040
C=O/C-N/N-H from polyamide	2212	2222
C=O-O from polysaccharides	2100	2110
C Cl from chlorinated hydrocarbons	1860	1870
C-H aryl	2188	2198
C-H aryl	2206	2216
C-H aryl	2294	2304
C-H aryl	2352	2362
C-H aryl	2387	2397
C-H aryl	2407	2417
C-H aryl	2440	2450
C-H aryl	2444	2454
C-H aryl	2469	2479
C-H aryl	2513	2523
C-H aryl	2525	2535
C-H aryl	2540	2550
C-H combination from lipids, aliphatic compounds	2470	2480
C-H from acetylene or methyne C-H	1530	1540
C-H from acrylate	1621	1631
C-H from alcohols as diols	1664	1674
C-H from aliphatic C-H	1703	1713
C-H from aliphatic hydrocarbon combination	1390	1396
C-H from aliphatic hydrocarbons	1410	1420
C-H from aliphatic hydrocarbons	1631	1641
C-H from aliphatic hydrocarbons	1635	1645
C-H from aliphatic hydrocarbons	1693	1703
C-H from aliphatic hydrocarbons	1693	1703
C-H from aliphatic hydrocarbons	1694	1704
C-H from aliphatic hydrocarbons	1695	1705
C-H from aliphatic hydrocarbons	1704	1714
C-H from aliphatic hydrocarbons	1709	1719
C-H from aliphatic hydrocarbons	1727	1737
C-H from aliphatic hydrocarbons	1762	1772
C-H from aliphatic hydrocarbons	2308	2318
C-H from aliphatic hydrocarbons	2347	2357
C-H from aliphatic hydrocarbons	2363	2373
C-H from aliphatic hydrocarbons	2458	2468
C-H from aliphatic hydrocarbons	2470	2480

(Continued)

TABLE 4.6 (*Continued*)

NIR Overtone Band Positions – By Functional Group Alphabetical Order (Workman 2016, 2000, Workman and Weyer 2012)

NIR Overtone Band Positions (by Functional Group)	Start (nm)	End (nm)
C-H from aliphatic hydrocarbons	2318	2328
C-H from aliphatic hydrocarbons, combination	1440	1450
C-H from aliphatic hydrocarbons, methylene	1725	1735
C-H from alkenes	1620	1630
C-H from alkenes	2170	2180
C-H from alkenes or polyenes	1170	1180
C-H from amides	2300	2310
C-H from amines	1735	1745
C-H from aromatic hydrocarbons	1142	1152
C-H from aromatic hydrocarbons	1680	1690
C-H from aromatic hydrocarbons	1685	1695
C-H from aromatic hydrocarbons	2477	2487
C-H from aromatic hydrocarbons ($ArCH_3$)	1736	1746
C-H from aromatic hydrocarbons ($ArCH_3$)	1744	1754
C-H from aromatic hydrocarbons ($ArCH_3$)	1770	1780
C-H from aromatic hydrocarbons, combination	1360	1370
C-H from aromatic hydrocarbons, combination	1417	1427
C-H from aromatic hydrocarbons, combination	1446	1456
C-H from aryl	1689	1699
C-H from aryl groups	1671	1681
C-H from aryls	2148	2158
C-H from aryls	2154	2164
C-H from aryls	2167	2177
C-H from cellulose	1780	1790
C-H from ethers	1701	1711
C-H from ethers	1733	1743
C-H from halogenated (CH_3Br)	1650	1670
C-H from halogenated (CH_3Br)	1660	1670
C-H from halogenated (CH_3Cl)	1656	1667
C-H from halogenated (CH_3Cl)	1667	1677
C-H from halogenated (CH_3I)	1661	1671
C-H from halogenated methyl (CH_3Br)	1695	1705
C-H from halogenated methyl (CH_3Cl)	1700	1710
C-H from halogenated methyl (CH_3I)	1694	1704
C-H from halogenated methyl (CH_3I)	1711	1721
C-H from hydrocarbons, methyl group	1705	1715
C-H from hydrocarbons, methylene	1765	1775
C-H from ketones	1678	1688
C-H from ketones	1682	1692
C-H from ketones	1693	1703
C-H from ketones	1695	1705
C-H from lipids	2310	2320
C-H from lipids	2380	2390
C-H from nitro (CH_3) as (CH_3NO_2)	1654	1664

(Continued)

TABLE 4.6 (*Continued*)

NIR Overtone Band Positions – By Functional Group Alphabetical Order (Workman 2016, 2000, Workman and Weyer 2012)

NIR Overtone Band Positions (by Functional Group)	Start (nm)	End (nm)
C-H from nitro (CH_3) as (CH_3NO_2)	1702	1712
C-H from polysaccharides	2280	2290
C-H from polysaccharides	2322	2332
C-H from polysaccharides	2330	2340
C-H from polysaccharides	2335	2345
C-H from polysaccharides	2352	2362
C-H from silicone (dimethyl siloxane)	1748	1758
C-H from silicone (dimethyl siloxane)	2295	2305
C-H from vinyl and vinylidene as isoprene	1629	1639
C-H from vinyl as 1-ethenyl oxybutane (1-ethenyl-methyl-ethyl-ketone)	1613	1623
C-H from vinyl-associated group	1637	1647
C-H from vinyl on pyrrolidinone	1621	1631
C-H methyl from aliphatic hydrocarbons	1194	1200
C-H methylene from aliphatic hydrocarbons	1211	1222
C-H/C=O combination from carbohydrates	2200	2210
C-H/C=O from lipids	2140	2150
C-H/C-C combination from cellulose	2488	2498
C-H/C-C/C-O-C combination from polysaccharides	2500	2510
C-H/O-H from alcohols, diols	1688	1698
C-H/O-H from alcohols, diols	1701	1711
C-H/O-H from alcohols, diols	1732	1742
C-N-C from proteins	2470	2480
C-N-C from amide	2530	2540
$CONH_2$ from amide A and amide II, proteins (polypeptides)	2060	2070
$CONH_2$ from proteins	2174	2184
$CONH_2$ from proteins in aqueous solutions	2167	2177
N-H ($2\nu_S$) from ammonia in water	1534	1544
N-H ($3\nu_S$) for ammonia in water	2210	2220
N-H combination band from primary amides	1470	1480
N-H from amide/protein	1471	1481
N-H from amide/protein	1483	1493
N-H from amide/protein	1490	1500
N-H from amide/protein	1492	1502
N-H from amide/protein	1500	1510
N-H from amide/protein	1510	1520
N-H from amide/protein	1520	1530
N-H from amide/protein	1530	1540
N-H from amide/protein	1570	1580
N-H from amides/proteins	1463	1473
N-H from amides/proteins	1980	1990
N-H from amides/proteins	2050	2060
N-H from amides/proteins	2055	2065
N-H from amides/proteins	2060	2070

(*Continued*)

TABLE 4.6 (*Continued*)

NIR Overtone Band Positions – By Functional Group Alphabetical Order (Workman 2016, 2000, Workman and Weyer 2012)

NIR Overtone Band Positions (by Functional Group)	Start (nm)	End (nm)
N-H from ammonia in water	2000	2010
N-H from polyamide	1480	1490
N-H from polyamide	1515	1525
N-H from polyamide	1538	1548
N-H from polyamide	1570	1580
N-H from polyamides	1618	1628
N-H from polyamides	2053	2063
N-H from primary amides	1430	1440
N-H from primary amides	1490	1500
N-H from primary amides	1960	1970
N-H from primary amides	2010	2020
N-H from primary amides	2030	2040
N-H from primary amine	2205	2215
N-H from primary aromatic amine	1977	1987
N-H from primary aromatic amines	1963	1973
N-H from protein	1529	1600
N-H from protein	2060	2070
N-H from proteins/amino acids	2180	2190
N-H from urea	1460	1470
N-H from urea	1470	1480
N-H from urea	1485	1495
N-H from urea	1500	1510
N-H from urea	1990	2000
N-H from urea	2040	2050
N-H from urea	2070	2080
N-H from urea	2220	2230
N-H/C=O from polyamides	2127	2137
N-H/C=O from polyamides	2012	2022
N-H/C-N from urea	2080	2090
N-H/C-N/C=O from proteins	2145	2155
N-H/C-N/C=O from proteins	2180	2190
N-H/C-N/C=O from urea	2180	2190
N-H/C-N/N-H from polyamides	2183	2193
O-H bound water from amide group of ovalbumin	1550	1560
O-H from alcohols as R-C-O-H	1580	1590
O-H from alcohols or water	1583	1593
O-H from alcohols or water	1603	1613
O-H from alcohols or water	2096	2106
O-H from aliphatic hydrocarbons	1410	1420
O-H from alkyl alcohols (diols)	1415	1425
O-H from alkyl alcohols (diols)	1460	1470
O-H from alkyl alcohols with single O-H group	1389	1426
O-H from alkyl alcohols with single O-H group	1389	1426

(*Continued*)

TABLE 4.6 (*Continued*)

NIR Overtone Band Positions – By Functional Group Alphabetical Order (Workman 2016, 2000, Workman and Weyer 2012)

NIR Overtone Band Positions (by Functional Group)	Start (nm)	End (nm)
O-H from alkyl alcohols with single O-H group	1460	1603
O-H from aromatic hydrocarbons	1420	1430
O-H from cellulose	1820	1830
O-H from crystalline sucrose	1441	1451
O-H from liquid water – near boiling	1938	1948
O-H from liquid water – near freezing	1938	1948
O-H from liquid water – room temperature	1940	1950
O-H from methanol	2016	2026
O-H from methanol	2123	2133
O-H from methanol	2519	2529
O-H from methanol (non-hydrogen-bonded)	1470	1480
O-H from phenolic groups	1464	1474
O-H from phenolics	1401	1441
O-H from polyols, alcohols, water	1802	2198
O-H from polysaccharides	1960	1970
O-H from polysaccharides and carbohydrates	2100	2110
O-H from primary alcohols	1452	1462
O-H from secondary alcohols	1464	1474
O-H from starch/polymeric alcohol	1490	1500
O-H from starch/polymeric alcohols	1540	1550
O-H from starches/polymeric alcohols	1450	1460
O-H from tertiary alcohols	1468	1478
O-H from water	1790	1800
O-H from water and polyvinyl alcohol OH	1942	1952
O-H polymers	2090	2100
O-H water – near boiling	1425	1435
O-H water – near freezing	1453	1463
O-H water – room temperature	1450	1460
O-H/C-H from methanol	1965	1975
O-H/C-O combination from cellulose	2270	2280
O-H/C-O combination from glucose	2273	2283
P-OH/O-H from phosphate	1908	1918
Polysaccharides	1930	1940
S-H from thiols	1740	1750
Si-O from silicone (dimethyl siloxane)	1452	1462
Si-O-H stretch + Si-O-Si from silicone (dimethyl siloxane)	1933	1943
Traditional cellulose wavelength	2336	2346
Traditional lignin position	2270	2280
Traditional reference position	2230	2240
Traditional reference wavelength	1680	1690
Traditional wavelength for lipids/oils	2310	2320
Traditional wavelength for proteins	2180	2190

TABLE 4.7
NIR Overtone Band Positions – By Ascending Wavelength Order (Workman 2016, 2000, Workman and Weyer 2012)

NIR Overtone Band Positions (Group)	Peak Position (nm)
C-H from aromatic hydrocarbons	1147
C=O carbonyl from aliphatic hydrocarbons	1165
C-H from alkenes, or polyenes	1175
C-H methyl from aliphatic hydrocarbons	1197
C-H methylene from aliphatic hydrocarbons	1217
C-H from aromatic hydrocarbons, combination	1365
O-H from alkyl alcohols with single O-H group	1408
O-H from alkyl alcohols with single O-H group	1408
C-H from aliphatic hydrocarbon combination	1393
O-H from phenolics	1421
C-H from aliphatic hydrocarbons	1415
O-H from aliphatic hydrocarbons	1415
O-H from alkyl alcohols (diols)	1420
C-H from aromatic hydrocarbons, combination	1422
O-H from aromatic hydrocarbons	1425
O-H water – near boiling	1430
N-H from primary amides	1435
C-H from aliphatic hydrocarbons, combination	1445
O-H from crystalline sucrose	1446
C-H from aromatic hydrocarbons, combination	1451
C=O carbonyl from ketones & aldehydes	1455
O-H from starches/polymeric alcohols	1455
O-H water – room temperature	1455
O-H from primary alcohols	1457
Si-O from silicone (dimethyl siloxane)	1457
O-H water – near freezing	1458
N-H from urea	1465
O-H from alkyl alcohols (diols)	1465
O-H from alkyl alcohols with single O-H group	1532
N-H from amides/proteins	1468
O-H from phenolic groups	1469
O-H from secondary alcohols	1469
O-H from tertiary alcohols	1473
N-H combination band from primary amides	1475
N-H from urea	1475
O-H from methanol (non-hydrogen-bonded)	1475
N-H from amide/protein	1476
N-H from polyamide	1485
N-H from amide/protein	1488
N-H from urea	1490
N-H from amide/protein	1495
N-H from primary amides	1495
O-H from starch/polymeric alcohol	1495
N-H from amide/protein	1497

(*Continued*)

TABLE 4.7 (*Continued*)
NIR Overtone Band Positions – By Ascending Wavelength Order (Workman 2016, 2000, Workman and Weyer 2012)

NIR Overtone Band Positions (Group)	Peak Position (nm)
N-H from amide/protein	1505
N-H from urea	1505
N-H from amide/protein	1515
N-H from polyamide	1520
N-H from amide/protein	1525
N-H from protein	1565
C-H from acetylene or methyne C-H	1535
N-H from amide/protein	1535
N-H ($2\nu_S$) from ammonia in water	1539
N-H from polyamide	1543
O-H from starch/polymeric alcohols	1545
O-H bound water from amide group of ovalbumin	1555
N-H from amide/protein	1575
N-H from polyamide	1575
O-H from alcohols as R-C-O-H	1585
O-H from alcohols or water	1588
C=O + N-H from polyamides	1603
O-H from alcohols or water	1608
C-H from vinyl as 1-ethenyl oxybutane (1-ethenyl-methyl-ethyl-ketone)	1618
N-H from polyamides	1623
C-H from alkenes	1625
C-H from acrylate	1626
C-H from vinyl on pyrrolidinone	1626
C-H from vinyl and vinylidene as isoprene	1634
C-H from aliphatic hydrocarbons	1636
C-H from aliphatic hydrocarbons	1640
C-H from vinyl-associated group	1642
C-H from halogenated (CH_3Br)	1660
C-H from nitro (CH_3) as (CH_3NO_2)	1659
C-H from halogenated (CH_3Cl)	1662
C-H from halogenated (CH_3Br)	1665
C-H from halogenated (CH_3I)	1666
C-H from alcohols as diols	1669
C-H from halogenated (CH_3Cl)	1672
C-H from aryl groups	1676
C-H from ketones	1683
C-H from aromatic hydrocarbons	1685
Traditional reference wavelength	1685
C-H from ketones	1687
C-H from aromatic hydrocarbons	1690
C-H/O-H from alcohols, diols	1693
C-H from aryl	1694
C-H from aliphatic hydrocarbons	1698
C-H from aliphatic hydrocarbons	1698

(*Continued*)

TABLE 4.7 (*Continued*)
NIR Overtone Band Positions – By Ascending Wavelength Order (Workman 2016, 2000, Workman and Weyer 2012)

NIR Overtone Band Positions (Group)	Peak Position (nm)
C-H from ketones	1698
C-H from aliphatic hydrocarbons	1699
C-H from halogenated methyl (CH_3I)	1699
C-H from aliphatic hydrocarbons	1700
C-H from halogenated methyl (CH_3Br)	1700
C-H from ketones	1700
C-H from halogenated methyl (CH_3Cl)	1705
C-H from ethers	1706
C-H/O-H from alcohols, diols	1706
C-H from nitro (CH_3) as (CH_3NO_2)	1707
C-H from aliphatic C-H	1708
C-H from aliphatic hydrocarbons	1709
C-H from hydrocarbons, methyl group	1710
C-H from aliphatic hydrocarbons	1714
C-H from halogenated methyl (CH_3I)	1716
C-H from aliphatic hydrocarbons, methylene	1730
C-H from aliphatic hydrocarbons	1732
C-H/O-H from alcohols, diols	1737
C-H from ethers	1738
C-H from amines	1740
C-H from aromatic hydrocarbons ($ArCH_3$)	1741
S-H from thiols	1745
C-H from aromatic hydrocarbons ($ArCH_3$)	1749
C-H from silicone (dimethyl siloxane)	1753
C-H from aliphatic hydrocarbons	1767
C-H from hydrocarbons, methylene	1770
C-H from aromatic hydrocarbons ($ArCH_3$)	1775
C-H from cellulose	1785
O-H from water	1795
O-H from polyols, alcohols, water	1807
O-H from cellulose	1825
C-Cl from chlorinated hydrocarbons	1865
C=O and C=OOH from carboxylic acids	1905
P-OH/O-H from phosphate	1913
C=O ($3\nu_S$) and .C=ONH from amides	1925
Polysaccharides	1935
Si-O-H stretch + Si-O-Si from silicone (dimethyl siloxane)	1938
O-H from liquid water – near boiling	1943
O-H from liquid water – near freezing	1943
O-H from liquid water – room temperature	1945
O-H from water and polyvinyl alcohol OH	1947
C=O from acids and esters	1955
N-H from primary amides	1965
O-H from polysaccharides	1965

(*Continued*)

TABLE 4.7 (*Continued*)
NIR Overtone Band Positions – By Ascending Wavelength Order (Workman 2016, 2000, Workman and Weyer 2012)

NIR Overtone Band Positions (Group)	Peak Position (nm)
N-H from primary aromatic amines	1968
O-H/C-H from methanol	1970
N-H from primary aromatic amine	1982
N-H from amides/proteins	1985
N-H from urea	1995
N-H from ammonia in water	2005
N-H from primary amides	2015
N-H/C=O from polyamides	2017
O-H from methanol	2021
C=O from urea	2035
N-H from primary amides	2035
N-H from urea	2045
N-H from amides/proteins	2055
N-H from polyamides	2058
N-H from amides/proteins	2060
$CONH_2$ from amide A and amide II, proteins (polypeptides)	2065
N-H from amides/proteins	2065
N-H from protein	2065
N-H from urea	2075
N-H/C-N from urea	2085
O-H polymers	2090
O-H from alcohols or water	2101
C=O-O from polysaccharides	2105
O-H from polysaccharides and carbohydrates	2105
O-H from methanol	2128
N-H/C=O from polyamides	2132
C-H/C–O from lipids	2145
N-H/C-N/C=O from proteins	2150
C-H from aryls	2153
C-H from aryls	2159
C-H from aryls	2172
$CONH_2$ from proteins in aqueous solutions	2172
C-H from alkenes	2175
$CONH_2$ from proteins	2179
N-H from proteins/amino acids	2185
N-H/C-N/C=O from proteins	2180
N-H/C-N/C=O from urea	2185
Traditional wavelength for proteins	2185
N-H/C-N/N-H from polyamides	2183
C-H aryl	2193
C-H/C=O combination from carbohydrates	2205
N-H from primary amine	2210
C-H aryl	2211
N-H ($3\nu_S$) for ammonia in water	2210

(*Continued*)

TABLE 4.7 (*Continued*)
NIR Overtone Band Positions – By Ascending Wavelength Order (Workman 2016, 2000, Workman and Weyer 2012)

NIR Overtone Band Positions (Group)	Peak Position (nm)
C=O/C-N/N-H from polyamide	2217
N-H from urea	2225
Traditional reference position	2235
O-H/C-O combination from cellulose	2275
Traditional lignin position	2275
O-H/C-O combination from glucose	2278
C-H from polysaccharides	2285
C-H aryl	2299
C-H from silicone (dimethyl siloxane)	2300
C-H from amides	2305
C-H from aliphatic hydrocarbons	2313
C-H from lipids	2315
Traditional wavelength for lipids/oils	2315
C-H from aliphatic hydrocarbons	2323
C-H from polysaccharides	2327
C-H from polysaccharides	2335
C-H from polysaccharides	2340
Traditional cellulose wavelength	2341
C-H from aliphatic hydrocarbons	2352
C-H aryl	2357
C-H from polysaccharides	2357
C-H from aliphatic hydrocarbons	2368
C-H from lipids	2385
C-H aryl	2392
C-H aryl	2412
C-H aryl	2445
C-H aryl	2449
C-H from aliphatic hydrocarbons	2463
C-H aryl	2474
C-H combination from lipids, aliphatic compounds	2475
C-H from aliphatic hydrocarbons	2475
C-N-C from proteins	2475
C-H from aromatic hydrocarbons	2482
C-H/C-C combination from cellulose	2493
C-H/C-C/C-O-C combination from polysaccharides	2505
C-H aryl	2518
O-H from methanol	2524
C-H aryl	2530
C-N-C from amide	2535
C-H aryl	2545

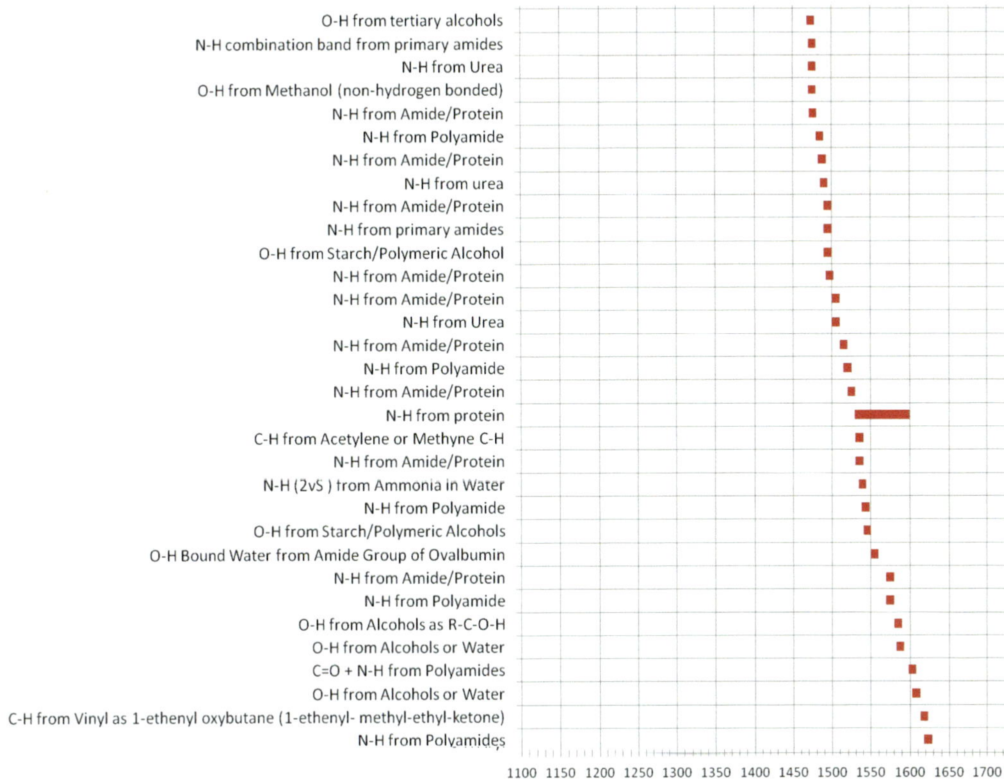

FIGURE 4.19 NIR overtone band positions – by ascending wavelength order.

REFERENCES

Fermi, E. 1931. Über den Ramaneffekt des kohlendioxyds. *Zeitschrift für Physik*, 71(3–4), 250–259.

Goddu, R.F. and D.A. Delker. 1960. Spectra structure correlations for near infrared. *Anal. Chem.* 32, 140–141, 1960.

Herschel, W.J. 1880. Skin furrows of the hand. *Nature*, 23(578), 76.

Mayo, D.W., Ed. 1994. *Infrared Spectroscopy: I. Instrumentation, and II. Instrumentation, Raman Spectra, Polymer Spectra, Sample Handling, and Computer-Assisted Spectroscopy* (Vol. 1), Bowdoin College, Brunswick, ME.

Workman Jr., J. 2000. *The Handbook of Organic Compounds, Three-Volume Set: NIR, IR, R, and UV-Vis Spectra Featuring Polymers and Surfactants*. Elsevier, Amsterdam.

Workman Jr., J. and Weyer, L. 2012. *Practical Guide and Spectral Atlas for Interpretive Near-Infrared Spectroscopy*. CRC Press, Boca Raton, FL.

Workman Jr., J.J. 1996. Interpretive spectroscopy for near infrared. *Appl. Spectrosc. Rev.*, 31(3), pp. 251–320.

Workman Jr., J.J. 2016. *Concise Handbook of Analytical Spectroscopy, The: Theory, Applications, and Reference Materials (In 5 Volumes)*. World Scientific, Singapore.

Part II

Instrumentation

5 Traditional NIR Instrumentation

*Howard Mark**
Mark Electronics

CONTENTS

5.1 HISTORICAL NOTE

Arguably, the first near-infrared (NIR) instrumentation was developed by William Herschel contemporaneously with his discovery of the NIR spectral region [1–3], if we accept his ingenious use of thermometers to detect the heat signature of NIR radiation as constituting an "instrument". In practical application terms, however, instruments for the detection and measurement of NIR radiation, consisting of a monochromator, sensitive detectors, and electronic amplification of the detector signal, were developed as byproducts of the use of NIR for military purposes in World War II. Even then, despite the work of Wilbur Kaye [4], NIR was long considered a "useless" region of the spectrum by chemists and spectroscopists because "there was no signal there".

* hlmark@nearinfrared.com.

At that time, and lasting until roughly the mid-1960s, there were no specific NIR instruments commercially available. NIR measurement was available as an adjunct of the "UV-Vis-NIR" instruments that were designed and constructed primarily for measurements in the ultraviolet (UV) and visible (Vis) regions of the electromagnetic spectrum. It was taught in "Chemical Instrumentation" classes that the spectral region between the strong signals in the visible region of the spectrum (electronic transitions) and the strong signals in the mid-IR region of the spectrum (fundamental molecular vibrations) was bereft of anything useful because any signals in the region were too weak to be useful or worth measuring.

That perception was partially due to the instruments available at that time. In keeping with the more-or-less "standard" setups used in other spectral regions, the interest was in determining the wavelengths at which various molecules and functional groups absorbed. To accomplish this, slits were closed down to optimize the spectral resolution of the instrument. This had the effect of greatly reducing the energy available at the detector, resulting in poor signal-to-noise performance, perceived as lack of a spectral signal.

Fortunately for the world, Karl Norris was an engineer, not a chemist, so he wasn't aware that "there was nothing in the NIR region worth measuring" or that you couldn't measure anything there, because there was nothing there to measure. Additionally, he knew that the absorption bands of the natural products he was interested in measuring were fairly broad, so he didn't need to close the slits to improve spectral resolution.

5.2 INTRODUCTION

When I was invited to contribute a chapter on instrumentation to this fourth edition of the handbook, I read what had been written by others in the previous editions. It was an interesting experience. The "same" chapter in the various editions was different from each other, reflecting both the varying viewpoints of the authors and the advances in the field as time went on, and new technologies were developed and applied. The first edition contained mainly a historical overview of the field, including the outgrowth of NIR from the much more widespread discipline and technology of mid-infrared spectroscopy. The second and third editions dealt less with the history and more with the advantages of each of the various technologies used for making NIR measurements.

My own predilections also lie in the direction of concentrating on the various technologies and how those effect the nature of the measurements that can be made. In a work such as this, the historical aspects cannot be completely ignored. This chapter will cover the technologies that are, to some extent, mainly of historical interest, although the intent is to bring in more of the recently developed technologies that have been applied to NIR measurements. All optical instruments must have (at least) three things:

1. A source of the light of appropriate wavelength(s),
2. A place to put the sample (internally or externally),
3. A detector (even if that's the human eye).

To be a spectrometer, the instrument must also have a method of measuring the light at different wavelengths so that a spectrum can be determined. This key component that distinguishes spectrometers from other types of optical instruments could be one that separates the wavelengths optically, electronically, or computationally. All these options give rise to an enormous variety of possibilities for how to organize it all into an instrument that will perform the planned measurements without being excessively expensive. Even listing them all is a daunting task, although previous editions of this handbook have included lists of the requirements.

We will forego repeating that listing, and for instruments that are commonly available, diving straight into discussing how the various key functions are implemented.

5.3 THE INSTRUMENT

Figure 5.1 shows block diagrams that represent the concepts for what the instrument has to do. Each configuration shown is better suited to particular applications, but all have the basic structures described above: source, sample, wavelength selector, and detector. Numerous variations exist that are not shown, for example, employing an integrating sphere in conjunction with reflectance measurements or configurations involving optical fibers. Since most of these figures illustrating the various parts of spectrometer are conceptual, auxiliary optics and other necessary parts of an actual instrument are not shown.

To calculate the sample's absorbance, most spectrometers have a provision for measuring the light intensity falling on the detector in the absence of a sample, and again when the sample is present. This is necessary to avoid confusing extraneous instrumental variations (noise, stray light, etc.) with changes in the sample properties. This is accomplished by replacing the sample with a stabl reference material (temporal separation), sometimes by using a dual-beam configuration similar to those used in UV or (mid)-IR spectrometers (spatial separation). No matter how it is accomplished,

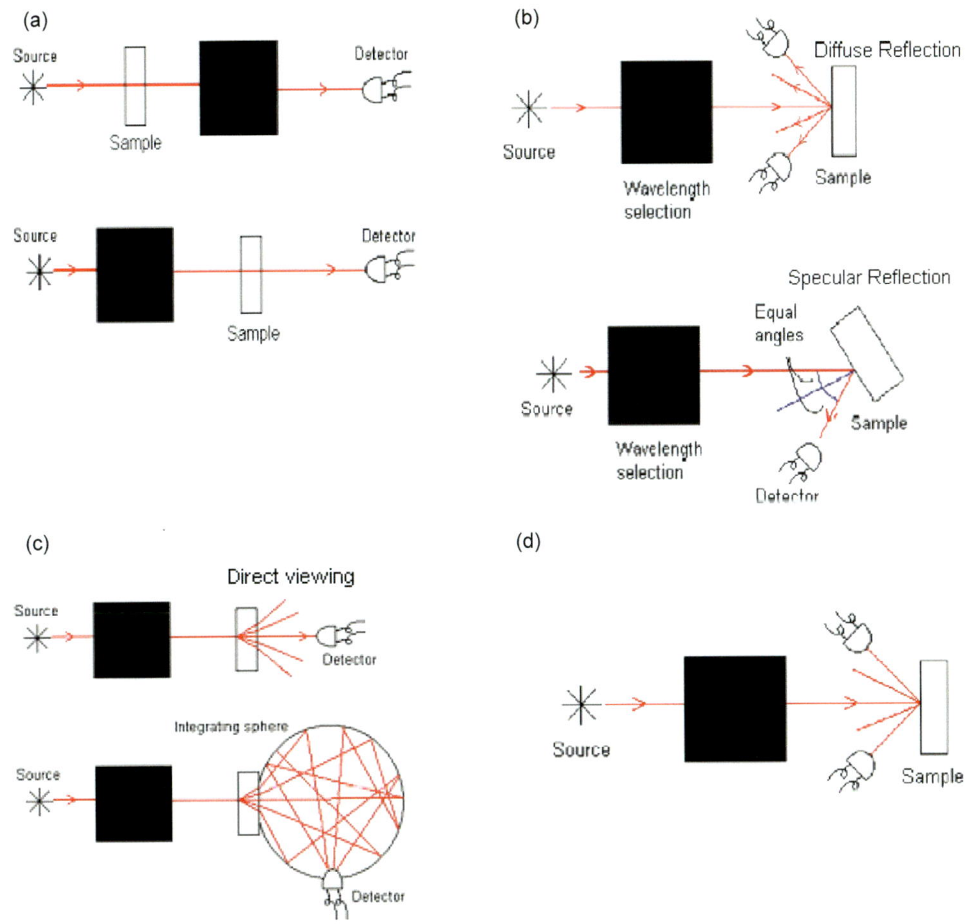

FIGURE 5.1 Block diagrams of generic instrument designs, showing how various types of NIR instruments are organized. In all parts of this figure, the "black box" represents the wavelength-selection part of the instrument. Parts (a) and (c) illustrate various types of transmission measurements. Parts (b) and (d) illustrate the reflection measurements. The examples shown here are by no means exhaustive of the possible configuration variations that have been used.

two readings of optical intensity are collected at each wavelength: one reading (I_o) representing the reading from a stable reference sample as a surrogate for the intensity of the light impinging on the sample and the other (I) representing the measurement of the sample's optical properties. The two readings at each wavelength are combined via the equation T (transmittance) or R (reflectance) $= I/I_o$, indicating the reflectance or transmittance of the sample at each wavelength. This is *ipso facto* the spectrum of that sample. A much more detailed discussion of various aspects of the measurement techniques can be found in the other chapters in this book, so this one will concentrate on the hardware.

Figure 5.2 shows some typical instruments of what might be called the "historical" era of instruments, and the types that were designed and made available to users based directly on the pioneering work of Karl Norris. These are collectively called "benchtop" instruments. Due to

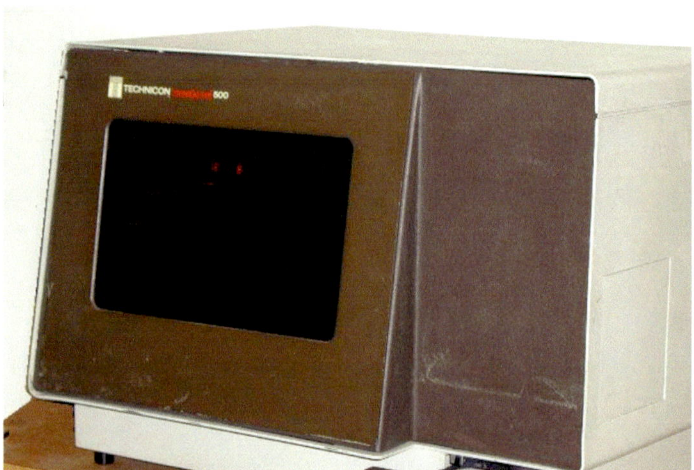

FIGURE 5.2 Early benchtop NIR instruments. (a) Technicon InfraAlyzer-500 benchtop diffraction grating-based reflectance spectrometer (circa 1978), shown with a sample cup in place. (b) Neotec Feed Quality Analyzer tilting filter-based reflectance spectrometer (FQA-51 – circa 1977). Shown with a sample cup in place, and various sample-handling implements. (Picture courtesy Robert Rosenthal, used with permission of Carole Rosenthal.)

their size, weight, and relative sensitivity to mechanical, electrical, optical, and environmental perturbations, their installation requires being mounted on rigid, rugged, and stable surfaces: normally laboratory benches.

We now segue into looking at the key components comprising all instruments, with some particular attention paid to ones available for and used in modern NIR instruments.

5.4 SOURCES

Most instruments need broadband sources in order to be able to measure spectra. By far, the most common source of optical radiation used for NIR instrumentation is a quartz-halogen (also called "tungsten-halogen") incandescent lamp (as shown in Figure 5.3). While nominally the same as incandescent bulbs used for general lighting, these quartz-halogen incandescent lamps differ in the materials used as well as their construction. In both types of lamps, the filament is made of tungsten (wolfram, W) and is heated electrically such that it reaches a temperature high enough to glow.

A major difference between the bulbs used for instruments and (for example) household bulbs is the envelope. In ordinary (household) bulbs, the envelope is made of glass and is large enough to keep it away from the filament, and therefore, the envelope is relatively cool. In bulbs for instruments, on the other hand, the envelope is made of fused quartz and is kept close to the filament. This proximity, plus the presence of a halogen gas (bromine or iodine under pressure as high as 5–8 atm) filling the inside of the bulb, allows the envelope to reach a much higher temperature. In an ordinary bulb, the high temperature of the filament causes some of the tungsten to evaporate; this then gets deposited on the envelope and darkens it. The presence of the halogen gas (e.g., bromine, Br) plus the high temperature of the envelope (~600°C) allow a chemical reaction to proceed: $W + Br \rightarrow WBr$. At the temperatures inside the bulb, the tungsten bromide has a high enough vapor pressure to evaporate, thus cleaning the envelope.

This process removes the deposit of tungsten from the bulb and prevents its darkening. When a WBr molecule migrates to the filament, the higher temperature of the filament causes the reverse reaction to occur: $WBr \rightarrow W + Br$. This reaction deposits the tungsten back on the filament and frees the halogen atom for another cleaning cycle. Thus, the halogen gas keeps the envelope clean, increases the light output, extends the lifetime of the bulb, and most importantly, keeps the spectral characteristics of the bulb emisson constant.

While we often think of voltage control when dealing with electronically controlled components, instrument power supplies for the sources are often current-controlled instead. The reason is in the inherent nature of the materials used for the construction of an incandescent source. Tungsten, like many other materials, increases its resistivity as its temperature increases. Bulbs used for household lighting share that characteristic. Although not always obvious, other devices on the same power line as a large incandescent bulb will sometimes flicker when a large, high-powered bulb is turned on. This is due to that same characteristic of the bulb. When power is suddenly applied to

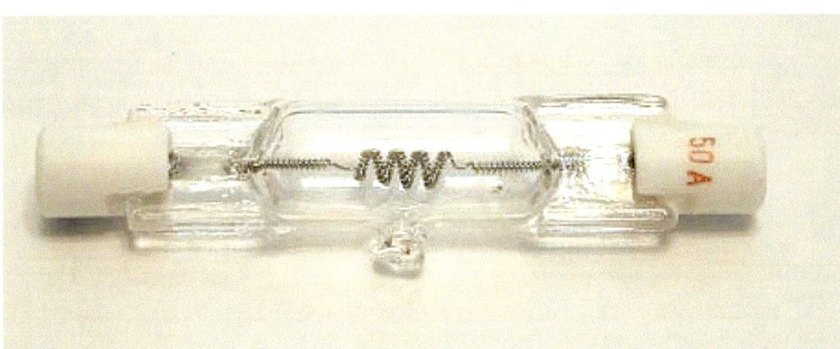

FIGURE 5.3 A representative example of a quartz-halogen incandescent instrumentation bulb.

a cold (~room-temperature) filament, the resistance is low and therefore the initial current is high. After the filament warms, its resistance increases and the start-up surge is reduced to its normal operating value.

What is seldom appreciated is that the resistance of the hot filament can be as high as five times the resistance of the room-temperature filament. In a household lightbulb, this is not a major issue, because the supply (the power line) has enormous reserve capability and can easily handle the surge the bulb requires. In an analytical instrument, however, the power supply often does not have a similar reserve capability. Therefore, to supply this start-up, surge would tax the capability of the power supply every time it was turned on, as well as putting strain on the bulb itself. This would lead to premature failure of the bulb and/or power supply. Using a controlled-current power supply prevents the surge from occurring and extends the life of the instrument (or at least extends the intervals needed before requiring a service call).

5.4.1 Other Sources

Other sources that have been used are, for the most part, specialty sources for idiosyncratic applications. These include sources as prosaic as the sun (for measurements of outdoor air pollution) and low-pressure mercury arc lamps (which emit some narrow atomic emission lines, useful for wavelength calibration purposes) and as exotic as the synchrotron at the National Synchrotron Light Source at Brookhaven (NY) National Laboratory (https://www.bnl.gov/ps/).

5.5 LASERs

One type of source, which shouldn't be neglected, are LASERs. The last several years have seen the advent of tunable lasers. These effectively combine both the source and the wavelength-determining functions into a single device. Several types of tunable lasers are known, even discounting those that are "tunable" only over the bandwidth of a single laser line, and those that "tune" by jumping from one laser mode to another. The bases for tunability of lasers that can cover a wavelength range of interest to spectroscopists are as follows:

1. MEMS (micro-electro-mechanical systems) structure that varies in the length of an external cavity. These can mode-hop between modes that correspond to wavelengths that are close enough together to be quasi-continuous.
2. External cavity lasers (ECL) use prisms and gratings similar to mode-hop.
3. Distributed feedback (DFB) lasers (that may be based on laser diode, quantum cascade laser, or optical fiber laser principles). The tuning is done thermally, by varying the temperature of the laser.
4. VCSEL (vertical cavity surface-enhanced laser) is a semiconductor diode laser with the laser cavity lying vertically, so that the beam comes out normally to the surface of the semiconductor.

These lasers share the basic characteristics such as narrow bandwidth, good collimation, and the high-intensity characteristic of all lasers. Most of them are also small and light, especially those based on semiconductor diode lasers. The main disadvantage is high cost.

5.5.1 Detectors

5.5.1.1 Phototubes

Historically, the first detectors for use in the visible and NIR region were outgrowths of earlier technologies that were available for electronic devices, namely, vacuum tubes. Phototubes and photomultipliers were built into evacuated glass housings, and the principle of operation was fairly simple (see Figure 5.4).

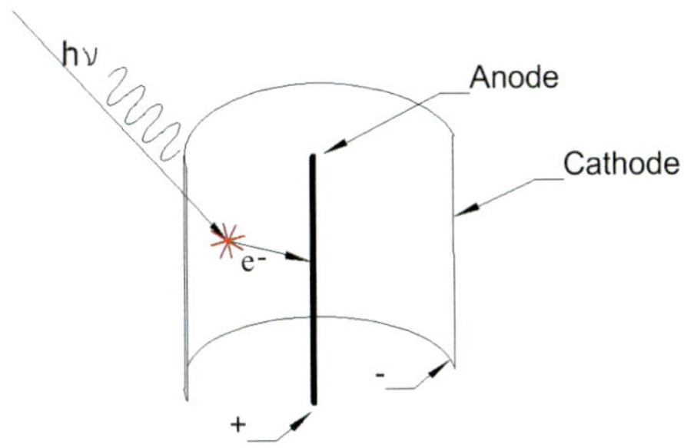

FIGURE 5.4 A schematic diagram of a phototube detector. The vacuum-tight enclosure is not shown. When an incoming photon strikes the photosensitive layer of the cathode, an electron is ejected. The electron is then attracted to the positively charged anode and thus contributes to the electrical current through the anode.

A photon hitting a cathode coated with a photosensitive layer like cesium causes the cathode to emit an electron. The electron is attracted to the positively charged anode; this creates a current in the circuit, and the current can then be measured. The more the emitted light, the more the electrons and the higher the current. The main limitation is that the photon has to have enough energy to knock the electron out of the surface of the cathode, which is called the "work function" and limits the responsivity of the phototube at the long-wavelength end of its range. Therefore, the cathodes in phototubes are typically coated with a material having weakly bound outer electrons, like cesium. Even so, most phototubes are limited to operation in the visible or very near-infrared region.

A photomultiplier operates on the same basic principle, except that more electrodes are inserted between the cathode and the anode. Each electrode has a successively higher voltage applied to them. These "extra" electrodes act as electron multipliers, which increase the sensitivity of the photomultiplier, to the point where individual photons can be detected. This part of the phototube is essentially the same as the electron multiplier used in mass spectrometers. While still in use in UV, X-ray, and gamma-ray spectroscopies, photomultipliers have been replaced by semiconductor-based detectors for NIR instruments.

5.5.2 Semiconductor Detectors

In most NIR applications, there is usually no shortage of photons, so the exquisite sensitivity of photomultipliers isn't needed. The advantages of semiconductor detectors such as smaller size, lighter weight, lower cost, and the obviation of the need for high voltages to operate the detectors overwhelm the need for extreme sensitivity. Therefore, modern instruments invariably use one or more of these semiconductor-based detectors (see Figure 5.5).

Table 5.1 presents the key characteristics of some semiconductor detectors commonly used for NIR spectrometers.

While there is no vacuum in a semiconductor detector, in some ways the operation is analogous to that of the phototube. Without delving deeply into semiconductor physics, Figure 5.5 shows that, as with all materials, a semiconductor's valence electrons are confined to the discrete energy levels dictated by quantum mechanics; these electrons are bound to their atoms. In addition, there is a conduction band which has a continuum of energy levels. Electrons in the conduction band are not confined to either a discrete energy or a single location. These electrons can move freely, and this provides the conductive capability of the semiconductor.

The valence band and the conduction band are separated by an energy gap: a "forbidden zone" (or "neutral zone", in Star Trek parlance), where no electrons can exist. However, an incoming photon carrying sufficient energy can give a valence electron enough energy to jump the gap and enter the conduction band. This electron can now contribute to the conductivity of the material, and add to the current flow, similarly to the way an electron does in a phototube, despite the differences in

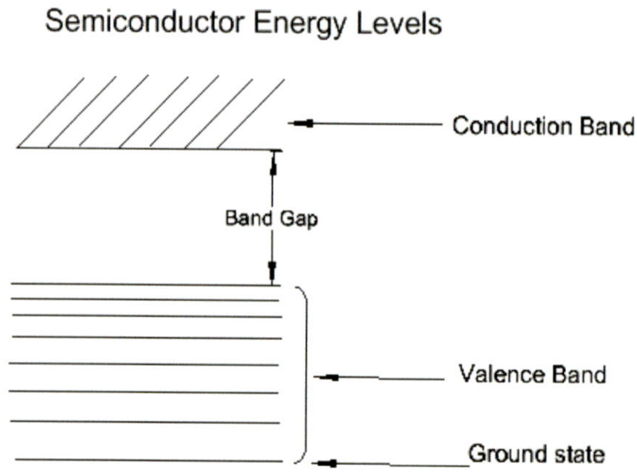

FIGURE 5.5 Electronic structure of semiconductor detectors.

TABLE 5.1
Detector properties

Detector Type	Wavelength Range (nm)	Speed	Sensitivity
Silicon (Si)	400–1100	Very high	High
Lead sulfide (PbS)	700–2500	Moderate	Very high
Indium gallium arsenide (InGaAs)	1100–1700	Very high	Varies depending on composition
Extended indium gallium arsenide (InGaAs)	1100–2300	Very high	Varies depending on composition

the physics involved. The nature of the semiconductor material determines the size of the bandgap and therefore the long-wavelength limit of the detector. While semiconductor detectors can be made very small, larger sizes can also be useful, especially when for reflection measurements. Figure 5.6a shows an unmounted lead sulfide (PbS) detector useful for this type of application.

Technology marches on: recent developments include a novel combination of solid-state physics and electron multipliers, forming a semiconductor electron multiplier [5].

5.5.3 Diode Arrays

A diode array is just what it sounds like: a group of semiconductor detectors mounted together in the same holder (indeed, usually built of the same block of substrate) and encapsulated in a block that can also be used as a mount, as shown in Figure 5.6b. Just about any type of semiconductor detector can be used for this. It can be used in conjunction with a diffraction grating (see Figure 5.6c) or with a linear variable filter (LVF, see Figure 5.9) so that each detector "sees" a different wavelength. Thus, it can measure an entire spectrum without having to mechanically move any parts of the spectrometer. This is a great advantage in portable and/or mobile operation, where small size, low weight, and ruggedness are critical factors.

5.6 WAVELENGTH SELECTION (OR, WHAT'S INSIDE THE BLACK BOX?)

Wavelength selection technologies are more varied than any other part of spectroscopic instruments. Many of them can be found to be in more-or-less common use in commercial instruments. Some of them are preferred, should low cost be the most important factor. Others are specified

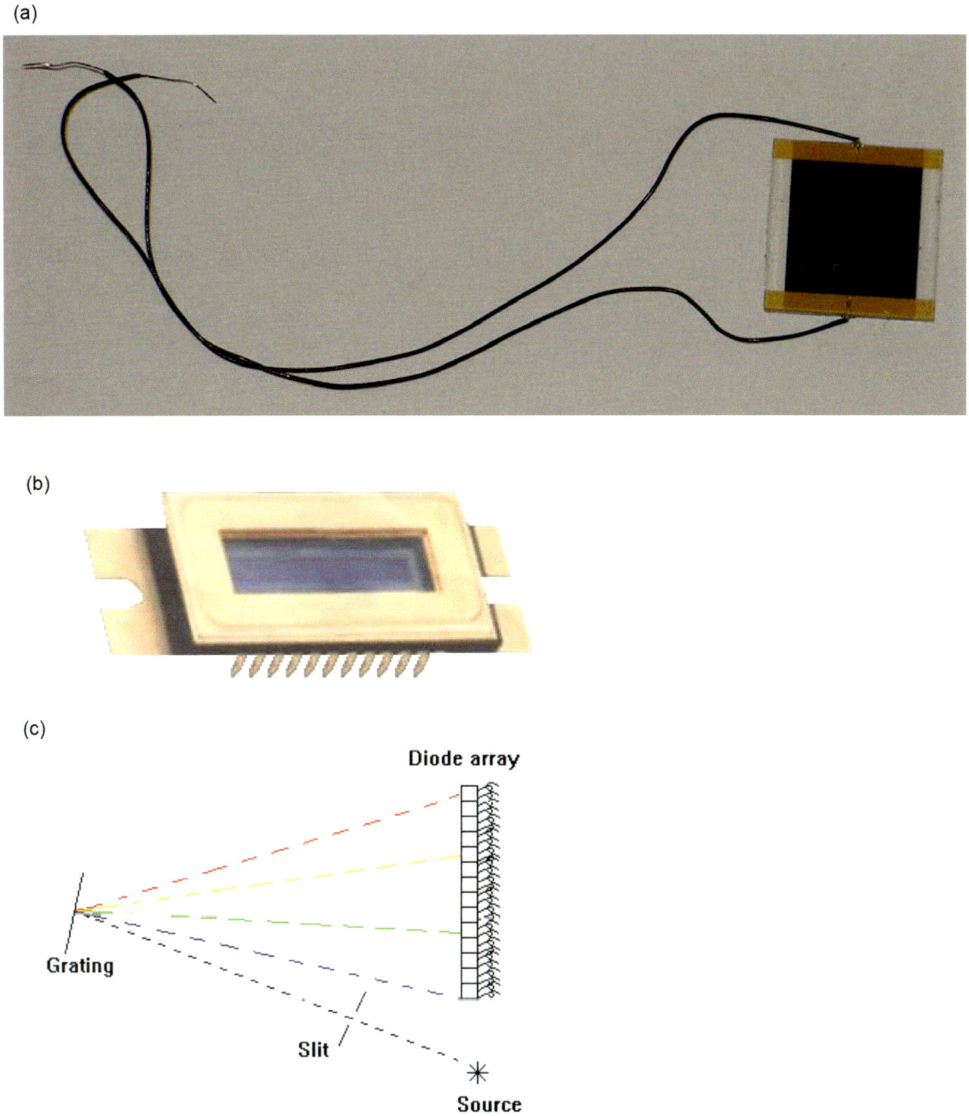

FIGURE 5.6 Pictures of semiconductor detectors and a schematic diagram of how an array detector can be used. (a) 1 × 1 cm PbS detector. The gold bands at the two sides of the detector are the contacts to the photoactive PbS layer. (b) Picture of a diode array detector. The active detectors are protected under an NIR-transmitting window. (c) Schematic of the use of a diode array detector in conjunction with a diffraction grating.

when the best spectral resolution is important. Still others are recommended if the instrument is to be used in mobile applications and ruggedness is critical. Here, we present and discuss some of the common, more widely used methods of selecting wavelengths from "white" light (NIR, actually).

5.6.1 Interference Filters

The simplest and arguably the most basic form of wavelength selector is an interference filter. It is basically a piece of glass, quartz, or other transparent material (commonly flat although focusing versions are available), upon which thin layers of various materials are deposited.

(a)

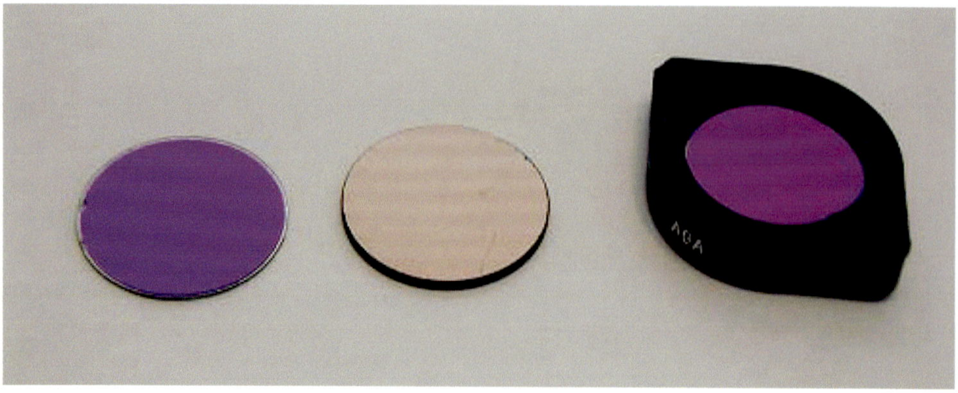

(b)

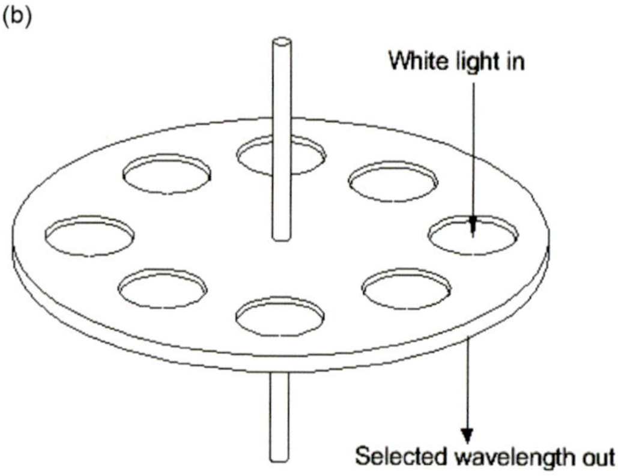

FIGURE 5.7 Interference filter usage. (a) This figure shows two bare (unmounted) filters and one filter mounted in a holder, ready for installation on a filter wheel for an instrument. The various colors reflect (no pun intended) the effect of the interference patterns of the filter, albeit in higher-order interference than the ones operative in the NIR. (b) Schematic filter wheel. Each filter allows a different wavelength to pass through, when the wheel rotates different filters into the beam.

Figure 5.7a shows some individual interference filters, while Figure 5.7b shows a typical filter wheel: a piece of round metal that the filters can be mounted on so that rotating the wheel moves different filters into the optical beam of the instrument, enabling measurements at different wavelengths.

So how does an interference filter work? Figure 5.8a shows how interference phenomena at a thin layer reinforce the energy at the selected wavelength. Figure 5.8a shows a thin film whose thickness equals one wavelength. The interference phenomenon repeats for thicknesses equal to integral numbers of one-half of the wavelength. In between those thicknesses where the transmission is maximum (constructive interference), the amount of light transmitted is reduced, and at one point reduced to zero (destructive interference), also because of the interference phenomenon. In addition, if monochromatic light of varying wavelength is shined on the filter, the intensity of the light passing throughout the range will undergo periodic brightening and dimming because of the way the intensity periodically goes to zero, as illustrated in Figure 5.8b. The net effect is that transmission maxima are at regularly spaced wavelengths, with corresponding regularly spaced wavelengths of zero transmission between the peaks (as shown in Figure 5.8a).

To obtain a single wavelength, these extra transmission peaks have to be removed. Typically, this is done by adding a second layer to the filter, so that the zeros of this layer correspond to every other

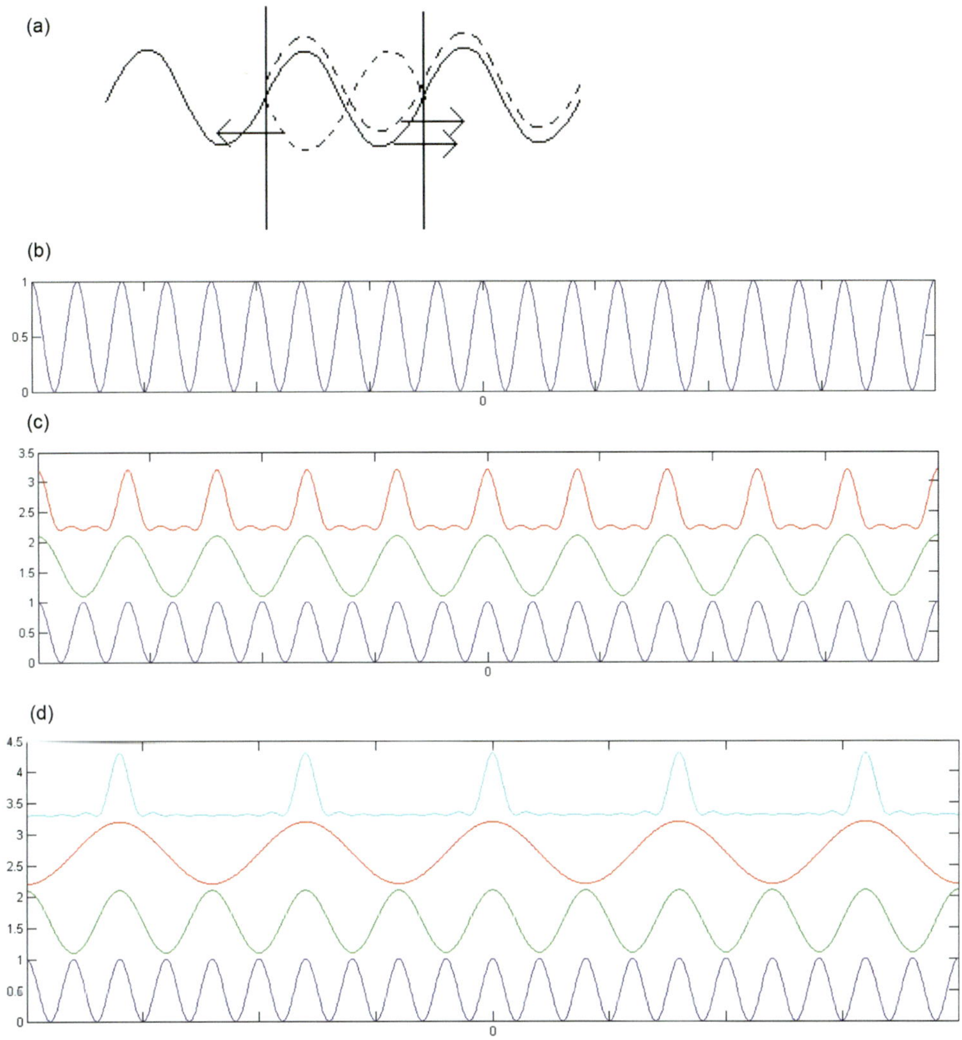

FIGURE 5.8 Behavior of interference filters. (a) Showing how interferences between waves reflected at the two surfaces of a thin layer create reinforcement of the light at a selected wavelength, increasing its brightness. (b) Sinusoidal interference pattern obtained over a wavelength range, as two beams go in and out of constructive interference. The peaks represent the filter's transmission maxima. The zero on the abscissa represents the intended (target) wavelength of the filter. (c) A sinusoid corresponding to a thickness of twice the target wavelength, created by a layer of calculated thickness where their zeros null some of the original peaks that are indicated in (b). The lower spectra indicate the individual layers' transmission behavior, while the resulting wavelength performance is shown in the top spectrum. Note that the zero levels of the top two spectra are offset for clarity. (d) Adding another layer to the filter eliminates half of the remaining transmission peaks, thus removing some of the spurious ones. The lower spectra indicate the individual layers' behavior, while the resulting wavelength performance is shown in the top spectrum. Note that the zero levels of the top three spectra are offset for clarity.

spurious peak of the main filter layer. As shown in Figure 5.8c, some spurious peaks still exist but their number is reduced. Figure 5.8d shows the result of adding still another layer: more of the spurious intensity peaks are removed. Only at the intended target wavelength do all the layers exhibit constructive interference. This process may be continued, but not indefinitely; it becomes more and more difficult to correctly match up the zeros of each new layer with the other existing layers.

Fortunately, that is not necessary. It is only necessary to continue the process to the point where the nearest spurious wavelength is far enough away from the desired wavelength, allowing the use of a material with an absorption edge to block all the wavelengths beyond that point.

5.6.2 Tilting Filters

All the above considerations are based on the assumption that the NIR beam strikes and passes through the filter normal (at an angle of 90°) to the surface of the filter; then, the beam passes through a distance *D* within the filter, as shown in Figure 5.9. If the filter is tilted so that the beam inside the filter is at an angle to the normal, then (also as shown in Figure 5.9) the distance the beam passes through the filter is increased by an amount of 1/cos (θ) and the wavelength of the NIR emerging from the filter is increased proportionately. The effective increase is even greater, which is because the higher refractive index of the filter material creates a further effective increase beyond the strictly geometric one. By rotating the filter while collecting data when the incoming NIR is at various angles to the filter, a limited amount of wavelength scanning can be achieved.

5.6.3 Linear Variable Filters (LVF)

Another way interference filters can be used to measure spectra is to incorporate LVF in an instrument.

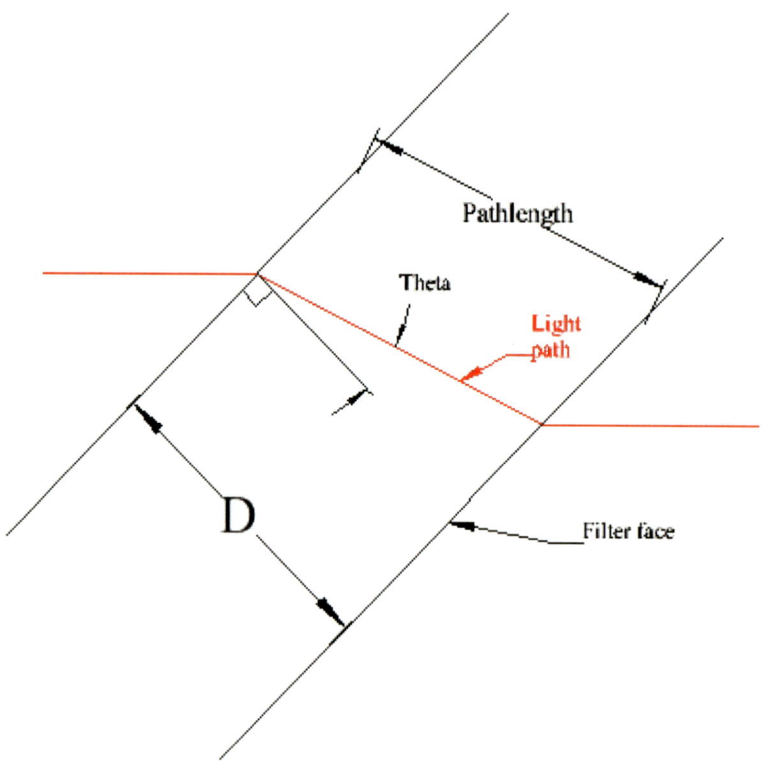

FIGURE 5.9 Showing the light path through a tilting filter. The red lines represent a light beam. When the filter is normal to the beam direction, the light passes through a distance *D* of the filter. When the filter is tilted to some arbitrary angle θ (theta), the light passes through a somewhat larger distance, due to both geometric and optical effects. That increase in path changes the wavelength at which the filter passes the optical radiation.

FIGURE 5.10 A schematic representation of a linear variable filter. This has a continuously varying thickness along its length, causing different parts of the filter to pass different wavelengths. The thickness of the filter, as well as the change in thickness, is greatly exaggerated for clarity.

As Figure 5.10 shows, the layers of a linear variable interference filter are manufactured so as to not be constant over their entire surface but vary in thickness along the length of the (relatively long, narrow) filter. Thus, different places along the filter length undergo the constructive interference phenomenon at different wavelengths.

5.6.4 Diffraction Gratings

Diffraction gratings are flat plates (or, like interference filters, they are sometimes curved so as to allow them to focus the light as well as to separate the different wavelengths) containing parallel lines or grooves on the surface. Historically, they were made mechanically (ruled), but in more recent times they are made optically by using lasers to create interference patterns to form the grooves (holographic gratings). The grooves are spaced apart a distance corresponding to the wavelength range of the radiation to be separated. For use in the NIR region, this corresponds to a groove density of 1000 grooves per mm; this is a very rough ballpark figure because so many factors affect it.

As Figure 5.11a shows, when the light is reflected from adjacent facets of the grating, it can again interfere constructively or destructively. In this case, however, the nature of the interference phenomenon depends on the angle at which the light comes off the grating surface. The net effect is that the light is separated by having the different wavelengths come off the grating at different angles. The grating can then be rotated so that these different wavelengths successively go in the direction to the detector, as shown in Figure 5.11b.

The details about how the light is spread out depend on many factors, all of which affect the design of the grating and of the instrument. The most important factor is the groove density, but other influential factors include the size of the grating (or rather, the area used), the angle at which the polychromatic light hits the grating, the solid angle subtended by the incoming light at the grating surface, whether the grating is flat or curved (either spherically or cylindrically), and the blaze angle of the lines on the grating (the blaze angle is the angle of the lines on the grating). Figure 5.11a shows these angles as being 45°, but other angles can be used. These will affect the properties of the grating and must be carefully considered in the design of the grating and the optical system.

5.7 FOURIER TRANSFORM SPECTROMETERS (FTIR)

5.7.1 Michelson Interferometer

Fourier transform spectrometers (FTIR) are based on the use of an interferometer, most commonly a Michelson interferometer, as shown in Figure 5.12. This interferometer consists of a beam splitter (a partially reflecting mirror), which splits the incoming radiation into two beams of equal intensity at 90° to each other. One of these beams goes to the fixed mirror, while the other to the movable mirror. The mirrors reflect the two beams back to the beam splitter, where they recombine.

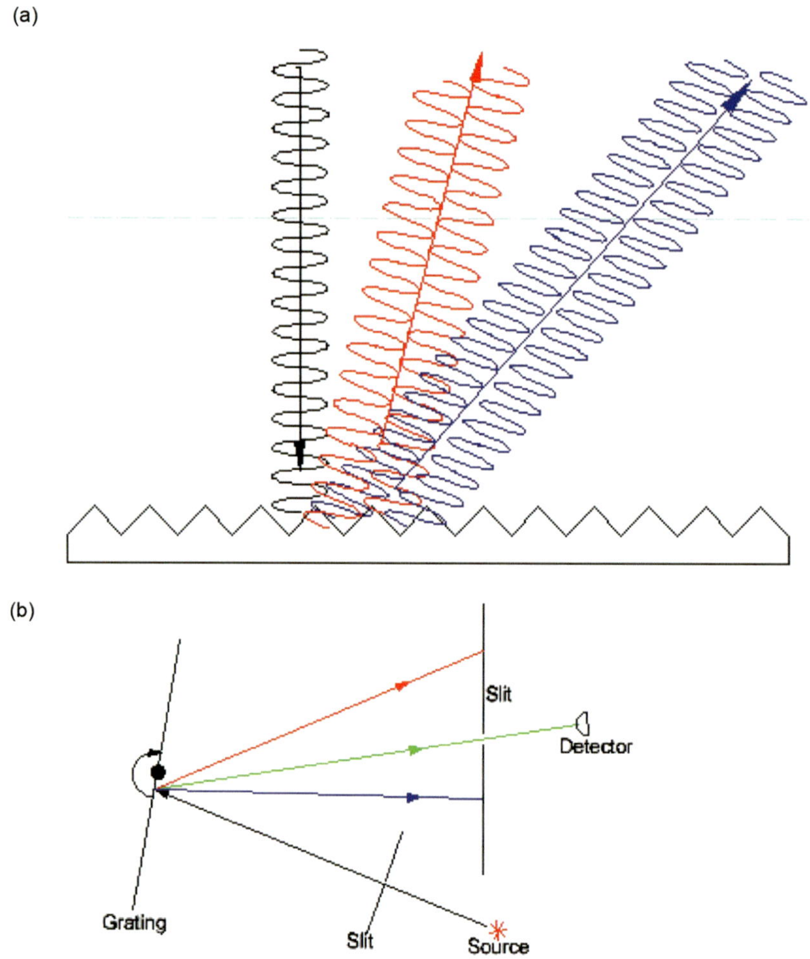

FIGURE 5.11 Operation of a diffraction grating. (a) A diffraction grating separates light of different wavelengths by virtue of the fact that when the light comes off the different facets of the grating, the different wavelengths will constructively interfere at different angles. This separates the wavelengths according to the angle of reflection from the grating. (b) By rotating the grating, different wavelengths are successively passed through the slit and sent to the detector.

As one of the mirrors is moved, it changes the phase relationships between the two beams, causing them to alternately cancel and reinforce. This causes the output beam to be alternately intensified and diminished, as shown in Figure 5.12. The frequency of the alternation between high and low intensity depends on the speed of the mirror and the wavelength of the light. To accomplish this, the key parts of the interferometer have to be manufactured and controlled with wavelength-of-light precision. This seems like a difficult criterion to meet, but with modern technology, it is now a routine operation.

The net effect of this is to code each wavelength in the incoming beam by modulating its intensity sinusoidally at a frequency inversely proportional to its wavelength. Thus, when the beam impinges on a detector, the electrical signal out of the detector also varies sinusoidally for each wavelength contained in the incoming NIR beam. The net output signal is thus the summation of all those sinusoids.

Typically, this signal (called an "interferogram") is recorded by digitally sampling it and saving the data in a computer memory. This raw interferogram signal does not resemble a spectrum but can

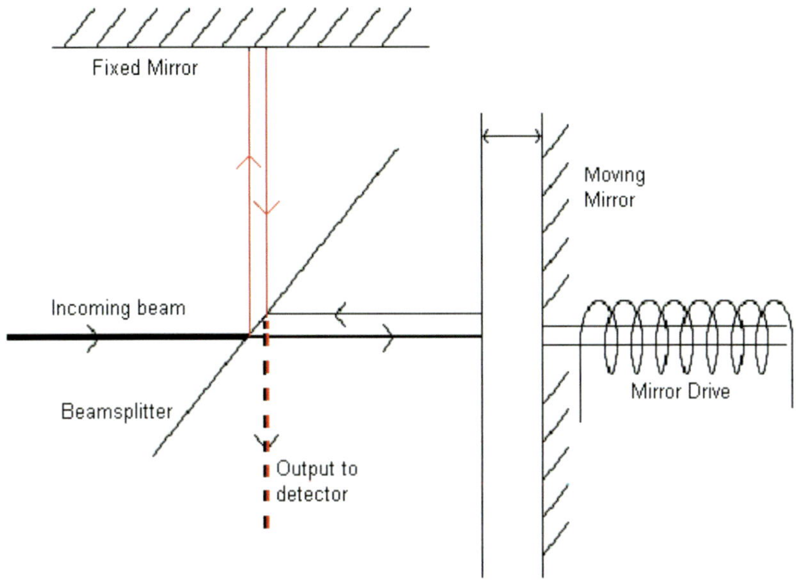

FIGURE 5.12 Schematic illustration of a Michelson interferometer. Fundamentally, it consists of three parts: a beam splitter, a fixed mirror, and a movable mirror. The beam splitter splits the incoming beam into two beams at 90° to each other. One of these goes to the fixed mirror, while the other goes to the movable mirror. The mirrors reflect the two beams back to the beam splitter. Here, the two incoming and returning beams are offset for clarity, although in an actual interferometer they are superimposed.

be converted into a spectrum by performing mathematical operations upon it. The chief operation needed is the mathematical operation known as a "Fourier transform". To prove this requires some heavy-duty math, so we will forego it here. But trust us: it works (not only in NIR, but it has a long history in mid-infrared spectroscopy, too)!

As with new abilities of solid-state devices such as in creating photomultipliers (see the "Detectors" section) so too are monolithic devices taking over what previously the exclusive territory of assembled spectrometers. A recent report reveals a MEMS version of a complete Michelson interferometer-on-a-chip [6].

The spectrum obtained by performing the Fourier transform on the data is the same as that obtainable from a diffraction grating-based instrument with one major difference: unless special steps are used to interpolate the data points, the resulting spectrum is found to be measured at equally spaced frequency intervals (conventionally called "wavenumbers") instead of at equally spaced wavelength intervals as a diffraction grating-based instrument produces. Thus, comparing spectra from the two types of instruments can sometimes be confusing, especially to inexperienced spectroscopists.

5.7.2 Polarization Interferometer

In the history of FTIR, many types of interferometers have been proposed and manufactured. Currently, the Michelson interferometer is the one most commonly used in NIR spectrometers. There is another type of interferometer, however, that is currently available in commercial NIR instruments: the polarization interferometer.

The polarization interferometer uses the same underlying optical principles as the Michelson interferometer, of having two beams interfere with each other, but accomplishes it differently. Instead of separating two beams spatially, the two beams take the same path through the interferometer but are distinguished by the polarization properties of the two beams.

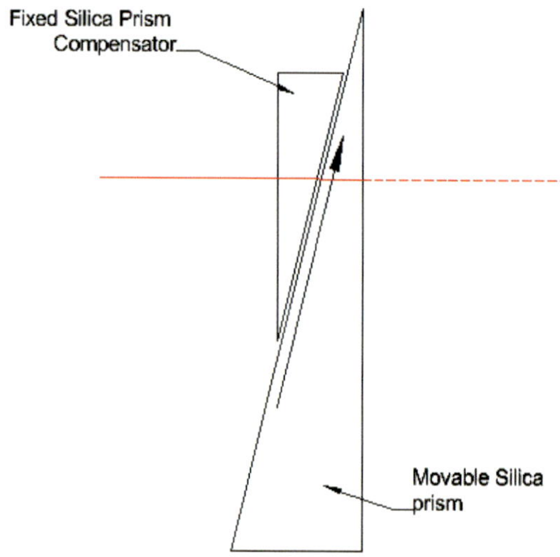

FIGURE 5.13 Polarization interferometer – see text for discussion of operation.

As shown in Figure 5.13, the interferometer consists of two prisms made from crystalline silica (quartz). Crystalline quartz is optically active, which means that the speed of light through a piece of quartz is different depending on the polarization of the light. Not shown in Figure 5.13 are auxiliary polarizers and other optical components, but the net result of passing through the quartz prism is that one polarization of the light is retarded compared to the other polarization. Since the main prism can be moved in the direction of the large arrow, the light will then pass through different thicknesses of quartz, and the amount of retardation of one beam compared to the other will change. This change in retardation of the light gives rise to an interference phenomenon with alternate reinforcement and cancelation of the two beams as the quartz prism is moved, even though they're in the same block of material. The interference phenomenon happens wavelength by wavelength, just the same as in the Michelson design, and the modulated light exits the prism from the opposite face from where it came in.

An advantage of the polarization interferometer compared to the Michelson design is that the interference effects are generated within the quartz block. Thus, since both beams travel through the same piece of quartz, vibrations or other mechanical phenomena cannot affect the interference phenomenon.

5.7.3 Acousto-Optic Tunable Filter Spectrometer (AOTF)

Figure 5.14 presents a schematic diagram of an acousto-optic transducer. At the heart of the transducer is a crystal with some special properties; one is that it is optically active. Another is that it exhibits Bragg's diffraction. While not the only type of crystal may be used, a common crystal useful for NIR spectroscopy is tellurium dioxide (TeO_2). This material is optically active, much as the crystalline quartz is. In addition, it exhibits Bragg's diffraction under the proper circumstances.

The key to using a suitable crystal to perform spectroscopy is that it can be excited with high-frequency sound waves. Then, a suitable transducer must be attached to the crystal, so that when high-frequency signals are applied to the transducer, it vibrates and emits sound waves into the crystal.

When the crystal is excited by a suitable source of sound waves, the structure of the crystal allows it to mediate an interaction between the sound waves and light waves in the crystal. A typical sound frequency useful for operation in the NIR range is ~90 to ~100 MHz. The non-quantum

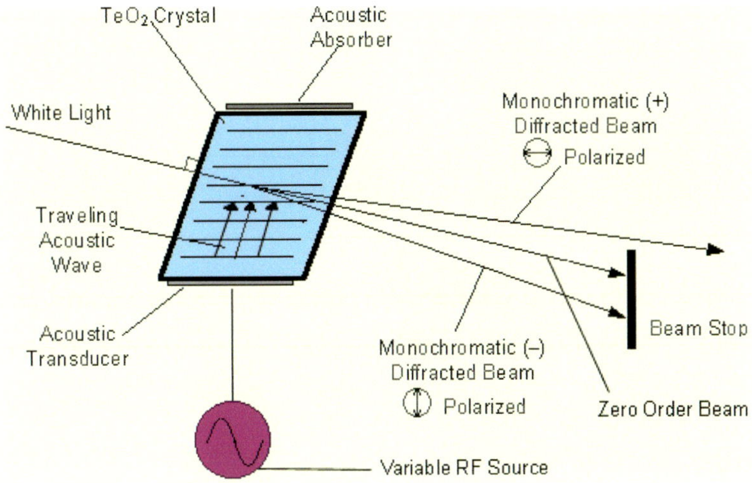

FIGURE 5.14 Acousto-optic prism – see text for details of operation. (Picture courtesy of Brimrose Corp of America.)

mechanical explanation for the interaction is that the sound waves compress and rarify the structure of the crystal, so that density differences exist in the crystal. When the size of the density differences matches the wavelength of light in the crystal, the crystal can act as a diffraction grating for that wavelength of light and cause it to deviate from the beam (this is the Bragg diffraction mentioned above).

By changing the frequency of the Radio Frequency (RF) signal, different wavelengths of light can be separated from the incoming beam of white light. As Figure 5.14 shows, two beams emerge from the crystal, one on either side of where the remainder of the white light exits. This dual diffraction is caused by the optical anisotropy of the crystal, which – since the crystal is optically active – causes the "ordinary" ray to emerge in one direction and the "extraordinary" ray to emerge in the opposite direction. Because the crystal is optically active, these two rays from nominally the same input wavelength have different frequencies, and therefore different wavelengths, within the crystal. The two rays also have slightly different wavelengths when they emerge from the crystal. Careful work requires this difference to be taken into account when analyzing any results from this type of measurement.

The factors determining which wavelength will be separated from the beam are as follows:

1. The crystal used,
2. The frequency of the RF signal,
3. The temperature of the crystal,
4. Whether the beam represents the ordinary ray or the extraordinary ray.

5.7.4 Micro-Mirrors

Figure 5.15 shows a schematic demonstration of the use of micro-mirror arrays to select one wavelength out of a spectrum. As the figure shows, the grating spreads the incoming white light such as to cover the micro-mirror array, giving each mirror a portion of the spectrum. Undesired wavelengths are directed away from the sample or detector, and one mirror at a time is activated so as to direct the wavelength of interest further down the optical system to the detector. A spectrum is generated by activating mirrors sequentially, so that a continuum of wavelengths is generated.

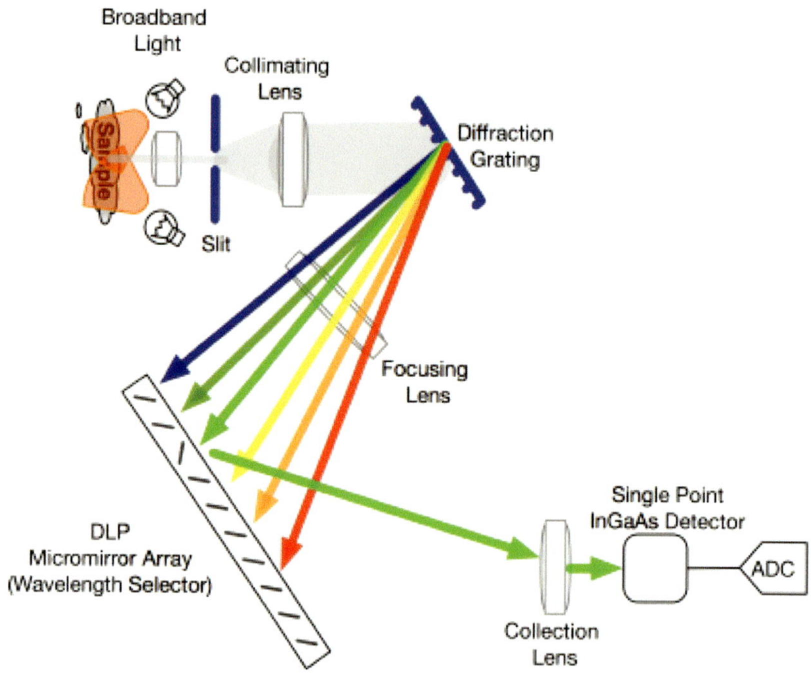

FIGURE 5.15 Schematic description of the use of a micro-mirror array. One mirror, which is flipped, directs its wavelengths' energy to pass through the optical system to the detector. All the other mirrors direct their corresponding wavelengths away from the active parts of the instrument and toward where they will be harmlessly dissipated. (Courtesy of Texas Instruments DLP® Products Traditional versus DLP technology-based spectrometer (2015).)

The use of micro-mirrors mimics a diode array, in that each mirror (or each detector in a diode array) controls whether a given wavelength will contribute to the measured spectrum. In a diode array, however, all pixels are active all the time, and while data from some might be ignored, they still require time and resources to measure. With micro-mirrors, on the other hand, setting a mirror to the "measure" position is independent of the status of any other mirror. This makes it possible, for example, to have two or more mirrors direct their corresponding wavelength energy to the detector and thus measure multiple wavelengths at once. Having these capabilities would have enabled Bill Fateley to conduct his Hadamard transform experiments in a near-ideal manner.

5.8 MODERN INSTRUMENTS

The above-described technologies, and some even more exotic and arcane, have been incorporated into instruments that add new capabilities to the repertoire of analyses that NIR can accomplish. These were developed over the 30 or so years since the days when there were only three manufacturers of commercially available modern NIR spectrometers on the market. Now there are well over a dozen providers of instruments, not counting those companies that make devices but don't necessarily call them "NIR spectrometers". A company could manufacture, market, and sell, for example, an "oil analyzer" based on NIR principles, but nobody would ever need be told about the technology being used. In the medical arena, for example, the pulse oximeter is widely used to measure blood oxygen content based on two carefully chosen NIR wavelengths, where filters are used to separate the important wavelengths from the spectrum, and data processing is used to eliminate errors due to pulse, stray light, and other known error sources.

5.9 THE TRANSITION TO MODERNITY

At the 1993 Pittsburgh Conference (PittCon), the 1600 aisle of the exhibit hall was completely blocked by throngs of conferees all trying to see (what was then) an amazing sight. Trying to push my way through the aisle reminded me of my undergraduate days at City College of New York (CCNY) when I had to push my way through the rush-hour crowds on the subway. What was the cause of all the excitement? A "new kid on the block" company (Ocean Optics) was exhibiting what was then a completely novel and unexpected type of instrument: a spectrometer that was built completely on a plug-in PC card (in the era when plug-in PC cards were the standard way to install accessories into the computer). While strictly speaking, that was a visible spectrometer rather than an NIR spectrometer, that device was the prototype for a completely disruptive technology for the time, and making the spectrometer simply another accessory for the computer, rather than a separate instrument, was the harbinger of instrumentation to come for NIR as well as all other types of spectrometers. See Figure 5.16 for a picture of that ground-breaking instrument.

Once it was shown that it could be done, manufacturers have evolved modern instruments in various ways. One major impetus has been to make them small and light so that can be brought to the sample (instead of the sample to the instrument). Here, we have various manufacturers' designs that move the technology in a modern direction. Figures 5.17–5.25 show (in alphabetical order by company name) some modern commercial instruments, incorporating the above-described technologies to create instruments that are small, light, and often portable. Another path of development was to create imaging instruments, named "hyperspectral imaging". These instruments measure spectra in an array of pixels; these allow the creation of images of optical behavior at any wavelength in the range of the spectrometer. Alternatively, if calibration model can be generated, an image of the concentration of any component in the sample can be displayed. This allows the user to directly detect heterogeneity and "hot spots" of excess concentration or "deserts" of low concentration or missing components.

There are some commonalities among the instruments. For example, a fiber-optic connector to bring the (NIR) light in is almost universally used. The portable instruments all require external battery power, so a supply of either power (e.g., from a USB port) or batteries must be arranged.

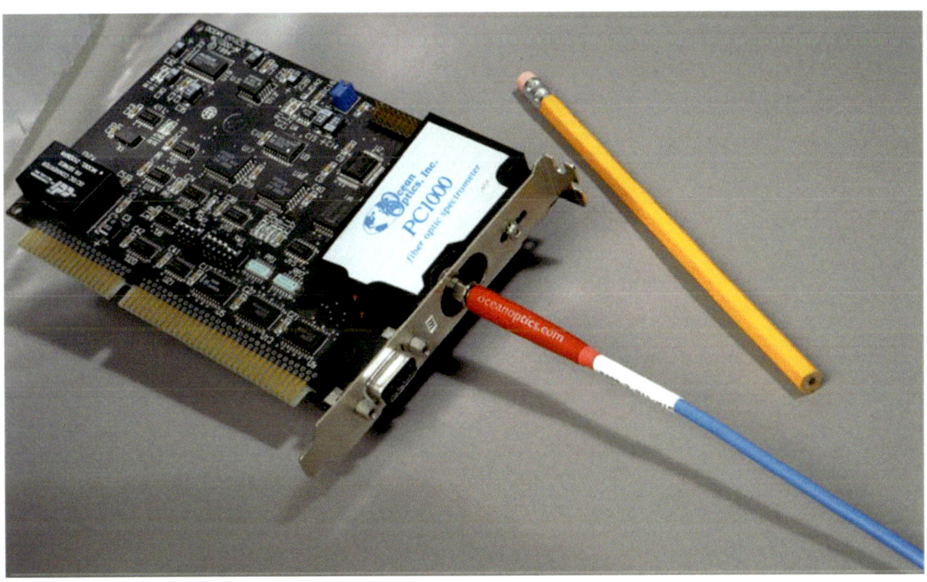

FIGURE 5.16 Ocean Optics' PC-1000 spectrometer on an IBM-PC ISA circuit board, which created a stir at the 1993 Pittcon Conference. (Picture used with permission from Ocean Optics.)

FIGURE 5.17 Bayspec SuperGamut Volume Phase Grating spectrometer. (Picture used with permission from Bayspec, Inc.)

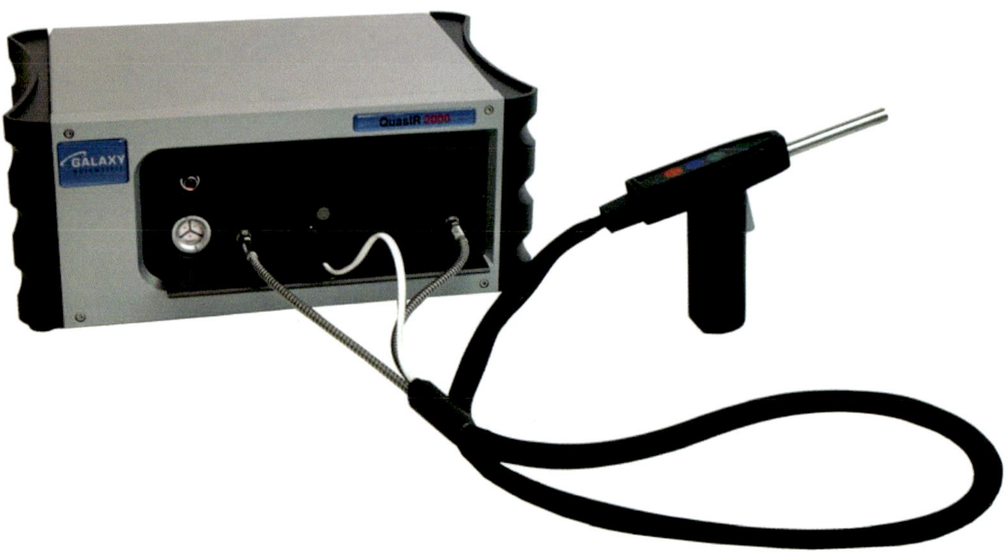

FIGURE 5.18 Galaxy Scientific's portable QuasIR-2000 interferometer-based spectrometer with hand-held probe. (Picture used with permission from Galaxy Scientific.)

Some argue that these are not "instruments" because they are not complete, and they are missing one (or more) of the key components mentioned in the beginning of this chapter. By requiring an external source of NIR radiation, they are, indeed, not complete spectrometers. However, I am not interested here in arguing terminology. Since the wavelength-separation capability distinguishes

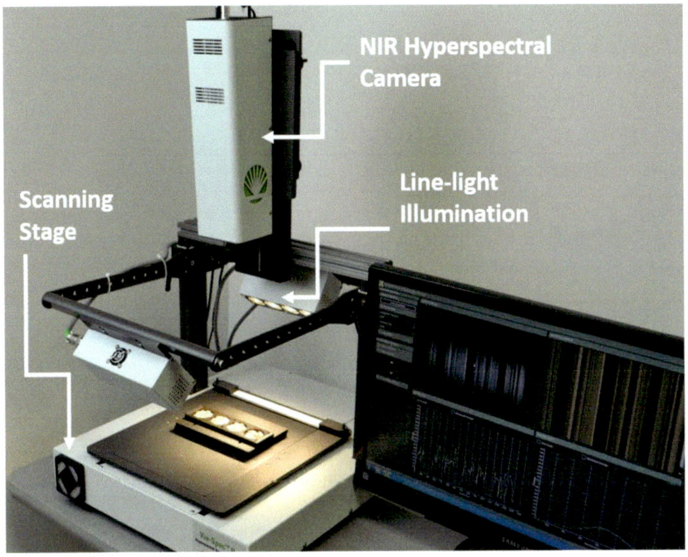

FIGURE 5.19 Middleton Hyperspectral Camera (imaging spectrometer) Model MSV-640 based on diffraction grating plus InGaAs diode-array detector, shown with a ViaSpec™-II scanning stage. (Picture used with permission from Middleton Instruments.)

FIGURE 5.20 This is the Ocean Optics' Ocean-FX diffraction grating-based high-speed spectrometer. (Picture used with permission of Ocean Optics Corp.)

FIGURE 5.21 Si-Ware NeoSpectra FTIR spectrometer on a smartphone, being used to measure a hamburger roll. (Picture used with permission of Si-Ware Corporation.)

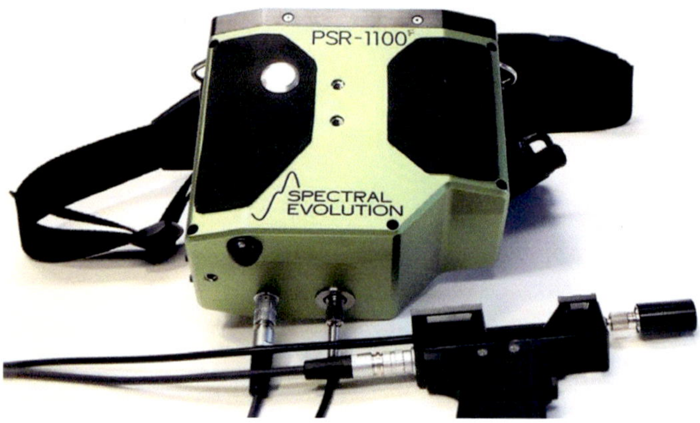

FIGURE 5.22 Spectral Evolution's PSR-1100 diode-array instrument, showing the portable spectrometer with fiber-optic probe and sample cell. (Picture used with permission from Spectral Evolution.)

FIGURE 5.23 This is the StellarNet Dwarf Star diffraction grating spectrometer. (Picture used with permission of Stellarnet Corp.)

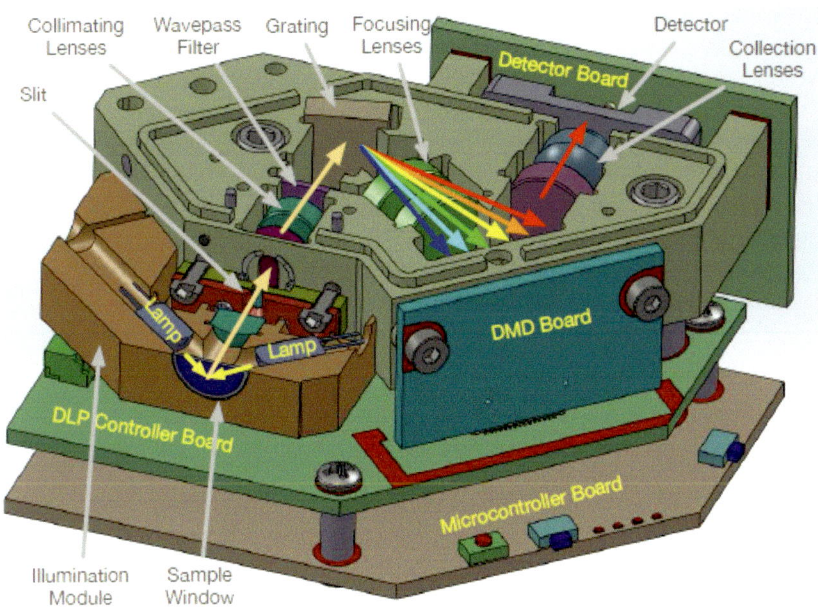

FIGURE 5.24 Texas Instruments' DLP NIRscan Nano micro-mirror-based spectrometer. (Courtesy of Texas Instruments DLP® Products DLP® NIRscan™ Nano optical engine (2015).)

FIGURE 5.25 Micro-NIR spectrometer from Viavi Corp. using linear variable filter and diode array detector. (Picture used with permission from Viavi Corp.)

these spectrometers from other types of optical instruments, for my purposes here I will accept the "spectral engine" (as some call it) as being the key part of a spectrometer, even if a source and detector and possibly other components have to be added to it to "complete" it.

And of course, technology does not repeal the laws of physics. Each of the instruments described above also has its limitations, dictated by the engineering trade-offs that must be made to achieve their main purpose for the instrument. A user must therefore carefully compare their analytical needs to the capabilities of any instrument under consideration and assure themselves that instrument can in fact solve their analytical problem.

REFERENCES

1. Herschel, W.; *Phil. Trans. Royal Soc. London*; 90 p. 255–283 (1800).
2. Herschel, W.; *Phil. Trans. Royal Soc. London*; 90 p. 284–292 (1800).
3. Herschel, W.; *Phil. Trans. Royal Soc. London*; 90 p. 293–329 (1800).
4. Kaye, W.; *Spectrochim. Acta*; 7 p. 181–204 (1955).
5. Patek, S., Hergert, E.; *Photonics Spectra*; 51 (10), p. 54–58 (2017).
6. Said, A. K.; *Photonics Spectra*; 51 (10), p. 48–52 (2017).

6 NIR
21st-Century Innovations

*Sean M. Christian**
Neolitics

Jess V. Ford
Weatherford

CONTENTS

* schristian@neolitics.com

6.1 A BRIEF HISTORY OF NIR SPECTROSCOPY (NIRS)

It is difficult to dispute the important role industry has played in the development of the modern-day near-infrared (NIR) spectrometer. The brief history on the development of NIRS that follows is provided to give an insight into the three-and-a-half centuries of innovation that have led to the development of the 21st-century NIRS.

6.1.1 The Development of the Spectroscope [1666–1896]

There is a long-standing and well-documented history of innovation related to spectroscopy. While the first known use of the word "spectroscopy" occurred in 1882 during a lecture by Arthur Schuster at the Royal Institution in London, England, the origins of spectroscopy predate Schuster's lecture by at least 226 years.

The first well-documented and systematic observation and measurement of electromagnetic radiation dates to 1666, when Sir Isaac Newton recorded the separation of sunlight into bands of colors using the sun, a small circular aperture, and a prism. In a 1672 publication entitled "A New Theory about Light and Colours" [1], Newton coined the word "spectrum" [def. Latin: apparition] to describe the colored images that were hidden within the light, which danced around like ghosts with the movement of his prism.

While one could argue that Newton's apparatus was the first spectroscope, having a source (the sun), a detector (the human eye), and a dispersive element (a prism), there are those that argue since he didn't have a slit or bandwidth limiter, it wasn't a spectroscope. In fact, it wasn't until 1802, close to a century and a half later, that William H. Wollaston, through the introduction of a slit, observed the first-line spectrum consisting of discrete lines emanating from the sun [2]. As a matter of principle, one must ask, should Wollaston, the creator of the first monochromator reduced to practice, be credited with the development of the spectroscope?

Unfortunately, Wollaston did not recognize the significance of his innovation nor the observations he made using it. Twelve years later, credit for the discovery of the sun's spectral lines went to Joseph von Fraunhofer, who it is said independently conceptualized the same innovation as Wollaston in 1814 and used it to catalog 576 narrow, dark spectral lines of the sun [3].

It is important to note that Fraunhofer's cataloging of the lines that still bear his name today, the "Fraunhofer lines", would never have been possible without another innovation, the diffraction grating, which he made from winding fine silver wire on a small frame.

David Rittenhouse of Philadelphia is credited with the development of the first grating [4], which was nothing more than hairs strung between two fine-pitched screws.

The world's first spectroscope was invented in 1859, by a physicist, Gustav Robert Kirchhoff, and a chemist, Robert Wilhelm Bunsen, whose innovation led to the development of the first optical source, the Bunsen burner. Oddly enough, the Bunsen burner produced very little "spectrum", but this is what made it ideal for Kirchhoff and Bunsen's research, which involved the measurement of spectral emission lines from common salts that possessed "the power of developing peculiar bright lines in the spectrum of a flame into which they were introduced" [5]. Some of the first data recorded using their spectrograph is shown in Figure 6.1 to the left of the solar spectrum first observed by Fraunhofer. In principle, it is important to note that Bunsen's burner wasn't the source which powered the first spectroscope. The source was the flame-ionized elements that were introduced to the Bunsen burner that enabled the first spectrograph.

Using their spectroscope, Kirchhoff and Bunsen set out to acquire data that resulted in a groundbreaking publication entitled "On Chemical Analysis by Spectrum-Observations" [6]. This seminal body of work not only laid the ground works for modern-day spectroscopy but also allowed the 19th-century scientists to explain Fraunhofer's lines – an explanation that had eluded them for half a century.

In order to perfect Kirchhoff and Bunsen's design as shown in Figure 6.2, they collaborated with an optician Carl August von Steinheil to engineer a commercial spectroscope, which allowed

FIGURE 6.1 An illustration of line spectra taken between 1859 and 1860 by a physicist, Gustav Robert Kirchhoff, and a chemist, Robert Wilhelm Bunsen, of six high-purity salts annexed besides the solar spectrum for comparison. Cleary shown are the solar "dark lines" first discovered by William H. Wollaston in 1802 and later catalogued by Joseph von Fraunhofer in 1814. Today, the dark lines are still referred to as "Fraunhofer lines", a testament to the importance of the early 19th-century discovery.

FIGURE 6.2 An illustration of the apparatus used by Kirchhoff and Bunsen between 1859 and 1860 to measure the line spectra of various high-purity salts. With the addition of the Bunsen burner as a source, it is recognized as the world's first spectroscope. The above spectroscope was the result of multiple innovations dating back to 1666 and is still being manufactured today.

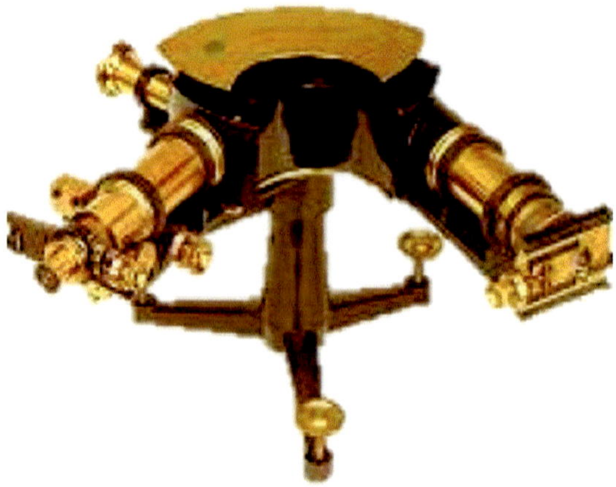

FIGURE 6.3 A photograph of an early 19th-century commercial spectroscope manufactured and distributed by C.A. Steinheil & Söhne of Munich Germany.

contemporaries of Kirchhoff and Bunsen to discover helium on the Sun 27 years before the element was discovered on the Earth and prior to the turn of the 20th century. Steinheil and Söhne's spectroscope as shown in Figure 6.3 latter led to the characterization and identification of several noble gases and elements that would eventually be predicted by Mendeleev's periodic law.

6.1.2 The Discovery of NIR Radiation

It is important to note that the early spectroscope relied on the human eye for the detection of spectral lines. It wasn't until 1800 that William Herschel noticed light passing through colored filters transferred different amounts of heat to each filter. In order to explore this phenomenon, Herschel set out to determine the "heat" energy of colors by placing a thermometer across multiple points within the sun's color spectrum. This led to an unexpected discovery. Specifically, Herschel noticed that the largest temperatures existed beyond red, where there was no visible light that could be seen by the human eye. He went on to determine that these "heat rays" could be transmitted, reflected, and absorbed just like visible light. This portion of the electromagnetic spectrum became known as **infrared** ("below red") radiation.

6.1.3 The First Spectrophotometer

It wasn't until Leopoldo Nobili and Macedonio Melloni exploited the thermoelectric effect discovered in 1822 by Thomas Seebeck to develop the first thermomultiplier that the modern-day spectroscope began to take shape. Their seminal work ultimately led Arnold O. Beckman to conceptualize the 20th-century spectrophotometer in 1939. The effort to commercialize Beckman's innovation took place at National Technologies Laboratories and was led by Howard H. Cary.

It is said that Bechman's DU Spectrophotometer reduced a chemical analysis process from weeks to minutes and improved the accuracy of the result from 25% to 99.99%. Within 12 months of the release of the Model A spectrophotometer, which utilized a glass prism, the Company released a Model B which utilized a quartz prism for improved UV wavelength performance, and later, a Model C with improved resolution. In 1941, it released the Model DU, which effectively remained unchanged until 1976 when it was discontinued. Bruce Merrifield, a Nobel Laureate who received the Nobel Prize for Chemistry in 1984 for the development of an ingenious method for synthesizing chains of amino acids or polypeptides in any predetermined order, once said that the spectrophotometer was "probably the most important instrument ever developed towards the advancement of bioscience".

In 1952, Karl Norris and Wade Brant, using a modified version of the Beckman DU, set out to develop a method for detecting blood in white-shell eggs for the US Department of Agriculture to grade and ultimately sort eggs [7]. While the NIR spectrum generated by the Beckman showed an early promise for the development of a real-time sorting method, it was later found to have no correlation with egg quality. It would be close to a decade later that Karl Norris would revisit the NIR region to measure the moisture content in seeds using a Perkin-Elmer UV/Vis/NIR scanning spectrophotometer [8]. Unfortunately, the instrument he utilized struggled with highly scattering media, so Karl proceeded to develop a custom spectrometer for analyzing moisture in grains and oilseeds [9]. It can be argued that the field of NIR spectroscopy (NIRS) truly took root after Norris published the results of these efforts.

6.1.4 Twentieth-Century Innovations

Over the course of the next 35 years, NIR instrumentation evolved. The advancements made during this time period are thoroughly covered in this chapter on traditional instrumentation. While extremely important to the field of NIRS and still prevalent throughout industry, filter, scanning monochromator, Fourier transform (FT), acousto-optic tunable filter (AOTF), and diode array-based NIR platforms are all considered the 20th-century innovations and will not be reviewed within this chapter.

6.1.5 Twenty-First-Century Innovations

Typically, technological advancements are driven by either necessity or curiosity. While Plato believed that necessity was the mother of all invention, there are those that would argue it is curiosity, which has led to some of the most significant innovations in history. Steve Jobs agreed, for he once said, "Much of what I stumbled into by following my curiosity and intuition turned out to be priceless later on". Why curiosity? Walt Disney may have hit the nail on the head when he stated, "When you're curious, you find lots of interesting things to do". Even Albert Einstein believed in the power of curiosity, for he once said "I have no special talents. I am only passionately curious".

When considering the historical development of the NIR spectrometer chronicled above, it is fair to say that innovations related to NIR instrumentation have been proportionately driven by both forces, especially when one examines the developments of the past 15 years, which have led to the creation of several MEMS-based NIR platforms.

To appreciate the current state of NIRS as it relates to these new platforms, a brief review of (1) the motivation behind the past 15 years of development and (2) optical MEMS is mandated.

6.1.5.1 The Driving Force for the 21st-Century Innovations

The world is currently experiencing the Fourth Industrial Revolution, appropriately titled Industry 4.0. It is the current automation trend in manufacturing that encompasses cyber-physical systems, the Internet of Things (IoT), and cloud computing, which are the critical attributes of any "smart factory".

Smart factories allow for a real-time process optimization, which is critical for enabling the economic viability of many consumer products. Low yields, poor quality, and reduced throughput limit the commercial practicality of many products.

The smart factory movement is so pervasive throughout industry that a special term, Industrial IoT (I_2oT), has been coined to describe the market servicing the requirements of today's smart factories. Last year, I_2oT garnered over \$200B of the \$1.2T in investment that poured into the IoT sector, making I_2oT the largest segment of the IoT sector to date.

Since the mid-1980s, the NIR spectrometer has been a mainstay analytical tool for the process analytical chemist who is responsible for converting raw materials into finished goods. In support of this statement, multiple sources estimated the NIRS market is expanding at a compound annual growth rate (CAGR) approaching 10%, second only to Raman spectroscopy. While this growth rate is exciting to all instrument manufactures, what is more exciting to NIR spectrometer manufacturers is the fact that these same sources claim that only 25% of the NIR analyzers in the marketplace are being utilized in a real-time industrial setting. In other words, over 75% of the NIR technology in the marketplace is being utilized either at-line, de-coupled from the live process, or offline within a laboratory environment, which is the antithesis of a smart factory solution. This dichotomy has created a real and apparent opportunity for NIR instrument manufacturers and has clearly been the primary driver spurring the development of MEMS-based 21st-century NIR platforms.

While the size, expense, and complexity of exiting in-line industrial solutions have received a valid criticism from industry experts, the primary factors limiting the deployment of NIR within industrial automation strategies have been twofold:

1. Existing NIR solutions require a significant manual interaction (i.e., labor) to maintain the technology in an industrial in-line or at-line setting,
2. Manufactures struggle to scale successful NIR automation strategies within their manufacturing environments due primarily to the instrument-to-instrument variability.

While industry has been investing aggressively in connected devices, cloud computing, advanced analytics, and machine learning, in parallel, NIR instrument manufacturers have invested in smaller, lower-cost, more robust, but perhaps most importantly, ultra-reliable I_2oT NIR platforms. Why? These manufacturers understand that at the core of an I_2oT solution is the connected device (i.e., the NIR spectrometer). Further, they understand that the reliability of the device is paramount to the success of any automation strategy.

6.1.5.2 Optical MEMS and NIR (MOEMS-NIR)

In 1965, Nathanson and Wickstrom developed what many consider to be the first MEMS device, a frequency-selective micro-electromechanical system (Figure 6.4) [10]. In Nathanson's era, the technology was known to only a select few, but today, over half a century later, the acronym evokes visions of extremely small machines even in the minds of the common, nontechnical, consumer. What the common consumer might not know is that to be considered as a MEMS device, the microfabricated device must have both an electrical and a mechanical or nonelectrical component.

In 1993, Dr. M. Edward Motamedi coined a new term, "MOEMS" (micro-optical electromechanical systems), which he introduced during an invited talk at the SPIE Critical Reviews of Optical Science and Technology conference in San Diego (Figure 6.5) [11]. He described MOEMS as an integrated micro-optical electromechanical system.

For over a decade, Texas Instruments (TI) had been working on the development of micromirror devices. During his presentation, Dr. Motamedi reported on a laser scanner that utilized a programmable or adaptive optical element: a micromirror that would one day enable a plethora of adaptive optical applications.

The technology Dr. Motamedi spoke of would later become known as a "digital micromirror device" (DMD) and find commercial success in the consumer electronics industry, where it served as the core technology for the rear-projection TV (RPTV) market and later, the digital projector market.

Due to the breath of applications that were being developed around the DMD at the turn of the 21st century, TI appropriately branded the technology as the Digital Light Processor (DLP™), and it is fair to state that the DLP ushered in a new era for optical technologies. The era of adaptive optical systems, or more apropos, is the era of MOEMS.

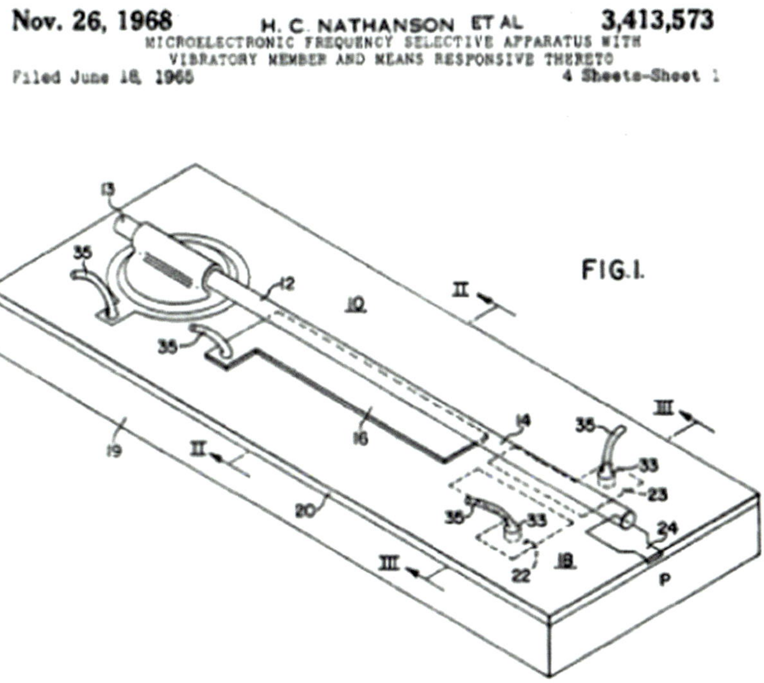

FIGURE 6.4 Illustration of a frequency-selective MEMS device that is considered the first MEMS device every patent. The patent was filed on June 18, 1965.

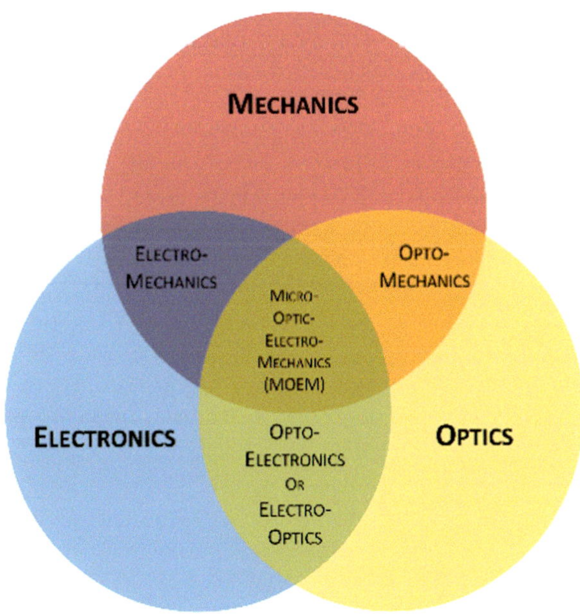

FIGURE 6.5 Reproduction of a figure first introduced by M. Edward Motamedi at an SPIE conference in San Deigo in 1993.

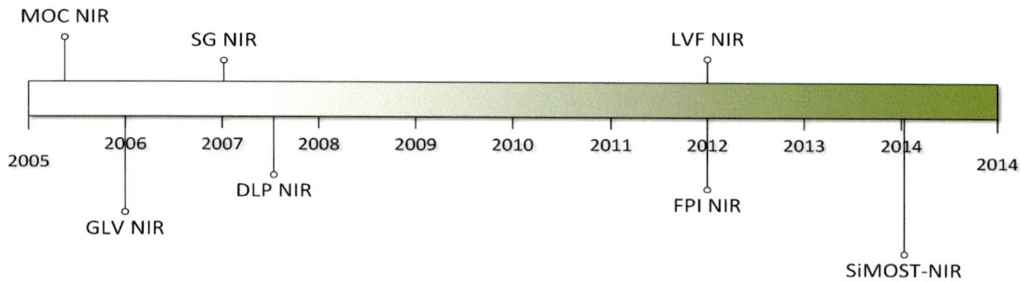

FIGURE 6.6 Timeline that chronicles the development of seven MOEMS-NIR innovations which have entered the commercial market since the turn of the 21st century with notable commercial success.

TABLE 6.1
Technology Platforms Developed in the 21st Century

Platform Technology	Monochromator Configuration
Multivariate optical computing (MOC)	Encoded filter
Grating light valve (GLV)	Bandwidth limiter – grating light valve
Scanning grating (SG)	Bandwidth limiter – grating-microscanner
Digital light projector (DLP™)	Bandwidth limiter – grating-micromirror array
Linear variable filter (LVF)	Variable filter
Fabry–Perot interferometer (FPI)	Micro-actuated optical cavity

Of the seven 21st-century MOEMS-based NIR platforms chronicled in Figure 6.6, it is interesting to note that the platforms leverage similar source and detector technologies. Where they differ, and where the true innovation lies, is within each platform's monochromator. Consequently, the following reviews will focus on the monochromators in each platform and exclude any discussion of integration of the monochromator with either a source or a detector, unless the innovation is intimately coupled with one of these elements. Table 6.1 summarizes each platform's monochromator innovation.

As shown in Figure 6.6, there was a flurry of activity in the mid-2000s that resulted in the commercialization of four MOEMS-based NIR platforms, followed by a 5-year draught of MOEMS-based NIRS innovations. Again in 2012, another wave of innovations resulted in the release of two additional inventions, followed by a third in 2014. Of particular note regarding Figure 6.6 is the fact that there have not been any notable commercial innovations since 2014. Section 6.3 will attempt to address the question of whether there appears to be another wave of technology innovations in the pipeline, which shows a commercial promise for NIRS.

In our estimation, the NIR industry CAGR won't expand by 9%–12%, as predicted by multiple industry analyst, it is in a position to expand much faster due to innovations spurred by the latest industrial revolution, Industry 4.0. Further, it is very probable that the innovations being development today, behind closed doors, will lead to a future in which NIRS is not only a mainstay technology in industry but also a pervasive technology within everyday consumer products.

6.2 TWENTY-FIRST-CENTURY PLATFORMS

Since each technology platform is unique, the reviews are ordered solely by the chronological date when the technology first entered the marketplace and by no other weighting. In reading the reviews that follow, one should come to appreciate that there is not a clear "leading" technology platform. It should also be noted that each platform could have multiple commercial options, and where necessary, each option will be presented.

For many technologists, this is a hard concept to grasp. How can there not be a leading technology platform? For the applied spectroscopist, it is easy to understand how two or more competing technologies can exist in the marketplace at the same time with no clear leader. The spectroscopist understands that commercial and/or industrial NIR applications each have their own unique set of application-specific requirements that dictate whether a platform technology is a viable solution to a particular application. Specifically, the spectroscopist must review the technology platform's technical attributes or specifications, and decide which mix of attributes are a best match to their application requirements. While this sounds difficult and potentially subjective, there are six key technology attributes (KTAs) that can be used to assess the viability of a given platform for each application.

The KTAs are as follows:

Integrated Drift Compensation (IDC):	Unit Price (Offline/On-line):
Environmental Sensitivity:	Dynamic Range/SNR:
Instrument Stability:	Spectral Range/Resolution:

6.2.1 Key Technology Attributes (KTAs)

In looking at the KTAs above, many may argue that scan rate is missing or is a critical factor in evaluating a technology platform and they would be correct in the instances where the fast spectral acquisition is a requirement. The screening of individual tablets produced by a tablet press in the pharmaceutical industry would be a perfect example. Unfortunately, the scan speed is a function of spectral range, spectral resolution, and even signal-to-noise ratio (SNR). Narrow the range, decrease the resolution, and/or degrade the SNR requirement, and every platform can scan faster. Further, while platforms exist that are capable of scanning even large ranges at relatively high resolution very quickly, many of these platforms require the acquisition and subsequent averaging of multiple scans to deliver practical SNRs. For these reasons, the scan speed has been ignored as a practical KTA for accessing the viability of a particular platform for a given application.

To enable the reader to assess the relative strengths and potential weaknesses of a technology platform as it relates to their specific application requirements, general values of Yes/No or Low/Average/High have been assigned to each KTA when applicable or known.

Further, due to the plethora of instrument configurations available, even within a given platform technology, and the impact even small changes to the spectral configuration can have on the KTAs, the authors have elected to assign KTAs for each platform to the instrument configuration that represents the largest spectral range the platform is capable of achieving.

A description of each KTA is as follows:

IDC	All NIR platforms are susceptible to source, mechanical, and electronic drift over time. Spectroscopists have developed techniques, such as the automated sampling of a reference material at periodic intervals to compensate for instrument drift; however, it is possible for a platform to include the integrated drift compensation (IDC), thereby eliminating or significantly delaying the need for automated drift correction strategies. This can also be called "real-time or dynamic referencing".

(Continued)

Environmental sensitivity	All NIR platforms exhibit some degree of sensitivity to changes in environmental parameters. Specifically, fluctuations in temperature, pressure, and humidity can lead to measurement inaccuracy if uncompensated. With the advent of MOEMS-based NIR platforms, it is common to encounter applications that require battery-operated systems. With batteries, the power level (i.e., the charge level) of the batteries and power quality will influence the measurement accuracy. With non-battery-powered devices, power quality impacts the measurement accuracy. Since manufactures have developed electronic strategies for the management of both power quality and level, and a device's sensitivity to humidity and pressure is typically a function of industrial packaging, the end user's calibration strategy, and the sample interface, this KTA only considers a platform's sensitivity to environmental temperature changes. Platforms with a LOW rating exhibit minimal changes in wavelength, amplitude response, or SNR over a significant temperature range. An AVERAGE rating is given to platforms with moderate temperature sensitivities that can be corrected for through advanced calibration strategies. Platforms with a HIGH rating must be held at a constant temperature due to their inability to tolerate temperature changes.
Instrument stability	Instrument stability could also be referred to as "ruggedness or robustness" towards mechanical shock and vibration. Specifically, if a device was dropped and experienced a large, instantaneous G-force, would it still be operational and if so, would the calibration still be valid. Further, if the device was installed in a high-vibration environment, can it remain calibrated and/or operable? A platform with HIGH instrument stability has both high shock and high-vibration tolerance. A platform with AVERAGE stability would need recalibration due to abnormally high shock or vibration. A LOW rating indicates that the device would be rendered inoperable due to any reasonable shock or vibration.
Unit price (offline/in-line)	Since one could argue that each of the 21st-century innovations reviewed below was developed with price point in mind, one would think that the unit price rating for each platform would be low, but this isn't necessarily the case as many of the platforms have expensive packaging and/or provisioning requirements if the intended use is for in-line, real-time, autonomous monitoring. Therefore, a unit price rating will be given for both offline and in-line solutions.
Dynamic range/SNR	Traditional spectroscopists define dynamic range as the difference between the largest photon flux an instrument can measure and the smallest photon flux the instrument can measure. With modern (e.g., digital) NIR spectrometers, dynamic range is defined by the ceiling of a spectrometer's analog-to-digital converter (ADC) and the ADC's floor, which is defined by the system's dark current. Signal-to-noise ratio (SNR) represents an instrument's ability to resolve small changes in flux within the systems' operable dynamic range. SNR is calculated by dividing the mean signal level by the RMS noise level. Many manufacturers will measure the instantaneous noise on the ADC in a high-flux environment to calculate SNR. However, true SNR is measured by calculating the full-spectrum shot-to-shot RMS noise across dozens of spectra. While an SNR value can be calculated for each spectral channel within the instrument's spectral range, it is common to specify SNR for the spectral channel with the highest throughput or spectral amplitude. It is also common to specify the wavelength for which the SNR is stated. Typically, for a given instrument configuration, an AVERAGE NIR spectrometer can measure between two and three decades of photon flux and has a true SNR between 1000 and 7000.
Spectral range/resolution	As a rule, spectral range and resolution are closely coupled parameters. While it is rare for a platform to have both high spectral range and high resolution, it is possible. For the purposes of this review, any system with a range approaching one optical order (e.g., 1250–2500 nm) will be rated high. A range less than 500 nm will be rated low, and anything in between will be considered average. With respect to resolution, for the purposes of this review, resolution refers to full-width half-max (FWHM) optical resolution and not wavelength precision of the platform's monochromator. Any platform with a FWHM resolution exceeding 0.5% of the system's spectral range will be rated high. A platform with a FWHM resolution worse than 2% will be rated low, and anything in between will be rated average.

6.2.2 The Utility of KTAs

What follows are two examples that should serve to highlight the utility of the KTA rating system as a tool for selecting the right technology platform for a given application.

Example 1

If a client has a price-sensitive application in a relatively nontechnical field where the technology is being utilized by unskilled users to classify, not characterize, a substance in a nonhostile environment (i.e., a first responder required to check an unknown substance for drug classification), then a platform that has the following KTA ratings is a viable candidate since classification, as opposed to quantification, is all that would be required. Further, the sampling is performed manually, which allows for the frequent background sampling to correct for calibration drift or offsets due to changing environmental parameters.

IDC:	No	Unit Price (Offline/In-line):	Low/NA
Environmental Sensitivity:	High	Dynamic Range/SNR:	Low/Low
Instrument Stability:	Average	Spectral Range/Resolution:	Average/Average

Basically, in this example, the system would need average stability to survive within a mobile environment, it would need to be relatively low cost to be in the hands of first responders, it would need average range and resolution to cover the plethora of substances that may need identification, and last, since the end user could routinely scan a background, even systems with high environmental sensitivity could be utilized. With respect to dynamic range and SNR, since the application only requires classification, it is likely that a system which is rated low for both would still work as long as the spectral range and resolution were sufficient for the substance tested.

Example 2

On the contrary, if the application required the characterization of the actual concentration of the unknown drug, then it is likely a different platform with the following attributes would be required.

IDC:	No	Unit Price (Offline/In-line):	Low/NA
Environmental Sensitivity:	Average	Dynamic Range/SNR:	Average/Average
Instrument Stability:	High	Spectral Range/Resolution:	High/Average

In this case, since quantification is needed, the application would benefit from a platform with enhanced spectral range, dynamic range, SNR, environmental sensitivity, and stability. In this example, IDC could be beneficial but would not be required.

6.2.3 Multivariate Optical Computing (MOC) Platform

IDC:	No	Unit Price (Offline/In-line):	Average/Average
Environmental Sensitivity:	Low	Dynamic Range/SNR:	Average/Low
Instrument Stability:	High	Spectral Range/Resolution:	NA/NA

FIGURE 6.7 Multivariate optical computing (MOC) multichannel wheel with 20 MOE elements and their corresponding reference filters. Figure reprinted courtesy of Sensors (MPDI Open Access) [13].

The MOC has the distinguished privilege of being not only the first 21st-century NIRS platform reviewed below but also the most controversial. In fact, some would argue that the MOC: (1) isn't a 21st-century platform since it was first conceptualized by Dr. Michael Myrick at the University of South Carolina in the late 1990s [12] and (2) doesn't classify as a MOEMS technology. In response to the first observation, it is important to note that the MOC was not commercialized until 2005, and in response to the latter, while its true the first iterations of the technology utilized bulk optical elements and large mechanical systems, the latter iterations integrated optically encoded filters fabricated using MEMS fabrication techniques and micro-optical electromechanical elements and represent potentially the smallest and most rugged NIRS platforms available on the market today.

The downside of the MOC is it is a fit-for-purpose solution and not a traditional NIRS platform that allows for the acquisition of an NIR spectrum. Specifically, it is a compressed sensing technique that utilizes an a priori selection of discrete wavelengths within the NIR spectrum that have been encoded into a multivariate optical element (MOE) to measure a singular analyte or parameter of interest. Specifically, an MOC with a single MOE is not a multianalyte or multiparameter NIRS platform. Further, the nonrecurring engineering (NRE) expense required to develop the a priori knowledge required to calculate and ultimately encode the MOE within the MOC for the measurement of a particular parameter may outweigh the return on investment for some niche or specialty low-volume applications.

The upside of the MOC is that with a careful application development work prior to the construction of an MOE, the MOC can measure a single analyte or parameter of interest very well. Further, due to the size of the MOE, it's possible to integrate multiple MOEs into a single MOC for the measurement of multiple parameters. In fact, the MOC as shown in Figure 6.7 includes 20 MOEs and their corresponding reference filters, each of which has been designed to measure a particular chemical or physical parameter of interest.

To understand the MOC, one must acquire a detailed understanding of the MOE that sits at the core of the MOC. Further, since MOC predictions are radiometric, it is important to note that an MOC requires the acquisition of transmission or reflectance spectra, as opposed to absorbance spectra.

6.2.3.1 The MOE

The MOE is a multilayer optical encoded interference filter designed to generate a unique signal response on a single-element photodetector, which is predicative of the parameter of interest.

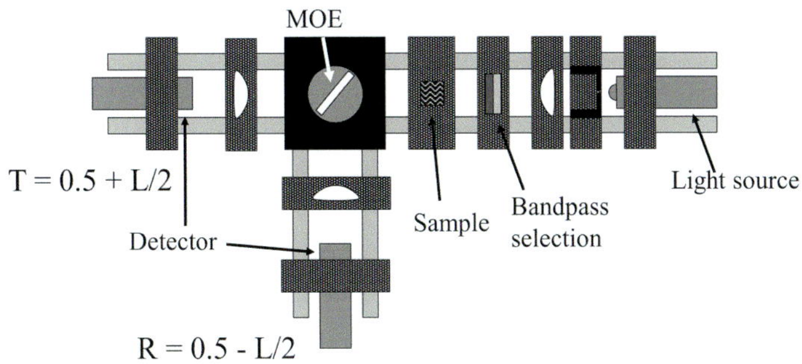

FIGURE 6.8 Schematic diagram of a simple single-element system for the prediction of chemical properties based on the transmission spectroscopy of samples in cuvettes. In this design, MOE is a multivariate optical element, the bandpass selection filters are simple colored glass filters that roughly limit the spectral window, the detectors are matched, T and R stand for transmittance and reflectance on the scale of zero to one, and L is the scaled regression vector of the MOE [14].

Specifically, the MOE circumvents the post-chemometric analysis of a traditional NIRS platform by encoding a regression vector into an optical transmission filter, the MOE, which not only senses but optically computes in a simultaneous step, the parameter of interest.

With any MOE, the spectral response related to a variation in a parameter of interest is typically designed using alternating layers of different materials to emulate a regression vector that correlates the variation with the response function. The classical regression of the dot product with the regression vector and the sample spectrum is then performed by placing the filter directly in the optical path [15], as shown in Figure 6.8, which illustrates a simple measurement arrangement [14] of a basic MOC. The convolved output intensity of the source, sample, and MOE measured by a detector can then be directly related to a physical or chemical material property. Since a spectrum is never collected, it can be argued that an MOC is not a spectroscopic technique, but rather falls into the category of "photometric analysis". That is not the opinion of the authors who consider MOC as a one of seven commercially viable 21st-century NIRS innovations.

It must be noted that the KPAs of spectral range and resolution are not applicable (NA) to MOCs. Further, while it is possible to define an SNR for the MOC, it is more accurate to discuss the standard error of prediction (SEP), which is a chemometric term as opposed to an instrument specification, when defining the accuracy of an MOC. SEP is determined experimentally by testing a validation set of samples with known chemical or physical properties derived from standard laboratory methods against the MOC measurement, and thus, SEP can never be better than the accuracy of the standard method [14].

There are currently two approaches for generating an MOE:

1. Principle component regression (PCR);
2. Vector relaxation, which requires a correlated reference measurement.

Effectively, PCR allows for the creation of a regression vector that serves as the basis function for generating an MOE. Scaling of the basis function is arbitrary but should yield realistic transmittance/reflectance values. The second method, vector relaxation, is a regressive method. A random number generator produces a random multilayer structure, which serves as the starting point for iterating a proprietary algorithm that seeks an optimal MOE [16,17]. Interestingly, while the two approaches yield very different MOE configurations, the resulting predictions and SEPs are typically comparable. However, since the vector relaxation method results in a thinner filter, it appears to have become the preferred method for generating MOEs.

6.2.3.2 MOC KTA Summary

For a detailed review of MOCs, the reader is encouraged to read the references listed within Section 6.6.1 and prior to 2019, would have been referred to Cirtemo for more information on MOEs. However, Cirtemo was acquired by Thorlabs in May of 2019. This new division of Thorlabs still operates out of South Carolina under the name Thorlabs Spectral Works, and information on MOEs can be found on Thorlabs website.

When considering the KTAs for MOCs, the high instrument stability and low environmental sensitivity are both critical attributes for harsh environments, such as the downhole oil and gas industry, where MOCs found early commercial success.

While the unit price is listed as average for both offline and in-line implementations, this was a difficult assessment to make, as the unit price is extremely sensitive to sales volume. In fact, the high NRE associated with the design of the MOE potentially has an insignificant impact on the unit price for high-volume applications. Unfortunately, it limits the commercial viability for low-volume applications, unless the value of the information outweighs the cost of development. Again, the oil and gas industry is a good example. There aren't a lot of downhole oil and gas tools operating in the market yet a downhole MOC still makes sense. Even when you consider the low SNR or relatively high SEPs common with MOCs, they still make sense for the downhole market. This is because the value of the information, even low-resolution information from a potentially expensive sensor, is used to develop oil field exploration strategies.

To summarize, if you are working in a harsh environment where the value of information is at a premium, or you have a high-volume application with low quantitative or simply qualitative requirements, an MOC-based NIRS may be the ideal platform for you.

6.2.4 Grating Light Valve (GLV) Platform

IDC:	No	Unit Price (Offline/In-line):	Average/High
Environmental Sensitivity:	Low	Dynamic Range/SNR:	Average/Average
Instrument Stability:	High	Spectral Range/Resolution:	Average/Average

Silicon Light Machines (SLM), in Sunnyvale CA, markets and licenses Grating Light Valve (GLV™) technology [18]. The GLV is a "microprojection" technology that utilizes a programmable diffraction grating. It competes with other light valve technologies such as Digital Light Processing (DLP™) chip and liquid crystal on silicon (LCoS) chip for implementation in video projector technologies. In comparing the GLV to DLP or LCOS, it is interesting to note that the actuation of the GLV can occur in ~20 ns, which is a million times faster than LCoS, and a thousand times faster than the DLP. This high speed is attributed to the low mass and extremely small submicron displacement requirements of the GLV ribbons. Effectively, the ribbons are massless due to their small size, and since there is no physical contact between moving elements, GLV lifespans exceeding 210 billion switching cycles have been documented. At standard video rates, this translates into a 15-year lifespan under continuous operation. With mil-spec shock survivability exceeding 30,000 g and equally impressive vibration results, the GLV is an extremely robust platform technology.

6.2.4.1 The GLV

The GLV, a dynamic diffraction grating that utilizes an array of extremely small parallel rows of highly reflective microribbons suspended above an air gap over a silicon substrate, was originally conceptualized by Dr. David M. Bloom and his colleagues in the electrical engineering department at Stanford University [19]. Bloom's array of analog-controlled reflective ribbons that when actuated, in a concerted fashion, created a series of diffractive elements or pixels led Dr. Bloom, in 1994, to found Silicon Light Machines to commercialize the GLV in the projection industry.

Within 2 years, a group from MIT, in 1996, independently developed an alternative GLV chip, in collaboration with the Defense Advanced Research Projects Agency (DARPA). The goal of the project was to develop a GLV-based device called the "polychromator", a programmable diffraction grating that could generate synthetic chemical spectra for use as a component in a mid-infrared correlation spectrometer.

The GLV chip consists of an array of flat-parallel mirror elements. Each array of six elements represents an analog-controlled diffractive pixel capable of attenuating incident light. In the actuated state, light is diffracted if the mirror displacement is exactly one-quarter the wavelength of the incident light. In the unactuated state, light striking the mirror elements at normal incidence is completely reflected. The precise displacement to create diffraction depends on the wavelength and bandwidth of the light within each diffractive pixel, and it is important to note that generating the precise displacement is only possible due to the analog nature of the GLV. Another element of the polychromator that proved critical to the commercialization of polychromator-based device was the double-layer interlaced architecture of static and dynamic elements, where the static upper mirror elements remained flat and parallel to the substrate during actuation, enabling each diffractive pixel to induce a precise optical phase shift with significantly improved diffraction characteristics. Effectively, an incident narrow-band optical beam of light could be turned fully-on and fully-off or attenuated to any intensity in between in a consistent and reliable manner.

The DARPA-funded polychromator was first commercialized in 2000 by Polychromix for NIR and C-band telecommunications applications. The optical design was very similar to a dispersive spectrometer in that light entered from a fiber was collimated and then hit a fixed diffraction grating splitting the light along one axis into different wavelengths. Subsequently, the light was passed through intermediate optics and was focused on the polychromator, a diffractive MOEMS chip, at the near-normal incidence. The reflected light passed back through the system and was focused to another fiber. All diffracted light was trapped and prevented from traveling back through the system. Along the length of the MOEMS chip, the diffractive elements were grouped into 100 six-element diffractive pixels, each representing a discrete optical channel within the C-band. Each channel was under analog control, enabling a variable transmission, including a complete attenuation. This allowed the telecommunications industry to correct the nonuniform gain curves of erbium-doped fiber amplifiers used to boost signals in long-haul communication networks. Effectively, with analog control of each optical channel, every channel within the C-band could be acted upon so that all channels entering the communication network processed the same relative intensity.

With the knowledge above, it is easy to envision the basic system architecture of the first GLV or polychromator-based spectrometer. All Polychromix, the first company to commercialize a GLV-based spectrometer, had to do was to replace the output fiber of the C-band device with a detector.

Further, it is easy to envision the system acting as a traditional monochromator that individually scans across a given wavelength range by addressing each channel independently in series, to acquire an optical spectrum.

However, the polychromator was capable of much more than raster scanning through optical spectrum. It wasn't like traditional monochromators due to the fact that the GLV had the ability to dynamically address all 100 individual optical channels simultaneously. Polychromix took an advantage of this capability to enable what later came to be known as "Digital Transform Spectroscopy" (DTS™).

Given the programmability and speed of the MEMS chip, it was possible to implement a technique similar to a Hadamard transform. Performing the Hadamard transform of this signal resulted in the calculated spectrum as shown in the bottom of the figure. The implementation required less than 10 ms for each mask such that the overall time to acquire a spectrum was ~1 second.

At the time, the only competitor that could compete with the speed of Polychromix's DTS™ spectrometer was the diode array spectrometer; however, diode arrays for the NIR spectrum were significantly more expensive than the single-element photodiodes required by the polychromator. But more importantly, the diode array spectrometer struggled with the elimination of stray light and temperature-sensitive dark current.

6.2.4.2 GLV KTA Summary

Since the DTS™ spectrometer was capable of individually modulating each diffractive pixel, which represents an individual optical channel within the dispersed optical path, the DTS™ spectrometer could effectively reject stray light and eliminate dark current effects on the measured optical spectrum. Its insensitivity to stray light was a significant design advantage that when coupled with the cost savings of its single-element detector, gave the DTS™ spectrometer a significant price-to-performance ratio relative to competitive systems.

The Polychromix spectrometer was commercialized as a small benchtop spectrometer in 2004 and later miniaturized and released as a handheld spectrometer in 2006. Polychromix's Phazir™ was likely the first handheld commercial analyzer to incorporate an NIR spectrometer, a light source, a probe, an onboard computer, a color LCD display, and rechargeable batteries. In 2009, Polychromix designed a space-hardened version of the DTS™ spectrometer that was flown on the NASA LCROSS mission and successfully detected water on the moon. In addition to operating in space for the duration of the mission, the MEMS chip was also capable of high-temperature operation and was validated for downhole measurements in petrochemical applications. In 2010, Polychromix was purchased by Thermo Fisher Scientific and marketed by their Portable Optical Analysis division alongside Ahura and Niton, completing their handheld portfolio.

6.2.5 Scanning Grating (SG) Platform

IDC:	No	Unit Price (Offline/In-line):	Low/High
Environmental Sensitivity:	Unknown	Dynamic Range/SNR:	Average/Average
Instrument Stability:	Unknown	Spectral Range/Resolution:	High/Average

At the turn of the 21st century, researchers at the Fraunhofer Institute began exploring the idea of replacing the tradition tilted grating in a Czerny–Turner spectrometer with a MEMS grating structure [4,5]. Their goal was the creation of a compact, 10 mm × 10 mm × 5 mm spectrometer. By 2004, the team had filed their first patents [20,21].

Their first challenge involved the development of an efficient grating structure. Initially, they etched lines and spaces in a silicon substrate, creating rectangular grating structures. Unfortunately, their rectangular gratings had extremely low efficiency (4%–5%), which were insufficient for the real-world applications. Using anisotropic wet etching, a common MOEMS fabrication process, the team was able to achieve grating efficiencies in the 40%–50% range over a broad wavelength range. Figure 6.9 shows the surface electron microscopic (SEM) images of three grating structures: a simple V-grove (top), a sinusoidal structure realized by glass filling of V-grooves followed by metallization (middle), and lastly, a tilted crystal sphere geometry that could serve as a master for blaze-grating creation [22]. Fortunately, the simple V-groove met the developmental goals of a high-efficiency grating capable of operation of the spectral range of 900–2500 nm [23].

6.2.5.1 The Scanning Grating

Realization of a viable grating structure upon a scanning MEMS element allowed for the creation of a miniaturized spectrometer with a configuration resembling the classical Czerny–Turner system. Figure 6.10 compares the traditional configuration with the MEMS scanning grating

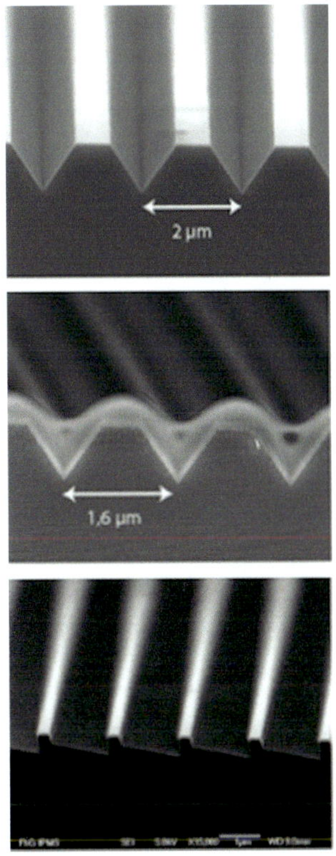

FIGURE 6.9 SEM micrographs of different grating structures realized by anisotropic silicon etching. This section is based on a briefing paper written by Heinrich Grüger, Jens Knobbe, and Tino Pügner from Fraunhofer IPMS. The authors appreciate their willingness to supply background material for this section.

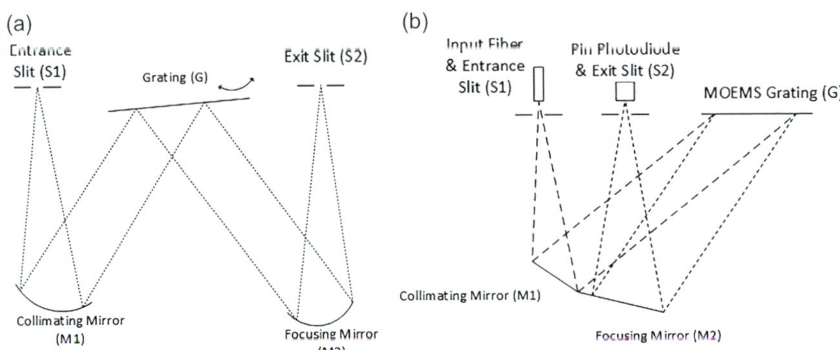

FIGURE 6.10 Czerny–Turner monochromator design in two different configurations: (a) classical setup with tilted grating and (b) modified version with an in-plane scanning grating microassembly, G: the grating; M1: collimating mirror; M2: refocusing mirror; S1: entrance slit; S2: exit slit.

configuration. After entering the entrance slit (S1) proximal to the distal end of an optical fiber, the incoming light is collimated via a high-precision micromachined static mirror (M1). The collimated light is then incident on the scanning grating where it is diffracted and subsequently captured by a second high-precision refocusing mirror (M2). M2 captures the diffracted light of

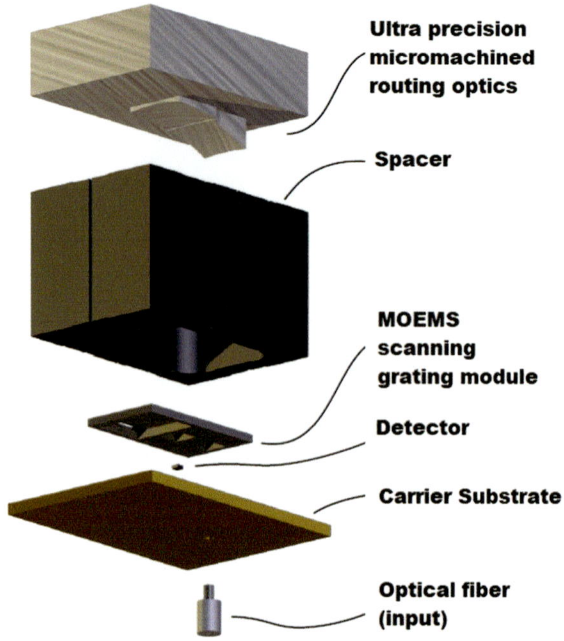

FIGURE 6.11 Exploded view of the scanning grating (SG) platform Graphic provided courtesy of the authors [26].

a given wavelength interval that is correlated with the grating's active position, focusing onto a detector behind the exit slit (S2). It must be noted that extreme care had to be taken in the design of the system to account for even-order aberrations, resulting in complex surfaces for both M1 and M2 [23].

The scanning grating assembly utilizes an electrostatic resonant drive (ESD) that is co-fabricated with the grating assembly in a single MOEMS unit in order to scan the grating. The ESD oscillates the grating at unique resonances that can be physically modeled as a mass-spring system. The grating size determines the range of deflection angle and the optimal resonance frequency. As an example, a larger grating area allows for a high number of ruled lines, leading to a higher overall throughput. The larger mass of the larger grating results in a lower resonance frequency, allowing for longer integration times at the detector and ultimately, improved SNRs. The resonant working principle results in high-stability scanning, and the correlation of the wavelength scale with the magnitude of the deflection [23,24].

Figure 6.11 shows an exploded view of a scanning grating spectrometer module. Clearly visible are the ultra-precision micromachined routing optics [25]. The microfabrication process allows for the integration of the following items within a single monolithic device:

- The entrance and exit slits;
- The grating plate;
- The springs attached to the grating plate;
- The comb structures of the ESD;
- The position measurement transducers.

6.2.5.2 Scanning Grating KTA Summary

The SG platform of Figure 6.11, with associated control and signal processing electronics, is commercially available from HiperScan in a variety of configurations. Unfortunately, the authors were unable to determine the sensitivity of the SG platform to environmental stimuli or the overall stability of the platform. Due to the resonate nature of the drive mechanism, an elevated temperature operation is likely to degrade the spectral range, resolution, and SNR. As a MOEMS platform, the SG is an elegant solution. If your application requires a small platform technology with average dynamic range, resolution, and SNR over a large spectral range – and your environmental conditions are relatively mild – the SG platform is likely a viable option.

6.2.6 Micro-Mirror Array (DLP ™) Platform

IDC:	Yes	Unit Price (Offline/In-line):	Average/Average
Environmental Sensitivity:	Low	Dynamic Range/SNR:	High/High
Instrument Stability:	High	Spectral Range/Resolution:	High/Average

The micromirror array, a MOEMS component consisting of millions of hinged microscopic aluminum mirrors on a silicon chip, was developed by Dr. Larry Hornbeck of Texas Instruments (TI) in 1987, after struggling for close to a decade to perfect an analog-controlled MEMS mirror. Over the course of the next 15 years, Dr. Hornbeck and his colleagues at TI were awarded a series of patents that formed the foundation for micromirror array technology and established the practical methods for manufacturing high-density arrays of micromirrors on integrated circuits. US patent no. 4,710,732 entitled "Spatial Light Modulator and Method", which was granted in 1987, is considered the seminal patent to potentially the most successful MOEMS technology developed to date. The secret to Hornbeck's success and the successful commercialization of the micromirror away was a move from analog to digital control and the addition of a mechanical hard stop to control the micromirror's displacement, creating a bistable mirror and enabling the repeated and consistent actuation of millions of mirrors in concert [27].

Six years later, in 1993, TI formed its Digital Imaging Division, which was tasked with commercializing Dr. Hornbeck's invention. While it is true, the digital projection industry was the first to capitalize on the micromirror array as a core technology, the television (TV) industry soon followed with Mitsubishi's release in 2001 of the RPTV. Eventually, TI's Digital Imaging Division rebranded the micromirror array as a digital light processing chip or DLP® chip, and the future of DLP® technology was established.

Unfortunately, with the release of the first LED TV by Sony in 2005, by 2008, Mitsubishi was the only remaining manufacturer of RPTVs, and by 2012, they too ceased production. Fortunately, with billions invested in the development of the DLP® chip and moreover, the availability of a proven technology and established supply chain, at the turn of the 21st century, researchers around the world began looking into other opportunities for the core technology [28].

The DLP® chip (Figure 6.12), with its tiny digitally controlled mirrors tilting thousands of times a second, has led to the creation of advanced 3D imaging systems, tiny pico projectors embedded in cell phones, in addition to the state-of-the-art digital cinema projectors. Further, the DLP® chip has found its way into a variety of nondisplay applications such as optical switching, holographic data storage, stereolithography, programmable lighting, hyperspectral or multipoint imaging, and programmable spectroscopy [29].

While the term "programmable spectrometer" was likely coined to describe the grating light valve [GLV] spectrometer, reviewed above, it is fair to say it has become synonymous with the DLP spectrometer thanks to TI and the spectroscopy group inside TI that was created in 2011 to facilitate the development and commercialization of DLP-based spectral products.

The terminology stuck because the programmable nature of the DLP® technology allowed for the development of new and/or advanced spectral methods previously not possible with conventional spectrometers [30,31].

FIGURE 6.12 A rendering of TI's DLP™ Chip, which is considered an adaptive optical element. The chip shown consists of over a million tiny mirrors aligned in columns and rows. The mirror diagonals are approximately 8 times smaller than the average diameter of a human hair and can move between two stable positions at kHz rates. The mirrors are so small; they are effectively massless, allowing the chip to be considered a solid-state component.

6.2.6.1 The DLP

In 2006, Christian et al. [32] conceptualized a spectrometer built around a DLP™ chip that could route and encode spectrally dispersed light between a sample and the reference path within a NIR spectrometer designed to go downhole to characterize oil and gas within a reservoir. The spectrometer had to be capable of operating over a 200°C temperature range in a high shock and vibration environment. Further, the spectrometer had to be capable of mitigating both temperature- and time-dependent measurement (i.e., calibration) drift (Figure 6.13).

The DLP platform was chosen due to its ability to not only route light repeatedly and accurately between two independent light paths but also encode the routed light at frequencies well above the electromagnetic frequencies common within an oil and gas operating environment. Further, due to the natural variability in the composition of crude oil, the device required a relatively large dynamic range and high SNR.

If one envisions the light within the spectrometer dispersed across the DLP™ chip such that columns are wavelengths and rows represent the intensity, an understanding of how the DLP spectrometer works begins to take shape. By turning one or more columns at a time, it is possible to scan through wavelengths, effectively capturing an optical spectrum. By turning one column at a time, resolution is maximized but throughput and scan time are impacted. By attenuating rows, which

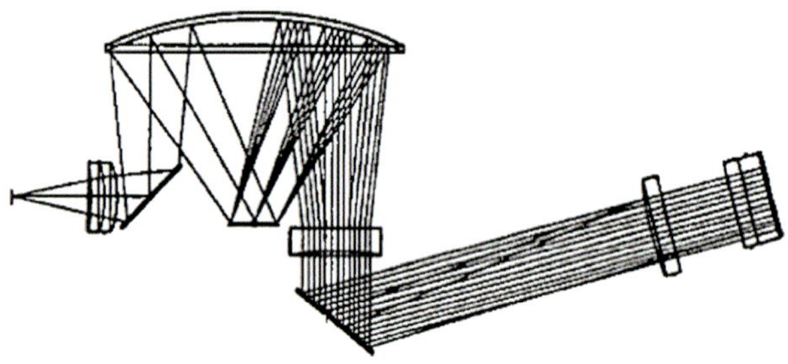

FIGURE 6.13 A ray trace of an Offner relay DLP-based NIR spectrometer first conceptualized by Christian et al. [32]. The component at the far right of the ray trace is a DLP™ chip. On the far left were three input fibers. A center fiber for bringing light into the relay and two fibers on opposite sides of the input fiber for receiving light that had been encoded and routed by the DLP™ chip.

represent amplitude, throughput is impacted, as is dynamic range. The beauty of the DLP is in its flexibility, its tunability, or better stated, its programmability. Next, imagine being able to modulate a predetermined series of columns and rows to deliver a specific number of photons within a limited spectral range to a sample and then lock-in on the modulated signal after it is interacted with that sample. All of this and more is possible with a DLP-based spectrometer.

In fact, the programmable nature of the DLP™ chip allows for other forms of encoding such as Hadamard encoding, enabling the DLP spectrometer to continually operate under a high-flux condition, maximizing SNR across the full spectral range of the spectrometer, and maximizing dynamic range.

The importance of the above capabilities is likely apparent to a spectroscopist, but what might not be apparent is:

1. The importance of the DLP's ability to route light accurately and repeatedly between two independent light paths. In the case of the chip as shown in Figure 6.13, the two light paths are separated by ±12° normal to the DLP surface. The innovation of Christian et al. in 2006 centered around the realization that either of the two independent states could be utilized for the internal full spectrum drift and/or environmental correction;
2. The programmability of the DLP, which allows instrument manufacturers, for the first time, to clone NIRS analyzers. Since NIRS instruments require complex multivariate methods to produce rational outputs, the importance of spectral clones with respect to the scaling of applications within an industrial environment should not be underestimated;
3. The DLP platform's ability to only scan the wavelengths of interest, allowing for extremely fast methods if an application can be reduced to just a handful of relevant wavelengths.

While DLP-based NIRS can be purchased direct from multiple vendors, Neolitics, Inc., a company in the process automation industry, retains the rights to using the traditional off-state of the DLP for IDC.

6.2.6.2 DLP KTA Summary

To the authors' knowledge, the DLP platform is the only platform capable of offering true full-spectrum integrated drift correction or IDC. Further, the low environmental sensitivity and high instrument stability are attractive KTAs for industrial clients or systems integrators. The average offline unit price may not work for low-end, low-cost applications but the average unit price for in-line process control applications is very attractive. With the high dynamic range, spectral range, and SNR achievable with the DLP platform, it could be the right technology for your application. The only consideration is the average resolution, but like diode array-based spectrometers, with the DLP spectrometer, you can sacrifice a range for resolution.

6.2.7 Linear Variable Filter (LVF) Platform

IDC:	No	Unit Price (Offline/In-line):	Low/High
Environmental Sensitivity:	High	Dynamic Range/SNR:	Average/Low
Instrument Stability:	High	Spectral Range/Resolution:	Average/Average

The LVF platform, like the MOE, is not technically a MOEMS device in the strictest sense, since it is difficult to identify an element with mechanical functionality. It is, however, enabled by micro-fabrication techniques, and MEMS developers have come to agree that MEMS devices with static micromechanical features are still MEMS devices [33]. Armed with this understanding, the LVF, with its array of photodiodes placed with micron mechanical precision and further, optically filtered with nanometer accuracy using microfabrication techniques, is without a doubt, a MOEMS platform technology.

6.2.7.1 The LVF

After decades of demonstrated success of filtered photodiodes in the field of photometric analysis, it is difficult to understand why the LVF spectrometer wasn't commercially available prior to 2012 [34]. The answer to this question can be found by examining the complexity of the linear variable filter (LVF), also known as the "linear variable bandpass filter" (LVBF) or the "linear variable optical filter" (LVOF).

As shown in Figure 6.14, the basic architecture of the LVF consists of an optical substrate upon which a first dielectric mirror stack, a precision-tapered optical cavity, and a second dielectric mirror stack are layered. Effectively, the wedge separation between the dielectric layers creates a Fabry–Perot (FP) cavity with variable thickness laterally, enabling a linearly progressive series of wavelength-selective FP filters. The relatively modern ability to create such precision coatings makes this platform a powerful approach for designing the fit-for-purpose LVF platforms.

The LVF element can be considered equivalent to a Fabry–Perot interferometer (FPI), except spatial coordinates as opposed to temporal coordinates define the passband.

To create a spectrometer, the LVF must be precisely integrated with a linear detector array. For NIRS, the preferred detector is an InGaAs linear photodiode array. While the LVF design can limit the spectral range and resolution of the integrated platform, the physical specifications of the detector array typically determine the spectral resolution and spectral range of the LVF platform. Table 6.2 summarizes the operational specifications of the Viavi MicroNIR 1700 ES for two different InGaAs photodiode arrays: a standard range detector array that covers 950–1650 nm and an extended range detector array that covers 1150–2150 nm. An image of the LVF platform specified by Table 6.2 can be found in Figure 6.15 [35–37].

6.2.7.2 LVF KTA Summary

Figure 6.15 clearly illustrates an attractive attribute of the LVF platform, its size. Clearly visible in the image are the two integrated sources that allow the MicroNIR 1700ES to perform reflectance spectroscopy. It should be noted, however, that a platform capable of transmission spectroscopy is also available. The package shown, which is solid state in nature, is a complete NIRS system with integrated electronics.

While the LVF itself is extremely robust with respect to shock and vibration, unfortunately, InGaAs photodiode arrays are sensitive to temperature fluctuations and moreover, struggle with dynamic range and SNR at elevated temperatures. In fact, the recommended maximum operating

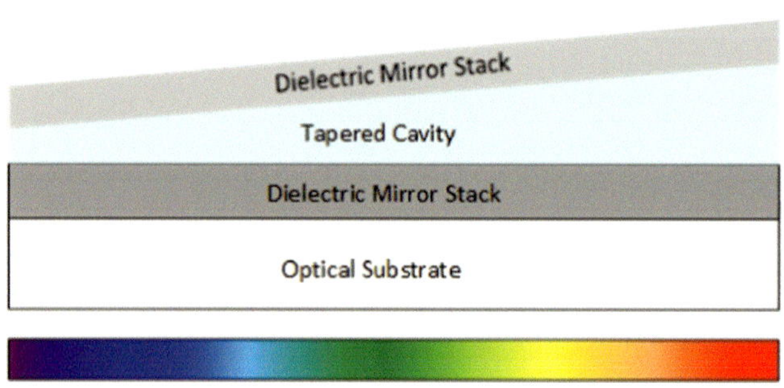

FIGURE 6.14 Schematic of an LVF. Illumination occurs from the top of the element, and the resulting color spectrum is displayed.

TABLE 6.2
VIAVI MicroNIR specifications

Detector array	Standard: 950–1650 nm Extended: 1150–2150 nm
Geometric resolution (pixel size)	Standard: 6.25 nm Extended: 8 nm
Optical resolution (spectroscopy)	1.25% of center wavelength @ 1000 nm, resolution = 12.5 nm @ 2000 nm, resolution = 25 nm
Wavelength accuracy	<3 nm (NIST SRM-2036)
Wavelength repeatability	<1 nm (NIST SRM-2036)

FIGURE 6.15 MicroNIR 1700ES from VIAVI. Graphic provided courtesy of VIAVI Solution, Inc.

temperature for an InGaAs photodiode array is 50°C. To mitigate the influence of environmental temperature on the performance of InGaAs photodiode arrays, manufacturers integrated single- or dual-stage thermoelectric coolers into the InGaAs package but doing so adds to the cost of the arrays and increases the electrical power requirements of the overall platform. In all, you have an application within a relatively low and stable thermal environment, and if you are concerned about shock and vibration or size, the LVF platform may be a viable option.

6.2.8 Fabry–Perot Interferometer (FPI) Platform

IDC:	No	Unit Price (Offline/In-line):	Low/High
Environmental Sensitivity:	High	Dynamic Range/SNR:	Average/Low
Instrument Stability:	Low	Spectral Range/Resolution:	Low/Average

In terms of size, it's difficult to compete with today's FPI NIR platform. There are two ways to integrate an FPI to form the monochromator: (1) integrated with the photodetector and (2) integrated with the light source. While both are available commercially, a MEMS-FPI integrated with a photosensor is the most common approach. This section will examine both options, realizing that the limitations and advantages of the FPI are not dependent on where integration occurs. Regarding size, fully integrated FPI NIR modules, inclusive of electronics, are commercially available in the range of 25 mm^3. Further, OEM modules without associated electronics can be packaged in TO-5 cans.

6.2.8.1 The FPI

Figure 6.16 shows an exploded view of the FPI integrated with a photodetector, which for the NIR spectral region is typically an InGaAs pin photodiode (Hamamatsu MEMS-FPI technical note). The FPI consists of an air gap between upper and lower mirror structures. The air gap is defined by physical spacers, also shown in this figure. The application of an electrostatic voltage to the upper mirror membrane changes the gap shifting the transmission wavelength of the assembly. Increasing the applied voltage reduces the gap, blue shifting the transmission. The voltage step size corresponds to the spectral resolution. The FPI substrate is silicon, and the mirrors are common multilayer dielectric coatings (silicone dioxide, silicon nitride, and poly-silicon). Since the FPI element is voltage-driven, the device is electrostatically sensitive. The FPI element may also be driven with piezoelectric actuators. Piezo-driven systems have significantly larger apertures [38,39].

In this configuration, the FPI module requires an axial collimated illumination of the assembly. The bandpass filter sets the wavelength range of the assembly and supplies high out of band rejection. The FPI filter allows for scanning across the wavelength range within the pass band. Typically, these types of FPI modules have limited bandwidths (250–450 nm) and 12–28 nm measurement resolution. Consequently, several assemblies are necessary to span the full NIR region.

Since the FPI platform supplies the scanning capability for the spectrometer, it is important to understand how scanning occurs and the underlying control required. FPI scan rates can be very fast, in the 50–100 µs range; however, averaging is needed at fast scan rates to improve SNR. Further,

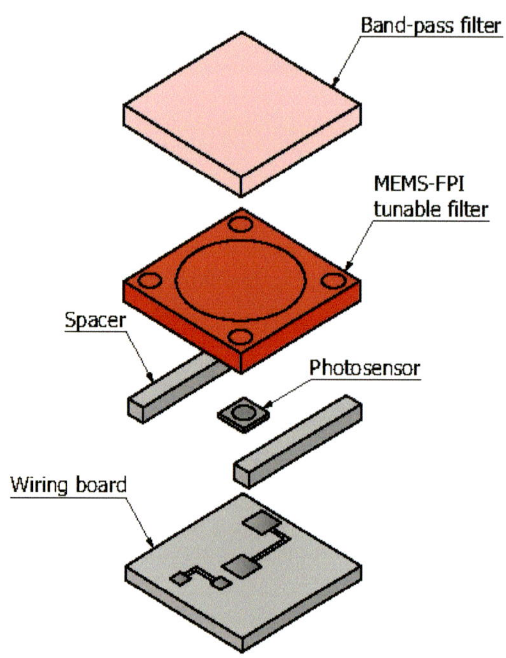

FIGURE 6.16 An exploded view of a Fabry–Perot Interferometer-based tunable filter integrated with a photosensor. Image courtesy of Hamamatsu Photonics [39].

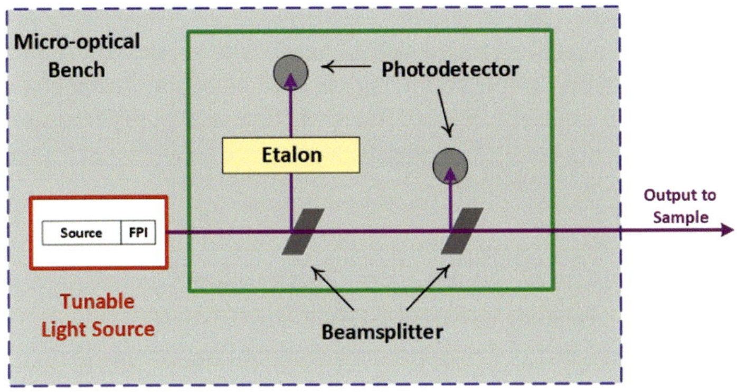

FIGURE 6.17 Block diagram of the Axsun Technologies integrated micro-optical bench. The diagram illustrates the integrated source, wavelength referencing, and source power referencing schemes. Output to the sample occurs via secondary optics or fiber optic cable.

both bandwidth and spectral resolution are impacted by the scan rate. Longer scan periods minimize the need for averaging multiple scans, with similar SNR levels and improved optical resolution.

The allowed wavelength range is determined by the maximum deflection of the FPI. Since the filter response is nonlinear and temperature-dependent, suppliers of FPI assemblies provide a detailed information on the drive voltage requirements as a function of temperature.

In addition, the upper mirror is sensitive to both shock and vibration. In a laboratory environment, these sensitivities are less problematic because the instruments typically experience a minimal thermal variation or shock and vibration. However, in process applications, care must be taken with respect to packaging, and complex electronic control systems are required to make the FPI a viable platform. For example, a careful characterization of the wavelength response versus temperature generates an additional correction factor that changes the applied voltage profile to keep the transmission wavelength constant and eliminate the thermal drift. Typically, this is accomplished with on-board thermal measurement either in the control electronics or in the FPI module. When properly characterized and controlled, the FPI assembly keeps center wavelength across the temperature range of interest to ±1.4 nm, which is acceptable for many NIR applications. Further, it must be noted that if the temperature range is sufficiently wide, the tunable spectral range of an FPI could be limited.

As noted above, a second integration point for an FPI is at the light source (Figure 6.17). This creates a variable wavelength light source, a predispersive NIR spectrometer, with the photodetection occurring post-sample interaction. The widest application of this approach is for optical coherence tomography (OCT) [40,41].

Commercially, the FPI is integrated with either a superluminescent LED (Axsun IntegraSpec XL) [42] or a solid-state laser (Axsun IntegraSpec2) [43]. The laser-based approach is by far the most common. Both these light sources have limited bandwidth ranges resulting in a spectrometer with high resolution (~1 nm) or a narrow spectral range. The commercial FPI light source is a complex module and needs a bit of unpacking [44].

First, a discussion of light sources is made. Solid-state light sources are small in size and offer brightness levels 1000 times or greater than traditional broadband light sources. The small sizes make an integration on micro-optical benches significantly simpler. Additionally, both options have small coherent beams, minimizing the need for beam-shaping optics. The FPI in this assembly has all the same requirements for scanning control as described above. However, the light source itself sets the scanning range, not an integrated filter on the FPI. Two added source control elements are integrated on the optical bench: real-time wavelength referencing and real-time source power referencing. Axsun has branded this integrated module as WARM (wavelength and amplitude reference module). The entire module or micro-optical bench is environmentally controlled allowing the better overall operational control, minimizing the environmental fluctuations, and enhancing the output stability. Optical coupling between integrated elements uses free-space micro-optics. Wavelength referencing occurs via the fringe pattern analysis, supplying exact wavelength values under changing environmental conditions.

6.2.8.2 FPI KTA Summary

From the review above, it can be seen that there are a variety of FPI platforms available within the market for NIRS. Further, depending on the application requirements, there may be more than one implementation that is acceptable. If size is a concern and the application environment is relatively benign, FPI could be the right solution. Just keep in mind that FPI platforms have a limited spectral range.

With this said, it can be argued that Axsun's Warm module with its amplitude reference arm is a form of IDC, allowing WARM-enabled micro-benches to maintain a 24-hour baseline stability on the order of ±0.5%, across the operational temperature range of 0°C–55°C.

6.3 THE FUTURE OF NIR & I_2OT

The authors are of the opinion that the future of NIR is bright, pun-intended. We envision a future where MOEMS-based NIR platforms are small, easily installed by a client, and require virtually no maintenance after installation. A future where the TOC of an in-line NIR I2oT strategy is literally the cost of the technology itself, the CAPEX, and the traditional OPEX associated with the deployment of the technology is virtually zero. In this future, due to the ease of use and affordability of MOEMS-based NIR platforms, the distributed process monitoring is enabled, which subsequently enables distributed process analytics (DPA). Distributed units acting in concert to optimize throughput and yield, ensure quality, and enhance safety will usher in a new era in process visualization and optimization, a new era for smart factories. In our opinion, the NIR industry is not expanding by an estimated 9%–12% CAGR. It is expanding much, much faster.

REFERENCES

1. I. Newton, Phil Trans. No. 80, 3075–3087 (1671/72). *Facsimile reproduction in Isaac Newton's Papers and Letters on Natural Philosophy*, 2nd ed., I. B. Cohen, Ed. (Harvard University Press, Cambridge, MA, 1978), pp. 47–59.
2. W.H. Wollaston, A method of examining refractive and dispersive powers, by prismatic reflection, *Phil. Trans. Royal Soc.* 92 (1802) 365–380.
3. J. Fraunhofer, Bestimmung des Brechungs-und des Farbenzerstreungs-Vermögens verschiedener Glasarten, in Bezug auf die Vervollkommnung achromatischer Fernröhre, *Annalen der Physik* 56 (1817) 264.
4. G. R. Harrison and E. G. Loewen, Ruled gratings and wavelength tables, *Appl. Opt.* 15 (1976) 1744–1747.
5. G. Kirchhoff und R. Bunsen, *Poggendorff's Ann. Bd.* Ex.8 161.
6. G. Kirchhoff und R. Bunsen, *Poggendorff's Ann. Der Physik und Chemie* 110 (1860) 161–189.
7. K.H. Norris, History of NIR, *J. Near Infrared Spectrosc.* 4 (1996) 31–37.
8. J.R. Hart, K.H. Norris and C. Golumbic, Determination of the moisture content of seeds by near-infrared spectrophotometry of their methanol extracts, *Cereal Chem.* 39 (2) (1962) 94.
9. K.H. Norris and J.R. Hart, Direct spectrophotometric determination of moisture content of grain and seeds, in *Proceedings 1963 International Symposium on Humidity and Moisture, Principles and Methods of Measuring Moisture in Liquids and Solids* Vol. 4, pp. 19–25, Reinhold Publishing, New York (1965).
10. H.C. Nathanson and R.A. Wickstrom, US Patent 3,413,573 (1968).

11. M.E. Motamedi, Merging micro-optics and micromechanics: Micro-opto-electro-mechanical (MOEM) devices, in *Proceedings of SPIE: Diffractive and Miniaturized Optics: A Critical Review*, p. 10271G, (1993). doi: 10.1117/12.170189.
12. M.P. Nelson, J.F. Aust, J.A. Dobrowolski, P. G. Verly and M. L. Myrick, Multivariate optical computation for predictive spectroscopy, *Anal. Chem.* 70 (1998) 73–82. doi: 10.1021/ac970791w.
13. C.M. Jones, B. Dai, J. Price, J. Li, M. Pearl, B. Soltmann, M. L. Myrick, A new multivariate optical computing microelement and miniature sensor for spectroscopic chemical sensing in harsh environments: Design, fabrication, and testing, *Sensors* 19 (2019) 701. doi: 10.3390/s19030701.
14. M.L. Myrick, O.O. Soyemi, M.V. Schiza, R. Farr, F. Haibach, A. Greer, H. Li, R. Priore, Application of multivariate optical computing to simple near-infrared point measurements, in *Proceedings of SPIE: Instrumentation for Air Pollution and Global Atmospheric Monitoring*, p. 4574 (2002). doi: 10.1117/12.455161.
15. O. Soyemi, D. Eastwood, L. Zhang, J. Li, P. Karunamuni, R. Gemperline, A. Synowicki and M. L. Myrick, Design and testing of a multivariate optical element: The first demonstration of multivariate optical computing for predictive spectroscopy, *Anal. Chem.* 73 (6) (2019) 1069–1079. doi: 10.1021/ac0012896.
16. O.O. Soyemi, P.J. Gemperline, L. Zhang, D. Eastwood, H. Li, M. L. Myrick, Novel filter design algorithm for multivariate optical computing, in *Proceedings of the SPIE 4205: Advanced Environmental and Chemical Sensing Technology*, pp. 288–299, (2001). doi: 10.1117/12.417462.
17. O.O. Soyemi, F.G. Haibach, P.J. Gemperline and M.L. Myrick, Nonlinear optimization algorithm for multivariate optical element design, *Appl. Spectrosc.* 56 (4) (2019) 477–487.
18. D. T. Amm, R. W. Corrigan, GLV technology: Update and novel applications, *Society for Information Display Symposium*, Anaheim, CA (May 19, 1998).
19. S.D. Senturia, M. Ramani, Y. Geller, Diffractive MEMS: A versatile tool, Laser Focus World, 2004. https://www.laserfocusworld.com/optics/article/16549809/diffractive-mems-a-versatile-tool
20. H. Schenk and H. Grueger, Spectrometer, US Patent 7,07,152 (2006).
21. H. Schenk and H. Grueger, Spectrometer, US Patent 7,034,936 (2006).
22. F. Zimmer, H. Grüger, A. Heberer, et al. Development of a NIR micro-spectrometer based on a MOEMS scanning grating, in *SPIE Proceedings* Volume 5455 MEMS, MOEMS and Micromachining, Photonics Europe, Strasbourg (August 16, 2004).
23. T. Pügner, J. Knobbe and H. Grüger, Near-infrared grating spectrometer for mobile phone applications, *Appl. Spectrosc.* 70 (5) (2016) 734–745. doi: 10.1177/0003702816638277.
24. T. Pügner, J. Knobbe and H. Lakner, Basic angles in microelectromechanical system scanning grating spectrometers, *Appl. Opt.* 50 (2011) 4894–4902. doi: 10.1364/AO.50.004894.
25. T. Pügner, J. Knobbe, H. Grüger and H. Schenk, Realization of a hybrid-integrated MEMS scanning grating spectrometer, in *Proceeding of SPIE: Next Generation Spectroscopic Technologies*, 8374 (2012). doi: 10.1117/12.919068.
26. H. Grüger, J. Knobbe, T. Pügner, et al., MEMS based systems and their applications in NIR spectroscopy for materials analysis, Presented at OCM 2013 – Optical Characterization of Materials (March 6, 2013).
27. L. J. Hornbeck, Digital light processing and MEMS: timely convergence for a bright future, in *Proceedings of Micromachining and Microfabrication Process Technology*, K.W. Markus, ed., SPIE 2639, 2, Austin, USA (1995).
28. D. Dudley, W. M. Duncan and J. Slaughter Emerging digital micromirror device (DMD) applications, Proc. SPIE 4985, MOEMS Display and Imaging Systems (20 January 2003). doi: 10.1117/12.480761.
29. B. Rose, M. Rasmussen, N. Herholdt-Rasmussen and O. Jespersen Programmable spectroscopy enabled by DLP, Proc. SPIE 9376, Emerging Digital Micromirror Device Based Systems and Applications VII, 93760I (10 March 2015). doi: 10.1117/12.2175803.
30. R. A. Crocombe Handheld spectrometers in 2018 and beyond: MOEMS, photonics, and smartphones, Proc. SPIE 10545, MOEMS and Miniaturized Systems XVII, 105450C (22 February 2018). doi: 10.1117/12.2286492.
31. E. Pruett Techniques and applications of programmable spectral pattern coding in Texas Instruments DLP spectroscopy, Proc. SPIE 9376, Emerging Digital Micromirror Device Based Systems and Applications VII, 93760H (10 March 2015). doi: 10.1117/12.2083863.
32. Christian, et al., US Patent Application no. US7440098B2.
33. MEMS and Nanotechnology Exchange. https://www.mems-exchange.org.

34. E. H. Wu, G. de Graaf and R. Wolffenbuttel, Design and implementation of a sub-nm resolution microspectrometer based on a linear-variable optical filter, *Opt. Express* 20 (1) (2012), 489–507. doi: 10.1364/OE.20.000489.
35. R.R. McLeod and T. Honda, Improving the spectral resolution of wedged etalons and linear variable filters with incidence angle, *Opt. Lett.* 30 (19) (2005) 2647–2649. doi: 10.1364/OL.30.002647.
36. C.G. Pederson, D.M. Friedrich, C. Hsiung, et. al., "Pocket-size near-infrared spectrometer for narcotic materials identification," Proceeding of SPIE: Next-Generation Spectroscopic Technologies VII, 9101, 2014, 91010O. https://doi.org/10.1117/12.2050019
37. N.A. O'Brien, C.A. Hulse, D.M. Friedrich, et.al., Miniature near-infrared (NIR) spectrometer engine for handheld applications, Proceeding of the SPIE 8374: Next-Generation Spectroscopic Technologies V, p. 837404 (2012). doi: 10.1117/12.917983.
38. Spectral Engines, Whitepaper: NIROne: Next generation spectral device and performance testing. https://www.spectralengines.com/learn-more/whitepapers.
39. Hamamatsu, Technical Note: MEMS-FPI spectrum sensor. https://www.hamamatsu.com/us/en/product/type/C13272-02/index.html.
40. B. Lee, W. Choi, J. Liu, C. Lu, J.S. Schuman, G. Wollstein, et al., Cardiac-gated en face doppler measurement of retinal blood flow using swept source optical coherence tomography at 100,000 axial scans per second, *Invest. Ophthal. Vis. Sci.*, 56(4) (2015) 2522–2530. doi:10.1167/iovs.14–16119.
41. K. Karnowski, I. Grulkowski, N. Mohan, I. Cox and M. Wojtkowski, Quantitative optical inspection of contact lenses immersed in wet cell using swept source OCT, *Opt. Lett.*, 39(16) (2014) 4727–4730. doi:10.1364/OL.39.004727.
42. Axsun Technologies, IntegraSpec XL spectrometers for process-control applications, www.axsun.com.
43. Axsun Technologies, IntegraSpec2 spectrometers for process-control applications, www.axsun.com.
44. Axsun Technologies, Technical Note: Designing a Miniature Spectrometer (2005), www.axsun.com.

ADDITIONAL READING

M. Alcalà, M. Blanco, D. Moyano, N.W. Broad, N.A. O'Brien, D. Friedrich, F. Pfeifer, H.W. Siesler, "Qualitative and Quantitative Pharmaceutical Analysis with a Novel Hand-Held Miniature near Infrared Spectrometer", *Journal of Near Infrared Spectroscopy*, 2013, 21 (6), 445–457. https://doi.org/10.1255/jnirs.1084

R.A. Crocombe, "MEMS technology moves process spectroscopy into a new dimension", *Spectroscopy Europe*, 2004, 16–19.

R.A. Crocombe, "Portable Spectroscopy", *Applied Spectroscopy*, 2018, 72 (7), 1701–1751. doi: 10.1177/0003702818809719

B. Dai, C.M. Jones, M. Pearl, M. Pelletier, M.L. Myrick, "Hydrogen Sulfide Gas Detection via Multivariate Optical Computing", *Sensors*, 2018, 18, 2006. doi: 10.3390/s18072006

D. Eastwood, O.O. Soyemi, J. Karunamuni, L. Zhang, H. Li, M.L. Myrick, "Field applications of stand-off sensing using visible/NIR multivariate optical computing", *Proceedings of SPIE: Water, Ground, and Air Pollution Monitoring and Remediation*, 2001, 4199. doi: 10.1117/12.417366

D.M. Friedrich, C.A. Hulse, M. von Gunten, E.P. Williamson, C.G. Pederson, N.A. O'Brien, "Miniature near-infrared spectrometer for point-of-use chemical analysis," *Proceedings of SPIE*: Photonic Instrumentation Engineering, 2014, 8992, 899203. doi: 10.1117/12.2040669

S.I. Hintschich, T. Pügner, J. Knobbe, J, J. Schröder, P. Reinig, H. Grüger, H. Schenk. "MEMS-based miniature near-infrared spectrometer for application in environmental and food monitoring". *Proceedings of the 8th International Conference on Sensing Technology*, September 2–4, 2014, Liverpool, UK.

C.M. Jones, J. Price, B. Dai, J. Li, D.L. Perkins, M.L. Myrick, "In-Situ Methane Determination in Petroleum at High Temperature and High Pressure with Multivariate Optical Computing", *Analytical Chemistry*, 2019, 91(24), 15617–15624. doi: 10.1021/acs.analchem.9b03715

M.L. Myrick, O.O. Soyemi, H. Li, L Zhang, D Eastwood, "Spectral tolerance determination for multivariate optical element design", *Fresenius' Journal of Analytical Chemistry* (Analytical and Bioanalytical Chemistry), 2001, 369, 351–355. doi: 10.1007/s002160000642

N. O'Brien, C.A, Hulse, F. Pfeifer, and H.W. Siesler, "Near Infrared Spectroscopic Authentication of Seafood", *Journal of Near Infrared Spectroscopy*, 2013, 21 (4), 299–305. https://doi.org/10.1255/jnirs.1063

T. Pügner, J. Knobbe, and H. Grüger, "Near-Infrared Grating Spectrometer for Mobile Phone Applications". *Applied Spectroscopy*, 2016, 70, 734–735.

A. Rady, J. Fischer, S. Reeves, B. Logan and N.J. Watson, "The Effect of Light Intensity, Sensor Height, and Spectral Pre-Processing Methods When Using NIR Spectroscopy to Identify Different Allergen-Containing Powdered Foods", *Sensors*, 2020, 20, 230–253. doi: 10.3390/s20010230

O.O. Soyemi, F.G. Haibach, P.J. Gemperline, and M.L. Myrick, "Design of angle-tolerant multivariate optical elements or chemical imaging", *Applied Optics*, 2002, 41 (10), 1936–1941.

Spectral Engines, *White Paper: Comparison of Miniaturized Near Infrared Spectral Sensing Technologies.* https://www.spectralengines.com/learn-more/whitepapers

Y. Wan, J. Carlson, B. Kesler, et al., "Compact characterization of liquid absorption and emission spectra using linear variable filters integrated with a CMOS imaging camera", *Scientific Reports*, 2016, 6, 29117. doi: 10.1038/srep29117

H. Yan and H.W. Siesler, "Hand-held near-infrared spectrometers: State-of-the-art instrumentation and practical applications", *NIR News*, 2018, 29 (7), 8–12. doi: 10.1177/0960336018796391

7 NIR Hyperspectral Imaging

Aoife A. Gowen[*]
University College Dublin

CONTENTS

7.1 INTRODUCTION

Hyperspectral imaging is an optical technique that enables the investigation of the spatial distribution of biochemical components on or within a sample. It can be considered as an extension of traditional point spectroscopy where, instead of obtaining a single spectrum for a sample, many spatially adjacent spectra are obtained over a surface, which can be visualized as an image. Equally, hyperspectral imaging can be viewed as an extension of conventional imaging into the spectral domain where, instead of obtaining an image of a sample at a single wavelength (as in grayscale imaging) or at three wavelengths (as in red-green-blue (RGB) color imaging), images are typically obtained at 100 or more contiguous wavelengths. A hyperspectral image can thus be thought of as many spectra arranged spatially to form an image, or many images stacked together to make a three dimensional structure, commonly known as a hypercube. This structure is perhaps more easily understood by consideration of the schematic diagram in Figure 7.1, which shows how the spatial (x, y) and spectral (λ) components of a hyperspectral image are linked. This schematic includes an NIR hyperspectral image of mixed packaging materials (described in more detail in Section 7.2). Four different materials are present in the image, identifiable by their shape. For instance, the squares are high-density polyethylene (HDPE). The reflectance spectrum of one pixel of the HDPE sample is plotted, and as can be seen in Figure 7.1, it exhibits an absorption feature at 1216 nm. A single slice of the

[*] aoife.gowen@ucd.ie.

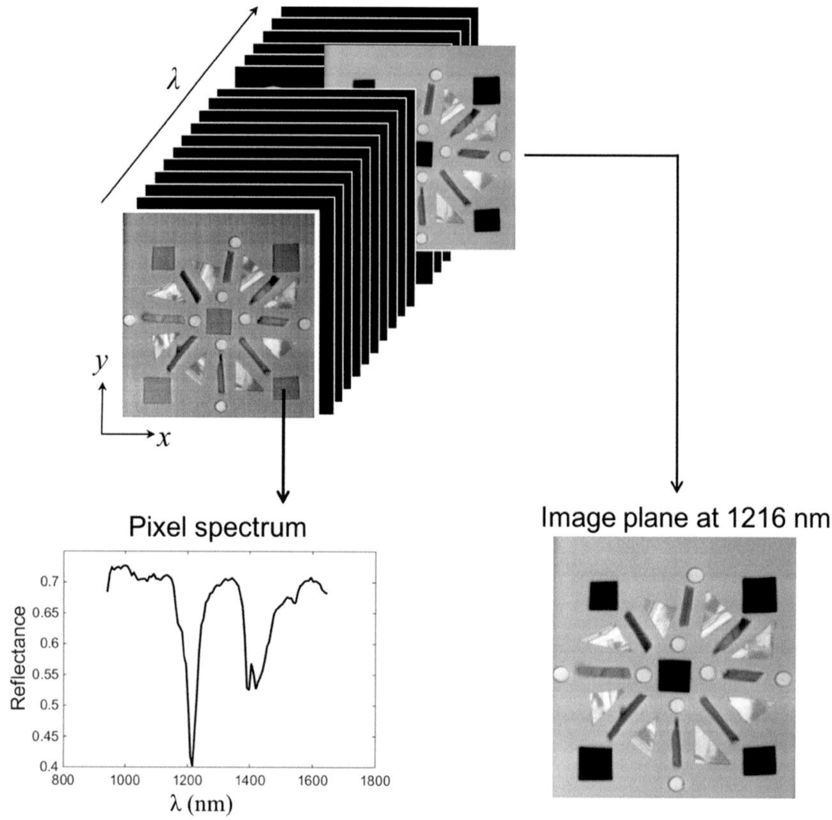

FIGURE 7.1 Schematic of hyperspectral image structure: x and y represent spatial dimensions, while λ represents the wavelength dimension. Hyperspectral image data for the "packaging" dataset (described in Section 7.2) is presented. A single-pixel reflectance spectrum of the HDPE sample is shown on the lower left-hand side. The largest absorbance peak occurs at 1216 nm. On the lower right-hand side, a single-wavelength image at 1216 nm is displayed. The HDPE samples (squares in the image) appear darker than the other materials in the image due to their relatively higher absorbance of light at 1216 nm.

hyperspectral image at 1216 nm is also shown. In this image, the HDPE samples appear darker than the other ones because they absorb more light at this specific wavelength.

"Hyperspectral imaging" is a general term that covers a wide variety of spectroscopic and spectrometry techniques (e.g., fluorescence, vibrational, Raman spectroscopy; mass spectrometry), wavelength ranges (e.g., UV, Vis, NIR, MIR), and spatial resolutions (from the nano-scale to remote sensing scale). The term originated in the mid-1980s from the remote sensing community with the development of the Airborne Imaging Spectrometer (AIS) at NASA's jet propulsion laboratory, an airborne instrument capable of imaging large regions of the earth in the NIR wavelength range (1200–2400 nm) (Chiou, 1985; Goetz et al., 1985). The subsequent development of NASA's Airborne Visible/Infrared Imaging Spectrometer (AVIRIS) expanded the application of spectral imaging to a wide range of tasks, from vegetation monitoring to mineral mapping on the earth's surface. The term "imaging spectroscopy" is now preferred over "hyperspectral imaging" by NASA (https://aviris.jpl.nasa.gov); however, use of the term "hyperspectral imaging" has persisted and grown, both in the literature and in common use; indeed, a recent paper title search in Web of Science indicates that in excess of three times more articles have been published in the past 10 years with the words "hyperspectral imaging" in the title than with "imaging spectroscopy" (1557 vs. 472). Despite this ambiguity in terminology, NIR hyperspectral imaging has in recent years become the topic of intense research in highly applied areas such as food quality analysis (Feng and Sun, 2012), pharmaceutical quality control (Gowen et al., 2008), and agricultural analysis (Adao et al., 2017). Regardless of the applications, many common considerations are shared by potential users of this technique, such as instrument selection, sample presentation, and data analysis. With the aim of introducing the reader to the general concepts involved in NIR hyperspectral imaging, this chapter is divided into two main sections: Instrumentation and Data Analysis, with several illustrative examples provided throughout.

7.2 DESCRIPTION OF DATASETS USED IN THIS CHAPTER

In order to illuminate the main points of this chapter, different samples and NIR imaging instruments were employed as described below.

7.2.1 NIR Hyperspectral Instrumentation

Instrument 1 (Macroscopic Dispersive NIR Imaging system): This pushbroom NIR hyperspectral imaging system (DV Optics, Padua, Italy) operates in the wavelength range 870–1720 nm, with reflectance values measured every 7 nm (further details of the instrument can be found in Gowen et al., 2011). The spatial resolution obtained using this instrument was approximately 320 μm.

Instrument 2 (Microscopic Dispersive NIR Imaging system): This pushbroom NIR hyperspectral imaging system (Cytotviva Ltd, US) also operates in the wavelength range 870–1720 nm, where reflectance is measured every 6 nm. The spatial resolution obtained using this instrument is approximately 3.2 μm at 10× magnification.

Instrument 3 (Microscopic Fourier transform infrared (FT-NIR) imaging system): This pushbroom system (Nicolet iN10 MX Imaging system by Thermo Scientific (Madison, WI, USA)) employs an mercury cadmium telluride (MCT) array detector, which was liquid nitrogen-cooled. Images were obtained in the wavelength range 1333–2500 nm at a spectral resolution of approximately 5 nm and a spatial resolution of approximately 25 μm.

7.2.2 Sample Description

Packaging: NIR hyperspectral images of four different packaging materials HDPE, polystyrene (PS), cardboard, and polyethylene terephthalate (PET) – were obtained in diffuse reflectance using Instrument 1. Single images of each material were obtained as well as a mixed image, including all materials (entitled "Packaging-mix"). In the mixed image, the materials were cut into different shapes to enable their identification, as follows: squares (HDPE), rectangles (PS), circles (cardboard), and triangles (PET).

Water in Oil: Approximately 400 μL of olive oil was placed into a 1-mm pathlength quartz cuvette, and a drop of water of approximate volume 80 μL was injected into the oil. This sample was imaged using Instrument 1 (in transflectance), Instrument 2 (in transflectance and transmittance at 10 × magnification), and Instrument 3 (in transmittance).

White Reference: A piece of Spectralon was imaged in reflectance using Instrument 1 and Instrument 2.

Chickpeas: Dry chickpeas (*Cicer arietinum* L.) were hydrated to different moisture contents by immersion in 1 L of tap water at 40°C ± 1°C. Ten chickpeas were removed from the water after the following soaking times: 30 min, 1 h, 2 h, 3 h, 4 h, 5 h, and 6 h. Their surface was lightly dried using tissue paper, and the samples were allowed to equilibrate at room temperature for 20 min. Each chickpea was subsequently sectioned along its central axis using a blade, and the inner surface of the chickpea was imaged using Instrument 1. The experiment was repeated twice, with the first repetition designated for model calibration and the second for model testing. Moisture content was determined by oven drying (Gowen et al., 2014).

7.3 NIR HYPERSPECTRAL IMAGING INSTRUMENTATION

Many different instrumental setups exist for the acquisition of NIR hyperspectral images. These can vary in terms of modality (e.g., reflectance, transmission), components (e.g., light sources, detectors), and in how the hyperspectral image is built up. Quartz tungsten-halogen (QTH) bulbs are often used as light sources in NIR imaging. These sources have the advantage of being cheap while covering a wide wavelength range, from 400 to 2500 nm; however, they generate significant amounts of heat, and this may alter or even burn a sample. To overcome this issue, fiber optic lines are often used to transmit the light. The detector employed in an NIR hyperspectral imaging system is generally optimized for wavelength range. For example, HSI systems operating in the visible–very near-infrared (Vis–vNIR (400–1000 nm)) range typically use charge-coupled device (CCD) or complementary metal oxide semiconductor (CMOS) sensors. Both types of detectors are sensitive in the wavelength range 300–1000 nm, but experience a sharp decline in quantum efficiency below 400 and above 900 nm. Indium Gallium Arsenide (InGaAs), or MCT, detectors are more commonly used for NIR hyperspectral imaging in the 1000–2500 nm range. However, these detectors are more expensive than CCDs and require cooling.

7.3.1 HYPERSPECTRAL IMAGE ACQUISITION

Three common ways in which a hyperspectral image can be acquired are shown schematically in Figure 7.2. Each technique has its own advantages and disadvantages as discussed below.

Point mapping is perhaps the most basic extension of conventional point spectroscopy for acquiring a hyperspectral image. It involves acquisition of a spectrum at one spatial position, followed by the movement of either the sample or the spectrometer to an adjacent position for spectral acquisition. This is continued until an entire image of a sample is built up. This methodology is commonly used in FT-NIR microscopic imaging. Important parameters to control include aperture size (this controls the spatial region on the sample represented by each pixel; typically, higher aperture sizes allow more light to reach the detector, and this improves the signal-to-noise ratio – however, this is at the expense of lower spatial resolution) and spatial step size (this should be equal to the aperture width (when stepping in the x-direction) or length (when stepping in the y-direction) to provide spatially contiguous spectra). This technique has the advantage of being simple and low cost, and can provide high-quality spectra, requiring relatively minor modification of a conventional spectrometer with a single detector, typically the addition of an x-y translation stage for the movement of the sample underneath a spectrometer. The disadvantage of this approach mainly lies in the amount of time it takes to build up the image. In the case of a 100×100 pixel region and a single spectrum acquisition time of 20 s, for example, it would take over 50 h for complete image acquisition. To overcome this limitation, spatially non-contiguous pixel spectra can be obtained; for example, selecting every tenth spectrum in the 100×100 pixel region would be reduced to a 10×10 region, reducing the acquisition time just under 1 h; however, the resultant spatial map would be missing information between each pixel.

"Pushbroom" or line scanning refers to acquisition where the hyperspectral image is built up line by line through the use of an array detector. This type of detector is used to acquire sequential spectral images of a line on the sample. Typically, remote sensing or proximal pushbroom systems are made with dispersive optical components. In these systems, the light from each spatial position along the line passes through a spectrograph and is split into its component wavelengths, resulting in a spectrum. The resultant two-dimensional image is captured by the array detector (see Figure 7.2). The hyperspectral image is built up either through moving the sample underneath the detector or, as is the case for remote sensing, through moving the detector above the sample. FT-NIR microscopic systems, on the other hand, do not require a spectrograph, relying instead on an interferometer. In those systems, a linear or matrix array of detectors can be used to obtain pushbroom images with speeds up to x times faster than point mapping, where x is the number of detectors in the array.

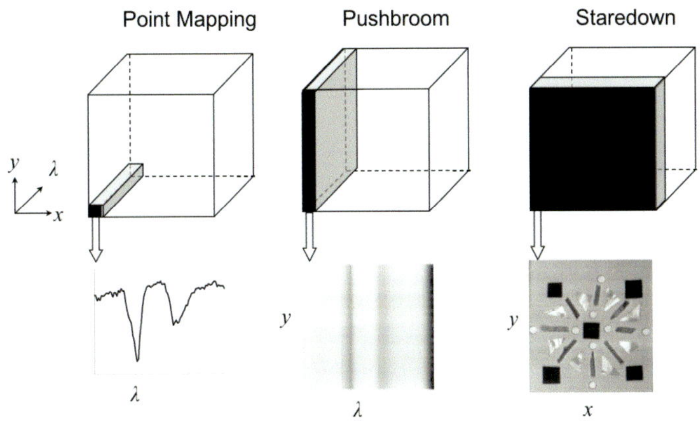

FIGURE 7.2 Schematic displaying three major ways in which a hyperspectral image can be acquired: x and y represent the spatial dimensions, while λ represents the wavelength dimension.

The pushbroom method thus enables a rapid hyperspectral imaging acquisition and is well suited to conveyor belt-type applications, such as food quality control. Downsides of this technique include the relatively high cost of InGaAs or MCT sensors used in array detectors, relatively low spectral resolution, and the presence of image and spectral artifacts, as will be discussed later in Section 7.3.

"Staredown" hyperspectral imaging systems acquire the hyperspectral image wavelength slice by wavelength slice by sequentially obtaining single-wavelength images. This, the only "true" imaging technique of the three methods as shown in Figure 7.2, can be achieved through the use of a tunable filter to select only light at a specific wavelength or through the use of a tunable light source that only shines light of a specific wavelength on a sample. A major advantage of this technique is that no sample movement is required. Using a tunable light source removes the need for a spectrograph and enables the use of higher-intensity light at each wavelength than would be the case for a broadband source and thus increased signal-to-noise ratio. However, producing homogeneous illumination over a large area can be challenging.

More recently, "one-shot" hyperspectral imaging has emerged, where the entire hyperspectral image can be obtained simultaneously in one acquisition. This can be achieved through the use of prisms and waveplates or mirrors to project simultaneously different wavelength images of a sample onto a detector array (Takatani et al., 2017) or by integrating spectral filters on top of detector elements (Gonzalez et al., 2015). So far, these systems have been commercialized for the Vis–vNIR (400–1000 nm) wavelength range, but it is expected that they will become more versatile and widespread in years to come. With the dual advantages of low cost and fast acquisition times, one-shot hyperspectral cameras stand to revolutionize the next phase of hyperspectral imaging in both industrial and consumer applications.

As is the case for conventional NIR spectroscopy, NIR hyperspectral images can be obtained using different modalities, such as transmission, reflectance, transflectance, or interactance. Most pushbroom instruments are designed for reflectance or interactance measurements only. However, it is straightforward to modify such an instrument to enable transmittance, simply by moving the light source underneath the sample. The selection of modality is usually sample dependent, and it is important to consider the optical properties of the sample. For example, transmission imaging is generally preferable to reflectance for materials that absorb a lot of light in the NIR, such as aqueous or high water content samples. In order to enable imaging of highly absorbing samples, it is advisable either to work with thin samples or to work in transmission mode. If this is not possible, the use of a highly reflective substrate can also improve transflectance measurements, although the effectiveness of this approach depends on sample thickness. For example, in Figure 7.3, two NIR hyperspectral images are shown for the "water-in-oil" sample. Images of the same field of view were

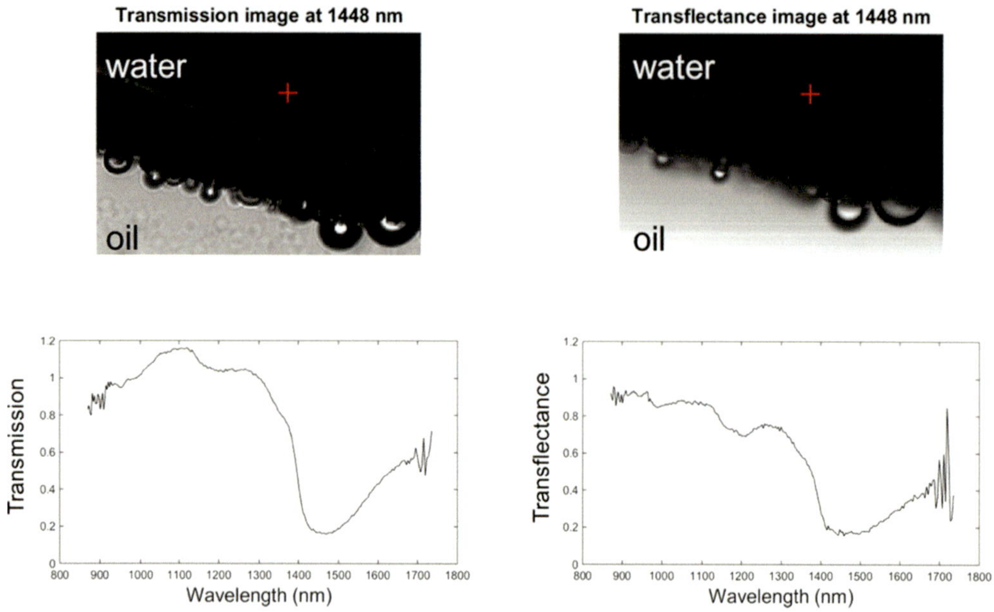

FIGURE 7.3 Transmission and transflectance hyperspectral imaging data for water-in-oil sample obtained using Instrument 2 (described in Section 7.2). Single-wavelength images at 1448 nm display the higher absorption of light by water compared to oil. A single-pixel spectrum of the water region is plotted in the lower panel from the spatial region represented by a red cross in each image.

obtained by the same instrument at the same magnification (10 ×) but using different modalities, transmission, and transflectance. For the transflectance measurements, a gold substrate was placed underneath the sample. It can be seen that the transmission image is sharper than the transflectance one, while some line artifacts appear in the transflectance image. A single-pixel spectrum of water is also shown, and there is a clear difference in the spectra profile for each modality. For the transflectance setup, the spectrum is noisier and the profile of the water peak appears to be broader than it is in the transmittance images. This is because of near-signal saturation due to the double-pass of light through the sample to the gold and back again to the detector, at least doubling the pathlength and reducing the amount of light reflected back to the detector.

7.3.2 Instrument Calibration

When light impinges on a detector in an NIR HSI system, the signal is recorded in raw intensity counts, ranging from 0 to $2^x - 1$, where x is the bit depth of the sensor. This is known as the dynamic range of the sensor. For example, an 8-bit sensor has a dynamic range of 0–255. The intensity detected by the sensor depends on a number of factors apart from the light interaction with a sample. These include dark current of the detector (i.e., stochastic signals arising from thermal energy), interelement variations in sensitivity on the array detector, wavelength dependence of the light source, stability of the light source, and any optical aberrations in the system. In order to account for these variations, and enable comparisons of NIR HSI data obtained from the same instrument at different points in time, it is necessary to perform instrument calibration. This should be carried out on a daily basis at a minimum. The measured signal intensity (I) is typically calibrated with respect to the measurement of a dark current (D) and white reference measurement (W), resulting in the calculation of relative reflectance (R), as shown in Equation 7.1.

$$R(\lambda) = \big(I(\lambda) - D(\lambda)\big) / \big(W(\lambda) - D(\lambda)\big) \tag{7.1}$$

The dark current (D) is usually measured by blocking the light entering the system (e.g., covering the lens) and measuring the resultant signal intensity at each element of the detector array, while W is usually measured using a highly reflective, nonabsorbing calibration sample. Materials commonly used for this purpose include spectralon, white ceramic tiles, and gold. The calibration sample is required to be large enough to fill the entire field of view of the imaging system to capture the response of each sensor element in the array. For this reason, gold is more commonly used in microscopic systems, while white tiles, being cost-effective and easy to clean, are commonly used in pushbroom imaging. If the NIR reflectance spectrum of the calibration sample is known (through measurement with an independent spectrometer), then the relative reflectance (R) in Equation 7.1 is usually multiplied by this spectrum to enable the comparison of spectral data between multiple instruments. In the case of transmission imaging, W is usually measured as the light passing directly from the light source to the detector with no sample in between. After calibration, it is advisable to take a hyperspectral image of the reference sample. This can be used to investigate the signal-to-noise characteristics of the system and stability over time.

7.3.3 Signal-to-Noise Estimation

Due to the wavelength dependence, both of light sources and of detectors in NIR imaging, it is important to characterize the signal-to-noise ratio across the spectral range. This can be estimated with the reference material used in calibration by measuring a region of the calibration tile, assumed to be homogeneous, and subsequently inspecting the variability in spectra. For example, a comparison of the response of two different NIR hyperspectral imaging systems is shown in Figure 7.4. Instrument 1 is a macroscopic system (pixel size approx. 320 μm), while Instrument 2 is a microscopic one (pixel size approx. 3.2 μm). Both systems are pushbroom, operate in the 950–1700 wavelength range, are illuminated with tungsten-halogen lightbulbs, and have InGaAs detectors. White spectralon was used as a calibration reference spectrum, and a region of 10×10 pixels of the spectralon was then imaged by each system. The mean image of each 10×10 pixel hypercube is shown in Figure 7.4. The first thing to notice is that the mean image for the microscopic system exhibits some spatial variability when compared with the macroscopic system, indicating that the spectralon sample is not uniformly homogeneous at the microscopic scale. Inspection of the pixel spectra from each instrument indicates that considerable noise is present at the ends of the spectral range – i.e., below 950 nm and above 1650 nm, mainly due to a decrease in detector sensitivity in those wavelengths. The coefficient of variation (% standard deviation/mean) at each wavelength is a useful estimator of the noise/signal ratio, as plotted in the bottom panel of Figure 7.4. It can be seen to increase at extreme edges of the wavelength range. It is important to remove such noisy regions from the hyperspectral image before carrying out subsequent data analysis, since the noisy wavelengths will impact any subsequent multivariate analysis (as is shown later in Section 7.4). This could be achieved by setting a threshold, for example, 5%, and exclude wavelength regions with a %CV higher than this threshold.

7.4 DATA ANALYSIS

Hyperspectral image analysis comprises a variety of pretreatment, chemometric, and image analysis steps, and the selection of these is generally goal dependent – for example, whether the intention is to explore, classify, or quantify the data. While all chemometric methods used in traditional point NIR spectroscopy can be applied to hyperspectral imaging data, certain steps need to be carried out to format the data for analysis. In addition, some issues regularly found in hyperspectral images are not common to traditional NIR spectroscopy and require special consideration. The most obvious is the three-dimensional structure of the hyperspectral image. In order to apply conventional data pretreatments and analysis to a hyperspectral image, the first step is usually to unfold the hyperspectral image, as shown in Figure 7.5. Unfolding a hyperspectral image simply means stacking each spectrum one on top of the other to make a two-dimensional matrix.

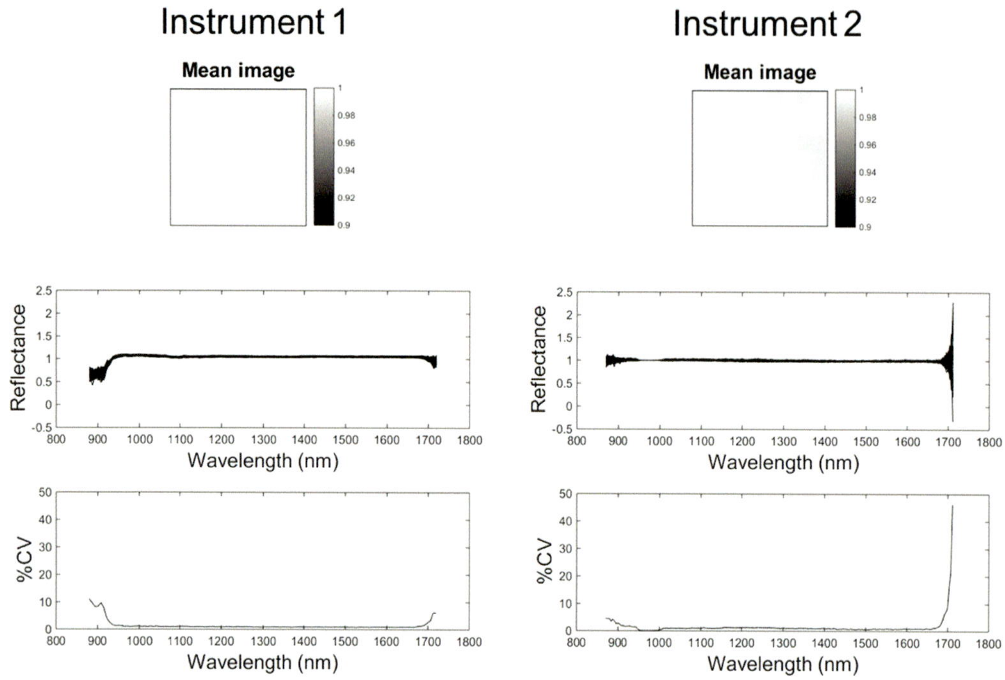

FIGURE 7.4 Reflectance hyperspectral imaging data for white spectralon sample obtained using Instruments 1 and 2 (described in Section 7.2). Mean wavelength images were obtained by averaging over all wavelengths. All pixel spectra within the 10×10 pixel region displayed are plotted in the middle row. The lower row shows the percentage coefficient of variation (%CV) calculated for each wavelength, demonstrating high noise at the extremes of the spectral range.

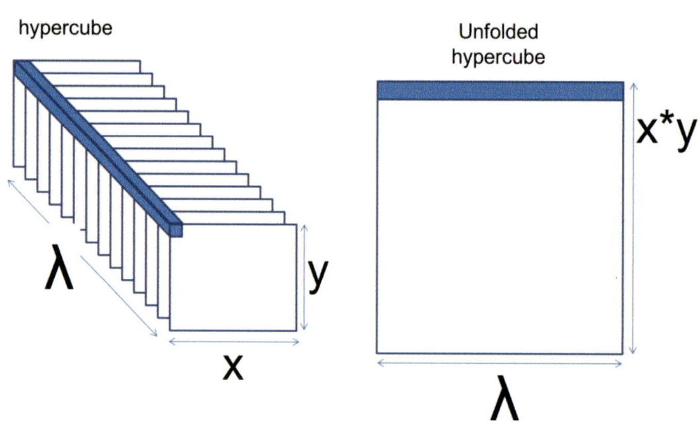

FIGURE 7.5 Schematic showing unfolding process: x and y represent spatial dimensions, while λ represents the wavelength dimension.

7.4.1 Data Pretreatments

Data pretreatments are typically required to remove undesired spectral or spatial variability from hyperspectral image data. This variability can arise due to the instrumentation (e.g., bad pixels, nonuniform lighting), environment (e.g., changes in temperature or shadows in the case of remote sensing), or even the sample itself (e.g., sample morphology). Since these sources of variability are not usually informative in the subsequent classification or interpretation of a sample, it is important to remove them prior to data analysis.

7.4.1.1 Background Removal

In NIR imaging of discrete objects, it is usually necessary to remove background (i.e., nonsample) pixels from the image. This is an important step in subsequent analysis where background pixels are not of interest. This can be achieved by thresholding the hyperspectral image at a single wavelength. However, to be able to do this, prior to image acquisition, it is beneficial to select a background material that is spectrally flat and distinct from the spectrum of the sample. For example, an NIR hyperspectral image of a PS sample is shown in Figure 7.6. The sample's spectrum exhibits a peak at 1146 nm, not present in the spectrum of the background material (a white tile). The histogram of the image at 1146 nm shows the distribution of pixels at different reflectance values. There are two main distributions: a unimodal distribution centered at 0.7, representing the background, and another trimodal distribution centered at 0.42, representing the sample. The image at 1146 nm can be thresholded by setting all pixels with an intensity less than 0.55–1 and all other pixels to 0. This results in a "mask" image in which all pixels of the background are set to zero.

7.4.1.2 Bad Pixel Removal

Detectors used in NIR hyperspectral imaging are prone to the occurrence of "bad" pixels, i.e., pixels that generate erroneous readings. These can materialize as "dead" pixels, where no response is recorded, or "spikes", where an unusual spike is present in a spectrum (Burger, 2009). Bad pixels can appear in fixed positions on a detector array or occur intermittently in different positions on a detector. They are a common scourge in NIR hyperspectral imaging and tend to increase in

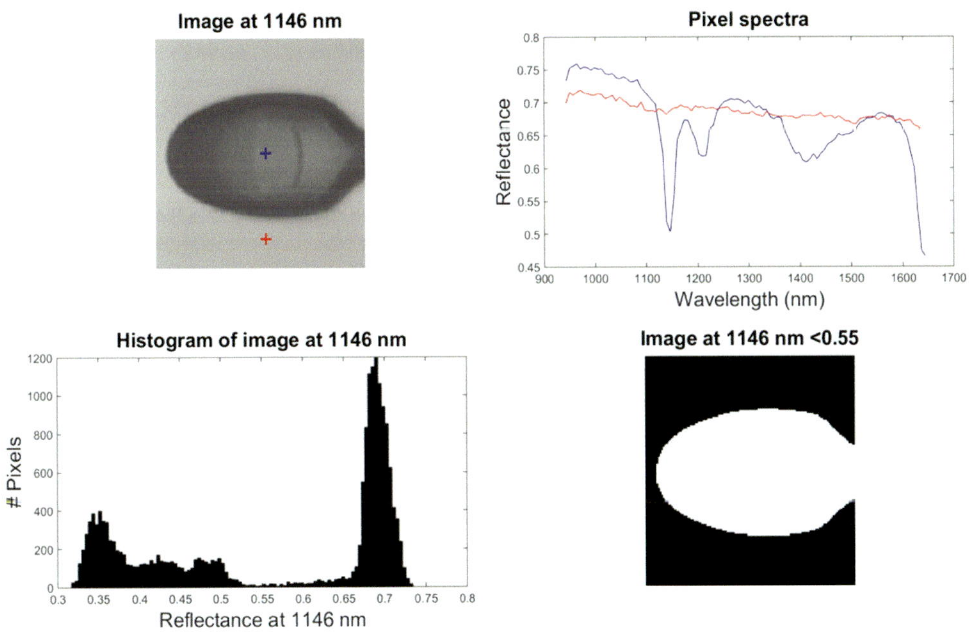

FIGURE 7.6 Removal of background from a hyperspectral image using a thresholding approach. An NIR reflectance image of a polystyrene sample is shown at 1146 nm (see "packaging" dataset, described in Section 7.2). Spectra of the background and sample are plotted on the upper-right panel, with their spatial coordinates marked on the adjacent image by a red and blue cross. The histogram of the image at 1146 nm shows two main distributions representing the foreground and background. Thresholding the image at 1146 nm results in a mask image shown in the lower-right panel.

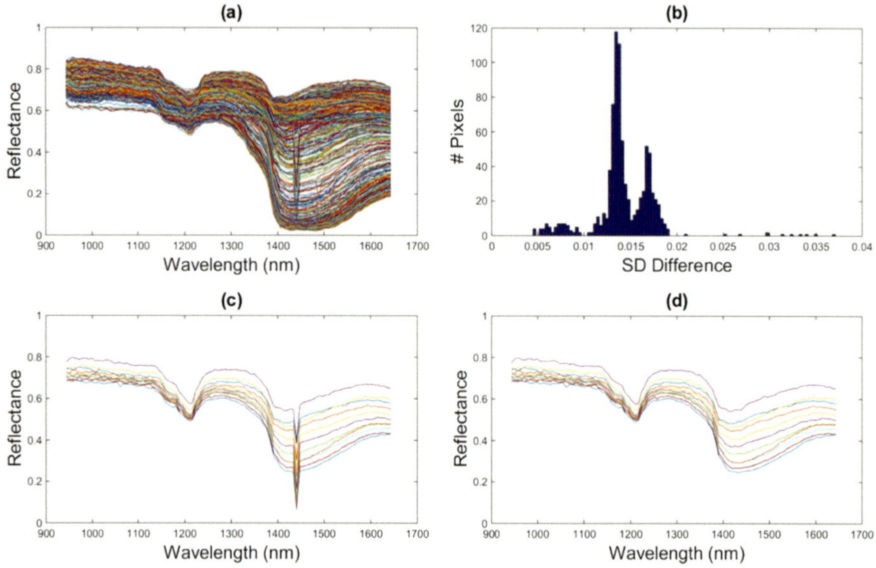

FIGURE 7.7 Identification and correction of bad pixels in hyperspectral imaging data. Pixel spectra from the "water-in-oil" sample imaged with Instrument 1 (as detailed in Section 7.2) are plotted in panel (a). The histogram of the standard deviation of difference spectra ("SD difference", obtained for each spectrum individually by subtracting consecutive wavelength reflectance data from each other) is shown in panel (b). The spectra with high values of SD difference (greater than 0.02) are plotted in panel (c) and are identified as bad pixels. These are replaced by median filtering around the spike and plotted in panel (d).

population over the lifetime of a detector. Bad pixels, although typically low in number compared to the total number of spectra in a hyperspectral image, can have a large influence on subsequent data analysis. Therefore, it is important to identify and replace them prior to further data analysis. One straightforward technique, described in Dorrepaal et al. (2016), involves calculating a difference spectrum for each spectrum by subtracting reflectance at subsequent wavelengths from each other. An example is provided in Figure 7.7, in which pixel spectra from the "water-in-oil" dataset are plotted. It is somewhat difficult to see, but there are some bad pixels presenting as spikes at around 1447 nm. The standard deviation of this difference spectrum ("SD difference") tends to be high for spectra harboring bad pixels. For example, the histogram of SD difference is plotted in Figure 7.7b. The spectra of pixels outside of the main distribution, with SD difference values greater than 0.02, are plotted in Figure 7.7c, where the bad pixels at 1447 nm are clearly evident. The bad pixels were removed by making a window around the bad pixel wavelength, five wavelengths in length, and replacing the bad pixel with the median value within the window. The resultant corrected spectra are plotted in Figure 7.7d.

7.4.1.3 Removal of Spatial Artifacts

As previously mentioned in Section 7.3, hyperspectral image artifacts in the spatial domain are also commonplace. Pushbroom systems are prone to the so-called striping artifacts, i.e., line artifacts in the spatial domain which can obscure features of interest. For example, an FT-NIR image of the "water-in-oil" sample is shown in Figure 7.8. The image at 1447 nm exhibits striping artifacts in the spatial domain which obscure the spatial features of the oil–water interface. Two spectra from the water region of the image are shown, where it can be seen that a nonlinear baseline is present. A method recently developed by Xu et al. (2017) uses extended multiplicative signal correction (EMSC) in the spatial domain to remove the effect of these kinds of artifacts. The corrected image and spectra are shown on the right-hand side of Figure 7.8. The method not only removes the

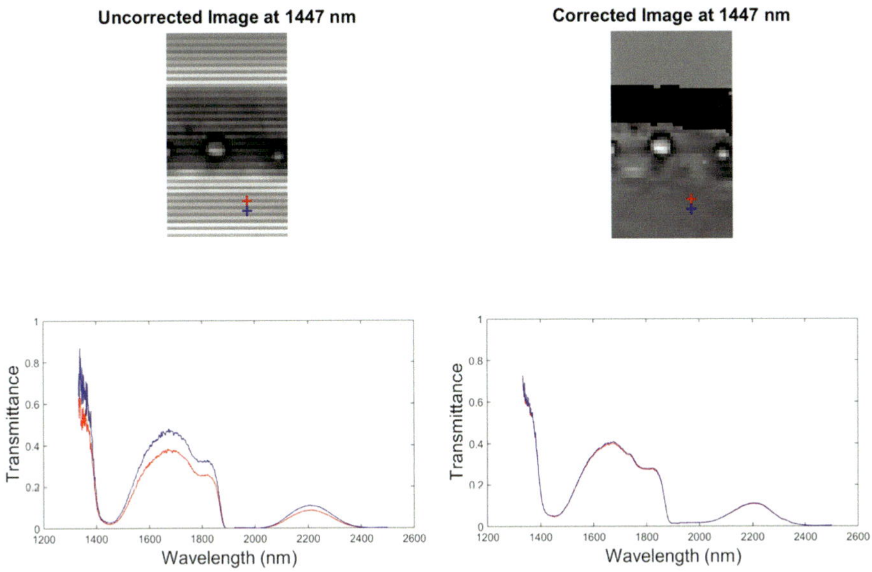

FIGURE 7.8 Correction of line artifacts in FT-NIR imaging data for "water-in-oil" sample measured in transmission using Instrument 3 (see Section 7.2). Line artifacts evident in single-wavelength images result in nonlinear shifting of the spectrum as shown in the lower-left panel. The spatial positions of the spectra shown are identified by crosses in the upper-left image. After image correction, the line artifacts and baseline effects on the spectra are removed (see right-hand side of image).

striping in the spatial domain but also removes the associated spectral variability, enabling a more accurate analysis of the data. The technique has the added benefit of smoothing noisy regions of the spectra.

7.4.1.4 Removal of Sample Morphology-Related Variability

Spectral pretreatments like EMSC are also useful for the removal of variations in spectra due to changes in the laboratory environment (Gowen et al., 2015) or sample morphology (Esquerre et al., 2011). For example, an NIR image of the PS sample is shown in Figure 7.9. Due to the nonflat nature of this object, a significant spectral variability is noticeable over the surface. Spectral pretreatments such as SNV (standard normal variate), MSC (multiplicative scatter correction), and EMSC can be employed to reduce this spectral variability, as shown in Figure 7.9. EMSC works better than SNV or MSC, resulting in lower subsequent spectral variability over the sample, but requires the input of interferent spectra, while MSC requires the provision of a target spectrum. In this example, the target spectrum chosen for MSC was the mean spectrum of the sample and the interferent spectrum chosen for EMSC was the first PC loading of the sample's spectra. This approach is reasonable for images in which only one type of sample is present; however, care must be taken in applying the same method to a mixed image if the object cannot be segmented because there is a risk of inducing new spectral features (using EMSC) or producing outlier spectra (using EMSC or MSC) (Fearn et al., 2009). Therefore, as a compromise, simpler row-wise pretreatments like SNV are often preferred.

7.4.2 Exploratory Analysis

Exploratory analysis is often carried out to locate the main sources of variability in a hyperspectral image or a set of hyperspectral images in an unsupervised way. Exploratory methods like principal components analysis (PCA) can be applied directly to an unfolded hyperspectral image (X),

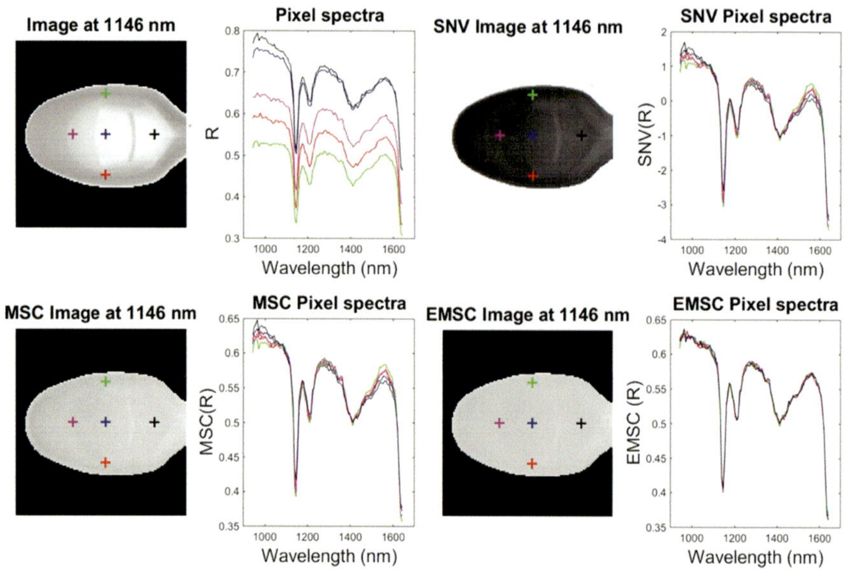

FIGURE 7.9 Effect of spectral pretreatments on pixel spectra of homogeneous sample with curved morphology. An NIR reflectance image of a polystyrene spoon is shown at 1146 nm (see "packaging" dataset, described in Section 7.2). In the top-right panel, the effect of sample morphology is pronounced, resulting in shifting amplitude of the reflectance spectra. Standard normal variate (SNV), multiplicative scatter correction (MSC), and extended multiplicative signal correction (EMSC) reduce the variability both in the image plane and in the spectra.

to decompose it into scores (S) and loadings (L) matrices as shown in Figure 7.10. PCA enables the typically >100 wavelength channels of a hyperspectral image to be reduced down to a smaller number (p) of principal components (PCs). The resultant scores matrix can be refolded into a stack of score images. This distinct advantage of hyperspectral image enables the analyst to interpret the spatial distribution of score values. For example, PCA is applied to the "packaging-mix" NIR hyperspectral image as shown in Figure 7.10. PCA is sensitive to any data pretreatments applied to the hyperspectral image, such as selection of wavelength range, background removal, and pretreatments. As previously mentioned in Section 7.3, the extreme wavelength regions of the detector are very noisy and should be removed from the data prior to analysis. In order to demonstrate the influence of including such regions on subsequent exploratory analysis, PCA was applied first to the hyperspectral image with the full wavelength range and then to the hyperspectral image with a reduced wavelength range of 943–1643 nm. The resultant PC score images are shown in Figure 7.10b and c, respectively. It is clear that including the entire wavelength range produced noise features in the score images, manifested as lines. This noise, a source of variability identified by PCA, is present in all of the PC images as shown in Figure 7.10b. By removing the noisy parts of the spectrum, this noise is removed from the score images (see Figure 7.10c). Upon closer examination of Figure 7.10c, it can be seen that PCA, an unsupervised technique, is capable of summarizing the main spectral variability between samples from 100's of wavelength channels down to just 4 or 5 PCs. In the first PC score image, the HDPE and PS samples are brighter than the paper and PET samples. In the second PC score image, the HDPE is darker than any of the other samples; in the third PC score image, the cardboard samples are darker than the rest; in the fourth PC score image, the PS and PET samples are darker than the rest; and in the fifth PC score image of the sample, there is little discriminatory information. When the background was removed from the image, the PCA score images were modified again, since the variability of the background was not included in the analysis (see Figure 7.10d). When SNV was applied to the data, most of the variability was moved

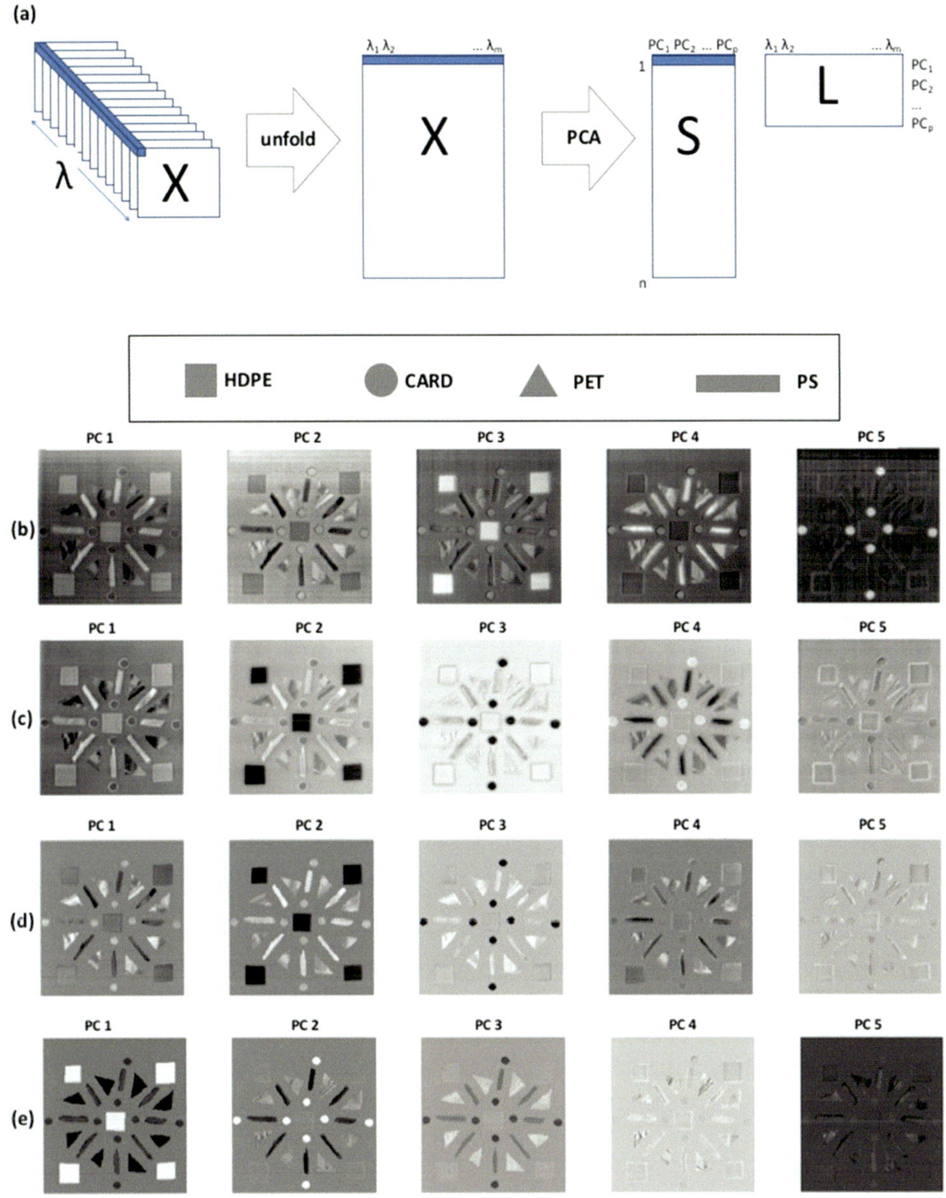

FIGURE 7.10 Principal component analysis (PCA) applied to hyperspectral imaging data. The top row (a) is a schematic showing how PCA can be applied to a hyperspectral image (*X*), producing scores (*S*) and loadings (*L*). PCA scores can be refolded to make PC score images, as shown in (b)–(e) for the "packaging" dataset (see Section 7.2 for details). PCA was applied to the image with: no background removal and full wavelength range (b); no background removal and reduced wavelength range (c); background removal (d); and background removal and standard normal variate pretreatment (e).

up to the first three PCs. In addition, spatial variability in the PET samples (triangles) due to their morphology was largely removed from the data (see Figure 7.10e). PCA can also be performed on sets of hyperspectral images, by unfolding them and stacking together. This enables the examination of interimage variability and the identification of common materials among images (for further information, see Dorrepaal et al., 2016).

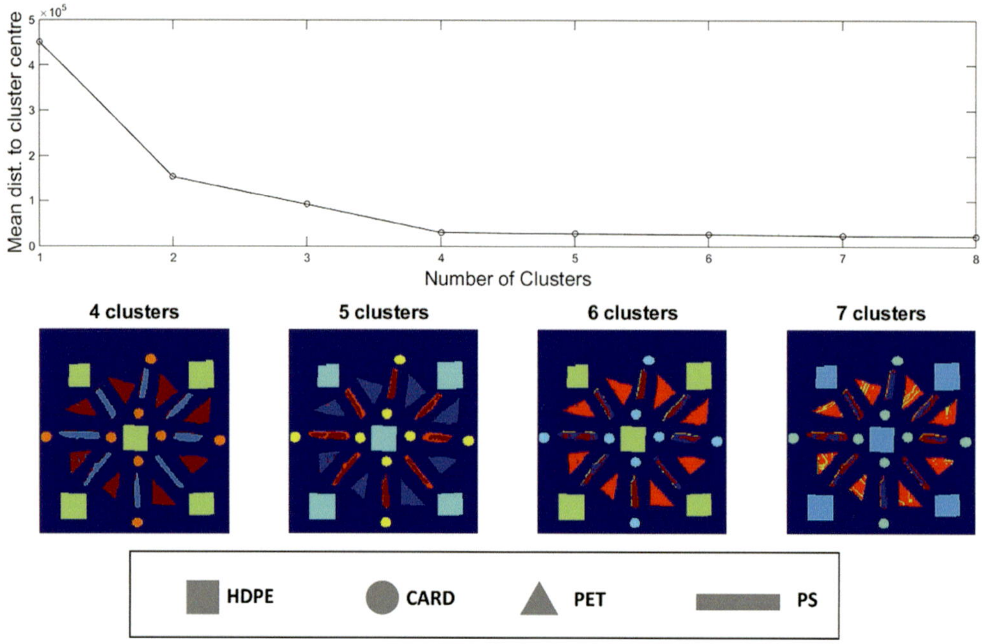

FIGURE 7.11 *K*-means clustering applied to "mix" image (see Section 7.2 for details). Objective function of the mean distance of each pixel to its cluster center is plotted, indicating four clusters are appropriate for this dataset. Clustering results for 4–7 clusters also shown. As cluster number is increased above 4, different regions within the PS and PET samples become clustered into new groups due to their morphology.

Data clustering techniques can also be considered as exploratory analysis tools. For example, *k*-means clustering can be used to estimate the number of classes in a particular image or set of images. Using again the "packaging-mix" dataset as an example, *k*-means was applied to the SNV-pretreated, non-background spectra. In this example, it is already known that there are four materials in the image, so four clusters should be the appropriate value of *k* in the *k*-means analysis. However, this is not always known in advance. In those cases, it is useful to calculate the mean distance of each point in a cluster to its center, averaged over all clusters, for each value of *k*. This is plotted in the top panel of Figure 7.11, and it can be seen to reach a plateau at four clusters. The *k*-means clustering results can also be visualized as an image, as shown in the lower panel of Figure 7.11. It can be seen that with $k=4$ clusters, most all of the material classes are correctly clustered. When the five-cluster image is considered, it can be seen that the PS sample is divided into two clusters, representing the center and edge regions of this material. As the number of clusters is increased, new clusters appear mainly near sample edges. These are not "real" clusters – they arise due to differences in reflected light intensity at sample edges. The color assigned to each cluster is random: Each time *k*-means is applied and the initial clusters are selected from the data by random seeding at the start of the algorithm. Therefore, this approach cannot be used for classification directly. However, the cluster center spectra of the clusters selected by *k* means could be used as inputs to a subsequent classification model.

7.4.3 Classification and Quantitation

Classification of hyperspectral images, i.e., identifying each pixel as belonging to a certain category, can be achieved using the plethora of multivariate classification techniques available in the literature (e.g., LDA, SIMCA, PLS-DA). As with the standard point spectroscopy, it is first necessary to create a calibration dataset of spectra representing each modeled class. This can be done by manual selection of a region of interest in an image representing a given class, or through random or stratified selection of spectra from an image region. The selected spectra are then combined in a two-dimensional matrix with a corresponding label vector, where each row of the matrix is labeled according to its class. When a calibration model is developed, it can then be applied to a new image for pixel classification resulting in a pixel classification map, as shown schematically in Figure 7.12.

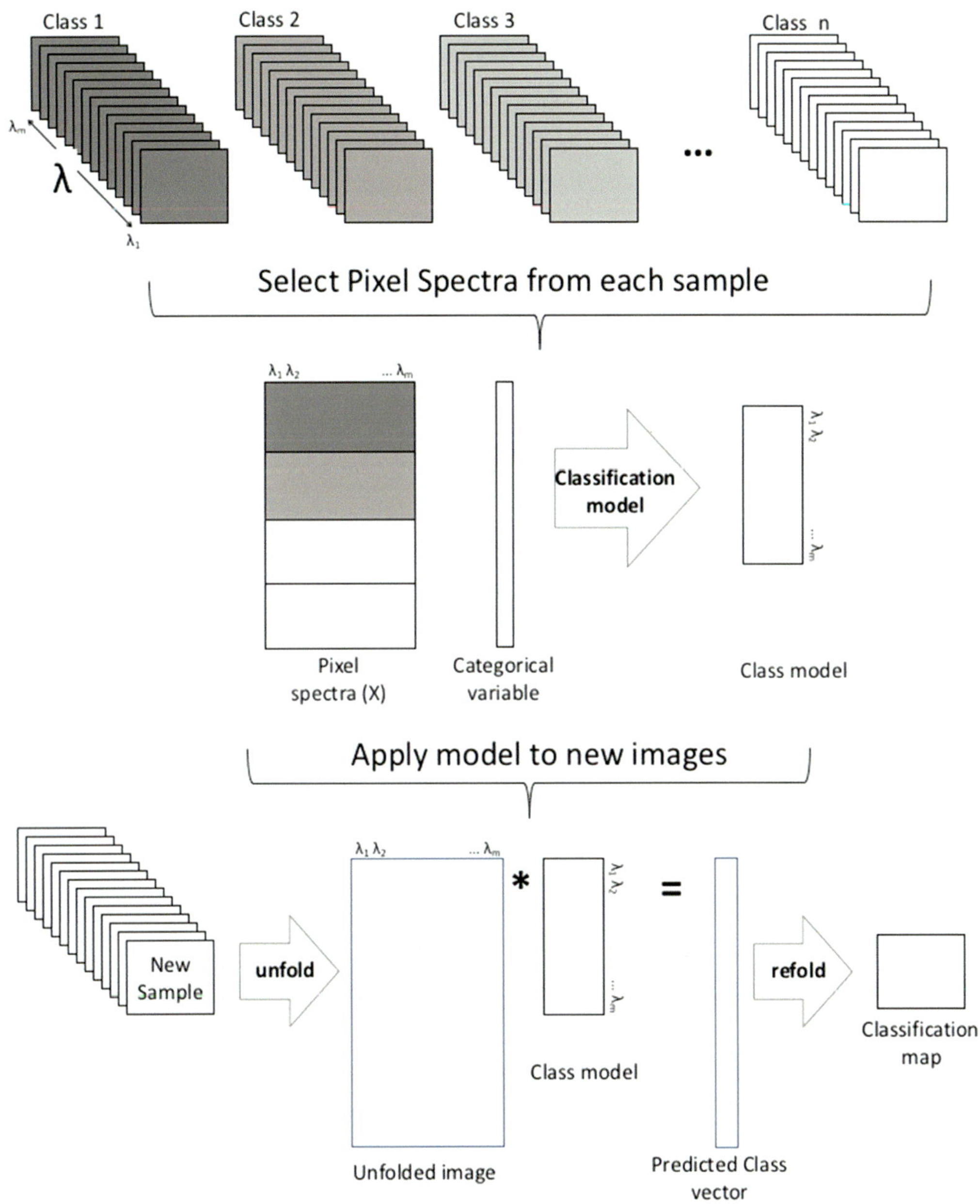

FIGURE 7.12 Schematic showing how classification models can be built from a set of hyperspectral images: λ represents the wavelength dimension, m is the number of wavelength variables, while n is the number of images.

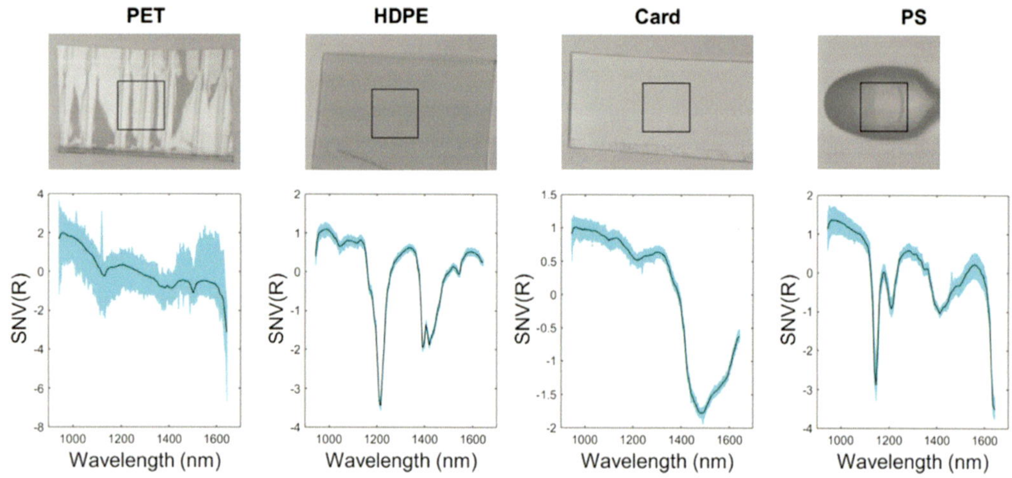

FIGURE 7.13 Mean NIR images of materials imaged in "packaging" dataset (see Section 7.2 for more details). Box represents a region of interest of 50×50 pixels in size. Lower panel displays the pixel and mean spectra from the region of interest.

As an example, calibration spectra were selected from a user-defined rectangular region of interest of 50×50 pixels in size (i.e., 250 spectra per class) from images of each of the materials in the "packaging" dataset (i.e., PET, HDPE, cardboard, and PS). The mean images of each sample and corresponding regions of interest and spectra are shown in Figure 7.13. The PET sample showed the most variability due to the morphology of the sample. These spectra were combined to make a matrix of 1000 spectra in length, and a corresponding label vector was created where each material was assigned a number label (PET = 1, HDPE = 2, card = 3, PS = 4).

An *n* class partial least squares discriminant analysis (PLS-DA) model was constructed to classify the spectra in the calibration set. The model classification error is plotted for each number of latent variables (LV), as shown in Figure 7.14. This plot reaches a minimum of 0 at 4 LV and does not change after that, indicating a perfect classification on the calibration dataset at four LV. This model was then applied to the image of mixed packaging materials, as shown in the bottom panel of Figure 7.14. Classification maps for different numbers of LV are also shown – it can be observed that at 2 LV, the model is underfitted, with classification errors occurring between the HDPE, cardboard, and PS classes, while the higher LV models also experience misclassification at the edges of the PS sample. Clearly, the 4 LV model is optimal for the data considered. After classification modeling has been optimized, the classification map can be further probed, for example, by calculating the number of pixels of each class present in an image. As an example application, in waste recycling at present, PS is often regarded as a contaminant as it is not widely recycled. Waste streams are characterized by the percentage of contaminant contained in them and are rejected if the level of contamination is above a certain percentage, for example, 10%. The percentage of a given class of material in a classification map can easily be quantified by counting the number of pixels equal to that class and dividing by the total number of non-background pixels. In this example, approximately 20% of the pixels represented in the image are PS. However, it can be noted that some of the PS pixels are misclassified at the edges probably due to pixel mixing with the background at the edges. This kind of misclassification could be reduced by image post-processing such as erosion.

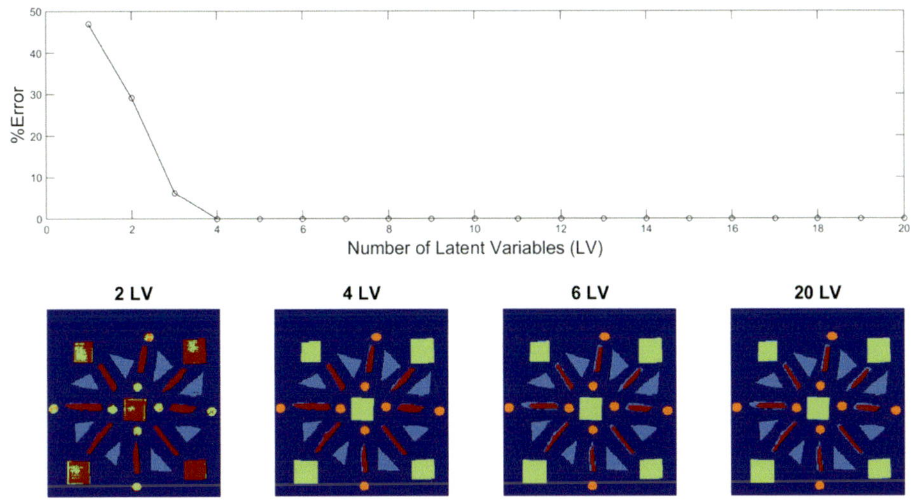

FIGURE 7.14 Partial least squares discriminant analysis (PLS-DA) applied to "packaging" dataset. Spectra as shown in Figure 7.13 were combined and labeled to make a calibration set. N-PLSDA was used to build a classification model. The percentage classification error for each number of latent variables in the model (up to 20) is shown on the top panel, indicating the appropriateness of a 4 LV model. Prediction maps for 2, 4, 6, and 20 LV are shown on the bottom panel. The number of pixels misclassified increases when the model is underfitted (at 2 LV) or overfitted (at 6 or 20 LV).

7.4.4 Regression Analysis

As in conventional NIR spectroscopy, regression techniques like partial least squares regression (PLSR) can be used to develop quantitative calibration models from hyperspectral imaging data to predict the concentration of a property of interest in a sample (e.g., moisture content, protein). However, since it is often impossible to measure the quantity of a given analyte at a pixel level, gross quantities like moisture content are usually measured for a sample, and then the mean or median spectrum of its hyperspectral image is used for PLSR modeling. This approach is shown schematically in Figure 7.15. After mean or median spectra are extracted from each image in a set of hyperspectral images, they can be combined into a calibration set (*X*) the same length as the set of measured quantities (*Y*), i.e., the number of samples (*n*). A regression model can then be built to relate the *X* and *Y*. This model can be applied to new images to predict quantities at a pixel level.

As an example of this approach, a PLSR model was built to predict moisture content in the "chickpea" dataset. The leave-one-out root mean-squared error of cross-validation (RMSECV) for the model, plotted as a function of number of LV as shown in Figure 7.16, exhibits an elbow at 3 LV and a global minimum at 10 LV. It is quite difficult to select the optimal number of LV for the model using the RMSECV plot alone. However, inspection of the prediction maps for different numbers of LV shows that at 3 LV, the model is underfitted and at 10 LV the model is overfitted. From the process of hydration, it is expected that the center of the sample should have a lower moisture content than the surrounding region – this is not the case in the prediction maps for LV 3 or LV 8 or higher. The noise prediction maps can be used to aid in estimating the correct numbers of LV for a PLSR model, by inspecting the changing spatial patterns in the prediction maps as the number of LV changes, or by calculating the prediction map error

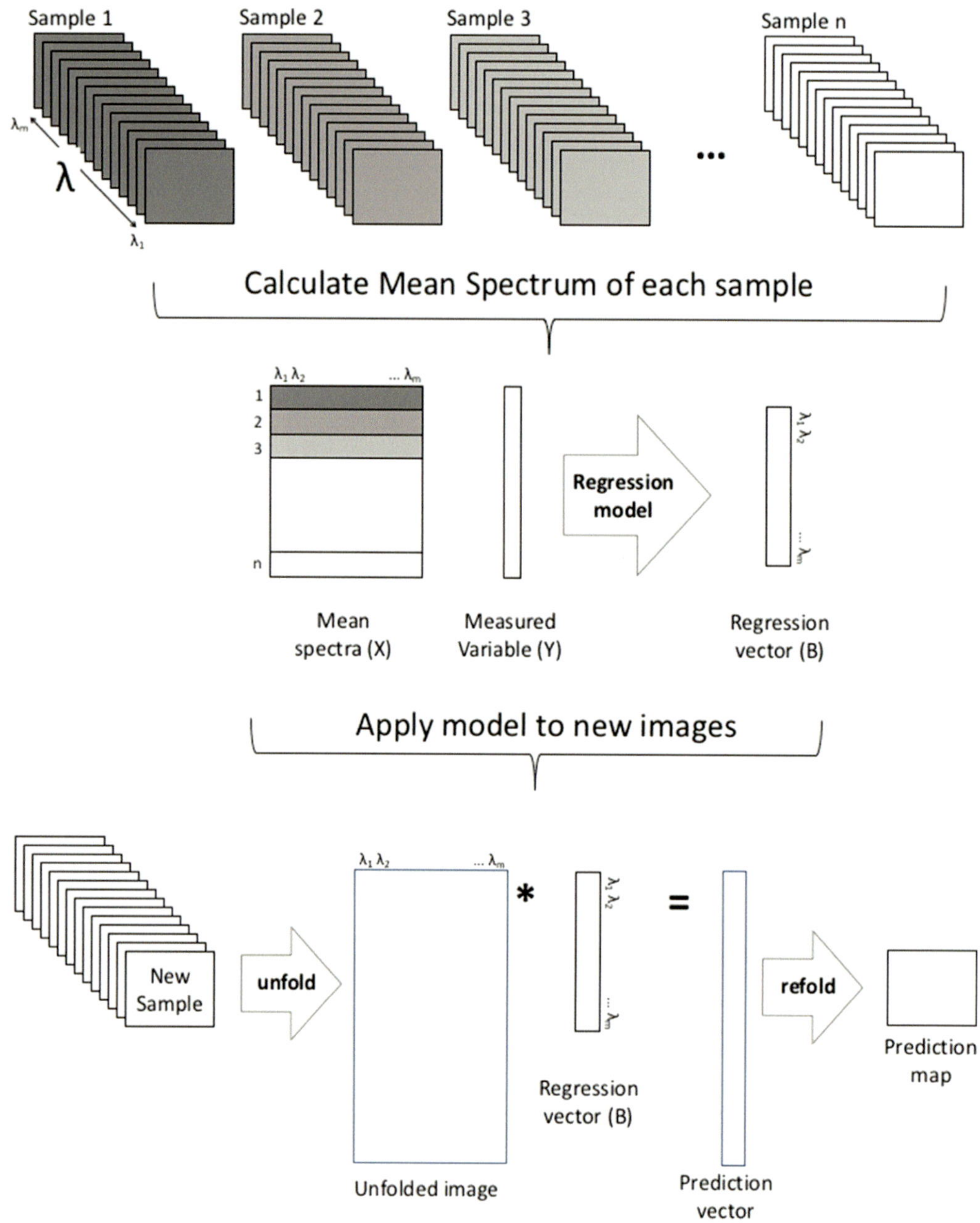

FIGURE 7.15 Schematic showing how regression models can be built from a set of hyperspectral images: λ represents the wavelength dimension, m is the number of wavelength variables, while n is the number of images.

($RMSEP_{Im}$) (Gowen et al., 2014). The $RMSEP_{Im}$ exhibits a global minimum at 7 LV, which corresponds with visual analysis of the prediction map indicating overfitting after 7 LV. This final example demonstrates the added advantage of hyperspectral imaging to visualize the spatial distribution of predicted values as a prediction map.

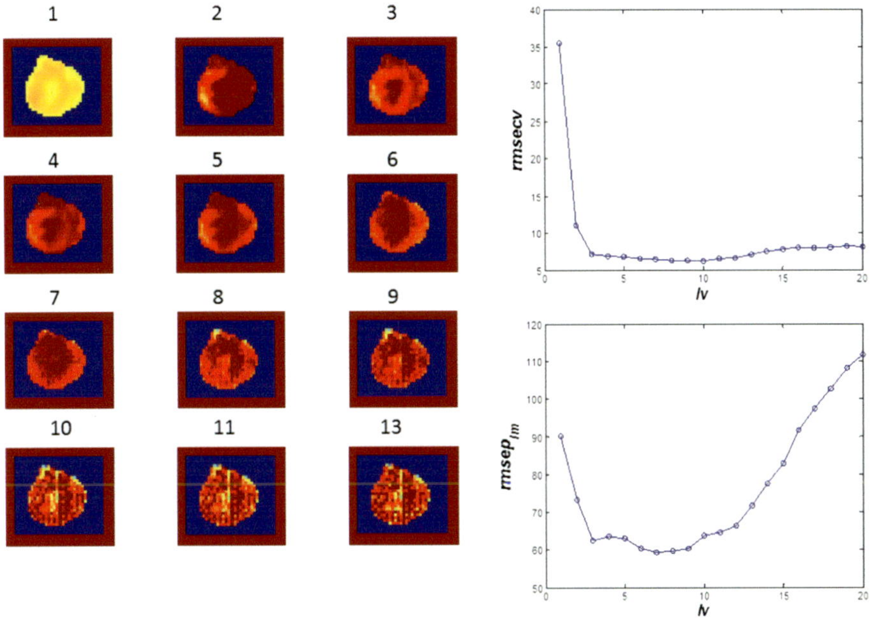

FIGURE 7.16 Prediction maps showing the prediction of moisture content (%) for different numbers of latent variables (LV, shown above each prediction map) for a sample hypercube selected from the chickpeas dataset (see Section 7.2). Border surrounding each prediction map represents the target color corresponding to the expected value of moisture content in each image. All images are scaled according to the maximum and mini mum stated moisture content in the dataset. Model evaluation metrics leave-one-out root mean-squared error of prediction (RMSECV) and root mean-squared error of prediction from the prediction images ($RMSEP_{Im}$) are also shown. While the RMSECV indicates a model with 3 or 10 LV to be optimal, the $RMSEP_{Im}$ and inspection of prediction maps indicate the optimal model performance at 7 LV (Gowen et al., 2014).

ACKNOWLEDGMENTS

I would like to acknowledge my first teacher in hyperspectral imaging, Dr. Jim Burger. Jim brought life into hyperspectral imaging through his unyielding curiosity, enthusiasm, and positivity.

I would also like to acknowledge funding for my research group provided by the European Research Council (ERC) under the starting grant program ERC-2013-StG call – Proposal No. 335508 and Science Foundation Ireland under the Investigators Program Grant – Proposal No. 15/IA/2984.

Dr. Junli Xu is acknowledged for preparation of the water-in-oil sample.

REFERENCES

Adao, T., Hruska, J., Padua, L., et al. 2017. Hyperspectral Imaging: A Review on UAV-Based Sensors, Data Processing and Applications for Agriculture and Forestry. *Remote Sensing* 9:11.

Burger, J. 2009. Replacement of Hyperspectral Image Bad Pixels. *NIR News* 20:7.

Chiou, W. 1985. NASA Image-Based Geological Expert System Development Project for Hyperspectral Image Analysis. *Applied Optics* 24:14.

Dorrepaal, R., Malegori, C., and Gowen, A. 2016. Tutorial: Time Series Hyperspectral Image Analysis. *Journal of Near Infrared Spectroscopy* 24:2.

Esquerre, C., Gowen, A., O'Donnell, C.P., and Downey, G. 2011. Suppressing Sample Morphology in Near Infrared Spectral Imaging of Agriculture Products by Chemometric Pre-Treatments. *Chemometrics and Intelligent Laboratory Systems* 117: 129–137.

Fearn, T., Riccioli, C., Garrido-Varo, A., and Guerrero-Ginel, J.E. 2009. On the Geometry of SNV and MSC. *Chemometrics and Intelligent Laboratory Systems* 96:22.

Feng, Y., and Sun, D. 2012. Application of Hyperspectral Imaging in Food Safety Inspection and Control: A Review. *Critical Reviews in Food Science and Nutrition* 52:11.

Goetz, A., Vane, G., Solomon, J., et al. 1985. Imaging Spectrometry for Earth Remote Sensing. *Science* 228:4704.

Gonzalez, P., Geelen, B., Blanch, C., Tack, K., and Lambrechts, A. 2015. A CMOS-Compatible, Monolithically Integrated Snapshot-Mosaic Multispectral Imager. *NIR News* 26:4.

Gowen, A., O'Donnell, C., Cullen, P., et al. 2008. Recent Applications of Chemical Imaging to Pharmaceutical Process Monitoring and Quality Control. *European Journal of Pharmaceutics and Biopharmaceutics* 69:1.

Gowen, A., Marini, F., Esquerre, C., O'Donnell, C., Downey, G., and Burger, J. 2011. Time Series Hyperspectral Chemical Imaging Data: Challenges, Solutions and Applications. *Analytica Chimica Acta* 705:272.

Gowen, A., Burger, J., Esquerre, C., Downey, G., and O'Donnell, C. 2014. Hyperspectral Image Analysis: Methods for Evaluating Regression Model Performance and Preventing Overfitting. *Journal of Near Infrared Spectroscopy* 22:.261.

Gowen, A., Marini, F., Tsuchisaka, Y., et al. 2015. On the Feasibility of Near Infrared Spectroscopy to Detect Contaminants in Water Using Single Salt Solutions as Model Systems. *Talanta* 131:609.

Takatani, T., Aoto, T., and Mukaigawa, Y. 2017. *One-Shot Hyperspectral Imaging Using Faced Reflectors*, IEEE Conference on Computer Vision and Pattern Recognition (CVPR) 2692–2700.

Xu, J., Sun, D., and Gowen, A. 2017. Chemometric Methods Applied in the Image Plane to Correct Striping Noise in Hyperspectral Chemical Images of Biomaterials. *Journal of Chemometrics* e2986.

8 NIR Instrument Sampling Configurations

*Jerome Workman, Jr.**
MJH Life Sciences

CONTENTS

8.1 INTRODUCTION

The quality, repeatability, and reproducibility of near-infrared (NIR) spectra depend upon the consistency in sample preparation and the precise alignment of the optical interface between the sample and the spectrophotometer. A variety of sampling accessories have been offered

* jworkman2002@yahoo.com.

commercially to optimize measurements of NIR spectral data. The most common NIR sampling techniques include reflectance measurements of powders, slurries, and liquids; transmittance of clear liquids; and fiber optic sampling for remote, quality control, hazardous, or multiple measurement head analysis.

There are no guarantees that a specific recommended sampling method will be efficacious as individual samples often present unique challenges. Issues include oil-in-water or water-in-oil emulsions, variance in particle size, high absorptivity, high crystallinity, high scattering material content, colorants or dye content, sample layering, low analyte concentration (i.e., below 0.5 wt%), small-range change in concentration, contaminants, and variable excipient quantities within a formulation. Other occasional problems with NIR measurement accuracy may include unusual reference tests that are not directly related to spectroscopic measurements, e.g., digestibility or calories; total oven volatiles used as a reference for moisture measurements; and where no direct signal exists for the selected analyte, e.g., sulfur in petroleum, or minerals in forages. In addition, one must remember that spectroscopy measures the molar concentration as weight per volume or moles per volume and not as weight percent.

For analytical thoroughness in sampling, one should document the sample, as well as any unusual issues encountered during sample preparation. For NIR, most solid or opaque slurry samples are measured using a diffuse reflection measurement technique. A transflectance or transmission method is optimum for water clear liquids. NIR is useful for quantitative analysis, qualitative identification, or spectral matching, or for conformity or quality estimation screening for finished products or ingredients. Initially, the unprocessed spectral data is collected. These spectra may then be preprocessed using baseline or scatter correction, smoothing, derivatives, or normalization.

8.2 GENERAL CONSIDERATIONS FOR SAMPLING

Near-infrared is a form of electromagnetic radiation nominally extending from 780 to 2500 nm or even 2600 nm. This spectral region is designated as 0.78–2.5 μm, or 12,820–4000 cm^{-1}.

NIR measurements for powders may be made using reflection directly from the sample surface using a variety of powdered sample cups (Figures 8.1 and 8.2). A transflection measurement is made when energy passes through the sample, reflects from a diffuse reflecting background, and then passes through the sample again to the detector (Figure 8.3). Since the absorptivity for molecules in

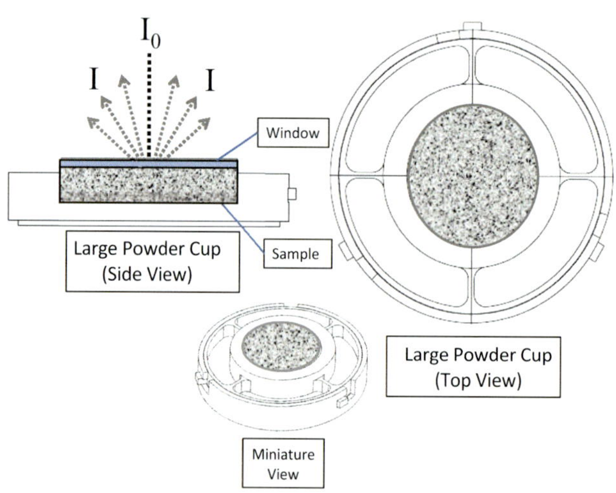

FIGURE 8.1 Traditional diffuse reflectance sample cup.

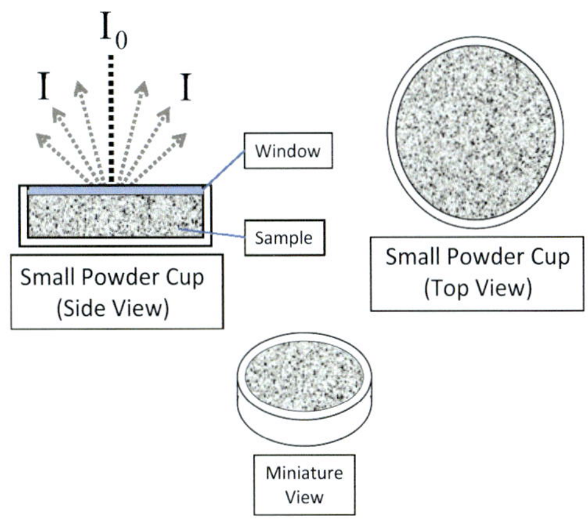

FIGURE 8.2 Small powder cup for diffuse reflectance measurements.

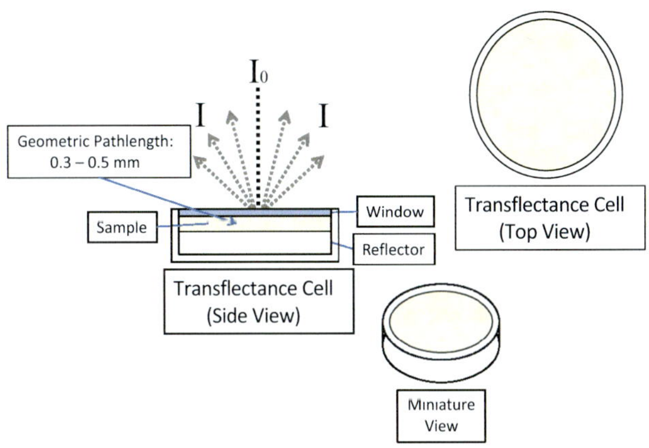

FIGURE 8.3 Sample cup illustration for diffuse transflectance measurements.

the NIR region is much less than in the IR region, a larger pathlength is required to obtain a useful spectrum. For a transflectance cell, the pathlength may nominally be 0.30 mm, and for powdered materials, a sample pathlength may nominally be approximately 7 mm compressed to 5 mm. Note that the required pathlength for appropriate analyte signal, the specific application, and type of sample dictate which method should be used for measuring a spectrum.

For NIR measurements, the minimum wavelength measured is determined by the sampling method, the sample type, and the spectrophotometer optics. For the visible spectral region, the most common detector is silicon-based with a detectivity range of 300 to 1100 nm (0.30–1.1 μm). On the higher end of the NIR spectrum, the limitation is from Indium Gallium Arsenide (InGaAs) with a working maximum typically near 2750 nm for the extended types (i.e., 2.7 μm). Alternative detectors with a slower response include lead sulfide (PbS) with a maximum response wavelength near 2800 nm (2.8 μm). NIR as a technique has been treated separately (Burns and Ciurczak 2008). The NIR spectral region is composed of molecular absorptions as a result of vibrational overtones and combination absorption bands. The absorptivity associated with these bands is less, by one or more orders of magnitude, than the fundamental infrared absorptions. For example, the absorptivity of the first overtone of C-H stretching in the NIR is 100 times less than the fundamental; the second overtone is 1000 times less; and the third overtone is 10,000 times less than the infrared fundamental absorptivity. As noted earlier, the magnitude of the absorption may be used to define the best procedure for pathlength and sampling. NIR tends to be less rigorous for sample preparation than infrared spectroscopy. Sampling for NIR analysis involves transmission or reflectance (normally diffuse reflectance) or a combination of both. For a general overview of the technique and its related sampling procedures, the reader is referred to a standard text on the subject (Burns and Ciurczak 2008).

In many cases, factors such as price, convenience of operation, optical performance/efficiency, and final application may be the final arbiter in the selection of a sampling method. The actual application can be an important deciding factor. Both NIR and mid-IR are used for factory-based measurements during manufacturing and processing of synthetic chemical materials and natural products. Sampling techniques may be manual or automated, at-line or "offline" for batch-oriented processes, or integrated on-line for continuous process monitoring. For additional reading, one is referred to references (Workman 2016, Chalmers and Griffiths 2002, Williams and Norris 1987, Roberts et al. 2004, Wiberley et al. 1990).

8.3 SAMPLE TYPE VERSUS SAMPLING METHOD

NIR spectroscopy is most often applied to powdered samples and measured using diffuse reflectance. It is not particularly useful for gas phase or dark solids or slurries. It also gives inconsistent results to scattering slurries and variable layered materials with alternating scattering absorption layers. This is because diffuse reflection has limited capabilities of layered sample types. Fine, ground material is the optimum sample type for the NIR diffuse measurements. Slurries and liquids may be measured using transflectance or diffuse reflectance. As will be mentioned throughout this text, it is essential that the analyst carefully document all experimental procedures, sampling methods, and conditions, and conscientiously note any abnormalities discovered during sample preparation. These should be recorded in appropriate lab analysis notebooks, either hard copy or electronic.

If the sample (solution or mixture) is known to contain water, then not only should this information be recorded but also extreme care must be taken to ensure that the sampling method is compatible with water being present. Comments are provided in Section 8.3.3 regarding the handling of samples containing emulsions and water that can interfere with proper analysis.

8.3.1 Liquids

NIR is compatible with low-OH (i.e., low-water) quartz, glass, and silicate cell windows. NIR strongly absorbs water in glass near 1453 and 1938 nm. Other much less intense absorptions occur near 770, 847, 979, 1200, and 1780 nm (Workman and Weyer 2012). Traditionally, liquids have been considered to be relatively easy except for low-molecular-weight compounds, which often suffer from volatility problems and extremely high absorptivity that typically require extremely small optical pathlengths (most often 0.50–1.0 mm). The main method for liquids is transflectance in a sealed transflectance cell (Figure 8.3, as above). The transflectance cell should have a temperature range from 10°C to 65°C for most sample types from volatile chemicals, aqueous solutions, or natural or synthetic oils. The temperature control for accurate multivariate calibrations should be ±0.1°C with a pathlength variation of no more than 3%–5% relative. For liquid analysis, the sample type and special characteristics, especially if a mixture, should also be noted.

8.3.2 Viscous Materials

A simple and convenient method for measuring viscous materials using NIR is to utilize an open slurry cup (or more simply a slurry cup). This cup is open and used in diffuse reflectance mode and has a depth so as to provide infinite optical thickness (Figure 8.4).

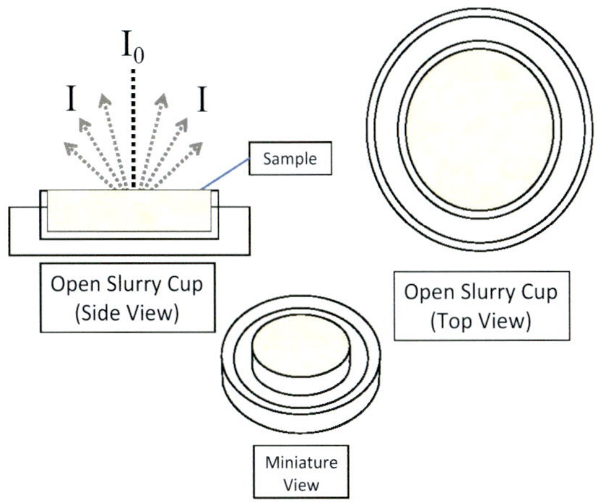

FIGURE 8.4 Open slurry cup for diffuse reflectance measurement of viscous samples, pastes, emulsions, and slurries.

8.3.3 Pastes, Emulsions, and Slurries

Depending on the constituents, pastes and slurries generally may be handled in a similar manner as proposed for viscous liquids (Figure 8.5). Always be aware of potential homogeneity problems. A slurry cup may be used for sample measurement, but if inhomogeneity is a problem, then the sample must be homogenized or carefully remixed to provide a more uniform suspension or emulsion (Figure 8.6). With multiple-phase emulsions or suspensions, be certain that separation does not occur if using a cup with an optical window. At the window interface, the polar nature of the optical material may cause segregation to occur. Common examples are emulsions, such as mayonnaise or margarine, and protein-containing materials; in this latter case, the protein may actually form a coating on the optical window. If an open cup is used, make certain that the separation of layers does not occur as the light layer will be measured and the heavy layer at the bottom of the cup will not be measured in proportion to the mixture composition. A disposable slurry cup may be used for rapid sample measurements when high numbers of samples must be measured and when measured samples are not retained (Figure 8.7).

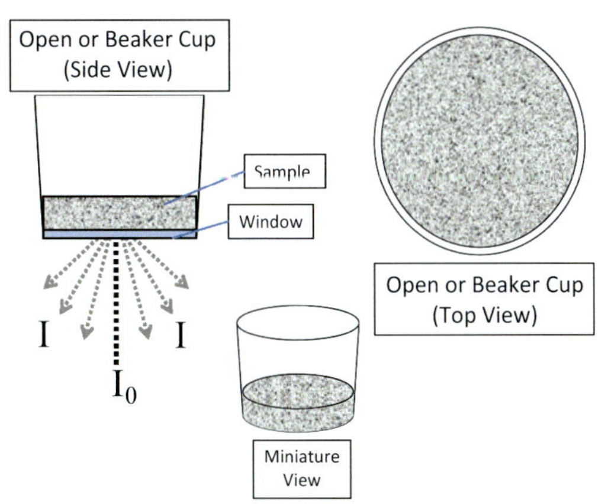

FIGURE 8.5 Open or beaker-type slurry cup for diffuse reflectance measurements.

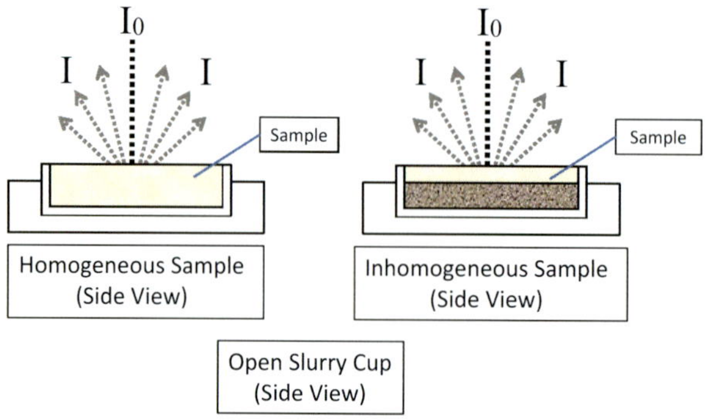

FIGURE 8.6 Illustration of homogeneous and inhomogeneous samples using a slurry cup for diffuse reflectance measurements.

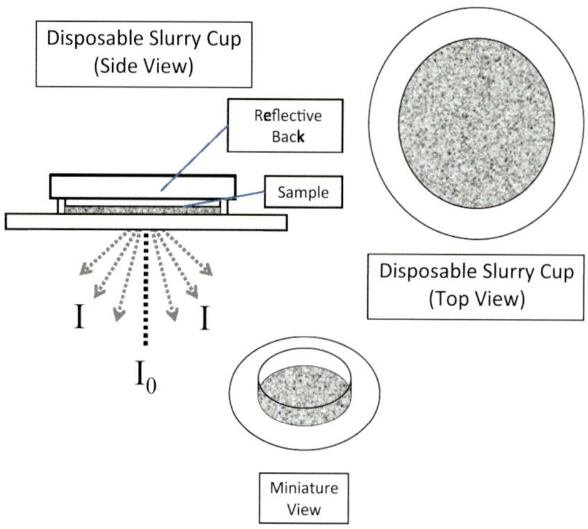

FIGURE 8.7 Disposable slurry cup for diffuse reflectance measurements.

8.3.4 Solids

Solids for NIR measurements may be solid blocks of material or are ground materials to pass a 1.0-mm screen (i.e., that are finer than 1.0 mm diameter particles). These samples may be measured "as-received" with high water concentration (above 30 vol%) or may be dried using microwave or forced air drying methods to near 0% water. The use of microwave drying carries with it the potential for damaging the sample due to high heat in various portions of the sample. Areas of particularly high heat occur due to the uneven field generation found in many commercial microwave heating systems. It also may occur due to bound or trapped water in microenvironments (i.e., pockets) of the sample. Note that traditional microwave ovens operate in a 1200 W field of 2450 MHz microwave energy. The purpose of the microwave field is to rotate the water molecules and cause surrounding molecules, based on the heat of friction, to rise in temperature. These molecules are rotating at 2450 million rotations per second to generate heat. In making spectroscopic measurements, one must attempt to mitigate spectral artifacts caused by refractive index, scattering, and polarization effects; and how to sample reproducibly and in a manner that is representative of the sample. Solid polymer samples of sufficient thickness may also be used for direct measurements using NIR. Typically, NIR will penetrate many polymer samples to 10 mm or more, and one should be aware that the NIR energy is penetrating the sample and the thickness will affect both shape and intensity of spectral bands.

8.3.4.1 Sampling Powders

Unlike the many methods used to prepare powders for measurement using infrared spectra, for NIR measurements the samples are generally dried and milled using a cyclone mill with a 1.0-mm screen, to make particles less than 1 mm in diameter. The powdered samples are placed in a variety of sample cups specifically designed for dried and ground material. Such cups include the locking powder cup (Figures 8.1 and 8.2) and open powder cups (Figure 8.5). There are a myriad of variations of this theme of presenting an infinite thickness of powdered material for the NIR measurement. Sample thickness for these cups may vary from 5 to 30 mm or more. The typical active window of the sample for measurement using different cup types is from 35 mm up to 5 inches or approximately 125 mm. When the measurement beam size is far less than the sample cup diameter, the sample may be rotated off-axis to measure multiple replicates of the sample during a single rotation.

8.3.4.2 Sampling Amorphous Materials (Organic), Continuous Sheets, and Polymeric Film

Polymer sheets may be measured in transmittance geometry or in diffuse reflectance geometry. A film holder may be used for transmittance (Figure 8.8). For diffuse reflectance, a standard reflectance cup with a diffuse back reflector may be used. For this application, a piece of the sheet or film is cut to the size of the sample cup and a diffuse reflectance measurement is made. It is important to remember that for a sample with limited thickness, the effective pathlength of a measurement is more than twice the physical pathlength of the sample. If the film sample has a metalized coating, it will act as an etalon and give interference fringes in addition to the molecular spectrum. These fringes are due to multiple reflections in the film and have a periodicity proportional to the film thickness. See film thickness measurement in the index of this set of volumes. Other methods for handling polymers and films may be used from infrared sampling techniques. The reader is referred to the index on thin films and polymer sampling as well as the chapter within the infrared text volume on the subject of sample preparation.

8.3.4.3 Sampling Lumps, Granules, or Pellets

This class of solids is an extension of the sample types already discussed, and many of the procedures already highlighted may be used here. These materials may be used in various diffuse reflectance cups in bulk form or ground and mixed for sample presentation using powder cups.

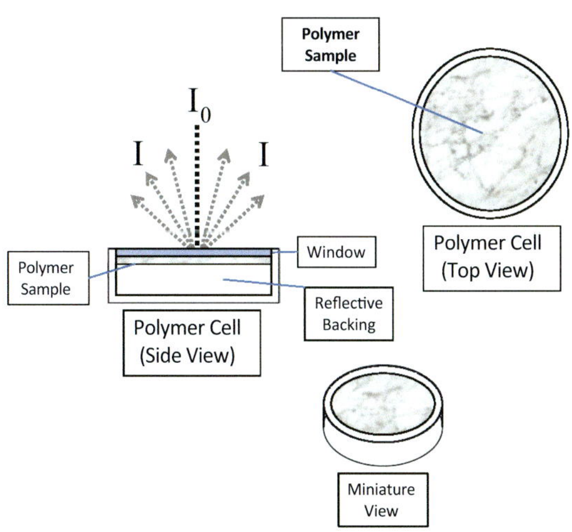

FIGURE 8.8 Film holder for transmittance measurements of films.

8.3.4.4 Sampling Intractable Materials

Sample with high absorbance (or less than 1% reflectance) may not provide enough signal for many commercial NIR spectrometers. If such samples must be measured routinely, one may try a spectrometer with much higher dynamic range. The measuring instrument must have at least one absorbance unity higher dynamic range than the typical samples measured.

8.4 SAMPLING METHOD VERSUS SAMPLE TYPE

This section provides similar information and cross-referenced information to those of the previous section, but in this case, it is more procedure-oriented. Hints and tips are provided for the actual method rather than focusing on the sample and its properties.

8.4.1 Transmittance Windows

The most basic sampling device for NIR includes a sampling system with a NIR transmissive window; low O-H-containing special glasses should transmit more than 90% of energy across the entire NIR region.

8.4.2 Transmittance Cells (Liquid Handling)

Liquids are generally measured using a fixed transflectance geometry with many commercial NIR instruments (Figure 8.3). Some instruments enable configuring either reflectance or transmittance geometry, and so transmittance measurements are permitted. Typical transmittance cells may be rounded (cylindrical) or rectangular, as illustrated in Figure 8.9. Rectangular cuvets may be square-shaped or be described as having an elongated rectangular cross section with a thin pathlength (of approximately 1.0 mm) and a width several (i.e., 5–10) times the pathlength. These are designed for ease of handling or when the beam size has a relatively large cross section.

8.4.3 Gas Cells

Gas cells are considered a form of transmission cells. Infrared is a technique often used for measuring gases due to the requirements for high resolution and high sensitivity. NIR is 100–1000 times less sensitive than infrared to the same functional groups, and thus, NIR is generally not considered to be adequate as a laboratory or process gas measurement technique. It could be applied to the identification or quantification of high-pressure gases if there is a simple matrix and high resolution is not a requirement.

8.4.4 Self-Supporting Films

Measuring thin polymer films has been previously mentioned. Adequate sample for NIR measurement of polymers generally requires a 1 mm thickness of material. Thin films of less than a few hundred microns in thickness are generally measured using infrared spectroscopy due to the increased sensitivity of infrared and the higher resolution available in most commercial infrared spectrometers.

8.4.5 Specular Reflectance Measurements

Specular reflectance (or mirror reflectance) represents an elastic photon reflection where only the direction of the incoming photon is changed; therefore, the original energy (i.e., frequency or wavelength) is identical after the reflection. The most common applications of specular reflectance are for the direct measurement of surfaces for contaminants or for the measurement of optical properties of

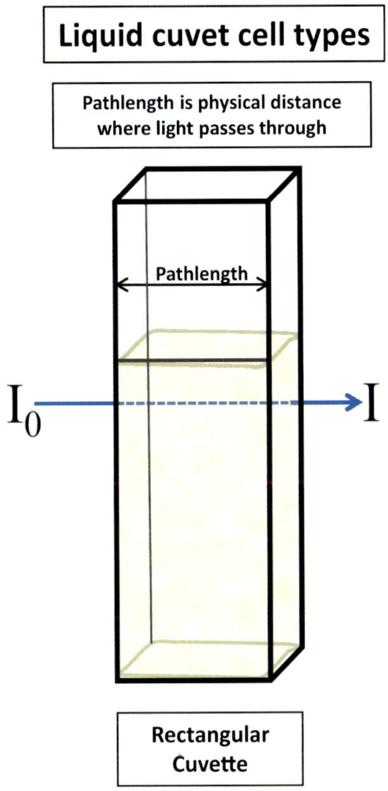

FIGURE 8.9 Illustration of rectangular cuvet type.

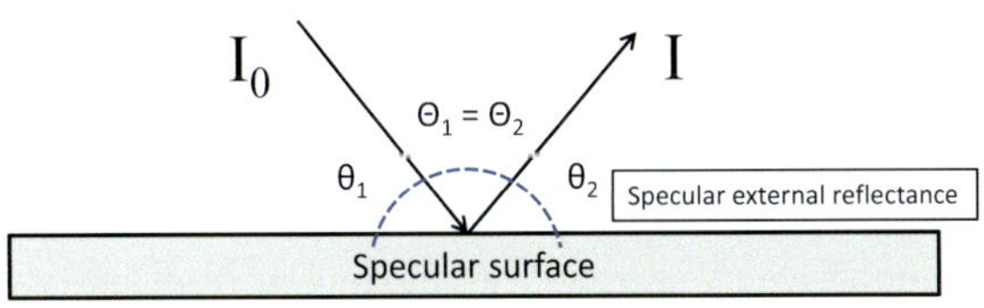

FIGURE 8.10 Illustration of specular surface reflection – the incident and reflection angles are identical.

materials in the NIR spectral region – e.g., certain optical mirror or lens materials, paints, or coatings. For the NIR region, these measurements are referred to as transflectance or reflection/absorption (Figure 8.3, as above). As may be observed in Figure 8.10, for specular reflection measurements, the incident angle and reflection measurement angle are the same. If the incident angle and the reflection measurement angle are different, other reflective interactions will be measured, such as diffuse reflection, interactance, or transflectance. In these types of measurements, an inelastic photon interaction occurs and an absorbance due to interaction with the sample is the most likely result.

8.4.6 Diffuse Reflectance Measurements

Diffuse reflectance is the most common of all measurement techniques using the NIR spectroscopy. The method is used when a Lambertian (or light-diffusing material), such as a fine powder, is measured. Optical effects which primarily occur on the surface of the sample, such as specular

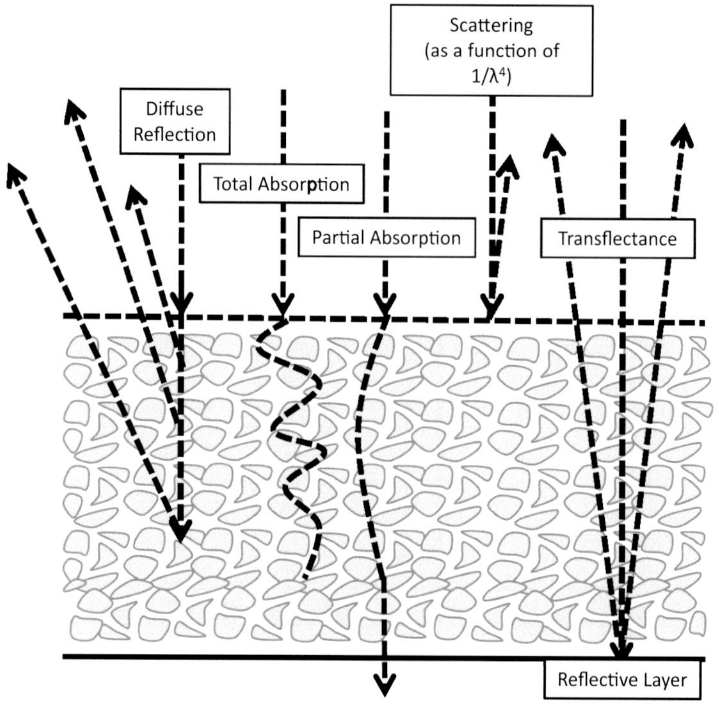

FIGURE 8.11 Reflection modes encountered for solids and powders.

reflection, Mie or Rayleigh scattering, and diffuse reflection (Figure 8.11), can impact the quality and characteristics of the final measured spectrum. A conventional sampling cup and measurement geometry for systems designed for diffuse reflectance measurements are shown in Figures 8.3, 8.13, and 8.14.

For the NIR measurements, uniform particle size is critical. Reducing the particle size by grinding will often improve the quality of the final spectrum. Also, a uniform particle size or a narrow particle size distribution is beneficial. Nonpowdered samples can be studied directly from the sample surface as long as the surface is irregular (granular or diffuse) and not polished. This chapter covering infrared diffuse reflection measurements contains several useful recommendations for dilution with KBr.

The spectrum produced from a diffuse reflectance measurement does not bear a direct numerical relationship between peak intensity and composition, in contrast to traditional transmission measurements in which the absorbance form of the data is directly proportional to concentration. With diffuse reflectance, spectral distortions are generated by a constantly varying effective pathlength, which is defined by the depth of penetration of the beam into the sample. This is dependent on the absorptivity of the sample at each particular wavelength (frequency), and the measured intensity throughout the spectrum varies accordingly. A correction that can be applied to the diffuse reflectance spectrum, which can help linearize the data, is the Kubelka–Munk function (Equation 8.1):

$$f(R_\infty) = \frac{(1 - R_\infty)^2}{2R_\infty}, \tag{8.1}$$

where $f(R_\infty)$ is the corrected spectrum and R_∞ is the ratio of the sample diffuse reflectance spectrum to a nonabsorbent reference (normally KBr or KCl), both measured at infinite depth. In the mid-IR region, infinite depth corresponds to approximately 3 mm.

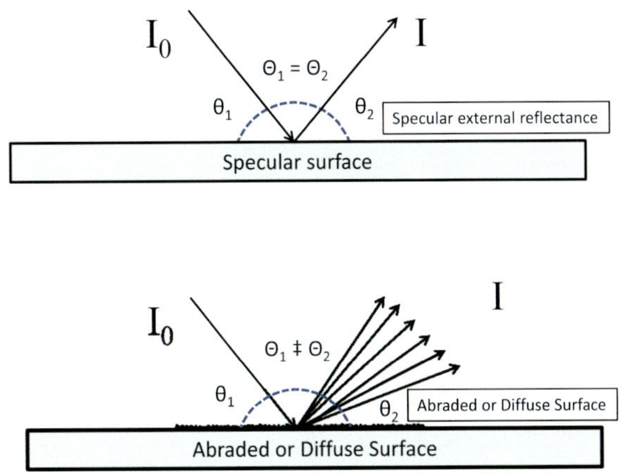

FIGURE 8.12 An abraded specular reflectance sample produces diffuse reflectance surface regimes.

The consequence of the $1 - R_{\infty}$ is to invert the spectrum and provide an absorbance-like format. It will be noted from the previous expressions that this corrected function can be correlated with concentration. Note that this function is normally only applied to undiluted (or concentrated) powdered samples.

8.4.7 Abrasion Measurement Techniques

Very hard surfaces can be studied by the abrasion of the surface with a sheet of abrasive material, such as silicon carbide or carborundum paper. At this point, a number of different methods may be used to analyze the abraded material. Following a roughening of the surface, diffuse reflectance may be used to measure the material (Figure 8.12).

8.4.8 Polyethylene Bag (1 mil Thickness)

Samples may be placed in thin but strong 1-mil-thick polyethylene bags for disposable sampling. These bags will leave a small, generally insignificant spectral signature, which must be included during the calibration phase to eliminate the bias in predicted values caused by the positive photometric offset and additional bands. The difference between reflectance spectra measured using the bag and standard window solid cup is minimal.

8.5 OPTICAL CONFIGURATIONS FOR NIR MEASUREMENTS

Most NIR instruments either exhibit a 0/45° geometry or use an integrating sphere detector. The look-up and look-down configurations are illustrated in Figures 8.13 and 8.14, respectively (Wendlandt and Hecht 1966, Kortüm 2012).

The integrating sphere provides a more uniform sampling of the reflection from samples, whereas the detector plane tolerances in non-integrating sphere measurements cause greater variation between instruments. They may be compensated for by precisely adjusting the height and the angle of detectors with respect to the sample measurement plane.

The integrating sphere is most often coated with barium sulfate or halon coating, or is machined from Spectralon™ or Fluorilon to provide a highly uniform and reflective light integration chamber (Figure 8.15).

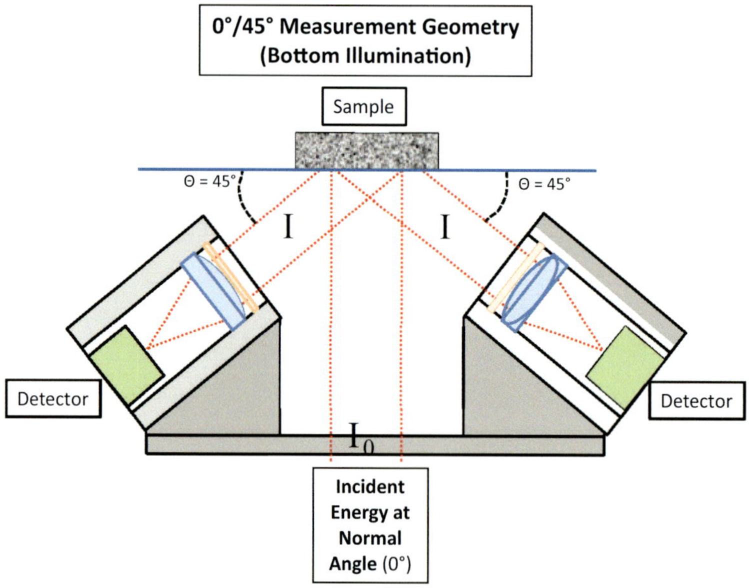

FIGURE 8.13 An illustration of a look-up 0/45° sampling geometry.

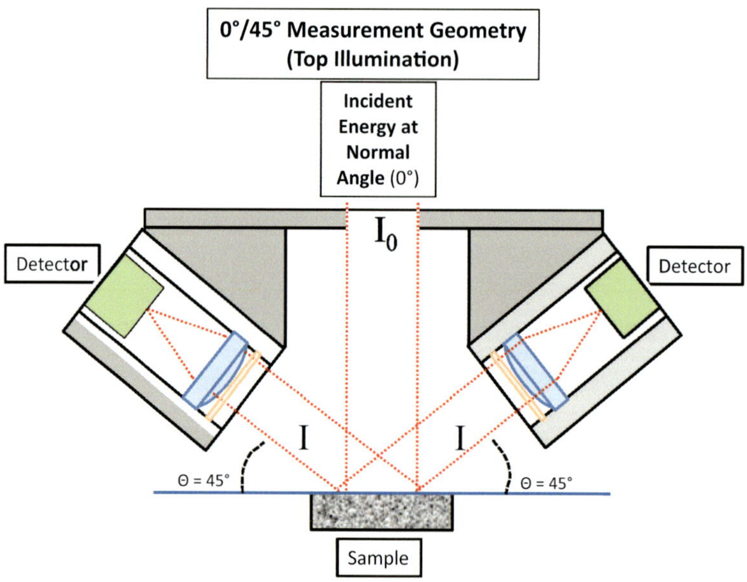

FIGURE 8.14 An illustration of a look-down 0/45° sampling geometry.

The advantages of an integrating sphere become obvious when it is realized that the image pattern at the detector varies with the characteristics of the samples presented to the instrument. When the beam angle is scattered or refracted off-axis, or is diverted from normal angle, the beam image area can vary and a normal 0/45° detector will lose some of the energy not captured. The integrating sphere is able to capture such diverging beam energy and direct it toward the detector surface. A standard measurement configuration for reference reflectance standards is referred to as the d/8° measurement geometry. This configuration is used to calibrate the reflectance of photometric reference material standards (Figure 8.16).

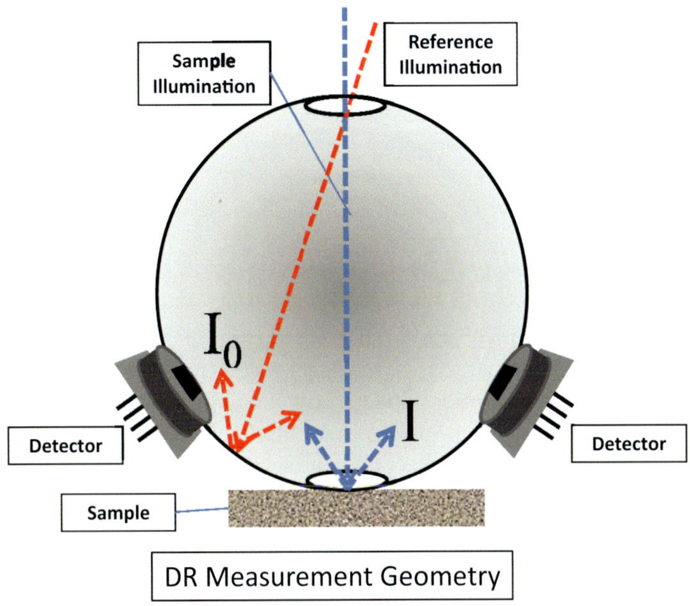

FIGURE 8.15 Illustration of standard integrating sphere reflectance measurement configuration.

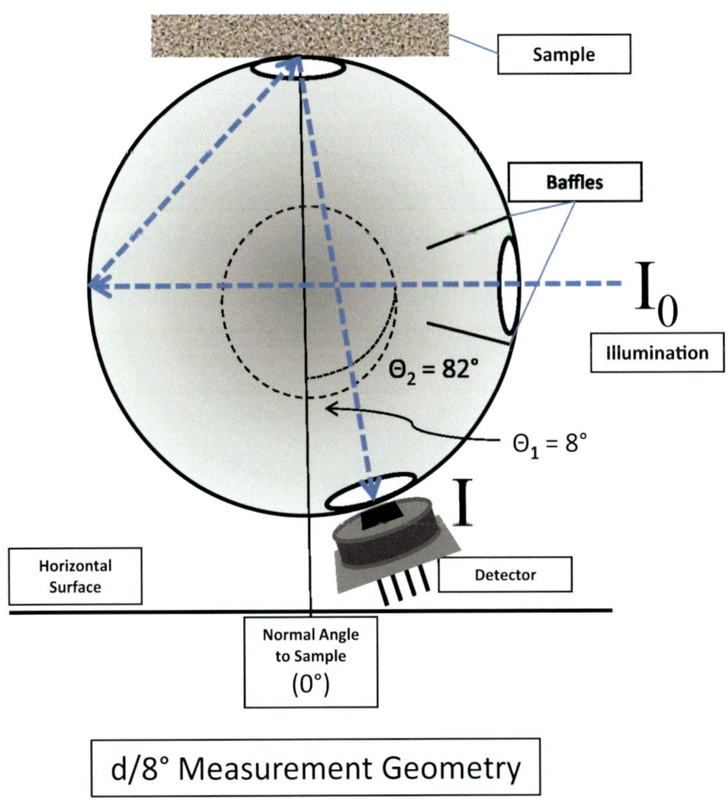

FIGURE 8.16 Illustration of d/8° integrating sphere reflectance measurement configuration.

The various specialized reflectance or transmittance measurements are used to determine the precise sample characteristics desired. Specular reflection and diffuse reflection represent the major reflectance techniques using an integrating sphere. Specular reflection is mirror reflection and consists in elastic collisions of light photons from the immediate surface of the sample. Specular reflected light energy contains only information related to the surface characteristics of the sample. Specular reflection allows surface properties such as refractive index and thin film thickness to be determined. The technique also permits the measurement of atomic or molecular surface structure for film monolayers.

The use of linearly polarized light in specular reflectance is provided by the Glan–Taylor prism polarizer or similar polarizing elements. The linearly polarized light is required to provide calculations for the most optical constants. The use of polarized light makes these measurements possible due to the requirement that the quantity of reflected light from a sample depends upon the state of polarization of the incident light. These relationships are explained by the Snell and Fresnel law equations.

Grating spectrophotometers always produce partially polarized light, which can become a disadvantage for quantitative specular reflectance measurements. The nature of reflectance spectra depends upon the angle of incoming light as well as the polarization properties of this light. Diffusely reflected light is light that has penetrated into a sample where it is exposed to the absorption properties of the sample. Thus, the light is attenuated based upon the atomic or molecular interaction of the sample with the incident light. If a sample absorbs no energy, such as in elastic collisions, total reflection occurs. If a sample absorbs all of the incident energy, such as in the case of a light trap, no reflection occurs. Most real diffuse samples reflect some measureable quantity of light. The light that interacts with the sample in an inelastic manner yields chemical information about that sample.

Both transmittance and reflectance data are used for qualitative or quantitative work. In quantitative analysis work, both transmittance and reflectance data are related to concentration with the assumption that Beer's law applies to both cases. In transmittance (T) measurements, we relate concentration directly to absorbance as:

$$\text{Absorbance} = -\log_{10} T. \tag{8.2}$$

In reflectance (R) measurements, we relate concentration directly to absorbance using:

$$\text{Absorbance} = -\log_{10} R. \tag{8.3}$$

On occasion, we see Kubelka–Munk transformed data also used to relate concentration to reflectance measurements as:

$$\text{Kubelka-Munk Reflectance} = \frac{(1-R)^2}{2 \cdot R}. \tag{8.4}$$

The Kubelka–Munk and absorbance equations for reflectance require a near-infinite thickness for the sample; this requirement is met if the sample thickness is sufficient to absorb or reflect all the impinging energy.

The refractive index of materials changes the angle of refraction for light impinging upon a sample surface. This angle of refraction from the front surface of a sample follows Snell's law as:

$$n_1 \cdot \sin(\phi_1) = n_2 \cdot \sin(\phi_2), \tag{8.5}$$

where n_1 is the refractive index of the initial medium the light is passing through, n_2 is the refractive index of the second medium, ϕ_1 is the incident angle of the light, and ϕ_2 is the angle of refraction of the light.

The amount of reflection from the surface of a material for different polarizations of light is computed using the following formulas. Snell's law is used to compute the angle of refraction from the angle of incidence of light reflection from a surface as light as it passes through two materials of differing refractive indices. The reflection of s-polarized (i.e., R_s, parallel ($R_{\parallel}$) or 0° polarized light) is computed using the following:

$$R_S = \frac{\sin^2(\phi_1 - \phi_2)}{\sin^2(\phi_1 + \phi_2)}. \tag{8.6}$$

The symbols are those as used in the Snell's law example, where n_1 is the refractive index of the initial medium the light is passing through, n_2 is the refractive index of the second medium, ϕ_1 is the incident angle of the light, and ϕ_2 is the angle of refraction of the light.

Furthermore, the reflection of p-polarized (i.e., R_p, perpendicular ($R_{\perp}$) or 90° polarized light) is computed using the following:

$$R_p = \frac{\tan^2(\phi_1 - \phi_2)}{\tan^2(\phi_1 + \phi_2)}. \tag{8.7}$$

8.6 THE PROPERTIES OF FIBER OPTICS

8.6.1 Typical Fiber Optic Applications

Optical fibers are useful for remote or process measurements using optical spectroscopy, particularly NIR, visible, and the higher-frequency infrared (typically above 1000 cm^{-1}). The basic properties of fiber optics include materials, construction, dimensions, and overall optical properties (Newport Corp. 2018, Ganz et al. 1995, Fiberguide Industries 2018).

NIR fiber optic materials consist of low-OH silica (SiO_2) with minimal inclusions, like entrapped air. A fiber bundle consists in cladding layer(s) and a coating layer for protection.

8.6.2 Materials for Optical Fibers

A variety of materials are used for the construction of fiber optics. Although many more materials are useful for communication-grade fibers, fibers used for spectroscopy require low attenuation or high transmission of energy over the wavelength or frequency of interest. The main materials used for spectroscopic fibers include low-OH silica (visible and NIR up to 2.5 μm), fluoride fiber (long-wavelength NIR into the infrared region to 4 μm), and chalcogenide (for the infrared region above extending to above 10 μm).

8.6.3 Construction of Optical Fibers

The basic components of spectroscopic-grade fiber optic cables and standard cable connectors are shown in Figure 8.17; they include several key elements as follows. The **core**, composed of low-hydroxyl-containing glass or plastic, is the light-transmitting portion of the optical fiber. The larger the core diameter, the greater the light transmitted, which is determined by the area function of $A = \pi \cdot r^2$. Surrounding the core is the cladding. The cladding is the boundary zone for the light transmitted down the core. Its purpose is to provide an interface of material at a different refractive index to cause reflection of the light within the core so that the light waves are transmitted down the fiber; the cladding is of *lower* refractive index than the core. The coating surrounds both the core and the cladding. Coatings are for protection of the basic optical cable and are generally multilayers of plastic meant to absorb some mechanical shock and preserve the

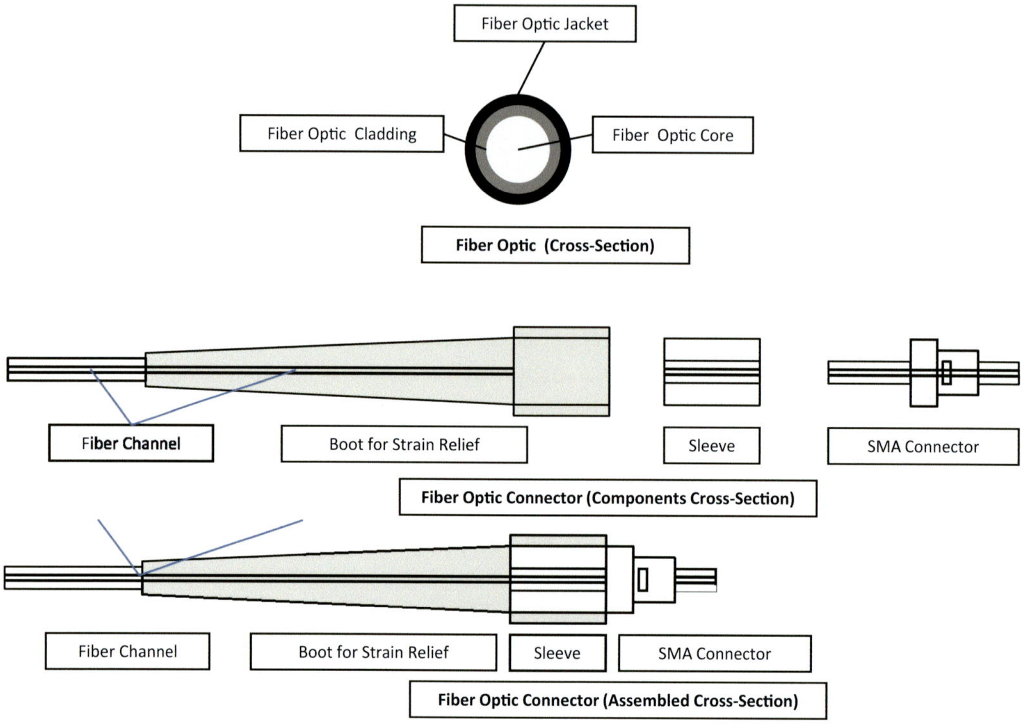

FIGURE 8.17 Cross section of typical optical fiber and SMA fiber optic connector components.

TABLE 8.1
Typical Optical Specifications for an Optical Fiber System

Core	Low hydroxyl (OH) silica	Diameter ± Tolerance
Cladding	Doped silica (SiO_2)	Diameter ± Tolerance
Buffer	Polyimide	Diameter ± Tolerance
Jacket	Material X	Diameter ± Tolerance
Numerical aperture	0.25 ± 0.05	
Attenuation	≤7 dB km^{-1} maximum @ 912 nm (i.e., 4 dB km^{-1} typical)	

fiber. The coatings are either mechanically or chemically strippable. Tables 8.1 and 8.2 show the typical characteristics of NIR fibers.

8.6.4 Dimensions of Optical Fibers

Fiber size for spectroscopic use varies from 200 to 600 μm, with 600 μm the most common. Fiber types include single mode, graded index, and step index. Single-mode fibers are generally 5–10 μm in diameter and are used exclusively for video or telecommunications. The single-mode fiber allows only one mode or light ray to be transmitted down the core. Graded index fibers consist of a core surrounded by a number of glass cladding material layers, each with a successively smaller refractive index as you move away from the center of the fiber. Most graded index fibers are used for communications and are typically found in 50, 62.5, 85, and 100 μm sizes. Step index fiber consists of a core surrounded by a single cladding layer of lower refractive index. The step index fiber is most common in spectroscopy, and it is found in the 200–600 μm sizes.

TABLE 8.2
Typical Optical Fiber for Vis-NIR

Characteristic	Specification
Type	Multimodal with step index
Core	Fused silica
Cladding	Fluorine-doped silica
Core size ranges	50–1500 μm
Wavelength range (UV-Vis)	190–1250 nm
Wavelength range (Vis-NIR)	300–2400 nm (low OH)
NA ranges	0.11–0.28
Coatings types available	Acrylate
	Polyimide
	Nylon
	Tefzel
	Aluminum
	Gold

8.6.5 Fiber Optic Connections

Figure 8.17 demonstrates the typical fiber connector type, often referred to as the SMA connector. This connector provides both strain protection and a fast-locking system for rapid locking and unlocking of the fiber connections. The fiber connector terminates the fiber as it is connected to a transmittance device, a source lamp, a spectrometer, or a detector. The connectors allow a quick connect/disconnect and align the core elements of the fiber ends to minimize attenuation. The alternative to the use of connectors is to splice the fibers, both time-consuming and inaccurate. In addition to the quick connect/disconnect SMA connectors, there are many threaded types which rely on a screwing mechanism for connection and disconnect. There are dozens of different connector types commercially available.

8.6.6 Properties of Optical Fibers

8.6.6.1 Light Transmittance Losses in Fiber Optics (Attenuation)

Losses in fiber optic transmitted light are due to absorption, reflection, scattering, misalignment, and microbending; the losses can vary with temperature, wavelength, and fiber condition. The greatest causes of attenuation include OH absorption at 960 and 1950 nm – and some attenuation from 1800 to 2500 nm – and Raleigh scattering in the 200–600 nm range. The quantity of optical loss is expressed as an attenuation rate in decibels (dB) of optical power per kilometer (km). Typical losses result from launch optics, temperature variations, fiber couplings in optical path, aging (solarization), optical alignment, preparation at exposed surfaces, normal attenuation due to fiber length, coupling at detector, and repairs/splices.

The losses in energy transmitted through a fiber can be calculated by using Beer's law. Beer's law states that the irradiance of energy through an absorbing medium falls exponentially with the distance of transmission following the relationship:

$$I_d = I_0 \cdot 10^{\alpha d/10}, \tag{8.8}$$

where I_d is the irradiance at distance (d) from the source, I_0 is the source irradiance at $d=0$, α is the attenuation (absorption) coefficient in units of dB/unit length (e.g., meters or kilometers), and d is the fiber length in unit length (meters or kilometers).

Attenuation losses are wavelength dependent; thus, the value for α is a function of the incident wavelength (λ) (Table 4.8). During fiber manufacture, thermal variations are present when the glass is in a molten state; the result is a refractive index variation throughout the fiber. As light passes through these variations, a phenomenon termed Rayleigh scattering occurs; this scattering reduces the fiber throughput and is proportional to λ^{-4}, where λ is the wavelength of the incident light. At longer wavelengths, the O-H present as an impurity in the SiO_2 (glass) causes absorption losses at 950, 1250, 1380, 1920, and above 2200 nm. The overall quality of a fiber is sometimes standardized to the Rayleigh scattering limit of 0.17 dB km^{-1} at 1550 nm.

8.6.6.2 Determination of Signal-to-Noise for Single Fiber

The theoretical total signal (SN) from an optical system (fiber) can be given by

$$\mathrm{SN} = R_S \cdot B_\lambda \cdot \theta \cdot \tau \cdot Q, \tag{8.9}$$

where R_S is the source spectral radiance, B_λ is the spectral bandwidth, θ is the etendue (throughput), τ is the transmittance loss, and Q is the quantum efficiency (electrons per second).

The measured signal-to-noise can be calculated as the full transmitted signal divided by the RMS noise (in transmittance units).

REFERENCES

Burns, D.A. and E.W. Ciurczak. 2008. *Handbook of Near-Infrared Analysis*. Boca Raton: CRC Press. doi: 10.1201/9781420007374

Chalmers, J. and P. Griffiths. 2002. *Handbook of Vibrational Spectroscopy*.

Fiberguide Industries. 2018. Optical Fibers. http://www.fiberguide.com/product/optical-fibers/.

Ganz, A. et al. 1995. Personal Communication.

Kortüm, G. 2012. *Reflectance Spectroscopy: Principles, Methods, Applications*. Berlin Heidelberg: Springer Science & Business Media. 10.J.

Newport Corporation. 2018. Projects in Single-Mode Fiber Optics.

Roberts, C.A., J. Workman and J.B. Reeves, eds. 2004. *Near-Infrared Spectroscopy in Agriculture* (Vol. 44). Madison: American Society of Agronomy.

Wendlandt, W.W. and H.G. Hecht. 1966. *Reflectance Spectroscopy*. New York: Interscience.

Wiberley, S.E., N.B. Colthup, and L.H. Daly. 1990. *Introduction to Infrared and Raman Spectroscopy*, ed. N.B. Colthup and L.H. Daly, San Diego: Academic Press, Inc. pp. 75–108.

Williams, P. and K. Norris. 1987. *Near-Infrared Technology in the Agricultural and Food Industries*. St. Paul: American Association of Cereal Chemists, Inc.

Workman, J., Jr and L. Weyer. 2012. *Practical Guide and Spectral Atlas for Interpretive Near-Infrared Spectroscopy*. CRC Press.

Workman, J.J, Jr. 2016. *Concise Handbook of Analytical Spectroscopy: The Theory, Applications, and Reference Materials* (In 5 Volumes). World Scientific.

9 Instrument Performance Calibration

*Jerome Workman, Jr.**
MJH Life Sciences

CONTENTS

9.1 INTRODUCTION

One of the essential aspects for determining the efficacy and quality of spectrometers is to directly measure multiple characteristics of its measurement capabilities and properties. There are many standard tests that are used to determine alikeness between spectrophotometer instruments, and several of those tests will be described in this chapter. The tests described are most applicable to the near-infrared (NIR) region, although the tests and standards may be varied to become applicable to spectrometers using different wavelength or frequency ranges. Eight basic tests and a summary evaluation are described here as a standard method to determine instrument measurement

* jworkman2002@yahoo.com.

performance, namely, wavelength accuracy, wavenumber repeatability, absorbance/response accuracy, absorbance/response repeatability, photometric linearity, photometric noise, signal averaging (noise) tests, and the instrument line shape (ILS) test. If carefully conducted, these experiments provide specific information for diagnosing mechanical, optical, and electronic variations associated with design or manufacturing limitations. They provide objective data for correcting and refining instrumentation quality and alikeness (Workman and Springsteen (1998), ASTM E275–08 (2013), Workman and Mark (2013a, 2013b), Workman and Mark (2014a, 2014b), Feudale et al. (2002), Abookasis and Workman (2011), Shenk and Westerhaus (1991), Shenk and Westerhaus (1989), Bland and Altman (1986), Mark and Workman (2010)).

9.2 INSTRUMENT OPTICAL QUALITY PERFORMANCE TESTS

The following series of tests is used to qualify instrument performance and to determine which issues are problematic due to deficiencies in instrument design features. These tests are related to alikeness in measurement performance between instruments, and to overall accuracy and precision (as repeatability and reproducibility). Note that the terms "optical density" (O.D.) and "absorbance units" (AU) are synonyms. (Note that "optical density" is found in historical documents and is still used in physics, and biomedical and optical engineering, but not often in analytical chemistry.)

9.2.1 Wavenumber Accuracy Test

Verify the wavenumber accuracy of the spectrophotometer using a suitable reference standard. The results must be consistent with the model performance specifications for the application in use. *Note: For transmittance measurements of a specific reference material, use a highly crystalline polystyrene standard polymer filter with 1 mm thickness; for reflectance measurements of the polystyrene, use a backing of highly reflective material, with a measured reflectance of greater than 95%.*

For this test, repeat measurements of the same polystyrene filter by placing it in the sample beam and not mechanically moving the sample over a normal measurement cycle for the instrument. For example, a 30-s measurement period and 15-s reference spectrum may be typical. Then, calculate the first derivative of each of the background-corrected replicate spectra – compute the inflection or zero-crossing positions for the center band at the polystyrene absorbance peak near the reference wavenumber position (ν_{ref}) specified (~5940 cm^{-1}). The ratioed spectrum is computed for each scan (i.e., scan-to-scan for each sample) and for the mean spectrum over the full 30-s measurement period. Therefore, calculate the standard deviation of difference of the wavenumber positions for the zero-crossings for scan-to-scan within (n) replicate samples, and the mean spectrum position $(\bar{\nu}_i)$ for the measured (ν_{ij}) versus reference (ν_{ref}) wavenumber values. The standard deviation (s_i) is calculated as:

$$s_i = \sqrt{\frac{\sum_{j=1}^{n}\left(\nu_{ij} - \nu_{\text{ref}}\right)^2}{n}} \tag{9.1}$$

The mean difference for wavelength accuracy is determined by:

$$\text{Wavenumber Accuracy (as mean difference)} = \bar{\nu}_i - \nu_{\text{ref}}, \tag{9.2}$$

where $(\bar{\nu}_i)$ is the average wavenumber for the scan-to-scan set, and (ν_{ref}) is the reference wavenumber position for the polystyrene filter near $5940\,cm^{-1}$. The results are reported as: Wavenumber (cm^{-1})

precision and accuracy as shown. This may be accomplished for multiple wavelengths depending upon the standard reference material measured.

9.2.2 Wavenumber Repeatability Test

Verify the wavelength repeatability of the spectrophotometer using a suitable reference standard, like a highly crystalline polystyrene sample of 1 mm thickness. The standard deviation of the selected wavenumber must be consistent with the model performance specifications for the application in use. *Note: For transmittance measurements of a specific reference material, use a highly crystalline polystyrene standard polymer filter with 1 mm thickness; for reflectance measurements of the polystyrene, use a backing of highly reflective material, with a measured reflectance of greater than 95%.*

For this test, repeat measurements of the same polystyrene filter by placing it in the sample beam and not mechanically moving the sample over a normal measurement cycle for the instrument. For example, a 30-s measurement period and 15-s reference spectrum may be typical. Then, calculate the first derivative of each of the background replicate spectra – compute the inflection or zero-crossing positions for the center band at the polystyrene absorbance peak near 5940 cm^{-1} for each scan (i.e., scan-to-scan for each sample) and the mean spectrum wavenumber position ($\bar{x}_i$) for the measured (x_{ij}). The standard deviation is calculated as:

$$s_i = \sqrt{\frac{\sum_{j=1}^{n} \left(v_{ij} - \bar{v}_i\right)^2}{n-1}}, \tag{9.3}$$

where (s_i) is the standard deviation for the scan-to-scan wavelength precision or repeatability for scan-to-scan measurements, and where (v_{ij}) are the individual wavenumber shifts of the zero-crossover for sample i and scan-to-scan number j; $(\bar{v}_i)$ is the average value for the scan-to-scan set; n is the number of replicate measurements (pool all scan-to-scan data).

The mean spectrum wavenumber position $(\bar{v}_i)$ is calculated as:

$$\bar{v}_i = \frac{\sum_{i}^{n} v_i}{n}. \tag{9.4}$$

The results are reported as: Wavenumber (cm^{-1}) Repeatability

9.2.3 Absorbance/Response Accuracy Test

Verify the response accuracy of the spectrophotometer using a suitable standard, e.g., a prespecified reference neutral density (ND) filter or reflectance standard with nominal absorbance units (AU) of approximately 1.0 AU (i.e., 10% transmittance or reflectance). This reference standard must be provided with reference measurements of at least two separated wavenumber positions, e.g., at 7000 cm^{-1} (1429 nm) and 4500 cm^{-1} (2222 nm). The standard deviation of the optical response must be consistent with the model performance specifications for the application in use.

For this test, repeat measurements of the same photometric standard by placing it in the sample beam and not mechanically moving the sample over a normal measurement cycle for the measurement period. For example, a 30-s measurement period and 15-s reference spectrum may be typical. Then, calculate the absorbance units (AU) for the entire spectrum and specifically at the two measured reference points of, e.g., 7000 and 4500 cm^{-1}. The reference AU for each wavenumber position ($A_{7000\,cm^{-1}}$ and $A_{4500\,cm^{-1}}$) is measured for each scan for each sample (n) and for the mean spectrum ($\bar{A}_i$) at each wavenumber location over the measurement period. The statistics are calculated as:

$$s_{7000\,cm^{-1}} = \sqrt{\frac{\sum_{j=1}^{n}\left(A_{ij} - A_{7000\,cm^{-1}}\right)^2}{n}}, \text{ and} \tag{9.5}$$

$$s_{4500\,cm^{-1}} = \sqrt{\frac{\sum_{j=1}^{n}\left(A_{ij} - A_{4500\,cm^{-1}}\right)^2}{n}}, \tag{9.6}$$

where $\left(s_{cm^{-1}}\right)$ is the standard deviation (precision) for the scan-to-scan absorbance units at a specified wavenumber (AU) accuracy for the set of measurements, and where $\left(A_{ij}\right)$ are the individual measurements of the absorbance units for sample i through scan-to-scan replicate measurement number j; $\left(A_{cm^{-1}}\right)$ are the reference values for the reference material at each of 7000 and 4500 cm^{-1}; n is the replicate measurement number. (Remember to pool all scan-to-scan data.)

The mean difference for photometric accuracy at each wavenumber position is determined by:

$$\text{Photometric Accuracy (as mean difference)} = \bar{A}_{cm^{-1}} - A_{cm^{-1}}, \tag{9.7}$$

where ($\bar{A}_{cm^{-1}}$) is the average photometric value for the scan-to-scan set, and ($A_{cm^{-1}}$) is the reference photometric value at each of 7000 and 4500 cm^{-1}. The results are reported as: Absorbance/Response (Photometric) Accuracy (in AU). These can be computed for multiple calibrated wavenumber or wavelength positions if desired.

9.2.4 Absorbance/Response Repeatability Test

Verify the response repeatability of the spectrophotometer using a suitable standard, e.g., a prespecified reference sample with nominal absorbance units (AU) of 1.0 AU (i.e., 10% transmittance or reflectance). The standard deviation of the optical response must be consistent with the model performance specifications for the application in use.

For this test, repeat measurements of the same photometric standard by placing it in the sample beam and not mechanically moving the sample over a normal measurement cycle for the measurement period. For example, a 30-s measurement period and 15-s reference spectrum may be typical. Then, calculate the absorbance units for the entire spectrum and specifically at the two measured reference points of 7000 and 4500 cm^{-1}. The absorbance unit is measured for each scan (i.e., scan-to-scan for each sample). Calculate the mean and standard deviation of the AU at the two wavenumber positions for scan-to-scan (within-replicate samples). This statistic is calculated for both 7000 and 4500 cm^{-1} wavenumbers as:

$$s_i = \sqrt{\frac{\sum_{j=1}^{n}\left(A_{ij} - \bar{A}_i\right)^2}{n-1}}, \tag{9.8}$$

where $\left(s_i\right)$ is the standard deviation for the scan-to-scan absorbance units (AU) repeatability for the scan-to-scan measurements; $\left(A_{ij}\right)$ are the individual measurements of the absorbance for sample i

and scan-to-scan number j; $(\bar{A}_i)$ are the mean measured values for the reference sample at each of 7000 and 4500 cm^{-1}; n is the replicate number of spectra (pool all scan-to-scan data). The results are reported as follows. This can be repeated using a different calibrated set of wavelengths or wavenumbers as needed.

9.2.5 Photometric Linearity Test

Verify the photometric linearity of the spectrophotometer by using a set of reference ND filters or suitable reflectance standards. Plot the observed response against the expected response. The slope of the line for reference (x) versus measured (y) data should be 1.00 ± 0.05 and the intercept 0.00 ± 0.05. Calculate the slope and intercept using the reference material measurements of 5%, 10%, 20%, 40%, 60%, and 80% reflectance, or 1.3, 1.0, 0.70, 0.40, 0.22, and 0.10 AU, respectively. Measure a 15-s background and 30-s sample run for each reference standard. *Note the reference value used can be the mean value for each reference standard measured on a minimum of 4 (four) instruments used for initial calibration modeling.*

The results are reported as a graph of the measured linear response against the expected response at the two measured wavenumbers, i.e., 7000 and 4500 cm^{-1}; record full spectral data; and include the slope and intercept for linearity at 7000 and 4500 cm^{-1}. Other wavenumbers or wavelengths may additionally be selected as needed. Note the slope (b) and intercept (a) for the data set of x_i (actual) and y_i (measured) pairs of measurements for each wavenumber position are given as:

$$b = \frac{\sum_i \left\{ (x_i - \bar{x})(y_i - \bar{y}) \right\}}{\sum_i (x_i - \bar{x})^2}, \text{ and} \tag{9.9}$$

$$a = \bar{y} - b\bar{x}. \tag{9.10}$$

9.2.6 Photometric Noise Test

Determine the photometric noise by using a reference ND filter or reflectance standard at 1 AU (10% R or T). Repeat measurements of the same reference standard by placing it in the sample beam and not mechanically moving the sample over a 30-s period. For the measurements, use a 15-s open beam reference spectrum followed by a 30-s sample run. Calculate the photometric noise, peak-to-peak, over the entire spectrum as the standard deviation of the spectrum over the measurement region, *excluding* trim areas. The photometric noise is computed as the standard deviation of the spectral response, and must be consistent with the model performance specifications for the application in use. This statistic is calculated for a single spectrum averaged over a standard measurement period (e.g., 30 s) as:

$$s_i = \sqrt{\frac{\sum_{j=1}^{n} \left(A_{ij} - \bar{A}_i \right)^2}{n - 1}}, \tag{9.11}$$

where (s_i) is the standard deviation (noise) for the *averaged* spectrum comprising a number of scan-to-scan measurements for each wavenumber (30 s), (A_{ij}) are the individual absorbance measurements i for the averaged spectrum at wavenumber j, $(\bar{A}_i)$ is the average absorbance units value for the averaged spectrum, and n is the number of data points (in wavenumbers). This can be reported at one or more wavenumbers or over the entire measurement region.

9.2.7 Signal Averaging Test

Determine the photometric noise by measuring a reference ND filter or suitable reflectance material at 10% transmittance (~1.0 AU), and *report results in transmittance or reflectance.* Repeat measurements of the same reference standard by placing it in the sample beam and not mechanically moving the sample over the entire measurement period. For the measurements, use a 30-s open beam reference spectrum at the start of the measurement and at the conclusion of the measurement period. Calculate the photometric noise (peak-to-peak) over multiple sets of scan-to-scan spectra as the standard deviation of the ratioed transmittance spectrum for each measurement (average the before and after background measurements before ratioing the spectra).

This signal averaging test is to be completed using three methods.

9.2.8 Random Noise Test

This test excludes short-, medium-, and long-term drift; slope; and background curvature with time using measurements of alternating background and sample measurement spectra. This test simulates "dual-beam" conditions and excludes most of the impacts from the longer-term periodic instrument drift, e.g., for $n = 2$, measure background, then measure sample, then background, then sample; reference each spectrum, then average the two referenced spectra – repeat this sequence for the appropriate number of co-added spectra: 1, 2, 4, 16, 64, 256, etc.; compute the background-corrected spectra by referencing alternate (i.e., sandwiched) spectra for averaged scans, then compute the standard deviation using Equation 9.8.

9.2.9 Medium-Term Drift Test

Take background measurements of the same number of scans as sample measurement used for co-added result, e.g., measure 1, 2, 4, 16, 64, 256, etc. as alternate background co-added set and then sample co-added set, e.g., for $n = 2$, measure 2 scans as background followed in sequence by 2 scans of the reference sample, ratio these and calculate the standard deviation; for $n = 4$, measure 4 background and then 4 sample spectra, average these, and ratio as a single spectrum, continue this sequence, and calculate the standard deviation using Equation 9.8.

9.2.10 Long-Term Drift Test

Measure background at the start of run and then measure samples in sequence using only the original background. Thus, generate average spectra from r number of scans, across the entire number of scans available, e.g., for $n = 4$: average scans 1–4, 5–8, etc.; for $n = 16$: average scans 1–16, 17–32, etc. Then, calculate the standard deviation across the averaged spectra using Equation 9.8.

Report noise results as Table 9.1.

TABLE 9.1
Results Reporting for Noise Tests (a)–(c)

Signal Averaging Test	Spectral Region	Actual Results (RMS)
(a) Random noise test		
(b) Noise test (including medium- or short-term drift)		
(c) Noise test (including long-term drift)		

9.2.11 Signal Averaging Test

Signal Averaging Test – Obtain a series of replicate scan-to-scan spectra in transmittance or reflectance mode, and compute a subset of replicate scans and process as described below. Do this for the following number of scans: 1, 4, 16, 64, 256, 1024, 4096, 16,384, etc., up to the maximum measurement time of interest. Ratio each pair of sample and background spectra, and calculate the noise level using equation 9.8 at 7500 cm^{-1} (1333 nm), 7000 cm^{-1} (1429 nm), 6000 cm^{-1} (1667 nm), 5000 cm^{-1} (2000 nm), 4500 cm^{-1} (2222 nm), and 4000 cm^{-1} (2500 nm). The noise level should be reduced by a factor of 2 for each successive ratioed spectrum; for example, if 1 scan gave a noise level of 1, 4 scans would give 1/2, 16 would give 1/4, 64 would give 1/8, and so on until signal averaging fails. The percent noise level for each successive ratioed spectrum should be a factor of 2 lower: e.g., 1, 1/2, 1/4, 1/8, 1/16, 1/32, 1/64, 1/128, etc. (see results reporting table below).

Failure of Signal Averaging – Report the number of scans and the measurement time for each set of scan-to-scan data used in the particular ratioed spectrum and the noise level. Report a failure when the computed/measured noise level is a minimum of twice (two times) that of the expected noise reduction. All spectrometers have a limit to their practical signal averaging capability, often set by residual interference fringing by optical components, by the apodization-determined feet of the moisture interferences, by electronic noise floor due to amplifier and detector performance, or mechanical spectrometer alignment or servo errors.

The results are reported as: Photometric noise signal averaging test at each wavenumber with a table reported for each of 7500 cm^{-1} (1333 nm), 7000 cm^{-1} (1429 nm), 6000 cm^{-1} (1667 nm), 5000 cm^{-1} (2000 nm), 4500 cm^{-1} (2222 nm), and 4000 cm^{-1} (2500 nm) (Table 9.2).

9.2.12 Instrument Line Shape (ILS) Test

The criterion for the laser emission band or etalon is that the FWHM (full-width at half peak maximum) of the emission line or transmittance band be less than one-tenth of the maximum resolution (nominally 10 nm for many NIR instruments) of the spectrometer (or less than 1 nm). The beam diameter should be larger than the collecting optics; the radial intensity profile of the beam in the plane of the field stop should be comparable to that of the standard source beam; the frequency of the source should not drift significantly during the period of a full-resolution scan (30 s) and the emission band should lie in one of the routinely measured spectral regions. A stabilized gas or semiconductor laser is an optimum choice for its narrow line width (limited by acoustic vibrations to ~10^{-5} cm^{-1}), typical for cylindrical HeNe laser cavity systems usually available in randomly polarized or linearly polarized versions. Frequency-stabilized lasers are ideal for this measurement

TABLE 9.2
Method for Reporting Signal Averaging Test

Number of Scans	Expected Noise Reduction Factor	Measured Photometric Noise (s_i)	Measured Noise Reduction Factor
1	1		
4	1/2		
16	1/4		
64	1/8		
256	1/16		
1024	1/32		
4096	1/64		
16,384	1/128		
65,536	1/256		

with a frequency stability of ±3 MHz or $\pm 10^{-10}$ cm^{-1} (a stability improvement of 10^{-5} or better over nonstabilized lasers).

Use of a laser system might consist of a 1.5231 μm (6565.99 cm^{-1}) helium:neon laser (ideally with optics to expand the beam, flatten its radial intensity profile, and monitor its power). The NIR (1.5 μm) laser is enclosed in an insulated box to prevent temperature fluctuations, which would, in turn, cause the laser frequency to drift. If these precautions are not available, it should be noted with the reported test results. The beam from the NIR laser is positioned as the standard NIR source. The output from the laser is amplified and displayed on an oscilloscope to indicate the laser power. The beam is directed through the interferometer or optical train of the spectrometer as if it represents the source energy.

9.2.13 Measurements of Laser Line

> Note that most spectroscopists will not have access to the equipment required to either measure or modify their instrument line shape. It is discussed here for informational purposes only.

After temperature stabilization, the temperature telemetry data is recorded from the instrument over the 30-s run periods. The laser cavity temperature for non-frequency-stabilized lasers requires a maximum temperature drift of not more than ±0.1°C in any measurement (30-s) interval (duration of one scan at maximum resolution), with the long-term drift of ±0.06°C h^{-1} enabling repeatable measurements of the ILS to be obtained. For low laser line drift rates (less than one part in 10^6), the FWHM of the measured laser line will not change significantly (see #3, and blue solid line in Figure 9.1), but the band will become asymmetric. Asymmetry of the measured ILS is defined as the ratio of the negative lobe depths of the minima immediately on either side of the central peak. This ratio is expressed as a percentage. For example, Figure 9.1 shows two theoretical spectra taken some period apart, say 1 h. The asymmetry in relative intensity of the lines is given by the absolute value of |#2 depth/#1 depth|×(100), or approximately |−0.3/−0.9| (100) = 30%. In addition, the drift in the laser frequency (#4) is measured during the course of replicate scans. From theoretical models, it can be demonstrated that laser drift rate should produce an asymmetry of approximately <10%, which is significantly less than the 30% asymmetry illustrated in Figure 9.1. If the asymmetry is greater than about 10%, there is generally some instrumental effect producing excess asymmetry, which must be corrected before calibration transfer can be effective.

9.2.14 Summary and Test Data Analysis Considerations

An idealized or reference line shape is required for comparison purposes. This *idealized* instrument line shape (iILS) can be calculated from theory, or a *reference* instrument line shape (rILS) can be

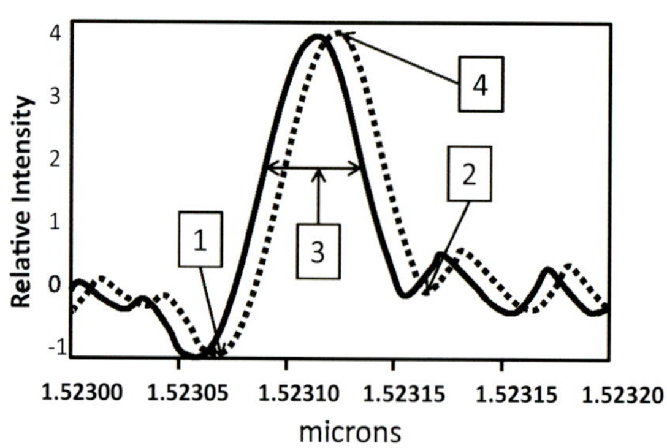

FIGURE 9.1 Replicate laser lines recorded at two different time intervals (solid versus dotted lines)

measured by averaging the laser or using an etalon measurement for multiple well-functioning spectrophotometers *used for the initial calibration modeling* and then by using the mean profile of these instruments as the comparison line shape reference. The challenge then for quality assurance and calibration transfer is to replicate this reference line shape (rILS) for subsequent instrument systems to which the original calibration will be transferred. Acceptable ILS tolerances are then used to test and adjust instrument line shape during manufacturing. ILS tolerances are established by testing the calibration model against synthetic (virtual) data where the ILS has been purposefully modified.

Other critical parameters available when measuring ILS are center wavenumber position, FWHM, and asymmetry (see reporting table below). The tolerances for center wavenumber position are small for spectrometers subject to multivariate applications. This can be measured by center of mass position changes for bands over time, or by the measured shifting of the zero-crossing of the laser band first derivative over time. The effects of wavenumber shift on calibration performance are tested using artificial data shifts and measuring prediction performance. FWHM is an indication of spectral resolution and should be recorded at time of manufacturing. This parameter is recorded in wavenumbers as the width of a band at one-half the total band height. Finally, the asymmetry can be recorded for a laser line as described above. These measurements are retained for historical performance and for tracking and diagnosing instrument performance issues. Compute a table, including (1) the reference center wavelength, (2) the measured center wavelength, (3) the measured FWHM, and (4) the measured percent asymmetry.

9.2.15 Summary Specifications for Instrument "Alikeness" Testing

General specifications for comparing instruments should meet the basic minimum pre-established criteria. These criteria are based on the actual use of the spectrophotometer and the accuracy required for measurements for a high signal-to-noise medical application. General requirements for each spectrometer should depend on the use, application, and overall performance requirements. The specifications here are used for illustrative purposes only. If extremely high performance and "alikeness" are required, then comparison statistics are more important. Here are some basic performance criteria that indicate a reasonable alikeness between two or more instruments. Specific calibrations and prediction performance criteria will vary and so then will the comparative instrument performance metrics (Table 9.3).

9.3 OPTICAL STANDARDS FOR ALIGNING THE *X*-AXIS

Alignment of the *x*-axis (i.e., the wavelength or wavenumber axis) is assumed for all spectra used for the purposes of quantitative and qualitative work. Note that the assumption for regression modeling as well as spectral comparison algorithms is that the data channels, or *x*-axis scales of the spectroscopic data, are precisely aligned and that only amplitude in the *y*-axis changes with respect to concentration in a predefined mathematical relationship. This section provides a discussion and set of spectra to illustrate the use of reference materials to align the *x*-axis for the NIR spectrophotometers.

The alignment of the *x*-axis (i.e., the wavelength or wavenumber axis) for spectroscopic data is essential to the basic use of these spectroscopic techniques. The primary assumption of all quantitative regression and qualitative comparison methods is that the data channels, as wavelength or frequency, are precisely aligned for comparison. Without such a condition, no spectroscopic methods are valid. Thus, the alignment of the *x*-axis requires a known standard material that may be measured to either align the axis or verify whether the axis is aligned. Multiple standards are available for each spectroscopic region. The typical standards used for spectrometer alignment or testing will be illustrated within this chapter (Workman (2016a, 2016b), Feudale et al. (2002), Workman and Mark (2014a, 2014b)).

The NIR spectral region is generally considered to be from 780 to 2500 nm.

TABLE 9.3
Typical Performance Specifications for Commercial NIR Spectrometers

9.2.1 Wavenumber accuracy	Absolute deviation must be less than ±0.8 cm^{-1} versus NIST reference following calibration using NIST traceable reference measurements for 5447.42 and 4333.51 cm^{-1} bands of crystalline polystyrene (1.0 mm thickness). Accuracy must be ±0.1 cm^{-1} in agreement with calibration (reference) as absolute maximum deviation.
9.2.2 Wavenumber repeatability	< 0.01 cm^{-1} (1 σ)
9.2.3 Absorbance/response accuracy	Absolute deviation must be less than ±0.02%*T* versus NIST traceable measurement of 0.09%*T* (3.046 AU) to 0.10%*T* (3.000 AU) for specific reference standards at ~1333 nm (7500 cm^{-1}) and ~2222 nm (4500 cm^{-1}), respectively. Accuracy must be ±0.01%*T* (±0.02 AU) in agreement with calibration (reference) parent instrument (absolute maximum deviation)
9.2.4 Absorbance/response repeatability	<0.001 AU (1 σ)
9.2.5 Photometric linearity	Slope: 1.00 ± 0.02; intercept: ±0.02 (absolute) for method
9.2.6 Photometric noise	RMS <0.001 AU for 0.1%*T* ND filter
9.2.7 Signal averaging test	
9.2.8 Random noise test	Must pass method to 1024 scans
9.2.9 Noise test (including medium-or short-term drift)	Must pass method to 500 scans
9.2.10 Noise test (including long-term drift)	Must pass method to 480 scans
9.2.11 Signal averaging test	
9.2.12 Instrument line shape (ILS) test	
(a) Center wavelength measured	Accuracy must be ±0.05 cm^{-1} in agreement with calibration (reference) line using interpolation or center of mass peak picking method
(b) FWHM measured (% deviation)	<5% deviation from calibration (reference) parent instrument (absolute maximum deviation)
(c) Asymmetry measured (% deviation)	<5% deviation as per prescribed method from calibration (reference) parent instrument (absolute maximum deviation)

An excellent summary paper is given related to reference standards in the NIR region (Weidner 1986). Similar NIST research includes details of the following standards: high-pressure mercury arc, water vapor, gas standards SRM 2517, SRM 2519. Also covered are solid materials SRM 1920, SRM 2035, SRM 2036; McCrone M-27 and M-42; and polystyrene. Liquids described include 1,2,4 trichlorobenzene and methylene chloride. For this chapter, we will describe polystyrene and SRM 1920 – rare earth oxide (REO) standards.

9.3.1 Polystyrene Reference Standard

Polystyrene in a highly crystalline structure is a useful wavelength/wavenumber standard for the NIR spectrophotometers. It provides a number of useful bands to verify or align the wavelength axis for all types of NIR instrumentation. The specific wavelengths and wavenumbers used for verification and calibration of the spectrophotometer *x*-axis measurements using polystyrene are illustrated in Figures 9.2 and 9.3, and described with resolution details in Tables 9.4 and 9.5 (Avian Technologies 2018).

For Figures 9.2 and 9.3 and Tables 9.4 and 9.5, 1.0-mm-thick polystyrene samples of 28 mm diameter were measured using a Perkin-Elmer Lambda-9/19 UV-Vis-NIR Spectrometer with traceability

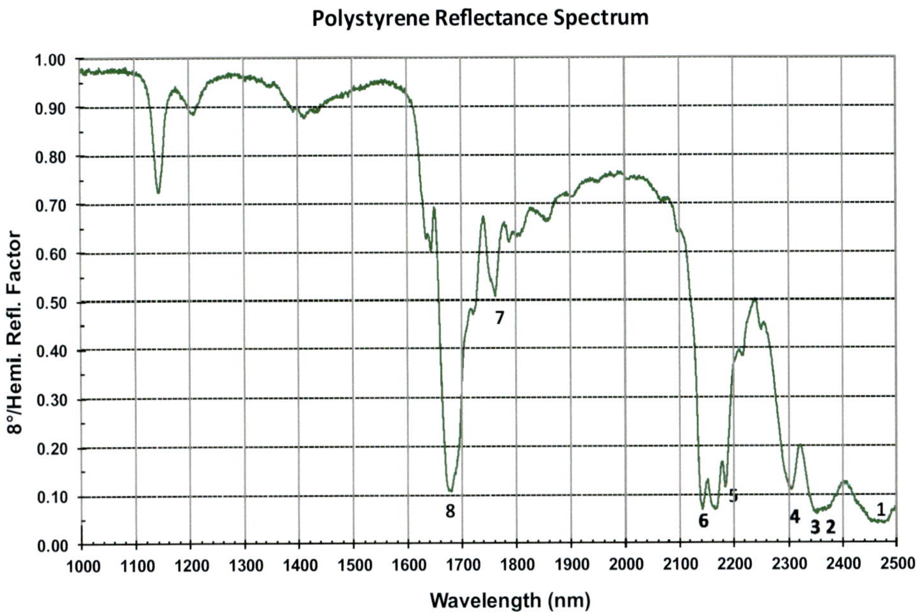

FIGURE 9.2 Polystyrene (crystalline form) as a reflectance spectrum, 1.0 mm thickness. (Spectrum used with permission from Avian Technologies, New London, NH.)

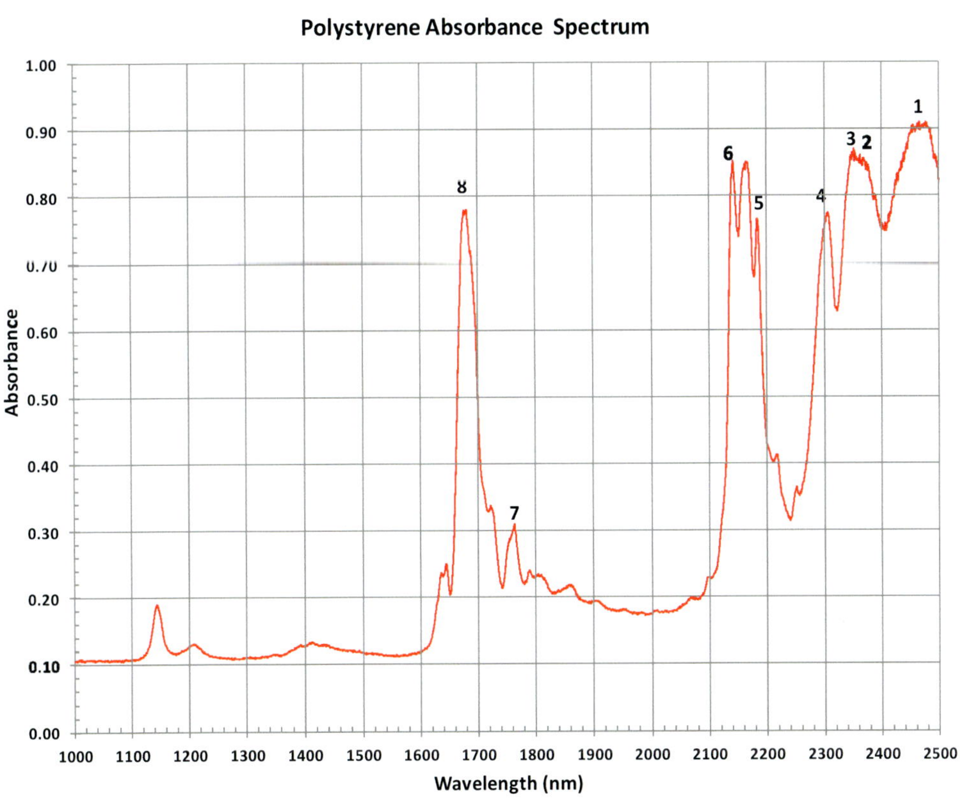

FIGURE 9.3 Polystyrene (crystalline form) as an absorbance from a reflectance measurement, 1.0 mm thickness. (Spectrum used with permission from Avian Technologies, New London, NH.)

TABLE 9.4
Polystyrene Nominal Wavelengths from NIST Traceable Instrument

Peak Designation Number	Literature Peak Wavelength[a]	Transmittance (Two Samples)	Bandpass at Peak-*T* (nm)	Reflectance w/FW99 Backing	Bandpass at Peak-*R* (nm)
1	2464.0	2475.5	2.6	2366.4	8.3
2	2362.0	*****	n/a	*****	n/a
3	2353.0	2352.2	2.1	2353.3	*6.3*
4	2308.0	2307.1	2.1	2308.2	*5.8*
5	2188.0	2186.6	1.5	2185.7	*4.7*
6	2145.0	2143.7	1.4	2143.1	*4.5*
7	1766.0	1764.0	1.0	1764.2	*2.9*
8	1681.0	1680.7	1.0	1681.1	2.7

Source: Spectrum used with permission from Avian Technologies, New London, NH.
[a] 32 cm^{-1} resolution. Polystyrene (crystalline form) Measurement, 1.0 mm thickness.
***** – No Values measured

TABLE 9.5
Polystyrene (Crystalline Form) as Absorbance (Transmittance) Spectrum from a NIST-Calibrated Double-Monochromator Instrument

Measured PS Peak Positions on NIST-Calibrated Dispersive Spectrometer at Two Specified Resolutions: Polystyrene Filters – Three Randomly Selected Filters			
Bandpass Approximately 5 nm at 2500 nm, 2.0 nm at 1500 nm, and 2.5 nm at 1300 nm		**Bandpass Approximately 3.5 nm at 2500 nm, 1.5 nm at 1500 nm, and 2.0 nm at 1300 nm**	
(nm)	**(cm^{-1})**	**(nm)**	**(cm^{-1})**
2476.1 ± 0.1	4038.6090	2476.5 ± 0.1	4037.9570
2353.0 ± 0.2	4249.8940	2353.1 ± 0.1	4249.7130
2307.6 ± 0.1	4333.5070	2307.4 ± 0.1	4333.8820
2187.3 ± 0.0	4571.8470	2187.3 ± 0.0	4571.8470
2170.0 ± 0.0	4608.2950	2170.1 ± 0.0	4608.0830
2144.2 ± 0.1	4664.1790	2144.2 ± 0.0	4663.7440
1790.7 ± 0.1	5584.4080	1790.7 ± 0.1	5584.4080
1764.2 ± 0.2	5668.2920	1764.3 ± 0.1	5667.9700
1681.4 ± 0.1	5947.4250	1681.6 ± 0.1	5946.7170
1646.4 ± 0.1	6073.8600	1646.4 ± 0.1	6073.8580
1414.5 ± 0.3	7069.6360	1415.7 ± 0.5	7063.6430

Source: Data used with permission from Avian Technologies (2018), New London, NH.
A set of three samples are averaged for band positions.

of measurements to ASTM Test Method E 1331-96, ASTM Test Method E903-96, NRC Certificate PO1948 (Didymium Oxide Glass), NRC Certificate Cal PAR-2008-2614 (Holmium Oxide Filter), NIST SRM-1920a, and NIST SRM-2036. Traceability to NIST is achieved through NORAMET, the North American Metrological Agreement (Avian Technologies (2018)).

Measurement conditions are the mean temperature of 21.9°C and relative humidity at 45%. The instrumental parameters are as follows: bandpass: n/a NIR sensitivity corresponding to a bandpass (approximately 1.5 nm from 1000–2500 nm); reporting interval: 1 nm; recording interval: 1 nm; scan speed: 60 nm min^{-1}; integration time: 1 s; number of measurements averaged: 6;

measurement range: 1000–2500 nm. The instrument was set up in 8° hemispherical geometry with the sample positioned normally to the beam. NIR sensitivity of 3 and scan speed of 1 nm s^{-1} were used with peak picking at 0.1-nm interval over the wavelength range of 1000–2500 nm.

9.3.2 SRM 1920a Former Reference Standard

SRM 1920a is a discontinued Standard Reference Material® that was used for verification and calibration of the NIR wavelength scale from 740 to 2000 nm. It was applicable for reflectance spectrophotometers or reflectometers using hemispherical geometry. It consisted of a mixture of three REOs such as dysprosium oxide (Dy_2O_3), erbium oxide (Er_2O_3), and holmium oxide (Ho_2O_3) that was pressed into a cylindrical cavity with an NIR-transmitting sapphire widow. The outer dimensions of the standard are 51 mm in diameter and 12 mm in height. The inner dimensions of the powder-holding cavity are 30 mm in diameter by 6 mm in depth. The SRM samples were calibrated using the NIST high-precision spectrophotometer. The spectra of a replicate mixture are shown in Figures 9.4 and 9.5, with the verified wavelengths for various bandwidths given in Tables 9.6 and 9.7 (Avian Technologies (2018), NIST (SRM)® 1920a (2004)).

9.3.3 Summary

A variety of materials exist for the important characteristic of aligning the wavelength or wavenumber axis for spectrophotometers from the ultraviolet through infrared regions, and Raman. *x*-axis alignment and maintenance is a critical step required for the reasonable application of quantitative

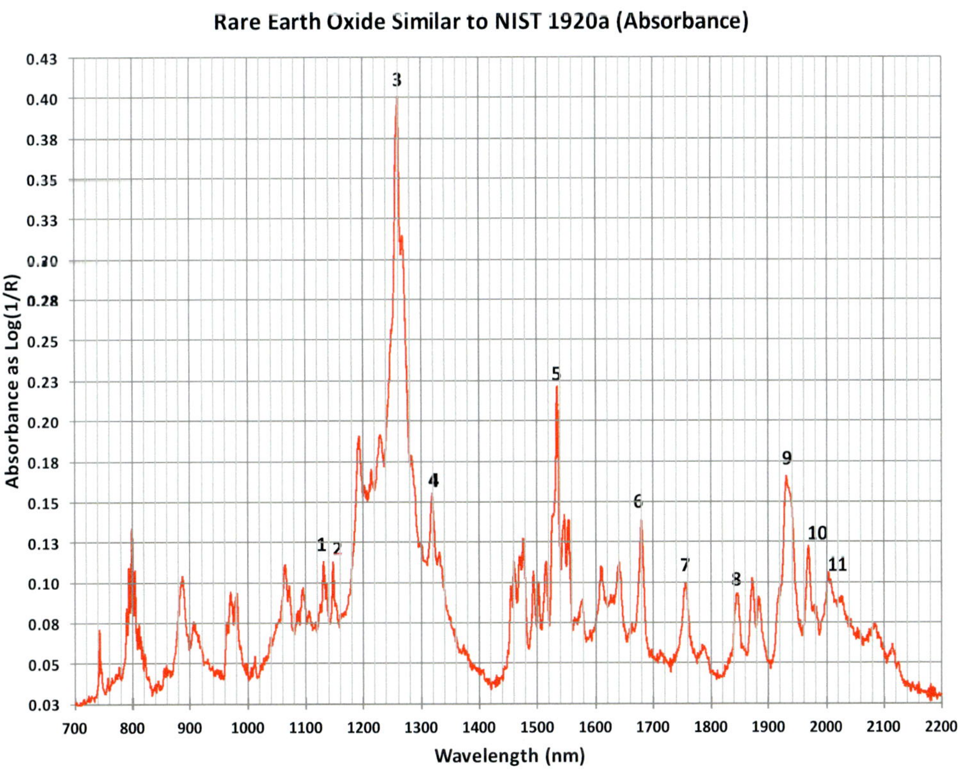

FIGURE 9.4 Rare earth oxide similar to NIST SRM 1920a as absorbance (reflectance) spectrum from a NIST-calibrated double-monochromator instrument. (Data used with permission from Avian Technologies, New London, NH.)

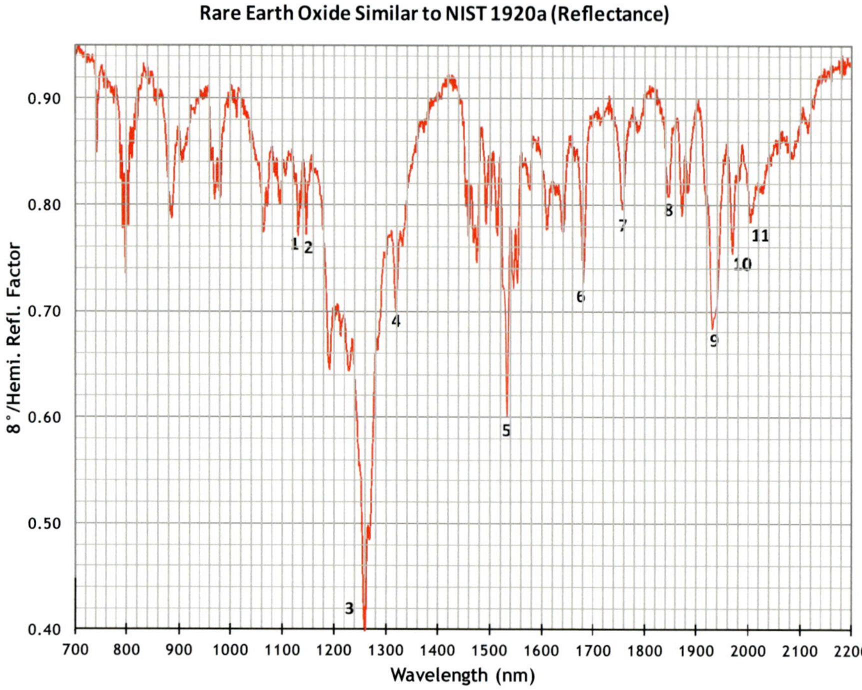

FIGURE 9.5 Rare earth oxide similar to NIST SRM 1920a as a reflectance spectrum from a NIST-calibrated double-monochromator instrument. (Data used with permission from Avian Technologies, New London, NH.)

TABLE 9.6

Certified Band Positions of NIST 1920a at Various Bandwidths for 11 Wavelengths from 1133 to 2006 nm

Peak Designation Number	Band	2 nm	3 nm	4 nm	5 nm	10 nm	(Variable)[a]	FT[b]	FT-10 nm
1	1133	1132.2	1132.3	1132.4	1132.9	1132.9	1132.3	1193.24	60.34
2	1148	1148.1	1148.4	1148.5	1148.7	1148.6	1147.9	1230.50	81.90
3	1261	1261.0	1260.9	1260.8	1260.8	1261.8	1261.5	1391.68	129.88
4	1321	1320.7	1320.7	1320.8	1320.7	1320.2	1320.7	1397.92	77.72
5	1536	1535.5	1535.6	1536.5	1535.4	1534.6	1535.6	1535.20	0.60
6	1682	1682.6	1682.3	1682.2	1682.2	1681.4	1681.7	1682.70	1.30
7	1757	1756.5	1757.8	1757.9	1757.8	1757.6	1757.1	1758.18	0.58
8	1847	1847.5	1847.3	1846.9	1847.0	1847.3	1847.1	1847.75	0.45
9	1933	1930.9	1931.6	1932.2	1932.5	1935.5	1932.6	1931.61	−3.89
10	1971	***	1970.6	1970.7	1970.8	1970.8	1970.8	1971.13	0.33
11	2006	***	2004.5	2005.9	2005.8	2006.3	2005.1	2005.43	−0.87

Source: Used with permission from NIST (NIST (SRM)® 1920a (2004), Workman (2016a), Weidner et al. (1986), NIST SRM website (2018), Isaksson et al. (2003)), column 8 data is from reference Isaksson et al. (2003).

[a] Variable bandpass in NIR. Appx. 2 nm from 1100 to 1850 nm; appx 8 nm below 850–1050 nm. ±2 nm accuracy in NIR.

[b] Isaksson, T., H. Yang, G.J. Kemeny, R.S. Jackson, Q. Wang, M.K. Alam, and P.R. Griffiths. 2003. Accurate wavelength measurements of a putative standard for near-infrared diffuse reflection spectrometry. *Applied Spectroscopy*, *57*(2), pp. 176–185.

*** – Not Measured

TABLE 9.7
Certified Bands of NIST 1920a at Various Bandwidths

Band Number	2 nm	3 nm	4 nm	5 nm	10 nm	Variable Bandpass[b]	FT Measured
1	***	743.0	743.4	743.4	****	743.3/752.0[a]	****
2	***	799.0	799.0	798.6	798.0	799.3	****
3	887.2	886.9	886.7	886.7	886.5	888.3	****
4	906.3	906.8	907.5	907.3	****	907.8	****
5	970.6	971.3	971.6	971.6	****	971.6	****
6	979.6	980.8	980.8	980.0	****	980.7	****
7	1012.9	1013.2	1012.9	1012.8	****	1013.3	1013.27
8	1064.7	1065.0	1065.0	1064.9	****	1064.8	1064.92
9	1095.6	1096.0	1096.2	1096.4	****	1089.0[a]	1095.98
10	1132.2	1132.3	1132.4	1132.9	1132.9	1134.6[a]	1132.21
11	1148.1	1148.4	1148.5	1148.7	1148.6	1148.9	1148.13
12	1192.2	1192.7	1192.9	1192.9	****	1193.5	1193.24
13	1230.2	1230.2	1230.2	1230.3	****	1232.4	1230.50
14	1261.0	1260.9	1260.8	1260.8	1261.8	1261.7	1261.87
15a						1301.1[a]	
15	1320.7	1320.7	1320.8	1320.7	1320.2	1320.6	1320.60
16	1456.2	1456.4	1456.7	****	****	1456.7	1456.84
17	1461.7	1461.9	1462.2	****	****	1462.2	1462.18
18	1471.2	1471.6	1471.6	****	****	1472.1	1470.50
19	1477.4	1477.6	1477.5	****	****	1478.2	1478.07
20	1494.8	1494.9	1495.0	1495.0	****	1495.4	1495.62
21	1503.4	1503.5	1503.5	****	****	1504.1	1504.14
22	1516.0	1516.0	1515.9	1515.7	****	1516.8	1516.60
23	1535.5	1535.6	1536.5	1535.4	1534.6	1535.4	1535.20
24	1544.7	***	****	****	****	1544.9	1545.23
25	1548.1	***	****	****	****	1548.3	1548.49
26	1555.1	1555.0	1554.8	****	****	1555.0	1556.07
27	1577.1	1577.2	1577.2	****	****	1578.2	1578.60
28	1611.8	1611.7	1611.7	1611.9	****	1612.5	1612.49
29	1642.7	1642.7	1642.5	1642.5	****	1643.1	1643.34
30	1682.6	1682.3	1682.2	1682.2	1681.4	1682.3	1682.70
31	1756.5	1757.8	1757.9	1757.8	1757.6	1758.1	1758.18
32	1847.5	1847.3	1846.9	1847.0	1847.3	1846.2	1847.75
33	1874.3	1874.0	1873.8	1874.0	****	****	1875.00
34	1885.0	1885.3	1885.7	1885.5	****	1888.6[a]	1885.37
35	1930.9	1931.6	1932.2	1932.5	1935.5	1936.2	1931.61
36	***	1970.6	1970.7	1970.8	1970.8	1970.6	1971.13
37	***	2004.5	2005.9	2005.8	2006.3	****	2005.43
37a						2020.2[a]	

Source: Data used with permission from NIST (NIST (SRM)® 1920a (2004), Workman (2016a), Weidner et al. (1986), NIST SRM website (2018), Isaksson et al. (2003)), column 8 data is from (Isaksson et al. 2003).

[a] Wavelengths that are appreciably different from packed powder standards.

[b] Variable bandpass in NIR. Appx. 2 nm from 1100 to 1850 nm; approx. 5 nm between 850 and 1050 nm. Claim is ±2 nm accuracy in NIR.

**** – Not measured.

and qualitative analyses using spectroscopic methods. The assumption for all analytical methods using spectrometers is that the *x*-axis, representing unique data channels, be constant with only the amplitude changing with respect to concentration. All chemometrics relies on the assumption of a precisely aligned *x*-axis for analytical techniques in order to compare data for calibration purposes.

9.4 INTRODUCTION TO PHOTOMETRIC ACCURACY

Alignment of the *y*-axis is a helpful alignment tool for all spectra when used for both quantitative and qualitative measurements using spectroscopy. The assumption of regression is that the data channels or *x*-axis of data is aligned and that only amplitude in the *y*-axis changes with concentration in some mathematically defined manner. An aligned *y*-axis provides an improved method to compare spectra and to relate amplitude to concentration. The correction of *y*-axis (amplitude) variation provides a minimum bias between instruments and provides a platform for improved calibration transfer between instruments. This chapter provides a discussion and presentation for using photometric reference materials to align the *y*-axis for ultraviolet, visible, NIR, infrared, and Raman intensity spectroscopic methods. A general method for correcting each type of absorption-based spectrophotometer is given as well as the types of reference standards used for calibration or verification of the photometric scale. For Raman spectroscopy, a method is described for intensity correction.

Photometric accuracy is a formal term used to describe how well a measuring device (spectrophotometer or illuminance meter) is able to determine the total energy flux transmitted through (or reflected from) a Standard Reference Material (SRM)®. Volume 14.01 of the American Society for Testing and Materials (ASTM) International document (ASTM E275-08, 2013) describes the measurement of the photometric accuracy for an instrument as follows (Sections 20.1 and 20.3):

> Select the appropriate Standard Reference Material and obtain ten successive readings of the apparent absorbance or transmittance at the specified wavelength. Average the ten readings. The photometric accuracy is the difference between the true absorbance or transmittance value and the average observed value.

It is further stated that the following applies to the reporting of photometric accuracy.

> Report the photometric accuracy in the following order: reference material, wavelength, true absorbance or transmittance, observed absorbance or transmittance plus or minus the standard deviation.

Example of Reporting Photometric Accuracy

As an example, the stated specifications for photometric accuracy for a test instrument might be written as:

Photometric Accuracy

±0.002 AU (for a range of 0.0 to 0.5 absorbance units)
±0.004 AU (for a range of 0.5 to 1.0 absorbance units)
±0.3%*T*
SRM: Measured with NIST 930D filters.

Here is an evaluation of these specifications in greater detail. A photometric accuracy of "0.002 AU at 0–0.5 AU" indicates that after the instrument has been set to zero, a standard sample which transmits between 30.1%*T* (i.e., 0.5 AU) and 100%*T* (i.e., 0.0 AU) measures within ±0.002 AU as compared to the specified value for that reference standard material at any wavelength position.

The tolerance for each of the 930d filters is stated by NIST as ±0.5% relative. Thus, a 10%*T* filter (which corresponds to 0.1 T or 1.0 AU) ND 930D-type reference standard would be expected to be within the range of 9.95%*T*–10.05%*T*. These values are equivalent to 1.0022–0.9978 AU. The difference in AU is 1.0022 − 0.9978 AU equals 0.0044 AU. The standard tolerance in AU is 1.0000 ± 0.0022 AU, rounded to 1.000 ± 0.002 AU. Conservatively for transmittance, we state 10.00%*T* ± 0.03%*T*, and related to 0.0 AU or 100%*T*; we specify 100.0%*T* ± 0.3%*T*.

If the stated tolerance for the NIST 930D filter is ±0.002 AU at 1.0 AU, we are stating that our maximum deviation in photometric accuracy for our instrument at 1.0 AU is also ±0.002 AU. For the worst-case measurement for the NIST filter, we would have the filter error of ±0.002 added to the instrument error of ±0.002, to result in a total maximum variation ±0.004 AU error. (This is our stated specification for the example given.)

And one can readily calculate the reflectance of transmittance of any measurement given the absorbance of that measurement as follows. For conversion of absorbance to transmittance, we use Equations 9.12–9.15:

$$A = -\log_{10}\left(\frac{I}{I_0}\right) = \log_{10}\left(\frac{I_0}{I}\right) \tag{9.12}$$

$$\frac{I}{I_0} = 10^{-A} \Rightarrow \frac{I_0}{I} = \frac{1}{10^{-A}} \tag{9.13}$$

$$\frac{I}{I_0} = 10^{-A} = T \tag{9.14}$$

$$\%T = T \times 100. \tag{9.15}$$

To convert absorbance to reflectance, we use Equations 9.16–9.19:

$$A = -\log_{10}\left(\frac{I}{I_0}\right) = \log_{10}\left(\frac{I_0}{I}\right) \tag{9.16}$$

$$\frac{I}{I_0} = 10^{-A} \Rightarrow \frac{I_0}{I} - \frac{1}{10^{-A}} \tag{9.17}$$

$$\frac{I}{I_0} = 10^{-A} = R \tag{9.18}$$

$$\%R = R \times 100. \tag{9.19}$$

TABLE 9.8
Relationship of %*T* and Absorbance Unit (AU) Values

T (as ppm)	%*T*	*T*	1/*T*	AU
10,000,000	1000	10	$0.1 = 10^{1}$	−1.0
1,000,000	100	1	$1.0 = 10^{0}$	0.0
301,000	30.1	0.301	3.01	0.5
100,000	10.0	0.1	$10 = 10^{1}$	1.0
10,000	1	0.01	$100 = 10^{2}$	2.0
1000	0.1	0.001	$1000 = 10^{3}$	3.0
100	0.01	0.0001	$10{,}000 = 10^{4}$	4.0
10	0.001	0.00001	$100{,}000 = 10^{5}$	5.0
1	0.0001	0.000001	$1{,}000{,}000 = 10^{6}$	6.0

9.4.1 Photometric Correction for Absorbance-Based Spectrophotometers

For making photometric calibrations using the known reference materials, the ratioed replicate spectral data are retained, along with the replicate spectra of a reference sample and dark background, or dark sample. Equation 9.20 represents Beer's law for taking measurements using a spectrophotometer. To convert the light reflected or transmitted from a sample (I) ratioed to the incident energy (I_0) to absorbance (A), as a linearized estimate of the spectral response as related to analyte concentration, the following Equations 9.20–9.23 may be applied.

If only an internal reference is used for spectral collection, then the absorbance spectrum with respect to wavelength is computed as Equation 9.20:

$$A = -\log_{10}\left(\frac{I}{I_0}\right) = \log_{10}\left(\frac{I_0}{I}\right) = \log_{10}\left[\left(\frac{R_I - D_{R_I}}{S - D_S}\right)\right] \tag{9.20}$$

If an external standard material is used to calibrate an automated internal reference material, then the absorbance spectrum with respect to wavelength is computed as Equations 9.21 and 9.22:

$$A = -\log_{10}\left(\frac{I}{I_0}\right) = \log_{10}\left(\frac{I_0}{I}\right) = \log_{10}\left[\left(\frac{R_I + R_\Delta - D_{R_I}}{S - D_S}\right)\right], \tag{9.21}$$

where

$$R_\Delta = \left(R_E - D_{R_E}\right)/\left(R_I - D_{R_I}\right), \tag{9.22}$$

where for Equations 9.20–9.23, S is the sample measurement, D_S is the Dark measurement for the Sample, R_I is the internal Reference Measurement, D_{R_I} is the Dark Measurement for the Internal Reference Sample, R_E is the measurement of the external standard reference material for calibration, D_{R_E} is the Dark Measurement for the external standard reference material, and R_Δ is the ratio between the dark-corrected R_E and R_I. The final spectrum is simplified to the ratio of (Internal Reference corrected by External Reference minus Dark) divided by (Sample – Dark). Note that Dark is a measurement where no energy from the source is allowed to the detector (it is blocked) or where a totally dark sample or light trap is measured. The Dark measurement represents the dark current changes in the instrument during a measurement that relate to electronic noise where the detection electronics, computational electronics, and control electronics perform the basic measurement functions with no imputed energy reaching the detector (i.e., the source is turned off or blocked using a shutter closure during the Dark measurement).

When a comprehensive method of photometric correction includes a light trap measurement to determine the dark response to correct for optical and window scatter, we use equation 9.23:

$$A = \log_{10}\left[\frac{\left(R_I - D_{R_I}\right)\cdot\left(\dfrac{\left(R_E - D_{R_E}\right)-\left(I_W - D_W\right)}{R_I' - D_{R_I}'}\right)}{S - D_S}\right], \tag{9.23}$$

where the variable terms used include the following:

R_I' is the photometric correction calibration measurement of the internal reference sample,
D_{R_I}' is the photometric correction calibration measurement of the dark signal intensity for the internal reference sample,

I_W is the intensity measurement of the light trap to determine the window scatter,
D_W is the measurement of the dark signal intensity for the light trap measurement.

equation 9.23 is simplified for understanding by removing the dark correction terms as equation 9.24:

$$A = \log_{10}\left[\frac{R_I \cdot \left(\dfrac{R_E - I_W}{R_I'}\right)}{S}\right]. \quad (9.24)$$

9.4.2 Near-Infrared (NIR) Reflectance Photometric Standards

The NIR spectral region is generally considered to be from 780 to 2500 nm. Two types of available reflectance standards are shown, namely, R50 (50% reflectance) and sintered Fluorilon (PTFE) (R99 or 99% reflectance), which are used for calibrating or verifying photometric alignment of NIR spectrophotometers. Photometric values for these standards are illustrated in Figures 9.6 and 9.7, and the reference reflectance values are given in the accompanying Tables 9.9 and 9.10. Materials are noted when provided by courtesy by Avian Technologies New London, NH. Note that in transmittance measurement mode, various solvents or dry air may be used to calibrate or verify the transmittance accuracy.

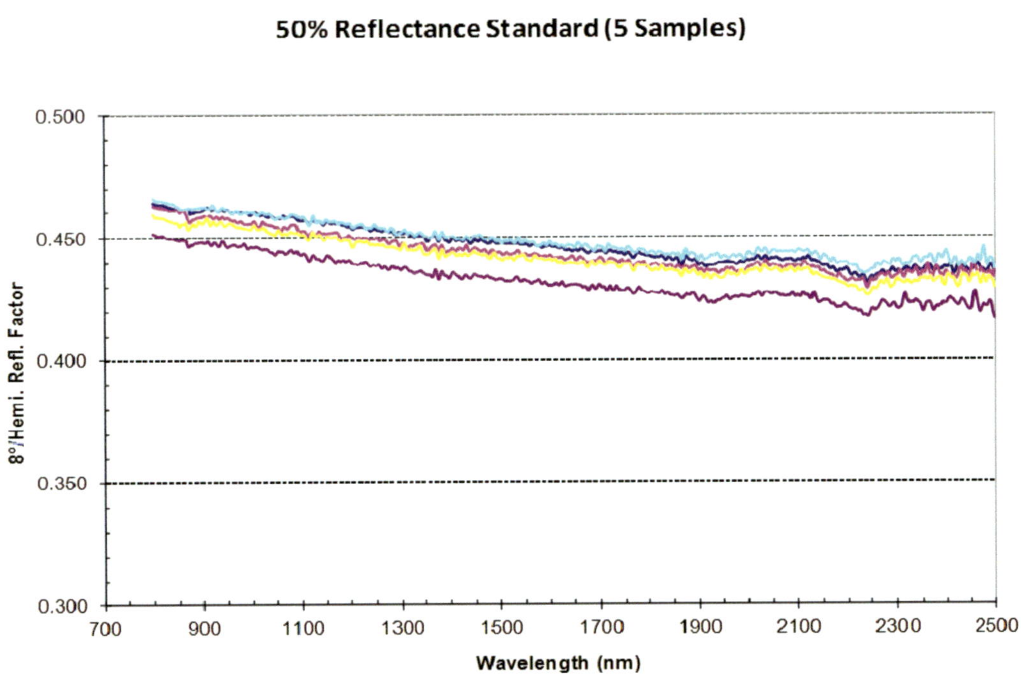

FIGURE 9.6 Fluorilon and carbon black mixture as a 50% reflectance standard. (Spectrum used with permission from Avian Technologies, New London, NH.)

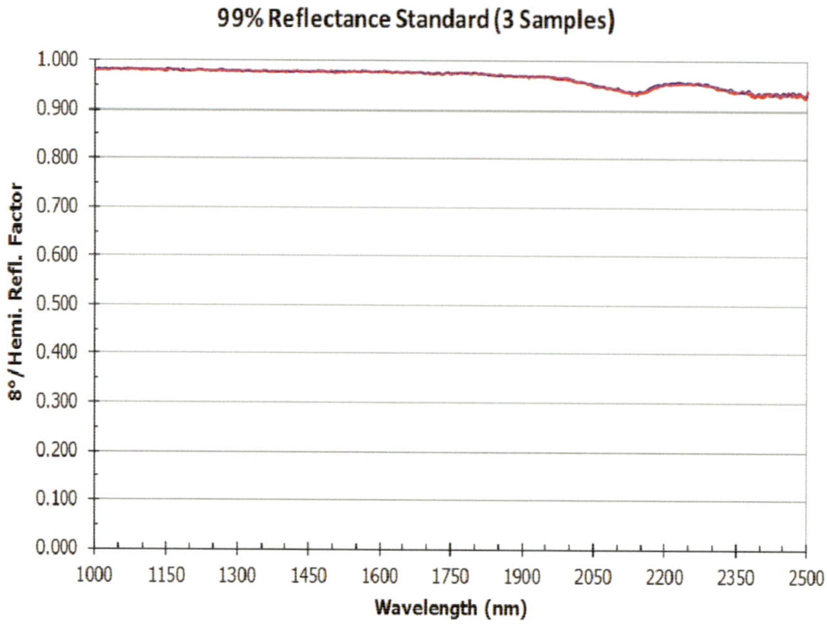

FIGURE 9.7 Fluorilon as a 99% reflectance standard. (Spectrum used with permission from Avian Technologies, New London, NH.)

TABLE 9.9
R50 Fluorilon Reflection Data from NIST Traceable Instrument

Wavelength (nm)	R50 Refl.	Wavelength (nm)	R50 Refl.
800	0.464	1625	0.446
825	0.463	1650	0.445
850	0.461	1675	0.444
875	0.460	1700	0.444
900	0.461	1725	0.444
925	0.461	1750	0.443
950	0.462	1775	0.443
975	0.461	1800	0.442
1000	0.459	1825	0.442
1025	0.460	1850	0.441
1050	0.458	1875	0.442
1075	0.458	1900	0.439
1100	0.457	1925	0.439
1125	0.457	1950	0.440
1150	0.456	1975	0.439
1175	0.456	2000	0.441
1200	0.454	2025	0.441
1225	0.454	2050	0.441
1250	0.453	2075	0.441
1275	0.452	2100	0.441
1300	0.452	2125	0.442
1325	0.450	2150	0.438

(Continued)

TABLE 9.9 (*Continued*)
R50 Fluorilon Reflection Data from NIST Traceable Instrument

Wavelength (nm)	R50 Refl.	Wavelength (nm)	R50 Refl.
1350	0.450	2175	0.437
1375	0.449	2200	0.437
1400	0.448	2225	0.434
1425	0.448	2250	0.434
1450	0.449	2275	0.436
1475	0.448	2300	0.436
1500	0.447	2325	0.436
1525	0.447	2350	0.436
1550	0.447	2375	0.436
1575	0.446	2400	0.438
1600	0.447	2425	0.438
		2450	0.436
		2475	0.437
		2500	0.435

TABLE 9.10
R99 Fluorilon Reflection Data from NIST Traceable Instrument (Three Samples)

Wavelength (nm)	Calibrated (#1) R99 Refl.	Checked (#2) R99 Refl.	Checked (#3) R99 Refl.
1000	0.981	0.981	0.981
1025	0.980	0.983	0.980
1050	0.982	0.981	0.982
1075	0.982	0.982	0.981
1100	0.980	0.983	0.981
1125	0.980	0.981	0.980
1150	0.978	0.984	0.980
1175	0.980	0.980	0.982
1200	0.979	0.981	0.980
1225	0.980	0.981	0.981
1250	0.979	0.980	0.979
1275	0.979	0.981	0.980
1300	0.978	0.981	0.980
1325	0.978	0.978	0.979
1350	0.977	0.979	0.978
1375	0.976	0.979	0.978
1400	0.975	0.979	0.977
1425	0.978	0.979	0.979
1450	0.977	0.978	0.978
1475	0.977	0.980	0.977
1500	0.976	0.978	0.978
1525	0.976	0.978	0.977
1550	0.977	0.978	0.977
1575	0.977	0.979	0.977

(*Continued*)

TABLE 9.10 (*Continued*)
R99 Fluorilon Reflection Data from NIST Traceable Instrument (Three Samples)

Wavelength (nm)	Calibrated (#1) R99 Refl.	Checked (#2) R99 Refl.	Checked (#3) R99 Refl.
1600	0.977	0.979	0.978
1625	0.975	0.977	0.976
1650	0.975	0.977	0.977
1675	0.975	0.976	0.975
1700	0.974	0.975	0.976
1725	0.974	0.975	0.976
1750	0.972	0.976	0.973
1775	0.974	0.976	0.976
1800	0.975	0.976	0.976
1825	0.972	0.971	0.973
1850	0.971	0.969	0.971
1875	0.969	0.971	0.970
1900	0.967	0.970	0.969
1925	0.967	0.969	0.968
1950	0.967	0.968	0.968
1975	0.964	0.966	0.966
2000	0.962	0.965	0.964
2025	0.957	0.959	0.961
2050	0.948	0.954	0.950
2075	0.945	0.949	0.949
2100	0.941	0.948	0.944
2125	0.934	0.940	0.938
2150	0.935	0.940	0.939
2175	0.946	0.951	0.948
2200	0.954	0.958	0.956
2225	0.954	0.960	0.957
2250	0.955	0.959	0.958
2275	0.954	0.954	0.956
2300	0.948	0.950	0.952
2325	0.943	0.943	0.946
2350	0.937	0.940	0.940
2375	0.934	0.943	0.939
2400	0.929	0.935	0.935
2425	0.933	0.933	0.935
2450	0.933	0.938	0.934
2475	0.930	0.936	0.932
2500	0.939	0.931	0.937

REFERENCES

Abookasis, D., and J.J. Workman. 2011. Direct measurements of blood glucose concentration in the presence of saccharide interferences using slope and bias orthogonal signal correction and Fourier transform near-infrared spectroscopy. *Journal of Biomedical Optics*, *16*(2), p.027001.6.

American Society for Testing and Materials. 2013. *Standard Practice for Describing and Measuring Performance of Ultraviolet and Visible Spectrophotometers* (ASTM Method E275-08), ASTM International, West Conshohocken, PA.

Bland, J.M., and D. Altman. 1986. Statistical methods for assessing agreement between two methods of clinical measurement. *The Lancet*, *327*(8476), pp.307–310.
Feudale, R.N., N.A. Woody, H. Tan, A.J. Myles, S.D. Brown, and J. Ferré. 2002. Transfer of multivariate calibration models: A review. *Chemometrics and Intelligent Laboratory Systems*, *64*(2), pp.181–192.
Isaksson, T., H. Yang, G.J. Kemeny, R.S. Jackson, Q. Wang, M.K. Alam, and P.R. Griffiths. 2003. Accurate wavelength measurements of a putative standard for near-infrared diffuse reflection spectrometry. *Applied Spectroscopy*, *57*(2), pp. 176–185.
Mark, H., and J. Workman. 2010. *Chemometrics in Spectroscopy*. Elsevier, Amsterdam.
NIST SRM website. 2018. https://www.nist.gov/srm.
NIST Standard Reference Material (SRM)® 1920a Certificate. 2004. *Near Infrared Reflectance Wavelength Standard from 740nm to 2000nm*. National Institute of Standards and Technology, Gaithersburg, MD, 6 pages.
Shenk, J.S., and M.O. Westerhaus. 1989. *Optical Instrument Calibration System*. U.S. Patent 4,866,644.
Shenk, J.S., and M.O. Westerhaus. 1991. New standardization and calibration procedures for NIRS analytical systems. *Crop Science*, *31*(6), pp.1694–1696.
Spectrum and data used with permission from Avian Technologies, 2018, New London, NH.
Weidner, V.R., P.Y. Barnes, and K.L. Eckerle. 1986. A wavelength standard for the near infrared based on the reflectance of rare-earth oxides. *Journal of Research of the National Bureau of Standards*, *91*(5), p.243.
Workman, J., Jr., and A. Springsteen. 1998. *Applied Spectroscopy: A Compact Reference for Practitioners*. San Diego: Academic Press.
Workman, J., Jr., and H. Mark. 2013a. Chemometrics in spectroscopy calibration transfer, part III: The mathematical aspects. *Spectroscopy*, *28*(6), pp.28–35.
Workman, J., Jr., and H. Mark. 2013b. Calibration transfer, Part IV: Measuring the agreement between instruments following calibration transfer. *Spectroscopy*, *28*(10), pp.24–33.
Workman, J., Jr., and H. Mark. 2014a. Calibration transfer, Part V: The mathematics of wavelength standards used for spectroscopy. *Spectroscopy*, *29*(6), pp.18–27.
Workman, J., Jr., and H. Mark. 2014b. Calibration transfer, Part VI: The mathematics of photometric standards used for spectroscopy. *Spectroscopy*, *29*(11), pp.14–21.
Workman, J.J., Jr. 2016a. *Concise Handbook of Analytical Spectroscopy: The Theory, Applications, and Reference Materials* (In 5 Volumes). World Scientific. Volume 1, pp. 163–214; Volume 2, pp. 217–278; Volume 3, pp. 363–424; Volume 4, pp. 323–360; Volume 5, pp. 229–250. ISBN-13: 978-9814508056. (Permission from NIST for data and spectra related to SRM materials).
Workman, J., Jr. 2016b. First principles of instrument calibration. *NIR News*, *27*(3), pp. 12–15.

10 Standards for Near-Infrared Analysis

Emil W. Ciurczak, Doramaxx Consulting,
*and Gabor J. Kemeny**
Middleton Spectral Vision

CONTENTS

10.1 INTRODUCTION

The need for specific and standardized reference materials for near-infrared spectroscopy (NIRS) arises from both technical and legal sources:

1. Unlike "normal" (or "traditional") spectroscopic measurements, where a sample is destroyed and the analyte of interest is extracted with a solvent chosen to not interfere with the analysis (as per Beer's law), NIRS routinely determines the analyte or property *in situ* (whole corn, intact tablets, milk and butter, etc.). This demands an instrument be capable of generating reproducible, accurate, and precise spectra. This can only be accomplished by checking/calibrating the instrument quite often (daily or more often during a longer process run), using materials agreed upon in the industry and certified by a responsible agency (NIST, USP, AOAC, etc.).
2. Unlike standard or "normal" spectroscopic techniques, there are no reference samples of the materials to be analyzed: no synthetic corn, no reference pharmaceutical tablets, and so forth. This necessitates a parallel analytical method be used to confirm the material/characteristic being sought in the NIR analysis. This places more of an onus on the instrument to be more precise and accurate than a technique wherein a "known" is run in parallel and any instrument deviations from the norm are mitigated. The response of the instrument to the sample(s) being run must be shown to be as close to exact (precise and accurate) as possible, each time the samples are run.

* gabor@middletonspectral.com

3. From the legal/pedagogical standpoint, a determination of some parameter (chemical or physical) by NIRS is called a "prediction," not an "assay." An assay is defined as a test wherein a certified material is run concurrently through the same process (solvent, pathlength, and, in the case of High Performance Liquid Chromatography (HPLC), both before and after the "unknown") at the time of determination. Since there are no certified (or synthetic) samples (in almost all cases) of, for example, a pharmaceutical tablet, ear of corn, leaf of tobacco, or loaf of bread, NIRS is run differently. Representative samples are scanned by the appropriate instrument, then analyzed for the analyte (chemical or physical) via recognized compendial methods. The results of the analysis and spectra are married into a mathematical equation (via Chemometrics: see chapters within this text for more information). Since there are no outside references of the sample being analyzed, the analyst needs to be assured that the instrument is performing as it was when the equation was generated: thus, reference standards *for the instrument.*

A comprehensive set of articles demonstrating both wavelength and photometric standards for NIR (as well as ultraviolet, visible, infrared, and Raman standards) for both *x*-axis (wavelength or wavenumber) and *y*-axis (intensity) is given by reference (Workman and Mark 2019a, b)

10.2 STANDARD REFERENCE MATERIALS (SRMs)

Since NIR is used in gas, liquid, and solid samples, reference standards need to come in all three forms. In addition, the nature of NIRS is that it is mainly used for complex samples (food, textiles, medicines, detergent mixes). As a consequence, it is used in both transmission and reflection modes. The light reflected from a solid (surface) can either be peculiarly or diffusely reflected. Little information is gleaned from specular light, but there are standards available. The majority of SRMs (Standard Reference Materials), however, are based on diffuse reflection and "normal" and diffuse transmission (see Table 10.1).

Since NIRS is used widely for pharmaceutical and medical applications, government agencies such as the US Food and Drug Administration (USFDA) and the European Medicines Agency (EMA) demand much more stringent calibrations and validations than for, say, moisture in wheat by the US Department of Agriculture (USDA) or chain length of a polymer through the American Society for Testing and Materials (ASTM). As a consequence, numerous agencies (i.e., FDA, EMA, ASTM, and ICH – International Conference on Harmonisation) have all generated both "suggested" and mandatory guidelines for performing NIR predictions.

The United States Pharmacopeia (USP), for example, has established the scientific rationale and recommendations for the technologies used in pharmaceutical analysis and manufacturing. The USP is one of the guiding documents for optical instrumentation as to what performance needs to be proven. The chapter on the NIR analysis establishes that wavelength scale, linearity, repeatability/noise, and robustness of calibration have to be verified. Optical standards and their operating procedures need to support the determination of these factors.

Optical standards are also used in non-pharmaceutical areas such as in the legal system (forensic evidence in court cases), environmental enforcement, veterinary medicine, food analysis, Federal Aviation Administration (FAA) supervised manufacturing procedures, etc.…

Another aspect of standards that are used in regulatory environment is that an expiration date has to be established for all standards (Standard Reference Materials or SRMs). Some standards may be very stable and durable, while others may break down chemically or from ambient UV radiation, may get contaminated or scratched during repeated use. This is why standards need to be recertified from time to time, preferably at intervals scientifically established by the end user or by intervals recommended by the manufacturer (Springsteen 1997, 2002).

TABLE 10.1
Types of NIR Standards

Three principle documents for used wavelength calibration		
European Pharmacopeia #2.2.40: NIR Spectrophotometry	**US Pharmacopeia General Chapter 1119: NIR Spectrophotometry**	**ASTM 275-01 Standard Practice for Describing and Measuring Performance of UV, Vis, and NIR Spectrophotometers**
Transmission:	A rare earth oxide (REO) in glass transmission wavelength standard (SRM 2035), certified by NIST	High-pressure Hg arc
Methylene chloride at an optical pathlength of 1.0 mm		
Other suitable standards may be used	Holmium oxide solution NIST SRM 2034 (240–650 nm) may be used in the 650-11 nm region, though not NIST certified for that	1,2,4-Trichlorobenzene at 1 mm pathlength
		A REO in glass transmission wavelength standard, SRM 2035, has been certified by NIST
Reflection and Transflection:	A mixture of dysprosium, holmium, and erbium oxides, NIST SRM 1920a	
A mixture of dysprosium, holmium, and other certified materials may be used.		
A suspension of 1.2 g of titanium dioxide in ~4 mL of methylene chloride		

Even stable SRMs need to be checked for several reasons:

1. Moisture can adsorb on surfaces of transmission windows or reflection surfaces. In the IR or NIR, moisture bands quickly overwhelm the spectral bands of even the most stable reference materials. If there is a communication with the supplier (and they don't lose money by NOT having to recertify the materials), they might instruct you how to repair the materials without harming them. (Although, if the analyst does regenerate the standards, it would be recommended that every several years, the SRMs be sent for recertification.)
2. There are some SRMs that are sealed and have an inert atmosphere (USP NIR standard, for example) for protection. The matrix for the REO standards is finely ground Teflon®, while not susceptible to moisture damage, it, also, could adsorb water (or organic) vapors and affect the NIR spectrum (Figure 10.1 and Table 10.3). (Again, even a sealed standard should be checked every few years, as even a "sealed" cell can leak, and AR is not visible. And, if the noble gas can escape, ambient air can enter.)

10.3 RECERTIFICATION REQUIREMENTS

There are a number of "reference" materials for NIRS (see Table 10.1). In most cases, they are inorganic and stable materials, such as rare earth oxides or aluminum oxide disks. Fortunately, some are usable in different wavelength regions (e.g., ultraviolet) and, in addition to saving the expense of duplicate references, it can be cross-checked in more than one range for stability and cleanliness.

It is advisable that a majority of materials used for reference be recertified or checked on a routine basis, since other than rare earth oxides, protected by an inert atmosphere, can either deteriorate or,

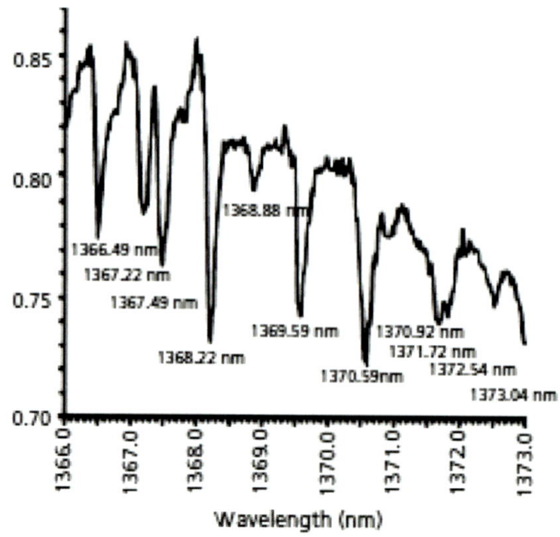

FIGURE 10.1 Water vapor spectrum (FT-NIR; 2 cm^{-1} resolution; 1366–1373 nm).

at a minimum, gather adsorbed moisture or organic vapors which can cause spurious peaks or cover needed reference peaks.

Something as simple as a porcelain (aluminum silicate) disk reference, extensively used in the 1970s and 1980s as a "blank" or "100% reference" (although it seldom reflected more than 90% of any wavelength) for diffuse reflection work, can easily attract moisture. The resultant scan of NIR radiation slowly dries the surface, such that multiple, so-added scans can vary in spectral integrity. (As an example, using a slightly moist porcelain disk for reference can give the appearance of "noise" as the water evaporates since most "noise" measurement routines merely report the peak-to-peak levels or repeat scans. As the water evaporates, the "noise" grows. Co-adding more scans would merely show and increase in "noise," further confusing matters.)

10.4 USE OF STANDARD REFERENCE MATERIALS (SRMs)

Periodic testing (i.e., yearly, monthly) is spelled out in the company operating procedures (SOP's) supplier's manuals. Some instrument parameters need daily testing, such as noise level and wavelength accuracy, and are easily performed, often with algorithms and standards built into the software and hardware, and need only a few keystrokes to initiate. Others, such as linearity and noise at high and low flux, need operator intervention (e.g., placing Teflon® powder, laced with various levels of carbon black in the light path for a linearity check). These are seldom performed more than once per instrument use (assuming that the instrument will not be used some days).

The actual scientific methods used for calibration are based on "traceable" materials. That is, they have been certified by a recognized authority, in the case of wavelength standards (see Tables 10.1 and 10.4). Traceability of materials is critical (similar to "chain of custody" in law enforcement evidence). A recognized source, whether it is a reputable manufacturer of chemicals or an agency, is tasked with certification (e.g., The National Institute of Standards and Technology or N.I.S.T., formerly the National Bureau of Standards); also, there needs to be a "paper trail" or certification (Certificate of Analysis or CoA) as to the identity and purity of any standard.

The International Standards Organization (ISO) issued a set of criteria to accredit laboratories and organizations performing the recertifications of standards. ISO 17025 and its newest 2017 release establishes the requirements to become a certified laboratory or to be able to follow the ISO 17,025 guidelines.

TABLE 10.2
Some Standards for Various Techniques

Type of Standards	UV–Visible Range	Near-Infrared Range	Mid-Infrared Range
Wavelength standards	Holmium oxide solution Holmium oxide glass	Rare earth oxide mixture Glass doped with rare earth oxides Thick polystyrene window	Polystyrene film
Linearity standards	Solution set Thin film set	Sintered PTFE with carbon black Powder mixture in cups with windows	NG-11 Glass
Dark reference standards	Light exclusion	Dark paint or material Electrolytically deposited black Deep cup	Light exclusion
Specular reflectance standards	Polished steel	Polished germanium disk	Polished germanium disk
Attenuated total reflectance standards	N/A	N/A	Flexible silicone element Polystyrene film or thick polystyrene disk

TABLE 10.3
Select Absorption Bands of Water in the Vapor Phase (*in vacuo* in cm^{-1}, Corrected for Refractive Index in Nanometers)

Wavenumber *in vacuo* cm^{-1}	Corrected for Refractive Index nm
7281.0812	1373.04
7283.7311	1372.54
7288.0910	1371.72
7292.3739	1370.92
7294.1231	1370.59
7299.4311	1369.59
7306.7398	1368.22
7310.6779	1367.49
7312.1974	1367.20
7315.7091	1366.55
7315.9877	1366.49

Some of the optical standards used in UV–visible, NIR, and mid-infrared wavelength ranges are shown in Table 10.2. Standards used for other techniques, Raman, fluorescence, and other optical measurements are beyond the scope of this chapter.

10.4.1 Wavelength Standards

The wavelength scale is one of the most important spectroscopic parameters for performing spectral measurements, both qualitative and quantitative. Instruments using different optical designs will

TABLE 10.4
Some Wavelength Standards for NIR Calibration

Standard	Type	Mode	Range(nm)/#Peaks
Mercury (atomic)	Discharge tube	Transmission	670–2325 (19)
Acetylene	Gas SRM 2517)	T	1510–1545 (>50)
Water vapor	Gas	T	1355–1373 (11)
Hydrogen cyanide	Gas (SRM 2519)	T	1525–1665 (>50)
REO glass	Solid (SRM 2035)	T	976–1946 (7)
REO glass	Solid (SRM 2065)	Reflection	976–1946 (7)
YAG-doped crystal	Solid (Nelson M-42)	T	1000–1645 (>50)
REO mix	Solid (SRM 1920)	R	740–2000 (37)
REO glass	Solid (SRM 2036)	R	976–1846 (7)
1,2,4-Trichlorobenzene	Liquid (1-mm pathlength)	T	1660–2543 (7)
Methylene chloride	Liquid (1-mm pathlength)	T	1155–2245 (8)
Methylene chloride	Liquid (w/TiO_2)	T	1155–2245 (8)

have to produce the same scale at different times, independent of environmental conditions. Some instruments, such as Fourier transform instruments, are inherently more (wavelength/wavenumber) accurate because their wavelength scale is tied to very accurate laser frequencies; others are tied to opto-mechanical accuracy and linearity of reflective gratings, and therefore, frequent, periodic wavelength calibration is more important.

10.4.2 Near-Infrared Reflection Standards

For NIR reflection, a very stable blend of rare earth oxides, erbium oxide (Er_2O_3), dysprosium oxide (Dy_2O_3), and holmium oxide (Ho_2O_3), was the original NIST (National Institute of Standards and Technology) standard. Sealed in a metal cup behind an optical window, the diffuse reflection spectrum has sharp and very stable absorptions due to the long-wave electronic transitions of the large electron clouds of the rare earth oxides. In an improved version of the same standard, an additional stable mineral, talc, was added in equal quantity to extend the wavelength coverage to 2500 nm, usually available on a commercial NIR instrumentation. Talc added four well-determined sharp bands in the regions void of rare earth oxide bands.

The band centers were determined in a multilaboratory Fourier transform infrared study (Griffiths et al. 2003) even more accurately than using the original dispersive instruments of the Bureau of National Standards (Weidner et al. 1986). The reflectance spectrum and comparison of the band center are shown in Figure 10.1 and Table 10.2, respectively.

SRM-2035x (where $x=a$, b, etc.) "UV-Vis-NIR Wavelength/Wavenumber Transmission Standard" (Figures 10.2 and 10.3). This SRM supersedes SRMs 2035 and 2065. It is a certified transfer standard intended for the verification and calibration of the wavelength/wavenumber scale of ultraviolet (UV)–visible–NIR spectrometers operating in transmission mode. SRM 2035x is certified for the location of the seven absorbance bands in the spectral region from 10,300 to 5 130 cm^{-1} at 8 cm^{-1} resolution in vacuum wavenumber. The same seven bands in the spectral region from 970 to 1946 nm are certified in wavelength (air) at 1 and 3 nm spectral bandwidth. In addition, 13 transmittance peaks spanning the spectral region from 334 to 805 nm are certified for peak location at 1 nm and 3 nm spectral bandwidth. The optical filter is 25 mm in diameter and 1.5 mm thick. The combination of constituent rare earth oxide concentrations and filter thickness yields absorption bands between 0.1 and 0.6 absorbance units in the NIR and 5%–60% transmittance in the UV–visible.

Another reflectance wavelength reference is the NIST standard, SRM 2036 (Choquette et al., 2005). The reference is a thin holmium oxide-containing glass, mounted on the surface of an ideal

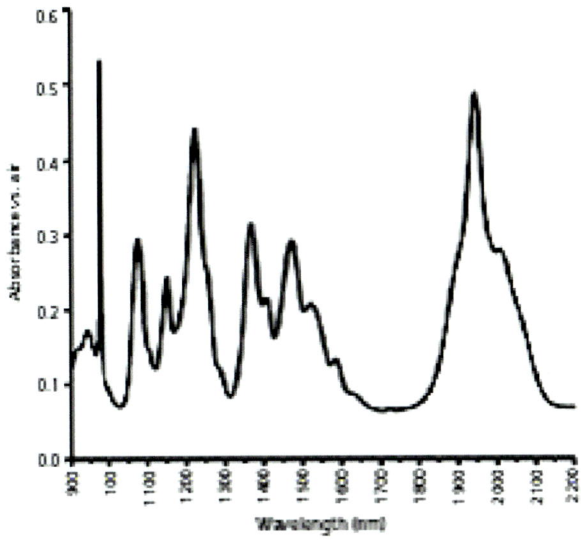

FIGURE 10.2 SRM 2035 FT-NIR spectrum at 8 cm^{-1} resolution.

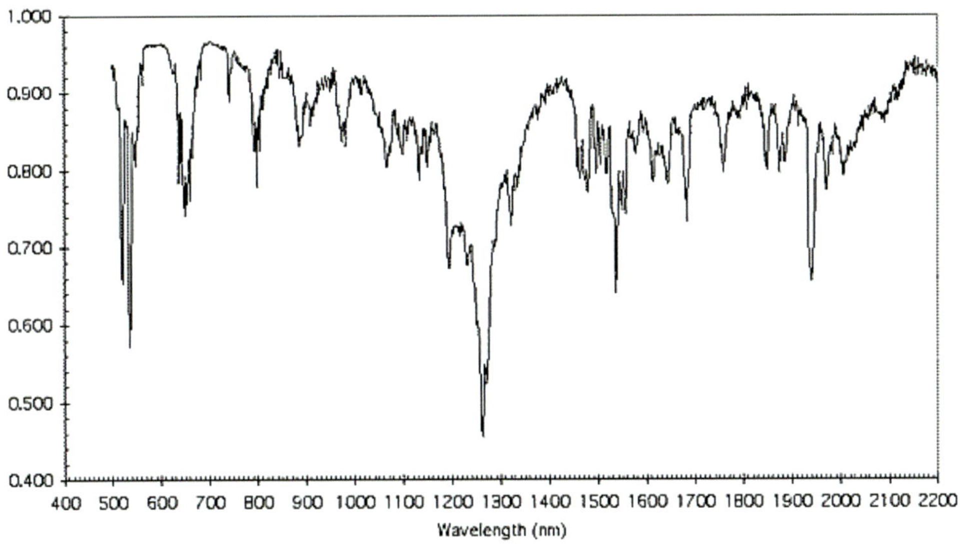

FIGURE 10.3 Transmission spectrum of NIST SRM 2035 in wavelength space (nm).

diffuse reflectance disk, made of sintered polytetrafluoroethylene (PTFE). In this case, the illuminating light transmits through the glass, is reflected by the diffuse scattering material, and transmits again through the glass. The band centers of the double transmission spectra were established using the centroid method.

10.4.2.1 Holmium Oxide

One of the most studied wavelength standard materials is the holmium oxide glass (David 2007). Over the four decades it was used, there have been no recorded changes in the certified bands, even in the face of humidity and temperature. D. W. Allen of NIST concluded that based on the extensive experience of robust behavior of holmium oxide, NIST no longer recommends recertification of these standards.

10.4.2.2 Linearity Standards

Since NIR spectra, due to massive spectral overlaps and hydrogen-bond shifts (as the temperature and concentrations of the materials change), can vary, it is imperative that the instrument *not* vary outside of very tightly defined parameters. (Remember that, in NIRS, we use predictions, not assays, so, without reference materials (e.g., synthetic corn), the instrument is the source of reproducibility and needs to be tightly controlled). The most common too; for rapid linearity checks on the instrument is a series of Spectralon (a non-absorbing matrix, basically Teflon powder) samples with varying amounts of powdered carbon. These "known" standards are scanned and the Absorbances graphed. Linearity is either demonstrated or not.

10.4.2.3 Specular Reflectance Standards

Specular reflection standards are used with special optical experiments and measurements as well as specular reflectance accessories (Zwinkels et al. 1994). In the NIR, materials such as polished stainless steel or other mirror-like artifacts can be used. These need to be validated with an external material with known reflectance. One standard relies on the absolute reflectance of the material/element germanium. The highly polished top surface of the germanium has a known specular reflectance controlled by the Fresnel equations that is only dependent on the very well-known index of refraction of the material and the angle of incidence. The bottom surface of the material, which is otherwise transparent to infrared radiation, is treated to fully absorb the transmitted component. This way the apparent reflectance is only controlled by the front surface and can be calculated from the independent and very accurately determined index of refraction of germanium. Other available standards usually rely on the degree of polish and the material the reflective standard, like stainless steel, is made from.

With these many factors, for standardization, most parameters need to be kept the same, for example, the type of crystals, such as zinc selenide (ZnSe), diamond, germanium (Ge), silicon (Si) have to be specified, the angle of incidence kept the same, even the temperature that influences the air-crystal index difference has to be the same (Icenogle et al., 1976). Due to the hard-to-control conditions of the measurements, standards organizations so far stayed away from issuing absolute measurements, so only a few standards are available to be used with this important type of optical measurement.

On the other hand, testing of the proper functioning of the ATR accessory is very important because the crystal is fragile and sensitive, and the crystal part can get dislodged or delaminated. Even if different instruments and accessories are different, the out-of-spec condition of the accessory can be detected with known and reproducible samples and reproducible sampling techniques. Once a baseline measurement has been established, unless the alignment, the crystal type has been changed, the spectral measurements are the same within established tolerance levels. Several materials have been used for this purpose, such as thick polystyrene, polystyrene films, mineral oil, and flexible silicone-based samples

REFERENCES

Choquette, S.J., Duewer, D.L., Hanssen, L.M., Early, E.E., SRM 2036 a reflection wavelength standard, *Applied Spectroscopy*, **59** (4), pp. 496–504 (2005).

David W.A., Holmium Oxide Glass Wavelength Standards, *Journal of Research of the National Institute of Standards and Technology* **112**, 303–306 (2007).

Isaksson, T., Yang, H., Kemeny, G.J., Jackson, R.S., Wang, Q., Alam, M.K., Griffiths, P.R., Accurate Wavelength Measurements of a Putative Standard for Near-Infrared Diffuse Reflection Spectrometry, *Applied Spectroscopy*, **57**(2), pp. 176–185 (2003).

Icenogle, H.W., Platt, B.C., Wolfe, W.L., Refractive indexes and temperature coefficients of germanium and silicon, *Applied Optics*, **15** (10), pp. 2348–2351 (1976).

Springsteen, A., "Photometric standards for near-infrared spectrometry" in *Handbook of Vibrational Spectroscopy*, J. Chalmers and P. Griffiths, eds. John Wiley & Sons, Chichester (2002).

Springsteen, A., Techniques for the measurement of specular reflectance, *Spectroscopy Magazine* **12** (8), 24–28 (1997).
Standard Practice for Describing and Measuring Performance of Ultraviolet, Visible, and Near-Infrared Spectrophotometers, ASTM E 275-93 (1993), Annual Book of ASTM Standards, 03.06, 708–718 (1997), West Conshohocken, PA, 10.
Weidner, V.R., Barnes, P.Y., Eckerle, K.L., A wavelength standard for the near infrared based on the reflectance of rare-earth oxides, *Journal of Research of the National Institute of Standards and Technology*, **91**, 243 (1986).
Workman Jr., J., Mark, H., Using Reference Materials, Part I: Standards for Aligning the X-Axis, *Spectroscopy*, **34**(2), 36 (2019a).
Workman Jr., J., Mark, H., Using Reference Materials, Part II: Photometric Standards, *Spectroscopy*, **34**(10), 18–28 (2019b).
Zwinkels, J.C., Noël, M., Dodd, C.X., Procedures and standards for accurate spectrophotometric measurements of specular reflectance, *Applied Optics* **33**, 7933–7944 (1994).

Part III

Modeling/Chemometrics

CALIBRATE – TRANSITIVE VERB

Meanings:

1. To determine, rectify, or mark the graduations of (something, such as a thermometer tube)
2. To standardize (something, such as a measuring instrument) by determining the deviation from a standard so as to ascertain the proper correction factor
3. (see https://www.merriam-webster.com/dictionary/calibrate)

Other meanings include

4. To check, adjust, or determine by comparison with a standard, the gradations of a quantitative measuring instrument: e.g., to calibrate a thermometer
   (see https://www.thefreedictionary.com/calibrate).

Thus, if we want to calibrate a thermometer, we need to have a way to provide an appropriate signal (temperature, in this case) with an accurately known value or values. As shown in Figure 1, we can measure that signal with the thermometer to be calibrated and know that a certain reading corresponds to that temperature. Figure 1 shows a stylized thermometer, which can be placed in an ice bath. When the reading stabilizes, we mark the thermometer such that the end of the liquid column corresponds to 0°. We repeat that process for at least one more (different) value of the temperature, e.g., a steam bath, and mark that as 100°

In olden days (predigital age), we would write down the instrument readings corresponding to the two temperatures on graph paper and draw a line between the points. That constituted the "calibration." To measure an unknown temperature, we would couple our thermometer to the sample whose temperature is to be measured, get a reading, and use our calibration graph, interpolating (or extrapolating, depending on how confident we are about the linearity of the measurement process) to determine what the temperature of the sample is.

In the digital age, we no longer need write down the readings. The concept is the same, but the details are all done automatically, and a computer calculates the slope and intercept of the graph, which constitutes the "calibration," which is now represented by an equation:

$$\text{Temperature} = \text{Slope} \times (\text{reading}) + \text{Intercept}.$$

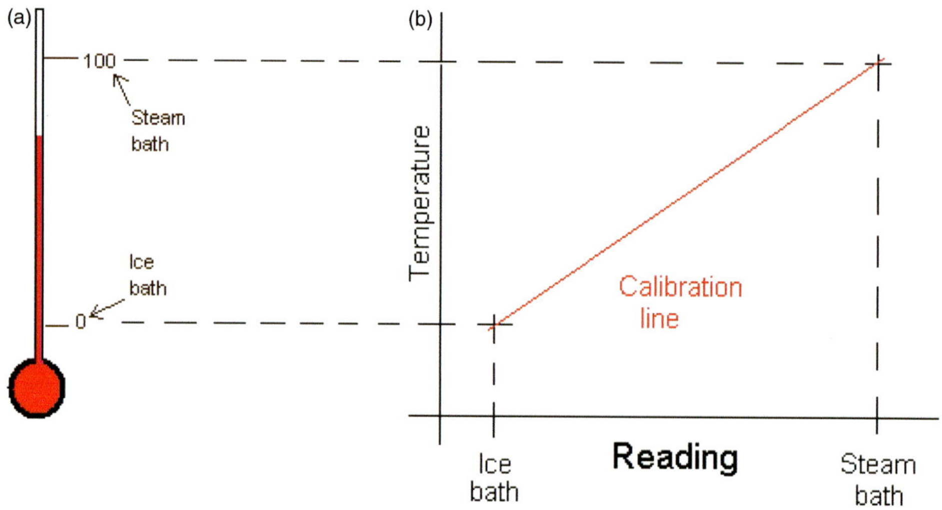

FIGURE 1 (a) A stylized thermometer, showing marks for temperatures corresponding to ice water (0°C) and steam (100°C); (b) A graph of the calibration shows the equivalence between thermometer readings and the actual temperature.

To measure an unknown temperature, the thermometer reading is plugged into the equation and the value of the temperature is computed.

The same thing is done to calibrate NIR spectrometers, but, in addition to the need for multiple NIR readings (at different wavelengths) for each sample, we have a problem. For most types of samples, and especially when dealing with natural products, by definition, we cannot create artificial samples with controlled and accurately known values of the desired analyte, to use to calibrate the instrument. We cannot, e.g., create a leaf, or piece of beef with known amounts of moisture or protein, or a soybean with controlled amounts of protein and oil, or make any of them have the same optical properties as an actual leaf, piece of beef, or soybean.

So what do we do? Since we cannot create samples to order, we have to use actual samples and determine the amount of our analyte in them by some other means. What are those "other means?" Generally, those "other means" are the analytical methods currently used to analyze their samples; very often this is the method that the NIR analysis will be replacing. The values produced by this alternate analytical method are the values that will be used along with the NIR readings, to produce the calibration.

In the context of NIR analysis, these alternate analytical methods are generally called "reference methods" and the values they produce are called "reference values." The need to use alternate methods, producing reference values to calibrate instruments, introduces difficulties of their own. Not the least of these is that the reference values are generally not nearly as accurately known as, say, the temperature used to calibrate a thermometer. Depending on the sample and the nature of the analyte, reference methods vary in quality from highly accurate analysis produced on a strictly scientific basis to the equivalent of "throwing the sample at a wall and seeing how long it sticks."

The overall process for calibrating an NIR instrument follows the same paradigm as for calibrating a thermometer (outlined in Figure 2, a simplified presentation of the calibration process and use of NIR spectrometers). Much of the rest of this book deals with the details of how it all is accomplished, problems you are liable to run into when trying to implement this very basis scheme, and recommendations for how you can deal with those difficulties.

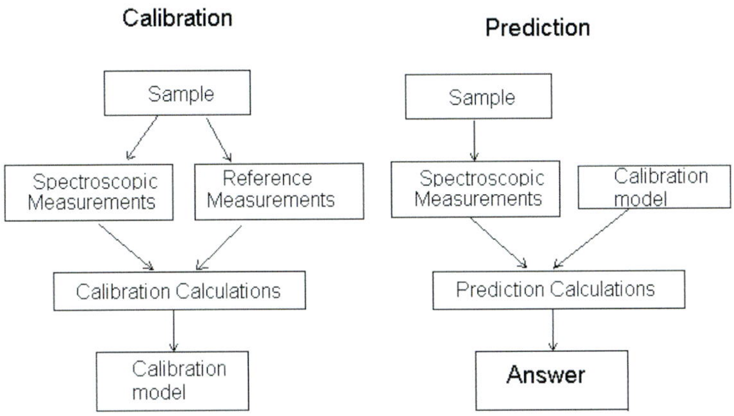

FIGURE 2 A simplified block diagram showing data flow through the calibration and analysis (prediction) processes.

Thus, to use NIR for performing analysis, you need to follow this procedure, summarized in Figure 2 which gives the "Reader's Digest condensed version":

1. Collect samples
2. Get reference values for the analyte
3. Measure the spectra of the samples
4. Perform the calibration calculations
5. Select a model for routine analysis
6. Validate that model

and finally:

7. Use the model in conjunction with the instrument to perform routine analysis

The rest of this book gives the details for how to do all that.

11 Pattern Recognition Applied to the Classification of Near-Infrared Spectra

Barry K. Lavine and Francis Kwofie
Oklahoma State University

CONTENTS

11.1 INTRODUCTION

Many relationships in spectral data cannot be expressed in quantitative terms. These relationships are better expressed in terms of categories or classes of samples which are correlated with specific physical and chemical properties of the samples. Just as in a calibration, this problem can be divided into two steps. First, the known samples are combined to separate information from noise in the data and to build a predictive model (that is also known as a discriminant). Second, the predictive model is used to determine the class membership of new samples. To develop a relationship suitable for identifying and isolating classes from multivariate spectral data, analytical chemists have turned to pattern recognition methods, which are a collection of methods (mathematical, statistical, and numerical) that have been developed to categorize samples on the basis of observed regularities (also known as patterns) in data (Kowalski and Bender 1972, Tou and Gonzalez 1974, Kowalski and Wold 1982).

In pattern recognition studies involving near-infrared (NIR) spectral data, samples are classified according to a specific property (which is difficult to measure or cannot be measured directly) using spectral measurements that are indirectly related to the property of interest. A discriminant or classification rule is developed from a set of samples for which the property of interest and the measurements are known. The set of samples is called the training set. The classification rule is used to predict the desired property in samples that are not part of the training set. The property of interest may be the geographic origin of raw materials used to formulate a pharmaceutical tablet, and the measurements are the absorbances at specific wavelengths directly obtained from a diffuse reflectance NIR spectrum of the tablet (Rostaing et al. 1988). The concept of an indirect relation between a spectrum and the property of a substance—first proposed by Hirschfeld and Martens (Martens 1983, Honigs et al. 1984, and Martens and Martens 1986)—is feasible because the physical and chemical properties of many substances (such as pharmaceutical tablets) can be determined by their chemical composition, which is often captured by their NIR spectra.

In this chapter, pattern recognition methods routinely used in NIR spectroscopy are discussed in the context of analyzing large NIR data sets generated in monitoring and profiling studies. The data preprocessing steps that must be implemented in order to apply these techniques to the real-world problems are also enumerated. The foci of the discussion in this chapter are principal components analysis (Wold et al. 1987, Jackson 1991), classification methods utilizing the Mahalanobis distance, and multivariate curve resolution (MCR) as these techniques are the most frequently cited in the published NIR literature. Examples of chemical analysis problems that employ NIR spectroscopy and pattern recognition techniques discussed in this chapter (as part of a brief survey of the literature) include the detection of adulterants in food, discrimination of plant genotypes, authentication of excipient and active ingredients in pharmaceutical tablets, polymorphism, and geographic and varietal origins of raw materials. Applications of NIR spectroscopy and pattern recognition methods to problems in pharmaceutical validation have been previously reviewed (Raggo et al. 2007). In the preparation of this chapter, an attempt has been made to strike a balance between applications and fundamentals.

11.2 PRINCIPAL COMPONENT ANALYSIS

Principal component analysis (PCA) is the best known of the unsupervised techniques in multivariate data analysis. The overall goal of PCA is dimensionality reduction, while retaining the information that is present in the original data. Dimensionality reduction is possible with PCA because measurement variables are usually correlated. Consider a set of samples (Mandel 1985) that are characterized by two measurement variables, X_l and X_2 (see Figure 11.1). X_l appears to be correlated with X_2 because fixing the values of X_l limits the range of values for X_2 in the vector space defined by these two variables. If X_1 and X_2 were uncorrelated, the entire rectangle shown in Figure 11.1 would be fully populated by data points. Correlations between measurement variables restrict the data to a smaller region of the measurement space. When the measurement variables are highly correlated, the data will reside in a subspace.

Correlations between the measurement variables are also an indication that a new set of basis vectors exist that are better at conveying the information present in the data than axes defined by the original measurement variables. This new basis set, which is linked to a variation, is used to develop a new coordinate system for displaying the data (see Figure 11.2). The variance-based axes of the new coordinate system are the principal components of the data. Each principal component is a linear combination of the original measurement variables. The first or the largest principal component is formed by determining the direction of maximum variance and modeling it by fitting a line that passes through the center of the data using the technique of linear least squares. The second largest principal component also passes through the center of the data and defines the direction of

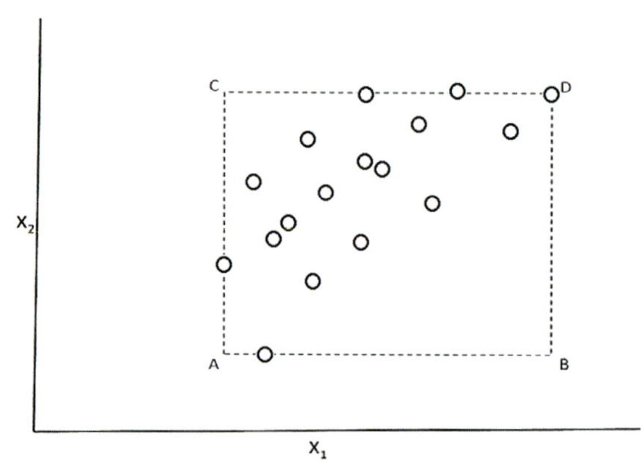

FIGURE 11.1 Seventeen samples plotted in a two-dimensional space defined by the measurement variables X_1 and X_2. A, B, C, and D represent the vertices of a rectangle that represents the smallest and largest values of X_1 and X_2. (Adapted from J. Mandel, "The regression analysis of collinear data," *J. Res. NBS*. 90(6), 465–476, and printed 1985, National Bureau of Standards.)

the next largest variance. The third largest principal component again passes through the center of the data and lies in the direction of the next largest variance. All principal components are mutually orthogonal (Figure 11.2). This constraint ensures that each principal component describes a different source of information (i.e., variance) in the original data.

In PCA, an eigenvalue decomposition of the data matrix **X** ($n \times p$) is performed after the data is mean-centered or auto-scaled. Each principal component is associated with an eigenvalue (λ), which reveals the degree of variation in the data captured by the principal component. The first principal component of the data has the largest eigenvalue followed by the second principal component, which has the second largest eigenvalue and so forth. PCA decomposes **X** ($n \times p$) into a score matrix **T** ($n \times f$), loading matrix **P** ($f \times p$), and residual matrix **E** ($n \times p$), where n is the number of spectra in the data set, p is the number of points or features in each spectrum, and f is the number of principal components necessary to represent the spectral data. Typically, f is smaller than p due to correlations among the measurement variables. The decomposition of **X** into the score, loading, and residual matrices is shown in Equation 11.1, where 1 is a column vector ($n \times 1$) of ones and $\mathbf{x}_{mean}$ is a ($1 \times p$) row vector representing the mean or centroid of the data.

$$X = 1x_{mean} + TP' + E \tag{11.1}$$

The relationships between the original measurement variables (wave numbers) and the principal components of the data are provided by the loading matrix. The coordinates of the samples in the principal components space are defined by the score matrix. Both the score matrix and the loading matrix describe the signal in the data. The residual matrix represents the noise in the data. By plotting the columns of the score matrix against each other, a representation of the distribution of the NIR spectra in the p-dimensional multivariate space can be obtained. The number of principal components used to represent the spectral data is equal to the number of columns in the score matrix. Often, most of the variance in a data set can be represented by the two or three largest principal components of the data because of correlations among the measurements variables. By analyzing a data set using principal component (score) plots, it is possible to find relationships in the data, e.g., detect similarities and dissimilarities among groups of samples, and identify the presence of outliers (discordant observations) in the data.

PCA is often used as a mapping and display technique for exploratory analysis to visualize relationships among samples in a data set. Dimensionality reduction is achieved by transforming the original measurement variables of the data matrix (i.e., the columns of the matrix) into principal

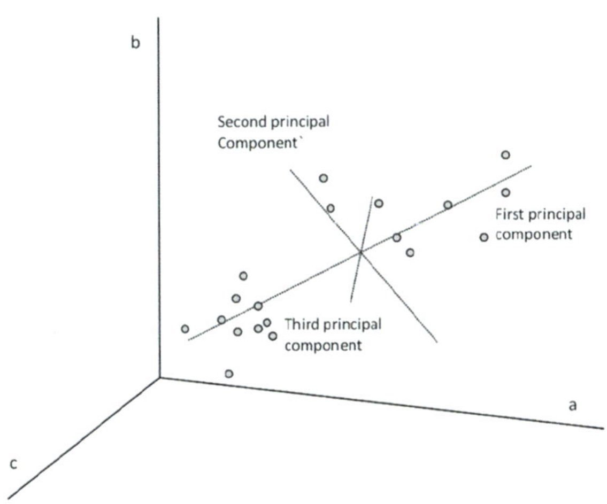

FIGURE 11.2 Principal components developed from the measurement variables: a, b, and c. (Courtesy of S. D., Brown. 1995. Chemical systems under indirect observation: Latent properties and chemometrics. *Appl. Spec.* 49, 14A–30A).

components. Examples of PCA applied to NIR spectral data include the detection of adulterants in food (Ellis et al. 2012), classification of wine and cheese (Cen and He 2007), discrimination of cereals such as wheat and barley from different geographical origins (Cozzolino 2014), and tablet identification for pharmaceutical quality control (Alvarenga et al. 2008). Preprocessing of the NIR spectra prior to the application of PCA using either multiplicative scatter correction (Isaksson and Naes 1988) or signal normal variate (Barnes et al. 1989) is crucial for minimizing baseline shifts (including offsets) and scattering effects that otherwise would prevent the detection of sample clustering in a PC score plot of the data, whereas the use of derivatives improves the resolution of overlapping bands by enhancing the importance of peaks in the spectra.

11.3 CLASSIFICATION

Pattern recognition techniques were originally developed to solve the class membership problem. In this section, classification methods, which utilize information about class membership to categorize the spectra in the training set, are presented. The discussion is restricted to discriminant analysis (linear discriminant analysis (LDA) and quadratic discriminant analysis (QDA), SIMCA, and partial least squares (PLS) discriminant analysis) as these techniques are the most popular. Each NIR spectrum is treated as a data vector $x=(x_1, x_2, x_3,\ldots x_j, \ldots x_p)$, where component x_j is the absorbance of the jth wavelength. A spectrum can be considered as a point in a high-dimensional measurement space. A basic assumption inherent in all pattern recognition studies is that distances between samples (i.e., points) in the measurement (pattern) space are inversely related to their degree of similarity. Thus, points representing samples from one class will tend to cluster in the same region of this space, while samples from the other class will cluster elsewhere. Pattern recognition methods seek to optimize a discriminant that will separate data possessing this type of structure. The class membership of a sample is ascertained by examining the class labels of the data points nearest to it or by determining the identity of the class centroid which is closest to the sample in the measurement space.

Discriminant analysis is frequently used by workers as it can provide insights into relationships contained within sets of spectral measurements used to develop a spectrum–property relationship. LDA generates a classification rule using the mean vector and covariance matrix of each class in the data set (James 1985). LDA is probably the best known example of a minimum distance classifier. A sample is assigned to the class with the lowest discriminant score, see Equation 11.2, where $d_k(x)$ is the discriminant score for class k, x is the data vector representing the sample, and C_p^{-1} is the inverse of the pooled or average class covariance matrix of the data. Since LDA assumes that all class covariance matrices are equal, the classification rule is a Mahalanobis distance, which is similar to a squared Euclidean distance with a correction term added, which is the inverse of the covariance matrix, to take into account correlations among the measurement variables when computing the distance between a sample and each class centroid.

$$\min\left[d_k(x)=(x-m_k)^T C_k^{-1}(x-m_k)\right] \tag{11.2}$$

An obvious drawback of LDA is that it does not take into account differences in the degree to which the samples comprising each class are dispersed in the high-dimensional measurement space as equal-class covariance matrices are assumed. Samples comprising one class may be more dispersed in the measurement space than samples comprising the other class. For the dispersed class, a relatively large distance between a sample and the class mean is not as significant as in the case of a compact class. In these circumstances, QDA, which does not assume equal-class covariance matrices, is preferred (Lachenbruch 1975, McLachlan 1992). The classification rule for QDA is given by Equation 11.3, where C_k^{-1} is the inverse of the covariance matrix for each class and $\ln|C_k|$ is the determinant of the class covariance matrix, which corresponds to the volume of space occupied by the sample data points representing class k.

$$\min\left[d_k(x) = (x - m_k)^T C_k^{-1} (x - m_k)\right] + \ln|C_k| \tag{11.3}$$

Examples of LDA and QDA applied to NIR spectral data include the determination of the origin of fishmeal using different species of fish for the purposes of quality control (Cozzolino et al. 2005), contamination of fishmeal by meat and bone meal (Murray et al. 2001), evaluation of the quality of wood for the construction of bows in stringed instruments (Casale et al. 2010), adulteration of minced beef by turkey meat (Alamprese et al. 2013), prediction of the origin of ciprofloxacin tablets as part of a surveillance program to detect counterfeit drugs (Fuenffinger et al. 2017), and the adulteration of orange juice (Twomey et al. 1995). In some studies, LDA and QDA were performed using the scores of the largest principal components of the data, whereas in other studies variables were selected based on their individual relevance as determined by metrics like Fisher weights to identify a small subset of wavelengths for QDA or LDA. Preprocessing of the NIR spectra in these studies was undertaken to reduce a baseline variation and eliminate scattering effects using a variety of methods, including a Savitzky–Golay first- or second-derivative filter with a first-order or second-order polynomial for smoothing followed by either the standard normal variate (with or without detrending) or multiplicative scatter correction (standard, weighted, and inverse) and then mean centering or auto-scaling of the data.

For LDA and QDA, the number of samples in the training set should be at least five times greater than the number of variables in order to develop a reliable estimate of the inverse of the covariance matrix. When the sample-to-variable ratio (or object-to-descriptor ratio) is less than five, LDA or QDA has been successfully performed using the scores of the nonzero principal components as variables. An alternative approach is to develop a better estimate of the inverse of the covariance matrix using a method like SIMCA (Wold 1976, Wold and Sjostrom 1977). For each class in the data set, the inverse of the covariance matrix is approximated by the so-called secondary principal components (eigenvectors) of the data, i.e., those with smaller eigenvalues. The inverse of the covariance matrix for each class is computed using spectral decomposition, see Equation 11.4, where ν_{jk} is the jth principal component of the inverse of the class k covariance matrix C_k^{-1}, λ_{jk} is the eigenvalue corresponding to the jth principal component, and p is the dimensionality of the data. As it is the smaller eigenvalues (not the larger ones) that are important for reconstructing C_k^{-1}, developing better estimates of the secondary eigenvalues is crucial. By taking the average of the smaller eigenvalues, more reliable estimates of them and the inverse of the covariance matrix for each class can be obtained, see Equation 11.5, where A is equal to the number of principal components used to describe each class in the data set (Frank and Friedman 1989).

The value of A for each class is directly determined from the data by cross-validation (Wold 1978). For cross-validation, the data from each class is divided into training set/prediction set pairs. For each training set/prediction set pair, segments of the data are omitted during PCA. The omitted data (prediction set) is predicted using the principal component models developed from the remaining data (training set) for one, two, three, etc. principal components. The predicted data is then compared to the actual data. This procedure is repeated until every data element has been omitted once in this round robin. The number of principal components for each class that yields the minimum prediction error for the omitted data is retained and is equal to the value of A.

$$C_k^{-1} = \sum_{j=1}^{p} \left[\nu_{jk} \nu_{jk}^T\right] / \lambda_{jk} \tag{11.4}$$

$$C_k^{-1} = \left[\sum_{j=A+1}^{p} \left[\nu_{jk} \nu_{jk}^T\right] / \sum_{J=A+1}^{p} \lambda_{jk}\right] \tag{11.5}$$

The classification rule for SIMCA is similar to the rule used in QDA (see Equation 11.3) except that a stabilized inverse of the class covariance matrix is substituted for C_k^{-1} and the determinant of the class covariance matrix is omitted. Examples of applications of SIMCA to NIR spectroscopy include the adulteration of camellia oil by soybean oil (Wang et al. 2006), detection of counterfeit pharmaceuticals (Paris et al. 2010), and identification of pharmaceutical excipients (Candolfi et al. 1999). In these three studies, data preprocessing was performed using the signal normal variate and the first- or second-derivative function to minimize baseline effects followed by auto-scaling of the data.

In recent years, PLS discriminant analysis (Brereton 2009, Stahle and Wold 1987) has been used as an alternative to SIMCA for the classification of NIR spectra. PLS is similar to PCA except that the eigenvector projection of the data is implemented using the Y-block (sample class membership) as well as the X-block (spectral measurements). Each PLS component is a linear combination of the original measurement variables that are rotated to ensure a better correlation with the class label of the samples comprising the training set. As in the case of PLS regression, a PLS discriminant can be used to predict the class membership of samples not part of the original training set. According to Rayens (Barker and Rayens 2003), the success of discriminant PLS to capture the class structure of the data is due to PLS being similar to LDA. In fact, if all nonzero PLS components are retained for discriminant analysis, the two methods are identical. As pointed out by Brereton and Lloyd (2014), when there are many more variables than samples, the results from PLS can be misleading. Brereton has provided the following example to demonstrate this deficiency, which is summarized here. Consider a data set consisting of 40 samples with each sample described by 200 random numbers generated using a uniform distribution. Twenty samples are assigned to class 1, and the other 20 samples are assigned to class 2. As there are no significant differences between these two sets of samples, it should not come as a surprise that no particular distinction exists between these two classes in a plot of the two largest principal components of the data. However, a plot of the two largest PLS components of this data would show an excellent separation between these two classes. Discriminant PLS can exploit chance correlations (which occur due to the large number of variables and the small number of samples comprising the training set) when developing the loadings for the more significant PLS components. According to Brereton, PLS discriminant analysis should be regarded as a first step in a full classification procedure as it can provide an insight into variables useful for classification through the weights and loadings computed by the PLS algorithm.

11.4 MULTIVARIATE CURVE RESOLUTION

The goal of MCR is to perform a mathematical decomposition of NIR spectral data of a mixture into its contributions by each component of the mixture (de Juan and Tauler 2000, Malik et al. 2015). Data obtained from an image or from a process monitored as a function of time is arranged in the form of a data matrix D ($n \times p$), where the n rows represent each spectrum recorded and the p columns represent the absorbance value of each wavelength comprising the spectra. MCR decomposition of D is performed using Equation 11.6, where each row of C is the concentration profile of a constituent, each column of S^T represents the spectrum of each constituent, and each row of E (residual matrix) describes the error associated with the data not explained by the product of C and S^T (Garrido, Rius et al. 2008).

$$D = CS^T + E \tag{11.6}$$

The matrix, D, can be decomposed using noniterative or iterative MCR methods. Examples of noniterative methods are window factor analysis (Malinowski 1992), heuristic evolving latent projections (Liang and Kvalheim 1992), and subwindow factor analysis (Manne et al. 1999). Examples of iterative methods are iterative target transformation factor analysis (Vandeginst et al. 1985) and alternating least squares (ALS) (Tauler et al. 1993). Iterative methods are used more frequently because of their flexibility, and fewer assumptions about the data are required. Of the iterative methods used, ALS is the most popular. ALS works by taking initial estimates of the concentration or spectral profiles of the constituents (using the appropriate boundary conditions) to develop better estimates of the concentration and spectral profiles during each iteration until convergence as defined by a specific threshold value for a user-specified criterion indicative of improvement to the fit has been met (de Juan et al. 2014). The actual values selected for these initial estimates are crucial to ensure the success of ALS. Evolving factor analysis (Maeder and de Juan 2009), which is a local rank method that detects the emergence and decay of the components in the data, has been used to develop initial estimates for ALS. Simple-to-use interactive self-modeling mixture analysis (Windig et al. 2007) is another method frequently used to select initial estimates for ALS. As for the boundary conditions in ALS selected by the user, these constraints work towards defining the shape of each concentration and spectral profiles in the bilinear model imposed on the data (see Equation 11.6). Examples of constraints frequently used in ALS include non-negativity, unimodality, closure, known pure spectra and/or concentration profiles, and local rank (Tauler et al. 2009).

An important issue in the successful implementation of ALS is assigning the appropriate number of components responsible for the variability in the data (Tauler and Maeder 2009). The rank of the data can exceed the number of components due to instrumental or other data-related artifacts or to interactions among the constituents present in the data. Alternatively, the data can be rank-deficient because of closure, the constituents sharing the same concentration or spectral profile or some constituents not being linearly independent of each other. Rank deficiency has been addressed through judicious preprocessing or column-wise augmentation of the data matrix (Saurina et al. 1998, Amrheim et al. 1999, Garrido, Lazaro et al. 2004).

Examples of MCR applied to NIR spectral data include monitoring the concentration of reactants and products in an aza-Michael reaction using aniline and a modified fatty acid ester with α, β-unsaturated ketone groups as reactants at 95°C (del Rio et al. 2009) by modeling mixtures of pure gasoline and ethanol in a distillation process with the concentration profile of each constituent used as seed information to formulate multivariate statistical process control models (de Oliveira et al. 2017); estimating the reaction profile for the esterification of a mixture of caprylic and capric acids with glycerol (Blanco et al. 2006); and resolving the spectrum and concentration profiles of the reactants, products, and intermediates encountered in the curing of an epoxy resin (Garrido, Lazaro et al. 2004). In these studies, the problem of rank deficiency was solved through the column-wise augmentation of the data matrix. Evolving factor analysis or SIMPLISMA provided the initial estimates of the spectra for ALS. NIR spectroscopy and ALS have also been used in the study of polymorphic transformations in pharmaceutical formulations expressed as tablets (Terra and Poppi 2014, Brondi et al. 2014, Blanco, Alcala et al. 2006). Other applications of NIR and ALS include the quantification and identification of adulterants in biodiesel-diesel blends (Camara et al. 2017), oxidation of edible oils (Wojcicki et al. 2015), detection of document fraud (Silva et al. 2014), and characterization of perturbations of hydrogen bonding in aqueous solutions by salts differing in their contribution to the ionic strength of the aqueous solution and their Lewis acidity (Gowen et al. 2013).

11.5 CONCLUSION

There is a substantial benefit to be gained by applying pattern recognition methods to NIR spectral data. A brief survey of the literature provided in this chapter demonstrates the breadth of applications. The application of multivariate methods to NIR spectral data is expected to increase in the near future due to the availability of software and training courses, both commercial and academic.

REFERENCES

Alamprese, C., M. Casale, N. Sinelli, S. Lanteri, and E. Casiraghi. 2013. Detection of minced beef adulteration with turkey meat by UV-visible, NIR, and MIR spectroscopy. *LWT – Food Sci. Technol.* 53, 225–232.

Alvarenga, L., D. Ferreira, D. Altekruse, J.C. Menezes, and D. Lochman. 2008. Tablet identification using near-infrared spectroscopy (NIRS) for pharmaceutical quality control. *J. Pharm. Biomed. Anal.* 48, 62–69.

Amrheim, M., B. Srinivasan, D. Bonvin, and M.M. Schumacher. 1996. On the rank deficiency and rank augmentation of the spectral measurement matrix. *Chemom. Intell. Lab. Syst.* 33(1), 17–33.

Barker, M. and W.S. Rayens. 2003. Partial least squares for discrimination. *J. Chemom.* 17, 166–173.

Barnes, R.J., M.S. Dhanoa, and S.J. Lister. 1989. Standard normal variate transformation and de-trending of near infrared diffuse reflectance spectra. *Appl. Spec.* 43, 772–777.

Blanco, M., M. Alcala, J. M. Gonzalez and E. Torras. 2006. Near infrared spectroscopy in the study of polymorphic transformation. *Anal. Chim. Acta* 567, 262–268.

Blanco, M., M. Castillo, A. Peinado and R. Beneyto. 2006. Application of multivariate curve resolution to chemical process control of an esterification reaction monitored by near-infrared spectroscopy. *Appl. Spec.* 60, 641–648.

Brereton, R.G. 2009. *Chemometrics for Pattern Recognition.* John Wiley & Sons, Chichester.

Brereton, R.G. and G.R. Lloyd. 2014. Partial least squares discriminant analysis: Taking the magic away. *J. Chemom.* 28, 213–225.

Brondi, A.M., L.A. Terra, G.P. Sabin, J.S. Garcia, R.J. Poppi and M.G. Trevisan. 2014. Mapping the polymorphic forms of fexofenadine in pharmaceutical tablets using near infrared chemical imaging. *J. Near Infrared Spectrosc.* 22, 211–220.

Camara, A.B.F., L.S. de Carvalho, C.L.M. de Morais, L.A.S. de Lima, H.O.M. de Araujo, F.M. de Oliveira, K.M.G. de Lima. 2017. MCR-ALS and PLS coupled to NIR/MIR spectroscopies for quantification and identification of adulterant in biodiesel-diesel blends. *Fuel.* 210, 497–506.

Candolfi, A., R. De Maesschalck, D.L. Massart, P.A. Hailey, and A.C.E. Harrington. 1999. Identification of pharmaceutical excipients using NIR spectroscopy and SIMCA. *J. Pharm. Biomed. Anal.* 19, 923–935.

Casale, M., L.R. Schimleck, and C. Espey. 2010. Classification of pernambuco (caesalpinia echinata lam) wood quality by near infrared spectroscopy and linear discriminant analysis. *J. Near Infrared Spectrosc.* 18, 435–442.

Cen, H. and Y. He. 2007. Theory and application of near infrared reflectance spectroscopy in determination of food quality. *Trends Food Sci. Technol.* 18, 72–83.

Cozzolino, D. 2014. An overview of the use of infrared spectroscopy and chemometrics in authenticity and traceability of cereals. *Food Res. Intern.* 60, 262–265.

Cozzolino, D., A. Chree, J.R. Scaife, and I. Murray. 2005. Usefulness of near-infrared reflectance (NIR) spectroscopy and chemometrics to discriminate fishmeal batches made with different fish species. *J. Agric. Food Chem.* 53, 4459–4463.

Ellis, D.I., V.L. Brewster, W.B. Dunn, J.W. Allwood, A.P. Golovanov and R. Goodacre. 2012. Fingerprinting food: Current technologies for the detection of food adulteration and contamination. *Chem. Soc. Rev.* 41, 5706–5727.

Frank, I.E. and J. Friedman. 1989. Classification: Oldtimers and newcomers. *J. Chemom.* 3(3), 463–476.

Fuenffinger, N., S. Arzhantsev, and C. Gryniewicz-Ruzicka. 2017. Classification of ciprofloxacin tablets using near-infrared spectroscopy and chemometric modeling. *Appl. Spec.* 71, 1927–1937.

Garrido, M., I. Lazaro, M.S. Larrechi, and F.X. Rius. 2004. Multivariate resolution of rank-deficient near infrared spectroscopy data from the reaction of curing epoxy resins using the rank augmentation strategy and multivariate curve resolution alternating least squares approach. *Anal. Chim. Acta* 515(1), 65–73.

Garrido, M.R., F.X. Rius, and M.S. Larrechi. 2008. Multivariate curve resolution – alternating least squares (MCR-ALS) applied to spectroscopic data from monitoring chemical reaction processes. *Anal. Bioanal. Chem.* 390, 2059–2066.

Gowen, A.A., J.M. Amigo and R. Tsenkova. 2013. Characterization of hydrogen bond perturbation in aqueous systems using aquaphotomics and multivariate curve resolution – alternating least squares. *Anal. Chim. Acta* 759, 8–20.

Honigs, D.E., G.M. Hieftje, and T. Hirschfeld. 1984. A new method for obtaining individual component spectra from those of complex mixtures. *Appl. Spec.* 38, 317–322.

Isaksson, T. and T. Næs. 1988. The effect of multiplicative scatter correction (MSC) and linearity improvement in NIR spectroscopy. *Appl. Spec.* 42, 1273–1284.

Jackson, J.E. 1991. *A User's Guide to Principal Component Analysis*, John Wiley & Sons, New York.

James, M. 1985. *Classification Algorithms.* John Wiley & Sons, New York.

de Juan, A., J. Jaumot and R. Tauler. 2014. Multivariate curve resolution (MCR). Solving the mixture analysis problem. *Anal. Methods* 6(14), 4964–4976.

de Juan, A. and R. Tauler. 2000. Soft modeling of analytical data. In R. A. Meyers (ed.), *Encyclopedia of Analytical Chemistry: Instrumentation and Applications*, John Wiley, New York.

Kowalski, B.R. and S. Wold. 1982. Pattern recognition in chemistry. In P. R. Krishnaiah and L. N. Kanal (eds.), *Classification, Pattern Recognition and Reduction of Dimensionality*, North Holland, Amsterdam.

Kowalski, B.R. and C.F. Bender. 1972. Pattern recognition: A powerful approach for interpreting chemical data. *J. Am. Chem. Soc.* 94, 5632–5640.

Lachenbruch, P.A. 1975. *Discriminant Analysis*, Hafner Press, New York.

Liang, Y. and O.M. Kvalheim. 1992. Heuristic evolving latent projections: Resolving two-way multicomponent data. *Anal. Chem.* 64(8), 936–946.

Maeder, M. and A. de Juan. 2009. Two-way data analysis: Evolving factor analysis. In S. D. Brown, R. Tauler, and B. Walczak (eds.), *Comprehensive Chemometrics*, Elsevier, Oxford, vol. 2, pp. 261–274.

Malik, A., A. de Juan and R. Tauler. 2015. Multivariate curve resolution: A different way to examine chemical data. In B. K. Lavine, S. D. Brown and K. Booksh, *40 Years of Chemometrics – From Bruce Kowalski to the Future*, ACS Symposium Series, Oxford University Press, Oxford, vol. 1199.

Malinowski, E.R. 1992. Window factor analysis: Theoretical derivation and application to flow injection analysis data. *J. Chemom.* 6(1), 29–40.

Mandel, J. 1985. The regression analysis of collinear data. *J. Res. NBS* 90(6), 465–476.

Manne, R., H. Shen, and Y. Liang. 1999. Sub-window factor analysis. *Chemom. Intell. Lab. Syst.* 45(1, 2), 171–176.

Martens, H. 1983. In H. Martens and H. Russwurm (eds.), *Food Research and Data Analysis*, Applied Science Publishers, London, 1983, pp. 5–38.

Martens, M. and H. Martens. 1986. Near infrared reflectance determination of sensory quality of peas. *Appl. Spec.* 40, 303–310.

McLachlan, G. L. 1992. *Discriminant Analysis and Statistical Pattern Recognition*, John Wiley & Sons, New York.

Murray, I., L.S. Aucottb and I.H. Pikec. 2001. Use of discriminant analysis on visible and near infrared reflectance spectra to detect adulteration of fishmeal with meat and bone meal. *J. Near Infrared Spectrosc.* 9, 297–311.

de Oliveira, R.R., R.H.P. Pedroza, A.O. Sousa, and K.M.G. Lima. 2017. Process monitoring and control applied to real-time monitoring of distillation processes by near-infrared spectroscopy. *Anal. Chim. Acta* 985, 41–53.

Paris, I.S., H. Riebere, M. Matoga, C. Civade, P.A. Bonnet, M.H. Tissier, and P.H. Chaminade. 2010. Challenging near infrared spectroscopy discriminating ability for counterfeit pharmaceuticals detection. *Anal. Chim. Acta* 658, 163–174.

Raggo, Y., P. Chalus, L. Maurer, C. Lema-Martinez, A. Edmond. 2007. A review of near infrared spectroscopy and chemometrics in pharmaceutical technologies. *J. Pharm. Biomed. Anal.* 44, 683–700.

del Rio, V., M.P. Callo, M.S. Larrechi, and L.M. de Espinosa. 2009. Chemometric resolution of nir spectra data of a model aza-Michael reaction with a combination of local rank exploratory analysis and multivariate curve resolution-alternating least squares (MCR-ALS) method. *Anal. Chim. Acta* 642, 148–154.

Rostaing, B., P. Delaquis, and Y. Roche. 1988. Automation of identification process of excipients and pharmaceuticals. fourier transform infrared spectroscopy and near-infrared diffuse reflectance spectroscopy applied to rapid in process control of pharmaceutical raw material, *S. T. P. Pharma* 4(6), 509–515.

Saurina, S., R. Hernandez-Cassou, R. Tauler, and A. Izquierdo-Ridorsa. 1998. Multivariate resolution of rank-deficient spectrophotometric data from the first order kinetic decomposition reaction, *J. Chemom.* 1293, 183–203.

Silva, C.S., M. F. Pimentel, R. S. Honorato, C. Pasquini, J. M. Prats-Montaban, and A. Ferrer. 2014. Near infrared hyperspectral imaging for forensic analysis of document forgery. *Analyst* 139, 5176–5184.

Stahle, L. and S. Wold. 1987. Partial least squares analysis with cross validation for the two class problem: A monte carlo study. *J. Chemomet.* 1, 185–196.

Tauler, R., B.R. Kowalski, and S. Fleming. 1993. Multivariate curve resolution applied to spectral data from multiple runs of an industrial process. *Anal. Chem.* 65(15), 2040–2047.

Tauler, R., M. Maeder and A. de Juan. 2009. Two-way extended curve resolution. In S. D. Brown, R. Tauler, and B. Walczak (eds.), *Comprehensive Chemometrics*, Elsevier, Oxford, vol. 2, pp. 473–505.

Tauler, R. and M. Maeder. 2009. Two-way data analysis: Multivariate curve resolution – error in curve resolution. In S. D. Brown, R. Tauler, and B. Walczak (eds.), *Comprehensive Chemometrics*, Elsevier, Oxford, 2009, vol. 2, 345–363.

Terra, L.A. and R.J. Poppi. 2014. Monitoring the polymorphic transformation on the surface of carbamazepine tablets generated by heating using near-infrared chemical imaging and chemometric methodologies. *Chemom. Intell. Lab. Syst.* 130, 91–97.

Tou, J.T. and R.C. Gonzalez. 1974. *Pattern Recognition Principles*, Addison Wesley Publishing Company, Boston, MA.

Twomey, M., G. Downey, and P.B. McNulty. 1995. The potential of NIR spectroscopy for the detection of the adulteration of orange juice. *J. Sci. Food Agric.* 67, 77–84.

Vandeginst, B.G.M., W. Derks, and G. Kateman. 1985. Multicomponent self-modeling curve resolution in high performance liquid chromatography by iterative target transformation analysis. *Anal. Chim. Acta* 173, 253–264.

Wang, L., F. S. C. Lee, X. Wang, and Y. He. 2006. Feasibility study of quantifying and discriminating soybean oil adulteration in camellia oils by attenuated total reflectance MIR and fiber optic diffuse reflectance NIR. *Food Chem.* 95, 529–536.

Windig, W., S. Markel, and P.M. Thompson. 2007. Self-modeling image analysis with SIMPLISMA. In H. F. Grahn and P. Geladi (eds.), *Hyperspectral Image Analysis*, John Wiley & Sons, New York.

Wojcicki, K., I. Khmelinskii, M. Sikorski, and E. Sikorska. 2015. Near and mid infrared spectroscopy and multivariate data analysis in studies of oxidation of edible oils. *Food Chem.* 187, 416–423.

Wold, S. 1976. Pattern recognition by means of disjoint principal component models. *Pattern Recogn.* 8, 127–139.

Wold, S. 1978. Cross validatory estimation of the number of components in factor and principal components models. *Technometrics* 20, 397–406.

Wold, S., K. Esbensen, and P. Geladi 1987. Principal component analysis. *Chemom. Intell. Lab. Syst.* 2, 37–57.

Wold, S. and M. Sjostrom. 1977. SIMCA, a method for analyzing chemical data in terms of similarity and analogy. In B. R. Kowalski (ed.) *Chemometrics, Theory and Application*, American Chemical Society, Washington, DC, vol. 52.

12 Supervised Methods
MLR, PCR, and MCR

Howard Mark[*]
Mark Electronics

CONTENTS

I have contributed chapters about calibration methodology and algorithms to the previous editions of this Handbook of NIR Spectroscopy. In them, I usually discuss the various aspects of near-infrared theory where little or no information was available. To some extent, this is still true, but since the previous (third) edition, some progress has been made. Under other circumstances, this would be unremarkable. In the context of NIR, where little or no attention has been paid to the underlying physical phenomena except for optical scattering (with little to show for it even there), or even the underlying mathematical knowledge, for over 30 years, this little bit of progress is tantamount to a major breakthrough.

The majority of discussions about how to calibrate a NIRS analysis pretty much cover the same ground: Algorithms include multiple linear regression (MLR), principal component regression (PCR), partial least squares (PLS), and data transforms (numerous types). Test statistics include standard error of estimate (SEE), root mean square error of calibration (RMSEC), correlation coefficient (R), correlation coefficient-squared (R^2), root mean squared standard error of prediction (RMSEP), predicted residual error sum of squares (PRESS), and other assorted types. They include recommendations for dealing with outliers, choosing algorithms and their parameters, and a few other global topics. In this book, much advice about using those tools is described in the many chapters exploring the various applications that NIR analysis is one of the premier analytical methods for. It seemed to me that adding more of the same to the literature would not be particularly beneficial either to the literature of NIR or to the readers of this book.

I decided, therefore, that a better use of this space available would lie in an exploration of more obscure corners of NIR and the mathematics upon which it is based. Those corners are not commonly covered in the various explanations available, so rather than write another rehash of the standard topics, I would deal with these obscure topics. These "other corners" include discussions

[*] hlmark@nearinfrared.com

of both the historical and mathematical backgrounds that the algorithms we use rest upon with some emphasis on aspects that are not usually discussed and also presentations of some techniques that are well known in other areas of the applications of the statistical methodology, but largely unknown to the NIR community. I hope that by emphasizing some of these lesser-known tools, at least some of our readers will start using them, even if not on a regular basis, then by writing the software to support the methodology, or clamoring for the instrument and third-party software companies to include those capabilities in their software packages.

12.1 LEAST SQUARES

The principle of least squares is one of the three fundamental principles underlying the science of statistics [1]. Gauss developed this mathematical theory over the period 1794–1798 [2], so even for a genius like Gauss, it was not easy! Nowadays, we consider "least squares" an almost trivial approach to calibration, especially those of us who are facile with multivariate mathematics and can sling the multivariate equations with ease. Unfortunately, I am not among those; while I can sometimes follow the explanations couched in the language of multivariate math, I often do not understand what those equations are telling us. I need to simplify them down to the point where they are simple enough that I can think about them.

I suspect that many people are in the same situation. We accept the equations that define the algorithms we use because, after all, they are *math* so they *must* be correct. And besides, all the experts vouch for them. But, how many of us really understand what is going on "under the hood" of those equations?

Some of them are relatively simple, so let's go back to basics. Everyone (everyone reading this book, at least), I am sure, has had occasion to draw a graph in some high school or college course on chemical analysis or physical chemistry, if nowhere else. Sooner or later, we all have to draw a calibration line (or possibly a curve) through some data that we collected. I dug out my old book on Experimental Physical Chemistry [3]. The extent of its discussion of data analysis, which I take as representative of what I expect most chemists are exposed to, consists of five pages on spectrophotometry and one entire chapter consisting of 19 pages on "Treatment of Experimental Data," of which one page was devoted to "Method of Least Squares." This was, of course, only univariate least squares (regression). Students were referred to another book, one on "Higher Mathematics for Engineers and Physicists," for the treatment of data involving more than two variables. For the univariate case, they presented the algebraic equations for the least squares calculations, but those were presented after having been solved explicitly for the slope (m) and intercept (b) of the line representing the (univariate) calibration model:

$$m = \frac{(n)\sum y_i x_i - \sum x_i \sum y_i}{(n)\sum x_i^2 - \left(\sum x_i\right)^2} \tag{12.1a}$$

$$b = \frac{(n)\sum x_i^2 \sum y_i - \sum x_i \sum x_i y_i}{(n)\sum x_i^2 - \left(\sum x_i\right)^2} \tag{12.1b}$$

The authors of that book also included a recommended table for recording the intermediate results and then noted (recall that this was at a time before even pocket calculators were generally available) "Calculations by the Least Square method are time-consuming and are, therefore, carried out only for the most precise data." For a univariate set of data, to evaluate Equations 12.1a and 12.1b and calculate the slope and intercept of a line for even a 5-sample experiment would require 32 multiplications/divisions and 46 additions/subtractions. That is a lot of computation, especially when done by hand, but it is what is needed just to compute coefficients of the one-variable equation, never mind error estimates or any of the fancy test statistics we all enjoy today! Personally, I do not know anybody who ever hand-calculated any of that (except, perhaps, textbook authors, and I am not even sure of them).

So, what did scientists of that day do? The data were simply plotted on a graph paper. If there was a theory to describe the physics involved (e.g., the Nernst equation for electrochemistry, the Clausius–Clapeyron equation for gas-law or vapor-pressure problems, or Beer's law for optical absorption measurements, for example), then the data could be linearized (transformed, in modern terminology) by plugging the data into the appropriate equation and plotting the transformed data values. If the data followed the theory, then the transformed data points would fall on a straight line and a line would be fitted to the data "by eye" with a straightedge or by calculating the slope of the line as $\Delta Y/\Delta X$, where ΔY and ΔX were computed as the differences between the two extreme values of the data. Using the extreme values that way was to minimize the effect of data errors on the slope and intercept of the line. If there was no theory and the data were not linear, then a tool called "French curve" was used to empirically fit a curve to the data and was also performed piecewise "by eye." Yes, we have come a long way!

But those tools only worked for univariate equations. For even a three-variable case (two independent and one dependent variable), data could be plotted as a family of lines (or curves) with values of one of the variables serving as a parameter for the calculation of the relationship between the other two variables. A scientist would be hard put to determine the values of the dependent variable if the value of the "parameter" variable fell between the plotted lines.

Ah, the good old daze!!

Those manual methods were, at best, slow, imprecise, labor-intensive, and subject to human error and interpretation. One scientist could fit one line to a set of data; then, another scientist could use the same data, fit a line, and get a completely different answer (that is the imprecision). The scientists of old were obviously up against the same problems. And surely, if they had any tools (straightedges and such) at all, then they were not that much better than the ones we can create today (and yes, I do know about the Egyptian capabilities of laying out their pyramids).

It is bad enough that the original data collection was subject to physical errors, but when the tools used to analyze the data added further error, something had to be done. Fortunately, there was a cavalry to call on to save the day, in the form of the mathematical community. I propose that we call what they did the First Digital Revolution.

Nevertheless, the world had to wait for the development of more capable tools, in both software (mathematics) and hardware (computers) domains for significant progress to be made.

12.2 MULTIPLE LINEAR REGRESSION

What did mathematicians of old do? Writing and counting had long since been invented, as was the concept of plotting data as graphs. So, their concept was to decide, for any data measurement, what the right answer to a problem was, and then WRITE THAT DOWN. Once reduced to written (digitized) form, the value would be constant. It would not be misread or misinterpreted. Whether it was right or wrong, another scientist seeing, for example, a "4" would not think it was some other number, would not misread it as "4.01," for example, and thereby introduce another layer of error.

The hardest part was to decide *what* the right answer to any given problem was. It turned out that there was a tool available for doing that, too. A lot of this stuff was going on more or less in parallel, over the years, at that time. Gerolamo Cardano had invented what turned out to be the origins of probability theory [1,2], which he applied to the analysis of games of chance (gambling). Once the word spread, it was not a great leap for scientists to realize that for any given problem, the answer that was of highest probability was the one most likely to be correct, too. And so, it was written.

But probability theory has its own problems. High on the list is the amount of computation involved. If you thought the amount of computation needed for least squares calculations was massive, you do not want to even *think* about the amount of computation involved in computing a maximum-likelihood result. The amount of computation a given problem requires increases factorially with the size of the problem. Again, this was at a time when not only modern pocket calculators were unknown, but even Charles Babbage was hundreds of years in their future. So, the scientists and mathematicians of the day knew there was an answer to their needs, but they could not compute it.

Enter Carl Friedrich Gauss, who was not alone in trying to find a solution to this conundrum. There are some opinions that Adrien-Marie Legendre also discovered the use of the least squares method (see Ref. 4) contemporaneously with Gauss. Certainly, others of his contemporaries were also working on the underlying problem. Gauss just had the luck, or the smarts, or whatever, to find an answer...and to be first (very important to be first!). Despite the historical controversy over priority, in modern times, Gauss gets the credit.

Gauss' big breakthrough was to be able to show that, when the data met certain prescribed conditions, the model obtained by fitting a model to the data using least squares calculations was the same model as would be obtained from using maximum-likelihood calculations. And it took much less computational effort. In case you ever wondered why least squares is so widely used for fitting models to data, you just read the justification for that.

What often gets lost in the practical application of least squares modeling for near-infrared analysis, however, is Gauss' conditions for the equivalence of least squares and maximum likelihood. There are four of them, based on the meaning of "error" as being the difference between an actual measured value for the analyte (or, more generally, the actual Y-value representing the quantity being modeled) and the value computed by the model. Then, Gauss requires that, for each sample:

1. Errors affect only the Y-variable.
2. There is no error in the X-values (the predictor variable(s)).
3. The error in the Y-values follows a normal (Gaussian, in Gauss' honor) distribution: $\exp(-(y-y_0)^2)$.
4. The model is linear in the coefficients.

In practice, condition 4 is usually adhered to, because today it is built into the software that is used to analyze the data. Condition 2, on the other hand, is almost never met. The values of the X-variables are all assumed to be known exactly, but in reality, it is virtually impossible to remove all error from a measurement. In practice, the assumption is made (whether warranted or not) that errors of the X-variable are at least "small enough" so as to not affect the computations, compared to a truly "error-free" set of data. But this is an assumption and is rarely checked in the routine use of NIR analysis. Similarly, condition 3 is also rarely, if ever, checked. One wonders how much of the difficulties encountered in routine use of NIR analysis is caused by failure of the data we use to conform to the requirements Gauss specified all those hundreds of years ago? How many analysts, barely able to push the right button on their computer, are even aware of what Gauss specified about the nature of the data to be used? Does anybody ever check their data against the assumptions Gauss made?

As an aside, some statisticians argue that there are even more fundamental assumptions that data must meet in order to apply the least squares concepts. Daniel [4] has recommended that meeting the following three conditions is the fundamental requirement for all computational methods for fitting data:

1. We know, or have chosen, the correct form of the equation, i.e., that the relationships are, in fact, linear.
2. The data we have are typical, i.e., are a representative sample of data to which we will apply the results of our data-fitting operations.
3. The Y-values (what we call the reference laboratory measurements) are statistically uncorrelated.

Daniel expands on these assumptions to explain them in detail, but you will have to read his discussion on page 7 of Ref. [4] in order to get his full argument. This is the end of the aside.

De facto, Gauss invented regression analysis, now called MLR when applied to calibration of near-infrared spectrometers. In this application, the X-values are the readings (a "reading" usually being the absorbance at a given wavelength) at various wavelengths, and the Y-values are the known properties of the samples (as determined by the reference laboratory measurements) for which the calibration is being developed.

An interesting fact about least squares and regression analysis is that our common usage of it is not the only possible way that Gauss' theory, and the restrictions specified, can be applied. This point is probably a good place to introduce the equations describing the MLR approach. The most common expression of MLR for application to NIR analysis is the very straightforward equation for absorbances:

$$A_\lambda = A_{\lambda 1} + A_{\lambda 2} + A_{\lambda 3} + \ldots \tag{12.2}$$

where A_λ is the total absorbance of a sample at a specified wavelength and $A_{\lambda i}$ is the absorbance of the ith component of the sample at that wavelength.

Since the absorbance of any component of a sample (which obeys Beer's law) depends on ε (the inherent absorption strength of the molecule of interest) and c (the concentration of the given component in the sample), we can rewrite Equation 12.2 as:

$$A_\lambda = c_1\varepsilon_{\lambda 1} + c_2\varepsilon_{\lambda 2} + c_3\varepsilon_{\lambda 3} + \ldots \tag{12.3}$$

(we ignore the fact that the absorbance also depends on the pathlength of light through the sample, because the pathlength is, for all practical purposes, constant and the same for all measurements of interest to us here; certainly, all the chemical components in any given sample are measured at the same pathlength. So, we leave the constant that would otherwise be included in the equations to represent the pathlength, out of our considerations completely). Here, we also ignore the true relationship between concentration of a given species in the mixture and the absorbance of that species in the mixture, because we have discussed that point elsewhere in this book (Chapter 20). So, for our current purpose, we accept that the relationship is described by Equation 12.3.

Given all these conditions, we find that for multiple samples measured at multiple wavelengths, we can set up and solve a set of equations similar to Equation 12.2 for the concentrations of the components (see Ref. [5] for the details of how one set of equations is converted to the other). One scan of a sample provides a set of readings at the various wavelengths, and the mathematics give us the coefficients needed to convert those data values into an analytical answer:

$$c = B_0 + B_1A_1 + B_2A_2 + \ldots \tag{12.4}$$

where c represents the concentration of the analyte of interest, B_i are the (empirically determined) coefficients needed to convert the absorbance at a given wavelength into the contribution of the optical data at that wavelength to the final computed value representing the concentration of the sample, and A_i is the absorbance of the sample at a given (ith) wavelength.

The calculations needed to convert the equations as expressed in the form of Equation 12.3 into the form expressed as in Equation 12.4 are themselves lengthy. As we noted above, even a univariate problem requires six multiplications/divisions and eight additions/subtractions **per sample** to compute the slope and intercept of a line representing the least squares values. When multivariate conditions are involved, the amount of computation increases as roughly the square of the number of variables in the problem; two variables would require 24 multiplications/divisions and 32 additions/subtractions **per sample**. Besides the actual amount of computation and the time needed to perform it, even the notation expressed in Equation 12.1 quickly becomes inadequate to describe the computations that need to be performed, much less to have any hope of displaying them in closed-form explicit expressions such as Equation 12.1 use.

12.3 MATRIX ALGEBRA

Mathematicians again came to the rescue, this time by invention of a new form of algebra, called matrix algebra. In it, complex computations could be described by equations that look, on the surface, very much like the ordinary algebraic equations normally used. Discussions of matrix algebra have been included in several books on chemometrics; the one I will pick on here is in the first few chapters of Ref. [6]. However, the definitions of the operations that those new types of equations deal with allow the expression of multiple algebraic operations to be described in a very compact notation. That did not affect the need for enormous amounts of computation, but it at least made it possible to write down problems and solutions on no more than a few sheets of paper.

We will start with a definition for symbolism of matrices: A matrix corresponding to a matrix variable A will be symbolized as $[A]$. The brackets do no more than indicate that the quantities inside the brackets are matrices and are subject to treatment with the operations appropriate for matrices, rather than as ordinary numbers.

The four major arithmetic operations (addition, subtraction, multiplication, and division) have their counterpart matrix operations, although some of them are not so easily recognizable as such. Addition and subtraction are pretty straightforward; the sum (or difference) of two matrices is a matrix of the same size as the operand matrices, and where each term is the sum or difference of the corresponding term of the operand matrices. Brereton [7] has presented a brief but clear exposition of the elementary matrix operations, as have several other authors of books and articles on chemometric analysis [6,8–15]. There are also operations that have no equivalent in ordinary algebra. One such operation, which is very common and very important, is the transpose of a matrix. That means to change the positions of all the entries in the matrix so that a number in the ith row and jth column is moved to the jth row and ith column. The transpose of a matrix $[A]$ is commonly written as $[A]^T$, although some authors use the notation A' as the symbol for a matrix transpose.

Multiplication is a bit more complicated. The result of the multiplication of two matrices is not even necessarily the same size as either operand, but does form a product that is sensibly related to the multiplication of ordinary numbers. We will not go further into this right now.

Division is even more complicated. In fact, division *per se* is not even a defined operation in matrix algebra. However, a set of corresponding definitions for matrices, taken together, serve the same function as the division operation does in ordinary algebra. There is a unit matrix, symbolized by [1], that has the same property as the number 1: Multiplying any matrix by a unit matrix returns the same matrix as the original matrix, just as multiplying any number by the number 1 gives the original number as the result.

Having defined a unit matrix, we can now define the inverse of a matrix. Just as multiplying any number by its inverse gives the number 1 as the answer, so too does multiplying any matrix by its inverse give a unit matrix as the answer. The notation for the inverse of a matrix $[A]$ is $[A]^{-1}$. So, while the division operation is not defined within matrix algebra, multiplying a matrix by the inverse of another matrix performs an equivalent function to what the division operation would perform. For example, as we mentioned above, multiplying any matrix by its own inverse gives the unit matrix as the answer; e.g., $[A]\times[A]^{-1}$ gives the unit matrix as the answer, i.e.,

$$[A]\times[A]^{-1}=[1] \tag{12.5}$$

Comparing all these operations with the equivalent operation of an inverse in ordinary algebra,

$$A\times A^{-1}=A\times 1/A=1 \tag{12.6}$$

reveals the equivalency of all these corresponding terms and definitions for the two cases.

There is an important difference, however: Any given nonzero number has an inverse, but that is not always true for matrices. Not all matrices can be inverted, for any of several reasons. We will not pursue this discussion here, but when we later on include a caveat "assuming the matrix is invertible," you will understand why.

12.4 BRIEF DERIVATION OF MLR

To convert Gauss' specifications into something that a computer program can be based on is relatively straightforward. We do this here for an example using a three-variable example (three is large enough for all practical effects of importance to be operative, but not so large as to make writing it out be overwhelming, and also, we understand that when dealing with larger amounts of data, everything follows the same pattern):

First, we write out the equations representing data from several samples of the type we will want to create a calibration model for:

$$C_1 = b_0 + b_1A_{11} + b_2A_{12} + b_3A_{13} + \text{error} \tag{12.7a}$$

$$C_2 = b_0 + b_1A_{21} + b_2A_{22} + b_3A_{23} + \text{error} \tag{12.7b}$$

$$\ldots$$

$$C_n = b_0 + b_1A_{n1} + b_2A_{n2} + b_3A_{n3} + \text{error} \tag{12.7c}$$

where C_i represents the value of the reference laboratory analysis for the ith sample, A_{ij} represents the measured absorbance value of the ith sample at the jth wavelength, and b_j represents the coefficient (to be calculated) for the jth wavelength.

The above equations represent the first to the last equations of the set containing the n samples we will be dealing with. The use of MLR signifies that we do not expect the model to fit the data exactly, so the "error" represents the difference between the measured data point and the value that the model will produce.

We want to compute the values of the coefficients, the b_i that will give the smallest possible value of the sum of squares of the errors, when the errors from each sample are squared and then all added together. So, we first solve Equations 12.7a–12.7c for the errors:

$$\text{error}_1 = C_1 - \left(b_0 + b_1A_{11} + b_2A_{12} + b_3A_{13}\right) \tag{12.8a}$$

$$\ldots$$

$$\text{error}_n = C_n - \left(b_0 + b_1A_{n1} + b_2A_{n2} + b_3A_{n3}\right) \tag{12.8b}$$

To generate the sum of squared errors we wish to minimize, we square the errors and add them together over the sample set (replacing all the individual errors with the symbol E_i):

$$\Sigma E_i^2 = \Sigma_i\left(C_i - \left(b_0 + b_1A_{i1} + b_2A_{i2} + b_3A_{i3}\right)\right)^2 \tag{12.9}$$

Now that we have the expression for the quantity we want to minimize, we can minimize the sum-squared error by finding the values of the b_i that provide that condition. We do that by applying the well-known technique we learned in elementary calculus: Take the derivative and set it equal to zero. Since there are three values for the various wavelengths b_i (plus b_0), we have to take four derivatives and end up with a set of four equations, all based on Equation 12.9.

$$\partial/\partial b_0\Sigma E^2 = \partial/\partial b_0\Sigma_i\left(C_i - \left(b_0 + b_1A_{i1} + b_2A_{i2} + b_3A_{i3}\right)\right)^2 \tag{12.10a}$$

$$\partial / \partial b_1 \Sigma E^2 = \partial / \partial b_1 \Sigma_i \left(C_i - \left(b_0 + b_1 A_{i1} + b_2 A_{i2} + b_3 A_{i3}\right)\right)^2 \tag{12.10b}$$

$$\partial / \partial b_2 \Sigma E^2 = \partial / \partial b_2 \Sigma_i \left(C_i - \left(b_0 + b_1 A_{i1} + b_2 A_{i2} + b_3 A_{i3}\right)\right)^2 \tag{12.10c}$$

$$\partial / \partial b_3 \Sigma E^2 = \partial / \partial b_3 \Sigma_i \left(C_i - \left(b_0 + b_1 A_{i1} + b_2 A_{i2} + b_3 A_{i3}\right)\right)^2 \tag{12.10d}$$

Taking the derivatives:

$$\partial / \partial b_0 \Sigma E^2 = \Sigma_i 2\left(C_i - \left(b_0 + b_1 A_{i1} + b_2 A_{i2} + b_3 A_{i3}\right)\right) \partial / \partial b_0 \left(b_0 + b_1 A_{i1} + b_2 A_{i2} + b_3 A_{i3}\right) \tag{12.11a}$$

$$\partial / \partial b_1 \Sigma E^2 = \Sigma_i 2\left(C_i - \left(b_0 + b_1 A_{i1} + b_2 A_{i2} + b_3 A_{i3}\right)\right) \partial / \partial b_1 \left(b_1 A_{i1} + b_2 A_{i2} + b_3 A_{i3}\right) \tag{12.11b}$$

$$\partial / \partial b_2 \Sigma E^2 = \Sigma_i 2\left(C_i - \left(b_0 + b_1 A_{i1} + b_2 A_{i2} + b_3 A_{i3}\right)\right) \partial / \partial b_2 \left(b_1 A_{i1} + b_2 A_{i2} + b_3 A_{i3}\right) \tag{12.11c}$$

$$\partial / \partial b_3 \Sigma E^2 = \Sigma_i 2\left(C_i - \left(b_0 + b_1 A_{i1} + b_2 A_{i2} + b_3 A_{i3}\right)\right) \partial / \partial b_3 \left(b_1 A_{i1} + b_2 A_{i2} + b_3 A_{i3}\right) \tag{12.11d}$$

Then, we need to set the derivatives to zero and do some algebra to put the equations into a neater-looking and more useful form. The interested reader can work through the details as an "exercise for the reader" (note that a majority of the partial derivative terms are zero) or can check out the details in Draper & Smith [16]. The result of doing that is a set of four equations in four unknowns, the b_i are the unknowns in the equations, and everything else represents constants that are determined by computations based on the data.

$$\Sigma_i C_i = b_0 + b_1 \Sigma A_1 + b_2 \Sigma A_2 + b_3 \Sigma A_3 \tag{12.12a}$$

$$\Sigma_i A_{1i} C_i = b_0 + b_1 \Sigma A_1^2 + b_2 \Sigma A_1 A_2 + b_3 \Sigma A_1 A_3 \tag{12.12b}$$

$$\Sigma_i A_{2i} C_i = b_0 + b_1 \Sigma A_2 A_1 + b_2 \Sigma A_2^2 + b_3 \Sigma A_2 A_3 \tag{12.12c}$$

$$\Sigma_i A_{3i} C_i = b_0 + b_1 \Sigma A_3 A_1 + b_1 \Sigma A_3 A_2 + b_2 \Sigma A_3{}^2 \tag{12.12d}$$

Equations 12.12 are called the normal equations. These equations can be solved for the values of b using techniques learned in high school for solving simultaneous equations, since once all the measured data (the A_i for the spectroscopic readings and the concentrations of the analyte in the samples) are known, all the summations evaluate to numbers (which are constants), then the values of b can be found as the solution to the four simultaneous equations.

However, to connect this to our previous discussion about matrix algebra, we convert Equations 12.10 into their equivalent in matrix notation:

$$\begin{bmatrix} \Sigma C \\ \Sigma A_1 C \\ \Sigma A_2 C \\ \Sigma A_3 C \end{bmatrix} = \begin{bmatrix} b_0 & b_1 \Sigma A_1 & b_2 \Sigma A_2 & b_3 \Sigma A_3 \\ b_0 & b_1 \Sigma A_1^2 & b_2 \Sigma A_2 A_1 & b_3 \Sigma A_3 A_1 \\ b_0 & b_1 \Sigma A_1 A_2 & b_2 \Sigma A_2^2 & b_3 \Sigma A_3 A_2 \\ b_0 & b_1 \Sigma A_1 A_3 & b_2 \Sigma A_2 A_3 & b_3 \Sigma A_3^2 \end{bmatrix} \tag{12.13}$$

From which the individual variables can be factored (note that while it may not be obvious that Equation 12.13 is the same as Equations 12.12, it is relatively straightforward to perform the indicated operations on Equation 12.13 and verify that Equations 12.12 are recovered), similarly for the subsequent few lines that follow. Factoring Equation 12.13, we obtain:

$$\begin{bmatrix} \Sigma C \\ \Sigma A_1 C \\ \Sigma A_2 C \\ \Sigma A_3 C \end{bmatrix} = \begin{bmatrix} 1 & \Sigma A_1 & \Sigma A_2 & \Sigma A_3 \\ 1 & \Sigma A_1^2 & \Sigma A_1 A_2 & \Sigma A_1 A_3 \\ 1 & \Sigma A_1 A_2 & \Sigma A_2^2 & \Sigma A_2 A_3 \\ 1 & \Sigma A_1 A_3 & \Sigma A_3 A_2 & \Sigma A_3^2 \end{bmatrix} \times \begin{bmatrix} b_0 & b_1 & b_2 & b_3 \\ b_0 & b_1 & b_2 & b_3 \\ b_0 & b_1 & b_2 & b_3 \\ b_0 & b_1 & b_2 & b_3 \end{bmatrix} \quad (12.14)$$

and then broken down into the separate vectors, matrices, and matrix operations:

$$\begin{bmatrix} A_{11} & A_{12} & A_{13} & \cdots & A_{1n} \\ A_{21} & A_{22} & A_{23} & \cdots & A_{2n} \\ A_{31} & A_{32} & A_{33} & & A_{3n} \end{bmatrix} \begin{bmatrix} C_1 \\ C_2 \\ C_3 \\ C_n \end{bmatrix}$$

$$= \begin{bmatrix} 1 & 1 & 1 & & 1 \\ A_{11} & A_{12} & A_{13} & & A_{1n} \\ A_{21} & A_{22} & A_{23} & \cdots & A_{2n} \\ A_{31} & A_{32} & A_{33} & & A_{3n} \end{bmatrix}^T \begin{bmatrix} 1 & 1 & 1 & & 1 \\ A_{11} & A_{12} & A_{13} & & A_{1n} \\ A_{21} & A_{22} & A_{23} & \cdots & A_{2n} \\ A_{31} & A_{32} & A_{33} & & A_{3n} \end{bmatrix}$$

$$\times \begin{bmatrix} b_0 & b_1 & b_2 & b_3 \\ b_0 & b_1 & b_2 & b_3 \\ b_0 & b_1 & b_2 & b_3 \\ b_0 & b_1 & b_2 & b_3 \end{bmatrix} \quad (12.15)$$

We can then put Equation 12.15 into the shorthand notation for matrices:

$$A^T C = A^T A b \quad (12.16)$$

Having done that, it becomes clear that solving Equation 12.16 for **b** by multiplying both sides of Equation 12.14 by $[A^T A]^{-1}$, the set of coefficients is:

$$b = \left[A^T A\right]^{-1} A^T C \quad (12.17)$$

which gives the same result, in matrix notation, as solving the algebraic equations (Equations 12.12).

The above discussion and the associated derivations tell us how to calculate the line (or plane, or hyperplane) that best fits the data. There is much more that needs to be computed about that best-fitting line in order to make it useful. These auxiliary computations are important for letting the users know how good a fit that best-fitting line gives is. There are many of these auxiliary statistics, but, in general, every statistician has their favorite set. For NIR calibrations, some of the more common ones are those that have been developed by statisticians over many years. These are the coefficient of multiple correlation (R), the standard error of estimate (SEE), and the F for regression (F). We have described and explained these statistics in the third edition of this handbook. As described in the introduction to this chapter, therefore we forebear to repeat those descriptions, in favor using the space for some less commonly discussed characteristics of regression.

12.4.1 Pseudo-Inverse

The new matrix notation brings with it the possibility of definitions for new operations, including some that are not easily expressed in ordinary algebraic language. We have already encountered one of these: the transpose operation. In particular, there is an operation called the pseudo-inverse. One point not stated in the above explanation of the matrix inverse is the fact that, in order to compute the inverse of a matrix, the matrix must be square (i.e., contain the same number of rows as columns). Sometimes, however, it is necessary or desirable to "divide" a matrix by another matrix when that second matrix is not square. In order to deal with that situation, the concept of the "pseudo-inverse" of a matrix was created.

The pseudo-inverse of a matrix A is defined as $([A][A]^T)^{-1} \times [A^T]$ (see Appendix A.1.2.3 in Ref. [7]). One property of matrix multiplication is that if any matrix is multiplied by its own transpose, then the resulting matrix will be square. Hence, multiplying the matrix $[A]$ by its own transpose gives a square matrix that can then be inverted. But, since the original matrix is not directly a square matrix, the result of inverting the product of those two matrices is termed the pseudo-inverse of the original matrix.

That is all very well and good, but why go to all that trouble? Is there anything special about the pseudo-inverse that makes it something more than a nice little bit of formalism that fits into the structure of matrix algebra? To answer that question, as I am fond of saying, "Of course there is or I wouldn't have brought it up (and certainly not with such a long lead-in)!"

So, what other useful property does the pseudo-inverse have that makes it important enough to spend all this much time and effort on it? To answer that, let's compare it with the ordinary inverse described above.

If we have a result that is a product of two matrices, as in Equation 12.18,

$$[A] = [B] \times [C] \tag{12.18}$$

and we know the matrices $[A]$ and $[B]$ but we do not know $[C]$.

Then, to find $[C]$, we can multiply both sides of Equation 12.18 by $[B]^{-1}$ (assuming that $[B]$ is square and invertible):

$$[B]^{-1} \times [A] = [B]^{-1} \times [B] \times [C] \tag{12.19}$$

Noting that $[B]^{-1}$ $[B]$ results in a unit matrix (as discussed above), we find that we have solved the equation for $[C]$:

$$[B]^{-1} \times [A] = [C] \tag{12.20}$$

In order to do this, in equation 12.19 we had to know that the matrix $[B]$ was square, in order that we can invert it. Suppose it wasn't square? Is there a way we can solve Equation 12.19 for C? Again, "Of course there is…"

It is what we have been talking about for the past few paragraphs: Instead of computing the inverse of $[B]$, we need to compute the pseudo-inverse of $[B]$ and use that pseudo-inverse instead of the inverse of Equation 12.19.

$$\left([B] \times [B]^T\right)^{-1} \times \left[B^T\right] \times [A] = [C] \tag{12.21}$$

So now a very important question arises: Are the answers for the matrix $[C]$ from Equation 12.20 the same as the answers in the matrix $[C]$ from Equation 12.21?

The answer is no, not only because of the differences in the equations themselves, but also because of the nature of the data to which each applies. Equation 12.20, which applies when $[B]$ is a square matrix, is the matrix expression for a solution by simultaneous equations, which has an exact solution and is not subject to any variability. There are neither excess samples nor excess variables in the data, and the equations are exactly solvable just like you learned in high school (barring any pathological properties of the data itself such as exact linear dependence).

Equation 12.21, on the other hand, applies when [*B*] is non-square, which typically means that there are more samples than variables in the data, so that Equation 12.21 is the matrix expression for a solution by least squares. In general, however, much the pseudo-inverse is an expression of matrix algebra formalism, and the presence and use of the pseudo-inverse is a sign that, as shown above, the problem under consideration is going to be solved using least squares calculations. In fact, while we did not make a point of it at the time, this construction appears in Equation 12.17 above, demonstrating that the pseudo-inverse is, in fact, the result of imposing the least squares condition on the result of solving the equations for regression analysis.

12.4.2 Variations on a Theme

At first glance, it would seem that, once the concept of least squares is expanded into the multivariate domain, we have exhausted the range of possibilities for how the concept of least squares could be applied to analyze a set of data. But that is by no means the case. Such thoughts reveal only the limitations of our own thinking, brought about to some extent by the fact that few of those expanded concepts have yet been applied to NIR analysis. The standard computation of a least squares fit as described above has some desirable properties. Among those properties is that the resulting calibration model is unbiased (in a defined statistical sense that has nothing to do with "bias" as commonly used in NIR spectroscopy). All other alternate methods that we describe below result in models that are biased in one way or another. This bias is not necessarily a bad thing, because in exchange for having bias in the model, we can gain some useful other properties for the calibration. Here, we describe some alternate ways in which a least squares calibration can be performed:

12.4.2.1 Variation 1: Zero Intercept

The first, and simplest, alternate condition we will consider is the possibility that b_0 is exactly zero: the model passes through the origin. This is unusual in ordinary NIR applications of regression, since it means that the calibration model passes exactly through the zero intercept. However, in certain applications, this is a known effect. One example is where the theory indicates that the model should pass through the zero intercept.

Another example is when multiple regression, as is often applied to data from designed experiments, includes factorial and mixture designs. In these applications, the target values for the various factors, or for the values of the components of a mixture, are the X-values for the several variables of the regression, and the measured response or property of interest is used for the Y-value. This is common and, in fact, in general is probably more common than the method we use in NIR analysis, i.e., using instrument readings for the X-values. In an experimental design, the design specifies the target values for the factors, and with a good design, the factors can be made orthogonal, creating a near-ideal mathematical situation.

When the design is a mixture design, the situation is not so salutary. One characteristic of multiple regression that is not often discussed in elementary treatments is the effect of intercorrelation of the X-variables, although more advanced discussions [16] treat it at length. Without going into too great detail here, let us just note that the results are not good: The model has increased sensitivity to noise, greater instability and error, than could be obtained from an optimal model. As discussed above, an ideal situation will have orthogonal X-variables, which means no intercorrelation among them. In a mixture design, just the opposite condition obtains: The intercorrelation is total, by virtue of the fact that for each sample, the total amount of all the different materials must equal 100%. This condition will ordinarily prevent the regression from being performed because in the course of the calculations, a "divide by zero" error will occur, and the computer will not complete the calculations.

One solution is to eliminate one of the ingredients in the sample from the calculations, thus keeping it from being used in the model. That removes the term corresponding to that ingredient, reducing the intercorrelation and allowing the calculations to proceed, creating a model based on the remaining information. This solution will work, but raises a couple of problems: one practical and one of principle. The practical problem is very straightforward: How do you choose which ingredient's data you should eliminate? If you have the data to try this out with, you will quickly learn that your results depend fairly strongly on the choices you make.

The problem of principle is similar: The original mixture design is presumably balanced, so that removing any ingredient alters the characteristics of the whole design. The statistics community eventually realized that rather than removing any individual subset of data by removing the corresponding variable from the calculations, the better approach is to remove the constant (the b_0) from the model; this can be accomplished by modifying the above equations suitably. By leaving the constant term out of the model shown in Equation 12.7, we are left with the model:

$$C_1 = b_1 A_{11} + b_2 A_{12} + b_3 A_{13} + \text{error} \tag{12.22}$$

We can now proceed with the derivation for this somewhat simpler case in the same way we did above:

1. Calculate the model for all the samples in a data set as we did in Equation 12.7:

$$C_1 = b_1 A_{11} + b_2 A_{12} + b_3 A_{13} + \text{error} \tag{12.23a}$$

$$\ldots$$

$$C_n = b_1 A_{n1} + b_2 A_{n2} + b_3 A_{n3} + \text{error} \tag{12.23b}$$

2. Then, calculate the errors as we did, using Equation 12.8:

$$\text{error}_1 = C_1 - (b_1 A_{11} + b_2 A_{12} + b_3 A_{13}) \tag{12.24a}$$

$$\ldots$$

$$\text{error}_n = C_n - (b_1 A_{n1} + b_2 A_{n2} + b_3 A_{n3}) \tag{12.24b}$$

3. Calculate the sum-squared errors:

$$\Sigma E_i^2 = \Sigma_i \left(C_i - (b_1 A_{i1} + b_2 A_{i2} + b_3 A_{i3})\right)^2 \tag{12.25}$$

4. Take the derivatives and set them equal to zero as previously:

$$\Sigma A_1 C = 0 = b_1\, \Sigma A_1{}^2 + b_2\, \Sigma A_1 A_2 + b_3\, \Sigma A_1 A_3 \tag{12.26a}$$

$$\Sigma A_2 C = 0 = b_1\, \Sigma A_2 A_1 + b_2\, \Sigma A_2^2 + b_3\, \Sigma A_2 A_3 \tag{12.26b}$$

$$\Sigma A_3 C = 0 = b_1\, \Sigma A_2 A_1 + b_2\, \Sigma A_2^2 + b_3\, \Sigma A_2 A_3 \tag{12.26c}$$

5. Writing the equations in matrix notation gives us the same expression as Equation 12.17. The difference is not in the modeling, but in the data that the model is applied to; normally, though, the code implementing the regression calculations has to be modified, to not include the constant term of the model.

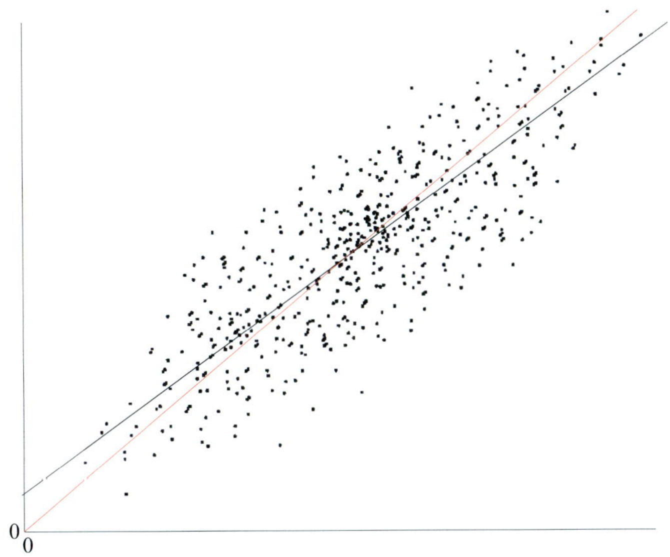

FIGURE 12.1 Stylized data showing a typical ordinary regression line (black) and a regression line forced through the origin (red). Both lines pass through the mean of the data set, so that neither of them is "biased" in the sense that NIR practitioners use the term.

$$A^T C = bA^T A \tag{12.17}$$

Figure 12.1 (graphically) shows the relationship between an ordinary calibration and one where intercept is forced to be zero. Here, we begin to see the effect of the statistician's meaning of "bias." Overall, both the biased calibration model and the unbiased model give the same average value for the entire data set and the average predicted value will be equal to the average reference lab value. The biased model will give predictions for the lower part of the range which are lower than they would otherwise be, while the predictions for the upper part of the range will be higher than they would otherwise be.

12.4.2.2 Variation 2: Coefficients that Sum to Zero

In the early days of NIR analysis, attempts were made to counteract the effect of the physical/optical variations (e.g., particle size) on the analytical readings, and that continues to this day. It was clear from the beginnings of the technology that the optical effects, loosely called "particle size effects," although they could have causes other than actual changes in the particle size of powdered solid samples, would affect the final analytical result. The effect on the absorbance readings was to create a systematic offset in the values of the spectral measurements (see page 56 in Ref. [5]). This offset affected the predicted analytical value of the sample component being measured. Attempts to modify the data to reduce this effect included using derivatives of the spectra ($dA/d\lambda$), spectral ratios (A_i/A_j) [5], and Fourier transforms (see Chapter 6 in Ref. [17]).

The amount of change of the predicted value depended on the model being used as well as the nature of the data. A measure of the sensitivity of any given model to this type of systematic variation in the data was termed the index of systematic error (ISE), which is computed as the sum of the calibration coefficients (except B0) (see page 57 in Ref. [5], or page 58 in the second edition of this text). The ISE could be used to compare models to see which was the least sensitive to these systematic variations of the spectra; the smaller the ISE, the less sensitive the model would be to the optical changes [5]. Ideally, if the ISE was zero, there should be no effect of the optical changes on the final values.

In practice, the variations at the different wavelengths for powdered solid samples are usually not exactly the same, but very often they are sufficiently similar that the assumption that the changes are substantially equal can serve as a good approximation. In these cases, the ISE is a useful guide as to how stable the value predicted by a given model will be, in the face of these optical changes due to particle size effects.

That leads to another question: Can we create a model where the ISE is exactly zero? And, again, the answer to that question is "Of course we can…" The required algorithm is variously called "Lagrange's method," or "the method of undetermined multipliers." It is an application of the calculus of variations, invented by the Italian mathematician Joseph-Louis Lagrange. The calculus of variations provides tools for describing and solving problems that have constraints superimposed on them. Many explanations of the algorithm are available, both in standard texts [16,18] and online [19,20]. There are tutorials and fundamental studies; tutorials deal with very generalized mathematical formulations of both the concept and its implementation. For our purposes here, we will demonstrate it strictly in and for the much narrower context of creating a calibration model with the constraint that the sum of the coefficients must add to zero.

The development of this algorithm follows very closely the development of the MLR algorithm (Equations 12.7–12.14).

12.4.2.2.1 Lagrange's Method

We start with Equation 12.27, which is the same as before, and as we recall, we limit ourselves to a three-wavelength case for simplicity in writing out the equations, as we did when discussing MLR (above):

$$C_{1i} = b_0 + b_1 A_{i1} + b_2 A_{i2} + b_3 A_3 + \text{error} \tag{12.27}$$

We calculate the error (substituting E for the error):

$$E = C_{1i} - \left(b_0 + b_1 A_1 + b_2 A_2 + b_3 A_3\right) \tag{12.28}$$

Then, as we did before, we create the sums of squares to be minimized, analogously to Equation 12.9, and recalling that all summations are taken over the samples (which are indexed by i when present, and which are inferred when not present):

$$\Sigma E^2 = \Sigma\left(C_1 - \left(b_0 + b_1 A_1 + b_2 A_2 + b_3 A_3\right)\right)^2 \tag{12.29}$$

where b_i represents the ith calibration coefficient.

We want to add the constraint that:

$$b_1 + b_2 + b_3 = 0. \tag{12.30}$$

We add the constraint to Equation 12.29, by multiplying the constraint by an undetermined multiplier; conventionally, λ is used to represent the multiplier and is termed the Lagrange multiplier in this context.

$$\Sigma E^2 = \Sigma\left(C_i - \left(b_0 + b_1 A_1 + b_2 A_2 + b_3 A_3\right)\right)^2 + \lambda \cdot [0] \tag{12.31}$$

The constraint that λ multiplies must equal zero in order to prevent the additional term from affecting the value of the equations. In our case, since the constraint is that the sum of coefficients is zero, we need only add the constraint itself to the equations. Here is the equation with the Lagrange multiplier added:

$$\Sigma E^2 = \Sigma\left(C_i - \left(b_0 + b_1 A_1 + b_2 A_2 + b_3 A_3\right)\right)^2 + \lambda \cdot \left[b_1 + b_2 + b_3\right] \tag{12.32}$$

For the time being, we will leave the value of the constraint in the equation, to keep the expressions simple. Expanding the squared term (in steps):

$$\Sigma E^2 = \Sigma C_i^2 - 2\Sigma C_i\left(b_0 + b_1A_1 + b_2A_2 + b_3A_3\right) + \Sigma\left(b_0 + b_1A_1 + b_2A_2 + b_3A_3\right)^2 + \lambda\cdot[0] \tag{12.33}$$

Expanding further:

$$\begin{aligned}\Sigma E_i^2 = &\sum_{i=1}^{n} C_i^2 - \sum_{i=1}^{n} 2C_ib_0 - \sum_{i=1}^{n} 2C_ib_1A_1 - \sum_{i=1}^{n} 2C_ib_2A_2 - \sum_{i=1}^{n} 2C_ib_3A_3\\ &+\sum_{i=1}^{n} b_0^2 + \sum_{i=1}^{n} b_0b_1A_1 + \sum_{i=1}^{n} b_0b_2A_2 + \sum_{i=1}^{n} b_0b_3A_3\\ &+\sum_{i=1}^{n} b_1b_0A_1 + \sum_{i=1}^{n} b_1^2A_1^2 + \sum_{i=1}^{n} b_1b_2A_1A_{i,2} + \sum_{i=1}^{n} b_1b_3A_1A_{i,3}\\ &+\sum_{i=1}^{n} b_2b_0A_2 + \sum_{i=1}^{n} b_2b_1A_{i,1}A_{i,2} + \sum_{i=1}^{n} \Sigma b_2^2A_2^2 + \sum_{i=1}^{n} b_2b_3A_2A_3\\ &+\sum_{i=1}^{n} b_3b_0A_3 + \Sigma b_3b_1A_{i,3}A_{i,1} + \sum_{i=1}^{n} b_3b_2A_{i,3}A_{i,2} + \sum_{i=1}^{n} b_3^2A_{i,\,3}^2 - \lambda\cdot[0]\end{aligned} \tag{12.34}$$

Substituting the constraint values from Equation 12.30 $[0-b_1-b_2-b_3]$ for [0], we obtain:

$$\begin{aligned}\Sigma E_i^2 = &\sum_{i=1}^{n} C_i^2 - \sum_{i=1}^{n} 2C_ib_0 - \sum_{i=1}^{n} 2C_ib_1A_{i1} - \sum_{i=1}^{n} 2C_ib_2A_{i2} - \sum_{i=1}^{n} 2C_ib_3A_{i3}\\ &+\sum_{i=1}^{n} b_0^2 + \sum_{i=1}^{n} b_0b_1A_1 + \sum_{i=1}^{n} b_0b_2A_2 + \sum_{i=1}^{n} b_0b_3A_3\\ &+\sum_{i=1}^{n} b_1b_0A_1 + \sum_{i=1}^{n} b_1^2A_1^2 + \sum_{i=1}^{n} b_1b_2A_1A_{i,2} + \sum_{i=1}^{n} b_1b_3A_1A_{i,3}\\ &+\sum_{i=1}^{n} b_2b_0A_2 + \sum_{i=1}^{n} b_2b_1A_1A_2 + \Sigma b_2^2A_2^2 + \sum_{i=1}^{n} b_2b_3A_2A_3\\ &+\sum_{i=1}^{n} b_3b_0A_3 + \Sigma b_3b_1A_3A_1 + \sum_{i=1}^{n} b_3b_2A_3A_2 + \sum_{i=1}^{n} b_3^2A_3^2 - \lambda\cdot\left[0-b_1-b_2-b_3\right]\end{aligned} \tag{12.35}$$

Now, we can take the derivatives (with respect to all the b_i and with respect to λ) of Equation 12.35:

$$\begin{aligned}\partial\left(\Sigma E_i^2\right)/\partial b_0 = &\,0 - \sum_{i=1}^{n} b_0 2C_i - 0 - 0 - 0 + \sum_{i=1}^{n} 2 + \sum_{i=1}^{n} b_1A_1 + \sum_{i=1}^{n} b_2A_2 + \sum_{i=1}^{n} b_3A_3\\ &+\sum_{i-1}^{n} b_1A_1 + 0 + 0 + 0 + \sum_{i=1}^{n} b_2A_2 + 0 + 0 + 0\\ &+\sum_{i=1}^{n} b_3A_3 + 0 + 0 + 0 - \lambda\cdot\left[0-b_1-b_2-b_3\right]\end{aligned} \tag{12.36a}$$

$$\begin{aligned}\partial\left(\Sigma E_i^2\right)/\partial b_1 = &\,0 - 0 - \sum_{i=1}^{n} 2C_iA_1 - 0 - 0 + 0 + \sum_{i=1}^{n} b_0A_1 + 0 + 0\\ &+\sum_{i=1}^{n} b_0A_1 + \sum_{i=1}^{n} 2A_1^2 + \sum_{i=1}^{n} b_2A_1A_2 + \sum_{i=1}^{n} b_3A_1A_3 + 0\\ &+\sum_{i=1}^{n} b_2A_1A_2 + 0 + 0 + 0 + \Sigma b_3A_3A_1 + 0 + 0 - \lambda\cdot\left[0-b_1-b_2-b_3\right]\end{aligned} \tag{12.36b}$$

$$\partial\left(SE_i^2\right)/\partial b_2 = 0-0-0-\sum_{i=1}^{n} 2C_1A_2 - 0+0+0+\sum_{i=1}^{n} b_0A_2 + 0$$
$$+0+0+\sum_{i=1}^{n} b_1A_1A_{i,2} + 0 + \sum_{i=1}^{n} b_0A_2 + \sum_{i=1}^{n} b_1A_1A_2 + \Sigma 2A_2^2$$
$$+\sum_{i=1}^{n} b_3A_2A_3 + 0 + 0 + \sum_{i=1}^{n} b_3A_3A_{i,2} + 0 - \lambda\cdot\left[0-b_1-b_2-b_3\right] \quad (12.36c)$$

$$\partial\left(\Sigma E_i^{\,2}\right)/\partial b_3 = 0-0-0-0-\sum_{i=1}^{n} 2C_iA_3 + 0+0+0+\sum_{i=1}^{n} b_0A_3$$
$$+0+0+0+\sum_{i=1}^{n} b_1A_1A_3 + 0+0+0+\sum_{i=1}^{n} b_2A_2A_3 + \sum_{i=1}^{n} b_0A_3$$
$$+\Sigma b_1A_3A_1 + \sum_{i=1}^{n} b_2A_3A_2 + \sum_{i=1}^{n} 2b_3A_3^2 - \lambda\cdot\left[0-b_1-b_2-b_3\right] \quad (12.36d)$$

$$\partial\left(\Sigma E_i^2\right)/\partial\lambda = 0-b_1-b_2-b_3 \quad (12.36e)$$

Removing the superfluous zeros:

$$\partial\left(SE_i^{\,2}\right)/\partial b_0 = -2\sum_{i=1}^{n} C_i + 2\sum_{i=1}^{n} 1 + \sum_{i=1}^{n} b_1A_1 + \sum_{i=1}^{n} b_2A_2 + \sum_{i=1}^{n} b_3A_3$$
$$+\sum_{i=1}^{n} b_1A_1 + \sum_{i=1}^{n} b_2A_2 + \sum_{i=1}^{n} b_3A_3 - \lambda\cdot 0 \quad (12.37a)$$

$$\partial\left(\Sigma E_i^{\,2}\right)/\partial b_1 = -2\sum_{i=1}^{n} C_iA_1 + \sum_{i=1}^{n} b_0A_1 + \sum_{i=1}^{n} b_0A_1 + 2\sum_{i=1}^{n} A_1^2 + \sum_{i=1}^{n} b_2A_1A_2$$
$$+\sum_{i=1}^{n} b_3A_1A_3 + \sum_{i=1}^{n} b_2A_1A_2 + \Sigma b_3A_3A_1 - \lambda\cdot 1 \quad (12.37b)$$

$$\partial\left(\Sigma E_i^{\,2}\right)/\partial b_2 = -2\sum_{i=1}^{n} C_1A_2 + \sum_{i=1}^{n} b_0A_2 + \sum_{i=1}^{n} b_1A_1A_2 + \sum_{i=1}^{n} b_0A_2$$
$$+\sum_{i=1}^{n} b_1A_1A_2 + 2\Sigma A_2^2 + \sum_{i=1}^{n} b_3A_2A_3 + \sum_{i=1}^{n} b_3A_3A_2 - \lambda\cdot 1 \quad (12.37c)$$

$$\partial\left(\Sigma E_i^{\,2}\right)/\partial b_3 = -2\sum_{i=1}^{n} C_iA_3 + \sum_{i=1}^{n} b_0A_3 + \sum_{i=1}^{n} b_1A_1A_3 + \sum_{i=1}^{n} b_2A_2A_3$$
$$+\sum_{i=1}^{n} b_0A_3 + \Sigma b_1A_3A_1 + \sum_{i=1}^{n} b_2A_3A_2 + 2\sum_{i=1}^{n} b_3A_3^2 - \lambda\cdot 1 \quad (12.37d)$$

$$\partial\left(\Sigma E_i^2\right)/\partial\lambda = 0-b_1-b_2-b_3+0 \quad (12.37e)$$

Collect like terms and set the derivatives equal to zero:

$$0 = -2\sum_{i=1}^{n} C_i + 2\sum_{i=1}^{n} 1 + 2\sum_{i=1}^{n} b_1A_1 + 2\sum_{i=1}^{n} b_2A_2 + 2\sum_{i=1}^{n} b_3A_3 - \lambda\cdot 0 \quad (12.38a)$$

$$0 = -2\sum_{i=1}^{n} C_iA_1 + 2\sum_{i=1}^{n} b_0A_1 + 2\sum_{i=1}^{n} b_1A_1^2 + 2\sum_{i=1}^{n} b_2A_1A_2 + 2\sum_{i=1}^{n} b_3A_1A_3 - \lambda\cdot 1 \quad (12.38b)$$

$$0 = -2\sum_{i=1}^{n} C_1 A_2 + \sum_{i=1}^{n} b_0 A_2 + 2\sum_{i=1}^{n} b_1 A_1 A_2 + 2\, b_2 \Sigma A_2^2 + 2\sum_{i=1}^{n} b_3 A_2 A_3 - \lambda \cdot 1 \quad (12.38c)$$

$$0 = -2\sum_{i=1}^{n} C_i A_3 + 2\sum_{i=1}^{n} b_0 A_3 + 2\sum_{i=1}^{n} b_1 A_1 A_3 + 2\sum_{i=1}^{n} b_2 A_2 A_3 + 2\sum_{i=1}^{n} b_3 A_3^2 - \lambda \cdot 1 \quad (12.38d)$$

$$0 = 0 - b_1 - b_2 - b_3 + 0 \quad (12.38e)$$

Divide each equation through by 2:

$$0 = -\sum_{i=1}^{n} C_i + \sum_{i=1}^{n} 1 + \sum_{i=1}^{n} b_1 A_1 + \sum_{i=1}^{n} b_2 A_2 + \sum_{i=1}^{n} b_3 A_3 + \lambda \cdot 0 \quad (12.39a)$$

$$0 = -\sum_{i=1}^{n} C_i A_1 + \sum_{i=1}^{n} b_0 A_1 + \sum_{i=1}^{n} b_1 A_1^2 + \sum_{i=1}^{n} b_2 A_1 A_2 + \sum_{i=1}^{n} b_3 A_1 A_3 - \lambda \cdot 1 \quad (12.39b)$$

$$0 = -\sum_{i=1}^{n} C_1 A_2 + \sum_{i=1}^{n} b_0 A_2 + \sum_{i=1}^{n} b_1 A_1 A_2 + b_2 \Sigma A_2^2 + \sum_{i=1}^{n} b_3 A_2 A_3 - \lambda \cdot 1 \quad (12.39c)$$

$$0 = -\sum_{i=1}^{n} C_i A_3 + \sum_{i=1}^{n} b_0 A_3 + \sum_{i=1}^{n} b_1 A_1 A_3 + \sum_{i=1}^{n} b_2 A_2 A_3 + \sum_{i=1}^{n} b_3 A_3^2 - \lambda \cdot 1 \quad (12.39d)$$

$$0 = 0 - b_1 - b_2 - b_3 + [0] \quad (12.39e)$$

The equations can be rearranged thusly:

$$\sum_{i=1}^{n} C_i = b_0 \sum_{i=1}^{n} (1) + b_1 \sum_{i=1}^{n} A_1 + b_2 \sum_{i=1}^{n} A_2 + b_3 \sum_{i=1}^{n} A_3 - \lambda \cdot 0 \quad (12.40a)$$

$$-\sum_{i=1}^{n} C_i A_1 = b_0 \sum_{i=1}^{n} A_1 + b_1 \sum_{i=1}^{n} A_1^2 + b_2 \sum_{i=1}^{n} A_1 A_2 + b_3 \sum_{i=1}^{n} A_1 A_3 - \lambda \cdot 1 \quad (12.40b)$$

$$-\sum_{i=1}^{n} C_1 A_2 = +b_0 \sum_{i=1}^{n} A_2 + b_1 \sum_{i=1}^{n} A_1 A_2 + b_2 \Sigma A_2^2 + b_3 \sum_{i=1}^{n} A_2 A_3 - \lambda \cdot 1 \quad (12.40c)$$

$$-\sum_{i=1}^{n} C_i A_3 = \sum_{i=1}^{n} b_0 A_3 + \sum_{i=1}^{n} b_1 A_1 A_3 + \sum_{i=1}^{n} b_2 A_2 A_3 + \sum_{i=1}^{n} b_3 A_3^2 - \lambda \cdot 1 \quad (12.40d)$$

$$0 = 0 - b_1 - b_2 - b_3 + 0 \quad (12.40e)$$

We now have five equations in five unknowns: four for the regression coefficients (b_0 and $b_1 - b_3$) and one coefficient for λ. If we compare these equations with Equations 12.12, we see that Equations 12.40 are the same as the normal equations with some additions:

1. The presence of a term including λ on the RHS of each equation. This comprises another variable to be solved. This is necessary, since adding the constraint to the normal equations created an extra equation, and we therefore need an extra variable in order to have a set of equations that are solvable algebraically
2. An extra equation specifying the constraint. This equation added to the normal equations ensures that the coefficients arrived at will not only be the ones that minimize the sum-squared error, but since all five equations must be satisfied, the coefficients must also have the property that their sum of the coefficients equals zero.

We are almost there.

We are now ready to invoke the magic of a matrix algebra approach to multivariate analysis. We write out Equations 12.40 in matrix format:

$$\begin{bmatrix} \sum C_i \\ \sum C_i A_1 \\ \sum C_i A_2 \\ \sum C_i A_3 \\ 0 \end{bmatrix} = \begin{bmatrix} b_0 \sum 1 & b_1 \sum A_1 & b_2 \sum A_2 & b_3 \sum A_3 & 0 \cdot \lambda \\ b_0 \sum A_1 & b_1 \sum A_1^2 & b_2 \sum A_1 A_2 & b_3 \sum A_1 A_3 & 1 \cdot \lambda \\ b_0 \sum A_2 & b_1 \sum A_1 A_2 & b_2 \sum A_2^2 & b_3 \sum A_2 A_3 & 1 \cdot \lambda \\ b_0 \sum A_3 & b_1 \sum A_1 A_3 & b_2 \sum A_2 A_3 & b_3 \sum A_3^2 & 1 \cdot \lambda \\ 0 \cdot b_0 & 1 \cdot b_1 & 1 \cdot b_2 & 1 \cdot b_3 & 0 \cdot \lambda \end{bmatrix} \tag{12.41}$$

Factor the RHS:

$$\begin{bmatrix} \sum C_i \\ \sum C_i A_1 \\ \sum C_i A_2 \\ \sum C_i A_3 \\ 0 \end{bmatrix} = \begin{bmatrix} \sum 1 & \sum A_1 & \sum A_2 & \sum A_3 & 0 \\ \sum A_1 & \sum A_1^2 & \sum A_1 A_2 & \sum A_1 A_3 & 1 \\ \sum A_2 & \sum A_1 A_2 & \sum A_2^2 & \sum A_2 A_3 & 1 \\ \sum A_3 & \sum A_1 A_3 & \sum A_2 A_3 & \sum A_3^2 & 1 \\ 0 & 1 & 1 & 1 & 0 \end{bmatrix} \begin{bmatrix} b_0 \\ b_1 \\ b_2 \\ b_3 \\ \lambda \end{bmatrix} \tag{12.42}$$

In the shorthand notation of the matrices, Equation 12.42 is the same as Equation 12.16, except that now, matrix **[A]** contains variables beyond the absorbances Equation 12.16 contains (i.e., it also includes λ and an extra row and column). Thus, in the shorthand notation of a matrix, Equation 12.42 becomes the same as Equation 12.16:

$$A^T C = A^T A \cdot b \tag{12.43}$$

Unsurprisingly, therefore, the solution is the same as for Equation 12.16:

$$b = [A^T A]^{-1} \cdot A^T C \tag{12.44}$$

The matrix calculations specified here are the same as with ordinary regression, and the difference is in the nature of the matrices that are used. The presence of the b_i in the normal equations specified by Equation 12.42 means that the algebraic solution to the normal equations must be such that the sum of the coefficients equals zero. And now, the vector of least squares coefficients, b, additionally has the property that the sum of the coefficients (not including b_0) equals zero.

12.4.2.3 Variation 3: Ridge Regression

Ridge regression is an interesting variation on the basic theme of calibration which has been sporadically reported for NIR calibration. It was another attempt to remove the effect of the optical variations due to particle size and scattering effects, from the calibration model. The normal equations described above in Equations 12.12 are what statisticians call "unbiased" (in a carefully defined sense that has nothing to do with the term "bias" as used in NIR), but are sensitive to the noise variations of the data. Calibration models calculated from data that are highly intercorrelated, as NIR spectra from powdered solids tend to be, are highly sensitive to noise more than similar data would be in the absence of the intercorrelations. One definition of ridge regression (RR) is that it is a way to create a parsimonious model when the number of predictor variables in a set exceeds the number of observations, and/or when a data set has multicollinearity (correlations between predictor variables) [21].

The calibration model and the computations involved are the same as for "ordinary" regression; the difference lies in a post-computation calculation. There are many Web sites that describe the calculation needed and the rationale for these calculations to see this small sampling of sites [21–25] or perform a simple online search for "ridge regression" which will turn up many more explanations.

As a matrix equation, the post-calculation has a very simple description:

$$\left[X^T X + \gamma[1]\right]^{-1} \tag{12.45}$$

which is substituted for $[X^T X]$ in Equation 12.17.

Operationally, this has a very simple meaning: After computing the sum-of-products matrix $[\mathbf{X}^T\mathbf{X}]$ to form the normal equations, the main diagonal of that matrix is modified by adding the constant $\boldsymbol{\gamma}$ to each element of the main diagonal in the algebraic development. When the matrix is inverted and the rest of the computations carried out in the same manner as they normally otherwise would, the resulting calibration model is less sensitive to the vagaries of the noise in the data. This is by virtue of the fact that the regression coefficients are smaller than they would be if the unmodified $[\mathbf{X}^T\mathbf{X}]$ matrix was used.

This calibration model is no longer an "unbiased" model as described above. It is a biased model and can be expected to be somewhat less accurate than the original unbiased model. The value of $\boldsymbol{\gamma}$ to use is determined empirically and can be set to optimize the performance, keeping in mind that larger values of $\boldsymbol{\gamma}$ correspond to models with less noise sensitivity but potentially more overall error.

The Internet is a rich source of information about ridge regression. Users can perform their own search or visit the Web sites listed above to learn more about ridge regression. Ref. [24] recommends that if the variance inflation factor (VIF) of a calibration model exceeds 10, then that is an indication that the data set is a possible candidate for ridge regression.

12.5 PRINCIPAL COMPONENT REGRESSION (PCR)

In practice, the MLR approach is the conceptually simplest one in routine use for calibrating NIR spectrometers. Even with the need for multiple wavelengths and least squares computations, the amount of calculation needed is less than other algorithms in current use or that we describe above. At least, that is true when you know the wavelengths to use to specify the data you will use to perform the calculations.

The "fatal flaw" in the application of MLR to generate calibration models for NIR analysis is the need to select the wavelengths that specify the absorbance data to use. In the earliest days of NIRS, several commercial instruments used interference filters to separate the individual wavelengths at which the absorbance data were measured (at least one is still in business and selling interference filter-based instruments). Some instruments used as few as six filters, which were enough to create calibration models for the critical components (oil, protein, and moisture) in the applications that were important (mainly for agricultural products: wheat, barley, corn, soybeans, tobacco). Even with as few as six wavelengths to choose from, problems arose in trying to select the "best" set to use for a given application:

1. Simply using all the wavelengths resulted in calibrations that were excessively sensitive to random noise, since the intercorrelations were very high and therefore the calibrations were unstable.

2. Because of this instability, trying to recreate a model by using a separate set of calibration samples resulted in two different models, with no clear way to determine which to choose.
3. Trial-and-error work revealed that using less than the full set of data would improve the stability of the model generation process, but usually at the expense of accuracy of the analytical values obtained from the resulting models.
4. Even culling wavelengths did not solve the wavelength-selection problem, because often more than one combination of wavelengths could produce calibration models of similar accuracy and it was still not clear whether any one was "best."
5. When scanning spectrometers or even instruments with more than six interference filters became common, the number of possible wavelength combinations increased factorially, making empirical wavelength selections unfeasibly lengthy.
6. Automated computer algorithms for searching through the possible combinations were developed, but could not overcome the factorial increase in computation needed to find the "best" set of wavelengths. Some search algorithms were developed that limited the amount of searching to be done, to make the search process become feasible. However, those algorithms inevitably were limited to inspect only a small fraction of the potential wavelength combinations that a given data set could provide, and therefore, the user could never be sure that the best combination was included in the search.

The problem of wavelength selection was never completely solved satisfactorily, although, in practice, several algorithms were developed that produced results in the form of calibration models that were "good enough." Simultaneously, the time for a search was reduced so as to be "short enough." Improvements in hardware, in the form of faster computers, also helped.

But what really brought about demise of MLR was the advent of new calibration algorithms that were able to utilize the entire spectrum without running afoul of the intercorrelation effects that plague MLR. We now know these algorithms as PCR (principal component regression) and PLS (partial least squares). We will consider primarily PCR here.

PLS is a relatively minor variation of the algorithm, although considerably more complicated to understand and describe. Instead of trying to fight the intercorrelation, the PCR algorithm uses that very characteristic to create new "spectra." These "spectra" are empirical mathematical functions, computed from the spectral data, that are exactly designed to use the correlations within the data set, and put them into the empirical functions that are being developed.

A key distinguishing characteristic of the empirical functions that are created is that each is the least squares estimator of the data set from which it is derived, analogously to the way a regression line is the least squares estimator of the data to which it is fitted. An example is shown in Figure 12.2, which presents four more or less similar spectra, shown as black lines. A principal

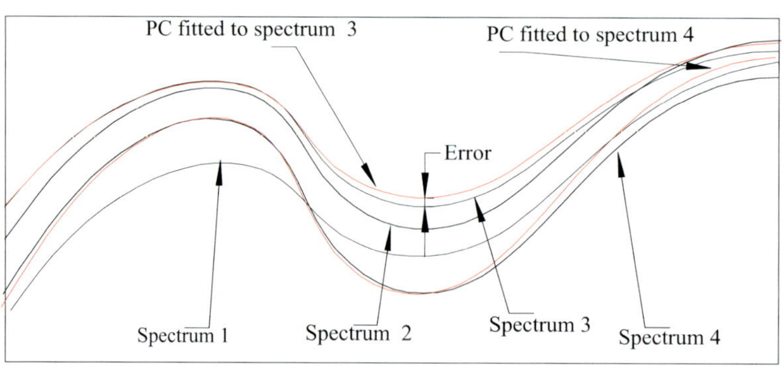

FIGURE 12.2 This figure shows four (stylized hypothetical) spectra (in black) and (for clarity) only one principal component (in red) fitted to two of the spectra.

component is shown fitted to two of the data spectra (to show more of them quickly made the diagram become too "busy" and confusing). The error between the data and the fitting function (PC) is shown at one wavelength for one of the spectra. However, after the PC is fitted to any given spectrum, the error is calculated for all wavelengths in the spectrum and the sum-squared error calculated. This process is repeated for all the spectra in the data set and the total sum-squared error calculated. The exact shape of the PC and the ability for it to be fitted to the individual data spectra are optimized, so that the total sum-squared error is the least possible value that can be achieved for that data set and that PC.

The mathematics of how this is accomplished is similar to the development of MLR calibration equations with coefficients that sum to zero, as described above: Errors (as shown in Figure 12.2) are squared and summed, the Lagrange multiplier is introduced, and then, derivatives are taken and set equal to zero. A key difference between computing principal components and MLR calibrations with coefficients that sum to zero is that whereas the MLR calibration required only a matrix inversion to solve for the calibration coefficients, computing principal components is more intricate and requires solving an eigenvalue problem. It is a bit lengthy to include the entire derivation here, but the details of that can be found in a series of Spectroscopy columns [26–30] and also in Chapters 82–86 in Ref. [31].

All this, however, tells us how to compute one principal component. Having done all that, then what do we do?

The answer to that question is as follows: The principal component we have computed is, by virtue of the way it is defined and computed, the empirical mathematical function that accounts for the maximum possible amount of variance in the data set which any mathematical function (empirical or theoretical) can do, that is, any single mathematical function. But that does not mean that it has removed all the variance, or that there is no more variance left to remove. But we need something else to do it; the one we just computed will have done all it could.

So, the next step is to actually do it: that is, remove the variance that this PC, for which the first PC (simply because we calculated it first) accounts. How do we do that? Compared to what we just went through, it is really pretty simple and straightforward. Figure 12.3 provides a block diagram of the procedure. We fit the PC we have computed, one by one, to the spectra in the original data set. That gives us a set of coefficients, called the scores, that represents the amount of variance that

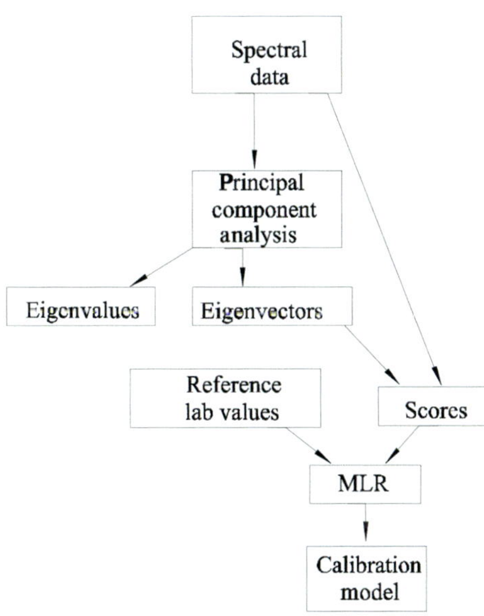

FIGURE 12.3 Block diagram of the PC-based calibration modeling algorithm.

the specified PC has contributed to each spectrum. Then, we multiply the principal component, wavelength by wavelength, by its score and subtract each set of products from the corresponding data spectrum. This results in the complete removal of the contribution of that PC from the set of data spectra. Repeating that for each data spectrum leaves us with a set of remainders, or residual spectra.

Going back to the beginning, we do the whole thing all over again, but using the set of residual spectra as the data, computing a second (and third…) PC, and replacing the residual spectra with a set of new residual spectra. This allows us to calculate the second PC.

We repeat this process as many times as we care to, or as many times as we think is needed to account for all the variance in the set of original data spectra, each time computing one more principal component. In actual fact, here we are explaining concepts, and in practice, there are shortcuts that are normally applied to reduce the amount of computation needed.

By virtue of the fact that, at each step, the contribution to the variance of the computed PC is removed by removing the contribution of that PC to the spectra, as described above. That means that all further PCs that are computed from that data set are necessarily orthogonal to the PCs that were already computed. It could not be otherwise; if any given PC had a contribution that was the same as a previously computed PC, then that contribution would have been computed, and its contribution removed, along with the contribution of the previous PC.

Even so, computing principal components is only part of the story. What we have been describing up to this point is called principal component analysis (PCA): the separation of the variance of a data set into discrete pieces, which have the following primary properties:

1. They are all mutually orthogonal (uncorrelated).
2. Each one is the least squares estimator of the data set left after removing the contribution of all the previous principal components.
3. Each one removed the maximum possible amount of variance from the data set, from the variance that was left after computing the previous PCs.
   There are also some secondary properties:
4. No more than one of them can have all positive or all negative values. Usually, that is the first one.

Having calculated the principal components, the next step is to relate those derived spectra to the composition of the samples. This is what is known as PCR (principal component regression). The procedure is summarized in the block diagram of Figure 12.3.

The product of the eigenvectors resulting from the PCA, together with the spectral data, gives the scores for each eigenvector for each sample. These scores are then used to replace the spectral data in an MLR calculation. There are typically many fewer PCA scores than there are wavelength points in the original data, making the process of selecting the correct factors much simpler and easier than to use the original wavelength data. The fact that the procedure used to calculate the PCA eigenvectors and eigenvalues inherently does so in decreasing order of the importance of the eigenvectors that are calculated is also helpful in selecting the scores to use for the finding the relationship between the optical data and the constituent concentrations.

REFERENCES

1. Mark, H.; Statistics for spectroscopy; in *Encyclopedia of Spectroscopy and Spectrometry*; Linden, J. C., Tranter, G. E., Koppenaal, D. W.: Eds.; Elsevier; Oxford; Vol. 4; pp. 250–260 (2017).
2. Kruskal, W., Tanur, J.; *International Encyclopedia of Statistics*, The Free Press, Division of MacMillan Publishing Co., Inc.; New York; Vol. 2 (1978).
3. Daniels, F., Mathews, J. H., Williams, J. W., Bender, P., Alberty, R. A.; *Experimental Physical Chemistry*; Fifth ed; McGraw-Hill Book Company, Inc.; New York (1956)

4. Daniel, C., Wood, F.; *Fitting Equations to Data - Computer Analysis of Multifactor Data for Scientists and Engineers*; First ed; John Wiley & Sons; New York (1971).
5. Mark, H.; *Principles and Practice of Spectroscopic Calibration*; John Wiley & Sons; New York (1991).
6. Mark, H., Workman, J.; *Chemometrics in Spectroscopy*; Elsevier; Amsterdam, New York (2007).
7. Brereton, R. G.; *Chemometrics: Data Analysis for the Laboratory and Chemical Plant*; Wiley; West Sussex (2003).
8. Sharaf, M., Illman, D., Kowalski, B.; *Chemometrics*; John Wiley & Sons; New York (1986).
9. Cooley, W. W., Lohnes, P. R.; *Multivariate Data Analysis*; John Wiley & Sons; New York (1971).
10. Mark, H., Workman, J.; Chemometrics in Spectroscopy - Part 2 - Elementary Matrix Algebra, Part 1. *Spectroscopy*; 8 (7), pp. 16–19 (1993).
11. Workman, J., Mark, H.; Chemometrics in Spectroscopy - Part 3 - Elementary Matrix Algebra, Part 2. *Spectroscopy*; 8 (8), pp. 16–18 (1993).
12. Workman, J., Mark, H.; Chemometrics in Spectroscopy - Part 4 - Matrix Algebra and Multiple Linear Regression, Part 1. *Spectroscopy*; 8 (9), pp. 16–20 (1993).
13. Workman, J., Mark, H.; Chemometrics in Spectroscopy - Part 5 - Matrix Algebra and Multiple Linear Regression, Part 2. *Spectroscopy*; 9 (1), pp. 16–29 (1994).
14. Workman, J., Mark, H.; Chemometrics in Spectroscopy - Part 6 - Matrix Algebra and Multiple Linear Regression, Part 3. *Spectroscopy*; 9 (4), pp. 18–19 (1994).
15. Workman, J., Mark, H.; Chemometrics in Spectroscopy - Part 7 - Matrix Algebra and Multiple Linear Regression: Conclusion. *Spectroscopy*; 9 (5), pp. 22–23 (1994).
16. Draper, N., Smith, H.; *Applied Regression Analysis*; Third ed; John Wiley & Sons; New York (1998).
17. McClure, F.; Analysis using Fourier transforms; in *Handbook of Near-Infrared Analysis*, Third edition; Burns, D., Ciurczak, E.: Eds.; Taylor & Francis Group; New York; p. 808 (1992).
18. Taylor, A. E.; *Advanced Calculus*; Ginn and Company; New York (1955).
19. https://www.khanacademy.org/math/multivariable-calculus/applications-of-multivariable-derivatives/constrained-optimization/a/lagrange-multipliers-examples.
20. http://tutorial.math.lamar.edu/Classes/CalcIII/LagrangeMultipliers.aspx.
21. www.statisticshowto.com/ridge-regression/.
22. https://onlinecourses.science.psu.edu/stat857/node/155p.
23. http://web.as.uky.edu/statistics/users/pbreheny/764-f11/notes/9-1.pdf.
24. http://ncss.wpengine.netdna-cdn.com/wp-content/themes/ncss/pdf/Procedures/NCSS/Ridge_Regression.pdf.
25. https://www.analyticsvidhya.com/blog/ 2017/06/a-comprehensive-guide-for-linear-ridge-and-lasso-regression/.
26. Mark, H., Workman, J.; *Spectroscopy*; 23 (2), pp. 30–37 (2008).
27. Mark, H., Workman, J.; *Spectroscopy*; 23 (5), pp. 14–17 (2008).
28. Mark, H., Workman, J.; *Spectroscopy*; 23 (6), pp. 22–24 (2008).
29. Mark, H., Workman, J.; *Spectroscopy*; 23 (10), pp. 24–29 (2008).
30. Mark, H., Workman, J.; *Spectroscopy*; 24 (2), pp. 16–26 (2008).
31. Mark, H., Workman, J.; *Chemometrics in Spectroscopy*, 2nd Edition; Elsevier; Amsterdam (2018).

13 Bayesian Regression for Chemical Calibrations of NIR Spectra

Steven D. Brown[*]
University of Delaware
and
IRSTEA

CONTENTS

The recent emphasis on data science has led to increasing interest in Bayesian methods for data analysis. Philosophically, the Bayesian approach to data analysis is close to that employed by practitioners in traditional measurement science, but the implementation of Bayesian methods often seems foreign to researchers who are more comfortable with conventional univariate and multivariate calibration. The flexibility offered by Bayesian methods is coupled with some mathematics that differs from the usual matrix algebra needed for the conventional regression modeling used in multivariate calibration of spectral mixtures. A Bayesian approach to regression modeling generally requires more background in statistics and computation than that needed for conventional regression modeling, and often the analyses require longer computation times, but the return gained from a Bayesian approach can be substantial, both in performance and in interpretability of the results.

[*] sdb@udel.edu

Because Bayesian methods are still rather obscure in chemometrics, and especially in analysis of spectral data, this chapter begins with a brief tutorial on some basic Bayesian approaches to regression. It then illustrates some ways in which Bayesian methods have been applied to problems in multivariate calibration of near-infrared spectral data and assesses the advantages and drawbacks of Bayesian methods for routine use.

13.1 A VERY BRIEF INTRODUCTION TO BAYESIAN STATISTICS

Before considering Bayesian regression, it is necessary to discuss what is meant by a "Bayesian" analysis and how that approach differs from an analysis using conventional statistical methods. These conventional methods include ordinary classical least squares regression, inverse least squares regression, principal components regression (PCR), partial least squares (PLS) regression, and other methods, linear or not, that are all based on the same basic idea of how the data behave and how to extract information from the data.

Most researchers performing multivariate calibration of spectrometric data ask "what do my data show?" There is the implicit understanding that the data carry information and that a set of unknown, true model calibration parameters summarize that information. In conventional calibration, there is an emphasis on finding estimates of those model parameters, such as regression coefficients, from the experimental data. Most commonly, a particular model form is assumed, and the calibration data, presumed to contain noise, are then fit to that form to find estimates for this set of fixed, true model parameters. From these estimates of the true parameters, a prediction of the value of the dependent variable y_{unk} in new data is estimated from the model parameters and the new predictor vector, $\mathbf{x}_{\text{unk}}$. The usual approach is to regard the predictor variables $\mathbf{X}$ as noise-free, and the dependent variable(s) y as corrupted by a source of homoscedastic, independent, additive noise. The model parameters, here the coefficients in the regression model, are always taken as unknown, noise-free constants. Probability distributions, if they are considered at all, generally are discussed in terms of means and variances of the estimated y_{unk} as reflected in the root mean squared error of many predictions (RMSEP), which ultimately connect to the uncertainty in the responses in the calibration set $\mathbf{y}$ and the reliability of the computation, exemplified in classical least squares regression by the accuracy of the matrix inverse $(\mathbf{X}^{\mathrm{T}}\mathbf{X})^{-1}$ (Mardia, et al., 1979).

This analysis has an implicit probability distribution for $\mathbf{y}$ that is Gaussian, centered on the average y value and with a variance specified by the noise in y. This distribution is convenient for predictions of properties for data that fall in the center of the calibration set because that is where the y values most frequently occur.

Data analysts who use Bayesian methods—Bayesians for short—call this modeling approach "frequentist" because the results are based on the frequency distribution of a sample of noisy $\mathbf{y}$ data provided for the calibration. Under this view, the data are a sample from a larger population and the data consist of a deterministic part and a random, noisy, part. What matters in the frequentist approach to modeling is the histogram of the $\mathbf{y}$ values used in the calibration.

The Bayesian approach to calibration modeling takes a completely different view of the problem of finding a calibration model and using it to predict the amount unknown in new samples. In the Bayesian approach, the emphasis is again on determining the calibration model parameters and predictions of the property for new data, but as a consequence of the new information contained in the calibration data we collect. Unlike the frequentist approach described above, the Bayesian approach relies on a fundamentally different view of the data and the parameters in the model. Bayesian methods are based on probability distributions because these methods presume that the set of model parameters θ are stochastic quantities, depending on what data are observed. The Bayesian approach also presumes that the observed data $\mathbf{D}$ are deterministic. This view of data analysis contrasts with the conventional (frequentist) view that the parameters are fixed, unknown constants and the data $\mathbf{D}$ are made of a stable, deterministic part affected by additive, stochastic error, so each observed datum is a sample drawn from some distribution with a frequency f.

The parameters in a Bayesian model are taken as stochastic because these are derived from data, and they are not observed. The data—both the predictors **X** and the response variable **y**—are taken as deterministic and without error in a Bayesian analysis because these are what we observed—they are not considered a sample from some population of possible data values, as is done in the frequentist approach. Because the distributions of stochastic variables in the model are based on our assessment of the possible value of a variable, not on the number of times a value for that variable occurs—its relative frequency—we call these belief distributions, reflecting our assessment of the value based on what we know about it. A belief distribution usually has the properties of a probability distribution.

A Bayesian approach has another major difference from the usual frequentist modeling: It relies on *two* models. One model incorporates any information on the desired parameters *before* data become available. This model is the prior distribution, often called the prior for short, and it specifies the prior belief in the stochastic parameters in the model. The prior distribution is specified by its form (e.g., Gaussian) and its parameters (usually the mean and variance). Some authors also give the prior values for the mean and variance as part of the definition, so a fully specified prior for some parameter β might be represented as $p(\beta)=G(\mu;\ 0,\ \sigma^2)$, indicating that the prior distribution for the stochastic parameter β is believed to be a Gaussian distribution with mean=0 and variance set by the value of the parameter σ^2. This prior indicates that the parameter β is probably zero but with some uncertainty. The value for the variance of β is an indirectly observed parameter, so it could also be set in the definition of the prior model, and then adjusted as information obtained from new data requires, or it might be specified as a hyperparameter, which can be either a preset, fixed, deterministic value that is not adjusted with information available from new data or a stochastic value that is modeled and adjusted based on observed data through another, "hierarchical" Bayesian analysis (Gelman, et al., 2013). Either approach to the prior provides a route to input information about the system being modeled before data are available, a capability that is not possible in a frequentist analysis.

Then, there is the model for new data. This is called the likelihood, and it relates the observed experimental data to the model parameters. The likelihood describes how the stochastic parameters in the model are related to the observed experimental data. This relationship is often mathematically similar to the linear algebraic model relating parameters to predictors used in the usual frequentist approach to modeling, but it, too, is a distribution in the Bayesian formalism.

The prior and likelihood distributions are then combined to give an estimate of the belief distribution for the stochastic parameters updated to consider the new information found in the data. This is the posterior distribution, often just called the posterior. It describes what is known about the stochastic model parameters after incorporation of any new information about those parameters provided in the experimental data.

In the Bayesian approach, then, the two models are each a probability distribution, and they combine to produce a posterior model probability distribution. This distribution embeds the prior information on the model system as updated by experimental observations. This posterior distribution can be queried for the value of a parameter in the model, and a calculation is known to statisticians as inference. Inference on the posterior might involve calculation of means and variances of the parameters, quantities sometimes obtained in the conventional frequentist approach, but often much more is available. The emphasis on probability distributions for all model parameters, on combining probability distributions, and on inferring values of model parameters from those distributions is what distinguishes a Bayesian method. The reliance on distributions and the need to extract parameters from those distributions make Bayesian analysis computationally demanding and difficult to learn for those with limited background in statistics.

While this introduction gives a sense of the approach and defines some of the terms common in a Bayesian analysis, those seeking more extensive background might want to start with general texts on Bayesian methods (Gelman, et al., 2013; Bolstead, 2010). There is also a very good introduction to Bayesian methods in regression in the text by Bishop (2006). For those readers comfortable with

R software, there are an enormous number of current, public-domain software packages implementing Bayesian methods at the CRAN repository (https://cran.r-project.org/). There is also a "task view," updated frequently, that summarizes much of the basic Bayesian material available in R and in related, special-purpose languages at https://cran.r-project.org/web/views/Bayesian.html.

13.1.1 Marginal and Conditional Probabilities

Bayesian methods make heavy use of marginal and conditional probabilities. A marginal probability distribution can be regarded as a part of a larger distribution, for example, a multidimensional probability distribution that varies over values for all three model parameters. Each of these parameters has its own marginal probability distribution where the probability is dependent only on values for that one parameter. For statistically independent parameters, these marginal distributions are independent of the other parameters; the product of the three marginal distributions is the full distribution, and each marginal distribution can easily be examined separately. If the parameters are not independent, getting a marginal distribution for one parameter from the full distribution requires integrating out—marginalizing—the contributions from all of the other parameters in the full distribution.

A conditional probability distribution reflects the probability distribution of some variable conditioned on the value(s) of another quantity or quantities; these conditional probabilities are indicated by a vertical bar, so, for example, $p(y| X, \beta)$ is the probability distribution of y conditioned on specific values of X and β. The likelihood function $p(D|\theta)$ is a conditional probability distribution that conditions the probability of observing some value of the data D on specific values of the model parameters θ.

Given the full prior probability distribution $p(\theta)$ for all of the model parameters θ and the overall likelihood function, $p(D|\theta)$, a conditional distribution describing how the data D are affected by the model parameters θ, the goal is to find the posterior conditional distribution $p(\theta|D)$, the conditional distribution that describes how the parameters θ are affected by the data D. This distribution describes the full distribution of the set of model parameters after the information present in the data is considered. This posterior conditional distribution summarizes results of the modeling. Standard statistical inference is then needed to extract information (e.g., the mean and variance) from this distribution, or from marginalized parts of this distribution, as appropriate.

13.1.2 Bayes' Theorem

Given the probabilistic models for the likelihood conditional dependence of the data on the parameters, $p(D|\theta)$, and the prior belief $p(\theta)$, Rev. Thomas Bayes was first to show that assuming statistical independence of the prior and the likelihood distributions and using conditional probability lead to $p(\theta|D)$ directly. His simple approach was formalized and promoted by Laplace, but the formula is known as Bayes' theorem. Bayes' theorem uses the principles of conditional probability to combine the prior distribution $p(\theta)$ and the likelihood distribution function $p(D|\theta)$ to permit the estimation of the conditional posterior, $p(\theta|D)$, and from it, the dependence of the parameters describing the model on the data D

$$p(\theta \mid D) = p(\theta)\, p(D \mid \theta)/p(D) \tag{13.1}$$

where $p(D)$, the probability distribution of the observed data, is usually unknown and often is integrated out and taken as simply a constant of normalization, so that the value of $p(\theta|D)$ integrates to 1, as required for a probability distribution.

Once the conditional posterior is obtained by using Equation 13.1, it is possible to make statistical inference on the posterior distribution to obtain the model parameters, including estimating their optimal values.

Thus, the Bayesian modeling approach involves four steps:

1. Establishing a full prior probability model for the model parameters, using all available information;
2. Establishing a likelihood model conditioning values of the parameters on data;
3. Establishing a full posterior probability model by conditioning the likelihood model on new data using Bayes' theorem; and
4. Assessing the fit of the model to the data and evaluating the optimal predictions from the posterior distribution by appropriate inference on the conditional posterior.

This sequence used in a Bayesian regression is illustrated in Figure 13.1.

The literature on Bayes' modeling agrees on these basic steps. However, there are several details in the Bayes' theorem calculation that cause difficulty in implementing those steps.

First, a choice of a prior distribution for each parameter θ in the model must be made. There are many ways to parameterize the model (what to use for θ) and what to use for the prior distribution, $p(\theta)$, for each θ in the model. There is also the need to decide which of the θ will be allowed to vary and which will be specified by fixed hyperparameters. The choice of priors in the Bayesian modeling is very flexible; in addition, what is used for the prior often has a significant impact on the model. Sometimes, the prior distribution $p(\theta)$ can be taken as uniform and unbounded; this is called an "improper" prior because the integral of $p(\theta)$ over an unbounded θ is infinite, and $p(\theta)$ is therefore not a "proper" probability distribution with an integral of 1. Choosing a "flat," "data-dominated," or "non-informative" prior model means that the analyst believes that there is no prior information available—as is implied in the usual frequentist analysis of the data. The flat prior can be used for Bayes' regression, and often is selected because it is simplest to implement and because it should give results very close to those from a frequentist regression using maximum-likelihood estimation.

Sometimes the form of a prior is selected to ease the computation. When a closed-form "analytical" solution is desired for a Bayesian analysis, as in a text or in an example, a conjugate prior is usually selected; the equation defining this prior comes from the same mathematical family of distributions as that of the distribution used for the likelihood model. This prior simplifies the calculations used to obtain the posterior distribution. For example, a multivariate Gaussian prior can

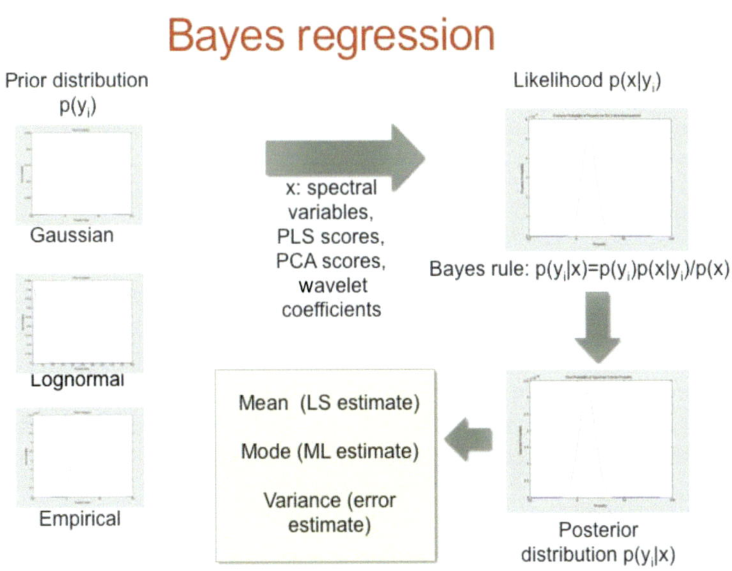

FIGURE 13.1 Workflow for a Bayesian regression.

be imposed on the parameters; this prior is conjugate to a multivariate Gaussian likelihood distribution. The value of the covariance of the multivariate Gaussian prior distribution can be set as a hyperparameter—usually taking the variances as fixed and homoscedastic—or the covariance can itself be regarded as a stochastic model parameter and allowed to adjust according to experimental data. Using either of these Gaussian priors with a Gaussian likelihood leads to a Gaussian posterior, and the computation (and the mathematics to derive and to integrate the conditional marginal posteriors) is greatly facilitated. For many situations, for example, in modeling with non-Gaussian likelihoods, using a conjugate prior is often is not possible, however. Fortunately, computational methods have improved greatly so that a numerical solution can be obtained, as is necessary in many of the applications of Bayesian regression methods to spectral calibration considered here; having convenient math is not always essential.

Once the posterior distribution is obtained, some sort of calculation must be used to find an optimal value for the θ parameter(s) in the model. To make that calculation, a decision is made on what is meant by $L\left(\theta,\widehat{\theta}\right)$ "optimal." Optimality must be specified through choice of a loss function that determines the consequences of any difference between the true model parameter value θ and its estimate $\widehat{\theta}$ (Berger, 2005). Much of the Bayesian literature often uses the maximum *a priori* (MAP) "0–1" loss function, where only an estimate made at the true value is considered useful, and all other values are considered worthless, regardless of their proximity to that true value. Choice of this loss function leads to using the mode of the posterior distribution as the optimal estimate of the parameter θ. When the prior distribution $p(\theta)$ is uniform, the use of this loss function results in a conventional maximum-likelihood estimate (MLE) $\widehat{\theta}$. Clearly, this choice is best used with unimodal conditional posterior distributions; not all posterior distributions are, in fact, unimodal, especially in regressions.

It is also possible to base optimality on minimizing the usual sum of squared differences between the true model parameter value θ and its estimate $\widehat{\theta}$ (the conventional "squared" (L_2) loss function). This choice leads to the usual minimum-variance (least squares) estimate using the mean $m(\theta)$ of the posterior distribution $p(\theta|D)$, where

$$m(\theta)=\sum_{k}\theta p(\theta|D) \tag{13.2}$$

is the least squares estimate for the mean of parameter θ. This estimate does not rely on unimodality of the posterior and is commonly used when the posterior is multimodal. Thus, how the posterior distribution is interrogated matters, and different ways of interrogation used in inference often give rise to different estimates of the parameters θ.

13.1.3 Belief and Priors

The many choices involved in setting up and interrogating a Bayesian model permit great flexibility in the modeling. This flexibility is a significant advantage in use, but it poses a major impediment to the newcomer to Bayesian analyses. The prior model can be structured to capture any information on the model that is available before data are observed. For example, the prior may reflect the belief that an intercept should be zero in a Beer's law model, but unlike the usual frequentist approach to modeling in which the intercept term is often dropped, the Bayesian approach puts a prior mean value of zero on this parameter, reflecting the belief that the value of the intercept parameter is usually zero, and then allows the available data to alter that value, or not, as appropriate. In a more complex multivariate model, priors may be set on all of the parameters, reflecting the belief that some terms in the regression model are more important than others in the modeling. Here, too, the data then provide information that alters these priors, reflecting the contribution of each term in the model based on the data and the loss function employed.

Because the choice of priors is subjective, reflecting the skill and prior knowledge of the researcher, any probability statement about the parameters resulting from the analysis is interpreted as the "degree of belief." This seemingly arbitrary selection of prior belief is balanced by the fact that Bayes' theorem provides the only consistent way to modify belief about parameters *given observed data*.

13.1.4 Establishing the Likelihood

The likelihood function also permits a flexibility not seen in the usual frequentist approach. Because the likelihood is a conditional probability distribution, and not a simple matrix algebraic relationship, it is possible to encode complex relationships between model parameters and data. Even empirical relationships, as discussed later, can be encoded in the likelihood. There is a cost involved with this flexibility: The full conditional distribution defining the likelihood must be defined over its region of applicability. While a frequentist relation relates data to parameters by a deterministic, algebraic equation, the Bayesian likelihood distribution is defined by the measured data, and fully defining the likelihood can sometimes require a substantial amount of data. The Bayesian model can use a deterministic likelihood structure defining what explanatory variables are included and how they are related to the response variable and to one another, as in a frequentist analysis, but the joint probability distribution for all observable and unobservable quantities in the problem must also be defined in the full probability model for the likelihood.

13.1.5 Obtaining and Assessing the Posterior

With the data defining the likelihood distribution, the appropriate posterior distribution can be computed from Equation 13.1. Getting the distribution $p(D)$ needed in Equation 13.1 is usually not an easy task, so either this quantity is conventionally omitted or it is treated as a normalization constant, and given a value so that the posterior probabilities integrate to 1 over the region of applicability.

The posterior distribution that results from Equation 13.1 is a joint probability distribution of all parameters, conditioned on the observed data. The multiple integration that is needed to estimate the marginal conditional posteriors for each of the parameters can be difficult; making this step easier is the main reason for choosing a conjugate prior. If the prior is not conjugate to the likelihood, integrating out the marginal conditional posterior distributions is challenging even when the number of parameters is small, and is almost always done numerically. However, even numerical integration may not be simple or quick when the number of parameters in the Bayesian model is large. In some cases, it is possible to simulate the posterior and then use sampling methods to estimate distributional parameters, but with "large" numbers of parameters in the model (more than about 30), even a sampling scheme as illustrated below can sometimes take substantial time, and approximate computational methods must be employed.

The numerical integrations and the significant computational burden they impose have been the largest impediment to use of Bayesian methods. Early workers in Bayesian multivariate regression used a Cray X-MP supercomputer for a Bayesian cross-validation approach to calculate contributions of only 10 predictors to a regression model (Mitchell and Beauchamp, 1988). The computational burden that these methods can impose has been substantially reduced by improvements in computation and by new approaches to extracting the marginal posteriors. These newer approaches permit the use of Bayesian regression on the large numbers of predictor variables typical in NIR spectral data, with results available in a few minutes' time, even on a laptop.

And last, once the marginal conditional posteriors are available, inference is performed on the posterior distributions: The parameters in the model are estimated, an assessment is made of how well the model fits the data, and predictions are made from the model by adding new data x to D and determining y from the joint posterior distribution. In almost all cases, a numerical approach is used because deciding on what is included in the model and deciding on the priors for those parameters and suitable values for any hyperparameters may require multiple passes of model building.

13.1.6 Simulation of the Posterior Distributions

By far, the most popular way of getting a numerical solution for a marginal conditional posterior from a Bayesian regression involves the use of a simulation. Most of these simulations are based on Markov chain Monte Carlo (MCMC) methods. An MCMC method relies on random draws (that's the Monte Carlo part), where the next step depends only on the current status, with no memory of historical status information (the Markov chain part). The aim is to simulate the target posterior distribution by sampling, with the distribution of any draw dependent on only the last sample drawn, so that the approximate distributions improve at each step. Eventually, the simulated distribution converges on the true target posterior distribution. MCMC simulation is sometimes implemented by the Metropolis–Hastings algorithm, a kind of random walk with an acceptance/rejection rule that is used to converge on a specific target distribution. This algorithm uses a jumping distribution in which the mean value of the distribution either stays where it is currently or moves to a new value, with the probability of movement depending on the ratio of probability densities between the new location and the current one. Steps that increase the density ratio are always accepted, but some steps that decrease density are, too, to avoid any local maxima. The Gibbs sampling algorithm is an alternative to Metropolis–Hastings that makes a movement based on distributions of each variable, conditioned on the values of all other variables. More details on these numerical simulation methods go beyond the scope of this review, but can be found in texts (e.g., Robert and Casella, 2004) and in the literature.

MCMC modeling using either a Metropolis–Hastings algorithm or Gibbs sampler depends on two important practical considerations. One is the number of trials needed to remove any significant effect from initial guesses used to start the simulation—the "burn-in" period. Once the burn-in period is reached, the shape of the estimated posterior distribution stabilizes and the parameter estimates from that distribution are improved considerably. Often, the burn-in for simulation of the posterior is long, and results are sometimes discarded from about 1/2 of the Markov chain, which might be thousands or more trial calculations, so burn-in for MCMC simulation can require significant computation time. And, second, the Markov chain optimization usually shows strong serial correlation in successive guesses, regardless of the algorithm used to choose the guess at each step. Successive guess values give little new information on the shape of the posterior, so these are usually "thinned" by choosing and keeping only one result from a set number of trials to reduce effects of serial correlation. The frequency of accepted points from the Markov chain is called the thinning rate, and it involves even more computation. Because of the need for burn-in and thinning to get an accurate simulation of the marginal posterior distributions, MCMC estimation of the posterior can be quite computationally demanding; this computation limits what can be done in a Bayesian analysis of highly multivariate data. Because the posterior is very high dimensional, many of parameters in the model are not statistically independent, and a marginal posterior might be required for each of the parameters in the multivariate model.

Given the issues mentioned above, it might seem that a Bayesian analysis is quite demanding to reduce to useable computer code, but there are now several packages that can be used to implement a typical Bayesian regression easily, and with minimal coding. The software in such free packages as Bayesian inference Using Gibbs Sampling (BUGS) and its variants RBUGS and WinBUGS, in STAN, in JAGS (Just Another Gibbs Sampler), and in other packages, now makes much of the computational side of a Bayesian regression analysis very straightforward, with user-selectable priors and built-in MCMC sampling. These packages generally run as an add-on to basic *R* or python software and in some cases to MATLAB, too. Some, like the WinBUGS package, are stand-alone. As the BUGS site notes in their "health notice," though, MCMC simulation is very effective, but it is not as robust as a closed-form solution, and there is no oversight against a careless application or misuse of the software, regardless of the package used.

13.2 BAYESIAN REGRESSION

The use of a prior distribution for the parameters in the model is an important main distinguishing characteristic of a Bayesian analysis. Even in the simplest case, the choice of the prior model makes a significant difference in the results from a Bayesian regression, but for the moment, the focus will be placed on the likelihood model. For example, consider the case where the data $\mathbf{D}=[\mathbf{X};\mathbf{y}]$ are modeled with a classical least squares (CLS) regression. This regression relationship defines a likelihood model where the deterministic predictor variables $\mathbf{X}$ relate to the stochastic response data $\mathbf{y}$:

$$\mathbf{y} = \boldsymbol{\beta}^T \mathbf{X} + \varepsilon \tag{13.3}$$

Here, the random fluctuation in the dependent variable is set by the error ε for each predictor variable. Using the conventional shorthand, the distribution for the error ε is taken as IID$(0,\sigma^2\,\mathbf{I})$ for each y, so the observed responses $\mathbf{y}$ in the likelihood are stochastic, unlike the deterministic, observed data in $\mathbf{X}$. Now, Equation 13.3 defines the explicit relationship between the model parameters $\boldsymbol{\beta}$ and the observed data in $\mathbf{X}$ and $\mathbf{y}$, but this relationship is expressed as a distribution. The error term ε is Gaussian-distributed with mean 0 and fixed variance σ^2, and since Equation 13.3 defines the relation between the observed data and the parameters, the likelihood expresses that relation as a multivariate Gaussian distribution with mean $\boldsymbol{\beta}^T\mathbf{X}$ and variance σ^2, giving the likelihood distribution $p(y|\boldsymbol{\beta})=G(y;\,\boldsymbol{\beta}^T\mathbf{X},\,\sigma^2)$. Here, the notation $G(a,\,b)$ indicates the likelihood distribution is Gaussian in shape with mean a and variance b. The semicolon indicates that the mean for y is set by the prior to be at the value $\boldsymbol{\beta}^T\mathbf{X}$.

In this model, there is no error in the deterministic predictor variable $\mathbf{X}$, but in the Bayesian view, the model parameter $\boldsymbol{\beta}$ is stochastic because it is not an observed quantity. The Bayesian model must specify its prior probability distribution, $p(\boldsymbol{\beta})$.

In this Bayesian model, the variance σ^2 is a hyperparameter; it is a deterministic parameter with a value set outside of the Bayesian modeling, so, while the value of σ^2 affects the performance of the model, as a hyperparameter, it is not assigned a prior and is not adjusted by the experimental data.

It is convenient to use a zero-mean Gaussian prior $p(\boldsymbol{\beta})$ for the regression parameter $\boldsymbol{\beta}$ because this prior distribution is conjugate to the Gaussian likelihood distribution, so a zero-mean distribution with fixed variance is selected:

$$p(\boldsymbol{\beta} \mid \alpha) = G\left(\boldsymbol{\beta}; 0, \alpha^{-1}\mathbf{I}\right) \tag{13.4}$$

conditioned on the hyperparameter α, a measure of the spread of the multivariate Gaussian distribution defined in Equation 13.4 above. Note that the quantity $\alpha^{-1}\mathbf{I}$ indicates that the covariance term is diagonal in this model, so the prior defined in Equation 13.4 assumes statistical independence of the parameters in the $\boldsymbol{\beta}$ vector.

Then, after experimental data are used to populate the likelihood distribution $p(\mathbf{y}|\mathbf{X},\,\beta)$, from application of Bayes' theorem, the un-normalized, conditional posterior in Equation 13.5 is obtained directly:

$$p\left(\beta|\mathbf{X},\mathbf{y}\right)\left(\beta|\right) \sim p\left(\mathbf{y}|\mathbf{X},\boldsymbol{\beta}\right)p\left(\boldsymbol{\beta}|\alpha\right) = G\left(\boldsymbol{\beta};\mathbf{m},\mathbf{S}\right) \tag{13.5}$$

where the posterior mean is

$$\mathbf{m} = \alpha^{-2}\mathbf{S}\mathbf{X}^T\mathbf{y} \tag{13.6}$$

and the posterior variance is

$$\mathbf{S} = \left(\alpha\mathbf{I} + \sigma^{-2}\mathbf{X}^T\mathbf{X}\right)^{-1} \tag{13.7}$$

The proportionality ~ in Equation 13.5 arises because division by $p(\mathbf{X})$ has been omitted from the product of likelihood and prior, so the conditional posterior is un-normalized, and the area under the distribution defined in Equation 13.5 will not be 1.

The solutions to Equation 13.5 differ, depending on what value of the hyperparameter α is used in the conditional prior distribution for the regression parameters **β**. If $\alpha=0$, the prior $p(\beta|\alpha)$ is a uniform distribution (often called a "flat" distribution). If, however, either an inverse gamma distribution conjugate prior is used for the prior distribution of variance $p(\sigma^2)$ or the variance hyperparameter is taken as fixed value ~ 0, the conditional posterior solution that results from using a squared loss function is the familiar maximum-likelihood estimate:

$$\boldsymbol{\beta} = \left(\mathbf{X}^T\mathbf{X}\right)^{-1}\mathbf{X}^T\mathbf{y} \tag{13.8}$$

the same result as obtained from the usual frequentist solution to Equation 13.3. This agreement occurs because both maximum-likelihood regression (and its special case of least squares regression) and the Bayesian regression model developed here follow the likelihood principle. Since these models both share the same likelihood and prior, the inferences from them must be identical. By providing a flat prior distribution to the Bayesian regression, no information is available on **β** before the data are obtained, and it is reasonable that the "usual" (frequentist) approach—which also admits no prior information—and the Bayesian approach with no useful prior information should agree.

However, if $\alpha >> 0$, there is information about the model parameters in the prior. In this case, the prior distribution of **β**, $p(\boldsymbol{\beta}|\alpha)$, is $G(\mathbf{0},1/\alpha)$, a normal distribution with mean vector **0** and variance $1/\alpha$, and now the MLE estimate that results from the Gaussian posterior is

$$\boldsymbol{\beta} = \sigma^{-2}\left(\alpha\mathbf{I} + \sigma^{-2}\mathbf{X}^T\mathbf{X}\right)^{-1}\mathbf{X}^T y \tag{13.9}$$

Equation 13.9 has the form of a conventional ridge regression with ridge parameter $\alpha\sigma^2$, a function of the variance $1/\alpha$ of the prior distribution $p(\beta|\alpha)$ and the variance σ^2 of the stochastic parameters in the model for **y**. The ridge solution in Equation 13.9 is a sparse solution to the regression given in Equation 13.3. This solution expects that most of the terms in the regression parameter **β** are nearly zero because the information provided in the Gaussian prior indicates that these coefficients are, on average, zero. In this case, a relationship results that is different from the usual frequentist model, reflecting the differing contributions of the prior, the prior belief in how close each of the parameters is to zero (as measured by the hyperparameter α), the hyperparameter describing variance in the y data (σ^2), and the conditional posterior probability that results from these choices.

The posterior has the form of a ridge regression in this case, but unlike conventional ridge regression, the ridge parameter is not determined by a cross-validation procedure, where multiple regression models are created over portions of the data and the root mean squared error of prediction for the withheld portion of the data is determined as a function of the ridge parameter. There are two common ways to get appropriate estimates for hyperparameters in a Bayesian regression. One is to maximize the marginal likelihood distribution as a function of the hyperparameter over the entire dataset, a method known in the literature as evidence approximation (Bishop, 2006). It is also possible to form another, higher-level Bayesian model where the hyperparameters from the lower-level model are the stochastic parameter components of a higher-level model, giving what is called a hierarchical model. The hyperparameters are then modeled with a prior and likelihood, and estimated from their posterior, just as in any other Bayesian analysis (Gelman, et al., 2013).

Prediction of the y value given a new, observed x done using any of these Bayesian models is also somewhat different than that done with a frequentist model. Once the posterior distribution of β is determined, a prediction of a new x_{unk} to yield y_{pred} is accomplished by integration over the posterior as shown in Equation 13.10:

$$p\left(y_{\text{amk}}\middle|y,\alpha,\sigma^2\right) = \int p\left(y_{\text{amk}}\middle|\beta,\sigma^2\right)p\left(\beta\middle|y,\alpha,\sigma^2\right)d\beta = G\left(y_{\text{amk}};m^T X_{\text{amk}},\sigma^2\left(x_{\text{amk}}\right)\right) \tag{13.10}$$

where

$$\sigma^2(\mathbf{x}_{\text{unk}}) = \sigma^2 + \mathbf{x}_{\text{unk}}^{\mathrm{T}}\mathbf{S}\mathbf{x}_{\text{unk}} \tag{13.11}$$

Because the prior is conjugate to the likelihood distribution, the conditional marginal posterior is another Gaussian distribution, and its mean provides the maximum-likelihood estimate for the unknown y value. A benefit not seen in a frequentist solution is that the uncertainty in a *single prediction*—obtained from inference on the variance of the posterior after updating for the new data in x_{unk}—is readily available, too. Not surprisingly, the uncertainty in the predicted property y_{unk} depends on the uncertainty of the calibration data and the uncertainty in the estimates of the model parameters, but also on the new data $\mathbf{x}_{\text{unk}}$.

From the Bayesian perspective, the solutions obtained by using classical least squares and ridge regression reflect the different prior distributions chosen and our belief as to the correct values of the regression parameters. The "flat" prior distribution indicates both a lack of prior information and an openness to any value for the regression coefficients in $\boldsymbol{\beta}$. Proponents of Bayesian approaches consider the flexibility that results from this approach quite useful, but it is worth noting that, assuming Gaussian priors for each parameter, two hyperparameters (mean and variance) must be specified in these prior distributions for each of the terms in $\boldsymbol{\beta}$, reflecting prior knowledge or a best guess of each value and of its uncertainty.

When the multivariate Gaussian distribution is used as a prior, the choice of its mean location reflects an opinion as to the likely value of each term in β, and its variances reflect the strength of that prior information—the degree of "belief" in each value. When the mean of the multivariate Gaussian prior is not zero, there is a bias introduced in the regression in the strict "frequentist" sense. In the Bayesian analysis, some of the possible values for model parameters are regarded as more likely than others, not in the sense of their frequency of occurrence, but in the sense that these values are more consistent with what is known about this system—there is a higher degree of belief in these values. This is similar to the dropping of the intercept term in fitting Beer's law models to spectral responses as a function of concentration.

The prior belief provided in a Bayesian analysis interacts with the new data in a sensible way, too. Limited new data generally have a small effect on the posterior distribution unless a flat prior is used; the information provided in the prior dominates in the posterior because any information in the small amount of new data, even if different from that provided in the prior, is not sufficient to substantially change the information about the model provided as part of the prior. Bayesian methods are not as useful when applied to small datasets because they are so sensitive to the specifics of the prior in these cases. The practitioner should keep this aspect of Bayesian methods in mind. A poor choice of prior, including providing a flat prior, in connection with having limited data, is not a good strategy to obtain a useful Bayesian analysis.

However, providing a prior that is highly consistent with information in the new data also results in a very small change in the posterior, even with a large amount of new data. Change in the posterior distribution from that of the prior distribution occurs only when there is a large enough difference between information provided in the prior and information extracted from the incoming data. The degree of belief provided in the variance hyperparameters that are part of the prior determines what difference is "large enough" to cause the posterior distribution to differ from that of the prior.

So, while consistency between the prior and posterior distributions can be a sign of either consistency of prior and new information or woefully insufficient data, when a large amount of new data is provided, and when the information in the new data is *not* consistent with that in the prior, the information in these new data dominates in the posterior, and the model is updated by the new information. In this case, the specific prior that is used may not matter! With enough new data, the prior is forgotten in the analysis. This updating of the model by extensive new information is now commonly used in chemical process modeling and in other applications of Bayesian multivariate time-series modeling (Nounou, et al., 2002a).

13.2.1 An Example of Simple Bayesian Regression

Unlike a conventional frequentist regression, a Bayesian regression does not give a set of fitted parameters as output, but it gives a posterior distribution. Getting the information needed from that distribution will require some work.

Two examples illustrate the steps needed.

First, consider a simple two-parameter linear regression using the model in Equation 13.3. Estimates of the model parameters b_0 and b_1 are sought, given some data (x_i, y_i) shown in Figure 13.2.

We presume that the dependent variable y is stochastic and is normally distributed about a mean y value where

$$\bar{y} = \beta_0 + \beta_1\bar{x} \tag{13.12}$$

with variance φ, so that the conditional distribution

$$p\left(y_i \mid \beta_0, \beta_1, \varphi\right) \sim N\left(\beta_0 + \beta_1 x_i, \varphi\right) \tag{13.13}$$

defines the likelihood model.

The prior model includes our belief on the distributions of each parameter in the model. Again, this model is specified by distributions:

$$\beta_0 \sim N\left(m_{0,}t_0\right)$$

$$\beta_1 \sim N\left(m_{1,}t_1\right)$$

$$\Phi \sim \text{IG}\left(\text{shape} = a, \text{ rate} = \gamma\right)$$

Here, we assume that each of the beta values is independently and normally distributed, each with a pre-specified, but possibly different mean m and variance t, and that the uncertainty in y is inverse gamma (IG)-distributed with mean a and variance γ. The distributions and their set of values $(m_0, m_1, a, t_0, t_1, \gamma)$ that determine this prior model are the hyperparameters for this analysis. All are adjustable, but these are fixed at specific, preset values for the Bayesian analysis that follows. Hyperparameters are taken as deterministic in a basic Bayesian analysis.

This set of models has the non-normalized (to within a constant) posterior

$$p\left(\beta_0, \beta_1, \varphi \middle| y\right) \sim \left[\prod_{i=1}^{n} p\left(y_2 \middle| \beta_0, \beta_1, \varphi\right)\right] p\left(\beta_0 \middle| m_0, t_0\right) p\left(\beta_1 \middle| m_1, t_1\right) p\left(\phi \middle| a, \gamma\right) \tag{13.14}$$

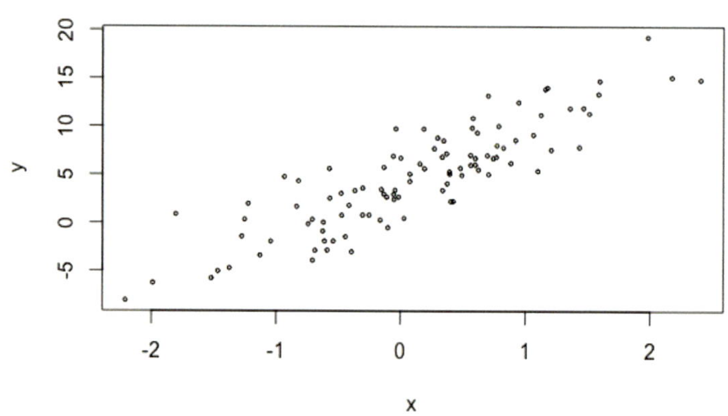

FIGURE 13.2 Noisy data for the example Bayesian regression.

because Equation 13.14 does not include division by $p(x)$. It happens that, because the priors are Gaussian and inverse gamma, and the likelihood is also Gaussian, the priors are conjugate to the likelihood, and the joint posterior distribution in this special case is closed form, the parameters are all independently distributed and can be evaluated analytically. Here, though, the posterior is determined numerically. The posterior for φ is statistically independent of those for β, so it can be considered separately, without the need to marginalize it from the joint posterior of the other parameters. This posterior is another inverse gamma distribution, with

$$\text{shape} = a + n/2 \tag{13.15}$$

and

$$\text{rate} = \frac{1}{2\varphi}\sum_{i=1}^{n}\left(y-\left(\beta_0+\beta_1 x_i\right)\right)^2 + \gamma \tag{13.16}$$

so just the joint posterior for β is considered here. The Gibbs sampler is used to find the marginal conditional posterior distributions of the β terms.

First, the un-normalized, marginal, conditional posterior is numerically evaluated for each of the β parameters in this model at specific values of β_0 and β_1. Then, a grid is set up of probability density over these possible values. A kernel smoother is sometimes used to increase the resolution of entries on the grid. Then, the Gibbs sampler is used to sample from the conditional posterior at each grid point. Approximation of the joint posterior by sampling from a pre-calculated grid is a time-consuming way to avoid the algebra associated with a closed-form solution but is very useful when the solution is not closed form. The sampling is not required here, but it is usually necessary in larger problems because the conditional posteriors cannot generally be obtained in closed form.

There are a few other issues that arise in using a grid. One is that the span must be selected to include all "reasonable" values that can be expected for the parameters. Deciding on the appropriate span of parameter values may take some trial and error. Second, because an un-normalized posterior is evaluated to get conditional marginal posteriors, it is possible to encounter numbers that exceed the software's/computer's capability to represent them. To avoid problems with underflow or overflow, the logarithm of the conditional probability is normally evaluated from the grid and then converted later.

For this simple regression, done on 100 samples of synthetic data, the Gibbs sampler, using only 1000 draws from the gridded conditional posterior and a burn-in period of 100 samples with no thinning used, provides very good estimates that come close to the true values for β_0, β_1, and φ.

Figure 13.3 shows the MCMC traces for the Gibbs sampling used to estimate the conditional posterior in this regression. The sampling has stabilized and the fluctuations of the distribution means, corresponding to MLE estimates of β_0 and β_1, are generally about the true values (indicated in the figure by the horizontal lines). The estimates for the variance φ also are stable, though there is more fluctuation. These estimates are all positive because variance estimates cannot be negative.

The estimates of the model parameters are shown in Figure 13.4. The multivariate normal relationship of the parameters is apparent. Inference here using the mean (a maximum-likelihood estimate) gives good agreement for the slope but only fair agreement for the intercept, a consequence of the considerable additive noise in the data. The Bayesian regression gives b_0=4.29±0.46, b_1=4.91±-0.62. A conventional least squares regression gives b_0=4.11±-0.26 and b_1=5.21±0.29.

Even this small analysis takes some time because of the creation of the likelihood grid and the sampling of the marginal conditional posteriors. This Bayesian analysis takes about 2 minutes to complete on a recent Intel processor, while the conventional frequentist version of this regression is essentially instantaneous.

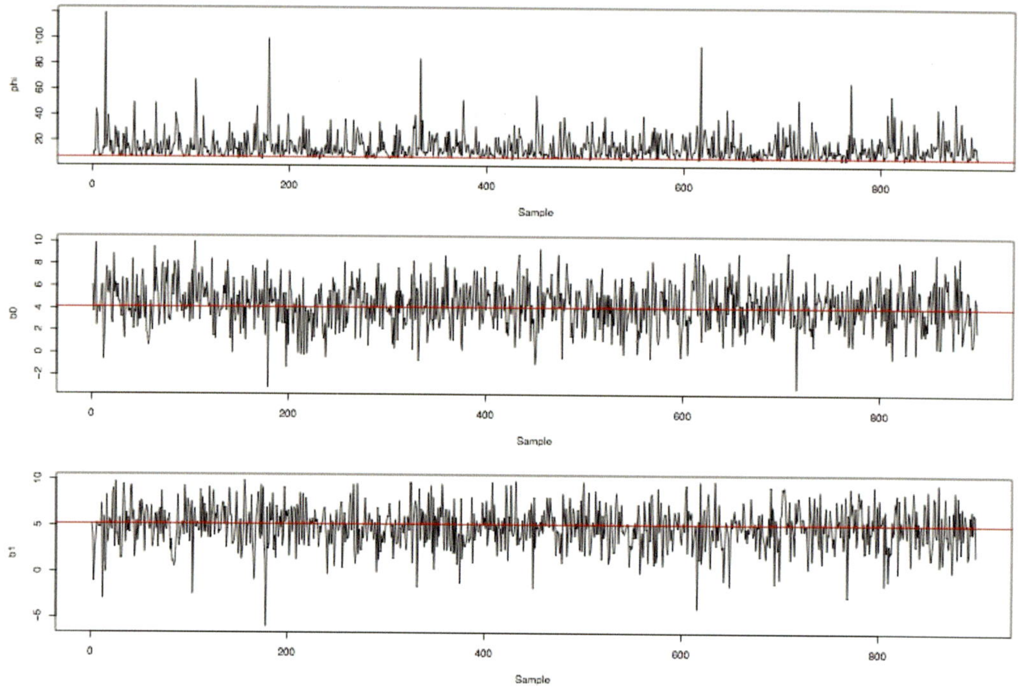

FIGURE 13.3 Markov chain Monte Carlo sequences from the Gibbs sampling for the simple Bayes regression. Only samples after the 100-point burn-in period are shown here.

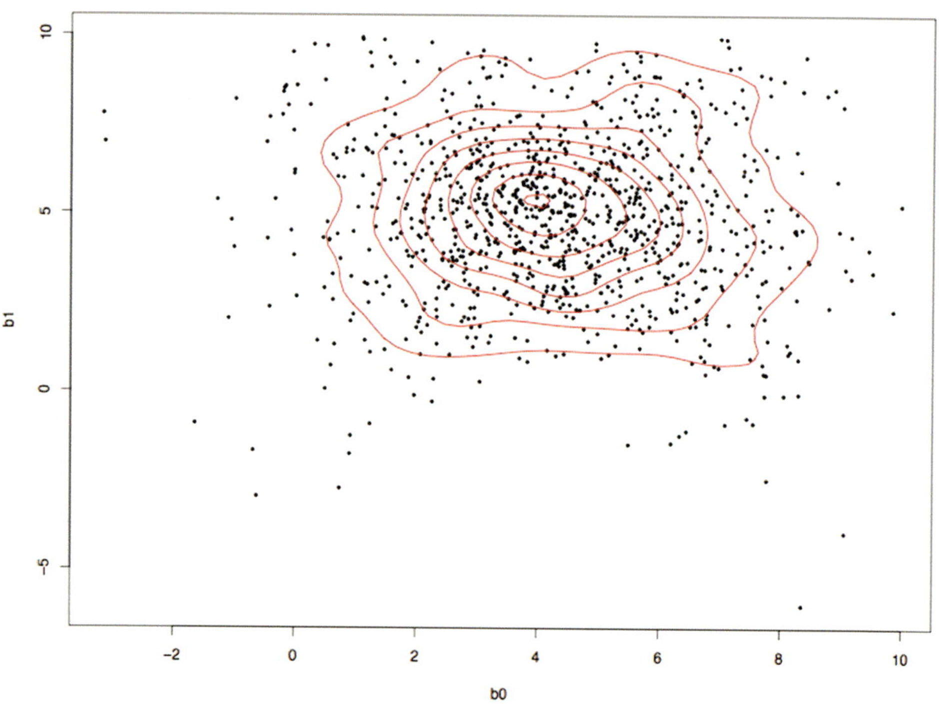

FIGURE 13.4 Posterior joint probability for the regression parameter estimates from the simple Bayesian regression. The lines show iso-probability levels.

TABLE 13.1
Estimator Evaluation from Four-Parameter Bayesian Regression

Parameter	True Value	Estimator Mean	Estimator Variance	Bias	Percent Bias (of Truth)	Coverage of 95% CI
b0	1000	1000.55	26,358	0.55	0.06	0.95
b1	50	50.53	348	0.53	1.06	0.95
b2	–50	–50.05	2.06	–0.05	0.10	0.94
b3	10	9.99	2.64	–0.01	–0.11	0.96
Phi	10,000	10,401.96	40,97,173	401.96	4.02	0.96

As the regression task becomes more complex, the computation time increases. A 5-parameter regression model ($y_i = b_0 + b_1x_i + b_2x_i + b_3x_i + \varphi$) is fit to the data; here, the model parameters are not assumed to be independently distributed. Using a flat, improper prior for the beta terms and an inverse-gamma-distributed variance term for the dependent variable y to model the observed data with this 5-parameter model takes about 7 minutes to run, fitting to only 50 observations, using 500,000 samples in the simulation (done here as block Gibbs sampling, where a vector draw is performed to both speed the calculation and to account for any interactions between the four model parameters), with a 100,000-sample burn-in and a trim rate of 1 in 50. Table 13.1 shows that the longer simulations and high trim rate pay off in the estimates. Other than the variance of the noise term, all parameter estimates are very close to the true values.

13.3 BAYESIAN CALIBRATION OF NEAR-INFRARED SPECTRA

These examples are very simple, and it is unlikely that Bayesian regression would be used to solve such simple modeling problems because the benefits will be minor and the effort more significant than that needed for the usual frequentist approach. When more complex models are fitted to large amounts of data, a Bayesian approach can show significant benefit over a frequentist model, however. Bayesian regression methods have become more common as the computational resources needed to implement them have become available. To date, however, much of the published work using Bayesian regression has been carried out by statisticians. A significant fraction of their published material has considered Bayesian regression on near-infrared (NIR) spectrometric data.

The complexity of the modeling and the computational demands make a Bayesian regression on near-infrared spectra a challenge. The benefits of performing the regression arise in three main areas: (1) selecting wavelength variables important to the modeling; (2) obtaining one-sample estimates of error in a result; and (3) modeling nonlinear and nonparametric relationships in spectral data. Each of these is discussed below.

13.3.1 Direct Variable Selection in Calibration using Bayesian Regression

One common motivation for Bayesian regression is a desire to reduce variables in some systematic way or to discover those variables most important to a model when there are many variables. That information is part of the posterior.

Selecting variables can be done by putting a prior distribution on each regression coefficient β. Using Bayesian methods to identify important variables in multivariate data by using a prior distribution was first reported by Mitchell and Beauchamp (1988). They selected variables using a non-informative prior on the variance in y, and a "spike-and-slab" prior distribution for each of the regression parameters β_j. A spike-and-slab distribution is a proper uniform distribution between two limits $-f_j$ *and* $+f_j$—the slab—but with extra density concentrated at 0—the spike—if the variable is vulnerable to deletion. The parameter γ_j relates the probability density at the "spike" to the height of the "slab" of uniform probability density. To exempt the regression coefficient corresponding to variable j from removal, the prior for the parameter is set so that $\gamma_j=0$, ensuring no prior density for the jth variable at the value $\beta_j=0$. The spike-and-slab prior defined sets of possible variables in a number of submodels. To find suitable values for γ, they used a Bayesian cross-validation, in which a predictive error is calculated in a leave-one-out sense, but unlike conventional, frequentist cross-validation, the prediction of a given sample $\mathbf{x}$ is not determined by estimating the response and then comparing it with the known value; instead, the posterior conditional distribution over the whole set is used in connection with the known response to estimate the error associated with estimating the response for each withheld sample directly, just s in a prediction of new data. Variable selection was investigated for synthetic data made with known models with and without predictor collinearity in a non-spectral dataset with ten related variables, and on a real, energy-conservation dataset. The posterior resulting from the spike-and-slab prior identified sets of submodels with different retained variables. Studies of synthetic data using this spike-and-slab prior showed that the Bayesian approach easily identified the terms in the data associated with larger regression coefficients, but variables with smaller coefficients and with zero coefficients in the true model often were misidentified as present, not zeroed out, with high posterior probability in the submodels. When the variables were made collinear in the synthetic datasets, the posterior density of several possible submodel choices was close, and choosing a single "best" submodel proved challenging, though competitive submodels with similar posterior probabilities tended to identify many, but not all, of the same predictor variables.

Even though this study did not attempt to examine spectra, because it is one of the first attempts to use the prior, as adjusted by experimental data, to decide on retention of variables in a regression model, it is important to the use of Bayesian methods in modeling spectral data. The datasets were small in size because of the computational burden imposed by the Bayesian regression and the cross-validation scheme they employed. Indeed, all of their work was done on a supercomputer, though it should be noted that in terms of computational capability, the processing power of the 1987-era Cray X-MP they used is slightly inferior to that of the 2012 Apple iPad 2 tablet (Larabel, 2012) on the LINPACK library, a conventional test!

It is also of interest to compare the model averaging done in this Bayesian approach with another common form of ensemble model averaging that does not use a Bayesian approach: stacking (Breiman, 1996; Tan and Brown, 2003). In stacking, weights are not determined from posterior probabilities of submodels, but they are determined by the errors from conventional cross-validation. Clarke has shown that when the true model is present in the set of submodels subject to Bayesian averaging, the results of the Bayesian regression are never worse than those from stacking, and may be significantly better. However, when terms in the Bayes' submodel set are outlying to the behavior of the data because of bias or noise effects in the data, stacking performs better; overall, stacking seems more robust than Bayesian model averaging (Clarke, 2003).

The first work aimed at using Bayesian regression to perform modeling and variable identification on real near-infrared spectral data examined two sets: One set of NIR data was collected on sucrose, glucose, and fructose in aqueous solution, a set originally created to test modeling by frequentist regressions using inverse and classical least squares regression. This early work by Brown (1982) considered conventional multivariate calibration of the NIR spectrometric responses to the amounts of the sugars in aqueous solution. The calibration set consisted of 125 transmission spectra collected on designed mixtures of the 3 sugar components and measured over the range 1100–2498 nm in increments of 2 nm. Another 21 samples were provided as a prediction set. This set was constructed to pose a challenge in multivariate prediction; many of the test samples had levels of sugar that exceeded the range used in calibration. The sugar data are available from Vannucci (http://marina.blogs.rice.edu/software/).

The second NIR dataset was first studied by Osborne et al. (1993). This dataset is also available online (http://marina.blogs.rice.edu/software/). NIR spectrometry was used to measure the composition of formed, but unbaked, pieces of cookie dough from their reflectance spectra. This set is also multivariate in response, with four constituents (fat, sucrose, dry flour, and water) subjected to calibration by varying a standard cookie recipe over a wide range. A total of 39 cookie dough samples were available in the calibration set, and another 39 samples were used for the validation set. All were measured by NIR reflectance from 1100 to 2498 nm with 2 nm resolution. The earliest work on this set made use of stepwise multiple linear regression to select wavelengths for a least squares calibration (Osborne, et al., 1993). This study treated the four target responses separately, using four stepwise, single-property classical regressions and finding different variable sets for each of the calibrated components.

Many of the published studies reporting Bayesian regression have examined one of the two datasets described above. These two sets have many of the issues typically seen in near-infrared spectra, and because they are spectral sets with far fewer samples than variables, they offer a challenge to Bayesian modeling. However, they also offer opportunities well suited to Bayesian analysis and highlight why a Bayesian approach has advantages in modeling near-infrared data. The most immediate problem with most NIR spectral datasets investigated by calibrations based on Bayesian regression lies in the large number of variables in the regression model. Both the sugar spectral data and the cookie dough spectral data have 700 variables, far too many to use in a Bayesian analysis of the sort illustrated above because of limitations in computer memory and because of the computational time needed to complete a full analysis. In addition, many of the spectral variables do not contribute independent information to the modeling. Near-infrared spectra commonly are relatively featureless and have strong correlations among the spectral variables. A third, related issue is that the number of samples collected for an analysis is less, sometimes far less, than the number of spectral variables.

For a Bayesian regression analysis of data with this many variables, variable reduction has to be performed. Thus, one focus of the many reports on Bayesian methods in analysis of NIR data has been on methods to reduce the size of the data by eliminating "unimportant" spectral variables from the data.

Brown revisited the sugar set with Fearn and Vannucci (Brown, et al., 1998) using a Bayesian approach. They began by arbitrarily reducing the number of variables from the original 700 variables available from the spectrometer to a set of 160 equally spaced, new variables located by linear interpolation of the original data. In addition to the usual priors on the regression parameters in the model, they added an additional mixture prior indicating whether or not an interpolated wavelength channel in the model was useful in modeling. The prior was implemented as a Bernoulli distribution on each of the 160 interpolated variable channels, so only a value of 0 or 1 was possible for each. In this way, the variance term in the prior model could be used to exclude a variable channel from consideration by setting both the mean and variance of the channel to 0 in the prior. Even with the reduction in variables from the original 700 wavelengths to 160 interpolated variable channels, the Bayesian regression analysis with conjugate priors done in closed form took far too long to compute on computers available then. Their estimate was that a closed-form solution to estimate conditional posteriors could be run with at most 20 spectral variable channels on the computational equipment then available. Even a version that used a Gray-code scheme to calculate the marginal conditional probabilities in an effort to speed calculations was not computationally feasible with more than about 25 spectral variables (Brown et al. 2002). On the other hand, the Gibbs sampling approach outlined above could be used with the entire reduced set of 160 variable channels. A prior that initially selected 20 channels from these data with wide latitude on admitting more (or eliminating) variable channels was implemented. Five modeling runs were performed, each with a different starting point for the analysis: all selected wavelength channels initialized in the prior as 1 in the Bernoulli distribution; half of the selected wavelength channels initialized as 1; a random set of 20 of the wavelength channels initialized as 1; all but 20 channels initialized as 1; and the first 20 wavelength channels initialized as 1. In all runs except the last, the set of 160 possible wavelength channels were randomized prior to the run in an attempt to break up the strong serial correlations present in the NIR spectra.

Within about 200 samples, the MCMC traces settled to select about 10 channels of the 160 possible. Fairly consistent sets of wavelength channels were selected over these five runs, despite the widely differing priors used to start the analysis, an indication of the information present in even these greatly reduced spectra. Of the 160 possible channels, a total 23 wavelength channels were selected from the 5 runs, where each run independently selected 10. The average of these five Bayesian models accounted for 55% of the probability examined and gave mean squared errors of 0.116, 0.361, and 0.351 for the prediction of sucrose, glucose, and fructose, respectively. Using only the 10 wavelength channels selected most often, and no model averaging, accounted for 20% of the visited probability and gave somewhat worse prediction mean squared errors of 0.210, 0.446, and 0.510. Considering the nature of the prediction set, these are fairly good results, especially given that the wavelength channels selected are not the ones that maximize the usual cross-validated error, since no cross-validation at all was performed, and all predictions were generated from integration of the marginal posterior, as discussed above.

Chen and Martin (Chen, et al., 2007c; Chen and Martin, 2009) also used a Bayesian regression to select variables from NIR. Like Brown et al. (1998), they used a Gaussian prior on the regression parameters in $\boldsymbol{\beta}$, but they considered the multivariate case with M multiple, statistically independent response variables in **Y**. While the covariance of the **Y** is diagonal, the analysis does not reduce to a set of separate Bayesian regression models developed independently on univariate responses because the prior distribution $p(\boldsymbol{\beta}|\alpha^{-1})$ for each set of parameters is based on the same hyperparameter α^{-1}, reflecting Gaussian distributions with the same prior mean but with different variances for the regression parameters in the M response models. They took the four responses as independent and assumed independence of the regression coefficients for each response, so

$$p(\beta|\alpha) = \prod_{i=1}^{M} G(\beta_i; 0, \alpha_i^{-1}) \tag{13.17}$$

from which **S**, the vector of posterior variances of $\boldsymbol{\beta}$, is obtained

$$S = (A + \sigma^{-2} X^T X)^{-1} \tag{13.18}$$

where $\mathbf{A} = \mathrm{diag}(\alpha_1, \alpha_2, \alpha_3 \ldots, \alpha_M)$. Evidence approximation (Bishop, 2006) is then used to find maximum-likelihood estimates for α and σ. The hyperparameter α_i is used to decide on retention of variables: When the value of α_i falls below a threshold, the corresponding β_i can only be close to zero, and the *i*th predictor has little impact on the regression, effectively removing it from the regression model. The authors claim that the size of the threshold is not important to the predictive performance of the model and to the variables that are ultimately excluded, so long as the threshold is set high enough. This approach to identifying variables in a Bayesian regression is known as automatic relevance determination (ARD) (MacKay, 1992; Neal, 1996), and it is increasingly used in imaging NIR spectrometry and NIR tomography (Miyamoto, et al., 2015).

The variable selection was used in developing calibrations in two datasets based on NIR spectra. The first set studied was the wheat kernel dataset (Pedersen, et al., 2002), available from http://www.models.life.ku.dk/wheat_kernels as a MATLAB "mat" file. This set consists of a training set of 415 samples of wheat measured by reflectance over 100 wavelengths from 850 to 1050 nm. A test set of 108 additional samples is also included. The calibration related the percentage of protein to the NIR spectra. This set is challenging to calibrate because of the large amount of variation in the spectral data caused by the use of different varieties of wheat grown at different locations, and the delay in measurement of the test samples relative to the calibration samples.

The second study reported results for calibration of the four response variables moisture, oil, protein, and starch in the extensively studied Cargill cornmeal dataset, available as a MATLAB "mat" file from Eigenvector Technologies (http://www.eigenvector.com/data/Corn/). This heavily studied dataset consists of near-infrared spectra of 80 cornmeal samples, each measured in reflectance on 3 instruments; one instrument (labeled m5) was a NIRSystems 6500 spectrometer and the other two (labeled mp5 and mp6) were different NIRSystems 5500 spectrometers. All spectra were originally measured over the wavelength range 1100–2498 nm at 2 nm resolution.

Both studies compared the predictive performance of Bayesian regression with that of conventional partial least squares (PLS) regression. Full Bayesian regression, using all variables, and regression with the adaptive selection of variables discussed above were compared with PLS regression using the wheat spectra. The root mean squared error of prediction (RMSEP) from PLS regression was 0.70%. Bayesian regression using the Gaussian prior, but without variable selection, gave an RMSEP of 0.62% on the same data. When the adaptive variable selection discussed above was employed, the Bayesian regression gave an RMSEP of 0.55%, using only 16 of the predictors. The sparse set of 16 predictors used in the adaptive model generally agreed with those wavelength regions having larger posterior means for the regression parameters found without adaptive variable selection.

Both separate and multivariate Bayesian regression models were developed on the data for the three NIR instruments in the cornmeal dataset. The results from the four univariate response Bayesian regressions were compared to those from univariate PLS-1 regression models. Results from the multivariate Bayesian regression were compared to those from a multivariate PLS-2 regression model. The PLS models generally underperformed the Bayesian models in all cases. Table 13.2 summarizes their results. The m5 dataset is of notably higher quality, and many regression methods do well on predictions of properties using these spectral data, so it was not surprising that on this set, the univariate Bayesian regression for moisture using adaptive variable selection had a very low RMSEP, about 10% of that reported from PLS-1 regression. It was also noteworthy that moisture was so well predicted by a very sparse model with only two spectral variables, at about 1900 and 2100 nm. The PLS-2 model for the cornmeal produced mixed results: Predictions of protein were much superior to those from a PLS-1 model built on only protein, but predictions were worse for the other responses than those obtained from the corresponding univariate PLS-1 models. The multivariate Bayesian regression with adaptive variable selection also underperformed the four

TABLE 13.2
Prediction Results[a] (Reported as RMSEP) from Several PLS and Bayesian Regression Models Built on the Cargill Corn Dataset

Dataset	Model	Moisture	Oil	Protein	Starch
m5	PLS-1	0.036	0.154	0.317	0.261
m5	Bayes	0.018	0.140	0.127	0.230
m5	Bayes-VS	0.004(2)	0.101(23)	0.072(27)	0.211(21)
mp5	PLS-1	0.184	0.207	0.365	0.698
mp5	Bayes	0.190	0.172	0.297	0.556
mp5	Bayes-VS	0.164(20)	0.153(32)	0.237(20)	0.531(22)
mp6	PLS-1	0.280	0.196	0.262	0.836
mp6	Bayes	0.211	0.192	0.218	0.559
mp6	Bayes-VS	0.194(18)	0.180(17)	0.198(30)	0.508(23)
m5	PLS-2	0.046	0.206	0.144	0.408
m5	mBayes	0.038	0.142	0.143	0.330
m5	mBayes-VS (10)	0.005	0.127	0.130	0.311

[a] The number in parentheses indicates the number of wavelengths used in the sparse Bayesian model. mBayes is the Bayes regression using multivariate response variables. VS indicates variable selection by adaptive modeling. Data adapted from Chen, et al. (2007c).

univariate adaptive Bayesian regressions, no doubt because the number of wavelengths chosen for inclusion in the multivariate regression was a clear compromise: larger by far than those needed by a univariate Bayesian model for moisture, but smaller by far than univariate Bayesian models for the remaining three response variables. The mp5 and mp6 data again showed that the predictions from univariate Bayesian regressions were mostly superior to those from PLS-1 and that the adaptive variable selection further lowered prediction errors, though not to the levels seen from the analysis of the m5 data.

The cookie dough dataset has also been examined using variable selection based on a spike-and-slab approach (Hérnandez-Lobato, et al., 2015) in conjunction with an expectation propagation (Minka, 2001) to approximate the posterior distribution by a simpler parametric distribution, chosen to permit fast calculation. This combination gives RMSEP errors for the dry flour and moisture responses in the cookie dough data that were comparable to those found by other means, but speeds calculations by a factor of about 2000x or more.

The studies cited above demonstrate that some form of Bayesian regression using a prior on the variables included in the regression model is clearly an effective way to discover a set of variables that are highly informative about the regression relationship. While the set of informative variables has often been generated as a kind of by-product in the Bayesian analysis of NIR spectra where the main goal is reduced model size to reduce computational demands, the relative success of Bayesian variable selection has not yet been systematically compared to variable selection achieved by using other sparse modeling methods. Nott and Leng (2010) focused on studying the success of a Bayesian regression at identifying variables for inclusion in a regression model. As part of this work, they considered the situation where there were many more variables than samples, and, as in the studies on NIR spectra cited above, they found that the posterior distribution has reliable, though not always definitive, information on the active and inactive variables in the true model. Using synthetic data, they employed that a Bayesian version of the lasso, a sparse regression where, instead of the usual L_2 loss function based on the sum of squared differences, the L_1 loss function based on the sum of absolute differences is minimized (Tibshirani, 1996). This change in the loss function better concentrates the posterior probability on those variables present in the true model underlying the data. Lasso methods have been reported in Bayesian regressions on NIR spectra, and a lasso regression using empirically selected variables from the 700 available in the sugar dataset performed about as well as the much more computationally demanding method of Brown discussed above, but one using generalized cross-validation selected too many predictors and failed to predict the test data well (Yuan and Lin, 2005).

13.3.2 Bayesian Calibration on Transformed Spectral Data

Instead of performing interpolation to downsample spectra to data with far fewer channels, reducing the set of variables in a typical set of NIR spectra can also be done by use of a linear transform on the spectral data. Several linear transforms are often employed to reduce variables.

13.3.2.1 Bayesian Regression on Wavelet-Transformed Spectral Data

One linear transform is the discrete wavelet transform (DWT). Like its relative the discrete (fast) Fourier transform, the discrete wavelet transform converts data from the time domain to the frequency (time^{-1}) domain, but because the wavelet transform is performed with a set of orthogonal basis functions of differing frequency ranges, called scales in the wavelet terminology, the wavelet transform converts a conventional time-domain spectrum into a series of wavelet coefficients for each scale. A wavelet transform is a dyadic decomposition of the original signal, meaning that each new scale representation of the input data includes progressively 1/2 fewer wavelet coefficients. Thus, the DWT of an n-point NIR spectrum is represented by a set of l scales, with the highest scale frequencies in the spectral data contained in the first scale, expressed by $n/2$ wavelet coefficients. A lower set of scale frequencies is expressed by the second scale, using $n/4$ wavelet coefficients, a still lower set of scale frequencies is expressed in the third scale, by $n/8$ wavelet coefficients, and so on down to the last scale, which describes the lowest scale frequency with 1 wavelet coefficient. The scales generated by the DWT depend on the wavelet basis set used to define the wavelet filter function. The wavelet basis is defined by the so-called "mother" wavelet, a filter function that generates the orthonormal set of wavelet scale basis functions that transform the data. Popular wavelet bases for transforming NIR spectral data include the Daubechies 4, Symlet 4, and Symlet 8 wavelet filters; here, the numbers refer to the number of "vanishing moments" in the wavelet filter function, a value that represents the highest order polynomial that can be described exactly by the wavelet filter. The results of the wavelet transform filtering are a set of coefficients that describe the behavior of the spectral data in different frequency ranges, from the highest frequencies spanned by the first scale, down to DC (0) frequency in the last scale.

Using the full set of scales, the spectral data can be reconstructed perfectly from their wavelet representation. However, because the higher frequencies often describe spectral noise in the NIR spectral data, these can sometimes be omitted without much loss of true signal. It also may be that the lower frequencies describe baseline effects in the NIR data, and these also may be omitted. A 700-variable NIR spectrum, padded to a total of 1024 variables to make the length of the signal a power of 2 and to avoid signal "leakage" issues at the ends of the spectrum from application of the wavelet filtering, and then expressed by a total of 1024 wavelet coefficients in all, might be well represented by only 100 or fewer wavelet coefficients. Note that these wavelet coefficients are not spectral wavelengths, nor are they readily converted to represent discrete wavelengths (note, though, that conversion is possible using a particular type of wavelet reconstruction; Tan and Brown, 2003), so wavelet representation of the data is not like a conventional wavelength selection scheme. Readers seeking understanding of the wavelet transform might want to start with two excellent tutorials (Graps, 1995; and Polikar, 2006) before tackling textbooks (e.g., Mallat, 1999) or the literature concerned with wavelet signal analysis.

The wavelet transform can be applied to selection of portions of NIR spectra included in a calibration model in a manner analogous to that used to select channels. In this case, the usual regression model is used as the likelihood, and the prior distributions of the parameters for the intercept, the regression coefficients, and the covariance matrix are specified, with an additional normal prior indicating whether or not the wavelet coefficient is used in the Bayesian model. In this way, the posterior incorporates the probability that a wavelet coefficient is selected (or not) and has the benefit of highly weighting subsets of the spectra that account for large variation in the response variable while, because of the prior, penalizing choices with many wavelet coefficients. An 8-scale Daubechies 4 wavelet decomposition was used to reduce NIR spectral data after some data at the ends of the spectra were removed as non-informative because of noise effects (Brown, et al., 2001). Of the 10 terms in the best 8-scale predictive model, 37% of the wavelet coefficients in the mid-range scale (scale 4) were included, along with a few terms from higher frequencies (6%, 6%, 1%, and 2% of scales 5–8, respectively), but no wavelet coefficients from the low-frequency scales (1–3) appeared in the posterior model, even though these would be important in any reconstruction of spectra from the coefficients. Similar findings are reported for other calibrations of near-infrared data based on wavelets (Tan and Brown, 2003). Inverse wavelet transformation of the coefficients in the model showed the spectral information in the posterior to be concentrated in two main bands: one at 1400–1450 nm, and the second at 1650–1730 nm, a region known to be active in spectra of fats and oils.

It is also possible to build an adaptive Bayesian model using families of submodels based on different wavelet decompositions, with different mother wavelets, and different numbers of decomposition levels, with basis functions employing different numbers of vanishing moments, and set Gaussian priors on the weights for each submodel. The posterior weights of these submodels are adapted in an MCMC search, as outlined above. This adaptive search produced a regression model where different wavelet bases contributed over different portions of the spectrum. The Bayesian wavelet model was compared to an approach based on stacking of the sets of wavelet submodels, with the stacking weights set by a bootstrap procedure (Breiman, 1996). These two modeling methods were applied to the biscuit dough dataset discussed above (Donald, et al., 2011). All forms of the Bayesian regression produced results substantially better than those from conventional stepwise multiple linear regression, principal components regression (PCR), or partial least squares (PLS) regression applied to the spectra of dough, as shown in Table 13.2, but the stacked ensemble of Bayesian models generally showed superior predictive performance, in agreement with theory (Clarke, 2003). The stacked models consistently produced superior errors of prediction, but resampled stacking with different wavelet types performed worse than when only a single wavelet type was used, possibly because of the very large number of models used in the ensembles used to model, according to the authors.

13.3.2.2 Bayesian Calibration on Principal Component and Latent Variable Scores of Spectra

Another way to reduce the number of variables in a spectral dataset is to perform a singular value decomposition (SVD) on the spectral data to obtain the principal components of the data. If many of the singular values of the data are zero, or can be taken as zero, the number of loadings retained after the SVD is often far fewer, or can be taken as far fewer, than the number of original variables in the data. This sort of variable reduction is well established in chemometrics.

The first report of a Bayesian regression performed on the SVD loadings [32] parameterized the Bayesian model in terms of the regression coefficients and the SVD loadings, so that, if the usual SVD is written as

$$\mathbf{X} = \mathbf{U}\mathbf{S}\mathbf{V}^T = \mathbf{T}\mathbf{P}^T \tag{13.19}$$

with scores matrix $\mathbf{T}$ and loadings matrix $\mathbf{P}^T$, the principal components regression (PCR) model is then

$$y = \mathbf{X}\boldsymbol{\beta} = \mathbf{T}\boldsymbol{\theta} \tag{13.20}$$

where, for k "significant" loadings, the $1 \times k$ regression vector $\boldsymbol{\theta}$ is expressed as

$$y = \mathbf{X}\boldsymbol{\beta} = \mathbf{T}\boldsymbol{\theta} \tag{13.21}$$

Now, under these transforms, observing the response $\mathbf{y}$ provides information only on the loadings regression parameters $\boldsymbol{\theta}$, so the Bayesian model is specified in the loadings space, not in terms of the original data. This is an unusual Bayesian model because the loadings space is data-dependent; since the loadings span independent components of variance in the data, the loadings space often depends on the experimental design used to obtain the data and on the specific data $\mathbf{X}$ included in the model. Because of this, the principal components model changes as the number of singular values taken to be nonzero changes. This model is one that will have priors and posteriors on θ, not β, so there is a need to establish the inverse relationship between θ and β, since knowing what wavelengths contribute significantly is an important part of the modeling. Using the full 700 points of the biscuit dough data and the first 16 loadings, and employing 100,000-sample MCMC estimation of the posterior distributions with a burn-in of 1000 samples and a thinning rate of 1/5 gave results very similar to those produced by the wavelet decomposition of the reduced spectra. Bayesian regressions based on wavelet and SVD data compression both showed that wavelengths near 1718 nm were especially important in the determination of the properties of the dough. The SVD analysis also found significant wavelengths at about 1520 and 2350 nm. Interestingly, the number of loadings here was not a parameter in the Bayesian model, and so the number of loadings retained in the model was not subject to setting a prior and to adjustment by information in the data, and as a consequence, the spectral data seem overfit, judging from the regression vector plotted in the paper (West, 2003). In the SVD-based analysis reported here, an amazing 99.995% of the variance in the spectral data was incorporated in the Bayesian model, and the number of components was set "consistent with the inherent smoothness of the predictor variable."

The construction of a Bayesian model for a variable reduction using SVD was also considered (Nounou, et al., 2002a,b; Chen, et al., 2007a,b). This chapter provides a clear introduction to Bayesian modeling and considers the many complex issues arising from a Bayesian principal component analysis (Nounou, et al., 2002a). Their treatment illustrates the complexity of this approach; unfortunately, their results do not show significant predictive improvement over other approaches used on the same data.

The usual way to perform a principal components regression by singular value decomposition is to first decompose the spectral data in **X**, decide on an appropriate number of principal component scores to retain in the model by cross-validation or by examining the predictive performance of the various models on a validation set, and then use the model to predict the property value y for new data. That approach cannot be used in a Bayesian setting because the cross-validation step is computationally expensive and because it assumes equivalence of models. A different approach is necessary.

For a principal components decomposition of **X**, Equation 13.19 applies, but here both **T** (the scores of **X**) and $\mathbf{P}^T$ (the *x*-loadings) are taken as stochastic, because they are estimated from **X**, which is deterministic, *but with additive error.* This approach differs completely from that mentioned earlier, and it highlights an issue with the great flexibility of a Bayesian approach: Deciding how to parameterize a model, what is and what is not stochastic, what the distributions of the stochastic variables are, and what are suitable hyperparameters is not always clear, nor is there a ready set of guidelines to follow.

Under these authors' view of the decomposition, a simple principal components regression (PCR), where principal components decomposition of the spectral data **X** is combined with a regression step, the Bayesian analysis becomes much more complex because of the need to specify all the distributions needed. Here, **T**, **P**, and the regression coefficient vector $\boldsymbol{\beta}$ are all treated as stochastic, and so each must have a prior distribution specified. In addition, conditional distributions for **X** and **y** must be set. There is also the question of statistical independence of these distributions, because it is conceivable that some stochastic parameters depend on other parameters. So far, no one has checked to see what the consequences of a badly selected distribution are on the result of this sort of modeling, and only the effects of noise have been reported (Chen, et al., 2007a,b).

A Bayesian approach to latent variables assumes that both the predictor variables **X** and the response variable *y* are contaminated by independent, Gaussian noise $\varepsilon_{x_i} - N(0,R)$ and $\varepsilon_{Y_i} - N(0,R_y)$, where $\mathbf{R}_x$ is the covariance matrix of the set of predictor variables and R_y is the variance of the noise in the univariate response variable. Measurement noise in the predictors is presumed independent of the measurement noise in the response variable. Since some principal component directions in the data are dropped, the observed data are regarded as containing noise and therefore stochastic, since

$$X_i = t_i p^T + \varepsilon_{x_i} = \bar{x}_i + \varepsilon_x \tag{13.22}$$

and

$$y_i = \beta t_i + \varepsilon_{y_i} = \bar{y}_t + \varepsilon_x \tag{13.23}$$

where the tilde in the equations above indicates a noise-free quantity.

Then, the regression can be implemented with the usual PCR model:

$$\tilde{X} = TP^T \tag{13.24}$$

and

$$\tilde{y} = T\beta \tag{13.25}$$

The wavelength–space regression parameters **b** normally reported for a NIR calibration are:

$$b = P\beta \tag{13.26}$$

Using this set of relations as the model, the marginal likelihoods for X and y are:

$$p(x_i \mid \mathbf{P}, t_i) \sim N(\mathbf{P}t_i, \mathbf{R}_x) \tag{13.27}$$

and

$$p(y_i \mid \boldsymbol{\beta}, t_i) \sim N(\boldsymbol{\beta}^T t_i, R_y) \tag{13.28}$$

and because these are statistically independent, the full likelihood distribution is just the product of Equations 13.27 and 13.28.

Prior information must be provided for **b** and $\tilde{\mathbf{X}}$ because we observe the noisy **X.** These priors can be non-informative or Gaussian. The Gaussian priors are $p(\mathbf{b}) \sim N(\boldsymbol{\mu}_b, \mathbf{Q}_b)$ and $p(\tilde{x}_i) \sim N(\boldsymbol{\mu}_x, \mathbf{Q}_x)$, where $\boldsymbol{\mu}_x$ is the prior mean of the noisy **X** because the noise in **X** is assumed zero-mean. The prior for the response is just $p(y_i) \sim N(\mu_y, R_y)$, and the prior for β is $p(\beta) \sim N(\mathbf{P}^T\boldsymbol{\mu}_b, \mathbf{P}^T\mathbf{Q}_b\mathbf{P})$.

Because all priors are conjugate to the likelihoods here, once all priors and likelihoods are specified, application of Bayes' theorem leads to a closed-form conditional posterior probability expression describing Bayesian PCR:

$$p(\mathbf{P},\mathbf{T},\boldsymbol{\beta} \mid \mathbf{X},\mathbf{y}) \sim p(\mathbf{X},\mathbf{y} \mid \mathbf{P},\mathbf{T},\boldsymbol{\beta}) * p(\mathbf{P},\mathbf{T},\boldsymbol{\beta}) \sim p(\mathbf{X} \mid \mathbf{P},\mathbf{T}) * p(\mathbf{y} \mid \mathbf{T},\boldsymbol{\beta}) * p(\mathbf{T} \mid \mathbf{P}) * p(\boldsymbol{\beta} \mid \mathbf{P}) \tag{13.29}$$

presuming that the prior $p(\mathbf{P})$ is uniform and that the rank of **X** is a fixed hyperparameter. This solution is only proportional to the full posterior because $p(\mathbf{X})$ is not readily calculated and is omitted from the calculation defined in Equation 13.1.

Equation 13.29 describes a constrained least squares problem that is not easily or quickly solved for large numbers of variables or observations, so MCMC methods and Gibbs sampling were employed to sample the full posterior (Chen, et al., 2007a). After extensive algebra, the uniform prior led to the full conditional posterior

$$\beta \sim N\left(\left(T^T T\right)^{-1} y, \left(T' T\right)^{-1} \otimes R_y\right) \tag{13.30}$$

where $\otimes$ is the Kronecker product. This is a maximum-likelihood solution to PCR, as is expected from the use of a uniform prior.

Algebra using the Gaussian prior on Equation 13.29 led to the full conditional posterior

$$p(\beta \mid y, P, T) \sim N(m, v) \tag{13.31}$$

where

$$\text{the mean is } m = P^\tau \mu_b + P^T Q_b P T^\tau \left(T P^T Q_b P T^\tau + I \otimes R\right)^{-1} \left(y - T P^\tau \mu_b\right) \tag{13.32}$$

$$\text{and variance is } v = P^T Q_b P \left(I - T^T \left(T P^T Q_b P T^T + I \otimes R_r\right)^{-1} T P^T Q_b P\right) \tag{13.33}$$

If the prior mean $\mathbf{P}^T\boldsymbol{\mu}_b = \mathbf{0}$ and the prior covariance $P^T Q_b P = I \otimes k$, the mean of this distribution is a ridge regression solution with ridge parameter $\lambda = R_y/k$, again as expected from the use of a Gaussian prior.

Gibbs sampling was applied to these conditional posteriors. Because of the orthogonality constraint on the principal component loadings, it is not possible to use the MCMC simulation to draw samples to simulate the posterior distributions. The solution was to use a closed-form solution for loadings obtained from the PCA or PLS method, then use those latent variable loadings to perform MCMC draws, and then update the Bayesian model from these to update the loadings.

This Bayesian PCR modeling approach was applied to calibrate NIR data from the wheat flour dataset (Kalivas, 1997). This dataset is no longer available at the Web site indicated in the published paper, but is included with the OHPL package available on CRAN (https://cran.r-project.org/) and can be obtained from there. The set consists of the NIR spectra of 100 wheat samples over the range 1100–2498 nm, measured in increments of 2 nm. The response variables are moisture and protein. There are 50 samples in the calibration set and 20 samples in each of the two validation sets. Here, every fifth wavelength measurement was used in this analysis to reduce the computational burden. For PCR and PLS calibrations, 10-fold cross-validation was used, in which 90% of the data were used in training and 10% for testing.

The Bayesian SVD method (Nounou, 2002b; Chen, et al., 2007a) requires an estimate of the covariance of measurement noise over the spectral variables used. That was not available from the published data, so, based on older literature, the noise variance in the spectral data was assumed to be piecewise linearly dependent on the wavelength of the NIR spectrum, with a minimum noise value set at 1800 nm and symmetric, linear increases in the noise variance below and above 1800 nm using a literature value for the standard deviation of noise over the NIR spectra (Williams and Norris, 2001). Estimates of the uncertainty in the laboratory values for protein and moisture are also needed and were again not reported with the dataset. Estimates for these were obtained from other published work (Centner, et al., 2000).

This approach can be extended to one using PLS latent variables, and a Bayesian partial least squares regression can be developed along the same lines as the Bayesian principal components regression described above. For convenience, the Bayesian regression is called a Bayesian regression on latent variables (BRLV) below.

As shown in Table 13.3, the results of Bayesian regression on latent variables reflect the amount of information provided in the prior. The "*u*" refers to the uniform, non-informative prior, and the "*g*" refers to the Gaussian (normal) prior.

Uninformative priors led to BRLV models with very little improvement over that obtained from the usual latent variable regressions with principal components regression (PCR) and partial least squares (PLS) regression. This result is no surprise: As discussed above, a uniform prior provides

TABLE 13.3
Mean Squared Error of Prediction for Several Regression Methods Applied to the Biscuit Dough Dataset (Osborne, et al., 1993)

Regression Method	Fat	Sugar	Flour	Moisture
Stepwise MLR	0.044	1.188	0.722	0.221
Bayesian regression	0.076	0.566	0.265	0.176
Wavelet Bayesian regression	0.059	0.466	0.351	0.047
Wavelet Bayesian regression (best performing model)	0.063	0.449	0.348	0.050
PLS regression	0.151	0.583	0.375	0.105
PCR	0.160	0.614	0.388	0.106
Stacked wavelet regression[a]	0.043	0.340	0.182	0.037

Adapted from Donald et al. (2011).

[a] Results from using a stacked ensemble of Bayesian wavelet regression models built using a Coiflet wavelet with 1 level of decomposition and one vanishing moment. Other forms of the Coiflet wavelet performed less well on this dataset.

no additional prior information, equivalent to that used in a frequentist regression, so BLVR modeling with a uniform prior provides the most direct comparison with PCR and PLS regression. The comparison is especially easy here because the PCR used the same loadings as the Bayesian regression in this case. Results are similar, but slightly worse, for a Bayesian regression on latent variables using the uniform prior as compared to those from conventional PLS regression for both moisture and protein. Because both approaches used the same 15 latent variables to develop the regressions used to calibrate, the main difference in the results may be a consequence of the somewhat arbitrary spectral and laboratory uncertainty estimates used in the Bayesian regression. This result shows the challenge of performing a Bayesian regression when the prior information provided is omitted from the data collection and must be guessed from other sources.

When the Gaussian prior is used, the BLVR modeling method uses fewer latent variables than conventional PLS for both calibrated responses. BLVR with the Gaussian prior predicts slightly better than PLS for moisture, but worse for protein. Overall, the difference between the results of conventional and BLVR predictive modeling is slight.

There is a third approach to Bayesian regression, the one not yet considered. This makes more extensive use of prior information, but the approach taken here is to use expert knowledge about the spectral responses to set the prior. Unlike other approaches in which the wavelengths were selected in the Bayesian analysis by allowing data to adjust prior weights, the approach taken here was to attempt to focus the model on a small set of wavelengths selected from previous experience with the calibration. This is an "informed" prior, labeled with "*i*" in Tables 13.4 and 13.5. In this case, the prior on wavelength participation in the regression is set with a hyperparameter vector to reflect known bands for hydroxyl, which get high weights in the prior for the regression vector **b**, and to exclude those portions of the spectrum deemed irrelevant to this calibration, which get very small weights in **b**. The wavelength regions associated with protein absorption are less well established

TABLE 13.4
Validation Error for Moisture in Wheat Data (Kalivas, 1997)

	PCR	PLS	BLVR (u)	BLVR (g)	BLVR (i)
Rank of spectral model	23	12	12	10	11
MSE for validation set 1	0.0667	0.0689	0.0729	0.0434	0.0371
MSE for validation set 2	0.1322	0.1195	0.1231	0.1003	0.0817
Overall MSE	0.0995	0.	0.0980	0.0718	0.0594

Adapted from data provided in Chen et al. (2007).

TABLE 13.5
Validation Error for Protein in Wheat Data (Kalivas, 1997)

	PCR	PLS	BLVR (u)	BLVR (g)	BLVR (i)
Rank of spectral model	24	15	15	14	16
MSE for validation set 1	0.1850	0.1503	0.1542	0.1728	0.1621
MSE for validation set 2	0.1401	0.1350	0.1418	0.1442	0.1177
Overall MSE	0.1625	0.1427	0.1480	0.1585	0.1399

Adapted from data provided in Chen et al. (2007b).

than those for moisture, so the priors used here were informed by looking at the size of the regression coefficients that resulted from the use of the regression with the uniform prior. The larger regression coefficients were deemed relevant to the protein calibration, and smaller values were deemed irrelevant. The portions considered relevant to protein in these empirically selected wavelength regions were again weighted highly. This approach requires confidence in the regressions with uninformed priors, since it biases the model strongly, but as Table 13.3 shows, the additional information available on hydroxyl in the prior led to much improved predictions. Now, the MSEP from the Bayesian latent variable regression is much lower than that of PLS regression, and with a far more parsimonious model. Even though the information about hydroxyl absorptions is well known, there is no easy way to take advantage of this knowledge in PLS, other than to rebuild a PLS model using only these wavelengths, because of the implied flat prior in frequentist regression. When the information provided is less well known and more broadly dispersed in the spectrum, however, as in the case of the regions corresponding to protein absorption, the gains are more modest, as shown in Tables 13.4 and 13.5, but even here, the Bayesian regression proves superior to PLS modeling in predictions. It would be interesting to compare PLS models developed using the same set of wavelengths with the Bayesian approach to determine if the wavelength selection is truly the reason for the improvement in these models.

Several different types of Bayesian regression were compared against PLS-1 regression for the calibration of fatty acids in milk (Ferragina, et al., 2015). The authors collected 1264 samples of milk from 85 herds and measured the spectra of the milk samples in the near-infrared and infrared, from 925 to 5011 cm^{-1}. Four target fatty acids in the milk were measured by gas chromatography; then, an individual model cheese was produced from each milk sample. The spectra were then related to the fatty acid levels in the milk and various properties of the cheese and the whey. Calibration of the spectra to these properties was then attempted using conventional PLS-1 regression, the WinISI modified PLS (MPLS) regression, and three Bayesian regression methods from the BGLR package available at the CRAN repository (https://cran.r-project.org/web/packages/BGLR/index.html). The three Bayesian regressions employed here were similar, and all used the same likelihood distribution, but they employed different priors. Bayesian ridge regression used a Gaussian prior, with no wavelength selection, similar to that done in the Gaussian process regression methods discussed above. Bayes A regression used a prior distribution for the regression parameters based on a Student t-distribution, with more density at zero and with heavier tails than the Gaussian distribution. This prior distribution induces some shrinkage of the regression parameters, resulting in higher posterior density in parameters with higher association with the property, and lower density in the posterior for parameters with little association. Bayes B regression used a variant of the spike-and-slab prior mentioned earlier. This prior combines a density spike at zero with a t-distribution for the parameter to enhance the shrinkage in the posterior of any regression parameters without strong correlations with the property. Here, the data must provide strong evidence that the term is clearly nonzero to overcome the prior's high probability that the parameter is zero. This prior tends to create sparse models with only variables highly correlated with the property.

While MPLS and the three Bayesian regression methods all outperformed conventional PLS-1 on these data, the three Bayesian methods all had higher R^2 values and lower residual variance than MPLS for all of the many properties calibrated. Bayes A and Bayes B regressions gave the most accurate results, except for the fat property, where MPLS regression and Bayes B regression produced the same results. Bayes A and B regressions had the best lines of concordance for all properties, with intercepts nearest 0 and slopes nearest 1.0 and with lower bias than PLS or MPLS. Bayes B regression used very few wavelengths, suppressing all of the usual water bands, while the far less selective Bayes ridge regression had some posterior density at almost all wavelengths. The authors mention, but did not compare these Bayes results with those from, PLS regression combined with a genetic algorithm, but they note the high information content in the wavelengths selected in the Bayes A and B regressions.

13.3.3 Nonlinear Bayesian Calibration of Near-Infrared Spectra

The studies discussed so far have focused on variable reduction and on the use of the prior to enhance modeling power in a calibration of a multivariate response as found in spectra. Another area of research in Bayesian regression modeling concerns ways of constructing the likelihood. The likelihood distribution in a Bayesian regression is not constrained to be linear. While it is much easier to map the density as a Gaussian based on some predetermined relation developed from linear matrix algebra, and to set the variance from an estimate of error, how the likelihood is created is flexible.

One place where the flexibility of a Bayesian approach offers a unique advantage over the usual, frequentist approach is in nonlinear regression applied to spectral data. A conventional, frequentist nonlinear regression is not possible when the number of predictor variables is much larger than the number of samples available, the case that arises for most data based on near-infrared spectrometry. However, nonlinear regression based on a reproducing kernel Hilbert space (Cristianini and Shawe-Taylor, 2000) can be used. This approach provides automatic dimension reduction of the predictor variables. The Bayesian formulation of this approach for a dataset that has n samples and p predictors and a q-dimensional multivariate response y begins with the nonlinear relationship

$$y = f(X) + \varepsilon \tag{13.34}$$

where f refers to some unknown, nonlinear function and ε is random error. The prediction of multivariate y can be structured as the minimization of some loss function L:

$$\min\left[\sum_{i=1}^{n} L\left(y_t, f(x)\right) + \lambda J(f)\right] \tag{13.35}$$

where $J(f)$ is a penalty function, λ is the regularization parameter, and n is the number of samples available for the modeling. The kernel function K is a solution to this optimization where

$$f_\tau(X) = \sum_{i=1}^{n} \beta_m K\left(X_i, X_j | \psi\right), r = 1, \ldots, q \tag{13.36}$$

where ψ, the kernel parameter, controls the shape of the kernel. Several possible kernels can be used, but for any kernel, the relation

$$y_t = BK_i + e_i \tag{13.37}$$

describes the regression. The kernel has converted the problem from a p-dimensional task to an n-dimensional task, since the number of calibrated components q in the mixture should be far less than n. A polynomial kernel function was selected in one paper using this approach (Chakraborty, 2012; Chakraborty, et al., 2012). Putting this relationship into a Bayesian regression leads to a complex likelihood, requiring the use of Bayesian latent variables to speed the Gibbs sampling. Proper, diffuse priors were used, each with its own hyperparameter. In this study, the regression is structured hierarchically, and the set of hyperparameters used in the regression were subject to a higher-level Bayes analysis. This hierarchical structure provided for estimates of the uncertainty in the regression parameters β and the variance of the priors.

The resulting kernel model permitted a complex relationship between the predictor variables and the multiple responses in y, but unlike earlier studies using linear Bayesian regression, no variable selection was performed because of the projection of the predictors into the kernel space. Since the posterior model was not closed form, Gibbs sampling was necessary, and the MCMC fitting of the hierarchy required two independent MCMC chains of 20000 iterations, with a burn-in of 10,000 samples and a thinning rate of 5 (Chakraborty, 2012).

This modeling approach was applied to the biscuit dough data discussed above. While this dataset has been studied by many other groups, unlike the other studies reported for this dataset, the calibration set used here was developed from randomly selecting 52 of the 78 dough samples available, with the remaining 26 samples used as the test set, and with this modeling repeated 100 times. Also, unlike most of the other studies, multivariate response modeling was employed in the kernel regressions, and all four properties were predicted from the same model; the pairwise correlations of the four response properties were fairly high. The modeling led to a posterior estimate of the kernel parameter ψ of 3 in over 50% of the runs and 2 in over 30%, strongly supporting the idea that the NIR spectral data are nonlinearly related to the properties in the cookie dough data. While results from these studies were not exactly comparable to those from other attempts made to model the biscuit dough, the RMSEP errors for all four properties reported in this study were generally lower than those from other Bayesian methods, including those from multivariate Gaussian process regression, and they were considerably lower than those from kernel PLS regression (Kuhn and Johnson, 2013) or from conventional, linear PCR. The predictive errors in the Bayesian results were also far lower than those from methods using random forest regression and support vector machines using either a polynomial or Gaussian kernel function (Rosipal and Trejo, 2001).

The NIR wheat data were also modeled using the same multivariate Bayesian regression approach. Here, 66 samples, again randomly selected, were used to train, with the remaining 34 used in the test set. The training and predictions were again repeated 100 times. Here, too, the polynomial kernel parameter ψ was found to be 3 in about 25% of the runs and 2 in over 60%, again strongly supporting the idea that the near-infrared spectral data are nonlinearly related to the wheat properties. As with the biscuit dough data, errors of prediction for moisture and protein were substantially lower than those from the support vector machine or the random forest regression. The predictive errors from the multivariate kernel Bayesian regression were also substantially lower than those from Gaussian process regression and kernel PLS regression, but here, surprisingly, the multivariate kernel Bayesian regression gave predictive errors only slightly lower than those from principal components regression (Chakraborty, 2012). The time required for these calibrations and predictions to run is not given in either of these papers, but it is noteworthy that, even though the projection to the kernel space greatly reduced the size of the model, the Bayesian regression runs were made on a computer cluster system.

Cui and Fearn (2017) also compared the performance of nonlinear Bayesian methods, as well as other linear and nonlinear regression methods, on near-infrared data. They developed a Gaussian likelihood function for the univariate property in terms of three hyperparameters: the signal and length scale of the radial basis function kernel distribution, and the regularization parameter λ. The Gaussian process regression used in this study was implemented using the GPML package available in the MATLAB programming language (Rasmussen and Nickisch, 2010). Rather than employ hierarchical Bayesian modeling to estimate the hyperparameters in the Bayesian model, they used evidence approximation methods (Bishop, 2006) to determine appropriate values, which were then held constant. A near-infrared dataset taken on 1240 agricultural seeds measured by a hyperspectral, near-infrared instrument and calibrated to the percent nitrogen was used to compare the regression methods in this study. The spectra were measured using 239 wavelengths with a resolution of 6.26 nm over the range 993.0–2488.4 nm. The predictive error from kernel Gaussian process regression (Gramacy and Lee, 2008) was compared to that from support vector machine (SVM) regression and conventional partial least squares regression. Like the kernel Gaussian process regression, the SVM regression also used projection to a kernel space based on a radial basis function. The hyperparameters in the SVM modeling were set by cross-validation using an L_2 norm. In all cases, both of the nonlinear methods easily outperformed PLS predictive modeling on the near-infrared data. No substantial difference was found here between the support vector machine and the Gaussian process regression modeling on this dataset, as might be expected from the use of the same kernel function and the weakly informative prior. The benefit of getting full posterior information for each prediction from the Bayesian regression, which can help with outlier detection, balanced against the difficulty of setting suitable values of hyperparameters by evidence approximation in smaller datasets with relatively few degrees of freedom.

Gaussian process regression modeling has also been used in modeling moisture content in three freeze-dried pharmaceutical compounds. A fourth compound was used as the external validation set in testing the global models for moisture (Clavaud, et al., 2017). The authors compared many linear and nonlinear regression methods on near-infrared data of 3822 samples taken on two purportedly identical spectrometers. Because these freeze-dried samples all had residual humidity levels within a very narrow range centered near 1.00%, a range of moisture levels was created in the samples by carefully controlled humidification or by drying of samples from the three compounds used for calibration, but not the samples of the test set. Spectral measurements were made in triplicate through the glass storage vial in reflectance mode on each of the two Fourier transform NIR spectrometers by averaging 32 scans made over a spectral range of 4000–10,000 cm^{-1}, giving a total of 3822 spectra of the 637 vials for the calibration. Another 168 spectra of the fourth compound were collected in the same manner. The authors tested the usual PLS-1 regression models for moisture in each of the three compounds separately as well as a global PLS-1 model, where all three pharmaceuticals were included in the calibration for moisture. They focused modeling efforts on well-known water bands near 5000 cm^{-1} (2000 nm) and 7000 cm^{-1} (1450 nm). In this study, support vector regression with a radial basis kernel outperformed linear Bayesian Gaussian process regression modeling. Bayesian ridge regression was implemented using Bayesian routines taken from the Anaconda python software package, a conventional, linear Bayes likelihood, ridge-regression model with a Gaussian prior. It is not at all surprising that the linear Bayesian regression method was outperformed by the nonlinear regression in this study. Plots of predicted vs. measured moisture for both calibration and validation all appear slightly curved downward for PLS-1 regression, but appear strictly linear for the SVR model, lending strong support to the claim of a nonlinear relationship between the moisture content and the spectral response (Clavaud, et al., 2017).

13.3.4 Nonparametric and Empirical Bayesian Spectral Calibration Methods

The linear Bayes regression with Gaussian prior was also outperformed slightly by a nonparametric prediction method. This method is based on choosing the property associated with the K-nearest calibration samples, where the unknown is predicted from a calibration set by finding the calibration sample closest to the spectrum of the unknown, or in some cases by calculating the average property value from a group of K samples that comprise the set of K-nearest neighbors in spectral space to the unknown spectrum (Clavaud, et al., 2017). Prediction from a K-nearest neighbor calibration requires no calibration relationship other than a distance, but it is of limited use without a large number of calibration samples fully spanning the space of the analysis (Pérez-Marín, et al., 2005). With limited or insufficiently distributed data, an unknown spectrum may fall in a region far from calibration samples, and the K-nearest neighbor algorithm may decline to estimate its property because the distances to its nearest neighbors are too large. When there are sufficient numbers of sufficiently well-dispersed data, a local calibration based on nearest neighbor spectra offers real advantages over conventional linear and even nonlinear approaches to modeling because the distance-based calibration permits variably nonlinear, flexible calibration relationships that are otherwise very difficult to model.

A Bayesian approach to K-nearest neighbor calibration, where the SVD scores of the NIR spectra of individual calibration samples are converted to a likelihood distribution by application of a simple Gaussian kernel function, has also been reported (Fearn, et al., 2010). This approach to distance-based calibration has the advantage of smoothing the calibration distribution in score space so that the likelihood distribution is defined over the space of the calibration. This permits calibration on smaller datasets where gaps may be present in the calibration data. The use of a likelihood also has the advantage over conventional, local smoothing in that all of the calibration data contribute to any prediction, unlike the K-nearest neighbor approach, where only K samples participate. In the Bayesian approach, samples contributing to likelihood density nearer the unknown contribute more heavily to the prediction. And, because it is a Bayesian approach, it also makes use of prior information, as in any other Bayesian analysis.

The weak point in any kernelization of data, as is done here, is deciding on the nature of the kernel function and setting its hyperparameters. While there is no need for cross-validation in a Bayesian model, validation using a tuning set is necessary to determine the optimal tuning hyperparameters for the kernel by testing the kernel values against predictive performance (Pérez-Marín, et al., 2012). Study of the same NIR feedstuff data and same likelihood kernel as used in earlier work (Fearn, et al., 2010) but with a different tuning hyperparameter gave different results; apparently, small values for the tuning hyperparameter in the Bayesian kernel regression resulted in lower SEP values, but produced occasional outlier results, while larger tuning hyperparameter values led to more consistent results, but with larger (worse) SEP values, and a robustness–performance tradeoff resulted. Using either the K-nearest neighbor model or the Bayes kernel regression on scores of the preprocessed spectral data, predictive results on NIR spectra of the feedstuff data were consistently better than those from PLS-1 regression in 24 of the 26 reported properties analyzed, reducing the SEP values by about 50% on average, with some variation between ingredients in the feedstuffs. Here, the K-nearest neighbor method slightly outperforms the Bayes nonparametric regression in terms of SEP, but those slightly better SEP results come at a significant cost in reliability. The K-nearest neighbor calibration approach fails to predict on 10–40% of samples, depending on the calibrated ingredient, while the Bayes regression predicts on all samples, and gives realistic credible intervals—the Bayesian equivalent of a confidence interval (see Gelman, et al., 2013 for more on this subject)—from the analysis.

Fearn et al. (2010) also demonstrate, using a Bayesian formulation, how an empirical prior can be used to improve calibrations. Somewhat akin to the informed prior used in selecting wavelengths for a regression model, the prior for the property can be set to represent the distribution of property data actually used in a calibration. This approach, called empirical Bayes calibration (Lwin and Maritz, 1980; Brown, 1982), relies on the distribution of property values present in the training set being a good description of future data. Like the K-nearest neighbor calibration, large gaps in property values or a lack of representation over the full range of calibration will lead to very disappointing predictive results. In many cases where inverse regression models (such as those in PCR and PLS-1) are used to form the likelihood, a normal prior will generally produce the best results because this prior often accurately describes the data used in calibration. However, when a calibration set is not well represented by a normal distribution, as is the case in some trace analyses, using the distribution that better reflects the calibration data will lead to superior results (Fearn, et al., 2010). For example, calibration of NIR data for a property with a lognormal distribution, as occurs in measurements of organic carbon in soils (Bellon-Maurel, et al., 2010), benefits from the use of the empirical, approximately lognormal prior. Both the bias and the SEP are reduced compared to the usual PLS-1 methods. An empirical distribution extends that benefit to data where the form of the distribution of the property value is not easily determined as lognormal or some other well-known function, or where the distribution may be mixed.

13.4 CONCLUSIONS

As this chapter demonstrates, Bayesian methods offer many advantages in multivariate calibration of near-infrared spectra. The great flexibility of the Bayesian modeling permits easy examination of linear modeling of univariate properties and linear modeling of multivariate properties, like many chemometric methods, but also permits nonlinear modeling and even nonparametric modeling with minimal changes to the likelihood model. This flexibility is a real advantage for accurate modeling of NIR spectra where nonlinear relationships to the property may be present. The prior offers a way to select variables and to provide information about the data and about parameters that is not easily incorporated in the conventional measurement model. Together, the two models allow an unusually capable approach to multivariate calibration of spectra.

While Bayesian regression is best done on larger datasets, the main difficulties associated with Bayesian regression have been associated with its substantial computational burden, the need to write complex software incorporating a different perspective on data analysis, and the need for practitioners to develop a background in probability and statistics that goes beyond basic linear algebra. The great advances in computational power available on small computer systems and the increasingly common multi-core, parallel computational systems make running most Bayesian regressions possible on small computer systems, even on laptop computers. Improvements in software have greatly reduced the need for user-written Bayesian regression software. Public-domain packages and public-domain Bayesian regression software are now easily obtained and used. The remaining hurdle is the training of users of this computational power and readily available software in Bayesian methods. It seems that this, too, is occurring, if gradually. The use of Bayesian regression in spectroscopic calibrations is much like the use of PLS regression was in 1987: The capability is there, some software is there, and computers can run the analyses in reasonable times, but reports are somewhat scarce and users have not yet seen a clear reason to invest in the methods. This chapter offers motivation for spectroscopists to make that investment and provides key references, links to software, and background reading to get the motivated user ready to try Bayesian regression methods.

ACKNOWLEDGMENTS

This work was supported by Grant 1506853 from the United States National Science Foundation.

Initial work on this chapter was carried out when SDB was a visitor at IRSTEA in Montpellier, France. The author acknowledges, with gratitude, the assistance of Dr. Jean-Michel Roger and Dr. Jean-Claude Boullet. He also thanks Dr. Javier Haddoux and others at IRSTEA for their hospitality.

REFERENCES

Bellon-Maurel, V., E. Fernandez-Ahumada, B. Palagos, J.-M. Roger, and A. McBratney. 2010. Critical review of chemometric indicators commonly used for assessing the quality of the prediction of soil attributes by NIR spectroscopy, *Trends. Analyt. Chem.*, 29, 1073–1081.

Berger, J. O. 2005. *Statistical Decision Theory and Bayesian Analysis*, 2nd Ed., Springer-Verlag, New York.

Bishop, C. M. 2006. *Pattern Recognition and Machine Learning*, Springer-Verlag, New York.

Bolstead, W. E. 2010. *Understanding Computational Bayesian Statistics*, John Wiley and Sons, Hoboken, NJ.

Breiman, L. 1996. Stacked regressions, *Mach. Learn.*, 24, 49–64.

Brown, P. J. 1982. Multivariate calibration, *J. Roy. Stat. Soc.*, B44, 287–321 (with discussion).

Brown, P. J., M. Vannucci, and T. Fearn. 1998. Multivariate Bayesian variable selection, *J. Roy. Stat. Soc.*, B60, 627–641.

Brown, P.J., T. Fearn, and M. Vannucci. 2001. Bayesian wavelet regression on curves with application to a spectroscopic calibration problem. *J. Am. stat. Assoc.*, 96, 398–408.

Brown, P. J., M. Vannucci, and T. Fearn. 2002. Bayes model averaging with selection of regressors, *J. Roy. Stat. Soc.*, B64(part 3), 519–536.

Centner, V., J. Verdu-Andres, B. Walczak, D. Jouan-Rimbaud, F. Despagne, L. Pasti, R. Poppi, D. Massart, and O. de Noord. 2000. Comparison of multivariate calibration techniques applied to experimental NIR data sets, *Appl. Spectrosc.*, 54, 608–623.

Chakraborty, S. 2012. Bayesian multiple response kernel regression model for high dimensional data and its practical applications in near infrared spectroscopy, *Comput. Stat. Data Anal.*, 56, 2742–2755.

Chakraborty, S., M. Ghosh, and B. K. Mallick. 2012. Bayesian nonlinear regression for large p small n problems, *J. Multivar. Anal.*, 108, 28–40.

Chen, H., B. R. Bakshi, and P. K. Goel. 2007a. Toward Bayesian chemometrics - a tutorial on some recent advances, *Analytica Chimica Acta*, 602, 1–16.

Chen, H., B. R. Bakshi, and P. K. Goel. 2007b. Bayesian latent variable regression via Gibbs sampling: methodology and practical aspects, *J. Chemom.*, 21, 578–591.

Chen, T., J. Morris, and E. Martin. 2007c. Gaussian process regression for multivariate spectroscopic calibration, *Chemom. Intell. Lab. Syst.*, 87, 59–71.

Chen, T. and E. Martin. 2009. Bayesian linear regression and variable selection for spectroscopic calibration, *Analytica Chimica Acta*, 631, 13–21.

Clarke, B. 2003. Comparing Bayes model averaging and stacking when model approximation error cannot be ignored, *J. Mach. Learn. Res.*, 4, 683–712.

Clavaud, M., Y. Roggo, K. Dégardin, P.-Y. Sacré, P. Hubert, and E. Ziemons. 2017. Global regression model for moisture content determination using near-infrared spectroscopy, *Eur. J. Pharm. Biopharm.*, 119, 343–352,

Cristianini, N. and Shawe-Taylor, J. 2000. *An Introduction to Support Vector Machines and Other Kernel-Based Learning Methods*, Cambridge University Press, Cambridge.

Cui, C. and T. Fearn. 2017. Comparison of partial least squares regression, least squares support vector machines and Gaussian process regression for a near-infrared calibration, *J. NIR Spectrosc.*, 25, 5–14.

Donald, D., D. Coomans, and Y. Everingham. 2011. Joint multiple adaptive wavelet regression ensembles, *Chemom. Intell. Lab. Syst.*, 108, 133–141.

Fearn, T., D. Pérez-Marín, and A. Garrido-Varo. 2010. Inverse, classical, empirical and non-parametric calibrations in a Bayesian framework, *J. NIRS*, 18, 27–38.

Ferragina, A., G. de los Campos, A .I. Vasquez, A. Cecchinato, and G. Bittante. 2015. Bayesian regression models outperform partial least squares methods for predicting milk components and technological properties using infrared spectral data, *J. Dairy Sci.*, 98, 8133–8151.

Gelman, A., Carlin, J. B., Stern, H. S., Dunson, D. B., Vehtari, A., and Rubin, D. B. 2013. *Bayesian Data Analysis*, 3rd Ed. Chapman & Hall/CRC, London.

Gramacy, R. B. and H. K. H. Lee. 2008. Gaussian processes and limiting linear models, *Comput. Stat. Data Anal.*, 53, 123–136.

Graps, A. 1995. An introduction to wavelets, *IEEE Comput. Sci. Eng.*, 2 50–61.

Hernàndez-Lobato, J. M., D. Hernàndez-Lobato, and A. Suarez. 2015. Expectation propagation in linear regression models with spike and slab priors, *Mach. Learn.*, 99, 437–487.

Kalivas, J. H. 1997. Two data sets of near-infrared spectra, *Chemom. Intell. Lab. Syst.*, 37, 255–259.

Kuhn, M. and Johnson, K. 2013. *Applied Predictive Modeling*, Springer-Verlag, New York

Larabel, M. 2012. Apple iPad 2 as fast as the Cray-2 super computer. https://www.phoronix.com/scan.php?page=news_item&px=MTE4NjU (Accessed 25 Jan 2019).

Lwin, T. and J. S. Maritz. 1980. A note on the problem of statistical calibration, *J. Roy. Stat. Soc.*, C29, 135–141.

MacKay, D. J. C. 1992. Bayesian interpolation, *Neural Comp.*, 4, 415–447.

Mallat, S. 1999. *A Wavelet Tour of Signal Processing*, 2nd ed., Academic Press, Cambridge, MA.

Mardia, K. V., Kent, J. T., and Bibby, J. M. 1979. *Multivariate Analysis*, Academic Press, Cambridge, MA.

Minka, T. 2001. A family of algorithms for approximate Bayesian inference, *PhD Dissertation*, Massachusetts Institute of Technology.

Mitchell, T. J. and J. J. Beauchamp. 1988. Bayesian variable selection in linear regression, *J. Am. Stat. Assoc.*, 83, 1023–1032.

Miyamoto, A., K. Watanabe, K. Ikeda, and M.-A. Sato. 2015. Variational inference with ARD prior for NIRS diffuse optical tomography, *IEEE Trans. Neural. Net.*, 26, 1109–1114.

Neal, R. M. 1996. *Bayesian Learning for Neural Networks*, Springer-Verlag, New York.

Nott, D. J. and C. Leng. 2010. Bayesian projection approaches to variable selection in generalized linear models, *Comp. Stats. Data Anal.*, 54, 3227–3241, 2010.

Nounou, M., B. R. Bakshi, P. K. Goel, and X. Shen. 2002a. Process modeling by Bayesian latent variable regression, *AIChE J.*, 48, 1775–1793.

Nounou, M., B. R. Bakshi, P. K. Goel, and X. Shen. 2002b. Bayesian principal component analysis, *J. Chemom.*, 16, 576–595.

Osborne, B.G., Fearn, T., and Hindle, P.H. 1993. *Practical NIR Spectroscopy*, Harlow, Essex.

Pedersen, D.K., H. Martens, J. P. Nielsen, and S. B. Engelsen. 2002. Near-infrared absorption and scattering separated by extended inverted signal correction (EISC): Analysis of near-infrared transmittance spectra of single wheat seeds, *Appl. Spectrosc.*, 56, 1206–1214.

Pérez-Marín, D., A. Garrido-Varo, and J. E. Guerrero. 2005. Implementation of LOCAL algorithm with near-Infrared spectroscopy for compliance assurance in compound feedingstuffs, *Appl. Spectrosc.*, 59, 69–77.

Pérez-Marín, D., T. Fearn, J. E. Guerrero, and A. Garrido-Varo. 2012. Improving NIRS predictions of ingredient composition in compound feedingstuffs using Bayesian non-parametric calibrations, *Chemom. Intell. Lab. Syst.*, 110, 108–112.

Polikar, R. 2006. The Wavelet Tutorial, http://cseweb.ucsd.edu/~baden/Doc/wavelets/polikar_wavelets.pdf (Accessed 25 Jan 2019)

Rasmussen, C. E. and H. Nickisch. 2010. Gaussian processes for machine learning (GPML) toolbox, *J. Machine Learn. Res.*, 11, 3011–3015.

Robert, C. P. and Casella, G. 2004. *Monte Carlo Statistical Methods*, 2nd Ed., Springer, New York.

Rosipal, R. and L. J. Trejo. 2001. Kernel partial least squares regression in reproducing Kernel Hilbert space, *J. Mach. Learn. Res.*, 2, 97–123.

Tan, H.-W. and S. D. Brown. 2003. Wavelet multiscale regression analysis for spectral calibration models, *J. Chemom.*, 17, 111–122.

Tibshirani, R. 1996. Regression shrinkage and selection via the lasso, *J. Roy. Stat. Soc.*, B58, 267–288.

West, M. 2003. Bayesian factor regression models in the "Large p, small n" paradigm, *Bayesian Stat.*, 7, 733–742.

Williams, P. and K. Norris, Eds. 2001. *Near-Infrared Technology in the Agricultural and Food Industries*, 2nd Ed., American Association of Cereal Chemists, Saint Paul, MN.

Yuan, M. and Y. Lin. 2005. Efficient empirical Bayes variable selection and estimation in linear models, *J. Am. Stat. Soc.*, 100, 1215–1225.

14 Derivative Quotient Method Regression, or the Norris Regression

*David W. Hopkins**
NIR Consultant

CONTENTS

14.1 INTRODUCTION

In this chapter, I present the work of Karl Norris in developing the method of using quotients of derivative terms of spectroscopic data using multiple linear regression (MLR). Because the original Fortran program is no longer supported by modern Windows DOS, I worked with Norris to develop a program to utilize his methods to study and analyze spectral data. Finally, I will present how the fourth derivative can be used to estimate the spectral noise of one or more spectra. Some of this work has already been described in a NIR News article (Hopkins, 2016).

William Hruschka wrote the latest version of the Fortran program called MULR.EXE for multi-term regression in 1998. MULR was one of a suite of programs for data pretreatment, regression, and utilities for conversion and display, together called spectral analysis programs (SPAN). Hruschka also wrote an accompanying user manual (Hruschka, 1998). The program was developed over 20 years, and Karl was arguably its most proficient user. I have used it and found it difficult to master. It is no longer supported under the DOS window of Windows 7 or later systems. Therefore, I asked Karl if I could work with him to develop a program in MATLAB (MATLAB, version R2011a; MathWorks, Inc.) that would enable the use of his methods. We tested the MATLAB program and obtained the same results on these data sets as Hruschka reported in the user manual. The program also includes new utilities I have written to extend his work, such as the determination of spectral noise.

14.2 PROGRAM OVERVIEW

The MATLAB program was written as a series of tasks to implement Karl's derivative quotient math and therefore named the DQM program. These tasks are listed with brief descriptions in

* nir.consultant@gmail.com.

TABLE 14.1
DQM Program Tasks

Task	Description
Menu	Display various tasks available
dqm1	Regression, search pattern #1, save models in files (*.modl) for later use
dqm2	Regression, search pattern #2, save models in files (*.modl) for later use
dqm3	Make various plots to investigate the data
dqmPredict	Use the current (or any other selected) calibration model to predict a selected data set
dqmDat	Make a data file (*.dat) from text files of the spectral data, reference values, ids, and wavelengths measured, with operator-entered information
sdd	Calculate the SDD for the predictions in (the current validation or other) file with duplicate or other replicate readings.
Noise	Calculate the noise in one or more selected spectra
Info	Present a more detailed description of the tasks listed above

Table 14.1. The derivatives are formed by the gap method defined in Equation 14.1 (Hruschka, 1998 and Norris, 2001):

$$\frac{\mathrm{d}(f(x))}{\mathrm{d}x} = \frac{f(x+\text{gap}) - f(x-\text{gap})}{\text{Norm}} \tag{14.1}$$

where Norm may be 1 or Norm $= 2\times\text{gap}$

Karl often used Norm of 1, so the derivatives were not normalized and the magnitude of the derivatives increased with increasing the gap. The gap is defined as the number of intervals from the center point of the derivative interval to the points at the ends of the intervals. The Norm of twice the gap gives the change in absorbance units (or log ($1/R$) units) per wavelength interval. This allows the direct comparison to other gap derivatives or derivatives such as those calculated by the Savitzky–Golay method (Savitzky and Golay, 1964). Of course, if regression subsequently utilizes the ratio of any two derivatives, the values of the Norm factors are taken into account by the regression coefficient, and the method of normalization is immaterial. Higher derivatives are calculated using the same gap and sequential applications of the first derivative.

The gap derivatives may be smoothed using a "boxcar" average of absorbance values over an interval of $(2\times\text{smt}+1)$ points. Here, smt is the smoothing gap, which may be any integer of 0 (no smoothing) or greater. Therefore, twice the smoothing gaps cover the range used for smoothing, and twice the derivative gaps cover the range used to estimate the derivative.

Multiple linear regression is performed using readings at wavelengths for numerator and denominator in all possible combinations of wavelengths, rather than the readings of individual wavelengths as usually performed in MLR or full-spectrum methods such as principal component regression (PCR) or partial least squares regression (PLSR). A set of full spectra is used for training, and a similar set is employed for validation. The MULR program is totally command-based, so the user chooses the wavelength range for the searches of the center wavelengths, the range of derivative gaps (hereafter simply called a gap), and the range of the smoothing gaps (hereafter called a smt) for the numerator (with no denominator). Then, with that best numerator wavelength, gap, and smt, a search is conducted for the best wavelength for the denominator, gap, and smt, tabulating the corresponding SEC, r, and R^2 values for the regressions. Plots of the r or R^2 values vs the wavelengths allow the user to visualize the spectral relationships, for comparison with the absorption or derivative spectra of samples. Thereafter, the search for the numerator may be reiterated, because usually the selection of the denominator impacts significantly the regression performance. The user continues the iteration of selecting the best numerator or denominator or may elect to vary the smt for the numerator or denominator. The user controls the optimization of the calibration parameters, and stops when the calibration performance cannot be improved.

The prediction equation takes the form shown in Equation 14.2:

$$y_{\text{pred}} = B_0 + B_1 X \frac{\text{Num}_1}{\text{Den}_1} + B_2 X \frac{\text{Num}_2}{\text{Den}_2} + ... + B_n X \frac{\text{Num}}{\text{Den}_n} \quad (14.2)$$

where the B coefficients are determined by MLR regression, Num and Den are the derivative values at the optimized wavelengths, gaps, and smts, and n indicates the number of terms in the calibration.

The use of derivatives rather than absorbance values has two major advantages. First, the baseline zero is determined, so the ratios of spectral values are not affected by offsets that typically are found in absorbance spectra of diffusely scattering samples. Secondly, each higher derivative has more bands, and they are narrower than the bands in the original spectra. This makes it easier for the regression to find and utilize bands that are less influenced by interfering components in the samples. A first derivative of an isolated absorbance peak has an inverted S-shape with two peaks (a positive and negative peak) which passes through zero at the peak wavelength of the original spectrum (Figure 14.1). One can learn to interpret the band positions of first derivative spectra as readily as the original absorbance spectra.

The DQM program performs tasks that are listed in Table 14.1. The tasks dqm1 and dqm2 are designed so that the operator has no interaction after setting up the initial ranges for the wavelength searches and the searches of gap and smt. We used two different tasks to employ two individual patterns for the optimization of the gap and smt parameters. Task dqm1 follows the pattern (Figure 14.2) that was most used by Karl and which was illustrated in the user manual (Hruschka, 1998). Task dqm2 allows the optimization of the smt parameters immediately after finding the gaps (Figure 14.3). The MULR program optimized the regressions using the maximum R^2 and displayed the R^2 vs wavelength for the results. The DQM program optimizes the regression results using the minimum SEC and displays the SEC, r, or R^2 vs wavelength.

Task dqm3 performs many of the functions we found useful for effectively using dqm1 and dqm2. Table 14.2 enumerates the tasks that are presently performed using dqm3. In addition to putting together data files, possibly the most important function is allowing the user to become familiar with the appearance of derivatives with various choices of gaps and smts. Derivatives using small gaps and smts allow the observation of lesser shoulders and bands. Using gaps or smts that are larger, broader peaks are enhanced while the narrower bands may be quite deformed (see Norris, 2001, and Ritchie and Mark, 2002a). When the dqm3 MSC plot shows parallel or nearly parallel lines, that it is hopethe spectra are very similar. On the other hand, spectra that are highly divergent are likely different and may indicate problems or outliers (Ritchie and Mark, 2002b). Another dqm3 task allows the operator to plot the correlation between a derivative evaluated at a selected wavelength with a particular reference analysis and obtain the correlation coefficient r and the multiple correlation coefficient R^2. This allows the demonstration of the connection between the spectroscopy and the chemistry of the samples. A new noise determination method is presented in Section 14.7.

Tasks dqm1 and dqm2 validate the final model on a second data set and report the performance statistics. If the calibration process is repeated very often to obtain better performance on the calibration and validation sets, it is good practice to test a final, "best" calibration on another data set. The task dqmPredict predicts any data set with any model and allows the omission of selected samples identified as outliers. The option to delete samples may be useful if the user wants to remove spectra that do not have the same number of repeat scans prior to using the sdd task to calculate the standard deviation of differences (SDD) of the NIR predictions.

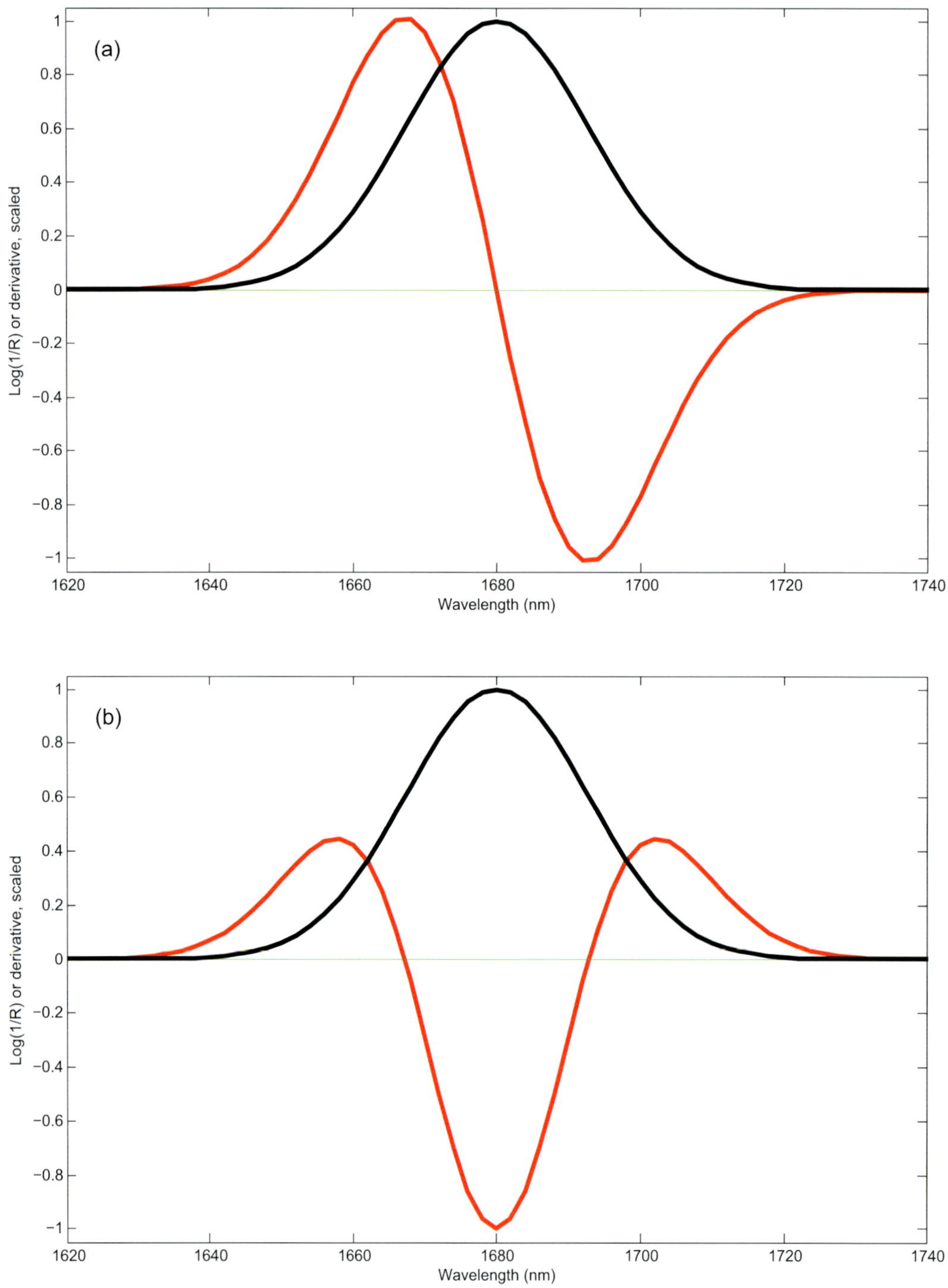

FIGURE 14.1 An isolated Gaussian absorbance band (black), with a peak at 1680 nm of 1.00 absorbance units and half-band width of 30 nm. (a) The first derivative (red) scale factor is 10.7. (b) The second derivative (red) scale factor is 41.6. (c) The third derivative (red) scale factor is 197. (d) The fourth derivative (red) scale factor is 595. The scale factors indicate how much the calculation of successive derivatives decreases signal (or height) of the derivative spectra.

(Continued)

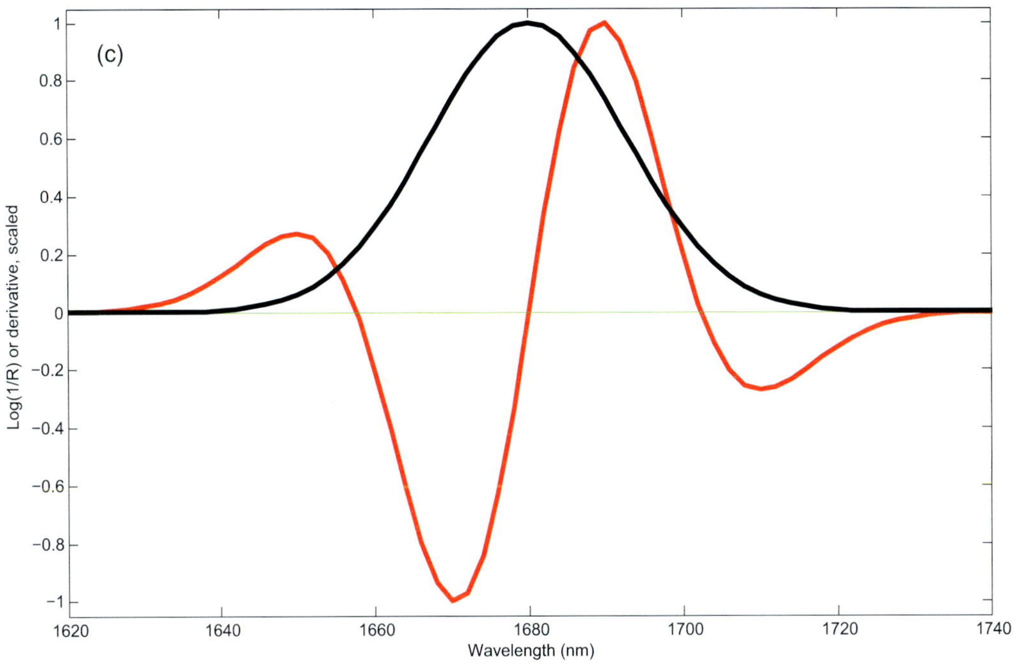

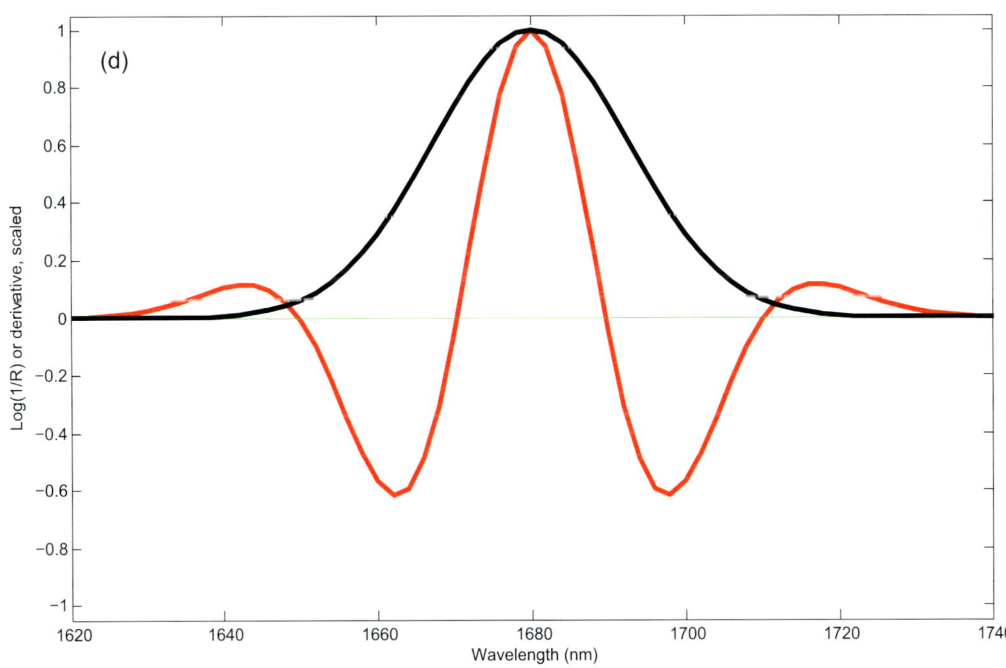

FIGURE 14.1 ***(CONTINUED)*** An isolated Gaussian absorbance band (black), with a peak at 1680 nm of 1.00 absorbance units and half-band width of 30 nm. (a) The first derivative (red) scale factor is 10.7. (b) The second derivative (red) scale factor is 41.6. (c) The third derivative (red) scale factor is 197. (d) The fourth derivative (red) scale factor is 595. The scale factors indicate how much the calculation of successive derivatives decreases signal (or height) of the derivative spectra.

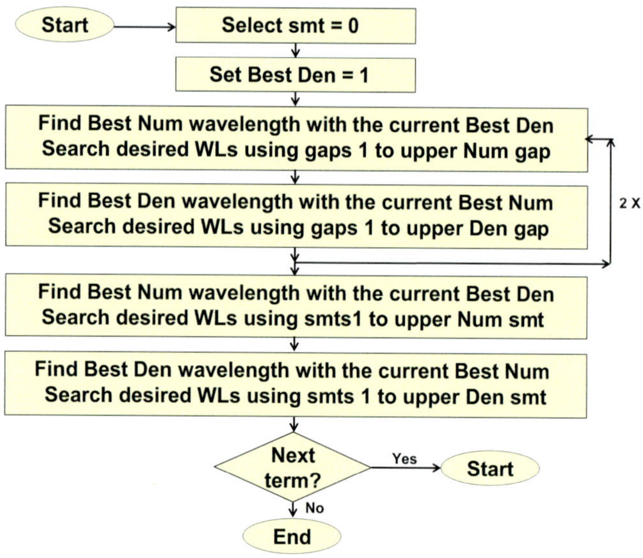

FIGURE 14.2 Workflow used in task dqm1. The order of the derivatives, wavelength ranges for the searches, and the ranges of gaps or smts are individually selectable for the numerator and denominator.

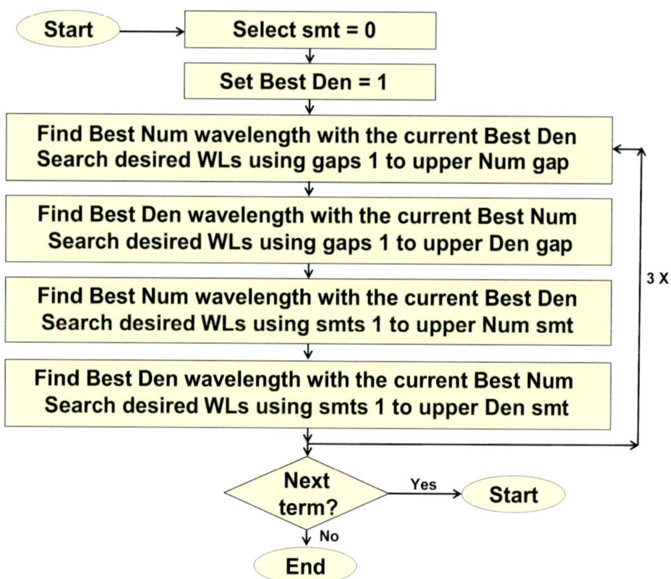

FIGURE 14.3 Workflow used in task dqm2. The order of the derivatives, wavelength ranges for the searches, and the ranges of gaps or smts are individually selectable for the numerator and denominator.

14.3 CONVOLUTION FUNCTIONS

The DQM program uses convolution functions for calculating all derivatives and smoothing operations. I first encountered convolution functions in the classic paper by Savitzky and Golay (1964), and details are found there. Briefly, a convolution function (CF) is not a function at all in the mathematical sense. Rather, it is a set of ordered numbers, usually represented by a set of integers and a normalization integer, that may be aligned with successive points in a spectrum, so that the CF

TABLE 14.2
Tasks for Each dqm3 Menu Entry

Task	Description
0	Display menu
1	Load data as *.dat file
2	Plot raw spectra for all samples
3	Plot derivatives for all samples
4	Plot constituents
5	Calculate correlations of constituents
6	Print ids and constituents for selected samples
7	Scroll through raw spectra
8	Scroll through derivatives
9	Plot selected raw spectra
10	Plot selected derivatives
11	Plot multiplicative scatter plots of selected spectra
12	Plot multiplicative scatter plots of selected derivatives
13	Plot selected difference spectra
14	Plot derivatives of selected difference spectra
15	Plot simple correlation of selected constituent with a derivative at a chosen wavelength
16	Reset axes of a selected plot
17	Enter new title for a selected plot
18 and 19	Enter new x- and y-axis labels
20–26	Enter new text and pointers to particular spectral features
99	Exit dqm3

values are multiplied by the corresponding spectral values and the products are all summed to give a single value that is divided by the normalization integer and then assigned to the center of the convolution interval in the convolved result. The CF of n factors is positioned at the start of the spectrum, and the first point of the result is at point $(n+1)/2$. Then, the CF is positioned at the next point of the spectrum, to determine the following point of the result, and the process is continued, moving the convolution interval along the spectrum until all of the original spectrum has been convolved. The process leaves $(n-1)/2$ points at the beginning and end of the convolved spectrum undefined, and these are usually set to zero. The magical thing (in my opinion) is that different CFs will yield either smoothed spectra or derivatives of various orders, depending on the CF factors chosen, while the mathematical operation is the same. Savitzky and Golay (1964) provided tables of CF integers and the normalization integers, but software packages today calculate the CF factors as needed (Madden, 1978), and the necessary calculations employ full precision multiplication, so if the CF are obtained at all, it is sometimes difficult to recognize integer values for the CF and normalization constant. There is an error in the third derivative, quantic or sexic fitting function, that has been corrected in the DQM program (H. Mark, personal communication, 2003).

While a derivative and smoothing of a spectrum may be accomplished with separate CFs, successive convolutions may always be accomplished by using a single new CF. In the DQM program, each search employs a CF based on the order of the derivative, and the gap and smt values desired. With modern computers, and the MATLAB programs optimized for vector math, the calculations for 50 samples at 1000 wavelengths, for 15 values of gaps, and 15 values of smt require less than 10 s. It now takes more time for the analyst to analyze the results than to perform the calculations.

The choices of gap and smt and the order of the derivative determine the length of the convolution interval and the CF factors. Importantly, the square root of the sum of the squares of the factors or the index of random errors (IRE) (Mark, 1991) gives the reduction in the random error, or noise reduction, that the application of the CF causes in the spectra. The IRE is given, along with the number of points and length (nm or cm^{-1}) of the convolution interval for each derivative calculated in the DQM program.

14.4 INSTRUMENT AND SAMPLE INFORMATION

Karl Norris routinely used a Cary 14 double-beam spectrophotometer custom-interfaced to a computer for the collection and analysis of spectra of samples. The instrument was operated with a slit width giving spectra with a half-band width of 3 nm (personal communication). With a wavelength interval of 1.60 nm, the range from 1000 to 2598.4 nm therefore contained 1000 wavelength points per spectrum. The spectra of a sample and a reference ceramic disk were collected in reflection mode and converted to log(1/*R*) values. The hard red winter wheat samples were finely ground in a cyclone mill and were subsequently analyzed for protein using the Kjeldahl digestion in the laboratory of Phil Williams at the Canadian Grain Commission. Each sample was analyzed 16 times, and the average result was used as the reference value. It was estimated (Hruschka, 1998) that the standard error of these values was 0.14% protein (dry basis). Moisture results for these samples were determined by NIR readings with a calibration that is not known at this writing. Two sets of 50 scans of different wheat samples were provided by Karl Norris in the original Fortran formats, WhelgA.r4d and WhelgB.r4d. These were converted to use with the DQM program and named WhelgA4.dat and WhelgB4.dat. The statistics of the protein values (reported on a dry basis) for the two sets are similar (Table 14.3).

14.5 RESULTS

With the DQM program, the user obtains information which is helpful in understanding the regression results and which is useful in guiding subsequent models. It is good to start by looking at the raw spectra and a selection of derivatives using dqm3. Figure 14.4 shows the original spectra. The main features are the moisture band at 1934 nm, a broad peak due the starch hydroxyl groups at 2100 nm, a slight shoulder at 2178 nm, and a sharp band at 2286 nm associated with CH and CH_2 bands of protein and starch as well as the small amount of lipids present.

The first derivative with a gap of only 2 points and no smoothing (Figure 14.5) elucidates the spectral details more clearly. The protein band centered at 2178 nm is resolved, with a peak of the derivative at 2160 nm. The inverted-S feature is located below the zero line, because the protein band is actually a shoulder on a descending region of the original spectra (compare Figure 14.4). In subsequent regression, the values at 2160 nm are used to estimate the protein content, even though the actual peak at 2160 is not visible with the smoothing produced a gap of 14 points and smt of 2 points. The smoothing spreads the protein information over a wide wavelength region, but it is interesting that the protein information is still maximal at 2160 nm.

It is useful to observe some of the plots of *r* and SEC vs wavelength as the searches proceed for the best numerator and denominator in task dqm1. Table 14.4 shows a summary of the program output for the searches. It is notable that the optimal wavelengths selected are not constant, but tend

TABLE 14.3
Statistics of the Protein Content (Percent by Weight on a Dry Basis) for the Calibration Set WhelgA4 and Validation Set WhelgB4

File Name	Minimum	Maximum	Average	Standard Deviation	Number of Scans
WhelgA4.dat	9.66	17.71	13.74	2.11	50
WhelgB4.dat	10.13	18.75	14.15	2.35	50

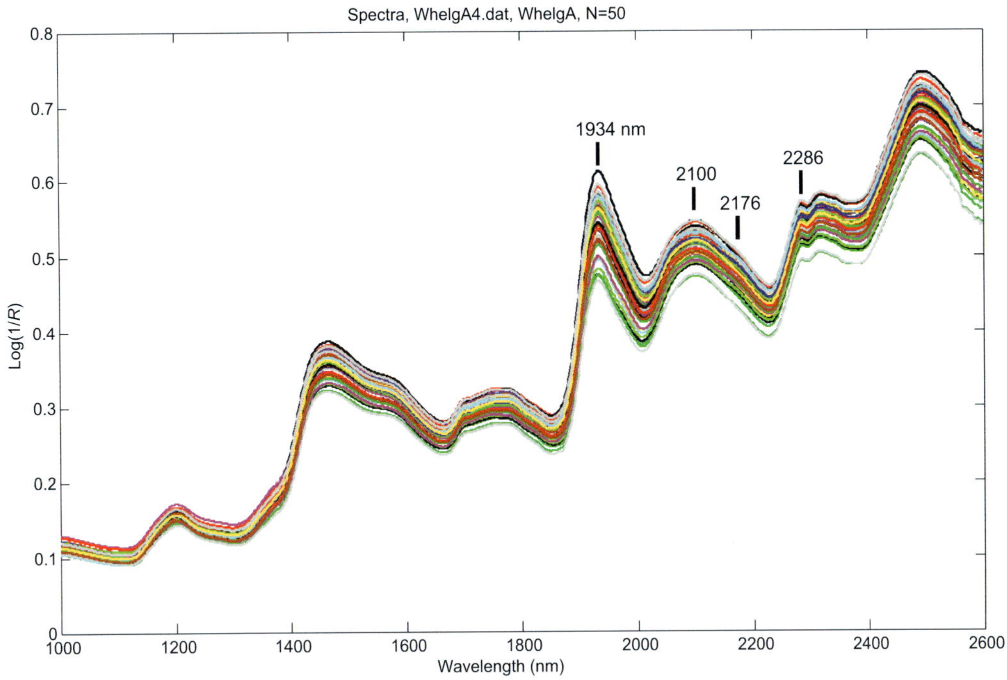

FIGURE 14.4 Original reflection spectra of ground samples of hard red winter wheat in WhelgA4.dat. The assignments of the major absorption bands are presented in the text.

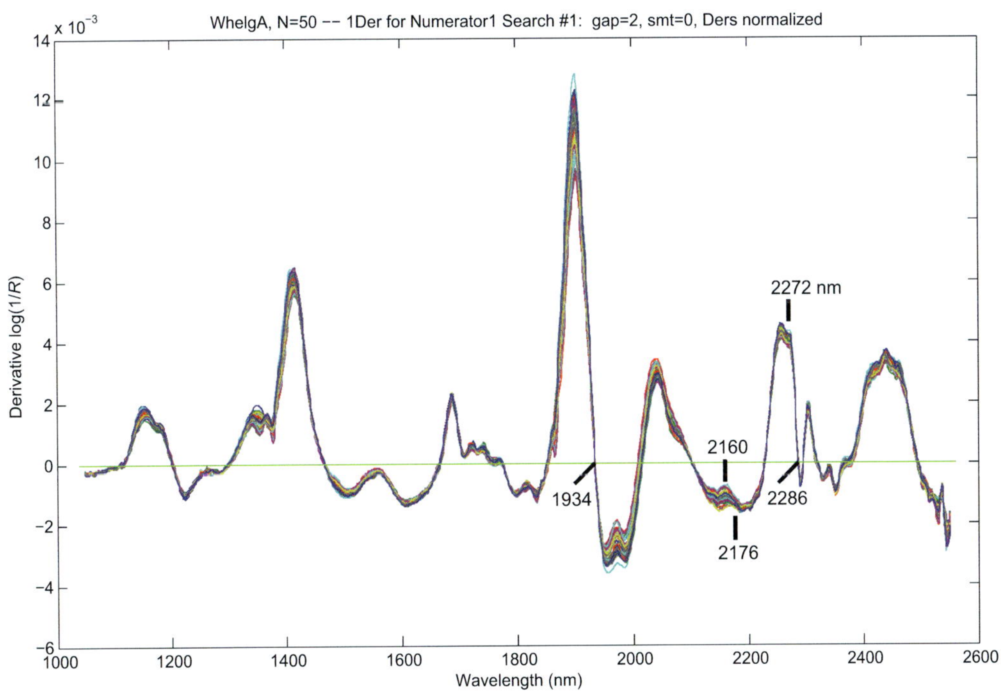

FIGURE 14.5 First derivative spectra (gap = 2, smt = 0) show more clearly the details about the components of the samples. Some important features identified in the course of this calibration study are indicated (see text). The derivatives cross the zero line at 1938 and 2286 nm, indicating peaks in the original spectra.

TABLE 14.4
Example of Search Output, Method 1

		Numerator1 Search #1		
Gap	Smt	SEC	RSQ	WL
1	0	0.40119*	0.9646	2158.4
2	0	0.36273*	0.9711	2156.8
3	0	0.36394	0.9709	2156.8
		Denominator1 Search #1		
Gap	Smt	SEC	RSQ	WL
1	0	0.26193*	0.9849	2265.6
2	0	0.25997*	0.9851	2272
3	0	0.25268*	0.9860	2268.8
4	0	0.25178*	0.9861	2270.4
5	0	0.24914*	0.9864	2268.8
6	0	0.24917	0.9863	2267.2
		Numerator1 Search #2		
Gap	smt	SEC	RSQ	WL
1	0	0.31769*	0.9778	2160
2	0	0.24798*	0.9865	2155.2
3	0	0.20707*	0.9906	2148.8
4	0	0.19965*	0.9912	2147.2
5	0	0.19571*	0.9916	2152
6	0	0.18530*	0.9924	2144
7	0	0.18080*	0.9928	2150.4
8	0	0.17420*	0.9933	2156.8
9	0	0.16619*	0.9939	2155.2
10	0	0.16274*	0.9942	2153.6
11	0	0.16049*	0.9943	2161.6
12	0	0.15755*	0.9945	2161.6
13	0	0.15059*	0.9950	2160
14	0	0.14939*	0.9951	2161.6
15	0	0.15071	0.9950	2161.6
		Numerator1 Search #4		
Gap	smt	SEC	RSQ	WL
15	0	0.13969*	0.9957	2158.4
15	1	0.13927*	0.9957	2160
15	2	0.13855*	0.9958	2160
15	3	0.13957	0.9957	2161.6
		Denominator1 Search #4		
Gap	smt	SEC	RSQ	WL
7	0	0.13855*	0.9958	2272
7	1	0.14080	0.9956	2272
7	2	0.14256	0.9955	2272

Summary of dqm1 Regression on Whelg4.dat, *N* = 50 Using 1 Term

Constituent: protein (db).	Omit samples: none.	
With set V, WhelgB4.dat, *N* = 50.	Omit samples: none.	
Term #1 Num	1-Der Search gap to 15, smt to 15	WLs 1048–2550.4 nm
Term #1 Den	1-Der Search gap to 15, smt to 15	WLs 1048–2550.4 nm

(Continued)

TABLE 14.4 (*Continued*)
Example of Search Output, Method 1

Gap	smt	SEC	RSQ	WL	Description
2	0	0.36273	0.9711	2156.8	Num1 gap Search #1
5	0	0.24914	0.9864	2268.8	Den1 gap Search #1
14	0	0.14939	0.9951	2161.6	Num1 gap Search #2
7	0	0.14008	0.9957	2272	Den1 gap Search #2
15	0	0.13969	0.9957	2158.4	Num1 gap Search #3
7	0	0.13969	0.9957	2272	Den1 gap Search #3
15	2	0.13855	0.9958	2160	Num1 smt Search #4
7	0	0.13855	0.9958	2272	Den1 smt Search #4

SEC = 0.13855 **RSQ = 0.9958** ***N* = 50**

RMSEP = 0.13485 **SEP = 0.13296** **Bias = 0.029** **RSQ = 0.9970** ***N* = 50**

One-term model: dave44-m1-dqm.modl

Term #1: (1D 2160 nm, gap = 15, smt = 2)/(1D 2272 nm, gap = 7, smt = 0)

Coefficients B0, B1, B2, … = 34.8914 63.9514

Term	Der Order	gap	Smt	CF Numb Pts	CF Width (nm)	IRE
Num1	1	15	2	35	54.4	0.0211
Den1	1	7	0	15	22.4	0.1010

First, the details of each search are presented, and for brevity, not all of the results are listed below. The asterisks in the SEC column indicate that each successive step of the search resulted in a decrease in the SEC; sometimes an increase in the gap or smt results in a loss of calibration performance. Second, the summary of the calibration is presented, with the details on the CFs including the IRE values.

to stay in the same narrow region as different amounts of smoothing are selected for the numerator and denominator treatments.

Search #1 for the numerator finds the MLR equivalent of a single-term calibration using the first derivative of log(1/*R*) values, and Figure 14.6 (and Table 14.4) shows that a wavelength of 2156.8 nm gives a SEC of 0.36% protein. The plot of the *r* values (Figure 14.7) shows basically two regions of good correlation with the protein content, at 1656–1780 nm and 2050–2200 nm. The inverse S figures for the protein amide bands with zero-crossings at 2054 and 2197 nm are evident, and two bands in the second overtone region at 1693 and 1731 nm are also apparent. The S-curves are significantly flattened in the positive and negative peaks, because the wavelengths across the peaks are nearly as effective in predicting the protein content. The maximum *r* is observed at 2156.8 nm, which necessarily agrees with the minimum SEC in Figure 14.6. The zero-crossing occurs at 2197 nm; we might expect this zero-crossing to be closer to 2176 nm (see Figure 14.5), but possibly neighboring bands influence the position of this feature in the *r* spectrum.

The program proceeds to search for the best denominator, using 2156.8 nm (gap = 2, smt = 0) just selected for the numerator. Figure 14.8 shows the SEC spectrum with the optimal wavelength of 2268.8 nm (gap = 5, smt =0). Often there are a number of wavelength regions with nearly as effective denominator choices, but the 2268.8 nm selection improves the SEC to 0.25% protein, a respectable value. In other applications, limiting the denominator searches to alternative wavelength regions indicated in the SEC spectra can lead to different calibrations with possibly improved performance. This may be the case, if the search selects an accidental correlation at a wavelength in the spectral noise at the end of the spectral, for example.

Figure 14.9 shows the SEC spectrum for Search #2 of the numerator, which is now paired with the denominator from the last search, at 2268.8 nm (gap = 5, smt = 0). A SEC of 0.15% protein is achieved. The SEC spectrum is rather different from the first search (Figure 14.6) because the regressions are now using ratio terms. Figure 14.9 shows that a numerator of 2161.6 nm (gap = 14, smt = 0) with the

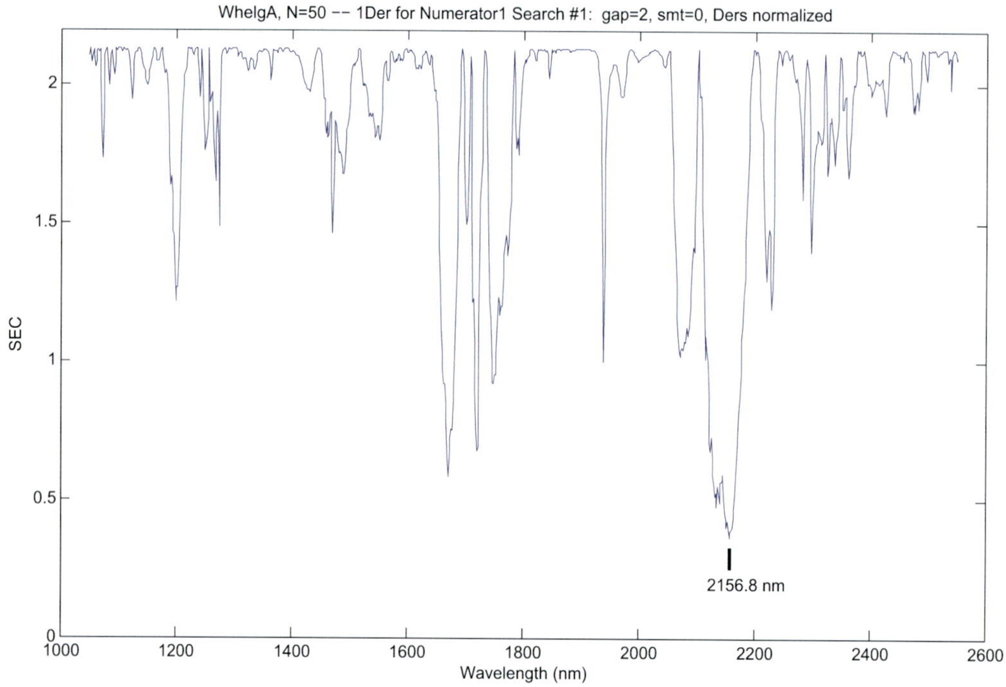

FIGURE 14.6 SEC spectrum for the first search of the numerator. Two major regions of correlation with protein content are 1656–1780 and 2050–2200 nm, and the best wavelength is 2156.8 nm, with gap = 2 and smt = 0.

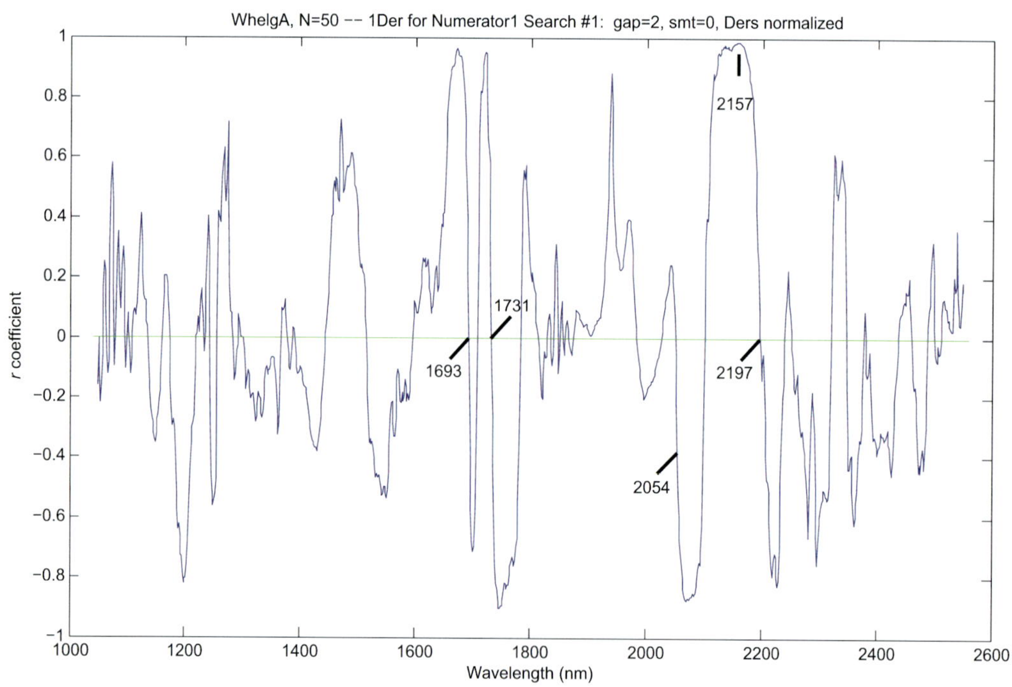

FIGURE 14.7 The *r* coefficient spectrum for the first search for the numerator. The two major regions of correlation with protein content also are seen in Figure 14.6.

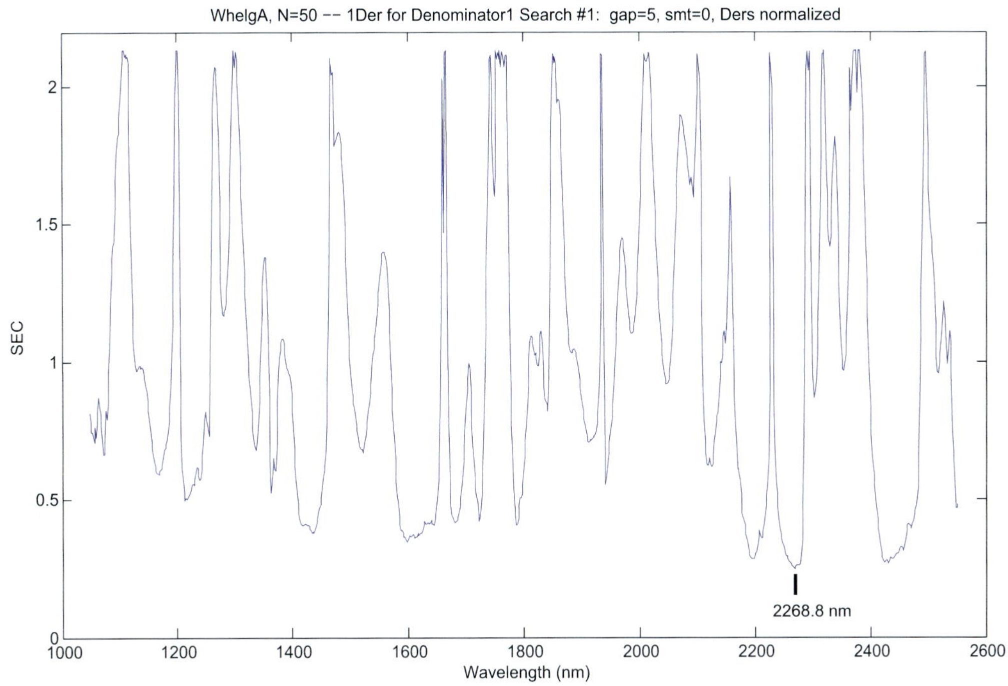

FIGURE 14.8 SEC spectrum for the first search of the denominator, which picks the best wavelength at 2268.8 nm, gap = 5, and smt = 0, with numerator = 2156.8 nm, gap = 2, and smt = 0.

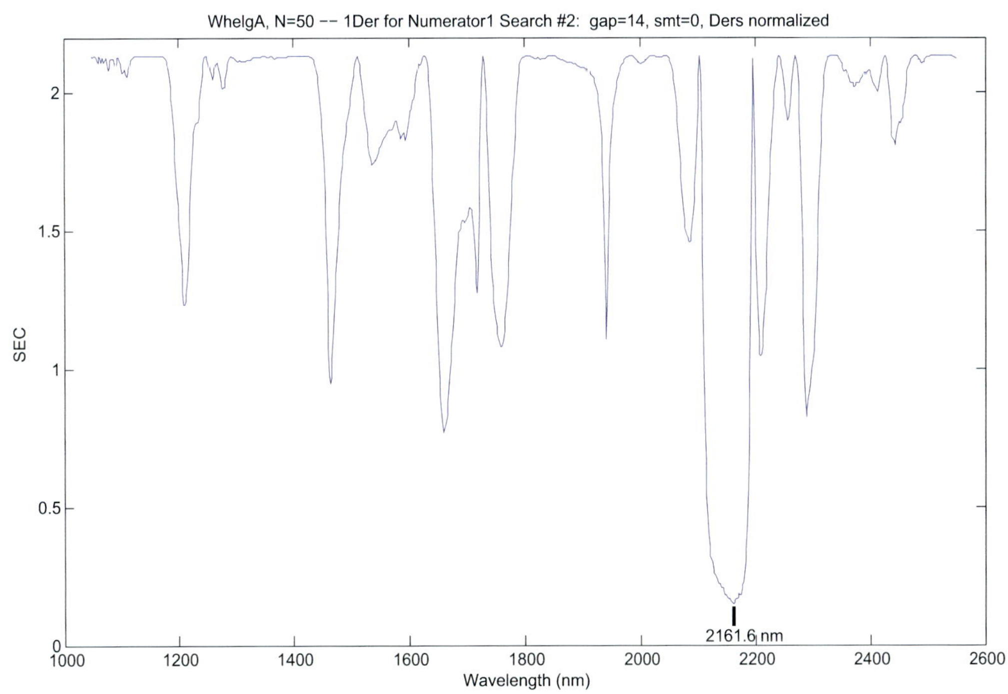

FIGURE 14.9 SEC spectrum for the second search of the numerator, which indicates the best wavelength is 2161.6 nm, gap = 14, and smt = 0, with denominator = 2268.8 nm, gap = 5, and smt = 0.

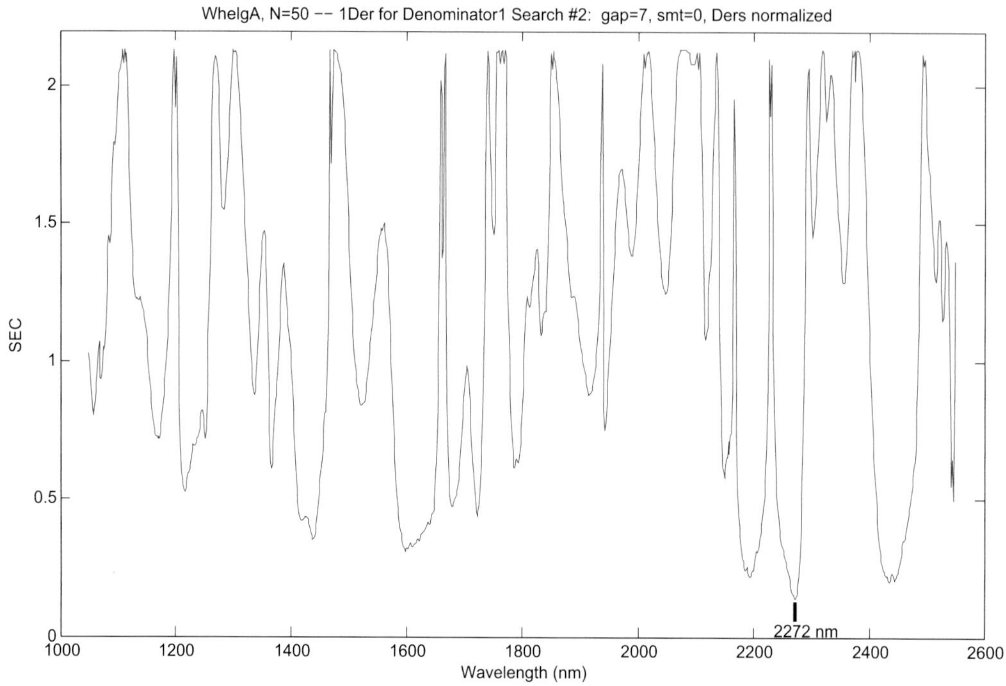

FIGURE 14.10 SEC spectrum for the second search of the denominator, which indicates the best wavelength at 2272.0 nm, gap = 7, and smt = 0, with the numerator = 2161.6 nm, gap = 14, and smt = 0.

denominator of 2268.8 nm is clearly the best choice. The improvement in SEC from 0.25% to 0.15% protein demonstrates the efficacy of this ratio of derivative terms in the model.

A bit more smoothing of the derivatives is indicated in Search #2 for the denominator (Figure 14.10), with a gap of 7 at the best wavelength at 2272.0 nm giving a SEC of 0.14% protein. In the third search of the numerator, more smoothing is found for the numerator than the previous search, with a gap of 15, but the SEC is still 0.14% protein (Table 14.4). The denominator shows no change in the optimal wavelength, or in the SEC, gap, and smt parameters in the third search. In the fourth searches of the numerator and denominator, the gaps of the derivatives are kept at the values found in the previous rounds. The wavelengths are searched as all the smt values are evaluated to find the optimum combination gap, smt, and wavelength selection. This final round found that the optimal wavelength is at 2160.0 nm, with gap of 15 points and smt of 2 points, while the denominator with those numerator values shows no change. The SEC values should only be reported with a precision of two significant figures, but the tasks compare SEC values at full precision and suggest an improvement even for a slight decrease in SEC. The SEC values are tabulated to five decimal places, so the operator can observe the progress as the searches proceed.

The optimized calibration from dqm1 is summarized in Table 14.5, with a SEC and RMSEP of 0.14% protein (dry basis). This SEC is about as low as we can expect, as it should not be less than the standard error of the reference method. A value of SEC less than the standard error of the reference method and RMSEP (for the validation set) greater than the SEC indicates over-fitting of the data. Calibration with dqm2 gives essentially the same performance (see Table 14.5), but the smts tend to be larger and the gaps smaller with dqm2 than with dqm1. I prefer the calibrations produced using dqm1, because I prefer smoothing by increased gaps over smoothing by a "boxcar" average. Table 14.5 also shows something I have noted over the years of a friendly competition with Karl, that the DQM approach generally gives calibrations with somewhat lower SEC and RMSEP values than can be achieved with PLS. Here, I used pretreatments that I have found optimal in removing

TABLE 14.5
Best Calibrations for Wheat Protein (db)

Regression Method	Calibration SEC	Validation RMSEP	SEP	Bias
A	0.14	0.14	0.13	0.03
B	0.14	0.14	0.14	0.03
C	0.17	0.18	0.18	0.01

A: Calibration using dqm1, (1D 2160 nm, gap = 15, smt = 2)/(1D 2272 nm, gap = 7, smt = 0) with B0 = 34.89 and B1 = 63.951. B: Calibration using dqm2, (1D 2160 nm, gap = 2, smt = 14)/(1D 2273.6 nm, gap = 5, smt = 4) with B0 = 34.79 and B1 = 60.935. The Num CF had 33 points spanning 51.2 nm, with IRE of 0.0244; the Den CF had 19 points spanning 28.8 nm, with IRE of 0.0471. C: Calibration using PLS in The Unscrambler 10.3, 4 factors, 1200–2432 nm, with data pretreatments of Savitzky–Golay 1 Derivative 15-point quadratic fit followed by multiplicative scatter correction.

multiplicative effects in PLS calibration, the first derivative plus multiplicative scatter correction (MSC) (Dhanoa et al., 1994). The difference in performance of the models is small but evident. It is suggested that including a wide spectral region for the PLS calibration forces the inclusion of wavelengths that are poorly correlated with the protein content, slightly degrading the performance that can be obtained.

14.6 DISCUSSION

The final prediction equation from dqm1 is a single-term quotient of first derivative terms:

$$Y_{\text{pred}} = 34.89 + 63.951\ X\ \frac{(\text{1D 2160 nm, gap} = 15,\ \text{smt} = 2)}{(\text{1D 2272 nm, gap} = 7,\ \ \text{smt} = 0)} \tag{14.3}$$

The first derivative at 2160 nm is located on the left lobe of the first derivative for the amide N-H combination band at 2176 nm of protein and may be taken as a measure of the protein content of the samples. Similarly, the first derivative at 2272 nm is near the left lobe of the derivative for the CH or CH_2 combination band at 2286 nm that can be taken as a measure of the total content of all constituents containing CH or CH_2 groups, or the total mass of the wheat. All the first derivative values at 2160 nm are negative and all the values at 2272 nm are positive, so that the ratio terms cause a subtraction from the B0 value (34.89) to arrive at the protein prediction. Even though the quotients are negative, they are *positively* correlated with the protein content. Thus, the derivative ratio is a direct measure of the protein concentration in the samples, as Karl Norris has argued the ratio should perform (Ritchie and Mark, 2002b). The deviations in the numerator derivative values due to the differences in light scattering in the sample spectra are corrected by the division by the denominator derivative values, and this allows the ratios to obey Beer's law (Norris, 2001).

14.7 NOISE AND THE FOURTH DERIVATIVE

In 2011, Karl Norris showed that the fourth derivative calculated with a small gap could detect a very narrow spectral band and the random noise of the spectrum, but wider bands were lost (not detected) in the fourth derivative (Norris and Davies, 2011). A good fourth derivative is shown in Figure 14.11, with gap = 3.5 and smt = 3. The positive peaks illustrate the richness of the fourth derivative bands we have already discussed in the first derivatives. There are two peaks evident

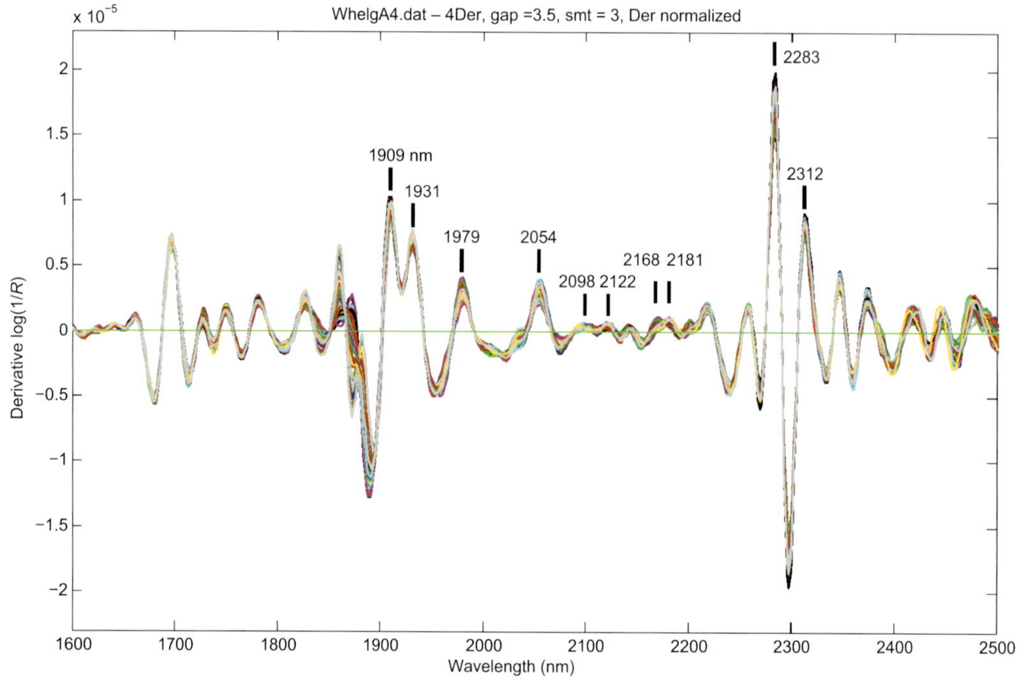

FIGURE 14.11 Fourth derivative of wheat spectra, gap = 3.5 and smt = 3. Bands corresponding to protein amide absorption, water, and starch CH and CH_2 are evident.

for the water combination bands at 1909 and 1933 nm. The amide bands appear at 1979, 2054, and 2168–2181 nm. The starch absorbance band at 2100 nm is so broad that two very low-intensity bands are resolved at 2098 and 2122 nm in the fourth derivative. This convolution function is 35 points long (the CF interval is 54.4 nm), IRE of 0.00132, giving a large reduction of the noise compared with the original spectrum. Figure 14.1d suggests that the signal is reduced by 1/595 = 0.0017 by a fourth derivative, and the IRE suggests that the noise is reduced by 0.0013, so the signal/noise ratio (SNR) is enhanced by 0.0017/0.0013 or 1.3. Thus, derivatives do not always increase the SNR, and it is a matter of optimizing the convolution function to obtain the desired IRE.

On the other hand, using gap = 0.5 and smt =0 gives fourth derivative spectra that display no bands, and simply noise (Figure 14.12). In fact, the IRE for this convolution function, [1 -4 6 -4 1], is 8.37; i.e., the noise in this fourth derivative is 8.37 times the noise in the original spectra. Therefore, this 5-factor convolution function can be used with a normalization factor of 8.37 to determine the noise in the original spectrum.

A task called noise is included in DQM, and it plots the original spectra overlaid with the noise spectrum calculated using the standard deviation (SD) of a moving window of 21 points across the entire spectrum. An example of the results is shown in Figure 14.13. The SD of 21 adjacent wavelength points in this case appears to give a noise level somewhat higher than measuring the SD of a number of repeated scans, as the noise of 15 micro-log units over 1200–1800 nm is higher than the level of 1 micro-log unit that I would expect from the Cary monochromator for scans of actual samples, or repeats of a reference ceramic disk. The increase in noise at wavelengths greater than 2500 nm is real, but the cause is not known; a similar effect is not observed with WhelgB4.dat. It is nice to have a technique to measure the noise in one spectrum, when none of the samples have been scanned with two or more replicates, which is the normal way to measure noise (Mark, 1991). We expect to see the spectral noise increase at higher absorbance levels, as peaks occur in the spectrum, and this method gives a tool to demonstrate the actual noise throughout the spectrum.

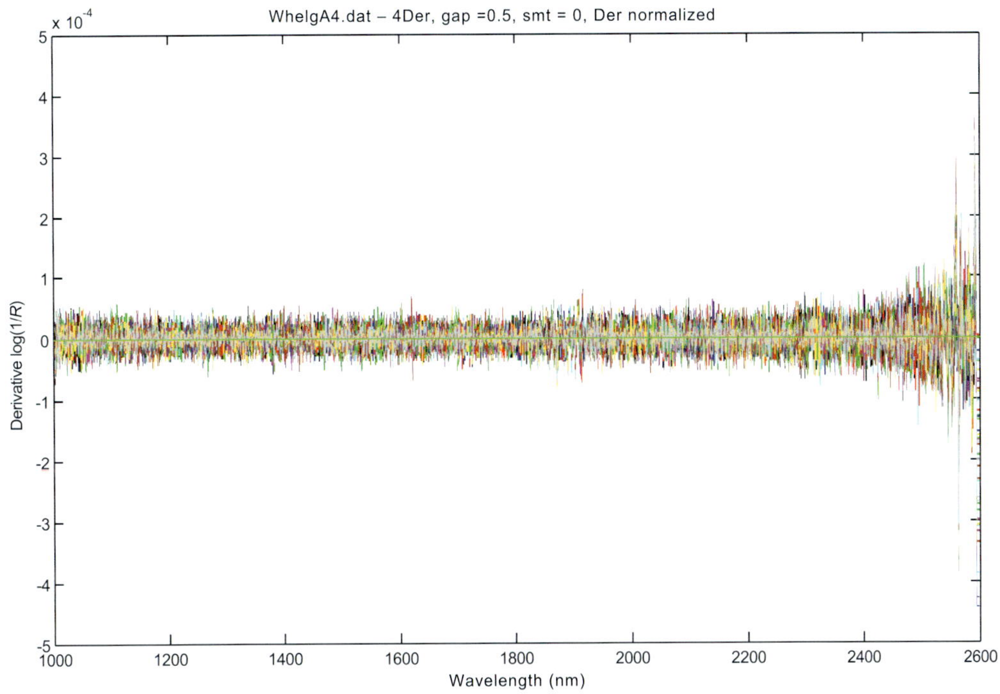

FIGURE 14.12 Fourth derivative of wheat spectra, gap = 0.5 and smt = 0, showing only noise.

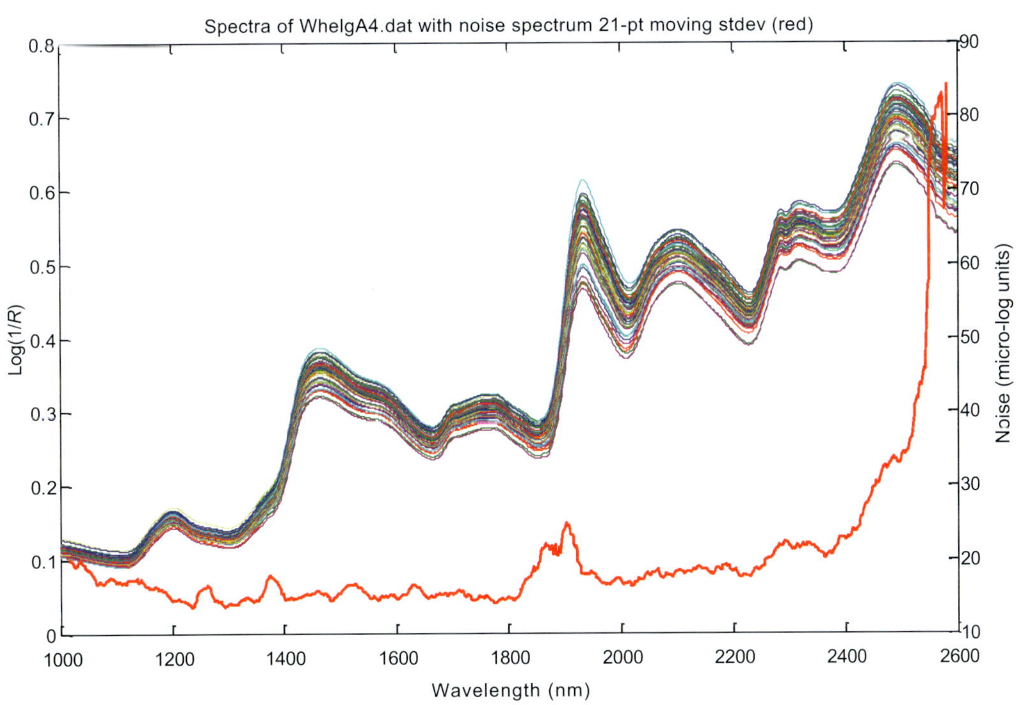

FIGURE 14.13 Absorbance spectra of wheat with noise spectrum.

It is important to notice that the definition of a "Norris" derivative has changed over the years, so care must be used to determine the method, if one uses a "Norris" derivative. The Unscrambler is the only software I know that presently employs a gap or gap–segment method of calculating derivatives (CAMO Software AS, Oslo, Norway, 2017), which was modeled after the early NSAS software from NIR Systems (which became Foss, and presently is Metrohm; the method is no longer available) (NSAS, NIR Spectral Analysis software, NIR Systems, 1993). For the first and third derivatives, gaps may be 1, 3, 5, ... and segments may be 1, 2, 3, ... points; for the second and fourth derivatives, gaps may be 1, 2, 3, ... and segments may be 1, 3, 5, ... points, where segments of 1 indicate no smoothing is used. The DQM programs use gaps defined differently, so that values of multiples of 0.5 are sometimes acceptable, and the smoothing gap is entirely different. For the first and third derivatives, gaps may be 1, 2, 3, ... and smt gaps may be 0, 1, 2, ...points; for the second and fourth derivatives, gaps may be 0.5, 1, 1.5, 2, ... and smt gaps may be 0, 1, 2, ... points, where smt gaps of 0 indicate no smoothing is used. It is hoped that the this new dqm program will allow widespread use of the innovative regression method using gaps and smts developed by Karl Norris.

14.8 SUMMARY

Karl Norris noted:

> The lack of available computer software for performing the derivative ratio regression has limited the application to a few workers in spectroscopy. The technique does require care and understanding of spectroscopy in selecting choices for calibration. However, the results obtained with these data sets should spur others to explore other data sets to understand the scatter effects in diffuse spectroscopy more fully.
>
> *Norris (2001)*

I would add that using this technique leads to an increased understanding of the chemistry underlying the spectroscopy. The calibrations appear to be very specific for the analyte of interest.

Dr. David W. Hopkins passed away during the production of the Handbook. The Editors would like to honor his memory and his contributions to the field of Near-Infrared Spectroscopy and to the Council for Near Infrared Spectroscopy (CNIRS).

Dr. David W. Hopkins, Ph.D. 1943–2021
NIR Consultant, Battle Creek, MI, USA
President, CNIRS (1994–1996)
CNIRS Awards, Chair (1997–2016)

Dr. David W. Hopkins was born in Oak Park, Illinois March 26, 1943; graduated from Carleton College in 1965 with a degree in chemistry and from UCSD in 1971 with a PhD in biology. A series of post docs took him to Harvard, the Biology Institute in Freiburg, Germany and the Smithsonian in Rockville, MD. He joined DICKEY-John in 1978 as Lab Supervisor responsible for lab determinations of moisture, protein, fat and fiber, and the determination of NIR calibrations. By producing calibrations for wheat, barley, corn, soybeans and many other food and feed products, he helped make NIR widely used in the food and agricultural sectors. In 1986 he joined Kellogg Co., where he led the NIR Analytical Group, producing calibrations for many finished food and raw materials. In 1992 he formed his own consulting company to bring his experience and expertise to companies engaged in the development and manufacture of NIR equipment and companies wishing to employ NIR technology effectively. He has worked in the Pharmaceutical, Food, Chemical, and Petrochemical sectors. David has published and presented many short courses in

NIR Spectroscopy, Chemometrics and the use of chemometric software at international meetings and at diverse corporate sites. He made it his specialty of studying and presenting the effective use of spectral pre-treatments.

He was a dedicated scientist and caring mentor to many students. His friendship and warm personality will be missed.

REFERENCES

Dhanoa, M. S., S. J. Lister, R. Sanderson and R. J. Barnes. 1994. The Link between Multiplicative Scatter Correction (MSC) and Standard Normal Variate (SNV) Transformations of NIR spectra. *J. Near Infrared Spectrosc.* 2 43–47.

Hopkins, David W. 2016. CNIRS Corner: Revisiting the Norris Derivative Quotient Math in Regression. *NIR News* 27(7), 23–28. doi: 10.1255/nirn.1643

Hruschka, W. 1998. SPANUSER.DOC Manual for Spectral Analysis Programs (SPAN).

Madden, H. 1978. Comments on the Savitzky-Golay Convolution Method for Least-Squares Fit Smoothing and Differentiation of Digital Data. *Anal. Chem.* 50(9), 1383–1386.

Mark, Howard. 1991. *Principles and Practice of Spectroscopic Calibration.* John Wiley & Sons, Inc, New York, ISBN 0-471-54614-3.

MATLAB, Version R2011a. 2011. MathWorks, Inc, Natick, MA.

Norris, Karl H. 2001. Applying Norris Derivatives. Understanding and Correcting the Factors Which Affect Diffuse Transmittance Spectra. *NIR News* 12(3) 6–9.

Norris, K.H. and A.M.C. Davies. 2011. Examining Diffuse Reflection and Transmission Spectra More Thoroughly: Part 1. Instrument Noise. *Spectr. Eur.* 23(6) 30–33.

NSAS. 1993. NIR Spectral Analysis Software Version 3, NIR Systems, Inc., Silver Spring, MD.

Ritchie, G.E. and H. Mark. 2002a. I Didn't See the Cows. Investigating NIR Transmittance Measurements through the Use of the Norris Regression (NR) Algorithm: Part 1: How Do We Come to "Norris Regression"? *NIR News* 13 (1) 4–6.

Ritchie, G.E. and H. Mark. 2002b. I Didn't See the Cows. Investigating NIR Transmittance Measurements through the Use of the Norris Regression (NR) Algorithm: Part 2: Theoretical Considerations. *NIR News* 13 (2) 3–5.

Savitzky, A., and M.J.E. Golay. 1964. Smoothing and Differentiation of Data by Simplified Least Squares Procedures. *Anal. Chem.* 36(8) 1627–1639.

The Unscrambler X10.3. 2017. CAMO Software AS, Oslo, Norway.

15 Preprocessing Methods in NIR Spectroscopy

Charles E. Miller[*]
Camo Analytics

Benoît Igne
GlaxoSmithKline

CONTENTS

* chuck.miller@camo.com.

15.1 OBJECTIVE

Preprocessing has long been a standard practice in NIR spectroscopy, likely because it was a key component to the earliest applications of NIR technology in the food and agricultural industry in the 1960s (Williams and Norris 1987). As applied NIR technology and enabling technologies have matured over the past several decades, so have the scope and complexity of preprocessing options that are at the method developer's disposal. Much like a customer shopping for a solution to a specific problem, this expansion is a double-edged sword for the NIR method developer:

- The presence of more options improves the chances of finding a more effective solution to one's specific problem, ...but
- It also challenges the method developer to better understand and differentiate the attributes of all of these options.

With the above in mind, the objective of this chapter is the following: to allow NIR method developers to better understand, and thus to better assess, the value of the ever-expanding repertoire of preprocessing solutions.

15.2 BACKGROUND

15.2.1 Definitions and Scope

The first step toward the above objective is to articulate operational definitions and scope of "preprocessing", within the specific context of NIR spectroscopy methods.

First, we will provide definitions of key terms to be used in this discussion:

- **Method:** This refers to the overall analytical method, which includes sampling protocol, settings of instrument and sampling hardware, and processing of data to generate results. **Example:** "Offline NIR reflectance method for determining % moisture content of wheat flour, which uses a manual sample presentation method, and a partial least squares (PLS) calibration model to generate % moisture values from NIR spectral data".
- **Model:** For applicable methods, this refers to the *algorithm* and the *parameters* used to transform a measured spectrum to a method output, which is typically a quality attribute. **Example:** "A PLS regression model used to convert a preprocessed NIR reflectance spectrum to a % moisture value".
- **Application:** This refers to the specific usage of the method. **Example:** "At-line determination of moisture content of wheat flour".
- **Objective:** This refers to the specific *purpose* of the application. **Examples:** quantitation, exploration, monitoring, detection, and classification.
- **Use Scenario:** This refers to a phase of the method life cycle during which the preprocessing is to be applied. **Examples:** data aggregation, model development, and online algorithm application.

With the above definitions in mind, a somewhat intuitive definition of preprocessing might be something like the following:

> **Preprocessing**: a mathematical treatment of raw analytical data that is done in preparation for model generation, or for algorithm application.

This seems like an intuitive and concise definition, but one could still argue that there are some ambiguities that need to be addressed, specifically:

1. What exactly does *raw analytical data* mean?
2. Is there a limit to the level of sophistication of the *mathematical treatment*?
3. Are there any constraints on the *algorithms* that can be used after preprocessing?
4. Which preprocessing usage *scenarios* will we cover?

To add some clarity and scope to this chapter, we will now address each of these possible ambiguities in order below:

1. For the purposes of this work, *raw spectral data* refers to the array data produced by the instrument as the "sample" spectrum, associated with the material being analyzed in real time, and after all relevant instrumental- and sampling-based spectral referencing is done. It is important to note that for NIR instruments used for online applications (process analytical technologies – or PAT), there can be several referencing operations (e.g., "dark scan", "white reference", "wavelength calibration") that can be considered part of the inner workings of the instrument hardware and optics. Because the configurations of such operations are typically not accessible to the method developer or are considered to be more *instrument*-specific rather than *application*-specific, they will be considered **out of scope** for preprocessing.
2. This discussion will avoid any limitations in the mathematical complexity of possible preprocessing methods. Keep in mind, however, that this will expose the reader to a wider array of methodologies, the more complex of which will begin to "look like" multivariate modeling methods themselves. In fact, an astute argument has been made that some preprocessing methods, such as extended multiplicative signal correction, can even preclude the need to subsequent multivariate modeling (Martens 2014). Furthermore, we will also consider preprocessing schemes that involve more than one sequential treatment: for example, baseline offset correction followed by vector normalization.
3. Similarly to (2) above, the scope will also be open to all types of subsequently applied modeling methods, including multivariate and univariate methods, as well as qualitative and quantitative methods. With that said, however, it is worth noting that the optimal selection of a preprocessing scheme can be impacted by the type of modeling method that is to be used afterward.
4. For typical online and offline NIR spectroscopy and PAT applications, there are only two usage scenarios for preprocessing: (1) to support method development and (2) to apply during method execution. Therefore, this work will focus only on these two usage scenarios. However, it will also explain how these scenarios can interact with each another.

15.2.2 Purposes of Preprocessing

Although the specific purposes of preprocessing can vary between applications, there are three purposes that are most frequently cited:

1. Reduction of irrelevant variance
2. Enhancement of relevant variance
3. Linearization of multivariate relationships

To better illustrate these purposes, Figure 15.1 shows a high-level process schematic of a typical process spectroscopy application that uses a multivariate calibration model. In typical NIR applications, the main purpose of the multivariate model (red box) is to impart sufficient specificity to the method, in the midst of irrelevant variance coming from several sources: the process (gray box), instrument (blue boxes), and sampling (green boxes) portions of the method. Note that spectral preprocessing is done immediately after the sample spectrum is generated by the instrument, and

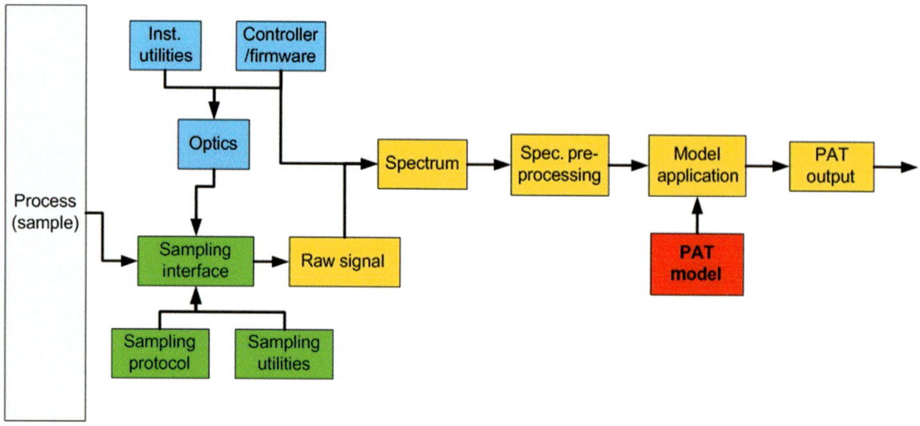

FIGURE 15.1 High-level process schematic of a typical process spectroscopy application that uses a multivariate calibration model.

immediately before the calibration model is applied to generate the PAT output. With this in mind, it becomes apparent that a key purpose of preprocessing is to "minimize the burden" that the model must bear to impart sufficient specificity onto the method, and this is the driver behind purposes (1) and (2) above.

Regarding purpose (3), it is generally accepted that latent variable modeling methods such as principal component analysis (PCA) and PLS are more efficient if the data presented to the model are "linearized" beforehand. In this context, linearization refers to both the relationships between different *X*-variables in the multivariate data array (spectrum) and the relationships between each of these *X*-variables and the PAT output variable(s), commonly called the "Y-variable(s)" (Næs and Isaksson 1993).

Of course, the "linearization mission" of preprocessing in NIR analysis is strongly linked to the fact that practical NIR analysis often requires that measurements be done under conditions that are spectroscopically non-ideal. Fundamental understanding of the non-linear effects in the various sampling modalities of NIR analysis (including transmission, diffuse transmittance, and diffuse reflectance) has been improved steadily over the past several years (Birth and Hecht 1987, Dahm and Dahm 2001, Griffiths and Dahm 2008), and as a result, preprocessing schemes' effectiveness has been accordingly adjusted and improved as well. Detailed discussion of this topic is dedicated to other chapters in this book.

15.2.3 Classes of Preprocessing Methods

After providing some background, scope, and definitions above, it is now time to dive into the various preprocessing methodologies that are at our disposal. For the purposes of clarity, these methods will be separated into the following five classes, with some examples listed (Wise et al. 2006, Miller 2010):

- **Centering:** mean-centering, median-centering, class-specific centering
- **Scaling:** autoscaling, group scaling, standard normal variate (SNV)
- **Transform:** log transform, wavelet transform, Fourier transform
- **Filtering:** Savitzky–Golay filtering (including derivatives), baseline filtering, moving average, Kalman filtering
- **Model-Based:** multiplicative signal correction (MSC), extended multiplicative scatter correction (EMSC), orthogonal signal correction (OSC)

As anyone who has done supervised or unsupervised cluster analysis of real data can attest, the above classes are not strictly exclusive, as some preprocessing methods could plausibly be assigned to more than one class. For example, autoscaling does contain a mean-centering component, and multiplicative signal correction (MSC) does typically include a scaling element. However, in general the methods within each class tend to have similar properties when applied to NIR spectral data – which leads us to the next section.

15.2.4 Critical Preprocessing Attributes

Although it could be considered a shameless repurposing of Quality by Design (QbD) terminology that has been well used in the life sciences industries, the concept of a "critical *preprocessing* attribute" (CPA) would seem to be quite relevant in the context of multivariate method development, in that it provides a means by which different preprocessing methods can be classified. When considering such attributes, one must consider that there are at least two main actions involved with every preprocessing method: (1) determination of the correction factors and (2) application of the correction factors to the input data.

The following can be considered key attributes of preprocessing methods, during both method development and method execution:

- **Execution Mode:** Is the preprocessing correction executed *row-wise* (one spectrum at a time), *column-wise* (one variable at a time), or *cell-wise* (one data value at a time)?
- **Dependence on Variable Subset:** Does the preprocessing outcome depend on the subset of variables (spectral wavelengths)?
- **Dependence on Object (Sample) Subset:** Does the preprocessing outcome depend on the subset of objects (samples)?
- **Mathematical Complexity:** How simple is the preprocessing? Does its algorithm involve more than two basic steps? How easy would it be to execute this preprocessing in real time?
- **Sensitivity to Outliers:** How much are the preprocessing results impacted by the presence of variable outliers, or sample outliers?
- **Relevance:** Does the preprocessing address known interfering source(s) of variation, or non-linear effect(s), in the input data?

The first three attributes (Execution Mode, Dependence on Variable Subset, and Dependence on Object Subset) are especially important when considering the use of multiple "chained" preprocessing steps, or when using variable selection or sample selection algorithms during method development. When chaining multiple preprocessing steps, it is generally more effective to execute row-wise methods before column-wise methods. This is the case because the row-wise methods generally exploit prior knowledge of the spectroscopy to minimize irrelevant *spectral* artifacts, and the column-wise methods are often considered a "finishing step" to allow the subsequent modeling math to focus on spectral variance instead of spectral magnitude.

Mathematical complexity is a key consideration both during model development and when determining real-time model execution strategies. Many preprocessing methods require no specification of parameters, but several require the user to select one or more parameters and judicious selection of these values requires some domain knowledge to avoid unwanted results. A good example of the latter case is Savitzky–Golay derivatives, which require the specification of a window width, the selection of which requires some understanding of the nominal resolution of the spectra obtained by the instrument, as well as its signal-to-noise properties (Savitzky and Golay 1964).

The sensitivity of preprocessing methods to outliers is most relevant during the early stages of data analysis and model development, when outliers are most likely to be present in the dataset. The impact of outliers depends on the number of objects or variables in the dataset, as well as the complexity of the preprocessing method. The final CPA, *Relevance*, refers to the specific underlying spectral phenomena that the preprocessing method is meant to address. The preprocessing used for a given application should address interfering spectral phenomena that are expected to be present in the data: both during method development and during method usage. Conversely, the use of a preprocessing method that reduces spectral variance that is relevant for the objective of the method can be quite counterproductive.

Table 15.1 summarizes the attributes of several commonly used preprocessing methods in NIR spectroscopy.

15.3 NIR DATASETS AND SOFTWARE

Two publicly available datasets will be used to demonstrate the effects of the most commonly used preprocessing methods on spectral data. As the main purpose of this chapter is to provide clarity on these preprocessing methods, these demonstrations will be limited to displays and comparisons of the raw and preprocessed spectral profiles.

15.3.1 Corn Dataset

This dataset, generated by Cargill (Eigenvector Research, 2005a), consists of the NIR diffuse reflectance spectra of 80 corn kernel samples, measured over the NIR spectral range of 1100–2498 nm, at a fixed interval of 2 nm per spectral variable. As observed in Figure 15.2a, each of the raw corn spectra contains rather broad absorption bands, with the different spectra having varying baseline offset and slope properties. Such properties are quite common in NIR diffuse reflectance analysis, where differences in the presentation, nominal particle size, particle size distribution, and bulk density of the solid material result in varying degrees of NIR light absorption and scattering as a function of wavelength. Refer to the chapters discussing the physics of NIR spectroscopy in this book.

15.3.2 Diesel Fuel Dataset

This dataset, generated by the Southwest Research Institute (Eigenvector Research, 2005b), consists of 784 spectra of diesel fuel samples, measured over the NIR spectral range of 750–1550 nm, at a fixed interval of 2 nm per spectral variable. As observed in Figure 15.2b, the raw diesel fuel spectra have sharper absorption bands than the corn spectra in Figure 15.2a. Although these spectra also exhibit some variation in baseline offset, the baselines appear to be better defined, in that there appear to be some baseline-resolved spectral regions (i.e., where no species have absorption bands). As these spectra were collected using a fixed-pathlength liquid sample cell, the variance in baseline offset is considerably less than that of the corn spectra, albeit still present.

15.3.3 Software

All plots and preprocessing were performed using either the PLS_Toolbox for MATLAB v. 8.6.2 (Eigenvector Research, Wenatchee, WA, USA), with MATLAB R2018b (The MathWorks, Natick, MA, USA), or The Unscrambler X version v. 10.5.1 (Camo Analytics AS, Oslo, Norway).

15.4 CONVENTIONS

When explaining the various pretreatment methods, a dataset of spectral data to be used for method development will be denoted as an (n by m) matrix $\boldsymbol{X}$, where n refers to the number of samples in the dataset and m refers to the number of spectral variables in the digitized spectrum. Using this convention, *columns* in $\boldsymbol{X}$ will refer to individual *spectral variables*, and *rows* in $\boldsymbol{X}$ will refer to individual *samples*.

TABLE 15.1
Attributes of Several Common Preprocessing Methods

Preprocessing Method	Preprocessing Class	Execution Mode	Dependence on Variable Subset	Dependence on Sample Subset	Mathematical Complexity	Sensitivity to Outliers (Severity)	Preprocessing Parameters	Relevance for *Spectral X* Data
Mean-centering	Centering	Column-wise	No	Yes	Low	Objects (medium)	None	Enables focus on spectral variance
Median-centering	Centering	Column-wise	No	Yes	Low	Objects (low)	None	Enables focus on spectral variance
Autoscaling (column normalization)	Scaling	Column-wise	No	Yes	Low	Objects (medium)	None	Enables focus on spectral variance; all variables have equal variance before modeling
A-priori variable scaling	Scaling	Column-wise	No	Yes	Medium	Objects (medium)	Scaling factors	Enables focus on spectral variance, exploits prior knowledge of spectral noise properties
SNV (row normalization)	Scaling	Row-wise	Yes	No	Low	Variables (medium)	None	Reduces multiplicative spectral effects (e.g., varying effective pathlengths); improves linearity
Log transform	Transform	Element-wise	No	No	Low	N/A	None	Improves linearity
Fourier transform	Transform	Row-wise	Yes	No	Medium	Variables (medium)	Frequency range	Reduces low-frequency (baseline) spectral components
Savitzky–Golay smoothing	Filtering	Row-wise	No[a]	No	Medium	Variables (medium)	Window width, polynomial order	Exploits collinearity of spectral variables; reduces spectral noise
Savitzky–Golay derivatives	Filtering	Row-wise	No[a]	No	Medium	Variables (medium)	Window width, polynomial order, derivative order	Exploits collinearity; reduces spectral noise; unambiguously reduces baseline variations
Fixed baseline removal	Filtering	Row-wise	Yes	No	Low/medium	N/A, or variable (medium)	Baseline wavelength(s)	Reduces baseline spectral variations; preserves spectral band shapes

(*Continued*)

TABLE 15.1 (*Continued*)
Attributes of Several Common Preprocessing Methods

Preprocessing Method	Preprocessing Class	Execution Mode	Dependence on Variable Subset	Dependence on Sample Subset	Mathematical Complexity	Sensitivity to Outliers (Severity)	Preprocessing Parameters	Relevance for *Spectral X* Data
Detrend	Filtering	Row-wise	Yes	No	Low/medium	Variables (medium)	Polynomial order, wavelength range	Reduces baseline spectral variations; preserves spectral band shapes
Weighted least squares (WLS) baseline removal	Filtering	Row-wise	Yes	No	High	Variable (low)	Polynomial order, wavelength range	Reduces baseline spectral variations; preserves spectral band shapes
MSC	Model-based	Row-wise	Yes	Yes/no	Medium	Variables (medium), and possibly objects	MSC reference spectrum, wavelength range, model equation	Reduces baseline and multiplicative spectral variations; preserves spectral band shapes
EMSC	Model-based	Row-wise	Yes	Yes/no	Medium/high	Objects and variables (medium)	Reference spectrum, basis vectors, wavelength range	Reduces baseline and multiplicative spectral variations and filters known interferences; preserves spectral band shapes
OSC	Model-based	Row-wise	Yes	Yes	Medium/high	Objects and Variables (medium)[b]	*Y* data, wavelength range, number of OSC factors	Reduces variance in *X* that is orthogonal to *Y*

[a] Preprocessed values at the edges of the spectral range(s) depend on the edge-filling algorithm.
[b] Also impacted by outliers in the *Y* (property) data.

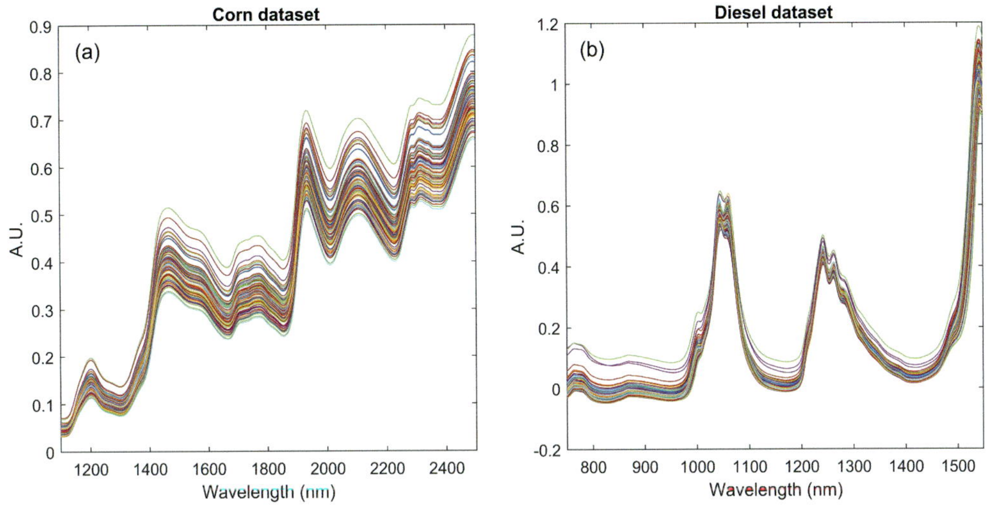

FIGURE 15.2 Raw corn dataset (a) and raw diesel dataset (b).

Regarding spectral variables, it will be assumed that these will be defined either as wavelength units (in nanometers) or as wave number units (in reciprocal centimeters). It will also be assumed that for a given application, the intervals between all adjacent spectral variables are equal in the unit being used. This is almost always the case in today's commercial NIR spectrometer systems. In rare cases where the intervals are not equal, it is still possible to use many of the pretreatments discussed. However, it should be noted that some smoothing and derivative methods assume equal intervals, and thus, non-intended results could be obtained in these cases.

15.5 CENTERING METHODS

In general, centering methods are used as a final preprocessing step, to allow the subsequent modeling method(s) to focus on spectral **variability** about a center, instead of spectral **magnitude**. Furthermore, the center is most typically a very meaningful spectral profile and is most commonly the mean spectrum in a specified dataset, in which case the operation is called **mean-centering**:

$$\boldsymbol{X}_{\mathrm{mc}} = \boldsymbol{X} - \bar{\boldsymbol{x}} \tag{15.1}$$

where $\bar{\boldsymbol{x}}$ is the array of mean spectral intensities at each wavelength in the spectrum, across all samples in the population. A robust variant of mean-centering is **median-centering**, where the median values at each spectral wavelength are used instead of the mean, which can be used in cases when one wishes to avoid the influence of sample outliers on the centering operation. Figure 15.3 shows the effect of mean-centering on the corn reflectance and diesel transmittance data.

From a spectroscopist's perspective, the most obvious effect of mean-centering is that the spectra now have positive- and negative-pointing bands. What might be less obvious, however, is that these positive- and negative-facing bands are predominantly at the same wavelengths as the original bands in the raw spectral data, especially for the corn diffuse reflectance data. Also, note that some of the centered spectra in both cases continue to have significant baseline offset or baseline slope artifacts. These observations underscore an important limitation of mean-centering: Although it allows the subsequent modeling method to focus more on spectral variance, it does not address baseline offset and multiplicative variations between spectral profiles: variations that are quite common in practical NIR spectroscopy.

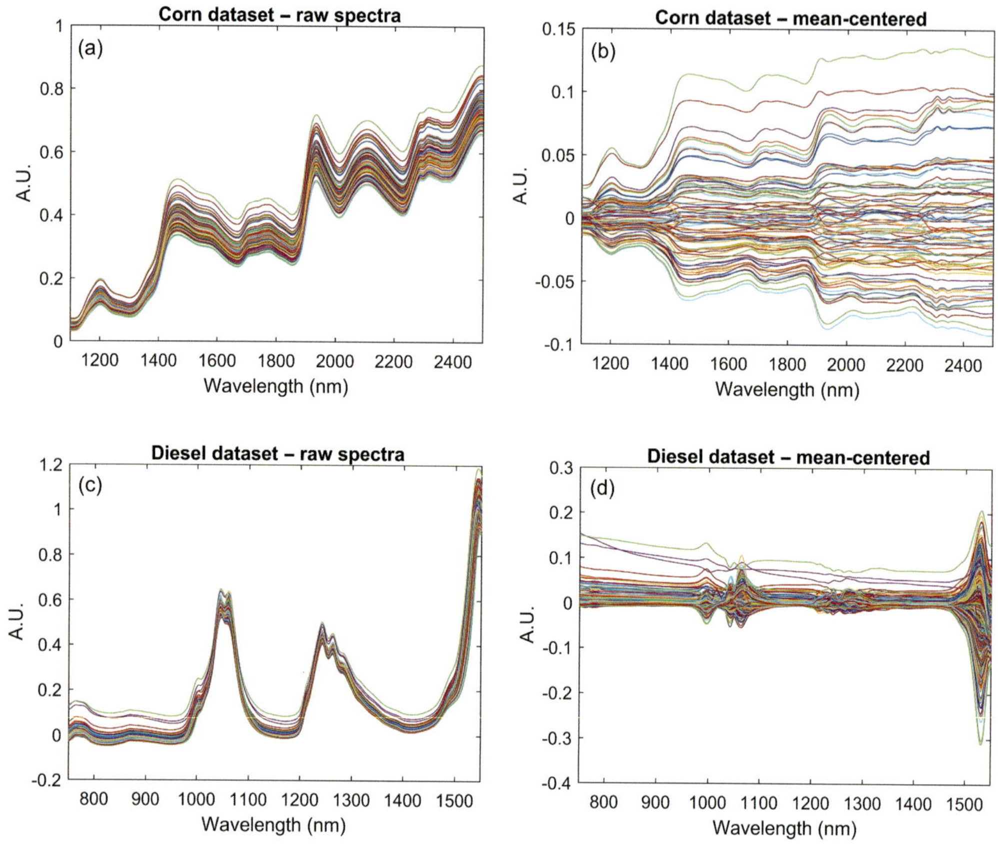

FIGURE 15.3 Effect of mean-centering on the corn (a and b) and diesel (c and d) datasets.

In almost all NIR spectroscopy applications, the method development objective is more efficiently attained by focusing on spectral variability, and thus, centering is almost always applied to the data before modeling. However, it is worth noting that there are plausible use cases of NIR spectroscopy where absolute spectral magnitude does have meaning, where centering might not be useful. One example of such a case could be traditional transmission spectroscopy, where there are fully resolved spectral bands corresponding to pure constituents in the sample, and where baseline artifacts are at a minimum. In such cases, one can be more confident in the relevance of raw spectral intensities, and thus, centering could inadvertently reduce or obscure relevant information.

Finally, it is important to note that there are row-wise (sample-wise) centering equivalents to mean-centering, which can be used to reduce row-wise variations in a spectral dataset. These include the SNV and baseline filtering methods, which are discussed in the sections below.

15.6 SCALING METHODS

Scaling methods involve a multiplicative transformation of the input data. Probably the most common form of scaling is called **autoscaling**, or unit variance scaling. For this method, the column-wise means and standard deviations are first determined. Then, each of the columns of the data in $\boldsymbol{X}$ is mean-centered and divided by the standard deviation:

$$\boldsymbol{X}_{\text{as}} = \boldsymbol{X}_{\text{mc}} \times \boldsymbol{S} \tag{15.2}$$

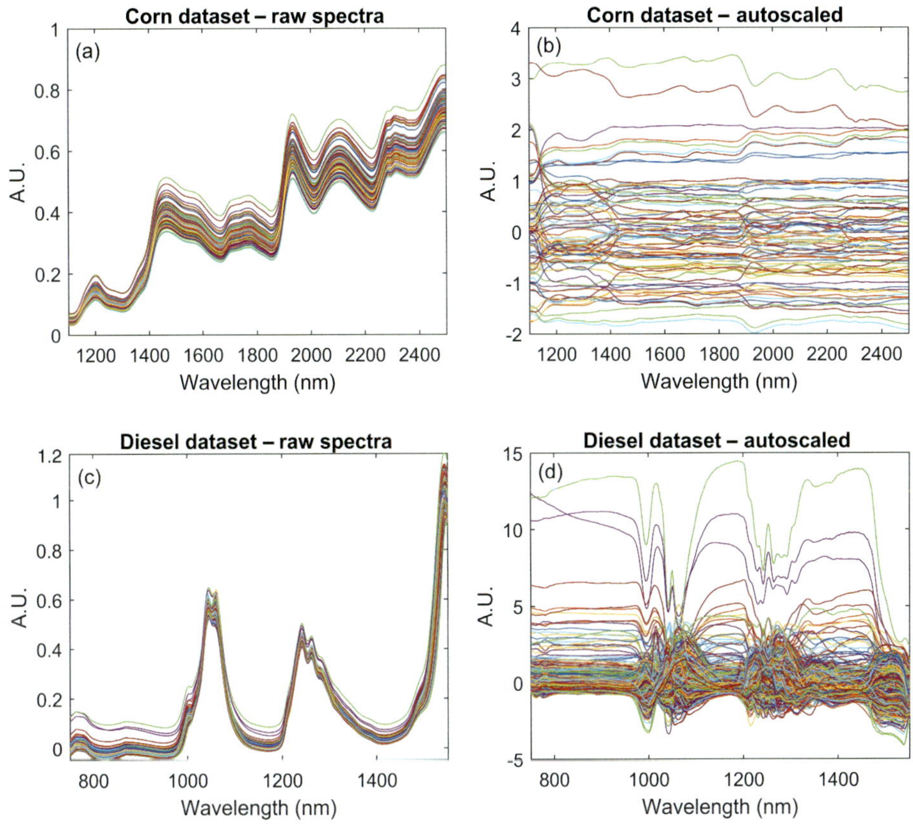

FIGURE 15.4 Effect of autoscaling on the corn (a and b) and diesel (c and d) datasets.

where $\boldsymbol{S}$ is a diagonal matrix in which the diagonal elements contain the inverses of the standard deviations of each spectral variable. Autoscaling results in data where the variance of each variable in the dataset is equal to 1. As many modeling methods are focused on finding and characterizing variance, this preprocessing method essentially puts all the spectral variables on equal footing before modeling math is applied. Figure 15.4 shows the effect of autoscaling on the corn and diesel fuel spectra.

Of course, there are many alterative column-wise (spectral variable-wise) scaling strategies that one could consider: for example, custom variable scaling based on *a priori* knowledge of the signal to noise of each spectral variable, or scaling based on known spectral regions where interferences or instrumental artifacts are suspected to be present. Such scaling simply involves the usage of a priori defined scaling factors, instead of inverse standard deviations, in the diagonal elements of $\boldsymbol{S}$ in the equation above.

Note that the autoscaled data in both cases contain localized band features at approximately the same wavelengths as the bands in the raw spectral data, but also show significant baseline offset and slope effects. Like mean-centering, autoscaling is also a column-wise preprocessing method, and thus cannot address row-wise (sample-wise) variations in spectral baselines.

Fortunately, there are plenty of row-wise (sample-wise) preprocessing methods that can be used to address such variations. One example, the **standard normal variate** (**SNV**) (Barnes, Dhanoa, and Lister 1989), is simply the row-wise equivalent of autoscaling: where the mean intensity over all variables is subtracted from each row's data, followed by division by the row-wise standard deviation:

$$x_{i,\,\mathrm{SNV}} = \frac{x_i - \bar{x}_i}{s_i} \tag{15.3}$$

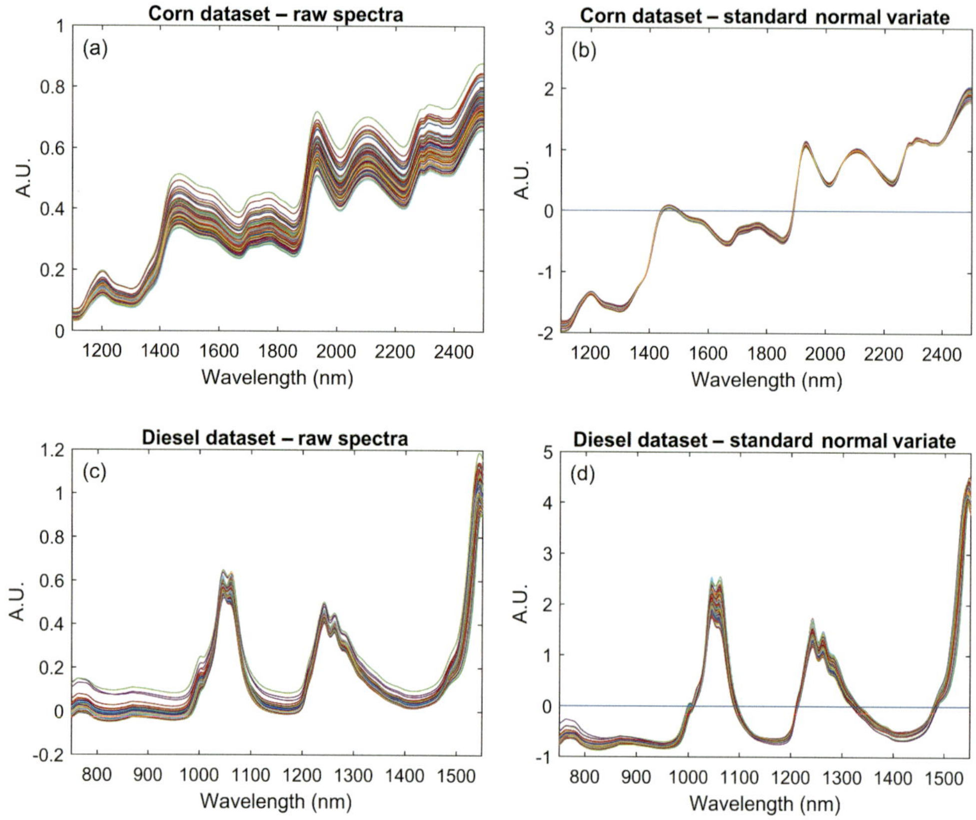

FIGURE 15.5 Effect of standard normal variate on the corn (a and b) and diesel (c and d) datasets.

where $x_{i,\,\mathrm{SNV}}$ is the SNV-corrected spectral intensity value for the ith sample, x_i are the spectral intensities for sample I, and $\bar{x}_i$ and s_i are the mean and standard deviation of the spectral intensities for sample i.

Similar to autoscaling, SNV involves two distinct operations: (1) offset of each spectrum by the mean and (2) scaling of each spectrum by its standard deviation. The former operation results in a spectral profile where the Y-axis zero passes through the spectrum, and the latter operation "expands" or contracts the amplitude of the spectral profile about the Y-axis. Figure 15.5 shows the result of SNV preprocessing on both of the datasets.

With the resulting spectral profiles crossing the Y-axis zero, SNV-transformed data might seem rather non-intuitive for spectroscopists. However, this preprocessing method does address two known effects in NIR spectroscopy: the baseline offset effect (through subtraction of the row-wise mean) and the multiplicative pathlength effect (through division by the row-wise standard deviation). Furthermore, most modeling algorithms don't really "care" where the Y-axis zero is located – only whether relevant spectral variance is observed. This, combined with the simplicity of the method, has led to SNV being a commonly used preprocessing method in practical NIR applications.

15.7 TRANSFORM METHODS

Several methods in this class aim to linearize the spectral responses to component concentrations, or another property of interest, based on knowledge of the sampling geometry or scattering physics of the application. For NIR reflectance and diffuse reflectance spectroscopy, the Kubelka–Munk theory (Kubelka and Munk 1931, Birth and Hecht 1987) was developed many years ago to provide a mechanistic basis for the reflectance signals obtained from partially absorbing and scattering materials. In principle, this theory can also be used for practical preprocessing of reflectance data, although such use requires *a priori* determination of absorption (K) and scattering (S) coefficients for the material being analyzed.

Other transform methods include the log(1/R) transform, which was empirically found to result in more linear spectral responses for diffuse reflectance spectroscopy, and the Beer–Lambert equation for transmission spectroscopy, or $-\log(I/I_0)$, where I_0 refers to the intensity obtained from a non-absorbing material, or reference standard.

Other row-wise transform methods are based on the premise that one wishes to filter out either high-frequency (noise) components or low-frequency (baseline offset, sloped baseline) components in a spectral data array. The Fourier transform, which changes a spectral response from the wavelength or wave number domain to a frequency domain, has been used to identify and subsequently filter out such components (McClure et al. 1984, Igne and Hurburgh 2010) and wavelet transform, which converts a spectral response to another domain that retains some position (wavelength) information and some frequency information (Tan and Brown 2001, Fearn and Davies 2003).

15.8 FILTERING METHODS

This family of methods exploits collinearity in adjacent responses to reduce high- or low-frequency interferences in the data. In NIR spectroscopy, the most commonly used filtering methods are row-wise, such that they take advantage of the collinearity of the spectral responses at adjacent wavelengths. Four types of filtering methods will be discussed in this chapter, all of which involve row-wise filtering: baseline removal, detrend, smoothing, and derivatives.

It is worth noting that when time-series data are used, there is an entire class of *column-wise* filtering methods that can be employed to take advantage of collinearity of responses at adjacent times, as well as expected time-dependent response behavior. However, such methods, including Kalman filtering and exponential filtering, are not considered to be *spectral* preprocessing methods, and thus will not be covered here.

15.8.1 Fixed Baseline Offset or Slope

In its simplest form, fixed baseline removal corresponds to subtracting an offset value from each spectrum. That value can be an intensity at a particular wavelength, or a mean intensity over a selected variable range or set of ranges.

In the case of the corn data, removing a constant offset would not be of much value, as there are no obvious spectral regions that can be definitively identified as baseline-resolved. However, for the diesel data, there are several regions that appear to be baseline-resolved, and thus could be useful for reducing baseline offset effects from the spectra. Alternatively, for the corn data, a baseline correction method based on a sloped baseline might be more effective. For such a correction, the offset and slope of each baseline could be calculated by determining the sloped line of best fit as a function of wavelength for each spectrum, using least squares regression.

Figure 15.6 shows the result of two different baseline removal methods for the two datasets. For the corn data, a sloped baseline correction was applied, based on the line defined by the slope and intercept of the regression between the spectrum and the wavelength values. For the diesel data, a constant baseline offset correction was applied based on the mean of absorbance values at three wavelengths (950, 1160, and 1414 nm). Note that in both cases, the varying baseline offsets and slopes have been reduced, thus making the absorption band features more visible.

It should be noted that baseline removal options often require the user to specify several parameters, including specific wavelengths or regions of wavelengths that are presumed to express only baseline offset or baseline slope, and no spectral absorption features. As a result, there are many possible options for such methods, and in many cases, the overall objective of the data analysis will dictate how the baseline should be removed.

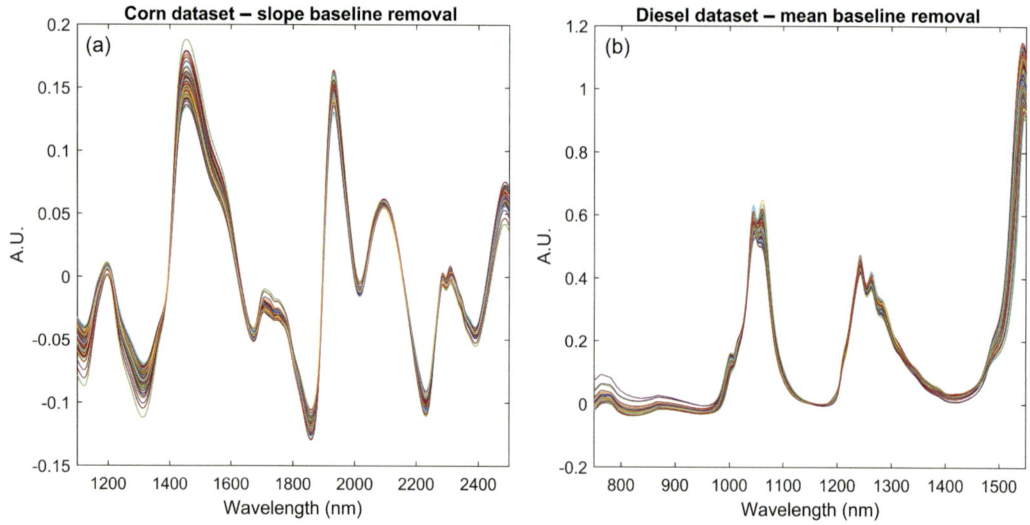

FIGURE 15.6 Corn dataset after a line removal defined by the slope and intercept of the raw spectra as a function of wavelength (a); diesel dataset after baseline correction by the mean of the absorbance values at 950, 1160, and 1414 nm (b).

15.8.2 Detrend

Detrend is a generalization of the sloped baseline removal method performed in Figure 15.6a above. Introduced by Barnes, Dhanoa, and Lister (1989), its objective is to remove the "increasing level of the log 1/*R* reflectance values over the range 1100 to 2500 nm". In contrast to the sloped baseline correction discussed above, the original version of the detrend method used a least squares fit to a second-degree polynomial, to better remove the curvilinear effect of particle size and density. In practice, the degree of the polynomial used for the baseline fit can be changed to better suit the data. Figure 15.7 shows the corn data treated with detrend preprocessing using a second-order polynomial. Both the slope baseline-corrected spectra shown earlier (Figure 15.6a) and the second-order detrended spectra show greatly reduced baseline variance as compared to the uncorrected corn diffuse reflectance spectra (Figure 15.2a), thus allowing more visibility of variance in the wavelength-localized absorbance bands in the spectra.

At this point, it is worth noting that all baseline reduction preprocessing methods (and most spectral preprocessing methods in general) make inherent assumptions about the underlying shape and variation of the spectral baselines in real data. The extent to which these assumptions are accurate can have two key impacts on the preprocessing method for any given application: (1) the efficiency of removing the *true* baseline variability, and (2) the potential to *add irrelevant variability*. Furthermore, it is a general rule that impact (2) can become more problematic as the complexity of the preprocessing method increases. Relating back to the baseline filtering methods discussed above, it should be clear that a simple baseline offset correction based on a single-wavelength intensity that is presumed to be baseline-resolved might be less efficient at removing true underlying baseline variability, but less likely to introduce irrelevant variability to the data. Conversely, a more complex second- or third-order detrend correction might be slightly more efficient at removing irrelevant baseline variance, but poses more of a risk of adding irrelevant variance or artifacts to the data. As always, the optimal selection and usage of preprocessing methods depends greatly on the specific context and objectives of the analysis.

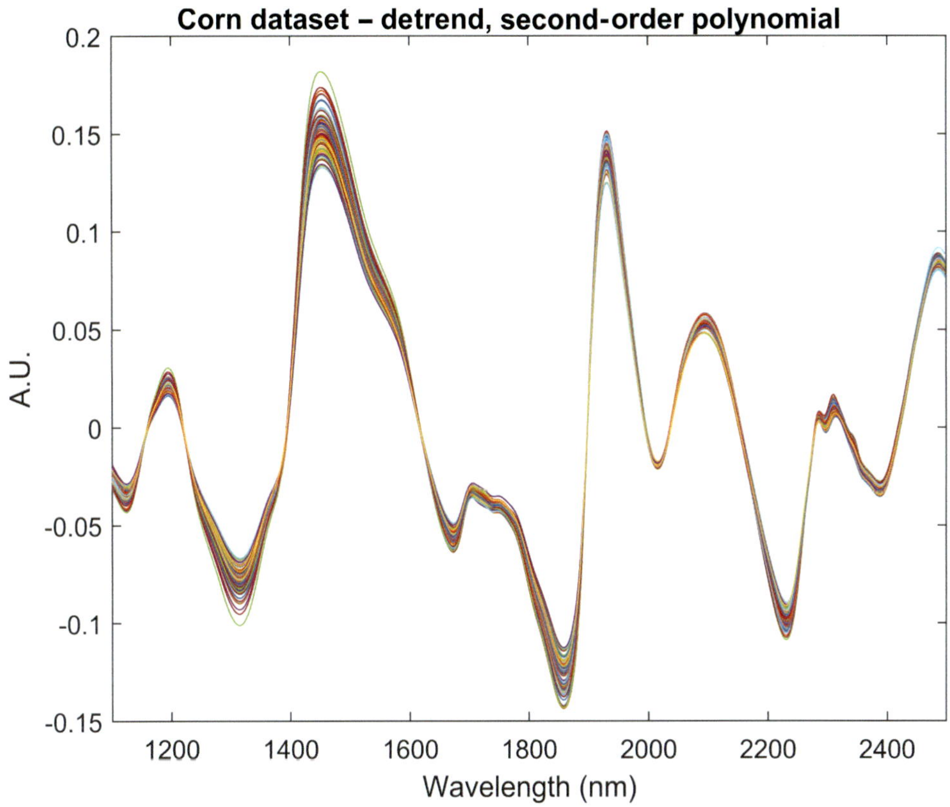

FIGURE 15.7 Result of applying a detrend preprocessing with a second-order polynomial to the corn spectra.

15.8.3 Weighted Least Squares (WLS)

As mentioned above, there are cases where it is safe to presume that there are some baseline-resolved wavelengths, but less safe to presume specifically *which wavelengths* are baseline-resolved. In such cases, there is a risk that the specification of incorrect wavelength could create artifacts at different regions of the spectrum, thus adding irrelevant variability to the data. If so, then the process of identifying appropriate baseline-resolved wavelengths can be automated by using a weighted least squares (WLS) approach. For the WLS approach, a low-order polynomial is used to fit the data by iteratively searching which points are most likely to belong to the baseline, based on their variance properties. It should be cautioned that if the order of the polynomial used for WLS is too high, then unwanted artifacts can be created in the processed spectral data. Figure 15.8 shows the result of applying a WLS baseline correction approach using a second-order polynomial to the diesel fuel NIR data.

Compared to the baseline offset-corrected data shown in Figure 15.6b, the shapes of the absorption peaks remained unaffected, but the baseline removal around 1400 nm was significantly improved. Like other baseline correction methods, WLS requires an *a priori* assumption regarding the order of the baseline to be removed, and thus, the usage of higher-order polynomial baselines can inadvertently remove useful variance that is detected as baseline variance. As for any preprocessing method, the practitioner must apply their prior knowledge of the chemistry and physics of the application in order to use such method optimally. With this in mind, it should be noted that some versions of the WLS algorithm allow one to specify reference spectra that could be used to guide the baseline removal algorithm toward an optimal result.

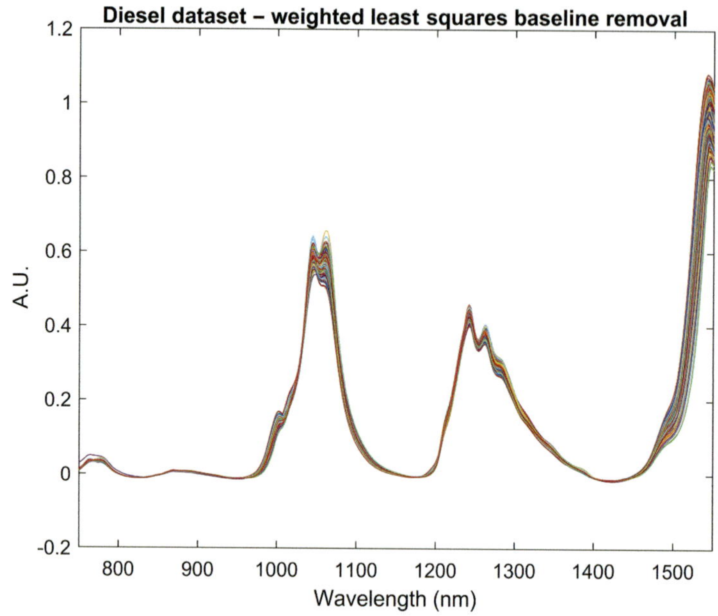

FIGURE 15.8 Application of a weighted least squares (WLS) second-order polynomial baseline correction to the NIR diesel data.

15.9 SMOOTHING AND DERIVATIVES

Smoothing and derivative preprocessing methods are often discussed together because they are applied in a very similar manner, and because they both exploit the expected covariance of adjacent variables in spectral data. Specifically, these methods work by applying a fixed spectral variable-width filter sequentially across the full spectral range. The work of Savitzky and Golay (1964) demonstrated that for digitized spectral data with wavelength intervals that are equally spaced, one can generate either a smoothed version of the spectrum, or an estimate of the analytical derivative of the spectrum, by simply applying a fixed-width filter to a spectrum in this manner. The specific values in the fixed Savitzky–Golay filter are determined using formulas provided in the original paper, and these based on three factors:

1. the order of the derivative, *D*, which equals 0 if only smoothing is to be done,
2. the width of the filter, *W*, defined in terms of the number of wavelength variable points, and
3. the polynomial order, *P*, referring to the assumed shape of all the spectral profiles within each of the filter windows.

In this context, Savitzky–Golay smoothing is simply a special case of Savitzky–Golay filtering where the derivative order is equal to zero ($D=0$).

Although Savitzky–Golay filtering has gained much traction in NIR spectroscopy, and analytical spectroscopy in general, there are other options for smoothing and derivatization of spectral data that will be briefly mentioned below.

15.9.1 Smoothing

The simplest approach to smoothing is to use a moving average, in which spectral intensity values within each wavelength window, each containing *W* spectral variable points, are averaged, and the result is used to represent the smoothed intensity for the wavelength at the center of the window.

The window is then moved by one wavelength point, and a second average is calculated to represent the smoothed intensity at the center of that new window. This process is repeated until the whole spectral range has been covered. The advantage of the moving average approach to smoothing is that it is simple (each of the W values in the filter is identical) and thus easy to understand. However, one concern with this approach is the possibility that subtle localized spectral features could be lost, even in cases where the window width is deemed appropriate for the spectral features of interest.

To help resolve this concern, and to preserve subtle spectral features while smoothing, Savitzky and Golay (1964) published coefficients of modified spectral filters that corresponded to the fitting of a polynomial function to each of the wavelength-localized spectral windows. The Savitzky–Golay filtering algorithm then proceeds as in the moving average case: The window is shifted one wavelength point and applied again, repeating until the entire spectrum has been covered. As in the moving average case, there are W numeric values in the filter, but they are *not identical*: Their values are determined by the assumed order of the polynomial (P) to be fitted in each window. In their original article, Savitzky and Golay published tables of coefficients corresponding to different values of W and P, to facilitate incorporation of these filters into software packages.

15.9.2 Derivatives

Derivatives are arguably the most used spectral pretreatment in NIR spectroscopy. This is likely the case because not only do they provide a non-ambiguous means of removing baseline offset and slope effects in spectra, but they can also be used to enhance localized spectral features. As discussed in the chapter on Norris regression in this book, derivative filters have been used since the earliest days of quantitative NIR reflectance spectroscopy, due to this dual functionality.

The simplest way to perform a first derivative of a spectrum would be to calculate the differences in spectral intensities of adjacent spectral variables, corresponding to a window width of $W=2$. The second derivative could then be performed by repeating this operation on the result, and so on for higher-order derivatives. Unfortunately, derivatives calculated in this way do not take advantage of the expected covariance of adjacent spectral variables and tend to enhance high-frequency noise in the spectra. As a result, such derivative filtering of spectra is of limited practical use.

Similar to the case of smoothing discussed above, Savitzky and Golay (1964) also developed derivative filters that both exploit the covariance of adjacent spectral variables and add smoothing functionality to avoid amplification of noise. The use of Savitzky Golay derivative filters requires also the specification of three parameters, D, W, and P. Regarding the derivative order, D, the optimal value for this parameter depends on the type(s) of irrelevant spectral variance that one needs to remove. For example, a first derivative filter ($D=1$) will remove baseline offset differences between the spectra, a second derivative filter ($D=2$) will remove baseline offset *and linear baseline slope variations* between the spectra, and higher-order derivatives will remove variations in more complex baseline components in the spectra. Also, as the derivative order increases, more localized band structures are generated in the spectra: A spectrum containing a single Gaussian band will have one positive and one negative band in its first derivative, two positive bands and one negative band in its second derivative, two positive bands and two negative bands in its third derivative, and so on. As in many other applications, it is best to use the lowest order of derivative that is needed to remove irrelevant variance and to enhance local spectral features.

Figure 15.9 shows the result of applying first, second, and third derivative filters on the corn data, using a window width of 7 points ($W=7$, corresponding to a range of 12 nm) and a polynomial order (P) of 2. Note that the first derivative (plot B) results in much less variance between the spectra, and the second derivative (plot C) reduces this variance even further, likely due to elimination of varying baseline slopes in the original data. However, advancing to the third derivative (plot D) clearly amplifies spectral noise, even though there are more localized spectral features produced versus the lower-order derivatives. In this specific case, a derivative order of either 1 or 2 would appear to be optimal, based on the objective of the application.

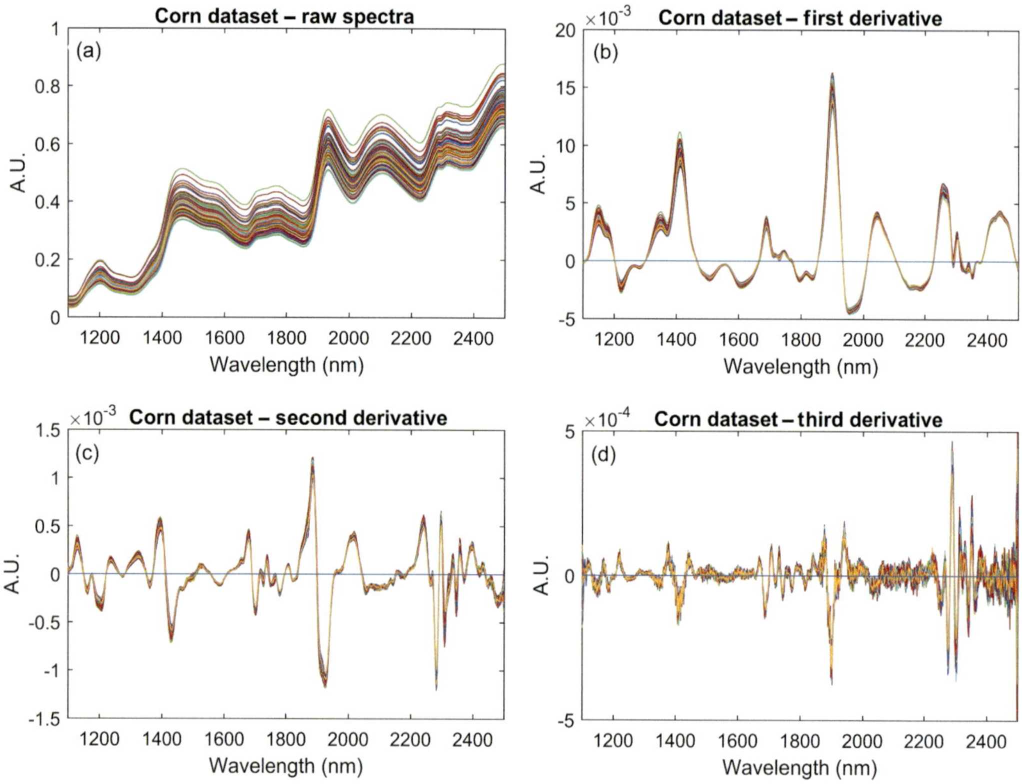

FIGURE 15.9 Effect of derivative order (parameter *D*) on Savitzky–Golay derivative filtering of the corn data: (a) raw data, (b) first derivative, (c): second derivative, and (d) third derivative. In all cases, polynomial order $(P)=2$ and window width $(W)=7$.

Regarding the window width (W) parameter, a general rule of thumb is to select a value that is no greater than the width of the sharpest absorption band in the spectrum, in order to avoid "smoothing over" finer spectral band features. Figure 15.10 shows the effect of different window widths on a localized spectral region of the diesel fuel spectra. From visual inspection of the raw spectra in this region (Figure 15.10a), a rough estimate of the width of the three bands at the top of the main absorption feature would be about 20–25 nm. Figure 15.10b–d shows the results of Savitzky–Golay first derivative filters using a second-order polynomial ($P=2$) and W values of 5, 11, and 21, respectively (corresponding to window widths of 8, 20, and 40 nm, respectively). In all three cases of derivative filtering, the spectral variance from baseline offset differences has been removed, allowing more visibility of localized spectral band variations. However, it is clear that the latter case of $W=21$ (window width of 40 nm) smooths out the finer, more localized spectral features, and this could have a negative impact on certain applications that require high specificity. In this specific case, it appears that the use of narrower windows of 8 and 20 nm (Figure 15.10b and c) preserves the localized spectral features while still maintaining some of the smoothing benefits. In general, the usage of narrower windows allows retention of localized spectral features while limiting any smoothing benefits, whereas the usage of wider windows improves the signal-to-noise properties of the data, at the possible expense of lost spectral specificity.

Regarding the polynomial order (P) parameter, a good general rule is to start with $P=2$, corresponding to the fitting of local second-order polynomials to each of the spectral segments. For most practical uses of NIR spectroscopy, this setting is sufficient, as its underlying assumption is that the spectral profiles *within each window* can be approximated by a second-order polynomial. The use of P values greater than 2 might enable improved retention of finer spectral features in the

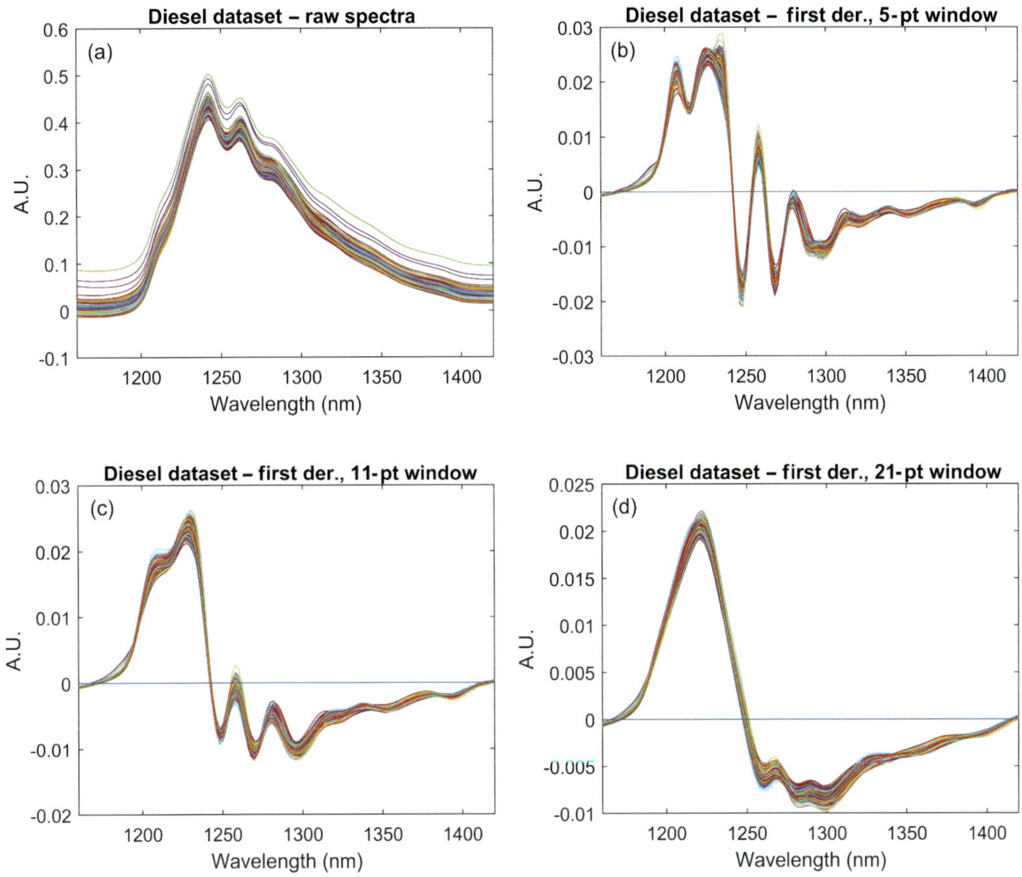

FIGURE 15.10 Effect of window width (W) on Savitzky–Golay first derivative filtering in a selected spectral subregion of the diesel fuel data (1160–1420 nm): (a) raw data, (b) $W=5$ (8-nm window), (c) $W=11$ (20-nm window), and (d) $W=21$ (40-nm window). In all cases, polynomial order (P) = 2 and derivative order (D) = 1.

derivative, but it might also constitute overfitting, especially for low window widths. Figure 15.11 shows the effect of changing the polynomial order from 2 to 5 for a Savitzky–Golay first derivative filter with $W=11$ (20 nm), for the selected spectral subregion of the diesel fuel data.

It is important to note that another method for calculating derivatives of NIR spectral data, often referred to as the "gap–segment" method, was introduced by Karl Norris during the earliest analytical applications of NIR diffuse reflectance analysis (Norris 2001, Hopkins 2016). This method is similar to the Savitzky–Golay derivative filters discussed above, with the adjustable parameters (named "segment" and "gap") *roughly* equivalent to the Savitzky–Golay W and P parameters discussed above. In the early years of NIR diffuse reflectance analysis, during which computing power was at a premium, stepwise MLR regression algorithms for calibration model development were often preferred over more computationally intensive PLS algorithms. In this scenario, the segment–gap derivative was used in conjunction with single-wavelength MLR searching algorithms in order to efficiently find single spectral variables to use in an MLR calibration model that effectively rejects both interfering species and irrelevant scattering artifacts. Refer to the chapter on Norris derivative in this book for more details.

A final topic relating to smoothing and derivative filters is "edge effect" associated with these methods. Specifically, when applying a smoothing or derivative filter of width W on a spectral array, the filter cannot be applied to the first $W-1$ and last $W-1$ variables in the array, and thus, it is not clear what the filtered values for these variables should be. There are several ways to handle this

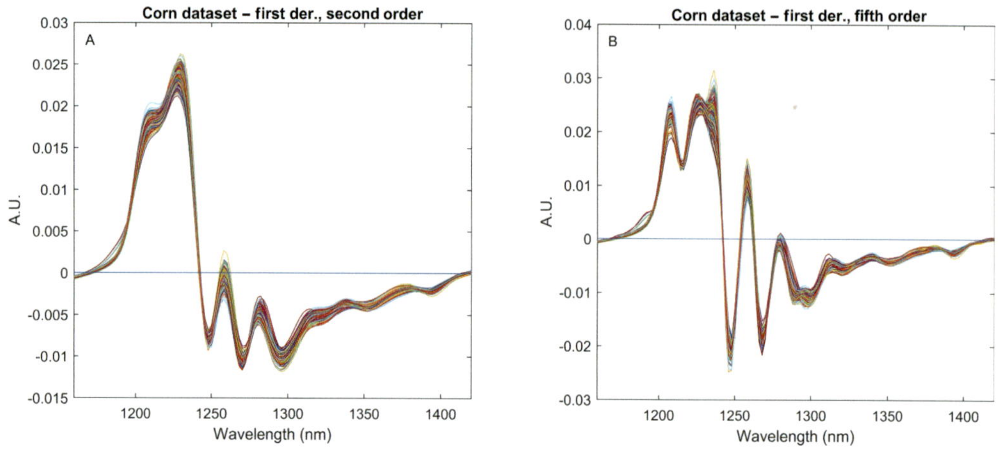

FIGURE 15.11 Effect of polynomial order (P) on Savitzky–Golay first derivative filtering in a selected spectral subregion of the diesel fuel data (1160–1420 nm): (a) $P=2$ and (b) $P=5$. In both cases, the window width $(W)=11$, corresponding to a 20-nm-wide window.

issue, including assigning each of these values as a missing value or to zero in the filtered array, assigning them to values corresponding to a filter using lower window width, or to extrapolated values. In any case, it is important for the method developer to understand how the specific algorithm to be used handles these spectral edges, in order to avoid unexpected results.

15.10 MODEL-BASED METHODS

This family of methods differs from those discussed above, in that they are each based on an explicit mathematical model of the spectral data.

15.10.1 Multiplicative Signal Correction (MSC)

Like the SNV method discussed above, MSC is a sample-wise scaling method that can effectively address multiplicative variations in the spectral data (Geladi, MacDougall, and Martens 1985). However, unlike SNV, the offset and multiplicative correction factors for each spectrum are calculated quite differently: by performing a linear least squares fit of a "reference spectrum" $\boldsymbol{x}_{\text{ref}}$ to each spectrum, using the following model:

$$\boldsymbol{x} = a\boldsymbol{x}_{\text{ref}} + b\mathbf{1} \tag{15.4}$$

where $\boldsymbol{x}$ is the spectrum to be corrected, a is a multiplicative correction factor, b is an additive correction factor, and $\mathbf{1}$ is a vector of ones of the same dimensionality as $\boldsymbol{x}_{\text{ref}}$. For most applications, $\boldsymbol{x}_{\text{ref}}$ is simply assigned to be the mean spectrum of the dataset, but it can be assigned to any spectrum that is deemed representative of the application. For example, if a scatter-free spectrum of the material of interest is available, it could be a more appropriate reference than the dataset mean spectrum.

The MSC model equation above reveals the key assumption behind this method: that any sample spectrum can be simply estimated as a multiple of the reference spectrum, plus a baseline offset. At first, this might seem counterintuitive, as it implies that there is no chemical-based spectral variation present. Although this is certainly not the case (otherwise, there would be no basis for establishing an analytical method using these spectra!), this is a good example of the famous quote from George Box: "All Models are Wrong, but some models are useful". In this case, despite the non-intuitive model, MSC can be quite effective when the offset and multiplicative spectral effects in the spectra are much larger in magnitude than the localized spectral effects from changes in sample chemistry. However, the converse is also true: The MSC method can be counterproductive when the opposite is the case.

Figure 15.12a and b shows the effect of MSC correction on the corn data, where the dataset mean is used as the MSC reference, and the full spectral range is used. Note the vast decrease in the baseline offset and multiplicative variations between the spectra. Also, note that, unlike derivative filtered spectra, MSC-corrected spectra retain the shape and directionality of the original absorption bands, as well as a spectroscopically meaningful *Y*-axis scale. For classical spectroscopists, this is particularly appealing because it allows easier interpretation of the subtle changes in band position and intensity from chemical-based changes in the samples.

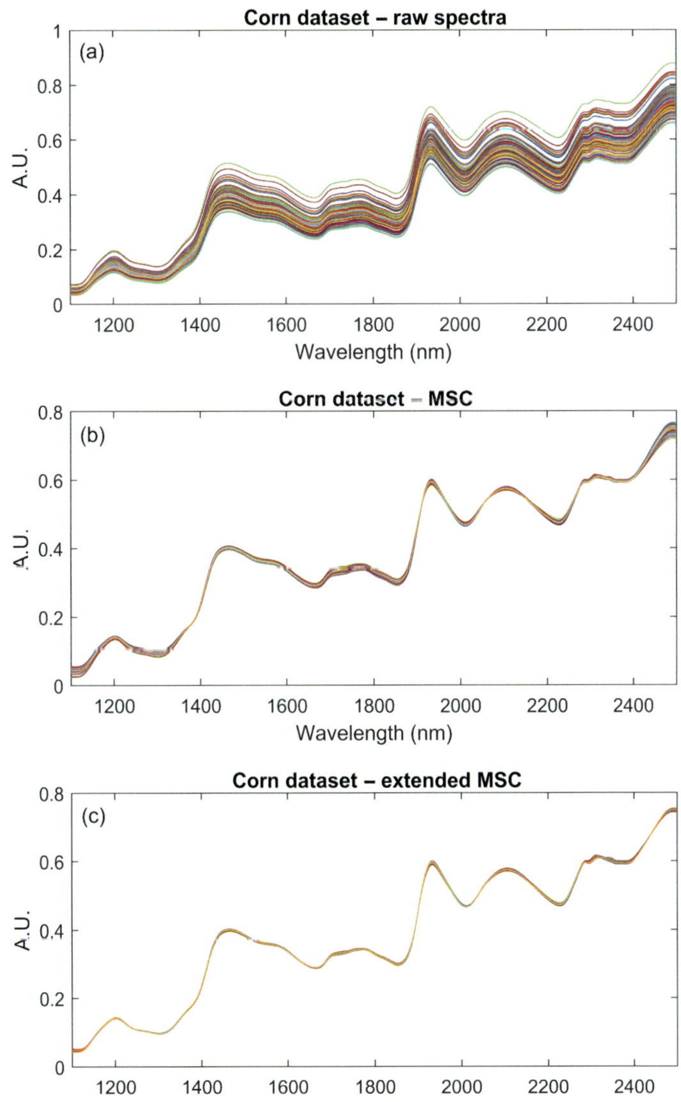

FIGURE 15.12 MSC and EMSC pretreatment of the corn spectra: (a) raw corn spectra, (b) MSC-corrected corn spectra, using the dataset mean as the reference and all 700 wavelength variables, and (c) EMSC-corrected corn spectra, using the dataset mean as the reference spectrum, along with both a linear and a quadratic baseline component.

MSC and SNV preprocessing are often compared to one another, as both of them address both baseline offset and multiplicative effects in spectral data. Figure 15.13 below shows the relationship between MSC and SNV results for the first spectrum of the corn dataset. The strong linear relationship between these results illustrates the equivalence of the two methods, at least from an analytical perspective. This linear relationship was further explored by Dhanoa et al. (1994), who found that both methods are interconvertible in cases when the dataset mean is used in MSC. The differences between these methods are more practical in nature: SNV does not require specification of a reference spectrum, and therefore is easier to use and to transfer to an online prediction engine. MSC does require specification of a reference spectrum, which needs to be registered and maintained with the method when it is put into use. In this way, SNV is somewhat easier to use than MSC, but MSC provides the opportunity to "extend" the model to provide more relevant spectral preprocessing, namely via extended multiplicative scatter correction (EMSC) (Martens and Stark 1991).

15.10.2 Extended Multiplicative Signal Correction (EMSC)

As mentioned in the previous section, the MSC method is expected to perform poorly in cases where spectral offset and multiplicative variations are small relative to chemistry-based variations. Fortunately, if sufficient a priori information about these variations is available, the MSC model equation (Equation 15.4) can be easily extended to account for these variations during estimation of the offset and multiplicative factors $\boldsymbol{a}$ and $\boldsymbol{b}$. This is the basis of the method known as extended

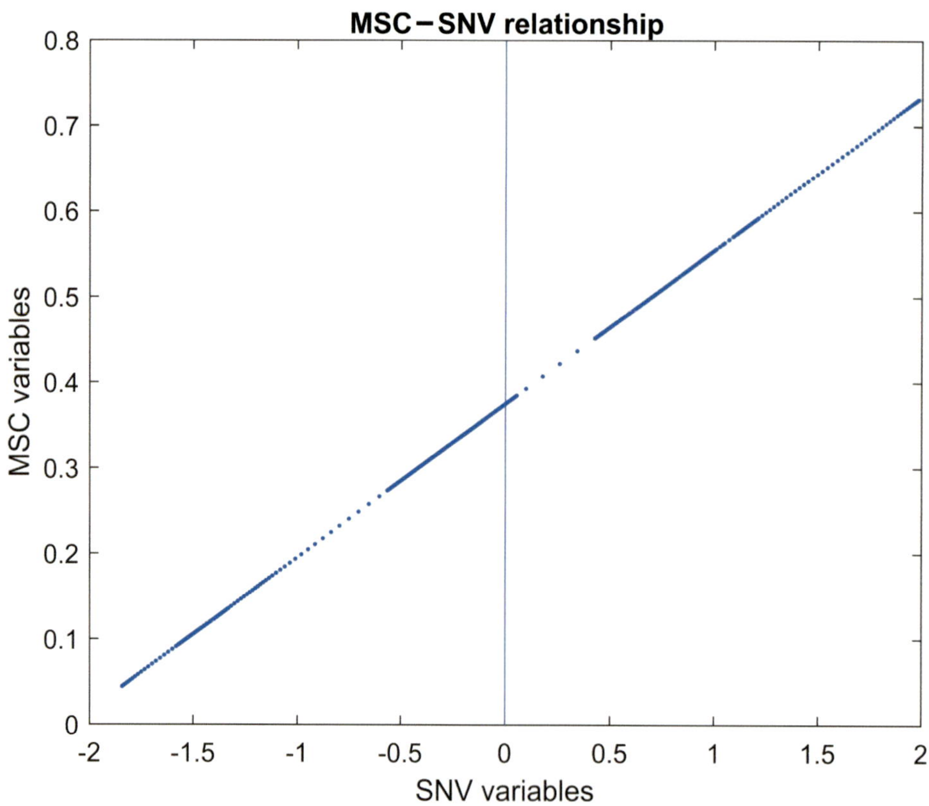

FIGURE 15.13 Scatter plot of the 700 MSC-corrected vs. the SNV-corrected spectral intensities for the first sample in the corn dataset. MSC-corrected spectral profiles are in Figure 15.12b, and SNV-corrected profiles are in Figure 15.5b.

multiplicative scatter correction (EMSC) (Martens and Stark 1991, Martens, Nielsen, and Engelsen 2003). One form of the EMSC model is given below:

$$x = ax_{\text{ref}} + b_H H + c_S S + d_P P \tag{15.5}$$

where $\boldsymbol{H}$ is a matrix containing different baseline profiles to be filtered out of the spectra (e.g., baseline offset, linear baseline slope, and/or curved baseline components), $\boldsymbol{S}$ contains spectral profiles of components or properties to retain in the filtered data, and $\boldsymbol{P}$ contains spectral profiles of components or properties to remove from the filtered data. Note that this model collapses to the MSC model (Equation 15.4) in special cases where no profiles are specified in $\boldsymbol{S}$, no profiles are specified in $\boldsymbol{P}$, and only the baseline offset profile (vector $\mathbf{1}_m$) is specified in $\boldsymbol{H}$. However, the benefit of EMSC over MSC is the ability to explicitly use prior knowledge to both (*a*) provide better estimates of the baseline offset and multiplicative effects and (*b*) enable the removal of spectral effects that are irrelevant to the problem. It is important to note that although Equation 15.5 is a common model form for EMSC, other model forms are possible, for example, where the same multiplicative factor a is distributed across all of the basis vectors $[\boldsymbol{x}_{\text{ref}}; \boldsymbol{H}; \boldsymbol{S}; \boldsymbol{P}]$.

For practical applications, it is often challenging to specify sufficiently relevant spectral profiles to use in $\boldsymbol{H}$, $\boldsymbol{P}$, and $\boldsymbol{S}$. For example, $\boldsymbol{S}$ and $\boldsymbol{P}$ can be populated with measured or "library" spectra of relevant pure components and suspected interferences, respectively. $\boldsymbol{H}$ can be populated with various arbitrary baseline functions, or a baseline profile calculated from measured "blank" samples taken on the spectrometer. With careful selection of these profiles, however, EMSC can be a very effective preprocessing method (Gallagher, Blake, and Gassman 2005). Figure 15.12c shows the result of a relatively simple EMSC preprocessing case of the corn spectra, where both a linear and a quadratic baseline component are included in $\boldsymbol{H}$, and $\boldsymbol{S}$ and $\boldsymbol{P}$ are not specified. From a visual perspective, one can see a further decrease in spectral variance versus the MSC-corrected spectra (Figure 15.12b).

15.10.3 Orthogonal Methods

These constitute an entire additional class of spectral preprocessing methods that operate by identifying and reducing variability in $\boldsymbol{X}$ that is deemed to be irrelevant to the property of interest ($\boldsymbol{y}$). Such methods are based on the premise that the spectral data $\boldsymbol{X}$ span two subspaces: one capturing the relevant information and one capturing the interfering signal. These methods operate by first identifying subspaces in $\boldsymbol{X}$ that are mathematically orthogonal to the property of interest, thus allowing subsequent modeling to focus on relevant spectral information. One method, orthogonal signal correction (OSC), was introduced by Wold et al. (1998) as a preprocessing method that would aim to identify the subspace of $\boldsymbol{X}$ that is orthogonal to the reference property data $\boldsymbol{y}$ to be used for PLS regression. For more information about orthogonal methods, readers should refer to the dedicated chapter in this book.

15.11 SPECIAL PREPROCESSING CONSIDERATIONS

This section discusses miscellaneous, yet very important, considerations that are frequently encountered when selecting and applying preprocessing methods.

15.11.1 Spectral Subregion Basis

In a perfect world, one could treat preprocessing and variable/feature selection as completely independent activities. Unfortunately, for most practical applications, there is a strong interaction between them. Note that in Table 15.1 above, one of the key preprocessing attributes is "Dependence on Variable Subset", which indicates whether the preprocessing result is dependent on the specific variable region(s) used in the spectral input. For those methods with a "yes" in that column, one must be sure to specify the exact set of spectral variables used for the preprocessing step. In addition, it is often observed in practical NIR spectroscopy that the inherent baseline offset or multiplicative effects can vary across the spectral range, in which case more wavelength-localized preprocessing might provide more effective results.

15.11.2 Fringe Benefits

What is often lost on users of chemometrics software is that many of the preprocessing methods discussed above not only transform the raw spectra, but also capture sample-wise coefficients from the filtering that are a convenient expression of the effects that were filtered out of the spectra. A good example of this is the set of mean and standard deviation coefficients ($\bar{x}_i$, s_i) captured from SNV preprocessing of raw NIR diffuse reflectance spectra (Equation 15.3). These express the baseline offset and multiplicative spectral amplitude for an entire dataset and can thus be quite useful for exploratory or diagnostic purposes.

15.11.3 Preprocessing Registration

Earlier, it was mentioned that the two key usage scenarios for spectral preprocessing are to support exploratory method development and to apply in an automated manner during method execution. Typically, the workflows for these two cases are quite different: The former one involves more exploratory usage of various preprocessing methods in conjunction with other method development tools in an attempt to find the optimal calibration model, whereas the latter one involves the repeated and automated execution of one of these schemes to a steady flow of raw spectral data. For any given application, there is one key element that links the two scenarios: the *registration* of the preprocessing scheme to be used in the deployed method.

For the purpose of this chapter, registration refers to the identification of the specific preprocessing methods to be applied, their associated parameters (if needed), and their sequence of application. Two of the attributes in Table 15.1, "Mathematical Complexity" and "Preprocessing Parameters", address the registration requirements for the preprocessing methods discussed in this work. It should be noted that the complexity of a method is not necessarily disadvantageous: It simply influences the requirements of the system on which the method is to be executed, as well as the scope of method management work that will be required.

15.12 SUMMARY

It is hoped that this work has provided you with an improved understanding and appreciation of the various spectral preprocessing methods that are used in NIR analysis. The spectral preprocessing scheme is often a key attribute for defining the effectiveness of any NIR application, and thus, it is critical to choose and "tune" these methods to obtain optimal results. Preprocessing also provides the method developer with great opportunities to apply prior knowledge of the spectroscopy, sampling, chemistry, process dynamics, and other factors, in order to generate the most relevant and robust NIR methods.

REFERENCES

Barnes, R. J., M. S. Dhanoa, and S. J. Lister. 1989. "Standard Normal Variate Transformation and De-trending of Near-Infrared Diffuse Reflectance Spectra." *Applied Spectroscopy* 43 (5):772–777.

Birth, G. S., and H. G. Hecht. 1987. "The Physics of Near Infrared Reflectance." In *Near-Infrared Technology in the Agricultural and Food Industries*, edited by P. Williams and K. Norris. 1–15. St. Paul, MN: AACC.

Dahm, D. J., and K. D. Dahm. 2001. "The Physics of Near-Infrared Scattering." In *Near-Infrared Technology in the Agricultural and Food Industries*, edited by P. Williams and K. Norris, 1–18. St. Paul, MN: AACC.

Eigenvector Research. 2005a. "NIR of Corn Samples for Standardization Benchmarking." http://www.eigenvector.com/data/Corn/index.html.

Eigenvector Research. 2005b. "Near Infrared Spectra of Diesel Fuels." http://www.eigenvector.com/data/SWRI/index.html.

Fearn, T., and A. M. C. Davies. 2003. "A Comparison of Fourier and Wavelet Transforms in the Processing of Near Infrared Spectroscopic Data: Part 1. Data Compression." *Journal of Near Infrared Spectroscopy* 11 (1):3–15.

Gallagher, N. B., T. A. Blake, and P. L. Gassman. 2005. "Application of Extended Inverse Scatter Correction to Mid-Infrared Reflectance Spectra of Soil." *Journal of Chemometrics* 19 (5–7):271–281. doi:10.1002/cem.929.

Geladi, P., D. MacDougall, and H. Martens. 1985. "Linearization and Scatter-Correction for Near-Infrared Reflectance Spectra of Meat." *Applied Spectroscopy* 39 (3):491–500.

Griffiths, P., and D. J. Dahm. 2008. "Continuum and Discontinuum Theories of Diffuse Reflection." In *Handbook of Near-Infrared Analysis*, edited by D. A. Burns and E. W. Ciurcsak, 21–64. Boca Raton, FL: Taylor & Francis.

Hopkins, D. W. 2016. "Revisiting the Norris Derivative Quotient Math in Regression." *NIR News* 27 (7):23–28. doi:10.1255/nirn.1643.

Igne, B., and C. R. Hurburgh Jr. 2010. "Using the Frequency Components of Near Infrared Spectra: Optimising Calibration and Standardisation Processes." *Journal of Near Infrared Spectroscopy* 18 (1):39–47. doi:10.1255/jnirs.865.

Kubelka, P., and F. Munk. 1931. "Ein Beitrag Zur Optik Der Farbanstriche." *Zeitschrift für Technische Physik* 12:593–601.

Martens, H. 2014. "EMSC as mathematical approach to Scattering." *International Diffuse Reflectance Conference*, Chambersburg, PA.

Martens, H., and E. Stark. 1991. "Extended Multiplicative Signal Correction and Spectral Interference Subtraction: New Preprocessing Methods for Near Infrared Spectroscopy." *Journal of Pharmaceutical and Biomedical Analysis* 9:625–635.

Martens, H., J. P. Nielsen, and S. B. Engelsen. 2003. "Light Scattering and Light Absorbance Separated by Extended Multiplicative Signal Correction. Application to Near-Infrared Transmission Analysis of Powder Mixtures." *Analytical Chemistry* 75 (3):394–404. doi:10.1021/ac020194w.

McClure, W. F., A. Hamid, F. G. Giesbrecht, and W. W. Weeks. 1984. "Fourier Analysis Enhances NIR Diffuse Reflectance Spectroscopy." *Applied Spectroscopy* 38 (3):322–329.

Miller, C. E. 2010. "Chemometrics in Process Analytical Technology (PAT)." In *Process Analytical Technology*, edited by K. Bakeev, 353–438. Chichester, UK: Wiley.

Næs, T., and T. Isaksson. 1993. "Non-Linearity Problems in NIR Spectroscopy." *NIR News* 4 (3):14–15. doi:10.1255/nirn.193.

Norris, K. H. 2001. "Understanding and Correcting the Factors Which Affect Diffuse Transmittance Spectra." *NIR News* 12 (3):6–9. doi:10.1255/nirn.613.

Savitzky, A., and M. J. E. Golay. 1964. "Smoothing and Differentiation of Data by Simplified Least-Squares Procedures." *Analytical Chemistry* 36:1627–1639.

Tan, Hu-Wei, and S. D. Brown. 2001. "Wavelet Hybrid Direct Standardization of Near-Infrared Multivariate Calibrations." *Journal of Chemometrics* 15 (8):647–663. doi:10.1002/cem.660.

Williams, P., and K. Norris, eds. 1987. *Near-Infrared Technology in the Agricultural and Food Industries*. St. Paul, MN: AACC Inc.

Wise, B. M., N. B. Gallagher, R. Bro, J. M. Shaver, W. Windig, and R. S. Kosh. 2006. "Chemometrics Tutorial for PLS_Toolbox and Solo." Eigenvector Research.

Wold, S., H. Antti, F. Lindgren, and J. Öhman. 1998. "Orthogonal Signal Correction of Near-Infrared Spectra." *Chemometrics and Intelligent Laboratory Systems* 44 (1–2):175–185. doi:10.1016/S0169-7439(98)00109-9.

16 Pretreatments by Orthogonal Projections

Jean-Michel Roger and Jean-Claude Boulet*
University of Montpellier

CONTENTS

16.1 INTRODUCTION

Whereas the term *orthogonal projection* refers to a generic mathematical expression widely used in chemometrics, the term *pretreatments (or preprocessings) by orthogonal projections* refers to a few recent methods. In this chapter, our first goal is to show that several preprocessings not usually associated with orthogonal projections *are* orthogonal projections. Our second goal is to provide a global view of these methods, for the reader to understand and to use them at best.

What is an orthogonal projection? Of course, we will enter lately into mathematical considerations. But, to begin with, let us present orthogonal projections as simply as possible.

The painting in Figure 16.1 represents shadow shapes, which can also be seen as orthogonal projections into our three-dimensional (3D) space. The axis light-hands is orthogonal to the wall. It is equivalent to say that the hands are projected *orthogonally onto* the wall or that the hands are

* jean-michel.roger@inrae.fr.

FIGURE 16.1 Shadow shapes by Ferdinand Loyen du Puigaudeau.

projected *orthogonally to* the axis hands-light. The result is a shadow that represents a rabbit head, not visible on the hands.

This simple example illustrates the main characteristics of orthogonal projections:

- an object is projected – the hands;
- dimensions are dropped after orthogonal projection, e.g., the axis hands-light;
- The same orthogonal projection can be defined by the remaining dimensions after orthogonal projection (e.g., the wall, 2D) or by the dimensions to drop (e.g., the axis hands-light, 1D);
- the result of the orthogonal projection reveals information that was not previously observed; e.g., the shadow shape of the hands represents a rabbit.

16.2 ORTHOGONAL PROJECTIONS, PRETREATMENTS SUITED FOR NIR SPECTROSCOPY

Let us compare the outputs from two methods of analytical chemistry: high-performance liquid chromatography (HPLC) and NIR spectroscopy. Both outputs are multivariate, and we will seek the variables that contain useful information, i.e., information about a given chemical compound $\mathscr{Y}$.

In an HPLC chromatogram (Figure 16.2a), the useful information is located within a few variables. All the other variables can be dropped without loss of information. The situation is very different for a NIR spectrum (Figure 16.2b): The useful information is located within a larger set of variables (sometimes, all). To extract this information, one has to take all the spectra into account.

The same considerations stand for the detrimental information, i.e., the information which jeopardizes the calibration; the detrimental information also impacts all the NIR spectra. A classical example in NIR reflectance spectroscopy is scattering, illustrated by Figure 16.3. The detrimental information is represented by a baseline with an important shift, with the shape of

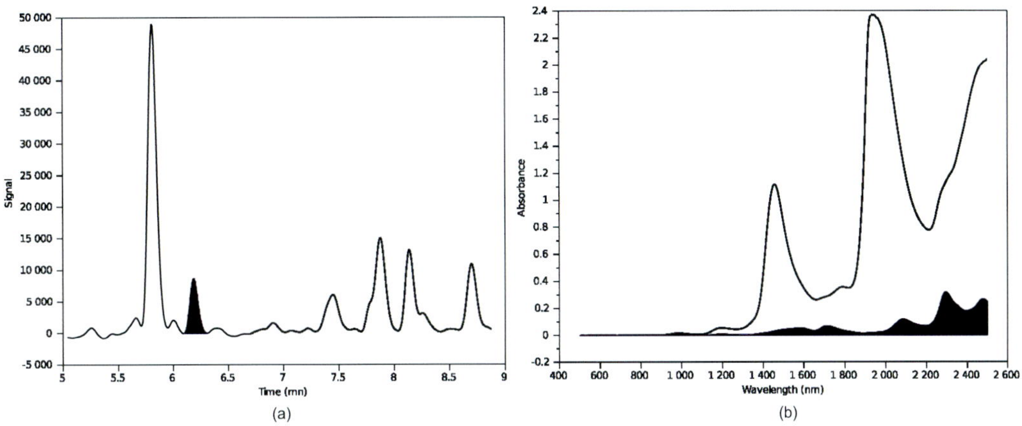

FIGURE 16.2 Comparison of a chromatogram (a) vs. an NIR spectrum (b); the black area corresponds to the contribution of one compound.

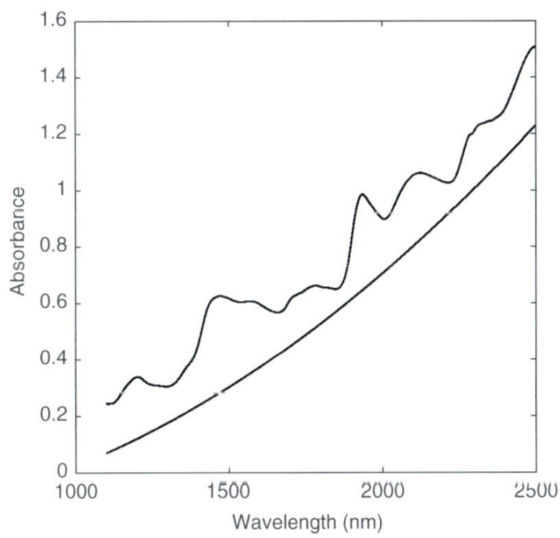

FIGURE 16.3 Example of a typical NIR spectrum measured in reflection. The baseline deformation represented as a curve leads to an important shift of the spectrum in the higher wavelengths.

a curve. As a consequence, the observed spectrum carries this shift, especially in the higher wavelengths.

Pretreatments based on orthogonal projections can remove such spectral alterations.

A set of real spectra is shown in Figure 16.4a. These spectra were acquired in reflection and then converted into pseudo-absorbance by $-\log(R)$. As previously, baselines cause severe alterations, which increase with the wavelengths. To simplify, these alterations will be approximated to straight lines, which mathematically are first-order polynomials of the wavelengths. Two orthogonal projections will be applied to these spectra: an orthogonal projection *onto* polynomials of order 1 (Figure 16.4b) and an orthogonal projection *to* polynomials of order 1 (Figure 16.4c). As a result, each spectrum of (a) has been split up into the sum of a spectrum of (b) that contains the baseline, i.e., the detrimental information, and a spectrum of (c) that contains the chemical, i.e., the useful information. When replacing the original spectra of (a) by the ones of (c), spectra of (b) are dropped and so baselines are removed.

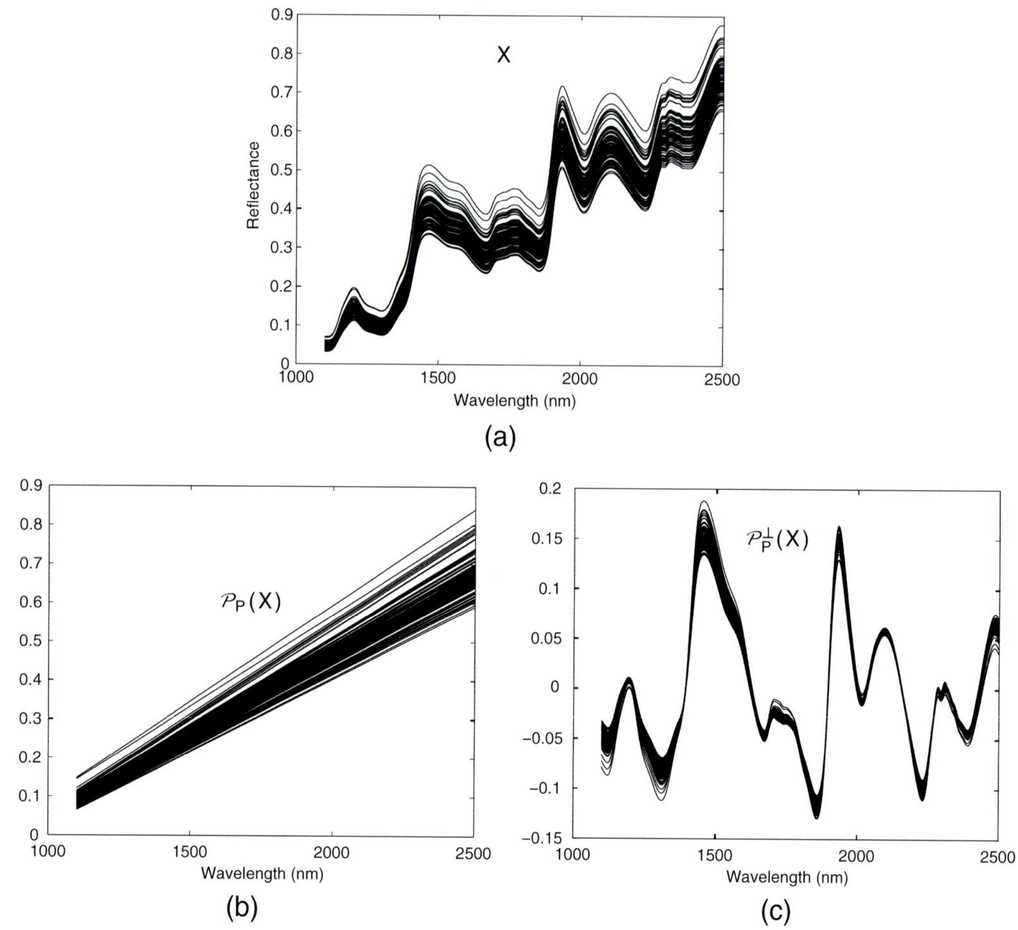

FIGURE 16.4 Example of orthogonal projection. (a) NIR reflectance spectra of corn; (b) orthogonal projection of the spectra on linear baselines; and (c) orthogonal projection of the spectra to linear baselines.

This example illustrates the practical interest of orthogonal projections as pretreatments. Before presenting the different methods, let us focus on the mathematical and geometrical aspects of orthogonal projections.

16.3 MATHEMATICS OF ORTHOGONAL PROJECTIONS

Orthogonal projections have been previously introduced with an example. Now, we will introduce mathematical considerations, in order to answer the question: How do orthogonal projections work?

16.3.1 Mathematical Writing of Orthogonal Projections

Let **X** of dimensions $(N \times Q)$ and **P** of dimensions $(Q \times A)$ be two matrices. Then, the projection of **X** *onto* the subspace spanned by the column vectors of **P** is:

$$\mathbf{X}_P = \mathbf{X}\mathbf{P}\left(\mathbf{P}^{\mathrm{T}}\mathbf{P}\right)^{-1}\mathbf{P}^{\mathrm{T}} \tag{16.1}$$

and the projection of $\mathbf{X}$ *orthogonal to* the subspace spanned by the column vectors of $\mathbf{P}$ is:

$$\mathbf{X}_P^{\perp} = \mathbf{X}\left(\mathbf{I}_Q - \mathbf{P}\left(\mathbf{P}^{\mathrm{T}}\mathbf{P}\right)^{-1}\mathbf{P}^{\mathrm{T}}\right) \tag{16.2}$$

with $\mathbf{I}_Q$ the identity matrix. If $\mathbf{P}^{\mathrm{T}}\mathbf{P}$ is not invertible, then it can be replaced by its Moore–Penrose pseudo-inverse $\left(\mathbf{P}^{\mathrm{T}}\mathbf{P}\right)^{+}$.

Then,

$$\mathbf{X} = \mathbf{X}_P + \mathbf{X}_P^{\perp} \tag{16.3}$$

Orthogonal projections are widely used in chemometrics. But when applied to pretreatments, then $\mathbf{X}$-lines are spectra, and $\mathbf{P}$-columns can be spectra or loadings obtained after calculation such as PCA.

From Equation 16.3, any spectrum can be decomposed in the sum of two spectra that contain different information. Depending on the cases, either $\mathbf{X}_P$ or $\mathbf{X}_P^{\perp}$ will be used. Of course, compared to $\mathbf{X}$, there will be a loss of information, respectively $\mathbf{X}_P$ if $\mathbf{X}_P^{\perp}$ is used or $\mathbf{X}_P^{\perp}$ if $\mathbf{X}_P$ is used. But by a smart choice for $\mathbf{P}$, the dropped information will be detrimental, and the remaining information will be useful, resulting in an improvement of the signal/noise ratio, as illustrated in Figures 16.1 and 16.4c.

16.3.2 Geometrical Approach of Pretreatments Based on Orthogonal Projections

Let us consider six individuals {a...f} characterized by two variables. The individuals could be six spectra from Figure 16.4a and the variables two of the wavelengths. The points representing the individuals in Figure 16.5 are aligned.

Imagine that an interference causes a shift in the direction of a vector $\mathbf{p}$ and that, consequently, the measured points shift from {a...f} to {A...F}. The effect is different for each point, but all share

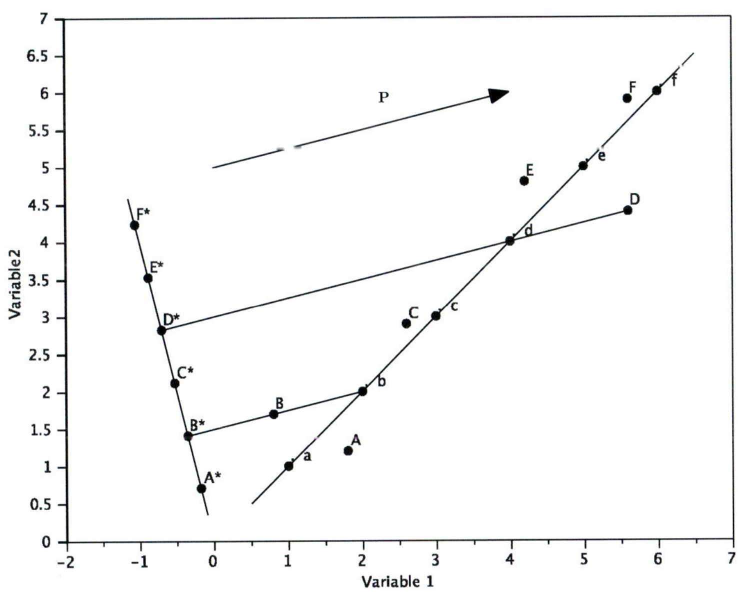

FIGURE 16.5 Illustration of a projection. The original values of observations {a...f} are aligned. An interfering factor represented by the direction of the arrow $\mathbf{p}$ shifts these points to {A...F}. The points {A*...F*} are the result of the projection of {A...F} on a subspace orthogonal to the direction represented by the arrow. This projection is illustrated by a dotted line for B and D.

the same direction. Points {A...F} are no longer aligned. A projection orthogonal to **p** yields the six corrected points {A*...F*}. Some properties of {a...f} have been lost: e.g., the initial direction of the point alignment and the original distances between the points. However, other properties of {a...f} that were lost in {A...F} due to the interfering factor are restored in {A*...F*}: the alignment along a straight line and the relative positions of the six points.

16.3.3 How Can Orthogonal Projection Improve Models?

At present, we have seen that orthogonal projections can remove detrimental information. But are they really useful to improve calibration or classification models?

The measured spectra are rarely formed only by the interference of the response to be estimated. Other chemicals and some physical factors such as product temperature or light scattering participate in the formation of the measured spectra. The useful space $\mathcal{E}^U$ contains the useful information. For methods such as PLSR, $\mathcal{E}^U$ is spanned by the latent variables. Every factor spanning a subspace that does not intersect $\mathcal{E}^U$ has no detrimental effect on the model. Every factor spanning a subspace that does intersect $\mathcal{E}^U$ potentially has a detrimental effect. The union of these subspaces is the detrimental space $\mathcal{E}^D$. Typically, the regression vector **b** should lie in $\mathcal{E}^U$ but not in $\mathcal{E}^D$.

When a linear model is applied to a spectrum, the estimated response is given by $\hat{y} = \mathbf{x}^T\mathbf{b}$. Let $\delta\mathbf{x}$ be an alteration added to the spectrum, due to an influence factor variation. The error induced on the estimation is given by $\delta\hat{y} = \delta\mathbf{x}^T\mathbf{b}$. Then,

$$|\delta\hat{y}| = \|\delta\mathbf{x}\| \times \|\mathbf{b}\| \times |\cos(\delta\mathbf{x}, \mathbf{b})| \tag{16.4}$$

From Equation 16.4, the absolute error caused by the influence factor variation is the product of three positive terms. Lowering any of these terms should lower the error. Reducing $\|\delta\mathbf{x}\|$ is the aim of generic preprocessings as derivative or normalization. Reducing $\|\mathbf{b}\|$ belongs to the good modeling practices (Seasholtz and Kowalski, 1993). The principle of the orthogonal projection (OP) methods is to lower the term $|\cos(\delta\mathbf{x}, \mathbf{b})|$, to constrain the model to be as orthogonal as possible to the spectral perturbation.

16.4 PRETREATMENTS BASED ON ORTHOGONAL PROJECTION

The parameter $\delta\mathbf{x}$ is not constant; $\delta\mathbf{x}$ varies inside the detrimental space $\mathcal{E}^D$. The parameter **b** belongs to the useful space $\mathcal{E}^U$. The orthogonality between **b** and $\delta\mathbf{x}$ is obtained by forcing $\mathcal{E}^U$ to be orthogonal to $\mathcal{E}^D$. This is done by removing from $\mathcal{E}^U$ the part which is collinear to $\mathcal{E}^D$, using an orthogonal projection. This operation is performed in three steps.

The first step is to identify $\mathcal{E}^D$. This is done by selecting as much detrimental information as possible and gathering it into a matrix called **D**. The lines of **D** and the spectra have the same dimension Q.

The second step is to compress the detrimental information of **D** into a few vectors representative of $\mathcal{E}^D$. Ideally, if all the relevant information has been put into **D**, as expected, these vectors should be a basis of $\mathcal{E}^D$. They are gathered and yield a matrix **P**.

The third step is to perform the orthogonal projection according to Equation 16.2. It is a simple calculation, but the result strongly depends on the **P** matrix, which depends on the **D** matrix. Moreover, the different pretreatments based on OP will differ based on how **D** is identified.

16.4.1 Detrimental Information Collection (D Matrix)

In order to identify **D**, one must first identify all the sources of detrimental information in relation to the different situations that are encountered. The knowledge of the samples, the conditions of the data acquisition, and the variable to be predicted is of great importance when applying an orthogonal projection. Then, the matrix **D** will be obtained by merging the detrimental information of all the identified sources.

Some sources, linked to *expert knowledge*, are provided by the spectroscopists when they can describe the nature of the detrimental information.

A detrimental influence due to a chemical compound is easily characterized by its pure spectrum. Orthogonal subspace projection (OSP; Harsanyi and Chang, 1994, Chang, 2003) is based on this principle. Pure spectra can be obtained from databases and put into **D**. In other situations, detrimental influences can be modeled as geometrical curves. For example, light scattering affects NIR spectra with baseline deformations consisting of an offset, a slope, and curvatures (see Figure 16.4). The deformations look like polynomials and thus were modeled so. Moreover, the same polynomial baseline corrections can be performed using a simplified Vandermonde matrix $\mathbf{V}$. For a second-order polynomial, $\mathbf{V}$ contains three columns obtained by raising the vector $\mathbf{v} = [1;2;3;...Q]'$ to powers 0, 1, and 2. These three vectors represent baseline examples spanning the subspace of all the second-order baselines. They are put into **D**. Pretreatments based on polynomials are detrend (Barnes et al., 1989, 1993) and MSC improved in extended multiplicative signal correction (EMSC; Martens et al., 2003).

But in most cases, all the sources are not known, and spectroscopists are not able to describe all the detrimental information, so alternative methods have been developed. They are based on *experimental knowledge* because they lie on experimental spectra containing detrimental information.

The task is easier when the experimental spectra do not contain any useful information. This occurs; for example, when the useful information is a contaminant, spectra without the contaminant are easy to obtain. Independent interference reduction (IIR) uses these spectra directly to build **D** (Hansen, 2001). But when the experimental spectra contain also some useful information, the previous approach cannot apply. So, smart strategies have been proposed to extract the detrimental information.

Most approaches are based on experimental designs. Spectral acquisitions are performed on the same set of samples and for several levels of a detrimental influence. The reference values are not necessary and not processed. Two approaches have been proposed, differing on the detrimental level management and on the way the useful information is removed.

The detrimental levels can be managed: They are set to K levels thanks to an experimental design (Figure 16.6). At each level, spectra are acquired for the same set of L samples, and a mean spectrum is calculated. All the K mean spectra form a matrix that is centered. Centering removes the useful information brought by the samples while preserving the detrimental information. Because of a loss of one dimension due to the centering, one of the resulting spectra is dropped and **D** is obtained. Moreover, the number of column vectors in **D** cannot exceed $K-1$. The method is external parameter orthogonalization (EPO; Roger et al., 2003).

The detrimental levels can also not be managed: L samples are measured at different levels of a detrimental factor (Figure 16.6). The number of levels can be different between two samples. Spectra are gathered by sample. The corresponding submatrices are centered, thus removing the useful information. For sample i, a matrix $\mathbf{D}_i$ is obtained. In error removal by orthogonal subtraction (EROS; Zhu et al., 2008), all the $\mathbf{D}_i^T\mathbf{D}_i$ matrices are calculated and added to obtain **D**. Merging of all the $\mathbf{D}_i$ to yield **D** is equivalent.

The two approaches present some differences. Whereas the former needs a balanced experimental design, the latter can be applied in most situations. Usually, the dimension of the **D** matrix is lower in the former than in the latter approach, which has the capability of preserving more detrimental information.

The following method has been designed for cases where it is not possible to measure the same product in normal and altered conditions. It is the case in process monitoring, when an unknown detrimental source appears. Another example is the season effect that appears between 2 years on

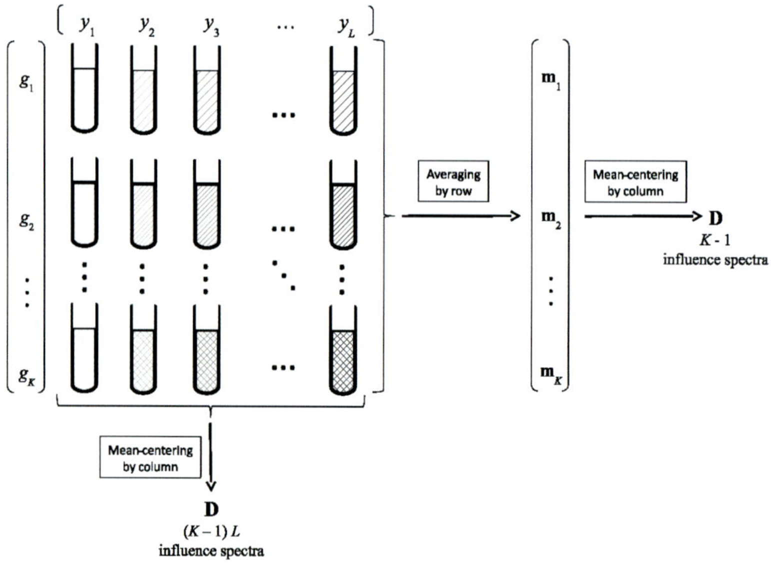

FIGURE 16.6 An experimental design for OP methods with K detrimental levels and L samples. Processings according to the horizontal or the vertical arrows yield EPO and EROS, respectively. Some of the $K \times L$ samples can be missing for EROS, not for EPO.

agricultural products. It is impossible to measure the same product under both conditions. Suppose, however, that we know the reference measurement y_c in correspondence with an altered spectrum $\mathbf{x}_c$. The goal is to calculate the spectrum $\hat{\mathbf{x}}_c$ that would have been acquired without the detrimental effects. This spectrum is estimated from the calibration database. For each individual i of this database, a weight w_i is calculated based on the difference between its y_i and y_c. The closer the y_i and y_c, the higher is the weight. The ideal spectrum is computed as the average of the calibration spectra weighted by the w_i. Then, the detrimental information $(\hat{\mathbf{x}}_c - \mathbf{x}_c)$ is put into **D**. This is the principle of dynamic orthogonal projection (DOP; Zeaiter et al., 2006), also applied successfully in calibration transfer (Igne et al., 2009) or vintage changes (Roger et al., 2008).

16.4.2 Detrimental Subspace Estimation (P matrix)

Once the matrix **D** is obtained, a basis **P** of the detrimental subspace must be calculated. The matrix **P** contains R vectors of $\mathbb{R}^Q$.

The proposed method to identify **P** is PCA (Roger et al., 2003, Andrew and Fearn, 2004). The first loadings of a PCA on **D** correspond to the main directions of **D** observed in the space of spectra, i.e., of $\mathscr{E}^D$. To avoid losing the first main direction, the **D** matrix *must not* be centered before PCA calculation. Special attention will be paid with commercial software packages, for which centering is often a default option. Alternately, users could also fill **P** with the eigenvectors of $\mathbf{D}^\mathrm{T}\mathbf{D}$, for example, by means of a singular value decomposition (SVD). Using the eigenvectors of $\mathbf{D}^\mathrm{T}\mathbf{D}$ (from PCA or SVD) produces orthonormal vectors for **P**, so that the expression of the OP in Equation 16.2 becomes simpler:

$$\mathbf{X}_P^{\perp} = \mathbf{X}\left(\mathbf{I}_Q - \mathbf{P}\mathbf{P}^\mathrm{T}\right).$$

The value R corresponds to the dimension of the subspace removed by the orthogonal projection. If R is too small, not all influences are removed, and the correction is not complete. If R is too big, some useful dimensions can be removed, and consequently, the model becomes poorer. Thus, finding the optimal value R_{opt} is a key point in the OP setup.

16.4.2.1 Adjusting R_{opt} from the Eigenvalue Evolution

The spectral influences to be removed are structured, and they lie in a very small subspace. So, it is natural to expect that only a few eigenvectors of $\mathbf{D}$ or $\mathbf{D}^T\mathbf{D}$ are necessary to represent $\mathcal{E}^D$. A simple approach for choosing R_{opt} is to search for a break point in the evolution of the eigenvalues of $\mathbf{D}^T\mathbf{D}$, sorted in decreasing order. Some statistical tests have been proposed to detect the break points for these kinds of curves (Jackson, 1991). However, as this curve is always decreasing, it is difficult to make an objective decision.

16.4.2.2 Adjusting R_{opt} with Cross-Validation

For each value of R between 0 – no projection – and an *a priori* large value R_{max}, a correction using R dimensions is performed on the calibration database and a regression is cross-validated for 1 to A_{max} latent variables. The RMSECVs of all these models provide a matrix of dimensions $\left(A_{max} \times \left(R_{max} + 1\right)\right)$. Because the calibration database is not supposed to contain any example of influence to be corrected, the positive effect of the projection cannot be observed plainly. A detrimental effect can, however, be observed when too many dimensions are removed. So, this approach helps to determine a new maximum value R_{max2} with $R_{max2} < R_{max}$. It will not give the optimum value R_{opt} that is between 0 and R_{max2}.

16.4.2.3 Adjusting R_{opt} from the Wilks Lambda Evolution

When a set of observations is structured in groups, the Wilks lambda (Λ) is defined as the ratio of the between-group variance and the total variance. This criterion varies between 0 and 1; the better the group separation, the higher the Λ (Lebart et al., 1984). Consider a data regrouped by samples. The between-group variance is carried by $\mathcal{E}^U$, while the within-group variance is carried by $\mathcal{E}^D$. If there were no external influences and if the spectra were perfectly acquired, the spectra of each group would match, yielding a Λ of 1. Thus, as far as the OP removes detrimental information, Λ increases with R. However, the dimensions that are removed can also carry some information related to $\mathcal{Y}$. So, removing too many dimensions can make Λ decrease. Hence, plotting Λ as a function of R produces a curve that should present a first increase, followed by a plateau or a decrease. R_{opt} corresponds to the maximum or to the break point of this curve.

The Wilks Λ can also be calculated on the prediction of the corrected model, as a function of R and of the number of latent variables (A). Varying R between 0 – no projection – and an *a priori* large value R_{max} and A between 1 and A_{max} provides a matrix of Wilks' Λ of dimensions $\left(A_{max} \times \left(R_{max} + 1\right)\right)$. Examining this matrix, e.g., as a 3D surface allows us to find the best dimensions for both the OP correction and the PLS regression.

16.4.2.4 Adjusting R_{opt} for Spectral Image Analysis

In spectral images, each spectrum corresponds to a pixel, the exact composition of which often is inaccessible. In this case, the methods based on cross-validation or test are not feasible. However, the extra information yielded by the spatial structure of the image can be used (Jaillais et al., 2012). Let us assume that an area $\mathcal{A}$ of the image can be considered as homogeneous regarding the response $\mathcal{Y}$. The spectra of $\mathcal{A}$ are gathered into a matrix $\mathbf{X}_a$. Let $\mathbf{k}$ be the pure spectrum of $\mathcal{Y}$. If $\mathbf{k}$ is unavailable, it could be replaced by the mean spectrum of $\mathbf{X}_a$ (Jaillais et al., 2012). Let $\mathbf{x}$ be a spectrum of $\mathbf{X}_a$ and $\mathcal{P}_R$ the projector using $\mathbf{P}$ with R dimensions; $\mathcal{P}_R = \mathbf{I}_Q - \mathbf{P}\mathbf{P}^T$. If the detrimental subspace were correctly removed, then $\mathbf{x}^T\mathcal{P}_R$ and $\mathbf{k}^T\mathcal{P}_R$ would be almost collinear and their squared cosine would be close to 1. The squared cosines are then computed between each spectrum of $\mathcal{A}$ after OP and $\mathbf{k}^T\mathcal{P}_R$, and their average values are plotted as a function of R. The optimum value R_{opt} is chosen according to a break point or a maximum of this curve.

16.4.3 Orthogonal Signal Correction

Orthogonal signal correction (OSC) was introduced by Wold et al. (1998) and used for calibration transfer by Sjöblom et al. (1998). This filter aims at removing from **X** dimensions that carry a large part of the variance and are orthogonal to **y**. OSC leads to computation problems because the orthogonality to **y** is defined in the space of the samples, and the correction must be realized in the space of the spectra, in order to be applied on new spectra. A discussion on this problem can be found in Boulet and Roger (2012). OSC effectiveness is controversial. Indeed, its objective is strictly the dual of PLS, which aims at finding directions that both carry variance and are correlated with **y**. Thus, when associated with PLS, OSC lowers the number of latent variables, but does not improve the predictive ability of the model (Fearn, 2000, Kemsley and Tapp, 2009, Verron et al., 2004).

Nonetheless, when used on appropriately designed data, OSC algorithm allows us to subtract from spectra the detrimental information while preserving the dimensions related to **y**. In Wold et al. (1998), OSC has been applied to spectra acquired according to an experimental design merging the response vector **y** and the influence factors (Preys et al., 2006, Boulet et al., 2007). The output of OSC algorithm with K factors is a matrix of loadings **P** and a matrix of weights **W**, both $(Q \times K)$. The OSC correction of a new spectrum **x** is made by the following oblique projection:

$$\mathbf{x}_{\text{osc}} = \mathbf{x}\left(\mathbf{I}_Q - \mathbf{W}\left(\mathbf{P}^{\mathrm{T}}\mathbf{W}\right)^{-1}\mathbf{P}^{\mathrm{T}}\right)$$

Note that the orthogonal projection is a particular case of the oblique projection.

16.5 ORTHOGONAL PROJECTIONS ILLUSTRATED BY EXAMPLES

Some practical examples are used here to illustrate the methods presented above and to focus on a few points of attention.

16.5.1 Apple Sugar Content and Fruit Temperature

This example comes from Roger et al. (2003), where all the experimental details can be found. The available data are sufficiently abundant to illustrate different ways to identify the harmful dimensions. Vis–NIR spectra (310–1150 nm, $Q = 256$ equally spaced wavelengths) and Brix degrees were measured on apples. A calibration set $(\mathbf{X}_0, \mathbf{y}_0)$ of 80 fruits was acquired at ambient room temperature (≈25°C). A second set of ten apples measured at eight different temperatures, varying from 5°C to 40°C, yielded a matrix of spectra $\mathbf{X}_1$ and of Brix $\mathbf{y}_1$. A third set of ten apples measured in the same conditions yielded a test set $(\mathbf{X}_2, \mathbf{y}_2)$, allowing us to calculate the prediction errors. A PLSR model realized on $(\mathbf{X}_0, \mathbf{y}_0)$ exhibited a cross-validation error (RMSECV) of about 0.8 Brix. Once applied on the test set, the model exhibited a global prediction error (RMSEP) of about 4.5 Brix. This error was mainly due to a bias dependent on the temperature.

Let us examine how to set up the correction of the temperature effect by applying an orthogonal projection on $\mathbf{X}_0$. As a reminder, the three steps are (1) to build a matrix **D** containing the spectral influences of the temperature, (2) to calculate a basis **P** of the subspace spanned by **D**, (3) to project $\mathbf{X}_0$ orthogonally to d to recalibrate the model. This example will illustrate two ways to build the **D** matrix and to examine the results on the test set.

The $\mathbf{X}_1$ matrix can be considered as $(8 \times 10 \times 256)$. Given the (8×10) individuals of $\mathbf{X}_1$, described by the 256 variables of the spectra, several sources of variations can be considered: the variation due to the difference between the ten apples, the variation due to the eight temperatures, and the variation due to all the uncontrolled phenomena, which is expressed differently for each acquisition. On this basis, the matrix **D** can be constructed in two different ways:

- To capture the variations solely induced by the temperature, we average $\mathbf{X}_1$ along its second dimension; i.e., we average all the spectra of all the apples measured at the same temperature. So, all other variations are minimized. This leads to eight averaged spectra $\{\mathbf{m}_1,\ldots\mathbf{m}_8\}$. The variation between these spectra is mainly due to temperature, and very little to repeatability, which has been reduced by the averages. We put in **D** the seven spectra of the differences between the eight $\mathbf{m}_i$ spectra. Equivalently, we can put into **D** the mean-centered matrix containing the $\mathbf{m}_i$ spectra. In this case, **D** contains seven independent spectra. Let us call this matrix $\mathbf{D}_1$.
- We can also adopt another point of view. We can decide to put in **D** all useless variations for the model, i.e., all variations non-related to apples. In doing so, we expect to correct the variations due to temperature and repeatability. This is done by mean-centering the $\mathbf{X}_1$ matrix along its first dimension, i.e., by removing to each spectrum acquired on an apple i the mean spectrum of all the spectra acquired on the apple i. In this case, **D** contains 70 independent spectra. Let us call this matrix $\mathbf{D}_2$.

Using these two matrices, two corrections are performed, including the building of **P**, the orthogonal projection, and the recalibration. These two corrections are performed for a number of OP dimensions varying from 1 to 10, and the recalibration is carried out using 1–20 latent variables. Figure 16.7a and b shows the error of prediction on $\mathbf{X}_2$, when using $\mathbf{D}_1$ or $\mathbf{D}_2$, respectively. These two part figures are drawn using the same scale, same angles, and same colormap. In Figure 16.7a, the number of OP dimensions is limited to 7, which is the rank of $\mathbf{D}_1$. In Figure 16.7b, all the numbers of OP dimensions are reported from 1 to 10, since the rank of $\mathbf{D}_2$ is 70. First, we can observe that the best performances of the two models are very similar; indeed, they are close to 0.5 Brix. Nevertheless, these performances are not obtained for the same combination of number of OP dimensions (R) and latent variables (LV). In Figure 16.7a, we can observe that for low values of R, the PLS model gives good predictions, but for a rather high number of latent variables ($R = 1 \text{ and } \text{LV} = 16$). As R increases, the number of optimal latent variables decreases. An optimum can be found for $R = 4$ and $\text{LV} = 12$. This illustrates that the orthogonal projections clean the space and make the calibration simpler. In Figure 16.7b, we can observe a different behavior. The orthogonal projection does not seem to have any effect for small values of $R(<3)$. This is due to the fact that $\mathbf{D}_2$ contains variations other than the ones due to temperature and that they carry a large

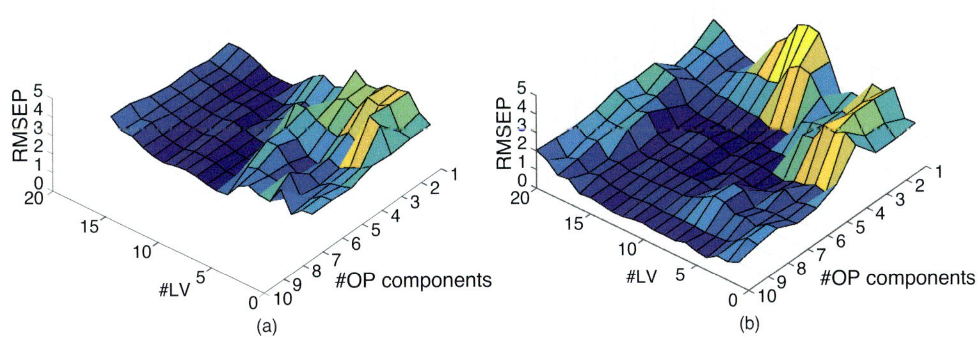

FIGURE 16.7 Prediction error (RMSEP) as a function of the number of dimensions removed by the OP and of the number of latent variables used by the PLS, using $\mathbf{D}_1$ (a) or $\mathbf{D}_2$ (b).

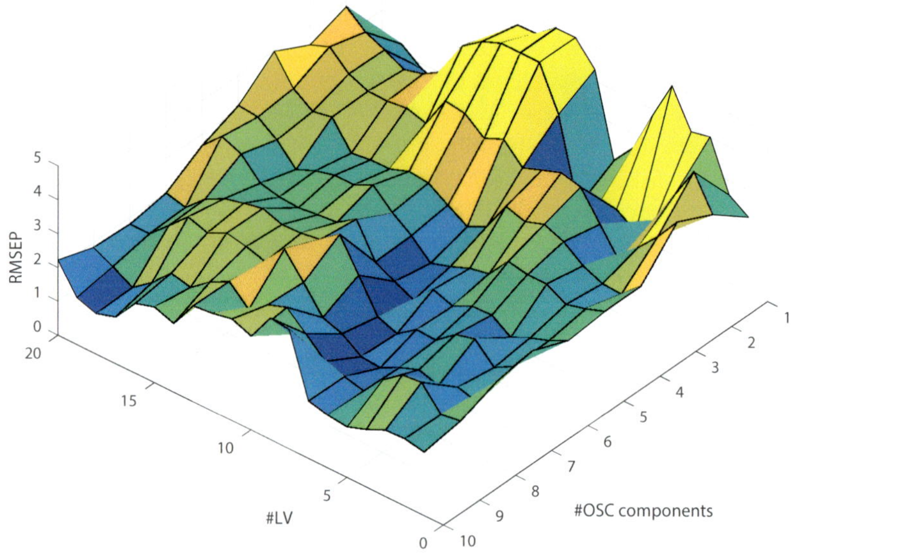

FIGURE 16.8 Prediction error (RMSEP) as a function of the number of dimensions removed by an OSC and of the number of latent variables used by the PLS.

variance. The first dimensions of $\mathbf{P}$ are not linked to temperature, and their deletion does not have a positive effect on the robustness of the model. An optimal zone, however, appears for $3 \leq R \leq 6$ and $8 \leq \text{LV} \leq 14$. This zone is, however, rather chaotic, which seems to indicate a greater sensitivity to the parameters when using $\mathbf{D}_2$ than using $\mathbf{D}_1$.

A correction was also made using an OSC calculated on $\mathbf{X}_1, \mathbf{y}_1$ and applied to $\mathbf{X}_0$ and $\mathbf{X}_2$, as explained in Section 16.4.3. Figure 16.8 shows the results, using the same scale and colormap as the previous figures. We can notice that, as for an OP using $\mathbf{D}_2$ (Figure 16.7b), the correction based on the OSC can delete more than seven dimensions. Indeed, the dimensions identified by the OSC are determined by the orthogonality to $\mathbf{y}_1$, so can go up to 79. One can also notice a zone of minimal error for a number of OSC components between 5 and 9 and a corresponding number of PLS latent variables of 11 and 7, respectively. The lowest RMSEP is about 1.18 Brix. The correction provided by the OSC is therefore much less efficient than that provided by the OP method. This difference in performance is due to the fact that the subspace useful to the model $\mathcal{E}^U$ and the subspace harmful to the model $\mathcal{E}^D$ have a significant intersection $\mathcal{E}^I$, as illustrated in Figure 16.9. In other words, the sugar content and the temperature of the fruits have some common influences on the spectra. The OP correction does not take into account the useful space and can therefore remove all the harmful space, including $\mathcal{E}^I$. On the contrary, correction by OSC does not remove any dimension of $\mathcal{E}^U$, since the dimensions that it withdraws are orthogonal to $\mathcal{Y}$. When $\mathcal{E}^I$ is big, the OSC-based correction is incomplete.

16.5.2 Wheat Protein Content and Several Influence Factors

This example comes from a chemometric contest proposed in the framework of the French Chimiométrie 2007 conference. It will allow us to examine how to combine several OP corrections. It concerned the calibration of protein content of ground wheat samples from their NIR spectra (1300–2398 nm, $Q = 550$ equally spaced wavelengths). All the details can be found in Pierna et al.

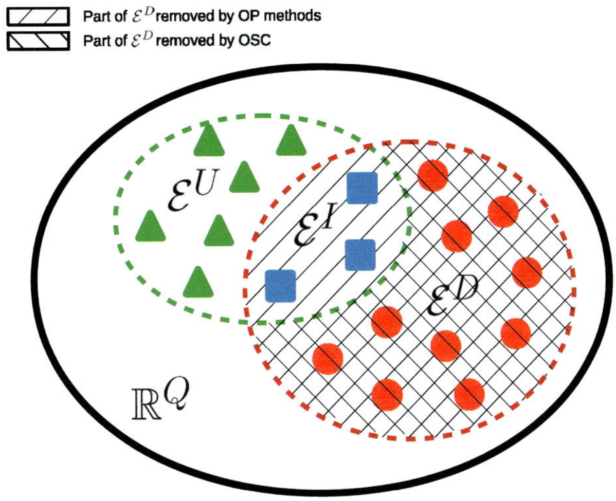

FIGURE 16.9 Illustration of how OP pretreatments and OSC pretreatments manage useful and detrimental subspaces.

(2011). The aim was to make an existing calibration robust versus five influence factors: particle size, temperature, moisture, repetition, and spectrometer changes. A calibration set $(\mathbf{X}_0, \mathbf{y}_0)$ of ten samples was provided. These samples were appropriately drawn from a bigger database of samples acquired at standard conditions of particle size, temperature, and moisture on a single spectrometer. The spectra of a second set of 11 wheat samples were measured according to an experimental design at two levels of particle size, two levels of moisture, and three levels of temperature and in duplicate. This provided a matrix of spectra $\mathbf{X}_1(11 \times 2 \times 2 \times 3 \times 2 \times 550)$. Additional spectra were supplied from one set (A) of 10 sealed cells scanned on 31 different instruments of the same type and from a second set (B) of 10 sealed cells scanned on 17 instruments. This provided two matrices of spectra $\mathbf{X}_2(10 \times 31 \times 550)$ and $\mathbf{X}_3(10 \times 17 \times 550)$. The protein content corresponding to $\mathbf{X}_1$, $\mathbf{X}_2$, and $\mathbf{X}_3$ was unknown. The spectra of a test set containing 2000 samples measured in varying conditions of particle size, temperature, and moisture and on different instruments were provided as a matrix $\mathbf{X}_4$. The protein contents of these test samples $(\mathbf{y}_4)$ were obviously unknown in the contest. They will be used here to calculate the prediction errors. A PLSR model realized on $(\mathbf{X}_0, \mathbf{y}_0)$ exhibited a cross-validation error (RMSECV) of about 2 g kg^{-1}.

Averaging the dimensions 3, 4, and 5 of $\mathbf{X}_1$ produces a matrix $\mathbf{Z}_a(11 \times 2 \times 550)$ containing the spectra of the 11 samples duplicated on two levels of particle size. Subtracting the 11 spectra of one level from the 11 other ones produces a matrix $\mathbf{D}_a$ that contains 11 examples of spectral influences due to particle size. Doing similarly with the other dimensions finally yields four matrices $\mathbf{Z}_a(11 \times 2 \times 550)$, $\mathbf{Z}_b(11 \times 2 \times 550)$, $\mathbf{Z}_c(11 \times 3 \times 550)$, and $\mathbf{Z}_d(11 \times 2 \times 550)$ and four matrices $\mathbf{D}_a(11 \times 550)$, $\mathbf{D}_b(11 \times 550)$, $\mathbf{D}_c(22 \times 550)$, and $\mathbf{D}_d(11 \times 550)$, respectively, related to particle size, moisture, temperature, and repetition. Mean-centering $\mathbf{X}_2$ along its second dimension yields a matrix $\mathbf{Z}_e(10 \times 31 \times 550)$, containing spectral variations mostly related to spectrometer changes. This matrix is unfolded to provide a matrix $\mathbf{D}_e(310 \times 550)$. Doing the same on $\mathbf{X}_3$ leads to a matrix $\mathbf{X}_f(10 \times 17 \times 550)$ and a matrix $\mathbf{D}_f(170 \times 550)$.

The management of several $\mathbf{D}$ matrices, related to several influence factors, can be done in two ways. Either each factor is individually studied, yielding individual projection, and all projections are merged, or all the matrices are merged and a global projection is made. Let's look at both strategies by example.

16.5.2.1 Individual Setup

Each of the five matrices $(\mathbf{D}_a, \mathbf{D}_b, \cdots, \mathbf{D}_e)$ yields a matrix of loadings **P**, the dimension of which must be determined. Since no reference values were given in association with the experimental design, let us use the method of the Wilks Λ, as described in Section 16.4.2.3. Each matrix $\mathbf{Z}_x$ is projected orthogonally to R_x dimensions of its corresponding $\mathbf{D}_x$, and the Wilks Λ is calculated considering the first dimension as the group variation (between variance) and the second one as the detrimental variation (within variance). Table 16.1 reports the values obtained for the five matrices and for R varying from 0 (no projection) to 10. The following conclusions can be drawn from the table:

- Observing the Wilks Λ at $R = 0$, it can be noticed that not all factors have the same detrimental effect. Repetition is the less detrimental factor, not surprisingly. Particle size appears as the most impacting factor. This is not surprising for measurement done in reflection mode. Spectrometer change has also an impact, irrespective of the experimental design (A or B). Temperature and humidity have a moderate impact.
- Observing the Wilks Λ evolution, it can be noticed that the orthogonal projection is very efficient for particle size, moisture, temperature, and repetition from the low values of R. For the spectrometer change, things are more complicated, since the Wilks Λ increases slowly. This is probably due to the complexity of the spectral deformations induced by the diversity of such numerous devices (31 and 17).

On the basis of these observations, let us retain 3, 3, 2, 3, 8, and 8 dimensions for the matrices $(\mathbf{P}_a, \mathbf{P}_b, \cdots, \mathbf{P}_f)$, respectively. These six matrices yield six orthogonal projections, which can be applied successively to $\mathbf{X}_0$:

$$\mathbf{X}_0^\perp = \mathbf{X}_0\left(\mathbf{I}_Q - \mathbf{P}_a\mathbf{P}_a^\mathrm{T}\right)\left(\mathbf{I}_Q - \mathbf{P}_b\mathbf{P}_b^\mathrm{T}\right)\left(\mathbf{I}_Q - \mathbf{P}_f\mathbf{P}_f^\mathrm{T}\right)\ldots$$

But doing that is not optimal, because the subspaces spanned by the $\mathbf{P}_x$ matrices are not orthogonal. Thus, applying the second projection destroys part of the effect of the first, and so on for the others. The application of this method produces an RMSEP of about 10 g kg^{-1}, which is far from the optimal value reported in Pierna et al. (2011).

Instead of piling up the projections, the recommended approach is to identify a unique detrimental subspace, prior to the orthogonal projection. This is done by merging all the $\mathbf{P}_x$ matrices into a $\mathbf{P}_{\text{all}}$ matrix and by carrying out a PCA on it. One must then determine again the optimal number R of PC to retain to form a new **P** matrix. Since we already know that $\mathbf{P}_{\text{all}}$ optimally describes the detrimental effects, it is now important to choose R in order to preserve the calibration model. Thus, several values of R are tested by cross-validation on $(\mathbf{X}_0, \mathbf{y}_0)$, as described in Section 16.4.2.2. Figure 16.10a

TABLE 16.1
Wilks' Λ Evolution for Each Influence Factor, According to *R*, the Number of Dimensions Removed by the Orthogonal Projection

Factor	0	1	2	3	4	5	6	7	8	9	10
Particle size	0.347	0.982	0.989	0.991	0.994	0.994	0.996	0.997	0.998	0.999	0.999
Moisture	0.817	0.982	0.986	0.993	0.995	0.996	0.997	0.996	0.997	0.999	0.999
Temperature	0.872	0.969	0.994	0.995	0.994	0.995	0.996	0.996	0.997	0.997	0.997
Repetition	0.996	0.997	0.998	0.998	0.999	0.999	0.999	1.000	1.000	1.000	1.000
Spectrometer change A	0.496	0.569	0.634	0.670	0.692	0.707	0.726	0.724	0.741	0.754	0.762
Spectrometer change B	0.524	0.607	0.659	0.714	0.744	0.777	0.801	0.819	0.825	0.838	0.849

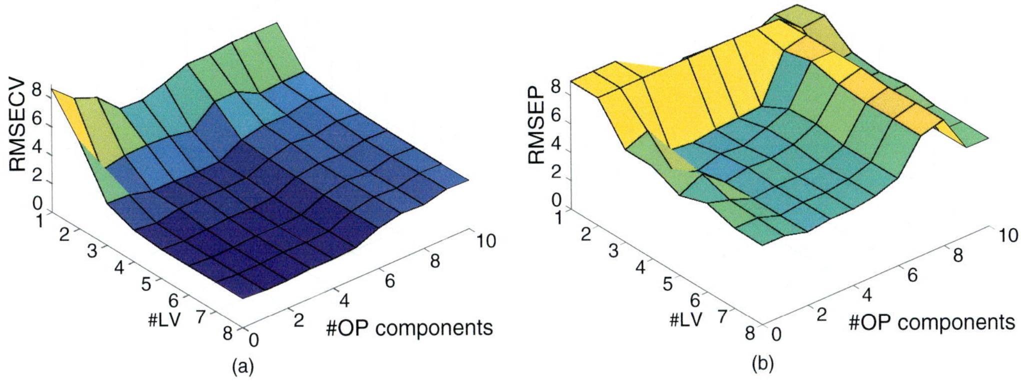

FIGURE 16.10 Errors as a function of the number of dimensions removed by the OP and of the number of latent variables used by the PLS: RMSECV (a) and RMSEP (b).

shows the evolution of the RMSECV as a function of R and of the number of latent variables of the model. One can reasonably adopt a value of about 3 or 4 for R and about 6 or 7 for the number of latent variables. This choice corresponds to the optimal zone in terms of RMSEP, presented in Figure 16.10b. The final corrected model exhibits an RMSEP of about 5 g kg^{-1}.

16.5.2.2 Global Setup

This option is much simpler than the previous one, but provides less knowledge about the data. It consists of merging all the $\mathbf{D}_x$ into a single matrix $\mathbf{D}_{\text{all}}$ and carrying out an orthogonal projection as if there was a single influence factor.

Let us illustrate the method explained in Section 16.4.2.3. In our example, $\mathbf{D}_{\text{all}}$ contains 535 lines, each of them being an example of an influence spectrum $\delta\mathbf{x}$ induced by variations in particle size, moisture or temperature, the repetition, or the spectrometer change. An orthogonal projection is carried out on $\mathbf{X}_0$, using $\mathbf{D}_{\text{all}}$, with a variable number R of dimensions, and a PLS regression is calibrated on the corrected matrix with a variable number A of latent variables. The model is applied on the matrices $\mathbf{X}_1$, $\mathbf{X}_2$, and $\mathbf{X}_3$, yielding a vector of 744 $(11\times(2+2+3+2)+10\times 31+10\times 17)$ predictions. However, these predictions are related to 31 $(11+10+10)$ samples. Figure 16.11a represents the evolution of the Wilks Λ calculated on the 744 predictions according to the 31 groups, as a function of R and A. We can notice a relatively flat and high zone for $R\geq 3$ and $A\geq 2$. Let us adopt the most parsimonious model, with $R=3$ and $A=2$. In Figure 16.11b, we can notice that it corresponds to a minimum of the RMSEP, about 4.5 g kg^{-1}.

16.5.3 Fermentation Monitoring and Temperature

This example comes from Zeaiter et al. (2006), where all the experimental details can be found. It will illustrate how to perform an orthogonal projection without experimental design, as explained for the DOP method in Section 16.4.1. NIR spectra ($1404-1546$ nm, $Q=71$ equally spaced wavelengths) and ethanol degree were measured during wine fermentation processes. A calibration set $(\mathbf{X}_0,\mathbf{y}_0)$ of 300 samples was acquired during a fermentation at constant laboratory temperature (25°C). A second fermentation was realized with the same material and conditions, except that the temperature increased from 25°C to 34°C during the first half of the process and decreased from 34°C to 25°C during the second half (see Figure 16.12). A test set $(\mathbf{X}_0,\mathbf{y}_0)$ of 1411 samples was acquired during this process. As shown in Figure 16.12, five control points were distributed over the first half of the process, yielding $(\mathbf{X}_c,\mathbf{y}_c)$. The first two points are taken during the isothermal phase; the three others are distributed over the temperature rise phase.

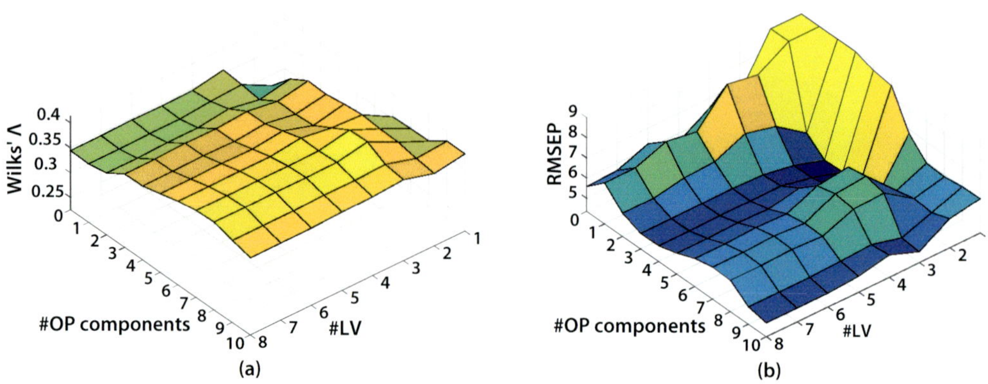

FIGURE 16.11 (a) Evolution of the Wilks Λ, expressing the grouping of the predictions of same samples, as a function of the dimension of the OP correction (R) and the number of latent variables of the PLSR (A). (b) RMSEP of the corresponding models.

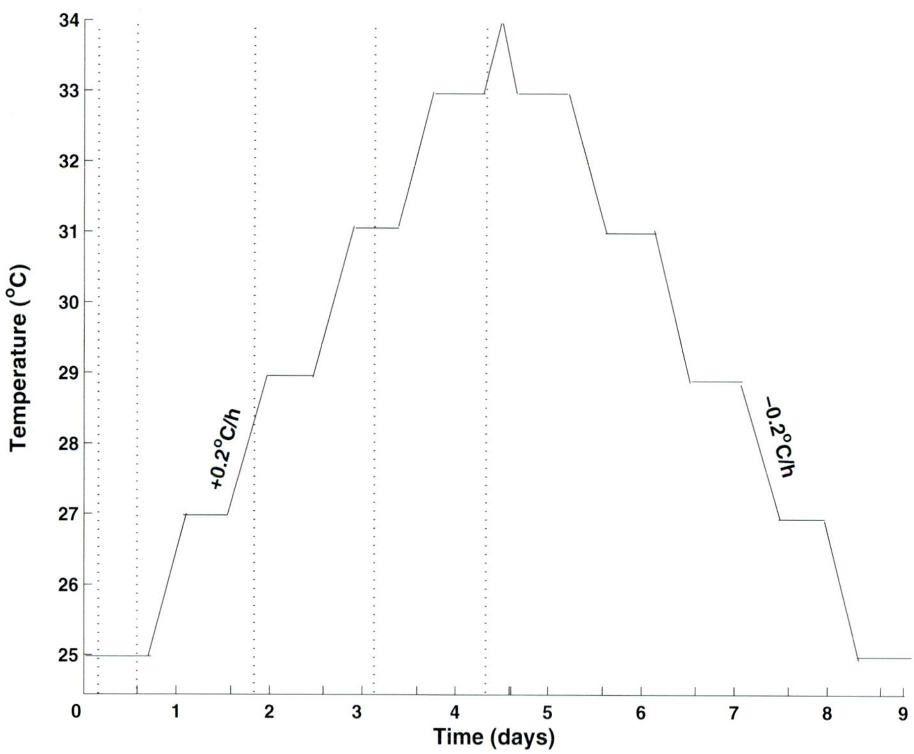

FIGURE 16.12 Evolution of the temperature during the fermentation process taken as test set. Vertical dash lines mark the control points.

At a given control time t_i, we can use the i first couples $(\mathbf{x}_c, y_c)$ to correct the model for these deviations. The most usual way for doing that is to compute a bias/slope (BS) correction. Another way is to perform an OP correction, following the DOP method described in Section 16.4.1. Figure 16.13 shows the different predictions of the degree of alcohol in superimposition to the reference one. At the beginning of the fermentation, the predictions present a negative bias, which is not due to temperature, but more probably to changes in physical properties, such

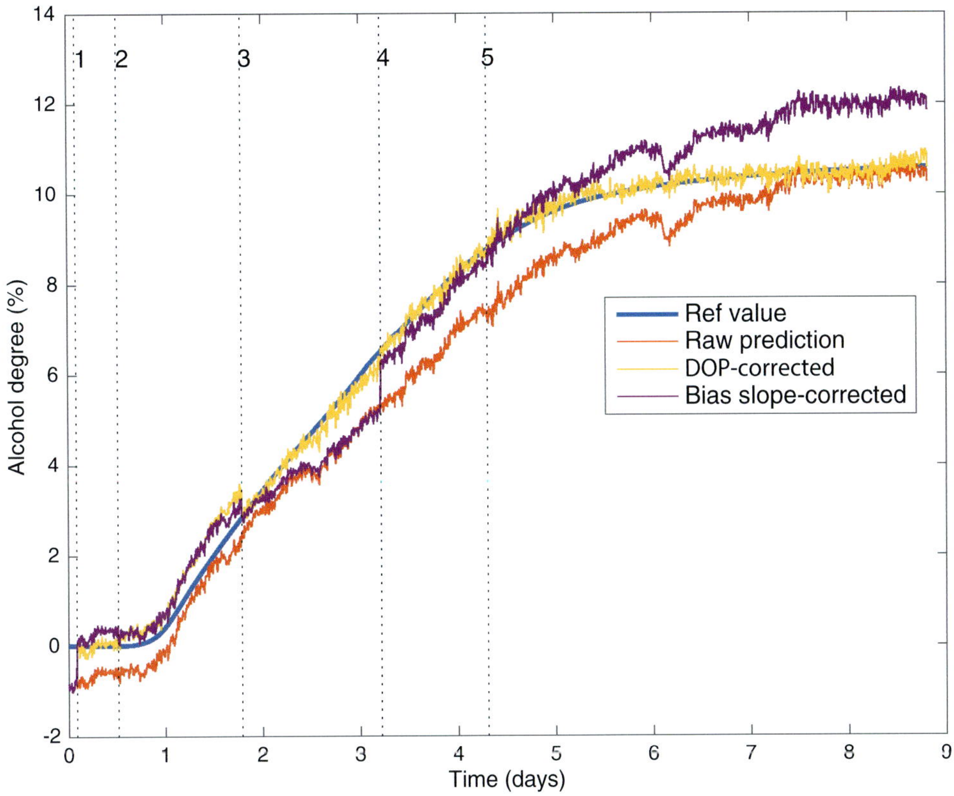

FIGURE 16.13 Evolution of the degree of alcohol: measured with reference laboratory, predicted by the PLS model, corrected by bias/slope, and corrected by DOP. Both corrections are performed using the control points marked by the vertical lines.

as turbidity. Up to the third control point, both BS and DOP corrections act similarly by correcting (imperfectly) the bias. At control point 3, both methods perfectly correct the prediction. But after this point, as the temperature continues to increase, the BS correction drifts severely as the DOP-corrected prediction remains on the reference curve. The same phenomenon can be noticed after control point 5; since the temperature decreases, the BS correction no longer acts in the correct direction, while the OP correction remains active. This is because OP has removed the subspace affected by the temperature, regardless of its level. Thus, at the end of the fermentation, when the temperature returns to the same value as the one of the calibration dataset, the raw prediction and the OP-corrected prediction are correct, while BS wrongly corrects an influence that has disappeared.

Let us examine the effect of the OP correction on the spectra. Figure 16.14 shows two spectra corresponding to two contrasted alcohol degrees (4% and 8%), in isothermal and anisothermal conditions. In Figure 16.14a, before any correction, one can see that the temperature causes a severe shift of the minimum. In Figures 16.12 and 16.13, one can notice that the anisothermal 8% spectrum corresponds to day 4, while the 4% one corresponds to day 2. Consequently, the difference of temperature for the 8% spectrum is about 8°C (25–33) and for the 4% spectrum is about 3°C (25–28). This is why the shift of the minimum is more important for the 8% spectrum than for the 4% one. In Figure 16.14b, after the OP correction, one can see that the isothermal and anisothermal spectra match perfectly. As explained in Section 16.3.2, the OP correction changes the shape of the spectra, but re-establishes the properties related to the useful information.

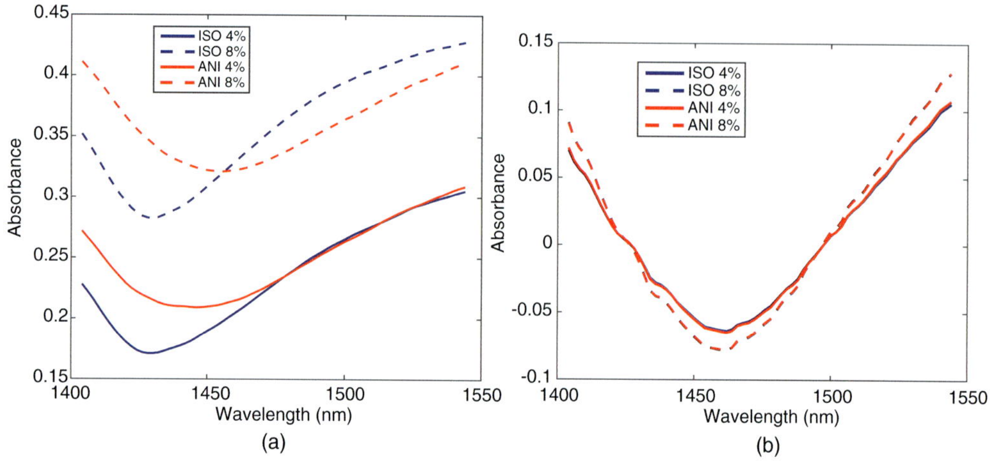

FIGURE 16.14 Isothermal and anisothermal spectra corresponding to 4% and 8% of alcohol: (a) without correction; and (b) after DOP correction.

16.6 CONCLUDING REMARKS

Orthogonal projections are useful tools to correct some effects on existing datasets which are unexpectedly confronted to new detrimental influences. This chapter has shown some practical examples on the setup of these orthogonal projections. It mainly depends on the available data. It has also shown the efficiency of orthogonal projections, in diverse situations. However, orthogonal projections cannot correct for multiplicative effects.

It must be understood that orthogonal projections are not pretreatments like the others. Even if they rely on calculations on the spectra, they make the model orthogonal to the perturbations. Thus, it is more correct to consider them as part of modeling. This property has two consequences: Firstly, there is no need to apply the orthogonal projection on the new spectra; and secondly, orthogonal projections should be made after all other pretreatments.

When several sources of detrimental information must be corrected, it is not recommended to correct each of them independently and thus to associate several projections. Indeed, the association of two orthogonal projections is generally not an orthogonal projection. It is preferable to concatenate all the detrimental information in a unique matrix **D** and to perform a global projection. This is the case when the orthogonal projections are used to maintain the robustness of a model over time, or over a network of spectrometers.

REFERENCES

A. Andrew and T. Fearn (2004). Transfer by orthogonal projection: Making near infra-red calibrations robust to between-instrument variation. *Chemom. Intell. Lab. Syst.*, 72:51–56.

B. Igne, J.M. Roger, S. Roussel, V. Bellon-Maurel, and C.R. Hurburgh (2009). Improving the transfer of near infrared prediction models by orthogonal methods. *Chemom. Intell. Lab. Syst.*, 99:57–65.

B. Jaillais, J.C. Boulet, J.M. Roger, F. Balfourier, P. Berbezy, and D. Bertrand (2012). Application of direct calibration in multivariate image analysis of heterogeneous materials. *Anal. Chim. Acta*, 734:45–53.

C.I. Chang (2003). *Hyperspectral imaging, techniques for spectral detection and classification*. Kluwer Academic / Plenum, New York.

E.K. Kemsley and H.S. Tapp (2009). OPLS filtered data can be obtained from non-orthogonalized PLS1. *J. Chemom.*, 23:263–264.

H. Martens, J.P. Nielsen, and S.B. Engelsen (2003). Light scattering and light absorbance separated by extended multiplicative signal correction. Application to near-infrared transmission analysis of powder mixtures. *Anal. Chem.*, 75:394–404.

J. Sjöblom, O. Svensson, M. Josefson, H. Kullberg, and S. Wold (1998). An evaluation of orthogonal signal correction applied to calibration transfer of near infrared spectra. *Chemom. Intell. Lab. Syst.*, 44:229–244.

J.A.F. Pierna, F. Chauchard, S. Preys, J.M. Roger, O. Galtier, V. Baeten, and P. Dardenne (2011). How to build a robust model against perturbation factors with only a few reference values: A chemometric challenge at 'Chimiométrie 2007'. *Chemom. Intell. Lab. Syst.*, 106:152–159.

J.C. Boulet and J.M. Roger (2012). Pretreatments by means of orthogonal projections. *Chemom. Intell. Lab. Syst.*, 117:61–69.

J.C. Boulet, T. Doco, and J.M. Roger (2007). Improvement of calibration models using two successive orthogonal projection methods, application to quantification of wine mannoproteins. *Chemom. Intell. Lab. Syst.*, 87:295–302.

J.C. Harsanyi and C.I. Chang (1994). Hyperspectral image classification and dimensionality reduction: An orthogonal subspace projection approach. *IEEE-Trans. Geosci. Remote Sens.*, 32(4):779–785.

J.E. Jackson (1991). *A user's guide to principal components.* Wiley & Sons.

J.M. Roger, F. Chauchard, and P. Williams (2008). Removing the block effects in calibration by means of dynamic orthogonal projection. Application to the year effect correction for wheat protein prediction. *J. Near Infrared Spectrosc.*, 16:311–315.

J.M. Roger, F. Chauchard, and V. Bellon-Maurel (2003). EPO-PLS external parameter orthogonalisation of PLS, application to temperature-independent measurement of sugar contents in fruits. *Chemom. Intell. Lab. Syst.*, 66:191–204.

L. Lebart, A. Morineau, and K.M. Warwick (1984). *Multivariate descriptive statistical analysis.* Wiley & Sons, New York.

M. Zeaiter, J.M. Roger, and V. Bellon-Maurel (2006). Dynamic orthogonal projection, a new method to maintain the on-line robustness of multivariate calibration, application to NIR-based monitoring of wine fermentations. *Chemom. Intell. Lab. Syst.*, 80:227–235.

M.B. Seasholtz and B.R. Kowalski (1993). The parsimony principle applied to multivariate calibration. *Anal. Chim. Acta*, 277:165–177.

P.W. Hansen (2001). Pre-processing method minimizing the need for reference analysis. *J. Chemom.*, 15:123–131.

R.J. Barnes, M.S. Dhanoa, and S.J. Lister (1989). Standard normal variate transformation and de-trending of near-infrared diffuse reflectance spectra. *Appl. Spectrosc.*, 43(5):772–777.

R.J. Barnes, M.S. Dhanoa, and S.J. Lister (1993). Correction to the description of standard normal variate (SNV) and de-trend (DT) transformations in practical spectroscopy with applications in food and beverage analysis, 2nd edition. *J. Near Infrared Spectrosc.*, 1:185–186.

S. Preys, J.M. Roger, and J.C. Boulet (2006). Robust calibration using orthogonal projection and experimental design, application to the correction of the light scattering effect on turbid NIR spectra. *Chemom. Intell. Lab. Syst.*, 91:28–33.

S. Wold, H. Antti, F. Lindgren, and J. Ohman (1998). Orthogonal signal correction of near infra-red spectra. *Chemom. Intell. Lab. Syst.*, 44:175–185.

T. Fearn (2000). On orthogonal signal correction. *Chemom. Intell. Lab. Syst.*, 50:47–52.

T. Verron, R. Sabatier, and R. Joffre (2004). Some theoretical properties of the O-PLS method. *J. Chemom.*, 18:62–68.

Y. Zhu, T. Fearn, D. Samuel, A. Dhar, O. Hameed, S.G. Brown, and L.B. Lovat (2008). Error removal by orthogonal subtraction (EROS): A customised pre-treatment for spectroscopic data. *J. Chemom.*, 22:130–134.

17 Calibration Transfer

*Benoît Igne**
GlaxoSmithKline

Jerome Workman, Jr.
MJH Life Sciences

CONTENTS

* igneben@gmail.com.

17.1 INTRODUCTION

Calibration transfer in near-infrared spectroscopy is a topic that has been extensively studied and documented in the scientific literature. A number of reviews are available, and readers interested in exploring the area fully should refer to them (de Noord 1994, Bouveresse and Massart 1996b, Fearn 2001, Feudale et al. 2002, Workman 2018). A number of book chapters have also discussed the subject (Dardenne 1992, Brown 2009, Workman 2015, Workman 2016). Finally, many publications compared multiple methods on the same calibration transfer problem. Some are as follows: Greensill et al. (2001) compared nine different methods for the transfer for models for melon soluble solids content; Lima and Borges (2002) compared 13 methods for the transfer of diesel oil property models; Pereira et al. (2008) compared nine methods for the prediction of naphthene and octane numbers; and Igne and Hurburgh (2008) compared methods for grain property model transfers.

The reason why calibration transfer, also referred to as standardization, is so important is that building a model and testing or validating it are just the beginning of an application. Once a method is in use, its life cycle management must be planned. Usually, a method is built on one instrument, in one location. What happens when that instrument changes (replace lamps, replace other optical parts)? When that instrument is decommissioned? When the application is moved to another location? When that model needs to be deployed as part of a network of instrument? What should be done when a company decides to use a different brand/sampling geometry/technology? Etc. Needless to say, these events should be as much as possible planned as part of the method development and method life cycle. As an example, a model should be made robust to lamp changes and potentially built on more than one instrument. However, that is not always possible and cannot always be anticipated.

The core of the problem of calibration transfer is that a same homogeneous sample scanned on different instruments will present differences in their spectra. These differences can be due to a number of factors: differences in the electronic and optical components that may meet the same specifications but present slightly different outputs; differences in the methods used to calibrate the instruments hardware (i.e., wavelength alignment); differences in signal processing (smoothing of the raw data performed at the firmware level, signal interpolation, etc.); differences in the conditions of use (embedded systems in harsh environments may provide a different signal than a benchtop system); etc.

While some of these elements are method-specific, others are dealt with at the instrument level. Standards exist to provide instrument vendors with guidelines that their instruments should meet. That is the case of the United States Pharmacopeia <1119> or the European Pharmacopoeia (Ph.Eur. 2.2.40) standards that specify criteria in linearity, noise, and wavelength accuracy when using calibrated stable standards. However, these standards are not used across the whole industry. Some instruments have internal controls to ensure the stability of the system. That is the case of Fourier transform-based systems that continuously use a laser at a particular wavelength to align the *X*-axis.

Nevertheless, once a model has been built on a particular system, historically called the master or parent unit but that will be referred to moving forward as primary unit, and needs to be transferred to one or more systems (often called slave or child units but that will be referred to as secondary units), if there is a non-acceptable difference in the predicted values for a same group of sample, something will have to be done. Note that in a similar way that a model error must be fit for purpose when compared to the error of the reference method, a transfer error must also be fit for purpose and it is necessary to understand beforehand the error that the transferred method must have when compared to the predictions on the primary unit but also against the reference method.

It is at this time important to differentiate model transfer from model update/model maintenance linked to the sample or the process. While both are part of the method life cycle management, they usually entail different approaches and the nature of the problem is different (readers should refer to Chapter 23 in this book). Method maintenance has to do with the fact that a model is no longer able to accurately predict a sample because the variability of that sample falls outside of the space from which it was built. On the other hand, method transfer deals with changing spectrometer conditions. While there is some overlap (models should be made robust to lamp changes, for instance), the techniques used in method calibration are usually addressing different fundamental questions:

- How can instruments be aligned or corrected to give the same spectra?
- How can the spectra collected on secondary units be transformed to look like the ones of the primary unit?
- How can the predictions on the secondary units be transformed to match those on the primary unit?
- How should a model be adapted so that it is able to predict a same sample with the same accuracy when collected on different units?
- How are all these issues addressed as instrumentation or calibration samples change with time.

Techniques developed to address these issues will be presented in the following sections.

17.2 INSTRUMENT ALIGNMENT AND CORRECTION

It is well understood, and widely accepted, that following the transfer of multivariate calibrations from one instrument to another, there is a bias adjustment required to the predicted results in order to conform to the reference values from a set of reference transfer samples. This will be discussed in detail in a subsequent section.

There have been many attempts to reduce this bias or intercept requirement, but the fact remains that intercept and slope are routinely used for calibration transfer from one instrument to another. There are many technical issues associated with making instruments alike enough to eliminate the bias and slope problem. The main differences between instruments that cause the slope and intercept challenges related to calibration transfer are wavelength shift (*X*-axis), photometric shift (*Y*-axis), and linewidth or lineshape (spectral resolution and shape).

17.2.1 Wavelength Axis Correction (*X*-Axis)

Wavelength repeatability and reproducibility for an individual instrument or multiple instruments are important parameters for quantifying spectrophotometer performance related to calibration transfer. If any single instrument is subject to measurement drift, the performance of calibrations developed using that instrument will begin to degrade at a rate proportional to the instrument drift rate. If one instrument is used to develop calibrations and those calibrations are transferred to one or more additional instruments, the repeatability and reproducibility of the predicted results will depend on how alike the instrument measurement performance is.

To transfer calibrations between multiple instruments over time, it is essential that the wavelength axis be stable within instruments and as nearly identical as possible across different instruments over varying measurement conditions. This is a difficult task for the basic commercial instrument designer and manufacturers in terms of making spectrometers highly precise and accurate for the wavelength axis. The requirements for engineering tolerances are high, and this equates to greater expense for mechanical, electronic, and optical components and assembly techniques and staff. In addition to building instruments that have nearly identical signals, it is essential to create and follow exact procedures to properly calibrate the wavelength axis of the spectrometer based on measurements of known reference standards (Wang, Veltkamp, and Kowalski 1991, Workman and Mark 2013).

17.2.1.1 Alignment of the Wavelength Axis

The task of selecting appropriate standards for certified peak wavelength positions is critical for improved *X*-axis (wavelength) alignment and calibration of spectrometers. Wavelength alignment technology is dependent on selecting a suitable set of such reference peak wavelengths and stable materials to align each spectrometer system using the best available measurement practices. If wavelength precision must approach ±0.05 nm and accuracy across the entire spectral region must be better than ±0.2 nm for 95% confidence interval, a series of technological improvements must be purposefully implemented.

If a primary instrument, with its own unique spectral signature, is used as a reference standard, many instruments may be made similarly to the master, but each instrument and the instruments as a whole may deviate from the true values corresponding to traceable standards. Early methods for wavelength correction involved measuring a paired reference standard with the primary instrument and aligning the newly manufactured instrument to the set of wavelength peak position values found using the primary instrument. This approach was used in the early development of near-infrared instruments in an attempt to make them more similar (Shenk, Westerhaus, and Templeton 1985, Nordqvist 1990, Shenk and Westerhaus 1991). The problem with this approach is that the entire world may have a different set of nominal wavelength values (measurands) than the primary instrument. Spectrometers change over time and conditions and wavelength alignment drifts. A common spectrometer may demonstrate inherent non-linearity in its wavelength axis, which does not allow accurate alignment throughout the complete spectral range. So, to consistently align the wavelength axis with other instrument types and even to align a single instrument over time require not only repeatability, but also accuracy indicated by agreement with reference standard peak positions (measurands). These reference peak positions are provided by a set of known physical standards (physical in the sense of primary physical lines that are universally reproducible). Such standards may be primary emission standards (i.e., atomic lines), physical standards of mixed rare earth glasses, or even organic liquids at standard pathlength and temperature. The measurands for calibrating wavelength scales would optimally be obtained using first principles whenever possible as this approach is universal and based on sound physical science and metrology, which by definition is fixed to a specific known uncertainty. The national laboratories around the globe use this approach as much as possible (Taylor and Kuyatt 1994, Springsteen 2006). There are other materials that may be used as physical standards, such as etalons. The use of etalons for wavelength alignment was successfully implemented for a process diode array spectrometer (Coates, Davidson, and McDermott 1992, Workman and Coates 1993, Tracy, Hoult, and Ganz 1994). The instrument was demonstrated to be both extremely accurate and precise.

Using a first-principles approach allows one to standardize each spectrophotometer based on known wavelength positioning at any time. First-principles wavelength alignment may be defined electronically, and thus, no physical instrument or object is required to transfer the technology required for wavelength axis calibration. Repeatable and well-characterized peak wavelength standards are required for this approach, and either emission sources, pure compounds, or solid rare earth oxide glasses with known peak wavelength positions, stability, and well-defined uncertainty may be used. Requirements are for glasses, crystalline polymers, and pure chemical solutions with an uncertainty of better than ±0.1 nm for multiple wavelengths. If a spectrometer is repeatable, but not reproducible (accurate), then database and calibration transfer becomes an issue of trying to align the wavelength axis to some unknown or poorly defined system.

Standard Reference Materials (SRMs) exhibit inherent uncertainties (Burgess and Hammond 2007). The uncertainties inherent in SRMs may be deemed by some to be too large for practical use to align modern spectrometers. As an example, National Institute of Standards and Technology (NIST) SRM 1920a has a published uncertainty of ±1 nm (Weidner, Barnes, and Eckerle 1986, SRM 1920a 2006). With large uncertainties in published wavelength peak positions, referred to as *measurands*, is it practical to use such samples for instrument alignment? The end goal is enhanced repeatability and reproducibility within and between instruments for spectral profiles.

Rare earth oxide mixtures are most often used for wavelength reference standards in near infrared (Weidner, Barnes, and Eckerle 1986). NIST SRM 1920a consists of a mixture of rare earth oxides (REOs). This SRM was reported to have stability problems related to variation in wavelength scale, and as such, it has been replaced with SRM 2036 glass (Standard Reference Material (SRM)® 2036 2013). SRM 1920a has reported wavelength registration changes based on temperature (±0.15 nm). In the certification of SRM 1920a, NIST also experienced challenges with their secondary Cary Model spectrometer and their manual method of peak selection resulting in the large reported uncertainty. In calibrating this SRM, NIST reportedly used the precise National Bureau of Standards/NIST spectrometer to measure three wavelengths (viz. 1012.9, 1260.7, and 1535.5 nm). The Cary instrument was reported to exhibit a deviation of ±1 nm during the test measurement period. For the NIST spectrometer, emission reference lamps included the use of Ne (neon) at 703.24 nm, Xe (xenon) at 881.94 and 979.97 nm, Hg (mercury) at 1013.97 and 1529.58 nm, and Kr (krypton) at 1816.73 nm. The use of such atomic line spectra for wavelength registration produces the lowest uncertainty (National Institute of Standards and Technology 2018).

NIST used 1,2,4-trichlorobenzene for the higher near-infrared wavelength region, to locate the 2152.60-nm position (Busch et al. 2000). Toluene was used for infrared locations at 3290.8, 3422.0,

3484.0, 5147.2, 5381.8, and 5549.0 nm. Carbon disulfide was used for 2224.0 nm and has been recommended for NIR wavelength calibration in early published work (Acquista and Plyler 1952, Plyler, Blaine, and Nowak 1957).

17.2.1.2 Measuring Wavelength Accuracy

The wavelength accuracy of any spectrophotometer may be computed by using a suitable reference standard. For this test, measurements of the same reference material are repeated by placing it in the sample beam and not mechanically moving the sample over a normal measurement cycle for the instrument. For example, a 30-s measurement period with 15-s reference measurement may be typical. The absorbance peak positions are computed near the reference wavelength position λ_{ref} specified. The spectrum is measured for each scan (i.e., scan-to-scan for each sample) and for the mean spectrum over the full 30-s measurement period. Then, the standard deviation of difference of the wavelength positions is computed for scan-to-scan readings within (n) replicate samples, and the mean spectrum position $\bar{\lambda}_i$ for the measured λ_{ij} versus reference (λ_{ref}) wavelength values. The standard deviation (s_i) as wavelength precision from reference value is calculated as:

$$s_i = \sqrt{\frac{\sum_{j=1}^{n} \left(\lambda_{ij} - \lambda_{\text{ref}}\right)^2}{n}} \tag{17.1}$$

The mean difference as wavelength accuracy is determined by:

$$\text{Wavelength Accuracy (as mean difference)} = \bar{\lambda}_i - \lambda_{\text{ref}} \tag{17.2}$$

where $\bar{\lambda}_i$ is the average wavelength for the scan-to-scan set and λ_{ref} is the reference wavelength position for the reference standard near the reference wavelength. The results are reported as follows: (17.1) wavelength (nm) precision from reference value and (17.2) accuracy. This may be accomplished for multiple wavelengths depending upon the standard reference material measured.

17.2.1.3 Measuring Wavelength Repeatability (Precision)

Measuring the wavelength repeatability of the spectrophotometer is accomplished using a suitable reference standard. For this test, repeat measurements of the same wavelength reference material are made. Then, the peak position for each scan is recorded and the mean spectrum wavelength position ($\bar{\lambda}_i$) is computed for all the measured spectra (λ_{ij}). The standard deviation for wavelength as precision is calculated as:

$$s_i = \sqrt{\frac{\sum_{j=1}^{n} \left(\lambda_{ij} - \bar{\lambda}_i\right)^2}{n-1}} \tag{17.3}$$

where s_i is the standard deviation for the scan-to-scan wavelength precision or repeatability for scan-to-scan measurements; λ_{ij} are individual wavelength shifts of the peak positions for sample i and scan number j; $\bar{\lambda}_i$ is the average value for the scan-to-scan set; and (n) is the number of replicate measurements (pool all scan-to-scan data).

The mean wavelength position ($\bar{\lambda}_i$) is calculated as:

$$\bar{\lambda}_i = \frac{\sum_{i}^{n} \lambda_i}{n} \tag{17.4}$$

The results are reported as wavelength (nm) precision/repeatability.

17.2.1.4 The NIST Uncertainty Number for SRMs

NIST uncertainty is a measure of uncertainty unlike the typical manufacturer report as either precision or accuracy. For example, NIST uncertainty has been calculated from Equation 17.5 (Weidner, Barnes, and Eckerle 1986):

$$U = \sqrt{A^2 + B^2 + C^2} \tag{17.5}$$

where A is twice the largest standard deviation of measurement of multiple emission lines (measurands) versus the spectrometer measured values over a period of 31 days (or some designated period); B is twice the standard deviation of the uncertainty in the peak selection method used); and C is the maximum variation in the standard due to temperature changes over a specified range and conditions. (Note: The number 2 (twice) is referred to as the k value, also known as the *coverage factor*; see Taylor and Kuyatt 1994.)

In testing SRM 1920a, the reported error for multiple wavelengths using emission lines for the NIST spectrometer was ±0.265 nm for all the data points measured over 31 days. The NIST actual reported uncertainty for SRM 1920a is equal to Equation 17.6:

$$U = \sqrt{0.57^2 + 0.41^2 + 0.15^2} = 0.72\,\text{nm} \tag{17.6}$$

This value was rounded up to ±1 nm considering the added uncertainty of the Cary spectrometer measurements and material stability over the lifetime of the SRM (Weidner et al. 1986).

A more detailed description of uncertainty computations and rationale with respect to wavelength SRMs is found in Standard Reference Material (SRM)® 2036 (2013), Taylor and Kuyatt (1994), and Choquette et al. (2005). There are significant complexities to the question of uncertainty in reference materials and so the reader is referred directly to these three citations. One may also obtain more detailed information on the history of the international metrical committees responsible for defining the technical details for uncertainty associated with measurement results; the reader is referred to Appendices C and D in Taylor and Kuyatt (1994) for historical details.

17.2.1.5 Using a NIST-Like Uncertainty Measurement

The data from a set of commercial NIR spectrophotometers exhibited a pooled standard deviation of ±1.303 nm that for a typical commercial instrument one may estimate the uncertainty with respect to wavelength accuracy (or agreement between measured wavelength peaks and measurand peak locations). Therefore, using the same uncertainty formula as NIST, a commercial instrument measured uncertainty for SRM 1920a is closer to the results shown in Equation 17.7 (Workman and Mark 2014a):

$$U = \sqrt{2.60^2 + 0.41^2 + 0.15^2} = 2.63\,\text{nm} \tag{17.7}$$

So, one must be careful when using the term uncertainty; NIST uncertainty is very well defined and does not relate to a simple measurement and standard deviation computation. Although the within-instrument wavelength repeatability or precision might be considered reasonable for commercial NIR spectrometers, the uncertainty of the wavelength registration as measured using solid reference materials and comparing those results to measurand values leaves much room for improvement (Workman and Mark 2014a).

17.2.2 Photometric Correction (*Y*-axis)

Photometric precision and accuracy, when measured with a certain protocol, are termed photometric repeatability and reproducibility. The photometric stability within a single instrument and between multiple instruments over time is important for seamless transfer of multivariate calibrations, and for the unbiased application of qualitative search libraries. In a previous chapter, we had discussed the importance of the stability in the wavelength axis of spectrophotometers and compared several leading instruments for differences (Workman and Mark 2014a). Data integrity depends on the accuracy and precision of the *X* (wavelength)- and *Y* (photometric)-axes. There are multiple factors affecting the photometric stability of spectrophotometers; these include the following: alignment and mechanical tolerances of optical elements; detector noise characteristics and amplitude, and detector linearity; detector electronics, including gain and offset settings; signal amplifier noise; and sample presentation repeatability and reproducibility (Wang, Veltkamp, and Kowalski 1991, Workman and Mark 2013).

Accurate and precise photometric axis alignment is a complex issue for commercial instrument manufacturers and design engineers. As always, the quality of manufacturing, design tolerances, quality of components, and maintenance of stringent system calibration protocols are required for photometric stability. As is the theme of this series of chapters, the appropriate use of primary and secondary reference standards and materials is also essential to superior metrical performance using spectroscopy.

17.2.2.1 Different Approaches to Alignment of the Photometric Axis

Approaches to aligning the photometric accuracy of any spectrophotometer may involve measuring a standard reference sample and correcting the instrument photometric values to those values obtained by measuring the same sample on a highly accurate and traceable spectrophotometer. This is basically a sound approach, but the question becomes: How often should the recalibration be performed, and how much difference between measured and reference standard values is acceptable before one recalibrates the photometric axis? Other questions are as follows: What values are to be used for the photometric reference standard and who is to decide what the correct reference (measurand) values should be? Are the values measured from a primary instrument to be used? Should the values measured from a typical production instrument be more appropriate to align all instruments such that production instruments are alike? This would be a viable strategy if all instruments were alike for all time. However, since this is not so, one must look to calibrate to a permanent set of references, such as pure atomic materials, metals, pure molecules, optical materials, or emission sources. If one uses these materials with known photometric values and spectral shapes, then instruments may always be corrected to a known reference standard, and this will always be true even in the future. For this physical reference approach, there is only dependency on natural laws and physics, not unstable materials and instrumentation.

A protocol to correct photometric values measured using a spectrophotometer would ideally provide high precision with universal accuracy so that expensive and time-consuming databases created over time are aligned with any instrument technology, which could be carried reproducibly into the future. In addition, using a first-principles approach allows one to standardize the data based on known photometric values at any time. First-principles photometric calibration may be defined electronically, and thus, no physical instrument or object is required to transfer the technology required for photometric adjustments or calibration.

To create a first-principles photometric calibration protocol, one requires repeatable and well-characterized photometric material standards, with known stability and with well-defined uncertainty. To have minimal effects for calibration transfer, absorbance or photometric response accuracy should have an absolute deviation of less than ±0.02%*T* (or *R*) versus a NIST traceable (or calibrated) spectrophotometer measurement at 0.09% (or *R*) (3.046 AU) to 0.10%*T* (or *R*) (3.000 AU) for specific reference standards at ~1333 nm (7500 cm^{-1}) and ~2222 nm (4500 cm^{-1}), respectively. Photometric accuracy must be ±0.01%*T* (or *R*) (±0.02 AU) in agreement with calibration (reference) instrument (absolute maximum deviation) across all wavelengths. Absorbance/response repeatability of ±<0.001 AU (1σ) would be optimal.

17.2.2.2 The NIST Uncertainty Number for SRMs

NIST uncertainty is a measure of uncertainty unlike the typical manufacturer report as either precision or accuracy. For example, NIST uncertainty has been calculated from Equation 17.8 (Weidner et al. 1986):

$$U = \sqrt{A^2 + B^2 + C^2} \tag{17.8}$$

where A is twice the largest standard deviation of measurement of multiple wavelength photometric values (measurands) versus the spectrometer measured values over a period of 1 month (or other designated period of several weeks' duration); B is twice the standard deviation of the uncertainty in the photometric data measurement method used (note that the noise on any photometric measurement adds uncertainty to the absolute photometric values); and C is the maximum variation in the reference standard due to humidity, material instability, photometric bleaching, or temperature changes over a specified range and conditions. (Note: The number 2 (twice) is referred to as the k value also known as the *coverage factor*; see Taylor and Kuyatt 1994.) As mentioned in a past chapter, one may also obtain more detailed information on the history of the international metrical committees responsible for defining the technical details for uncertainty associated with measurement results; the reader is referred to Appendices C and D in Taylor and Kuyatt (1994) for historical details.

17.2.2.3 Measurement of Photometric Accuracy as Absorbance

A comparison of the leading near-infrared spectrophotometer performance characteristics is very revealing as to the state of the art for wavelength and photometric accuracy and precision in modern instrumentation (Workman and Mark 2014a, 2014b). An identical sample of Fluorilon R99 was measured across four different instrument models and manufacturers using the same sample in the same sample holder (Avian Technologies LLC, New London, New Hampshire, USA). The identical sample was measured on a double-monochromator system carefully calibrated using NIST Standard Reference Materials and other calibration protocols to obtain the reference (or measurand) values.

The study indicated that data collected on any of the instrument models tested will result in significant photometric differences that must be compensated for when using databases collected across the various instrument models and manufacturers.

The mean difference between the measured and measurand values for photometric "accuracy" is determined by Equation 17.9:

$$\text{Photometric mean diff} = \bar{A}_i - \bar{A}_{\text{ref}} \tag{17.9}$$

where $\bar{A}_i$ is the average measured reference material photometric value for the test instrument in absorbance units (AU) over the entire 1100–2500 nm wavelength range and $\bar{A}_{\text{ref}}$ is the average measured reference material photometric values for the calibrated reference instrument in absorbance units (AU) over the entire 1100–2500 nm wavelength range. The results are reported as mean difference or "accuracy."

Table 17.1 demonstrates the average accuracy for each of the four commercial instruments as Models A through D over the wavelength range of 1100–2500 nm as compared to a reference calibrated spectrophotometer (Perkin-Elmer Lambda 9/19 UV-Vis-NIR Spectrophotometer). Note that instrument B is within a minimum specified goal of ±0.02 AU across the spectral range. This speaks well of calibrating each instrument based on a known photometric standard (Workman and Mark 2014b).

TABLE 17.1
Average Photometric Values for Commercial Instruments A through D as Compared to a Reference Spectrophotometer across a Range of 1100–2500 nm (Workman and Mark 2014b)

Instrument	Ave. Measured	Ave. Reference	Mean Difference
A	0.9950	0.9662	0.0288
B	0.9620	0.9662	–0.0042
C	0.7490	0.9662	–0.2172
D	0.8363	0.9662	–0.1300

17.2.2.4 Measuring Photometric Precision (Noise)

The RMS noise is computed for each wavelength for the three replicate measurements of Fluorilon R99. The computation is completed using Equation 17.10 as the standard deviation of the absorbance values at each wavelength for the sample set of three Fluorilon R99 spectra measured serially. In this equation, the mean spectrum photometric (absorbance) value is computed at each wavelength i as $\bar{A}_i$ and compared to the measured photometric values at each wavelength and for each of the replicates as A_{ij}. The photometric repeatability is then computed using Equation 17.10:

$$s_i = \sqrt{\frac{\sum_{j=1}^{n}\left(A_{ij} - \bar{A}_i\right)^2}{n-1}} \tag{17.10}$$

where s_i is the standard deviation for the scan-to-scan photometric values (as RMS or root mean square precision or repeatability) for the instrument under the specific measurement conditions; A_{ij} are individual photometric values (as absorbance) at wavelength position for sample i and replicate or scan number j; $\bar{A}_i$ is the average photometric (or absorbance) value for the replicate scan set; and n is the number of replicate measurements (i.e., three serial replicates of Fluorilon R99 in this case).

The mean spectrum photometric value at each wavelength position $\bar{A}_i$ is calculated using Equation 17.11:

$$\bar{A}_i = \frac{\sum_{i}^{n} A_i}{n} \tag{17.11}$$

17.2.2.5 Measuring Photometric Accuracy as Reflectance

The pooled accuracy for each instrument is determined using the standard deviation of error Rs_i between the measured and measurand (reference) reflectance values for the Fluorilon R99. We may compute the pooled standard deviation of the combined reflectance values at each wavelength for each instrument for the measured R_{ij} versus the measurand or reference R_{ref} wavelength values. Therefore, the pooled standard deviation of reflectance values for each instrument Rs_i is calculated using Equation 17.12:

$$Rs_i = \sqrt{\frac{\sum_{j=1}^{n}\left(R_{ij} - R_{\text{ref}}\right)^2}{n}} \tag{17.12}$$

17.2.3 Linewidth Correction (Bandwidth)

Instruments can be made more alike or precisely calibrated using first-principles physical standards with a pre-defined linewidth and lineshape.

TABLE 17.2
Predicted Value Results for Six Calibration Samples with Varying Linewidths

16.4nm	16.8nm	17.4nm	17.6nm	17.9nm	18.2nm
10.00	7.38	5.43	3.90	2.82	1.93
12.00	9.25	7.20	5.60	4.46	3.53
14.00	11.12	8.97	7.29	6.11	5.13
16.00	12.99	10.74	8.99	7.75	6.72
18.00	14.86	12.51	10.68	9.39	8.32
20.00	16.73	14.28	12.38	11.03	9.92

Linewidths are shown in the top row (in nm). Original reference values are given in the first column. Original calibration was completed with 16.4nm linewidth.

17.2.3.1 Changing the Linewidth or Lineshape

A changing linewidth as full width at half-maximum will have dramatic effects on calibration transfer prediction results from one instrument to another (Table 17.2). A single absorption band has been evaluated with different bandwidths (Workman and Mark 2015, 2017). In Table 17.2, the effects of linewidth broadening on prediction values for an example calibration developed at 16.4nm are shown. To match bandwidths across instruments, measurements are taken at a higher resolution (i.e., smaller linewidth) than desired, and then, a variety of smoothing techniques are applied to broaden the linewidth to a desired target value. Note that Table 17.3 demonstrates a summary of regression results for standard error of prediction (SEP), slope, and intercept following broadening in linewidth parameters.

17.3 CALIBRATION TRANSFER: COMPUTING THE CONFIDENCE LIMITS FOR EACH STATISTICAL PARAMETER

The equations for computing the confidence limits for each statistical parameter (i.e., predicted values, slope, and intercept) are given in Equations 17.13–17.17. If the transfer calibration indicates differences that are significant, there is a difference in the transferred calibration analytical performance from the primary to the secondary instrument.

17.3.1 Confidence Limits for Predicted Values from a Regression

To compute the upper and lower confidence limits for any predicted value, Equations 17.13 and 17.14 are used. First, we compute the SEC or standard error of calibration. The SEC is computed using Equation 17.13:

$$\text{SEC} = \sqrt{\frac{\sum_{i=1}^{n} (y_i - \hat{y}_i)^2}{n - k - 1}} \tag{17.13}$$

For this equation, n is the number of calibration samples; k is the number of wavelengths (or factors) used for the initial calibration; y_i is the reference laboratory results for each ith sample; and $\hat{y}_i$ is the predicted results for each ith sample.

The confidence limit for the predicted value is then computed using Equation 17.14 (ASTM E1655-05 2012):

$$\text{CL}_{\hat{Y}} = \hat{Y} \pm t.\text{SEC}.\sqrt{1 + D^2} \tag{17.14}$$

where $CL_{\hat{Y}}$ indicates the confidence limits for the predicted value based on the standard error of calibration, the t-value is for $n-k-1$ degrees of freedom, and the Mahalanobis distance (D) of the predicted sample is used. For this simulation, one may substitute 0 or 1 for the D value. The 95% confidence for a two-tailed $\alpha = 0.05$ test.

17.3.2 Confidence Limits for Slope from a Regression

The confidence limits for the slope of a regression are computed using Equations 17.15 and 17.16. First, we compute the standard error of the slope using Equation 17.15:

$$SE_{slope} = \frac{\sqrt{\frac{\sum_{i=1}^{n}(y_i - \hat{y}_i)^2}{n-2}}}{\sqrt{(x_i - \bar{x})^2}} \quad (17.15)$$

For this equation, n is the number of calibration samples; x_i is the reference laboratory results for each ith sample; and $\bar{x}$ is the mean reference value for all samples, i.e., the mean concentration value. So, y_i is the dependent variable; x_i is the independent variable; and n is the number of samples.

The confidence limit for the slope of the regression line is then computed using Equation 17.16:

$$CL_{slope} = S \pm t \cdot SE_{slope} \quad (17.16)$$

where CL_{slope} gives the confidence limits for the slope (S) value based on the standard error of the slope and the t-value is for $n-2$ degrees of freedom. The 95% confidence for a two-tailed $\alpha = 0.05$ test.

17.3.3 Confidence Limits for Intercept/Bias from a Regression

The confidence limits for the intercept or bias of a regression are computed using Equations 17.13 and 17.17. First, we compute the standard error of a validation set or alternatively the standard error of the calibration using Equation 17.13. The confidence limit for the intercept or bias value is then computed using Equation 17.17:

$$CL_{intercept} = I \pm \frac{(t \cdot SEC)}{\sqrt{n}} \quad (17.17)$$

where $CL_{intercept}$ gives the confidence limits for the intercept (I) value based on the standard error of calibration or prediction, and the t-value is for $n-2$ degrees of freedom. The 95% confidence for a two-tailed $\alpha = 0.05$ test.

17.4 ASSESSING THE EFFECT OF INSTRUMENT DIFFERENCES ON CALIBRATION TRANSFER

In practice, calibration transfer for every product and constituent calibration requires bias or slope adjustments following transfer. This phenomenon is primarily due to four factors: reference laboratory differences, the drift in product chemistry/spectroscopy requiring continuous updating of calibrations, the drift or changes in spectral characteristics from a single spectrophotometer making measurements over time, and the consistent differences in spectral characteristics between spectra measured from multiple spectrophotometer types.

Reference laboratory differences are due to a true bias in the reference chemistry measurement values caused by different analysts, different laboratories, or fundamental differences in analysis methods. The drift in product chemistry/spectroscopy is caused by changes in raw materials, manufacturing processes, or drifting ecotype expression in natural products. This product drift is best accommodated for by performing recalibration rather than bias changes. However, if the user has robust and well-characterized calibration models with consistent and audited reference laboratory results, the primary cause of bias is brought about by within- or between-instrument differences in measured spectra and their resulting prediction results will thus vary as compared to those spectra used for calibration. The key instrument factors requiring continuous biasing of prediction results following calibration transfer are wavelength registration differences, photometric offset, and linewidth or spectral shape differences.

The main issues associated with single instrument changes over time are minimal, mostly due to slight mechanical, electronic, or throughput (etendue) variations within a single instrument. These differences should be small and random for a properly designed and functional spectrophotometer. *The bias for each product and constituent prediction is caused by the relatively consistent differences between spectra measured from different instruments. The bias represents a zero-order correction in an attempt to accommodate for one or several consistent differences between instruments.* As long as the differences between instruments are relatively consistent, the bias changes will perform a very basic and functional correction for predicted values.

17.4.1 Differences between Instruments

For a single spectral band, one can illustrate a definite relationship between SEP, bias, and slope related to common changes in wavelength registration, photometric levels, and linewidth or spectral shape variation (Workman and Mark 2015, 2017). These are the usual changes in instrumentation between different serial numbers of the identical instrument design and manufacturer. One can actually establish a set of experiments for any product and constituent univariate or multivariate model to measure the relationship between the specific type of instrument variation and its effects on the prediction results. One may observe how the usual changes in instrument spectra cause significant changes in prediction results and alter the slope, bias, and SEP.

To illustrate these principles, a univariate case is described in Workman and Mark (2015, 2017). From this example, the typical differences in wavelength, photometric registration, and linewidth demonstrate specific trends in the variation of prediction results as SEP, bias, and slope following calibration transfer. In this example, constituent concentration values were between 10 and 20 units (or percent), exhibiting an initial SEP of 0.01. Using these data, one may estimate the magnitude of the variations in predicted results caused by typical instrument differences, including spectral linewidth. The analyte band absorbance for the referenced example is shown within the following data table (Table 17.3).

TABLE 17.3
Summary of Regression Results for Standard Error of Prediction (SEP), Slope, and Intercept Following Changes in Linewidth Parameters

Amount of Linewidth Change (nm)	Standard Error (SEP)	Conf. Limit (SEP) ±	Significant Difference Yes/No	Slope	Slope Sign. ±	Significant Difference Yes/No	Intercept	Intercept Sign. ±	Significant Difference Yes/No
0.00	0.01	0.04	No	0.0000	0.003	No	0.0000	0.011	No
0.40	3.62	0.04	Yes	0.9345	0.003	Yes	−1.9650	0.011	Yes
1.00	6.32	0.04	Yes	0.8856	0.003	Yes	−3.4308	0.011	Yes
1.25	8.43	0.04	Yes	0.8475	0.003	Yes	−4.5746	0.011	Yes
1.50	9.92	0.04	Yes	0.8206	0.003	Yes	−5.3830	0.011	Yes
1.80	11.15	0.04	Yes	0.7983	0.003	Yes	−6.0505	0.011	Yes

A univariate regression model is applied to a single spectral band for these comparisons.

17.4.1.1 Instrument Differences versus SEP

The SEP continues to increase as the linewidth changes relative to the calibration spectral data. One might use several methods to "correct" the prediction results, such as bias or slope, but making instruments alike with precise tracking of predicted results would require an actual bandwidth correction for the spectra.

17.4.1.2 Instrument Differences versus Bias

From the data table, one observes that a change in linewidth of +1.8 nm, where the original linewidth of spectra are changed from 16.4 to 18.2 nm, causes a bias of approximately −6.0 concentration units for a dataset averaging 15 concentration units.

17.4.1.3 Instrument Differences versus Slope

For prediction value slope change, the following results were reported in Workman and Mark (2017). Offset had no effect on slope, but wavelength and linewidth differences did have a significant effect on slope. Wavelength was reported to have a much lesser effect on slope than linewidth. Note that differences in wavelength registration of ±1.0 nm change slope minimally from 0.991 to 0.970. However, one may observe from the data table that a change in linewidth of +1.8 nm, where the original linewidth of 16.4 nm is broadened to 18.2 nm, causes a slope change of from 1.0 (no change) to approximately 0.8.

17.4.1.4 Summary of Instrument Variation on Calibration Transfer

Changes in linewidth will generally have a significant effect on calibration transfer slope, bias, and SEP values.

17.5 CALIBRATION TRANSFER BY OPTICAL STANDARDIZATION

Optical standardization methods aim to modify the spectra collected on secondary units so that a model built on a primary unit will provide an accurate prediction. Figure 17.1 shows for a sample scanned on three different instruments the magnitude of differences that can be experienced.

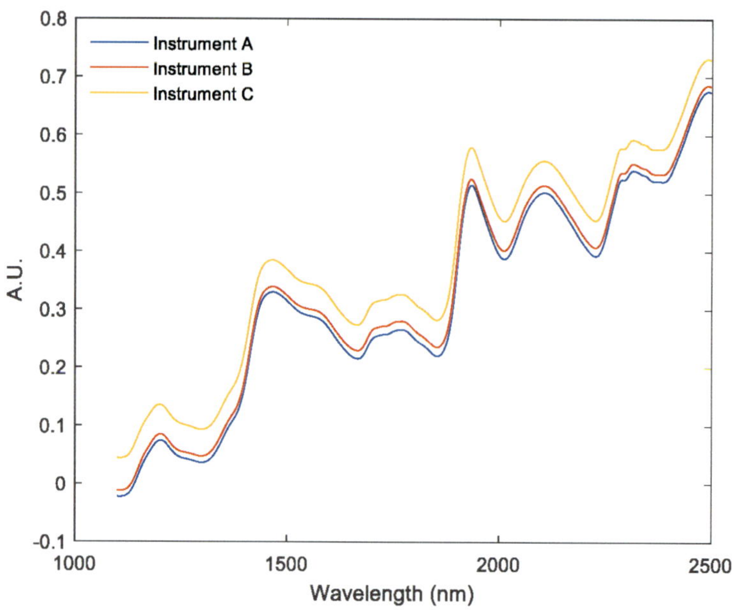

FIGURE 17.1 Spectra of a same sample on three different instruments

A large number of methods have been developed to address this problem. They can be characterized into three main categories: techniques using a transfer function to transform the spectra of the secondary units so they match the primary unit, and techniques removing the information different between the units. These methods often use a small set of samples called transfer or standardization samples.

17.5.1 Transfer Function-Based Methods

Transfer functions are mathematical manipulations used to alter spectra or calibration equations from a secondary or child instrument to have similar characteristics or predicted values to those measured on the primary or parent instrument, as spectra from the parent instrument have generally been used for calibration development.

17.5.1.1 Single-Wavelength Standardization

This method, developed by Shenk, Westerhaus, and Templeton (1985), consists of correcting, for each wavelength, the absorbance shifts. It is achieved by regressing the absorbance values of a set of samples scanned on the primary unit against the one obtained when the same samples are scanned on the secondary unit. A slope and an offset are obtained, and the secondary unit is corrected. This simplistic technique is applicable only when no shift in the wavelength axis exists. Figure 17.2a shows the relationship for 20 samples of the absorbance values of instrument A compared to instrument B at $\lambda = 1500$ nm. A slope and an offset can be obtained for each wavelength and used to correct the spectra from the secondary unit (Figure 17.2b). This algorithm assumes that no shift in the wavelength axis is present.

The same authors presented an updated version of the single-wavelength standardization that included the possibility to correct for shifts in the *X*-axis. Bouveresse, Massart, and Dardenne (1994) described the functioning of the algorithm.

It is divided into two steps. The first step is called wavelength index correction and consists in correcting the *X*-axis shifts, while the second part corrects the intensity differences. Standardization spectra are preprocessed using a first derivative treatment. This method is adapted to many standardization situations involving similar instruments. When facing instruments with

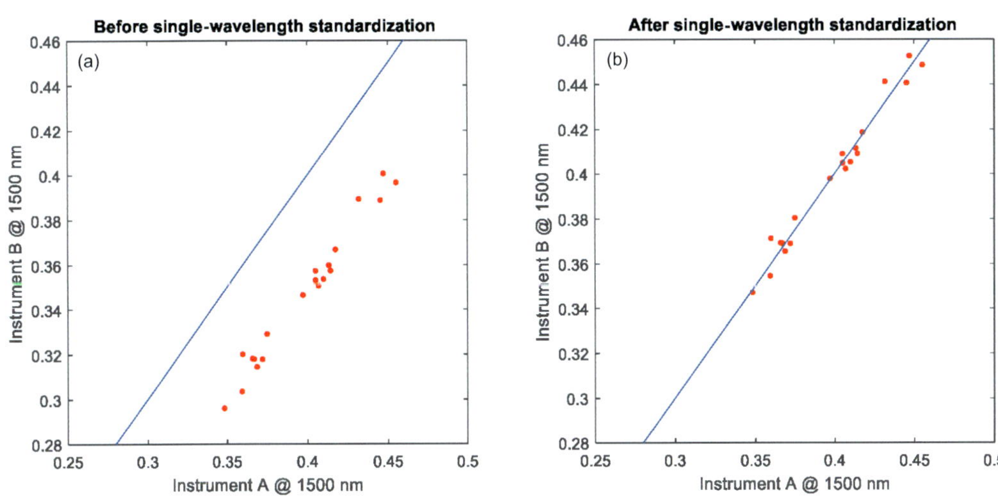

FIGURE 17.2 Spectra of instrument A, which can be approximated with the spectra of instrument B using the following equation: $B_{\text{corrected}} = B*1.0103 + 0.0479$. (a) Difference between raw spectra at 1500 nm. (b) Results of the correction.

more complex differences like a peak broadening, the patented algorithm is not sufficient because it "assumes that no relationship exists between neighboring correction models" (Feudale et al. 2002). An improvement to the algorithm was proposed by Bouveresse, Massart, and Dardenne (1995) that used locally weighted regression instead of a linear regression to correct for the intensity differences.

17.5.1.2 Direct Standardization (DS) and Piecewise Direct Standardization (PDS)

Wang, Veltkamp, and Kowalski (1991) introduced the concept of a transfer function that would modify the spectra of a spectrometer to make it match another. It is described by the following equation where R_1 represents the spectra from the primary unit and R_2 those from a secondary unit for the same samples:

$$R_1 = R_2 F \quad (17.18)$$

F is a square matrix of the dimension of the wavelength axis. The authors proposed that since there is no concentration information involved in this transformation, stable standards could be used to perform the standardization. However, most of the applications of direst standardization (DS) use a small set of samples scanned on the instruments to standardize. The F-matrix is calculated by the following equation where R_2^+ is the pseudo-inverse of the spectra from the standardization set of the primary unit:

$$F = R_2^+ R_1 \quad (17.19)$$

In direct standardization, the calculation of F assumes that the difference at each wavelength of the primary unit is due to variations across all the variables of the secondary unit(s). Because a limited number of samples are used, DS is subject to overfitting and so F is usually calculated by a singular value decomposition.

A local version of DS takes the form of relating only a subset of secondary instrument variables to each primary instrument variable. This is piecewise direct standardization (PDS). The advantage of PDS over DS is that changes between instruments may be localized over a short range of wavelengths. Thus, instead of being a square matrix, F now takes the form of a diagonal matrix whose parameters are evaluated by principal component regression (PCR) or partial least squares (PLS). The number of wavelengths of the secondary unit used to model each variable of the primary unit is usually subject to an optimization.

DS and PDS have been successfully used by a large number of authors. In early 2018, a search yielded over 150 documents discussing the topic in many fields. A few of these are cited here as examples: Wang, Lysaght, and Kowalski (1992), Wang and Kowalski (1993), Dreassi et al. (1998), Despagne, Walczak, and Massart (1998), Bergman et al. (2006), Liu, Han, and Yang (2011), and Nouri et al. (2017).

Further investigation into the performance of PDS was conducted by Bouveresse and Massart (1996a). The authors investigated the influence of the standardization samples on the performance of the transfer by PDS. They concluded that it is important to select the best samples for standardization by choosing those with the most leverage (as identified by a method such as the Kennard–Stone method; Kennard and Stone 1969). The authors also investigated how the selection of the model rank could affect the results.

While DS and PDS have mostly been used onto the spectra (raw or pretreated), some work has been done on deploying PDS onto spectra transforms. Most of that work has been with wavelet (Walczak, Bouveresse, and Massart 1997, Tan and Brown 2001). Wavelet decomposition allows the analysis of the signal at different levels of resolution. DS and PDS have been applied to these different levels, applying different transfer functions to different resolution levels.

17.5.1.3 Other Transfer Function Techniques

A number of other techniques have been used determine transfer functions in the context of calibration transfer. The following section will cover some but is far from complete. A more comprehensive review was provided by Workman (2018).

- Reverse standardization/piecewise reverse standardization

  Reverse standardization and piecewise reverse standardization are based on the same method as direct standardization and PDS. It simply transforms the spectra from the primary instrument to match the spectra of the secondary unit. The method was described by Lima and Borges (2002). The method requires that a new calibration model be developed.

  The authors showed that the reverse approach provided better results than the direct methods based on the fact that a new model was built and tested on the transformed data. That ensured that the model was best suited for the secondary unit.
- Canonical transfer

  As indicated by its name, this approach uses canonical correlation analysis to find a space into which the correlation between the primary and secondary instruments is maximized. This technique was proven to outperform PDS in situations where sufficient information about between-instrument variability is available (Fan et al. 2008).
- Finite impulse response

  The concept of finite impulse response was applied to calibration transfer by Tan, Sum, and Brown (2002). A finite impulse response is a filter that in the present situation of calibration transfer transforms the spectra of the secondary unit, so the model developed on the primary unit performs as intended. The filter, performed in $2p + 1$ spectral point window, is created by regressing the mean centered variables in each window to relate the primary and the secondary unit. The regression coefficients are then used to transform the secondary unit spectra. This technique was successfully used by Cogdill, Anderson, and Drennen (2005).
- Spectral space transformation

  Spectral space transformation uses spectral decomposition by singular value decomposition and classical least squares to perform the calibration transfer (Du et al. 2011). It starts by decomposing a matrix combining standardization samples scanned onto two different instruments, X_{comb}, using a singular value decomposition to obtain the following equation decomposing the spectral information and the noise:

  $$X_{\text{comb}} = T_S\left[P_1^T, P_2^T\right] + E \tag{17.20}$$

  where T_s are the scaled scored values and P the loadings for each instrument for the spectral information. E corresponds to the decomposition of X_{comb} for the noise. X_{comb} can also be decomposed using the classical least squares approach by:

  $$X_{\text{comb}} = C\left[S_1^T, S_2^T\right] + E \tag{17.21}$$

  where C is the concentration matrix and S the pure component signals on the primary and secondary instruments. The two equations above can be combined as follows:

  $$T_S P_1^T = C S_1^T \text{ and } T_S P_2^T = C S_2^T \tag{17.22}$$

  These can further be expanded to express the pure spectral signal for each instrument as follows:

  $$S_1^T = \left(C^T C\right)^{-1} C^T T_S P_1^T \text{ or } S_1^T = R P_1^T \tag{17.23}$$

$$S_2^T = \left(C^T C\right)^{-1} C^T T_S P_2^T \text{ or } S_1^T = RP_2^T \quad (17.24)$$

With these relationships established, the transfer function between instruments can now be derived. For a spectrum scanned on the secondary instruments, its concentration vector can be estimated by $\hat{c} = x_{\text{secondary}}\left(S_2^T\right)^+$. The difference between $x_{\text{secondary}}$ and $x_{\text{secondary Transform}}$ spectrum as if it had been scanned on the primary unit can be expressed as $\hat{c}\left(S_1^T - S_2^T\right)$. Thus, $x_{\text{secondary Transform}}$ can be calculated as follows:

$$x_{\text{secondary Transform}} = x_{\text{secondary}}\left(P_2^T\right)^+ P_1^T + x_{\text{secondary}} - x_{\text{secondary}}\left(P_2^T\right)^+ P_2^T \quad (17.25)$$

$x_{\text{secondary Transform}}$ can now be predicted by the calibration model built on the primary unit.

17.6 CALIBRATION TRANSFER BY POST-REGRESSION CORRECTION

This technique adjusts prediction values made on the secondary unit by the model developed on the primary, parent, or master unit. Post-regression correction techniques are easily implemented by every embedded software or firmware. Often called slope and bias correction, this technique is widely used among very similar instruments (one can note the misuse of the term bias that in reality corresponds to the term intercept). In this technique, a standardization set is scanned on all instruments (the secondary units as well as the primary unit) and predictions are compared to the reference values (Bouveresse et al. 1996). A linear regression is implemented, and the slope and intercept of the trend line are determined. These parameters are then used to systematically correct predictions from the secondary instrument using the prediction equation developed on a master unit. This is a very simple approach that requires only that the range of the standardization samples cover the range of the calibration samples.

A special case of slope and intercept correction is bias-only correction. Bias, as defined as the average difference between predicted and reference values, is calculated from standardization samples predictions and reference values and is added to the future predictions. This situation is suitable only when the slope is not significant. This signifies that the difference between the two units is mainly based on absorption intensities differences and not a shift in the ***X***-axis. The slope in post-regression correction appears when the wavelength alignment between the instruments is not correct.

As an example, predictions of a standardization set from a protein model developed on a primary unit and predicted from the spectra collected on a secondary unit are compared with the reference measurements. Figure 17.3 shows the relationship between the two units as well as the slope and intercept necessary to implement slope and bias correction and the bias for bias-only correction.

In this situation, using slope and bias correction, the prediction on the secondary unit should be corrected as follows:

$$\hat{X}_{\text{Cor}}^{\text{Sec}} = 0.982 \times \hat{X}_{\text{Raw}}^{\text{Sec}} + 0.595 \quad (17.26)$$

With bias-only correction, it would look like:

$$\hat{X}_{\text{Cor}}^{\text{Sec}} = \hat{X}_{\text{Raw}}^{\text{Sec}} + (-0.059) \quad (17.27)$$

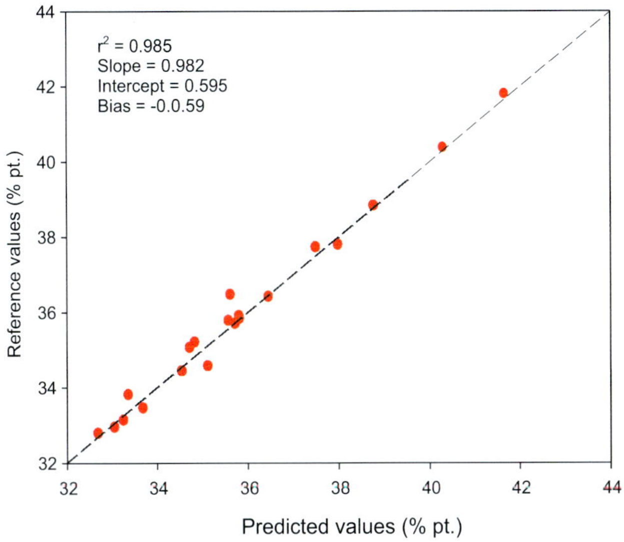

FIGURE 17.3 Post-regression correction results for protein. Slope and intercept or bias-only correction can be used to correct the measurements.

17.6.1 Bias or Slope Adjustments of Predicted Results across Primary and Secondary Instruments

A significant bias between primary and secondary NIR predicted values results mainly from instrumental differences. Other sources of significant bias changes between reference values and spectroscopy-based predicted values are due to chemical or spectral interferences. These cause significant bias in the measured analyte concentration due to the effect of another component, property of the sample, or analytical measurement anomaly (Interference Testing in Clinical Chemistry, 2005). Prediction bias as related to instrument differences is the subject of this discussion.

17.6.1.1 Bias (Means) One-Sample *t*-Test between Primary and Secondary Instruments

The concept of mean differences (or bias) requires a test of significant to determine whether a bias adjustment should be made. An appropriate statistical test will indicate whether the variation in the mean values (bias) between sample sets of predicted values is within expected random fluctuation for a normally distributed population of measurements. The larger the sample set used to test calibration transfer bias, the more accurate is the estimate of the true bias value. Thus, using 20 test samples would give a more accurate estimate of the bias than 10 samples. The smaller the standard error of the mean, the greater the confidence will be of the true bias value. The standard error of the mean is given as the SEM (Equation 17.2). If 20 rather than 10 samples are used for bias determination, a more powerful estimate of the true bias is indicated by a factor of $\frac{\sqrt{20}}{\sqrt{10}} = 1.41$. This is true only if there is no slope difference. Note that it is acceptable to make the statistically significant bias correction even if there is a meaningful slope difference.

The statistical test used to determine bias significance is a simple parametric one-sample *t*-test. For this test, we compute the average predicted mean value for a test set of samples on the primary instrument, and the set of predicted values for the same sample set on the secondary instrument following calibration transfer. In the case of NIR data, the mean NIR value is designated for the set of reference samples measured on the primary (i.e., calibration) instrument as $\bar{X}_{\text{Primary}}$ (for this *t*-test, only the mean value is known). The mean NIR value is computed for the same set of reference samples measured on the secondary (i.e., transfer) instrument as $\bar{X}_{\text{Secondary}}$. For comparison of the primary and secondary instrument biases, the test hypotheses are as H_0 (i.e., the mean values or biases are statistically the same) or H_A (i.e., the mean values or biases are statistically different). This is shown as follows:

$$H_0 = \overline{X}_{\text{Primary}} = \overline{X}_{\text{Secondary}}$$

$$H_A = \overline{X}_{\text{Primary}} \neq \overline{X}_{\text{Secondary}}$$

Next, the standard deviation for the secondary instrument ($s_{\text{Secondary}}$) is computed. The NIR predicted data on the measured set of samples are as follows:

$$s_{\text{Secondary}} = \sqrt{\frac{\sum_{i=1}^{n} (y_i - \bar{y}_i)^2}{n-1}} \tag{17.28}$$

where $\bar{y}_i$ is the mean predicted value for the sample set for the secondary instrument and y_i are the individual predicted values for the set of test samples for the secondary instrument.

Then, we compute the standard error of the mean (SEM) for the secondary instrument NIR data as follows:

$$\text{SEM} = \frac{s_{\text{Secondary}}}{\sqrt{n}} \tag{17.29}$$

These computations are used to test whether the predicted value mean is statistically the same for the primary and secondary instruments. This test determines whether the average predicted values are statistically the same for the test set used. Note that the reference values are not assumed to be known for the test set. For this t-test statistic, one is able to compute the t-test for mean bias significance as follows:

$$t = \frac{\overline{X}_{\text{Secondary}} - \overline{X}_{\text{Primary}}}{\text{SEM}} \tag{17.30}$$

If this resulting t-test is greater than the t critical value for $n-1$ degrees of freedom, then one knows the bias is significant and it should be changed. If the t-value computed is less than the critical value of t, it is not significant and should not be changed. If the bias is significant, one can test the difference between the slopes of the two lines.

17.6.1.2 Bias (Means) Two-Sample *t*-Test between Primary and Secondary Instruments

If the predicted values for the set of test samples are known for both the primary and secondary instruments, a parametric two-sample t-test may be used to compare biases. Note that an identical number of samples are used for each bias test, so the sample sizes for each test of the primary and secondary instruments are identical. It is not assumed that we have the reference values for the set of test samples. For this test, one computes the mean NIR value for the set of reference samples measured on the primary (i.e., calibration) instrument as $\overline{X}_{\text{Primary}}$. Then, the mean NIR value for the same set of reference samples measured on the secondary instrument is computed as $\overline{X}_{\text{Secondary}}$. For this test, our test hypotheses are identical to the one-sample t-test as follows:

$$H_0 = \overline{X}_{\text{Primary}} = \overline{X}_{\text{Secondary}}$$

$$H_A = \overline{X}_{\text{Primary}} \neq \overline{X}_{\text{Secondary}}$$

Then, we compute the standard deviations for both the primary (s_{Primary}) and secondary ($s_{\text{Secondary}}$) instruments. The NIR predicted data on the measured set of samples are as follows:

$$s_{\text{Primary}} = \sqrt{\frac{\sum_{i=1}^{n}\left(x_i - \bar{x}_i\right)^2}{n-1}} \tag{17.31}$$

$$s_{\text{Secondary}} = \sqrt{\frac{\sum_{i=1}^{n}\left(y_i - \bar{y}_i\right)^2}{n-1}} \tag{17.32}$$

where $\bar{x}_i$ and $\bar{y}_i$ are the mean predicted values for the sample set for each instrument and x_i and y_i are the individual predicted values for the set of test samples for each of the primary and secondary instruments, respectively.

Then, we compute the t-test for mean bias significance as follows:

$$t = \frac{\bar{X}_{\text{Secondary}} - \bar{X}_{\text{Primary}}}{\sqrt{\frac{s_{\text{Secondary}}}{n_c} + \frac{s_{\text{Primary}}}{n_p}}} \tag{17.33}$$

If this t-test value is greater than the t critical value for $n_C + n_p - 2$ degrees of freedom, one knows the bias to be significant (accept H_A); thus, the bias should be changed. If the t-value computed is less than the critical value of t, it should not be changed and is not significant (accept H_0). If the bias is significant, one knows to test the difference between the slopes of the two lines or to compare their correlation coefficients.

17.6.1.3 Slope Significance Limit Test between Primary and Secondary Instruments

If a secondary instrument is expected to perform identically to the primary instrument for predictions, following calibration transfer, one should compute the confidence limits for comparison. These limits may be computed between the primary and secondary predicted values first using criteria computed exclusively from the primary instrument. For a slope significance test, the slope change between the primary instrument-predicted values for a set of test samples over time is computed and the results retained. To accomplish this, we compute the predicted values for a set of 20 test samples measured on the primary instrument 1 week apart. Thus, we would have two sets of predicted values designated by different measurement times as $x_i$for time 0 (the reference values), and for the same set of samples measured 1 week later designated as (y_i). (Note: These test samples must be chemically stable over the 1-week period.)

For the computation of the acceptable slope confidence interval (C.I.) for the secondary instrument to be considered alike to the primary instrument, we compute three basic sets of numbers. These are (1) the set of primary predicted values at time 0 (x_i), (2) the set of primary predicted values at week 1 (y_i), and (3) the set of predicted values from the simple linear regression between these x_i and y_i values, designated as (y_i). Note that n is the number of x_i, y_i pairs (i.e., 20 for this example). From these three sets of values, we can compute the standard deviation of the residuals for the predicted values, from the regression ($\hat{y}_i = b + mx_i$), as follows:

$$s_{y/x} = \sqrt{\frac{\sum_{i=1}^{n}\left(y_i - \hat{y}_i\right)^2}{n-2}} \tag{17.34}$$

Then, we compute the standard deviation for the desired slope as follows:

$$s_m = \frac{s_{y/x}}{\sqrt{\frac{\sum_{i=1}^{n}\left(x_i - \hat{x}_i\right)^2}{n-2}}} \tag{17.35}$$

And, the confidence limit for the slope is given as:

$$\text{C.I.}_m = m \pm t \cdot s_m \tag{17.36}$$

where t is the critical value of the t-distribution, two-tailed test, at $\alpha = 0.05$ and with degrees of freedom of $n-2 = 18$. This value is 2.10 and can be found in any table of the critical values of the t-distribution.

Therefore, the test criteria for testing a secondary instrument slope following calibration transfer would be within the confidence limits of the test measured on the same primary instrument over this time interval. To complete this primary and secondary comparison, one would designate the primary instrument results as x_i and the secondary measured results as y_i. Then, the slope would be computed for the regression and must fall within the computed confidence limits for the time 0 and week 1 primary tests. This is a test of equivalence in slope between primary and secondary predicted values; note that Mark (1991) also includes such computations.

17.7 STANDARDIZATION SAMPLES

Both optical standardization and post-regression correction methods require the use of samples to estimate and correct the difference between units. Whether they use real samples or synthetic samples, care must be taken in their selection. In this section, the different types of standardization samples are presented and the main criteria for selecting them are discussed, namely the stability and representativity of the standardization samples.

17.7.1 Stability and Representativity of the Standardization Samples

Stability: The physical and chemical stability of the standardization samples is crucial for a successful calibration transfer. If changes occur in the samples, a part of the spectral differences will be due to these changes, and it will not be possible to differentiate these differences from those coming from the instruments. The resulting standardization parameters will therefore correct the instrumental differences but will also incorporate the differences due to the instability of the samples. Such standardization parameters will yield erroneous results when used for predictions in routine practice.

Representativity: Standardization samples should be as representative as possible of the samples to which the standardization parameters will be applied. If the samples are not representative, the differences seen in their use will usually not be germane to the intended use. Thus, erroneous correction factors may be derived: factors that will cause future predictions to be erroneous.

17.7.2 Approaches for Selecting Standardization Samples

Two main approaches can be distinguished: selection of a representative subset of samples from the calibration samples and the use of independent standardization samples. Selection of a representative subset consists of selecting a subset of representative samples among the numerous calibration samples collected in the calibration step and measuring the samples from this standardization subset in the prediction step. The main advantage of the standardization subset is the high representativity of the samples, provided a good subset selection method is used. Different subset selection methods have been proposed in the literature (Wang, Veltkamp, and Kowalski 1991, Kennard and Stone 1969). A comparative study (Bouveresse and Massart 1996a) showed that the smallest and most representative subset is obtained by the Kennard–Stone algorithm. One important consideration is to ensure samples in the subset cover the entire range of values.

Selection of independent standardization samples consists of measuring a set of independent samples on both instruments. These independent samples are not used for model development. Samples close to the physical and chemical structure of the analyzed samples or very different samples such as generic standards can be used. The main advantage of this approach is that not special storage conditions need to be developed.

17.7.3 Selection of the Most Suitable Approach

The selection of the most suitable approach consists of making the best compromise between representativity and stability for a given situation if both cannot be perfectly met. The use of a representative subset of calibration samples can yield representative standardization samples, but the stability of the samples may be a problem (e.g., for food products).

By using independent standardization samples, it is possible to choose samples stable enough between both calibration and prediction steps, but these samples may be less representative. This compromise can be summarized by the plot shown in Figure 17.4.

When the calibration samples are very stable or can be kept stable over time (e.g., by protecting samples from the light, temperature effects, and loss of volatile components, and by storing them under inert atmosphere), the subset selection approach should be used, because it enables one to obtain stable and representative standardization samples. This situation is ideal because there is no real compromise to be made between stability and representativity.

During the measurement of the calibration samples, a set of independent standardization samples has to be simultaneously measured to check the stability of the instrumental response. For the standardization of a single instrument over time, the stability is the most critical point since

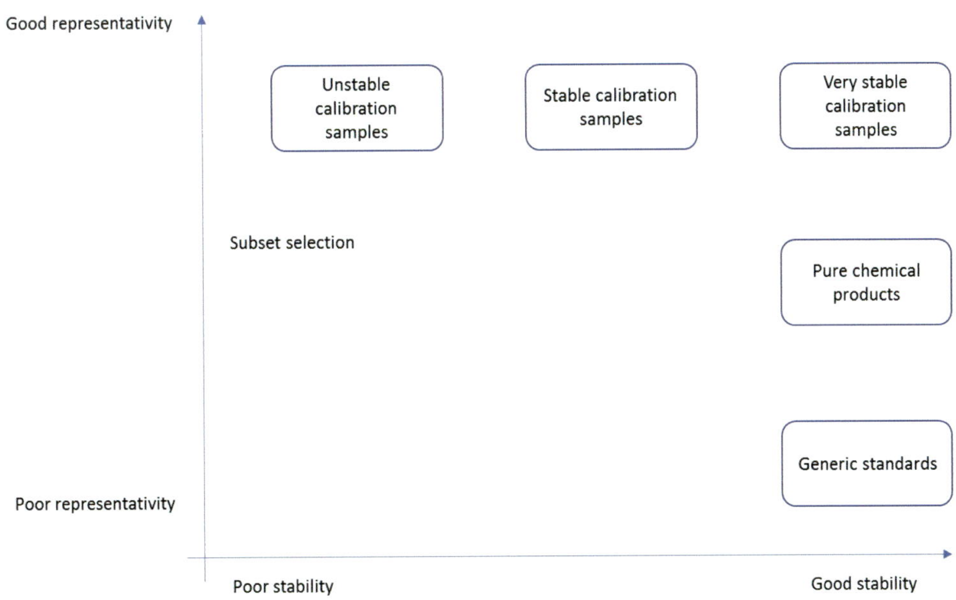

FIGURE 17.4 Stability–representativity relationship.

the same samples should be used over very long periods of time. In such a case, only perfectly stable samples can be used as standardization samples. If the analyzed samples can be kept perfectly stable, a representative subset can be selected among the calibration samples. However, if they cannot be kept stable over time, it is necessary to select independent stable standardization samples. These should be as representative of the samples to be analyzed as possible. The most stable standardization samples are generic standards such as halons (Dardenne 1992) and polystyrene (Wang and Kowalski 1992). However, these samples are usually so different from the analyzed samples that the obtained results may not be satisfactory. It has been suggested that the use of these standards should be limited to the monitoring of instrumental parameters over time (Bouveresse et al. 1997, Bouveresse, Casolino, and Massart 1998). A more suitable way of obtaining a stable and representative standardization set is to use a set of mixtures containing well-chosen pure chemicals similar to the analyzed samples (Bouveresse, Massart, and Dardenne 1995). However, even this may lead to error, as slight, unknown amounts of impurities can differ between bottles of the same production lot. One such example is the presence of particulates in containers used to store the chemicals. Some work has been done with virtual standards by which are mathematically derived from pure materials (Cooper, Larkin, and Abdelkader 2011, da Silva et al. 2017).

It is important to note that time and geography dictate the criticality of the stability of the samples. When NIR instruments must be standardized, the standardization samples only need to be stable between the moment at which they are measured on the primary instrument and the moment at which they are measured on the secondary instrument(s). This period may not exceed a few minutes if instruments are located at the same place, or may take several days or weeks if both instruments are far from each other. Therefore, stability requirements are paramount to the practical needs.

17.7.4 Number of Standardization Samples

A sufficient number of samples must be included in the standardization set in order to contain enough information concerning the instrumental differences between both calibration and prediction steps. Too few samples will yield an incomplete standardization. Too many samples represent a loss of time. There are no absolute rules to decide how many samples should be used, but this number is strongly conditioned by two factors, namely the standardization method used and the complexity of the spectral differences. The choice of how many standardization samples often depends on the standardization method and complexity of the spectra. A discussion of this point was provided by Bouveresse and Massart (1996a).

17.8 CALIBRATION TRANSFER BY GLOBAL MODELING

Various methods have been proposed to produce the universal model or a calibration that is mostly robust against standard instrument differences or changes with time as are common to commercial instruments of today. These have been referred to as robust models, or global models.

For computing a robust model, various experimental designs have been constructed to better represent the product, reference values, and instrument calibration space and to include typical changes and interferents that should be included within the model for it to be broadly applicable. Using this approach, one might design a factorial experiment for the composition of the calibration set to include multiple variations typically encountered during routine analysis. A list of some of these variations may consist of differences in sample pathlength, sample cup, sample temperature, sample moisture content, flow rate, particle size, interferent content, instrument type, constituent ratios, sampling parameters, and others (Abookasis and Workman 2011). These approaches will work for a period of time until the instrument drifts or the product or constituent chemistry changes. These types of changes are expected, and thus, routine recalibration (i.e., model updating) would be required as a standard procedure if the chemistry of instrument changes is considered significant.

A method for selection of specific robust wavelengths in MLR models that are more forgiving toward wavelength differences in interference filter-based instruments has been demonstrated to be effective (Mark and Workman 1988).

17.8.1 Calibration Augmentation over Time

Often called model update (MU) in the calibration transfer literature, it consists in adding to the calibration set samples representing new sources of variability. This approach will often include samples from the secondary unit(s) to ensure that the model will perform accurately once transferred. This technique is often successful, but there often exists a trade-off between robustness and model accuracy. As the calibration set becomes too large and starts encompassing different subpopulations, its performance may start degrading on some instruments or for some sources of variability. In these situations, it is necessary to revisit the membership of the calibration set.

This in effect uses the calibration transfer as a temporary solution or bridge to building an accurate model on the secondary instrument. Such a practice is often used when the instrument differences are too significant for direct and accurate calibration transfer.

A number of approaches have been investigated to efficiently update models in situations of calibration transfer. Methods not requiring the need for reference values were introduced. The use of spectral differences for a same sample without the need for reference values was proposed by Westerhaus (1991). Samples without the material of interest (placebo samples) were used by Kramer and Small (2007). As stated before, virtual standards, stable generic standards, and stable chemical standards were investigated by Cooper, Larkin, and Abdelkader (2011), da Silva et al. (2017), Wang and Kowalski (1992), Dardenne (1992), and Bouveresse, Massart, and Dardenne (1995).

Nevertheless, as more and more variability is included in models, an optimization of the spectral range and preprocessing methods may be necessary. A number of authors have proposed using spectral preprocessing methods to remove instrument variability. This includes multiplicative signal correction-based methods (Kramer, Morris, and Rose-Pehrsson 2008) and corrections in the frequency domain (Igne and Hurburgh 2010b).

17.8.2 Sample Selection

In situations where model performance is affected by the large spread of variability brought by samples coming from a large range of variance, an optimization of the sample membership may be necessary. There are a number of ways to optimize the samples: random subset selection to reduce the number of samples for all sources of variability, manual subset selection when it is possible to separate groups of samples, selection by highest leverage using the Kennard–Stone algorithm, etc.

Sample selection can also be achieved as part of the regression step through the use of local regression methods. In a local regression, a large pool of calibration samples is created and only the most relevant are selected for building the calibration. This can be particularly useful when subpopulations exist in the calibration set. There are a number of approaches to local regressions: for example, locally weighted regression (Cleveland and Devlin 1988, Naes, Isaksson, and Kowalski 1990), LOCAL (Shenk, Westerhaus, and Berzaghi 1997), and CARNAC (Davies et al. 1988, Davies and Fearn 2006); but they all work on the same principle. The sample to be predicted is compared to all the samples in the calibration set, and the "closest" samples are selected. A model is then built on these samples and used to predict the sample of interest. The various methods differ based on how the closest samples are selected. The application of local methods to calibration transfer was investigated by Igne and Hurburgh (2010a).

17.8.3 Tikhonov Regularization

Kalivas et al. (2009) introduced the use of the Tikhonov regularization in the context of multivariate model maintenance and transfer. The method belongs to the family of optimization methods such as LASSO regression and is based on identifying the regression coefficients b, in the context of spectroscopic data:

$$\min\left[\|Xb - y\|_2^2 + \lambda^2\|Lb - y_L\|_2^2\right] \tag{17.37}$$

where $\boldsymbol{X}$ are the calibration data, y are the target data, $\frac{2}{2}$ represents vector 2-norms, L is a regularization parameter, λ is a weighting factor for the second term, and y and y_L are the target values for X and L, respectively. In that context, the regression coefficients can be calculated as follows:

$$\hat{b} = \left(X^T X + \lambda^2 L^T L\right)^{-1}\left(X^T y + \lambda^2 L^T y_L\right) \tag{17.38}$$

In the situation of calibration transfer, $\boldsymbol{L}$ represents difference spectra from the secondary instrument(s) and y_L is 0. Thus, the equation above can be modified as follows:

$$\min\left[\|Xb - y\|_2^2 + \lambda^2\|Lb\|_2^2\right] \tag{17.39}$$

$$\hat{b} = \left(X^T X + \lambda^2 L^T L\right)^{-1}\left(X^T y\right) \tag{17.40}$$

Kalivas et al. (2009) provided further stabilization steps to the optimization equations above related to the issues generated by the high collinearity of the spectral data during matrix inversion. Readers should refer to the article for more details.

This optimization method aims to desensitize the regression vectors to the spectral artifacts brought by the new sources of variability as well as simultaneously minimizing the least squares error and reducing the norm of the regression vector to limit overfitting.

17.9 ORTHOGONALIZATION-BASED METHODS

These methods are based on the statistical theory that the column space of the $\boldsymbol{X}$-matrix (set of all possible linear combination of column vectors) is "the sum of two subspaces, among which only one contains information useful for the model "(Roger, Chauchard, and Bellon-Maurel 2003). Thus, by performing the appropriate projection, one can develop a model based only on the "adequate" $\boldsymbol{X}$-matrix for the model of interest. There exist two ways of estimating the "parasitic" subspace. The first consists of finding the space orthogonal to y. This is what orthogonal signal projection (OSC) and orthogonal projections to latent structures (O-PLS) are doing. In the context of calibration transfer, OSC has successfully been used for calibration transfer. The second approach estimates the space in which the external factor occurs and includes techniques such as external parameter orthogonalization (EPO) (Roger, Chauchard, and Bellon-Maurel 2003), independent interference reduction (IIR) (Hansen 2000), transfer by orthogonal projection (TOP) (Andrew and Fearn 2004), dynamic orthogonal projection (DOP) (Zeaiter, Roger, and Bellon-Maurel 2006), and error removal by orthogonal subtraction (EROS) (Zhu et al. 2008).

Orthogonal signal correction, DOP, TOP, and EROS will be discussed here. The subtraction of the influential subspace from the original X matrix (X_0) is performed by

$$X_0^* = X_0 - C \tag{17.41}$$

where $\boldsymbol{C}$ is a correction matrix. X_0^* is then used to develop robust models. The difference between the various orthogonal techniques is based on how $\boldsymbol{C}$ is calculated.

17.9.1 Orthogonal Signal Correction (OSC)

OSC was introduced by Wold et al. (1998). The algorithm decomposes X into scores using the NIPALS method. These scores are orthogonalized to **y** to obtain t^*. A PLS is then used to make Xw as close to t^*. NIPALS is the used on X to provide an updated t. That score vector is then orthogonalized to y. After convergence, the loading vectors corresponding to each round of orthogonalization can be calculated to form the correction matrix $C = tp^T$. The operation can be repeated until the desired number of corrections has been applied to X_0.

In the context of calibration transfer, the calibration set should contain spectra collected on the different instruments and the number of OSC factors to remove is optimized so that the performance of the model is similar on the various spectrometers involved.

17.9.2 Transfer by Orthogonal Projection (TOP)

Andrew and Fearn (2004) published an orthogonal technique specifically aimed at removing the interfering components arising from the difference in spectrometers from an $\boldsymbol{X}$-matrix. Standardization samples are measured on both primary and secondary units. A difference spectrum is calculated by subtracting the spectra from both units, and a PCA is performed on the difference matrix $\boldsymbol{D}$. The first few loadings of $\boldsymbol{D}$ are then used to form the matrix $\boldsymbol{P}$ representing the direction of main variation between units. $\boldsymbol{P}$ is orthogonalized on the $\boldsymbol{X}$-matrix to obtain a corrected $\boldsymbol{X}$-matrix ($\boldsymbol{X}_{\mathbf{corr}}$) for between-instrument differences using the following equation:

$$X_{\text{corr}} = X\left(I - PP^T\right) \tag{17.42}$$

where $\boldsymbol{I}$ is the identity matrix. $\boldsymbol{X}_{\mathbf{corr}}$ is then used to develop calibration models.

TOP requires the use of standardization samples. But in certain situations, collection of samples is not possible and orthogonalization is performed using the DOP method.

17.9.3 Dynamic Orthogonal Projection (DOP)

In DOP (Zeaiter, Roger, and Bellon-Maurel 2006), standardization samples are replaced by virtual standards. These standards are chosen to represent a variation to remove. In calibration transfer situations, standardization samples collected from the secondary instruments are used to create virtual standards using a kernel function based on the calibration set (X_0 and y_0 matrices) and the reference value of samples collected on secondary unit (Y_t matrix). Virtual standards are created as per the following equation:

$$\hat{X}_t = AX_t \text{ with } a_{ij} = F_{y_{ti}}\left(y_{0j}\right) \tag{17.43}$$

where $\hat{X}_t$ are the virtual standards and $F_{y_{ti}}$ is a Gaussian kernel function centered on y_{ti} for the ith sample and the jth variable. A difference matrix $\boldsymbol{D}$ is calculated ($D = \hat{X}_t - X_t$), and similarly to TOP, the first few loadings from a PCA are selected to form a loading matrix $\boldsymbol{P}$. The $\boldsymbol{X}$-matrix is then orthogonalized similarly as in TOP to obtain an $\boldsymbol{X}_{\mathbf{corr}}$ matrix that can be used for model development.

By not requiring the same samples run on both master and secondary units, DOP is more flexible, especially for online applications.

17.9.4 Error Removal by Orthogonal Subtraction

EROS (Zhu et al. 2008) is based on the same principles as TOP and DOP. The difference arises from the way the ***P*** matrix is calculated. In EROS, the difference matrix ***D*** is mean-centered and a new matrix ***W*** representing the difference between measurements is calculated as follows:

$$W = \sum_{i=1}^{m} \frac{D_i D_i^T}{(r-m)} \tag{17.44}$$

where ***r*** is the total number of spectra in the ***D*** matrices and m is the number of samples (also the number of ***D*** matrices). ***W*** represents the pooled within-sample covariance matrix of the spectra. The first few PCA loadings of ***W*** are selected and used to form the matrix ***P***, used to orthogonalize the ***X***-matrix.

Orthogonal methods need to be tuned. The number of times orthogonal components are removed for OSC needs to be optimized. Similarly, TOP, DOP, and EROS require selecting the right number of latent variables to remove from ***X***. Refer to Woody et al. (2004) and Igne et al. (2009) for examples of the use of these methods in the context of calibration transfer.

17.10 PERFORMANCE EVALUATION POST-CALIBRATION TRANSFER: WHAT TO COMPARE WHEN TRANSFERRING CALIBRATIONS?

For spectroscopic-based measurements using multivariate calibrations, one may compare the standard or reference concentrations for a set of samples to the spectroscopic-based predicted values. One may also compare the response of the primary instrument to that of the secondary instrument. In making these comparisons, one may perform statistical tests for bias, correlation, and slope. A statistically significant difference in bias should result in a change of the bias. A statistically significant result in correlation or slope should result in a basic multivariate recalibration, unless one can demonstrate that the differences between the compared values have some real slope variation between them due to fundamental scientific principles.

17.10.1 Comparing Instrument Predictions Following Calibration Transfer

The main question when comparing primary to secondary instrument predictions, or a reference laboratory method to an instrument prediction, or results from two completely different reference methods, is how to know if the differences are meaningful or significant and when they are not. There is always some difference expected, since an imperfect world allows for a certain amount of "natural" and expected statistical (or random) variation. However, when are those differences considered statistically significant differences, or when are the differences too great to be acceptable? Although many methods for analysis comparisons are proposed, one of the most powerful sets of tests for comparison of analytical methods is found in Youden and Steiner (1975). This reference describes details regarding collaborative laboratory tests, ranking of laboratories for accuracy, outlier determination, ruggedness tests for methods, and diagnosing the various types of errors in analytical results.

17.10.2 Bland–Altman Method for "Limits of Agreement"

One of the definitive papers used for analytical method comparisons within the highly regulated clinical sciences is named the Bland–Altman method. The Bland–Altman method comparison is the standard for clinical data where there are stringent requirements for accurate analytical results. There is an entire series of publications by these statisticians, and this is still taught and used for clinical analysis criteria.

The paper describing this method is entitled "Statistical Methods for Assessing Agreement Between Two Methods of Clinical Measurement," by J. Martin Bland and Douglas G. Altman (Bland and Altman 1986). This paper describes the errors often made when comparing methods. The authors summarize the contents of this paper as follows:

> In clinical measurement comparison of a new measurement technique with an established one is often needed to see whether they agree sufficiently for the new to replace the old. Such investigations are often analyzed inappropriately, notably by using correlation coefficients. The use of correlation is misleading. An alternative approach, based on graphical techniques and simple calculations, is described....

When analytical results are compared following calibration, one attempts to assess the degree of agreement between them. Bland and Altman discount completely the use of correlation as a useful parameter to assess analytical agreement. A summary of their arguments is given in the following brief discussion.

In using Bland–Altman as a method comparison, two single measurements are taken for each sample (1 and 2) on each instrument (A and B), given as an example as A1 and A2 and B1 and B2. The first measurement is used for comparison, and the second (replicate) is used for a repeatability computation. For this method of comparison, a line-of-equality plot is made to compare two methods or two instruments. The various *X*,*Y* data points are plotted against a perfect straight line of equality. The authors make the point that correlation (*r*) measures the strength of the relationship between two variables, but it does not measure the agreement between them (Figure 17.5). Perfect agreement is indicated by the data points lying directly on the line of equality. A perfect correlation is indicated if the points lie along any straight line. The authors emphasize that (1) correlation indicates the strength of a relationship between variables – not that the analytical results agree; (2) a change in scale does not affect correlation, but drastically affects agreement; (3) correlation depends upon the range of the true quantity (analyte) in the sample; (4) tests of significance are mostly irrelevant between two similar analytical methods; and (5) data in poor agreement analytically can be highly correlated (Figure 17.6). This figure shows three analytical sets with perfect correlation but poor agreement.

A Bland–Altman plot (Figure 17.7), extremely familiar to clinical analysts, demonstrates a good visual comparison technique to evaluate the agreement between two methods or instruments. The *X*-axis (abscissa) for each sample is represented by the average value for each sample obtained from the comparison results (using two methods or two instruments). The *Y*-axis (ordinate) for each

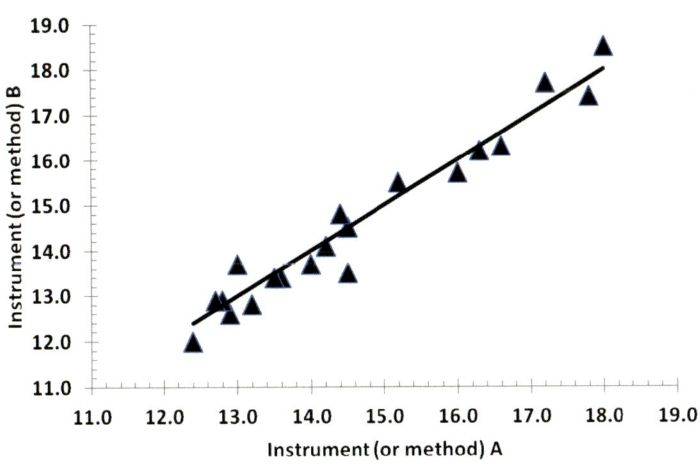

FIGURE 17.5 Instrument or method A1 (*X*-axis) as compared to instrument or method B1 (*Y*-axis), with data points compared to perfect line of equality.

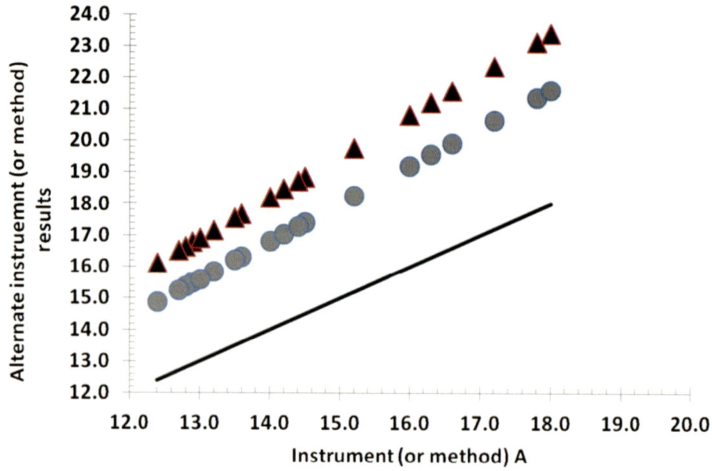

FIGURE 17.6 Instrument or method A1 (*X*-axis) as compared to instruments (or methods) C or D (*Y*-axis), showing line of equality for perfect agreement as well as other perfect correlations that are not in analytical agreement. This indicates correlation as an imperfect representation of agreement between methods.

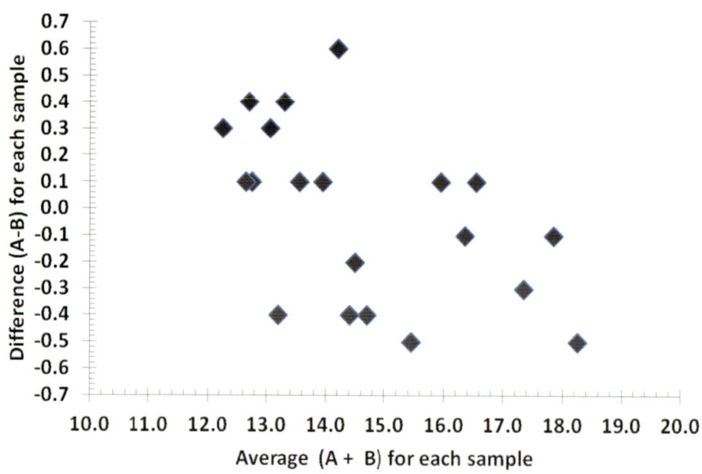

FIGURE 17.7 The Bland–Altman plot indicating the difference plotted against the mean for each sample for instruments (or methods) A1 and B1.

sample is represented by the difference between one method and the second method (or instruments A and B in our example) for each sample. Such a plot uses the mean and plus or minus two standard deviations as the upper and lower comparison thresholds.

To assess whether the data are in close enough agreement for analytical equivalence purposes between A1 and B1, the bias or mean difference $(\bar{d})$, the standard deviation of the differences (*s* or SD), and the expected "limits of agreement" are computed. The comparison test is expressed as $\hat{d} \pm 2 \cdot s$ for a 95% confidence level.

The mean difference is computed as the average of all the differences between the comparative instruments, for each sample, as follows:

$$\bar{d}_i = \frac{\sum_{i=1}^{n} (A_i - B_i)}{n} \tag{17.45}$$

The standard deviation for this comparison for the set of samples is computed as follows:

$$s_i = \sqrt{\frac{\sum_{i=1}^{n} (A_i - B_i)^2}{2n}} = \sqrt{\frac{\sum_{i=1}^{n} D^2}{2n}} \tag{17.46}$$

The following results are computed for each comparison.

Bias is $\bar{d}$, standard deviation for all samples is s, and the 95% confidence level is $(\bar{d} + 2s)$ and $(\bar{d} - 2s)$. If this number is considered too large for a 95% confidence of the result agreement, then these methods are not considered equivalent. On the other hand, if these limits of agreement are acceptable for the application where they are used, this is acceptable. In a clinical situation, a physician determines the level of accuracy or agreement required for critical intervention decision making; this would be analogous to a process control supervisor or analytical scientist assessing the acceptable level of agreement between comparative methods in order to use the alternate method or instrument as a substitute.

17.10.3 Conclusion

This classic and well-accepted method has been successfully applied to analytical results for research, commercial, and clinical laboratory analyses.

17.11 HARDWARE AND STANDARD REFERENCE MATERIAL CONSIDERATIONS: INSTRUMENT COMPARISON AND EVALUATION METHODS

One of the essential aspects for determining the efficacy and quality of calibration transfer is to make certain the spectrometer instrumentation is essentially the same, or as similar as possible, prior to the calibration transfer experiment. There are many standard tests that are used to determine alikeness between spectrophotometer instruments, and several of those tests using standard or calibrated reference materials are described in Chapter 9. The tests described are most applicable to the near-infrared region, although the tests and standards may be varied to become applicable to spectrometers using different wavelength or frequency ranges. Eight basic tests and a summary evaluation are generally used as a standard method to determine instrument measurement performance, including wavelength accuracy, wave number repeatability, absorbance/response accuracy, absorbance/response repeatability, photometric linearity, photometric noise, signal-averaging (noise) tests, and the instrument lineshape (ILS) tests. If carefully conducted, these experiments provide specific information for diagnosing mechanical, optical, and electronic variations associated with design or manufacturing limitations. They provide objective data for correcting and refining instrumentation quality and alikeness.

17.11.1 Instrument Optical Quality Performance Tests

The following series of tests is used to qualify instrument performance and to determine which issues are problematic due to deficiencies in instrument design features. These tests are related to alikeness in measurement performance between instruments and to overall accuracy, and precision (as repeatability and reproducibility). Note that the terms optical density (O.D.) and absorbance units (AU) are synonyms. See Chapter 9 for details of the following tests, which utilize standard or calibrated reference materials to perform each test:

1. Wave number accuracy test.
2. Wave number repeatability test.
3. Absorbance/response accuracy test.

4. Absorbance/response repeatability test.
5. Photometric linearity test.
6. Photometric noise test.
7. Signal-averaging test.
   a. Random noise test.
   b. Noise test (including medium- or short-term drift).
   c. Noise test (including long-term drift).
   d. Signal-averaging test.
8. Instrument lineshape (ILS) test.
   a. Center wavelength measured
   b. Full-width half-max (FWHM) measured (% deviation)
   c. Asymmetry measured (% deviation)

17.11.2 Summary Specifications for Instrument "Alikeness" Testing

General specifications for comparing instruments should meet basic minimum pre-established criteria. These criteria are based on the actual use of the spectrophotometer and the accuracy required for measurements for a high signal-to-noise application. General requirements for each spectrometer should depend on the use, application, and overall performance requirements. The specifications in Table 17.4 are used for illustrative purposes only. If extremely high performance and "alikeness" are required, then comparison statistics are more important. Here are some basic

TABLE 17.4
General Performance of Spectrophotometers Required for Optimum Calibration Transfer

1. Wave number accuracy	Absolute deviation must be less than ±0.8 cm^{-1} versus NIST reference following calibration using NIST traceable reference measurements for 5447.42 and 4333.51 cm^{-1} bands of crystalline polystyrene (1.0 mm thickness). Accuracy must be ±0.1 cm–1 in agreement with calibration (reference) as absolute maximum deviation.
2. Wave number repeatability	< 0.01 cm^{-1} (1σ)
3. Absorbance/response accuracy	Absolute deviation must be less than ±0.02%*T* versus NIST traceable measurement of 0.09%*T* (3.046 AU) to 0.10%*T* (3.000 AU) for specific reference standards at ~1333 nm (7500 cm^{-1}) and ~2222 nm (4500 cm^{-1}), respectively. Accuracy must be ±0.01%*T* (±0.02 AU) in agreement with calibration (reference) primary instrument (absolute maximum deviation)
4. Absorbance/response repeatability	<0.001 AU (1σ)
5. Photometric linearity	Slope: 1.00±0.02; intercept: ±0.02 (absolute) for method
6. Photometric noise	RMS <0.001 AU for 0.1%*T* ND filter
7. Signal averaging test	
7.1. Random noise test	Must pass method to 1024 scans
7.2 Noise test (including medium- or short-term drift)	Must pass method to 500 scans
7.3 Noise test (including long-term drift)	Must pass method to 480 scans
8. Instrument lineshape (ILS) test	
8.1. Center wavelength measured	Accuracy must be ±0.05 cm^{-1} in agreement with calibration (reference) line using interpolation or center of mass peak picking method
8.2. FWHM measured (% deviation)	<5% deviation from calibration (reference) primary instrument (absolute maximum deviation)
8.3. Asymmetry measured (% deviation)	<5% deviation as per prescribed method from calibration (reference) primary instrument (absolute maximum deviation)

performance criteria that indicate a reasonable alikeness between two or more instruments. Specific calibrations and prediction performance criteria will vary and so then will the comparative instrument performance metrics.

17.11.3 Conclusions

This chapter reviewed the most popular methods to perform calibration transfer. It is far from being exhaustive, and readers should refer to the various reviews presented in the introduction for additional details. Whether performed at the instrument level, spectral level, model level, or prediction level, calibration transfer remains a challenging activity, as complex as the method development level. It should be as carefully designed and tested as a calibration model and routine performance checks after calibration transfer should be applied. Nevertheless, as instrument manufacturers standardize their equipment and calibration protocols, it is expected that models should be more and more easily transferable over large networks of instruments.

ACKNOWLEDGMENT

The authors acknowledge the contributions from the following source, which comprises a portion of this revised and updated chapter: "Bouveresse and Campbell (2008)."

REFERENCES

Abookasis, D., and J. Workman. 2011. "Direct measurements of blood glucose concentration in the presence of saccharide interferences using slope and bias orthogonal signal correction and Fourier transform near-infrared spectroscopy." *J Biomed Opt* 16 (2)

Acquista, N., and E. K. Plyler. 1952. "Calibrating wavelengths in the region from 0.6 to 2.6 microns." *Journal of Research of the National Bureau of Standards (United States)* 49 (1):4.

Andrew, A., and T. Fearn. 2004. "Transfer by orthogonal projection: Making near-infrared calibrations robust to between-instrument variation." *Chemometrics and Intelligent Laboratory Systems* 72 (1):51–56. doi: 10.1016/j.chemolab.2004.02.004.

ASTM International. 2012. "Standard Practice for Infrared Multivariate Quantitative Analysis (ASTM E1655-05)." West Conshohocken, PA.

Bergman, E. L., H. Brage, M. Josefson, O. Svensson, and A. Sparen. 2006. "Transfer of NIR calibrations for pharmaceutical formulations between different instruments." *J Pharm Biomed Anal* 41 (1):89–98. doi: 10.1016/j.jpba.2005.10.042.

Bland, M., and D. Altman. 1986. "Statistical methods for assessing agreement between two methods of clinical measurement." *The Lancet* 327 (8476):307–310. doi: 10.1016/S0140-6736(86)90837-8.

Bouveresse, E. and B. Campbell. 2008. "Transfer of multivariate calibration models based on near-infrared spectroscopy." In *Practical Spectroscopy Series, Chapter 11, 35*, 231. Boca Raton, FL: CRC Press.

Bouveresse, E., C. Casolino, and D. L. Massart. 1998. "Assessing the validity of near-infrared monochromator calibrations over time." *Applied Spectroscopy* 52 (4):604–612. doi: 10.1366/0003702981943941.

Bouveresse, E., C. Hartmann, D. L. Massart, I. R. Last, and K. A. Prebble. 1996. "Standardization of near-infrared spectrometric instruments." *Analytical Chemistry* 68 (6):982–990. doi: 10.1021/ac9510595.

Bouveresse, E., and D. L. Massart. 1996a. "Improvement of the piecewise direct standardisation procedure for the transfer of NIR spectra for multivariate calibration." *Chemometrics and Intelligent Laboratory Systems* 32 (2):201–213. doi: 10.1016/0169-7439(95)00074-7.

Bouveresse, E., and D. L. Massart. 1996b. "Standardisation of near-infrared spectrometric instruments: A review." *Vibrational Spectroscopy* 11 (1 SPEC. ISS.):3–15. doi: 10.1016/0924-2031(95)00055-0.

Bouveresse, E., D. L. Massart, and P. Dardenne. 1994. "Calibration transfer across near-infrared spectrometric instruments using Shenk's algorithm: Effects of different standardisation samples." *Analytica Chimica Acta* 297 (3):405–416. doi: 10.1016/0003-2670(94)00237-1.

Bouveresse, E., D. L. Massart, and P. Dardenne. 1995. "Modified algorithm for standardization of near-infrared spectrometric instruments." *Analytical Chemistry* 67 (8):1381–1389. doi: 10.1021/ac00104a013.

Bouveresse, E., S. C. Rutan, Y. Vander Heyden, W. Penninckx, and D. L. Massart. 1997. "Detection, interpretation and correction of changes in the instrumental response of near-infrared monochromator instruments over time." *Analytica Chimica Acta* 348 (1–3):283–301. doi: 10.1016/S0003-2670(97)00017-2.

Brown, S. D. 2009. "Transfer of multivariate calibration models." In *Comprehensive Chemometrics: Chemical and Biochemical Data Analysis*, edited by S. D. Brown, Tauler, R., Walczak, B., 345–377. Elsevier.

Burgess, C., and J. Hammond. 2007. "Wavelength standards for the near-infrared spectral region." *Spectroscopy (Santa Monica)* 22 (4):40–48.

Busch, K. W., O. Soyemi, D. Rabbe, K. Humphrey, B. Dundee, and M. A. Busch. 2000. "Wavelength calibration of a dispersive near-infrared spectrometer using trichloromethane as a calibration standard." *Applied Spectroscopy* 54 (9):1321–1326. doi: 10.1366/0003702001950995.

Choquette, S. J., D. L. Duewer, L. M. Hanssen, and E. A. Early. 2005. "Standard reference material 2036 near-infrared reflection wavelength standard." *Applied Spectroscopy* 59 (4):496–504. doi: 10.1366/0003702053641414.

Cleveland, W. S., and S. J. Devlin. 1988. "Locally weighted regression: An approach to regression analysis by local fitting." *Journal of the American Statistical Association* 83 (403):596–610. doi: 10.1080/01621459.1988.10478639.

Clinical and Laboratory Standards Institute. 2005. "Interference Testing in Clinical Chemistry; Approved Guideline – Second edition, CLSI document EP7-A2." Wayne, PA.

Coates, J., T. Davidson, and L. McDermott. 1992. "The design and application of spectrometric analyzers for the chemical process industry." *Spectroscopy* 7 (9):40–49.

Cogdill, R. P., C. A. Anderson, and J. K. Drennen. 2005. "Process analytical technology case study, part III: Calibration monitoring and transfer." *AAPS PharmSciTech* 6 (2):E284–E297. doi: 10.1208/pt060239.

Cooper, J. B., C. M. Larkin, and M. F. Abdelkader. 2011. "Calibration transfer of near-IR partial least squares property models of fuels using virtual standards." *Journal of Chemometrics* 25 (9):496–505. doi: 10.1002/cem.1395.

da Silva, N. C., C. J. Cavalcanti, F. A. Honorato, J. M. Amigo, and M. F. Pimentel. 2017. "Standardization from a benchtop to a handheld NIR spectrometer using mathematically mixed NIR spectra to determine fuel quality parameters." *Analytica Chimica Acta* 954:32–42. doi: 10.1016/j.aca.2016.12.018.

Dardenne, P., R. Biston, and G. Simmaeve. 1992. "Calibration transferability across NIR instruments." In *Near Infra-Red Spectroscopy*, edited by K. I. Hildrum, Isaksson T., Næs, T., Tandberg, A., 453–458. Ellis Horwood, Chichester.

Davies, A. M. C., H. V. Britcher, J. G. Franklin, S. M. Ring, A. Grant, and W. F. McClure. 1988. "The application of Fourier-transformed near-infrared spectra to quantitative analysis by comparison of similarity indices (CARNAC)." *Microchimica Acta* 94 (1):61–64. doi: 10.1007/BF01205839.

Davies, A. M. C., and T. Fearn. 2006. "Quantitative analysis via near infrared databases: Comparison analysis using restructured near infrared and constituent data-deux (CARNAC-D)." *Journal of Near Infrared Spectroscopy* 14 (6):403–411. doi: 10.1255/jnirs.712.

de Noord, Onno E. 1994. "Multivariate calibration standardization." *Chemometrics and Intelligent Laboratory Systems* 25 (2):85–97. doi: 10.1016/0169-7439(94)85037-2.

Despagne, F., B. Walczak, and D. L. Massart. 1998. "Transfer of calibrations of near-infrared spectra using neural networks." *Applied Spectroscopy* 52 (5):732–745. doi: 10.1366/0003702981944157.

Dreassi, E., G. Ceramelli, P. L. Perruccio, and P. Corti. 1998. "Transfer of calibration in near-infrared reflectance spectrometry." *Analyst* 123 (6):1259–1264. doi: 10.1039/a706694e.

Du, W., Z. P. Chen, L. J. Zhong, S. X. Wang, R. Q. Yu, A. Nordon, D. Littlejohn, and M. Holden. 2011. "Maintaining the predictive abilities of multivariate calibration models by spectral space transformation." *Analytica Chimica Acta* 690 (1):64–70. doi: 10.1016/j.aca.2011.02.014.

European Pharmacopoeia (Ph.Eur. 2.2.40). "Near-Infrared Spectroscopy." EUROPEAN PHARMACOPOEIA 9.0.

Fan, W., Y. Liang, D. Yuan, and J. Wang. 2008. "Calibration model transfer for near-infrared spectra based on canonical correlation analysis." *Analytica Chimica Acta* 623 (1):22–29. doi: 10.1016/j.aca.2008.05.072.

Fearn, T. 2001. "Standardisation and calibration transfer for near infrared instruments: A review." *Journal of Near Infrared Spectroscopy* 9 (4):229–244.

Feudale, R. N., N. A. Woody, H. Tan, A. J. Myles, S. D. Brown, and J. Ferré. 2002. "Transfer of multivariate calibration models: A review." *Chemometrics and Intelligent Laboratory Systems* 64 (2):181–192. doi: 10.1016/S0169-7439(02)00085-0.

Greensill, C. V., P. J. Wolfs, C. H. Spiegelman, and K. B. Walsh. 2001. "Calibration transfer between PDA-based NIR spectrometers in the NIR assessment of melon soluble solids content." *Applied Spectroscopy* 55 (5):647–653. doi: 10.1366/0003702011952280.

Hansen, P. W. 2000. "Pre-processing method minimizing the need for reference analyses." *Journal of Chemometrics* 15 (2):123–131. doi: 10.1002/1099-128X(200102)15:2<123::AID-CEM636>3.0.CO;2-8.

Igne, B., and C. R. Hurburgh Jr. 2008. "Standardisation of near infrared spectrometers: Evaluation of some common techniques for intra- and inter-brand calibration transfer." *Journal of Near Infrared Spectroscopy* 16 (6):539–550. doi: 10.1255/jnirs.819.

Igne, B., and C. R. Hurburgh Jr. 2010a. "Local chemometrics for samples and variables: Optimizing calibration and standardization processes." *Journal of Chemometrics* 24 (2):75–86. doi: 10.1002/cem.1274.

Igne, B., and C. R. Hurburgh Jr. 2010b. "Using the frequency components of near infrared spectra: Optimising calibration and standardisation processes." *Journal of Near Infrared Spectroscopy* 18 (1):39–47. doi: 10.1255/jnirs.865.

Igne, B., J. M. Roger, S. Roussel, V. Bellon-Maurel, and C. R. Hurburgh. 2009. "Improving the transfer of near infrared prediction models by orthogonal methods." *Chemometrics and Intelligent Laboratory Systems* 99 (1):57–65. doi: 10.1016/j.chemolab.2009.07.007.

Kalivas, J. H., G. G. Siano, E. Andries, and H. C. Goicoechea. 2009. "Calibration maintenance and transfer using Tikhonov regularization approaches." *Applied Spectroscopy* 63 (7):800–809. doi: 10.1366/000370209788701206.

Kennard, R. W., and L. A. Stone. 1969. "Computer aided design of experiments." *Technometrics* 11 (1):137–148. doi: 10.2307/1266770.

Kramer, K. E., R. E. Morris, and S. L. Rose-Pehrsson. 2008. "Comparison of two multiplicative signal correction strategies for calibration transfer without standards." *Chemometrics and Intelligent Laboratory Systems* 92 (1):33–43. doi: 10.1016/j.chemolab.2007.11.009.

Kramer, Kirsten E., and Gary W. Small. 2007. "Blank augmentation protocol for improving the robustness of multivariate calibrations." *Applied Spectroscopy* 61 (5):497–506.

Lima, F. S. G., and L. E. P. Borges. 2002. "Evaluation of standardisation methods of near infrared calibration models." *Journal of Near Infrared Spectroscopy* 10 (4):269–278.

Liu, X., L. J. Han, and Z. L. Yang. 2011. "Transfer of near infrared spectrometric models for silage crude protein detection between different instruments." *Journal of Dairy Science* 94 (11):5599–5610. doi: 10.3168/jds.2011-4375.

Mark, H., and J. Workman. 1988. "A new approach to generating transferable calibrations for quantitative near-infrared spectroscopy." *Spectroscopy* 3 (11):28–36.

Mark, H. 1991. *Principles and Practice of Spectroscopic Calibration*. Wiley.

Naes, T., T. Isaksson, and B. Kowalski. 1990. "Locally weighted regression and scatter correction for near-infrared reflectance data." *Analytical Chemistry* 62 (7):664–673. doi: 10.1021/ac00206a003.

National Institute of Standards and Technology. (2018). "Basic Atomic Spectroscopic Data." http://physics.nist.gov/PhysRefData/Handbook/element_name.htm

Nordqvist, K. G. 1990. "Method of Reducing the Susceptibility to Interference of a Measuring Instrument." U.S. Patent.

Nouri, M., C. Gomez, N. Gorretta, and J. M. Roger. 2017. "Clay content mapping from airborne hyperspectral Vis-NIR data by transferring a laboratory regression model." *Geoderma* 298:54–66. doi: 10.1016/j.geoderma.2017.03.011.

Pereira, C. F., M. F. Pimentel, R. K. Galvao, F. A. Honorato, L. Stragevitch, and M. N. Martins. 2008. "A comparative study of calibration transfer methods for determination of gasoline quality parameters in three different near infrared spectrometers." *Anal Chim Acta* 611 (1):41–47. doi: 10.1016/j.aca.2008.01.071.

Plyler, E. K., Blaine, L. R., and Nowak, M. 1957. "Reference wavelengths for calibrating prism spectrometers." *Journal of Research of the National Bureau of Standards (United States)* 58 (4):5.

Roger, J. M., F. Chauchard, and V. Bellon-Maurel. 2003. "EPO-PLS external parameter orthogonalisation of PLS application to temperature-independent measurement of sugar content of intact fruits." *Chemometrics and Intelligent Laboratory Systems* 66 (2):191–204. doi: 10.1016/S0169-7439(03)00051-0.

Shenk, J. S., and M. O. Westerhaus. 1991. "New standardization and calibration procedures for NIRS analytical systems." *Crop Science* 31 (6):1694–1696. doi: 10.2135/cropsci1991.0011183X003100060064x.

Shenk, J. S., M. O. Westerhaus, and P. Berzaghi. 1997. "Investigation of a LOCAL calibration procedure for near infrared instruments." *Journal of Near Infrared Spectroscopy* 5 (4):223–232.

Shenk, J. S., M. O. Westerhaus, and W. C. Templeton. 1985. "Calibration transfer between near infrared reflectance spectrophotometers." *Crop Science* 25:159–161. doi: 10.2135/cropsci1985.0011183X002500010038x.

Springsteen, A. 2006. "Molecular spectroscopy workbench an ode to traceability and calibration interval." *Spectroscopy (Santa Monica)* 21 (11):14–16.

Standard Reference Material (SRM) ® 1920a Certificate of Analysis. (2 February 2006). "Near-Infrared Reflectance Wavelength Standard from 740 nm to 2000 nm." National Institute of Standards and Technology, Gaithersburg, MD, 7 pages.

Standard Reference Material (SRM)® 2036 Certificate. (24 January 2013). "Near-Infrared Wavelength/Wavenumber Reflection Standard, National Institute of Standards and Technology, Gaithersburg, MD, 7 pages.

Tan, H.-W., and S. D. Brown. 2001. "Wavelet hybrid direct standardization of near-infrared multivariate calibrations." *Journal of Chemometrics* 15 (8):647–663. doi: 10.1002/cem.660.

Tan, H.-W., S. T. Sum, and S. D. Brown. 2002. "Improvement of a standard-free method for near-infrared calibration transfer." *Applied Spectroscopy* 56 (8):1098–1106. doi: 10.1366/000370202321275015.

Taylor, B. N., and C. E. Kuyatt. 1994. "Guidelines for Evaluating and Expressing the Uncertainty of NIST Measurement Results." NIST Technical Note 1297.

Tracy, D., R. Hoult, and A. Ganz. 1994. Standardizing and Calibrating a Spectrometric Instrument. US Patent.

United States Pharmacopeia <1119>, Near infrared spectrophotometry, *Pharmacopeial Forum* 30 (6):2137.

Walczak, B., E. Bouveresse, and D. L. Massart. 1997. "Standardization of near-infrared spectra in the wavelet domain." *Chemometrics and Intelligent Laboratory Systems* 36 (1):41–51. doi: 10.1016/S0169-7439(96)00075-5.

Wang, Y., and B. R. Kowalski. 1992. "Calibration transfer and measurement stability of near-infrared spectrometers." *Applied Spectroscopy* 46 (5):764–771.

Wang, Y., and B. R. Kowalski. 1993. "Temperature-compensating calibration transfer for near-infrared filter instruments." *Analytical Chemistry* 65 (9):1301–1303. doi: 10.1021/ac00057a031.

Wang, Y., D. J. Veltkamp, and B. R. Kowalski. 1991. "Multivariate instrument standardization." *Analytical Chemistry* 63 (23):2750–2756. doi: 10.1021/ac00023a016.

Wang, Y., M. J. Lysaght, and B. R. Kowalski. 1992. "Improvement of multivariate calibration through instrument standardization." *Analytical Chemistry* 64 (5):562–564. doi: 10.1021/ac00029a021.

Weidner, V. R., P. Y. Barnes, and K. L. Eckerle. 1986. "Wavelength standard for the near infrared based on the reflectance of rare-earth oxides." *Journal of Research of the National Bureau of Standards (United States)* 91 (5):243–253.

Westerhaus, M. O. 1991. "Improving repeatability of NIR calibrations across instruments." In *Proceedings of the Third International Conference on Near Infrared Spectroscopy*, edited by R. Biston, Bartiaux-Thill, N., 671–674. Gembloux, Belgium: Agricultural Research Centre Publishing.

Wold, S., H. Antti, F. Lindgren, and J. Öhman. 1998. "Orthogonal signal correction of near-infrared spectra." *Chemometrics and Intelligent Laboratory Systems* 44 (1–2):175–185. doi: 10.1016/S0169-7439(98)00109-9.

Woody, N. A., R. N. Feudale, A. J. Myles, and S. D. Brown. 2004. "Transfer of multivariate calibrations between four near-infrared spectrometers using orthogonal signal correction." *Analytical Chemistry* 76 (9):2595–2600. doi: 10.1021/ac035382g.

Workman, J. J., Jr. 2015. "The essential aspects of multivariate calibration transfer." In *40 Years of Chemometrics – from Bruce Kowalski to the Future*, 257–282. Washington, DC: American Chemical Society.

Workman, J. J., Jr. 2016. "Calibration transfer." In *The concise Handbook of Analytical Spectroscopy: Physical Foundations, Techniques, Instrumentation, and Data Analysis*, edited by J. J. Workman, Jr., Vol. 3, pp. 185–188. Singapore: World Scientific Publishing–Imperial College Press.

Workman, J. J., Jr. 2018. "A review of calibration transfer practices and instrument differences in spectroscopy." *Applied Spectroscopy* 72 (3):340–365. doi: 10.1177/0003702817736064.

Workman, J., and J. Coates. 1993. "The importance of standardizing instrumentation." *Spectroscopy* 8 (9):36–42.

Workman, J., and H. Mark. 2013. "Calibration transfer, part IV: Measuring the agreement between instruments following calibration transfer." *Spectroscopy (Santa Monica)* 28 (9).

Workman, J., and H. Mark. 2014a. "Calibration transfer, part V: The mathematics of wavelength standards used for spectroscopy." *Spectroscopy (Santa Monica)* 29 (6).

Workman, J., and H. Mark. 2014b. "Calibration transfer, Part VI: The mathematics of photometric standards used for spectroscopy." *Spectroscopy (Santa Monica)* 29 (10):14–21.

Workman, J., Jr., and H. Mark. 2015. "Optimizing the regression model: The challenge of intercept/bias and slope 'correction.'" *Spectroscopy (Santa Monica)* 30:7.

Workman, J., Jr., and H. Mark. 2017. "Bias and slope correction." *Spectroscopy (Santa Monica)* 32 (2): 24–30.

Youden, W. J., and E. H. Steiner. 1975. *Statistical Manual of the Association of Official Analytical Chemists*. Washington: The Association.

Zeaiter, M., J. M. Roger, and V. Bellon-Maurel. 2006. "Dynamic orthogonal projection. A new method to maintain the on-line robustness of multivariate calibrations. Application to NIR-based monitoring of wine fermentations." *Chemometrics and Intelligent Laboratory Systems* 80 (2):227–235. doi: 10.1016/j.chemolab.2005.06.011.

Zhu, Y., T. Fearn, D. Samuel, A. Dhar, O. Hameed, S. G. Bown, and L. B. Lovat. 2008. "Error removal by orthogonal subtraction (EROS): A customised pre-treatment for spectroscopic data." *Journal of Chemometrics* 22 (2):130–134. doi: 10.1002/cem.1117.

Part IV

Method Development

18 Practical Aspects of Sampling for NIRS Analysis

Phil Williams[*]
PDK Projects, Inc.

CONTENTS

18.1 INTRODUCTION, AND THE PHILOSOPHY OF SAMPLING FOR NIRS ANALYSIS

Sampling is a subject that commands a monograph of its own. Sampling and sample preparation are two aspects of analytical work by NIRS (near-infrared spectroscopy) or reference methods where research has apparently lagged behind other developments in the technology. The main reason for this is that the requirements of sampling in industry have been recognized for many years. The research essential to effective and efficient sampling has been accomplished and documented accordingly. In most industrial operations involving large volumes of bulk materials, sampling is automated. The precision and reliability of the systems have been scrupulously established, and procedures are specified in detailed manuals. The physical act of collecting the sample is simply that of removing the sample from the sampling system and transferring it to the location where it will be tested by the chosen method of analysis, or even carrying out NIRS testing on-site.

This chapter is all about sampling, but it is about sampling for NIRS analysis. The tendency for users of NIRS is to think of the material in terms of the reference data, for the prediction of which the technology is to be used. The concept and principle of NIRS analysis differ from those of the classical chemical analysis in that it is exclusively based on the spectra. The spectra are, of course, derived from the chemical and physical makeup of the material, but spectra can differ among

[*] philwilliams@pdkgrain.com.

materials with the same chemical composition [1]. For example, consider the spectra of samples of wheat of the same variety and therefore genetic constitution when grown in the northern wheat-growing areas of Alberta and southern Saskatchewan. The samples may be reported to contain the same protein and moisture contents, but the different growing conditions can affect the molecular arrangement in space and, as a consequence, the relationship between the spectra and the reported chemical composition that results from the calibration process. This is the reason underlying the recommendation for replication of samples of agricultural materials with the same reference data. It is also at least a partial explanation for the occurrence of outliers.

Avogadro's number gives the number of molecules in a gram molecule as 6.022×10^{23}. A single kernel of hard red spring wheat, weighing about 35 mg, with 12% moisture would contain approximately 1.4×10^{20} molecules of water. Even when chemical analysis indicates that the composition of two samples of a material like wheat is identical, with differences that are usually reported to only two places of decimals, such differences result in differences in the order of billions of molecules, undetectable by reference chemistry, but spectrally detectable by spectrometers, and all with the potential for hydrogen bonding. Errors in reference testing affect the statistics of calibration development, but the molecular composition of the sample will not change. The molecules are all in motion in the ground state, until irradiation by a light source causes a change in energy levels, detectable in the form of absorbance. Hydrogen bonding can change the shapes of larger molecules, like proteins in space, and can cause association between constituents, such as protein and oil. Such associations change the shapes of the molecules and consequently the spectral features, as well as the functionality. The results for (in this case) protein content as predicted by NIRS are probably more reliable than those of the reference method.

The application of NIRS in analysis differs fundamentally from those of chemical reference methods. The technique is directly applied to the material, whether it is intact, or has been prepared for analysis by a suitable sample preparation system. It is nondestructive of the material to be tested, whereas practically all reference methods used in industry require sample preparation, which changes, and usually destroys, the sample. Any factor that affects the spectra will influence the effectiveness of NIRS application, but may not affect the process of reference analysis by wet-chemistry methods. This chapter encompasses the factors involved in the setup and establishment of a sampling system, which is an essential part of any industrial process that calls for analytical work. Six things are absolutely essential for the successful application of NIRS. These are (1) accurate and (2) reproducible reference analysis, (3) excellent spectral precision, (4) comprehensive sample selection, (5) assembly of samples that embody all sources of variance, and (6) efficient sample presentation to the instrument. No matter how refined is the chemometric software, it will not fully compensate for imperfections in any of these prerequisites. This chapter will address factors 3, 4, 5, and 6.

The two main categories of sampling are sampling for feasibility studies and sampling for use in industry. This chapter will focus on sampling for industrial-level NIRS analysis, and the hazards and necessities that must be overcome and provided in order to understand what is happening during calibration for, and application of, NIRS. The main difference between feasibility and industrial applications is that there can be no outliers in an industrial application, because everything that is presented to the instrument has to be analyzed. The most successful industrial applications of NIRS involve the prediction of one or more important constituents, such as protein, moisture, sugar, fiber (cellulose), or fat/oil contents, for which abundant absorption bands are present in the NIR region. Some functionality characteristics, such as flour water absorption and flour ash, can be predicted with confidence by NIRS because these attributes are strongly affected by constituents, like protein content, for which absorbers are plentiful. In the case of ash, there is no ash in flour until it has been burned, and what is being measured is related to the cellulosic components of the flour, for which absorbers are also clear-cut [2]. In general, because they share absorbers with major constituents, which will tend to "mask" their spectra, the smaller the amount of a component present, or the further a parameter to be predicted is remote from actual constituents, e.g., a flavor component, the less trustworthy is the application likely to be, and the less likely is it to be reliable for use in industry. The impact of error on the commercial aspect in industry, and the consequences of erroneous diagnosis in the medical profession are both features that call for careful consideration before the adoption of NIRS as a replacement for the existing methods of analysis in either field.

When an operation is considering the use of NIRS, one of the first things considered in budgeting for sampling, calibration development, and subsequent NIRS analysis is the cost. The costs may appear to be exorbitant, but they should always be considered in relation to the value of their analysis to the operation. From this aspect, the initial cost of a NIRS instrument is a particularly important factor at the outset, but the instruments are durable for several years, so the expense is not ongoing. Costs for assembly of a set of samples for a trustworthy calibration may amount to thousands of dollars, but the costs of the instrument, sampling, and calibration, must be assessed by comparison with the value of the material to which the analysis is to be applied, which is usually many times higher than the costs of sampling and analysis, and the convenience of instantaneous results when needed.

Once the sampling system has been established, collection of samples for calibration development includes transportation to and from the sampling site, containers for the samples, transportation by mail or other methods, storage before analysis, costs of the reference analysis, and other items. These costs should be considered in perspective to the cost of testing once the calibrations have been developed, when samples are always available. A salary of $50,000 for a technician translates into 42.7 cents a minute for a working week of 37.5 h. A NIRS test takes about 2 min, during which several constituents can be measured simultaneously, so the economic advantage of conversion from reference methods to the NIRS method is obvious, to say nothing of the equally important advantage of instantaneous results when needed, as in quality control. Additional benefits are that a lot of daily routine testing, such as protein and moisture contents of whole grain, can be carried out by staff with no analytical experience, and NIRS analysis is absolutely safe.

The accuracy and reproducibility of the analytical methods will vary among the laboratories responsible for the reference analysis used in the calibration development and monitoring aspects of NIRS analysis, but these are the laboratories to which assessment will be made as to the continued effectiveness of the NIRS testing. It is the responsibility of individual operations to determine whether they are acceptable. Laboratories vary in their degree of reproducibility. Provided that the accuracy and reproducibility are acceptable to the operation and its clients, useful NIRS calibrations can be achieved where the statistics of the reference laboratory performance would not meet the most stringent statistical evaluation. Because of the excellent spectral precision of NIRS instruments, the reproducibility of NIRS analysis is often superior to that of the reference laboratory.

The text will digress from time to time to introduce some apparently nonsampling issues, but the objective of such digressions is to enable the operator to get the best possible results, by indicating items that affect the results of application of NIRS. While it is proficient to achieve fastidious statistics, the most accustomed reasons for adopting NIRS in industry are to improve the efficiency of the operation by taking advantage of the speed of the testing, and to reduce costs. Although it is not the aspect of NIRS that is most frequently described, the chief quality feature of NIRS is the precision of the spectra, which is attributable to the efforts of the engineers of instrument companies. When an operation decides to adopt NIRS as an analytical technique, management will tend to compare the NIRS results to those of the tried-and-trusted methods that they had hitherto employed. But, as well as its significant impact on the speed and availability of results, NIRS is usually more reliable than wet-chemistry methods. The most practical advice to an operation is "If you have verified that the NIRS method works for you, use it" even if the calibration statistics do not conform to perfection.

This is the philosophy of sampling for NIRS analysis!

But why is sampling so important? The success of both NIRS and reference testing depends on the integrity of the sample, and if this is compromised, the results of any testing will be misleading, with the associated consequences. The main factors affecting the accuracy and precision of NIR

TABLE 18.1
Sample-Associated Factors in NIRS Analysis

Sampling	Samples
Type of sampling system	Type of material
Location of sampling system	Composition (oil, moisture, fiber)
Material to be sampled	Foreign material
Foreign material	Physical texture
Sample preparation	Grinder type
Size of sample	Particle size
Flow characteristics	Blending
Sample transfer method	Bulk density
Blending	Storage method
Storage	
Identification	
Heterogeneity of population	
Frequency of sampling	
Sampling method	
Sample selection (calibration)	

analysis are attributable to sampling, samples, or sample presentation. These are summarized in Table 18.1 (not in order of importance).

These will all be addressed in this chapter. Sample presentation to the instrument is a key factor in the effectiveness of NIRS analysis, but this is principally affected by the operator and the instrument, and the details are not considered to be within the scope of this chapter.

There are a number of further things that the NIRS user needs to know about samples, sampling, and sample preparation. These include the following:

1. What is a sample?
2. What is the purpose of the sample?
3. What is meant by sampling?
4. What affects the accuracy of sampling?
5. How can the accuracy of sampling be determined?
6. What is the significance of a "truly representative" sample?
7. How should the sample be taken?
8. Sample assembly
9. How should samples be prepared for analysis?
10. How should samples be stored?
11. How the sample is presented to the instrument?

18.2 WHAT IS A SAMPLE?

Statisticians and analytical chemists look upon a sample in different ways. To a statistician, a sample is a small quantity of a population that is intended to represent, reliably, the entire population in its original appearance; its physical, chemical, and physicochemical form; and its functionality. This may be referred to as "the first law of sampling". To the analytical chemist, a sample is what is contained in the envelope, or whatever container arrives in the laboratory for analysis. It is usually not the responsibility of the analyst to take the sample. But if the results differ significantly from the results that experience has shown to be typical of the analysis of that particular material, it behooves the analyst to determine the source and method of sampling, in case the material has to be re-sampled. To the staff of a country grain elevator, the sample is what is taken from the back of a truck, upon which the status and value of the farmer's delivery is assessed, and the person that takes the sample may also perform the NIRS test. Liquids with viscosities ranging from thinner than water to thick slurries; plastic solids like plastics themselves; or something of the consistency of cheese, a piece of cloth, soil, wood, or wool, an area of skin, a liter of river water, or a bag of manure all present individual precepts for either bringing the material into contact with the NIRS instrument, or, more recently, bringing the NIRS instrument to the material.

18.3 WHAT IS MEANT BY SAMPLING?

Sampling is the act of withdrawing a discrete amount of material from the bulk of material (called "the population"). Sampling systems vary widely, depending on the physical state of the material, e.g., whether it is intact or powdered solid, fibrous, slurry, liquid, and individual operations will identify or develop an appropriate system and equipment for withdrawing the sample. The purpose of the sampling process is to provide a suitable amount of the material, for analysis. The objective of any chemical or physicochemical analysis is to determine the potential functional properties of the material being analyzed, which in turn are applied to determine the value and price of the material [3].

18.4 WHAT IS THE PURPOSE OF THE SAMPLE?

The main reasons for sampling are (1) to determine the reproducibility (precision) of laboratory testing; (2) to determine the degree of heterogeneity of the material; (3) to determine the mean composition of a quantity of material, such as the contents of an elevator bin, or a field of bales of a forage; (4) to assemble a set of samples for calibration development; (5) to monitor the consistency of a manufacturing process; (6) to identify a particular type or class of material; and (7) to determine the effect of use of the material as it is on the environment, like spreading of liquid manure on a field.

The first reason for sampling is to determine the reproducibility of the analytical method(s) used in the testing process. The reproducibility of the reference method is a component of the overall error and is consequently a primary element in the determination of sampling error. Its reproducibility and repeatability should be determined at the outset, because it is the ultimate criterion underlying calibration, monitoring, and application of NIRS. Determination of the heterogeneity of the material is also fundamental to future analysis and use of the material in industry, but it can only be assessed once the reproducibility of the sampling procedure itself has been established. This, in turn, can only be verified by analysis for one or more constituents or parameters.

Reproducibility can be determined by the repeated application of the method to a single well-blended check sample, using all steps, including subsampling for analysis, sample preparation, and all of the analytical steps. It also includes all of the operators likely to use the method(s). The standard deviation (SD) of the results is the reproducibility of the method. It can best be determined and monitored by means of a check sample, or samples. The check sample is the most effective system for monitoring and maintaining laboratory performance. A check sample should be prepared from the material to which the calibration is to be applied, and should be assembled from samples that are routinely tested by the laboratory. These samples should incorporate the main variables likely to affect the application of the NIRS testing. They should be combined and very thoroughly blended. The final blended check sample should be of a size sufficient to last for a period of at least a few weeks. It is not practicable to prepare check samples of fresh material, such as fresh forages or silage, because of the physical nature of the material, its susceptibility to moisture changes, and its difficulty in storage.

The results of testing check samples cannot monitor reproducibility if the sample itself is changing. Check samples should be stored in heavy-duty plastic pails and refrigerated. Sufficient sample should be brought out from time to time to last for a few days, and should be allowed to equilibrate to room conditions before scanning by NIRS. For prolonged storage (1 week or more in warm climates), samples should be stored at 4°C–5°C. It is not advisable to freeze grains and other cellular materials because the expansion that accompanies the change from water to ice may affect the spectra. This is particularly important for fresh samples, which should be scanned at the time of withdrawal, or as soon as possible thereafter. When the check sample is becoming low in volume, a new check sample should be prepared and tested side-by-side with the original sample, before the original sample is exhausted.

The second reason for sampling is to determine the degree of heterogeneity of the population of material. Heterogeneity means variability among samples of the same material. The maximum variance among samples of the same material is the variance sought in samples assembled for NIRS calibration development. Deliveries of the same type of grain to country elevators by farmers rank next on the scale of sample-to-sample variability. The degree of variance will be defined by the maxima and minima in composition, or other characteristics that experience will have shown to exist in the material. For example, in western Canadian hard red spring wheat, the long-time mean protein content lies at 13.6% (13.5% moisture basis), but experience has shown that the range can extend from less than 9% to over 18% in any year. In other wheat-growing areas, the mean protein content may display a lower level, with a commensurate change in range. Samples withdrawn for quality control will exhibit much less variability, while at the other extreme of variability lie samples taken to assure the reproducibility of laboratory analysis.

To determine the heterogeneity, a useful system is to withdraw a number of samples (again at least 10), retain and analyze each of the samples individually, and make a composite sample using a constant amount, or volume, as each of the individual subsamples is added. Each of the progressive composites is analyzed by the reference method (or by the NIRS method, assuming a reliable calibration. It is quicker and cheaper, and just as accurate). The mean of the individual samples, which are subsamples of the final composite sample, should agree with the result obtained on the final composite sample. The SD of the results of the individual subsamples shows the variance and heterogeneity of the material. It will progressively decrease, until it stabilizes, at which stage no further samples need to be taken. "Normal" values for the coefficient of variance (CV) for wheat protein content should be between 2% and 3%. The CV for samples of grains like oats can be expected to be higher. The CV of soil, manures, forages, and more complicated materials will be even higher, but should be determined, and should be less than 10%. Frequency of sampling and sample size will be determined by the operator, using data derived from the determination of the heterogeneity.

The third reason for sampling is to determine the composition of a material, e.g. by analyzing the contents of a large bin of grain before shipment. Over the years, chemists and technologists in many fields have striven to develop tests which will enable the prediction of the end-product utilization potential of a substance, from a knowledge of its composition. The relationship of the composition of the material to its functionality will have been determined by earlier research. The best-known example is probably the protein test for wheat, which assists the cereal chemist in estimating bread and cookie baking potential. Moisture tests on grain are carried out to determine whether the grain is safe to store or whether it should be dried, and also to determine how much water, for which the buyer is paying the full grain price. For example, large-scale buyers of wheat often specify a maximum moisture content. An error of 1% in moisture content for a cargo of 35,000 tons of grain (a typical ship load) at $200.00 per ton can translate into the purchaser buying $70,000.00 worth of water. But during the 45 or so years that has passed since its first introduction, NIRS has transcended the confines of grain technology. Animal feed manufacturers and nutritionists need information on energy, digestibility, and other parameters. Pharmaceutical companies need to know that they are shipping the correct products with the correct composition. The NIRS technology has expanded into most fields of industry and commerce, and has made its presence felt in the areas of medical diagnosis [3] and environmental science.

The delivery/unloading of a farm truckload takes only a few minutes. Prior to the adoption of NIRS during the mid-1970s, the grain could not be tested at the time of delivery because of the time of testing, but the introduction of NIRS enabled testing for protein and moisture contents at the time of unload. This allowed the elevator to pay the farmer on the basis of the protein content, as well as the weight of grain delivered, and has generated millions of dollars for wheat growers. The mean protein content of the grain in the entire silo can be computed using the weighted mean system. The weighted mean is used to calculate the average of a series of observations, like protein content, based on, e.g., deliveries of different quantities of grain of different protein contents that may be added to a silo. It is calculated by multiplying the weight of each increment (e.g., farmers' deliveries to a country elevator) by the protein content, then dividing the total (weight×protein) by the total (weight). Table 18.2 gives an example of calculation of the weighted mean.

The fourth reason for sampling is the assembly of samples for calibration development. All NIRS instruments require calibration, whether it is carried out by the user or the factory. Sample selection is really a factor only in calibration development and validation, because for NIRS day-to-day analysis, the operator has no choice in the samples presented for routine analysis. Sample selection and assembly for calibration development is the most complicated, arduous, time-consuming, and expensive aspect of calibration development for industrial application, especially when the costs of reference analysis have to be included.

The fifth reason for sampling is that of quality control, to ensure that a product that is being manufactured is maintaining the specifications stipulated by the purchaser. For example, flour mills produce flour to comply with the requirements of bakers. To ensure that the flour is conforming to the specifications, the main flour stream is periodically sampled at intervals that range from 1 to 3 h, depending on the quality control strategy of individual mills, and tested for constituents like the ash (mineral matter) contents. The ash test has long been accepted as a criterion of the efficiency of milling. The reason for the use of ash as a criterion is that the ratio of mineral matter in the bran to that in the endosperm is approximately 29:1, which means that a very small increase in the bran content of the flour will impact the ash content. The ash test takes a minimum of 2 h, during which a large mill can produce over 100 tons of flour. If the flour is not meeting specifications, due to a change in the equipment or procedure, the mill is able to address the issue immediately, on the basis

TABLE 18.2
Example of Calculation of the Weighted Mean

N	Protein (%)	Weight[a]	Weight × Protein %
1	16.2	14.6	236.52
2	13.6	21.2	288.32
3	13.1	20.3	265.93
4	13.5	44.6	602.1
5	13.2	28.7	378.84
6	13.7	16.9	231.53
7	12.9	33.3	429.57
8	14.4	24.9	358.56
9	12.3	18.3	225.09
10	13.4	26.6	356.44
11	11.3	37.5	423.75
12	13.3	29.6	393.68
Sum	-	316.5	4190.33

Weighted mean protein content = sum (weight×protein)÷total weight = 4190.33 ÷ 316.5 = 13.24%.

[a] Weight of delivery in tons.

of the NIRS ash result, thereby avoiding the consequences of flours not meeting specifications. Demands for quality factors such as Farinograph and Alveograph parameters are increasing in the flour milling industry. The use of NIRS for determination of constituents such as protein and ash contents allows the laboratory to carry out the more time-consuming and technically demanding tests, and NIRS has become widely used in quality control.

Animal feed mills produce different feeds to comply with the requirements of farmers. Animal feed mills produce a wide range of different feeds to comply with these needs. Unlike flour mills, feed mills use a range of materials of plant and animal origin in making their products, and their undertaking is complicated by the fact that animal feeds contain several ingredients that differ widely in physical characteristics, as well as in composition. The sources of major constituents of the feeds may also change. To ensure that the feeds will conform to the recipes formulated by the mill nutritionists, the focus is on the ingredients used. The ingredients are sampled and tested for constituents such as moisture, protein, fiber, and ash (mineral matter), and the feeds are made up by weight. NIRS is increasingly used by feed mills. The wet-chemistry fiber test takes several hours, during which a large mill can produce 100s of tons. If the product is not meeting specifications, due to a change in the equipment or procedure, the mill is again able to address the issue immediately, on the basis of the NIRS results.

Industrial plants, including animal feed mills, can generate calibration samples by the creation of samples with a range in terms of composition of ingredients and physical characteristics wide enough to induce and accommodate the variance in spectral characteristics necessary for the development of stable calibrations. Calibrations for parameters like metabolizable energy are based on reference data that is estimated from constituents such as starch and fat. When calibrations are developed on these types of industrially prepared samples, fewer samples are required, but changes in composition, inclusion of different mill streams, sources of major ingredients, etc. should be replicated at least three times and preferably more frequently.

The sixth reason for sampling is for classification, or discriminant analysis. Classification applications involve the assembly of a training set. This set should consist of samples of the classes among which discrimination is required. The next two figures (Figures 18.1 and 18.2) illustrate the concept. The samples of each group to be assembled should represent all of the identified physical and chemical variables that are likely to affect the spectra. Calibrations are set up as "Teaching"

FIGURE 18.1 Three classes of Canadian hard red spring wheat. Left-to-right: CWRS, CWES, CPSR.

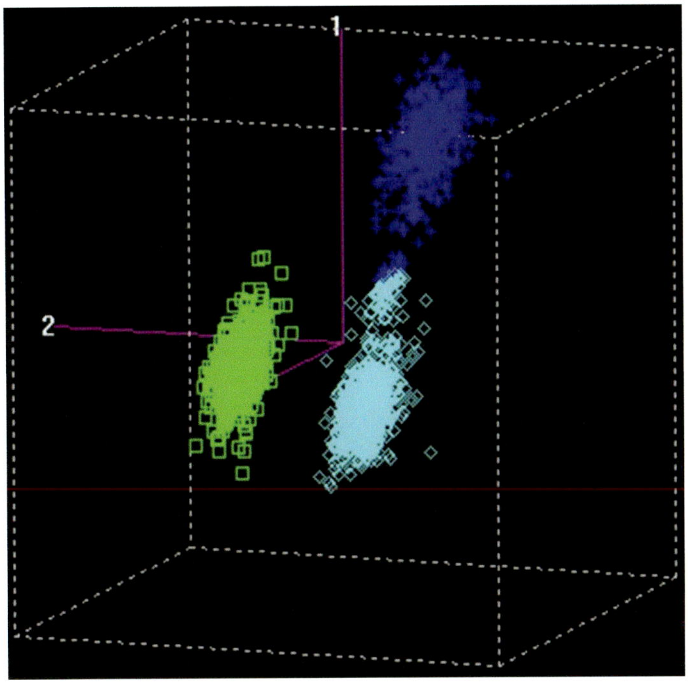

FIGURE 18.2 NIRS classification of the three wheats of Figure 18.9. Top-right CWRS, bottom-left CWES, bottom-right CPSR.

sets, using samples of classes that are the most pertinent to the operation. There is no requirement for specific reference data for every individual sample, other than to include samples that represent the ranges in reference data that are likely to be encountered. New samples are then recognized as whether they do or do not belong, and can accordingly be accepted or rejected. In setting up the teaching set, guidelines must be established to allow for a certain degree of uncertainty. The teaching set can be set up simply on the basis of sample type or class (e.g., different classes of wheat), as illustrated. Within a commodity, like a certain class of wheat, it is conceivable to set up a system for discriminating among levels of composition, such as protein and/or moisture contents with cutoffs, or banding levels for pricing purposes.

The seventh reason is that of analysis of material that is to be applied to agricultural land to determine the possible impact of its application. This includes the analysis of composts and manures, as well as the soil itself. For example, large-scale hog farming involves the accommodation of large numbers of animals in barns. The manure is stored in pits below the floors, or in external lagoons, each of which has to be emptied periodically, and the liquid manure pumped onto the adjacent farmland. The composition of liquid hog manure changes during pump-out, but it can be monitored accurately and continuously during pumping using NIRS. Continuous on-line analysis of liquid manure as it is being applied enables mapping of the amounts of nutrients being added to the land. Some areas of a field will receive too little, and some areas far too much, in terms of nutrients applied in the manure. Continuous analysis by NIRS during application can provide valuable information to farmers and to precision agriculture by improving the efficiency of fertilizer use. Figure 18.3 illustrates the type of information that can be achieved. Calibration development for on-line NIRS analysis of such materials calls for withdrawing samples at intervals, to obtain samples with variance in physical and chemical compositions, as well as variance in the prevailing conditions during the operation, which could take several days. Temperature during periods of continuous NIRS on-line analysis of liquid manure during application has ranged from 2°C to over 30°C. Samples should periodically be withdrawn during the application of the calibration, for monitoring.

Lb. applied to area at 6,000 gallons per acre. Early pumping. DM = 1.0% P = 16 N =125	Lb. applied to area at 6,000 gallons per acre. Later pumping. DM = 2.5% P = 29 N = 159
Lb. applied to area at 6,000 gallons per acre. Later pumping. DM = 5.0% P = 51 N = 215	Lb. applied to area at 6,000 gallons per acre. Advanced pumping. DM = 10.0% P = 95 N = 328

FIGURE 18.3 Mapping of a field for nutrient application.

18.5 HOW SHOULD A SAMPLE BE TAKEN?

The basic sampling or sample withdrawal systems are manual and automatic. Manual sampling of grain can be carried out by the specialized hand-held sampling devices, such as the Ellis cup and Pelican sampler, or probes. An array of such manual sampling equipment can be accessed from companies such as Seedburo and Gamet. Sampling of farm trucks at country elevators may be carried out simply by diverting a stream of grain falling from a truck into a pail by hand (Figure 18.4). This method is effective, provided that the sampling during the unloading of the truck continues throughout the unloading. Experience will determine how often should subsamples be taken.

At modern country primary elevators, truckloads are mainly sampled by means of hollow probes that are inserted into the bulk of grain and used to withdraw grain at all levels of the bulk. The subsamples are pneumatically delivered to the testing office (Figure 18.5), and composited before analysis. In the case of open trucks and trailers, at least six probes should be taken at areas that cover the full surface and depth of the bulk of grain in the truck or trailer. The subsamples should then be thoroughly blended before analysis. This can best be done by using a blending device like a Boerner sample divider (Figure 18.6). Many modern primary elevators that receive grain directly from farms are equipped with diverter samplers, which are used when grain is being taken for shipment.

Frequency of sampling and sample size will be determined by the operator, based on the physical nature of the material and the degree of heterogeneity. The shorter the interval between sampling actions, the more representative will it be. Because of different densities, sample size is a matter of

FIGURE 18.4 Hand-sampling a farm truck.

FIGURE 18.5 Pneumatic sampler for farm trucks.

volume, rather than weight. The heterogeneity of the material affects sample size, and in general, the higher the degree of heterogeneity, the bigger should be the sample, and the more frequently should the samples be taken. A further factor is that the larger the particles of material (e.g., whole wheat grains, chopped fresh forage), the larger should be the sample taken. The size of the sample in an industrial operation is also affected by the type of sampling system. There is usually a large bulk of material, and the larger the sample, the more likely is it to be representative of the population. For practical purposes, sample size is often limited to the size and type of container that will be used to transport the sample to where it will be tested. This applies to samples that are taken for calibration development, and monitoring, as well as regular testing. Day-to-day analysis by NIRS is made on-site, at the time that the sample is withdrawn.

The frequency of sampling also depends on the type of operation and the material to be sampled. It can vary from a single composite sample, made up from several hand-withdrawn subsamples such as the sampling of a farmer's truck at the time of unload, or bales of forage in a field, through periodical sampling, such as the loading of a vessel at a grain terminal, which can take several days, depending on the size of the ship. In exporting terminals, sampling is automated and continuous, and the timing of withdrawal has to be established. In a large terminal, grain may be loaded at the rate of 1000 tons h^{-1}. The diverter sampler takes a pass every 45 s or so, and the samples are combined hourly to give a composite sample for the bin, hold, or entire cargo. Automated sampling systems usually provide far too much grain, and stream splitters can be installed to divert a small percentage of the stream into the area where the NIRS testing is carried out.

FIGURE 18.6 The Boerner sample divider.

To determine the frequency of manual sampling a static bulk of material, a large number of samples should initially be taken (at least 12, from different areas of the population). Each sample is analyzed for composition, and composite samples are prepared as each sample is taken. The number of such subsamples is reached when the mean and SD stabilize. The mean of all of the individual samples should agree with that of the final composite. Samples for the development of on-line calibrations are one-time samples that should be taken to encompass the range in composition of the material, and to incorporate changes in work-time operating conditions. Sampling for NIRS interactance analysis involves applying the sensing head to the surface of the material for a number of times to assure a reliable value for the analysis. The number of applications is determined by the operator, based on the SD of the results of individual scans. When the frequency of sampling has been determined, the entire exercise should ideally be repeated to verify the reliability. The economics of the frequency of sampling and testing should be determined, because the costs of the sampling system are part of overall operating costs, and have to be practicable. The consequences of not developing and practicing an accurate and adequate sampling system far outweigh the costs of sampling.

Bales of straw, wool, and similar materials are manually sampled by augers, and the nature of the material makes blending difficult. Sampling of bags of grain, pellets, flour, or any other type of bagged material can be carried out by probes that penetrate the bags. Forages are usually baled in the field before storage or use. Bales can be sampled by manual or mechanical probes that penetrate the bale and withdraw samples from throughout the bale (Figure 18.7). The location of bales in the field can be documented to allow an estimate to be made of the mean and variability of forage composition all over the field. Wool and cotton bales can be sampled in the same way. Materials such as fresh or composted animal manure are sampled by grab samples, which should be scanned

FIGURE 18.7 Sampling a bale of forage using an auger.

by NIRS at the time of, or soon after, sampling, and protected from moisture loss before reference analysis. Because of the nature of the material, it is difficult to prepare a blended composite sample of fibrous materials, and the mean of the NIRS results has to be accepted as the reality. Sample transfer simply means the transfer of the sample into the container in which it will be transported to the laboratory or instrument.

Hand sampling of soil in a field is usually carried out manually on-site, by means of augers (Figure 18.8). Soil sampling is complicated by the size of the fields, the topography, and the need to take samples from at least two depths of the soil. The top 10–15 cm will contain roots, which should be removed before analysis of the soil itself, unless the soil is to be sampled for organic matter, including carbon. To obtain a reliable estimate of the composition and condition of the soil of a field of 100 acres is a formidable task. Sampling of big plants, like whole corn (maize) plants, is also complicated, and operations that work with this type of material have to develop their own sampling methods, and evaluate their reliability.

In large elevators and most flour and feed mills, bulk grain is sampled by automatic samplers that continuously withdraw subsamples from the main streams of grain, flour, feed ingredients, and other types of products. The subsamples are pneumatically transported to the elevator testing laboratory. Usually, this method of sampling removes too much sample for practicable analysis, and the stream of sampled materials are passed through a secondary divider, which allows a certain percentage of the stream to be transferred to the laboratory, and redirects the remainder to the original conveyor. The samplers are usually at the end of belts or conveyors, where the grain is discharged into a hopper and conveyed into the testing area. Diverter ("cross-cut") samplers move across the stream. They are also located at the end of belts or conveyors, where the grain is being discharged into a hopper, and divert subsamples into the pneumatic transfer system. Cross-cut samplers can also be installed in a spout or chute in the grain-handling system. Cross-cut samplers have become the sampling method of choice throughout the grain industry.

Automatic sampling of grains and powders, like flours, can be influenced by flow rate. The flow characteristics are affected by factors such as moisture content, particle size, and the physical state

FIGURE 18.8 The Vertis soil sampling auger.

of the material. Cereal grains usually flow freely. Flour from hard wheat also flows easily, but the particle size of flours from very soft wheat, like the wheat used for making biscuit (cookie) and pastry flours, is much smaller, and because of physicochemical features, including van der Waals factors, particles are attracted to each other, which impedes flow rate, and can impact sample cell loading in NIRS analysis.

Once calibrations have been developed, on-line continuous analysis by NIRS eliminates the need for sampling, except for monitoring. Sampling systems may be custom-designed to serve specific purposes in industrial processing plants. Some processing plants monitor their product streams directly with NIRS. This calls for optimization of the positioning of the sensor. Flour mills can either install NIRS instruments on selected flour streams, like the final products, and monitor them continuously for, e.g., protein, moisture, and ash contents, or withdraw samples manually or automatically for the laboratory. Application of NIRS to liquids, such as liquid manures, or sewage sludges, can be carried out by continuous on-line analysis by attaching the instrument to the piping through which the liquids are pumped in the field. Suitable devices can be incorporated in the field to enable sampling for calibration development, and subsequent monitoring of the performance (Figure 18.9).

The efficiency of automated sampling is affected by the degree to which the location at which the sample is taken is truly representative of the population. If the material is stratified, the sampling must penetrate all of the strata. This becomes difficult, e.g., in the sampling of something like a lagoon of hog manure, where the surface may have total solids content of only 0.5%–1%, whereas the bottom layers will be much higher in total solids and nutrients, so that the lagoon or storage tank must be continuously agitated during pump-out. Stratification also affects temperature. Layering of liquids changes refractive indices. Temperature has an effect by changing the viscosity of slurries, and during out-of-door activities, temperatures may vary by up to 30°C. All of these factors affect the spectra and must be incorporated into samples assembled for calibration models for NIRS analysis of these types of materials. In the case of continuous on-line analysis of liquids, like liquid manures, the NIRS instrument should be located at the center of the piping that carries the liquid,

FIGURE 18.9 Field sampler for liquid manure.

to avoid frothing at the surface and sedimentation at the bottom of the piping, both of which affect spectral characteristics.

Manual sampling is carried out directly on farm trucks as they arrive at elevators, either from the back of the truck as it is being unloaded, or by a pneumatic truck probe, which will be located at a position convenient for parking the truck. Automatic samplers have to be located in such a way that the entire depth of the belt or conveyor is being sampled to avoid the error of stratification. Vibration of conveyor belts and other conveyors cause lighter components of the grain, such as weed seeds and chaff, to migrate to the surface due to the differences in bulk density, and can influence the results. Foreign material in grain cannot be accurately determined by NIRS. The determination of such foreign material forms an important segment of the pricing, so that accurate sampling for the determination of the foreign material is essential. An error of only 0.1% in measurement of foreign material translates into hundreds of thousands of dollars in a terminal that handles over a million tons of grain annually. The measurement of foreign material is done by weighing samples of the different types of foreign material produced by equipment like the Carter Dockage tester (Figure 18.10).

18.6 WHAT AFFECTS THE ACCURACY (EFFICIENCY) OF SAMPLING?

Accuracy can best be defined as "Closeness to the true composition" in terms of chemical composition. Absolute accuracy is theoretically impossible to achieve, because of the sampling error and of the error inherent to any chemical method. In the case of physical parameters, such as texture or viscosity, accuracy is difficult to define, except in the terms that are acceptable to establish the functionality of the material (the capacity for doing what it is supposed to do). The most practical approach for the evaluation of a NIRS application is to achieve the highest degree of precision, in terms of reproducibility. The mean value of the results of repeated testing of a check sample is then the value that must be accepted by the laboratory and operation management for the accuracy of testing future samples of the same material for that parameter.

The efficiency of sampling, as defined by the accuracy and precision, is mainly affected by eight factors. The first is the error of the test procedure used in the evaluation. The second is the type of material and its physical characteristics, such as whether it is solid, granular, powdered, fibrous, liquid, and slurry. The third factor is the chemical composition of the material, with particular regard to the moisture content. Fresh materials can contain up to over 90% water, and should be tested for moisture content at the time of sampling or on-site NIRS scanning. The fourth factor is the inherent heterogeneity of the material. This can only be determined once the reproducibility of

FIGURE 18.10 The Carter Dockage tester.

the test methods and sampling procedure have been established. The fifth factor is the location of the sampling system, which is particularly important with automatic sampling systems. The location of the sampling systems will be decided by the operator and by the operation. Grain, flour, feeds, and similar materials can be sampled from belts or spouting. Forages, wool and cotton, and similar materials can be sampled from bales. Soils, manures, and composts can be sampled at the site by suitable augurs and probes. The sixth and seventh factors are the method and frequency of withdrawal, respectively, and the eighth factor is the amount of sample.

In the case of the eighth factor, sample size depends on the heterogeneity of the material, including its physical nature. In an industrial operation, it is also affected by the type of sampling system. The higher the degree of heterogeneity, the bigger should be the sample, and the more frequently should samples be taken. Under industrial conditions, there is usually an abundance of the material to be analyzed. The larger the sample, the more likely is it to represent the population, but for practical purposes, a sample size of 500–1000 g is usually a convenient amount for NIRS analysis of grains. For composite samples that characterize, e.g., a 50,000-ton cargo of grain, a sample size of up to 20 kg is often assembled from increments taken at regular intervals during loading the vessel, with blending of each composite as a new increment is added. Fibrous materials, such as forages or wool, pose additional difficulties associated with their bulk, springiness, and low bulk density. Here, sample size is a question of volume, rather than weight.

The guidelines for sampling presented here apply particularly to solid materials. Liquids and slurries call for different methods of sampling. Water is the main constituent of most of these materials. Water contains no constituents except total solids, so the application of any analytical method, including NIRS, is for the determination of the amount and composition of the solid material present.

High-moisture materials are the most difficult to handle. High-moisture grain is difficult to grind, e.g., for reference analysis, and impossible to grind without variable and uncertain moisture loss. For reference analysis of fresh forages, silages, vegetables, fruits, meat, fish, soils, and manures, the sample preparation steps will differ. All have to be dried and processed before reference analysis. Fresh materials are more difficult to prepare than are high-moisture grains. For example, for scanning fresh forages and corn silage by NIRS, size reduction by hand chopping is usually necessary, because of the size of the original material. A useful system is to determine the moisture content and scan such materials the way they are at the time of sampling or receipt. Then subsample, dry and grind them for analysis and determine the moisture content of the material at the time of reference analysis. The calibration model is developed using the scans on the fresh material, and the reference data has to be converted to the original moisture content.

Whether or not results have to be reported on a constant moisture basis, like moisture-free, it is important to test the moisture content both before and after sample preparation. This minimizes the risk of inducing significant and variable biases in the reported results. A lot of confusion in relating NIRS to reference data can be avoided by reporting all results on a constant moisture basis. An example follows. It shows a bias of less than 0.1% for the determination of protein content when the results were reported on the constant moisture basis of 12%, whereas the apparent bias was over 0.3%, based on the reference analysis, had the results not been converted to the same moisture basis. Many of the discrepancies between NIRS and reference data can be traced to misinterpretation caused by differences in moisture contents between the samples used for NIRS and reference analysis.

Protein (NIRS) as-received = 13.62% (13.6)
Moisture (NIRS) as-received = 12.0%
Protein (Reference) = 13.97% (14.0)
Moisture (at the time of reference testing) = 10.2%
Protein (12.0% moisture basis) = 13.69% (13.7)

18.7 HOW CAN THE ACCURACY OF SAMPLING BE DETERMINED?

After a sampling system has been established, the sampling error should be determined. But before this step is undertaken, it is advisable to determine the degree of tolerance that will be acceptable in the application of future analysis. For example, if a tolerance provided by a standard error of prediction (SEP) of ±0.3% protein content will fulfill the requirements of the entire testing operation, there is no need to undergo the extra work and expense in striving for a SEP of 0.2%. The degree of tolerance is related to the sampling error in that, provided that the overall testing error, including sampling error, is within the degree of tolerance accepted by the operation, it is satisfactory.

The sampling error can be determined by the analysis of a number of samples of the material to be analyzed (at least 10), and by the SD and CV of the results. The samples should be taken by using all of the steps that will be used for ongoing testing by NIRS. But the determination of NIRS sampling error is complicated by the fact that the testing error is a critical element of the sampling error, so it is essential to determine the SD of the test method (SET) before attempting to evaluate the sampling error, which is a separate operation. To save the expense and time of testing, NIRS can be used to determine the sampling error, because NIRS is the method that will be used in the operation. The CV is a useful guideline. If the CV of the standard error of the results of repeated testing the samples is high, based on NIRS or reference analysis, but the CV of the SEL (SET) is low, it means that the sampling error is high. If this is the case, the reasons should be determined and corrected. Sampling error is part of the total error, and it is hard to run an efficient analytical system if the sampling error is high.

TABLE 18.3
Influence of SEL on Total Sampling Error

TSE	**0.341**	**0.341**	**0.341**	**0.341**
SEL	0.125	0.200	0.250	0.300
ASE	0.295	0.223	1.158	0.077
% ASE of TSE	86.6	65.6	46.2	22.6

The contribution of the SEL (SET) to the overall error can be calculated by the following formula:

$$\left((\text{TSE})^2 - (\text{SEL})^2\right) \div (\text{TSE})^2 \times 100 = \text{ASE}, \tag{18.1}$$

where TSE=overall error, SEL=error of laboratory testing, and ASE=actual sampling error corrected for SEL. Table 18.3 illustrates how the SEL can influence the overall sampling error.

No matter what chemometrics are employed in the development of the calibration, the instrument regards the calibration equation as a constant, and generates its results by multiplying the spectral data of any sample scanned by the equation. So the effectiveness of NIRS analysis is very dependent on the attributes of the spectra. If the spectral precision is poor, this will inevitably result in impaired reproducibility of analytical results. It is worthwhile to determine, as scanned, the spectral variance in the material for which the calibration is intended at the outset. For example, the variance of spectral data from chopped corn silage will be significantly higher than that of wheat flour.

No calibration is required for the determination of spectral precision. It can be determined by scanning one or more typical samples of the material 10 times, and recording the spectral data at two or more wavelengths. The sample should be re-loaded into the sample cell between each loading, and the recently scanned subsample should not be returned to the original sample. After the tenth scan, the sample should be left in the cell and re-scanned nine more times without disturbing the cell. The SD of the spectral log $1/R$ data from the first ten scans is the variability in spectral data that will be experienced during NIRS analysis of the material (reproducibility). The SD of the spectral data from the 10th–19th scans is the repeatability of the scanning, and is strictly a function of the instrument, because there will have been no variance contributed by the operator. Table 18.4 gives a typical data for spectral precision of whole wheat kernels at two wavelength points.

Spectral precision should be determined for the application of NIRS to any type of material. The CV can be expected to increase for more complex materials such as soil and forages, and especially for fresh materials, and high-moisture materials. Its determination is a useful contribution to understanding both the potential and the results of application of NIRS. Spectral precision cannot be determined unless access to the raw spectral data is possible, such as on-line applications. When a calibration has been developed and is declared suitable for application, the same system can be used to determine the reproducibility and repeatability of the NIRS application.

18.8 SAMPLE PREPARATION

In these days of direct NIRS testing of intact commodities like whole grain, the errors introduced by grinding and some other aspects of sample preparation are largely avoided. But sample preparation is as important as ever for the reference analysis that is to be used for monitoring instrument performance and for development and extension of calibrations, as well as for materials such as forages, composted manures, and soils. There are several steps in sample preparation. These include the following:

TABLE 18.4
Spectral Precision of Whole Wheat Kernels

Scan	1210 nm	Scan	1210 nm	Scan	2230 nm	Scan	2230 nm
1	0.4741	10	0.4666	1	0.9046	10	0.8738
2	0.4641	11	0.4633	2	0.8871	11	0.8852
3	0.4708	12	0.4702	3	0.9111	12	0.8758
4	0.4549	13	0.4641	4	0.8923	13	0.8711
5	0.4802	14	0.4687	5	0.8947	14	0.8773
6	0.4733	15	0.4674	6	0.6984	15	0.8792
7	0.4691	16	0.4638	7	0.8791	16	0.8823
8	0.4744	17	0.4683	8	0.8663	17	0.8738
9	0.4723	18	0.4662	9	0.8905	18	0.8751
10	0.4666	19	0.4675	10	0.8738	19	0.8756
Mean	0.46998	Mean	0.46661	Mean	0.88689	Mean	0.87691
SD	0.00693	SD	0.00228	SD	0.01478	SD	0.00425
CV %	1.47	CV%	0.49	CV %	1.67	CV %	0.48

1. Identification and labeling of samples at the time of sampling or receipt of the samples
2. Removal of foreign material, unless the commodity is to be evaluated as received
3. Thorough mixing/blending of the sample before any sample preparation or testing is attempted
4. Testing for moisture content as received
5. Moisture reduction, as necessary
6. Accurate subsampling, if the sample is too big for processing
7. Size reduction to the condition whereby the material can be subjected to reference testing
8. Blending the sample after size reduction because grinding usually induces stratification
9. Cleaning of the grinder between samples, where necessary
10. Testing for moisture content after sample preparation at the time of analysis
11. Storage of samples after preparation and before analysis, to avoid changes in composition
12. Labeling of the stored samples.

Sample preparation for NIRS analysis must be achieved in such a way that the sample presented for analysis has not been changed from its original state, to avoid changes in spectral characteristics. This is impossible in most situations for reference analysis. The three main steps in sample preparation for NIRS analysis of agricultural materials are removal of foreign material, reduction in moisture content of fresh materials, and size reduction by grinding or chopping. Foreign material includes large items, such as whole heads of cereals and stones, and small items, such as weed seeds and chaff. These items differ in composition from the original grain and can cause bias in the analysis.

Accurate documentation is the first priority. About 3% of all major errors in laboratory testing are attributable to analytical results being associated with the wrong sample. It is usually necessary to abbreviate the identity, so an appropriate system for identification must be set up. The system should include the year when prolonged storage is anticipated. The durability of markers should be checked, to make sure that the identification doesn't get rubbed off, or fades between sampling and analysis. The ideal marker is one which is stable to heat, time and smudging, non-water-soluble, but removable with a solvent such as alcohol, to facilitate reuse of the containers. Where samples are stored in plastic bags, it is good practice to include the identity inside the bag as well as outside the bag.

Cleaning of grain before analysis is an essential part of sample preparation. Heterogeneity is affected by the presence of foreign material. Large foreign materials, such as seed pods and stones, can interfere with NIRS spectroscopic measurements and should be removed before the analysis. Samples should be analyzed on arrival, with the foreign material included after the removal of the large material. Because the amount of foreign material affects the value and price of the commodity, it is important that the blending preserves the distribution of foreign material as it was in the original material as received. Large-scale cleaning of grains and seeds is carried out in grain handling and processing plants, and detailed descriptions of the equipment and methods involved are outside the scope of this chapter [4,5].

Throughout the early years of application of NIRS in most situations, samples were presented to the instrument in some form of sample cell, after sample preparation. This involved a reduction of the material to a form that could be loaded into the cell, usually by grinding. Different types of grinders are available for processing different materials [6]. The Christy/Norris 8″ hammer mill, the Cyclone sample mill, and the Perten KT 3100 mills have proven records for grinding grains, and the seeds of food legumes, and are widely used in grain sample preparation. The Christy/Norris mill is very robust and can be used to grind seeds such as faba beans and palm kernels, which are extremely hard. It is a direct-drive mill, and noisy, so ear protection equipment should be provided for operators. The Retsch centrifugal grinder has proved to be the most suitable for working with oilseeds. Forages can be ground in any of the Wiley series of mills, or in a Christy/Norris 8-inch hammer mill. For grinding fibrous material, such as forages and straws, the Wiley series of grinders is suitable. The Wiley No. 4 grinder is a large mill and is easy to clean. Grinders vary, and if more than one grinder is used, operators should check that spectral characteristics are not affected by differences in grinders. Figure 18.11 shows spectra of the same wheat sample ground on two Cyclone grinders. Differences of this order would cause significant biases and would affect slopes.

Other than particle size, the most common changes caused by grinding are changes in temperature and moisture content. Samples ground in high-speed impeller-action grinders can become warm during grinding. Materials, like wheat grain, can reach temperatures as high as 45°C while grinding a large number of samples (20 or more) in sequence. Under daily operating conditions, it

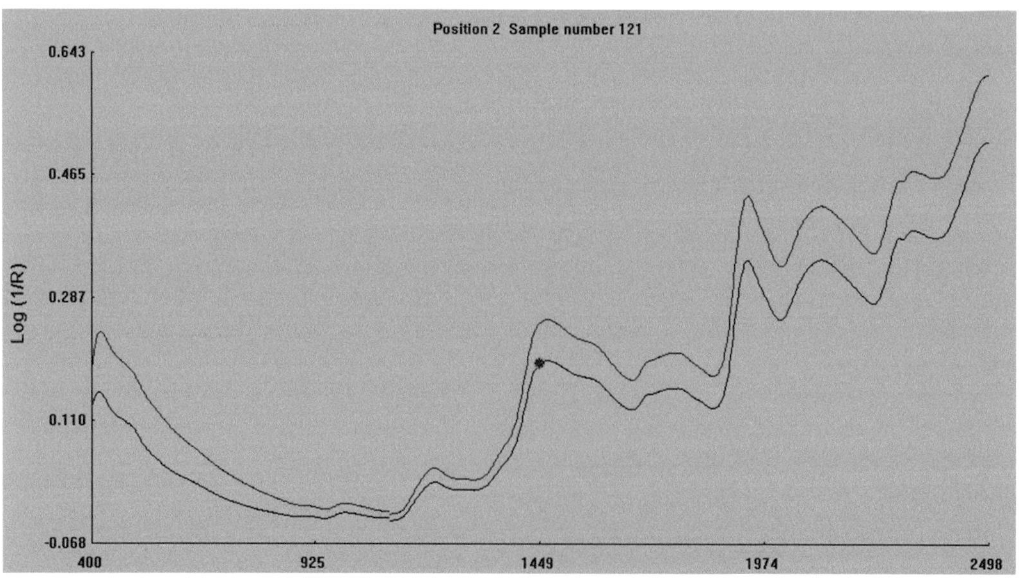

FIGURE 18.11 Spectra of the same sample of wheat ground on two Cyclone grinders.

is usually not practicable to allow samples to cool before testing, so in the case of samples that are to be ground, the samples used in the development of a calibration should also be ground one-at-a-time, and scanned right away.

Grinding the sample during sample preparation results in losses in moisture content that vary depending on the material, the original moisture content, the type of grinder, and the number of samples ground at the same time [7]. For accurate analysis by combustion, the sample should be ground to pass through a 1.0-mm screen in the grinder. Moisture loss for some grains during grinding by the Cyclone grinder is summarized in Table 18.5.

Flour milling, and many other types of processing in which the products are analyzed in-line, must be considered as a form of sample preparation, since the milling operation is the last thing that happens to the sample before analysis. In a flour mill, recently milled flour is at a temperature of up to 40°C. But it cools on standing, and unless it is tested right off the mill, calibration development should encompass a temperature range from normal room temperature up to 45°C–50°C.

All mills and grinders are subject to changes in performance due to aging, and wear of internal screens, which increases the size of the aperture. Cereal grains are usually ground to pass through a 1.0-mm screen. Soils are commonly ground to an average particle size of 2 mm using an automated version of the pestle and mortar, after air drying. Commodities like oilseeds are usually ground to pass through a 2.0-mm screen. Where grinding is necessary, all samples must be thoroughly blended after grinding before NIRS (or any other) analysis. This is usually done by mixing in the grinder receptacle, or in the sample container. Blending is also necessary when whole grains or seeds are to be tested by NIRS, to ensure an even distribution of physical and chemical compositions.

Blending of ground material is another essential to successful analysis. The different strata of ground grains have different composition. For example, after grinding wheat, the protein content of the finest particles can be 3%–4% (absolute) higher than the mean protein content obtained after thorough blending. In the preparation of dried forages, care has to be taken to preserve the ratio of leaf to stem during subsampling and sample preparation for reference analysis. Leaves are much higher in protein and lower in fiber contents than stems. Liquids must also be well blended and free from air bubbles or layers. Layers or strata in liquids are associated with changes in composition and/or changes in refractive index and density, all of which can affect spectra, and consequently accuracy and precision. Slurries and other viscous liquids must also be thoroughly mixed. Where practicable, liquids and slurries should be continuously agitated during scanning to prevent layering.

The efficiency of mixing and blending in flour and feed mills, and other processing plants, is sometimes tested by sending samples for the analysis of a minor constituent before and after the mixing/blending operation. This method is subject to the errors in the analytical tests used, as well as to delays in obtaining the analyses. The efficiency of mixing/blending can be determined simply and quickly by using NIRS spectral data with no chemical analysis. The first step is to determine what wavelengths to monitor. It is a good idea to monitor wavelength points that are known to be

TABLE 18.5
Moisture Loss during Grinding Grain

	Moisture Loss % Commodity		
Orig. Moisture %	Wheat	Durum	Barley
17	3.2	3.0	4.0
15	3.2	2.6	3.5
13	1.5	1.4	2.5
11	0.9	1.1	1.9
9	0.3	0.5	0.8

associated with the known constituents, such as water (960 and 1410 nm), oil (1210 and 2306 nm), and protein (1692 and 2054 nm).

Samples (at least 10) are taken and scanned after about 4–5 s of blending, and also after blending is believed to be complete. If the blending is efficient, the SD of the spectral data after 5 s of blending will be several times greater than the SD of the peak heights at the same wavelength points after blending is complete. Table 18.6 shows the influence of blending on the spectral data of wheat flour at two wavelength points. For brevity, only five replicates are shown. The variance in the log 1/*R* of the unblended series at both 1410 and 2306 nm was about seven times higher than that of the blended series. This showed that monitoring the variance in log 1/*R* provides an excellent estimate of the efficiency of blending.

For calibration purposes, the moisture content should be determined before and after preparation, and the analytical data reported on the basis of the original moisture content, except in cases where data have to be reported on a constant moisture basis. For the preparation of items such as feed mixes, or computation of total dietary intake by cattle or sheep, where a large proportion of the intake may be as fresh feed, the composition with respect to nutrients such as protein, fiber, and energy should be reported as a function of the actual composition of the feed as eaten.

The moisture content of fresh material should be determined at the time of sampling. The sample should be scanned at the time of sampling, or as soon as possible after sampling, and the sample protected from moisture loss before the preparation for reference analysis. The spectra recorded at the time of sampling will change accordingly, but composition may change to the extent of introducing error in calibration development and reporting of results. For example, a sample of corn silage may be assessed for protein content, to determine the supplementation required for the diet of the animal. The sample needs to be dried and ground for reference analysis. Dried fibrous material is very hygroscopic, and may pick up moisture after grinding, so the moisture content should be checked before reference analysis. The results for protein content of the sample as tested would not reflect the composition of the silage as feed, so the result used as a guideline for supplementation

TABLE 18.6
Changes in Precision of Log 1/*R* Values as a Result of Blending Flour

Replicate	Log 1/*R* at 1410 nm	Log 1/*R* at 2306 nm
1a[a]	0.3376	0.5318
1b	0.3364	0.5316
2a	0.3344	0.5277
2b	0.3369	0.5325
3a	0.3373	0.5329
3b	0.3369	0.5322
4a	0.3425	0.5423
4b	0.3376	0.5336
5a	0.3362	0.5303
5b	0.3371	0.5328
SDa	0.00301	0.00555
Mean log 1/*R*a	0.3376	0.5330
CVa	0.892	1.042
SDb	0.000432	0.00074
Mean log 1/*R*b	0.3370	0.5325
CVb	0.128	0.139
Ratio CVa/CVb	6.97	7.50

[a] Series a samples blended 5 s; series b samples were thoroughly blended.

would have to be adjusted to the original moisture content of the material. The protein content as eaten by the animal can be calculated by the following formula:

$$((100 - \mathrm{MS}) \div (100 - \mathrm{MT})) \times \mathrm{PT} = \mathrm{AP} \quad (18.2)$$

MS = moisture content at the time of sampling, MT = moisture content at the time of testing, PT = protein content as tested, and AP = actual protein content as eaten by the animal. An example is given for the NIRS analysis of a fresh forage:

Protein (PT)	= 21.8%
Moisture (MT)	= 3.8%
Moisture (MS)	= 73.4% (26.6% dry matter)
Protein as eaten (AP)	= 6.03%

The diet of the animal would need to be supplemented based on the actual protein content as eaten (6.0%) for the maximum productivity or weight gain. Application of NIRS enables the moisture content to be obtained directly from the fresh material at the time of sampling, once the appropriate calibrations have been developed and evaluated.

Drying is necessary in many high-moisture materials such as fresh corn, silages, and other substances, before grinding and before the analysis. Drying of grains and forages is extensively documented [e.g., Refs. [8,9]]. Fresh corn grain is often harvested at well over 30% moisture and must be reduced to 15% or lower before grinding or storage. This can be achieved by forced air drying, at temperatures of not higher than 40°C. Silages contain significant quantities of volatile fatty acids and esters, which can be lost on drying, with concomitant changes in nutritive value.

Changes in moisture content are dependent on the original moisture content of the sample and also on the relative humidity of the atmosphere. The laws of physics are inviolate and moisture exchange with the surrounding atmosphere is inevitable if the moisture content of the sample is not in equilibrium with the relative humidity of the atmosphere surrounding the sample. Meats and fish, and fresh vegetables and fruit all have moisture contents of 70% or higher, and should be analyzed by NIRS (or reference methods) ideally directly, or as soon as practicable after sampling. Other material with very high moisture content, i.e., above 20%, includes silages, fresh grass and leafy forages, peat soils, and manures, and the moisture content should be determined as soon as possible after sampling for the accurate computation of analytical data. An error of 1% or 2% in the reported moisture content will affect the accuracy of determination of all other constituents, and accurate moisture testing is probably the most important aspect of sample preparation for NIRS testing of fresh materials.

18.9 UNDETECTED MOISTURE

Oilseeds and other seeds, the results of which are reported on a moisture-free basis, are usually dried before analysis by reference methods, since the moisture interferes with the determination of oil by extraction. If the seeds are dried whole, then ground for analysis, the ground material will absorb water rapidly and may contain 2%–4% moisture by the time it is analyzed. This undetected moisture can induce significant errors to the reporting of reference results for oil or protein. Figure 18.12 illustrates the difference in the moisture band at 1930 nm for soybeans dried before and after grinding. The moisture content of the samples of "dried" soybean seed studied varied up to 3%–4%, enough to cause errors of up to 2% in protein content. The reason for the absorption of water is the presence of very hygroscopic cellulosic fiber in the oilseed testa–pericarp. If the oilseeds are to be analyzed by NIRS, there is no need for drying the samples before analysis.

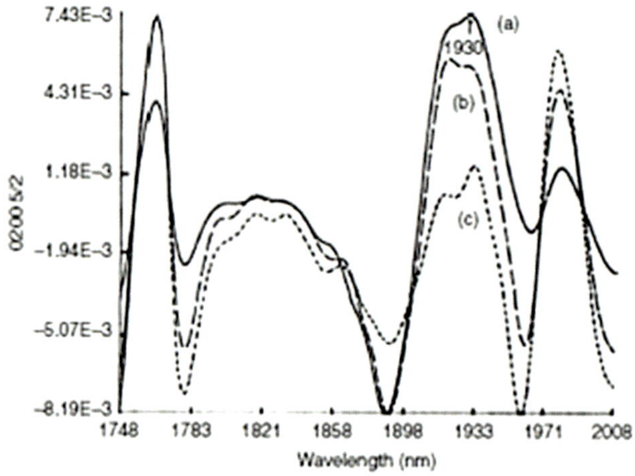

FIGURE 18.12 NIRS spectra of whole soybeans. (a) As-is beans. (b) Beans after drying and grinding. (c) Beans after grinding and drying. Note the reduction in the moisture band at 1930 nm and the commensurate increase in the protein band at 1978 nm.

Moisture can be determined directly by NIRS to enable reporting of the oil, protein, etc., results on a moisture-free basis.

18.10 SAMPLE ASSEMBLY

Before beginning the development of a calibration, it is essential to establish the sources of variance and to identify sources of samples that epitomize the variance factors. Samples of the material that represent variance that has not been included in the calibration will appear as outliers. The most important first step to take when an outlier has been identified is to verify that it is indeed an outlier. First check that all reference data have been correctly entered and that all sample identification is correct. Re-testing by the reference method may result in a significant change in the data. If this happens, the sample has to be re-tested again to determine which of the first two results is the correct one. If the difference between the results of re-testing is within the limits of the SET, the sample is probably a spectral outlier.

A re-scan may show the new spectrum to be significantly different from the earlier scan of the same sample. If the new scan is essentially identical to the earlier scan of the same sample, and the reference tests match each other, the sample is a spectral outlier, by definition. The outliers may be samples from a different origin, which was not considered as a source of variance during the initial assembly of samples for calibration development, and could be the forerunner of more samples of the same type. It is then necessary to seek more samples from the new origin. In the world of industry and commerce, there are no outliers – every sample presented to the instrument either at-line (in the testing area) or on-line has to be analyzed, and the result reported as recorded and used in the operation. Experience will show whether the result is obviously not in accordance with customary results from the same material, in which case the reason for the deviation should be researched.

Particularly in the agricultural and food industries, sources of variance are abundant in the material to be analyzed. The main sources of variance in agricultural materials are summarized in Table 18.7.

The main constituents that affect spectral data of agricultural materials such as grains, seed, and forages are the moisture, oil and fiber contents, and the physical texture. Interactions among these constituents, and their associated absorbers, combined with changes in physical texture, initiate changes in the spectra. Of these, the most influential constituent is water. At high moisture levels, all

TABLE 18.7
Sources of Variance in Agricultural Materials

Material	Variable
Grains and seeds	Type and variety, growing location and season, seed texture, composition, storage
Forages	Type, species, stage of growth, location and season, composition, terrain
Animal feeds	Animal, stage of maturity, diet, sources of ingredients, blending, composition
Foods	Type, texture, physical form, ingredients, composition
Manure	Animal, stage of maturity, diet, sources of ingredients, blending, composition
Soil	Type, location, terrain, time of year, farming practice, depth, composition, weather

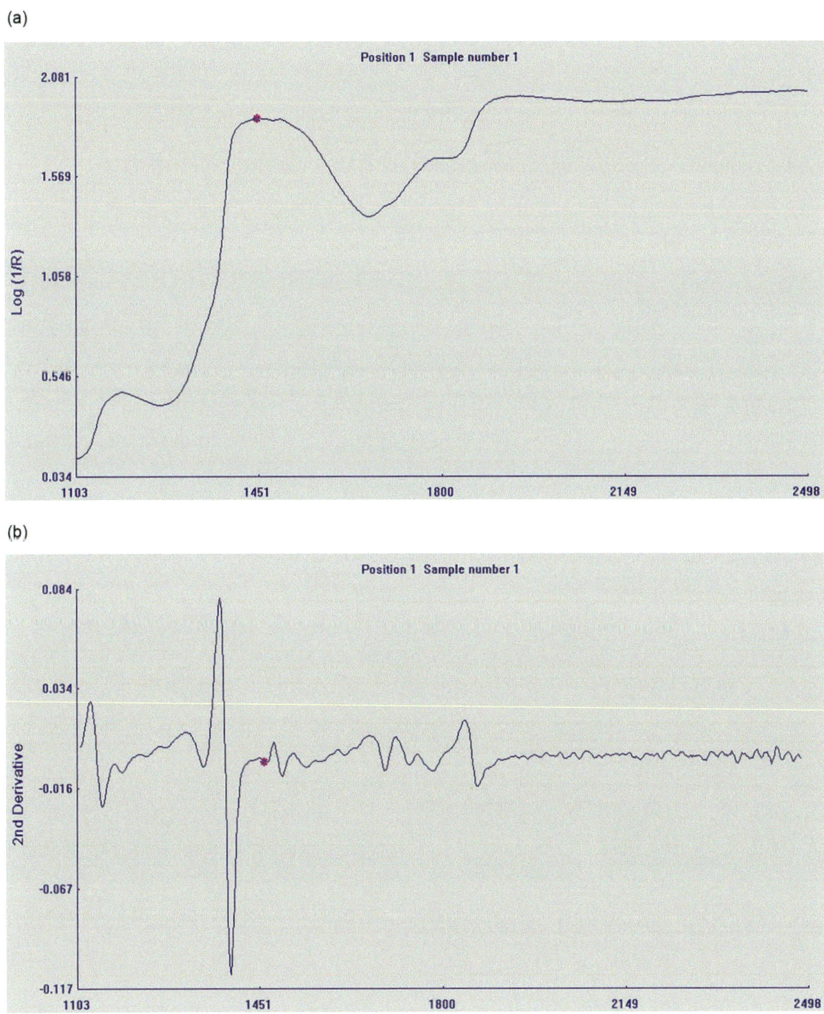

FIGURE 18.13 (a) High-moisture chicken breast meat: log $1/R$. (b) High-moisture chicken breast meat: D2OD 4/4.

absorbers above the widely quoted water band in the 1940-nm area become of no value, due to saturation of the detector. Figures 18.13 and 18.14, respectively, show the spectra of fresh chicken meat and liquid hog manure at high and low total solids contents. Oil affects the particle characteristics of oil seeds such as canola, sunflower, and flax seed after grinding, causing the ground material to

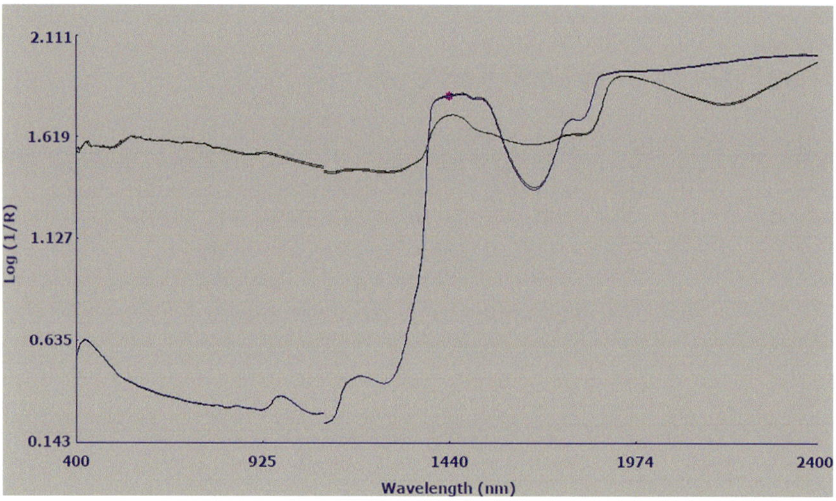

FIGURE 18.14 Liquid hog manure: upper spectrum 12% DM; lower spectrum 1% DM.

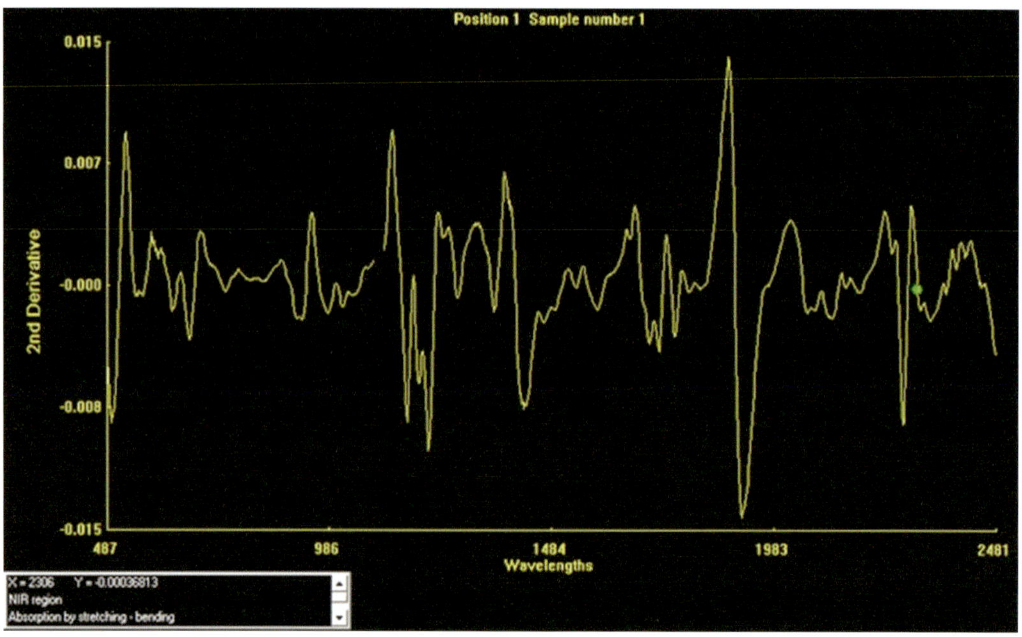

FIGURE 18.15 Whole sunflower seed: second derivative 4/4. The cursor is at 2306 nm.

paste, rather than remaining in discrete particle form. The physical form and texture of agricultural materials are further variables that have to be factored into sample assembly. Spectra are strongly influenced by the degree of hardness and softness in cereal grain, while the fiber content and the nature of the fiber affect the spectra of forages. Grain texture can be satisfactorily predicted by NIRS because of its effect on the spectra, which is independent of composition. A curious feature of intact sunflower and soybeans, even though sunflowers may contain 45%–50% oil, is that they may contain a constituent that inhibits the appearance of the familiar absorbance bands at 2306 and

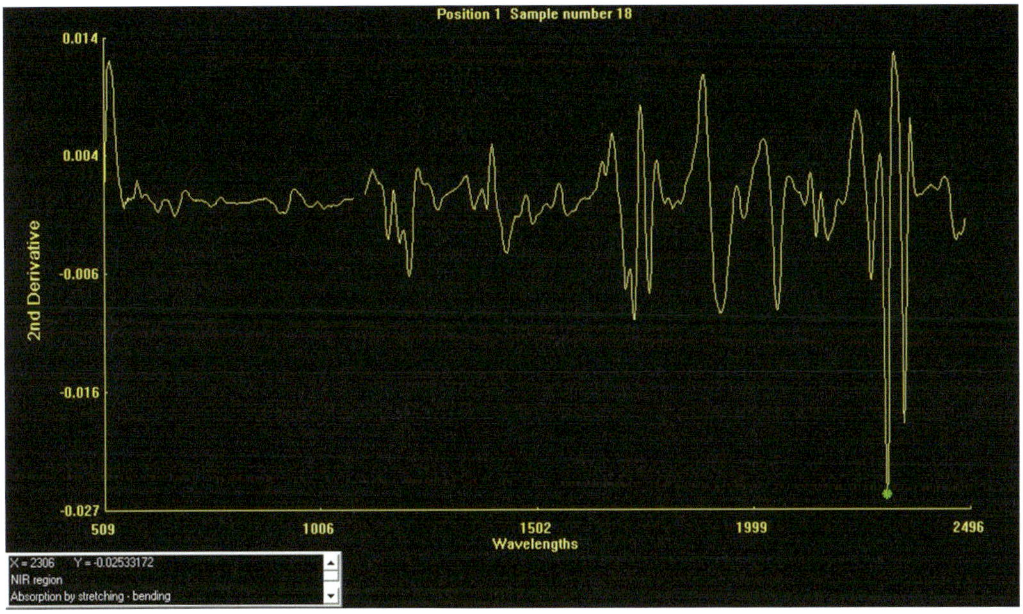

FIGURE 18.16 Ground sunflower seed: second derivative 4/4. The cursor is at 2306 nm.

2346 nm. Both bands are prominent in the same samples after grinding (Figures 18.15 and 18.16) and the oil and moisture contents can be determined accurately by NIRS in the intact seed.

It is difficult to assemble sufficient samples carrying the required range in moisture at the same time as the range in other constituents. Moisture ranges can be generated in wheat and barley and some other cereals by tempering, but grains such as corn, soybeans, and oilseeds cannot be tempered so effectively, while it is impossible to induce ranges of moisture artificially in most industrial materials. In terms of grains and forages, the best time to assemble grain samples for calibration development for moisture under natural conditions is at harvest time, when the moisture range is usually at its extremes. The labile nature of moisture content is a serious source of error in the development of calibration models for the accurate prediction of moisture by NIR or NIT spectroscopy.

Growing location is influenced by terrain, soil type, and farming practice, all of which affect spectra to varying degrees. Variance induced by the growing season is the most time-consuming factor in developing a stable calibration model in commercial grain crops, because it can take several years. Growing seasons appear to have a unique tendency to introduce spectral variance, mainly due to changes in texture. The best way to accommodate this is to monitor the first few (20–50) samples harvested in a new season, to detect whether the existing calibration can be used to test the new season's samples without bias. If bias does appear, the calibration should be strengthened by adding new samples early in the season until no bias is detected. After a few seasons (usually 3 or 4), the calibrations should be sufficiently stable to accommodate any subsequent seasonal variability. The ANN (artificial neural network) calibrations offered by some instrument companies for wheat protein and moisture have proved to be very stable to the seasonal effect, mainly because spectra from several seasons and many locations have been assimilated in their assembly. Because of environmental factors such as temperature and humidity, seasonal variance is applicable to many agricultural and biological materials, including those of animal origin.

The stage of growth is an important factor in developing calibrations for forages. Lignification that begins to develop as plants mature changes the texture of the material. Lignin is a complex polymer that cross-links with cellulose and adds strength to cell walls. This allows maturing plants to support the growing weight of seed development, and literally keeps the heads above ground. Lignification affects the particle characteristics of the ground sample that has been prepared for reference analysis, and affects the precision.

The most important factors affecting soils are the physical composition and moisture content. Soils vary in texture from very light sands and loams to heavy intractable clays, and this affects presentation of the sample to the instrument, as well as sample preparation for reference analysis. Long-term local variations in terrain such as elevated and low-lying areas affect soil texture and characteristics like fiber content. Drying manures and silages can cause losses of volatile constituents, such as ammonia and low-molecular-weight fatty acids. Drying and processing soils, manures, composts, and similar materials for reference analysis cause irreversible changes in texture and other physicochemical parameters, and these materials should be scanned in their natural state where practicable. Slurries like liquid manures vary in total solids content, which changes their diffuse reflectance or transmittance characteristics. Liquid manures and milk both tend to sediment very quickly on standing, and must be agitated thoroughly before scanning or reference analysis. Milk differs in that its fat globules sediment upward. Continuous on-line application of NIRS to liquids is becoming increasingly exploited, and largely obviates the need for sampling, sample preparation, and sample cell presentation.

Sources of variance in other industrial applications are usually less complicated than are those of agricultural materials and foods, where there are so many physical and chemical components that interact with each other. Operators of industries that use, or aspire to use, NIRS will identify their individual sources of variance, but the principle of acquiring samples for calibration will be the same as described.

Sources of variance include changes in the testing environment. Local conditions include altitude. Industrial conditions include daily fluctuations in temperature, relative humidity, dust intensity, and barometric pressure. Temperature is one of the most important of all variables that affect the application of NIRS. Even small changes affect the spectra. To accommodate variance factors arising from local operating conditions, samples for the development of calibration models should be scanned during normal operations, whether in the laboratory or under industrial environments, including out-of-door conditions. Normal operating conditions may include evening and night shifts, and samples for calibration should be assembled under all operating conditions. The atmosphere when the test area is entered for the first time in the day will change when people enter it due to changes in temperature, and relative humidity, and dust levels. Because in the working world analytical work will be carried out during the working day, including shifts, where practicable samples intended for calibration development should be collected all through the work-time. Out-of-door applications will unavoidably bring about bigger ranges in environmental conditions.

Earlier guidelines for assembly of samples for calibration development recommended assembly of equal numbers of samples at every level of composition, but if the samples are distributed in normal (Gaussian) fashion with respect to composition, the results of prediction will tend to regress toward the mean (Figure 18.17). Regression to the mean causes results at the high end to appear to be lower than they are, while results at the low end will appear to be higher. The amount by which the NIRS result of a given sample will differ from the true (actual reference) result on prediction depends on the distance of the predicted value of the sample from the population mean, and the coefficient of correlation between NIRS and reference data. The apparent deviations that can be expected can be calculated using the following formula:

$$\text{Apparent result}(x) = \bar{y} + (y1 - \bar{y}) \times r, \tag{18.3}$$

where $\bar{y}$ = population mean; $y1$ = reported result; and r = coefficient of correlation.

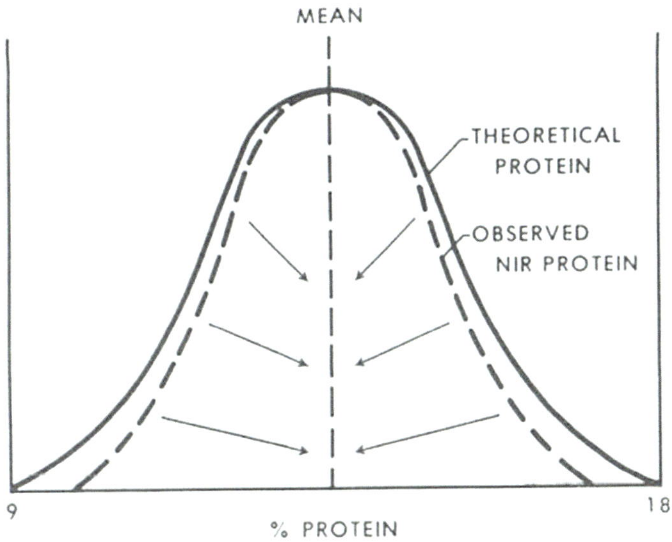

FIGURE 18.17 Regression to the mean.

Example 1

Stated result = 15.6; mean = 13.1; r = 0.98
Apparent result = 13.1 + (15.6 − 13.1) × 0.98 = 2.5 × 0.98 = 13.1 + 2.45 = 15.55

Example 2

Stated result = 15.6; mean = 13.1; r = 0.88
Apparent result = 13.1 + (15.6 − 13.1) × 0.88 = 2.5 × 0.88 = 13.1 + 2.20 = 15.30

Example 3

Stated result = 9.7; mean = 13.1: r = 0.98
Apparent result = 9.7 + (9.7 − 13.1) × 0.98 = −3.4 × 0.98 = 13.1 − 3.33 = 9.77

Example 4

Stated result = 9.7; mean = 13.1: r = 0.88
Apparent result = 9.7 + (9.7 − 13.1) × 0.88 = 3.4 × 0.88 − 13.1 − 3.06 = 10.11

For a given SEP, about 68% of all of the samples will be distributed within one SEP value of the mean result and 95% will occur within about twice the SEP value of the mean. The effect of regression to the mean is minimal if the coefficient of correlation is high, but becomes sizeable at lower values of r. At stated values close to the mean, the effect is also reduced, and an argument in favor of selecting samples on the basis of Gaussian distribution is that because the majority of the population will differ from the mean by only ±1 SD value, Gaussian selection will tend to favor the accuracy of the majority of samples to be tested. For day-to-day analytical work, where large volumes of commercial samples are to be processed and the price depends on accurate analytical data, Gaussian distribution is acceptable for sample assembly. The best way to minimize the effect of regression toward the mean is to ensure the highest r^2 values between reference and spectral data in calibration development.

Sample assembly becomes more complicated if the objective is to develop calibrations with replicated uniform distribution of composition, particularly if more than one constituent or parameter is required, and especially when the constituents/parameters are not correlated with one another, e.g., in the case of moisture and protein contents. When accuracy at extremes of reference data is important for some constituents, e.g., in plant breeding programs, uniform distribution across the range of reference data is helpful, even if it is only achievable for the most sought-after constituents. Assembly of samples with variance in all of the major factors expected to affect the spectra is the most demanding, time-consuming, and usually expensive part of development of a NIRS calibration model, especially where the costs of laboratory reference analysis have to be included.

In the case of NIRS analysis, sample selection and assembly is really a factor only in the development of calibrations and their validation, because the operator has no choice in the samples presented for routine analysis. The number of samples to be taken for calibration development depends on the number of variables that have been identified. The first step in the assembly of samples for calibration development for industrial use is to identify the sources of variance, and having achieved this, to protect from outliers a sample set for calibration development must include all of the sources of error, ideally replicated at least five times. This can result in the assembly of a sample set running into the hundreds. The reason for this apparent "overkill" is that material with the same reported chemical and even functional composition may exhibit substantial differences in spectral characteristics [1].

Some factors and constituents exert a strong influence on the spectra of the same material, whereas the same factors and constituents may have very little influence on the spectra, despite a wide difference in their presence. In Figure 18.18a, the kernel weights of the two similar wheat spectra were 27.4 and 46.3 g per 1000 kernels, but in Figure 18.18b, the kernel weights of the two widely different spectra were, respectively, 34.4 and 34.9 g. In Figure 18.19a and b, the kernel texture of the two wheat samples differed from a PSI (particle size index) of 70.3 (very soft) to 49.0 (hard), while the PSI of the wheat samples with the two very different spectra was identical at 49.8. These two examples are the important characteristics of wheat (kernel weight and kernel texture) for which no absorbers have been identified. Both phenomena can be attributed to a particle size effect. Wheat kernels can be considered as particles, albeit much bigger than the particles of ground wheat.

Turn to constituents for which absorbers are abundant. The two similar spectra of Figure 18.20a represent wheat samples with moisture contents of 10.6% and 17.1%, whereas in the two spectra of Figure 18.20b, both wheat samples had a moisture content of 13.3%. In Figure 18.21a, the two spectra had protein contents of 10.8% and 15.6%, while the wheat samples represented by the two spectra of Figure 18.21b both had a protein content of 10.3%. Sources of variance such as particle size and shape are common to several agricultural materials, such as grain, forages, and soil. Other sources of variance are unique to specific applications, which will be identified by the quality control staff.

Assembling samples for calibration for applications where uniformity in the end product is the goal is complicated by the lack of variance. The lower SD of the samples will affect the Relative Predictive Determinant (RPD). Flour mills can overcome this by using millstreams that differ in composition and particle characteristics. Industrial plants, like animal feed mills, can generate calibration samples by the creation of samples with a range in terms of composition of ingredients wide enough to induce and accommodate the variance in spectral characteristics necessary for the development of stable calibrations. When calibrations are developed on these types of industrially prepared samples, fewer samples are required, but changes in composition, inclusion of different mill

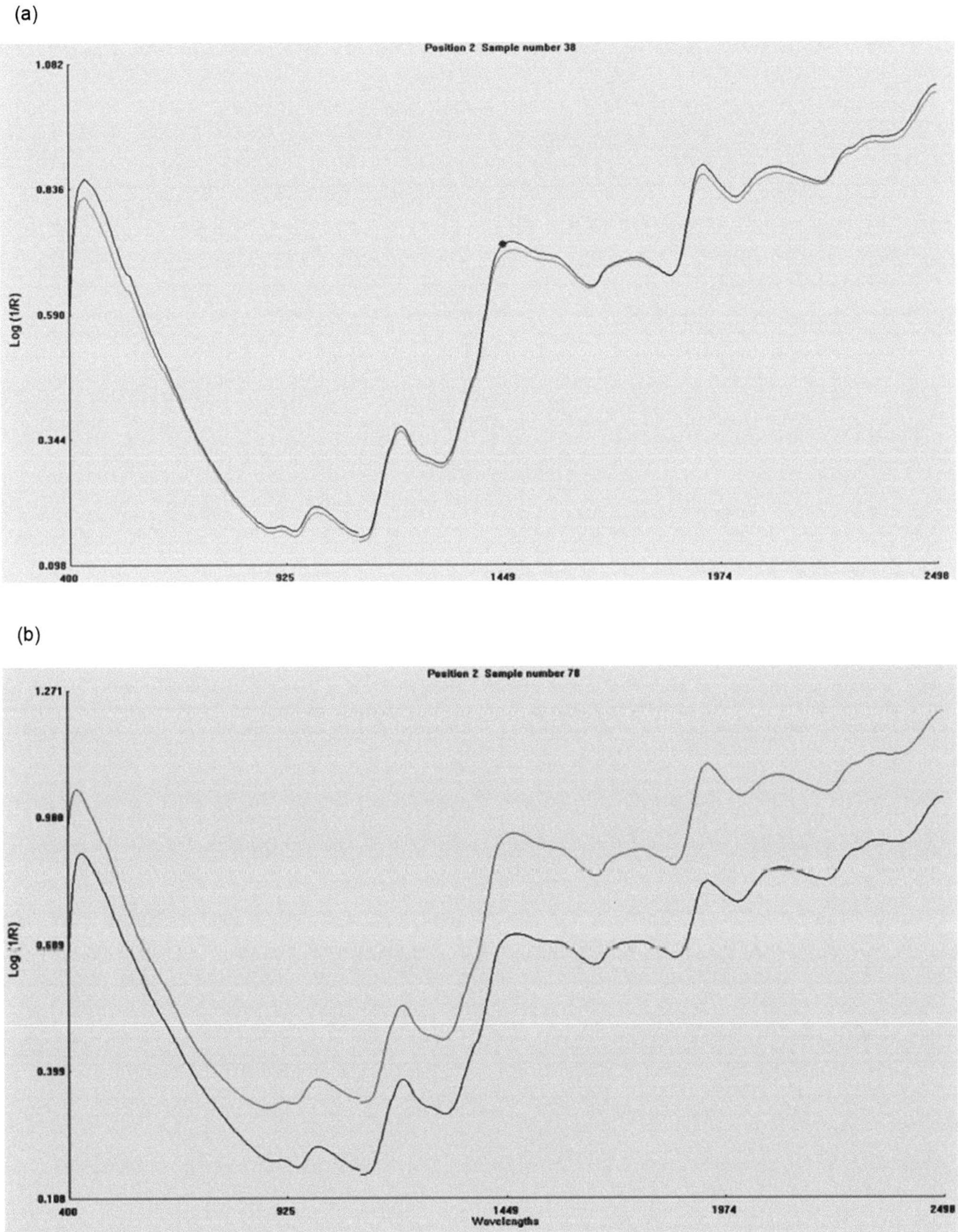

FIGURE 18.18 Influence of kernel size. (a) The spectra were very similar, but the 1000-kernel weights were 27.4 and 46.3 g. (b) The 1000-kernel weights were 34.4 (lower) and 34.9 g.

streams, sources of major ingredients, etc. should be replicated at least three times, and preferably more frequently. Samples for calibration of on-line instruments that are to be used to scan products or product mixes on moving belts or in piping or duct-work are calibrated with static samples, and the calibration is evaluated by subsequent removal and analysis of samples from the system.

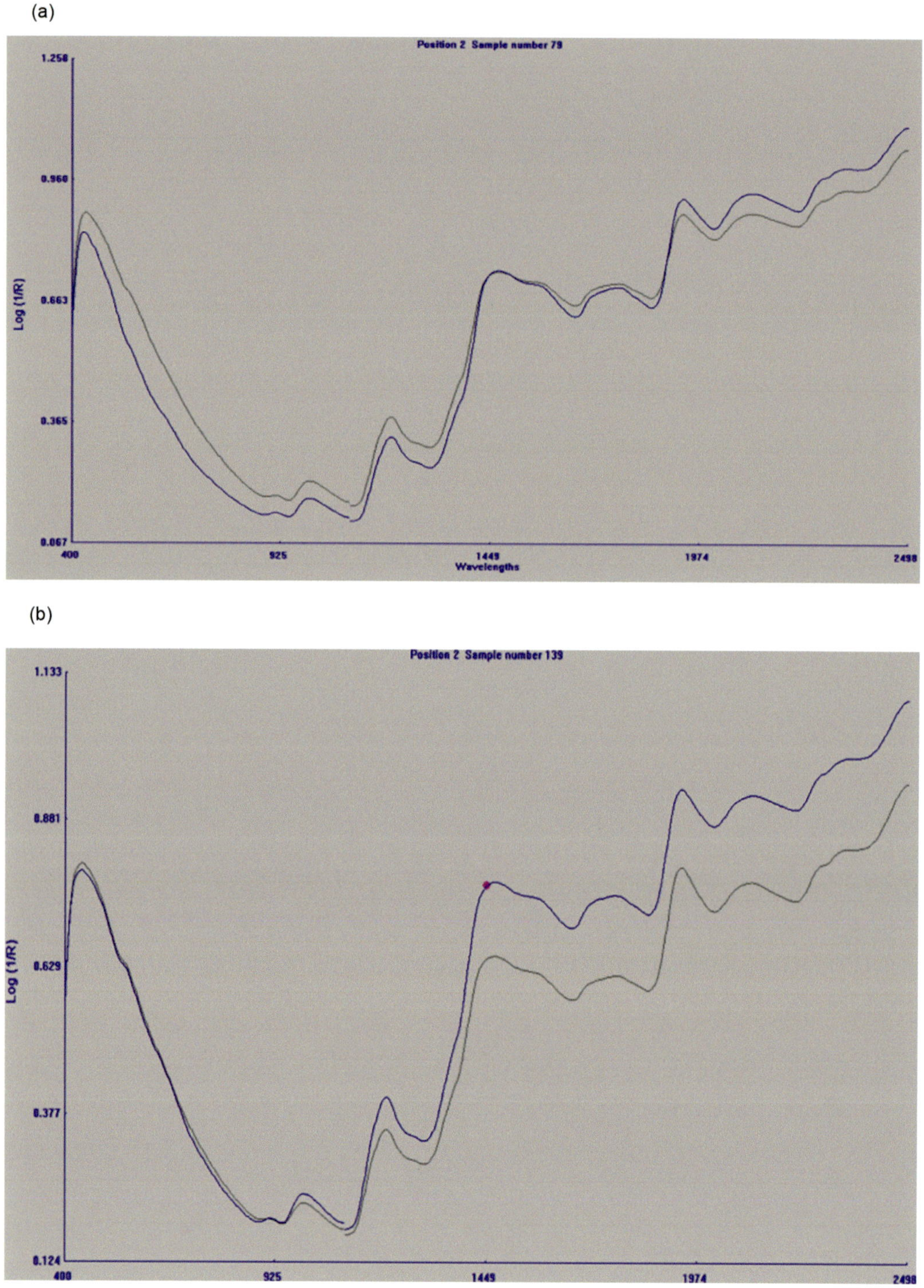

FIGURE 18.19 (a) Influence of kernel texture (particle size index). The respective PSI values of the two wheat samples were 70.3 (very soft) and 49.0 (hard). (b) The PSI of both wheats was 49.8.

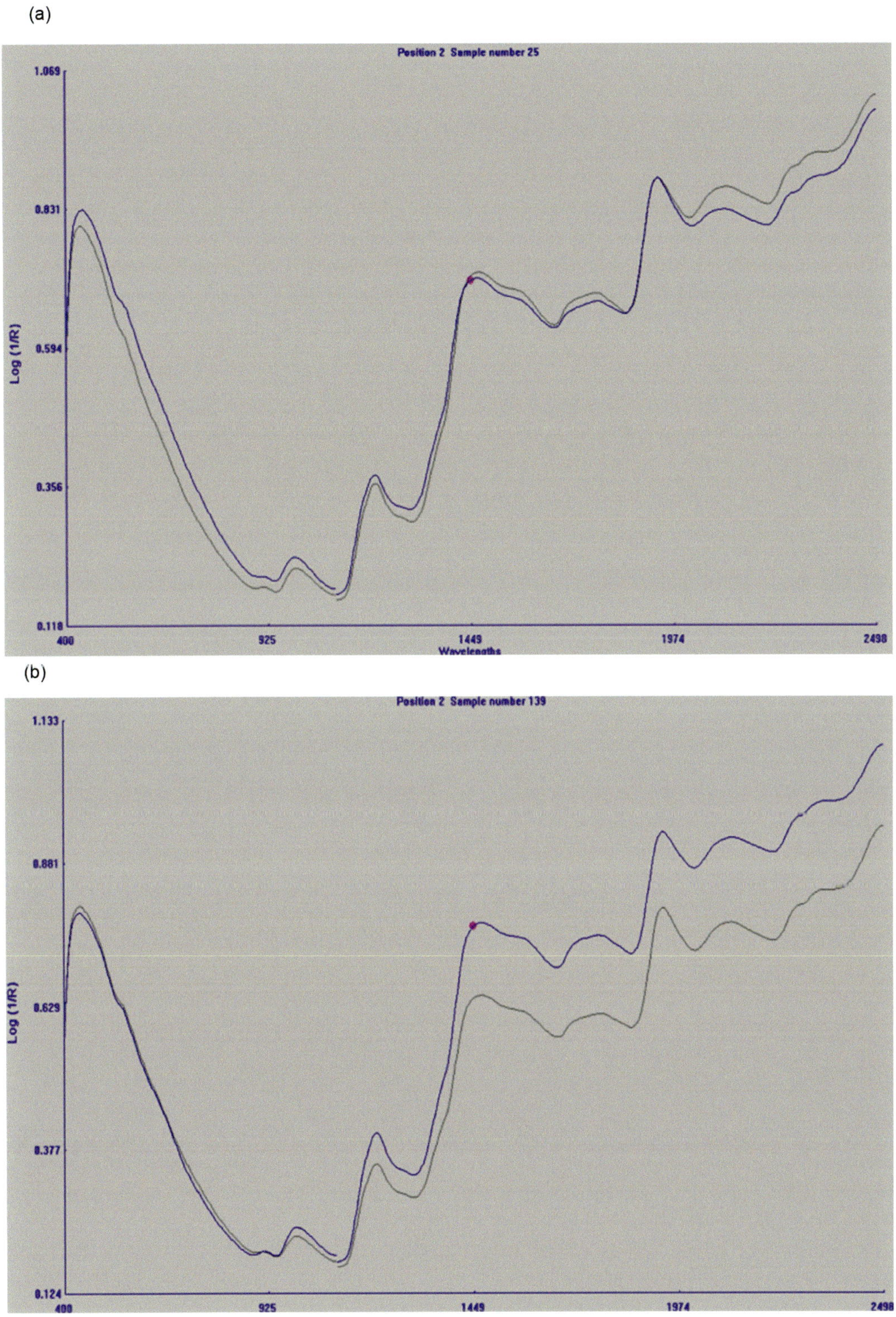

FIGURE 18.20 (a) Influence of water. In (a) with similar spectra, the respective moisture contents were 10.6% and 17.1%. (b) The moisture content of both wheat samples was 13.3%.

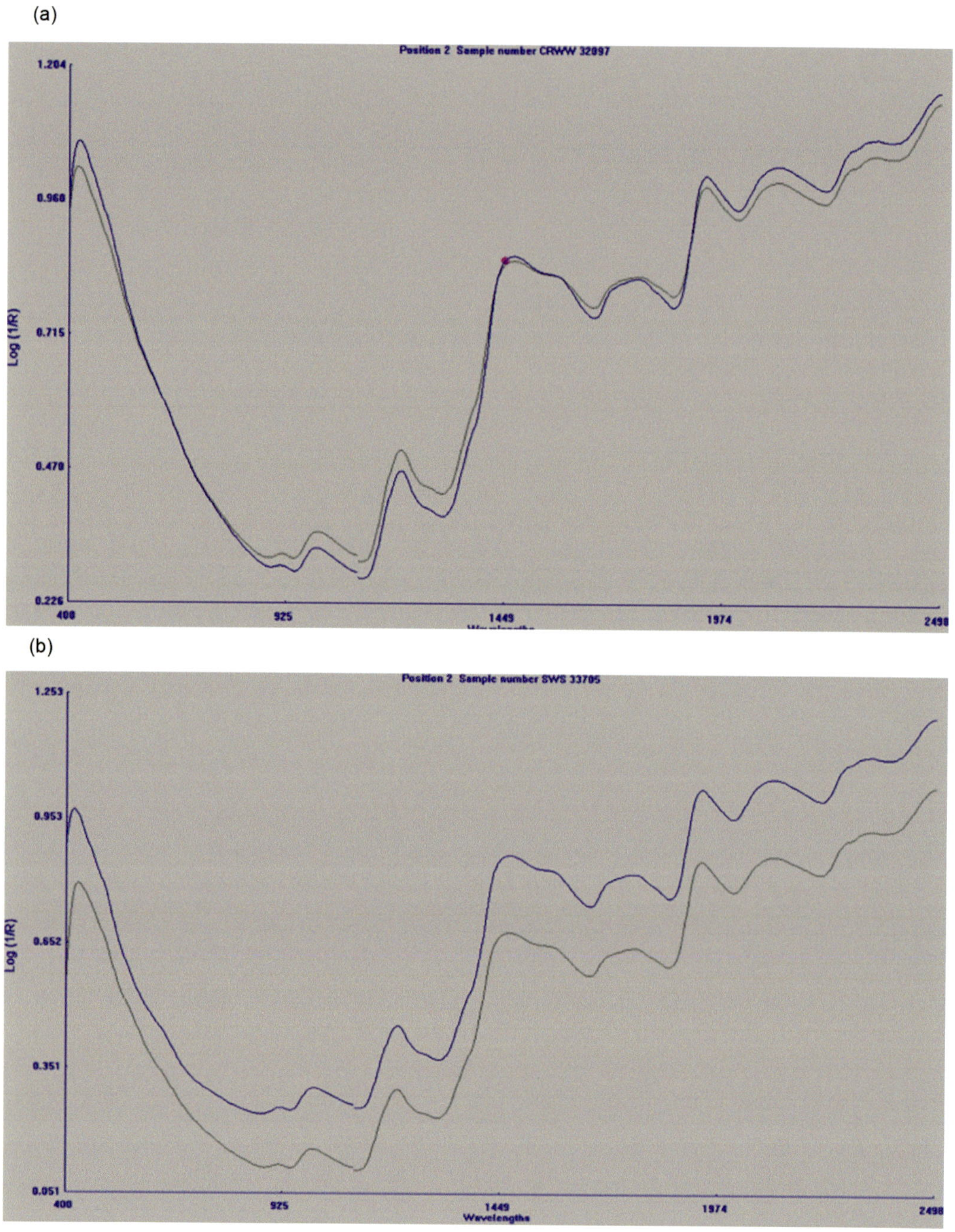

FIGURE 18.21 Influence of protein content. (a) The protein contents of the two wheat samples were 10.8% and 15.6%. (b) The protein contents of both wheat samples were 10.3%.

18.11 SAMPLE STORAGE

Unless the NIRS analysis is carried out directly on the material on-site, samples require protection from change between the sampling/sample preparation stages and analysis, and possible re-analysis, by NIRS or reference methods. Sample storage applies mainly to check samples, and samples assembled for reference analysis for calibration development, which should be scanned at the time of selection. In the most routine analysis situations, including those for quality control, storage of samples is not a factor because samples are either analyzed almost immediately or, in the case of on-site analysis, the sample is scanned, and there is no need or capacity for storage.

When sample storage is necessary, it is essential that the sample presented for NIRS analysis be unchanged from that which was taken in terms of composition and physical form. This includes storage of samples before preparation, and after preparation, but before analysis. In recent times, most grains and seeds are tested by NIRS without grinding, and can be stored and transported where necessary in heavy-duty plastic bags, such as the Ziploc bags of at least 4-mil thickness. Under a microscope, a plastic bag looks like a chicken fence, and water molecules can freely escape, at a rate that depends mainly on the moisture content and the surrounding relative humidity. The thicker the plastic, the more slowly does the moisture content change.

Bulk density of samples has to be taken into consideration when samples are to be stored. Samples of 500–1000 g of materials such as soil or grain have fairly small volume, and many samples can be stored in an area of a cubic meter or so. Samples of forages and similar light materials are bulkier and require more storage space.

18.12 WHAT IS THE SIGNIFICANCE OF THE TRULY REPRESENTATIVE SAMPLE?

Most analyses are carried out on what is purported to be a sample that is completely representative of the whole. It will enable the prediction of the likely functionality of the entire parcel of material represented by the sample in the process for which it is intended. It will signify the value of the sample and therefore the price. Composite samples of large bulks can mask the variance that may occur within the bulk of the material. For example, if a composite sample is taken during storage of farmers' deliveries of grain into a big silo, the composite may show that the moisture content is 12.9%, a safe value for storage. But the individual deliveries may have ranged from 11% to 16% in moisture content. The delivery of, e.g., 12 tonnes of grain at 16% moisture into a silo could cause heating, mold development, and spoilage within the silo, which would not become apparent until the grain was later moved out of the silo for marketing. The heated grain would then become dispersed throughout the grain as it moved out of the silo (Figure 18.22).

When a composite sample is going to be prepared from a series of deliveries, or subsamples of cargo loading, or field samples, the variance or the SD among the samples and the range in analytical data should be determined, e.g., to answer possible complaints. To obtain this important information, each individual subsample that goes to make up the composite should be tested before compositing. This is impracticable with reference testing, but it can be easily and cheaply achieved using NIRS by scanning every subsample as it is taken.

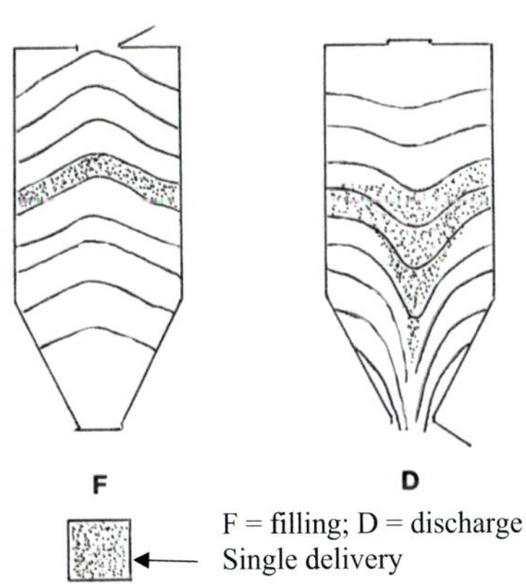

FIGURE 18.22 Grain blending during silo discharge.

18.13 NONSAMPLING

No samples are required for interactance analysis by NIRS, other than the samples (or rather the spectra) that are required for reference testing for calibration development, and subsequent monitoring. In the case of on-line NIRS systems, no samples are required because all of the materials are being continuously scanned. Modern diode array NIRS instruments take many scans per second over a wavelength range of 900–1700 nm. These instruments overcome the potential errors of nonuniform surface by taking and averaging multiple scans. Filter instruments work over the small wavelength range produced by a careful selection of discrete filters, and are equally applicable to on-line analysis for some constituents. Samples of liquids are presented for NIRS analysis usually by diode array on-line instruments by means of flow-through cells, where again many scans are taken and averaged. Flow-through cells intended for use with slurries and liquids that may contain abrasive materials, like sand, usually use sapphire windows, which are very hard and resistant to abrasion.

Interactance requires no sampling because the instrument is brought directly in contact with the material to be analyzed (Figure 18.23). The incoming irradiation is applied to the sample surface usually via the external cables, and the diffusely reflected signal is relayed back to the detectors via the internal cable. Interactance requires no sample preparation, other than cleaning the surface to which the sensing head is to be applied. Because of the small size of the sensor, it should be applied to the surface at least four times, and the spectra averaged. It is applicable to solid surfaces, and to materials such as meat, fish, fruits, and vegetables, which are difficult to sample, or load into sample cells.

18.14 FEASIBILITY STUDIES AND "GRAB" SAMPLES

This chapter has focused on sampling for industrial-level applications of NIRS, but the literature on application of NIRS has burgeoned to countless thousands of papers, including conference proceedings, most of which describe feasibility studies. There are two main types of feasibility study, and the sampling procedure differs accordingly. The first is that of an operation that is considering the adoption of NIRS to replace an existing method of analysis, and needs to determine whether NIRS is the most appropriate method. Sampling for such investigations is mainly concerned with calibration development, which has been discussed above. All major sources of variance should be represented, but the degree of replication need not be so rigorous. The adoption of NIRS is based on the statistics of the feasibility study and, if these are acceptable, final adoption of NIRS calls for further work, including assembly of a much more comprehensive set of samples under working conditions.

The second type of feasibility study is the motivation to present or submit a scientific paper. A high proportion of this type of study concerns more tenuous applications, using NIRS as a

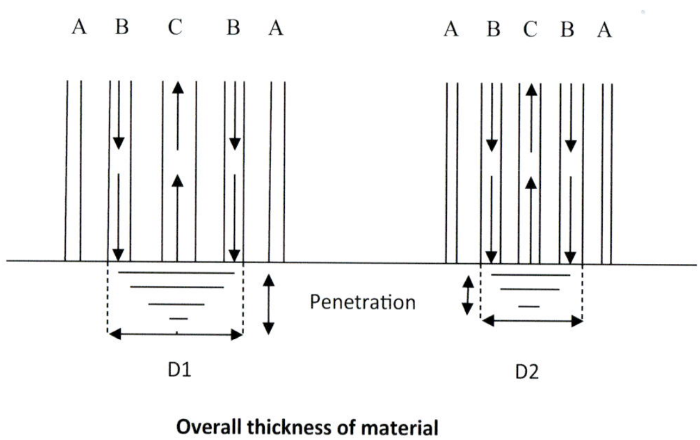

FIGURE 18.23 Illustrating the interactance principle. A, outer wall of cable; B, incoming radiation energy to surface; C, diffusely reflected energy to detector; D, area and depth of material scanned.

replacement for a traditional method, like visual appraisal, which has served the operation successfully over the years. The statistics of such studies are often not sufficiently acceptable to denote feasibility, or justify the replacement of a time-honored system by industry. Samples for such feasibility studies are often "grab" samples, assembled by no particular system. They may be assembled simply by buying them from a store.

Because of differences in the physical nature of materials, automated sampling methods may not be practicable, and in many areas of industry, some forms of grab samples are unavoidable. The practice of "grab" sampling is fairly widely used for NIRS testing in quality control. Provided that the reliability of the grab sample has been established, this is a rational procedure. The integrity of a grab sample taken from a bulk of material can be evaluated by testing for a constituent number of grab samples taken at the same time from different areas of the population, and computing the SD and CV of the results. The "population" could be a single bale of forage or cotton, but at least four samples should be taken from different areas of (in this case) the bale. The statistics will incorporate the error of the test method used. A common form of grab sampling is the sampling of farm truckloads of wheat for testing for protein and moisture contents at the time of delivery. The subsamples are emptied into a suitable container and thoroughly blended to give a composite sample before analysis, using equipment like a Boerner sample divider. This system is acceptable, provided that the sampling continues throughout the unloading and that the amount collected is thoroughly blended before analysis. For practical purposes, samples taken for quality control in flour mills are usually grab samples, taken by diverting flour from the stream of product. After all, two samples taken from a mill stream after an interval of only a few seconds would be truly different samples of the same material in a big mill. The precision of the NIRS testing will have been established during setup of the testing, and periodical monitoring by the laboratory.

Materials such as fresh forages and composted animal manure are usually sampled using the grab sample system, with the mean result for the analysis of the samples having to be accepted as the result, because of the difficulty of developing and blending of such material. Systems for sampling soils and fresh material in the field should be developed on-site by the concerned operation. The purpose of sampling forages is often to determine the degree to which the diets of the animals need to be supplemented to enable the optimum productivity or weight gain. Samples of crops taken in the early stages of growth can indicate the requirements for fertilizer application. Samples of fresh forage can be taken at different stages of growth, but the major factors will be the type of plant, its size, stage of development and moisture content, and the method of processing it to the stage where it can be scanned. For example, it is difficult, if not impossible, to devise an effective method for sampling, for NIRS analysis, whole corn (maize), or even wheat plants, in the field. Concerned operations develop their own system for sampling such materials, but the determination of the reproducibility of such methods, including spectral precision, is recommended.

18.15 SAMPLE PRESENTATION TO THE INSTRUMENT

This section refers to the actual presentation of the sample to the NIRS instrument. Some whole-grain analyzers have very simple sample access. The operator simply has to pour the sample into a hopper at the top of the instrument, the sample is scanned in about 40 s, the results are displayed, and the sample falls into a drawer at the bottom of the instrument. This is ideal for operators, who do not have time to load and clean a sample cell. Sample access is also simple in instruments that use an open sample cell revolving on a turntable and viewed from below. The lower surface is planar, and the sample can be poured into the cell and emptied out after analysis.

During the past three decades, applications of NIRS to intact material have increased, and several instruments present the samples in the form of moving cells, such as the cups offered by Bruker optics, Buchi Corporation, Perten, Foss, and Unity Scientific – all of which revolve on a turntable – and the vertically moving cell offered by Foss. These sample cells hold more material and reduce a lot of the sources of error associated with sample preparation. Samples for calibration

TABLE 18.8
Influence of Sample Cup Size on Sample Size and Instrument Precision

Cell Type	Area (cm^2)[a]	No. kernels	Wt. Grain[b]	Mean Log $1/R$	SD[c]	CV %
Spinning cup	4.9	45	1.51	0.3018	0.00987	3.27
Rectangular	33.8	235	7.87	0.3027	0.00366	1.21

[a] Area of surface presented to instrument.
[b] Estimated weight.
[c] $N=12$ scans.

development for instruments such as the Foss InfraTec Model 1241 and the Perten Model IM9500 can be directly collected from the grain that passes through the instruments during scanning, provided that sufficient variance in composition is represented, such as deliveries directly from farms at harvest time.

For instruments that present the sample in a sample cell, the cell design also has a significant influence on the amount of sample that the instrument scans, and therefore on potential sampling error. The area of sample scanned also affects the precision of scanning. This is summarized in Table 18.8. Note that the precision of the spectral data for the rectangular cell was nearly three times better than that of the spinning cup. The area presented to the instrument by a rotating sample cup of 12 cm in diameter is greater than that of the rectangular sample cell, with commensurate improvement in CV of spectral data.

As a result of interaction between customers and instrument manufacturers for specific applications, sample presentation cells occur in several shapes and sizes, although most companies retain the same sample cell for their off-the-shelf NIRS models. A successful sample cell should be easy to fill without stratification, should be easy to clean, and should provide a sample layer deep enough to ensure diffuse reflectance from the sample only, without the possibility of the signal reaching the detector consisting partly of reflectance from the cell cover. In loading sample cells with ground agricultural materials, the bulk density of the material affects the packing within the cell, and consequently the surface presented to the instrument, and the spectral characteristics. Table 18.9 indicates the bulk densities of some common materials. The data were derived from materials ground on grinders most suited for the material.

Static electricity is a factor when ground material is presented to the instrument in a sample cell with a window. Static electricity is caused when two surfaces contact each other. Electrons can be transferred. Most of the time the electric charges remain in the materials, but if a material carrying a static electricity charge comes into contact with another surface that has lower resistance to electrical conductivity, charges will be transferred. Static electricity can cause samples to become orientated in direction, and samples of ground grains and forages can become arranged in kaleidoscopic positions on the surface of the cell window, which is what the instrument scans. These differences can cause biases. This is the reason why sample cells should be rotated through at least 180° when they are being scanned for analysis. This source of error is eliminated when sample cells automatically rotate. Here, the Cyclone grinder suffers some disadvantage. Its plastic grinding chamber intensifies the presence of static electricity. Quartz windows are more susceptible to static electricity than glass windows. Oats are particularly prone to static electricity, and arguably, ground oats are the most difficult of all cereals in sample presentation.

Some instruments are designed so that the sample is presented in a cell that moves vertically or horizontally, rather than rotating. If the sample presentation system is designed so that the cell rotates on a turntable, the sample can be irradiated from either above or below. If the sample is viewed from above without a cell cover, the surface will be more or less uneven, but this is a variable that can be accommodated by the rotation of the cell. If the sample is viewed from below, the surface will be the bottom of the cell and will be planar (level). Provided the sample is viewed from

TABLE 18.9
Bulk Density of Some Ground Agricultural Materials

Material	Bulk Density (g cm^{-3})
Faba bean	0.639
Lentil	0.542
Wheat flour (hard)	0.514
Corn	0.5
Wheat (Hard Red Spring, HRS)	0.495
Soybean hulls	0.489
Barley	0.39
Soymeal	0.36
Wheat bran	0.349
Oats	0.345
De-fatted canola meal	0.338
Dried alfalfa	0.24
Mature barley straw	0.142

below, open cups can be used with the low-density fibrous materials such as wool and forages if a suitable weight is provided to compress the sample inside the cell. This is not practicable if the sample is viewed from above, and viewing the sample from below gives more flexibility to the sample presentation system. The same principles with regard to path length, etc., apply with regard to whether the sample is irradiated from above or below. Sample thickness should ensure complete diffuse reflectance. A sample thickness of 3 cm or more is adequate in the case of wheat and seeds of similar size. At least 5 cm thickness is recommended for maize and sunflower seeds (which have bigger kernels), or chopped forages, to ensure that light energy is diffusely reflected to the detectors, and is not dissipated, by passing right through the sample.

18.16 SOME FINAL WORDS

The theoretical aspects of sampling have been encompassed in comprehensive accounts, typical of which is Ref. [10]. This chapter has focused on sampling for analysis by NIRS and has drawn attention to the factors that are associated with the application of NIRS in industry in order to obtain the most reliable results. It has not detailed the actual methods for withdrawing samples. In practice, industrial operations that have decided to use NIRS will have developed calibrations and analytical quality control systems that are adequate for their own operation. The chief contribution of NIRS in industry has been the short time per test, which has facilitated on-line continuous analysis, and analysis of all components (subsamples) of composite samples, the individual analysis of which was formerly compromised by the prohibitive time of testing, and costs of application of reference methods to all of the subsamples. In quality control, NIRS has enabled the laboratory to test at more frequent intervals, and to extend its range of testing to accommodate additional quality factors, hitherto precluded by the time entailed. The excellent reproducibility of NIR spectra supports the premise that the NIRS technique is equivalent, if not superior, to most time-honored, but time-consuming and expensive reference methods.

REFERENCES

1. P. Williams. When is an outlier not an outlier? *NIR News* 30: No. 3, 9–11, 2019.
2. P. C. Williams, B. N. Thompson, D. L. Wetzel, G. McLay, and D. Loewen. Near-infrared instruments in flour mill quality control. *Cereal Foods World* 26: 234–237, 1981.

3. J. M. Conway, K. H. Norris, and C. E. Bodwell. A new approach for the estimation of body composition: infrared interactance. *American Journal of Clinical Nutrition* 40: 1123–1130, 1984.
4. R. C. Brook and G. H. Foster, Drying, cleaning and conditioning. In: *Handbook of Transportation and Marketing in Agriculture*, Vol. II. Field Crops, E. E. Finney, Jr. (ed.), CRC Press Inc., Boca Raton, FL, 1981, 63–110.
5. J. F. Lockwood. Ch. 14. The screen-room. In: *Flour Milling*, J. F Lockwood (ed.), The Northern Publishing Co., Ltd., Liverpool, England, 4th Ed., 1960, 209–219.
6. P. C. Williams. A study of grinders used for sample preparation in laboratory analysis of grains. *Cereal Foods World*, 29, 770–775, 1984.
7. P. C. Williams and J. T. Sigurdson. Implications of moisture loss in grains incurred during sample preparation. *Cereal Chemistry*, 55, 214–229, 1978.
8. G. H. Foster. Drying cereal grains. In: *Storage of Cereal Grains and Their Products*, C. M. Christensen (ed.), The American Association of Cereal Chemists, St. Paul, MN, 3rd Ed., 1982, 79–116.
9. P. E. Parker, G. R. Bauwin, and H. L. Ryan. Sampling, inspection and grading of grain. In: *Storage of Cereal Grains and Their Products*, C. M. Christensen, (ed.), The American Association of Cereal Chemists, St. Paul, MN, 3rd Ed., 1982, 1–35.
10. K. H. Esbensen and C. Wagner. Theory of sampling (TOS) versus measurement uncertainty. *TrAC Trends in Analytical Chemistry*, 57, 93–106, 2014.

19 A Framework for Representative Sampling for NIR Analysis – Theory of Sampling (TOS)

Kim H. Esbensen and Rodolfo J. Romañach*

CONTENTS

CHAPTER OVERVIEW

No analytical result can be validated in isolation; no analytical result is better than the quality of the aliquot, or the spectral signal upon which it is based. But the analytical horizon in itself is too narrow for proper validation. The relevant issue is *representativity* with respect to the primary target material, the lot, from which the analytical aliquot (or signal) ultimately originates. Aliquots/analytical signals are the end-products of a compound "lot-to-analysis" pathway in which *all* preceding sampling/signal acquisition operations must be representative in order for an analytical result to be a valid rendition of the original lot material. The attending sampling, subsampling, sample preparation, and sample presentation errors constituting the total sampling error (TSE) can typically be 10–25+ times larger than the NIRs (near-infrared spectroscopy) total analytical

* khe.consult@gmail.com

error alone (TAE_{NIR}); i.e., TSE dominates the relevant total measurement uncertainty (MU) budget. Focusing on analysis (TAE_{NIR}) alone is therefore a breach of due diligence when seen from the point of view of the user of analytical results. This chapter deals with all critical steps *before* NIR analysis, irrespective of whether the TSE contribution is large, intermediate, or small – all cases can be treated identically, including the widespread sensor-based solutions from the process and Process Analytical Technology (PAT) realm.

The key issue of "sampling" is lot heterogeneity and how to *counteract* its adverse influence on sampling/signal acquisition, subsampling, and sample preparation/ presentation processes, which all take place *before* analysis. The Theory of Sampling (TOS) is the guiding framework for valid analysis, including NIR spectroscopy, because this alone can help to eliminate or reduce maximally all such adverse influences. TOS emphasizes the Fundamental Sampling Principle (FSP), which states that all potential units from an original lot must have an equal probability of being sampled in practice, and that samples are not altered in any way after sampling. Units can be particles, particle fragments, or collections of units making up the practical sampling unit, termed "increments". In the realm of NIR analysis, compliance with FSP is often a hidden elephant in the room, far from always properly acknowledged, but here addressed heads on. In this light, there are a lot of potential pitfalls, especially regarding PAT, that must, and can, be avoided, which are dependent on the proper TOS involvement.

FSP is the first of six governing principles (GP) and four sampling unit operations (SUO), which must be honored in order to guarantee sampling-and-analysis representativity. This chapter presents all necessary sampling principles and practical operations in a unified, systematic framework. While it is often argued that the analyst is only responsible for TAE_{NIR}, someone else must then be responsible for controlling TSE. However, this is a most unfortunate division that positively invites serious sins-of-omission. It is infinitely better to view the full "lot-to-analysis" pathway as a unified whole, as a *common* responsibility that *includes* the NIR analyst (indeed also the data analyst, whether of chemometric or statistical inclination).

This chapter presents the minimum TOS competence needed to be scientifically, technically, and managerially responsible for guaranteeing relevant, meaningful, and reliable analytical results in all contexts, under all circumstances and for all types of materials that can be subjected to NIR analysis – illustrated by a set of generic sampling solutions and case examples from typical application fields within the NIR realm.

19.1 INTRODUCTION – THE CRITICAL ROLE OF SAMPLING

The aim of this chapter is to present the necessary and sufficient *preanalysis* principles and strategies for sampling and subsampling, sample preparation, and presentation needed to produce a representative aliquot. Quality control (QC) requires planning of the analytical methods involved as well as of the preceding sampling methods. Preanalysis handling of samples could, e.g., segregate a well-mixed pharmaceutical material, or a complex agricultural sample, leading to erroneous analytical results. Segregation or percolation of particles could, e.g., occur by mechanical transportation or flow towards an in-line analyzer, or could occur within the analytical laboratory by sample container transportation or other handling operations. NIRs (near-infrared spectroscopy) radiation would then interact with a more or less segregated material that would be *different* from the original aliquot, batch, or lot, which is the true target to be characterized. Such induced differentiation can in practice occur at all sampling stages.

But segregation is not the only factor inducing heterogeneity; many lots are significantly heterogeneous "from birth" – indeed, heterogeneity is a material characteristic that occurs at all scales, ranging from lot to aliquot. The cardinal task for the full "lot-to-aliquot" sampling responsibility is: How to provide an accurate estimate of the composition of the lot *in spite of* the heterogeneity encountered, intrinsic or induced, at different stages. It turns out that this task can only be accomplished by properly optimizing all sampling procedures involved, specifically designed to *counteract* lot heterogeneity.

The Theory of Sampling (TOS) to the fore.

The principles promulgated by TOS are *universal* and can be carried over to sampling all types of heterogeneous materials, for any lot size, shape, and form regardless of whether in a static state (typical in analytical laboratories in science, technology) or as a dynamic lot (the second dominating area of the applied NIR, mostly in process industry sectors). This chapter describes how the principles of TOS represent an opportunity, indeed a valuable empowerment, to understand and be in full command of the total measurement uncertainty (MU) for NIR spectroscopy and thereby close the loop for global, optimal analysis. Specifically, TOS' principles can become a valuable asset in the development of NIR methods (Romañach and Esbensen 2016). TOS is as necessary in NIR method development as it is, e.g., for multivariate data analysis, chemometrics (Esbensen and Swarbrick 2018).

TOS has an intricate relationship to MU in that it provides a strongly needed *complement* to $MU_{analysis}$ as shown in first forays in Heydorn and Esbensen (2004), and Esbensen and Heydorn (2004), augmented and completed in Esbensen and Wagner (2014a).

Within the NIR realm, very many approaches have been presented for extracting *increments*, or *samples* destined for analysis, but many of these are unfortunately far from being in compliance with the critical stipulations laid down by TOS; indeed, upon reflection, what results are often merely worthless "specimens" (see glossary for TOS definitions).

Sampling procedures can be characterized according to different criteria, e.g., sampling from *stationary lots or* from *dynamic lots* (moving lots); or sampling can be *manual, instrumental, or automated*, in which latter cases sampling can be an integral part of the analytical process (see further below regarding Process Analytical Technology). The acquisition of NIR spectra for a blending process, e.g., requires the prior decision of installing a sapphire window in the blender, but exactly where? (Igne et al. 2011, Corredor et al. 2015); the same issue is also relevant when analyzing powder flowing through a chute, for example (Vargas et al. 2017, 2018). Such preanalysis actions are part of *sampling* in the Process Analytical Technology (PAT) context whenever a NIR spectrometer is desired to monitor a process. It turns out that there are no PAT issues that do not also involve sampling considerations.

Sampling has very many *purposes*. Sampling may be carried out as a simple *statistical* random sampling from an *assumed* population of individual independent units (iid) – *or* it can be carried out as a TOS-guided sampling that is critically influenced, and aware, of the adverse lot (material) heterogeneity encountered. This latter context is quite different from the standard statistical population assumption as is explained in full below. Sampling may also be performed with a view of other external objectives, e.g., to map the spatial composition variation (e.g., 2-D material characterization of the distribution of an active pharmaceutical ingredient (API) across a tablet surface, for example), or with the aim of determining the mean concentration of an analyte and its variation throughout a lot in time and/or in space (Osorio et al. 2014, 2016). Or sampling may involve the identification of incoming raw materials by NIR at a pharmaceutical manufacturing site, or check for variations in the crystal form of an API (Romañach and Esbensen 2015). Or sampling may be performed to detect counterfeit products (Guillemain, Dégardin, and Roggo 2017). Importantly, sampling is very nearly always involved as a prerequisite to develop NIR calibration models (Romañach and Esbensen 2016, Vargas et al. 2018).

It is always dangerous to focus on a specific situation to develop what appears to be a *reasonable* "sampling solution" (but *without* the necessary TOS competence), and to proceed with the same approach to (all) other lots, materials, and processes with significantly different heterogeneities. Within applied NIR, a plethora of "sample interfaces" and "sample cells" that correspond with this description have been developed and published (Reshadat et al. 1999, Vargas et al. 2017) over time. In this context, there is a real danger for *illegitimate generalizations* as will be laid clear below.

This chapter argues that it is *futile* to try to convince the world of the virtue of universal "for-all-materials", "for-all-purposes" sampling tools/equipment, such as universal sampling valves and flow-through cells. It would likewise be futile to claim that such devices work equally well on *all* materials, and under *all* circumstances. For these reasons, this chapter does not contain sections on sampling of specific materials: grain, forages, food, feed, fibers, milk, wine, liquids, slurries, mixtures, or aggregates. Instead of such a piecewise approach, this chapter introduces a single set of *universal principles* and *unit operations* with which to undertake to design *appropriate* sampling tools, valves, or other equipment that will work on *all* materials under *all* circumstances. *Appropriate* shall be understood as meaning *commensurate* with the empirical heterogeneity of the target material (as estimated by a relevant assessment, a replication experiment (RE), or a variographic characterization, to be laid out in full below). This chapter will also present the specific sampling-for-NIR analysis solutions that are also meant to be generic guides for an even broader application potential.

19.2 CRITICAL SUCCESS FACTORS

The first critical success factor for reliable analysis (not just for NIR) is to acknowledge fully the need for professional and efficient *counteraction* of all of the potentially adverse issues introduced above.

1. All materials are *heterogeneous*[†] – it is only a matter of degree. NIR spectroscopy is highly effective in detecting heterogeneity in materials analyzed. If all samples, subsamples, and the final analytical aliquots are to be valid and reliable (below it is shown that an identical statement can be made regarding the material basis for spectral signals), there are specific demands governing *how to* sample representatively from heterogeneous materials and processes – and *how to* acquire valid spectral signals – and there are objective criteria for *how to* design, manufacture, operate, and maintain appropriate sampling equipment, or how to select and validate the already-existing equipment and procedures, all of which are encapsulated in TOS. This carries over identically to the PAT realm as well.
2. A key issue lies in realizing that there very nearly *always* is a *multistage* pathway "from lot to aliquot", which logically and physically is always *preceding* analysis. This point also applies to the PAT realm (see further below). An important bonus from TOS is that it is possible to focus on specific stages or part operations individually with a view of optimization, as there are no interactions between sampling stages elements in the full lot-to-analysis pathway.
3. Thus, analytical results do *not* spring into existence from analytical instrumentation alone. A comprehensive case for TOS in relation to the conventional focus on the total analytical error (TAE_{NIR}) can only be found in Esbensen and Wagner (2014a).

Case Example I

NIR spectroscopy is often used to analyze, e.g., complex agricultural products that are indeed very heterogeneous (Esteve-Agelet and Hurburgh 2010). In fact, NIR spectroscopy evolved from this type of highly heterogeneous materials (Williams and Norris 2001). For example, the effect of heterogeneity is clearly evident in a study to determine the quality traits of cashew apples (Ribeiro et al. 2016). A calibration set was carefully developed with 129 cashew apple samples from nine varieties, then randomly harvested, at different stages of maturation. The NIR spectra included in the calibration set were the average of three relevant individual fruit regions: apex, middle, and base.

[†] Heterogeneity is always dealt with conceptually, and counteracted in practice, by way of its two complementary components, constitution heterogeneity (CH) and distribution heterogeneity (DH), the former sometimes quantifying grain size heterogeneity as well when especially prominent.

The averaging of spectra was performed since previous studies indicated that the principal source of variation in cashew apple composition (the heterogeneity) occurred *along* the latitudinal direction, increasing from the apex to the base. The study carefully evaluated the results obtained from each of these three regions, to determine whether a significant inter-region variability was observed or not. This evaluation was conducted in hope of possibly identifying one specific region for measurement that would be able to *represent* the entire fruit. However, the study found a significant analytical variability between the three sampling regions selected.

The individual spectra collected at each of these regions were first used for prediction with a suitably calibrated multivariate model. These individual results were then compared with the results obtained for the three averaged spectra obtained at the apex, middle, and base. Bias was determined by comparing the predictions from the single region to those of the three averaged regions. This heterogeneity caused a significant *spatial sampling bias*, which completely dominated over the analytical bias, a situation that potentially could confuse analysts during the method development. Two of the four apple quality traits analyzed were in fact predicted with relatively low bias in the apex region, but the bias was significantly greater in the middle and base regions of the cashew apple, respectively. This study thus indicated that the *compound bias* defines a distinct spatial pattern for each property along the surface of the fruit.

This relatively simple example demonstrates the necessity of being able to discriminate between a sampling bias and an analytical bias. Below, it turns out that there is a fundamental difference between these two types of bias: One can be corrected for, while the other *cannot*, the latter with dramatic consequences if not fully understood. This type of spatial heterogeneity is the reason why *composite sampling* is a must (also) in NIR spectroscopy.

Unrecognized or uncompensated material heterogeneity is one of the main reasons for the occurrence of "outliers" that differ significantly from all other data results in multivariately calibrated NIR models. Because of this origin, such outliers cannot, in any way, be corrected for by any data analytical or statistical means.

As an example, NIR can consistently be measuring less than 5% (w/w) water in a mixture, but all of a sudden a result of, say, 30% (w/w) is observed. This high result would of course confound analysts and would require that a pertinent investigation be performed to confirm, to investigate *why* this occurred, or to *dismiss* the outlier (either outcome to be fully argued and substantiated of course). Multivariate data diagnostics tools may possibly be useful in identifying the root cause for the 30% (w/w) result; however, it may just as well be that the outlier reflects unexpected heterogeneity present in the target material.

This chapter presents an introduction to TOS, simplifying bewildering target material complexities and heterogeneities with a generic set of governing principles (GP) and sampling unit operations (SUO) with which to address all sampling issues, empowering NIR analysis. However, there are so many different situations, cases, lots, processes, materials in need of NIR analysis, that an overview can quickly be lost. This chapter will present the simplest, most efficient introduction to the operative TOS framework, sufficient to guarantee representative analytical results, complete with a plethora of literature from which to appreciate further in-depth understandings.

19.3 THEORY OF SAMPLING (TOS) – A MINIMUM SET OF DISTINCTIONS

The first operational distinction is that sampling/signal acquisition can take place either from *stationary lots or* from *dynamic lots* (flowing/moving/ducted streams of matter). While apparently distinctly different practical situations, the general principles governing representative sampling are in fact *identical*, the dynamic lot situation simply gives rise to two *additional* sampling errors, not met with for stationary lots.

In all sampling situations, there is one dominating characteristic that must be dealt with appropriately: *heterogeneity*. This is the sole reason why a simple "grab sample" is inadequate (although this does produce a very high mass reduction rate). But mass reduction does not by itself lead to representativity – on the contrary. Representative sampling must be viewed as mass reduction, successfully *counteracting* all adverse heterogeneity influences while keeping the essential characteristics of the lot material intact through a series of sampling and subsampling stages. "Counteracting" means that problem-specific knowledge is used to develop adequate sampling processes *commensurate* with the empirical degree of heterogeneity met. In some situations, the typical, expected heterogeneity is known (or well known) and does not demand any special precautions, while the interesting cases are those where standard application of solutions developed elsewhere are *not* appropriate, i.e., when NIR analysis is addressing fundamentally new materials or processes.

To get the right perspective on the demands that must be put on the full "lot-to-aliquot" pathway, the typical sampling rates are illustrated in Figure 19.1.

The uncertainty due to total sampling error effects (TSE) accumulated before the aliquot is delivered to the analytical instrumentation is *dominating*, typically of the order of 10–25× *larger* than the specific analytical errors (TAE), dependent upon the nature of the lot target material and/or whether the sampling procedures used are in compliance with TOS or not. This dictum applies identically to sampling of stationary as well as dynamic lots.

The sampling targets/materials/lots of interest in the NIR realm are innumerable in their kind (commercial lots, commodities, products) and can have almost any constitution, shape, form … and spanning three, six, or more orders-of-magnitude in size, e.g., from shiploads to single kernels. It is anything but trivial *how to* arrive at a representative aliquot of, say, a mass corresponding to the order-of-magnitude of *grams* typically with total sampling rates spanning 10^3–10^9 distributed over several sampling/subsampling stages. TOS is the *critical success factor* for being able to successfully span this gap between the macro-, meso-, and microscales indicated.

Similarly regarding acquisition of NIR spectra: How to optimize the way a NIR beam interacts with a stationary aliquot, e.g., in a sample cup holder? What is the effective volume represented by a NIR signal? These issues are traditionally considered as a strict analytical responsibility, but are in fact also an *optical sampling* issue. It is perhaps not always properly acknowledged that

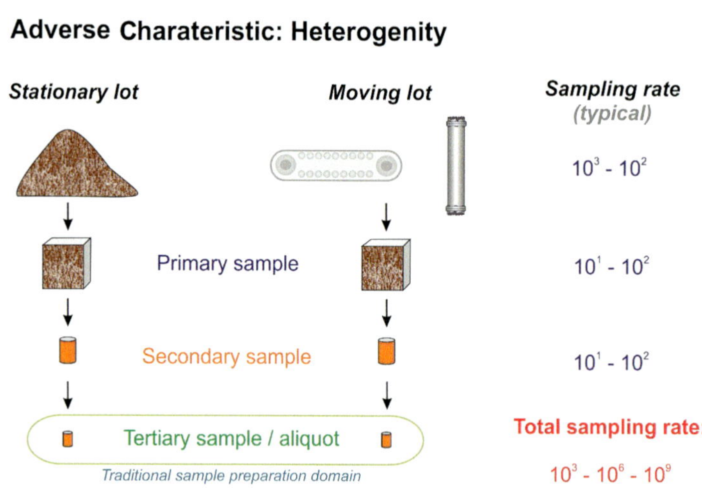

FIGURE 19.1 TOS' framework is identical for both stationary and dynamic lots, the latter indicated by cartoons of a conveyor belt and a pipeline. The principal issue is how to *counteract* the adverse influence from material heterogeneity over effective sampling ratios spanning 1: 10^3–10^6–10^9. Such large sampling rates are only possible by using a multistep sampling process "from lot to aliquot". Copyright KHE Consulting & HERZOG Maschinenfabrik GmbH & Co. KG, brought here with permission.

the design of a "*sample* cell" is in fact designing a "*sampling* cell". This distinction is subtle, yet absolutely crucial.[‡]

How does a NIR sensor interact with a flow of heterogeneous matter, either on a conveyor belt or within a pipeline (process sampling)? The *effective* field of view (FOV) is here actually the specific volume interrogated by the NIR beam for the duration of the spectral acquisition. This is not a simple question as the effective penetration depth varies according to the wavelength (Iyer, Morris, and Drennen III 2002, Bellamy, Nordon, and Littlejohn 2008, Ortega-Zuñiga et al. 2017). These are issues that are of course well known in the NIR analytical and technology realms, which make up the dominating part of the present handbook. Still there are here also crucial specific sampling issues that can only be fully optimized in the context of TOS. It is not only about TAE, but also about TSE+TAE. Sins-of-omission are committed in much of contemporary NIR analysis due to these issues not being paid sufficient attention, perhaps especially regarding process sampling: Inferior sampling procedures and equipment lead to *biased sampling*, which again lead to compromised analytical results suffering from inflated, uncontrollable total measurement errors, $MU_{total} := MU_{sampling} + MU_{analysis.}$ The sampling bias issue is illustrated further below.

There is, however, also an upside. Knowledge of TOS principles constitutes an opportunity to more fully understanding all sources of error in the results obtained from a NIR spectroscopic analysis. TOS offers complete pre-emptive solutions to these otherwise unavoidable sampling bias effects. Figure 19.2 illustrates the *critical success factor* for representative sampling in the case of moving streams/flows of matter. This is all about how the fundamental sampling *increment*

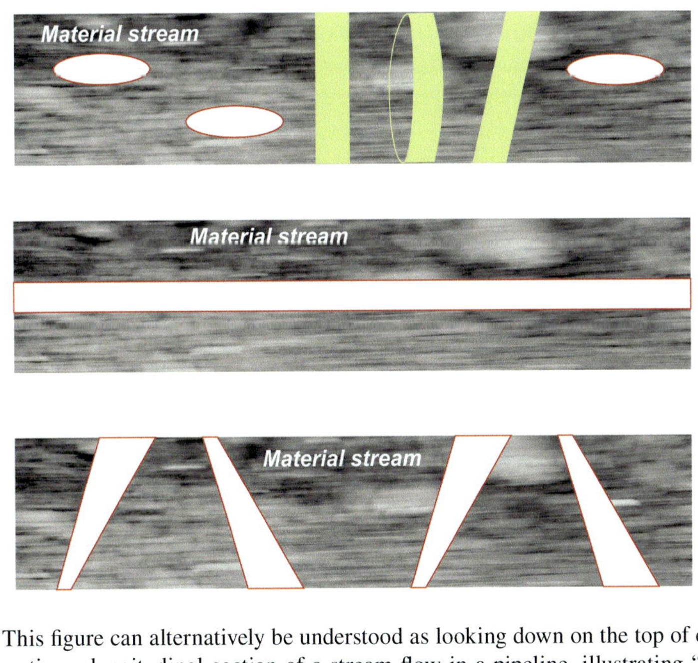

FIGURE 19.2 This figure can alternatively be understood as looking down on the top of conveyor belts, or viewed as representing a longitudinal section of a stream flow in a pipeline, illustrating "correct" delineation and extraction of cross-sectional increments or slices (green, top panel), critically needed in order for sampling to be *unbiased*. All other illustrated increments (white) are fraught with an uncontrollable *sampling bias*. Note incorrect single-unit extraction, in the form of "grab samples", either by "random scoop sampling" (ellipses in top panel), by a partial slice of the stream all of the time, or by unbalanced cross-slices. Grab samples, however, can *never* be representative, as none "cover" a complete cross-slice of the stream. Copyright KHE Consulting & HERZOG Maschinenfabrik GmbH & Co. KG, brought here with permission.

[‡] These reflections on NIR signal acquisition are much more general, indeed applying with equal meaning to all of the PAT; see Esbensen and Paasch-Mortensen (2010).

is defined, delineated, and extracted in practice. The only three "correct" (*unbiased*) increment delineations allowed are delineated with green in Figure 19.2. These are the only delineations that capture a *complete* cross section of the flowing matter, a complete *slice*. The lot material has the same possibility of being sampled regardless of whether it is positioned at the center or close to the edges in the correctly delineated slices pictured. Identical arguments apply for the third dimension, i.e., the thickness dimension. But NIR diffuse reflection spectra interact with approximately the first 1–2 mm top layers of a material only and thus do not analyze the entire cross section (Bellamy, Nordon, and Littlejohn 2008, Iyer, Morris, and Drennen III 2002, Ortega-Zuñiga et al. 2017). NIR transmission measurements therefore provide more comprehensive signal characterization than diffuse reflectance measurements (Sánchez-Paternina et al. 2016). Transmission methods offer the possibility of reducing the increment delimitation error influence (IDE) and increment extraction error (IEE) (see further below) by capturing the entire cross section of the material analyzed (Masithoh, Haff, and Kawano 2016). Note that Figure 19.2 can alternatively be viewed as portraying a longitudinal section through a pipeline – but all aspects of the conveyor belt comments above apply with exactly the same relevance. This exposé illustrates that sampling is often the principal source of analytical result uncertainty in NIR applications because the analytical method typically is under superior control, as witnessed by the complement of this handbook.

The key information in Figure 19.2 is that if increment delineation, extraction, and subsequent handling do not follow TOS' stipulations for "correctness" (see Glossary), sampling *will* be *biased*. This issue is compounded by the fact that a sampling bias is of a fundamentally different nature than the well-known analytical bias, which *can* be corrected for by a simple "bias correction". It may perhaps come as a surprise for many in the NIR arena (and elsewhere) that there is no similar correction possibility for a sampling bias (more on this critical issue below).

The most important lesson concerns the widespread (fatal) *mistake* that it is possible to extract one, singular "sample" that will be representative, if only materialized by the right piece of equipment and/or procedure (physical extraction), or by a "smart" NIR "sample cell/interface". This modus is the convenient "grab sampling" that unfortunately is extensively used in science, technology, or industry. TOS explains in very clear terms why this can *never* be the case – *because of the inherent material heterogeneity.*

19.4 HETEROGENEITY

Heterogeneity is the source of one of the two major classes of sampling error effects. If heterogeneity is not counteracted properly at the relevant scale, significant sampling errors *will* translate into unnecessarily inflated total measurement system uncertainties. Figure 19.3 shows a generic "significantly heterogeneous material", illustrating the essence of what the preanalysis process is up against, whether by physical sampling or by sensor signal acquisition.

The critical feature is the irregular spatial distribution of an analyte, the *distribution heterogeneity* (DH) (Esbensen and Wagner 2015, Danish-Standards-Foundation 2013). The average analyte lot concentration (illustrated by the black squares) is 9.85%; white squares represent the matrix, the part *not* carrying the analyte. The analyte can, e.g., be an API, another essential constituent of interest, or an impurity or a contamination.

Figure 19.3 renders an impression of the typical state of analyte heterogeneity in the overwhelming proportion of materials addressed by NIR. Note that the compound material is *not* homogeneous at the scale of the potential individual increment/sample volume (indicated by red rectangles). In TOS, homogeneous means that all the units of a material are strictly identical to one another (Gy 1995), while in other fields, this term is used to indicate that a high degree of *uniformity* has been obtained, e.g., as shown by a sampling variance below 2% relative standard deviation (Romañach 2015, Sánchez-Paternina et al. 2015). From numerous practical experiences across many application areas, it is well known that all mixed materials will in fact at best only mix to a degree commensurate with a material-dependent "minimum residual heterogeneity", which may

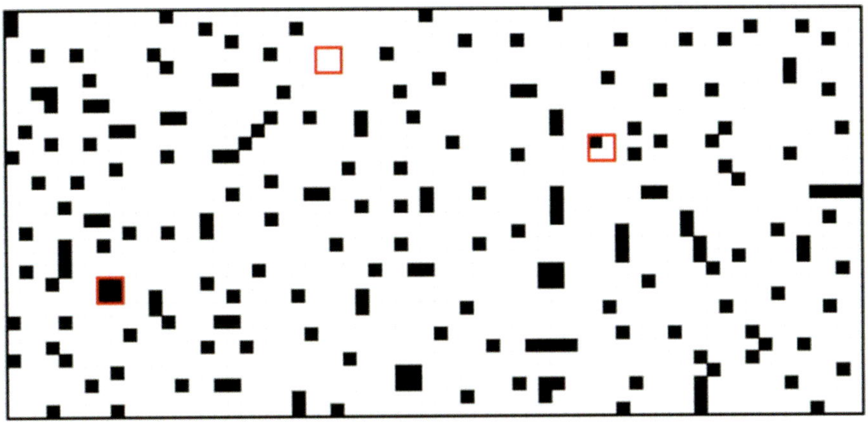

FIGURE 19.3 Principal illustration of the Fundamental Sampling Error (FSE) effect and its origin, heterogeneity. A model system contains 9.85% analyte (black squares), which has been mixed with the maximum efficiency, resulting in a minimum residual heterogeneity. When "sampled" (three grab samples are illustrated with red squares), the analytical results can still be significantly different from the lot average. The "samples" would be affected by the FSE regardless of whether a scoop, a spear, or a PAT sensor is used as the sampling tool. The FSE effect will always be present to some degree for other lot average concentrations, other heterogeneities, other grain sizes, and other sample sizes.

not seldom be higher than this threshold (2%). This issue is described in full detail in the pertinent TOS literature (DS3077 2013; Esbensen et al. 2016).

A single "random sample" of the lot *could* be obtained, e.g., as shown in the top part of Figure 19.3, by a grab sampling procedure. The illustration shows three such single-extraction samples, which (from right to left) turn out to have vastly different individual analyte concentrations, however (25%, 0%, and 100%). The bias, the deviation between each individual grab sample and the average analyte concentration (9.85%), is different for each grab sample. Other, equally realistic locations of similarly "randomly chosen samples" would lead to different quantitative manifestations of the above scenario, perhaps to a smaller degree, but will result in the exact same principal confusion: *Why do samples extracted by the same sampling method from the same* material end up with significantly *different* analytical concentrations? *What* goes on, *why*, and *what/who* is to blame? And *how* to improve the sampling and/or the analysis process?

Note that this situation is *identical* if the pictured grab sampling was performed through the use of a spectroscopic (PAT) method, the red squares now representing FOV area, or volume, of a NIR beam interacting with the lot material.

Consider the purpose of establishing a multivariate NIR calibration model. The spectral information, three rows in the **X**-matrix, would be related to the actual concentrations in the three virtual analytical aliquots involved, i.e., 25%, 0%, and 100%, still varying significantly around the commonly assigned y-level, the lot reference concentration (9.85%). Thus, heterogeneity can easily affect the development of NIR calibration models adversely, e.g., as laid out in detail in Romañach and Esbensen (2016). And there is a further subtlety here. Consider a PAT system successfully obtaining spectra corresponding to a specific FOV volume (**X**-spectra), while reference samples of the flowing lot material are sampled physically (for reference characterization in the laboratory, y-data) but with a relaxed attitude "as close to the point of analysis *as possible*".

It is clear that accepting such a *location mismatch* will introduce a structural flaw with respect to the calibration model, specifically because of the heterogeneity of the flowing material. Tacit, or unwitting, acceptance of this kind of error will unavoidably lead to inflated modeling and prediction errors of a kind that cannot be compensated for by any known chemometric (or statistical) correction approach. This may be unwelcome news but must nevertheless be confronted. There is no help to be had in the often-heard aspiration "more data ...." since these will be affected by the exact same affliction.

The *only* general solution to this issue is simple in principle, but not necessarily also simple to execute: It is *crucial* always to characterize the *same* aliquot, the *same* volume, and the *same* sample with both the analytical NIR spectra and the pertinent reference method. Taking a ruthlessly honest look at the plethora of currently installed "sampling interfaces" in ducted flow cases across the gamut of technology and industry reveals that this mismatch situation is unfortunately met with distressingly often – distressingly so because this *per force* establishes a highly significant structural sampling bias – but a sampling bias that is completely unnecessary.

In the general case of a significant lot heterogeneity compared to the individual increment/sample volume, there will always be a smaller-or-larger such *mismatch* between individual grab samples and the full lot analyte concentrations. Within the analytical realm alone (*which* analytical method is used, what is the *method* accuracy and precision, etc.), this situation is difficult, indeed impossible to understand. But from the perspective of proper understanding of spatial heterogeneity (TOS), this is but a trivial example of what is known as the Fundamental Sampling Error (FSE) at work (Figure 19.3); FSE is described and illustrated in more detail in Gy (1998), Pitard (1993), Danish-Standards-Foundation (2013), Romañach and Esbensen (2016), and Romañach (2017).

The illustrations in Figures 19.3 and 19.4 are purposefully simplistic, but highly relevant for this archetype situation. The basic FSE effect on the total MU has been demonstrated for all types of material systems for over 60 years, as described in the abundant TOS literature. The present exemplification also hints at how the fundamental principles of TOS apply equally to the more complex PAT or emerging technologies, a point more fully elaborated in Esbensen and Paasch-Mortensen (2010). The key thrust of TOS is oriented toward *how to* obtain representative samples (including NIR signals) from heterogeneous materials (Gy 1998; DS3077 2013). But TOS has so far been only very little discussed in textbooks on instrumental analysis and similar. TOS is very often also lacking in coverage in modern chemometrics textbooks as well, with one notable exception (Esbensen and Swarbrick 2018).

In the case illustrated by Figure 19.3, fingers are sometimes pointed at the analytical laboratory as the cause of such "inexplicable" significantly varying analytical values. However, this has absolutely nothing to do with the performance of the analytical method, the data, the analytical modeling, or the laboratory performance (it is very nearly a safe assumption that TAE is in full control and that chemometrics is applied properly), but is a clear sign of the FSE effect. FSE is always encountered to a larger or smaller degree when dealing with heterogeneous materials, i.e., with analytes subject to constitution - and/or distribution heterogeneity. This effect will be progressively more pronounced at lower and lower analyte concentration levels (trace concentrations) and/or for other marked heterogeneity manifestations, e.g., also involving grain size distributions. Scientists working with NIR methods have also observed the effects of FSE in the form of *short-term precision* instability, e.g., as determined in repeatability studies based on replicated aliquots.

Here is the universal truth: Single *grab samples* do not do a reasonable job *counteracting* the spatial heterogeneity! Before this is fully acknowledged, a futile question often crops up: "How *big* should a sample be ... to be representative?" It is clear that a sample would have to be of the order of quite unacceptable, very high proportions of the whole lot before it would stand a chance of being a useful vehicle for estimating the *average* overall concentration of a heterogeneous material, e.g., 10%, 30%, and 50% of the lot volume. Such primary sample sizes are obviously not of relevance in practice in science, technology, or industry – and would in any way just transfer the (sampling) problem to become somebody else's responsibility, namely, the laboratory's problem. But the analytical laboratory is neither well equipped (theoretically and in practice) to deal with such vastly *oversized* samples arriving at the delivery entrance; indeed, laboratories most often issue stringent stipulations as to a maximum primary sample size that will be accepted.

The solution is *composite sampling*, *not* an increase in grab sample size. The latter is a brute force approach, while the former is the intelligent solution.

19.5 COUNTERACTING HETEROGENEITY – COMPOSITE SAMPLING

Composite sampling is the main enabler for proper sampling for all material heterogeneities, at all sampling stages – but especially when procuring the all-important *primary sample*, which is usually further mass-reduced (subsampled/split) in several stages before ending up as an analytical aliquot (or is the all-important single sampling-and-analysis operation in PAT solutions). Composite sampling is equally the solution for accurate (unbiased) multivariate calibration models as extensively used for NIR instrument calibration as it is always possible to aggregate any number of increments of a lot to form a composite sample (Romañach and Esbensen 2016, Esbensen 2020).

The illustration in Figure 19.4 is again *generic*, a placeholder for all types of significantly heterogeneous materials, i.e., materials that often may aggregate, "clumpy materials" (displaying "hot zones"), or segregate irregularly at several scales (local vs. global segregation). With TOS' understanding, this type of irregular spatial heterogeneity brings about the grouping and segregation error (GSE), the effect of which constitutes the frontline problem that has to be counteracted by *all* practical sampling processes whether addressing stationary or moving lots. Graphically, it is obvious how composite sampling is designed to do just that. Figure 19.4 makes use of 40 *increments*, deployed randomly over the entire material volume (area is this rendition), which are subsequently aggregated to produce a *composite sample*. The number of increments aggregated is termed "Q". $Q=40$ in Figure 19.4, while $Q=1$ for grab sampling.

Aggregates of an increasing number of *increments* are able to compensate for the irregular spatial heterogeneity (DH) in an increasingly improved fashion, e.g., estimating the concentration as 9.38% in Figure 19.4 versus the true value of 9.85%. The "free parameter" of all composite sampling procedures is "Q", the number of increments one is willing to deploy. Should an analyst not be satisfied with a "too cautious" Q, the general rule for how to increase the *fit-for-purpose representativity* is simply to increase Q. This can either be done in a very direct way – try with an increased Q

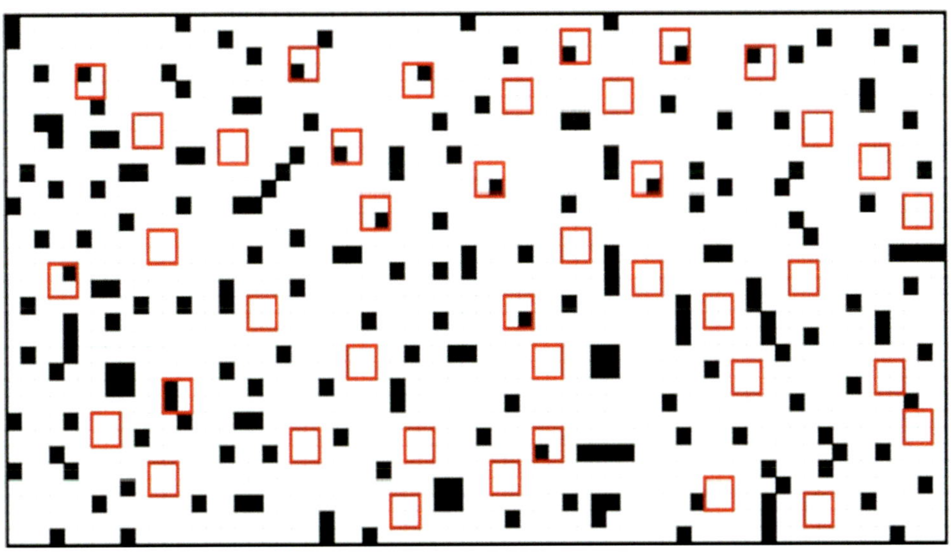

FIGURE 19.4 Extracting 40 randomly selected *increments* provides for an effective determination of the average concentration of the material shown in spite of adverse lot heterogeneity. Figure 19.4 illustrates how composite sampling provides a universal counteraction to both the grouping and segregation error (GSE) and the fundamental sampling error (FSE) effects. The lot with 9.85% analyte concentration can be sampled either using a grab sampling approach (individual red squares shown in Figure 19.3) or by deploying appropriate composite sampling procedures (the aggregate of all red squares in Figure 19.4), which provides an 9.38% analyte concentration. Composite sampling zooms in on the target concentration with efficiency and reliability, scaling directly with Q.

and test the total sampling+analysis error with a replication experiment (RE). Or one can deploy a perfunctory variographic experiment, in which one will be rewarded with a complete insight into the current process variability and the possibilities for complete management of *all* errors and uncertainty components. Both these facilities are described below.

Case Example II

NIR spectroscopy facilitates composite sampling, but there is a catch. Many NIR spectra can be obtained nondestructively in less than a second for any aliquot or spectral acquisition scenario employing a *stationary* sample holder, etc. But even though easily acquiring a high number of *scans*, averaging of course only pertains to a good estimation of the specific TAE. No other heterogeneity than that present in the aliquot (or aliquot volume) has been compensated for by averaging a number of spectra from an *identical* aliquot. But the case is necessarily different if composting spectral signals pertains to a series of replicated aliquots representing a *moving* material, for example.

NIR also offers opportunities for more problem-dependent composite sampling approaches. Here follows a two-step composite sampling case example. In the first step, a number of spectral *scans* are performed, resulting in an *average spectrum* (Romañach and Esbensen 2016), e.g., based on a total of 16 or 32 scans. This scan averaging improves the signal-to-noise ratio, i.e., the quality of the spectral characterization. In Fourier transform instruments, the signal-to-noise ratio will improve by approximately the square root of the number of scans obtained (Griffiths and de Haseth 2007). The second step consists of analyzing each of a relevant set of calibration samples using many locations (think of a NIR beam interacting with a rotating petri disk). This procedure is an optically driven composite sampling, covering more of the lot material than would otherwise be the case with just one location. Thus, a total of hundreds of spectra could easily be obtained for each of calibration samples.

Such a two-step composite approach has, e.g., been very effective in a number of applications where spectra were obtained of flowing powder mixtures (Mateo-Ortiz et al. 2014). Similar two-step approaches have been widely used in many pharmaceutical applications (Bakri et al. 2015, Durao et al. 2017, Martinez et al. 2013, Vanarase et al. 2010).

Specifically, NIR spectra may alternatively have to be obtained with powders in a static mode. A number of methods have been developed with static powders, which were subsequently used for the prediction of flowing powder mixtures (Cama-Moncunill et al. 2015, Vargas et al. 2017). A recent study compared a first approach where calibration spectra were obtained from 60 different sections of a longitudinal tray that contained calibration blends (Mateo-Ortiz et al. 2014). These 60 calibration spectra were obtained without movement of the powder. Subsequently, diffuse reflectance spectra were also obtained as the powder flowed by the spectrometer. Each of the "dynamic" spectra was the average of 12 spectra collected with an integration time of 6.6 ms, thus producing a time-composite sample, with the same effect as above rotating petri disk. The accuracy of the latter calibration model was improved by a factor of ~3 for the flowing powder system. These results reflect that as a calibration blend flows, new material is scanned ("replicate samples") and included as part of the pertinent composite sample for a given analyte composition. The number of effective increments used in the flow composite is very nearly always greater than for a static powder sample, improving the quality of a multivariate calibration model significantly.

19.6 PIERRE GY (1924–2015) – THE KEY CONCEPT OF SAMPLING ERRORS

Pierre Gy (1924–2015), the founder of the Theory of Sampling (TOS), single-handedly developed the Theory of Sampling from 1950 to 1975, and spent the following 25 years applying it in key industrial sectors (mining, minerals and metals processing, cement). In the course of his career, he

FIGURE 19.5 Pierre Gy (1924–2015), the founder of the Theory of Sampling (TOS), at one of his last public courses (Porsgrunn, Norway; 2005). Copyright KHE Consulting, brought here with permission.

wrote nine books and gave more than 250 international speeches on all subjects of sampling, including the all-persuasive aspect of hidden economic losses due to neglect of salient sampling issues. In addition to developing TOS, he also carried out a significant amount of practical R&D. But he never worked at a university; he was an independent researcher and a consultant for nearly his entire career – a remarkable scientific life. Gy's *oeuvre* is awe-inspiring. He wrote a five-paper personal scientific history published in 2004 (Gy 2004a, b, c, d, e). His scientific achievements are honored in a special issue of **TOS Forum**, no. 6 https://www.impublications.com/content/tos-forum-table-contents?issue=16_6 (open access) (Figure 19.5).

Gy's scientific breakthrough was to take on the overwhelmingly complex phenomenon of heterogeneity in full earnest. In this quest, he identified no less than eight potential sampling errors that represent everything that *can* go wrong in sampling, subsampling (sample mass reduction), sample preparation, sample presentation – due to heterogeneity and/or inferior sampling equipment design and performance. He meticulously worked out, over a period of 25 years, how to *avoid* committing such errors in the design, manufacturing, maintenance, and operation of sampling equipment, and elucidated how their adverse impact on the total accumulated uncertainty could be reduced as much as possible when doing sampling in practice. It was a monumental job, *ibid.* Along the way, he studied for and was awarded two PhDs (in mineral processing and statistics, respectively) in order to be adequately equipped to solve the complex theoretical and practical problems identified.

But although complex, TOS can in fact be made easily accessible. There are many systematic elements of the TOS, which makes mastering possible also from at a less in-depth theoretical level. For example, the eight sampling errors originate from only three sources: the *material*, the *sampling equipment*, and the *sampling process* – the latter bracketed by whether the lot is stationary or moving when sampling takes place, a distinction that is already well known within the NIR realm: "Sample only when the powder is moving". As it turned out, *some* sampling errors were even found to be able to be *eliminated* completely, which simplifies the sampling agenda considerably. But it is still necessary to know exactly *how to* identify these errors and exactly *how to* do this.

Recently, TOS has been presented in a fully axiomatic framework (Esbensen and Wagner 2014a). Figure 19.6 shows the systematics of all TOS' GP, SUO, and a graphic summary of the pertinent sampling error management rules. The relevance of this overview stems from the fact that all

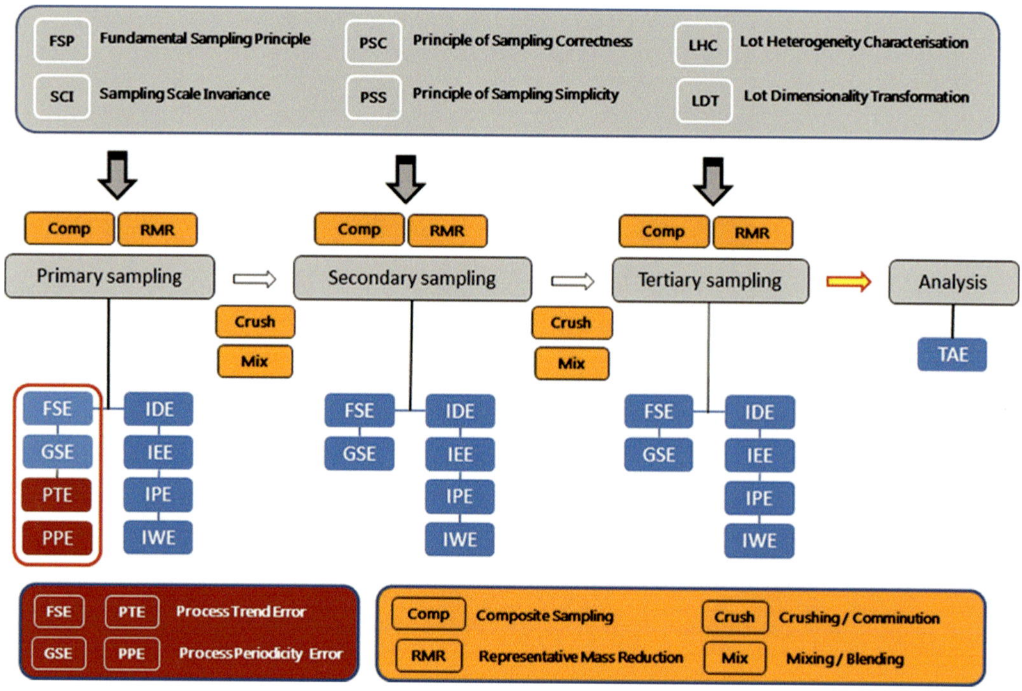

FIGURE 19.6 Theory of Sampling (TOS) – synoptic overview, shown here for three sampling stages. TOS is comprised of six governing principles (GP) (top gray panel), four sampling unit operations (bottom yellow panel), and eight sampling errors (blue/maroon). This diagram depicts a generic multistage sampling process whose sole aim is to deliver a representative analytical aliquot (yellow arrow). Sampling of stationary lots addresses how to treat six basic sampling errors (blue), while process sampling (sampling of dynamic lots) needs two more errors to be addressed in full (maroon). TOS logically demands that all prealiquot steps are collected under a unified sampling responsibility. For details of the interactions between TOS' elements, see DS 3077 (2013); Esbensen and Paasch-Mortensen (2010); Pitard (1993); Gy (1998); Esbensen and Wagner (2014a); Esbensen (2020); and other authoritative references herein. (Illustration copyright KHE Consulting, brought here with permission.)

principal elements of a guaranteed representative pathway to this goal are outlined here as a master guide and prelude to TOS in full.

The first issue on any sampling agenda is to eliminate the incorrect sampling errors (ISE), mainly an issue regarding the design, installation, operation, and maintenance of the sampling equipment. Subsequently what remains are the correct sampling errors (CSE), which can be dealt with by standard means, i.e., by scaling the number of composite sampling increments deployed, Q, with respect to the empirical heterogeneity encountered (always honoring the Fundamental Sampling Principle, FSP), and involving the pertinent GP's and other SUO's where needed.

Recent introductions to TOS in general, and in relation to chemometrics and multivariate data analysis in pharmaceutical production, can be found in Esbensen (2020), Esbensen and Swarbrick (2018) and Esbensen et al. (2018); these are complementary texts to the present review below.

19.7 A KEY DISTINCTION: ERROR VS. UNCERTAINTY

There will always be a definite difference, small or large, signifying a nonzero difference between the true, unknown content a_L of a lot L and the estimated (analyzed) content of a sample, a_S – even for a representative sample, because of material/lot heterogeneity. The analytical determination is fraught with a combined sampling-and-analysis *uncertainty*. This cannot be otherwise due to the massive mass reduction of heterogeneous material involved.

Tradition in many scientific fields has established the word "error" as common practice, though this term implies a mistake, an *error*, that could have been prevented, while statisticians prefer the word "uncertainty", which implies no culpability. However, TOS has demonstrated that there are *both* sampling errors and sampling uncertainties. The effects of some sampling errors *can* be preventatively minimized, or even eliminated (ISE), while some final sampling uncertainty for any sampling protocol is inevitable. The job is to minimize this remainder, CSE. Add to this that there is also an analytical error or uncertainty.

Because the word "uncertainty" is not strong enough in the real-world context of all significantly heterogeneous materials and processes, the term "error" has been retained in TOS (Gy's imperative), while making it clear that this does not necessarily imply culpability – but it *may*. The sampling errors which are subject to the possibility of elimination (ISE) shall and must be so! If not, *someone*, or *somebody*, is de facto responsible for committing an error resulting in a sampling bias (see further below), which unavoidably results in unnecessary inflation of MU_{total} – and this certainly constitutes an error for which a *legal person* is culpable (it could be a faulty standard, ditto procedural description, an ill-informed, or sloppy sampler, an incompetent supervisor, a CTO not in the know regarding TOS). The core of the understanding described here is from Pitard (1993 (p. 33); 2019) who informs that the basic arguments originate with Pierre Gy's fundamental analysis of the sampling process.

19.8 TOS IN PRACTICE: GOVERNING PRINCIPLES (GP) AND SAMPLING UNIT OPERATIONS (SUO)

Case Example III

As an example from current industrial approaches, note how TOS' multistage sampling process is in distinct opposition to the grab sampling approach common in the pharmaceutical industry, where powders traditionally are extracted with a *thief*, or *spear*, applied to a fixed set of locations, tacitly *assuming* that this rigid, codified approach is sufficient to ensure representative samples, which are then to be compared in order to estimate blend uniformity, etc.

However, this assumption has been criticized and shown to be a serious misunderstanding (Esbensen et al. 2016, Romañach 2015, Romañach and Esbensen 2015, Muzzio et al. 1997). But many companies are still using thieves (sampling spears); in fact, a total of 20 out of the 23 companies interviewed recently reported using thief sampling (Pinzon de la Rosa et al. 2017). This status quo is virtually unchanged from similar interviews in 2001 that showed a total of 25 of 28 companies used thief sampling then. Thief, or spear sampling, is in clear conflict with the FSP as argued forcefully in Romañach and Esbensen (2015). This issue is perhaps a disquieting new insight in pharma, e.g., but one that better be faced squarely.

The Fundamental Sampling Principle (FSP) applies with equal strength to PAT signal acquisition, as illustrated in Figure 19.7a, b. Strenuous efforts to design, install, and implement "smart sensors" notwithstanding, the lack of compliance with TOS' basic demand for increment representativity, Figures 19.2, 19.7a, b make the desired signal acquisition compromised to a degree that scales directly with the heterogeneity in the streaming flux of matter. In practice, only media of "uniform composition" will escape the danger of a significant process sampling bias. This may be a bitter pill to swallow for many NIR analytical Original Equipment Manufacturers (OEMs) and "whole solution" providers, but true nonetheless. There is no escape from the demand for *representative* primary signal acquisition.

Pointedly, the exact same sampling principles govern at all sampling stages, TOS' fundamental principles and unit operations are *scale invariant*. Thus, at all stages, *sampling errors* and their

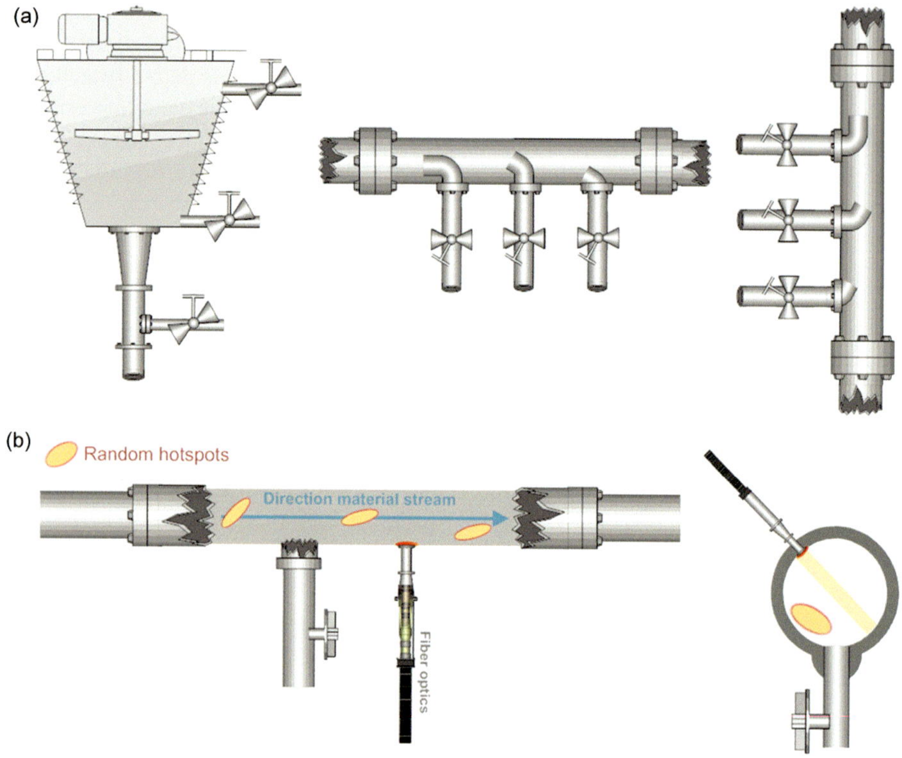

FIGURE 19.7 (a) The many ways to install a "sample outtake valve" in process technology. Upon reflection, *all* examples shown are not in compliance with TOS; see Figures 19.2, 19.7b and Esbensen and Paasch-Mortensen (2010) for details. (Copyright KHE Consulting & HERZOG Maschinenfabrik GmbH & Co. KG, brought here with permission.) (b) If/when PAT sensors are not "seeing" a volume corresponding to a full cross-sectional *slice* of a moving stream of matter, there is neither in this case TOS compliance; see Esbensen and Paasch-Mortensen (2010) for in-depth explanations. (Copyright KHE Consulting & HERZOG Maschinenfabrik GmbH & Co. KG, brought here with permission.)

uncertainty effects will impact significantly if not carefully managed and counteracted. Thus, the sampling responsibility is the same at all scales/stages. Thanks to TOS' overlying GP, this complex undertaking can in fact be carried out in a simplified, systematic fashion. TOS is put on the simplest possible axiomatic footing in Figure 19.6 and Table 19.1.

TABLE 19.1
Six Governing Principles (GP) and Four Sampling Unit Operations (SUO) Constitute the Necessary and Sufficient Framework for Representative Sampling (Danish-Standards-Foundation 2013, Esbensen and Wagner 2014a,b, Esbensen 2020).

Governing Principles	Sampling Unit Operations (SUO)
Fundamental Sampling Principle (FSP)	Composite sampling
Principle of Sampling Scale Invariance (SSI)	Comminution/crushing
Principle of Sampling Correctness (PSC)	Mixing/blending
Principle of Sampling Simplicity (PSS) (primary sampling + mass reduction)	Representative mass reduction (subsampling)
Principle of Lot Dimensionality Transformation (LDT)	
Lot Dimensionality Characterization (0-, 1-, 2-, 3-D)	

The system of GP offers a structured, rational competence needed to analyze, assess, evaluate, and suggest solutions for *any* sampling problem, no matter at what scale, for whatever type of material (only its heterogeneity is relevant). The four SUO are the *only* practical procedures at the disposition for *any* sampling process: (1) composite sampling, (2) comminution/crushing, (3) mixing/blending, (4) mass reduction, but not just *any* mass reduction – *only representative* mass reduction will do (Petersen, Dahl, and Esbensen 2004). This applies for all stages before analysis (in NIR spectroscopy, as well as for other analytical modalities).

Judicious application of SUO's are legion, suffice to mention but a few, e.g., comminution and drying of soil samples, which is often performed before obtaining NIR spectra to determine soil organic matter and carbon (Changkun and Xianzhang 2016), or grinding of sorghum wheat in order to be able to mix and subdivide this material (Junhui et al. 2015), or a homogenization (mixing/blending) step prior to analysis, before the determination of fat content in curry soup containing coconut milk (Ekkapong and Panmanas 2017). A number of biopharmaceutical applications also include mechanical agitation (mixing) of cultivations in bioreactors while NIR spectra are obtained (Sampaio et al. 2014). Application of one of more SUO's is also critical, e.g., in generating calibration samples with different degrees of moisture, or different particle size mixtures (Changkun and Xianzhang 2016, Zhou, Hines, and Borer 1998, Barajas et al. 2007).

Thus, while all parties in the NIR realm are familiar with crushing, mixing, blending, and subsampling in particular cases – exactly *how to* deploy these agents optimally when facing a *specific* heterogeneous material still needs the full complement of TOS' GP to succeed (even though not all GP are used for all situations/cases). It matters not that, e.g., crushing and mixing/blending are perhaps not widely recognized as *sampling unit operations*; what matters is that by viewing and designating them in this fashion opens up for a systematic way of addressing various hitherto compartmentalized procedures in the laboratory in a wider perspective with which negotiate all adverse issues in the full preanalysis realm with systematic rigor. There can be found all too many "locally smart" but "universally fatal" solutions in the NIR literature.

Within TOS, there exists an intimate interrelationship between the GSE and the FSE, expressed by the segregation factor, ζ, and the grouping factor, γ (Figure 19.8). These phenomenological factors are used in TOS' theoretical treatment of heterogeneity to represent segregation and local "grouping" in the more comprehensive equations originally derived by *Gy*. While it lies distinctly outside of the scope of the present review to present the pertinent theoretical treatments in full, one of the factors actually turns out to represent the operative size of the increment used for sampling, γ, for which reason it is easily appreciated how important a role γ plays for practical sampling. And the other, the segregation factor ζ, lends itself well to direct graphical illustration and understanding. Figure 19.8 thus shows with clarity the effect of reducing the segregation factor as a function of the SUO no. 3: mixing. Upon reflection, it will be appreciated how the GSE error effects can be reduced by reducing either one, two, or all three of the factors on the right-hand side of the equation given in Figure 19.8. Observe how only the segregation factor represents mixing, while the other represents a prerequisite

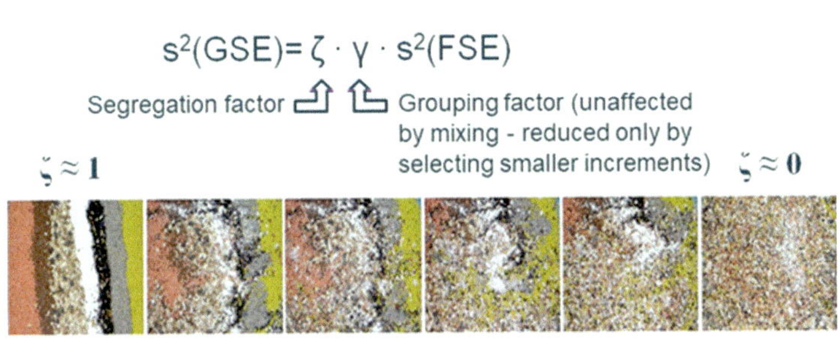

FIGURE 19.8 Illustration of the heterogeneity-counteracting effect of mixing (SUO no. 3) together with an indication of the underpinning relationships between FSE and GSE in the Theory of Sampling; see text for details. (Illustration copyright KHE Consulting, brought here with permission.)

for composite sampling, reducing the size of the operative increments. Without going into further details here, this key equation in fact allows the avid TOS theoretician to derive no less than three of the four SUO's, no.s 1,2,3: composite sampling, crushing (comminution), and mixing.

Should this sneak preview of some of the deeper levels of TOS have caught the interest of the reader, the relevant follow-up references can be found here: Gy (1988, 2004a, b, c, d), Pitard (1993, 2019), Esbensen & Wagner (2014a, b), Esbensen (2020), and DS 3077 (2013).

In 2013, the world's first *universal* standard for representative sampling was published, termed a "horizontal standard" (DS 3077: Danish-Standards-Foundation 2013)[§]. Horizontal means that the sampling principles presented avoid focusing on the bewildering diversity and the number of *types* of materials in science, technology, industry, and society, which are the target of sampling. The horizontal approach specifically only focuses on the effective material *heterogeneity*. Also, the *divide* between "the field" (the realm of primary sampling) and "the laboratory" (the realm of mass reduction and other relevant SUO operations) is *dissolved*; this artificial divide disappears when it is realized that all of TOS principles are *scale invariant*. Based on the synopsis presented in Figure 19.6, lots of all manifestations, forms, and sizes (containers, vessels, reactors, blenders, petri dishes …) can be treated in an identical, systematic fashion; *only* their heterogeneity differs.

Sampling's singular non-negotiable demand is to secure a documentable representative primary sample and that the relevant characteristics hereof (analyte concentration(s), grain size distribution) are also retained representatively along *all* further subsampling stages in the full pathway leading to the analytical aliquot. This can only be guaranteed if/when all sampling procedures follow TOS' principles and unit operations with no exceptions. Perhaps complex at first sight, the systematics of TOS (6 GP, 4 SUO) (Figure 19.6 and Table 19.1) makes this a much rational and easier undertaking, the backbone of which is presented in this chapter.

Case Example IV

As a major example, GP #4, the Principle of Sampling Simplicity (PSS), governs the multistage nature of *all* sampling-and-analysis processes, always starting with a primary sampling operation *followed* by a series of representative mass reductions (subsamplings) until the analytical aliquot has been produced. The latter tasks are covered by SUO #10 specifying that all successive mass reductions must be *representative*, no exceptions.

There is unfortunately an abundance of rule breaking taking place in many of today's analytical laboratories, e.g., "spatula subsample extraction" – which is nothing but grab sampling at the smallest scale of interest (Esbensen, 2015). From TOS' systematic point of view, subsampling is nothing but sampling with the specific purpose of achieving a maximum mass reduction. But sampling is all the same and therefore equally mandated to follow TOS' principles. This perspective of mass reduction, wherever it takes place, manually or automated, is a series of "self-similar" sampling operations at smaller and smaller scales but *all* must comply with all GP and SUO's involved. Petersen, Dahl, and Esbensen (2004) presented a comprehensive benchmark exposé of TOS applied for practically all prealiquot activities in the laboratory.

Also, as a PAT example, a sensor used to monitor a process will not necessarily interact with the complete cross section of the flowing stream of material produced (Figure 19.7). In this scenario, the effective FOV *volume* of a PAT signal is an example of *optical grab sampling*. Typically, across the many types of material analyzed by NIR, TSE can attain magnitudes anywhere 10–10^3 or higher, relative to TAE, much dependent on the heterogeneity.

The task of systematic application of TOS is to make *each* individual sampling operation representative (no chain is stronger than its weakest link). This is greatly helped by the fact that there is

[§] Note that DS 3077 (2013) only lists three GP, while the full complement of six (three were added subsequently) was first presented in Esbensen and Wagner (2014a,b), and fully elaborated in Esbensen (2020).

no *interaction* between any two sampling stages; indeed, all sampling issues can be addressed as a series of *individual* operations. All sampling and subprocesses can thus be optimized individually. This helps immensely when facing complex sampling problems.

Material heterogeneity cannot automatically be *assumed* to be discernible with the naked eye, a feature that makes proper sampling challenging. Figure 19.8 shows an example of the type of material that can be met with in very many application fields destined for NIR analysis, *powders*. The point is that here heterogeneity is invisible, but still very much present. It is always necessary to understand that material heterogeneity involves two aspects, compositional heterogeneity (CH) and distribution heterogeneity (spatial heterogeneity, DH). Esbensen (2015) (open access) gives a full outline of the complexities of compositional versus spatial heterogeneity CH/DH and the corresponding *appropriate* sampling modes (Figure 19.9).

The principal strategy is simply to *assume* that a significant material heterogeneity is *always* present. This is a rational approach completely in line with *never* to treat any material *as if* homogeneous; i.e., never to give in to spatula sampling at the laboratory scale if not having validated this approach and proved it produces fit-for-purpose subsampling¶.

FIGURE 19.9 The fallacy of relying on visual verification when assessing the degree of material heterogeneity. It is tempting to conclude that the leftmost powder lot is "homogeneous" and therefore allows spatula sampling (grab sampling) – while there is in fact a significant irregular spatial distribution of an *invisible* polluting analyte. In the right hand illustration, all particles larger than average diameter have been dyed blue, which reveals a highly irregular spatial distribution of the different particle sizes in this material (which indeed also *appeared* homogenous before dying). These complementary heterogeneity phenomena are considerably more complex if/when the analyte is present at different concentration levels in different particle size ranges. (Illustration copyright KHE Consulting, brought here with permission (original photos, courtesy erstwhile Ph.D. Peter Paasch-Mortensen).)

¶ Nevertheless, spatula sampling of powders at the ultimate laboratory aliquot mass scale is currently still often the preferred option. This *might* be acceptable but most emphatically only after a specific *validation*. The reasons for acceptance could, e.g., be because the material grain size has been so reduced, and thus having greatly facilitated effective mixing, that the effective subsampling variability has become so low so as to be below an *a priori* established fit-for-purpose threshold. There is obviously no generalization power from this powder fact to any other types of materials, at any higher scales. Grab sampling really is the *bête noir* at all scales unless rigorously validated by a perfunctory replication experiment (RE) or a variographic characterization.

The same applies to grab sampling at all larger scales (where matters quickly become significantly more difficult however because of the increasing lot size involved). A logical and rational stand is therefore to treat all materials as if they were *significantly heterogeneous*. This is codified in a mandatory manner as GP #6: heterogeneity characterization. By adhering to this mandate, rational planning for sampling becomes systematic and immensely powerful: All materials are simply to be treated in a *unified* fashion using *composite sampling*, SUO #7. The professional sampler does not need to switch the type of sampling approach nor the operational plan when addressing different materials, or different lot size, scale, or material configuration. Depending on the overall lot heterogeneity and its grain size distribution, a suitable basic increment tool size is selected (see, e.g., DS077 (2013)), but there is no other change of sampling procedures necessary (composite sampling). It is only necessary to modify the number of increments extracted (Q), while at all time honoring the FSP.

19.9 THE NIR PATHWAY FROM LOT TO ALIQUOT

NIR analysis is also influenced by how lots, samples, and smaller subsamples are *handled* before the aliquot is analyzed, like for any other analytical techniques, a dependence well recognized in prior publications (Williams 2001, Mark 1991). Williams listed about 30 sample-associated factors that affect the accuracy and precision of NIR analysis but only gave *some* indications of how to avoid many of these, and emphasized that: "*If [these] are not efficiently carried out, consistent and accurate NIR analysis is not possible, no matter how sophisticated the software*"(Williams 2001). This view is in complete agreement with the principles presented in this chapter. The issue was earlier that while well identified, many of these critical factors were *not* treated systematically from the point of view of *what-to-do-to-reduce-their-influence*. This chapter carries on and takes this evolution to its logical conclusion: TOS to the fore.

As an example, Figure 19.10 illustrates the analysis of a powder mixture by diffuse reflectance NIR spectroscopy. The NIR beam penetrates perhaps as much as 5 mm within the powder mixture. However, the radiation that returns to the near-infrared spectrometer is mostly from the top 2 mm (Popo et al. 2002, Iyer, Morris, and Drennen III 2002). The vial shown in Figure 19.10 could be up to 20 mm deep, say, but the NIR radiation is still only interacting with the top 1–2 mm. The powder in the vial could also be analyzed by a reference method, e.g., by high-performance liquid chromatography (HPLC), where a solvent is used to extract the analyte from the *entire* vial volume. If not carefully choreographed, a systematic support (volume) mismatch error would again be the result, because NIR and HPLC are not analyzing the same volume of material. This type of mismatch between the reference sample and the optical method was explained already many years ago (Mark

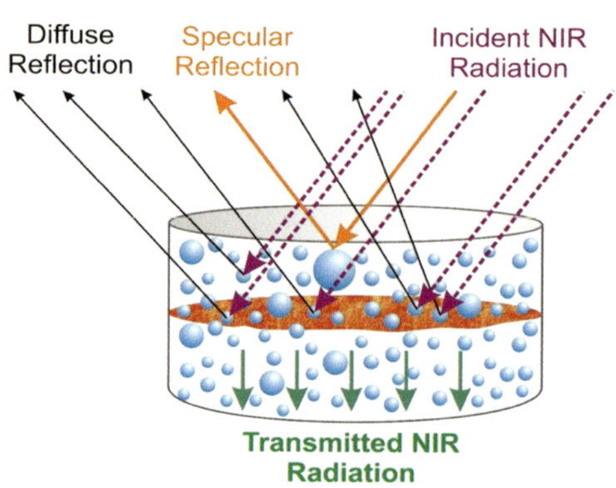

FIGURE 19.10 Illustration of an incurred incorrect sampling error (ISE), i.e., partial optical sampling/analysis (NIR signal) versus full-vial reference volume HPLC signal.

1991). A volume mismatch error, in the TOS parlance, is in fact a significant IDE, i.e., one of the four ISE involved in creating a sampling bias.

Case Example V

The NIR spectrometer indicated in Figure 19.10 is "sampling" a powder mixture, based on the interaction of the radiation with the mixture in the prepared sample holder. This operation is simultaneously the final sampling stage (*optical sampling*) as well as the physical basis for analysis through a multivariate calibration model, allowing to predict a specific property of the powder mixture (Esbensen and Swarbrick 2018, Romañach and Esbensen 2016). Predictions will be fraught with unnecessarily inflated prediction errors (prediction variances) however, and varyingly so due to the obvious aliquot volume mismatch.

It is sometimes argued that since this mismatch error is "identical" for all samples (calibration, validation, prediction samples), it will in effect create a "constant bias", which will influence calibration, validation, and prediction samples identically, and which can therefore be corrected for (as an analytical bias). This is a fatal *misunderstanding* however, first of all because the sampling bias is of a different nature than the analytical bias (see further below), but this understanding also tacitly presupposed that all materials, mixtures, and powders are *identical* when involved in NIR analysis, all differences in sample preparation (handling, pouring, compactification, presentation a.o.) notwithstanding; this is clearly not the case. Optical sampling is intricately related to the physics of diffuse reflection in NIR spectroscopy. As the top layers of a powder will receive more radiation than the bottom layers, less radiation reaches the bottom part of a powder blend on a volume-to-volume comparison. For a given material with a specific physical constitution (grain size distribution a.o.), the depth penetration also depends on the wavelength used. The lower wavelength radiations penetrate to greater depths at higher wavenumbers where molecules absorb less NIR radiations. Thus, the *effective sample mass* interacting with NIR radiation also *varies* according to wavelength (Dahm and Dahm 2007). These relationships cannot be resolved universally by the traditional route of demanding a "bigger sample". The solution is NIR radiation interacting with an increased analytical mass in a well-designed fashion so as to accomplish practical *composite sampling* and at the same time guarantee sampling correctness (elimination of sampling bias). This objective has been the holy grail for decades of development of suitable "sample cells" within the NIR realm, not all of which having reached this goal, however. While the specific analytical problems of this kind are well recognized, it takes more than a minimum of TOS competence to look through the immediate manifestations of many "sample interfaces", salient examples to appear below.

Such sampling errors cannot be corrected for by any mathematical correction algorithm, by physical pre- or post-treatment, or by any software approach. These issues will always affect adversely on the possibility of developing fully optimized, or even appropriate, calibration models. All NIR spectroscopic methods are similar to a multiple path-length cell where particles may interact with radiation more times than once (Johansson et al. 2002). The mass analyzed by the radiation may be estimated in some cases, but is generally not known in detail. Most analytical chemists know exactly the material that is analyzed. In NIR spectroscopy, the situation is rather the opposite; indeed, NIR spectroscopists are spectacularly the most successful analytical chemists that do not know the exact sample mass being analyzed!

19.10 ANALYTICAL IMPROVEMENTS – IMPRESSIVE, BUT NOT ENOUGH

Within the NIR realm, the term "sampling" is most often used in the analytical sense related to *how* NIR spectra are obtained, e.g., diffuse reflectance versus transmittance spectroscopy. There are scores of published papers with discussions about the many types of "sample cells" that have been developed over time. Upon reflection however, many of these are actually appropriate only within the perspective of "sample presentation" and are thus mostly (or exclusively) affecting TAE. It should be abundantly clear from the argumentations above, however, that the majority of cases are concerned with analysis of moving matter, in which the active aliquot volume delineation and the analytical signal generation are intimately related to the specifics of *physical*, as well as *optical sampling*. For example, if a material has been affected by *segregation* in a prior handling operation, or is actively segregating during transportation, the analytical uncertainty can very easily be affected by a *sampling bias* if the sampling/mass reduction/analysis intricacies outlined above have not been adequately counteracted. In a quite high number of current cases, this issue is actually unknown.

Case Example VI

A didactic example of how to cater for a greater effective analytical mass à la composite sampling, under fully controlled conditions, was presented recently (Sørensen, Larsen, and Esbensen 2014). With spectroscopic methods using a constant radiation beam aperture, the effective scanning footprint will be different for a stationary versus a spinning Petri dish, for a rolling bottle, and for a relatively newly developed "spiral sampler" configuration. These "sample presentation" modalities will impact differently on MU_{total} of a NIR analytical determination of heterogeneous materials. This study presented results evaluating the total sampling-and-analytical bias and precision characteristics for three alternative sample presentation approaches in a bench-top experiment using a mixture of two types of plastic polymer pellets as a test material with a significant heterogeneity, even though extensive mixing was applied. After removal of all ISE, there were still significantly CSE contributions, in the form of *FSE* and *GSE* uncertainties. The focus in this study was on the relative merits of the alternative "sample presentation" approaches related to the different effective scanning areas (FOV) relative to the potential full sample size (i.e., volume). This is basically a *FSE* issue, but also a residual heterogeneity concern; sampling should here address the effective sample volume and should also seek to minimize effects from *GSE*. The *effective* scanning volume is the sample surface area that is appropriately scanned *in depth* (see Figure 19.11). Typical proportions between the effectively scanned areas in this study were 18 cm², 30 cm², and 300 cm² for the Petri dish, the bottle spinner, and the spiral sampler, respectively, effective penetration depth being identical in the specific experimental design.

FIGURE 19.11 Illustration of alternative NIR "sample presentation" accessories, Petri sampler (left), bottle sampler (center), and spiral sampler (right). The effectively scanned areas are indicated (dependent on method/number of scans) as well as the maximum sample volume for the respective sample containers.

FIGURE 19.12 Illustration of competing aliquot/sample masses typically achieved with the Petri dish (left), rolling bottle (center), and spiral sampler (right). Typical analytical masses are 35 g, 50 g, and 600 g, respectively. N.B. The NIR system does not interact with the entire mass in the containers; see further in text.

These issues will always affect adversely on the possibility of developing fully appropriate calibration models. The CSE effects cannot be corrected for by any mathematical correction algorithm, by physical pre- or post-treatment, or by any software approach. This study has also a strong carrying-over possibility to address slurries and fluids (Figure 19.12).

For full description and details on materials, experimental design, and NIR analysis as well as detailed results for the three sample cell alternatives (accuracy and precision), see Sørensen, Larsen, and Esbensen (2014). The "winner" from this experiment, by a far margin, is clearly the spiral sample presentation *principle*, because of the dramatically increased aliquot mass available for analysis. It is usually claimed that this result can easily be generalized to many *similar* heterogeneous materials and systems... but is this true?

19.10.1 However, there is a Hidden Elephant in the Room

The critical issue is that with all three approaches, there is no concern for, far less optimization of, the *preceding* sampling stages. So, while it is possible to develop significant improvements in the sample preparation and analysis stage, in the full lot-to-aliquot pathway context, this will have only little effect with respect to the potentially 10–25 X larger *neglected* TSE incurred in the preceding stages. Analytical improvements are very welcome, and the NIR community has seen a plethora of significant and creative achievements of this kind – but they are no panacea by themselves. TOS' full framework presented in Figure 19.6 must reign comprehensively.

The NIR realm is still affected by this kind of analytical "tunnel vision" to a significant degree – beware of false "sampling" prophets in many clothings. However, it is also fair to state that matters are on the move and have been improving significantly in the past 10 years or so.

19.11 THE REPLICATION EXPERIMENT (RE)

RE is an established tool within TOS, which has been only very little used in NIR spectroscopy (Esbensen et al. 2016). RE represents a strong opportunity for increasing the impact of NIR spectroscopy on QC (Danish-Standards-Foundation 2013, Esbensen et al. 2013a, b). What follows constitutes the situation in which the NIR analyst/researcher must recognize *significant* sampling, handling, and preparation error effects in addition to TAE_{NIR}.

What is meant by "replicate samples" (upon which subsequently can be performed replicate analytical measurements as well)? The first thing to note is that replication may also be carried out at other levels, e.g., related to sampling of multiple batches, lots, or production units, sampling from different seasons or from different instruments, perhaps carried out by different operators a.o. There is often a marked conceptual imprecision when "replication" is invoked, which is seriously impairing comprehensive understanding.

Table 19.2 shows three different levels of an RE.

- The first step consists in deposition of about 250 grams of an acetaminophen (APAP) powder blend prepared with a concentration of 15% (w/w), onto a 3-m-long metal rig (see Esbensen et al. (2016)). After deposition, a total of 65 NIR spectra were obtained from one end of the rig to the other as it moved at 10 mm/sec. The deposition was in fact performed in a repeated fashion six times, giving a total of 390 spectra. Table 19.2 shows the predictions of APAP concentration in the six redepositions of 250 g that were analyzed. Table 19.2 also shows the results for two of the blending processes performed, a process with vibration and tumbling (VT) and a process with only tumbling (T2).
- The second replication level involved the acquisition of NIR spectra of a single blend, after it was deposited onto the same rig. NIR spectra were obtained from a one end of the blend to the other, a total of 10 times. A total of 647 spectra were obtained for a single deposition of the VT blend, and 650 spectra for the T2 process.
- The objective of the third level addressed the conventional analytical repeatability, consisting in acquisition of six consecutive spectra without moving the powder rig.

  For each of these replication scenarios, it is imperative that the reader is offered full disclosure of the stage, level, and intensity of "replication".

As the case in point then, what is meant by "replicated samples?" With reference to TOS, it will be appreciated that "replication" can concern (at least) the following alternative scenarios:

TABLE 19.2
Deposition of Acetaminophen Powder Blends Prepared at 15% (w/w) by Two Different Mixing Processes

	Deposition ($n=6$)		Replicate of Single Deposition $n=10$		Repeatability Study ($n=6$, at 10 points)	
	VT	T2	VT	T2	VT	T2
Average % (w/w)	14.93	16.39	15.21	16.24	15.78	15.54
Spectra (#)	390	390	647	650	36	60
SD % (w/w)	0.78	1.58	0.34	2.02	0.14	0.17
RSD (%)	5.20	9.62	2.23	12.46	1.3	1.12
MPE	0.04	0.7				
Sill	0.7					
SD blend*	0.75	1.34				
Inherent blend SD **			0.31	2.01		

Adapted from Esbensen et al. (2016).

Stage 1: Replication of the primary sampling process (followed by all the steps necessary to reach all the way to analysis), with due regard, and control over *possible* effects of time, sequence, raw material variation if/when present. The first stage in the above example involved the deposition of six different subsamples, and the acquisition of 65 different spectra for each. This first stage includes the variation in the deposition, the variation due to the heterogeneity of the blend, and finally the variation (random error) associated with the NIR measurements.

Stage 2: Replication starting at the secondary sampling stage (i.e., first mass reduction). This second stage included the variation due to the heterogeneity of the blend and the NIR measurements only in the case example. The deposition is now not included in this assessment, as the blend is already deposited over the rig, but the same blend is analyzed 10 times for a total of 650 determinations.

Stage 3: Replication starting with the tertiary sampling process (further mass reduction). In the RE described above, this consisted of six consecutive spectra without moving the powder mixture. This assessment only includes the uncertainty from the instrumental measurement. As expected, this corresponds to the absolute lowest variation. This situation corresponds to the term "replicate analysis".

The cited experiment is an example developed in an academic research laboratory to pioneer the development of the RE in NIR applications. However, other applications may require additional levels or stages. RE may of course be used with any analytical facility.

All three stages described above are important. When the analyst "presses the button" repeatedly, say six times, this corresponds to TAE *sensu stricto*. This is the minimum error that will be observed as it is only the exact same sample (vial, aliquot) that is being measured, repeatedly.

However, this measurement is not a good estimate of the *total* variance that may be expected in sampling and analysis of the powder mixture described above. The analyst may, e.g., consider it relevant to analyze the powder repeatedly, allowing normal temperature variations to influence TAE because this is a more *realistic* repetition of the between-samples situation than repeating measurements on one static sample only. This is a foray into "Taguchi thinking"[**], and is very much taking steps in the direction of more realistic validation. But to another analyst, it may perhaps appear equally reasonable also to include some, or all, of the "sample handling/preparation" variations in the replication scheme, for the same reason: to be more realistic. For which reason, such perturbations should then logically also be repeated.

Having opened up for this augmented replication context, it would seem an unavoidable logical step to follow up with further, equally relevant and realistic perturbations of the circumstances surrounding both sampling and subsampling in addition to analysis, i.e., also including sample preparation, and indeed the tertiary, secondary, and ultimately primary sampling errors in the replication concept. Following the full logic of TOS, it is now actually clear that the only complete "sampling-and-analysis" scenario which is guaranteed to include *all* possible uncertainty contributions to the total TSE+TAE uncertainty, is the one that starts with replication of the primary sampling, aka "replication from the top".

Repeating the primary sampling, say 10 times – each individual primary sample being subjected to the exact same protocol governing all the ensuing subsampling (mass reduction), sample handling, and preparation – is the only procedure that allows the full set of error effects to be reliably manifested as the total variability between the 10 analytical results. Table 3 shows that for the VT process, a RSD of 5.20% is obtained. This is the lowest variation observed for the two alternative mixing operations involved, and it indeed includes the variation related to the sampling as well as the TAE.

[**] Taguchi approach: http://en.wikipedia.org/wiki/Taguchi_methods

The replication from the top approach is the only *fully* realistic replication of all the elements in the sampling-and-analysis pathway commensurate with the routine workflow of typical laboratories; the reader is well served remembering that no aliquot materialized out of thin air at the door stop of the laboratory delivery backdoor. By contrast, starting at any other of the subsequent levels is guaranteed to result in inferior TSE+TAE estimations, which structurally unavoidably will be "too low". There is, however, no way one can estimate the magnitude of the missed-out variance.

RE very much goes beyond the usual efforts to validate an analytical method, and extends QC into the realm of validating and understanding the errors also associated with *all* the steps prior to analysis. There should *always* be an obligation for analysts, indeed also data analysts, to describe the rationale behind the choice of a specific replication scheme used, and to disclose publically exactly what was in fact replicated, else the user of the analytical data will be in the dark. The APAP powder blend example above shows three stages, or levels, all of which contribute to the variation observed in the final blend analytical results.

Undocumented, unexplained (and sometimes even ill-understood) application of the term "replication" has been the source of a significant amount of unnecessary confusion in analytical chemistry, statistics, and chemometrics. However, many times the problem boils down to that s^2(TAE) has been *misconstrued* to be identical to the (much) larger s^2(TSE+TAE), which is a grave *error*, for which *someone* must be responsible – but who? Who or what is the culprit in this context? And, more importantly, how can this issue be rectified?

The above scenario illustrates the unfortunate compartmentalization of responsibility, which rules the day in a wide swath of current laboratory, scientific, and industrial contexts. Defensive comments commonly heard are: "The analyst is not supposed to deal with sampling *outside* the laboratory"; " this department is *only* charged with the important task of reducing the primary sample to manageable proportions, as per the laboratory's instructions"; "sampling is *automated*, there is no sampling problem here"; "I am not responsible for sampling, I only analyze *the data*!" – and a legion similar excuses for not seeing, or wanting to deal with the complete "measurement uncertainty" issue. All too often the problem belongs to "somebody else".

If "excuses" like these continue to be allowed in the practical experimental work, in technical guidelines and reports, and in the scientific literature, there is a grave danger that this status quo will simply be perpetuated: "Replication" will then mostly still take its point of departure at stage 3 (maybe stage 2, but almost never from stage 1), the primary sampling stage. Add to this a certain lack of understanding on behalf of authors, reviewers, and editors to focus and crack down on this serious "replication" ambiguity. The issue is manifestly critical. Grave errors are still being committed, for which a universal remedy has not yet seen the light. This chapter a.o. intend to contribute to put an end to this unfortunate issue.

With reference to TOS, most aspects related to the vagueness concerning replication issues can be assigned to incomplete notions as to the role and influence of spatial heterogeneity DH_L of the analyte(s) in question. What is confusing, and frustrating, is also that in many cases the analysts (and a possible subsequent data analyst) often are, literally, miles away from the location where the issue originates. But it is not difficult, nor expensive to do something about this: RE can easily be carried out by the same personnel responsible for the primary sampling, for example.

Here is what to do – it is simplicity itself. As an example, consider a process sampler (see Figure 19.13), who, albeit employing a distinctly nonrepresentative manual procedure, nevertheless wants to estimate the total sampling-and-analysis sampling variability (in the vain *hope* that this may actually be low enough to pass muster – hope spring eternal, but the outcome in the illustrated situation is very likely doomed). Here is a way to grade any-and-all sampling procedure ... by simply *replicating* the primary sampling.

The quantitative effect of lot distributional heterogeneity (DH_L) *interacting* with a specific sampling process can be quantified by extracting and analyzing a statistically sufficient number of *replicate primary samples* with the objective to "cover the spatial geometry, the spatial heterogeneity, or the salient time span of the lot/process" as best possible (obeying FSP), and from this to calculate

FIGURE 19.13 A generic example of a specific primary sampling operation (in this case, a manual sampling of a process stream) that can easily be *replicated* (e.g., 10 times), which is all that is needed for a replication experiment, RE(**10**). NB. This is *not* an example of representative process sampling, but RE is applicable to all sampling processes, in order to pass objective judgment on their performance. Illustration copyright KHE Consulting, brought here with permission.

the empirical variance of the resulting analytical results a_S. This procedure is termed a "replication experiment", described in full in DS 3077 (2013) or Esbensen and Swarbrick (2018).

A relatively small number of primary samples may often suffice, though never less than 10 if possible, a standard statistical dictum proclaim. But the issue at hand is not only statistical, but also very much associated with the presumed level of lot heterogeneity. If DH_L is, or is suspected of being, significantly influential, it is senseless to be frugal with respect to the number of replicated primary samples for obvious reasons. It is suggested to use a *replication index*, **r**, associated with the replication experiment, RE(**r**). Thus, RE(**7**) can be meaningfully compared to RE(**12**), for example. The issue is not so much the exact **r** value; it is rather that the experimenter honors an obligation to report on what basis RE was physically carried out; e.g., in Figure 19.13, a manual stream sampling was subjected to a RE(**10**).

RE must be governed by a protocol that specifies precisely how all procedural elements are to be carried out. It is essential that primary sampling and *all* subsampling and mass reduction stages, sample preparation, etc. are replicated in a *completely identical* fashion.

Obviously, it is obligated for a RE(**r**) that all ISE have first been eliminated, i.e., that only *correct sampling* is employed (TOS), or the sampling bias will play havoc with this experiment (see below).

However, it is also *feasible* first to gauge an existing sampling-and-analysis procedure in which this requirement has not necessarily been fulfilled. RE will then *include* the pertinent error effects from these factors; i.e., include the adverse inconstant sampling bias effects a.o. (see further below). This will soon be seen as a major bonus, however, revealing the real practical power of RE(**r**).

It has been found useful to employ a standard statistic to characterize results from RE. The relative coefficient of variation, CV_{rel}, is an informative measure of the magnitude of the standard deviation (σ) in relation to the average ($\mathbf{X}_{avr}$) from a series of analytical results corresponding to the appropriately replicated primary samples, expressed in %:

$$CV_{rel} = [\sigma / \mathbf{X}_{avr}] \times 100 = \text{RSV [Relative Sampling Variability]} \quad (19.1)$$

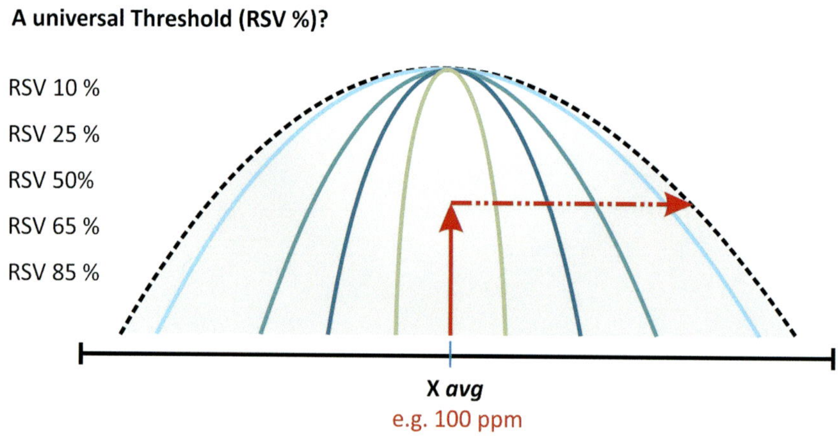

FIGURE 19.14 Schematic illustration of different empirical RSV magnitudes, expressed with respect to the pertinent average of the **r** analyses (e.g., 100 ppm is just for illustration). The larger the RSV magnitude [%], the larger the spread of the **r** analytical results realized through the RE (**r**).

When RSV is calculated from data originating from RE starting with the primary sampling, it includes *all* sampling and analy°tical errors as manifested **r** times as the **r** replicated primary samples all pass identically through the full "lot-to-analysis" pathway. RSV measures the total empirical sampling variance influenced by the specific heterogeneity of the lot material as expressed by the current sampling procedure. The core issue is that a properly designed RE is a reliable (TSE+TAE) estimator, and it simultaneously furnishes a quantitative measure of the effective heterogeneity of the lot, *as manifested by the sampling procedure in use*. A properly designed RE makes it possible to determine a quantitative specification for the heterogeneity of the material being sampled. The specific RSV magnitude can be seen to be directly proportional to the effective total heterogeneity of the targeted lot/material, because some form of sampling of the lot must be carried out in order to end up with analytical results. Indeed *what* could be more *relevant* as a proper quantitative characterization of any practical "total measurement situation" that includes all sampling errors incurred at all relevant stages?

Since all sampling errors, at all scales from lot to analysis, are included in the analytical result for each replicated primary sample, it is certain that both the inconstant sampling bias (if present) and all grouping and segregation (GSE)-induced variability effects (always present to some degree, as are the Fundamental Sampling Error effects, FSE) are allowed to manifest themselves **r** replicated times – as are all ISE that have not been eliminated. It follows that RSV provides a highly relevant expression of the effective MU_{total}. This has the desirable consequence that all sampling-plus-analysis procedures may easily be put to the test, *if* a specific threshold with which to compare is at hand (Figure 19.14).

From the TOS community, a general acceptance threshold has (reluctantly) been suggested as 20%. Thus, a RSV, which is higher than, say, 20%, signifies an unacceptably high sampling variability – with the mandate that the sampling procedure then *must* be *improved.*

However, such a general threshold is in reality passing a much too fleeting judgment over what is in reality a strongly problem-dependent issue. Material heterogeneity may very well be so that a higher threshold is warranted, but there also exist many types of materials for which a lower threshold than 20% is entirely appropriate[††]. Be all this as it may, it is sufficient to be conversant enough

[††] There has been an extensive debate within the international sampling community as to setting up a (completely) general *RSV* threshold. Opinions have slowly converged to accept a suggestion, which originates from the characteristics of the Poisson distribution (sampling can in a certain sense be likened to a Poisson selection process), that a *RSV* larger than 20% signifies that the average of repeated sampling and analysis is out of control. Be aware, however, that this is an attempt to characterize <u>all</u> the world's extremely different materials, lots, and processes with one singular threshold – a most dangerous simplification! More on this important issue can be found in DS 3077 [2] and Pitard [1].

with these matters to demand, that some form of RSV threshold must be available, else data analysis operates with a much too uncomfortable margin.

There is another very useful aspect of RE(**r**). Consider that an initial sampling procedure evaluation resulted in a RSV of, say, 127%. There is obviously something distinctly wrong here, 127% is way above 20%. This represents a situation in which one would very likely find a nonrepresentative operation *somewhere* in the sampling pathway. Remedial action (TOS) is of course unavoidable. It is possible to detect precisely where the culprit part process is to be found – for which reason the definition of full versus partial versus hierarchical REs is made first.

Full replication: By replication "from the top" (this is called a baseline RE), quantitative information as to the performance the *total* sampling + measurement procedure can be obtained. In the event of a RSV transgression, the alarm bell has been sounded – and remedying activity will be needed in order to lower the sampling variability below the pertinent threshold[‡‡]. In this fashion, a full RE(**r**) delivers immediate, highly relevant information on any sampling process currently in use. The "gamble" by starting out with a full RE(**r**) is that this *may* substantiate the current procedure, in which case no further action is necessary (the gambit was won). But the opposite side of this issue is that one *must* implement the relevant TOS-remedial actions in the case where RSV transgresses 20%–30%, no exceptions! There is only one remedy for locating where a too high RSV originates, and that is to perform a partial RE.

Partial replication: It is also possible to start RE at a "lower stage" in the replication hierarchy. Figure 19.15 illustrates the general setup for full versus partial REs.

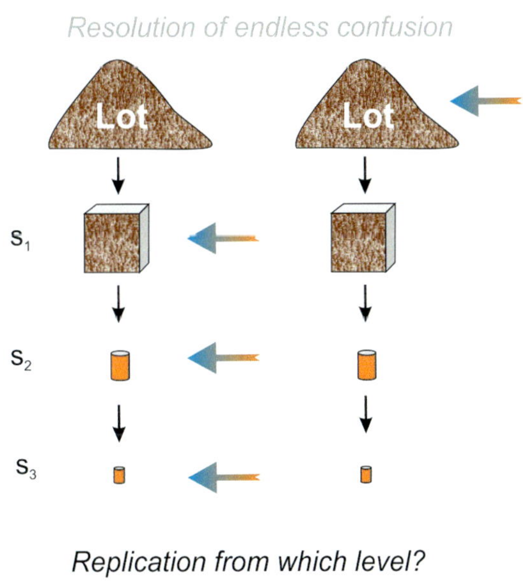

FIGURE 19.15 Illustrating full versus partial replication experiments, i.e., different starting levels (horizontal arrows). This illustration harks back to the hierarchical levels of "replication" delineated in the introduction of this section.

[‡‡]There exist many types of materials with significantly different heterogeneity levels. The general threshold for *RSV* (20% → 30%) refers to *significantly* heterogeneous materials, a very wide and diverse class of materials, e.g., mineralization, ores, pollutant effects, industrial aggregates (building materials, waste, biomass …), the list is exhaustive. There also exist many other types of materials which exhibit less heterogeneity, for which a lower threshold will be relevant, say at the 10%, or 5% level, or even lower. It is emphasized that the demand for a general *RSV* threshold is a demand which cannot be met universally. It is necessary to invoke common sense when making *RSV* operational in a distinct problem-dependent fashion.

FIGURE 19.16 Hierarchical replication setup. A separate RE(r) for each sampling stage allows to decompose the full RE(r) variance into individual-stage components. NB. This decomposition makes use of variances, $(std)^2$, as (std) is not an additive/subtractive statistical entity.

Hierarchical replication: A hierarchical replication consists of the full and all partial replication setup *combined*. This is complete replication at all levels. The situation is illustrated in Figures 19.15 and 19.16. In the latter, the magnitude of s^2 at each sampling level is represented by the length of a horizontal bar. As each lower-level variance is included in any of those pertaining to higher levels, the lower-level variances can be subtracted from those from all higher levels. By this simple subtraction, one can perform a complete *decomposition* of the level-specific s^2. For details, the reader is referred to:

https://www.spectroscopyeurope.com/system/files/pdf/Sampling%2028-1.pdf and Esbensen (2020).

From the above, it is clear that RE(**r**) furnishes the exact information needed to tackle all replication issues met with in current practice. RE(**r**) is an unambiguous quantitative index of the total sampling-and-analysis variability induced through the full lot-to-analysis pathway. In this sense, RE(**r**) is a "measure of everything", in the sense that RSV simultaneously grades the performance of a complete sampling-plus-analysis system and furthers a quantitative expression of the total material heterogeneity encountered.

Among other implications, this has a direct impact on the validation issues so important in multivariate calibration, especially regarding the aspect of "replicate measurements". Further discussions of these issues can be found in Chapter 9 in Esbensen and Swarbrick (2018) focusing on the development of multivariate calibrations for spectroscopic applications.

19.12 A POTENTIAL RSV[%] REVOLUTION

Imagine if all providers of sampling-plus-analysis results were *mandated* to disclose the pertinent RSV[%] for the specific combination of procedure and material. This would allow all *users* of analytical results a very simple and effective way to evaluate the relevant total quality of determination in the form of RSV[%] magnitudes, comparable at a glance: above or below 20% (or below or above any alternative threshold made public).

The didactic cases shown in Figures 19.17–19.20 illustrate how it will be possible to put an end to *all* uncertainty regarding the influence of the preanalysis pathway on the validity of analytical results (NIR, or otherwise): No more disingenious *sales talk* of the merit of this or that specific type, or brand of equipment; no more *opinionating* regarding this or that specific approach *claimed* to be superior – instead *just* the numerical facts RSV[%]. A scientific community decision to mandate RSV[%] would be a revolution for *objective* sampling-plus-analysis evaluation. What's keeping the NIR community from taking this quantum leap? See DS3077 (2013) and the above open access introduction for a full description of RE(r).

19.13 THE SAMPLING BIAS

A comparison between the analytical process and the preceding sampling process will make clear how all remaining ISE will inflate MU_{total} unnecessarily – but not only that. The sampling bias

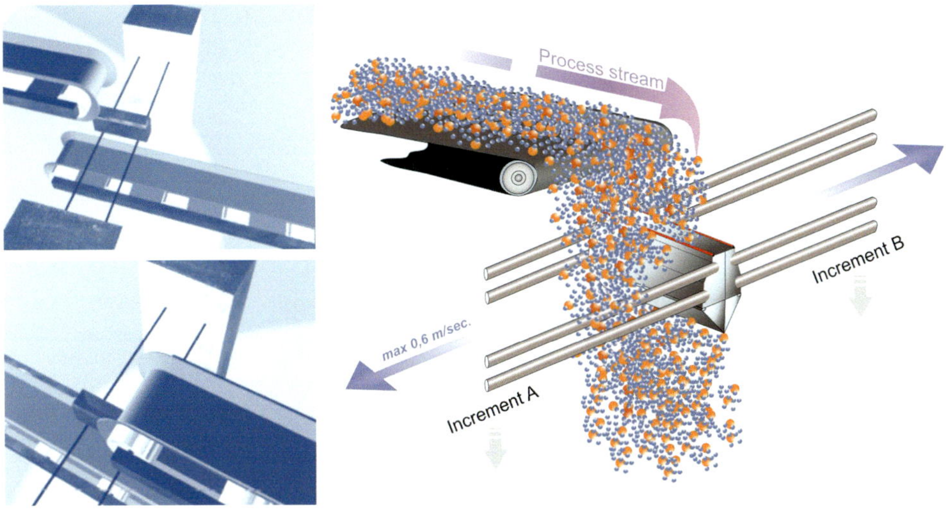

FIGURE 19.17 Schematic illustration of one of TOS' archetype primary samplers widely held always to be representative, the cross-stream sampler. If the performance characteristics are not known, e.g., because it is applied to a new material never tested or sampled before, it is easy to apply a RE(r) to establish the operative RSV[%] level (compare Figure 19.13 for a parallel application, but to a distinctly nonrepresentative approach). Illustration copyright KHE Consulting & HERZOG Maschinenfabrik GmbH & Co. KG, brought here with permission (leftmost illustrations, courtesy of erstwhile Ph.D. Hans Henrik Friis-Pedersen).

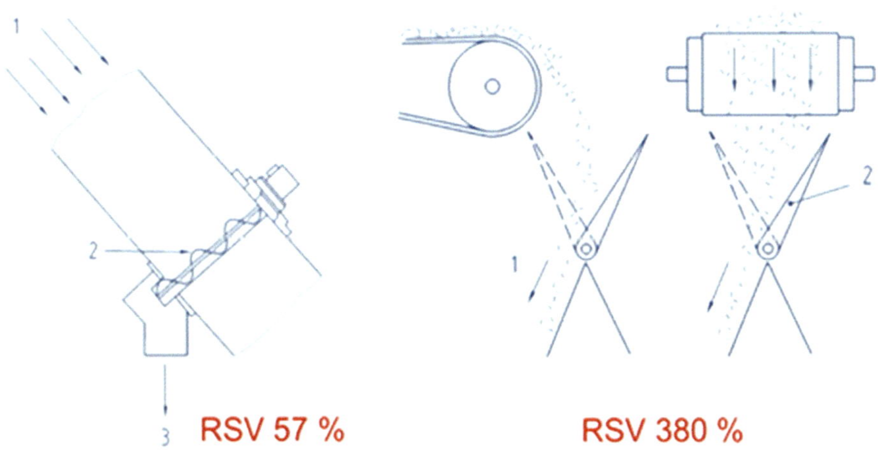

FIGURE 19.18 Examples of RSV[%] validation of two "ISO-recommended" process sampling solutions, the "air slide sampler" (left) and the "stream diverter" (right). Note that the flap valve design to the right may result in dramatically different RSV[%] levels as a result of the specific design and/or significant material heterogeneity differences. *Generic illustrations.*

defies characterization as it is *inconstant* (it is different every time one is trying to estimate it). This realization has not been presented in the NIR realm before. Acknowledging this will allow for a new perspective of the critical need to be reasonably conversant and competent in *all* salient sampling matters before analysis. There is but a small price to pay for acquiring the modicum of TOS knowledge needed, but the payback is enormous: 10–25+ times potential *reduction* in unnecessary sampling error effects. The specifics are as follows.

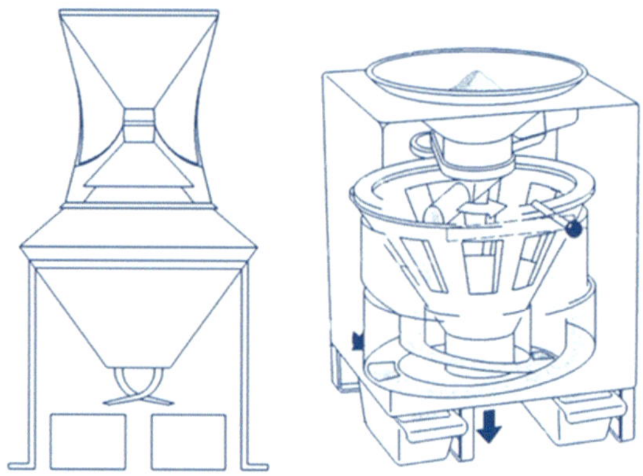

FIGURE 19.19 Examples of RSV[%] validation of two excellently designed laboratory subsampling tools, the "Börner divider" to the left and the "RAKO rotary splitter" to the right, tested on the same material. *Generic illustrations.*

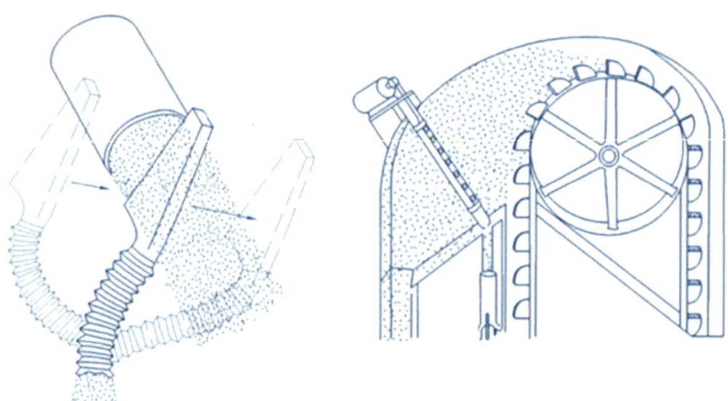

FIGURE 19.20 Two schematic examples of primary process sampling approaches applied to different materials. Note that a specific sampler can be certified for some materials, while failing for other materials (left). The right-hand illustration is related to the "air slide sampler" in Figure 19.18. *Generic illustrations.*

As is very well known, for an analytical process it is always possible to perform a rectifying *bias correction* if/when needed. It is less well known that a sampling bias is *not* amendable to a similar kind of correction (analytical, data analytical, statistical). This is one of TOS' most surprising and most penetrating understandings (Esbensen and Wagner 2014a, Gy 1998, Pitard 1993, 2019, Esbensen 2020). For all analytical methods, a fundamental credo is that if there is a systematic deviation between the true concentration and the average of a series of analytical determinations, this *bias* is constant. This fundamental assumption makes it possible to invoke a statistical bias correction, illustrated schematically in Figure 19.21 (top panel).

This is quite opposite with respect to a sampling process (strictly speaking a "sampling-plus-analysis" process). Try as one might, or as one would dearly *hope*, by replicating the full lot-to-aliquot pathway, the result of 60+ years of empirical and theoretical work within the realm of TOS has led to the understanding that the accuracy, and the bias, is *varying*, it is in fact *inconstant*. The empirical spread of repeated sampling-plus-analysis results will differ significantly among

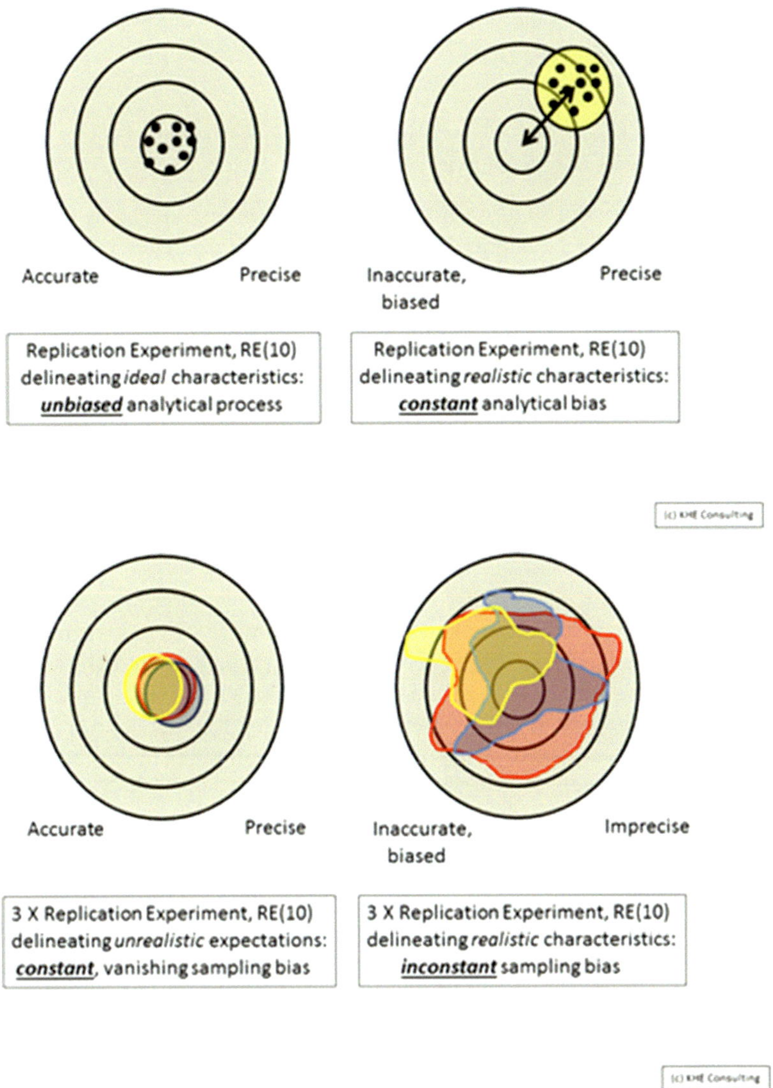

FIGURE 19.21 (a, b) The crucial difference in the nature of the analytical bias, which *can* be corrected for, and the sampling bias, which *cannot*, as revealed by the application of replication experiments (RE). The sampling bias is *inconstant*, because of material heterogeneity and sampling process defects (ISE). Illustration copyright KHE Consulting, brought here with permission.

themselves every time a new RE is carried out (upon which a similar bias correction *would* have had to be based). This fundamental result is illustrated in the lower panel of Figure 19.21.

Thus, it is *never* possible to carry out a sampling bias correction! This kind of sampling variability cannot be corrected for by any mathematical correction algorithm, by physical pre- or post-treatment, or by any software approach. WHAT to do in this case?

Dramatic at first sight, TOS makes the only logical, rational, and science-based conclusion.

Do not try to correct for sampling bias effects, it is but an *exercise in futility*. Instead, implement whatever is needed (TOS) in order that the sampling process eliminates *all* ISE, the bias-generating errors. While easy in principle, how to do this in practice involves a reasonable grasp of the key principles and tools in TOS. But the bonus is readily at hand; in fact, the major part of achieving sampling process representativity has been accomplished as soon as this stipulation has been implemented, GP #3, cfr. Table 19.1.

19.14 VARIOGRAPHIC ANALYSIS (QUALITY ASSESSMENT, MU_{TOTAL}, OF PROCESS SAMPLING)

Variographic analysis is a powerful tool with immense potential for advancing QC of process NIR spectroscopy and process understanding. The compact description below is followed by didactic examples of NIR application in the process industry. Full variographic initiation can be found in a plethora of TOS literature (Danish-Standards-Foundation 2013, Pitard 1993, 2019, Esbensen 2020, Gy 1998, Esbensen and Wagner 2014a; Esbensen and Paasch-Mortensen 2010).

Most experienced NIR spectroscopists have been in situations where production personnel have indicated: "The problem is with the NIR method, my process is stable and is working well". Variographic analysis provides a scientific approach to address this situation – the verdict *may* very well be different! Variographic analysis makes it possible to *discriminate* between true process variation and that part of the total *apparent* process data variation that actually stems from the combined sampling-and-analysis error instead; this latter is "the total measurement system error". This is a very powerful facility.

The basic setup for a variographic experiment, aka variographic data analysis, is an equidistant measurement series (Figure 19.22). As applied in the PAT regimen, TOS only requires a chronological series of analytical results. For example, process NIR spectra may be used to develop a variogram since the order in which the spectra were obtained is always known (Romañach, Sanchez-Paternina, and Esbensen 2018, Esbensen et al. 2016, Sierra-Vega et al. 2018). NIR spectra may be obtained, e.g., for granulates, powders or liquid, or slurries flowing through a duct or a chute, or from material moving on a conveyor belt. Following proper multivariate calibration and validation, new process monitoring NIR spectra are then used for prediction to obtain on-line analyte concentrations (or other, calibrated properties of interest). Such successive analyte concentration predictions form an ideal basis for calculating a variogram.

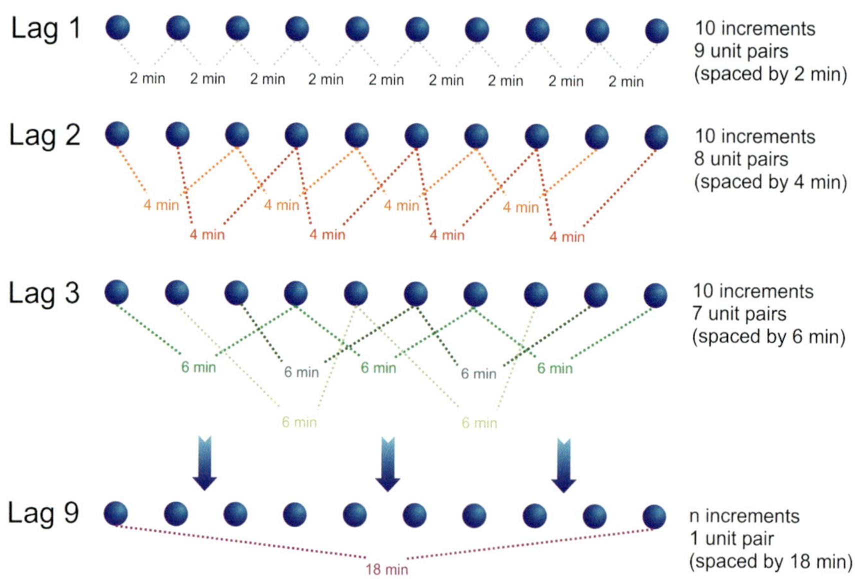

FIGURE 19.22 Basic setup for variographic data analysis. Any equidistant measurement series can be subjected to variographic analysis, but TOS is particularly aimed at characterizing chronological series of analytical results pertaining to process monitoring. The distance between pairs of increments/samples is termed "the lag, *j*".

The distance between *pairs* of process samples is termed "the lag, *j*". Variographic data analysis is used to characterize the average variability (variance) of all pairs of analytical results with an intersample distance equal to *j*, for *j* running from 1 to half the number of measurements, N_U, i.e., in the interval $[1; N_U/2]$.

The first objective of a variographic analysis is to get an overview of the *apparent* total process variance for all intersample distances *simultaneously*, and this is accomplished by plotting the series of $[j, V(j)]$ results (see Figures 19.23–19.24).

$$V(j) = \frac{1}{2(N_U - j)} \sum_m (h_{m+j} - h_m)^2$$

$$V(j) = \frac{1}{2(N_U - j)a_L^2} \sum_m [a_{m+j} - a_m]^2$$

V(j) = Variogram function [relative (h_m) or absolute (a_m)]

h_m = Increment heterogeneity contribution (mass prop. a_n)

$$h_n = \frac{a_n - a_L}{a_L} \times \frac{M_n}{M_n}$$

a_n: sample concentration
a_L: lot grade (process average)
M_n: sample mass

FIGURE 19.23 Using TOS' heterogeneity contribution transcription (h_n) of mass-weighed increment analytical results (a_n), the variogram function, *V(j,)* is a function of the lag distance between *pairs* of increment results. *V(j)* is expressed as standard statistical variances. When no confusion can arise, increments are sometimes referred to as samples.

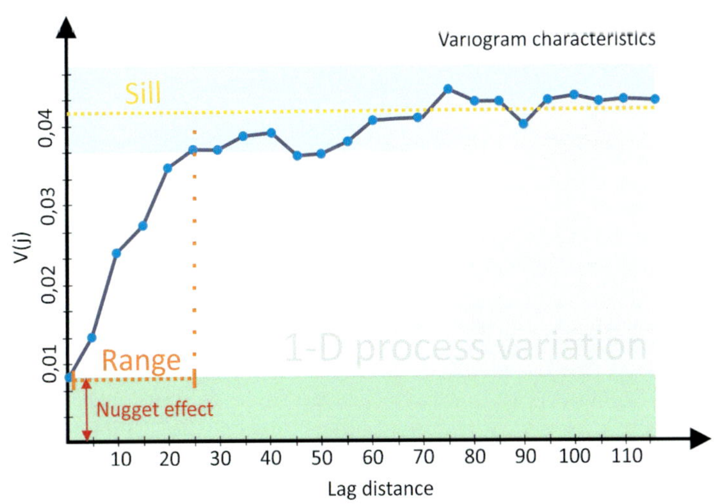

FIGURE 19.24 The "increasing variogram" with parameters: range, sill, nugget effect (n.e.), defined and described in the text. Estimation of the important nugget effect is usually a simple (3-, 4-, 5-point) back-extrapolation of a smoothed variogram curve. Technically, the sill is the *average V(j)* for all lags for which reason it lies asymptotically close to the maximum variance level as a "lower envelope", but is in practice often described as a "ceiling".

Based either on increment[§§] analytical results directly (a_n) or on the much more useful heterogeneity contribution transformation (h_n), the variogram function $V(j)$ is calculated for each lag (Figure 19.23), delivering the average variance corresponding to the contemporary lag, $V(j)$.

$$V(j) = \frac{1}{2(Q_{\text{total}} - j)} \sum_{q=1}^{Q_{\text{total}}-j} \left(h_{q+j} - h_q\right)^2. \tag{19.2}$$

Plotting the lag (horizontal X-axis) against $V(j)$ (plotted along the vertical Y-axis) gives a synoptic view of the lag-dependent variability of the original measurement series, notably for all scales (lags) of interest simultaneously. There is a wealth on process (or product) information residing in variograms (see, e.g., Esbensen et al. (2007) and Esbensen and Paasch-Mortensen (2014)).

Figure 19.24 shows the most often occurring variogram, called "the increasing variogram".

All variograms are characterized by three parameters (only): the *sill*, the *range*, and the *nugget effect* (n.e.).

- The *sill* represents the total variance of all measurement data if calculated without considering their mutual order; this is the maximum variance experienced by the data series. Technically, the sill is the asymptotic horizontal lower convolute corresponding to this maximum variance[¶¶].
- The *range* is the lag distance at which the increasing variogram flattens out and becomes well described by the sill (which applies to all lags greater than the range). The interval from lag=1 to the range reflects the discrete intersample distances at which *autocorrelation* is a characteristic feature of sample pairs. The autocorrelation is highest for lag=1 and disappears at lags greater than the range. The variogram is not defined for lag $j=0$, as this would correspond to the extraction of the exact same sample twice (which is a physical impossibility).
- The magnitude of the lag=0 variogram value, the *nugget effect*, is usually estimated by extrapolating a simple (3-, 4-, or 5-point) linear regression fit as a backward extension from a smooth variogram curve (Figure 19.23). There is a critical information hidden in the nugget effect estimate. The variogram function $V(0)$ is in fact an estimate of the *sum* of all error variances pertaining to ISE (to the degree these have not been completely eliminated)+CSE+TAE. Thus, the nugget effect furthers an estimate of *all* error manifestations of the *total* process measurement system (TPMS), $MU_{\text{sampling+analysis}}$ (Pitard 1993, Esbensen and Paasch-Mortensen 2010, Gy 1998, Esbensen 2020), which opens up for a breakthrough insight into the usefulness of variographic process characterization in general.

For very many purposes in process technology, industrial processing, and manufacturing, the most important parameter of an experimental variogram is the *nugget effect*. TOS' detailed theoretical analysis has led to the understanding that the nugget effect represents the sum total (TSE+TAE)=(ISE+CSE+TAE). This explanation of the "meaning" of the nugget effect is a very useful feature that has not been recognized before TOS (Pitard 1993, 2019, Danish-Standards-Foundation 2013, Gy 1998, Esbensen and Wagner 2014a, Esbensen and Paasch-Mortensen 2010,).

The nugget effect magnitude, expressed as a % of the sill level, turns out to be a *quality index* of the TPMS performance, a convenient feature of process data analysis (Figures 19.24–19.25). This feature is fully described in DS 3077 (2014), Esbensen et al. (2016), Pitard (2019), and many other

[§§] Depending on the specific case, either single samples or increments are extracted from the process (later destined to be aggregated to form a composite sample). The resulting variogram will then reflect the 1-D variability of samples or increments, respectively. In fact, the increment-based variogram is often used to determine the number of increments, Q, to be composited (see Esbensen et al. (2007), Esbensen (2020) and Esbensen and Paasch-Mortensen (2014) for full details).

[¶¶] Technically, the sill is defined as the average variance calculated for the entire measurement series.

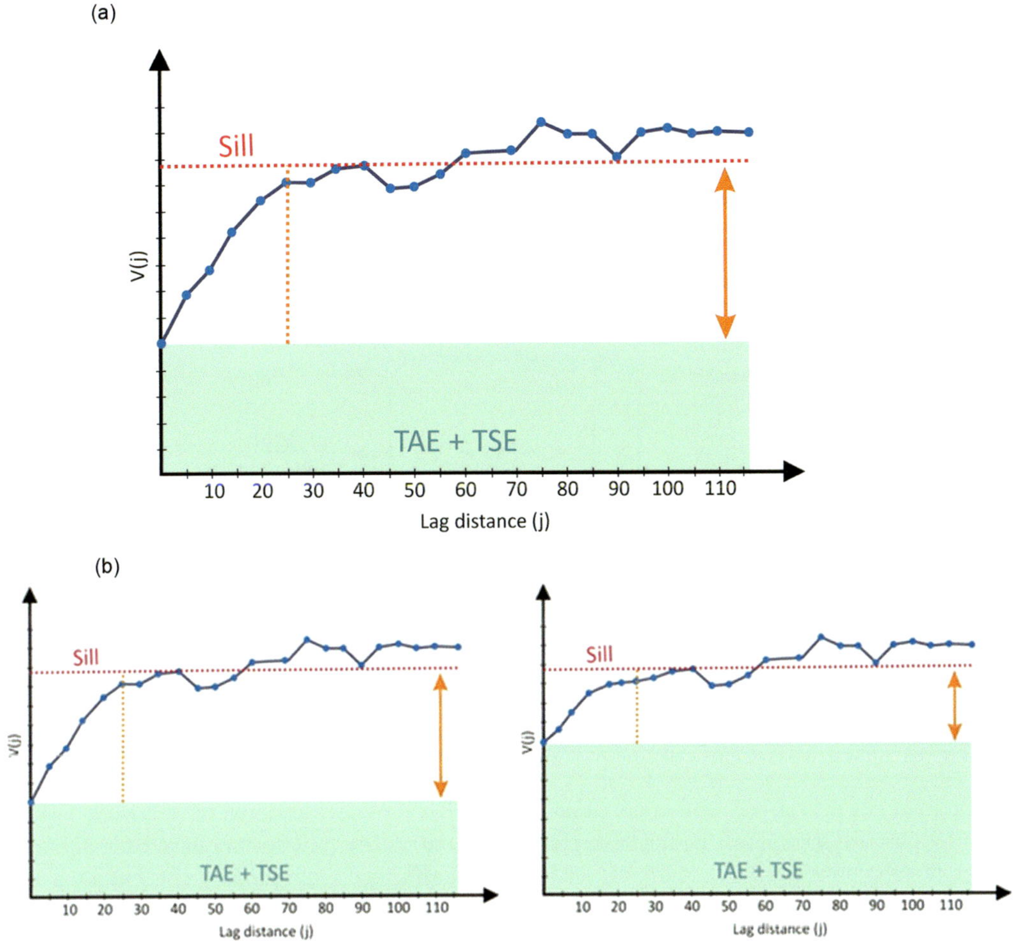

FIGURE 19.25 Variographic characterization of the total measurement system performance. The nugget effect (n.e.), expressed as a % of the sill level, is a quality index of the total process measurement system (TPMS) performance (see text for details). The left example shows a very risky TPMS, with a nugget-to-sill ratio of ~30 to 40%, while the TPMS to the right is hopelessly inadequate to reveal the true process variability, since ~70% of the total measurement series variance is nothing but sampling-and-measurement system noise.

TOS references. Suffice here to cite that the TOS community has agreed that any measurement system in which the nugget-to-sill ratio exceeds 30% is in general not satisfactory, *ibid*. However, in specific implementations, and for specific purposes, it is also acceptable to use another demarcation – but this is critically dependent upon *full disclosure* and *argumentation* for deviating from the standard 30% stipulation.

Case Example VII

Variographic analysis is specifically designed for characterization and monitoring of continuous processes in general, and will, e.g., be of particular support for continuous manufacturing of pharmaceutical products (Romañach 2017). In continuous manufacturing, materials and products are continuously charged and discharged from the system, throughout the duration of the process (Lee et al. 2015, Osorio et al. 2015).

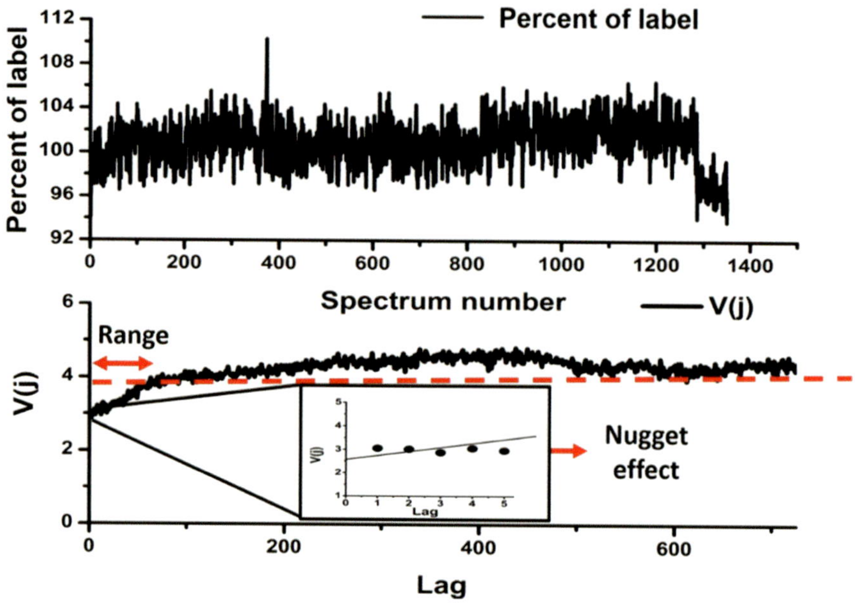

FIGURE 19.26 A continuous manufacturing example of process and measurement system monitoring based on variographic characterization. The TPMS, as measured by the nugget-to-sill ratio [%], is ~60% which points to a serious need for the sampling-plus-analysis system to be improved, which was the exact purpose with the feasibility experiment depicted.

The material discharged from a continuous blender is a perfect example of a moving 1-D lot as it flows toward a compressing machine (Esbensen et al. 2016). NIR spectra have been obtained for the flowing powder and variographic analysis was performed after the drug concentration was predicted by a calibrated partial least squares (PLS) model. In a "first" paper, the sill, nugget effect, and range have been determined for a commercial continuous manufacturing process that is monitored by NIR spectroscopy in pharmaceutics (Vargas et al. 2018, 2017). Figure 19.26 shows the drug concentration in terms of percent of label for NIR spectra obtained from a continuous manufacturing test facility. The consecutive drug concentration results are used to develop the variogram. Presentation in two plots is strongly recommended: One (the top plot) shows the time variation of drug concentration through the mixing process, while the other plot, the variogram, provides the complementary information needed to monitor and assess both the process variability and the decoupled total measurement system validation (n.e.). Figure 19.26 is based on almost 1400 determinations of drug concentration with NIR spectra obtained every 5 s. The sill for this data is ~4%, while the nugget effect is about 2.5% of the label claim making the n.e./sill ratio ~60%. This quality index indicates that there is a serious need for improvement of the current sampling and analytical system, which was a first principal feasibility design only. All details of this variographic data analysis are confidential.

This example shows the potential of variographic characterization for TPMS evaluation with a view of improvement, a didactic illustration that would not be as strong were the application showing an already perfected system. Instead of a simple RSV[%] test, which would only be able to show that something is not right (inflated sampling bias effect), the full variographic evaluation addresses the TPMS, which allows to account quantitatively for the % of the observable process variability that is in effect only a reflection of the TPMS performance only. This example can be a placeholder for a legion of other *blind* PAT valve/sensor installments in current process technology and process industrial sectors – "blind" because there has never been a proper variographic installment validation yet. It does not take more than a simple variographic evaluation to fully test or reject any TPMS – which can be carried out based on routine PAT data coming of the process anyway! Should older sampling installments have missed proper validation, the TPSM variographic test is available essentially without costs because of the vast amount of historical data often available.

From all of the above, it should now be clear that emphasis in many cases should not be placed on the TAE features, but at the very opposite end of the sampling-to-analysis pathway, i.e., the primary sampling stage. If the primary sample is not representative .... all is lost.

19.15 THE BY-PASS PRIMARY SAMPLING DEVICE – A MUCH-OVERLOOKED CRITICALITY

As a poignant example, consider a process by-pass, often hailed as a facility that furthers efficient, smart scaling down fluxes for on-line, in-line, or at-line process analysis. By-pass valves or ducts are in fact functioning as on-line process mass reduction facilitators, and as such, they must comply with the same demands for representativity as all other sampling devices. Generalizing the above scenario, Figure 19.27 presents five often met with types of by-pass valve implementation used

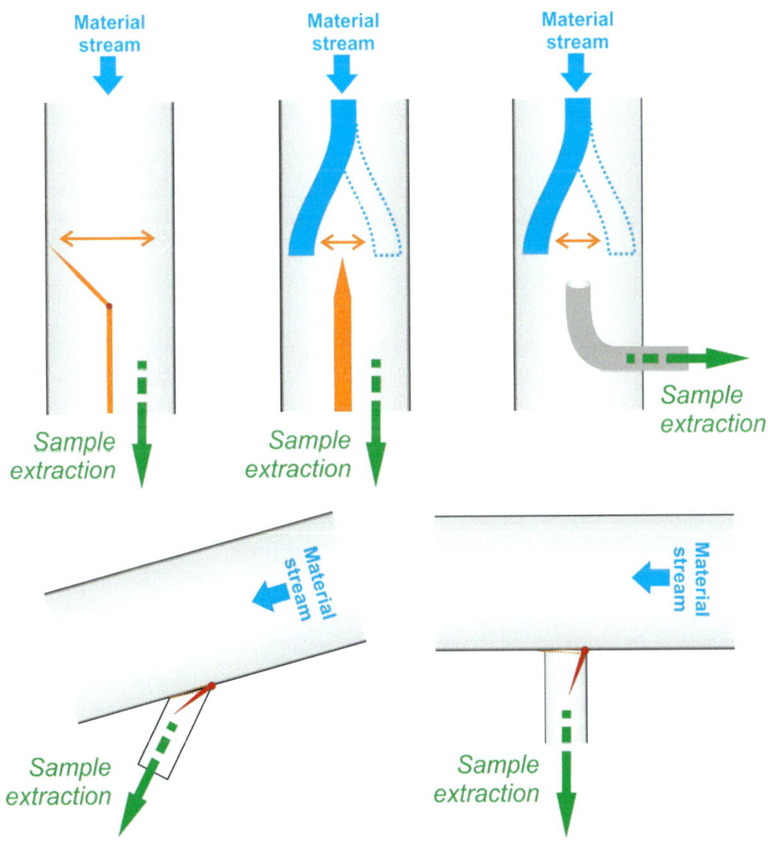

FIGURE 19.27 Five examples of "sample extraction valve" configurations. From the treatment of TOS above, the reader should be able to identify the *singular* correct extraction example; the remaining four operations result in worthless "specimenting". The adverse cases are known as variants of the "flap valve" implementation***. (Copyright KHE Consulting & HERZOG Maschinenfabrik GmbH & Co. KG, brought here with permission.)

*** Only the top rightmost illustration will result in a representative increment (or sample) outtake; see Esbensen and Paasch-Mortensen (2010) for details.

for "sample extraction" in many industry sectors (Esbensen and Paasch-Mortensen 2010), NIR no exception (see, e.g., also Figures 13.3–13.5 in Burns and Ciurczac (2008)).

The conclusions pertaining to Figure 19.27 is a serious warning of the consequences that will result if not paying sufficient attention to the primary sampling in the process regimen. The illustration may, e.g., relate to grain transportation or a host of similar aggregate, slurry, or powder transportation cases in which a "sample" is desired on the fly.

In the dedicated sampling literature, there is solid documentation of the difficulties involved in designing and operating by-pass ducts and valves properly, even when in the hands of sampling experts; see, e.g., Fouchee and Steinhaus (2015) from which can be quoted: *"This test work confirms that the implementation of split sampling does not yield reliable sampling results of 50% cross stream cuts, no matter how careful the design [....]. One must rather design the unit to accommodate a full cross stream cut for meaningful results [...]"*. A parallel view is presented in a general introduction to TOS: "TOS – pro et contra" (open access: https://www.spectroscopyeurope.com/system/files/pdf/Sampling_30-1.pdf).

Figure 19.27 thus again illustrates well the dangers of analytical tunnel vision.

Case Example VIII

Here follows a more advanced example of a sampling solution from process industry with strong relevance for much of current NIR process analytics.

Figure 19.28 presents a case in which it is necessary to obtain a reliable sample from the reactor vessel depicted at regular time intervals; this system could be a chemical reaction vessel or a biotechnological fermentation vessel, or similar. The point is that the material is significantly heterogeneous with a propensity for segregation if not mixed thoroughly at all times. Even so, it has been

FIGURE 19.28 How to sample representatively from a reactor or processing vessel? The bottom outlet (with outflow to the right) is traditionally also used for conventional sample extraction. Compare with Figure 19.7a. The innovative approach consists in implementing a recirculation loop to the vessel (left circuit). Experience shows that the volume below the lowest mixing impeller constitutes a *dead zone*, seldom not well enough mixed for proper sampling purposes. See text and Esbensen and Paasch-Mortensen (2010) for details. (Copyright KHE Consulting & HERZOG Maschinenfabrik GmbH & Co. KG, brought here with permission.)

observed that there exist several mixing *dead zones* in the vessel/ reactor, especially *below* the level of the lowest rotating impeller, which is supposed to be able to mix the entire lower-level volume satisfactorily, upon close scrutiny alas with distinctly nonoptimal results. But sampling is all the more critical for safe processing.

What to do?

Well, under no circumstances should one not even *try* sampling from the always less than satisfactorily mixed vessel volume! Instead, following the illustrations in Figures 19.28 and 19.29, an outtake valve facility is installed high up a vertical stretch of a *recurrent loop*, collocated with a PAT sensor. The function of the recurrent loop is to boost mixing of the entire vessel, by continuously placing increments from the most deviating compositions from the bottom dead zone (due to noneliminated segregation features) at the top of the entire vessel volume, where they will be subjected to the most effective mixing due to the two impellers shown. This feature has been shown to boost the full vessel mixing in a highly significant fashion, essentially after only a few full-volume circulations, but the recirculation looping is usually kept going continuously. So far, this is but a standard application of the SUO mixing. But it was *also* a design feature to allow for a maximum of *automixing* in the upward pipeline section brought about by a turbulent flow-front (with constantly varying front perturbations) working *against* a constant downward gravitational force; experiences show that a significant extra mixing boost inside the pipeline can be obtained this way.

It is furthermore a critical success factor to use a *small-bore* recirculation pipeline, for a specific sampling reason. By careful coordination of two internal butterfly valves detailed in Figure 19.29, it is possible to direct a ~cylindrical volume element (~ideal increment delineation) to become a *correctly* extracted reference sample. Observe that this volume has been subject to simultaneous PAT characterization by the colocated PAT sensor, making a support mismatch

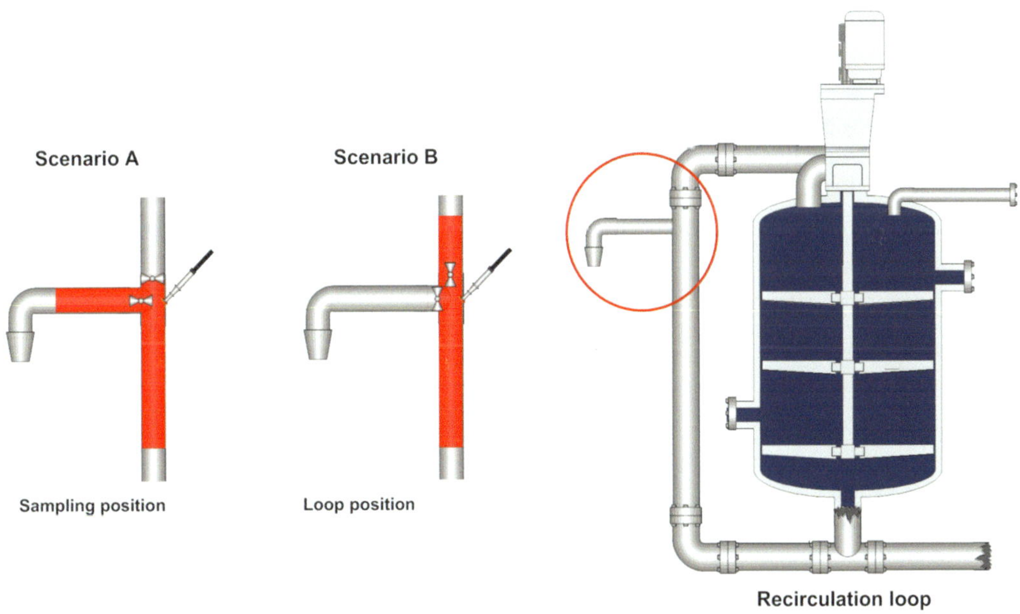

FIGURE 19.29 Details of a TOS solution for sampling via a recurrent loop. Two rotation-coordinated internal butterfly valves direct a ~perfect increment (cylinder) through an outtake valve, enabling to secure a reference sample which is subjected to simultaneous colocated PAT sensor characterization (scenario A). The same sensor will also be used as signal acquisition sensor with respect to the on-line PAT prediction (scenario B). The recurrent loop solution has effectively transformed an "impossible-to-sample" 3-D reactor into an easy 1-D process sampling issue. See text and Esbensen and Paasch-Mortensen (2010) for details. (Copyright KHE Consulting & HERZOG Maschinenfabrik GmbH & Co. KG, brought here with permission.)

error very small (easily "controllable" by making the sample cylinder length larger with respect to the recurrent loop diameter). A point of refinement is that both valves are designed to rotate in one direction only, counterclockwise in the illustrations shown, making the otherwise fatal "flap valve" issue disappear completely (compare Figure 19.17); full details can be found in Esbensen and Paasch-Mortensen (2010).

After multivariate calibration and proper validation, this TOS solution has the added advantage of already being fitted optimally for on-line prediction, in fact using the exact same PAT sensor deployment as for the calibration/validation.

Caveat. The observant reader will have noticed that this sensor positioning principally suffers from the same sampling error issue as was depicted in Figure 19.7b. However, the design criterium to use a *small-bore* recurrent loop pipeline diameter is specifically addressing this issue with an aim to reduce it quantitatively to below *fit-for-purpose* representativity status; a full description of this approach is found in Esbensen and Paasch-Mortensen (2010). This aspiration, needless to say, should be evaluated and verified by a proper variographic experiment.

For this type of solutions, and for *similar* cases, the final dictum is that all TSE issues must be resolved (TOS) before it is possible to take the proper full advantage of any-and-all TAE_{NIR} optimizations. Without this, TSE will dominate and swamp most TAE_{NIR} gains. TOS and NIR are complementary competences, neither of which can be neglected.

19.16 RELATIONSHIP BETWEEN TOS AND CHEMOMETRICS

The ultimate issue in this chapter relates to an alleged relationship between TOS and chemometrics, indicated by Figure 19.30. It has occasionally been claimed that the kind of "errors" so effectively under control in chemometric bilinear data modeling, the residual ε's in chemometric covariance data models, also take account of TOS' sampling errors (*first author's personal communications and observations over a full career in chemometrics, too numerous to single out, too embarrassing to identify*). *Were* this indeed the case, chemometric data analysis would resolve pretty much all the issues outlined in this chapter. What a potential power for chemometrics!

Alas, this is but wishful thinking (however, desirable). In fact, this is a completely unjustified assumption, as revealed by several attempts over the past two decades to generalize TOS' error

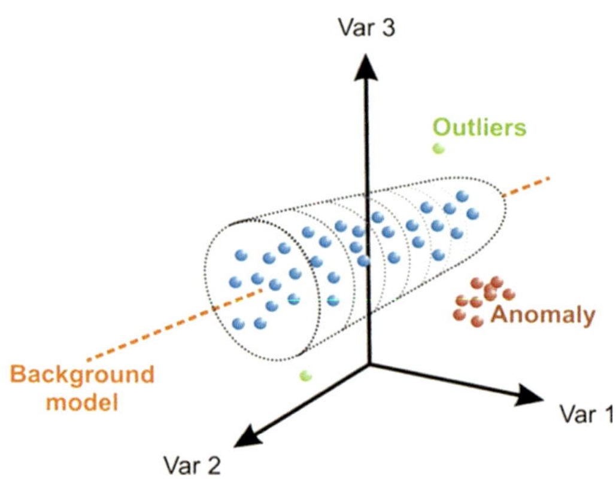

FIGURE 19.30 The claim from many chemometric stalwarts that multivariate "errors" so effectively screened away by bilinear data models, also take care of eliminating sampling errors, is sadly but wishful thinking. The relationship between TOS' distinctly univariate error concept and chemometrics' multivariate errors is much more complicated.

concept to the multivariate realm (KHE). While TOS optimization addresses only one variable at the time (usually directed at the one showing the largest variability, read heterogeneity), TSE errors unavoidably impact on the full multivariate model, affecting both loadings (loading weights) and the attendant model residuals. There are several underlying "dimensions" to TSE impacts (N.B. *not* in the form of classic data analysis components), originating from a plurality of analytes with their own, some correlated, other uncorrelated, variabilities due to disparate analytes residing in different mineral phases, some with correlated abundance, and others not. The issue is decidedly complex – the most prominent reason is that "data" cannot simply be viewed in splendid isolation ("Chemometricians only analyze the data") but rather the viewpoint must always be based on full TOS understanding, as shown in this Chapter with force.

The short story is that while chemometrics, so powerfully proven as the absolute champion data analytical modality over very many decades, is not *also* an omnipotent agent when it comes to dealing appropriately with preanalysis heterogeneity – only TOS is. A more fully developed framework for synergy between TOS and chemometrics is presented by Esbensen & Julius (2009), Esbensen and Geladi (2010), Esbensen and Swarbrick (2018).

ACKNOWLEDGMENTS

The authors are sincerely grateful to Martin Lischka, HERZOG Maschinenfabrik GmbH & Co. KG, for his enthusiastic contributions in the form of many new line drawings and other illustrations to this chapter. A similar heartfelt thank you is directed at Pedro A. Martinez, University of Puerto Rico, Recinto Mayagüez, for much valuable editorial help.

GLOSSARY

TOS – Theory of Sampling: Complement of governing sampling principles (GP), sampling unit operations (SUO), sampling error management rules, and normative practices needed to overcome all adverse effects of material heterogeneity and to design, implement, use, and maintain proper sampling processes in order to be able to guarantee *representativeness*.

DS 3077 (2013) (2. Ed): International standard codifying TOS' principles and practices. DS 3077 contains a complete TOS terminology. The key contents of this standard are augmented and completed in Esbensen and Paasch-Mortensen (2010), Esbensen and Wagner (2014a, b) and Esbensen (2020). http://www.ds.dk

FSP – Fundamental Sampling Principle: FSP states that all potential extraction (grains/particles/fragments) units from a lot must have an equal probability of being sampled, and that samples must not be altered in any way after sampling.

CH – Constitution Heterogeneity: The heterogeneity observed *between* individual units (grains/particles/fragments) of a material. CH is an intrinsic characteristic of the target material; CH can only be reduced by comminution or crushing of the material.

DH – Distribution Heterogeneity: The heterogeneity observed *between* different material groups-of-units, i.e., *between* all virtual and real sampling increments of a material. DH is expressing lot spatial heterogeneity. DH can be reduced by mixing and/or composite sampling; DH will also be reduced after comminution and mixing.

Fit-for-purpose representativeness: Characteristic of a sampling process which has reduced TSE+TAE to below a low threshold level, which has been set *a priori*.

FSE – Fundamental Sampling Error: The minimum sampling error that can be obtained only when all other errors (ISE, GSE) have been successfully eliminated.

GSE – Grouping and Segregation Error: GSE is a reflection of the spatial heterogeneity (DH) if not sampled optimally. This error occurs because of segregation and/or (local) grouping of material units. GSE can be reduced by mixing or by composite sampling, in which the effects stemming from GSE are progressively reduced as a function of an increased number of increments, Q.

Increment: The fundamental unit in practical sampling. Increments result from singular extraction operations, specifically with the intention to be aggregated to form *composite samples*.

Grab sample: "Samples" resulting from a single increment extraction only, intended to represent a complete lot. However, TOS shows that this is impossible for all significantly heterogeneous lots/materials. Single grab samples should be termed "specimens", in order not to be confused with representative samples.

Grab sampling: The process of extracting a singular "sample" (increment) in the faulty belief that this can adequately represent a whole lot.

ISE – Incorrect Sampling Errors: ISE occur due to structurally inferior or nonoptimized sampling procedures. ISE can only be eliminated when sampling complies with the principles of "correct sampling" as delineated in the Theory of Sampling (TOS). The presence of noneliminated ISE leads to a detrimental *sampling bias.*

CSE – Correct Sampling Errors: CSE are what remains after all ISE have been successfully eliminated, or reduced to below a fixed level characterized by the sampling process being *fit-for-purpose representative*. This level must be set before sampling process validation can take place.

Bias: The distinction between an analytical bias (which *can* be corrected for) and the sampling bias (which *cannot*) is crucial (see sampling bias).

Analytical bias: Systematic deviation between the true composition and the average of replicated analytical determinations. Under the statistical convention that the systematic deviation is *constant*, it is possible to make analytical bias corrections based on a number of repeated analytical determinations on the same aliquot.

Sampling bias: The Theory of Sampling (TOS) observes, from overwhelming empirical evidence that the sampling bias is *inconstant*, changing in magnitude for every estimation attempt. For this reason, it is not possible to perform a sampling bias correction in analogy with the well-known analytical bias, contrary to much unsubstantiated opinion. TOS draws the logical and scientifically correct conclusion that all influences producing a sampling bias must be *eliminated* from the sampling process, i.e., via radical ISE suppression (or complete elimination).

MU – Measurement Uncertainty: MU should always be understood as the *total* MU, MU_{total}. TOS terms this the global estimation error (GEE), which is composed of the total analytical error (TAE) and the total sampling error (TSE). The term "MU_{total}" is used to emphasize that measurement uncertainty does not solely depend on TAE.

$MU_{analysis}$: An estimate of the total analytical error (TAE) for a specific method of analysis, following the principles of MU.

$MU_{sampling}$: An estimate of the total sampling error (TSE) for a specific sampling-and-analysis method, following TOS.

GEE – Global Estimation Error: GEE=TSE+TAE; correspondingly: $MU_{total}=MU_{sampling}+MU_{analysis}$.

MSPC – Multivariate Statistical Process Control: Use of the multivariate latent variable space for statistical process control, to achieve, e.g., adequate monitoring, fault detection and diagnoses, or upset detection of processes. MSPC makes an extensive use of chemometrics.

PAT – Process Analytical Technology: A system based on relevant and adequate process understanding to design, analyze, monitor, and control manufacturing and processing through timely measurements of critical quality and key performance indicators of raw or in-process materials and of other relevant operational parameters, to ensure final product quality.

Q: Number of increments sampled to be *aggregated* to form a *composite sample*. It is critical that Q is *not* set by fiat, or by rote (like in continuously repeated standards via unreflected recitation). Q must specifically be linked to a proper lot heterogeneity characterization, e.g., by a RE (see below) or by *variographic* characterization.

RE – Replication Experiment: The *accumulated* variances from each step in a sampling procedure can be quantified by extracting and analyzing a number of *replicate primary samples* and sending these through the remaining laboratory procedures+analysis in a completely identical fashion. RE experiments shall "cover" the full spatial heterogeneity of the lot, by covering the full geometrical lot.

RSV – Relative Sampling Variability: The relative standard deviation obtained for the results of a replication experiment (RE). RSV is expressed as a relative coefficient of variation, CV [%].

Representativity; representativeness: It is not possible to ascertain the status of an individual sample by any means if removed from the context of its full sampling-and-analysis pathway. The characteristic "representative" can only be assigned to a specific *sampling process* that fulfills the proper demands specified by TOS.

Sampling (TOS): The practical ability of being able to produce a mass-reduced part of an original lot/material/target with guaranteed representativity with respect to both intrinsic features (e.g., composition, grain size distribution) and external attributes (e.g., moisture). The result of a TOS-compliant process is a *bona fide* representative sample. TOS is the only science-backed framework in existence that defines representativity in a rigid qualitative and quantitative fashion.

Sampling (trivial): If *not* based on TOS, any extraction operation from any lot *may* be called "sampling", but is manifestly unable to produce anything but a worthless, nonrepresentative "specimen" with no traceable relationship to the lot.

Sample: Term to be used only for a *bona fide* representative sample, i.e., a sample resulting from a representative (or a fit-for-purpose) sampling process.

Specimen: Any nonrepresentative part of a lot, extracted by an incorrect or a nonrepresentative process. Specimens are not worth analyzing.

Total Analytical Error (TAE): The sum of all error effects due to analytical measurements made on aliquots, $MU_{analysis}$ (aliquots may be representative, or not). Aliquots may be extracted as physical subsamples or may be *virtual samples*, e.g., a volume of a flux of matter interacting with a spectroscopic analytical instrument or similar (PAT). TAE only corresponds to the errors involved in analytical method validation. TAE is specifically singled out in TOS, so as to avoid overlapping, or confusion with respect to TSE.

Total Sampling Error (TSE): The sum of all sampling errors, TSE=ISE+CSE.

REFERENCES

Bakri, B., M. Weimer, G. Hauck, and G. Reich. 2015. "Assessment of powder blend uniformity: Comparison of real-time NIR blend monitoring with stratified sampling in combination with HPLC and at-line NIR Chemical Imaging." *European Journal of Pharmaceutics and Biopharmaceutics* 97:78–89. doi: 10.1016/j.ejpb.2015.10.002.

Barajas, M. J., A. R. Cassiani, W. Vargas, C. Conde, J. Ropero, J. Figueroa, and R. J. Romañach. 2007. "Near-infrared spectroscopic method for real-time monitoring of pharmaceutical powders during voiding." *Applied Spectroscopy* 61 (5):490–496.

Bellamy, L. J., A. Nordon, and D. Littlejohn. 2008. "Real-time monitoring of powder mixing in a convective blender using non-invasive reflectance NIR spectrometry." *Analyst* 133 (1):58–64. doi: 10.1039/b713919e.

Cama-Moncunill, R., M. Casado, Y. Dixit, D. Togashi, L. Alvarez-Jubete, P. Cullen, and C. Sullivan. 2015. "Moisture determination of static and in-motion powdered infant formula utilising multiprobe near infrared spectroscopy." *Journal of Near Infrared Spectroscopy* 23 (4):245–253.

Changkun, W., and P. Xianzhang. 2016. "Improving the prediction of soil organic matter using visible and near infrared spectroscopy of moist samples." *Journal of Near Infrared Spectroscopy* 24 (3):231–241. doi: 10.1255/jnirs.1184.

Corredor, C. C., R. Lozano, X. Bu, R. McCann, J. Dougherty, T. Stevens, D. Both, and P. Shah. 2015. "Analytical method quality by design for an on-line near-infrared method to monitor blend potency and uniformity." *Journal of Pharmaceutical Innovation* 10 (1):47–55. doi: 10.1007/s12247-014-9205-2.

Dahm, D. J. and Dahm, K. D. 2007. *Interpreting Diffuse Reflectance and Transmittance: A Theoretical Introduction to Absorption Spectroscopy of Scattering Materials.* NIR Publications, Chichester, West Sussex.

Danish-Standards-Foundation. 2013. DS 3077(2013). In Representative Sampling - *Horizontal Standard*: Danish Standards Foundation.

Durao, P., C. Fauteux-Lefebvre, J. M. Guay, N. Abatzoglou, and R. Gosselin. 2017. "Using multiple process analytical technology probes to monitor multivitamin blends in a tableting feed frame." *Talanta* 164:7–15 doi: 10.1016/j.talanta.2016.11.013.

Ekkapong, C., and S. Panmanas. 2017. "Rapid evaluation of fat content in curry soup containing coconut milk by using near infrared spectroscopy." *Journal of Near Infrared Spectroscopy* 26 (1):16–25. doi: 10.1177/0967033517746550.

Esbensen, K., H. (2020) Introduction to the Theory and Practice of Sampling. IMP Open, Chichester, UK (2020). ISBN: 978-1-906715-29-8. https://store.impopen.com/introduction-to-the-theory-andpractice-of-sampling.html

Esbensen, K. H., and P. Geladi. 2010. "Principles of proper validation: Use and abuse of re-sampling for validation." *Journal of Chemometrics* 24 (3–4):168–187. doi: 10.1002/cem.1310.

Esbensen, K. H., and K. Heydorn. 2004. "Metrology in sampling - a first foray (biological materials)." *Chemometrics and Intelligent Laboratory Systems* 74 (1):115–120. doi: 10.1016/j.chemolab.2004.04.010.

Esbensen, K. H., and L. P. Julius. 2009. "Representative sampling, data quality, validation - A necessary trinity in chemometrics." *Comprehensive Chemometrics: Chemical and Biochemical Data Analysis* 1–4:C1–C20.

Esbensen, K. H., and P. Paasch-Mortensen. 2010. "Process sampling: Theory of sampling – The missing link in process analytical technologies (PAT)." In K. A. Bakeev (ed.) *Process Analytical Technology*, 37–80. John Wiley & Sons, Ltd., New York.

Esbensen, K. H., and C. Wagner. 2014a. "Theory of sampling (TOS) versus measurement uncertainty (MU) - A call for integration." *Trac-Trends in Analytical Chemistry* 57:93–106 doi: 10.1016/j.trac.2014.02.007.

Esbensen, K., and C. Wagner. 2014b. "Why we need a theory of sampling." *The Analytical Scientist* 21:30–38.

Esbensen, K. E., and B. Swarbrick. 2018. *Multivariate Data Analysis – in Practice. An Introduction*, 6th ed., IMPublising, Chichester, West Sussex.

Esbensen, K. H., A. D. Román-Ospino, A. Sanchez, and R. J. Romañach. 2016. "Adequacy and verifiability of pharmaceutical mixtures and dose units by variographic analysis (Theory of Sampling) – A call for a regulatory paradigm shift." *International Journal of Pharmaceutics* 499 (1–2):156–174. doi: 10.1016/j.ijpharm.2015.12.038.

Esbensen, K. H., and C. Wagner. 2015. "Sampling is not Gambling! (exit grab sampling)." *Spectroscopy Europe* 27 (4):24–26.

Esbensen, K., P. Geladi, and A. Larsen. 2013a. "The replication myth 1." *NIR News* 24 (1):17–20.

Esbensen, K., P. Geladi, and Anders Larsen. 2013b. "The replication myth 2: Quantifying empirical sampling plus analysis variability." *NIR News* 24 (3):15–19.

Esbensen, K. H., H. H. Friis-Petersen, L. Petersen, J. B. Holm-Nielsen & P. P. Mortensen (2007). Representative process sampling – in practice: Variographic analysis and estimation of Total Sampling Errors (TSE). *Proceedings 5th Winter Symposium of Chemometrics (WSC-5)*, Samara 2006. *Chemometrics and Intelligent Laboratory Systems, Special Issue*, vol. 88, No.1, 41–49 (2007).

Esteve-Agelet, L., and C. R. Hurburgh. 2010. "A tutorial on near infrared spectroscopy and its calibration." *Critical Reviews in Analytical Chemistry* 40 (4):246–260. doi: 10.1080/10408347.2010.515468.

Fouchee, A., and R. C. Steinhaus. 2015. "Design advances and operational studies for the True Pipe® sampler: A symmetry based unit for reliable sampling of pressurised particulate streams." *TOS Forum* (5):151–156. doi: 10.1255/tosf.60.

Griffiths, P. R., and J. A. de Haseth. 2007. *Fourier Transform Infrared Spectrometry*. 2nd ed., Wiley, Hoboken, NJ.

Guillemain, A., K. Dégardin, and Y. Roggo. 2017. "Performance of NIR handheld spectrometers for the detection of counterfeit tablets." *Talanta* 165:632–640 doi: 10.1016/j.talanta.2016.12.063.

Gy, P. 1998. *Sampling for Analytical Purposes*. 1st ed., Wiley, New York.

Gy, P. 2004a. "Part IV: 50 years of sampling theory - a personal history." *Chemometrics and Intelligent Laboratory Systems* 74 (1):49–60. doi: 10.1016/j.chemolab.2004.05.014.

Gy, P. 2004b. "Part V: Annotated literature compilation of Pierre Gy." *Chemometrics and Intelligent Laboratory Systems* 74 (1):61–70. doi: 10.1016/j.chemolab.2004.05.010.

Gy, P. 2004c. "Sampling of discrete materials - a new introduction to the theory of sampling - I. Qualitative approach." *Chemometrics and Intelligent Laboratory Systems* 74 (1):7–24. doi:10.1016/j.chembolab.2004.05.012.

Gy, P. 2004d. "Sampling of discrete materials - II. Quantitative approach-sampling of zero-dimensional objects." *Chemometrics and Intelligent Laboratory Systems* 74 (1):25–38. doi: 10.1016/j.chemolab.2004.05.015.

Gy, P. 2004e. "Sampling of discrete materials - III. Quantitative approach-sampling of one-dimensional objects." *Chemometrics and Intelligent Laboratory Systems* 74 (1):39–47. doi: 10.1016/j.chemolab.2004.05.011.

Gy, P. M. 1995. "Introduction to the theory of sampling I. Heterogeneity of a population of uncorrelated units." *TrAC Trends in Analytical Chemistry* 14 (2):67–76. doi: 10.1016/0165-9936(95)91474-7.

Heydorn, K., and K. Esbensen. 2004. "Sampling and metrology." *Accreditation and Quality Assurance* 9 (7):391–396. doi: 10.1007/s00769-004-0808-z.

Igne, B., B. M. Zacour, Z. Q. Shi, S. Talwar, C. A. Anderson, and J. K. Drennen. 2011. "Online monitoring of pharmaceutical materials using multiple NIR sensors-part I: Blend homogeneity." *Journal of Pharmaceutical Innovation* 6 (1):47–59. doi: 10.1007/s12247-011-9099-1.

Iyer, M., H. Morris, and J. Drennen III. 2002. "Solid dosage form analysis by near infrared spectroscopy: Comparison of reflectance and transmittance measurements including the determination of effective sample mass." *Journal of Near Infrared Spectroscopy* 10 (4):233–245.

Johansson, J., S. Folestad, M. Josefson, A. Sparén, C. Abrahamsson, S. Andersson-Engels, and S. Svanberg. 2002. "Time-resolved NIR/Vis spectroscopy for analysis of solids: Pharmaceutical tablets." *Applied Spectroscopy* 56 (6):725–731.

Junhui, L., C. D. Mary-Grace, C. Shih-Fang, L. Song, S. Vijay, and J. B. Patrick. 2015. "Prediction of starch content and ethanol yields of sorghum grain using near infrared spectroscopy." *Journal of Near Infrared Spectroscopy* 23 (2):85–92. doi: 10.1255/jnirs.1146.

Lee, S. L., T. F. O'Connor, X. Yang, C. N. Cruz, S. Chatterjee, R. D. Madurawe, C. M. V. Moore, L. X. Yu, and J. Woodcock. 2015. "Modernizing pharmaceutical manufacturing: from batch to continuous production." *Journal of Pharmaceutical Innovation* 10 (3):191–199. doi: 10.1007/s12247-015-9215-8.

Martens, H., and T. Naes. 1992. *Multivariate Calibration*. Wiley, New York.

Martinez, L., A. Peinado, L. Liesum, and G. Betz. 2013. "Use of near-infrared spectroscopy to quantify drug content on a continuous blending process: Influence of mass flow and rotation speed variations." *European Journal of Pharmaceutics and Biopharmaceutics* 84 (3):606–615. doi: 10.1016/j.ejpb.2013.01.016.

Masithoh, R. E., R. Haff, and S. Kawano. 2016. "Determination of soluble solids content and titratable acidity of intact fruit and juice of satsuma mandarin using a hand-held near infrared instrument in transmittance mode." *Journal of Near Infrared Spectroscopy* 24 (1):83–88. doi: 10.1255/jnirs.1196.

Mateo-Ortiz, D., Y. Colon, R. J. Romanach, and R. Mendez. 2014. "Analysis of powder phenomena inside a Fette 3090 feed frame using in-line NIR spectroscopy." *Journal of Pharmaceutical and Biomedical Analysis*. 100:40–49. doi: 10.1016/j.jpba.2014.07.014.

Muzzio, F. J., P. Robinson, C. Wightman, and D. Brone. 1997. "Sampling practices in powder blending." *International Journal of Pharmaceutics* 155 (2):153–178. doi: 10.1016/s0378 5173(97)04865-5.

Ortega-Zuñiga, C., K. Reyes-Maldonado, R. Méndez, and R. J. Romañach. 2017. "Study of near infrared chemometric models with low heterogeneity films: The role of optical sampling and spectral preprocessing on partial least squares errors." *Journal of Near Infrared Spectroscopy* 25 (2):103–115. doi: 10.1177/0967033516686653.

Osorio, J. G., G. Stuessy, G. J. Kemeny, and F. J. Muzzio. 2014. "Characterization of pharmaceutical powder blends using in situ near-infrared chemical imaging." *Chem. Eng. Sci.* 108:244–257 doi: 10.1016/j.ces.2013.12.027.ss

Osorio, J. G., E. Hernández, R. J. Romañach, and F. J. Muzzio. 2016. "Characterization of resonant acoustic mixing using near-infrared chemical imaging." *Powder Technology* 297:349–356 doi: 10.1016/j.powtec.2016.04.035.

Osorio, J. G., A. U. Vanarase, R. J. Romañach, and F. J. Muzzio. 2015. "Continuous powder mixing." In P. J. Cullen, R. J. Romañach, C. D. Reilly, N. Abatzoglou (eds.) *Pharmaceutical Blending and Mixing*, 101–127. John Wiley & Sons, Ltd., New York.

Petersen, L., C. K. Dahl, and K. H. Esbensen. 2004. "Representative mass reduction in sampling - a critical survey of techniques and hardware." *Chemometrics and Intelligent Laboratory Systems* 74 (1):95–114. doi: 10.1016/j.chemolab.2004.03.020.

Pinzon de la Rosa, C., V. Rodriguez, M. L. Hormaza, and R. J. Romañach 2017. "TOS Meets the NSF I-Corps™ Program." *8th World Conference on Sampling and Blending*, Perth, Australia.

Pitard, F. F. 1993. *Pierre Gy's Sampling Theory and Sampling Practice. Heterogeneity, Sampling Correctness, and Statistical Process Control*. CRC Press, Boca Raton, FL.

Pitard, F. F. 2019. *The Theory of Sampling and Sampling Practise*. 3rd ed., CRC Press. ISBN-13: 978-1138476486.

Popo, M., S. Romero-Torres, C. Conde, and R. J. Romanach. 2002. "Blend uniformity analysis using stream sampling and near infrared spectroscopy." *AAPS PharmSciTech* 3 (3):E24. doi: 10.1208/pt030324.

Reshadat, R., S. Desa, S. Joseph, M. Mehra, N. Stoev, and S. T. Balke. 1999. "In-line near-infrared monitoring of polymer processing. Part I: Process/monitor interface development." *Applied Spectroscopy* 53 (11):1412–1418. doi: 10.1366/0003702991945768.

Ribeiro, L. P. D., A. P. M. da Silva, A. A. de Lima, E. de Oliveira Silva, Å. Rinnan, and C. Pasquini. 2016. "Non-destructive determination of quality traits of cashew apples (Anacardium Occidentale, L.) using a portable near infrared spectrophotometer." *Journal of Near Infrared Spectroscopy* 24 (1):77–82. doi: 10.1255/jnirs.1197.

Romañach, R. J. 2015. "Sampling and determination of adequacy of mixing." In P. J. Cullen, R. J. Romañach, C. D. Reilly, N. Abatzoglou (eds.) *Pharmaceutical Blending and Mixing*, 57–78. John Wiley & Sons, Ltd, New York.

Romañach, R. J. 2017. "Theory of sampling - from missing link to key enabler for process analytical technology (PAT)." *8th World Conference on Sampling and Blending*, Perth, Australia, May 9–11, 2017.

Romañach, R. J., and K. H. Esbensen. 2015. "Sampling in pharmaceutical manufacturing - Many opportunities to improve today's practice through the Theory of Sampling (TOS)." *TOS Forum* 4:5–9.

Romañach, R. J., and K. H. Esbensen. 2016. "Theory of sampling (TOS) - for development of spectroscopic calibration models" *American Pharmaceutical Review* 19 (6):138–139.

Romañach, R. J., A. Sanchez-Paternina, and K. H. Esbensen. 2018. "Variographic analysis of 1-D lots in pharmaceutical manufacturing (powder mixing)." *American Pharmaceutical Review* 21 (1):22–26.
Sampaio, P. N., K. C. Sales, F. O. Rosa, M. B. Lopes, and C. R. Calado. 2014. "In situ near infrared spectroscopy monitoring of cyprosin production by recombinant Saccharomyces cerevisiae strains." *Journal of Biotechnology* 188:148–157 doi: 10.1016/j.jbiotec.2014.07.454.
Sánchez-Paternina, A., A. Román-Ospino, C. Ortega-Zuñiga, B. Alvarado, K. Esbensen, and R. Romañach. 2015. "When "homogeneity" is expected—Theory of Sampling in pharmaceutical manufacturing." *TOS Forum* (5):67–70. doi: 10.1255/tosf.61.
Sánchez-Paternina, A., A. D. Román-Ospino, M. Martínez, J. Mercado, C. Alonso, and R. J. Romañach. 2016. "Near infrared spectroscopic transmittance measurements for pharmaceutical powder mixtures." *Journal of Pharmaceutical and Biomedical Analysis* 123:120–127 doi: 10.1016/j.jpba.2016.02.006.
Sierra-Vega, N. O., A. Sánchez-Paternina, N. Maldonado, V. Cárdenas, R. J. Romañach, and R. Méndez. 2018. "In line monitoring of the powder flow behavior and drug content in a Fette 3090 feed frame at different operating conditions using Near Infrared spectroscopy." *Journal of Pharmaceutical and Biomedical Analysis* 154:384–396 doi: 10.1016/j.jpba.2018.03.017.
Sørensen, M., A. Larsen, and K. Esbensen. 2014. "Visualisation of sampling error effects in near infrared analysis—comparison between Petri dish, roll bottle and spiral sampler." *TOS Forum* (2):22–26. doi: 10.1255/tosf.19.
Vanarase, A. U., M. Alcalà, J. I. Jerez Rozo, F. J. Muzzio, and R. J. Romañach. 2010. "Real-time monitoring of drug concentration in a continuous powder mixing process using NIR spectroscopy." *Chemical Engineering Science* 65 (21):5728–5733. doi: 10.1016/j.ces.2010.01.036.
Vargas, J. M., S. Nielsen, V. Cárdenas, A. Gonzalez, E. Y. Aymat, E. Almodovar, G. Classe, Y. Colón, E. Sanchez, and R. J. Romañach. 2018. "Process analytical technology in continuous manufacturing of a commercial pharmaceutical product." *International Journal of Pharmaceutics* 538 (1–2):167–178. doi: 10.1016/j.ijpharm.2018.01.003.
Vargas, J. M., A. D. Roman-Ospino, E. Sanchez, and R. J. Romañach. 2017. "Evaluation of analytical and sampling errors in the prediction of the active pharmaceutical ingredient concentration in blends from a continuous manufacturing process." *Journal of Pharmaceutical Innovation* 12 (2):155–167. doi: 10.1007/s12247-017-9273-1.
Williams, P. C., and K. H. Norris. 2001. *"Near-Infrared Technology: In the Agricultural and Food Industries."* American Association of Cereal Chemists, Saint Paul, MN, 19–37.
Williams, P. 2001. "Sampling, sample preparation, and sample selection." In E. Ciurczak, and D. A. Burns (eds.) *Handbook of Near-Infrared Analysis*, 2nd edn. CRC Press, Boca Raton, FL.
Zhou, X., P. Hines, and M. W. Borer. 1998. "Moisture determination in hygroscopic drug substances by near infrared spectroscopy." *Journal of Pharmaceutical and Biomedical Analysis* 17 (2):219–225. doi: 10.1016/S0731-7085(97)00182-9.

20 Volume Fraction vs. Weight Fraction

*Howard Mark**
Mark Electronics

CONTENTS

20.1 HISTORICAL BACKGROUND

Ever since Karl Norris assembled the three key components (use of the NIR spectral region, diffuse reflection measurements of powdered samples, and use of what we now call multiple linear regression—MLR, the prototype algorithm for chemometrics) that ultimately turned into modern NIRS analysis, NIRS has existed as a self-contained, albeit internally consistent, analytical method. There was little or no connection to other existing chemical or spectroscopic science. Since Karl was affiliated with the US Department of Agriculture (USDA), his concerns were with the measurement of samples of interest to the department: largely foodstuffs and nonfood agricultural products such as cotton and tobacco. These types of samples differed greatly from samples of concern to academicians in some key respects: they were largely nonhomogeneous powdered solids as opposed to the clear liquids and gases that the theory of classical spectroscopy was based upon (e.g., Beer's law, etc.).

It was generally not possible to separate the components of a sample (a kernel of wheat or a soybean, for example) so that the spectra of the "pure" components could be measured. It was at least as difficult to separate the components of something like a wheat kernel without changing them, as it would be to try to synthesize a wheat kernel from a collection of proteins, starches, vitamins, minerals, oils, and all the micro-ingredients that a natural wheat kernel contains. This was the task that Karl set out to address.

* hlmark@nearinfrared.com

I was working for Technicon Instruments Corp., one of the first three companies to commercialize the technology that Karl Norris had pioneered. When I joined the development group, the scientists already there were confused about how the technology was working. Compared with spectra in other spectral regions, the NIR spectra were anomalous. Baselines were nonexistent, absorbance peaks were at best broad and ill defined, standards were nonexistent, spectral interpretation was an unknown quantity, and what theory existed was known to be incomplete and likely somewhat incorrect. Since most of the samples of interest at that time were natural products (e.g., agricultural products such as wheat), spectra of pure constituents were unavailable; wheat "protein" could not be separated from a wheat kernel for example (at least, not and still remain recognizable as "wheat protein" in any way that could allow a meaningful measurement of a "wheat protein spectrum"). When we met scientists from the other NIR companies at conferences, or at the small "NIR meetings" that Karl would organize periodically, we found they shared our confusion and lack of understanding, as did Karl himself. Yet we all of us were well aware that from a practical, empirical, point of view, the technology "worked." By the application of multivariate mathematics, it could be used to reliably analyze various products for their compositions.

Newcomers to the field, having had a course or two in classical spectroscopy, invariably had the same reaction: "I learned about spectroscopy in school. But here there's no theory, no baseline, no reference beam, no standards, and no spectral peaks. Yet it works!! It must be magic!" All those missing pieces flew in the face of standard spectroscopic thought and practice. This distinguished and separated NIR analysis from the rest of science, making it appear to exist in a universe of its own. Yet, that "magic" made NIR an enormously successful commercial analytical method, being applied to many important industrial applications.

20.2 PROBLEM AREAS

The condition of existing in a universe of its own has persisted to this day. Very little progress (if any) has been made in connecting NIRS to the rest of the scientific universe. Despite its commercial success, many mysteries that were known back then still exist and still have no cogent (proven) explanations:

1. The apparent need for, and use of, more variables in calibrations than can reasonably be justified.
2. Algebra dictates that no more wavelengths or factors *should* be needed than variables present.
3. Difficulty in reproducing calibrations for the same constituents in the same type of samples.
4. Inability to reproduce wavelength selections (MLR models) in successive attempts.
5. Difficulty and/or inability in relating wavelengths chosen (for MLR) or prominent wavelength bands (for PCR/PLS) to spectral features.
6. Unexpected and/or unexplained "outliers."
7. The standard error of calibration (SEC) and standard error of prediction (SEP) *should* drop precipitously when the number of wavelengths/factors equals the number of variations in the samples.
8. Spectroscopic measurements should be accurate over the entire range of concentrations, not only for dilute solutions or a small range of values.
9. Calibrations should be extrapolatable, with a calculated reduced accuracy at the extremes.
10. Calibration transfer should be as easily and readily performed as comparing two mid-IR spectra.

In practice, all the above difficulties were observed, but none of the proposed or expected explanations were. And, quite frankly, nobody seemed to care. Through the magic of Chemometrics,

"IT WORKED!" so why question success? And yet those unexplained "mysteries" were signs of trouble looming. Many potential root causes of the problems have been proposed.

20.3 PROPOSED ROOT CAUSES

1. Sample effects
   a. Inhomogeneity
   b. Polymorphism
   c. Interactions
   d. Optical scatter
2. Reference lab error
3. Instrument noise and/or variability
4. Instrumental nonlinearities:
   i. Stray light
   ii. Detector saturation
      a. Electronic nonlinearities
5. Calibration issues
   i. "Wrong" transforms of spectral data
   ii. "Wrong" calibration algorithm
   iii. Incorrect calibration parameters (number of wavelengths/factors)
6. Unknown relation between spectrum and composition (data transforms)

Ad hoc empirical solutions abound, often following this template:

1. One or more of the suspected above-listed root causes is targeted.
2. A (spectral) data transform is proposed to address the suspected root cause.
3. A calibration transfer experiment is designed to verify the correctness of the solution.
4. The experiment is conducted, and the data analyzed. "Outliers" are deleted.
5. The SEP of the transferred calibration is compared to the SEC of the calibration data.
6. Success is declared!

Over the years, many scientists have tried to address one or more of the suspected problems. Except for these studies, the attack usually consisted of hypothesizing a cause of the problem and taking an empirical approach by applying an *ad hoc* solution, usually as part of the routine chemometric algorithm being used for the application. Surprisingly, this approach often "worked," insofar as the scientist was able to obtain satisfactory calibration and prediction performance for the analyte of interest in his samples and extrapolate that finding to the conclusion that his "fix" was actually correcting the situation. In contrast, even when that was the case, this approach usually gave little insight into the underlying (root) causes of the difficulties encountered. Even worse, the results obtained could rarely be generalized or extended to other applications of NIR or for other types of analysis. Every new application required a new research project, with the only common procedure being the application of chemometrics to the analytical situation.

Despite the practical usefulness of these *ad hoc* analytical methods, these exercises can hardly be called progress in the development of the technology. A given solution to an immediate analytical situation is often not applicable to another similar type of chemical analysis, even if the analyte and the sample types are the same, nor can they be applied to different types of analysis. No fundamental knowledge was developed through this work. The separation of an isolated method or calibration model from the rest of the scientific universe also isolates the method from further fundamental knowledge, so that there are no clues left as to which direction to make further explorations.

Often, the analytical "recipe" does not shed light on the fundamental science or processes underlying the particular analytical method used, nor does the analytical method used lend any insight into other potential applications of the analytical method, neither is the actual analytical method that is developed usable by other scientists. Indeed, not having explanations at that time for the effects seen, it made NIR analysis appear to be divorced from the mainstream of spectroscopy and existing in a universe of its own, separated from the universe of the rest of science.

The general NIR community, which at that time was quite small, shared the confusion over the behavior of NIR data. Attempts to explain this anomalous behavior went on for many years. Several of the common explanations that were proposed are listed above. Some of those common explanations have subcategories. For example, the use of the "wrong" calibration algorithm has, along with it, the use of an incorrect number of factors (for the principal components regression—PCR or partial least squares—PLS algorithm) or the wrong number of wavelengths (for the MLR algorithm). The use of MLR also carries with it the possibility of using the "wrong" wavelengths (even if the number of wavelengths is correct). Note: the equivalent is also true, for example, for PCR or PLS. That possibility usually goes unrecognized and unnoted, since most of the software packages available do not allow the user to select arbitrary subsets of the factors computed during the execution of the PCR or PLS algorithm.

For example, a user who decides to create a model that uses four principal components (PCs) is generally not free to select which four components to use, with all the common software packages for spectroscopic analysis, only the first four principal components may be used. This limitation is both a curse and a blessing. The curse is that the limitation prevents users from selecting just the minimum number of factors that are needed to optimally analyze the samples of interest. Instead, if a factor, say, the sixth, is important, the user cannot select four factors that include the sixth but are forced to include extra, unnecessary, factors such as the third and fifth, perhaps, that contribute to overfitting. The blessing is that a majority of users are already confused by needing to use complicated algorithms they are not comfortable dealing with and are happy and even relieved to not have another whole set of choices to make.

Matters remained thus for many years. Many successful applications of NIR and chemometrics were developed and published in the open literature, yet the "mysteries" remained. No general physical principles for determining a method to develop and optimize a calibration model, which could be applied by anyone, anywhere, were forthcoming. Only the general empirical methods seemed to "work" to develop a good calibration model.

Around 2010, a serendipitous discovery changed this picture [1]. It was found that the units in which the reference values for concentrations of an analyte are expressed actually matter. Use of the "wrong" units to express the analyte leads to large errors due to nonlinearity. Potential nonlinearity had previously been investigated [2], but the tools available at the time were inadequate, nor was attention directed to the right places. Nonlinearity can be covered up by the chemometrics, but it doesn't go away. It lays hidden until something changes (e.g., a different set of samples) and then pops up, often in the form of one or more "outliers". All the previous works for solving the problems of NIR analysis were directed toward investigations of problems with the spectral readings. Nobody had looked at the reference laboratory values except to disparage "bad reference laboratory results." Nobody had concerned themselves with the question of whether they were measuring the right quantity in the first place.

The fundamental problem is illustrated in Figure 20.1, which shows the relationship between weight fraction and volume fractions of two different mixtures. Figure 20.1a shows the volume fractions of each component of a binary mixture, acetone and dichloropropane, as the weight fraction of the dichloropropane varies from zero to one (0%–100%). Even in the absence of chemical interactions between the two mixture ingredients, we can see that there is nonlinearity in the relationship between the two quantities: weight fraction and volume fraction. This nonlinearity is caused by the difference in density of the two mixture ingredients, causing the density of the mixture to vary as more of either ingredient is added.

Another way to approach the problem is to note that, since acetone is less dense than dichloropropane, adding a given weight of acetone corresponds to a greater volume of acetone and therefore, a greater total volume, than adding the same weight of dichloropropane would cause.

For a two-component mixture, Figure 20.1a clearly demonstrates the problem. For mixtures with more ingredients, the situation is even more complicated. The complication is due to the fact that, plotting the weight percent vs. volume percent for a mixture containing more than two ingredients, the details of the nature of the curvature seen depends on the densities of all the ingredients of the mixture. Figure 20.1b shows the relationship between the weight fraction and the volume fraction of dichloropropane in a ternary mixture as the relative amounts of the other two ingredients (acetone and dichloromethane) vary.

As the mixture changes, so does the curvature of the lines. In this case, a single line in Figure 20.1a becomes a family of lines, as seen in Figure 20.1b. Another property of mixtures with more than two ingredients is that there is no longer a one-to-one relationship between the weight fraction and the volume fraction; the volume fraction corresponding to a given weight fraction depends on the amounts of the other ingredients in the mixture and vice versa, as can also be seen in Figure 20.1b.

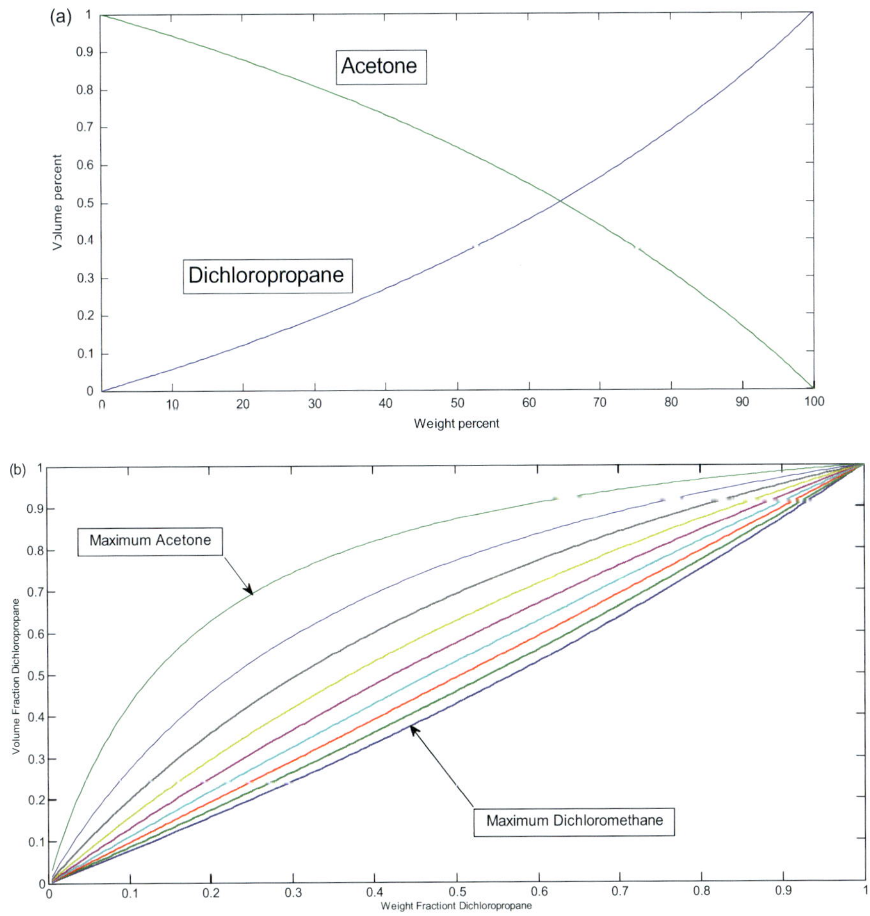

FIGURE 20.1 Volume percents versus weight percents of ingredients in various mixtures. (a) Volume percent of both ingredients of a binary mixture versus weight percent: dichlorpropane in acetone–dichloropropane mixtures. The *X*-axis represents the weight percent of dichloropropane in the mixtures. (b) Volume fraction versus weight fraction of one ingredient (dichloropropane) in a ternary mixture as the amounts of the other two ingredients (dichloromethane and acetone) vary (each curve represents a different, although constant, ratio of the other two ingredients).

Note that the relationships in Figure 20.1 have nothing to do with spectroscopy, it is purely a matter of the physical chemistry of mixtures [1]. What DOES have to do with the spectroscopy is a fact that physicists have known about at least since James Clerk Maxwell formulated his eponymous equations: matter absorbs light in proportion to the volume of the absorbing material. Physicists knew this, but the knowledge seems not to have made it into the chemical and spectroscopic arenas. To be sure, sporadic papers do appear in the literature where the authors specify measuring concentrations of the analyte in their products as volume fraction, but these are rare exceptions. The vast majority of publications dealing with quantitative analysis specify weight fractions (or percentages) as the units of concentration. This is due to their historical backgrounds; the development of analytical chemistry has led chemists (except in medical applications, where the weight of analyte per deciliter is common) to consider weight fractions the "standard" way to perform measurements.

In the past, when the standard of measurement was based on gravimetric and volumetric wet chemistry, this was eminently satisfactory. To specify steel, for example, by the weight percent of chromium (Cr), Tungsten (W), and other metals it contains was a satisfactory way to set specifications for the steel. Then similar volumetric and/or gravimetric methods were used to determine whether a steel sample at hand met those specifications. Even then, when optical spectroscopy was applied to the measurement, spectroscopists knew that UV measurements could only be applied to samples with sufficiently low concentrations of the analyte or the instrument response would become nonlinear with respect to the concentration. Modern multivariate mathematical methods of computation were unknown then, but a univariate method could still be used, even in the face of the nonlinearity. All that was needed was to plot the response of a set of standard samples of known concentration. The response of the ultraviolet (UV) measurement could be compared to that graph to find the analyte concentration, whether it was linearly related to the UV response or not.

Similar procedures were used for mid-range infrared (MIR) analysis when they were applied to measuring organic materials. In both cases, it was important that the sample preparation be identical when measuring the unknown samples as when measuring the samples used to create the calibration curve.

Those methods were not applied to NIR analysis. As described above, "standard" samples could not be created, neither could special "calibration" samples be prepared for a calibration curve, nor could spectra be reduced to a single wavelength for measurement. Furthermore, the best efforts of the physics community have failed, for over 100 years, to describe a theory for computing the reflectance of an inhomogeneous solid sample from other properties of the sample.

In applying the multivariate mathematics to NIR spectra, the linearity of the responses of the interferences as well as of the analyte was simply assumed. No thought was given to the relationships between the spectroscopic readings (which can only see one thing) and the various ways that "concentration" can be specified. Hence the calibration algorithms have to work very hard to fit the nonlinear data into the structure of what it believes is a linear relationship [3].

20.4 PREVIOUS WORK

The work described in the aforementioned paper [1] and, in considerably more detail, in a series of columns in the magazine spectroscopy [4–17] explain how an experiment intended for a completely different purpose shed light on the questions raised above. For the sake of completeness, we summarize that work here.

The intended purpose of the experiment was to describe and explain the operation of the classical least squares (CLS) algorithm. Toward this end, the experiment was set up to eliminate or minimize all the ROOT CAUSES described above. The sample effects (inhomogeneity, polymorphism, interactions, and optical scatter) were addressed by using completely miscible mixtures of clear organic liquids with minimum polarity as the samples.

Reference laboratory errors were reduced (nearly eliminated) by weighing the liquids; weighing is one of the most precise and accurate measurement methods available, with precisions of as much as 10^{-6} potentially available. No separate actual laboratory analyses were performed. Due care was exercised to prevent changes in the samples due to evaporation or other extraneous effects. Thus, we expect any errors from this source to be eliminated.

Instrumental effects were minimized by routinely checking the instrument characteristics and operating in a regime where the detector signal was neither too high (so as to saturate the detector) nor too low (where stray light or other extraneous effects might cause problems).

Calibration issues (incorrect transforms or number of PCR|PLS factors) were eliminated by using the CLS algorithm without any data transformations. CLS does not utilize any computed factors for performing the calibration; only the spectra of the pure materials comprising the samples are needed. Adherence to Beer's law was verified by ascertaining that mixture spectra reconstructed from the CLS coefficients and pure material spectra agreed with the actual spectra of the corresponding mixtures.

A set of 15 mixtures of toluene, *n*-heptane, and dichloromethane, specified by a statistical mixture design, was made using the design specifications to determine the weights of each ingredient to be added to each mixture. Their transmission spectra were measured. No transformations were applied to the data. No samples stood out as potential outliers and none were eliminated.

The results were surprising and disappointing. The computed concentrations of the ingredients in most of the mixtures showed large discrepancies (as much as 10%–15%, in some cases) from the specified values of the carefully made-up samples. A review of the experimental work revealed no blunders or other experimental discrepancies.

Since the reconstruction of the mixture spectra demonstrated that Beer's Law was operating as theory indicated, the only possible cause of the discrepancies was the use of an incorrect value for the actual mixture compositions. The care used in executing the experiment eliminated the possibility that the samples were made up incorrectly. The only other possibility that remained was that the optical results were being compared to the wrong property of the samples. As stated in that paper [1], "We expect the spectral values to correspond to some physical characteristic but there is no reason to expect that characteristic to be the weight of a sample component." Weight, after all, depends on the nuclei of the atoms, while the spectroscopic interactions depend on the behavior of the electrons in the atomic and molecular orbitals. It remained, then, to determine what characteristic(s) of the mixtures would provide values that agreed with the spectroscopically measured values.

Several possibilities were proposed and tested. The results from one particular sample with nominal weight percent of the various ingredients are shown in Table 20.1, showing the results from five different ways of expressing the concentrations of the various mixture ingredients.

Interestingly, the value for toluene comes pretty close to the same value as the spectroscopic measurement no matter how the "concentration" is expressed. Only the volume percent gives a value for all three ingredients that is close to the corresponding spectrally measured value.

TABLE 20.1

Results from a Sample with Nominal Composition: Toluene = 25%, Dichloromethane = 25%, *n*-Heptane = 50%. All Values are wt %

	Toluene	Dichloromethane	*n*-Heptane
Spectral values	23.01	17.48	56.92
Weight percent	23.86	26.45	49.68
Mole percent	24.29	29.21	46.50
Mole % of hydrogen	19.5	5.86	74.64
Mole % H corrected[a]	22.3	10.33	67.36
Volume percent	22.93	16.50	60.56

[a] Corrected for the density of the ingredient.

Table 20.1 presents the comparative information for one sample from the set of 15 used. In the original publication [1], the results from all 15 samples were examined and presented. The full set of results makes a compelling case that, just as with this single sample, volume percent is the only measurement unit that gives quantitative results for the concentrations of the various ingredients that consistently agree with the spectroscopic values.

What could not be done in the course of that work was to actually apply any of the modern calibration algorithms to the data used for that study, because the number of samples ($n = 15$) in the set available was simply too small.

20.5 NEW DEVELOPMENTS

Perusing the literature, a pair of papers about using spectroscopic analysis to examine clear liquid samples by Willem Windig attracted attention [18,19]. While Willem's interest in the dataset was for testing the Simplisma algorithm, the data itself appeared to have properties that would make it useful for the purposes of further examining the effects of different units on calibration, i.e., enabling tests of conventional calibration algorithms along with CLS analysis similar to the previous experiment [20].

20.6 EXPERIMENTAL

Details of the sample preparation and measurement methods are described in [18]. With the assistance of Tormod Naes, Prof. Windig was contacted, and he graciously agreed to supply the data he used for the Simplisma tests described above. Table 20.2 describes the properties of the sample set.

Spectra were supplied as absorbance values from 1100 to 2500 nm. The data was of very high quality, so no samples were deleted for any reason. Also, no data transformations were applied to the spectra.

The experimental design had an interesting characteristic. With five different ingredients making up the mixtures, there are ten possible ways they can be plotted pair-wise, as shown in Figure 20.2.

With the experimental design used, all ten pairs of ingredients produce essentially the same plot except for minor variations in the dispersion of the points around each concentration level. I have been unable to find a description of the general type of statistical experimental design that was used to create this pattern. (Prof. Windig, himself, had himself lost track of the source of the design.)

CLS calculations and plots were performed using custom software written in MATLAB (Mathworks, Natick, MA), in order to verify that the data followed Beer's Law and that results corresponding to those in the previous experiment [20] were achievable. CLS calculations were compared with the weight fraction values and with the (computed) volume fraction values as reference values; these results were plotted.

TABLE 20.2
Characteristics of the Dataset Described in [18,19] and Used for this Work

The sample set contained 70 mixtures, each containing five ingredients:

Acetone

2-Butanol

1,2-Dichloropropane

Methylene dichloride

Methanol

Concentrations were defined gravimetrically as weight fractions, and a statistical design was used to specify the concentration of each ingredient in each of the 70 mixtures. The concentration of each ingredient in each sample was set to one of the following concentrations: 10, 22.5, 35, 47.5, and 60 wt %.

The spectrum of each mixture was measured twice.

Key characteristics of the mixture design are described in Figure 20.2

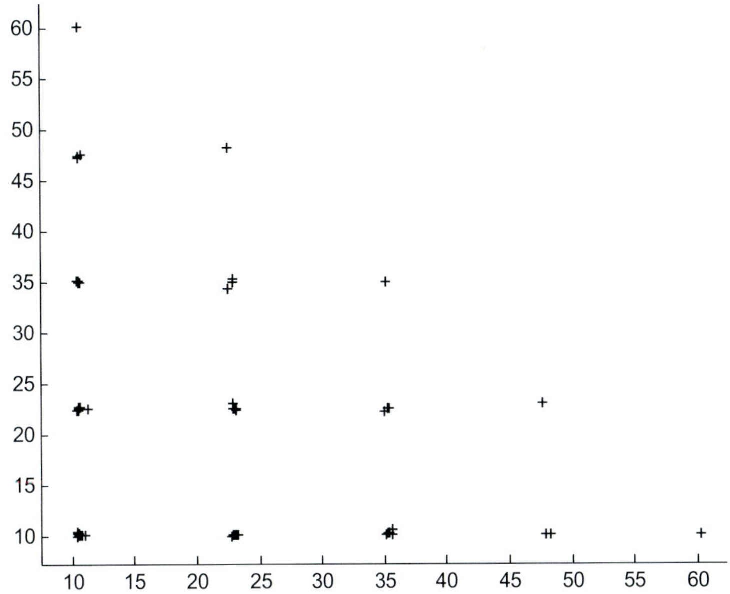

FIGURE 20.2 Overall characteristics of the orthogonal, balanced, symmetric statistical experimental mixture design used to specify the samples. The five different ingredients allow for 10 pairwise comparisons; each of the 10 possibilities provides a plot that looks like the one below.

PCA and PCR calculations and plots were performed using The Unscrambler (CAMO, Oslo, Norway) X 10.1. While as we will see, six factors were always enough to describe the mixture spectra, the PCA and PCR programs were allowed to compute up to as many as ten factors, to guard against potentially missing any interesting phenomena that might occur at higher numbers of factors.

20.7 RESULTS

CLS: spectral analysis without reference laboratory data

As described in the Introduction, our first goal was to verify that these spectra perform the same way that our own previous work [20] on a smaller sample set indicated. To this end, our initial investigations were to apply CLS analysis to the data to verify that they behave the same way as the previous, simpler data set behaved, before applying the more sophisticated calibration algorithms. This provides reassurance that the properties and behavior of the two data sets are similar, and that the conclusions based on the results of the first data set were general and not idiosyncratic to that data set, and therefore could be expected to apply to other data sets. The volume fractions of the five chemical components of the mixture were calculated using the known weights of the ingredients added to each mixture and the densities of the pure materials shown in Table 20.3 [21].

The CLS results were evaluated in two ways: First was the ability to reproduce the spectra of several of the mixtures in the dataset from their pure-component spectra and the CLS coefficients computed for that mixture, compared to the actual spectrum of the mixture. This is also a test of whether Beer's Law holds for these mixtures. Since the experimental design did not include the pure component spectra (although those were measured separately, and fortunately so, since they were needed for use with the CLS algorithm. The samples used as surrogates of the pure materials for this comparison were the ones containing each of the components at its maximum value (i.e., 60%) in the set of spectra, as surrogates for the pure component spectra.

These comparisons are shown in Figure 20.3, and we can see excellent agreement between the actual spectra of the various mixtures and the spectra computed from the pure components, which

TABLE 20.3
Densities of the Components Used for the Five-Component Mixtures

Component	Density
Acetone	0.792
2-Butanol	0.808
1,2-Dichloropropane	1.1593
Methanol	0.7928
Methylene dichloride	1.336

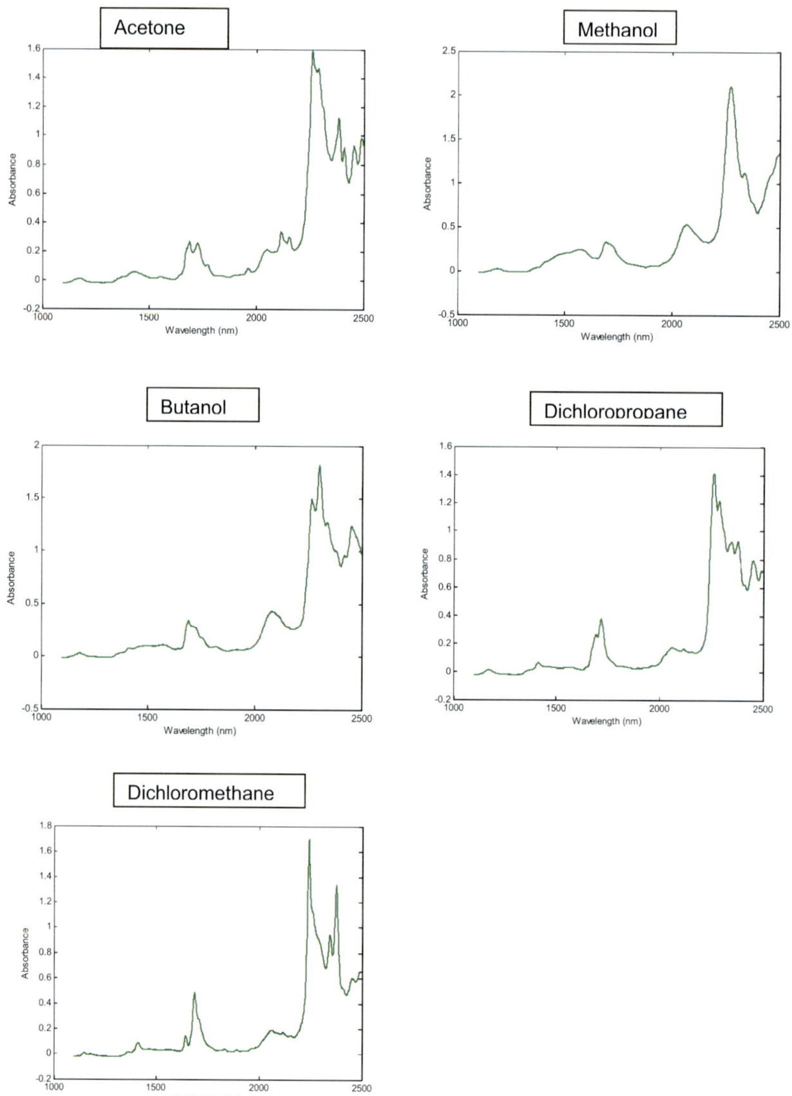

FIGURE 20.3 Comparison between actual spectra of mixtures (blue) and the corresponding spectra reconstructed from the pure components (green). The mixtures shown are the ones that have the maximum concentration (60 wt %) of the specified ingredient, with all other ingredients being at their minimum values (10 wt %).

verifies the applicability of Beer's law to the mixtures, and also the lack of any more than minor interactions between the any of the mixture components [22].

The second way of evaluating the results was to see whether they conform to the theory previously developed to explain the relationships between the spectroscopic analysis, as reconstructed from the CLS algorithm, and the different units (weight fraction and volume fraction) used to determine the compositions of the mixtures, as developed during the study of the smaller experiment. These comparisons are shown in Figure 20.4.

The left-hand column of plots in Figure 20.4 shows that the data points in each cluster, representing those samples with the same nominal weight fraction specified by the experimental design, lie on a vertical or near-vertical, line. However, this characteristic creates large differences between the nominal reference laboratory value and the spectrally determined value. When the reference values are converted to volume fractions, as in the right-hand column of plots in Figure 20.4, the clusters of data points no longer lie on a vertical line. The predictability of the volume fractions used as surrogates for the concentration causes individual clusters to spread out parallel to the horizontal axis as well, in a manner that makes them agree with the spectroscopic values and conform to a hypothetical line describing the overall behavior of the data, albeit some with increased random scatter. This behavior, which all five components of the mixtures exhibit, agrees with the corresponding results found in the previous experiment that used mixtures with only three components [20].

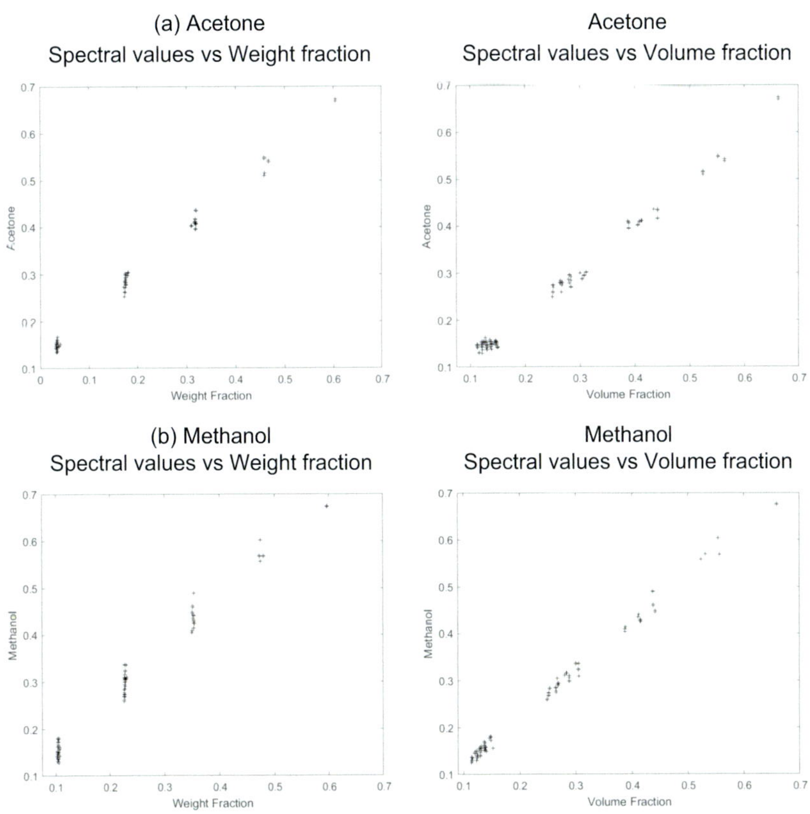

FIGURE 20.4 Plots of the spectroscopic measure of concentration from the CLS algorithm versus the weight fraction or volume fraction representing the known concentrations of the components of the mixtures.

(Continued)

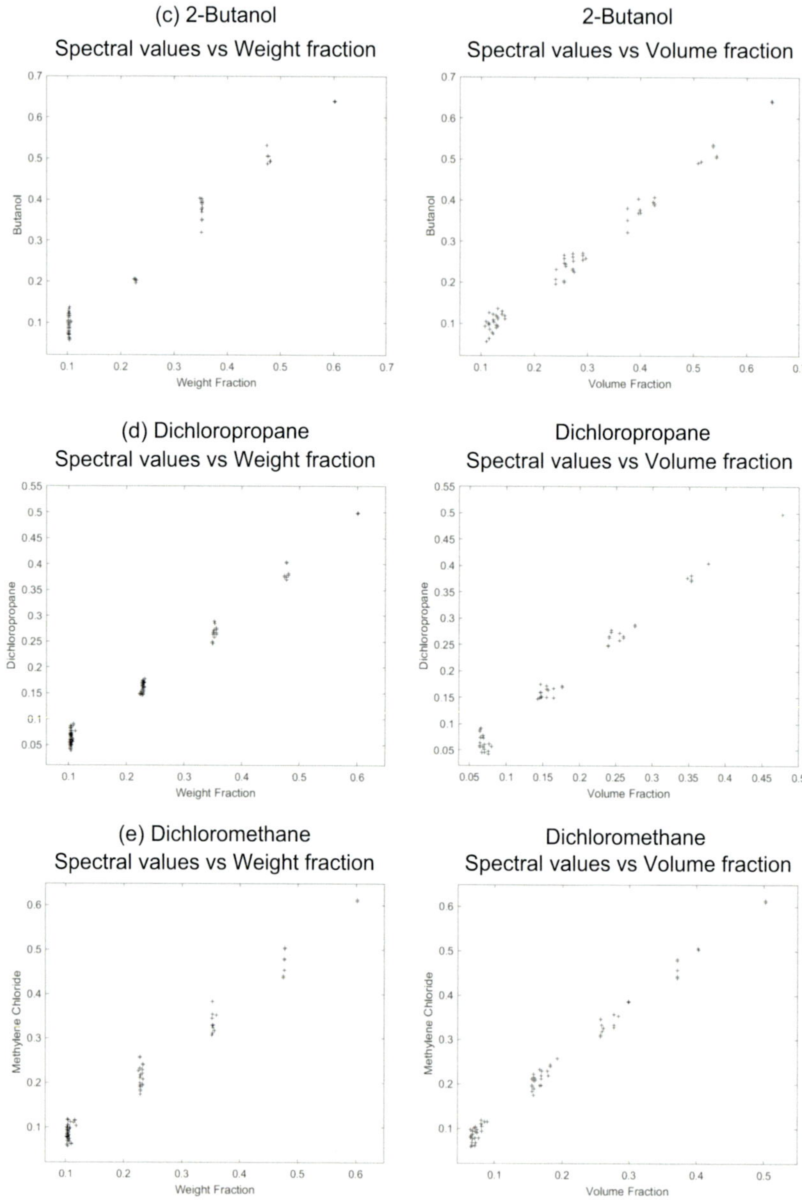

FIGURE 20.4 (*CONTINUED*) Plots of the spectroscopic measure of concentration from the CLS algorithm versus the weight fraction or volume fraction representing the known concentrations of the components of the mixtures.

20.8 SUMMARY OF CLS RESULTS

Preparatory to applying other calibration algorithms to the data, the CLS algorithm demonstrated that each of the five components of the mixtures behaved similarly to the way the data in the previous experiment performed. The ability to reproduce the spectra of mixtures from the spectra of the mixture components and the improved agreement between spectrally calculated concentration values and the volume fractions of the mixture components compared to the weight fractions constitute confirmation of the two major characteristics for the use of the CLS algorithm, which were previously found. These findings provide assurance that the theoretical underpinnings explaining the behavior of the spectral data are correct and apply to both data sets.

20.9 PRINCIPAL COMPONENT ANALYSIS OF THE SPECTRAL DATA—USING REFERENCE VALUES

The mixtures each contain five chemical components, which therefore bring with them five degrees of freedom. In a mixture, however, the concentration of each component is dependent on the other four since the sum of all five must equal unity (or 100%); this relationship reduces the total number of degrees of freedom in the data set to four. In theory, therefore, four principal components (PCs) should suffice to accommodate the variance introduced by the several chemical components of the mixtures.

Figure 20.5 shows the residual variance of the spectra as a function of the number of PCs computed and applied to the data. Contrary to what theory tells us, Figure 20.5 indicates that not all the variance of the spectral data is accounted for until seven principal components have been calculated and applied to the spectral data.

Figure 20.6 shows a set of plots, one for each of the chemical components of the mixtures, which shows the amount of residual variance from fitting the principal components of the spectra to the compositions of the mixtures for each ingredient. These graphs depend slightly on the reference values of the mixtures, but plots made using either weight fractions and volume fractions for the reference values are almost indistinguishable to the eye. Several points are notable:

For all ingredients, the total variance (e.g., variance for zero PCs applied to the data) is greater when concentration is expressed as volume fractions than for weight fractions. All ingredients enter the experimental design with the same set of values, so we expect that the behavior of one ingredient should reflect the behavior of all. We note, however, that three ingredients have densities less than unity, while only two have densities greater than unity. Since conversion from weight basis to volume basis is nonlinear, this implies that over the entire set of samples, there is a larger total numeric value for volumes than there were for weight and the variance scales correspondingly.

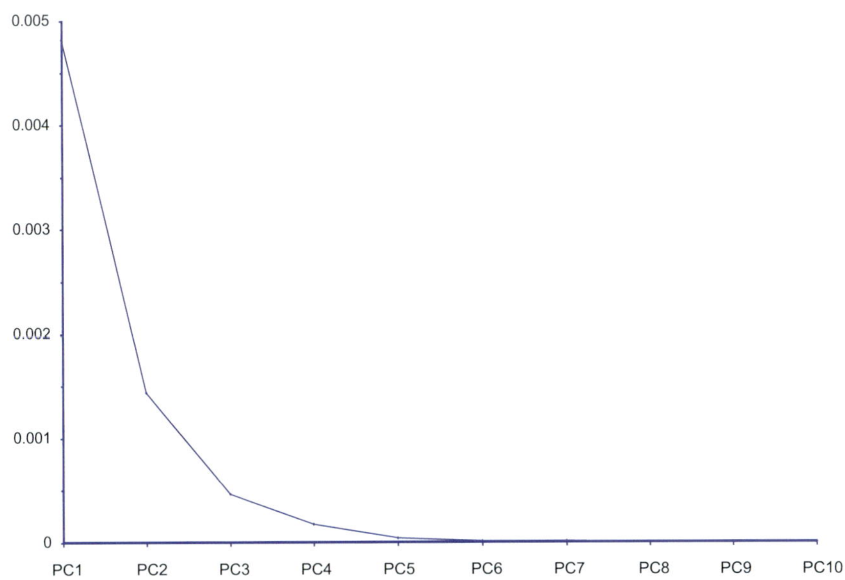

FIGURE 20.5 PCA residual variance of the calibration spectra as a function of the number of principal components used.

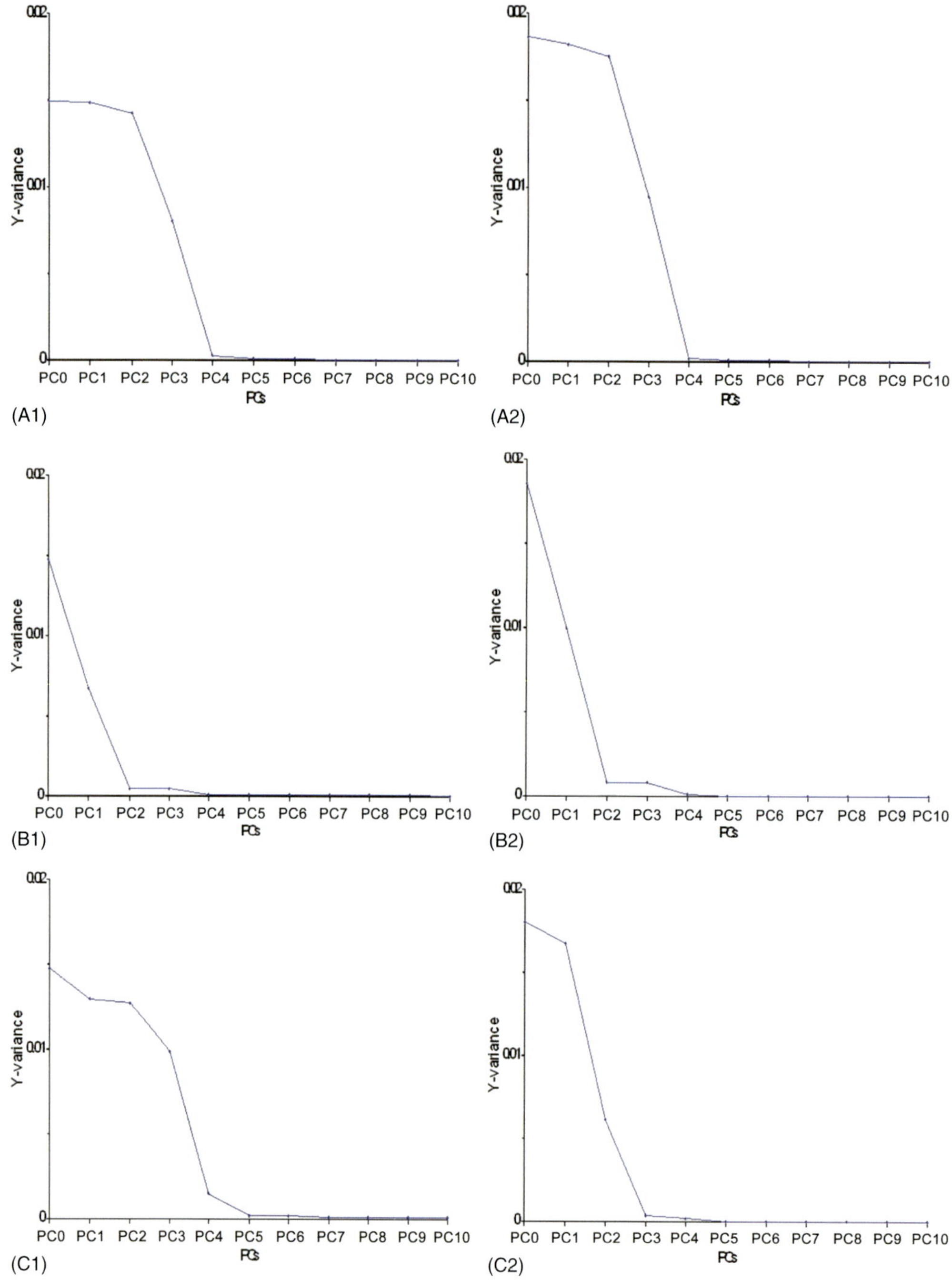

FIGURE 20.6 PCA residual variances of ingredient concentrations as a function of the number of principal components used. Each part of this Figure 20.6 shows the residual variance plot when the ingredient is expressed as weight fraction (Column 1) and volume fraction (Column 2). The several mixture ingredients are plotted in order of increasing density of the ingredient.

(Continued)

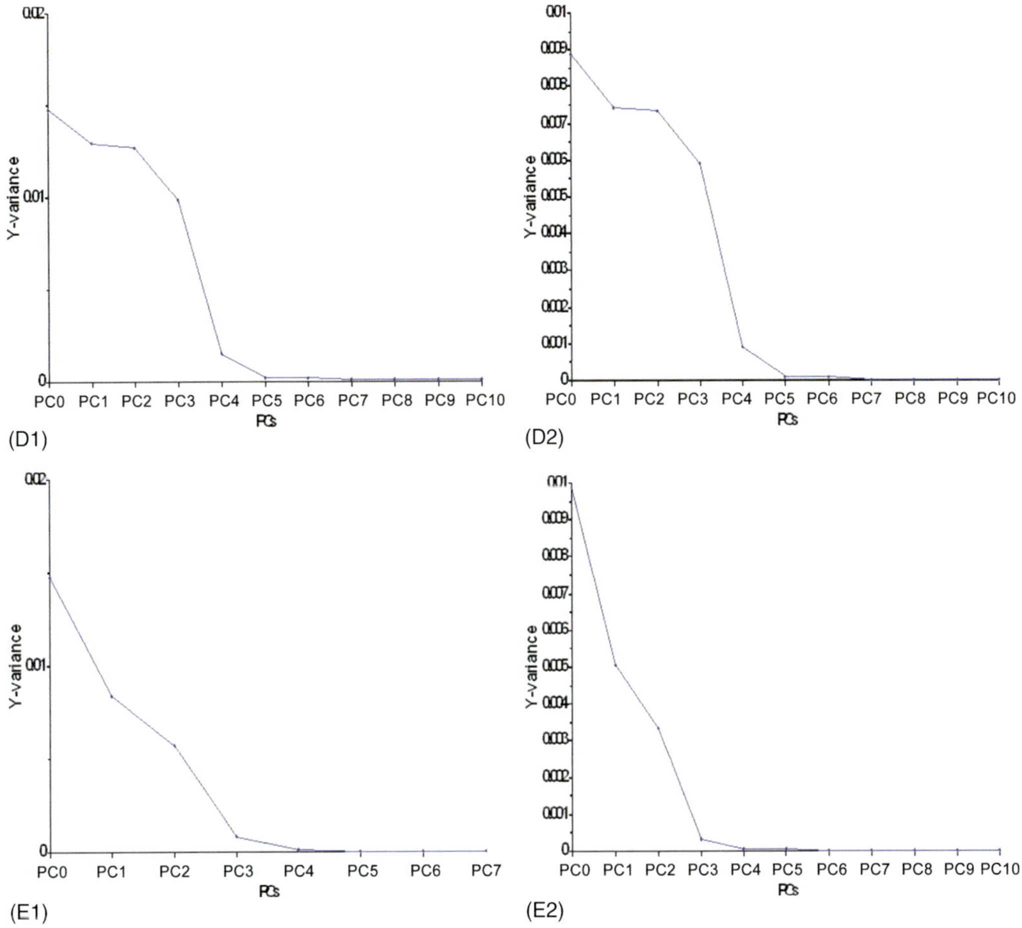

FIGURE 20.6 (*CONTINUED*) PCA residual variances of ingredient concentrations as a function of the number of principal components used. Each part of this Figure 20.6 shows the residual variance plot when the ingredient is expressed as weight fraction (Column 1) and volume fraction (Column 2). The several mixture ingredients are plotted in order of increasing density of the ingredient.

Since variances are not scaled by the computations, this trivial numerical relationship can be the cause of the larger total variance. The computation of principal components scales the results to the total variance of the corresponding dataset; this would then scale all the results linearly and make the plots appear to be essentially the same although they are in fact (inversely) density dependent. We will examine this behavior further, later in this paper.

Second, for each ingredient, the graphs of the behavior of the principal components to explain the ingredient data per Figure 20.6 appear to the eye to be the same whether the ingredient concentration is expressed as weight fraction or volume fraction. This is not very surprising, since the principal component loadings are calculated entirely from the spectral data and therefore are themselves the same regardless of the chemical component that is being fitted.

Third, several of the graphs (notably the ones for acetone, 2-butanol, dichloropropane, and, arguably, methanol) contain sections (between points representing successive principal components used in the calibration) where the graph is horizontal or nearly horizontal. This indicates that for the two points marking the horizontal sections, the second point of each pair, corresponding to the larger number of principal components, is not removing any variance due to that analyte, from the fit of that loading to the spectral data.

Finally, unexpected and highly interesting results were noted when the graphs in Figure 20.6 were compared between the different ingredients. Here we point out that the graphs for the various mixture ingredients in Figure 20.6 are plotted in order of increasing density of the ingredient. The following characteristics of the plots were noted: the least dense ingredients (acetone, methanol, methylene chloride, and 2-butanol) had their error variance reduced to zero with four or fewer principal components, the remaining one (dichloropropane) required more than four principal components. Furthermore, the mixture ingredients with the lowest densities had a tendency to utilize the fewest and lowest numbered principal components. Again, this appears to fly in the face of the theoretical prediction that no more than four PCs should be needed to fit the spectrum for any ingredient.

Another relationship between the PCs was then noted: those PCs where the plot contained horizontal sections are the ones that required excess (i.e., >4) PCs. This tendency is especially pronounced in the case of dichloropropane, where the horizontal section is the one between the first and second PCs. This indicates that the second PC is not contributing to the model for dichloropropane, and we see that for this ingredient, only PCs one through five are reducing the error variance. In other words, only four PCs are needed to explain the variance due to the dichloropropane, but those are not the FIRST four PCS.

The other ingredients behave similarly. Acetone and butanol also include PCs that are not reducing the variance of the respective ingredients, and thus, a calibration for those ingredients would be as good without including the corresponding PC (actually better, arguably, since such a calibration would avoid the loss of degrees of freedom associated with including the unnecessary PC).

In the light of the plots of methanol and acetone, where acetone uses PCs 3 and 4 while methanol uses PCs 1 and 2, the statement above: "the mixture ingredients with the lowest densities had a tendency to utilize the fewest and lowest numbers of Principal Components" seems incongruous. We should expect methanol to be plotted before acetone, in that sequence. We note another property of these data: methanol and acetone have almost identical densities. Then, we should expect them to have the same eigenvalues in the principal component decomposition. This property would cause these two ingredients to be degenerate in the experimental design used.

A known property of degenerate eigenvalues (a property used in, e.g., quantum mechanics) is that indistinguishable degenerate states mix to form distinguishable states. Thus, acetone and methanol, being the two materials that have the lowest, and almost identical, densities in the set of ingredients, are both degenerate with respect to each other and contribute the largest amount of variance to the total; they therefore have virtually all their variance accounted for by the first four principal components, and each of these ingredients requires two PCs to account for their variance. We can see this in the plots.

Degenerate eigenvectors are also orthogonal; of necessity, they account for different portions of the variance, albeit the same amount. Recall that two key properties of principal components are that they are all orthogonal and that the first one accounts for the source of maximum variance in the data. Thus, the analysis uses the first two PCs to explain all the methanol variance while acetone, being orthogonal, includes none of the variance of the first two PCs and is completely explained by the subsequent two PCs. All this can be seen in Figure 20.6.

Note a careful distinction here: while all the variance for acetone and methanol are explained by those four PCs, that does not mean that those PCs are in any way "used up" in the sense of not including any more variance that could be available to help explain the effects of other ingredients in the mixture, although not each other. Indeed, each of the other three ingredients have some of their variance accounted for by one or more of the first four PCs.

Another noteworthy point is that dichloropropane, which has the second highest density in the set of ingredients, has two sections of the residual variance plot that show no decrease in variance. This is also the ingredient that most clearly shows a requirement for more than four PCs to fully accommodate its contribution to the total variance of the data. Yet, if we ignore the two sections of the plot that show no decrease from the inclusion of the corresponding PCs into the model, we find that the dichloropropane also requires four PCs to accommodate its variance although those four are also not the first four that are computed from this data.

This explains two apparently contradictory observations: first, that mathematical theory tells us that no component should need no more than four factors (PCs) to create a calibration for any ingredient yet we find that we need seven for this set of data. The reason is that, as careful study of Figure 20.6 reveals, while no individual ingredient needs more than four PCs, those are not necessarily the first four, or even the same four, for all ingredients.

Not all the PCs contribute to the analysis of any particular ingredient. Horizontal or nearly horizontal sections of the residual variance plots of Figure 20.6 indicate that the higher-numbered PC from that section does not contribute to the analysis of the corresponding ingredient. Thus the first and second PCs do not contribute to analysis of acetone, the third PC to the analysis of methanol, the second and fifth PC to the analysis of dichloropropane, or the fourth PC to the analysis of methylene chloride. Because of that, sometimes it requires more than the first four PCs to be able to analyze all the ingredients in a given mixture, as it is here most clearly seen in the case of dichloropropane, which requires a total of five PCs in order to include the four actually needed to perform the analysis for that ingredient. Thus, an ingredient requiring N PCs for its calibration may still require more than N PCs to be computed, since some of those PCs may not be among the first "N" PCs.

20.10 EFFECTS OF DENSITY ON THE EXPERIMENTAL DESIGN

Figure 20.2 showed the diagrammatic experimental design that specified the weights, and therefore the weight fractions, of each of the ingredients. We noted above that the plot of the experimental design was essentially identical for any pair of ingredients that were plotted, regardless of the amount of the other materials in the mixture, and therefore all the data points for a given concentration pair all overlapped. This is not the case when the actual spectroscopically active units (volume fractions) are used to specify the design.

Figure 20.7a illustrates how the design is affected when volume fractions are plotted rather than weight fractions. In Figure 20.7a, volume fractions of butanol and methylene chloride are plotted against each other. It is clear that the volume fractions for mixtures with the same amount (by weight) are no longer the same, despite the same weight of each ingredient being plotted, because the volume fractions change depending on the other ingredients in the mixture. The groups of data points with the same amounts of butanol and methylene chloride have the same relationship to each other as they did when the weight fractions were being plotted, but within each group of data points, dispersion of the data has been introduced. The plot of each of the 10 combinations of ingredients has these same properties although the different common ingredient changes the variances of the data points in the plot. This is an important point.

An even more important effect is that the plots are no longer identical for the different ingredient pairs, as they were in Figure 20.2. While all the plots of the different pairs of ingredients are spread out as shown in Figure 20.7a, they are spread out differently and by different amounts due to the different variances induced in the data. This is another illustration of the same property presented earlier in the discussion of Figure 20.6, that less dense ingredients occupy larger volumes and therefore bring larger amounts of variance into the dataset; this is demonstrated in Figure 20.7b, where butanol volume fractions are plotted against volume fractions of the other four materials. Since the plot of butanol versus each of the other four ingredients would look very similar, it would be very difficult to distinguish them for comparison purposes. In order to be able to compare the different ingredient pairs, therefore, different colors are used to distinguish the four other ingredients from each other. We can now easily see that the plots from the different ingredients are of different sizes due to the differences in the variances of the different ingredients.

We also see now that the data points cover most of the area of the plot. More subtly, but more importantly, is that the extreme *Y* point for each of the other ingredients is at a different value with

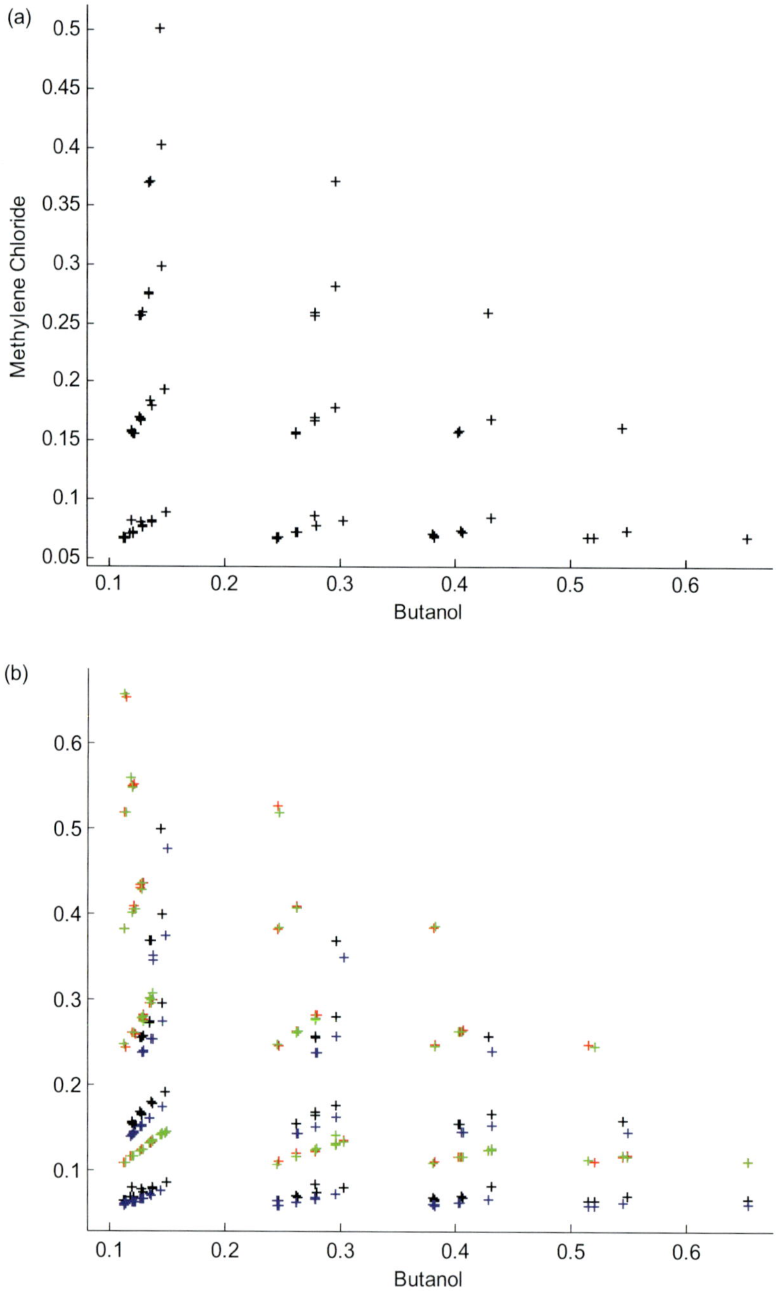

FIGURE 20.7 Experimental design when the volume fractions are used for the plot. (a) Volume fraction plot of butanol and methylene chloride (compare to Figure 20.3). (b) Volume fraction of butanol versus that of each of the other ingredients (black = methylene chloride, red = methanol, blue = dichloropropane, green = acetone).

the intermediate mixtures changing proportionately. This indicates that each ingredient introduces a different amount of variance into the design and, from the arguments above, we can now see that the low-density ingredients occupy more space on the plot (as they do in the samples). Since variance increases as the square of the amount of dispersion of the data, this effect magnifies the effect of the varying densities. While the weight variances of all the ingredients are the same, the different ingredients contribute a different amount of variance to the volume and to the spectroscopic response.

We also now see that, despite the use of a mixture statistical design that appears to be completely balanced for the five mixture ingredients, those ingredients with higher density add less variance to the dataset than ingredients with low density. Note that in Figure 20.7b, for the samples where the value of butanol = 0.1, the highest Y values for acetone and methanol (the two ingredients with the lowest density—red and green in Figure 20.7b), the highest spectral (Y) reading is roughly 0.65. The highest Y value for methylene chloride (the densest ingredient, black in Figure 20.7b) is roughly 0.5.

Thus, despite all good intentions, the design is NOT balanced among the five ingredients as intended. This effect is distorting the nature of the actual experimental design affecting the data being measured: what you see is NOT what you get when using weight fractions in the design for spectroscopic analysis. Compare Figure 20.2 with either part of Figure 20.7. When implementing the experimental design by weight, Figure 20.2 implies that all ingredients included with the same weight have the same influence on the results, but that is not the case when performing spectroscopic analysis. The actual design that is effectively controlling the experiment is the one presented in Figure 20.7. Interestingly, since the densities of the methanol and acetone are the same, their effects on the spectroscopic behavior are still subject to the degeneracy described above; note that in Figure 20.7b, the points representing acetone (green) and methanol (red) overlap each other. Also, if the design were plotted against an ingredient other than butanol as being common to the other four, then the entire design would appear to be different from Figure 20.7b.

We can extrapolate this finding to any experimental design used to specify design points for any experiment where spectroscopic readings are used for measurement. This discrepancy between different units to specify an experimental design needs to be taken into account when an experimental design is developed. The amount of the various ingredients in a mixture needs to be adjusted to ensure that specified volumes, rather than specified weights, of the different ingredients are used in creating the samples in order for the design to be properly balanced. Otherwise, the actual design implemented in the specification of the design points is not what it appears, or is intended, to be. Experimenters in the future, designing experiments in which spectroscopy plays a key role, need to take this effect of density of the ingredients into account in their experimental designs. One way this can be accomplished is by specifying the ingredient composition so as to use equal volumes of ingredients; the ingredients could even be measured and dispensed using classical volumetric equipment. Alternatively, if ingredients are specified by weight, this can be accomplished by specifying the ingredient weights that correspond to the needed volumes. Alternatively it can be accomplished post facto by using the various ingredient densities to correct the experiment to the actual ingredient volumes used.

20.11 SUMMARY

We summarize this as follows:

1. The experimental design is balanced (symmetric) in weight.
2. Densities differ, and therefore it is not balanced for volumes.
3. For the same weight, less dense liquids have greater volumes.
4. The design would therefore be distorted and differ for each ingredient pair.
5. Spectroscopy responds in proportion to volume and therefore is more sensitive to lower-density ingredients. This breaks the symmetry.
6. In the design, less dense ingredients have greater volume.
7. Acetone and methanol have the same density.
8. Acetone and methanol principal components are degenerate.

9. Less dense ingredients have greater optical variability (variance).
10. Principal components are sensitive to variance: first component represents the greatest variance and therefore the least dense ingredient.
11. Therefore, principal components are otherwise calculated in order of increasing density of the ingredient.

The various observations discussed above all lead to the following explanations: the theoretical mixture design was balanced, but the design and its properties were based on the use of the weight fractions to specify the various ingredients. Since the various materials have different densities, the specified design is NOT balanced with regard to the volume fractions of the various ingredients in the mixtures. Ingredients with the lowest densities (acetone and methanol) are present with higher volumes in the actual mixtures and therefore make a larger contribution to the total variance in the mixtures (and consequently the spectra) than the ingredients with higher densities. This can be deduced from the behavior of the principal components and their relations and can actually be seen in Figure 20.6.

This allows us to recognize that there is a direct connection between a physical property of the ingredients, in this case the density, and mathematical properties of the data set, in this case the order of calculation of the principal components. The PCs corresponding to the low-density ingredients are the ones that contribute the most to the variance to the data set and are consequently the ones computed first.

20.12 SUMMARY OF CLS RESULTS

Preparatory to applying other calibration algorithms to the data, the CLS algorithm demonstrated that each of the five components of the mixtures behaved similarly to the way the data in the previous experiment performed. The ability to reproduce the spectra of mixtures from the spectra of the mixture ingredients verifies the applicability of Beer's law, and the improved agreement between spectrally calculated concentration values and the volume fractions of the mixture components compared to the weight fractions constitutes the two major characteristics of the use of the CLS algorithm that were previously found. These findings provide assurance that the theoretical underpinnings explaining the behavior of the spectral data are correct.

20.13 SUMMARY OF PCA RESULTS

The intended purpose of the experimental design used was to ensure balance between the five ingredients' contribution to the variance of the data. This intent was subverted by the differences in the densities of the various components, the property affecting what the volume (the property the measurement is actually sensitive to [20]), which is not the same for the various ingredients in the mixture. Since the mixtures were made according to the weights of the various ingredients included, the corresponding volumes differed, and therefore the low-density ingredients affected the total variance more than the high-density ingredients did, effectively destroying the expected balance of the experiment. The residual variance plots of Figure 20.6 are in order of increasing ingredient density, so it becomes clear by scanning those plots that there is a distinct tendency for the PCs contributing to the spectrum of a given ingredient to move to the higher PCs as the density of the corresponding ingredient increases. This is in accord with theoretical expectations.

Not all the PCs contribute to the analysis of any particular ingredient. Horizontal, or nearly horizontal, sections of the residual variance plots of Figure 20.6 indicate that the higher-numbered PCs from that section do not contribute appreciably to the analysis of the corresponding ingredient. The first and second PCs do not contribute to the analysis of acetone, the third PC does not contribute to the analysis of methanol, nor do the second and fifth PCs to the analysis of dichloropropane, nor the fourth PC to the analysis of methylene chloride. Because of this effect and, despite the fact that each ingredient in the mixture requires only four PCs, it sometimes requires more than the first four PCs to be able to analyze all the ingredients in a given mixture. In the case of dichloropropane, we require a total of five PCs in order to include the four active ones actually needed to perform the analysis for that ingredient.

This explains the common observation that more total PCs often need to be computed for a calibration than theory would seem to indicate because a calibration using N PCs does not always find them among *the first* N PCs that are computed. The excess PCs that are often computed, which are not required for individual analyses, may be profitably deleted from the analysis for the corresponding ingredient. The previous conclusion [20] about the nonlinear relationship is now additionally seen to be a consequence of the direct relationship between the physical property (density) and mathematical properties (order of calculating PCs) of the spectroscopic data measured on samples affected by that physical property, and thus the nonlinearity found previously [20]) is only part of the explanation for the need for "extra" factors.

20.14 CONCLUSIONS

The principal component analysis of the data conforms to the theory, albeit in an unexpected way. Each of the five ingredients does in fact have their variance accounted for by four principal components although not necessarily the first four. Since PCA is performed on the spectra only, this is true whether correct (volume fractions) units or incorrect units are used to express analyte concentrations.

Due to the inverse relationship between ingredient density and the corresponding volume for equal weights of an ingredient, with everything else being equal, principal components were calculated in order of decreasing density, for the contribution of an ingredient to the variance of the spectral data.

When a statistical experimental design is implemented for spectroscopic measurement, the design parameters need to take into account the differences between specifications of the design in different units. Otherwise, analysis of the data will produce incorrect (or at least, unexpected) results for ingredients that are expected to have similar effects (i.e., due to equal weights) but in fact are affected by variables that have different spectroscopic activity. Corrections to the implementation of the design should be made to ensure that the design specifications represent the volume fractions of corresponding ingredients, so that the statistical analysis to mean what it is expected to mean.

The sensitivity of the spectroscopic measurement to the volume fraction of the ingredients of a mixture distorts the nature of an experimental design-based mixture made according to weight. Ingredients that might be expected to have effects of similar magnitudes because equal weights were added to the mixture will actually have different magnitudes of their effects. This is above and beyond any effects that could be attributed to the actual spectral differences, a question we have not considered.

REFERENCES

1. Mark, H., Rubinovitz, R., Dahm, D., Dahm, K., Heaps, D., Gemperline, P.; Pittcon, Atlanta, GA 2010(pub) (2011).
2. Mark, H.; *Applied Spectroscopy*; 42 (5), pp. 832–844 (1988).
3. Esbensen, K. H., Geladi, P., Larsen, A.; *NIR News*; 23 (5), pp. 16–18 (2012).
4. Mark, H., Workman, J.; *Spectroscopy*; 25 (5), pp. 16–21 (2010).
5. Mark, H., Workman, J.; *Spectroscopy*; 25 (6), pp. 20–25 (2010).
6. Mark, H., Workman, J.; *Spectroscopy*; 25 (10), pp. 22–31 (2010).
7. Mark, H., Workman, J.; *Spectroscopy*; 26 (2), pp. 26–33 (2011).
8. Mark, H., Workman, J.; *Spectroscopy*; 26 (5), pp. 12–22 (2011).
9. Mark, H., Workman, J.; *Spectroscopy*; 26 (6), pp. 22–28 (2011).

10. Mark, H., Workman, J.; *Spectroscopy*; 26 (10), pp. 24–31 (2011).
11. Mark, H., Workman, J.; *Spectroscopy*; 27 (2), pp. 22–34 (2012).
12. Mark, H., Workman, J.; *Spectroscopy*; 27 (5), pp. 14–19 (2012).
13. Mark, H., Workman, J.; *Spectroscopy*; 27 (6), pp. 28–35 (2012).
14. Mark, H., Workman, J.; *Spectroscopy*; 27 (10), pp. 12–17 (2012).
15. Mark, H., Workman, J.; *Spectroscopy*; 29 (2), pp. 24–37 (2014).
16. Mark, H., Workman, J.; *Spectroscopy*; 29 (9), pp. 26–31 (2014).
17. Mark, H., Workman, J.; *Spectroscopy*; 30 (2), pp. 24–33 (2015).
18. Windig, W., Stephenson, D. A.; *Anal. Chem.*; 64, pp. 2735–2742 (1992).
19. Windig, W.; *Chemomet. Intell. Lab. Syst.*; 36, p. 3–16 (1997).
20. Mark, H., Rubinovitz, R., Heaps, D., Gemperline, P., Dahm, D., Dahm, K.; *Appl. Spect.*; 64 (9), pp. 995–1006 (2010).
21. Hodgeman, C.; *"Handbook of Chemistry and Physics"*; 44 ed.; The Chemical Rubber Publishing Co., Cleveland, OH (1962).
22. Xiong, Z., Pfeifer, F., Siesler, H.; *Applied Spectroscopy*; 70 (4), pp. 635–644 (2016).

21 Process Measurements

*Emil W. Ciurczak**
Doramaxx Consulting

CONTENTS

21.1 INTRODUCTION

For decades, near-infrared spectroscopy (NIRS) has been a mainstay in food, agriculture, food, textiles, and, since the 1980s, pharmaceuticals. The popularity of the technology is based on physics, not marketing. NIRS utilizes strong sources and, when coupled with the newest quiet, sensitive detectors, can be used for rapid measurements. However, the vast majority of instruments (and applications) were based on available higher-end instruments, based in laboratories. In many locations, some of the more rugged lab-based instruments were placed near grain silos or pharmaceutical processes, but few were considered actual "process instruments."

Basic analytical chemistry theory would back the simple concept that more measurements usually mean more accurate understanding of the "actual value" of whichever parameter we are measuring: active ingredient, viscosity, color, pH, density, etc. (Statistically, were we to assay 100% of any batch of material, we would discover the "true" values… at the cost of the product, in the case of traditional, wet chemical methods, however.) Since the better values we produce, the more accurately we can control a process; clearly, spectrophotometric analyses are one answer. When considering sample preparation and/or sample destruction as time-consuming roadblocks to rapid evaluation, near-infrared spectroscopy becomes a top choice.

NIR process monitoring and analysis are becoming more important as more and more industries both become automated and, simultaneously, feel intense international competitive pressures. Clearly, better-controlled processes produce a better product, which, in turn, increased competitiveness and profitability. However, not until recently have all the pieces been available: speed, accuracy, and

* emil@ciurczak.com

software capable of computing as quickly as readings could be taken. Add to these attributes instrument ruggedness and wireless capability. Some of the newer instruments are low power (meant as hand-held) and do not need an enclosure to make them explosion proof. Many "hand-held" units are adaptable to holders on production equipment, allowing them to become part of the unit.

As we will see, "true" process instruments are often designed for process work, not modified from lab units. They almost always include fiber optic probes, allowing them to be used for both solids and liquids. Add to this faster and quieter detectors and incredible advances in computers and concomitant algorithms, and we have "the perfect storm" of process instrumentation. Since the last edition of this text (2008), much progress has been made in process equipment, computers, and understanding (experience) in rapid, accurate process monitoring and control.

21.2 TRADITIONAL (LAB-BASED) INSTRUMENTS

For several decades (1960s–1980s), the NIR spectrometer (usually grating based) was treated as any other analytical instrument: well-made, accurate and precise, and useful for lab work, but too sensitive to move from the laboratory bench to "where the chemistry was taking place" or to a warehouse to track incoming materials. Samples were gathered from the point of storage, delivery, or production; bagged; labeled; and shipped to the lab for analysis. The only different feature of a NIRS analysis ("prediction," actually) was that there was little or no need for sample preparation, making it quite a bit simpler and faster than, say, a gas or liquid chromatographic, "wet," or "traditional" spectroscopic method.

As is covered in other chapters, the popularity of the NIR instrument was based on its ability to measure "as is" samples. This was important in agricultural applications (where it was firmly established as an important method of analysis since the 1950s), since nearly every sample was heterogeneous and for which "standards," natural or synthetic, could not be produced (not easy to "make" an ear or kernel of corn synthetically). The paradigm of analysis was spun 90°… literally. The absorbance (better known because it was often precise/accurate to four digits) was on the *X*-axis and the chemical value appeared on the *Y*-axis, being the value with the greatest error/uncertainty. This led to simple (yet elegant) equations that calculated the analyte(s) of interest in the presence of the matrix (see all the chapters on MLR, PLS, etc.) Considering the amount of sample preparation time saved, the relatively slow analysis time of the NIR (several minutes) was negligible.

While there were newer additions to the instrument that allowed faster scans (i.e., diode array detectors for simultaneous wavelength measurement or acousto-optic tunable filters (AOTF) and interferometers for extremely fast scans) and more sensitive (with better signal-to-noise) detectors made from Indium Gallium Arsenide (InGaAs), these new-fangled additions came at a cost. I asked one of the leaders in NIR instrumentation (in 1990) what an InGaAs diode array cost him (cost to manufacturer, not retail to customer) and he simply said, "$10,000." When I asked about volume, he said "$10,000 each." It seems that InGaAs detectors were at the heart of the telecom industry. (They could accurately and rapidly receive and re-transmit voice and data streams, so were essential for long-distance (e.g., underwater) transmissions.)

As a consequence of these costs, a super-sensitive unit costs upward of $100,000 (in 1990 dollars) and wasn't considered something that was needed in great numbers. Also, the computers were *starting* to get better (1980's units often needed an additional math coprocessor to even compute something as simple as an MLR equation) and the times for the existing principal component and partial least squares programs (introduced commercially in 1987) were being measured in minutes and seconds, not in hours and days. (When I was the *beta* site for the first Mahalanobis distance program (written by H. Mark at Technicon), we would begin "crunching" the spectra for an equation on Friday afternoon and expect the completed equation when we returned on Monday.)

Process-hardened lab instruments could be produced (Bran+Leubbe, formerly Technicon) manufactured a polarized-interferometer-based instrument that could be placed on a cart and wheeled into production. It was mainly used for raw material qualification (having several orders of magnitude higher noise than a typical lab-based instrument, it was seldom used for any trace analysis). While somewhat primitive (in terms of a "process" instrument), it was a breakthrough in bringing the lab to the samples… but not truly portable, as far as convenience goes. That waited for some years and a number of technical advancements.

21.3 PORTABLE INSTRUMENTS

In the late 1980s, instrument manufacturers produced ruggedized versions of their lab instruments, capable of being placed on a wheeled cart and pushed to the warehouse. These were neither explosion proof nor easy to bring to the samples. They also still ran on 120/240 VAC and required extension cords to operate. That also precluded having them operate in any environment with powder or organic fumes in the air. That excludes many pharmaceutical operations, from API synthesis through tablet coating, grain elevators, flour milling, and many other industrial processes. So, "portable" is somewhat of a misnomer for these early instruments as they were tethered to electrical outlets, seated on carts. So, it is safer to call them "mobile" if not true portable instruments.

Some of the hazards of these converted lab instruments may be mitigated by placing them in "explosion-proof" enclosures. These NEMA (National Electrical Manufacturers Association) enclosures were basically a cabinet with a positive pressure that both kept fumes/dusts away from the electronics and simultaneously protected them from extremes in temperature. NIR units used for refinery control were air conditioned/heated against the 100°F (38°C) temperatures in Houston or the cold (–30°F or –34°C) in Alaska. However, aside from doubling (or more) the cost of the instrument, these are still "in-place" instruments, not in any way portable (except being carried to where they were installed).

21.4 HAND-HELD INSTRUMENTS

In the 1990s, some interesting events occurred (keep in mind that "interesting" may mean good or bad). One very notable event (although it was a drawn-out event) was the collapse of the (over-heated, over-extended) telecom industry. All the indium-gallium-arsenide (InGaAs) that had been hoarded by the suppliers was suddenly available to instrument manufacturers. And, in our capitalistic society, abundance often means price lowering. InGaAs diode-array detectors dropped in price by orders of magnitude, making them quite attractive to instrument companies, adding sensitivity and speed to their instruments.

Another victim of the telecom crash were the suppliers (many in the Boston area) of MEMS (Micro-Electro-Mechanical Systems) repeaters. These companies could either shutter their doors or find new applications for their equipment. Fortunately, the initial purpose of MEMS units was to sort through multiple frequencies of light striking them simultaneously and sort them into coherent signals. This is precisely what happens in the gratings, prisms, and interferometers in traditional spectrometers.

Simultaneously, the world of computing was evolving. We did not go from Univac to the iPhone overnight any more than we moved from the Wright brothers at Kitty Hawk to Boeing's B-777 airplane. The computer designations (largely from the controlling integrated circuits/memories) went 286, 385, Pentium, and so on, getting smaller, faster, and best yet, less expensive along the way.

So, with the addition of MEMS-based spectrometers, NIR instruments could truly become portable, often weighing no more than a hair drier. With existing on-board computing power, doing raw material IDs was reduced to a "point-and-shoot" approach. The operator merely had to open a container (usually powder) and scan the surface, and, within seconds, the material was identified and potentially qualified.

21.5 PROCESS INSTRUMENTS

The common denominators (non-negotiable parameters) needed for an instrument to be considered a process unit are obvious:

1. *Speed.* In a "process," a sample will either be changing with time (chemical or biological reaction), so a measurement needs to be a "snapshot," not a slowly developing, research-grade spectrum, or discreet units will be produced at a high rate of speed. With lab-based instruments, it is not uncommon for a NIR spectrum to take a minute or more to be generated. Except for long (usually biological) reactions, 2–72 h, for example, this is clearly not fast enough for a process instrument.

   For an instrument to be used in a production setting (whether pharmaceutical tablet/capsule production, blending of cake mixes, or an assay of olive oil squeezing), measurement times of less than 1 s are clearly required; in fact, millisecond measurements are preferred, where possible. This has only become possible with, for example, thermally cooled Indium Gallium Arsenide (InGaAs) or Lead Selenide (PbSe) detectors and concomitantly fast computers with software to make sense of the signals.
2. *Precision/accuracy.* Since in-process measurements are used to control the reaction (adding reagents for organic reactions, quenching reactions, adding nutrients to bioreactions, etc.), proper (accurate/precise) values need to be generated for timely "decisions" to be made, either manually or by a computer algorithm. In some cases (repetitive, small, discreet units, such as pharmaceutical tablets or filled vials), absolute accuracy may be compensated by the sheer number/rapidity of measurements, allowing real-time trend or out-of-specification monitoring.

   (By simply continuously averaging several units and observing their (moving) averages, a fairly good picture of the overall analysis may be made. A good analogy to this technique would be averaging multiple spectra (e.g., 32 or 64 scans) for a spectrometer to make up for baseline noise.)
3. *Sensitivity.* While many of the analytes or physical parameters sought by NIRS are fairly large (i.e., drug in tablets around 5%–10%, fat in milk between 0% and 3%, and so forth), some analytes might be small by comparison. In order for the instrument to achieve an accurate measurement of minor components, the speed of multiple measurements (point #2) allows averaging spectra to generate a signal-to-noise (S/N) ratio such that the "signal" or spectral peak is a minimum of four times as large as the baseline noise. In many cases, impurities that can ruin a reaction (chemical or biological) are present in low concentrations. Therefore, the process instrument needs the same sensitivity as a lab-based (research-grade) instrument while attaining a speed far beyond the typical R&D instrument.
4. *Reproducibility.* The fact that any NIR instrument is a single beam device, and it is optimal to check the zero absorbance or 100% transmission point for reference as often as possible. However, when a process instrument is required to run for a long period (up to 24 h for a single production batch, in some cases), it is not always simple to re-zero the baseline on the run. For processes that have been shown to change slowly over the course of the process, interrupting the measurements for a fast (internally run) baseline measurement every hour or so would not substantially affect the overall monitoring and control of the process.

   For example, a pharmaceutical tablet press could easily produce 100,000 tablets per hour. If the baseline realignment takes 1 min, the result would only be 1666 tablets or 1.67% of the product missed. Compared with the industry standard of randomly sampling and assaying 20 tablets post-production, this is less than insignificant. For industries such as paper and textiles, similar calculations may be made, showing that keeping the instrument calibrated distracts very little from "continuous" or 100% monitoring.

In a number of some very rapid instruments, a single, discreet sample may be measured several times at different points along its axis, giving a better picture of its internal homogeneity. Few instruments can achieve this (one example is Indatech, France), but, for complex samples (pharmaceutical tables, again), this spatial monitoring gives a better picture of the chemical (active ingredient) and physical properties (hardness, density) along the body of the tablet.

21.5.1 Converted Equipment

The majority of early process instruments were existing laboratory units, converted into process-compatible analyzers. One example was a unit made by NIRSystems (then Perstorp, later FOSS, then Metrohm), wherein their scanning (grating-based) unit (model 6500) was deconstructed and recombined inside a NEMA. The light was directed sequentially through a number of fiber optic probes (usually up to six) to the samples (most often liquids, such as biochemical processes) and back to the monochromator and lastly to the detector.

Technically, such devices were "process hardened" but were expensive and still as slow as lab instruments.

21.5.2 "New" Portable Equipment

Some very small instruments that use some of the following technologies and have an on-board battery, computer capacity, and a screen to view results have come onto the market (Figures 21.1 and 21.2 show two examples).

21.5.2.1 MEMS

MEMS (an acronym used in the United States, Microsystems Technology (MST) in Europe, and Micromachines in Japan) is a technology that can be defined as miniaturized mechanical and electromechanical elements (i.e., devices and structures) that are made using the techniques

FIGURE 21.1 Ahura hand-held NIR.

FIGURE 21.2 ASDI hand-held unit.

of microfabrication. The critical physical dimensions of MEMS devices can vary from well below one micron on the lower end of the dimensional spectrum, all the way to several millimeters. They can vary from relatively simple structures having no moving elements to extremely complex electromechanical systems with multiple moving elements under the control of integrated microelectronics. The one main criterion of MEMS is that there are at least some elements having some sort of mechanical functionality whether or not these elements can move. The term used to define MEMS varies in different parts of the world. In the United States, they are predominantly referred to as MEMS, while in some other parts of the world, they are called "Microsystems Technology" or "micromachined devices."

While the functional elements of MEMS are miniaturized structures, sensors, actuators, and microelectronics, the most notable (and perhaps most interesting) elements are the microsensors and micro-actuators. Microsensors and micro-actuators are appropriately categorized as "transducers," which are defined as devices that convert energy from one form to another. In the case of microsensors, the device typically converts a measured mechanical signal into an electrical signal.

In the context of NIRS, they are designed as a very, very small spectral engine. For example, the MEMS device could be a tiny Fabray–Perot interferometer with two parallel micro-lenses vibrating, rapidly changing the distance between the two lenses, effectively generating an interferogram.

21.5.2.2 MOEMS

Micro-opto-electro-mechanical systems (**MOEMSs**) are not a special class of micro-electro-mechanical systems (MEMS) but rather the combination of MEMS merged with micro-optics; this involves sensing or manipulating optical signals on a very small size scale using integrated mechanical, optical, and electrical systems. It includes a wide variety of devices including optical switch, optical cross-connect, tunable VCSEL (a type of semiconductor laser diode with laser beam emission perpendicular from the top surface—see Figure 21.3), microbolometers, among others. These devices are usually fabricated using micro-optics and standard micromachining technologies using materials such as silicon, silicon dioxide, silicon nitride, and gallium arsenide (Figure 21.4).

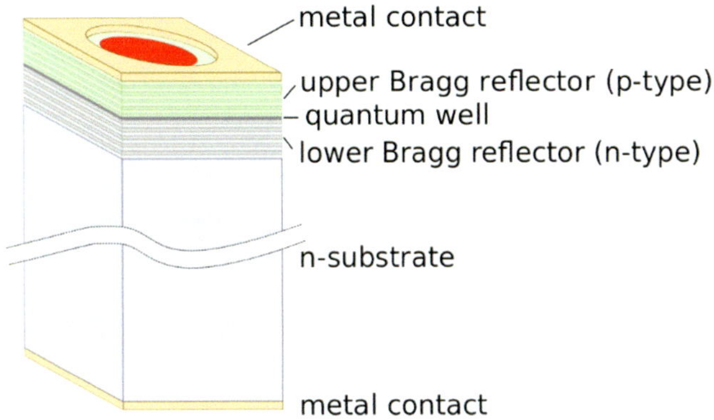

FIGURE 21.3 Diagram of a simple VCSEL structure.

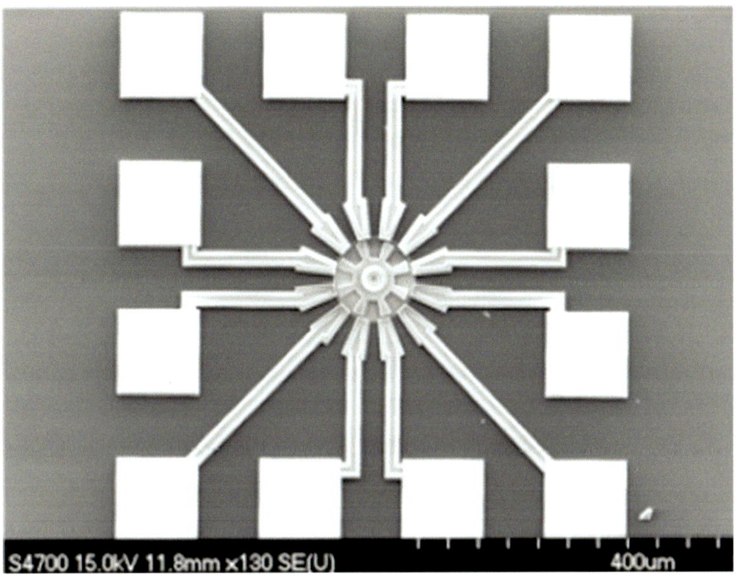

FIGURE 21.4 A surface micromachined electro-statically-actuated micromotor fabricated by the MNX. This device is an example of a MEMS-based microactuator.

21.5.3 Process Designed Equipment

21.5.3.1 AOTF

In general, acousto-optic effects are based on the change of the refractive index of a medium due to the presence of sound waves in that medium. Sound waves produce a refractive index grating in the material, and it is this grating that is "seen" by the light wave. These variations in the refractive index, due to the pressure fluctuations, may be detected optically by refraction, diffraction, and interference effects; reflection may also be used.

The acousto-optic effect is extensively used in the measurement and study of ultrasonic waves. However, the growing principal area of interest is in acousto-optical devices for the deflection, modulation, signal processing, and frequency shifting of light beams. This is due to the increasing availability and performance of lasers, which have made the acousto-optic effect easier to observe and measure. Technical progress in both crystal growth and high-frequency piezoelectric transducers has brought valuable benefits to acousto-optic components' improvements.

Along with the current applications, acousto-optics present interesting possible applications. It can be used in nondestructive testing, structural health monitoring, and biomedical applications, where optically generated and optical measurements of ultrasound give a noncontact method of imaging.

21.5.3.2 Diode Arrays

Diodes are simply semi-conductors that convert quanta of light to electrical current. They have a fixed electric potential and, when light strikes them, current runs. They are generally nonspecific in which wavelengths activate the current (there can be diodes specific to UV or visible or NIR, just not specific wavelengths), so to be useful as detectors, we have two choices: (1) the light is pre- or post-sample dispersed by a rapidly grating and shone on a single diode, generating a traditional spectrum or (2) the dispersed light (via a static grating) is shone on a collection of individual diodes (an array) to generate a spectrum.

Diode arrays are just what they sound like: a series of detectors (diodes) in an array. The number of diodes will determine the resolution of the spectrum, while the source and wavelength selection device will determine the wavelength range (and the material of the diodes, themselves). These are most often made from silicon (400–1100 nm), lead sulfide (PbS: 1100–2500 nm), and indium gallium arsenide (InGaAs: 1000–3500 nm, with cooling). The number is variable, but limited by simple geometry: 32 or 64 diodes take up less space than, say, 1024 diodes. Clearly, the resolution is higher for larger numbers of diodes, but, since (in general) NIR peaks are quite wide (low frequency), while "noise" is generally a higher frequency (narrow peaks), so there is a limit to how many diodes are needed, which means "diminishing returns" with many, many more diodes.

21.5.3.3 Chemical Imaging

Hyperspectral imaging is a combination of spectroscopy and image forming. It creates an electromagnetic profile of a two-dimensional (2D) scene by capturing the intensity of radiation emitted from each elementary spatial segment of the scene (pixel) at various wavelengths. The collected information is usually arranged into a hyperspectral image data-cube, which has two spatial dimensions and one spectral dimension. The data-cube can be seen as a batch of monochromatic images each corresponding to one spectral band. Hyperspectral imaging systems divide the radiation received from each pixel into many sections, that is, narrow spectral bands often only a few nanometers apart, before detecting them. Therefore, in these systems, the number of photons detectable at each wavelength is greatly reduced and hence the required sensor sensitivity or exposure time is increased in comparison with the conventional panchromatic or trichromatic imagers.

The spatial and spectral information provided by a hyperspectral image of a scene can be analyzed to detect, identify, or discriminate objects and patterns as well as the chemical composition of the material present at the scene. Hyperspectral imaging finds application in many areas such as computer vision, remote sensing, biomedicine, surveillance, precision agriculture, environmental studies, forensics, nanoparticle research, food science, mining, forestry, etc. [1–5]. Several techniques have been developed for hyperspectral imaging. They can roughly be categorized as whiskbroom (point scan), and Figures 21.5 and 21.6 show push-broom (line scan), tunable filter (wavelength scan), and snapshot.

For process work, the push-broom type is most amenable to rapidly moving samples (or stream of material, such as grain or paper or textiles), as the highest resolution data cube could take minutes to generate (and require terabytes of space to contain all the scans from a production run).

21.6 SAMPLING

One critical step in the calibration and validation (and application) of a process analysis is where and how the sample is. Over the years, instrumentation has been developed for grains and such and, in recent years, pharmaceutical powders and small solids (i.e., tablets, granules, or capsules).

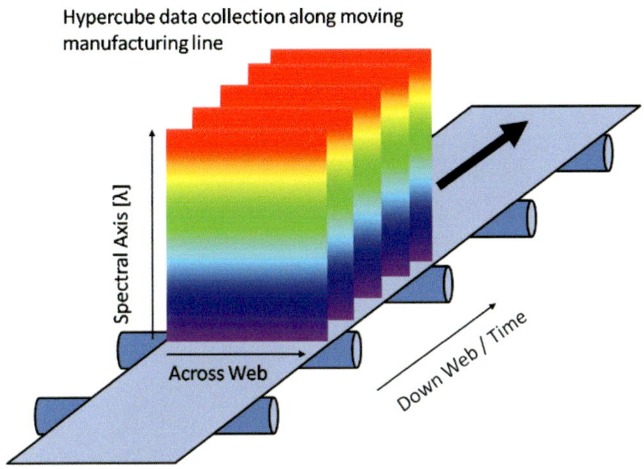

FIGURE 21.5 Push-broom imaging for qualitative and quantitative measurements in real time.

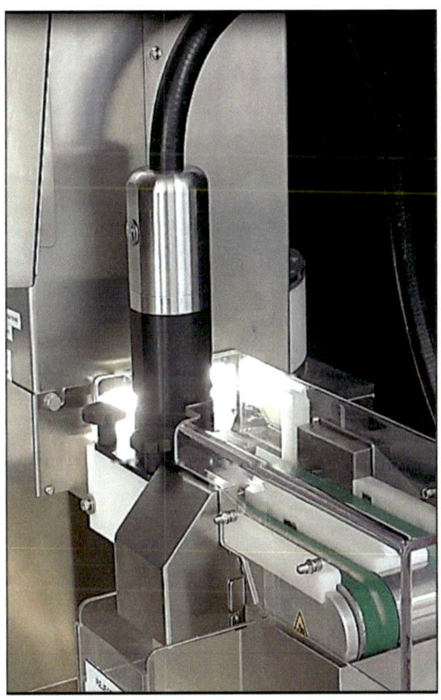

FIGURE 21.6 Push-broom imaging via a fiber optic probe for tablets (100,000$^+$ per hour/per channel).

Many texts have been authored, devoted to all manner of sampling, from railcars of ore to bins of mixed powders to containers of discreet units, such as pharmaceutical tablets or apples or eggs. Unfortunately, for our purposes, they are almost always based on static containers of materials. A process is a moving target.

As such, we need to "sample" a moving target, while scanning with the spectrometer is merely a matter of choosing a location for the light beam to strike the sample and an appropriate place for the detector to acquire the reflected or transmitted light. However, we need to calibrate our method with a referee (compendial) analytical method, so we need to procure a sample that has been measured by the NIR and submit THAT SAME sample for classic analysis.

In many cases, that is not a problem. "Finished" units, such as corn kernels or sealed vials, will not substantially change between sampling and assay. However, if we are examining a reaction, there is always the chance for a large difference between the material that is observed in the process and that which is delivered to the QC lab. Also, since continuous processes may drift (detergent synthesis, polymer manufacture, pharmaceutical tableting/encapsulation), discreet samples require a sampling method that is remote and robust, often by using fiber optics for both the illumination and collections optics (see for example Figures 21.7–21.9).

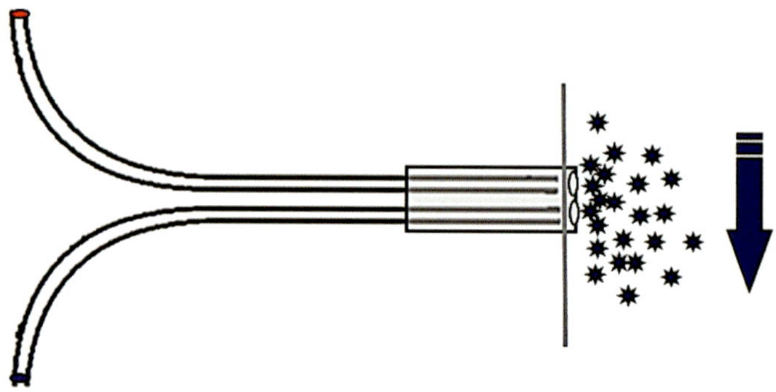

FIGURE 21.7 Diagram of a fiber optic sampling system with illumination and collection fibers and the sample shown as*.

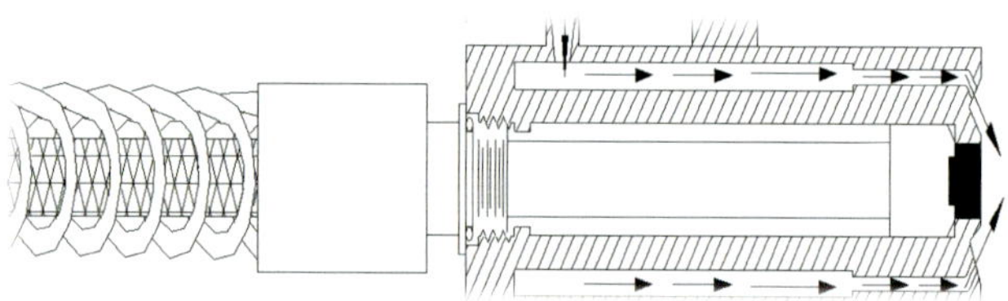

FIGURE 21.8 Cross-section of a fiber optic-based sampling system showing the direction of airflow that allows a sample-viewing area to be free from contamination.

FIGURE 21.9 Configuration of an on-line analysis of cheese curd.

21.7 CALIBRATION

When performing a "normal" near-infrared calibration, we have some advantages over a process model. We are under less time constraints, so we can repeat samples, we can use longer scan times, and we are able to construct a set of calibration samples with a wide set of analyte values (often from 80% to 120% of the expected/nominal value). These values are loosely based on HPLC (high-performance liquid chromatography), the most common reference analysis method (since NIR is not considered a primary method, needing a reference method for calibration).

With a large number of finished units (natural products or manufactured materials) to choose from, it is possible to build a set of calibration spectra with values across the expected range and more. (Since NIR is not as successful when extrapolated, so it is always preferable to calibrate beyond the expected range of values.) While this seems complicated and can take some time, it has proven successful for decades.

However, if a process is being monitored/measured, the logical approach would be to use process samples, not artificial samples, to perform calibrations. While a well-designed process *should* vary within only a few percent, "conventional wisdom" suggests a wide spread of values to give a more accurate value and help assure linearity.

21.8 VALIDATION

There are different levels of validation, depending on the product. If we are determining foodstuffs, tobacco, soft drinks, or candies, the validation is largely mathematical; when we analyze pharmaceutical/medical materials, we need to contend with regulatory agencies (e.g., the US Food and Drug Agency or the European Medicines Agency) as well as established statistical algorithms.

For a lab-based method, samples for validation may be gathered much like for calibration: lots of time and a large number of samples from which to choose. Capturing samples from a process line (see "sampling") and comparing the spectra with compendial methods can be more difficult.

REFERENCES

1. Lillesand, T. M., Kiefer, R. W., and Chipman, J. W., *Remote Sensing and Image Interpretation*, Wiley, New York (2008).
2. Chang, C.-I., *Hyperspectral Data Exploitation: Theory and Applications*, Wiley, New York (2007).
3. Grahn, H. and Geladi, P., *Techniques and Applications of Hyperspectral Image Analysis*, Wiley, New York (2007).
4. Schowengerdt, R. A., *Remote Sensing: Models and Methods for Image Processing*, Elsevier, Amsterdam (2007).
5. van der Meer, F. D. and de Jong, S. M., *Imaging Spectrometry: Basic Principles and Prospective Applications*, Springer, Berlin (2001).

22 Validation

*Gary E. Ritchie**
GER Compliance

CONTENTS

The first principle is that you must not fool yourself—and you are the easiest person to fool.

Richard Feynman, Cargo Cult Science, June 1974

22.1 INTRODUCTION

Suppose that measuring the amount of a substance contained in a clump of matter with invisible energy without physical or chemically affecting it, that the result would be the same if that clump of matter was analyzed chemically by destructive means? The answer to the question is not surprisingly answered with another question; how can the two results be compared in such a way as to make the two results equally valid? (The core question to any and all NIR measurements.)

Richard Feynman discussed in his lecture, *Cargo Cult Science* [1], on the integrity of experimental results and the scientists who generate them, "The first principle is that you must not fool yourself—and you are the easiest person to fool. So, you have to be very careful about that. After you've not fooled yourself, it's easy not to fool other scientists. You just have to be honest in a conventional way after that."

Validation is really nothing more than each scientist carefully experimenting so that they are first convinced of their own results and then others becoming convinced by convention. By building a body of facts from observation and experimentation (mostly serendipitously, by the way), the historic contribution by many scientists reinforces the data leading to knowledge and a record of how the science evolves from hypothesis to a theory and then to a scientific law. Near infrared is a fact

* gary.e.ritchie@gmail.com

of nature as evidenced by observation (Herschel), hypothesized from theory (Maxwell), measured by experimentation (Coblentz), and validated as a principle and a law (Franck-Condon & Hooke).

Validation, from Latin *validus* "strong," from *valere* "be strong," of course, has several different meanings, be it morally, legally, or analytically viewed. In this context, however, validation is being used in its analytical definition for providing irrefutable evidence of the equivalency of NIR spectral and analytical reference results. Just as important, if not more important, is that the criteria used to select the tests for validity are appropriate and valid in themselves.

22.2 FIGURES OF MERIT

Known as figures of merit (FOM), these criteria are numerical quantities based on one or more characteristics of a system, device, or method that represents a measure of efficiency or effectiveness. Dardenne [2] proposed a "standard" to report near-infrared (NIR) calibration results. The tabulated statistics of calibration results are, in fact, the FOM used to gauge the accuracy of a NIR calibration model (e.g., parameter, such as fat, for feed analysis, but virtually any parameter calibrated by NIR may be reported in this manner). The proposed FOMs for a valid NIR method are discussed in more detail in The Modern NIR Experiment section of this chapter and are presented in the summary section at the end of this chapter.

22.3 THE NIR EXPERIMENT

The basic NIR experiment is performed in three steps: NIR calibration, (approved) reference analysis, and NIR validation. In the first step, a number of samples (N) are measured, obtaining NIR spectra. For quantitative results, the samples may possess varying concentrations or other attributes that are being investigated. For qualitative results, the samples may be assessed for the identification or another qualitative physical attribute. In the second step, the (exact) same samples are measured by the reference method chemically and/or physically for the property being correlated to their NIR spectra. In the third step, the reference values obtained or property used for identification or other qualitative attributes are correlated by chemometric methods, most commonly: principal component analysis (PCA), principal component regression (PCR), multiple linear regression (MLR), and partial least squares (PLS). However, there are other statistical approaches used to relate the NIR spectral domain to the analytical reference values [3]. It is at this point that the consequence(s) of the NIR experiment, being a secondary correlation method, shows its primary effect from the results, that being bias or apparent calibration error. For quantitative NIR results, the accuracy of concentration or amount is what requires validation, and for qualitative NIR results, the identity, or other property of interest needs to be validated.

Defoggio [4,5] establishes, rather convincingly, that the sources of error between the calibration set's lab values (the reference value) and the NIR-fitted values are the apparent calibration errors. Both errors are responsible for the difference observed and therefore must be assessed, understood validated, and controlled. Furthermore, he makes the case that the three key attributes of NIR models (user requirements, spectral noise, and artifacts) propagate through a model and shows how lab error affects a model's coefficients:

$$Y = b_0 + b_1 X_1 + b_2 X_2 + \ldots + b_p X_p \quad (22.1)$$

where

b_0 is the regression coefficient for the intercept

b_i values are the regression coefficients (for variables 1 through p),

a models fitness for use (its validity) can be tested through a series of formalized mathematical and statistical FOMs for assessing an unknown as authentic, an outlier, or a residual.

It will be shown that the validation of a NIR spectroscopic results is based on both scientific and statistical tests that provide the necessary evidence and that the results are equal to that obtained by a suitable analytical reference method (when very strict and enforceable experimental controls are applied to minimize the sources of variation leading to error (bias) in the results). Through experience, it is known that these errors arise primarily from poor design of experiment, sample selection, NIR measurement, reference laboratory measurement, analyst variation, NIR instrument variability, software algorithm variability, model selection, and design and other sources to be cited in this paper.

Finally, the chapter will present an overview of regulatory, industrial, and novel approaches to validating a NIR model and results.

22.4 HISTORICAL PERSPECTIVES

NIR validation has its beginnings in 1800 with the observation by Herschel [6] of invisible solar rays "beyond the confines of red-coloured light." While many were convinced of his discovery, some were skeptical. Several critical advances occurred from 1822 to 1861 that validated Herschel's observations and established *a posteriori* the empirical evidence of invisible solar rays [7–11].

In ipsa experimenta inventa repetivit (discovery repeats itself in every experiment). It goes without saying, but on the discourse on the nature of validation in the context of the historical development of NIR (Table 22.1), it is appropriate to be reminded of the fact that it is crucial in the practice of science that for experimental results to be valid, no experiment can violate that which has been proven to be true before it or the whole science is invalid. The crux of the problem for validating experimental data then is to determine all sources of error, eliminate bad practices, minimize phenomenon that may interfere and contaminate observation and faithfully report all results (positive and negative) that are attributable, legible, contemporaneous, original, and accurate (ALCOA).

TABLE 22.1
NIR timeline

1666—Newton disperses white light into colors	1960—Robert Goddu constructs the "Goddu Chart"
1668—Mariotte finds heat disappears when glass inserted in beam	1965—First PDP-8 minicomputer is marketed
1704—Newton publishes Opticks, Spectroscopy Becomes Science	1965—Norris builds first computerized NIR spectrometer
1777—Scheele formulates the concept of radiant heat	1968—Kermit Whetsel published the first review of NIR spectroscopy
1781—Scheele confirms findings of Mariotte	1971—First microprocessor appears, the 4004
1800—Wm Herschel discovers the near infrared	1971—DICKEY-John & Neotec first reflectance NIR instruments
1801—T. Young writes on the interference of light	1971—Data General markets the Nova 2/10 minicomputer
1802—Wallaston observes Fraunhofer lines	1972—Choppin studies water structure with NIRS
1818—Fresnel explains polarization of light	1974—Neotec sells USDA the first Spectro-Computer
1822—Seebeck discovers thermoelectrics	1974—The Model 8080 (8-bit) microcomputer is sold
1825—Nobli expands thermopile to 27 junctions	1974—Technicon enters the NIR marketplace
1829—Niepce & Daguerre invent photographic plate	1976—Bill Gates & Paul Allen found Microsoft
1832—Ampere's theory says light and heat are the same phenomenon	1976—Wold publishes the PLS regression method
1840—J. Herschel confirms spectroscopic absorption	1977—Beckman published his first book on IR spectroscopy
1850—Melloni invents the first spectrometer	1978—The Zero IDRC held in Athens, GA

(*Continued*)

TABLE 22.1 (*Continued*)
NIR timeline

1852—Stokes gives a treatise on light and fluorescence	1978—USDA National Forage Network begins at six locations
1859—Kirchhoff & Bunsen say spectra unique to sample	1978—P. Williams writes about automatic digital analyzer—ADA
1869—Angstrom makes the first reflection grating	1978—NSAS becomes the first commercially available software
1869—Samuel Langley invents the bolometer	1978—Technicon introduces the InfraAlyzer 400
1881—Abney & Festing record spectra of organic liquids	1979—Forage Network takes delivery of six Neotec 6100s
1893—Angstrom formulates the wavelength of light	1980—Trebor introduces whole-grain analyzer
1897—JJ Thompson discovers the electron	1982—First IDRC held at Wilson College
1900—Max Planck discovers the quantum	1982—McClure proposes FT analysis of NIR data @ IDRC
1905—Einstein explains the photoelectric effect	1984—NIR Task Group Established in ASTM
1905—Coblentz makes the first detailed study of near infrared	1985—IUPAC defines NIR region to be 780–2500 nm
1925—JJ Collins is first to study NIR properties of water	1985—Ian Cowe was first to apply PCR to NIR data
1931—Kubelka & Munk publish Theory on Light Scattering	1985—Hirschfeld explained the saline/water calibration
1931—Hotelling publishes mathematics of principal components	1986—Tomas Hirschfeld Passes Away
1945—Hertzberger writes a thesis on quantum mechanics	1986—CNIRS founded by Ed Stark
1947—First European Conference on Molecular Spectroscopy held	1986—Norris receives AACC approval for NIR wheat-hardness test
1949—Coblentz publishes early history of near infrared	1989—Barton et al. granted AOAC approval for the ADF method
1951—Wilbur Kaye builds the first automatic recording spectrometer	1990—NIR news was founded
1954—The first Pittsburgh Conference held	1993—Journal of Near-Infrared founded
1954—Coblentz Society is formed	1995—Patent office authorizes wavelength patents
1958—Norris writes the first NIR paper on moisture in seeds	

Source: The NIR Spectrum, Special Issue, 2006, NIR Celebration History Timeline. W. Fred McClure, Professor, Editor-In Chief, The NIR Spectrum.

Validation then is more than just a mathematical and statistical proof of the current result; it is verification of the results that came before. The theories and laws that describe thermoelectric and electromagnetic radiation are the underpinnings of all NIR experiments that have followed since their discovery and are listed below. Table 22.2 presents several early key investigations that led to an empirical and theoretical understanding and ultimately the recording of the first near-infrared radiation spectrum.

Scientific laws of infrared radiation

I. Laws of thermoelectric circuits [12]
   - I Law of Homogeneous Materials
   - II Law of Intermediate Materials
   - III Law of Successive or Intermediate Temperatures

$$V = \alpha \Delta T \tag{22.2}$$

where
V = voltage,
α = Seebeck coefficient (S)

TABLE 22.2
Early Investigations

1822	1825	1831	1835	1840	1861	1881
Thomas Johann Seebeck—discovered the thermoelectric effect	Leopoldo Nobili—devised the Astatic galvanometer	Nobili and M Melloni, co-designed the thermoscope	Melloni—Refines the thermoscope into the thermopile	John Herschel confirms spectroscopic absorption	John Tyndall—experimented on the heat properties of gases by adapting the existing components of the setup as differential instruments	Abney, R. E., F. R. S., and Festing recorded the first NIR spectrum

$$S_{AB} = S_B - S_A = S_{AB} = S_B - S_A \; \frac{\Delta V_B}{\Delta T} - \frac{\Delta V_A}{\Delta T} \tag{22.3}$$

II. Oersted Law of Electromagnetism [13]

$$\vec{\nabla} \times \vec{\mathbf{B}} = \mu_0 \left(\vec{J} + \varepsilon_0 \frac{\partial \vec{E}}{\partial t} \right) \tag{22.4}$$

III. Maxwell's Electromagnetic equations [14]

I. Gauss's law for electric fields

$$\vec{\nabla} \cdot \vec{\mathbf{E}} = \frac{\rho}{\varepsilon} \tag{22.5}$$

II. Gauss's law for magnetic fields

$$\vec{\nabla} \cdot \vec{\mathbf{B}} = 0 \tag{22.6}$$

III. Faraday's law

$$\vec{\nabla} \times \vec{E} = \frac{\partial \vec{B}}{\partial t} \tag{22.7}$$

IV. Ampere–Maxwell law

$$\vec{\nabla} \times H = \vec{J} + \frac{\partial \vec{D}}{\partial t} \tag{22.8}$$

V. Franck – Condon Principle [15]

$$q_{\upsilon'\upsilon''} \left| \int \Psi_{\upsilon'} \Psi_{\upsilon''} dr \right|^2 \tag{22.9}$$

VI. Hooke's Law [16]

$$\tilde{v} = \frac{1}{2\pi c} \sqrt{\frac{k(m_1 + m_2)}{m_1 m_2}} \tag{22.10}$$

The first NIR spectrum was recorded by William de Wiveleslie Abney and Edward Robert Festing in 1881 [17,18].

While the laws necessary to support the validity of NIR spectroscopy have been thoroughly tested and verified, our focus is the validity of any subsequent measurements made on their basis. As such, in addition to the theorems and their tests accumulated over time after millions of experiments, we have been given empirical steps that when performed, making NIR experiments practical and trustworthy.

The basis of two validation parameters followed today, that is, the repeatability of the NIR measurement and wavelength uncertainty, can be traced to their historical beginnings to the very first experiments performed (John Herschel, 1840 and Abney and Festing, 1881). More importantly, Abney and Festing correctly deduced the origin of the NIR spectra from their studies. The following quotes are taken directly from their original published works.

On the Repeatability of Obtaining the Impress of Rays of Heat—1840

Note I. On the Distribution of the Calorific Rays in the Solar Spectrum

Received March 4, Read March 5, 1840.

"Since the communication to the Royal Society, of the foregoing pages, I have discovered a process by which the calorific rays in the solar spectrum are made to leave their impress on a surface properly prepared for the purpose, so as to form what may be called a thermograph of the spectrum, in which the intensity of the thermic ray of any given refrangibility is indicated by the degree of whiteness produced on a black or very dark ground, by the action of the ray at the point where it is received at the surface. From the nature of the process it is demonstrable, that only rays of heat, and not those which are usually called chemical, can be concerned in the effect produced."

[The following statement of dimensions, a mean of three experiments, will place this important proposition in the clearest evidence. The measures are, as usual, from the fiducial point or centre of the yellow ray]."

In another statement:

"The mean of a great number of measures, as precise as the nature of the observation admits, gives for the situation of the centres or maxima of these spots, or their coordinates along the axis of the spectrum, reckoning as usual from the fiducial yellow..."

and,

"The Fiducial Yellow line is referred to as 0.0, where this is observed to be the least refracted sensible termination."

On the Accuracy of Referring Some of the Principal Spectra to a Comparison with the Solar Spectrum -1881

In the words of Abney and Festing, "The accuracy of our plan of measurement might be doubted, but it is really trustworthy after a little experience. We should not have trusted to it, however, without referring some of the principal spectra to a comparison with the solar spectrum. The wavelengths of the Fraunhofer lines in the infra-red region were taken from the map furnished by one of us with the paper already referred to."

They go on and further elaborate:

"A test of the accuracy of the measurements was the fact that the same wavelengths of the absorption of any particular substance could always be obtained on several photographs. We invariably compared all absorption spectra with that of ethyl iodide, which furnishes almost entirely a line spectrum and, in fact, our measurements were referred to an empiric scale derived from the absorptions of this body, before we were fortunate enough to obtain a wave-length scale from strict comparison with the solar spectrum. We believe that the chloroform spectrum might be almost better to give a fiducial scale. The multiplicity of fine lines in both of these substances makes the mapping the spectra comparatively easy when once the position of those lines has been accurately fixed [19].

On the Origin of the Near-Infrared Spectrum

"A crucial test was to observe spectra containing hydrogen and chlorine, hydrogen and oxygen, and hydrogen and nitrogen.

We therefore tried hydrochloric acid and obtained a spectrum containing some few lines. Water gave lines, together with bands, two lines being coincident with those in the spectrum of hydrochloric acid.

In ammonia, nitric acid, and sulphuric acid we also obtained sharply-marked lines, coincidences in the different spectra being observed, and nearly every line mapped found its analogue in the chloroform spectrum, and usually in that of ethyl iodide. Benzine again gave a spectrum consisting principally of lines, and these were coincident with some lines also to be found in chloroform. It seems then that the hydrogen, which is common to all these different compounds, must be the cause of the linear spectrum.

In what manner the hydrogen annihilates the waves of radiation at these particular points is a question which is at present, at all events, an open one, but that the linear absorptions, common to the hydrocarbons and to those bodies in which hydrogen is in combination with other elements such as oxygen and nitrogen, is due to hydrogen there can be no manner of doubt."

As the development of the NIR technique continued to the end of the nineteenth and on into the beginning of the twentieth century, several scientific pronouncements cement NIR as a viable form of radiation that allows researchers to advance its utility and investigate materials separate from the ultraviolet, visible, and mid-infrared range.

1. Kirchhoff and Bunsen said spectra unique to sample (based first on inorganic element absorption and emission spectra and then observed in molecules)
2. Abney and Festing concluded from very strong experimental evidence that "the foundation of all [near-infrared] absorption line from these bodies arises primarily from the hydrogen atom in combination with carbon, nitrogen and oxygen."
3. NIR measurements are repeatable
4. NIR measurements are accurate where wavelengths are based on a known "fixed" value (standard)

22.5 THE BASIS OF THE MODERN NIR EXPERIMENT

Coblentz conducted an extensive study on the phenomenon of absorption and emission of select pure organic compounds culminating in his report in 1905. A supplement to that first report was subsequently published [19]. In this report, he provides insight into the degree of accuracy and the sources of error associated with obtaining the spectra. Table 22.3 lists both the observation on accuracy and errors as discussed by Coblentz.

Other considerations cited by Coblentz on the accuracy in obtaining the absorption and emission spectra of select organic compounds concerning the purity of the compounds are noted. He states, "Obviously it would have been impracticable to purify all the chemicals used. The alternative was to duplicate the work by examining the same compound manufactured by different firms." He goes on to further state, "due to the difference in the thickness of the films, which is of no consequence, rarely is there a difference in the spectra of the two samples. This means that the two samples of a compound were equally pure, or what amounts to the same thing, that they contained the same impurity if any was present." Another concern that was noted was the presence of water in minute amounts in the "pure" samples, which gives rise to its bands in the substances under investigation.

The significance of this work as it relates to validation falls into five categories as sources of error and their impact to variation of results:

1. Errors due to instrument (mechanical and electronic)
2. Errors due to computation
3. Errors due to sample positioning and measurement
4. Errors due to single determination from one "representative" sample and measurement
5. Errors due to the presence of moisture

TABLE 22.3
Coblentz Observations

Degree of Accuracy	Sources of Error
Agreement with that of previous investigations on absorption spectra of (CO_2)	Difference of emission lines due to the curvature of the slit
Estimation of the degree of accuracy of this band (CO_2) is based on the number of observers who have located it	
The accuracy of location is the 3.43 μ absorption band found in compounds containing CH_3 groups accurate to about 0.02 μ	Adjustment of the prism (mounting and remounting) (Allowing) for errors in computing, drawing, and reading wavelengths from the calibration curve, in adjusting the apparatus, in drawing the curves, etc.
In the present work, the adjustment was tested, by means of the sodium lines, at the beginning and conclusion of each series of measurements	The difficulty in keeping the apparatus in adjustment for a long period
Likely to occur with a gas flame but did not occur with the "heater" used on a storage-battery circuit	A source of error in determining the transmission The rigidity of the mounting of the cell in its carrier, which moved to a fixed height in a plane parallel to that of the spectrometer slit, in accurately fitting wooden ways upon the spectrometer arm, eliminated errors due to variation in the thickness of cell
On the question of the use of a prism kept at the minimum and at the nonminimum deviation	Any variation from the real minimum in the setting for the sodium lines will be magnified far out in the infrared when using the nonminimum deviation method
The difficulty in establishing the absolute value of any absorption band	Coblentz notes that on a marked band of carbon tetrachloride, found by Paschen at 6.45 μ would be an excellent check on the present work, it was examined and found to consist of two bands, of which this is the mean value

At this point, one must be reminded that the validation of NIR results (nondestructive) relative to those produced from chemical analysis (destructive) in the modern sense is not yet being considered. These early observations on near- and far-infrared made by Coblentz and his predecessors were generated using manually operated instruments by, in his words, "strings and shutters." Despite this fact, the case being made for accurate results was continually being improved upon. The observations made, not just by Coblentz, but other scientists as well, are quite relevant and real in as much as the ability to systematically and methodically obtain accurate and precise results, over time and on different instruments, were being considered in these early experiments. This last point is in fact, on the issue of calibration transfer, is still the single most important problem to be solved. More about this is in the section on calibration transfer.

22.6 AUTOMATED NIR SPECTROPHOTOMETERS

While the development of spectroscopy has been undergoing explosive growth since 1905, other developments in physics, technology, statistics, and material science were a few of the important areas of progress taking place making modern near-infrared spectroscopy into what it is today. In a summary of the history of spectroscopy up to 1927, Randall noted, "the infra-red region is not simply an extension of the spectrum at the long wave-length end. The data which it furnishes are not simply more data of the kind to be obtained in the other parts of the spectrum. The band spectra of the near infra-red region are different in type from those of the far infra-red and both are different in character from the bands in the photographic regions. The information to be obtained from these infra-red spectra is not to be found elsewhere" [20].

While these developments certainly accelerated the adoption and use of NIRS in many industries inside and outside of agriculture, it was not without problems. But, in fact, it was the problems and their subsequent solutions that became the pathway forward to the deployment and implementation of the modern diffuse reflection and transmission devices in use today. The following table lists the observations, the associated error(s), and solution(s) for each of the problems that have led to modern-day validation parameters for NIR spectroscopic measurements (Table 22.4).

TABLE 22.4
Modern Validation Parameters for NIR Spectroscopic Measurements

Number	Observation	Error	Solution	Source
1	Diffuse reflection rather than transmission to measure composition	Does not vary linearly with absorber concentration	Use of multiple regression techniques for calibration. Use sufficiently large population of samples with chemical analyses.	[21]
2	Factors affecting the accuracy of near-infrared reflectance spectroscopy (NIRS) are mean particle size, particle size distribution, temperature, and moisture	Sample-induced errors in NIR analysis can be caused by suboptimum wavelength selection and by increases in the variance of the NIR signal received as a result of factors associated with the sample surface	The incidence of such errors could be minimized by improving the methods of utilizing the energy signal received by the instrument. Normalized second-derivative treatment of data gave the best performance in predicting the protein content of samples varying widely in particle size.	[21]
3	Collected spectra as a function of time	Found an unknown absorber at 750 nm, which increased with age Results were repeatable on other specimens	Measuring the change in scattering properties of the shell as the moisture content of the shell stabilizes during the first hours after laying.	[22]
4	Organizing a set of samples and make the measurements without making errors	Most outlier data points can be traced to human error	Sampling of spectra should be performed by the same person entering the chemical data and operating the computer programs to do the analyses to minimize or eliminate human error.	[22]
5	Adequacy of signal to noise to make measurements to explore NIR diffuse reflectance to measure the composition of grains and oilseeds	Integrating sphere and all of the optical elements relating to the double-beam principle	Discarded the integrating sphere and all of the optical elements relating to the double-beam principle and designed a very simple single-beam reflectance unit.	[22]
6	Sampling problem using a cup for grain	Variation in spectra intensity due to ordinate shift	Use sealed cups with spring-loaded glass cover, which is filled to constant amount (later adoptions use a spinning cup to average out sampling variance).	[25]
7	The calibration problem of needing a large *N*, wide composition range, sample variance in environmental growing conditions (extremes in dry and wet conditions)	Effect on SEP, SEC, and bias (note on ground material sample size: smaller sample size is the variation resulting from systematic errors during weighing and transfer of sample. Small sample size increases subsampling error)	Recommended *N* is dependent on many factors requiring investigation into concentration linearity range and moisture content of samples.	[25]

(Continued)

TABLE 22.4 (*Continued*)
Modern Validation Parameters for NIR Spectroscopic Measurements

Number	Observation	Error	Solution	Source
8	Background scans in between sample could cause errors	Variation in spectra intensity due to ordinate shift	Build-in the optical standard for determining between sample and between instrument startup and shutdown to determine instrument stability.	[25]
9	Variation in environmental conditions during sample scanning	Variation in spectra intensity due to ordinate shift Appearance of bands due to excess moisture	Environmental controls should be monitored to minimize temperature, humidity, and particulate fluctuations. Care should be taken to ensure sample moisture is representative of population where moisture isn't a constituent of interest.	[25]
10	Variation in sample particle size	Variation in spectra intensity due to ordinate shift	Uniform 0.5 –1-mm particle size	[25]

In 1952, Norris was working with a Beckman DU spectrophotometer, measuring the transmittance properties of the shell, yolk, and white of the egg in the visible region (with overlap into the short-wave NIR (SWNIR) region 550–950 nm) [22]. The spectra obtained were the first modern-day automated recorded NIR spectra although the work went unreported. The key observation about this and other work provides significant insight into how experimentation on NIR in the future would proceed (Table 22.4).

There are several departures from Coblentz and Norris's investigations. First, the biggest difference between Coblentz and work performed during and after Norris was that the use of NIR went from studies on pure compounds to studies (in the modern era) by NIR for the composition of heterogeneous materials that were opaque. Eggs, fruits, vegetables, and seeds were among the first samples studied by NIR for their quality attributes. The second difference was that as work progressed with NIR on a Perkin–Elmer UV/VIS/NIR scanning spectrophotometer and then a modified Cary model 14 spectrometer, the mode of sampling changed from transmission to diffuse reflection due to the loss of sample preparation with the use of carbon tetrachloride (deemed a health hazard). Third, infrared radiation from pure compounds and their absorption and emission lines compared to the solar spectrum go from being photographed to being obtained point by point and, later, by automatic scanning.

By the time that Norris and others made use of infrared and near-infrared rays to study the composition of matter, the development of computer technology together with advances made in the statistical theory of chemometrics was available. Advances by Wold and Kowalski are credited with having the largest impacts on the advancement of NIRS in the modern era [23,24].

22.7 COMMERCIALIZATION OF NEAR-INFRARED INSTRUMENTATION

Following the studies by Norris and others at the United States Department of Agriculture (USDA), NIRS use gained favor for the analysis of forage, soil, and animal physiology. Soon after, its use expanded into other markets such as energy, food, textiles, beverage, pharmaceuticals, and more. Several publications appeared that codified the NIR experiment, most notably the USDA Handbook No. 64324 by Marten, Shenk, and Barton [25]. In this work, three steps for the validation of the NIR experiment are presented. In the history of NIR, this is a pivotal moment because key elements of the NIR experiment for obtaining meaningful results are given that allow the experimenter to separate

TABLE 22.5
Marten, Shenk, Barton, & Westerhaus Approach to the NIR Experiment Validation

1. Identification of two different types of outliers in the raw data selected for use in the sample calibration set.
   a. Laboratory values did not represent the samples at the time the scan was taken.
   b. Spectra are different from the spectra of the calibration samples.
2. Use the validation statistics to evaluate the calibration performance of the model.
   a. The standard error of prediction (SEP) is a true indication of the performance of the equation on unknowns from the same population (of samples used to build the model from the calibration set).
   b. Select the equation (model) with the smallest SEP. Unlike SEC, which must decrease with each additional term, SEP only decreases until overfitting becomes important and forces it to increase.
   c. The slope of the regression line relating the determinations to the laboratory values. Slopes should be near 1.0. If markedly different from 1.0, they indicate the high and low values will be consistently under or overestimated.
3. Validate the equation with unknown samples during routine analyses.
   a. Note a shift in the sample population from that used in the calibration. This shift can be detected easily as a spectral outlier, and it is different from the spectra of the calibration samples in its intensity, band positions along the abscissa, slope, combinations of any two or all three features upon visual inspection. The evaluation of the unknown spectra visually as well as the use of appropriate outlier statistics is always recommended.
   b. Differences observed in the unknown spectra and changes in the instrument have a large effect on the calibration set that may require recalibration. This is detected in bias between the average difference between the reference method and NIRS analytical values.

acceptable data from unacceptable data to produce valid results. Table 22.5 presents the Marten, Shenk, Barton, and Westerhaus approach to NIR experiment validation, extracted from the handbook.

The commercialization of NIR instruments, exclusively designed for the detection and analysis of diffuse reflection NIR spectral radiation, for identification and compositional analysis continued with Neotec Internal Quality Analyzer (1969) with limited success. This was followed by a device for high-speed apple sorting in 1972:

1. Neotec's Spectro Computer (1970),
2. Neotec's Grain Quality Analyzer (1971),
3. Dickey-John's Grain Analysis Computer (1971),
4. The Automatic Digital Analyzer installed at the Canadian Grain Commission (1974),
5. Neotec's Ground Meat Analyzer (1974),
6. Technicon's InfraAlyzer-400 (1976),
7. Dickey-John's GAC-III (1977), and
8. Neotec's GQA-31 (1973).

The first microprocessor-controlled NIR instrument was the Neotec GQA-41 (1974). The Neotec Feed Quality Analyzer, FQA-51 (1977), the Neotec Model-6300 High Energy Research Instrument (1978), and the Technicon InfraAlyzer-500 High Energy Research Instrument (1980) [26,27] soon followed.

At the time of this writing, the NIR industry has grown to many manufacturers, but the market is dominated by 13 key players providing everything from the transmission and diffuse reflection, dispersive, and Fourier transform (FT) NIR bench-top and process instruments to micro handheld devices.

The big suppliers are:

1. Thermo Fisher,
2. Bruker,
3. Buchi Labortechnik,
4. Agilent Technologies,
5. Foss A/S, Shimadzu,

6. PerkinElmer,
7. Sartorius,
8. Jasco,
9. Yokogawa Electric,
10. ABB,
11. Kett Electric,
12. ASD (PANalytical),
13. Unity Scientific.

Many other smaller, but nonetheless important, manufacturers of NIR devices occupy this market as well.

22.8 THE MODERN NIR EXPERIMENT

It is important to point out that modern NIR spectroscopy developed mainly from two schools simultaneously. One was from classical spectroscopy, which provided a firm theoretical basis for explaining and understanding the origin of the spectra and how to interpret and use them. The other was NIR instrument development. The development of the modern automated device spawned the adoption of NIR analysis for rapid, nondestructive, and noninvasive chemical and physical analysis in many agricultural and industrial applications. A key and early contributor to both the theory and the design of the modern NIR spectrometer was Wilbur Kay. He presented a historical review of NIR up to that time at the Conference on Analytical Chemistry and Applied Spectroscopy in Pittsburgh (PittCon), in Pennsylvania, March 3, 1954 [28]. He made several key observations that lay the foundation for the validation of the NIR experiment (Table 22.6).

Showing the development of NIR spectroscopy from first principals and tying the concept of validation to them secures the utility of the NIR experiment to the investigation of the physical and chemical properties of matter. The mechanics of the measurement devices used in those experiments, the

TABLE 22.6

Kay's Remarks on the NIR Experiment with Additional Comments

1. Spectral data arise primarily from hydrogenic (CH, NH, OH) stretching vibrations
2. Instruments should be assessed for their signal-to-noise ratio and wavelength/wavenumber resolution
3. Review the chemistry of the analyte of interest, selection rules, and anharmonicity corrections that are readily applicable to the near-infrared region
   a. Vibrations predominantly localized in two molecular groups within a given molecule do not produce intense combination bands unless they share a common atom or are connected through multiple bonds
4. Determine the limit of detectability
5. Determine precision in the range from 0.004% to 0.5%, which is possible when OH, NH, or CH groups can characterize the unknown
6. Record the temperature of NIR experiments. Near-infrared spectroscopy concerns intermolecular associations, the most familiar type being termed "hydrogen bonding." The result of this intermolecular interaction is association, and molecular groups strongly entering into the interaction are likely to exhibit marked perturbations. Such bonds will be temperature sensitive.
7. Application of near-infrared spectroscopy to any specific mixture requires the knowledge of the spectra of the components of the mixture. However, the knowledge of characteristic bands may facilitate the prediction of applications.
8. Two main questions concerning an [NIR] analysis are as follows:
   a. Can a band characteristic of the molecule in question be resolved?
   b. Is there sufficient difference in absorptivities between this characteristic band and the sum of absorptivities of all other components at this wavelength?

procedures followed in making the measurements, and the mathematical and statistical analysis applied to evaluate the results are all critical.

22.9 CROSS-VALIDATION

The development of statistical tools to build chemometric models from NIR spectra and laboratory reference values and evaluation of their results has advanced on several fronts:

Brereton *et al.* present an overview of the history of chemometrics [29]. With respect to the validation of chemometric methods, they make a critical and important observation on the use of an important statistical tool commonly employed in developing calibration models, which being cross-validation.

Cross-validation is traced first to Larson [30] to solve the well-known problem of the deviation of the coefficient of multiple correlation (R^2) obtained from predicted data as compared to the calibration equation derived from an independent set of data, where its R^2 value is inflated due to chance factors arising from the calibration. It is observed that the (predicted R^2) < (calibration R^2) by a significant amount. Larson investigated the source of this shrinkage and then showed that an optimal predicted R^2 can be obtained using a theoretical equation and then assessed the empirically obtained result against it.

Brereton *et al.* warn about the use of and relying solely on cross-validating multivariate (MVA) regression models for the problem stated above without additional rigor of the theory of sampling (TOS), developed and formalized by Gy [31], applied to sample selection for calibration and validation. Validation of the calibration model in statistical terms can arise from two ways: internal and external, where the former refers to the use of calibration samples for cross-validating the suitability of the prediction by leaving one or more blocks of samples out the calibration (statistical sampling or resampling) and the latter uses samples never used in the calibration.

They cite the Esbensen and Geladi [32] work on cross-validation within the framework of TOS and their conclusion that "the extensive and generalized use of cross-validation methods based on segmented procedures does not find a theoretical justification." We have arrived at the only solution left to the chemometrician who wants to test the validity of their calibration model, that is to use external samples and do not rely on cross-validation at all. In the words of Faber and Rajkó, "Test is Best." [33]. In their paper, "How to avoid over-fitting in multivariate calibration—The conventional validation approach and an alternative," they propose a randomization test that enables one to assess the statistical significance of each component that enters the model. The advantages of this approach used without and with cross-validation are discussed in the paper.

We come now to the essence of validation of a NIR MVA regression model, the figures of merit (FOM). In the following sections, we present and discuss the elements of validation as it applies to NIR spectroscopy in several fields of practice. Note that the emphasis on the scientific basis and instrumental aspect of NIR measurements begins to take a back seat to a focus on the mathematical and statistical nature of the multivariate NIR method. The evolution of scanning instruments as well as the computer technology (both hardware and software) jumpstarts research on chemometric applications in NIR spectroscopy.

22.10 REGULATORY VALIDATION SCHEMES

NIR validation protocols have existed in various forms from the very first NIR experiment up through its evolution as a modern analytical tool. In the modern era, we have seen that the use of NIR for the analysis of agricultural purposes, first codified in the USDA Agriculture Handbook No. 643. In the words of the authors, "This handbook is intended to provide an overview of the history of NIRS use for forage quality analysis, a summary of available equipment and computer software, recommended conventional and NIRS procedures for analysis of forage quality, an explanation of how and why NIRS functions, recommended NIRS instrument calibration (including equation development, validation, monitoring and transfer), progress in NIRS technology transfer,

TABLE 22.7
Validation Considerations from USDA Agriculture Handbook No. 643

Number	Handbook No. 643 Statement	Validation Element
1	NIRS use for forage quality analysis	Problem statement for NIRS use
2	A summary of available equipment and computer software	Provide the name, model, manufacturer, and serial no. of the instrument(s) and any associated attachments including equipment manual(s) and version no. Provide the name, manufacturer, and version of the software and any manual(s) and version no(s).
3	Recommended conventional and NIRS procedures for analysis of forage quality	Describe the NIRS procedure used: **Mode**—transmission, transflection, diffuse reflection, others as appropriate **Device type**—bench top, process analyzer, portable, handheld, others as appropriate **Format**—dispersive, Fourier transform (FT), charge-coupled device (CCD), others as appropriate **Population/sample description and properties** **Population/sample container and handling** N_{Cal}—Number of calibration samples N_{Test}—number of test samples N_{Val}—number of validation samples
4	An explanation of how and why NIRS functions	Statement of NIRS as a choice of analysis and sample suitability
5	Recommended NIRS instrument calibration (including equation development, validation, monitoring, and transfer)	Statement on equation development approach, validation, monitoring, and equation transfer
6	Progress in NIRS technology transfer	Statement of NIRS technology transfer
7	Consideration of future applications and certification procedures for NIRS technology	Statement of lifecycle management of NIRS technology as a solution to the problem statement

and consideration of future applications and certification procedures for NIRS technology." This carefully thought out and purposefully parsed statement provides instructions on what should be considered and reported when a NIR experiment for forage quality analysis is performed but may serve as a starting point for the basis of validating any NIRS experiment using the basic elements of method protocol development, method validation, and reporting the results.

In the handbook, Barton provides reasons why a NIR result (following a calibration) does not produce the same result on the analytical samples as they statistically should for similar samples in the calibration set and the factors that should be considered during validation [25]:

1. the NIR method is an empirical method and as such will never yield a "right" answer
2. the NIR spectrum and chemical analysis were not run on the same sample
3. the resultant value, predicted based on the calibration samples, may not have the same relationship to a nutritive entity (or any other analysis, the analyte of interest) as the calibration samples or the ones you are analyzing with NIRS.

He further elaborates the ways to mitigate the factors that confound the validation of NIR empirical results. Factors to be investigated should include analyst training and competency, the validity of the laboratory reference method, certification or standardization of the reference and NIR method, screening methods to ensure the sample is representative of the population from which the calibration samples were selected, analytical information about other constituents of the sample, the use of a known set of samples as alternates to determine fitting to the calibration equation, and the use of control or reference samples with known results to check that the correct equation has been used.

Since Norris' original publication on the determination of moisture in seeds, along with subsequent work done at the United States Department of Agriculture (USDA) and many other researchers around the world from many fields of applications, NIR methods of analysis have been firmly established as a valid means for determining the chemical and physical composition of materials [34]. Principles of validation have emerged that provide unequivocal evidence of NIR results being similar and equal to the laboratory reference method. For example, in the determination of the moisture content of seeds by NIRS of their methanol extracts [35], Hart *et al.* establish the use of standard deviation from the regression of the spectrophotometric determinations as a FOM for comparing the spectroscopic result to the standard Karl Fischer determination.

Further, the authors establish that results should be performed in duplicate and that they should differ by no more than 0.2%. The authors state that the use of the standard deviation from the regression of the NIR result and a limit for duplicate determinations establishes the accuracy of the spectroscopic method. On this basis, the NIR results are deemed equally valid to the Karl Fisher result. But further evaluation of the authors' report also provides even more critical steps that were carried out to ensure that the results are equivalent and valid. The authors maintain that in order that the NIR results are accepted, the NIR method must:

1. contain no compounds whose absorbances interfere (with water),
2. have no appreciable small variations in temperature (that may have an) effect on results, and
3. be completely free of suspended material.

A valid result, based on the author's requirements, also implies that a good experimental design be thought through and executed such that statistical criteria can be shown to be directly related to empirical evidence. This NIR experiment was based on transmission mode, and the authors took great care in advising the reader by providing explicit instructions to follow if equivalent and accurate results were to be obtained by the NIR method:

1. the analyte must be a known and documented near-infrared absorber or otherwise fully characterized as such
2. the method must be shown to be specific for the analyte (experiments were carried to determine whether substances were interfering with the determinations for water. Since the NIR results obtained by extraction experiments were in agreement with the corresponding results obtained by the Karl Fischer method, no interferences were found)
3. a suitable range should be established in order to determine the linearity of response
4. the critical experimental step (procedure) should be explored for retrieving the analyte
5. the measurement step (method) should be described clearly, concisely, and completely so that it can be performed repeatedly with the minimum variability
6. the instrument requirements are provided and instructions given for handling the sample during measurement (temperature effects NIR measurement, if reference and sample extract are not at the same temperature, serious error may result)
7. specific requirements as to the reagents used and the stipulation that reference and sample preparation solvent come from the same lot
8. a method for obtaining the spectral result is described (the clarity of the solutions for NIR transmission is stipulated as the variability of the result is impacted by suspended particles)

9. the units for the results are provided and expressed
10. a statement of proportionality to Beer's law is made since the data obtained from the NIR regression curve correlated to the Karl Fischer values resulted in a straight line

A statement of validity can be deduced from the reported study based on the study design and statistics evaluated. The method for the determination of the moisture content of seeds by near-infrared spectrophotometry of their methanol extracts can be shown to be accurate, precise, linear, specific, selective, sensitive, and robust.

22.11 INDUSTRIAL BEST PRACTICES

Since the publication of the USDA Handbook No. 64324, several more sources of standard practices have appeared around the world codifying the requirements for validating NIR measurements:

1. Starting with the publication of the Association of American Cereal Chemists (AACC) *Near-Infrared Technology in the Agricultural and Food Industries* by Williams and Norris in 1987 [36],
2. Standard NIR methods of analysis appeared over the next 27 years. The Association of Official Analytical Chemists by Barton and Windham in 1988 [37],
3. *Standard Practice for NIR Multivariate Quantitative Analysis* by the American Society for Testing and Materials (ASTM) by Workman and Brown in 1998 [38,39],
4. Near-Infrared Spectrometry in *European Pharmacopoeia* in 1997 [40],
5. Guidelines for the Development and Validation of Near-Infrared Spectroscopic Methods," Pharmaceutical Analytical Sciences Group in 2001 [41], and
6. USP *Near-infrared spectrophotometry, Chapter 1119* in 2003 [42].

In January 2014, the European Medicinal Agency (EMA) finalized their guidelines on the use of near-infrared spectroscopy by the pharmaceutical industry and the data requirements for new submissions and variations [43]. In March 2015, the United States Food and Drug Administration (USFDA) published a *Draft Guidance for Industry Development and Submission of Near-Infrared Analytical Procedures* [44].

Near-infrared spectroscopy has evolved to become a technique used in multiple industries on a global scale. So, it is not surprising that the validation requirements have evolved from various perspectives from commerce of agriculture, food, beverages, and petrochemical industries to provide a scientific basis of material quality and value and to medicine to provide noninvasive measurements on human and animal healthy and diseased states. As pointed out previously, the use of NIR, driven from both a theoretical construct to study the properties of matter and the engineered instruments to measure both spectral and hyperspectral images of materials using absorbance, transmission, and reflection modes, expands the work needed to validate the results generated from these various devices. Additional parameters are needed to validate process measurements obtained in real time, which is a departure from traditional laboratory measurements made using smaller sample aliquots. The use of fiber optic probes has allowed measurements to be made in environments beyond the controlled laboratory conditions of standard temperature and pressure (STP) and humidity (laboratory temperature, barometric pressure, and relative humidity) defined by the United States National Institute of Science and Technology (NIST) for a temperature of 20.0°C (293.15 K, 68.0°F) and an absolute pressure of 1 atmosphere (14.696 psi, 101.325 kPa) and relative humidity prescribed at 45%RH±5%RH.

For the purposes of this chapter, validation parameters and recommended FOMs will focus on laboratory diffuse reflection spectral measurements under standard conditions, and references will be provided for measurements made in all other modes, devices, processes, and conditions where appropriate.

Table 22.8 highlights key contributions from each standard introduced and its impact on the validation concept to the NIR analytical method.

TABLE 22.8
Key Contributions from NIR Standards

Validation Parameter	Impact on NIR Analytical Method	NIR Standard Reference
The ratio of performance to deviation (RPD) is the ratio of the standard error in prediction to the standard deviation of the samples for assessing the usefulness or goodness of fit of the calibration model.	RPD was initially used by Williams (1987) for assessing the goodness of fit for NIR calibration (in agricultural and food products). Batten (1998) wrote "Williams (pers. comm.) suggested that RPD values greater than 3 are useful for screening, values greater than 5 can be used for quality control, and values greater than 8 for any application."	[45]
• *Statistics for evaluation of the method for applicability to the determination of acid-detergent fiber and crude protein in any forage or feed sample included: • t-values • H-values • Standard error of calibration (SEC) • Coefficient of determination (R^2) • F-statistic of regression coefficients • Standard error of analysis (SEA). Note: is now referred to the standard error of prediction (SEP) • Bias	• A statement that "Accuracy of any chemometric procedure is limited only by validity of laboratory method used to measure desired quality parameter for all samples and selection of appropriate learning set." • A statement that "Equations selected from calibration statistics, which have been validated, are used to calc. acid-detergent fiber and crude protein content of feed and forage samples." • A statement that "Consistency in sample handling and preparation is crucial to successful use of NIRS technique." • A statement that "In practical terms, min. of 50 samples should be considered." • A statement that "Best equation for routine NIRS analysis is based on both superior calibration and validation statistics."	[37]
• Statistics for multivariate methods of (n k I) for mean centered models, and h is the leverage statistic • Bias • Precision of the model determined using $t \geq 3$ test samples and $r \geq 6$ replicate measurements per sample • Automatic preprocessing or postprocessing	• ASTM standard practice codified the modern NIR method along multivariate calibration and validation steps. • Model assessment for adherence to multivariate calibration and validation mathematical methodology	[38,39]
• Library data evaluation validated by correlation coefficient (R^2), the sum of squared residuals $SSR = \sum_{i=1}^{n} \left(Y_i - \hat{Y}_i\right)^2$, or the distance using cluster analysis (i.e., K-means cluster) must comply with the acceptance limits defined in the validation procedure. • Identification and characterization (qualitative analysis) validation parameters for specificity, robustness • Quantitative analysis based on multivariate methods of multiple linear regression (MLR). Principal component regression (PCR) or partial least squares (PLS-1)	• A statement that. "However, direct comparison of the spectrum obtained with the substance being examined Multiple Linear Regression (MLR). Principal Component Regression (PCR) or Partial Least Squares (PLS-1) • **Leverage statistic, h • Spectral Residual Outlier Statistic • Validation set $N > 20$ (Independent of calibration set) • Validation set must not include high leverage or spectral residuals • Validation samples $>4k$ if model not mean centered, or	[40]

(*Continued*)

TABLE 22.8 (*Continued*)
Key Contributions from NIR Standards

Validation Parameter	Impact on NIR Analytical Method	NIR Standard Reference
• Specificity—see requirements for qualitative analysis • Linearity—correlation of NIR results calculated from NIR responses within the used algorithms to reference method results distributed throughout the defined range of the calibration mode • Range—covered by the range of analyte reference values and the limit of quantitation • Accuracy—determined by assessing the standard error of prediction (SEP) $\text{SEP} = \sqrt{\frac{\sum_{i=1}^{n}\left(y_{V,i} - Y_{V,i}\right)^2}{n}}$ • Precision—for $n = 6$, measurements performed according to the developed analytical method. Precision may be considered at two levels, repeatability (replicate measurements of the same sample, with or without variation in sample positioning) and intermediate precision (replicate measurements by different analysts, different days of measurements). • Robustness—assessment of effects of variations of temperature, humidity, sample handling, and the influence of instrument changes. • Outliers—consideration is given to spectral and reference laboratory outliers relative to specification and retesting results. • Ongoing model evaluation—continuous monitoring of validation parameters, revalidation as new materials or samples are added to model or occurrences of process changes, compositional changes. • Transfer of databases—procedures and criteria must be applied to demonstrate that the model remains valid with the new database or new instrument. • Data storage—store the electronic NIR spectra, libraries, and data according to the current regulations.	• Validation > $4(k + 1)$ if model mean centered • Validation samples span 95% of the range of the calibration samples • For standard error of calibration (SEC), does 95 % of the results for the validation samples fall within $\pm t\cdot \text{SEC}\cdot\sqrt{1+h}$ of the reference values where t is the Studentized t value for $n-k$ degrees of freedom with a reference spectrum of a chemical reference substance, as used in infrared absorption spectrophotometry, is not appropriate. Suitable validated mathematical treatment of the data is required." • A statement that, "Measurements in the NIR region are influenced by many chemical and physical factors as described below; reproducibility and relevance of results depend on control of these factors and measurements are usually valid only for a defined calibration model." • A statement that, "Validation parameters Analytical performance characteristics to be considered for demonstrating the validation of NIR methods are similar to those required for any analytical procedure. Specific acceptance criteria for each validation parameter must be consistent with the intended use of the method." • A statement that, "Controls must be in place to ensure that results outside the validated range are not accepted."	
• Specificity—as required by ICH for identification/qualification/quantification (major components and impurities/minor components) • Linearity—as required by ICH for quantification (major components and impurities/minor components only)	• Guidelines for the Development and Validation of Near-infrared Spectroscopic Methods in the Pharmaceutical Industry, PASG NIR Subgroup, codifies NIR method by aligning with ICH validation parameters where applicable and defining NIR specific calibration and validation parameters	[41]

(*Continued*)

TABLE 22.8 (*Continued*)
Key Contributions from NIR Standards

Validation Parameter	Impact on NIR Analytical Method	NIR Standard Reference

• Range—as required by ICH for quantification (major components and impurities/minor components only)
• Accuracy—as required by ICH for quantification (major components and impurities/minor components only)
• Precision
  • Repeatability—as required by ICH for qualification/quantification (major components and impurities/minor components only)
  • Intermediate precision—as required by ICH for quantification (major components and impurities/minor components only
• Robustness—as required by ICH for identification/ qualification/quantification (major components and impurities/minor components
• Detection limit—not applicable
• Quantitation limit—as required by ICH for quantification (major components and impurities/ minor components
	• A statement that, "ICH guidelines Q2A and Q2B address traditional method validation requirements but do not address the unique and specific requirements for NIR method validation."
	• A statement that, "These PASG guidelines cover the particular NIR requirements whilst remaining complementary to ICH Q2A and Q2B."
	• A statement that, "Furthermore, revalidation may be necessary in the following circumstances:
	  • changes in the synthesis of the drug substance;
	  • changes in the composition of the finished product;
	  • changes in the finished product manufacturing process or sources/grades of ingredients;
	  • changes in the analytical procedure or the NIR instrumentation.
	• A statement that, "where test materials will be analyzed that are outside the model represented by the library, resulting in a "NIR non-conformance" against the NIR model. In these circumstances it is essential that the material is authenticated using appropriate alternative tests, prior to its acceptance and incorporation in the library	

• Method Validation parameters were not published in the first USP <1119> but appeared later following industry and regulatory comment on a proposed revision to the chapter in *PF 28(6)* In-Process Revision: <1119> NEAR-INFRARED PECTROPHOTOMETRY
• The proposed revision includes a section on Method Validation
• Method Validation parameters proposed are:
  • Specificity
  • Linearity
Note: The correlation coefficient, R^2, is not a true measure of linearity, but is rather a measure of the fraction of variation in the data that is adequately modeled by the equation. It is dependent on the standard error of the calibration equation (and hence the reference method) and on the range of the calibration data. Statistical tests such as Durbin–Watson are available for the determination of linearity. Other applicable statistics may be used as appropriate.
	• A statement that, "Validation of an analytical method requires establishment of the suitability through testing of each of these specifications. [Author Note: At the time of the first publication of USP <1119>, the chapter presented validation parameters for instrument suitability (referred to as specifications here] and not method validation]
	• A statement that, "validation is generally achieved through the assessment of specialized chemometric parameters, these parameters can still be related to the fundamental validation characteristics required for any analytical method."
		[42]

(*Continued*)

TABLE 22.8 (*Continued*)
Key Contributions from NIR Standards

Validation Parameter	Impact on NIR Analytical Method	NIR Standard Reference

• Range
• Accuracy
• Precision
  • Repeatability
  • Intermediate Precision
• Robustness
• A list of potential variables (not all inclusive) that may affect the spectral response should be considered:
  • the environment in which measurement takes place;
  • optical quality of the glassware;
  • cleanliness of the sample interface;
  • optical path length;
  • sample optical properties, temperature, thickness;
  • sample flow;
  • polymorphism;
  • particle size;
  • residual moisture and solvents;
  • homogeneity;
  • the age of the samples;
  • time of measurement and instrumental drift
Validation (qualitative procedures)
• Specificity
• Robustness
Validation (quantitative procedures)
• Specificity
• Linearity and range
• Accuracy
• Precision
• Robustness
• Limits of detection and quantification
NIRS transfer between NIRS instruments
	• A statement that, "NIRS method: describes the key elements, principally within the NIRS apparatus, which enable NIRS measurement of the analyte of interest;
	• NIRS model: describes how the NIRS spectral data measured using the NIRS method are related to the analyte of interest, generally employing chemometric software;
	• NIRS procedure: describes how the NIRS method and model are used for the intended purpose, within its defined scope.
	• A statement that, "This guideline should be read in conjunction with Directive 2001/82/EC and Directive 2001/83/EC [43].
	• This guideline should be read in conjunction with: Ph. Eur. Monograph 2.2.40 Near-infrared Spectroscopy and 2.9.47 demonstration of uniformity of dosage units using large sample sizes;
	• ICH Q2(R1) guideline on validation of analytical procedures (CPMP/ICH/381/95);
	• VICH guidelines GL1 & GL2 on validation of analytical procedures (CVMP/VICH/590/98 & CVMP/VICH/591/98); CHMP and CVMP guideline on process validation information and data to be provided in the regulatory submission (EMA/CHMP/CVMP/QWP/70278/2012-Rev 1); ICH Q8: guideline on pharmaceutical development; ICH Q9: guideline on quality risk management; ICH Q10: guideline on pharmaceutical quality system; ICH guideline Q8, Q9 and Q10 - questions and answers (CHMP/ICH/265145/2009)."
	• A statement that, "risk assessment should address the assumptions made in the development of the procedure and its scope, the variables affecting response, and sample presentation."
	• A statement that, "If the NIRS procedure is being presented in the initial registration dossier, validation data should also be presented for the reference analytical procedure, if used."	

(*Continued*)

TABLE 22.8 (*Continued*)
Key Contributions from NIR Standards

Validation Parameter	Impact on NIR Analytical Method	NIR Standard Reference
	• A statement that, "The applicant should indicate how they will manage the NIRS procedure lifecycle in the initial application, since this is essential for the efficient and effective implementation of post-approval changes to the NIRS procedure.	

• Qualitative NIR procedures based on identification libraries should be validated for specificity
• Quantitative procedures should be validated for:
  • Accuracy
  • Precision
  • Specificity
  • Linearity
  • Range
  • Detection and Quantitation Limits
  • Robustness [as per ICH Q2 (R1).
	• A statement that, "[The Draft Guidance also provides recommendations regarding how the concepts described in the International Conference on Harmonization (ICH) guidance for industry Q2 (R1) Validation of Analytical Procedures: Text and Methodology (ICH Q2 (R1)) and PAT — A Framework for Innovative Pharmaceutical Development, Manufacturing, and Quality Assurance can be applied to the development, validation, and submission of NIR analytical procedures."	[43]

* *t*-values —Large positive or negative *t*-value indicates that residual is 2.5 times the standard error of the difference between NIRS determination and the reference method determination and that laboratory values from the reference method were inaccurate or did not represent samples at time scan was taken (i.e., subsampling error). These samples were reanalyzed by a reference method. Moreover, output for samples that have large *H*-values was evaluated. Large *H*-statistic (>3) indicates that NIRS spectra used in calibration for that sample differ substantially from NIRS spectra of other samples. Calculation of *H* was mathematically derived from covariance matrix according to formula: $H = X\ (s^1x)^{-1}\ x^1$ (Landa. *Apparatus* (a)). High value on diagonal of *H* matrix indicates sample that is dissimilar to calibration set at wavelengths used in equation. Such samples were rescanned. If two scans agree and the sample belongs to the population, then retain the sample. If scans disagree, then the first scan was a mistake and should be discarded. The standard error of calibration (SEC) to determine fit of calibration samples: the lower the SEC, the better the fit. Equation with SEC about two times the laboratory repeatability standard deviation for acid-detergent fiber and crude protein reference method was selected. Coefficient of determination (R^2) to determine the proportion of variation in reference method values among samples explained by NIRS regression equation was examined. Low R^2 values often indicate that laboratory data from reference method are imprecise. If laboratory repeatability error from reference method is 1/4 of standard deviation, equation with $R^2 = 0.75$ was selected. *F*-statistic of regression coefficients was examined. High *F*-values indicate that the regression coefficient is significantly different from zero; small *F*-values indicate that the coefficient contributes little to the equation except to fit random errors. Probability that the observed *F*-value was obtained solely through chance does not follow standard *F* tables because *F* is selected as maximum of all wavelength combinations considered. As the number of choices increases, large *F*-values are needed to signify coefficient fitting more than just random errors. Reject equations with *F*-values <10.

** Leverage statistic, *h*, capability of detecting high leverage outliers. Capability to detect outliers via a statistic such as those based on spectral residuals.

As it has been shown in Table 22.8, the modern near-infrared spectroscopic method of analysis is a valid technique that has been firmly established as a scientific discipline for studying the chemical and physical properties of materials in many industries. From its core beginnings in the agricultural sciences, where it has been used to study soil to feces and everything in between [46], to the development, manufacturing, and quality of pharmaceuticals [47,48], NIR has become an integral analytical tool in the global commerce of goods. The trending of the same FOM selected for reporting the equivalence of the NIR spectra to the reference measurements among the standards developed throughout various agricultural and industrial practices is evident as one reviews Table 22.8 from the early to later standards.

One of the critical steps in the use of NIR is the ability to transfer the validated regression equation from a parent instrument where the original calibration was performed to two or more child instruments where the model can be deployed and used. The following discussion presents the issues and solutions for calibration transfer of NIR spectroscopic regression equations.

22.12 CALIBRATION TRANSFER

Calibration transfer (also referred to as instrument standardization) is a procedure that is used to apply a model (a chemometric algorithm that is the result of MLR, PCR, PLS, or other regression equation) constructed on a parent instrument to two or more child instruments. Very simply stated, a new model's slope and intercept should be statistically the same as the parent instrument within experimental error, and the new predicted results should be unbiased. Instrument variance within and between manufactured instruments identical in make and model are the factors that have the greatest effect on calibration transfer. The problem has been studied intensively, and multiple reviews have been published most recently by Workman [49]. While other factors are at the root of errors in bias, slope, and intercept, such as the analyst training, sampling, sample presentation, and environment, the fundamental problem of calibration transfer of NIR spectroscopic regression equations stems from the instrument itself.

Kazeminy et al. [50] showed that rather than transferring a regression equation, they would develop three separate models from the same calibration samples (n = 1224) to determine just how identical or different regression equations could be across different instruments and software packages. They demonstrated that, despite instrument and software differences, exact and accurate spectroscopic results can be achieved using NIR spectroscopy for discriminate analysis. Using the same samples, NIR method spectral range, spectral pretreatment parameters, and regression equation parameters, the same samples calibrated on three different instruments and controlled by three different software packages gave nearly the same multivariate models, despite gross instrumental and software differences, to accurately predict 100% the identity of pharmaceutical dosage forms comprising the validation set. This experiment demonstrated the primary reason why calibration transfer is desirable to eliminate the need to recalibrate and revalidate large data sets repeatedly across multiple instruments. Agricultural-based measurements in the pioneering days of NIR analysis developed calibration transfer methods using this approach and then used the coefficients obtained from the multiple instruments to normalize instrument responses as close as possible to the reference (parent) instrument.

As instruments were evolving into dispersive, FT, and other optical-detector arrangements, calibration transfer got more complex. The origin of the analyte signal from the reference instrument from which a model is based is composed of the radiation from the sample and variance from non-instrumental and instrumental effects. Griffiths et al. [51] showed that wavelength accuracy of NIR instruments and ultimately the accuracy of the NIR regression models are primarily affected by the following factors arising from the instrument:

- the spectral bandwidth (SBW), that is, the optical resolution of the instrument
- instrument line shape (ILS) (i.e., spectral slit function for grating spectrometers)
- band symmetry (due to the extent of overlap of the bands in the spectrum of the wavelength standard)
- the algorithm selected for the peak-picking algorithm used to assign peak maxima (i.e., NIST peak center of mass)
- vignetting of the beam by the diffuse reflection optics.

They state, "…the accuracy with which calibration data are transferred from instrument to instrument may be limited because of differences in the wavelength scale of the two instruments, thus precluding a broader use of NIR spectrometry."

Since no two NIR instruments are (yet) identical due to natural variation in design tolerances, in materials and process, calibration transfer solutions mainly address sampling and noninstrumental factors. The five factors previously listed will still have some impact on subsequent instrument measurements no matter the approach used, be it by manipulating sampling or addressing noninstrumental parameters (i.e., mathematical corrections), albeit minimal.

Workman provides via reference that calibration transfers are of three varieties where instrumental factors are not addressed directly:

1. Combining data from multiple instruments and by using spectral pretreatments of that data,
2. Modifying calibrations using bias and slope adjustments, and
3. By adjusting spectra (i.e., piecewise direct standardization (PDS)).

All three cases involve sampling and/or mathematical noninstrumental adjustments and not instrumental adjustments.

The aforementioned NIR standards also address calibration transfer, and the following table (Table 22.9) lists their remarks and considerations.

From the standard approaches listed in Tables 22.8 and 22.9, NIR method validation and calibration transfers can be summarized into several caveats on which a successful transfer hinges upon.

I. Instrument and noninstrument factors must be interrogated quantitatively to establish the degree of model likeness from pretransfer conditions and post-transfer results.
II. Standards differ as to the requirements to demonstrate the validity of models and equivalency of results.
III. While standardization of the NIR method appears to be harmonized globally across industries, NIR calibration transfer is given wide latitude on positions and recommendations provided by the standards for assessing the validity of transferred models and the equivalency of results from multiple instruments to the original.
IV. Regardless of the approach chosen for model calibration and calibration transfer, all official methods of NIR analysis (AACC, AOAC, ASTM, EMA, FDA, USDA, and USP) recommend, to various degrees, that model calibration and transfer validation statistics are necessary to assess the validity of the models and the equivalency of results.
V. Since all NIR methods are secondary methods, the basis of selecting the most appropriate validation statistics should always begin with assessing the level of risk for what the model is being used and to what degree of accuracy and precision the results must be to ensure quality that is equivalent or better to the reference method.

It has been shown how NIR spectroscopy has evolved from a curiosity of nature to a standard practice in analytical chemistry. NIR methods were able to advance mainly due to the advent of mathematical and statistical concepts that became chemometrics primarily attributed to Wold, Kowalski, and Massart and the keys to a modern analytical technique in chemistry in the hands of Norris, Williams, and others following them. According to Wold, the main focus of chemometrics is to "structuralize the chemical problem in mathematical terms," and that, rather than forming a separate branch of chemistry, chemometrics must "remain an integral part of all areas of chemistry" and Kowalski, "…chemometrics is aimed not at theoretical calculations, but at the extraction of useful chemical information from measured data". Given that this is the case, the objective of

TABLE 22.9
Calibration Transfer Considerations in NIR Standards

Calibration Transfer Statistics	Impact on NIR Analytical Method	NIR Standard Reference

• Considerations for calibration transfer from instrument to instrument
  • Wavelength uncertainty
  • Background reference
  • Sources of noise include lamp and lamp power supply, stray light, photometric nonlinearity, and surface reflectance.
  • Sources of photometric differences among instruments are stray light, specular component, nonlinearity in detector response and signal amplification, and small but important differences in the overall optical system.
  • A valid equation transfer program
• Calibration transfer validation Comparative statistics
  • mean,
  • standard deviation,
  • *N*,
  • bias,
  • the standard error of the difference (SED),
  • the standard error of the difference corrected for bias (SED(C)),
  • R^2,
  • the standard error of calibration (SEC), and the average of SED and SEC (AVE).
• A statistic summary that selects the best equation for each variable is printed also
	• A statement that, "The NIRS method of analysis requires very high wavelength reproducibility, typically within a standard deviation of less than 0.02 nm. The wavelength accuracy requirementswithin one instrument are not very important, and a tolerance of ±5 nm is acceptable. However, if calibrations are to be transferred from one instrument to another, corrections must be made for the wavelength differences."
	A statement that, "The reflectance of the reference must remain constant to maintain the calibration of an NIRS instrument. The absolute value of the reflectance is not of great consequence; but if calibrations are to be transferred from one instrument to another, any differences in the reference must be accounted for. It appears that the ceramic working standard is very stable and does not represent a significant source of error within one instrument, but between-instrument variability occurs because the position of the ceramic and its backing affect the measured reflectance."
		[25,51]

• In practically every case, the SED between instruments was less than or equal to the normal SEL of the chemical analysis within a laboratory. This means that by using the transfer process, errors between laboratories can be reduced to the level of those within laboratories (also known as Shenk'6989s law)	• A statement that, "errors become a more serious problem for multiple-instrument applications in which it is desirable to transfer calibrations from one instrument to another." • A statement that, "Although there are many potential problems with the transfer process, these procedures make it possible for commercial companies to provide laboratories with factory calibrated instruments. Only when precalibrated instruments are made available will this computerized technology begin to reach its potential.	
$N_{CAL} = 650$ $N_{VAL} = 94$ MLR Relative standard deviation for reproducibility (RSDR) Variance component for among-laboratory variation (coefficient of variation)	• Results confirm that it is possible to calibrate, validate, and transfer (NIRS) equations and data among laboratories. Adopted official first action, *J Assoc Off Anal Chem.* 1988 Nov-Dec;71(6):1162–7"	[37]

(*Continued*)

TABLE 22.9 (*Continued*)
Calibration Transfer Considerations in NIR Standards

Calibration Transfer Statistics	Impact on NIR Analytical Method	NIR Standard Reference
1. Validation of a calibration transfer is required. a. Validation set N b. Validation sample concentration range spans the expected population concentration range c. Uniform distribution d. Consideration is given to mixtures and complex mixtures. e. Outlier validation (if feasible) 2. Validation statistics c. Validation error $= v - v$ d. Variance of the validation error VAR_v e. Standard error of validation (SEV) f. Validation bias g. Standard deviation of validation errors (SDV) h. Significance of validation bias i. Validation of agreement between model and reference method j. Validation confidence limits k. Standard error of qualification	• ASTM E1655-00 • A statement that, "A complete description of calibration transfer methodologies is beyond the scope of these practices." • A statement that, "Instrument standardization is one means of achieving calibration transfer." • A statement that, "Calibration transfer may be required when maintenance is done to an instrument…" • Reports statistical techniques used for analytical method comparisons on two or more instruments • A statement that, "An alternative method can be used to demonstrate agreement between the model and the reference method."	[38,39]
• Makes a statement regarding the validity of a model when transferred but provides no recommendation on what to apply	• A statement that, "Transfer of databases. When databases are transferred to another instrument, spectral range, numbers of data points, spectral resolution and other parameters have to be taken into consideration. Further procedures and criteria must be applied to demonstrate that the model remains valid with the new database or new instrument."	[40]
	• A statement that, "The calibration model for a NIR method is developed, stored and applied in electronic form as part of an appropriate instrument/software package, and should not be transferred to another instrument unless procedures and criteria are applied to demonstrate that the model remains valid on the second instrument. Simpler means of transferring methods from one location to another include:" • Relocation of instrument and calibration model. • Use of the complete calibration and validation sample sets to regenerate the method on a second instrument. • In general, electronic calibration transfer is only recommended to another instrument of the same type and configuration.	[41]

(*Continued*)

TABLE 22.9 (*Continued*)
Calibration Transfer Considerations in NIR Standards

Calibration Transfer Statistics	Impact on NIR Analytical Method	NIR Standard Reference
• Makes a statement negating model transfer to ameliorate the need of validating a model on other instruments	• A statement that, "The need for method transfer can be planned into aNIR method at the development stage. This allows for an option whereby the calibration model is developedand validated for several instruments at the same time by including the instrument's variation in the calibration.By this means the calibration is valid on all instruments used in the development and the need for transfer between them is avoided. Furthermore, if a PLS calibration has been developed on a single instrument it can be supplemented to become a two-instrument calibration at a later time.	
• Makes a statement regarding the validity of a model when transferred but provides no recommendation on what to apply	• A statement that, "The model for an NIR method is developed, stored and applied in electronic form as part of an appropriate instrument/software package. • When a model is transferred to another instrument, procedures and criteria must be applied to demonstrate that the model remains valid on the second instrument. In general, electronic model transfer is only recommended for another instrument of the same type and configuration. A number of model transfer procedures exist and can be applied as appropriate. Procedures involve the use of various chemometric (mathematical and statistical) approaches with appropriate validation.	[42]

• Parameters essential to judge the similarity of instruments
  • hardware (e.g., identical spectrometer type and measuring setup)
  • software (including mathematical algorithm and how the spectra are treated in the calibration model);
  • interfaces (e.g., probes, waveguides and fiber optics)
• NIR method changes usually only require a bias and/or slope or vector correction
• In cases where mathematical compensation is insufficient, calibration and validation of the NIRS model should be repeated and confirmed on the additional instrument(s)
	• A statement that, "The following parameters are essential to judge the similarity of instruments."
	• A statement that, "Depending on the scope of the NIRS procedure (e.g., qualitative, trend analysis or quantitative) and the degree of similarity of the instruments involved, NIRS procedure transfer will be performed using option 1 or option 2 below:
	• 1. In cases for which there are NIRS method changes only
	• 2. In all other cases, for which mathematical compensation is insufficient
	• A statement that, "A convenient method for choosing samples is one that is based upon their good multivariate leverage. In this method, samples are selected that have a large influence on the calibration model. Depending on the complexity of the multivariate model, a smaller representative number of samples (in comparison with the number used to calibrate and validate the model originally) should be sufficient to support model transfer between instruments.
		[43]

(*Continued*)

TABLE 22.9 (*Continued*)
Calibration Transfer Considerations in NIR Standards

Calibration Transfer Statistics	Impact on NIR Analytical Method	NIR Standard Reference
	• A statement that, "In both cases (1) and (2), the transfer of an NIRS procedure to another instrument should be the subject of an appropriate comparability protocol to demonstrate successful transfer. • A statement that, "The protocol should include criteria that have been justified to be acceptable to demonstrate a satisfactory transfer.	

• Makes a statement regarding the validity of a model when transferred by suggesting showing acceptance limits are equivalent to those used for the original procedure	• A statement that, "Recognize circumstances that might warrant revision of the calibration model. Examples of such circumstances may include the following:
	  • Transfer of the NIR analytical procedure.
	• A statement that, "Examples of changes to an NIR analytical procedure that is used for release or stability testing of a drug product, a drug substance or in-process procedures used for RTRT:
	  • Transfer of an NIR procedure when the procedure is validated using acceptance limits equivalent to those used for the original procedure.
	• A statement that, "Examples of changes to an NIR analytical procedure that are used for testing raw materials and in-process material or intermediates:
	  • Transfer of NIR procedure provided that the procedure is validated using acceptance limits equivalent to those used for the original procedure.
		[44]

chemometrics in NIR methodology must be to produce spectroscopic results that can be statistically and, most importantly, chemically and physically related to the properties of the materials under study with a very high degree of certainty. This can only be achieved by validation.

22.13 NOVEL APPROACHES

Validation of NIR analytical methods has evolved from parametric-based models generated from several wavelengths and reference values or properties, where instrumental and noninstrumental factors must be considered in validating results from MLR regression models to validating results from multivariate regression and classification models (i.e., PCR, PLS, KNN), where a standard battery of tests is applied to the model for its predictability or as pharmaceutical regulations state, "fit for purpose." As spectral processing, regression algorithms, and graphical tools have been developed and adopted for extracting chemical and physical information from spectra-based models, the need for validating NIR models using approaches that lend themselves to greater interpretation of results and measuring the predictability of the models has significantly extended the capabilities of the NIR spectroscopist.

Chavez et al. [52] note in their critical review that NIR method validation approaches for pharmaceutical use fall mainly into four main decision methodologies:

1. Descriptive approach that use statistical descriptive parameters
2. Different approach using statistical hypothesis tests
3. Equivalence approach
4. Accuracy profile based on the statistical methodology of tolerance intervals

Classical NIR validation procedures were based on traditional chemometric parameters (i.e., RMSEC, RMSECV, R^2, RMSEP, bias, etc.). The basis of validation is to determine the predictability of a calibration model. Two approaches are used: internal and external validation. Internal validation refers to the use of various cross-validation techniques where a calibration model is assessed for its predictability by the use of iteratively removing calibration samples from the dataset and estimating the error of prediction called error of the cross-validation. External validation refers to the use of an independent dataset not used in the calibration set and estimating its error of prediction. With the adoption of NIR techniques by the pharmaceutical industry, standard analytical validation statistics from the International Conference on Harmonization (ICH) were added to the traditional validation statistics and also became standardized [53].

Much of the previous discussion centers on MLR and PLS regression analysis, which has become the *de facto* chemometric method for NIR in many industrial applications. The simple reason is that these methods tend to correlate variables that are linear. However, there are still many other approaches that should be considered since they are better suited for relating nonlinear variables such as Bayesian, Monte Carlo simulations, support vector machines, artificial neural networks (ANNs), and others [54,55].

An example of a completely different application for qualitative classification that introduces a new estimate called polar distance is called polar quality system (PQS). It is different from other data reduction methods previously discussed. Kaffka et al. [56] describe a method for transforming NIR spectra and representing them in a polar co-ordinate system. It is useful for qualitative analysis of NIR spectra for discriminating among very similar spectra. The spectra are transformed into their polar spectrum and coordinates. When a spectrum is transformed into a polar co-ordinate system, the radius is a function of the spectral value, while the angle is a function of the wavelength. As a result of the transformation, changes in the composition of the sample corresponding to chemical changes in wavelengths are exhibited as changes in angles in the polar coordinate and physical changes as changes in radii. Despite the radical transform of the NIR spectra, results are evaluated by evaluating polar distances from the center or quality points (similar to Euclidian or Mahalanobis distances in the rectangular spectral space). Validation is based on evaluating polar component plots, similar to principal component score plots for discriminating quality point clusters.

Many tests exist beyond the traditional and ICH statistics for assessing qualitative and quantitative model characteristics such as sensitivity, specificity, selectivity, correlation, permutation tests, and others for providing additional criteria for predictability and model selection. These are reviewed briefly below:

Net analyte signal (NAS) was first proposed by Lorber [57]. It has gained wide acceptance despite problems. However, its utility has been shown for both NIR and Raman spectroscopy [58]. NAS is defined as that part of the signal that is orthogonal to the spectra of the other components. As a FOM, it has the effect of increasing accuracy and precision by reducing the error of calibration and prediction. The benefit is that three other FOMs are assessed for model predictability, sensitivity (manifested as the slope of the regression), selectivity, and the limit of detection (LOD). The use of the successful application of NAS, as compared to other similar pure component projection methods, improved direct calibration (IDC), science-based calibration (SBC), and augmented classical least squares (ACLS) calibration [59].

Another significant step in ascribing the importance of utilizing nonunivariate FOM to multivariate validation was the introduction of orthogonal signal correction (OSC) [60,61]. As was mentioned previously, the use of multivariate FOM is primarily for understanding the attributes of a model with respect to the strength of the correlation, errors in the calibration, cross-validation, and prediction and bias. But what is most important in achieving a valid multivariate regression is establishing criteria for selecting a valid model as well as assessing the accuracy of the predictions by the model. In this respect, it is necessary to consider that the most appropriate model is one that is born out the appropriate criteria established to evaluate the model during the validation step (this step should be protocol driven) as well as ascertaining the degree of bias between the models predicted values and the reference values.

OSC was originally proposed on the premise that PLS modeling of NIR spectra, composed of the analyte of interest in combination with all other components comprising the sample matrix, could be shown to perform better by providing the methodology for improving predictions and interpretations. Wold *et al.* suggest that OSC, in combination with PLS, can be used to achieve a signal correction that is as close to orthogonal as possible to a given *Y*-vector or *Y*-matrix, without removing information from the spectra regarding *Y* (the measured response variable in multivariate calibration applications). Variants of this approach have emerged since then, and the applicability of the approach is undeniable since many of the early works reporting misuse and erroneous programing in software led to wrong conclusions about its utility.

A proposed use for an estimate of the accuracy of NIR models is the β-expectation tolerance interval (β-TI) [62]. β-TI defines an interval where an expected proportion of NIR results will fall with a defined probability β. Chavez et al. [63] propose the use of this validation criteria where they argue that a low RMSEP alone is not sufficient as acceptance criteria for the validity of model results.

Most other NIR validation FOM includes the use of decision rules for NIR hyperspectral imaging applications [64].

22.14 SUMMARY

This overview presented the concept of FOM for the development, validation, and use of NIR methods from multivariate analysis. In the background are the scientific advances that were occurring in the theoretical, instrument engineering, computer technology in both hardware and software, and mathematical and statistical revolution out of which the science of chemometrics was born. At this juncture, it has been proposed that a standard approach be adopted for reporting NIR models so that they can be reproducibly be reconstructed or transferred. Dardenne proposed such a standard [2]; however, having gone through the history, development, and current requirements of NIR analytical methods in various industries, this proposal can be improved upon to facilitate method reproducibility and transfer as well as to satisfy the regulators.

22.15 PROPOSED FIGURE OF MERITS

Some considerations about NIR spectroscopy [2]:

The table as presented as an idea is appropriate for consideration, but it requires augmentation so that the method can be useful for users and regulators who must use NIR experiments from any instrument in any industry.

TABLE 22.10
Statistics of Calibration Results

Parameters	Fat
Units	%Dry matter or "as is"
SEL—Reproducibility	0.25
N	65
Outliers	0
Min	6.39
Mean	10.49
Max	15.08
SD	1.59
SEC	0.33
R^2_{Cal}	0.93
SECV	0.38
R^2_{CV}	0.94
NIR repeatability	0.19
Number of terms	3
RPD_{Cal}	4.82
RPD_{CV}	4.18
Segments (LOO)	4
WL range/step	100–2498/2
Pretreatment(s)	SNV-DETREND-1st Derivative,4,4
Regression method	PLS

Clearly, from the previous discussions, much more information is needed for reporting a NIR analytical method. There are several components that must be taken into account to preserve the validity over time and different instruments. They are as follows:

A. Noninstrumental factors
 1. A description of the analytical problem
 2. An accurate description of the sample
 3. A description of the reference specification, method, procedure, acceptance criteria (instrument, sample preparation, standard preparation, measurement procedure, etc.)
 4. Date(s)
 5. Time(s), temperature, humidity
 6. Calibration location(s)
 7. Analyst(s)
 8. A statement of the NIR solution and selection justification
 9. NIR specification, method, procedure, acceptance criteria (instrument, sample preparation, standard preparation, measurement procedure, etc.)

B. Instrumental factors
 10. NIR mode of measurement
 11. NIR instrument makes and models number and a description of the optical components

C. Software factors
 12. Software name and version number
 13. Pretreatment(s) (standard nomenclature: derivative (D1, D2, D3, D4), gap number, segment number, (G#S#), standard normal variate (SNV), detrend (DT), multiplicative scatter correction (MSC), extended multiplicative scatter correction (EMSC), smoothing (SM), baseline normalization (BN)

14. Regression method and description of the algorithms used for handling spectral processing and modeling

D. Model factors

See Tables 22.1–22.10

REFERENCES

1. Feynman, R. P., *Cargo Cult Science*, June 1974.
2. Dardenne, P., "Some considerations about NIR spectroscopy: Closing speech" at NIR-2009, IM Publications.
3. Martens, H. and Naes, T. *Multivariate Calibration*, Wiley, New York, 1989.
4. Difoggio, R., "Examination of some misconceptions about near-infrared analysis", *Applied Spectroscopy,* 49(1), 67–75, 1995.
5. Difoggio, R., "Guidelines for applying chemometrics to spectra: feasibility and error propagation," *Applied Spectroscopy,* 54(3), 94A–113A, 2000.
6. Herschel, W., "XIV. Experiments on the refrangibility of the sun," *Philosophical Transactions of the Royal Society (London)*, 284, 1800.
7. Seebeck, T. J. "Magnetische Polarisation der Metalle und Erze durch Temperatur-Differenz". *Abh. Akad. Wiss. Berlin*, 289–346, 1820–1821, 1822.
8. Nobili, L., "Sur une nouvelle classe de phènomènes èlectrochimiques". *Annales de chimie*, 34, 280–297, 1827.
9. Nobili et Melloni, Recherches sur Plusieurs Phénomènes Calorifiques Entreprises au Moyen du Thermo-Multiplicateur. *Annales de Chimie et de Physique*, 48, 198–217, 1831.
10. Melloni, M., "Note sur la ŕflexion de la chaleur rayonnante," *Annales de chimie*, 60(402), 426, 1835.
11. Tyndall, J., "The Bakerian lecture: On the absorption and radiation of heat by gases and vapours, and on the physical connexion of radiation, absorption, and conduction", *Philosophical Transactions of the Royal Society of London*, 151, 1–36, 1861.
12. Burns, G. W. and Scroger, M. G., "The Calibration of Thermocouples and Thermocouple Materials, NIST Special Publication 250-35", U.S. Department of Commerce National Institute of Standards and Technology, June 12, 1989.
13. Oersted, H. C. "Experiments on the effect of a current of electricity on the magnetic needles", *Annals of Philosophy*, London: Baldwin, Craddock, Joy, 16, 273, 1820.
14. Fleisch. D., *A Student's Guide to Maxwell's Equations*, Wittenberg University, 2008.
15. Nicholls, R. W., "Franck-Condon factors to high vibrational quantum numbers I: N2 and N2+", *Journal of Research of the National Bureau of Standards - A. Physics and Chemistry*, 65A(5), September–October 1961.
16. Hooke, R., *De Potentia Restitutiva*, or of Spring, Explaining the Power of Springing Bodies, London, 1678.
17. Abney, W., Bakerian Lecture "On the photographic method of mapping the least refrangible end of the spectrum," *Philosophical Transactions of the Royal Society*, 1880.
18. Abney, R. E. and Festing, R. E., "On the influence of the atomic grouping in the molecules of organic bodies on their absorption in the infra-red region of the spectrum", *Philosophical Transactions of the Royal Society*, 172, 889, 1881.
19. Coblentz, W. W., "Supplementary Investigations of Infra-Red Spectra Part V Infra-Red Reflection Spectra, Part VI" "Infra-Red Transmission Spectra, Part VII Infra-Red, Emission Spectra", The Carnegie Institution of Washington, 1908.
20. Randall, H.M., "The American Association for the advancement of science: Infra-red spectroscopy", *Science*, LXV(1677), 1927.
21. Norris, K.H. and Williams P.C., "Optimization of mathematical treatments of raw near-infrared signal in the measurement of protein in hard red spring wheat. I. Influence of particle size", *Cereal Chemistry*, 61(2), 158–165, 1984.
22. Norris, K., "Early history of near infrared for agricultural applications", *NIR News*, 3(1), 12–13, 1992.
23. Kowalski, B.R., Jurs, P.C., Isenhour, T. L., Reilly, C.N. "Computerized learning machines applied to chemical problems-multi category pattern classification by least squares", *Analytical Chemistry*, 41, 695–700, 1969.
24. Wold, S. "Spline functions, a new tool in data-analysis", *Kem Tidskr*, 3, 34–37, 1972.

25. Marten, G. C., Shenk, J. S., Barton II, F. E. editors, *Near Infrared Reflectance Spectroscopy (NIRS): Analysis of Forage Quality*. U.S. Department of Agriculture, Agriculture Handbook No. 643 (revised with supplements), 110 p., 1989.
26. Barton, F. E., "Theory and principles of near infrared spectroscopy", *Spectroscopy Europe*, 14(1), 2002.
27. Ritchie, G. E., Mark, H., The 13th International Diffuse Reflectance Conference (IDRC) "Commemorating 25 Years of IDRC, Theory and Design", Sponsored by The Council for Near Infrared Spectroscopy (CNIRS), The Chambersburg Conference Wilson College, Chambersburg, PA, August 2006.
28. Kaye, W., "Near-infrared spectroscopy: A review. I. Spectral identification and analytical applications", *Spectrochimica Acta*, 6, 257–287, 1954.
29. Brereton, R. et al, *Analytical and Bioanalytical Chemistry*, 409(25), 5891–5899, 2017.
30. Larson, S.C., "The Shrinkage of the Coefficient of Multiple Correlation," *Journal of Educational Psychology*, 22, 45–55, 1931.
31. Gy, P., Heterogeneite, Echantillonnage, Homogeneisation (Heterogeneity, Sampling, Homogenizing), xiv+607 p (Masson: Paris), 1988.
32. Esbensen, K. H., Geladi, P., "Principles of proper validation: use and abuse of re-sampling for validation", *Journal of Chemometrics*, 24,168–187, 2010.
33. Faber, N.M., and Rajkó, R., "How to avoid over-fitting in multivariate calibration—How to avoid over-fitting in multivariate calibration—the conventional validation approach and an alternative", *Analytica Chimica Acta 595*, 98–106, 2007.
34. Pasquini, C., "Near infrared spectroscopy: Fundamentals, practical aspects and analytical applications", *Journal of the Brazilian Chemical Society*, 14(2), 198–219, 2003.
35. Hart, J.R., Norris, K.H., Golumbic, C., "Direct spectrophotometric determination of moisture content of grain and seeds", *Cereal Chemistry*, 39, 94, 1961.
36. Williams, P., Norris, K. "*Near-Infrared Technology in the Agricultural and Food Sciences; American Association of Cereal Chemists*", St. Paul, MN, 1987.
37. Barton, F.E., II and Windham, W.R., "Determination of acid detergent fibre and crude protein in forages by near infrared reflectance spectroscopy: Collaborative study", *Journal of the Association of Official Analytical Chemists*, 71, 1162–1167, 1988.
38. Workman Jr, J. and Brown, J. "A new standard practice for multivariate, quantitative infrared analysis-Part I", *Spectroscopy*, 11(2), 48, 1996.
39. Workman Jr, J. and Brown, J. "A new standard practice for multivariate, quantitative infrared analysis-Part II", *Spectroscopy*, 11(9), 24, 1996.
40. 'Near-infrared Spectrometry, in Methods of Analysis 2.2.40' in "European Pharmacopoeia", 3rd edition, Council of Europe, Strasbourg, 43, 1997.
41. "Guidelines for the Development and Validation of Near Infrared Spectroscopic Methods," *Pharmaceutical Analytical Sciences Group*, January, 2001.
42. USP Near-infrared spectrophotometry, Chapter 1119. United States Pharmacopoeia USP26 NF21, 2388–2391, 2003.
43. European Medicines Agency. "Guideline on the use of Near Infrared Spectroscopy (NIRS) by the pharmaceutical industry and the data requirements for new submissions and variations", 2014:28, 2014.
44. U.S. Department of Health and Human Services FDA. "Development and Submission of Near Infrared Analytical Procedures Guidance for Industry Draft Guidance", 2015.
45. Williams, P.C., "Variables affecting near-infrared reflectance spectroscopic analysis." *Near-Infrared Technology in the Agricultural and Food Industries* (eds P. Williams & K. Norris), pp. 143–167. American Association of Cereal Chemists Inc., Saint Paul, MN, 1987.
46. García-Sánchez, F., Galvez-Sola, L., Martínez-Nicolás, J. J., Muelas-Domingo, R. and Nieves, M., "Using Near-Infrared Spectroscopy in Agricultural Systems, Developments in Near-Infrared Spectroscopy" Konstantinos Kyprianidis, IntechOpen, DOI: 10.5772/67236. Available from: https://www.intechopen.com/books/developments-in-near-infrared-spectroscopy/using-near-infrared-spectroscopy-in-agricultural-systems Developments in Near-Infrared Spectroscopy, Using Near-Infrared Spectroscopy in Agricultural Systems, 2017.
47. Burns, D. A., Ciurczak, E. W., *Handbook of Near-Infrared Analysis*, Third Edition (Practical Spectroscopy), 2008.
48. Bakeev, K., *Process Analytical Technology: Spectroscopic Tools and Implementation Strategies for the Chemical and Pharmaceutical Industries*, 2010.
49. Workman, J., "A review of calibration transfer practices and instrument differences", *Spectroscopy Applied Spectroscopy*, 72(3) 340–365, 2018.

50. Kazeminy, A., Hashemi, S., Williams, R. L., Ritchie, G. E., Rubinovitz, R., Sen, S., "A comparison of near infrared method development approaches using a drug product on different spectrophotometers, and chemometric software algorithms", *Journal of NIRS*, 17(5), 2009.
51. Griffiths, P. R., Ritchie, G. E., Casay, G. A., Jones, B. J., "Stimuli to the revision process: Wavelengths of calibration standards for near-infrared diffuse reflection spectrometry", *Pharmacopeial Forum*, 38(1), 1–11, 2012.
52. Chavez, P.F., Bleye, C., Sacré, P.Y., Rozet, E., Hubert, P., Ziemons, E., "Validation methodologies of near infrared spectroscopy methods in pharmaceutical applications", *European Pharmaceutical Review*, 18(1), 2013.
53. International Conference on Harmonization (ICH) of technical Requirements for Registration of Pharmaceuticals for human Use, Topic Q2 (R1), "Validation of Analytical Procedures: Text and Methodology", Geneva, 2005.
54. Balabin, R. M., Safieva, R. Z., Lomakina, E. I., "Comparison of linear and nonlinear calibration models based on near infrared (NIR) spectroscopy data for gasoline properties prediction", *Chemometrics and Intelligent Laboratory Systems*, 88, 183–188, 2007.
55. Fearn, T., et al., "Inverse, classical, empirical and non-parametric calibrations in a bayesian framework", *Journal of Near Infrared Spectroscopy* 18, 27–38, 2010.
56. Kaffka, K.J., Gyarmati, L.S., "Qualitative (comparative) analysis by near-infrared spectroscopy", Biston, R. and Bartiaux-Thill, N. (Eds.) *Proceedings of the Third International Conference on Near Infrared Spectroscopy*, Agricultural Research Centre Publishing. Gembloux (Belgium) pp. 135–139, 1991.
57. Lorber, A., "Error propagation and figures of merit for quantification by solving matrix equations", *Analytical Chemistry*, 58, 1167–1172, 1986.
58. Short, S. M., Cogdill, R. T., Anderson, C. A., "Determination of figures of merit for near-infrared and Raman spectrometry by net analyte signal analysis for a 4-component solid dosage system", *AAPS PharmSciTech*; 8(4), 96, 2007.
59. Sharma S., Goodarzi M., Wynants L., Ramon H, Saeys W., "Efficient use of pure component and interferent spectra in multivariate calibration", *Analytica Chimica Acta*, 778, 15–23, 2013.
60. Wold, S., Antti, H., Lindgren, F., Ohman, J., "Orthogonal signal correction of near-infrared spectra", *Chemometrics and Intelligent Laboratory Systems*, 44, 175–185, 1998.
61. Trygg, J., Wold, S. "Orthogonal projections to latent structures, O-PLS", *Journal of Chemometrics*, 16, 119–128, 2002.
62. Rozet, E., Wascotte, V., Lecouturier, N., Préat, V., Dewé, W., Boulanger, B., et al., "Improvement of the decision efficiency of the accuracy profile by means of desirability function for analytical methods validation. Application to a diacetyl-monoxime colorimetric assay used for the determination of urea in transdermal iontophoretic extracts", *Analytica Chimica Acta*, 591, 239–247, 2007.
63. De Bleye, C., Chavez, P. F., Mantanus, J., Marini, R., Hubert, P. H., Rozet, E., Ziemons, E., "Critical Review of near infrared spectroscopic methods validations in pharmaceutical applications", *Journal of Pharmaceutical and Biomedical Analysis* 69, 125–132, 2012.
64. Vermeulen, P., Fernández Pierna, J.A., van Egmond, H.P., Zegers, J., Dardenne, P., Baeten, V., Validation and transferability study of a method based on near-infrared hyperspectral imaging for the detection and quantification of ergot bodies in cereals, *Analytical and Bioanalytical Chemistry*, 405(24), 7765–7772, 2013.

23 Method Development and Lifecycle

*Benoît Igne**
GlaxoSmithKline

Gary McGeorge
Bristol-Myers Squibb

CONTENTS

23.1 INTRODUCTION

Ensuring high-performance standards for analytical methods is an area of continuous interest. Governing bodies define best practices across a variety of industries, and for pharmaceutical testing practitioners, one follows a general set of guidelines published by the International Conference on Harmonization (ICH). More specifically, the ICH document Q2(R1) presents the general best scientific practices for analytical method validation for a variety of differing applications. ICH Q2(R1)

* igneben@gmail.com.

was last updated in 1994, and it does not explicitly discuss the use of spectroscopic techniques such as near-infrared spectroscopy (NIRS). However, the framework it introduces and the subsequent ICH documents (Q8 and Q11) are largely applicable to NIRS and will be discussed in this chapter. The process of validation essentially lays down the performance requirements for which the method must comply with in order to be deemed suitable for use. While mainly targeted towards the pharmaceutical industry, these principles are directly applicable to all fields where a model is developed, tested, validated as appropriate, and maintained through its lifecycle.

NIRS is such a wonderfully powerful technique in that it can analyze samples in their native form, with no or little sample preparation. Since these samples will both absorb and scatter radiation, it is possible to measure composition, crystal form, moisture content, particle size, or material density from a single measurement, if it is suitably designed to do so. However, since it is not possible to cover the specifics of each method, in this chapter, we will present the framework for the development of a NIRS method, which is applicable to all scenarios.

23.2 ANALYTICAL QUALITY BY DESIGN (AQbD)

AQbD arises from the concept of quality by design (QbD), introduced in ICH Q8 as "a systematic approach to development that begins with predefined objectives and emphasizes [...] understanding and [...] control, based on sound science and quality risk management." A process developed using the framework of QbD will be well understood and deliver quality product within a design space determined during method development. Applied to analytical methods, AQbD concepts allow for the development of a "well understood, fit for purpose, and robust method that consistently delivers the intended performance throughout its lifecycle" (Reid et al. 2013). Figure 23.1 presents a schematic of an AQbD workflow. Each of its components will be discussed below.

23.2.1 Method Feasibility

Before moving forward with a complex method development and testing/validation, it is often wise to ensure that the application that is being targeted is feasible within the limitations of the particular situation. These constraints could be the type of instrument selected, the nature of the samples available, the sampling method, the quality of the reference method, or the variability in process. Literature often provides examples of what has been shown to work in a particular situation, but it may be a business risk to engage in a complex program without at least an indication that it works as expected with a limited set of samples and conditions. Once a practitioner is comfortable with progressing, the method goals can then be more clearly defined.

23.2.2 Method Performance Requirements

It is often insufficient to simply define what need a method will address. It is important to define ahead of method development what criteria the method should meet. Adherence to the criteria will demonstrate that the method is performing against the intended purpose. It is also a critical time where the method developer and the method customer agree on what the expectations are. A method built by a highly skilled worker but not transferable to plant operators may not meet acceptance criteria, even if its performance is adequate; a method built in a lab but not working on plant equipment without recalibration may also not meet the customer demands. Thus, the voice of the customers should be taken into account, if applicable when listing criteria. These criteria can be figures of merits as outlined in ICH Q2(R1) such as accuracy, linearity, precision, specificity, robustness, and/or operation's requirements such as analysis speed and ease of use. After method development, these criteria are compared with what is obtained during testing and potentially validated to determine whether they were met. Criteria can be adjusted if proper rationale can be articulated or the method can be determined to be unfit for the intended use and not considered for deployment.

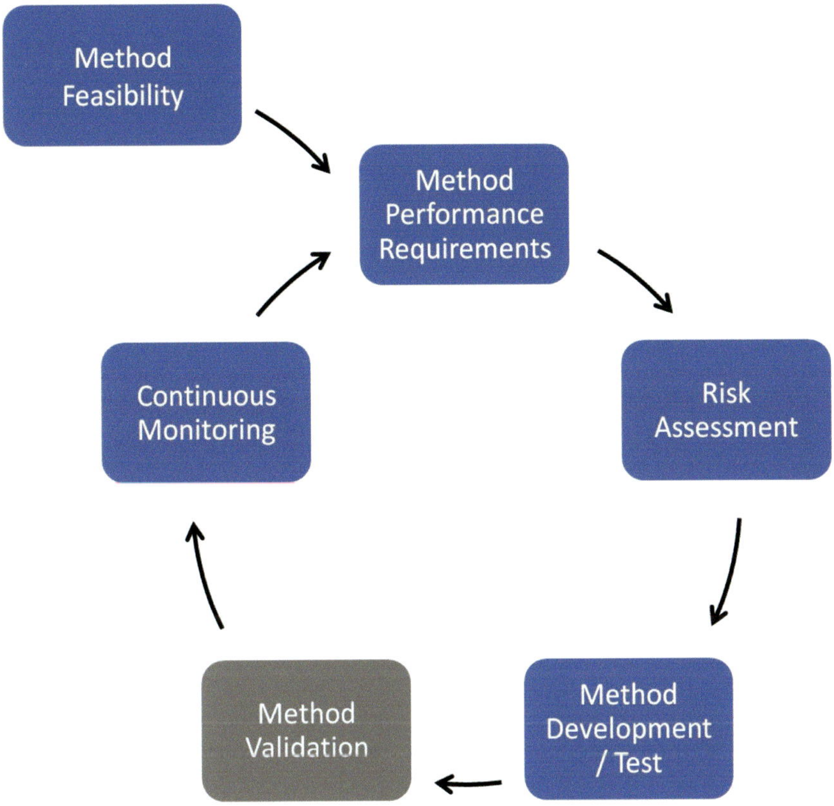

FIGURE 23.1 Analytical quality by design workflow.

23.2.3 Risk Assessment

The next step in the analytical QbD process will be to identify what can negatively affect the method performance. This step, often called a risk assessment, will try to list, score, and rank all the factors that could affect a model performance. Figure 23.2 shows an example of a risk identification process for a pharmaceutical method aimed at predicting the content of active ingredient. While far from being exhaustive, the risk map shows the categories that may need to be investigated. For natural samples, other risks will need to be identified (i.e., crop year representativeness, sample particle size (for whole or ground samples), etc.).

Risk scoring is then performed to determine the risks that will be accepted, avoided, or need to be mitigated. This is often done with a risk analysis method that will score the risk as per its

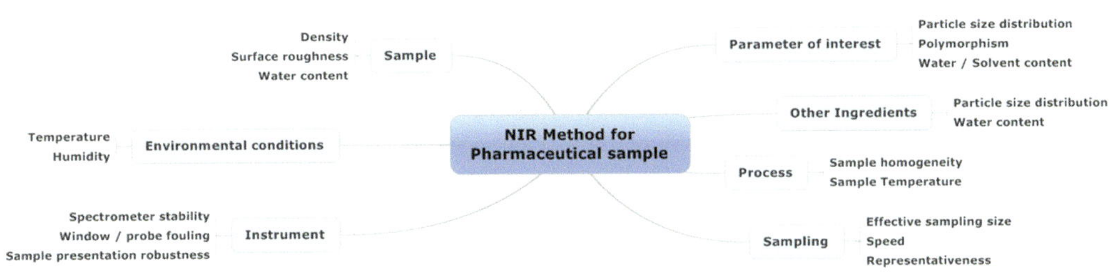

FIGURE 23.2 Examples of risks identified for a pharmaceutical method.

severity, its impact, and its detection. However, mitigating or avoiding all risks is often impossible. Time and resource constraints will force the model developer to make trade-offs to build a model that will be robust to most but not necessarily all sources of variability that it can encounter. Thus, in making any model, some risk of robustness will be passed onto the user who needs to accept the predetermined risk factors that may arise from future previously unseen sources of variability as well as factors that were simply not identified *a priori*. The assessment of these existing and any future risks needs to be handled through a method lifecycle management process. The design of such a process or system will be discussed later and will include outlier/trend detection tools to demonstrate that the test sample(s) are still applicable for the predefined model.

23.2.4 Model Development

The method development stage takes the learnings from the feasibility work, the risk assessment, and the performance requirements from the analytical target profile to inform a more complete method design. This could include the creation of the calibration and test sets via a suitable design of experiments (DoE) in concert with the data collection parameters of the spectrometer. It is in this step that spectral ranges and preprocessing options are optimized.

Some industries have the possibility to manually prepare "nontarget" laboratory samples in addition to using samples from the production line. This is unlike many fields where there is only one way to obtain samples (soil, grains, …). To ensure robustness, production samples should be included if most of the samples are made in the laboratory to ensure that the calibration samples are representative of the finished product. Such synthetic samples could be specifically designed to expand the range for a quantitative model since production samples will inherently have narrow range or to forcibly introduce extra variance to expand the design space such that model loadings will more accurately describe the analyte of interest. More generally, the sources of variability identified in the risk assessment should be included as integral components in a DoE for the calibration sample set. Alternately they should be evaluated as a component of method testing or validation.

The error of the laboratory reference method needs to be determined, at the latest, at this stage of development for methods that build a relationship between the reference method and the NIRS method. Such examples include a NIRS method for the quantitative determination of crystal form impurities in an active pharmaceutical ingredient, that is usually determined by powder X-ray diffraction, a NIRS method for the quantitative measure of grain or soil moisture using an oven-based method (AACC method 44-15A 2000), or a component of interest such as protein content analyzed by combustion (AACC method 46-30 2000). The method error of the reference will be convoluted with the NIRS-specific errors resulting in a larger total analytical error (TAE) and will influence what performance criteria can be expected from the method. For example, if an error of 2% is required but the reference method provides a 1.5% error, it only leaves a small error that the NIR method can generate and it may not be possible to obtain the relevant method error.

Finally, having the possibility to test the method performance not only by cross-validation but also with one or several test sets made from samples not included in the calibration set may help understand the true performance of the model.

23.2.5 Method Validation

Method validation is an exercise that may not be necessary depending on the field of application of the method. In the pharmaceutical industry, models that will be used to make a decision on the process or used to release a product to the market will need to undergo validation. It consists of demonstrating that the method meets the predetermined validation criteria often based on the principles outlined in the analytical target profile. It tests method figures of merit outlined in ICH Q2(R1), such as accuracy, linearity, and precision. Method validation is the topic of another chapter in this book and will not be further discussed here.

23.2.6 System Suitability

Just like with any analytical method procedure, it is critical to ensure that at the time of use, the instrument, sampling interface, and any sample preparation are functioning correctly such that high-quality data can be analyzed. The instrument would undergo routine performance verification by appropriately trained technical staff on a periodic basis, but this verification is insufficient in itself. Furthermore, suitable test standards should be tested at the time of use, and the results of the defined procedure should be compared against acceptance criteria and trended against historical values. The test standards are often NIST traceable rare earth oxide samples, except if the method developer has developed their own approach that could prove acceptable. This, for example, could leverage a stable set of standards that are unique and specific to the method, potentially a small subset of the validation sample set.

Outlier diagnostics will be discussed later, but in summary, they can also serve as a system suitability metric for each and every test article, much in the same way as a tailing factor in a chromatographic method. If, for example, the sample was not aligned correctly or the instrument wavelength accuracy was off, then the diagnostics could pick this up.

23.3 METHOD LIFECYCLE

The development and testing of a NIR spectroscopic model (when validation is not required) are usually the first step in the use of a method for practical use. As discussed in the previous section, the thoroughness of the amount of robustness that is included in the model will dictate its applicability beyond the samples with which the models were developed. The end result is that models may initially lack robustness to certain factors, whether known or unknown. The net result is a need to put in place a plan that will be used to manage any update to the model that may be necessary for ensuring it remains fit for purpose. The development of a plan such that it is embedded into the operational business model for the company's quality system is critical to the long-term success of the application. One clearly does not want to figure this out on the fly. A discussion by Takuya Miyano and colleagues provides an excellent perspective on how to design such a model maintenance procedure in a compliant manner for pharmaceutical companies (Miyano et al. 2015).

A second element to lifecycle management is to make sure there is a system to track that the model is still performing as expected. This can be done using model diagnostics. Statistical limits can be established on these diagnostic statistics to be used as warning and action limits where applicable. If any diagnostic metric was to exceed a predefined limit, an action could be triggered to understand the nature of the errant result, which may include a verification of the model performance using the primary reference method. In the event that the model was indeed in error, then the spectra and corresponding reference values could then be used for the model update as appropriate, which will be discussed later.

Broadly speaking, there are two main types of changes that will need to be managed: changes to the samples and changes to the instrument. For the first case, it is important that any change to the process (crop year, manufacturing process, field, ...) that was not captured in the initial model be managed. Any change should be documented and its impact on the model assessed. The second type of change considers modification to the analytical instrument (lamp, detector, ...) that may affect the model performance over time; this may include model transfer to another instrument. If a model update is required, similar processes to development and testing should be followed. Some pharmaceutical regulatory agencies have published guidance documents regarding what model lifecycle should consider (EMEA Guideline 2014, FDA Guidance 2015).

23.3.1 Robustness as Method Lifecycle Strategy

Several factors also need to be considered when trying to make a model as robust as possible.

- When robustness increases, it usually affects negatively the model accuracy. The regression algorithm now needs to be able to handle sources of variability that may not exist naturally and that will degrade its performance. This is also related to the fact that by including all possible sources of variability (determined at the time of the risk assessment), it may make the model span a large range in various factors (particle size, density, ...) that does not occur naturally. Thus, the model is potentially built to handle a space of variability that normal operating conditions may not cover.
- Anticipating all sources of variability at the time of model conception may not be possible. If the samples are to change due to a natural process or a change in manufacturing process of the samples or ingredients in the samples, it may not be possible to know how that change will affect the model performance. Moreover, large design of experiments can be expensive as they may require large sample sizes exploring several factors via synthetically prepared samples, and for each sample, a reference analysis that can be costly.

Such a strategy may not be possible for all applications but should be considered when model updates are more problematic than the original model roll-out.

In situations where the samples to be analyzed are not expected to change significantly over the period of utilization of the method, a comprehensive risk assessment and the corresponding design of experiments could be used to build the necessary robustness into the model. It could simplify the use of the model in practical implementations as it would require less periodic updates but would not negate the need to track the model performance through diagnostics.

23.3.2 Model Maintenance Procedures

23.3.2.1 Method Diagnostics—Outlier Trending/Tracking

For the purpose of outlier tracking and trending, we are referring to the outlier metrics for test samples and comparing their spectral characteristics to those of the calibration or validation sets. Two diagnostics are commonly employed that define the deviation from the sample response from the center of the model, either within the space defined by the model (Hotelling's T^2) or outside the model space (Q residuals). Upon tracking these metrics for a series of test samples/batches, one can ascertain what typically should be forecast with certain statistical power and limits would be placed on them. This commonly employs standard Western Electric univariate control charting. These control charts can be at the sample level or batch level depending on the nature of the method application. For example, with a tablet assay method, over a hundred tablets could be measured, generating a composite assay value describing the batch. Isolating a single tablet outlier has a very different impact compared to a whole batch worth of *suspect* data.

Chuck Miller succinctly captured that "It is very important to note that the term 'outlier' does not imply 'incorrect'" in the book *Process Analytical Technology* (Miller 2005). It is common to make an observation that an outlier statistic is noncharacteristic of the past and quickly assume "wrongness" of the NIR method. What should happen is a careful inspection of the NIR data, manufacturing process, and raw material characteristics to lead the practitioner to a logical explanation of the origin of the difference. To understand the potential reasons for any outlier observations, it is beneficial to routinely monitor the risk factors that may contribute to the deterioration of the method performance.

23.3.2.2 Continuous Monitoring of Risk Factors

Through the risk assessment, many factors would likely have been determined to have an impact on the model. These factors, if deemed important enough via a scoring or *risk prioritization assessment*, would be systematically monitored through the lifecycle of the method or until it can be confidently deprioritized and eliminated as a key risk. Alternately, if risks are realized for a specific factor, then it may be necessary to restrict the property ranges for the risk attribute. Alternately the learnings from the monitoring would drive a conversation regarding the need for a model revision (see later).

23.3.2.3 Periodic Assessment against Reference Method

One interesting area that has been quite a challenge is that of periodic method assessment. Upon a defined schedule, a pair-wise comparison of the NIR method against the reference method is applied to demonstrate that the method is still performing. This assessment has been employed as a last resort to ensure that the samples have not drifted from the model, in spite of meeting the diagnostic tests and continuous monitoring showing now concerns. In industrial processes, normal operating conditions will influence the fact that only samples at target processing conditions and composition may be available. Hence the testing will demonstrate that routine production data are aligned with the method requirements. There is very little guidance on what this test should look like, but applying an equivalency statistical test is a sensible choice. The null hypothesis for an equivalency test is that "the two tests are not equivalent," with a conclusion that can be drawn as "the NIR method is equivalent to the reference method." In essence, this testing approach determines that the two tests produce a *practically equivalent result*. With an equivalency test, one must define the size of the goal posts, that is, what level of difference is acceptable. This is for the practitioner to decide based on the specification margins for the method and the risk tolerance to the conclusion of the test. For example, if the limit is set at 2% difference between the two tests and the result is greater, hence demonstrating nonequivalence, then a systematic investigation into the study will proceed. If the threshold is too loose, then nonequivalent methods may not be identified, passing the risk onto the consumer. Alternately if the limits are too restrictive, the method sponsor would be constantly and unnecessarily performing investigations. Both considerations need detailed, careful assessment for the sponsor of the method. Limentani et al. (2005) and Chatfield and Borman (2009) provide good additional detailed discussion for those interested in equivalency testing for analytical methods.

Periodic performance assessment requires nevertheless that system suitability tests are performed so that any drift of the instrument would have been detected and addressed (refer to the system suitability section for further discussion).

23.3.3 Method Transfer as Part of Method Lifecycle

As part of the initial method design, the scope for potential routine deployment should be a key element. The deployment may be to a single site and instrument or potentially to a global network with ten or more analyzers. The strategy on how to manage this can be quite complex for NIR. While method transfer will be covered elsewhere in this book, we will discuss it briefly here since it has direct implications on method development.

23.3.3.1 Robust Modeling

As presented in the risk assessment section, instrumentation can play a significant role in either introducing risks or helping to mitigate them. Many analyzer designs are known to be variable across different builds, while others may be more consistent. The nature of instrument variability is therefore essential.

1. For stable, consistent analyzers, the idea of a single model that is transferred as-is to secondary instruments is possible. This model may have been built on several primary benches to generate a *global model* that is potentially more robust. This single global model is then transferred to all analyzers.
2. For more variable instrumentation, or for more exacting/sensitive methods, model revisions on each instrument may be required, possibly via bias and slope adjustments, or appropriate transfer strategies like piecewise direct standardization (PDS). Here robustness is challenging to build in the model but must be added during the transfer stage.

23.3.3.2 Method Transfer during Lifecycle

In the example (1) above, it is sensible to consider that the model/method transfer is simply a partial re-validation of the method to the receiving site on the new instrument. For case (2) with variable instruments, it is less likely that a simple transfer would pass the acceptance criteria. Hence the transfer strategy would look to identify the new characteristics of the receiving instrument/setting and attempt to align the previously designed model to work on the new unit. Alternately, it could be possible to simply re-build the model on the fly with the new instrument and then manage multiple models across the global network, which brings with it other challenges.

23.3.4 Model Revision

It should simply be expected that at some point in the analyzer network, and at some time, the model will simply not meet the original design requirements or that the spectrometer vendor changes analyzer specifications or that regulations will change. The end result is a necessity to update the model to ensure that it functions as required. Model revision is often one of the largest concerns that a company and regulators worry about since it seems so distinctly different from what the quality organizations are used to. From the regulators' perspective, there is the need to ensure that the model is valid under all prospective or registered conditions that it will be deployed, and that the model revision is not some constant activity that results in little confidence that the model is performing as it is purported to do. From the company perspective, it is a worrying idea that the models may need revision, resulting in significant regulatory scrutiny and oversight on model changes, significant delays and reduction in manufacturing efficiency, increased production costs, and variable staff resourcing.

Once the decision to modify the model has been made, it is important to ascertain a path forward. Some model revisions may require modification of the pretreatment options or changes to the number of latent variables. This option is quite straightforward and would occur in the situation where a new change has occurred that does not invalidate the original calibration samples/spectra. However, it should be expected that the revision could require new calibration standards to be prepared. Often this is presented as due to shifting raw material characteristics that were not seen previously. With the assumption that these new characteristics would be observed in the future, it is therefore essential to incorporate their features in the model. At this point, it is necessary to consider whether the old calibration data are still useful or is it too far away in spectral space and will never been seen again. Any prospective rebuilding of the model should justify the addition to or removal of samples from the calibration sample set. For instance, it may make sense to remove samples including a particular raw material when that same raw material is no longer sourced from that supplier; it may be necessary to remove samples from crop years presenting extreme conditions when the samples were chemically affected and do not represent a current year's expected production. For synthetic samples, this simple component has significant ramifications since the commercial site making the product may not have the ability to manufacture off-target pilot-scale batches that could be needed to ensure that appropriate leverage is built into the model with the right samples.

One scenario could play out where the company chooses to modify the model in a fashion that all the original calibration/validation samples need to be re-tested. This could happen in the event that the sampling system changed from a transmission tablet testing module to a novel automation solution. In such a case, one would like to leverage the previously prepared standards, and therefore ensuring that the "old" standards are still suitable for use is critical to the model building process.

23.3.5 Case Studies

To illustrate some of these concepts, three case studies will now be presented. In each scenario, the nature of the samples to be predicted changed due to year-to-year variability, modification in processing conditions, or material properties (different lots/vendors or physical conditions):

- Case Study 1: Sample variability linked to crop years detected by continuous monitoring of error statistics
- Case Study 2: Raw material variability detected by global diagnostics (Hotelling's T^2 and Q residual)
- Case Study 3: Model calibration via knowledge integration

23.3.5.1 Case Study 1: Sample Variability Linked to Crop Years

When external factors are responsible for sample variability, it can be challenging to develop a robust model. This is particularly true in the field of agriculture where year-to-year variability will affect the nature of the samples or from pharmaceutical ingredients that are derived from natural materials, for example, cellulose. To take as example the development of calibration for a grain property, different crop years will exhibit different climates and growing conditions that will affect the grain properties (chemical—their content in various properties of interest and physical—size and density). Additional external factors such as soil conditions, geographic origin, and disease effects can also create sources of variability that may only be seen once or that may be the norm for future crop years. Finally, a multitude of cultivars will often be available and change based on growing years. Thus, a useful calibration must be able to handle these sources of variability. Moreover, it is not uncommon to have networks of instruments mounted on harvesting equipment, at grain elevators and in testing centers all over the world. Thus, changing models every year may not be always feasible.

The consequences of these sources of variability are an increase in the identification of outliers or the appearance of subpopulations within a calibration set. An accurate and robust prediction model will require that this progressive variability be included through model updates. The problem can be accentuated if rapid genetic changes are occurring, such as during a breeding programs.

To illustrate this, the development of a triticale protein model across several crop years will be presented (Igne et al. 2007). Triticale grain samples harvested from 2002 to 2006 were harvested as part of a breeding trial managed by Iowa State University at five geographical locations. Winter and spring varieties were available in 2002–2004. Only winter lines were cultivated in 2005 and 2006. A total of 414 samples with associated protein reference values were available: 126 for 2002, 93 for 2003, 94 for 2004, 98 for 2005, and 93 for 2006. Their spectra were acquired by an Infratec 1241 (Foss North America, Eden Prairie, MN, USA) from 850 to 1048 nm at 2 nm increment with an 18 mm pathlength.

Partial least squares models were developed using two different approaches: (1) spectral addition where a model was first built with the 2002 data and then used to predict 2003; the 2003 data were then added to the 2002 spectra to predict the 2004 dataset; this was repeated until 2006 was used in prediction and (2) a spectral adaptation approach where the best combination of crop years was added to best predict the 2006 samples.

The spectra were pretreated with the second derivative and autoscaling. Cross-validation and independent set prediction statistics were used to evaluate the model performance.

Figure 23.3 shows the results of the spectral addition process. Model performance was calculated by cross-validation, by randomly taking out 25% of the calibration set and using it as a test

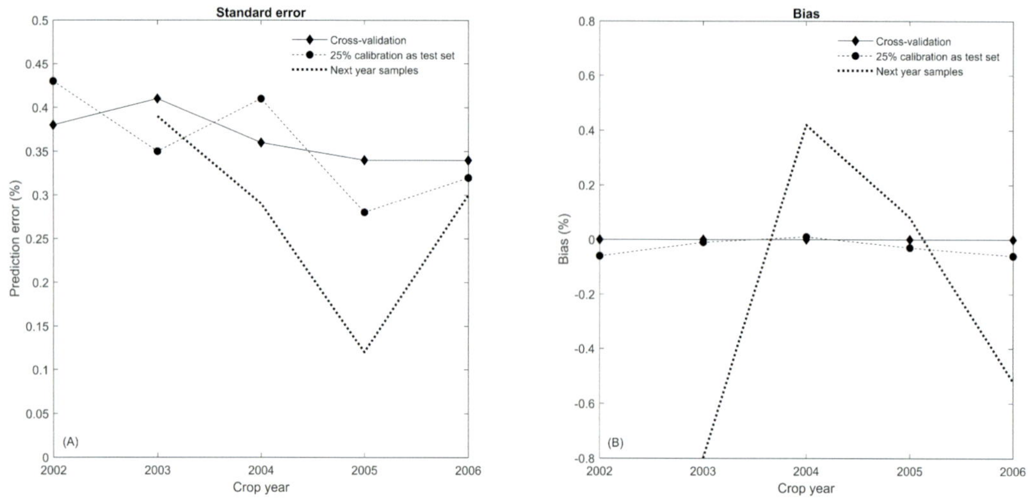

FIGURE 23.3 Model performance (standard error (a) and bias (b)) for the spectral addition approach.

set, and by predicting the samples from the following crop year. The standard error of prediction (Figure 23.3a) and bias (Figure 23.3b) were used to evaluate the model performance.

While the internal validation statistics (cross-validation and 25% of calibration as test) generally showed a progressive improvement of the model performance through the inclusion of more samples, there was no indication of the performance of the model for the next crop year. While this was not particularly clear from the standard error values, it was the bias value that was particularly affected by unknown variability. Thus, the 2003 samples predicted from the model built with the 2002 samples only exhibited the same precision information as the internal validation statistics but was incapable of accurate predictions. The situation only really improved when predicting the 2005 samples from a calibration set made of 2002–2004 samples. However, the 2006 samples were not well predicted by the model constituted of 2002–2005 samples.

Thus, in situations where a significant difference exists between the samples, it is imperative to collect enough relevant data to build a robust model and have diagnostics in place to identify when the model should not be trusted because a new sample falls outside of the space covered by the sample. This will be discussed by the next case study. However, it can be highly beneficial to try to understand the sources of variability and their impact on the model predictions. This is what is shown in Figure 23.4. Here, all combinations of samples available were used to build models for predicting the 2006 samples. $2^4 - 1 = 15$ was the total number of crop year combinations that could be evaluated. It is important to note that in 2006, only triticale winter lines were available. Thus, it is necessary to look at the model performance from models built with and without the spring lines.

Several interesting observations could be made. First, all but one bias (out of 16 results) were negative when including the crop year 2002 samples to predict 2006. Second, most combinations including 2003 showed higher biases than without it. Finally, 2005 appeared to represent the closest variability to 2006. The presence of spring lines when predicting winter lines appeared to affect model performances in some years but not others.

These results showed the importance of keeping track of the sample constituency of the calibration model not to create subpopulations that may deteriorate model robustness when they are not representative of the current and upcoming variabilities.

23.3.5.2 Case Study 2: Raw Material Variability

The previous case study showed how uncontrolled sources of variability can affect a model's ability to provide accurately predicted values. In this next example, the effect of deliberate variability onto

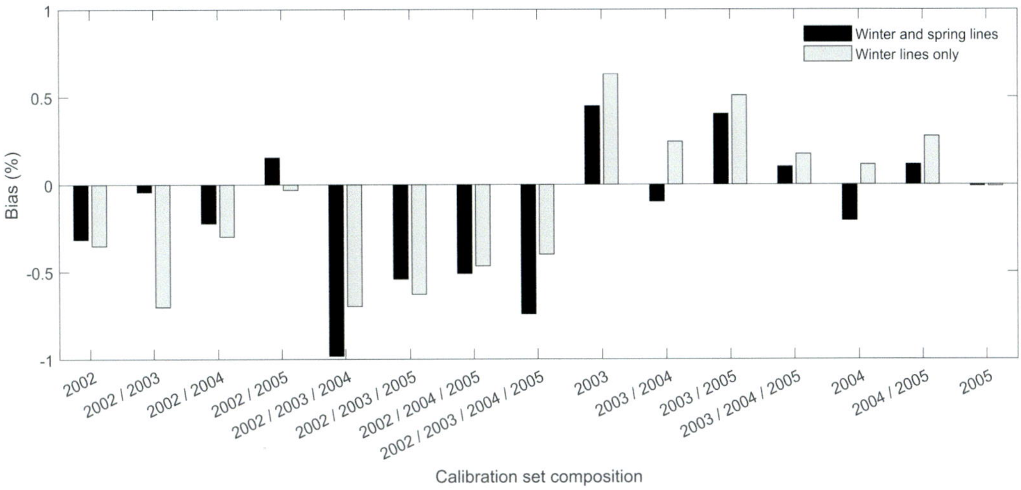

FIGURE 23.4 Prediction bias evolution as a function of calibration set composition.

prediction and model diagnostics will be investigated (Igne et al. 2014). A simple pharmaceutical formulation composed of anhydrous theophylline, lactose monohydrate, microcrystalline cellulose, and soluble starch was used to build a calibration set based on a fully balanced quaternary mixture design varying from 0% to 60% w/w. Tablets were made from the 28 formulations and scanned with a Foss NIRSystem 5000 bench-top spectrometer. An additional design point, close to the model center (25% w/w for all ingredients), was formulated with slightly different compositional changes as outlined in Table 23.1. The 12 conditions were used to make additional tablets and evaluate the PLS model (built from the 28 design points) ability to detect the effect of the sources of variability.

The effect of these changes to the calibration model was evaluated using a significance test on the prediction to answer the question: were the predicted values affected by the compositional change? The model diagnostics Hotelling's T^2 and Q residuals were also used to understand if the model could even detect these changes.

TABLE 23.1
Conditions Evaluated to Test Model Robustness and Detection Ability

Design	Theophylline	LAC Part Size (μm)	Starch
1	Anhydrous[a]	50	Vendor 1[a]
2	Monohydrate	50	Vendor 1[a]
3	Anhydrous[a]	100[a]	Vendor 1[a]
4	Monohydrate	100[a]	Vendor 1[a]
5	Anhydrous[a]	125	Vendor 1[a]
6	Monohydrate	125	Vendor 1[a]
7	Anhydrous[a]	50	Vendor 2
8	Monohydrate	50	Vendor 2
9	Anhydrous[a]	100[a]	Vendor 2
10	Monohydrate	100[a]	Vendor 2
11	Anhydrous[a]	125	Vendor 2
12	Monohydrate	125	Vendor 2

[a] The condition was part of the calibration set.

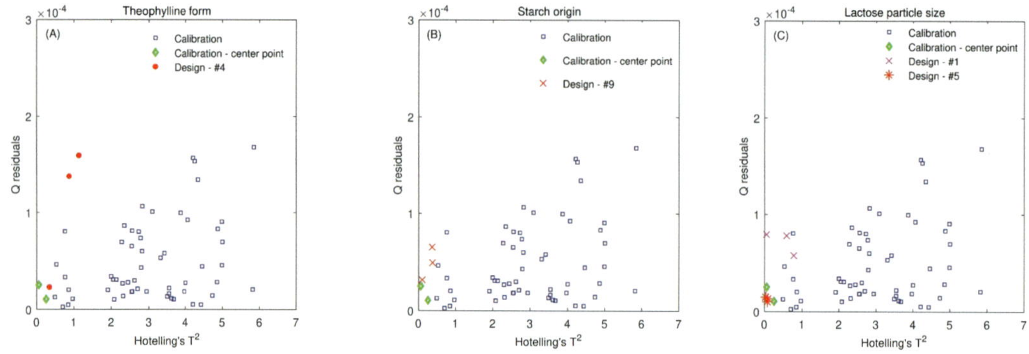

FIGURE 23.5 Hotelling's T^2 and Q-residual plots for the calibration (blue squares), center point (green diamond), and formulation changes for the active ingredient (a), the starch vendor (b), and lactose particle size (c).

Using an analysis of variance with $\alpha = 0.05$, only the effect of lactose particle sizes was significant. The change in the polymorphic form of the active ingredient or the vendor for the starch was not detected by the model. To be clear, the model was not built with the intent to detect these changes, and in a traditional model development process, if the possibility existed that the form of the active ingredient changes during manufacturing, a risk would be raised and the model specificity to the form changes would be investigated to ensure that it can identify when the right form is not present. This may be done using diagnostics as presented in Figure 23.5. Note that Figure 23.5 presents only the cases where only one parameter is changed. For the results of all 12 conditions, readers should refer to the original article (Igne et al. 2014).

If the test samples made by the various conditions in Table 23.1 were considered as identical to the calibration data for the inner and outer distances to the model represented by the Hotelling's T^2 and Q residuals, respectively, they should overlap with the samples represented by the green diamonds, which are samples made at the center point but with the same ingredients as the calibration set. The model diagnostics clearly showed that the samples made with the different forms of the theophylline (test condition 4) and the smaller lactose particle size (test condition 1) were different from those in calibration. However, it is not as straight forward for the starch manufacturer and the larger lactose particle size.

These results are interesting because they indicate that while predictions and diagnostics are useful tools, they don't tell the whole story and thus must be complemented by a robust calibration effort that will ensure the model will have seen the expected robustness (i.e., different particles of excipients and/or different vendors of excipients (if it is expected that the excipient supplier will change during the product lifecycle)). However, for the change in the form of the active, there may be little the NIR model can do, and it will be the robustness of the manufacturing process or a form identification method (see pharmaceutical analysis chapter) that will ensure that only the right active ingredient is present in the tablets.

23.3.5.3 Case Study 3: Process Variability

This last case study will extend the example of the change in raw materials to look at the impact of changing process conditions onto the model performance. More particularly, the effect of climate conditions will be investigated (Igne, Drennen, and Anderson 2014). A pharmaceutical formulation composed of acetaminophen, hypromellose, lactose, microcrystalline cellulose, and magnesium stearate was used to make tablets over a period of a calendar year. The manufacturing process included a granulation step. The climatic conditions where the work was done (North-East of the USA) exhibit humid summers and dry winters. To cope with these issues and build a robust model, a comprehensive method development process was initiated that included samples made at small and large scales during the summer and the winter and samples placed in environmental chambers where the relative humidity was varied from 11% to 75%. This comprehensive approach, or risk-based approach, was intended to include all the moisture variabilities that could affect the model performance. However, as outlined in Figure 23.6, it allowed us to split the data into two separate assessments and investigate less robust scenarios (risk associated with moisture variations and risk associated with process variations) and their impact on model performance.

Calibration models built with NIR data collected from these three scenarios were applied to the prediction of an independent set of samples manufactured from a granulation process manufactured in the summer (higher humidity). Figure 23.7 shows the PLS calibration and prediction plots. Based on a priori information from the scenarios, the following results should be expected:

- the risk-based model (including moisture variability and manufacturing scale variability) should be able to predict the new batch without any issue.
- the climate risk scenario should not be able to account for the difference in moisture between the calibration samples and those used in the test set.
- the process risk scenario where the batch-to-batch variability was not available to the model should not be expected to perform as well as in the risk-based scenario.

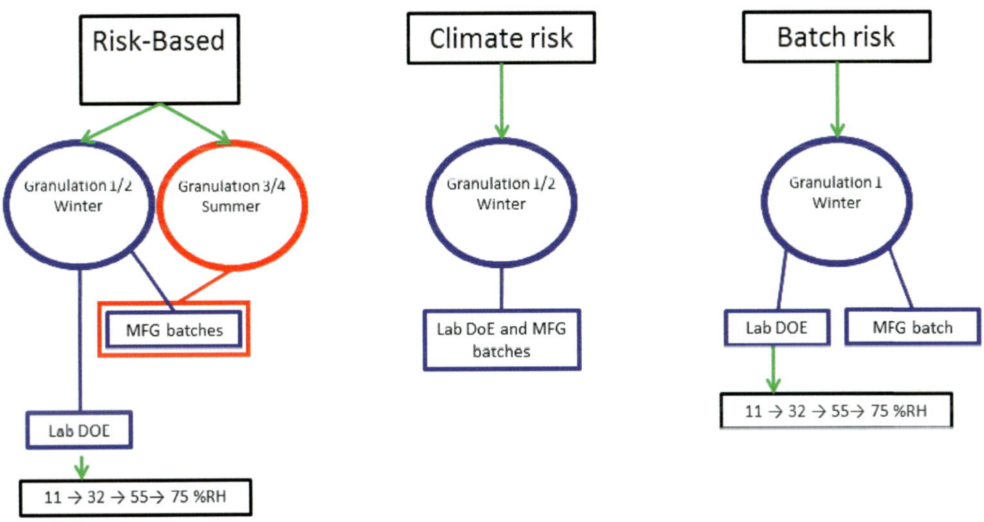

FIGURE 23.6 Calibration scenarios.

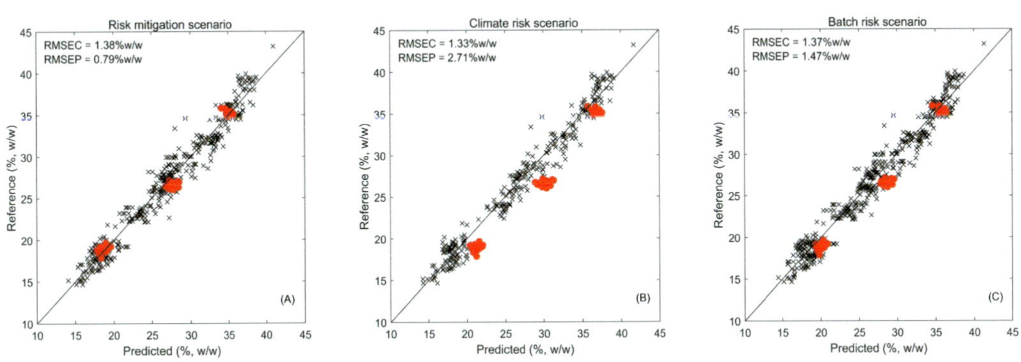

FIGURE 23.7 Calibration and prediction vs. reference plot for the prediction of active content in different modeling scenarios. (a) Risk mitigation scenario. (b) Climate risk scenario. (c) Batch risk scenario.

Results from Figure 23.7 show exactly what one could have expected. The variability due to the lack of moisture robustness is larger than the one linked to batch-to-batch variability, but it may not always be the case. The results differed in their linearity, with the climate risk and the batch risk scenarios exhibiting nonlinearity in the predicted values.

These three case studies showed very simple examples of risks that could affect the development of a calibration model and how they could be mitigated or monitored. The literature is full of such examples, and readers should refer to their particular fields for more applicable examples.

23.4 CONCLUSION

This chapter presented a high-level overview of the method development process and the maintenance of a model. Building a model is often the simpler step in the usage of the model. It can make or break the effort if a fit for purpose performance is not obtained. But when the model does what it is supposed to do, ensuring its performance in operational settings can be a lot more challenging. The analytical quality by design framework can help guide the process, and while it was mainly designed for pharmaceutical applications, the need to properly design a calibration model, test it, and possibly validate it (in the regulatory sense) is common to all fields. Also, to ensure that a model is valid when applied to new samples, model diagnostics should be used. Hotelling's T^2 and Q residuals are powerful statistics that can help explain why a method is not performing as intended or simply to confirm that it is doing its job.

ACKNOWLEDGEMENTS

Authors would like to thank Zhenqi (Pete) Shi from Ely Lilly and Company for his help and review of the manuscript.

REFERENCES

AACC International. 2000. *Approved Methods of the American Association of Cereal Chemists*, 10th Ed. Methods 44-15A.

AACC International. 2000. *Approved Methods of the American Association of Cereal Chemists*, 10th Ed. Methods 46-30.

Chatfield, M. J., and P. J. Borman. 2009. "Acceptance Criteria for Method Equivalency Assessments." *Analytical Chemistry* 81 (24):9841–9848. doi: 10.1021/ac901944t.

EMEA Guideline. Guideline on the Use of Near Infrared Spectroscopy by the Pharmaceutical Industry and the Data Requirements for New Submissions and Variations. http://www.ema.europa.eu/docs/en_GB/document_library/Scientific_guideline/2014/06/WC500167967.pdf.

FDA. Draft Guidance for Industry on Development and Submission of Near Infrared Analytical Procedures, March 2015. http://www.fda.gov/ucm/groups/fdagov-public/@fdagov-drugs-gen/documents/document/ucm440247.pdf.

ICH. Q2 (R1) Validation of Analytical Procedures: Text and Methodology, *International Conference on Harmonization (ICH).*

ICH. Q8 Pharmaceutical Development, *International Conference on Harmonization (ICH).*

ICH. Q11 Development and Manufacture of Drug Substances, *International Conference on Harmonization (ICH).*

Igne, B., J. K. Drennen, and C. A. Anderson. 2014. "Improving Near-Infrared Prediction Model Robustness with Support Vector Machine Regression: A Pharmaceutical Tablet Assay Example." *Applied Spectroscopy* 68 (12):1348–1356. doi: 10.1366/14-07486.

Igne, B., L. R. Gibson, G. R. Rippke, and C. R. Hurburgh Jr. 2007. "Influence of Yearly Variability of Agricultural Products on Calibration Process: A Triticale Example." *Cereal Chemistry* 84 (6):576–581. doi: 10.1094/CCHEM-84-6-0576.

Igne, B., Z. Shi, J. K. Drennen, and C. A. Anderson. 2014. "Effects and Detection of Raw Material Variability on the Performance of Near-Infrared Calibration Models for Pharmaceutical Products." *Journal of Pharmaceutical Sciences* 103 (2):545–556. doi: 10.1002/jps.23816.

Limentani, G. B., M. C. Ringo, F. Ye, M. L. Bergquist, and E. O. McSorley. 2005. "Beyond the t-Test: Statistical Equivalence Testing." *Analytical Chemistry* 77 (11):221 A–226 A. doi: 10.1021/ac053390m.

Miller, C. 2005. "Chemometrics in Process Analytical Technology (PAT)." In *Process Analytical Technology*, edited by K. Bakeev. John Wiley & Sons Ltd, Chichester, UK, pp. 353–434.

Miyano, T., H. Nakagawa, T. Watanabe, H. Minami, and H. Sugiyama. 2015. "Operationalizing Maintenance of Calibration Models Based on Near-Infrared Spectroscopy by Knowledge Integration." *Journal of Pharmaceutical Innovation* 10 (4):287–301. doi: 10.1007/s12247-015-9226-5.

Reid, G. L., J. Morgado, K. Barnett, B. Harrington, J. Wang, J. Harwood, and D. Fortin. 2013. "Analytical Quality by Design (AQbD) in Pharmaceutical Development." *American Pharmaceutical Review*.

Part V

Applications

24 Application of NIR for Analysis of Beverages

Lana R. Kington
Bristol-Myers Squibb

Tom M. Jones
Sanofi-Aventis

CONTENTS

24.1 INTRODUCTION

Important applications of near-infrared analysis (NIRA) are found in the beverage and nutritional formula industries. Technologies have been described for near-infrared (NIR) determinations of constituents in alcoholic beverages such as beer, wines, and distilled spirits; nonalcoholic beverages such as fruit juices, coffees, teas, and soft drinks; and other products such as infant and adult nutritional formulas. Some of these applications are described in this chapter.

Near-infrared (NIR) spectroscopy, both inline and online, since the late 1980s has been proven as a continuous monitoring and process control tool for product quality in both food and agriculture industries. A comprehensive review has been published that describes NIR spectroscopy for the process analysis and control of meat, fruit, grain, dairy products, and beverages. This review focuses on the literature published from 1998 to 2008. The review topics include techniques for process measurements, calibration using chemometric methods, and interpretation of results (Huang et al. 2008).

Near-infrared (NIR) spectroscopy was shown to exhibit rapid and efficient food and beverage analysis. A review paper reports published literature from 1998 to 2008 demonstrating applications of NIR spectroscopy for both raw and processed or factory prepared foods. NIR spectroscopy has been used extensively to measure food and beverage chemical composition, functional properties, quality attributes, and regional and varietal differences. This precise and relatively accurate information is essential for addressing consumer confidence and food safety issues (Woodcock et al. 2008). A review published in 2015 provides an updated and detailed description of data fusion as used in the field of food and beverage quality assessment and authentication (Borràs et al. 2015).

24.1.1 Sample Preparation

Liquid NIRA differs from NIR applications for ground or powdered samples in a few key ways. Rather than utilizing true reflectance measurements, liquid analysis involves transmission or a combination of transmission and reflectance sometimes referred to as transflectance (D. A. Burns, personal communication). Another way in which liquid measurement differs from the measurement of powders involves the introduction of the sample into the instrument. Preparation of a liquid sample may involve dilution, suspension, or homogenization prior to sample introduction. A different type of sample cell, usually a flow cell of some type, is then required.

24.1.2 Liquid Cell Design

Several general factors should be weighed when designing or choosing a liquid cell. The mode of detection-transmission or transflectance will influence the decision. Other factors to consider are the physical characteristics of the sample. A liquid sample could be a slurry, a suspension, a solution, or an emulsion. An additional consideration is the application environment in which the sample will be used. Remote applications could utilize a flow cell in which the sample is injected manually or is processed via a homogenizer and pumped into the cell automatically. Online applications would probably require the ability for the sample to flow continuously or alternately fill and empty the cell as needed during process control.

Liquid cell design often takes into account the effect of temperature on the reproducibility of measurements. Temperature effects on NIR protein are reported to be a factor for differences discovered between Federal Grain Inspection Service Laboratories in the early 1980s (Anon. 1981). Williams concluded that an inverse relationship exists between temperatures of ground samples and their corresponding apparent protein contents (Williams et al. 1982). Three suggestions for compensating for the differences are presented. One method is to select wavelengths that are minimally sensitive to temperature fluctuations. A second approach is to analyze samples at the same temperature at which the calibration was developed. The third suggestion is to develop a temperature compensation factor using a correlation chart. This sensitivity to temperature has also been noted by users of NIR technology for liquid analysis (Siddiqui and Jones 1981). Although the preferred tactic for addressing this potential problem is to select wavelengths that are not temperature sensitive, many available liquid cells are designed with mechanisms to maintain constant temperature. Note that transreflectance (i.e., transflectance) was used successfully for measuring rice wine spectra through glass bottles for alcohol content, titratable acidity, and pH, respectively (Yu et al. 2009).

24.2 ALCOHOLIC BEVERAGES

NIRA is widely utilized for the analysis of constituents in a variety of alcoholic beverages. These include beer, wine, and distilled spirits. Applications range from direct readings for alcohol content to calibrations for original gravity, a function of alcohol.

24.2.1 Beers

NIRA has been applied for determinations of alcohol and other constituents in beers. Coventry and Hunston have reported calibrations for alcohol in beers using transmission (Coventry and Hunston 1984). The most successful calibration for alcohol in a variety of light and dark beers utilized a

TABLE 24.1
Wavelengths Used in Calibrations to Determine Composition of Beers

Constituent	Wavelength (nm)
Alcohol (ref.)	2230
Alcohol	2310
Water reflectance	1445
Sugar	2100

Source: M. Dzwinczyk, *Proc. 8th Int. Symposium on Near Infrared Reflectance Analysis (NIRA)*, Technicon, Tarrytown, New York, 1985.

single wavelength, 1672 nm. Standard errors of the calibration ranged from 0.2 to 0.3 over a range of 0%–11% alcohol content.

The European brewing industry uses NIRA extensively to monitor the brewing process. Critical parameters measured by NIRA include alcohol; original gravity, which is a function of alcohol; real extract; apparent extract; sugars; and total solids. To ensure reproducibility of these measurements, attention is given to maintaining a fixed sample temperature. One method of achieving this result is to circulate water at the desired temperature through a modified solid sample compartment. A cell for this purpose has been developed by Technicon Europe. Approximately 20 mL of liquid is injected into the cell. Equilibration time to achieve the desired temperature is only about 30 s. Some key wavelengths for these types of calibrations described by Dzwinczyk are presented in Table 24.1 (Dzwinczyk 1985).

In a related application, the brewing industry has also extended the uses for NIRA to include analysis of raw material components. Total nitrogen determination in barley and malt and α-acid determination in hops are two such applications (Day 1984).

Fourier-transform near-infrared (FT-NIR) spectroscopy was evaluated for the monitoring of beer fermentation process parameters under different conditions. For the experiment, wort with two different strains of yeast underwent fermentation at different temperatures for 9 days. FT-NIR was used to test its analytical abilities for monitoring biomass, Brix, and pH. For the calibration work, locally weighted regression (LWR), partial least squares (PLS), and principal component analysis (PCA) were evaluated. It was concluded that the LWR calibration technique was a suitable quantitative method for monitoring these parameters in beer production (Grassi et al. 2014).

NIR was evaluated for usefulness in following the process of aging in beer. For this application,

- Successive projection algorithm (SPA) was used in combination with linear discriminant analysis (LDA) for aging classification models. For the experiment, 83 alcoholic and nonalcoholic beer samples from cans and bottles were examined. For aging, simulation samples were stored under dark conditions at 40°C for 10 and 20 days. NIR spectra over the range of 12,500–5405 cm^{-1} were compared to those of freshly brewed samples. Principal component analysis (PCA) results showed clear discrimination of fresh and 10- and 20-day aged samples. For the nonalcoholic samples, the discrimination method was not successful when using either PCA or soft independent modeling of class analogy (SIMCA) discriminant methods. The application of SPA-LDA did indicate successful classification when using a single wavenumber in each case by specifically 5550 cm^{-1} for nonalcoholic samples and 7228 cm^{-1} for alcoholic samples. Note that only one sample was misclassified (Ghasemi-Varnamkhasti et al. 2012).
- NIR and refractometry were combined for a new approach to the analysis of alcohol and real extract for a variety of beer samples. Calibration was accomplished using partial

least squares (PLS) regression between NIR spectra and refractometry with the alcohol and extract concentrations. Results of this work yielded a root mean squared errors of validation (RMSEV) for the validation of the constituents of alcohol and extract concentrations in mass percent (w/w)% as 0.23 Mas% (method A), 0.12 Mas% (method B), and 0.19 Mas% (method C) and 0.11 Mas% (method A), 0.11 Mas% (method B), and 0.11 Mas% (method C), respectively (Castritius et al. 2010).

24.2.2 Wines and Distilled Liquors

The NIR measurement of alcohol and sugars in wine and distilled liquors is used routinely in more than 50 sites in France (Day 1984). The concentration of alcohol in wine was determined by Kaffka and Norris as early as 1976 (Kaffka and Norris 1976). The accuracy of their calibration is reported as better than 0.1% by volume at 11%–17% alcohol using transmittance as the mode of detection. Sucrose in wine has also been investigated. An accuracy of 0.2% is achieved with a calibration for sucrose levels of 0.8%–8.8%. The same authors reported a determination of tartaric acid, another constituent of interest in the wine-making industry, over a range of 0.5%–8.2% with an accuracy of 0.1%.

Near-infrared (NIR) and visible (VIS) spectroscopy have been used to classify commercial Riesling wines from Australia, New Zealand, France, and Germany. For the experiment, commercial Riesling wines (n = 50) were measured over the spectral region of 400–2500 nm in transmission using a monochromator instrument. Principal component analysis (PCA), partial least squares discriminant analysis (PLS-DA), and stepwise linear discriminant analysis (SLDA) based on PCA scores were used to classify Riesling wines based on their country of origin. Full cross-validation (leave-one-out) was used as the validation method when classification models were developed. For the best results, PLS-DA models correctly classified 97.5%, 80%, and 70.5% of the Australian, New Zealand, and European (France and Germany) Riesling wines, respectively. It was recommended by the authors that a larger sample study be completed for validation and refinement of the method (Liu et al. 2008).

NIR was used for online analysis of rice wine using a combination of visible and near-infrared (Vis–NIR) spectra. The calibration method used was least squares support vector machines (LS-SVM). Transreflectance (i.e., transflectance) spectra of rice wine samples were taken using a fiber optic probe across sealed round brown glass bottles with labels removed. LS-SVM models demonstrated improved prediction statistics compared to partial least squares regression (PLSR) models. For the LS-SVM method, correlation coefficients for validation (r_{val}) were 0.915, 0.888, and 0.872, and the root mean square error of prediction (RMSEP) was 0.168 (%(V/V)), 0.146 (g L^{-1}), and 0.033 for alcohol content, titratable acidity, and pH, respectively. Based on the results, online wine quality determination using LS-SVM was reported to be a reliable method (Yu et al. 2009).

24.3 NONALCOHOLIC BEVERAGES

Constituents present in many nonalcoholic drinks have also been quantitated with NIRA, for example, moisture and caffeine in coffee (Day 1984). Other applications have been developed for constituents in teas, soft drinks, and fruit juices. Most of these applications are unpublished.

Coffee beverage sensory data were evaluated using NIR spectra of 51 Arabica roasted coffee samples. The experiment was intended to predict the quality of the coffee beverage based on the levels of acidity, bitterness, flavor, cleanliness, and flavor body. Partial least squares (PLS) combined with ordered predictor selection (OPS) was used to perform calibration using preselected significant spectral regions. The regions of the spectrum recognized as meaningful for sensory quality were related to the NIR spectra of pure caffeine, trigonelline, 5-caffeoylquinic acid, cellulose, coffee lipids, sucrose, and casein. It was reported that the specific chemistry, as measured using NIR, was related to the following coffee quality characteristics: (1) lipids and proteins are associated with the coffee attribute body; (2) caffeine and chlorogenic acids are related to bitter taste; (3) chlorogenic acids were correlated to acidity and flavor quality; (4) and finally, the cleanliness factor and overall evaluated coffee beverage quality are related to the content of caffeine, trigonelline, chlorogenic acid, polysaccharides, sucrose, and protein (Ribeiro et al. 2011).

24.3.1 Fruit Juices

Calibrations for degrees Brix and acidity in fruit juices have been performed utilizing transflectance and a 45°C liquid cell (McDermott 1988). The calibration parameters are presented in Tables 24.2–24.4 with the wavelengths utilized for each calibration.

Carbohydrate in fruit juices has been determined using NIRA transmittance with a 2.2-nm path-length cell (Osborne and Fearn 1986). Such calibrations have worked best when performed for each type of fruit and have achieved relative standard deviations of 0.25% in the range of 9.3%–11.3% for orange juice.

NIR is proving to be an objective, quantitative, and nondestructive technique for the quality assessment of citrus fruit, often used for juice products. A review paper discusses the use of near-infrared (NIR) spectroscopy (NIRS) for internal and external quality measurement of citrus fruit. The methods used for traditional NIRS for citrus analysis are discussed as well as the application of

TABLE 24.2
NIR Calibrations for Brix and Acid over Entire Range

	Brix (range = 4.4%–19%)	Acid (range = 0.3%–2.26%)
Wavelengths (cm)	1445	1828
	1778	2010
	1818	1576
	2190	2150
R	0.999	0.996
SEE	0.200	0.041
N	66	66
F ratio	7348	1779

Source: Courtesy of L. P. McDermott, personal communication.

TABLE 24.3
NIR Calibrations for Brix on the Basis of Percent Brix

	Brix Range (%)		
	4–8	9–14.4	14.5–19
Wavelengths (nm)	1436	1352	1436
	1450	1744	2206
	1506	2220	2388
R	0.990	0.995	0.997
SEE	0.161	0.146	0.127
N	15	29	22
F ratio	184	89	1072

Source: Courtesy of L. P. McDermott, personal communication.

TABLE 24.4
NIR Calibrations for Acid on the Basis of Percent Brix

	Brix Range (%)		
	4–8	**9–14.4**	**14.5–19**
Wavelengths (nm)	2276	1436	1352
	1954	1828	1534
	2052	1758	1800
R	0.999	0.990	0.960
SEE	0.023	0.038	0.024
N	15	29	22
F ratio	4822	405	70

Source: Courtesy of L. P. McDermott, personal communication.

hyperspectral imaging and optical coherence tomography for the modern assessment of fruit quality (Magwaza et al. 2012).

NIR has been applied to the analysis of biological fluids and beverages, in particular for commercial brands of orange juice. A study was undertaken to differentiate between commercial brands of orange juice and naturally derived juices. When a comparative experiment is performed, it was demonstrated that discriminating a pure natural juice from commercial brands was possible. The key spectral region of interest for discrimination was from 1461 to 1635 nm. The use of NIR spectroscopy showed that when natural juice was compared to six commercial brands, only one brand was spectrally close to the natural juice; the remaining five commercial brands were quite different spectrally from the natural juice (Alfouzan et al. 2017).

24.3.2 Glucose in Water Solutions

NIRA calibrations have also been developed for glucose solutions manufactured for pediatric use (Siddiqui and Jones 1981). These solutions typically range from 5% to 10% glucose by volume in water. The simplicity of preparing standard solutions directly from glucose raw material sources makes this an extremely rapid, accurate, and reproducible alternative to standard AOAC methodology.

24.4 FORMULA PRODUCTS

Infant formula products and other nutritional formulas directed to the needs of hospital patients provide a multibillion dollar annual market for several manufacturers. These liquid formulas are complex suspensions of nutrients, the levels of which are carefully controlled. Regulation of these products, especially infant formulas, dictates the need for a complex array of nutrient analyses. While these preparations may often resemble milk or dairy products, they are actually quite different and provide unique analytical problems. Several of the required tests, however, lend themselves well to spectroscopic techniques including NIRA.

24.4.1 Historical Background

The use of NIRA for monitoring constituent levels in infant formula products began in the late 1970s. Protein, fat, content, and total solids monitoring were the targets of initial efforts in this area (N. U. Siddiqui, personal communication). The early work paralleled the development of NIRA for dairy applications. Instrumentation was relatively simple at this point, utilizing a flow cell in the place of the powder sample holder. Samples were prewarmed and injected into the flow cell manually via syringe.

In the early 1980s, significant progress was made to expand the application of NIRA to other types of formula products. Calibrations for constituents in several soy-based, milk-based, and glucose/water formulations were developed (N. U. Siddiqui personal communication, T. M. Jones, personal communication).

These included infant- and hospital-directed products. Concurrently, instrumentation was greatly improved for dairy and liquid applications (Luchter 1983). Samples were now preheated, homogenized, and introduced into the instrument through the use of an external sample preparation module.

The evolution of formula applications involving NIRA extended to the process environment in 1986. Calibrations were developed for different process lines in which the protein, fat, and total solids levels were monitored at different process stages by laboratory personnel using benchtop equipment (Jones and Rademacher 1986). Wavelengths and regression statistics reported for these calibrations are produced in Tables 24.5 and 24.6. The next logical extension of NIRA development would be to perform this testing using online techniques.

TABLE 24.5
Soy Formula Calibration for Major Constituents

	Protein	Fat	Solids
Wavelengths (nm)	2100	1759	1818
	2180	2270	1982
		2310	2139
		2348	
R^2	0.9933	0.9989	0.9998
SEE	0.0770	0.0529	0.0751
N	31	31	31
F ratio	6316	19,378	172,592

Source: Courtesy of Bristol-Myers Squibb MJNG, T. M. Jones and L. R. Rademacher, *Proc. Rocky Mountain Conference*, Denver, CO, 1986.

TABLE 24.6
Milk-Base Formula Calibration for Major Constituents

	Protein	Fat	Solids
Wavelengths (nm)	1818	2270	1445
	2100	2310	2100
	2139	2336	1982
	2180		2270
R^2	0.9946	0.9993	0.9995
SEE	0.535	0.0477	0.1446
N	45	45	45
F ratio	5977	62,364	59,852

Source: Courtesy of Bristol-Myers Squibb MJNG, T. M. Jones and L. R. Rademacher, *Proc. Rocky Mountain Conference*, Denver, CO, 1986.

TABLE 24.7
Typical Accuracy and Precision of NIRA for Liquid Formula Analysis

	Total Solids	Fat/Oil	Protein
Precision (% RSD)	<0.5	<0.3	<0.5
Relative accuracy (%)	±1	±1	±3
N	6	6	6

Source: Courtesy of Bristol-Myers Squibb MJNG, L. R. Rademacher, *Proc. 4th Int. Symposium on Near Infrared Reflectance Analysis*, Technicon, Tarrytown, New York, 1983.

TABLE 24.8
Liquid Formula NIRA Applications

Application	Constituents
Whey-casein infant formula	Protein, fat, solids
Soy protein infant formula	Protein, fat, solids
Glucose/water solution (5% and 10%)	Glucose
Complete isotonic formula	Protein, fat, solids
Complete liquid diet	Protein, fat, solids

Source: Courtesy of Bristol-Myers Squibb MJNG.

24.4.2 Accuracy and Precision

Utilizing NIRA for formula analysis has proven to be cost effective and reliable [14]. The use of a careful check sample program ensures the continued accuracy and reliability of analytical data obtained by the technique over a long period(Rademacher 1983). Table 24.7 provides an indication of the accuracy and precision typical of formula applications.

A summary of specific liquid formula applications in which NIRA is used as a primary method of analysis is given in Table 24.8. As NIRA becomes more widely used in this industry, the list of applications will continue to expand.

24.5 CONCLUSIONS

The use of NIRA for beverage and infant formula applications is rapidly increasing. Typically involving high-volume markets with concurrently large sample volumes in quality control, these product types are ideal candidates for NIR technologies. Investments in instrumentation, calibration software, and operator training are easily and quickly recoverable. Relatively few NIR applications for beverage and formula products have been published to date, however, due to the highly competitive markets involved. This situation may well be resolved as the number of applications and participating laboratories continues to increase.

REFERENCES

Alfouzan, M., B. Al-Otaibi, K. Issa and S.A. Alshebeili. 2017, November. Near infra red (NIR)-based classification of orange juice. In *Electrical and Computing Technologies and Applications (ICECTA), 2017 International Conference on IEEE*, Ras Al Khaimah, UAE, 1–5.

Anon. 1981. *Mill. Bak. News*, *20*, 33.

Borràs, E., J. Ferré, R. Boqué, M. Mestres, L. Aceña and O. Busto. 2015. Data fusion methodologies for food and beverage authentication and quality assessment–A review. *Analytica Chimica Acta*, *891*, 1–14.

Castritius, S., Kron, A., Schäfer, T., Rädle, M. and Harms, D., 2010. Determination of alcohol and extract concentration in beer samples using a combined method of near-infrared (NIR) spectroscopy and refractometry. *Journal of Agricultural and Food Chemistry*, *58*(24), 12634–12641.

Coventry, A.G., and M.J. Hunston. 1984. Application of near-infrared spectroscopy to the analysis of beer samples. *Cereal Foods World*, *29*, 716–718.

Day, M. 1984. *Food Proc.*, *53*, 51.

Dzwinczyk, M. 1985. NIRA in the brewing industry. In *Proc. 8th Intl Symp. on Near Infrared Reflectance Analysis (NIRA)*, Technicon, Tarrytown, NY.

Ghasemi-Varnamkhasti, M., S.S. Mohtasebi, M.L. Rodriguez-Mendez, A.A. Gomes, M.C.U. Araújo, and R.K. Galvão. 2012. Screening analysis of beer ageing using near infrared spectroscopy and the Successive Projections Algorithm for variable selection. *Talanta*, *89*, 286–291.

Grassi, S., J.M. Amigo, C.B. Lyndgaard, R. Foschino, and E. Casiraghi. 2014. Beer fermentation: monitoring of process parameters by FT-NIR and multivariate data analysis. *Food Chemistry*, *155*, 279–286.

Huang, H., H. Yu, H. Xu, and Y. Ying. 2008. Near infrared spectroscopy for on/in-line monitoring of quality in foods and beverages: A review. *Journal of Food Engineering*, *87*(3), 303–313.

Jones, T.M. and L.R. Rademacher. 1986. A strategy for the use of NIRA in liquid formula quality control applications. In Proc. Rocky Mountain Conference, Denver, CO.

Kaffka, K.J. and K.H. Norris. 1976. Rapid instrumental analysis of composition of wine. *Acta Aliment*, 5, 267–279.

Liu, L., D. Cozzolino, W.U. Cynkar, R.G. Dambergs, L. Janik, B.K. O'neill, C.B. Colby, and M. Gishen. 2008. Preliminary study on the application of visible–near infrared spectroscopy and chemometrics to classify Riesling wines from different countries. *Food Chemistry*, *106*(2), 781–786.

Luchter, K. 1983. InfraAlyzer 400D evaluation. In *Proc. 4th Intl Symp. on Near Infrared Reflectance Analysis*, Technicon, Tarrytown, NY.

Magwaza, L.S., U.L. Opara, H. Nieuwoudt, P.J. Cronje, W. Saeys and B. Nicolaï. 2012. NIR spectroscopy applications for internal and external quality analysis of citrus fruit—a review. *Food and Bioprocess Technology*, *5*(2), 425–444.

Osborne, B.G. and T. Fearn. 1986. *Near Infrared Spectroscopy in Food Analysis*, Longman Higher Education, New York, p. 132. ISBN 10: 0582494893ISBN 13: 9780582494893.

Rademacher, L.R. 1983. Quantitative aspects of NIRA: Its value as a control tool. In *Proc.* 4th *Intl Symp. on Near Infrared Reflectance Analysis*, Technicon, Tarrytown, NY.

Ribeiro, J.S., M.M.C. Ferreira and T.J.G. Salva. 2011. Chemometric models for the quantitative descriptive sensory analysis of Arabica coffee beverages using near infrared spectroscopy. *Talanta*, *83*(5), 1352–1358.

Siddiqui, N.U. and T.M. Jones. 1981. Bristol-Myers, Evansville, IN, unpublished study.

Williams, P.C., K.H. Norris, and W.S. Zarowski. 1982. Influence of temperature on estimation of protein and moisture in wheat by near-infrared reflectance. *Cereal Chemistry*, *59*, 473–477.

Woodcock, T., G. Downey and C.P. O'Donnell. 2008. Better quality food and beverages: the role of near infrared spectroscopy. *Journal of Near Infrared Spectroscopy*, *16*(1), 1–29.

Yu, H.Y., X.Y. Niu, H.J. Lin, Y.B. Ying, B.B. Li, and X.X. Pan. 2009. A feasibility study on on-line determination of rice wine composition by Vis–NIR spectroscopy and least-squares support vector machines. *Food Chemistry*, *113*(1), 291–296.

25 Applications of NIR Spectroscopy in the Confectionary Industry

*Jerome Workman, Jr.**
MJH Life Sciences

W. Jeffrey Hurst
The Hershey Company

CONTENTS

25.1 INTRODUCTION

For the topic of applications of NIR spectrometry in the confectionary industry, a modest literature search using Google with the search term *NIR and confectionery* indicates the number of citations in the thousands. Furthermore, using data from Statistica, it can be gleaned that worldwide chocolate sales are 82 B $USD, while US consumption is approximately 22 B $USD and Western Europe at 39B $USD. Worldwide cocoa harvest is 4.55 million metric tons with the largest consumption being in Switzerland. This provides some perspectives on the use and potential use of analytical techniques for measuring the quality of raw and processed cocoa. NIR has broad applications in the confectionary industry segment, and some of the various applications with references will be given in this chapter. There will be a few terms that might create confusion such as *percent (%) cacao* used by some manufacturers and *percent (%) sum of cocoa solids and cocoa butter.* There are some who feel that the greater the *% cacao* in a product, the better it would be as a confectionary, but in an extreme case, there is a product called cocoa butter soap that approaches 100% cacao and contains no cocoa solids suitable for consumption. Note that cacao as a powder is made by cold-pressing (pressurized) unroasted cocoa beans. The result of this process is a powder retaining the natural living enzymes but with the cocoa fat (cacao butter) removed. Ingredients for chocolate: The following tables describe the common ingredients and ingredient names used for chocolate confectionary products (Tables 25.1 and 25.2).

* jworkman2002@yahoo.com.

TABLE 25.1
A List of Ingredients Present in Chocolate

Dark Chocolate: Cacao liquor, sugar, cacao butter, lecithin, and vanilla
Milk Chocolate: Cacao liquor, sugar, cacao butter, milk solids, milk fat, lecithin, vanilla
White Chocolate: Sugar, cacao butter, milk solids, milk fat, lecithin, vanilla
Chocolate Confections: Cacao liquor, milk solids, cacao butter, butter, vanilla, natural flavorings, cacao beans, cacao seeds, cocoa powder, nuts, and soy lecithin/glucose, natural flavors include real vanilla, vanilla beans, fresh herbs, fresh fruit, fruit purees, liquor

TABLE 25.2
Ingredient Names Common to Chocolate Confectionery Labels

Cocoa Related: Cocoa liquor, chocolate liquor, unsweetened chocolate, cacao mass, cocoa mass, chocolate fondant, cocoa beans, cacao beans, chocolate beans, cacao seeds, cocoa seeds, chocolate seeds, cacao butter, cocoa butter, cacao oil, cocoa oil, cocoa fat, cacao fat
Dairy Related: Milk, cream, whole milk, condensed milk, milk crumb, dry milk powder, dry cream powder, milk solids, dry milk solids, milk fat, butter oil, butter fat
Soybean Related: Soy lecithin, lecithin, soya lecithin
Vanilla Bean Related: Vanilla bean, real vanilla, ground vanilla beans, whole vanilla beans, vanilla bean extract

25.2 PARAMETERS MEASURED USING NIR

The application of NIR in the confectionery space has included the entire variety of available technologies, including filter-based instruments (tilting and discrete filters), monochromator or dispersive instruments (the most common), acousto-optic tunable filter (AOTF) systems, and Fourier transform-NIR (FT-NIR). Several manufacturers will provide basic starter calibrations, calibration services, networking, and even advanced calibrations. There are a variety of technologies and calibration approaches that are effective for measuring raw materials and processed final products.

Table 25.3 provides an example of a typical NIR instrument that has been adapted to measurements of cocoa-based products; such an instrument includes precalibrated applications, sample types, and typical data. The assays given in this table are typical for the industry.

In addition to the various lab-based NIR instruments, there are another group of instruments that have been developed to monitor parameters in real time and for potentially harsh plant process environments. One of the newer application areas that have been evolving over the last few years is the application in the detection of food adulteration. A consortium of industry, government, and

TABLE 25.3
Analysis of Chocolate and Cocoa Parameters Using NIR

Constituent or Parameter	Measured Range (s)
Moisture	0.2%–1.8%
Dry matter	10.2%–52.8%
Fat	0.1%–18.2%; 7.5%–46.0%; 27.3%–90.7%
Solid fat at 20°C	48.0%–72.2%
Protein	2.4%–4.6%; 2.0%–4.5%
Lactose	0.4%–18.1%
Sucrose	11.5%–53.3%
Theobromines (3,7-dimethyl-1H-purine-2,6-dione)	1.056%–4.246 mg kg^{-1}

instrument company scientists published a paper on the detection of the adulteration of skim milk powder by melamine resulting from one of the first incidents of adulteration (Moros et al. 2007). For additional information on the topic of detection and analysis of food fraud and adulterants analysis, refer to Ellis et al. (2015).

25.3 NIR FOR QUANTITATIVE CONFECTIONERY ANALYSIS

In an early research report, near-infrared (NIR) spectroscopy was described as a general method for measuring a variety of food types based on the absorption of electromagnetic radiation in the wavelength range of 780–2500nm (Osborne 2006). The NIR method is described as measuring overlapping absorption bands primarily composed of overtones and combinations of fundamental vibrational modes. The main molecular interactions involve O–H, C–H, and N–H. The measurement of the NIR spectrum, when combined with a variety of regression methods, where the reference chemical measurements are mathematically regressed against the NIR spectra, provides a method requiring from 15 to 90s for measuring primarily moisture, protein, fat, and carbohydrate constituents in ground samples. NIR when applied to chocolate liquor or ground cocoa beans or cocoa powder provides a rapid analysis quality tool for the chocolate industry.

25.3.1 Chocolate Quantitative Analysis

Calibration equations for analysis of sucrose, lactose, fat, and moisture in chocolate were computed using Fourier transform near-infrared (FT-NIR) spectra (Tarkošová and Čopíková 2000). These spectra were also used for estimating rheological properties as viscosity and yield of chocolate. The NIR-predicted results were compared with the standard reference techniques for calibration and validation. The spectra of 96 chocolate samples were measured over a spectral range of 910–2500nm using an FT-NIR Nicolet Avatar 360N spectrometer and Up diffuse reflection infrared Fourier transform (UpDRIFT) spectroscopy accessory. Mathematical preprocessing consisted of either first or second derivative was combined with a partial least squares (PLS) calibration algorithm. The models were validated using a cross-validation method. For the work, it was determined that fat, lactose, and sucrose could be predicted with suitable accuracy. However, the predictions for moisture, viscosity, and yield were not considered accurate enough at this time. Moisture is a difficult constituent to calibrate unless based on Karl Fischer moisture reference values since oven volatiles consist of many different chemical constituents. The NIR spectra are able to measure water quite accurately if the sample reference values are accurate for water content while spectra are measured.

Artificial neural networks (ANNs) have been combined with diffuse reflectance near-infrared Fourier transform spectroscopy (DRIFTS) for measuring nutritional parameters in chocolate (Moros et al. 2007). The reference methods to obtain this information are time consuming and laborious. For this study, 36 chocolate samples were analyzed for the main nutritional parameters in chocolate, such as fat, proteins, carbohydrates, cocoa content, and energetic value. For the NIR analysis, triplicate measurements of each sample were made. After spectra were collected, cluster hierarchical analysis (CHA) was used to select both calibration and validation sets. The calibration set comprised a limited set of 19 with a validation data set of 17 unique samples. ANNs were selected as a spectral data calibration method for developing a predictive model for the nutritional parameters, based on modeling the NIR spectra compared to the reference laboratory data. Results of the calibration work were reported as the root-mean-square error of prediction (RMSEP). The resulted prediction results obtained for carbohydrates, fat, energetic value, and cocoa were 1.0% (w/w), 1.0% (w/w), 50kJ $(100\,g)^{-1}$, and 1.4%, respectively. The mean difference (d_{x-y}) and standard deviation of mean differences (s_{x-y}) of the carbohydrates, fat, protein content, energetic value, and cocoa content were 0.9% and 2.4% (w/w), 0.2% and 1.0% (w/w), 9.1 and 50kJ $(100\,g)^{-1}$, and −0.5% and 1.4%, respectively. The maximum relative error for the prediction (QC) of any of these parameters for a new sample did not exceed 5.2%. Of course, the limited number of samples did not yield conclusive results but indicated some significant potential for using NIR to predict complex parameters in chocolate.

Near-infrared spectroscopy (NIRS) was evaluated as a rapid analytical method for the determination of sucrose content in chocolate mass (da Costa Filho 2009). A variety of chocolate mass types and recipes, with a wide range of composition, were used to develop a predictive model for sucrose composition. The more diverse sample set was purposed to provide improved model performance and compute a single comprehensive calibration. Calibration methods used included classical multiple linear regression (MLR) and PLS. GA-MLR models were developed using a variable selection method based on the regression coefficient of the PLS regression vector and by applying a genetic algorithm (GA). High correlation coefficients (r) resulted for calibrations as 0.998, 0.997, and 0.998 for PLS, MLR, and the GA-MLR methods, respectively, and low prediction errors were achieved. The results demonstrated that NIR is suitable as a rapid method to determine sucrose in chocolate mass for typical chocolate production.

Near-infrared (NIR) spectroscopy was reported to be a useful technique for the analysis of protein, fat, sugar, and water content in chocolate base using the spectral region of 4000–12,000 cm^{-1} when combined with chemometric calibration techniques (Stohner et al. 2012).

25.3.2 Processed Cocoa Quantitative Analysis

A study tested the use of near-infrared (NIR) spectroscopy for routine analytical prediction of biochemical quality parameters in processed cocoa. For most producers, quality control is achieved by visual inspection and sensory testing (Krähmer et al. 2015). Chromatographic methods have not proven practical for flavor evaluation due to laborious isolation and purification steps. Therefore, a rapid and reasonably accurate method would be of great benefit to the industry. The NIR method was evaluated for cocoa quality measurements, and the following results were achieved: phenolic substances (R^2 = 0.93), organic acids (R^2 = 0.88), epicatechin (R^2 = 0.93), lactic acid (R^2 = 0.87), fermentation time (R^2 = 0.92), and pH value (R^2 = 0.94).

25.3.3 Cocoa Bean Quantitative Analysis

Near-infrared (NIR) reflectance spectroscopy was shown to be useful for the prediction of procyanidins in cocoa beans (*Theobroma cacao*) (Whitacre et al. 2003). Procyanidins may be associated with potential health benefits. NIR spectra were used for computing a predictive model for the quantitation of procyanidin oligomers using high-performance liquid chromatography (HPLC) as reference data. The calibration set comprised cocoa beans from different seasons, geographical environments, and fermentation levels in order to provide a wide concentration range for procyanidins compared to naturally occurring levels. The predictive results for the NIR calibrations were as R^2 = 0.983 with standard deviation and standard error of cross-validation ratio (SD/SECV) of 5.68. A global version of the calibration resulted in an R^2 = 0.98 and a SD/SECV = 6.20 across 20 FOSS NIR instruments. The NIR predictive model is able to provide a rapid method for the analysis of procyanidins in cocoa liquors for quality assurance and product manufacturing.

The comparison of spectral regions, such as near-infrared (NIR) and Fourier-transform infrared (FTIR), provides additional information for cocoa powder analysis of major constituents. In one set of experiments, spectral information was compared and then combined for calibrations using 100 cocoa powder samples over the spectral ranges of 1100–2500 and 2500–16,667 nm, representing 9090–4000 and 4000–600 cm^{-1} (Veselá et al. 2007). Calibration models were created for the prediction of moisture (mean = 3.98%), fat (mean = 13.51%), and nitrogen (mean = 3.77%). The ranges for the calibration samples were as follows: moisture (1.60%–7.80%), fat (2.42%–22.00%), and nitrogen (0.88%–4.48%). For the NIR spectra, prediction results as *relative* root mean square error of cross-validation (RMSECV/mean) were as follows: 5.2% (R^2 = 0.94) for moisture, 7.0% (R^2 = 0.96) for fat, and 1.7% (R^2 = 0.98) for nitrogen. These correspond to computed RMSECV values of 0.21% for moisture, 0.95% for fat, and 0.065% for nitrogen, respectively. For FTIR, the *relative* (RMSECV/mean) was unacceptable for moisture, 10.4% (R^2 = 0.94) for fat, and 3.9% (R^2 = 0.95) for nitrogen. These correspond to computed RMSECV values of 1.41% for fat and 0.15% for nitrogen, respectively. The use of data fusion by combining the NIR with FTIR spectral regions allowed more detailed studies in order to characterize frequencies in one domain based on the information of the other domain. A second derivative using NIR spectra was determined to be the best overall approach.

FT-NIR spectroscopy was evaluated for nondestructive evaluation of cocoa bean quality. The feasibility study for using NIR for analysis consisted of developing calibration models for fermentation index, total polyphenol content, and pH (Sunoj et al. 2016). The cocoa pod samples were pretreated by storage for 0, 7, 14, and 21 days in order to test the effect on these quality parameter measurements. The set of test beans was then fermented using the heap method. For the test samples, the fermentation index range was 0.535–1.242 and the total polyphenol content ranged from 6.48 to 15.58 mg g^{-1}, with the sample pH ranging from 4.26 to 6.13. All test samples were measured over the spectral range of 12,500–3600 cm^{-1} (800–2778 nm). Traditional quality analysis results were compared to the NIR spectra using PLS regression. Spectra were preprocessed using vector normalization, multiplicative scatter correction (MSC), and first-derivative (1D) techniques. The calibration model results were as follows: for fermentation index and total polyphenols, the coefficient of determination as R^2 was greater than or equal to 0.80 and pH results as $R^2 < 0.80$. The proposed method was quite rapid for all constituent analysis at less than 1 min total analysis time. Conventional laboratory methods require ≈28 h. The conclusion was that FT-NIR spectroscopy produced rapid and accurate quality attributes of fermenting cocoa and could possibly be employed to similar food products.

Cocoa bean quality and value are often associated with the total fat content in the beans. For this reason, Fourier transform near-infrared (FT-NIR) spectroscopy was tested for total fat analysis in beans in order to evaluate its use as a rapid and accurate analysis method (Teye and Huang 2015). A unique set of calibration approaches was tested. These included the use of efficient spectra interval selection by synergy interval partial least squares (Si-PLS), support vector machine regression (SVMR), and synergy interval support vector machine regression (Si-SVMR). The calibration algorithms were compared using root-mean-square error of prediction (RMSEP) and correlation coefficient (R_{pre}) in the prediction set. The optimum calibration for this experiment was reported as the Si-SVMR model with an RMSEP = 0.015% and $R_{pre} = 0.9708$. This study reported that the total fat content in cocoa beans could rapidly be predicted by FT-NIR spectroscopy using the Si-SVMR calibration technique.

Near-infrared spectroscopy (NIRS) and electronic tongue (ET) technology were combined to test an analytical method for the determination of total polyphenol content (TPC) as a measure of phytochemicals in cocoa beans (Huang et al. 2014). For this experiment, 110 cocoa bean samples with different polyphenol content were measured with both NIRS and ET. Both NIR and ET were compared to the data fusion vector from NIR plus ET. The vectors from each method and the combined method were scaled using normalization and principal component analysis (PCA). Support vector machine regression (SVMR) was used to construct calibration models. Final model performance was evaluated using the correlation coefficient (R_{pre}), RMSEP, and bias in the prediction set. Compared with a single technique (NIRS or ET), the data fusion was superior for the determination of TPC in cocoa beans. The optimal data fusion model was achieved with $R_{pre} = 0.982$, RMSEP = 0.900 g/g, and a bias = 0.013 g/g in the prediction set. The overall results demonstrate combining NIRS and ET yielded improved prediction of TPC in cocoa beans over the individual techniques.

Near-infrared spectra and denaturing gradient gel electrophoresis (DGGE) profiles were both used to evaluate the potential for predicting the microbiological changes taking place during the fermentation of cocoa (Nielsen et al. 2008). For the experiment, cocoa bean samples were sampled at 24-h intervals, then dried, and analyzed using NIR. This represents the first study to microbiological changes during cocoa bean fermentation using DGGE as correlated to NIR using multivariate data analysis. The DGGE profiles were correlated to the NIR spectra using partial least squares regression models (PLS2). It was determined that a correlation (r) of 0.87 was obtained for the bacterial-derived DGGE profiles and $r = 0.81$ for the yeast-derived DGGE profiles. These results were considered promising for rapid analysis of fermentation using NIR.

Near-infrared (NIR) hyperspectral imaging (HSI) over a spectral range of 1000–2500 nm (10,000–4000 cm^{-1}) was used to predict fermentation index (FI), total polyphenols (TP), and antioxidant activity (AA) of individual dry fermented cocoa beans (Caporaso et al. 2018). These beans were scanned nondestructively on a single seed basis. Cocoa bean samples for testing were selected as 10 beans from each of 17 batches. Samples were scanned as single seeds using NIR-HIS, and calibrations were developed versus reference data using PLS regression. External validation performance was determined as $R^2 = 0.50$ (RMSEP = 0.27) for fermentation index (FI); $R^2 = 0.70$ (RMSEP = 34.1 mg ferulic acid per gram) for total polyphenols (TP); and $R^2 = 0.74$ (RMSEP = 60.0 mmol Trolog per kilogram) for antioxidant activity (AA). It was concluded by this study that the prediction performance was suitable for screening purposes useful for food quality control applications.

Fourier transform near-infrared spectroscopy (FT-NIRS) was evaluated as an analytical technique for measuring total fungi count (TFC) (Kutsanedzie et al. 2018). TFC is used as a quality indicator for cocoa beans, as the presence of various fungi within the beans can result in a food safety problem. A solution of cocoa beans was made for multiple samples, and these were measured using FT-NIRS and correlated to the reference laboratory TFC values using a variety of chemometric calibration algorithms. These calibration methods included PLS, Si-PLS, synergy interval-genetic algorithm-PLS (Si-GAPLS), ant colony optimization – PLS (ACO-PLS), and competitive-adaptive reweighted sampling-PLS (CARS-PLS). The calibration performance test parameters included the coefficients of the prediction (R_p), the RMSEP, and the ratio of the sample range standard deviation to RMSEP (RPD). The optimum results obtained were $0.951 \leq R_p \leq 0.975$ and $3.15 \leq \text{RPD} \leq 4.32$, using FT-NIRS combined with the Si-GAPLS calibration algorithm. The method was reported as being suitable for the quantification of TFC in cocoa beans for quality and safety monitoring.

25.4 NIR FOR QUANTITATIVE CONFECTIONERY ANALYSIS

Near-infrared spectra were compared to data from a trained sensory panel for a large set of raw and roasted cocoa beans, chocolate mass, and finished chocolate (Davies et al. 1991). The sensory panel data demonstrated clear differences between samples based on geographical origin and different processing. NIR was able to discriminate raw bean data better than finished chocolate; a correlation of $r = 0.86$ was demonstrated between the raw bean spectra and the sensory data on finished chocolate. For discrimination of high- versus low-quality raw beans, the NIR accurately identified 64% of the low-quality samples and rejected 20% of acceptable quality samples. This was considered a positive result for feasibility for the replacement of difficult and demanding sensory analysis.

Near-infrared (NIR) spectroscopy was tested for the rapid identification of cocoa bean varieties (Teye 2016). Five cocoa bean cultivars were evaluated for discrimination using the spectral region from 10,000 to 4000 cm^{-1} (1000–2500 nm). Support vector machine (SVM) and linear discriminant analysis (LDA) algorithms were compared for discrimination capabilities. The comparison of discrimination models was based on the principal component analysis (PCA) and verified by cross-validation. The research indicated that the SVM model outperformed the LDA model. The SVM model was based on five principal components and achieved a 100% correct identification for both training and prediction sets. The five cocoa bean cultivars identified were reported as IMC 85 × IMC 47, PA7 × PA150, PA150 × Pound7, Pd10 × Pd15, and T63/967 × T65/238.

Discriminant analysis was performed using FT-NIR to compare Ghana cocoa beans based on their geographical region of origin (Teye et al. 2013). For this study, NIR spectra were measured for 194 cocoa bean samples from seven cocoa-growing regions. Data analysis of the spectra included an initial PCA followed by the comparison of various algorithms for discrimination power. Four different discriminant analysis methods were compared. The compared discriminant analysis methods included K-nearest neighbor (KNN), linear discriminant analysis (LDA), support vector machine

(SVM), and back-propagation artificial neural network (BPANN). The algorithms tested as follows: the SVM model exhibited a discrimination success rate of 100% for both the training and prediction sets with mean centering (MC). BPANN showed a discrimination rate of 99.23% for the training set and 96.88% for the prediction set. The LDA model had 96.15% and 90.63% for the training and prediction sets, respectively. KNN model had 75.01% for the training set and 72.31% for the prediction set. This work demonstrated powerful discrimination for origin of beans even from a similar geographical region.

The specifics of the fermentation process determine the cocoa flavor and taste as derived from cocoa beans. To study the effects of fermentation on the measurable chemistry of cocoa beans, a combination of analytical methods was applied. The methods evaluated included the cut test, colorimetry, fluorescence spectroscopy, near-infrared spectroscopy, and gas chromatography–mass spectrometry (GC–MS). The data obtained using these analytical techniques were evaluated using chemometric methods to examine cocoa beans sampled at different durations of fermentation and samples representing fully fermented and dried beans from all cocoa-growing regions of Ghana (Aculey et al. 2010). The results of this study were as follows: changes were observed in measured values for colorimetry (higher a^* and b^* values related to length of fermentation), and the volatile compound content was related to fermentation time. For example, the formation of 2-phenylethyl acetate, propionic acid, and acetoin and the breakdown of diacetyl occurred as fermentation progressed.

Cocoa bean authentication was tested using FT-NIR spectra with SVM and Si-PLS algorithms (Teye et al. 2014a). SVM was applied to discriminate between fermented cocoa beans (FC), unfermented cocoa beans (UFC), and adulterated cocoa beans having 5–40 wt/wt% content of unfermented cocoa beans (UFC); the Si-PLS model was used to quantify the UFC content in the FC samples. The SVM was found to accurately discriminate the cocoa bean samples; the identification rate obtained was 100% for both the training set and the prediction set using three principal components (PCs). The Si-PLS calibration performed with an RMSEP of 1.68% and the coefficient of correlation in prediction (R_{pred}) = 0.98 for % UFC in FC. It was concluded that FT-NIR spectroscopy together with the tested algorithms was useful for both identification of fermented and unfermented cocoa beans and quantification of UFC down to 5% concentration in FC.

Data fusion combining near-infrared (NIR) spectra and electronic tongue (ET) data was tested together for the classification of five cocoa bean varieties (Teye et al. 2014b). PCA was used to fuse the data vectors by normalization, and SVM was used to develop the classification model. The results indicated that either single sensor (NIR or ET) was able to classify between 83% and 93% accuracy, while the SVM and data fusion (NIR plus ET) yielded a classification accuracy of 100% using three principal components for both training and prediction tests. It is concluded that ET combined with NIR data can accurately classify cocoa bean varieties. NIR has proven itself to be a cost-effective, rapid, and accurate analytical technique when properly applied for quantitative and qualitative analyses of cocoa-related products.

REFERENCES

Aculey, P.C., Snitkjaer, P., Owusu, M., Bassompiere, M., Takrama, J., Nørgaard, L., Petersen, M.A. and Nielsen, D.S., 2010. Ghanaian cocoa bean fermentation characterized by spectroscopic and chromatographic methods and chemometrics. *Journal of Food Science*, *75*(6), pp. S300–S307.

Caporaso, N., Whitworth, M.B., Fowler, M.S. and Fisk, I.D., 2018. Hyperspectral imaging for non-destructive prediction of fermentation index, polyphenol content and antioxidant activity in single cocoa beans. *Food Chemistry*, *258*, pp. 343–351.

da Costa Filho, P.A., 2009. Rapid determination of sucrose in chocolate mass using near infrared spectroscopy. *Analytica Chimica Acta*, *631*(2), pp. 206–211.

Davies, A.M.C., Franklin, J.G., Grant, A., Griffiths, N.M., Shepherd, R. and Fenwick, G.R., 1991. Prediction of chocolate quality from near-infrared spectroscopic measurements of the raw cocoa beans. *Vibrational Spectroscopy*, *2*(2–3), pp. 161–172.

Ellis, D.I., Muhamadali, H., Haughey, S.A., Elliott, C.T. and Goodacre, R., 2015. Point-and-shoot: rapid quantitative detection methods for on-site food fraud analysis–moving out of the laboratory and into the food supply chain. *Analytical Methods*, *7*(22), pp. 9401–9414.

Huang, X., Teye, E., Sam-Amoah, L.K., Han, F., Yao, L. and Tchabo, W., 2014. Rapid measurement of total polyphenols content in cocoa beans by data fusion of NIR spectroscopy and electronic tongue. *Analytical Methods*, *6*(14), pp. 5008–5015.

Krähmer, A., Engel, A., Kadow, D., Ali, N., Umaharan, P., Kroh, L.W. and Schulz, H., 2015. Fast and neat–determination of biochemical quality parameters in cocoa using near infrared spectroscopy. *Food Chemistry*, *181*, pp. 152–159.

Kutsanedzie, F.Y., Chen, Q., Hassan, M.M., Yang, M., Sun, H. and Rahman, M.H., 2018. Near infrared system coupled chemometric algorithms for enumeration of total fungi count in cocoa beans neat solution. *Food Chemistry*, *240*, pp. 231–238.

Moros, J., Iñón, F.A., Garrigues, S. and De la Guardia, M., 2007. Near-infrared diffuse reflectance spectroscopy and neural networks for measuring nutritional parameters in chocolate samples. *Analytica chimica acta*, *584*(1), pp. 215–222.

Nielsen, D.S., Snitkjaer, P. and van den Berg, F., 2008. Investigating the fermentation of cocoa by correlating denaturing gradient gel electrophoresis profiles and near infrared spectra. *International Journal of Food Microbiology*, *125*(2), pp. 133–140.

Osborne, B.G., 2006. Near-Infrared Spectroscopy in Food Analysis. Encyclopedia of Analytical Chemistry: Applications, Theory and Instrumentation. in *Encyclopedia of Analytical Chemistry*, R.A. Meyers and R.J. McGorrin eds., John Wiley & Sons, Ltd: Hoboken, New Jersey. doi: 10.1002/9780470027318.a1018

Stohner, J., Lukas, B., Suter, M., Zucchetti, B., Deuber, F. and Hobi, F., 2012. NIRS of chocolate and its chemometric analysis. *Newfood*, *15*(6), pp. 21–28.

Sunoj, S., Igathinathane, C. and Visvanathan, R., 2016. Nondestructive determination of cocoa bean quality using FT-NIR spectroscopy. *Computers and Electronics in Agriculture*, *124*, pp. 234–242.

Tarkošová, J. and Čopíková, J., 2000. Fourier transform near infrared spectroscopy applied to analysis of chocolate. *Journal of Near Infrared Spectroscopy*, *8*(4), pp. 251–257.

Teye, E., 2016. Nondestructive authentication of cocoa bean cultivars by FT-NIR spectroscopy and multivariate techniques. *Focus On Medical Sciences Journal*, *2*(3) pp. 1–10.

Teye, E., and Huang, X., 2015. Novel prediction of total fat content in cocoa beans by FT-NIR spectroscopy based on effective spectral selection multivariate regression. *Food Analytical Methods*, *8*(4), pp. 945–953.

Teye, E., Huang, X., Dai, H. and Chen, Q., 2013. Rapid differentiation of Ghana cocoa beans by FT-NIR spectroscopy coupled with multivariate classification. *Spectrochimica Acta Part A: Molecular and Biomolecular Spectroscopy*, *114*, pp. 183–189.

Teye, E., Huang, X.Y., Lei, W. and Dai, H., 2014a. Feasibility study on the use of Fourier transform near-infrared spectroscopy together with chemometrics to discriminate and quantify adulteration in cocoa beans. *Food Research International*, *55*, pp. 288–293.

Teye, E., Huang, X.Y., Takrama, J. and Haiyang, G., 2014b. Integrating NIR spectroscopy and electronic tongue together with chemometric analysis for accurate classification of cocoa bean varieties. *Journal of Food Process Engineering*, *37*(6), pp. 560–566.

Veselá, A., Barros, A.S., Synytsya, A., Delgadillo, I., Čopíková, J. and Coimbra, M.A., 2007. Infrared spectroscopy and outer product analysis for quantification of fat, nitrogen, and moisture of cocoa powder. *Analytica chimica acta*, *601*(1), pp. 77–86.

Whitacre, E., Liver, J.O., Van Den Broek, R., Van Engelen, P., Remers, B.K., Van Der Horst, B., Tewart, M.S. and Jansen-Beuvink, A., 2003. Predictive analysis of cocoa procyanidins using near-infrared spectroscopy techniques. *Journal of Food Science*, *68*(9), pp. 2618–2622.

26 NIR Analysis of Wheat Products

Brian G. Osborne[*]

CONTENTS

26.1 INTRODUCTION

The flour milling and baking industry was an early adopter of NIR technology. This was a consequence of a commercial need for the measurement of wheat protein content at the point of trade. Once NIR had been adopted for this purpose, its use was extended to the analysis of flour at critical control points in flour milling as well as in baking and snack food production. This chapter reviews the applications of NIR spectroscopy in the cereals industry for process control, compositional analysis of wheat-based products (bread, cookies, flour confectionery, snack foods and breakfast cereals) and cereal chemistry research.

26.2 FLOUR MILLING

An essential aspect of mill monitoring and control is the maintenance of product quality, which can be accomplished by NIR in the laboratory, at-line or on-line. On-line NIR technology has enabled

[*] bandgosborne@bigpond.com.

the analysis of intermediate and finished flours to be performed continuously. By careful selection of the sampling points, the miller can determine when and where adjustments should be made to ensure consistency of product quality. Many mills incorporate continuous flour blending to tailor-make products to each customer's specifications using a few base straight-run flours. This provides the maximum flexibility in responding to customer requirements.

26.2.1 Protein

The determination of protein in wheat flour by NIR is defined by AACC International Approved Method 39-11.01. The method utilizes NIR readings either at discrete nonadjoining wavelengths – usually 2180, 2100, 1680 nm (Osborne et al. 1982) – or in a continuous wavelength region with a standard error of prediction (SEP) of 0.2%. Furthermore, the accuracy of the analysis is stable over a long period without adjustment (Osborne et al. 1982) and is maintained when calibrations are transferred between instruments of the same model (Osborne and Fearn 1983).

The protein content of wheat flour is a key specification that determines end-product quality. It can be controlled in several different ways: (1) by blending of wheats, (2) by blending of flours and (3) by supplementation of flour with dried gluten. NIR on-line samplers were first developed to measure the protein content of flour. This provides an excellent example of a NIR feedback control system. Dried gluten can be used to replace wholly or partly the protein in flour, which would otherwise be derived from high-protein wheat in the grist. The success of on-line NIR for monitoring flour protein content therefore led to its incorporation into a closed-loop control system for gluten addition to flour (Cliffe 1988; Fearn and Maris 1991; Osborne 1986). A mixing screw is installed between the gluten feed and the NIR sampler station from which a feedback signal controls the gluten feeder. The system proved to be an efficient and accurate method of control of gluten addition to achieve a target protein content in the flour. However, it was established that differences in flour temperature caused a difference in protein readings of sufficient magnitude (+0.02% protein per °C rise in temperature) to require a correction for this effect. This was accomplished by installing a thermocouple probe projecting into the moving stream of flour that produced a signal in the form of a 4–20 mA current representing the temperature range 0°C–50°C. This signal was transmitted to the instrument's computer where it was used to correct the measured protein reading to a 20°C basis, as this was the temperature at which the calibration samples had been measured. The results of performance trials conducted using two different instruments in two different flour mills showed that using either system, the protein levels of gluten-supplemented flours could be controlled to a standard deviation of less than 0.1%.

26.2.2 Moisture

The moisture content of flour is set by the conditioning regime and may be measured by NIR analysis concurrently with protein with a SEP of 0.15% (Osborne et al. 1982).

26.2.3 Granularity

As NIR spectra are strongly affected by the scattering of light by a sample as a consequence of its particle size, any method to correct for the scattering effect can also be used to measure the particle size. The simplest method (and the only one available for use with a fixed filer instrument) is to use the log $1/R$ value at a wavelength at which there is no significant chemical absorption, for example, 2230 nm (Osborne et al. 1982). The particle size of flour can also be measured using methods that capture the effects of particle size on NIR spectra in a continuous wavelength region, for example, the first principal component or the scatter coefficient calculated using the multiplicative scatter correction (MSC) algorithm. Zhu et al. (2017) proposed the use of hyperspectral scattering for the determination of the particle size of flour.

Granularity is the particle size distribution of flour calculated as a cumulative percentage of mass. It is used as a basis for information about roll flute condition, gap setting and sifter efficiency. Pasikatan et al. (2001, 2002, 2003) have proposed a NIR on-line granulation sensor as a means of roller mill automation. The concept is to develop calibrations for the cumulative mass of sieved fractions to produce a NIR granulation curve as a basis for roll-gap optimization.

26.2.4 Starch Damage

The proportion of starch granules mechanically damaged during the milling process is an important quality factor in bread and cookie making because the baker requires consistent flour water absorption from one batch to another. Although starch damage is not the only factor that influences water absorption, it is one that the miller can control in a given grist by adjusting the reduction roll pressures. This can be accomplished automatically by means of computer-controlled actuators on each pair of rolls.

Mechanical damage to the starch granules in the flour is accompanied by cleavage of hydrogen bonds between starch molecules and between starch and bound water. Absorption of NIR energy in the region of 1100–2500 nm due to free and hydrogen-bonded hydroxyl bonds in starch and water, hence starch damage, can be measured using a monochromator in reflectance mode over the range of 1100–2500 nm (Osborne 1996). The first such NIR starch damage calibration was derived by Osborne and Douglas (1981), and then appropriate filters were obtained for this measurement to be performed on a fixed filter instrument (Osborne et al. 1982). The SEP for this calibration was 3.2 Farrand units, which is large in comparison with the range normally encountered in bread flours (20–40 Farrand units). Further research using commercial flour samples, together with samples produced on a 650 kg h^{-1} pilot mill, resulted in a calibration with an r^2 0.94 and SEP 0.41% (Osborne 1998a) (Figure 26.1). In this study, the improvement to the accuracy of starch damage calibration depended on a more precise reference method (Megazyme) than that used previously, the use of a monochromator instrument and modified partial least squares regression, and the control of sample temperature within ±0.5°C. This is a much tighter control than for protein as the effects of starch damage on the spectrum are more subtle, and, as a result, sample temperature must either be controlled very closely or be measured and a correction applied to the apparent NIR

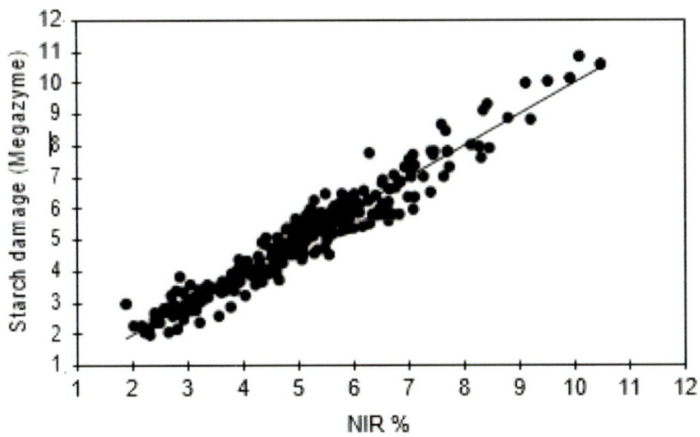

FIGURE 26.1 Plot of Megazyme versus NIR flour starch damage (Osborne 1996).

starch damage. The NIR calibration for starch damage was validated by using it to track changes that resulted from increases and decreases to reduction roll pressures during pilot milling trials. Random samples analyzed by the reference method confirmed that the NIR and reference starch damage data follow the same trends. Furthermore, the response to a series of roll pressure changes of a 50:50 blend of two wheats was close to that predicted from the NIR starch damage values for the two component wheats.

26.2.5 Water Absorption

Flour water absorption may be predicted from its protein, moisture and starch damage contents (Fearn and Osborne 1998). Since each of these can be determined by NIR, it would be expected that water absorption can also be determined by this method. The results of studies based on the use of a monochromator in reflectance mode over the range of 1100–2500 nm suggest that a SEP between NIR and Brabender Farinograph of approximately 1% may be achieved (Delwiche and Weaver 1994; Osborne et al. 1983a).

26.2.6 Qualitative Analysis

Rather than developing quantitative calibrations for a range of quality parameters, some researchers have proposed qualitative analysis to detect flours that are out of specification. Nielsen et al. (2001) employed a new discriminant approach to the identification of flours that are out of specification with respect to either particle size or composition. Similarly, Gergely et al. (2002) have used NIR discriminant analysis (polar qualification system) to separate, in quality space, flour fractions produced by two different mills. The discrimination focused on NIR wavelengths around 1700 and 2300 nm with the aim of distinguishing starch-rich and bran-rich fractions. Kumagai et al. (2002) have reported the discrimination of 15 kinds of noodle flour using principal component analysis, and González-Martín et al. (2014) have traced the origin of flours produced in different regions of Chile. Williams and Dexter (2002) recorded the standard deviation of optical density at 1978 nm as a measure of the efficiency of flour mixing. NIR can also discriminate flours milled from wheat, sago and taro grain species (Rachmawati et al. 2017).

26.2.7 Contaminants

Food manufacturers increasingly need to give attention to screening for components that may result in allergies or intolerances, and NIR methods have been developed for this purpose. Principal component analysis of hyperspectral images of flour with additions of ground peanut over the range of 1000–2200 nm allowed the detection of peanut traces down to 0.1% (Mishra et al. 2015). NIR has also been shown to predict the concentration of gluten content of flours and batters but was not sufficiently sensitive to detect trace amounts of gluten in gluten-free foods (Albanell et al. 2012).

26.2.8 Summary of Flour Applications

- Flour protein, moisture, granularity, starch damage and water absorption can be measured by NIR.
- On-line or at-line measurements can be used to adjust roll settings or to control flour blending or gluten addition.
- On-line or at-line measurements require correction for the effect of higher and more variable sample temperature compared to that in the climate-controlled laboratory in which the calibrations were performed.

26.3 BAKING PROCESSES

26.3.1 Dough Making

Baking begins with the blending of flour with water and other ingredients to produce a dough or batter, which is then baked in an oven to form the product. Detection of faults in the dough avoids energy wastage in baking an out-of-specification product and, for cookie doughs, may allow the dough to be recycled. NIR has been used for compositional analysis of intact cookie and bread dough pieces and for monitoring the progress of bread dough mixing.

During baking, the moisture content of dough is reduced from about 18%–2% for cookies and 60%–40% for bread. Monitoring the changes in the moisture content during baking allows the oven performance to be optimized. On-line NIR is commonly used to monitor the moisture contents of cookies on a conveyer as they leave the oven (Scott 1992). This measurement depends on a reproducible relationship between the NIR reading, which represents the surface moisture content in the freshly baked cookie, and the bulk moisture content of the fully equilibrated product in packet and on strategies to compensate for the discontinuous nature of discrete products on a conveyer. A fiber optic probe of 1 mm diameter was designed to measure the local water content inside bread dough during baking (Thorvaldsson and Skjøldebrand 1998). This sensor provides the potential for the measurement of the moisture content distribution and water transport rate in bread doughs as these factors influence crumb texture and crust formation.

The application of NIR for the monitoring of automatic metering equipment used for fat, sucrose, dry flour and water in short dough cookie products has been demonstrated (Osborne et al. 1984b). This study used an experimental design to minimize intercorrelations in which a standard recipe was varied to provide as great a range as possible for each of the four constituents under investigation while producing doughs of commercially acceptable consistency. To prevent high correlations between fat and sucrose, the recipe was randomized by varying independently the amounts of fat and sucrose. Then, for each recipe, the amounts of water and flour were calculated so that the dough produced would be capable of manipulation on a rotary molder. The dough pieces were presented intact to the NIR instrument. It can be seen from the spectra in Figure 26.2 that the absence of an ingredient could readily be detected from the NIR spectrum. Following multiplicative scatter correction, plots of correlation coefficient versus wavelength showed high correlations that corresponded exactly to the absorption maxima in the spectra of fat and sucrose (Osborne 1988a). The

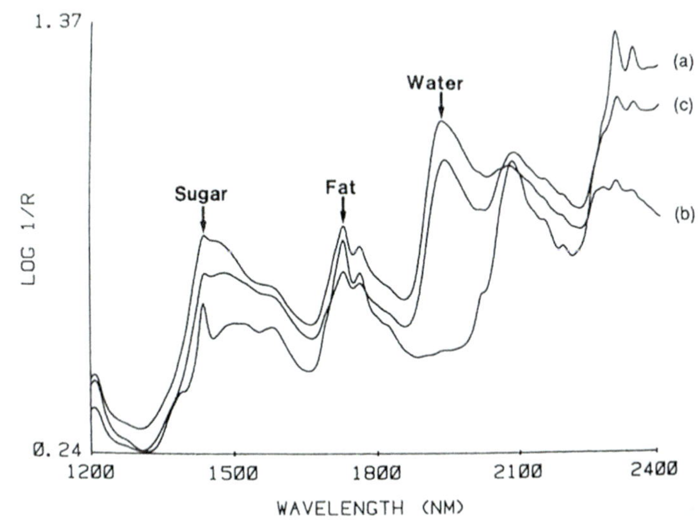

FIGURE 26.2 NIR spectra of mixtures of sugar/fat (a), sugar/fat/water (b) and sugar/fat/flour/water (c).

resulting calibrations enabled the simultaneous detection of errors in fat of 5% relative to the total amount of fat in the recipe and gross errors in the water, sucrose and flour percentages. Although this work was carried out using a monochromator, only eight wavelengths in total were required so that the application would be of practical value using a fixed filter instrument.

Dough mixing is a critical stage of the breadmaking process. During mixing, energy is imparted to the dough to develop the gluten matrix. This matrix enables the gas produced during fermentation to be retained during proving and baking. Poor mixing is one of the major single causes of bread faults. Craft bakers judge the optimum dough development by sight and touch, but in industrial bakeries, an automated approach is required. The availability of diode array instruments enabled the development of dynamic NIR spectroscopy (Wesley et al. 1998) and its patented application to the determination of the point of optimum dough mixing (Wesley et al. 2002). The diode array instrument provided sufficient speed of measurement on a moving sample to allow averaging of many spectra over the 3–5 min mixing time. In its original implementation (Figure 26.3), the instrument was placed above the mixer, and the reflected radiation from the moving dough mass was measured. The NIR absorption at 1160 nm decreases to the point of optimum dough mixing, as the water in the dough is bound to the protein and starch fractions and then increases as excess water is released. This provides a direct chemical measurement of the process that is independent of the mixer design and speed. It has been shown to be consistent with the changes to the quality of the bread baked at different stages of mixing. Subsequently, Wesley et al. (1999a) employed a fiber optic interactance probe to make a similar measurement using a 35-g mixograph. The use of NIR for real-time monitoring of dough mixing has been confirmed and extended by other research groups (Alava et al. 2000, 2001; Huang 2001; Huang et al. 2001). These groups have shown that the NIR method works

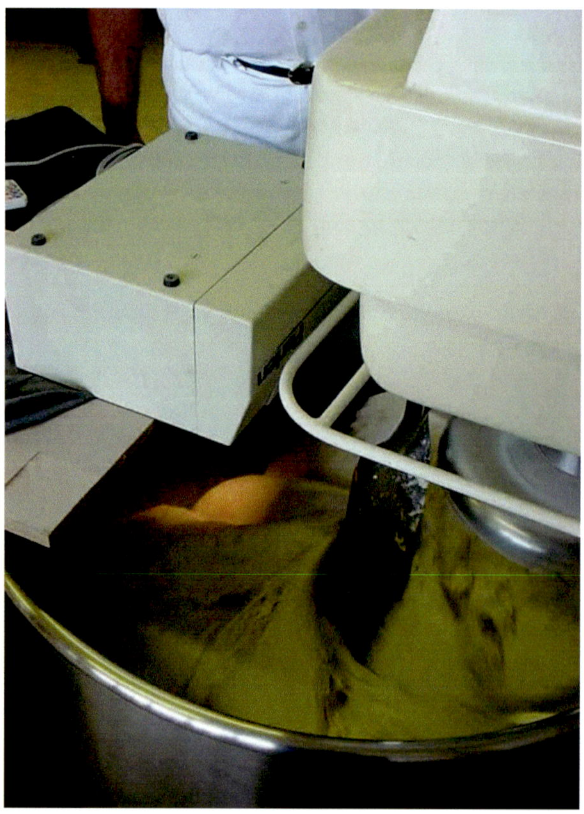

FIGURE 26.3 NIR monitoring of dough mixing.

for a variety of full formula doughs in commercial-scale mixers. Furthermore, NIR can be used to monitor dough mixing for different flour:water ratios, that is, crumbly dough (Aït Kaddour and Cuq 2009), bread dough (Aït Kaddour et al. 2007) and batter (Aït Kaddour et al. 2008). In the sponge and dough process, the dough (sponge) undergoes a second mixing with the addition of new ingredients. NIR measurements made at-line using a fiber optic probe together with a parallel factor analysis of the resulting spectra have been used to follow this process and determine the optimum leavening time (Ulrici et al. 2008).

26.3.2 Cookie Baking

In a parallel study to that described above for cookie doughs, whole cookies were presented to a NIR scanning monochromator with the writing and docker pin holes facing the light beam, as would be the case on a production line. NIR calibrations for fat, sucrose, dry flour and water in the whole cookies were obtained. However, although the calibrations had high correlation coefficients and acceptable precision compared with other methods, only that for fat was capable of sufficient accuracy for the detection of ingredient metering errors likely to be encountered in a commercial bakery.

26.3.3 Cake Production

Visible/NIR hyperspectral imaging has been used to follow and control the heat treatment of flours during cake production (Verdú et al. 2016).

26.3.4 Extrusion Cooking

Extrusion cooking is used to manufacture from cereals a wide range of products including breakfast cereals, snack foods and pet foods. In these products, texture and density are key quality requirements that are controlled by the processing conditions summarized by the term "degree of cook." Conventionally, control is attempted by measurement of inputs such as the moisture content, screw speed and barrel temperature, on the basis that the relationship between these and the product quality is understood. However, a more direct means of control is highly desirable. In contrast to breadmaking, where the product structure is largely controlled by protein development, in extrusion cooking, it is the starch component of the cereal-based raw material that is the chief functional ingredient. After crystalline regions of starch granules melt at high temperatures and pressures, the application of powerful shearing forces in the extruder softens granular structures and causes disruption and dispersal. A continuous phase of starch polymers containing bubbles of steam in the extrudate allows for an extensible foam-like structure to develop. The extent of transformation of the raw material, referred to as the degree of cook, is crucial to final product quality. Degree of cook increases when there is an increase in depolymerization of the starch molecules, which in turn results in an increase in the number of free hydroxyl bonds and a decrease in paste viscosity. The degree of cook of a powdered extrudate sample may be measured by means of a viscometer. However, viscometry can only be carried out on a sample that has already been processed and hence is not suitable for feedback process control. The first attempt to use NIR spectroscopy at-line for the measurement of degree of cook in real time involved the preparation of products from wheat, maize, oat and rice over a wide range of processing conditions and the extrudates freeze dried and ground to powders prior to NIR spectroscopy. Successful calibrations based on spectral features related to starch structure were obtained using the powdered samples (Guy et al. 1996; Millar et al. 1996) and at-line (Guy 1996). Subsequently, a fiber optic reflectance probe was installed in the die of the extruder, and the light was transmitted through the melt and reflected back from a polished steel pillar. The spectral characteristics of the different forms of starch found in the hot melts close to the die of the extruder were identical to those of the powdered extrudates (Guy 1996).

FIGURE 26.4 Fiber optic transmission probes in the extruder die.

At the temperature and pressure within the die of the extruder, the melt becomes a transparent fluid. Hence on-line NIR transmission spectroscopy using paired fiber optic probes (Figure 26.4) was used to monitor the physical state of starch polymers in the barrel melt as a means of measuring the degree of change, which has occurred under a given processing regime (Evans et al. 1999). A NIR measurement based on the band at 1430 nm identified as a result of the strategic research on wheat starch structure (Osborne 1996) was used to monitor changes, which resulted from adjustments to different process variables.

26.4 FOOD COMPOSITIONAL ANALYSIS

The compositional analysis of wheat-based foods is required for (1) the quality control of the finished product and (2) the nutritional labeling.

NIR has been used for the compositional analysis of bread. Calibrations have been derived for protein, fat, moisture, starch and total sugars. Two different approaches to sample presentation have been used: (1) drying and grinding the bread (Duvenage 1986; Groenewald 1984) and (2) placing intact slices, cut into a disc of appropriate diameter using a pastry cutter, into a standard cell (Osborne et al. 1984a; Suzuki et al. 1986). There was no difference in the analytical performance of the NIR method using either approach. In two independent studies, calibrations for the analysis of dried bread (Duvenage 1986) and bread discs (Suzuki et al. 1986) were found to be readily transferable between bread samples baked in different bakeries using widely different ingredients and processes.

The NIR calibrations for doughs, breads and cookies described previously were derived by means of designed experiments carried out under carefully controlled conditions in pilot bakeries. An example of NIR in use under commercial conditions is the analysis of dry cake mixes (Osborne et al. 1983b). The instrument used for this application was calibrated in the laboratory and then assessed under quality control conditions in the bakery. A single calibration for a range of different mixes (bread, scone, sponge, shortbread and shortcrust) was achieved for fat and sucrose. The calibration for fat gave a SEP of 1.7% in the range of 3%–26%, while that for sucrose gave a SEP of 2.7% in the range of 13%–41%. Although these SEP figures are much higher than those obtained between replicate determinations using the corresponding reference methods (fat by Soxhlet 0.38%

and sucrose by high-performance liquid chromatography 1.04%), they proved acceptable for quality control purposes in view of the speed with which NIR results could be obtained.

Baker et al. (1979) investigated the relationship between NIR measurements and neutral detergent fiber values of breakfast cereals. In a preliminary study (Baker et al. 1979), a correlation coefficient of 0.985 and an SEC of 1.36% over the range of 0%–31.7% was obtained using 213 samples of various cereals based on wheat, maize, oats and rice. These results were achieved by means of a two-term calibration equation in which the NIR data were treated as the first derivative of log $1/R$ divided by the first derivative of log $1/R$ at a reference wavelength (Norris regression). Further work with a different monochromator instrument (Baker 1983) resulted in a very similar equation with a SEC of 1.5%; testing this equation on independent samples gave a SEP of 1.8%. Calibration wavelengths for fiber (1352, 2204, 2448 and 2450 nm) had to be selected so as to avoid interference caused by the presence of crystalline sucrose in the samples. Because of the interference, it became apparent that it should be possible to determine the total sugar content of breakfast cereals by NIR. In an experiment designed to test this hypothesis (Baker and Norris 1985), 84 branded cereals, of which the total sugar content ranged from 0% to 54%, were used. These were divided into calibration and prediction sets of 40 and 44 samples, respectively, and a calibration equation was derived using the first-derivative ratios of 1432/1822 and 1692/1782 nm. This equation gave a SEP of 2.8% on the prediction samples against the reference gas–liquid chromatography method. The SEP is in very close agreement with that reported for sucrose in cake mixes (Osborne et al. 1983b) and confirms that NIR is applicable to the determination of sugars in cereal foods. Lin et al. (2014) used a handheld NIR instrument to determine the sucrose content of sugar-coated breakfast cereals.

Kays and Barton II reported calibrations for dietary fiber (Kays and Barton II 1998), protein (Kays and Barton II 2000) and energy content (Kays and Barton II 2002) in a wide range of dry-milled cereal products. In further work, this group derived acceptable calibrations in cereal products for lipid classes (saturated, monounsaturated and polyunsaturated) with SEP values of 0.76%–1.30% (Sohn et al. 2009) and cis- and trans-fats with a SEP of around 0.7% (Kim and Kays 2013).

Osborne (1988b) derived a single calibration for the determination of fat in bread, cookies and breakfast cereals. This calibration using log $1/R$ data of 1720 and 1700 nm after normalization by the MSC algorithm gave a SEP of 1.10% over a range of fat content of 1.2%–27.3%. It was, however, necessary to freeze dry the bread and to grind all the samples prior to NIR analysis. Therefore, although a universal calibration could be obtained, more sample preparation was required. Calibrations have also been reported for fiber and starch in a wide range of snack foods (Baker 1983, 1985; Baker and Norris 1985) and for protein, moisture, ash, starch, fiber and lipids in wheat bran (Hell et al. 2016). Cereal snack bars have been classified into regular, diet and light using NIR discriminant analysis (Brito et al. 2013). This means that NIR spectroscopy could be used for the compositional analysis of cereal products in a general purpose food analysis laboratory. The range of applications is summarized in Table 26.1.

TABLE 26.1
Summary of NIR Applications for Wheat-Based Foods

Constituent	Bread	Cookies	Dry Mixes	Snack Foods	Breakfast Cereals
Protein	X			X	X
Moisture	X				
Fat	X	X	X		X
Lipids		X		X	X
Starch	X			X	
Sugars	X	X	X		X
Fiber				X	X
Energy				X	X

26.5 CEREAL CHEMISTRY RESEARCH

26.5.1 Breadmaking Quality of Flour

There have been some reports in the literature of NIR calibrations for loaf volume or height but none for loaf quality. Williams et al. (1988) obtained a SEP between NIR measurements on flour and loaf volume of 48 and 30 mL, respectively, while Delwiche and Weaver (1994) reported a SEP of 0.11 cm for loaf height. These calibrations do not appear to have been adopted in routine use because the emphasis has transferred to the measurement of biochemical parameters since the high and low molecular weight glutenin protein fraction has been found to account for 50%–60% of the variation in dough properties relating to baking quality (Bhandari et al. 2000; Wesley et al. 1999b, 2001). The spectra of glutenin and gliadin are shown in Figure 26.5. A curve fitting method applied to the spectra of mixtures of glutenin, gliadin and starch over the region 2050–2220 nm resulted in an r^2 of 0.76 for glutenin content in the mixtures (Wesley et al. 1999b). Following this demonstration of the spectral basis, Wesley et al. (2001) reported partial least squares calibrations for glutenin and gliadin contents in flour. Importantly, the calibrations were shown to be independent of the total protein content by decoupling of the two properties (composition and content).

The development of new wheat cultivars that target low glycemic index requires specific starch characteristics such as low or zero amylose contents of partially waxy and waxy wheats, respectively. Therefore, the application of NIR to the determination of amylose was investigated (Wesley et al. 2003). Figure 26.6 shows the spectra of amylose and amylopectin derived by the same master curve deconvolution method that had been used for glutenin and gliadin. It is apparent that the spectra are only linearly independent on noncontinuous subsets of wavelengths. As a result, localization methods were applied to enable successful calibrations for amylose content.

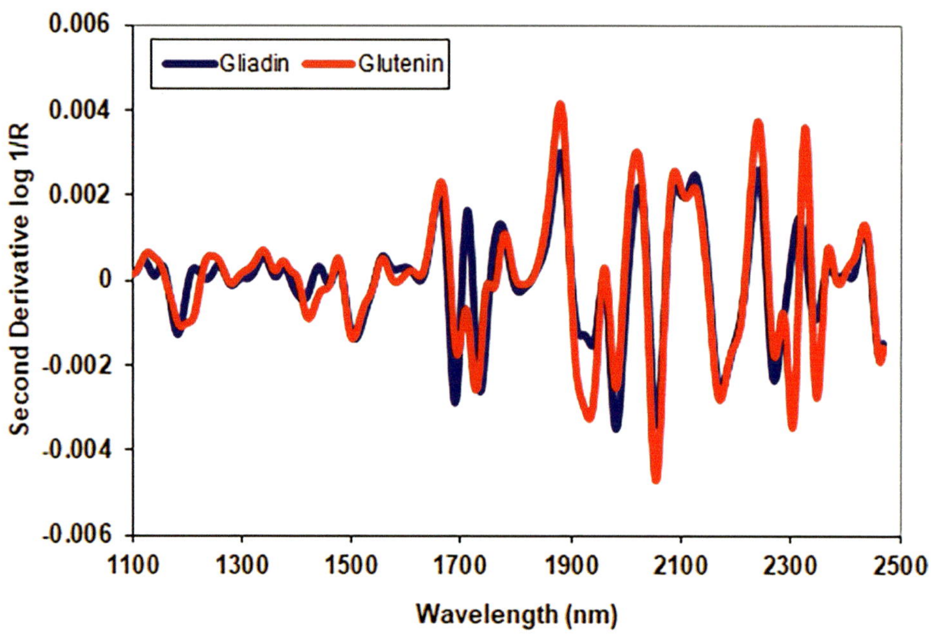

FIGURE 26.5 NIR spectra of flour protein fractions (Wesley et al. 1999b).

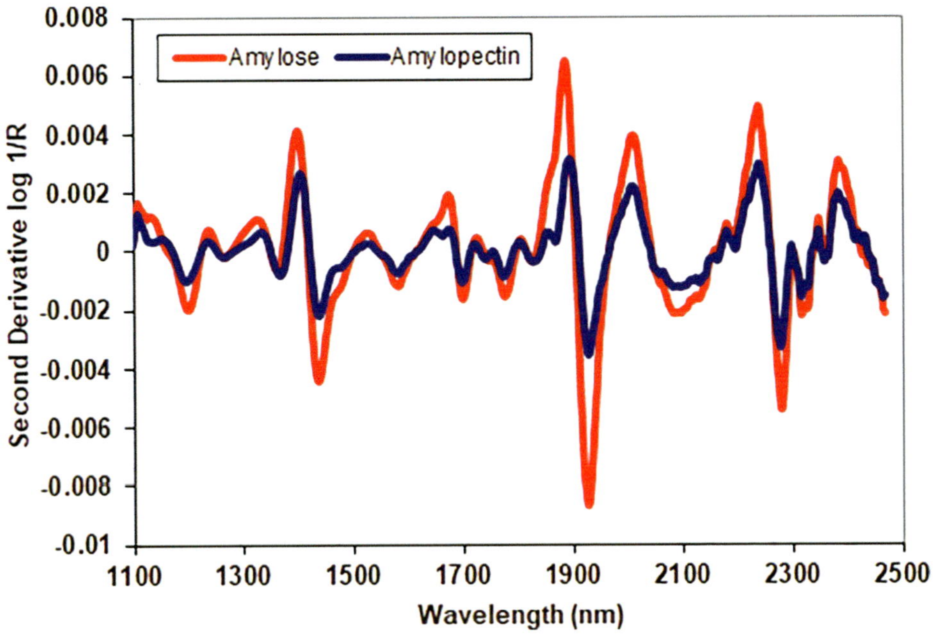

FIGURE 26.6 NIR spectra of flour starch fractions (Wesley et al. 2003).

26.5.2 Breadmaking Process

The use of NIR for real-time monitoring of dough mixing has been confirmed and extended by research groups who have shown that the NIR method works for a variety of full formula doughs in commercial-scale mixers and can be used for research into the relationships between the NIR mixing curves, flour quality, ingredient interactions, flour protein functionality and final bread quality (Aït Kaddour et al. 2007; Alava et al. 2000, 2001; Huang 2001; Huang et al. 2001). Verdú et al. (2015) have used NIR imaging to characterize flours on the basis of their physicochemical, rheological and fermentation kinetics properties.

26.5.3 Bread Staling

NIR spectroscopy has been applied in areas of research where the spectral measurements can be used directly to study biochemical processes. For example, if bread crumb is stored for a few hours or more, the process known as staling or retrogradation occurs. In one of the first reports of the use of NIR for the study of reaction kinetics in food systems, the second derivative data at 1414 and 1465 nm were used to calculate rate constants for the retrogradation process in bread (Osborne 1998b; Wilson et al. 1991). NIR spectra were recorded using discs sampled from bread slices taken from loaves during storage. The rate constants calculated from the second derivative at 1414 nm was 15.29 $kh^{-1} \times 10^3$ and at 1465 nm 14.16 $kh^{-1} \times 10^3$ compared with the value obtained by differential scanning calorimetry of 16.06 $kh^{-1} \times 10^3$. A generic spectroscopic rationale for this application was proposed by Osborne (1996): retrogradation occurs through the slow redevelopment of crystallinity in the amylopectin fraction of the starch. This crystallinity is a direct result of extensive hydrogen bonding, both intramolecularly and to water molecules, of the amylopectin molecules. It is these changes to the hydrogen bonding of the starch and water O–H bonds at 1410 and 1460 nm that are measured by the NIR spectra of the intact bread crumb. Further research showed that NIR had a high correlation to texture during bread storage and that NIR could follow bread staling more accurately than texture analysis (Xie et al. 2003). These workers went on to study the effects of starch, protein and temperature on bread staling using NIR (Xie et al. 2004). They concluded that NIR followed moisture and starch structure changes associated with amylopectin retrogradation. Two-dimensional correlation spectroscopy on near- and mid-infrared spectra of bread crumb during aging demonstrated that the major bread staling processes such as amylopectin retrogradation and water loss can be followed with both near- and mid-infrared spectroscopy (Ringsted et al. 2017).

Another example of the direct use of NIR spectra in research is the study of the relationships between the NIR dough mixing curves, flour quality, ingredient interactions, flour protein functionality and final bread quality (Aït Kaddour et al. 2007).

26.6 CONCLUSION

NIR is a mature technology for process control in flour milling, bread baking, cookie baking and extrusion cooking and for the compositional analysis of wheat-based foods. Samples such as sliced bread or cookies may be analyzed intact. However, if samples are ground (after drying if necessary), then a single calibration can be used for a wide variety of different foods. NIR has also been used for cereal chemistry research. In some cases, the NIR measurement does not require calibration against a reference method.

REFERENCES

AACC International. *Approved Methods of Analysis, 11th Ed. Method 39-11.01. Near-Infrared Reflectance Method for Protein Determination in Wheat Flour.* St. Paul, MN: AACC International.

Aït Kaddour, A., Barron, C., Morel, M.-H. and Cuq, B. 2007. Dynamic monitoring of dough mixing using near infrared spectroscopy: Physical and chemical outcomes. *Cereal Chem.* 84: 70–79.

Aït Kaddour, A. and Cuq, B. 2009. In line monitoring of wet agglomeration of wheat flour using near infrared spectroscopy. *Powder Technol.* 190: 10–18.

Aït Kaddour, A., Morel, M.-H. and Cuq, B. 2008. Description of batter dough mixing using near-infrared spectroscopy. *J. Cereal Sci.* 48: 698–708.

Alava, J. M., Millar, S. J. and Salmon, S. E. 2000. The assessment of dough development during mixing using near infrared spectroscopy. *Spec. Publ. R. Soc. Chem.* 261: 439–441.

Alava, J. M., Millar, S. J. and Salmon, S. E. 2001. The determination of wheat breadmaking performance and bread dough mixing time by NIR spectroscopy for high speed mixers. *J. Cereal Sci.* 33: 71–81.

Albanell, E., Miñarro, B., and Carrasco, N. 2012. Detection of low-level gluten content in flour and batter by near infrared reflectance spectroscopy (NIRS). *J. Cereal Sci.* 56: 490–495.

Baker, D. 1983. The determination of fiber in processed cereal foods by near-infrared reflectance spectroscopy. *Cereal Chem.* 60: 217–219.

Baker, D. 1985. The determination of fiber, starch and total carbohydrate content in snack foods by near-infrared reflectance spectroscopy. *Cereal Foods World* 30: 389–392.

Baker, D. and Norris, K. H. 1985. Near-infrared reflectance measurement of total sugar content of breakfast cereals. *Appl. Spectrosc.* 39: 618–621.

Baker, D., Norris K. H. and Li, B. W. 1979. Food Fiber Analysis: Advances in Methodology, in *Dietary Fibers, Chemistry and Nutrition*, ed. G. E. Inglett and S. I. Falkenhag, 67–78. New York: Academic Press. doi: 10.1016/B978-0-12-370950-9.50011-9

Bhandari, D. G., Millar, S. J. and Scotter, C. N. G. 2000. Prediction of wheat protein and HMW-glutenin contents by near infrared (NIR) spectroscopy. *Spec. Publ. R. Soc. Chem.* 261: 313–316.

Brito, A. L. B., Brito, L. R., Honorato, F. A., Pontes, M. J. C. and Pontes, L. F. B. L. 2013. Classification of cereal bars using near infrared spectroscopy and linear discriminant analysis. *Food Res. Int.* 51: 924–928.

Cliffe, D. 1988. Technical profile: Near infrared on-line sampler, analyser. *World Grain* 6(1): 37.

Delwiche, S. R. and Weaver, G. 1994. Bread quality of wheat flour by near-infrared spectrophotometry: Feasibility of modeling. *J. Food Sci.* 59: 410–415.

Duvenage, E. 1986. Measurement of fat and moisture in air-dried bread by near infrared spectrophotometry. *J. Sci. Food Agric.* 37: 384–386.

Evans, A. J., Huang, S., Osborne, B. G., Kotwal, Z. and Wesley, I. J. 1999. Near infrared on-line measurement of degree of cook in extrusion processing of wheat flour. *J. Near Infrared Spectrosc.* 7: 77–84.

Fearn, T. and Maris, P. I. 1991. An application of Box-Jenkins methodology to the control of gluten addition in a flour mill. *Appl. Stat.* 40: 477–484.

Fearn, T. and Osborne, B. G. 1998. Prediction of flour water absorption from other flour properties. In: *Proc. Australian Cereal Chemistry Conference*, ed. L. O'Brien, A. B. Blakeney, A. S., Ross, and C. W. Wrigley, 439–442, Melbourne: Royal Australian Chemical Institute.

Gergely, S., Handzel, L., Zoltan, A. and Salgo, A. 2002. Near infrared spectroscopy – a tool for the evaluation of milling procedures. In: *Near Infrared Spectrocopy: Proc. 10th International Conference*, ed. A. M. C. Davies and R. K. Cho, 33–37. Chichester: NIR Publications.

González-Martín, M. I., Wells Mocada, G., González-Pérez, C. et al. 2014. Chilean flour and wheat grain: Tracing their origin using near infrared spectroscopy and chemometrics. *Food Chem.* 145: 802–806.

Groenewald, C. A. 1984. New applications of near infrared spectroscopy in the food industry. *Suppl. SA Food Rev.* 59: 69–71.

Guy, R. C. E. 1996. Spectroscopic measurements of physical changes in starch caused by extrusion cooking. *Extrusion Communique*, 18–19.

Guy, R. C. E., Osborne, B. G. and Robert, P. 1996. The application of near infrared reflectance spectroscopy to measure the degree of processing in extrusion cooking processes. *J. Food Eng.* 27: 241–258.

Hell, J., Prückler, M., Danner, L. et al. 2016. A comparison between near-infrared (NIR) and mid-infrared (ATR-FTIR) spectroscopy for the multivariate determination of compositional properties in wheat bran samples. *Food Control* 60: 365–369.

Huang, W. N. 2001. Rheological dough and mixing behavior of food ingredients as studied by Labtron and NIR spectroscopy. *Paper 28 AACC Annual Meeting*, 14–18 October, Charlotte, NC.

Huang, W. N., Olewnik, M. C., Psotka, J. J. and Dempster, R. E. 2001. Measuring rheological dough and mixing properties of full formula ingredients and microingredients in real time using NIR. *Paper 133 AACC Annual Meeting*, 14–18 October, Charlotte, NC.

Kays, S. E. and Barton, F. E., II. 1998. The use of near infrared reflectance spectroscopy to predict the insoluble dietary fibre fraction of cereal products. *J. Near Infrared Spectrosc.* 6: 221–227.

Kays, S. E. and Barton, F. E., II. 2000. Predicting protein content by near infrared reflectance spectroscopy in diverse cereal food products. *J. Near Infrared Spectrosc.* 8: 35–43.

Kays, S. E. and Barton, F. E., II. 2002. Prediction of energy content in cereal food products by near infrared spectroscopy. In: *Near Infrared Spectrocopy: Proc. 10th International Conference*, ed. A. M. C. Davies and R. K. Cho, 217–220. Chichester: NIR Publications.

Kim, Y. and Kays, S. E. 2013. Fourier transform near infrared spectroscopy prediction of trans and cis fats in ground cereal products at different resolutions. *J. Near Infrared Spectrosc.* 21: 141–147.

Kumagai, M., Karube, K., Sato, T. et al. 2002. A near infrared spectroscopic discrimination of noodle flours using a principal-component analysis coupled with chemical information. *Analyt. Sci.* 18: 1145–1150.

Lin, C.-A., Ayvaz, H. and Rodriguez-Saona, L. E. 2014. Application of portable and handheld infrared spectrometers for determination of sucrose levels in infant cereals. *Food Anal. Methods* 7: 1407–1414.

Millar, S., Robert, P., Devaux, M. F., Guy, R. C. E and Maris, P. 1996. Near-infrared spectroscopic measurements of structural changes in starch-containing extruded products. *Appl. Spectrosc.* 50: 1134–1139.

Mishra, P., Herrero-Langreo, A., Barreiro, P. et al. 2015. Detection and quantification of peanut traces in wheat flour by near infrared hyperspectral imaging spectroscopy using principal-component analysis. *J. Near Infrared Spectrosc.* 23: 15–22.

Nielsen, J. P., Bertrand, D., Micklander, E., Courcoux, P. and Munck, L. 2001. Study of NIR spectra, particle size distributions and chemical parameters of wheat flours: a multi-way approach. *J. Near Infrared Spectrosc.* 9: 275–285.

Osborne, B. G. 1986. The use of near infrared reflectance spectroscopy in the control of flour protein content. *J. Chem. Technol. Biotechnol.* 36: 364–366.

Osborne, B. G. 1988a. Comparative study of methods of linearization and scatter correction in near infrared reflectance spectroscopy. *Analyst, Lond.* 113: 263–267.

Osborne, B.G. 1988b. Determination of fat in a variety of cereal foods using NIR spectroscopy. In: *Analytical Applications of Spectroscopy*, eds. C.S. Creaser and A.M.C. Davies, 68–71. London: Royal Society of Chemistry.

Osborne, B. G. 1996. Near infrared spectroscopic studies of starch and water in some processed cereal foods. *J. Near Infrared Spectrosc.* 4: 195–200.

Osborne, B. G. 1998a. Improved NIR prediction of flour starch damage. In *Proc. Australian Cereal Chemistry Conference*, ed. L. O'Brien, A. B. Blakeney, A. S., Ross, and C. W. Wrigley, 434–438. Melbourne: Royal Australian Chemical Institute.

Osborne, B. G. 1998b. NIR measurement of the development of crystallinity in stored bread crumb. *Analysis* 26: M39–M41.
Osborne, B. G., Barrett, G. M., Cauvain, S. P. and Fearn, T. 1984a. The determination of protein, fat and moisture in bread by near infrared reflectance spectroscopy. *J. Sci. Food Agric.* 35: 940–945.
Osborne, B. G. and Douglas, S. 1981. Measurement of the degree of starch damage in flour by near infrared reflectance analysis. *J. Sci. Food Agric.* 32: 328–332.
Osborne, B. G., Douglas, S. and Fearn, T. 1982. The application of near infrared reflectance analysis to rapid flour testing. *J. Food Technol.* 17: 355–363.
Osborne, B. G., Douglas, S. and Fearn, T. 1983a. Recent progress in the application of NIR to the measurement of quality parameters in flour. In: *Progress in Cereal Science and Technology*, ed. J. Holas and J. Kratochvil, 577–581. Amsterdam: Elsevier.
Osborne, B. G. and Fearn, T. 1983. Collaborative evaluation of universal calibrations for the measurement of protein and moisture in flour by near infrared reflectance. *J. Food Technol.* 18: 453–460.
Osborne, B. G., Fearn, T., Miller, A. R. and Douglas, S. 1984b. Application of near infrared reflectance spectroscopy to the compositional analysis of biscuits and biscuit doughs. *J. Sci. Food Agric.* 35: 99–105.
Osborne, B. G., Fearn, T. and Randall, P. G. 1983b. Measurement of fat and sucrose in dry cake mixes by near infrared reflectance spectroscopy. *J. Food Technol.* 18: 651–656.
Pasikatan, M. C., Haque, E., Spillman, C. K., Steele, J. L. and Milliken, G. A. 2003. Granulation sensing of first-break ground wheat using a near-infrared reflectance spectrometer: studies with soft red winter wheats. *J. Sci. Food Agric.* 83: 151–157.
Pasikatan, M. C., Haque, E., Steele, J. L., Spillman, C. K. and Milliken, G. A. 2001. Evaluation of a near-infrared reflectance spectrometer as a granulation sensor for first-break ground wheats: Studies with six wheat classes. *Cereal Chem.* 78: 730–736.
Pasikatan, M. C., Steele, J. L., Haque, E., Spillman, C. K. and Milliken, G. A. 2002. Evaluation of a near-infrared reflectance spectrometer as a granulation sensor for first-break ground wheats: Studies with hard red winter wheats. *Cereal Chem.* 79: 92–97.
Rachmawati, Rohaeti, E. and Rafi, M. 2017. Combination of near infrared spectroscopy and chemometrics for authentication of taro flour from wheat and sago flour. *J. Phys.: Conf. Ser.* 835 012011. doi: 10.1088/1742-6596/835/1/012011.
Ringsted, T., Siesler, H. W. and Engelsen, S. B. 2017. Monitoring the staling of wheat bread using 2D MIR-NIR correlation spectroscopy. *J. Cereal Sci.* 75: 92–99.
Scott, M. J. 1992. On-line and point measurement of moisture in food. *Food Technol. Int. Europe* 81–84.
Sohn, M., Kim, Y., Vines, L. L and Kays, S. E. 2009. Near infrared analysis of lipid classes in processed cereal products. *J. Near Infrared Spectrosc.* 17: 127–133.
Suzuki, K., McDonald, C. E. and D'Appolonia, B. L. 1986. Near-infrared reflectance analysis of bread. *Cereal Chem.* 63: 320–325.
Thorvaldsson, K. and Skjøldebrand, C. 1998. Water diffusion in bread during baking. *Food Sci. Technol.—LW & T.* 31: 7–8.
Ulrici, A., Vigni, M. L., Durante, C. et al. 2008. At-line monitoring of the leavening process in industrial bread making by near infrared spectroscopy. *J. Near Infrared Spectrosc.* 16: 223–231.
Verdú, S., Ivorra, E., Sánchez, A. J., Barat, J. M. and Grau, R. 2015. Study of high strength wheat flours considering their physicochemical and rheological characterisation as well as fermentation capacity using SW-NIR imaging. *J. Cereal Sci.* 62: 31–37.
Verdú, S., Ivorra, E., Sánchez, A. J., Barat, J. M. and Grau, R. 2016. Spectral study of heat treatment process of wheat flour by VIS/SW-NIR image system. *J. Cereal Sci.* 71: 99–107.
Wesley, I. J., Gras, P. W., Larsen, N., Osborne, B. G. and Skerritt, J. H. 1999a. A preliminary comparison of the 35g Mixograph and dynamic near infrared spectroscopy for studying dough development. In *Proc. Australian Cereal Chemistry Conference*, ed. L. O'Brien, A. B. Blakeney, A. S., Ross, and C. W. Wrigley, 447–450. Melbourne: Royal Australian Chemical Institute.
Wesley, I. J., Uthayakumaran, S. Anderssen, R. S., et al. 1999b. A curve-fitting approach to the near infrared reflectance measurement of wheat flour proteins which influence dough quality. *J. Near Infrared Spectrosc.* 7: 229–240.
Wesley, I. J., Larroque, O., Osborne, B. G., Azudin, N., Allen, H. and Skerritt, J. H. 2001. Measurement of gliadin and glutenin content of flour by NIR spectroscopy. *J. Cereal Sci.* 34: 125–133.
Wesley, I. J., Larsen, N., Osborne, B. G. and Skerritt, J. H. 1998. Non-invasive monitoring of dough mixing by near infrared spectroscopy. *J. Cereal Sci.* 27: 61–69.
Wesley, I. J., Larsen, N., Osborne, B. G. and Skerritt, J. H. 2002. Monitoring of dough properties. US Patent 6,342,259 B1, Washington, DC: United States Patent Office.

Wesley, I. J., Osborne, B. G. Anderssen, R. S., Delwiche, S. R. and Graybosch, R. A. 2003. Chemometric localization approach to NIR measurement of apparent amylose content of ground wheat. *Cereal Chem.* 80: 462–467.

Williams, P. C. and Dexter, J. E. 2002. Application of near-infrared raw spectral data to determine efficiency of flour mixing. *Paper 360, 87th AACC Annual Meeting*.

Williams, P. C., El-Haramein, F. J., Ortiz-Fereira, G. and Srivastava, J. P. 1988. Preliminary observations on the determination of wheat strength by near-infrared reflectance. *Cereal Chem.* 65: 109–114.

Wilson, R. H., Goodfellow, B. J., Belton, P. S., Osborne, B. G., Oliver, G. and Russell, P. L. 1991. Comparison of Fourier transform mid infrared reflectance spectroscopy with differential scanning calorimetry for the study of the staling of bread. *J. Sci. Food Agric.* 54: 471–483.

Xie, F., Dowell, F. E. and Sun, X. S. 2003. Comparison of near-infrared reflectance spectroscopy and texture analyzer for measuring wheat bread changes in storage. *Cereal Chem.* 80: 25–29.

Xie, F., Dowell, F. E. and Sun, X. S. 2004. Using visible and near-infrared reflectance spectroscopy and differential scanning calorimetry to study starch, protein and temperature effects on staling. *Cereal Chem.* 81: 249–254.

Zhu, Q., Xing, Y., Lu, R., Huang, M. and Ng, P. K. W. 2017. Visible/shortwave near infrared spectroscopy and hyperspectral scattering for determining bulk density and particle size of wheat flour. *J. Near Infrared Spectrosc.* 25: 116–126.

27 Application of NIR Spectroscopy to the Analysis of Forages

*Olivier Minet, Virginie Decruyenaere, Bruno Godin, and Vincent Baeten**
Walloon Agricultural Research Centre

CONTENTS

Preamble

The Walloon Agricultural Research Centre (CRA-W) located in Gembloux, Belgium, maintains numerous contacts with farmers and horticulturists, companies, universities, and Belgian and foreign research centers. CRA-W is involved in various European, national, and regional research projects. Among many other initiatives, CRA-W has focused on promoting expertise in near-infrared spectroscopy (NIRS) since the 1980s. This chapter is mainly based on the knowledge and experience accrued by CRA-W after almost 40 years of NIRS forage analyses. The content of this section is quite technical and aims to provide clear and practical information to anyone interested in forage analysis.

27.1 INTRODUCTION

Forages are the primary self-produced feed resources of ruminants and make a significant contribution to the feeding autonomy of the farm. Forages are cropped under different forms, such as grazing or harvesting to produce silages and haylages. It is obvious that in the current economical context (e.g., price volatility of main inputs and agricultural productions), it is essential to give utmost importance to the quality of forage production to minimize feeding costs and cover animal requirements while protecting the environment.

* v.baeten@cra.wallonie.be.

At the nutritional level, diet must bring all the vital nutriments (e.g., energy, proteins, fibers, minerals, vitamins) in the right proportions, based on the metabolic state of the animals (e.g., pregnancy, growing, fattening). On the basis of the average values of the forages, a farmer may either overestimate the nutritive value of feed and not cover his animals' needs or underestimate it, thus with the risk to produce manure that is too rich and could potentially pollute the environment. From the minerals that are taken from the soil by forages, it is also possible to calculate nutritional indices (based on the nitrogen content of the plant) and to put in place a reasoned system of soil fertilization.

This being the case, it is evident that the tools that will allow to determine the chemical composition and nutritive value of forage production also become essential not only to adapt the diets to the farm's animal requirements but also to manage the overall operation of the farm.

Near-infrared spectroscopy is one of these tools and presents many advantages:

- Speed of analysis: the spectral acquisition and the prediction step usually take less than 5 min
- Costs (≈10 × cheaper than reference methods), around 30 € for the determination of the main constituents by NIR spectroscopy
- Simultaneous quantification of different parameters and criteria
- Ecologically friendly (reduced use of reagents, especially after validations of the NIR calibrations)
- Simple sample preparation, which is generally limited to a drying and a grinding step at the laboratory level and requires practically no preparation if the analysis is performed at the farm
- Analysis is easy to implement

Therefore, near-infrared reflectance spectroscopy can contribute significantly to the best qualification of forages for animal nutrition.

27.2 TYPES OF FORAGES

Grasslands are the largest ecosystems in the world. They cover 26% of the world land area (FAOSTAT, 2009). At the European level, permanent grasslands cover 57 million of Mha. Temporary grasslands cover smaller areas (less than 9.7 Mha). Other fodder crops cover 9 Mha where maize silage accounts for 5 Mha (Huyghes et al., 2015).

Forages comprise all the fibrous materials consumed by herbivores. The most frequently encountered forages are grass (e.g., fresh, silage, pre-wilted silage, or haylage) from temperate or tropical origin, maize (e.g., whole plant silage, corn cob mix silage), sugar beet pulp, cabbage, etc. All these types of forages can be subjected to NIRS analyses for predicting chemical composition and digestibility.

In temperate regions, forages from grasslands are usually composed of a mix of grasses (ryegrass, cocksfoot, timothy, etc.) and legumes (alfalfa, white, and/or purple clovers). These are grazed or harvested to produce haylages and silages. Many other plant species are also used in a multispecies mixture for feeding herbivores. This is the case, for instance, of mixtures of cereals, grasses, legumes, and protein crops harvested immature as silage. Whole plant maize in the form of silage or sugar beet pulp is also included in the forage resources of a farm. From a nutritional point of view, the uniqueness of these forage resources is their heterogeneity.

Some resources such as corn silages and sugar beet pulp are energy rich. Others such as grass products provide protein and/or fibers to the food rations. Moreover, within a same category of forage, there is a strong variability, so it is essential that the producer is aware of the quality of the forage. Indeed, low-quality forages may lead to an undesirable economic performance and metabolic disorders (e.g., acidosis or ketosis) resulting from the distribution of unbalanced diets.

FIGURE 27.1 Grass field in Belgium.

The five classes of forage are generally described as follows:

- **Green forages (Figure 27.1)** are grazed grass, whole plant maize, immature cereals, and protein mixed crop.
- **Silage forages (Figure 27.2)** are composed of grass, whole plant maize, or beet pulp. Silage is a method of preserving wet forage through anaerobic lactic fermentation. This fermentation has the effect of stabilizing the product by inhibiting its degradation by acid production thus lowering the pH
- **Dry fodder (Figure 27.3):**
- **Artificially dehydrated fodder (Figure 27.4) like** alfalfa pellets, etc.
- **Straws and stalks** are products like straws of cereals, peas, or corn cob

27.3 FORAGE CONSTITUENTS

From a chemical point of view, forages are composed of two elements: water and dry matter (Figure 27.5). According to the type of forage, this proportion can be very different. For example, maize silage forage contains roughly less than 35% of dry matter versus hay forage with 85%.

Knowledge of water content is important for several reasons:

- It is related to the ability of the animals to ingest.
- It allows reporting the results of analyses to a fixed base and thus compares the different foods between them.
- It allows assessing the risks during conservation.

FIGURE 27.2 Maize silage from the experimental farm of CRA-W.

FIGURE 27.3 Haystack from the experimental farm of CRA-W.

FIGURE 27.4 Alfalfa pellets in a small cup from a Foss DS2500 in the NIR laboratory of the CRA-W.

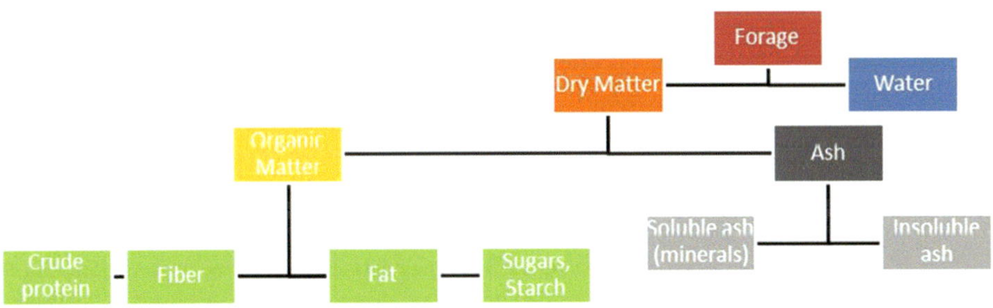

FIGURE 27.5 Major constituents of forage.

Dry matter is composed of an organic fraction (carbohydrates, proteins, starch, fibers, etc.) and an inorganic matter (i.e., minerals and oligo minerals) obtained by calcination.

- Organic matter is composed of
  - ***Crude proteins (CPs):*** they play a role in the formation of tissue, enzymatic, and hormonal activities and represent an energy source through various pathways of metabolic degradation, milk, and meat production. The metabolism of the protein is very complex especially in ruminants. That's why there are different ways of evaluating the quality of forage. Crude proteins are commonly determined by dosing total nitrogen (N) and multiplying the N value by 6.25. This represents a mixture of true protein nitrogen and nonprotein nitrogen.
  - **Fats:** they are present in small quantities in forage. Their roles are mainly energetic and act as a potential source of omega 3 fatty acids.

- **Carbohydrates:** they are the main sources of energy for rumen microbes, but they are also used directly by the animal. There are two groups of carbohydrates according to their functions: nonstructural and structural.
  - *Nonstructural carbohydrates* that include simple sugars or rapidly digestible compounds (i.e., glucose, fructose, sucrose, and maltose) and polysaccharides (i.e., starch, fructosan). Starch is the storage polysaccharide in legumes, while fructosan plays the same role in grasses.
  - *Structural carbohydrates* are the substances that form the cell walls and lend rigidity to the plant: they include cellulose, hemicellulose, and lignin. The main structural carbohydrate found in plants is cellulose.
    - Cellulose is a homopolymer consisting of glucose. It has a crystalline structure that makes it difficult for rumen microbes to degrade it enzymatically
    - Hemicelluloses are heteropolymers consisting of xylose, arabinose, galactose, mannose, and glucose. They do not have a crystalline structure, so they are easily digestible by rumen microbes
    - Lignin is a complex compound that settles on cellulose, hemicellulose, pectin, and structural protein compounds as the plant ages. It binds to the constituents of cell walls making them vulnerable to biotic and abiotic stresses

  Analysis of the fiber or cell wall present in forages is a major concern in ruminant nutrition because diets often contain large amounts of forage, and the fiber fraction affects both feed intake and animal performance.

  The complexity of its structure explains the difficulty of digesting it. The more cellulose present in forage, the harder to digest.

  The fibrous fraction that dissolves plant pectins, starch, fructans, proteins, sugars, and lipids, commonly known as NDF (neutral detergent fiber) and insoluble in a neutral detergent, contains hemicellulose, cellulose, and lignin but not pectin; it also includes fiber-bound protein and minerals, as well as heat-damaged protein (N-ADF†).

  The level of NDF in the animal ration influences the animal's intake of dry matter and the time of rumination. The concentration of NDF in forage is negatively correlated with energy concentration.

  The complex of lignin and cellulose is generally grouped under the term lignocellulose or nonsoluble fibers in an acid detergent (ADF fibers, "acid detergent fiber"). This fibrous component contains cellulose and lignin, as well as fiber-bound protein and minerals. It increases with the maturity of the plant. Hemicellulose is estimated by subtracting the ADF from the NDF. The proportions of ADF and NDF of forage are indicative of its food value. ADF fiber is generally related to digestibility and the energy value of the forage: the more ADF fiber in the forage, the lower the digestibility and the energy content.

  Treatment of the ADF residue with 72% sulfuric acid followed by calcination permits determining the fraction of crude lignin (ADL—acid detergent lignin). Lignin is totally resistant to enzymatic degradation in the digestive tract of ruminants; consequently, it has no nutritional value and, furthermore, adversely affects the digestion of other structural carbohydrates. With the aging of the plant, the cellulose chains surround themselves with lignin and hemicellulose, which leads to a decrease in the digestibility of cellulose and fodder. Cellulose can then be estimated by the difference between ADF and lignin plus ash.
- **Non-nitrogenous extracts** consist mainly of sugars.

† N-ADF refers to a contamination of nitrogen in ADF (acid detergent fiber).

- **Inorganic matter** is contained in soluble form in mineral substances (i.e., calcium, phosphorus, sodium, magnesium, sulfur, and trace elements) essential for the vital functions of the animals and in insoluble form in the physical constituents of soil (i.e., clays, silts, and sands), which is characteristic of forage contaminated by these soils. The role of minerals is multiple and varied according to the animal species and its physiological stage (e.g., pregnancy, sowing, fattening). Minerals intervene at the level of skeleton development (major role of calcium, phosphorus, and magnesium), cell activity of organisms (role of potassium), hormonal and enzymatic reactions (role of trace elements), and ruminal digestion of feed.

  Rumen microorganisms have specific requirements for major elements, such as phosphorus and sulfur and trace elements such as copper, zinc, cobalt, and manganese.

27.4 LABORATORY REFERENCE METHODS OF CHEMICAL ANALYSES

NIR spectroscopy is directly related to wet chemistry reference methods of analysis. Indeed, NIR calibrations are calculated from databases where spectra and the corresponding sample reference values match.

Reference methods are generally described under ISO regulations, and sometimes different official reference methods exist for the same parameters. That's why when a database is built or a validation step is carried out, it is always important to use a single reference method.

Below, a list of wet chemistry reference methods commonly used for forage analyses is provided.

- Water: drying the sample generally in an oven at 60°C for 2 days and at atmospheric pressure. The drying conditions may vary according to the nature of the fodder
- Analytical dry matter: drying the sample in an oven at 103°C for 4 h.

  Norm: REGULATION (EC) no. 152/2009—Methods of sampling and analysis for the official control of feed (E): NF ISO 6496: Determination of moisture and other volatile matter contents
- Total ash: obtained by calcining the dry matter in a muffle furnace at 550°C. They are then solubilized with nitric or hydrochloric acid on a sand bath. The soluble and insoluble ashes are separated by filtration.

  Norm: NF V18–101: Raw ash determination (R): REGULATION (EC) no. 152/2009—Methods of sampling and analysis for the official control of feed Annex III M

  A distinction is made between major elements (0.2%–1%) and minor or micro-nutrients (10–500 ppm or mg kg^{-1}), but this distinction is purely quantitative as the trace elements are as essential as the major elements. The major elements routinely analyzed are potassium (K), phosphorus (P), sodium (Na), magnesium (Mg), and calcium (Ca). The most often analyzed trace elements are iron (Fe), copper (Cu), zinc (Zn), and manganese (Mn). The identification of cations is based on atomic absorption using a spectrophotometer, while phosphorus is determined by colorimetry.
- Total proteins or crude proteins: the nitrogen content is determined by the chemical destruction of the organic matter it contains (Kjeldahl method). It is then multiplied by a factor of 6.25 to obtain the total protein material.—*Another method is based on the Dumas method*

  Norm: REGULATION (EC) no. 152/2009—Methods of sampling and analysis for the official control of feed (B): NF EN ISO 5983-2: Determination of nitrogen content and calculation of crude protein content—Part 2: Block digestion and steam distillation method (A): NF EN ISO 5983-1: Determination of nitrogen content and calculation of crude protein content—Part 1: Kjeldahl method NF V 18–100

  NF EN ISO 16634-1: Determination of the total nitrogen content by combustion according to the Dumas principle and calculation of the crude protein content—Part 1: Oilseeds and animal feeding stuffs

- Soluble sugars: the reference analysis method consists of an extraction in ethanolic medium followed by a Luff–Schoorl method prior to inversion (reduction of sugars) and following inversion (total sugars). The results are expressed as percentage of glucose.
  Norm: REGULATION (EC) no. 152/2009—Methods of sampling and analysis for the official control of feed
- Starch: based on the difference between the rotatory polarimetric measurement of a hydrolysate (hydrochloric acid) and that of an ethanol extract of the sample.—*Another method is based on an enzymatic reaction.*
  Norm: REGULATION (EC) no. 152/2009—Methods of sampling and analysis for the official control of feed • (N): NF ISO 6493
- Fibers: the Weende method is most commonly used to determine raw cellulose content. It is based on an acid attack (sulfuric acid) followed by an alkaline attack (sodium hydroxide) on the sample.
  Norm: REGULATION (EC) no. 152/2009—Methods of sampling and analysis for the official control of feed • (B): NF V03–040: Determination of raw cellulose—General method • (I): NF EN ISO 6865: Determination of crude fiber content—Method with intermediate filtration

  A more precise characterization of cell walls, based on the use of detergent solutions at different pHs, allows the quantification of three distinct fractions, namely:
  - Insoluble fibers after treatment in neutral pH, comprising hemicellulose, cellulose, and lignin, which corresponds to the totality of the cell walls (NDF for neutral detergent fiber)
  - Insoluble fibers after treatment in acid pH, comprising cellulose and lignin (ADF for acid detergent fiber). Hemicellulose is the result of the difference between NDF and ADF
  - Insoluble fibers after a 72% H_2SO_4 attack on lignin (ADL for acid detergent lignin). Cellulose is the result of the difference between ADF and ADL
  Norm: VF V18–122: animal foods: sequential determination of parietal components.
- Enzymatic digestibility of organic matter: the digestibility of organic matter (dMO) is the determining parameter in the prediction of the energy value of forage. Indeed, for a given substrate, the digestive energy and dMO are closely correlated. There are a large number of methods devised to measure dMO; the most easily standardized and routinely applied methods are enzymatic techniques using pepsin and cellulase.

27.5 SAMPLE PREPARATION AND NIR ANALYSIS

Due to the heterogeneity of forages (Figure 27.6a and b), sampling is a critical step of their analysis.

Samples must be as representative as possible of the whole stock of forages (bales or silo). When sampling, several portions must be taken (at least 10 subsamples of 50 g each) and collected in a bag that must be hermetically closed and stored under cool conditions until their arrival at the laboratory. In NIR analyses, sample preparation is relatively easy but critical (Reddersen et al., 2013). For forages, preparation usually considers a drying and a grinding step. Forage samples have to be properly dried (until constant weight), generally for 2 days at 60°C maximum to avoid Maillard reactions in particular (Figure 27.7).

The operator will carefully arrange the samples in a thin layer (or in a micro-perforated plastic bag) and turn them regularly to obtain a homogeneous drying that will prevent mold growth. The grinding step is also very important and can influence the result of a NIR analysis. The recommendations are to grind the samples using a 1-mm sieve. Sometimes pregrinding will be necessary. In any case, it is important to always use the same grinding protocol. The grinders commonly used for forage are the following:

(a)

(b)

FIGURE 27.6 Grass silage is a very heterogeneous product constituting a mix of leaves, stems, and grains (b). According to the configuration of the silo, the sampling collection is often very difficult to realize (a).

- A hammer mill for a rough grinding (Figure 27.8)
- A knife hammer used with a 1-mm sieve (Figure 27.9)

Ground samples are submitted to NIR analysis at least 30 min after grinding to allow time for the samples to reach the moisture level equilibrium.

Generally, roughly 10 g of samples is analyzed (Figure 27.10) so all the preliminary sampling steps are critical especially with heterogeneous products such as forages.

FIGURE 27.7 Oven for drying fodder (Chemistry lab at CRA-W). The recommended temperature is 60°C. The samples can be placed in a micro-perforated bag or in a tub until constant weight.

FIGURE 27.8 Retsch grinder. (CRA-W Laboratory.)

FIGURE 27.9 Cyclotec grinder. (CRA-W Laboratory.)

FIGURE 27.10 Grass sample in a quarter cup from a Foss XDS in the NIR laboratory of CRAW before NIR analysis.

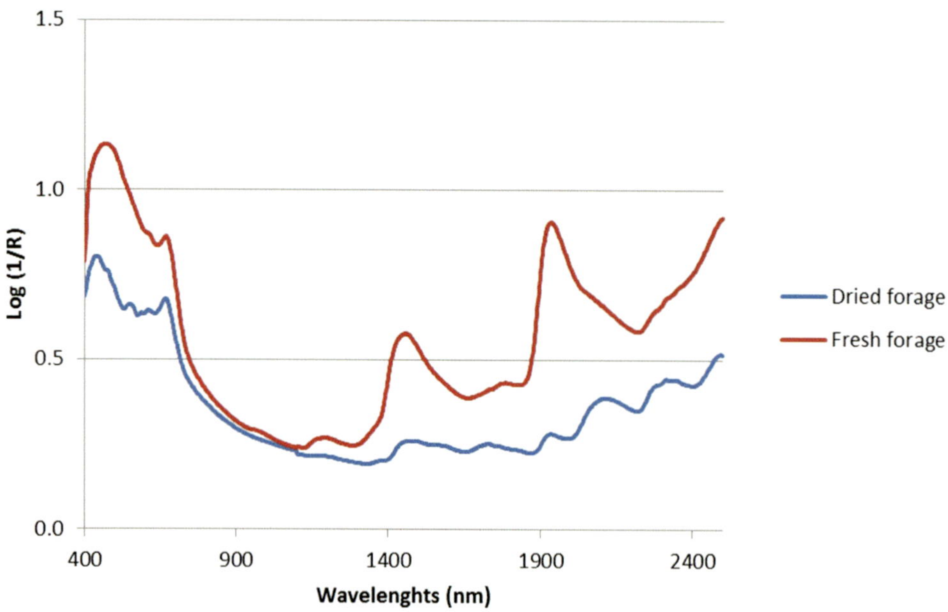

FIGURE 27.11 NIR spectra (dried and fresh) of a sample of maize forage on a Foss XDS.

The NIR spectrum (Figure 27.11) will vary especially according to the sample composition and its granulometry.

27.6 PERFORMANCE OF NIRS CALIBRATIONS FOR PREDICTING CHEMICAL COMPOSITION OF FORAGES

Studies on the potential of near-infrared spectroscopy (NIRS) for assessing feeding values of forages are numerous (Norris et al., 1976; Andueza et al., 2011; Deepa et al., 2016). A way to evaluate the accuracy of a calibration is to calculate the RPD (residual prediction deviation) value, which is the ratio of the standard deviation for a component, by the error of the corresponding NIR calibration ($RPD = SD_{REF}/SEC_{NIR}$). A calibration is generally considered as accurate for RPD values of 3 and higher.

In most cases, RPD ratios for forage calibrations are higher than 3 and confirm the good potential of near-infrared spectroscopy in forage analyses. More recent developments (Ward et al., 2011) used NIRS for a rapid assessment of the mineral concentration of forages. But the RPD ratio was lower, and the method could only be used for screening.

The databases for forages at CRA-W contain several thousand samples analyzed by wet chemistry. These databases present a very wide variability since the data were acquired during the last 40 years. The algorithm of calibrations commonly used is the partial least squares (PLS).

Depending on the application, the GLOBAL or the LOCAL methods will be used. In the GLOBAL system, all the spectra are used to build the calibrations. In the LOCAL approach, a new specific calibration is built automatically when a new sample is scanned. The system is set up in a way that the software can select in the database only the most similar spectra of the sample spectrum to be analyzed. Generally, the LOCAL approach gives better results (Sinnaeve et al., 1994; Tran et al., 2010).

Tables 27.1 and 27.2 give the characteristics and performance of NIR calibrations conducted by CRA-W. The parameters are expressed at 100% of dry matter (DM).

TABLE 27.1
Performance of NIR Calibrations for Dried Grass Forages (CRA-W Database)

Parameters	*N*	Min	Max	Mean	SD	R^2	SEC	RPD
Analytical dry matter	1877	88.84	97.49	93.16	1.44	0.78	0.68	2.1
Proteins % DM	1877	4.45	31.26	15.49	5.26	0.98	0.76	6.9
Fiber % DM	1465	11.27	41.10	26.18	4.97	0.95	1.11	4.5
Total ash % DM	1989	3.44	16.66	10.05	2.20	0.85	0.86	2.6
NDF % DM	806	26.00	70.18	48.09	7.36	0.95	1.61	4.6
ADF % DM	539	11.47	43.16	27.31	5.28	0.95	1.13	4.7
ADL % DM	530	0.00	5.88	2.76	1.04	0.87	0.37	2.8
Digestibility of dry matter (De Boever) % DM	1156	50.19	108.28	79.23	9.68	0.96	1.89	5.1
Digestibility of organic matter (De Boever) % DM	1291	46.02	108.06	77.04	10.34	0.96	1.97	5.3
Total soluble sugars % DM	629	0.12	36.12	11.47	8.22	0.97	1.35	6.1

Max, maximum value; Min, minimum value; N, number of spectra; R^2, coefficient of determination; SD, standard deviation; SEC, standard error of calibration.
RPD = SD/SEC.

TABLE 27.2
Performance of NIR Calibrations for Dried Maize Forages (CRA-W Database)

Parameters	*N*	Min	Max	Mean	SD	R^2	SEC	RPD
Analytical dry matter %	7197	88.54	98.15	93.34	1.60	0.90	0.52	3.1
Proteins % DM	6514	3.75	11.77	7.76	1.34	0.90	0.42	3.2
Fiber % DM	5749	8.86	32.68	20.77	3.97	0.95	0.93	4.3
Total ash % DM	7962	1.43	7.07	4.25	0.94	0.77	0.45	2.1
Starch % DM	6586	0.00	58.39	28.07	10.10	0.98	1.60	6.3
NDF % DM	4751	22.60	62.97	42.78	6.73	0.95	1.58	4.3
ADF % DM	3645	9.65	38.35	24.00	4.78	0.95	1.04	4.6
ADL % DM	3326	0.40	4.91	2.66	0.75	0.81	0.34	2.2
Digestibility of dry matter (De Boever) % DM	2217	53.89	92.55	73.22	6.44	0.92	1.83	3.5
Digestibility of organic matter (De Boever) % DM	2530	52.22	92.07	72.14	6.64	0.92	1.90	3.5
Digestibility of dry matter (Aufrère) % DM	3941	49.86	86.54	68.20	6.11	0.90	1.92	3.2
Digestibility of organic matter (Aufrère) % DM	3547	47.57	87.76	67.67	6.70	0.90	2.10	3.2
Residual soluble sugar % DM	4693	0.00	16.87	5.48	3.80	0.95	0.84	4.5
Total soluble sugars % DM	3046	0.00	23.02	7.47	5.18	0.99	0.63	8.3

Max, maximum value; Min, minimum value; N, number of spectra; R^2, coefficient of determination; SD, standard deviation; SEC, standard error of calibration.
RPD = SD/SEC.

27.7 ANALYSIS OF WET FORAGE

In addition to classical NIR analyses on dried and ground samples, calibrations are developed for the determination of the chemical composition and enzymatic digestibility of undried forage (Andueza et al., 2013). Indeed, it is most interesting for farmers or nutritionists to mount near-infrared devices on silage cutters or to analyze forage directly at the farm site using a portable spectrometer to obtain a first evaluation of the quality of the product (estimation of humidity, crude protein content). The main purpose is to secure a quick determination of the composition by skipping the drying and grinding stage thus saving a significant amount of time.

For example, the NIR fresh grass database built by CRA-W contains different types of samples harvested during the 2014 and 2015 growing seasons (Decruyenaere, 2006). Calibrations were developed for determining the dry matter content (DM 60°C) and the chemical composition: total ash, crude protein, crude cellulose, and enzymatic digestibility of organic matter (cellulase method).

In practice, before the NIR analysis, grass samples were cut into 8–10 cm long strands. Samples were then homogenized and analyzed on a Foss XDS equipped with a rectangular full cup (Figure 27.12). Every sample was scanned five times, and the absorbance of the samples was recorded as log (1/*R*).

At the time of creation of the database and in an effort to save time and money, fresh samples were not analyzed through wet chemistry but dried in a ventilated oven at 60°C for 48 h for dry matter determination and then ground through a 1-mm sieve using a cyclotec grinder. The next step consisted of subjecting the ground and dried sample to the same NIR analyzer (XDS). The measurements were made in duplicate in this case, and the classical chemical composition and enzymatic digestibility of organic matter were then predicted from existing and approved NIR calibrations. Since the NIR predictions are checked on a regular basis as part of the activities undertaken by interlaboratory studies (BIPEA, Germ services, REQUASUD), the level of confidence on dried grass forage is very high.

Compared to NIR calibrations developed with dried samples, the performances (Table 27.3) obtained from fresh samples are slightly lower.

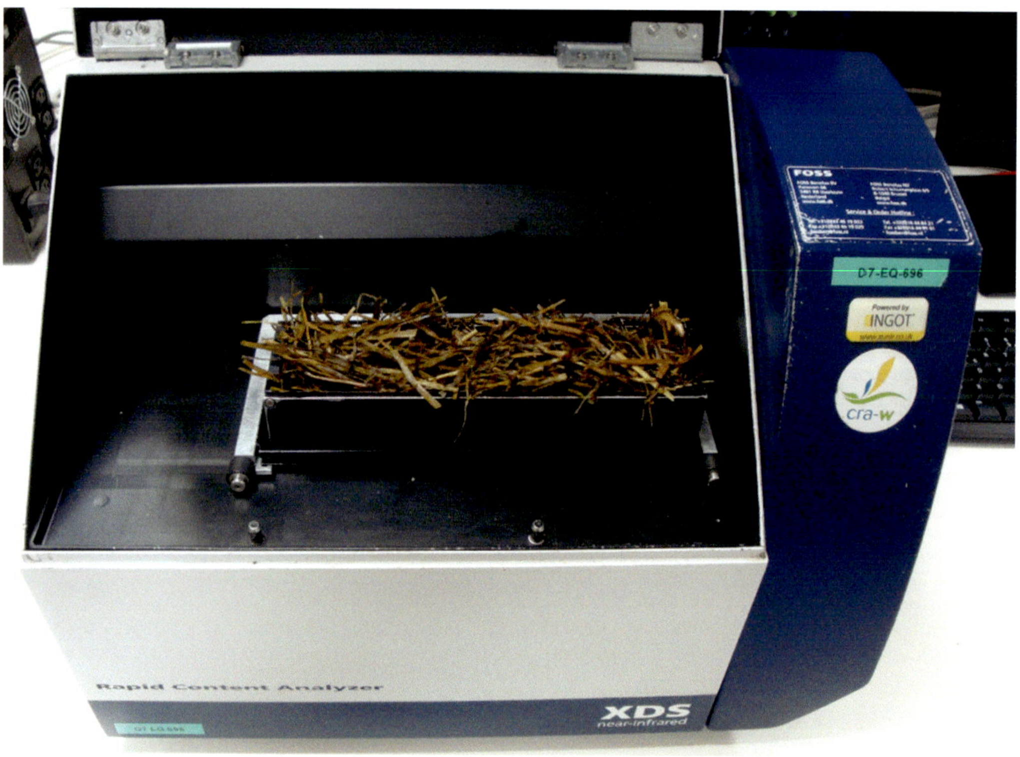

FIGURE 27.12 Analyze of wet forage with a Foss XDS spectrometer in the NIR laboratory of CRA-W. (The Foss XDS spectrometer is a monochromator NIR instrument working in reflectance and covering the range of 400–2498 nm.)

TABLE 27.3
Performance of NIR Calibrations for Wet Grass Forages (CRA-W Database)

Parameters	*N*	Min	Max	Mean	SD	R^2	SEC	RPD
Residual dry matter %	192	13.32	69.39	28.67	13.54	0.99	1.48	9.1
Protein % DM	222	2.76	29.64	14.79	5.43	0.93	1.85	2.9
Ash % DM	219	6.00	31.66	11.11	4.24	0.93	1.19	3.6
Cellulose % DM	218	18.50	34.69	24.82	3.48	0.90	1.38	2.5
Digestibility of organic matter % DM	218	44.80	93.17	72.96	9.26	0.93	3.08	3.0

Max, maximum value; Min, minimum value; N, number of spectra; R^2, coefficient of determination; SD, standard deviation; SEC, standard error of calibration.
RPD = SD/SEC.

The results are similar to those obtained by other research teams. Previous work (Sinneave et al., 1994; Park et al., 1998) has shown the potential of NIR analyses of fresh materials to determine the chemical composition and digestibility of grass silages. According to Cozzolino and Labandera (2002), the NIR fresh analyses can be used to predict the dry matter and crude protein of forages with errors of calibration of 1.2% and 1.9%, respectively. This potential has been confirmed by more recent studies. Thus, Park et al. (2014) have shown that moisture, proteins, and ash can be estimated with errors of calibration of 1.89%, 1.14%, and 1.47% DM, respectively.

NIR fresh analyses allow estimating the dry matter content and the chemical composition and digestibility of grass with precision. The range of variation of the samples used to develop the calibrations suggests that this precision is sufficient to differentiate the samples qualitatively. The estimation of dry matter content, a parameter of importance for the calculation of livestock diets, seems sufficiently accurate. The database should, however, be supplemented with new samples to consolidate the calibrations.

27.8 HAND-HELD NIR SPECTROMETERS

In addition to classical analyses of dried fodder, nutritionists and farmers are also interested in the analyses of fresh forage directly at the farm. Indeed, changes in nutriment in the silages can be important according to the season, the place where the samples are taken or in the way of how the silage has been done. Having a real-time analysis tool, even with a more limited accuracy can be very interesting to help the farmers feed their cattle the most efficient way. Miniaturized and mobile NIR instruments allow very frequent forage analysis with immediate results and for a lower cost.

There are various devices on the market that are more or less adapted to the measurement of fresh fodder. Each of them has its specificity and needs to be optimized. In 2018, CRA-W had the opportunity to test the FieldSpec of ASD (Figure 27.13), the microNIR 1700 of VIAVI (Figure 27.14), and the NIR Flame of Ocean Optics (Figure 27.15) on the forages of its experimental farm but also from 12 other farms in Wallonia. The materials tested were mainly maize and grass silage (Figure 27.16).

In this mini-study, the variability is of course very limited by the number of farms and because all the silages analyzed were produced in 2017 and 2018. That's why presenting RPD and SECV values at this stage of the project were not very significant. What emerges from this study is that analyses of wet samples at the farm are much more complicated than on dried samples in a laboratory, especially for maize.

The main reason is certainly the sample presentation (Figure 27.17) but also the restricted spectral ranges of most of the miniaturized spectrometers, the influence of the external environmental conditions (temperature, rain, and relative moisture), and the unconscious human behavior which

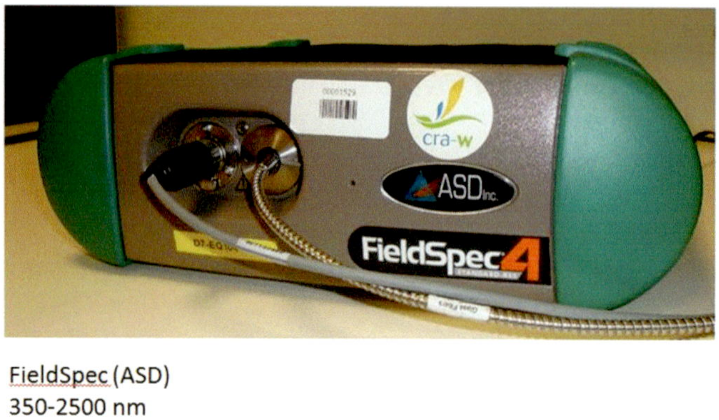

FIGURE 27.13 ASD spectrometer. (NIR Laboratory of CRA-W.)

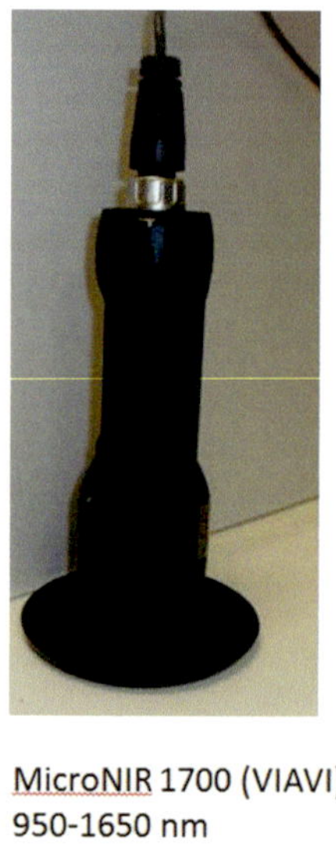

FIGURE 27.14 VIAVI spectrometer. (NIR Laboratory of CRA-W.)

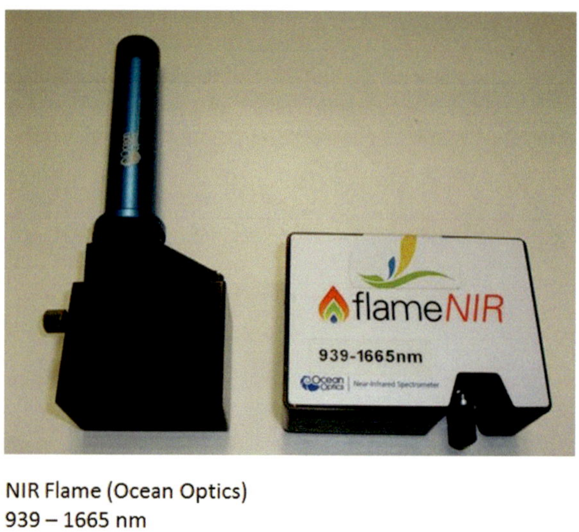

FIGURE 27.15 Flame NIR. (NIR Laboratory of CRA-W.)

FIGURE 27.16 Sample collecting on the silo at the experimental farm of CRA-W.

FIGURE 27.17 Analysis of fresh forage at the experimental farm of CRAW with the VIAVI.

unintentionally tends to place the spectrometers at the best place in the silage. Nevertheless, the first results seem to confirm that a dry matter determination is possible with a good accuracy in silage with such instruments.

For the other main parameters such as fibers, digestibility, ash, or protein, the level of accuracy seems to be more limited but acceptable in grass. In fresh maize silage, the sample presentation (unground grain, leaves, etc.) was probably the main limiting factor for a good NIR analysis, resulting in less accurate models as on grass silage.

27.9 REQUASUD: A LABORATORY NETWORK FOR NIR ANALYSIS IN BELGIUM

The Walloon Agricultural Research Center manages a network of laboratories called REQUASUD (www.requasud.be). Each of these laboratories owns a Foss XDS NIR spectrometer. The CRA-W is responsible for calibration updates, generating standardization procedures on an annual basis, and checking the results obtained by the organizations of Inter-Laboratories Studies (ILS) for different products (forage, cereals, food, etc.).

The network is driven by the MOSAIC (Foss) system, while the LOCAL methods are used for forage analysis. The principle consists of selecting from a vast database hundreds of spectra similar to the spectrum of the sample being predicted. Specific calibrations for each component are built automatically. The minimal and maximal number of samples selected in the database and the number of PLS factors used have been previously optimized.

These ring tests take place twice a year. The first test is usually conducted in March and focuses on grass (hay and silage), while the second test is usually run in September and targets maize (whole plant and silage). Two blind duplicate samples of each type (fermented and not fermented) are sent to each laboratory. Before being sent, homogeneity tests are conducted to ensure that each laboratory has received identical samples.

The samples are analyzed through wet chemistry procedures and NIR spectroscopy. The requested components are dry matter, crude proteins, cellulose, ash, and digestibility (including starch for maize only). The results are compiled, and a value is assigned to every parameter of each sample. This assigned value corresponds to the average of the results of all laboratories minus outliers. The outliers are identified by statistical tests:

- Grubbs: makes possible the detection of outliers in terms of average dispersion. The principle of this test is to compare absolute values vs. reduced deviations
- Cochran: makes possible the detection of outliers in terms of dispersion and is applied to the standard deviations of the measurements
- Z-score analysis: reflects the consistency between the analytical result and the assigned value. It is a measure of how many standard deviations below or above the population mean a raw score is.

27.10 NIR APPLIED TO FECES TO ASSESS THE QUALITY OF FORAGES

Methods for assessing the quantitative and qualitative characteristics of grass are numerous but difficult to apply in the case of grazing ruminants. Several studies have shown that NIRS spectra of feces can be correlated with the chemical or biological composition of samples in order to develop calibrations (Decruyenaere, 2015, Dixon and Coates, 2009, Boval et al., 2004).

Major studies show that NIRS has great potential for estimating in vivo digestibility and voluntary intake by grazing ruminants and that feces are a good indicator of ingested diets. Based on both large or small and varied databases, results suggest that NIRS spectral libraries could be developed for characterizing ruminant feed intake. The accuracy of the NIRS models in estimating in vivo digestibility and voluntary intake is similar to or better than that of other methods generally used to assess these parameters. NIRS could also be used to predict ruminants' diet composition in terms of plant species. However, these predictions should only be used for ranking, because of the current lack of accurate procedures for determining diet selection individually.

The main difficulty associated with this method is generating diet–fecal pairs as reliably as possible. NIRS calibrations for predicting in vivo diet characteristics are derivative calibrations. The samples analyzed for reference values (diet samples) differ from the samples submitted for NIRS analyses (feces). In connection with research on forages, in vivo trials with animals confined in pens or digestibility crates appear to be the best reference method for generating NIRS calibrations.

Future work will involve developing NIRS calibrations for predicting independent datasets and using them to create decision-support tools for improving diverse grazing management schemes. The focus should be on comparing different feeding strategies rather than obtaining an exact estimate of feed intake values. As a low-cost and rapid prediction technique, NIRS could contribute significantly to the development of a methodology that would help improve our knowledge of forage and animal variability.

27.11 CONCLUSION

In a world where everything needs to be optimized, NIR spectroscopy is a wonderful tool that can replace traditional physicochemical methods to determine the composition of forage and to evaluate its quality. The classical NIR analyses of dry and ground samples are well in place now. The future is in miniaturization and portability. That's why we are seeing on the market a new generation of hand-held spectrometers. This is only the beginning, and a lot of work has still to be done, but for sure in the future, at least a preliminary analysis of forage will occur at the farm level.

Forage analysis in the field is one of the elements of what we can call "smart farming" where the farms will be equipped with different types of sensors to control and optimize all its processes.

ACKNOWLEDGMENTS

The main authors want to acknowledge all the people who helped in writing this chapter and/or who contributed to the building of the spectral databases of the CRA-W over the years in particular Richard Agneessens, Bruno Godin, Vincent Baeten, Bernard Lecler, Claudine Clément, Sandrine Mauro, Stéphane Brichard, Nicolas Crasset, Eric Fontaine, Nicolas Chamberland, and Pierre Dardenne.

REFERENCES

Andueza D., Picard F., and Aufrère J. (2013). Ability of local and global fresh permanent grassland calibrations to predict the chemical composition and nutritive value of permanent grassland hays using NIRS, in *16th International Conference on Near Infrared Spectroscopy*, 2–7 June 2013, la Grande-Motte, France, pp 269–273.

Andueza D., Picard F., Jestin M, Andrieu J., and Baumont R., (2011). NIRS prediction of the feed value of temperate forages: efficacy of four calibration strategies. *Animal*, 5(7), 1002–1013. doi:10.1017/S1751731110002697.

Boval M., Coates D.B., Lecomte P., Decruyenaere V., and Archimède H., (2004). Faecal near infrared reflectance spectroscopy (NIRS) to assess chemical composition, *in vivo* digestibility and intake of tropical grass by Creole cattle. *Anim. Feed Sci. Technol.*, 114(1), 19–29.

Cozzolino D. and Labandera M., (2002). Determination of dry matter and crude protein contents of undried forages by near-infrared reflectance spectroscopy. *J. Sci. Food Agric.*, 82, 380–384.

Decruyenaere V., (2015). Estimation of diet digestibility and intake of grazing ruminants through near infrared reflectance spectroscopy analysis of faeces. Application in various contexts of livestock production (PhD Dissertation in English). Gembloux Agro-Bio Tech, University of Liège, Gembloux, Belgium, p. 163.

Decruyenaere V., Agneessens R., Toussaint B., Anceau C., Goffaux M-J. and Oger R. (2006). Qualité du fourrage en région wallonne, *REQUASUD*.

Deepa K., Senthilkumar S., Kalpana K., Suganya T., Sasikumar P., Thirumalaisamy G., Sureshkumar R., and Vasanthakumar P., (2016). IRS in animal sciences. *Int. J. Sci. Environ. Technol.*, 5(2), 605–610.

Dixon R. and Coates D. (2009). Review: Near infrared spectroscopy of faeces to evaluate the nutrition and physiology of herbivores. *J. Near Infrared Spectrosc.*, 17, 1–31.

FAOSTAT. 2009. Statistical Database 2007. Rome.

Huyghe C., Peeters A., De Vliegher A., 2015. La prairie en France et en Europe. In : Adapatation des prairies semées au changement climatique. Colloque INRA 16-17 novembre 2015, Poitiers, France, 13–23.

Hyung Soo Park, Sang Hoon Lee, Ki Choon Choi, Young Cheol Lim, Ji Hea Kim, Ki Won Lee and Gi Jun Choi, 2014. Prediction of the Chemical Composition and Fermentation Parameters of Winter Rye Silages by Near Infrared Spectroscopy. Journal of the Korean Society of Grassland and Forage Science 34(3): 209–2013.

Minet O. Ferber, F., Jacob, L., Lecler, B., Agneessens, R., Cugnon, T., Decruyenaere, V., Genot, V., Gofflot, S., Pitchugina, E., Planchon, V., Rennesson, M., Sinnaeve, G., Wavreille, J., Dardenne, P. & Baeten, V., (2016). la spectrométrie proche-infrarouge : une technologie rapide, précise et écologique pour déterminer la composition et la qualité des produits agricoles et alimentaires, Gembloux, ASBL Requasud, cellule de coordination, 32.

Norris K.H., Barnes R.F., Moore J.E., and Skenk J.S. (1976) Predicting forage quality by infrared reflectance spectroscopy. *J. Anim. Sci.*, 43, 889–897.

Park R.S., Agnew R.E., Gordon F.J., and Steen R.W.J., 1998. The use of near infrared reflectance spectroscopy (NIRS) on undried samples of grass silage to predict chemical composition and digestibility parameters. *Anim. Feed Sci. Technol.*, 72, 155–167.

Reddersen B., Fricke T., and Wachendorf M. (2013). Effects of sample preparation and measurement standardization on the NIRS calibration quality of nitrogen, ash and NDFom content in extensive experimental grassland biomass. *Anim. Feed Sci. Technol.*, 183, 77–85.

Sinnaeve G., Dardenne P., Agneessens R., and Biston R., 1994. The use of near infrared spectroscopy for the analysis of fresh grass silage. *J.Near Infrared Spectrosc.*, 2, 79–84.

Sinneave G., Dardenne P., and Agneessens R. (1994). Global or Local? A choice for NIR calibrations in analyses of forage quality. *J.Near Infrared Spectrosc.*, 2, 163–175.

Tran H., Salgado P., Tillard E., Dardenne P., Nguyen X.T., and Lecomte P., (2010). "Global" and "local" predictions of dairy diet nutritional quality using near infrared reflectance spectroscopy. *J. Dairy Sci.*, 93, 4961–4975.

Ward A., Nielsen A.L., and Moller H., (2011). Rapid Assessment of Mineral Concentration in Meadow Grasses by Near Infrared Reflectance Spectroscopy Rapid Assessment of Mineral Concentration in Meadow Grasses by Near Infrared Reflectance Spectroscopy. *Sensors*, 11, 4830–4839.

28 NIRS Is Ripe for Use in Horticulture

Nghia Nguyen-Do-Trong, Bart Nicolaï, and Wouter Saeys[*]
MeBioS Division

CONTENTS

28.1 NIRS AS A RAPID ALTERNATIVE FOR DESTRUCTIVE QUALITY ASSESSMENT

The quality of a batch of fruit and vegetables is typically evaluated through subjective visual inspection and destructive measurements on a small number of samples. The fast and nondestructive character of NIR spectroscopy makes it an appealing alternative for use in horticulture. As a result, its potential has been investigated for a wide range of applications in horticulture, as reported by Nicolaï et al. (2007). Since then, a vast number of articles have been published, which further expand its application to nondestructive evaluation of quality attributes of fruit and vegetables. Most of these applications are in five main groups of fruit and vegetables: pome fruit, citrus fruit, banana, leafy greens, and root vegetables. In Figure 28.1, some representative NIR spectra for fruit and vegetables from these five groups are presented as obtained in diffuse reflectance mode with an FT-NIR spectrophotometer (MPA, Bruker Optics, Ettlingen, Germany) equipped with an integrating sphere. It can be seen that the NIR spectra of these fruit and vegetables are very similar and are dominated by broad absorption peaks around 970, 1200, 1450, and 1940 nm. These correspond to the overtone and combination bands of the O–H group in water (Williams and Norris 2001). This is not surprising as water is the main constituent in fruit and vegetables (70%–96%) (Canet 1988). Two other more subtle absorption peaks at around 1800 and 2500 nm correspond to the overtone and combination bands of the C–H groups in the carbohydrates, which are the main component in the dry matter of fruit and vegetables (Williams and Norris 2001).

[*] wouter.saeys@kuleuven.be

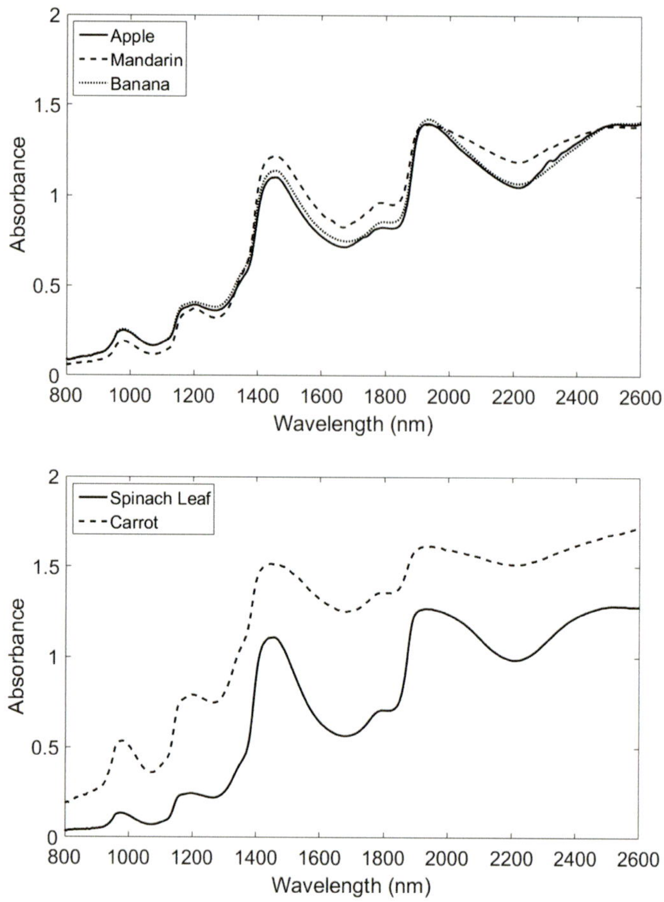

FIGURE 28.1 NIR absorbance spectra for representative fruit and vegetables from the five distinguishable groups as obtained in diffuse reflectance mode with an FT-NIR spectrophotometer.

28.1.1 Pome Fruit

Within this group, apple and pear are the fruit for which most reports on the use of NIRS technology for nondestructive quality evaluation can be found. In Table 28.1, an overview is given on the quality aspects of different apple and pear cultivars for which the potential of NIR spectroscopy has been evaluated. Taste (e.g., soluble solids content) and texture (e.g., firmness) are the most important attributes contributing to apple and pear overall quality besides their size, appearance, aroma, and safety attributes, and, hence, they have been the most targeted quality attributes so far (Nicolaï et al. 2007). Reflectance and interactance modes were used for spectral acquisition, because the relatively large size and high water content of the fruit make it challenging to perform a transmittance measurement through an intact apple or pear. Most studies were limited to the wavelength range below 1700 nm, because the high moisture content results in very low signals for longer wavelengths. In some studies, the visible range (400–780 nm) was also included to exploit the correlation between the concentrations of the biological chromophores (chlorophyll, carotenoids, anthocyanins) and the quality features (Nicolaï et al. 2007). For soluble solids content (SSC), successful predictions with an accuracy around 0.4% (RMSEP) have been reported for various apple and pear cultivars (Lu, Guyer, and Beaudry 2000; Nicolaï, Theron, and Lammertyn 2007; Fan et al. 2009; McGlone, Jordan, et al. 2003; Nicolaï et al. 2008). However, the results reported for the prediction of texture properties (e.g., firmness) of apple and pear were more variable and less convincing. Dry matter content for

TABLE 28.1
Overview of Applications of NIR Spectroscopy for Nondestructive Evaluation of Apple and Pear Quality.

Species	Attribute	Cultivar	Spectral Range (nm)	RMSEP/ (Classification Accuracy)	Reference
Apple (*Malus domestica Borkh.*)	SSC (soluble solids content) (°Brix)	Jonagold	380–1865	0.61 (SEP)	Lammertyn et al. 1998
		Golden Delicious, Jonagold	350–999	1.1	Ventura et al. 1998
		Empire, Golden and Red Delicious	800–1700	0.4–0.5	Lu, Guyer, and Beaudry 2000
		Various	380–2000	0.7	Peirs et al. 2001; Peirs, Scheerlinck, and Nicolaï 2003
		Various	1000–2000	0.7–0.9 (SEP)	Peirs et al. 2002
		Royal Gala	500–1100	0.5–0.7	McGlone, Jordan, and Martinsen 2002
			800–1000	0.3–0.5	McGlone, Jordan, et al. 2003
		Various	800–1000	0.2 (SEP)	Temma, Hanamatsu, and Shinoki 2002a
			680–1235	0.3–0.5 (SEP)	Temma, Hanamatsu, and Shinoki 2002b
		Fuji	300–1100	0.3	Walsh, Golic, and Greensill 2004
			812–2357	0.5	Liu and Ying 2005
		Golden Delicious	800–1690	0.4	Nicolaï, Theron, and Lammertyn 2007
		Red Fuji	500–1100	0.39 (SEP)	Fan et al. 2009
		Braeburn	500–1000	0.69	Nguyen Do Trong et al. 2014
			550–1000	0.57	Van Beers et al. 2015
		Fuji and Golden Delicious	350–2500	0.51–0.68	Paz et al. 2009
		Elshof	1000–2500	0.7; 0.39–0.46	Travers, Bertelsen, and Kucheryavskiy 2014; Møller et al. 2013
	Firmness (*N*)	Various	800–1100	–	Lovász, Merész, and Salgó 1994
		Various	1000–2000	8.26–10.94 (SEP)	Peirs et al. 2002
		Royal Gala	500–1100	7.0–7.5	McGlone, Jordan, and Martinsen 2002
		Elshof	1000–2500	–	Møller et al. 2013
		Fuji and Golden Delicious	350–2500	8.28–8.83	Paz et al. 2009
		Braeburn	500–1000	8.91	Nguyen Do Trong et al. 2014
			550–1000	7.36	Van Beers et al. 2015
		Red Fuji	500–1100	0.53	Fan et al. 2009
		Golden Delicious, Gala	400–1800	4.91–7.33 N (SEP)	Park et al. 2003

(Continued)

TABLE 28.1 *(Continued)*
Overview of Applications of NIR Spectroscopy for Nondestructive Evaluation of Apple and Pear Quality.

Species	Attribute	Cultivar	Spectral Range (nm)	RMSEP/ (Classification Accuracy)	Reference
	Stiffness	Jonagold	380–1650	2.49 (SEP)	Lammertyn et al. 1998
	Dry matter	Various	800–1100	–	Lovász, Merész, and Salgó 1994
		Royal Gala	800–1000	0.32%–0.51%	McGlone, Jordan, et al. 2003
		Elshof	1000–2500	0.7%; 0.37%–0.48%	Travers, Bertelsen, and Kucheryavskiy 2014; Møller et al. 2013
		Braeburn	650–950	0.43%–0.48%	McGlone and Martinsen 2004
	pH, acidity	Various	800–1100	–	Lovász, Merész, and Salgó 1994
		Fuji	812–2357	0.068%; 0.044% (SEP)	Liu and Ying 2005
		Jonagold	380–1650	0.068 (SEP)	Lammertyn et al. 1998
		Elshof	1000–2500	–	Møller et al. 2013
		Various	1000–2000	1.61–2.37 mL NaOH (SEP)	Peirs et al. 2002
		Royal Gala	500–1000	0.052%	McGlone, Jordan, and Martinsen 2002
	Optimal harvest date, Streif index	Various	380–2000	5–9 days; 0.18	Peirs et al. 2001; Peirs, Schenk, and Nicolaï 2005
	Starch pattern index, quantitative starch	Royal Gala	500–1100	0.95%; 0.17%	McGlone, Jordan, and Martinsen 2002
	Streif index, starch conversion value	Braeburn	550–1100	0.062; 0.785	Van Beers et al. 2015
	Brownheart disorder	Braeburn	300–1140	7.9%–15.4%	Clark, McGlone, and Jordan 2003
			650–950	4.1% (RMSECV)	McGlone et al. 2005
	Sensory attributes	Golden Delicious, Braeburn, Fuji	400–2500	0.29–0.54 (RMSE)	Mehinagic et al. 2004
	Bruises	Golden Delicious	400–1700	(16.3%)	Xing et al. 2006
		Golden Delicious	430–930[a]	–	S. Kim, R. Chen, and M. Mehl 2001
		Red and Golden Delicious	900–1700[a]	(59%–94%)	Lu 2003
		Jonagold	900–2500[a]	[98%]	Keresztes, Goodarzi, and Saeys 2016
		Jonagold, Kanzi, Joly Red	1000–2500[a]	(94.4%)	Keresztes et al. 2017

(Continued)

TABLE 28.1 (*Continued*)
Overview of Applications of NIR Spectroscopy for Nondestructive Evaluation of Apple and Pear Quality.

Species	Attribute	Cultivar	Spectral Range (nm)	RMSEP/ (Classification Accuracy)	Reference
		Various	400–2500[a]	(>90%)	Baranowski, Mazurek, and Pastuszka-Woźniak 2013
		Golden Delicious, Pink Lady	900–1700[a]	(92.42%; 94.04%)	Ferrari et al. 2015
	Fungal contamination	Golden Delicious	430–930[a]	–	S. Kim, R. Chen, and M. Mehl 2001
	Fecal contamination	Various	450–851[a]	–	S. Kim et al. 2002
	Surface defects/ contaminations	Various	424–899[a]	–	Mehl et al. 2004
	Surface defects	Honeycrisp, Redcort, Red Delicious	450–940[a]	(>82%)	Ariana, Guyer, and Shrestha 2006
Pear (*Pyrus communis L.*)	SSC (soluble solid content) (°Brix)	Conference	780–1700	0.44	Nicolaï et al. 2008
		Huangguan	1000–2500	0.28	Jiang and Zhu 2013
		Ya	930–2548	0.35	Li et al. 2016
		Ya	400–1000[a]	0.49	Fan et al. 2015
		Doyenne du Comice	450–1000[a]	1.1%	Itoh et al. 2013
	Firmness (*N*)	Conference	780–1700	–	Nicolaï et al. 2008
		Huangguan	1000–2500	0.6	Jiang and Zhu 2013
		Rocha	400–950	–	Cavaco et al. 2009
		Ya	400–1000[a]	0.72	Fan et al. 2015
		Doyenne du Comice	450–1000[a]	11.6	Itoh et al. 2013
	Bruises	Shingo	950–1650[a]	(92%)	Lee et al. 2014

[a]*Indicates hyperspectral imaging measurement.*

apple could also be predicted with an accuracy around 0.32% (RMSEP) (McGlone, Jordan, et al. 2003). Prediction performance for titratable acidity (TA) was too variable to be useful in practice. As the detection of superficial, internal defects, and surface contaminants requires spatial information, hyperspectral imaging has been used to detect these quality attributes (Keresztes et al. 2017; Keresztes, Goodarzi, and Saeys 2016; Lorente et al. 2012).

28.1.2 Citrus

In Table 28.2, an overview is given on the applications of NIR spectroscopy for quality evaluation of citrus fruit. While the skin of citrus fruit is thicker and more turbid than that of pome fruit, its flesh is more translucent. Transmittance measurements are, thus, more feasible for citrus fruit such as mandarins and oranges (Magwaza et al. 2012). This allows us to obtain spectroscopic information from the flesh deeper towards the fruit's core than is possible in pome fruit. In contrast to pome fruit, texture properties like firmness are of less interest in citrus fruit. Therefore, most applications of NIR spectroscopy on citrus fruit have focused on the prediction of the chemical composition of the fruit flesh relevant for its nutritional value and taste (SSC, TA, and vitamin C). The following prediction accuracies (RMSEP) have been reported for SSC, TA, and vitamin C, respectively, for mandarin: 0.32% (Kawano, Fujiwara, and Iwamoto 1993; McGlone, Fraser, et al. 2003), 0.09 (% citric acid) (Liu et al. 2010; Sun, Zhang, and Liu 2009), 2.7 (mg/100 g) (Sun, Zhang, and Liu 2009) and for orange: 0.51% (Cayuela 2008), 0.14 (% citric acid) (Magwaza et al. 2013), 3.9 (mg/100 g) (Xia et al. 2007). It has also been shown that various defects such as rot can be detected with hyperspectral imaging (Kim et al. 2004; Blasco et al. 2007, 2009).

28.1.3 Banana

Banana is the most important fruit in the world (Statista, 2014). An overview of the applications of NIRS for quality evaluation of banana is given in Table 28.3. In contrast to pome fruit, it has a thick skin, which is highly turbid. The flesh, which is the only part that is consumed, is also highly turbid, which makes the transmittance approach applied in citrus fruit quite challenging. Diffuse reflectance or interactance measurements are, therefore, preferred. However, as nondestructive measurements require that the skin remains intact, the measurement configuration should be designed such that sufficient information is collected from the flesh below the skin. As both the skin and flesh colors change during ripening, many researches have included the visible range (400–760 nm) to exploit this color information for nondestructive quality evaluation of the flesh tissue. However, as the color varies within and between different banana varieties, the multivariate calibration models should be made robust against this. The main quality attributes of banana which have been measured with NIRS are SSC and firmness of the flesh. Prediction accuracies of 0.79% for SSC (Akmalia, Saputro, and Handayani 2017) and 2.94 N for firmness (Xie, Chu, and He 2018) have been reported.

28.1.4 Leafy Greens

An overview of the application of NIRS for quality evaluation of leafy greens is given in Table 28.4. It is obvious that most researchers have focused on spinach and lettuce as these are commercially important leafy greens that are produced in large quantities. In contrast to the aforementioned fruit categories, leafy vegetables are typically thin and flat and have a moisture content of around 95%. Therefore, spectra are typically acquired in transmittance or transflectance mode. Their green color is a result of the absorption of red and blue light by chlorophyll, the natural chromophore involved in photosynthesis. Apart from chlorophyll, they may also contain other biological chromophores such as carotenoids and anthocyanins. Quality degradation in leafy greens is mostly related to chlorophyll degradation and water loss, subsequently leading to changes in the visual appearance and

TABLE 28.2
Overview of Applications of NIR Spectroscopy for Quality Evaluation of Citrus Fruits.

Species	Attribute	Cultivar	Spectral Range (nm)	RMSEP/ (Classification Accuracy)	Reference
Mandarin fruit (*Citrus reticulata* L.)	SSC (soluble solid content)	Imperial	306–1130	0.4 (RMSEC)	Guthrie et al. 2005; Guthrie, Reid, and Walsh 2005
		Nanfeng	400–1040	0.65	Sun, Zhang, and Liu 2009
			350–1040	0.65	(Liu et al. 2010
		Page tangelo	400–1000	0.45 (RMSECV)	Antonucci et al. 2011
		Satsuma	400–1000	0.59 (RMSECV)	Antonucci et al. 2011
			500–1100	0.32	McGlone, Fraser, et al. 2003
			350–2500	0.33	Gómez, He, and Pereira 2006
			800–1100	–	Tsuchikawa et al. 2003
			680–1235	0.32 (SEP)	Kawano, Fujiwara, and Iwamoto 1993
		Unshiu	636–1236	0.42 (SEP)	Lee et al. 2004
	Titratable acidity (TA)	Imperial	306–1130	0.2 (RMSEC)	Guthrie et al. 2005; Guthrie, Reid, and Walsh 2005
		Nanfeng	400–1040	0.09	Sun, Zhang, and Liu 2009
			350–1040	0.09	Liu et al. 2010
		Page tangelo	400–1000	0.09 (RMSECV)	Antonucci et al. 2011
		Satsuma	400–1000	0.07 (RMSECV)	Antonucci et al. 2011
			500–1100	0.15	McGlone, Fraser, et al. 2003
			800–1100	–	Tsuchikawa et al. 2003
	Dry matter	Imperial	306–1130	0.6 (RMSEC)	Guthrie et al. 2005; Guthrie, Reid, and Walsh 2005
	pH	Satsuma	350-2500	0.18	Gómez, He, and Pereira 2006
	Firmness	Satsuma	350–2500	0.53	Gómez, He, and Pereira 2006
	% Juice	Imperial	306–1130	5.0 (RMSEC)	Guthrie et al. 2005; Guthrie, Reid, and Walsh 2005
	Vitamin C	Nanfeng	400–1040	2.7	Sun, Zhang, and Liu 2009
	Color	Nanfeng	400–1040	0.81	Sun, Zhang, and Liu 2009
	Various defects	Various	400–1800[a]	(86%)	Blasco et al. 2007; Blasco et al. 2009
	Green mold rottenness	Clemenules	320–1100[b]	(91.1%)	Gómez-Sanchis et al. 2008
	Fungal infection	Clemenvilla	650–1700	(97.8%)	Lorente et al. 2015
	Maturity	Unshiu	650–955	(97%)	Kim et al. 2004
	Defects	Unshiu	650–955	(92.5%)	Kim et al. 2004
Orange fruit (*Citrus sinensis* L.)	SSC (soluble solid content)	Gannan	800–2500	0.75	Lu et al. 2006
		Navel	350–1800	0.68	Liu, Sun, and Ouyang 2010
			640–920	0.44	Wang and Xie 2014
		Various	350 2500	0.74	Cayuela and Weiland 2010
		Valencia Late	578–1850	0.51	Cayuela 2008
		Valencia	780–2500	0.58	Magwaza et al. 2013
	Vitamin C	Navel	800–2500	3.9 (RMSECV)	Xia et al. 2007
		Valencia	780–2500	8.01	Magwaza et al. 2013
	Titratable acidity (TA)	Various	350–2500	0.17	Cayuela and Weiland 2010
		Valencia Late	578–1850	0.33	Cayuela 2008
		Valencia	780–2500	0.14	Magwaza et al. 2013
	pH	Various	350–2500	0.15	Cayuela and Weiland 2010
		Valencia Late	578–1850	0.49	Cayuela 2008
	Various defects	Various	400–1800[a]	(86%)	Blasco et al. 2007; Blasco et al. 2009

[a,b] Indicates multispectral and hyperspectral imaging, respectively.

TABLE 28.3
Overview of Applications of NIR Spectroscopy for Quality Evaluation of Banana.

Species	Attribute	Cultivar	Spectral Range (nm)	RMSEP/ (Classification Accuracy)	Reference
Banana (*Musa acuminata*)	SSC (soluble solid content)	–	400–1000[b]	–	Rajkumar et al. 2012
		Ambon		0.79	Akmalia, Saputro, and Handayani 2017
		Cavendish	Color features*	0.85 (RMSECV)	Sanaeifar, Bakhshipour, and de la Guardia 2016
			532, 660, 830*	1.30 (RMSECV)	Hashim et al. 2018
		Berangan	1000–2500	3.91	ALI et al. 2018
	Moisture content	–	400–1000[b]	–	Rajkumar et al. 2012
		Raja		0.25%	Siregar, Handayani, and Saputro 2017
	Firmness	–	400–1000[b]	–	Rajkumar et al. 2012
		Cavendish	Color features[a]	4.91 (RMSECV)	Sanaeifar, Bakhshipour, and de la Guardia 2016
		AAA group, cv. Brazil	380–1023[b]	2.94	Xie, Chu, and He 2018
	Color (L^*, a^*, b^* values)	AAA group, cv. Brazil	280–1023[b]	2.09, 1.32, 2.89	Xie, Chu, and He 2018
	Titratable acidity (TA)	Cavendish	Color features[a]	0.02 (RMSECV)	Sanaeifar, Bakhshipour, and de la Guardia 2016
	pH	Cavendish	Color features[a]	0.12 (RMSECV)	Sanaeifar, Bakhshipour, and de la Guardia 2016
		Berangan	1000–2500	0.36	ALI et al. 2018
	Chilling injury	–	660, 785[a]	(99.33%)	Hashim et al. 2013
	Carbohydrate contents	AAA group, Cavendish subgroup	1100–2500	0.80%	Tarkosova and Copikova 2000
	Ripeness levels	AAA group, Cavendish subgroup	300–1000	–	Acharya et al. 2015
	Chlorophyll index	Cavendish	532, 660, 830[a]	0.333 (RMSECV)	Hashim et al. 2018
	Elasticity	Cavendish	532, 660, 830[a]	0.053 (N/mm) (RMSECV)	Hashim et al. 2018

[a,b] Indicates multispectral and hyperspectral imaging, respectively.

TABLE 28.4
Overview of Applications of NIR Spectroscopy for Quality Evaluation of Leafy Greens.

Species	Attribute	Spectral Range (nm)	RMSEP/(Classification Accuracy)	Reference
Bok Choi (*Brassica rapa subsp. chinensis*)	Nitrate ion content	514, 884, 902, 906	–	Ito and Idezawa 2006
	Chlorophyll content	350–2500	22.1 mg m^{-2}	Xue and Yang 2009
Broccoli (***Brassica oleracea L.***)	Total glucosinolate contents	400–2498	1.44 mmol kg^{-1} (SEP)	Hernández-Hierro et al. 2012
Chicory (***Cichorium intybus L.***)	Sensory attributes	350–2500	0.22–0.41 (RMSECV)	François et al. 2008
	Optimal harvesting date	400–1680	8.9–10.6 d	François et al. 2009
Chinese cabbage (*Brassica campestris L.* ssp. Pekinensis)	Nitrogen content	400–2500	4.0 g kg^{-1} (SEP)	Min et al. 2006
Chinese kale (*Brassica albograbra* Bailey)	Glucosinolate content	400–2500	2.87 mmol kg^{-1} (SEP)	Chen et al. 2014
Cabbage (*Brassica oleracea L.*)	Glucosinolate content	400–2500	2.38 mmol kg^{-1} (SECV)	Font, Del RÍO-Celestino, et al. 2005
Coriander (*Coriandrum sativum L.*)	Essential oil content and composition	1108–2490	0.29 mL/100 g (SECV)	Schulz et al. 1998
Dill (*Anethum graveolens L.*)	Essential oil content and composition	1108–2490	0.30 mL/100 g (SECV)	Schulz et al. 1998
Komatsuna leaves (*Brassica rapa* var. peruviridis)	Nitrate concentration distribution	607–967[b]	1446 ppm (WSE)	Itoh et al. 2010
Lamb's lettuce (*Valerianella locusta L.*)	Prior storage period	380 1690	3.7 d	Jacobs et al. 2016
	Freshness levels (pH, moisture, total phenol content)	400–1000	RMSECV: 0.13, 0.12%, 12.64 mg/100 g gallic acid	Beghi et al. 2014
	Shelf-life	400–1000	(74%–96%)	Giovenzana et al. 2014
Leaf rape (***Brassica napus ssp. pabularia***)	Glucosinolate contents	400–2500	3.72 mmol kg^{-1} (SECV)	Font, Río-Celestino, et al. 2005
Lettuce (*Lactuca sativa L.*)	Chlorophyll, carotenoid, and anthocyanin contents	500–1000	0.495; 0.105; 0.592 mg kg^{-1}	Steidle Neto et al. 2017
	Chlorophyll content	350–2500	22.1 mg m^{-2}	Xue and Yang 2009
	Anthocyanin content	RGB image[a]	–	Yang et al. 2016
Rocket (***Eruca vesicaria***)	Mineral composition	400–2500	Na, K, Ca, Mg: ≤ 9.5 g kg^{-1}; Fe, Cu, Mn, Zn: ≤21.3 mg kg^{-1} (RMSECV)	Villatoro-Pulido et al. 2012
Spinach (*Spinacia oleracea*)	Chlorophyll content	350–2500	22.1 mg m^{-2}	Xue and Yang 2009
	ATP content	610–950	–	Ye et al. 2009
	Chlorophyll and carotenoids contents	874–1734[b]	–	Zhang et al. 2017
	Quality grading	400–1000[b]	(>95%)	Diezma et al. 2013
	Shelf-life	400–1000[b]**	–	Lara et al. 2013
	Fecal contamination	456–950[b]	(>87%)	Everard, Kim, and Lee 2014
	Escherichia coli contamination	400–1000[b]**	0.038 log(CFU/g) (MSE)	Siripatrawan et al. 2011
	Quality deterioration	450, 680, 800	–	Lunadei et al. 2012
Wild rocket (*Diplotaxis tenuifolia L.*)	Color and textural quality	18 wavelengths (405–970 nm)[a]	–	Løkke et al. 2013

[a,b] Indicates multispectral and hyperspectral imaging, respectively.

texture of the leaves. Since chlorophyll and anthocyanins are spectrally active in the visible range, most researchers have included the visible range (400–780 nm) in their spectral measurements for nondestructive quality evaluation. Chlorophyll, carotenoids, and anthocyanins in lettuce could be predicted with accuracies of 0.495, 0.105, and 0.592 mg kg^{-1} (Steidle Neto et al. 2017). The content of glucosinolates in broccoli could be assessed with an accuracy of 1.44 mmol kg^{-1} (Hernández-Hierro et al. 2012).

28.1.5 Root Vegetables

As root vegetables are storage structures that have developed under the ground, they have higher dry matter content than leafy greens, ranging from 12% for carrots to 30% for sweet potatoes. They have a distinct skin, and their flesh is highly turbid, which makes diffuse reflectance or interactance measurements the preferred measurement modes. An overview of the applications of NIR spectroscopy for quality evaluation of root vegetables is given in Table 28.5. Potato and sweet potato have obtained the most attention as they are important food crops with a high starch content. For the measurement of the dry matter and starch content in potato and sweet potato, errors below 1% have been reported (Brunt and Drost 2010; Haase 2011; Hartmann and Büning-Pfaue 1998).

28.2 HURDLES AND PITFALLS

While the wide potential of NIRS in horticulture is obvious from the above, many challenges are faced in bringing this technology to practical applications. In this section, the challenges posed by the fruit anatomy, biological variability, and spatial heterogeneity will be discussed, and recommendations will be given on how to cope with these.

28.2.1 Fruit and Root Anatomy and Sample Presentation

The fact that fruit have a multilayered structure with an outer skin covering the internal flesh tissue creates a real challenge for applying NIRS technology. As the flesh is the main (or only) part that is consumed, the quality attributes of the flesh determine the consumption experience of the product. So, we are mainly interested in the information that can be obtained from the interaction of the NIR light with the flesh tissue. However, as a nondestructive measurement on an intact fruit or root requires that the light travels through the skin to reach the internal flesh, the acquired reflected or transmitted light is also influenced by the optical properties of the fruit skin. As the skin of most fruit and vegetables is covered with a waxy layer, it has a glossy appearance. As this specular reflection may dominate the acquired signals, it has been proposed to use cross-polarization to minimize these effects of specular reflection on the acquired diffuse reflectance spectra (Nguyen-Do-Trong et al. 2016) for hyperspectral imaging and appropriate sample presentation (0°/45° or interactance) for point spectroscopy measurements.

While we are typically interested in the variation in the flesh properties, the composition and structure of the skin can also vary. This adds unwanted variation to the spectra, which interferes with the information on the quality attributes of the flesh (Nicolaï et al. 2007). Saeys et al. (2008) quantified the differences in optical properties (absorption and scattering coefficients) of apple skin and flesh in the wavelength range of 350–2200 nm and concluded that the optical properties of the skin have a strong impact on the NIR spectra acquired on intact fruit. Vaudelle and L'Huillier (2015) also reported that variation in the apple skin thickness had a large impact on the apple flesh optical properties estimated from the acquired signals. Moreover, Van Beers et al. (2017) found that the optical properties of both the apple skin and the flesh change during maturation. So, the unspecific correlation between the optical changes in the skin and the quality changes in the flesh might be used by the multivariate calibration models to predict the latter, resulting in calibration models that are not robust. To minimize the effect of the skin on the properties or on the quality predictions for the

TABLE 28.5
Overview of Applications of NIR Spectroscopy for Quality Evaluation of Root Vegetables

Species	Attribute	Spectral Range (nm)	RMSEP/(Classification Accuracy)	Reference
Carrot (***Daucus carota L.***)	Carotenoid and sucrose contents	1108–2490	1.54; 0.74 mg/100 mL (SECV)	Schulz et al. 1998
	Sucrose, glucose, and fructose contents	800–1700	15.4%; 4.6%; 2.3% (SECV)	Zude et al. 2007
	Carotenoid content		<18% (SEP)	
Casava (*Manihot esculenta*)	Carotenoid, β-carotene, and dry matter contents	400–2500	SEP: 1.74 mg kg^{-1}; 1.28 mg kg^{-1}; 2.20%	Sánchez et al. 2014
	Total nitrogen, ash, sugar, and starch contents	350–2500	SEP: 0.36%; 0.15%; 0.7%; 2.07%	Lebot, Malapa, and Jung 2013; Lebot et al. 2009
Potato (*Solanum tuberosum L.*)	Dry matter	770–2498	1.3%–1.5% (SEP)	Scanlon, Pritchard, and Adam 1999
		760–1040	0.91% (RMSECV)	Helgerud et al. 2012; Helgerud et al. 2015
	Dry matter and starch contents	964–1655	1.5%; 1.4% (RMSECV)	Su and Sun 2017
	Dry matter, starch, and protein contents	1100–2500	0.5%; 0.63%, 0.06%	Brunt and Drost 2010
	Nitrogen concentration	9 wavelengths (1680–2310 nm)	0.092% (SEP)	Young, Mackerron, and Davies 1997
	Dry matter, starch, reducing sugar, sucrose, and total sugar contents	850–2500	SEP: 0.39%; 0.47%; 48.3; 106; 156 mg. kg^{-1}	Haase 2011
	Dry matter, starch, protein, and sugar contents	1100–2500	SEP: 0.19%; 0.28%; 0.06%; 0.061%	Hartmann and Büning-Pfaue 1998
	Sprouting capacity	400–2500	0.53–0.60 (SEP)	Jeong et al. 2008
	Glucose, sucrose, soluble solid contents, leaf primordia, density	446–1125 900–1685 400–1000[a]	Glucose: 0.052% Sucrose: 0.044% SSC: 0.400% Leaf primordia: 0.163 density: 0.012 g cm^{-3}	Rady et al. 2014
	Glucose and sucrose contents	400–1000**	0.026%; 0.027%	Rady, Guyer, and Lu 2015
		446–1125	0.019%; 0.108%	Rady and Guyer 2015
	Starch, soluble sugar, protein, and amino acid contents	400–1050	–	Kjær et al. 2016

(*Continued*)

TABLE 28.5 *(Continued)*

Overview of Applications of NIR Spectroscopy for Quality Evaluation of Root Vegetables

Species	Attribute	Spectral Range (nm)	RMSEP/(Classification Accuracy)	Reference
Radish (*Raphanus sativus L.*)	Sponginess degree	400–2500	–	Marcelis et al. 1995
Sweet potato (*Ipomoea batatas L.*)	Starch physiochemical quality and pasting properties	1100–2500	–	Lu, Huang, and Zhang 2006
	Starch, moisture, and sugar contents	680–1235	SEP: 1.91%; 2.00%; 1.21%	Katayama, Komaki, and Tamiya 1996
	Total nitrogen, ash, sugar, and starch contents	350–2500	SEP: 0.82%; 0.61%; 1.87%; 2.75%	Lebot, Malapa, and Jung 2013; Lebot et al. 2009
	Starch, sugar, protein, and mineral contents	350–2500	SEP: 3.13%; 2.31%; 0.44%; 0.73%	Lebot, Ndiaye, and Malapa 2011
	Protein content	400–2500	0.29 %	Magwaza et al. 2016
	Dry matter and starch contents	964–1655	1.5%; 1.4% (RMSECV)	Su and Sun 2017
Taro (***Colocasia esculenta L.***)	Starch, total sugar, protein, and mineral contents	350–2500	SEP: 2.14%; 1.50%; 0.57%; 0.44%	Lebot, Malapa, and Bourrieau 2011
	Total nitrogen, ash, sugar, and starch contents	350–2500	SEP: 0.67%; 0.74%; 1.51%; 2.98%	Lebot, Malapa, and Jung 2013; Lebot et al. 2009
Yam (***Dioscorea spp.***)	Total nitrogen, ash, sugar, and starch contents	350–2500	SEP: 0.89%; 0.90%; 1.97%; 2.79%	Lebot, Malapa, and Jung 2013; Lebot et al. 2009
	Starch, sugar, and protein contents	350–2500	SEP: 1.299%; 0.481%; 0.542%	Lebot and Malapa 2013
Yam bean (*Pachyrhizus* spp.)	Dry matter and protein contents	400–2500	SEP: 1.20%; 1.94%	Velasco and Grüneberg 1999

[a]Indicates hyperspectral imaging measurement.

flesh properties, it has been suggested to make sure that the captured photons have traveled deep enough into the fruit or root to have interacted with the flesh. This can be obtained by measuring in interactance mode at a source–detector distance that provides the maximal signal-to-noise ratio for the quality properties of the flesh (Nguyen Do Trong et al., 2014; Van Beers et al., 2015).

28.2.2 Biological Variability

As fruit and vegetables are typically grown under natural conditions, the spectra acquired from these products are subject to variation between cultivars, harvesting seasons, geographical locations, production sites, and postharvest storage conditions. Several researchers have quantified the impact of these sources of biological variation on the prediction performance for the quality features of interest (Peirs et al. 2003; Bobelyn et al. 2010; McGlone and Kawano 1998; Wedding et al. 2013; Louw and Theron 2010). Bobelyn et al. (2010) applied functional analysis of variance (FANOVA) to demonstrate the significance of the effects of cultivar, origin, shelf-life exposure time, and season on the NIR spectra acquired in the 800–1700 nm range for a very large number of apples. They also indicated their potential impact on the prediction performance. From their analysis, they concluded that a large data set, which encompasses all relevant variabilities that can be expected in future samples (e.g., season, orchard, cultivar, and origin effects), is required to build a robust calibration model.

28.2.3 Spatial Heterogeneity

Fruit and vegetables often exhibit large spatial variability in their quality features. For example, apples typically have a sun-exposed side and a shadow side that have different quality properties, while leek has white, light green, and dark green parts. To properly assess these spatial quality variations with NIRS, spectra should be acquired at various locations on the analyzed samples for predicting and mapping this variation in quality traits over the sample surface. Hyperspectral imaging, a technology combining point spectroscopic measurement and imaging, fulfills this requirement, and its potential has been reported for a wide range of applications in fruit and vegetables (Elmasry et al. 2012; Dale et al. 2013; Nicolaï et al. 2007; Magwaza et al. 2012; Su, He, and Sun 2017; Gowen et al. 2007; Lorente et al. 2012; Qin et al. 2013; Wang et al. 2016; Wu and Sun 2013; Pu, Feng, and Sun 2015).

28.3 NEW TRENDS

28.3.1 NIRS in the Orchard and Packhouse

While NIRS has already been successful in replacing many destructive reference methods in the laboratory, most applications still require sampling and transport of the samples to be scanned in the laboratory. As the scanned samples are typically not returned to the batch, the inherent potential to measure all products in a nondestructive way is not yet fully exploited. Thanks to the development of more robust and portable spectrophotometers, there is a trend to move the NIRS sensors to the packhouse and even to the orchard (dos Santos et al. 2013). Recent developments in micromechanical systems for wavelength scanning and on-chip deposition of Fabry–Perot filters have resulted in the commercialization of miniaturized spectrophotometers, which are more affordable and can easily be taken to the packhouse or orchard (Gutiérrez et al. 2015; Leo et al. 2009; Pérez-Marín et al. 2010; Sánchez et al. 2011; Torres et al. 2017). Some systems are even equipped with a global positioning system, which allows us to tag the measured spectra with the geodetic coordinates of the location where these have been measured. This would allow us to map the quality features throughout an orchard or region and provide valuable information for precision horticulture (Dambergs, Gishen, and Cozzolino 2015).

Several researchers have already tested the potential of such portable and hand-held spectrophotometers for applications on fruit and vegetables and reported prediction performances, which are close to those reported for benchtop instruments (dos Santos et al. 2013; Plans et al. 2013; Pan, Lu, et al. 2015; Pan, Zhu, et al. 2015; Das et al. 2016; Urraca et al. 2016; Choi et al. 2017; dos Santos Neto et al. 2017; Torres et al. 2017; Wang et al. 2017). However, it should be noted that most studies were still performed in laboratory conditions, while the measurement conditions in a packhouse or orchard are far more variable than those in laboratory conditions (e.g., fluctuations in temperature and humidity of the environment, fluctuations in sample temperature, interference from ambient light, …). So, the calibration models will have to be made robust against these sources of variation to obtain reliable quality measurements in practice (Nicolaï et al. 2007).

28.3.2 Separation of Scattering and Absorption

As mentioned in Section 28.2.1, fruit and vegetables consist of multiple turbid tissue layers. This makes that the acquired reflectance or transmittance spectra are influenced by the scattering and absorption properties of the different tissue layers. The absorption properties are the result of the interaction of the photons with the organic groups (O–H, C–H, N–H, …) and thus related to the chemical composition. In contrast, the scattering properties are the result of the spatial variation in the refractive index and thus related to the microstructure of the samples. As the microstructure has an impact on the macroscopic texture properties of the fruit and vegetables, new spectroscopic measurement techniques have been proposed to separate the absorption and scattering properties of the samples for predicting their quality attributes related to chemical composition and texture, respectively (Nicolaï et al. 2007).

In spatially resolved spectroscopy (SRS), the diffuse reflectance or transmittance is acquired at different source-detector distances. Several researchers have reported on the potential of SRS for applications in fruit and vegetables: apple (Cen et al. 2013; Huang and Lu 2010; Lu et al. 2010; Mendoza et al. 2011; Nguyen Do Trong et al. 2014; Qin, Lu, and Peng 2009; Zhu et al. 2016); peach (Lu and Peng 2006), tomato (Zhu et al. 2015), and various species (Qin and Lu 2008). The best prediction accuracy obtained for SSC of apples was 0.69% (Qin, Lu, and Peng 2009; Nguyen Do Trong et al. 2014), which is still slightly less accurate than that of conventional NIR spectroscopy measurements for apples. Prediction of apple firmness remains a challenge with varying prediction accuracies.

In time-resolved spectroscopy (TRS), a short light pulse (pico or femtoseconds) is pumped into the sample, and the diffuse reflectance or transmittance is measured as a function of time with nanosecond resolution at a fixed distance from the point of illumination. The potential of TRS for applications in fruit and vegetables has also been reported by different researchers: apple (Rinaldo et al. 2001; Rizzolo et al. 2010; Valero et al. 2005; Vanoli et al. 2009), pear (Nicolaï et al. 2008; Zerbini et al. 2002), nectarine (Rizzolo et al. 2009), cherry (Zude et al. 2011), and various fruits (Cubeddu et al. 2001).

While results comparable with those obtained with point spectroscopy have been reported for both SRS and TRS, the prediction of texture properties of apples from the scattering properties is still not very reliable. It is hypothesized that changes in texture properties of apples resulting from the enzymatic conversion of protopectin to pectin in the middle lamella and the solubilization of pectin in the cell wall may not result in changes in the scattering properties. Therefore, the potential for optical measurement of fruit firmness may be limited. However, good results have been obtained predicting the porosity of apples from the scattering properties (Wang et al. 2019).

28.4 CONCLUSION

From the vast number of reports on the successful application of NIRS for nondestructive quality evaluation of fruit and vegetables, it is clear that NIR spectroscopy is ripe for use in horticulture. Most of the successful reports are for chemical composition-related quality attributes (SSC, dry matter content, defects, or contaminants, …), while the prediction of texture-related quality attributes (e.g., firmness) and minor chemical components remains challenging. The multilayered fruit anatomy, the biological variability, and the spatial heterogeneity have been identified as the main challenges for application in practice, which demand for careful design of the optical configuration and the training of robust calibration models. Recent technical innovations make it possible to bring the NIRS equipment to the packhouse and even the orchard. This introduces even larger variability for which the calibration models have to be made robust. Combination of multiple spectroscopic measurements in space or time allows us to separate the information on the microstructure (e.g. porosity) from that on the chemical composition, but the success of these methods in predicting fruit firmness has so far been limited.

REFERENCES

Acharya, U.K., K.B. Walsh, P.P. Subedi, and W.B. McGlasson. 2015. Can visual reflectance indices be related to ripeness of banana fruit?

Akmalia, D., A.H. Saputro, and W. Handayani. 2017. A non-destruction measurement system based on hyperspectral imaging for sugar content in banana (Musa sp.). Paper read at 2017 International Seminar on Sensors, Instrumentation, Measurement and Metrology (ISSIMM), 25–26 Aug. 2017.

Ali, M.M., R.B. Janius, N.M. Nawi, and N. Hashim. 2018. "Prediction of total soluble solids and pH in banana using near-infrared spectroscopy." *Journal of Engineering Science and Technology* no. 13 (1):254–264.

Antonucci, F., F. Pallottino, G. Paglia, A. Palma, S. D'Aquino, and P. Menesatti. 2011. "Non-destructive estimation of mandarin maturity status through portable VIS-NIR spectrophotometer." *Food and Bioprocess Technology* no. 4 (5):809–813. doi: https://doi.org/10.1007/s11947-010-0414-5.

Ariana, D., D.E. Guyer, and B. Shrestha. 2006. "Integrating multispectral reflectance and fluorescence imaging for defect detection on apples." *Computers and Electronics in Agriculture* no. 50 (2):148–161. doi: https://doi.org/10.1016/j.compag.2005.10.002.

Baranowski, P., W. Mazurek, and J. Pastuszka-Woźniak. 2013. "Supervised classification of bruised apples with respect to the time after bruising on the basis of hyperspectral imaging data." *Postharvest Biology and Technology* no. 86:249–258. doi: https://doi.org/10.1016/j.postharvbio.2013.07.005.

Beghi, R., V. Giovenzana, R. Civelli, C. Malegori, S. Buratti, and R. Guidetti. 2014. "Setting-up of a simplified handheld optical device for decay detection in fresh-cut Valerianella locusta L." *Journal of Food Engineering* no. 127:10–15. doi: https://doi.org/10.1016/j.jfoodeng.2013.11.019.

Blasco, J., N. Aleixos, J. Gómez-Sanchís, and E. Moltó. 2009. "Recognition and classification of external skin damage in citrus fruits using multispectral data and morphological features." *Biosystems Engineering* no. 103 (2):137–145. doi: https://doi.org/10.1016/j.biosystemseng.2009.03.009.

Blasco, J., N. Aleixos, J. Gómez, and E. Moltó. 2007. "Citrus sorting by identification of the most common defects using multispectral computer vision." *Journal of Food Engineering* no. 83 (3):384–393. doi: https://doi.org/10.1016/j.jfoodeng.2007.03.027.

Bobelyn, E., A.-S. Serban, M. Nicu, J. Lammertyn, B.M. Nicolai, and W. Saeys. 2010. "Postharvest quality of apple predicted by NIR-spectroscopy---Study of the effect of biological variability on spectra and model performance." *Postharvest Biology and Technology* no. 55 (3):133–143. doi: https://doi.org/10.1016/j.postharvbio.2009.09.006.

Brunt, K., and W.C. Drost. 2010. "Design, construction, and testing of an automated NIR in-line analysis system for potatoes. Part I---off-line NIR feasibility study for the characterization of potato composition." *Potato Research* no. 53 (1):25–39. doi: https://doi.org/10.1007/s11540-010-9149-y.

Canet, W. 1988. "Determination of the moisture content of some fruits and vegetables by microwave heating." *Journal of Microwave Power and Electromagnetic Energy* no. 23 (4):231–235. doi: https://doi.org/10.1080/08327823.1988.11688062.

Cavaco, A.M., P. Pinto, M.D. Antunes, J.M. da Silva, and R. Guerra. 2009. "'Rocha' pear firmness predicted by a Vis/NIR segmented model." *Postharvest Biology and Technology* no. 51 (3):311–319. doi: https://doi.org/10.1016/j.postharvbio.2008.08.013.

Cayuela, J.A. 2008. "Vis/NIR soluble solids prediction in intact oranges (Citrus sinensis L.) cv. Valencia Late by reflectance." *Postharvest Biology and Technology* no. 47 (1):75–80. doi: https://doi.org/10.1016/j.postharvbio.2007.06.005.

Cayuela, J.A., and C. Weiland. 2010. "Intact orange quality prediction with two portable NIR spectrometers." *Postharvest Biology and Technology* no. 58 (2):113–120. doi: https://doi.org/10.1016/j.postharvbio.2010.06.001.

Cen, H., R. Lu, F. Mendoza, and R.M. Beaudry. 2013. "Relationship of the optical absorption and scattering properties with mechanical and structural properties of apple tissue." *Postharvest Biology and Technology* no. 85:30–38. doi: https://doi.org/10.1016/j.postharvbio.2013.04.014.

Chen, J., L. Li, S. Wang, X. Tao, Y. Wang, A. Sun, and H. He. 2014. "Assessment of glucosinolates in Chinese kale by near-infrared spectroscopy." *International Journal of Food Properties* no. 17 (8):1668–1679. doi: https://doi.org/10.1080/10942912.2012.678535.

Choi, J.-H., P.-A. Chen, B. Lee, S.-H. Yim, M.-S. Kim, Y.-S. Bae, D.-C. Lim, and H.-J. Seo. 2017. "Portable, non-destructive tester integrating VIS/NIR reflectance spectroscopy for the detection of sugar content in Asian pears." *Scientia Horticulturae* no. 220:147–153. doi: https://doi.org/10.1016/j.scienta.2017.03.050.

Clark, C.J., V.A. McGlone, and R.B. Jordan. 2003. "Detection of Brownheart in 'Braeburn' apple by transmission NIR spectroscopy." *Postharvest Biology and Technology* no. 28 (1):87–96. doi: https://doi.org/10.1016/S0925-5214(02)00122-9.

Cubeddu, R., C. D'Andrea, A. Pifferi, P. Taroni, A. Torricelli, G. Valentini, C. Dover, D. Johnson, M. Ruiz-Altisent, and C. Valero. 2001. "Nondestructive quantification of chemical and physical properties of fruits by time-resolved reflectance spectroscopy in the wavelength range 650–1000 nm." *Applied Optics* no. 40 (4):538–543. doi: https://doi.org/10.1364/ao.40.000538.

Dale, L.M., A. Thewis, C. Boudry, I. Rotar, P. Dardenne, V. Baeten, and J.A. Fernández Pierna. 2013. "Hyperspectral imaging applications in agriculture and agro-food product quality and safety control—A review." *Applied Spectroscopy Reviews* no. 48 (2):142–159. doi: https://doi.org/10.1080/05704928.2012.705800.

Dambergs, R., M. Gishen, and D. Cozzolino. 2015. "A Review of the state of the art, limitations, and perspectives of infrared spectroscopy for the analysis of wine grapes, must, and grapevine tissue." *Applied Spectroscopy Reviews* no. 50 (3):261–278. doi: https://doi.org/10.1080/05704928.2014.966380.

Das, A.J., A. Wahi, I. Kothari, and R. Raskar. 2016. "Ultra-portable, wireless smartphone spectrometer for rapid, non-destructive testing of fruit ripeness." *Scientific Reports* no. 6:32504. doi: https://doi.org/10.1038/srep32504. https://www.nature.com/articles/srep32504#supplementary-information.

Diezma, B., L. Lleó, J.M. Roger, A. Herrero-Langreo, L. Lunadei, and M. Ruiz-Altisent. 2013. "Examination of the quality of spinach leaves using hyperspectral imaging." *Postharvest Biology and Technology* no. 85:8–17 doi: https://doi.org/10.1016/j.postharvbio.2013.04.017.

dos Santos, C.A.T., M. Lopo, R.N.M.J. Páscoa, and J.A. Lopes. 2013. "A review on the applications of portable near-infrared spectrometers in the agro-food industry." *Applied Spectroscopy* no. 67 (11):1215–1233.

dos Santos Neto, J.P., M.W.D. de Assis, I.P. Casagrande, L.C. Cunha Jr., and G.H. de Almeida Teixeira. 2017. "Determination of 'Palmer' mango maturity indices using portable near infrared (VIS-NIR) spectrometer." *Postharvest Biology and Technology* no. 130:75–80. doi: https://doi.org/10.1016/j.postharvbio.2017.03.009.

Elmasry, G., M. Kamruzzaman, D.-W. Sun, and P. Allen. 2012. "Principles and applications of hyperspectral imaging in quality evaluation of agro-food products—A review." *Critical Reviews in Food Science and Nutrition* no. 52 (11):999–1023. doi: https://doi.org/10.1080/10408398.2010.543495.

Everard, C.D., M.S. Kim, and H. Lee. 2014. "A comparison of hyperspectral reflectance and fluorescence imaging techniques for detection of contaminants on spinach leaves." *Journal of Food Engineering* no. 143:139–145. doi: https://doi.org/10.1016/j.jfoodeng.2014.06.042.

Fan, G,, J. Zha, R. Du, and L. Gao. 2009. "Determination of soluble solids and firmness of apples by Vis/NIR transmittance." *Journal of Food Engineering* no. 93 (4):416–420. doi: https://doi.org/10.1016/j.jfoodeng.2009.02.006.

Fan, S., W. Huang, Z. Guo, B. Zhang, and C. Zhao. 2015. "Prediction of soluble solids content and firmness of pears using hyperspectral reflectance imaging." *Food Analytical Methods* no. 8 (8):1936–1946. doi: https://doi.org/10.1007/s12161-014-0079-1.

Ferrari, C., G. Foca, R. Calvini, and A. Ulrici. 2015. "Fast exploration and classification of large hyperspectral image datasets for early bruise detection on apples." *Chemometrics and Intelligent Laboratory Systems* no. 146:108–119 doi: https://doi.org/10.1016/j.chemolab.2015.05.016.

Font, R., M. Del RÍO-Celestino, E. Rosa, A. Aires, and A. De Haro-BailÓN. 2005. "Glucosinolate assessment in Brassica oleracea leaves by near-infrared spectroscopy." *The Journal of Agricultural Science* no. 143 (1):65–73. doi: 10.1017/s0021859605004806.

Font, R., M. del Río-Celestino, E. Cartea, and A. de Haro-Bailón. 2005. "Quantification of glucosinolates in leaves of leaf rape (Brassica napus ssp. pabularia) by near-infrared spectroscopy." *Phytochemistry* no. 66 (2):175–185. doi: https://doi.org/10.1016/j.phytochem.2004.11.011.

François, I.M., E. Mariën, K. Brijs, P. Coppin, and M. De Proft. 2009. "The use of Vis/NIR spectroscopy to predict the optimal root harvesting date of chicory (Cichorium intybus L.)." *Postharvest Biology and Technology* no. 53 (1):77–83. doi: https://doi.org/10.1016/j.postharvbio.2009.03.003.

François, I.M., H. Wins, S. Buysens, C. Godts, E. Van Pee, B. Nicolaï, and M. De Proft. 2008. "Predicting sensory attributes of different chicory hybrids using physico-chemical measurements and visible/near infrared spectroscopy." *Postharvest Biology and Technology* no. 49 (3):366–373. doi: https://doi.org/10.1016/j.postharvbio.2008.02.011.

Giovenzana, V., R. Beghi, S. Buratti, R. Civelli, and R. Guidetti. 2014. "Monitoring of fresh-cut Valerianella locusta Laterr. shelf life by electronic nose and VIS–NIR spectroscopy." *Talanta* no. 120:368–375. doi: https://doi.org/10.1016/j.talanta.2013.12.014.

Gómez-Sanchis, J., L. Gómez-Chova, N. Aleixos, G. Camps-Valls, C. Montesinos-Herrero, E. Moltó, and J. Blasco. 2008. "Hyperspectral system for early detection of rottenness caused by Penicillium digitatum in mandarins." *Journal of Food Engineering* no. 89 (1):80–86. doi: https://doi.org/10.1016/j.jfoodeng.2008.04.009.

Gómez, A.H., Y. He, and A.G. Pereira. 2006. "Non-destructive measurement of acidity, soluble solids and firmness of Satsuma mandarin using Vis/NIR-spectroscopy techniques." *Journal of Food Engineering* no. 77 (2):313–319. doi: https://doi.org/10.1016/j.jfoodeng.2005.06.036.

Gowen, A.A., C.P. O'Donnell, P.J. Cullen, G. Downey, and J.M. Frias. 2007. "Hyperspectral imaging – an emerging process analytical tool for food quality and safety control." *Trends in Food Science & Technology* no. 18 (12):590–598. doi: http://dx.doi.org/10.1016/j.tifs.2007.06.001.

Guthrie, J.A., D.J. Reid, and K.B. Walsh. 2005. "Assessment of internal quality attributes of mandarin fruit. 2. NIR calibration model robustness." *Australian Journal of Agricultural Research* no. 56 (4):417–426. doi: https://doi.org/10.1071/AR04299.

Guthrie, J.A., K.B. Walsh, D.J. Reid, and C.J. Liebenberg. 2005. "Assessment of internal quality attributes of mandarin fruit. 1. NIR calibration model development." *Australian Journal of Agricultural Research* no. 56 (4):405–416. doi: https://doi.org/10.1071/AR04257.

Gutiérrez, S., J. Tardaguila, J. Fernández-Novales, and M.P. Diago. 2015. "Support vector machine and artificial neural network models for the classification of grapevine varieties using a portable NIR spectrophotometer." *PLOS ONE* no. 10 (11):e0143197. doi: https://doi.org/10.1371/journal.pone.0143197.

Haase, N.U. 2011. "Prediction of potato processing quality by near infrared reflectance spectroscopy of ground raw tubers." *Journal of Near Infrared Spectroscopy* no. 19 (1):37–45. doi: https://doi.org/10.1255/jnirs.919.

Hartmann, R., and H. Büning-Pfaue. 1998. "NIR determination of potato constituents." *Potato Research* no. 41 (4):327–334. doi: https://doi.org/10.1007/bf02358965.

Hashim, N., S.E. Adebayo, K. Abdan, and M. Hanafi. 2018. "Comparative study of transform-based image texture analysis for the evaluation of banana quality using an optical backscattering system." *Postharvest Biology and Technology* no. 135:38–50. doi: https://doi.org/10.1016/j.postharvbio.2017.08.021.

Hashim, N., M. Pflanz, C. Regen, R.B. Janius, R.A. Rahman, A. Osman, M. Shitan, and M. Zude. 2013. "An approach for monitoring the chilling injury appearance in bananas by means of backscattering imaging." *Journal of Food Engineering* no. 116 (1):28–36. doi: https://doi.org/10.1016/j.jfoodeng.2012.11.018.

Helgerud, T., V.H. Segtnan, J.P. Wold, S. Ballance, S.H. Knutsen, E.O. Rukke, and N.K. Afseth. 2012. "Near-infrared spectroscopy for rapid estimation of dry matter content in whole unpeeled potato tubers." no. 1 (4). doi: https://doi.org/10.5539/jfr.v1n4p55.

Helgerud, T., J.P. Wold, M.B. Pedersen, K.H. Liland, S. Ballance, S.H. Knutsen, E.O. Rukke, and N.K. Afseth. 2015. "Towards on-line prediction of dry matter content in whole unpeeled potatoes using near-infrared spectroscopy." *Talanta* no. 143:138–144. doi: https://doi.org/10.1016/j.talanta.2015.05.037.

Hernández-Hierro, J.M., J. Valverde, S. Villacreces, K. Reilly, M. Gaffney, M.L. González-Miret, F.J. Heredia, and G. Downey. 2012. "Feasibility study on the use of visible–near-infrared spectroscopy for the screening of individual and total glucosinolate contents in broccoli." *Journal of Agricultural and Food Chemistry* no. 60 (30):7352–7358. doi: https://doi.org/10.1021/jf3018113.

Huang, M., and R. Lu. 2010. "Apple mealiness detection using hyperspectral scattering technique." *Postharvest Biology and Technology* no. 58 (3):168–175. doi: https://doi.org/10.1016/j.postharvbio.2010.08.002.

Ito, H., and F. Idezawa. 2006. Non-destructive determination of nitrate ion in leaf stalk of Qing Gin Cai using visible (Vis)-near infrared (Nir) spectroscopy.

Itoh, H., J. De Baerdemaeker, N.N. Do Trong, L. Wang, and W. Saeys. 2013. "Estimation of pear ripeness by hyperspectral laser scatter imaging." *IFAC Proceedings Volumes* no. 46 (4):160–165. doi: https://doi.org/10.3182/20130327-3-JP-3017.00037.

Itoh, H., S. Kanda, H. Matsuura, N. Shiraishi, K. Sakai, and A. Sasao. 2010. "Measurement of nitrate concentration distribution in vegetables by near-infrared hyperspectral imaging." *Environmental Control in Biology* no. 48 (2):37–49. doi: https://doi.org/10.2525/ecb.48.37.

Jacobs, B.A. J.G., Bert E. Verlinden, E. Bobelyn, A. Decombel, P. Bleyaert, J. Van Lommel, I. Vandevelde, W. Saeys, and B.M. Nicolai. 2016. "Estimation of the prior storage period of lamb's lettuce based on visible/near infrared reflectance spectroscopy." *Postharvest Biology and Technology* no. 113:95–105. doi: https://doi.org/10.1016/j.postharvbio.2015.11.007.

Jeong, J.-C., H.-C. Ok, O.-S. Hur, and C.-G. Kim. 2008. "Prediction of sprouting capacity using near-infrared spectroscopy in potato tubers." *American Journal of Potato Research* no. 85 (5):309–314. doi: https://doi.org/10.1007/s12230-008-9010-x.

Jiang, H., and W. Zhu. 2013. "Determination of pear internal quality attributes by fourier transform near infrared (FT-NIR) Spectroscopy and multivariate analysis." *Food Analytical Methods* no. 6 (2):569–577. doi: https://doi.org/10.1007/s12161-012-9480-9.

Katayama, K., K. Komaki, and S. Tamiya. 1996. "Prediction of starch, moisture, and sugar in sweetpotato by near infrared transmittance." *HortScience* no. 31 (6):1003–1006.

Kawano, S., T. Fujiwara, and M. Iwamoto. 1993. "Nondestructive determination of sugar content in satsuma mandarin using near infrared (NIR) transmittance." *Journal of the Japanese Society for Horticultural Science* no. 62 (2):465–470. doi: https://doi.org/10.2503/jjshs.62.465.

Keresztes, J.C., E. Diels, M. Goodarzi, N. Nguyen-Do-Trong, P. Goos, B. Nicolai, and W. Saeys. 2017. "Glare based apple sorting and iterative algorithm for bruise region detection using shortwave infrared hyperspectral imaging." *Postharvest Biology and Technology* no. 130:103–115. doi: https://doi.org/10.1016/j.postharvbio.2017.04.005.

Keresztes, J.C., M. Goodarzi, and W. Saeys. 2016. "Real-time pixel based early apple bruise detection using short wave infrared hyperspectral imaging in combination with calibration and glare correction techniques." *Food Control* no. 66:215–226. doi: http://dx.doi.org/10.1016/j.foodcont.2016.02.007.

Kim, G., K. Lee, K. Choi, J. Son, D. Choi, and S. Kang. 2004. "Defect and ripeness inspection of citrus using NIR transmission spectrum." *Key Engineering Materials* no. 270–273:1008–1013.

Kim, M.S., A.M. Lefcourt, K. Chao, Y.R. Chen, I. Kim, and D.E. Chan. 2002. "Multispectral detection of fecal contamination on apples based on hyperspectral imagery—Part I. Application of visible and near–infrared reflectance imaging." *Transactions of the ASAE* no. 45 (6):2027. doi: https://doi.org/10.13031/2013.11414.

Kim, M. S., Y. R. Chen, and P. M. Mehl. 2001. "Hyperspectral Reflectance and Fluorescence Imaging System for Food Quality and Safety." *Transactions of the ASAE* no. 44 (3):721. doi: https://doi.org/10.13031/2013.6099.

Kjær, A., G. Nielsen, S. Stærke, M.R. Clausen, M. Edelenbos, and B. Jørgensen. 2016. "Prediction of starch, soluble sugars and amino acids in potatoes (Solanum tuberosum L.) using hyperspectral imaging, dielectric and LF-NMR methodologies." *Potato Research* no. 59 (4):357–374. doi: https://doi.org/10.1007/s11540-017-9335-2.

Lammertyn, J., B. Nicolaï, K. Ooms, V. De Smedt, and J. De Baerdemaeker. 1998. "non-destructive measurement of acidity, soluble solids, and firmness of jonagold apples using NIR-spectroscopy." *Transactions of the ASAE* no. 41 (4):1089. doi: https://doi.org/10.13031/2013.17238.

Lara, M.A., L. Lleó, B. Diezma-Iglesias, J.M. Roger, and M. Ruiz-Altisent. 2013. "Monitoring spinach shelf-life with hyperspectral image through packaging films." *Journal of Food Engineering* no. 119 (2):353–361. doi: https://doi.org/10.1016/j.jfoodeng.2013.06.005.

Lebot, V., R. Malapa, and M. Jung. 2013. "Use of NIRS for the rapid prediction of total N, minerals, sugars and starch in tropical root and tuber crops." *New Zealand Journal of Crop and Horticultural Science* no. 41 (3):144–153. doi: https://doi.org/10.1080/01140671.2013.798335.

Lebot, V., A. Champagne, R. Malapa, and D. Shiley. 2009. "NIR determination of major constituents in tropical root and tuber crop flours." *Journal of Agricultural and Food Chemistry* no. 57 (22):10539–10547. doi: https://doi.org/10.1021/jf902675n.

Lebot, V., and R. Malapa. 2013. "Application of near infrared reflectance spectroscopy for the evaluation of yam (Dioscorea alata) germplasm and breeding lines." *Journal of the Science of Food and Agriculture* no. 93 (7):1788–1797. doi: https://doi.org/10.1002/jsfa.6002.

Lebot, V., R. Malapa, and M. Bourrieau. 2011. "Rapid estimation of taro (Colocasia esculenta) quality by near-infrared reflectance spectroscopy." *Journal of Agricultural and Food Chemistry* no. 59 (17):9327–9334. doi: https://doi.org/10.1021/jf202310w.

Lebot, V., A. Ndiaye, and R. Malapa. 2011. "Phenotypic characterization of sweet potato [Ipomoea batatas (L.) Lam.] genotypes in relation to prediction of chemical quality constituents by NIRS equations." *Plant Breeding* no. 130 (4):457–463. doi: https://doi.org/10.1111/j.1439–0523.2010.01840.x.

Lee, K., G. Kim, S. Kang, J. Son, D. Choi, and K. Choi. 2004. "Measurement of sugar contents in citrus using near infrared transmittance." *Key Engineering Materials* no. 270–273:1014–1019. doi: https://doi.org/10.4028/www.scientific.net/KEM.270-273.1014.

Lee, W.-H., M.S. Kim, H. Lee, S.R. Delwiche, H. Bae, D.-Y. Kim, and B.-K. Cho. 2014. "Hyperspectral near-infrared imaging for the detection of physical damages of pear." *Journal of Food Engineering* no. 130:1–7. doi: https://doi.org/10.1016/j.jfoodeng.2013.12.032.

Leo, P. Schuler, S. Milne Jason, M. Dell John, and F. Lorenzo. 2009. "MEMS-based microspectrometer technologies for NIR and MIR wavelengths." *Journal of Physics D—Applied Physics* no. 42 (13):133001.

Li, J., X. Tian, W. Huang, B. Zhang, and S. Fan. 2016. "Application of long-wave near infrared hyperspectral imaging for measurement of soluble solid content (SSC) in pear." *Food Analytical Methods* no. 9 (11):3087–3098. doi: https://doi.org/10.1007/s12161-016-0498-2.

Liu, Y., X. Sun, and A. Ouyang. 2010. "Nondestructive measurement of soluble solid content of navel orange fruit by visible–NIR spectrometric technique with PLSR and PCA-BPNN." *LWT - Food Science and Technology* no. 43 (4):602–607. doi: https://doi.org/10.1016/j.lwt.2009.10.008.

Liu, Y., X. Sun, H. Zhang, and O. Aiguo. 2010. "Nondestructive measurement of internal quality of Nanfeng mandarin fruit by charge coupled device near infrared spectroscopy." *Computers and Electronics in Agriculture* no. 71:S10–S14. doi: https://doi.org/10.1016/j.compag.2009.09.005.

Liu, Y., and Y. Ying. 2005. "Use of FT-NIR spectrometry in non-invasive measurements of internal quality of 'Fuji' apples." *Postharvest Biology and Technology* no. 37 (1):65–71. doi: https://doi.org/10.1016/j.postharvbio.2005.02.013.

Løkke, M.M., H.F. Seefeldt, T. Skov, and M. Edelenbos. 2013. "Color and textural quality of packaged wild rocket measured by multispectral imaging." *Postharvest Biology and Technology* no. 75:86–95. doi: https://doi.org/10.1016/j.postharvbio.2012.06.018.

Lorente, D., N. Aleixos, J. Gómez-Sanchis, S. Cubero, O.L. García-Navarrete, and J. Blasco. 2012. "Recent advances and applications of hyperspectral imaging for fruit and vegetable quality assessment." *Food and Bioprocess Technology* no. 5 (4):1121–1142. doi: https://doi.org/10.1007/s11947-011-0725-1.

Lorente, D., P. Escandell-Montero, S. Cubero, J. Gómez-Sanchis, and J. Blasco. 2015. "Visible–NIR reflectance spectroscopy and manifold learning methods applied to the detection of fungal infections on citrus fruit." *Journal of Food Engineering* no. 163:17–24 doi: https://doi.org/10.1016/j.jfoodeng.2015.04.010.

Louw, E.D., and K.I. Theron. 2010. "Robust prediction models for quality parameters in Japanese plums (Prunus salicina L.) using NIR spectroscopy." *Postharvest Biology and Technology* no. 58 (3):176–184. doi: https://doi.org/10.1016/j.postharvbio.2010.07.001.

Lovász, T., P. Merész, and A. Salgó. 1994. "Application of near infrared transmission spectroscopy for the determination of some quality parameters of apples." *Journal of Near Infrared Spectroscopy* no. 2 (4):213–221. doi: https://doi.org/10.1255/jnirs.47.

Lu, G., H. Huang, and D. Zhang. 2006. "Prediction of sweetpotato starch physiochemical quality and pasting properties using near-infrared reflectance spectroscopy." *Food Chemistry* no. 94 (4):632–639. doi: https://doi.org/10.1016/j.foodchem.2005.02.006.

Lu, H. H. Xu, Y. Ying, X. Fu, H. Yu, and H. Tian. 2006. "Application Fourier transform near infrared spectrometer in rapid estimation of soluble solids content of intact citrus fruits." *Journal of Zhejiang University. Science. B* no. 7 (10):794–799. doi: https://doi.org/10.1631/jzus.2006.B0794.

Lu, R. 2003. "Detection of bruises on apples using near–infrared hyperspectral imaging." *Transactions of the ASAE* no. 46 (2):523. doi: https://doi.org/10.13031/2013.12941.

Lu, R., H. Cen, M. Huang, and D.P. Ariana. 2010. "Spectral absorption and scattering properties of normal and bruised apple tissue." *Transactions of the ASABE* no. 53 (1):263. doi: https://doi.org/10.13031/2013.29491.

Lu, R., D.E. Guyer, and R.M. Beaudry. 2000. "Determination of firmness and sugar content of apples using near-infrared diffuse reflectance1." *Journal of Texture Studies* no. 31 (6):615–630. doi: https://doi.org/10.1111/j.1745-4603.2000.tb01024.x.

Lu, R., and Y. Peng. 2006. "Hyperspectral scattering for assessing peach fruit firmness." *Biosystems Engineering* no. 93 (2):161–171. doi: https://doi.org/10.1016/j.biosystemseng.2005.11.004.

Lunadei, L., B. Diezma, L. Lleó, L. Ruiz-Garcia, S. Cantalapiedra, and M. Ruiz-Altisent. 2012. "Monitoring of fresh-cut spinach leaves through a multispectral vision system." *Postharvest Biology and Technology* no. 63 (1):74–84. doi: https://doi.org/10.1016/j.postharvbio.2011.08.004.

Magwaza, L.S., U.L. Opara, H. Nieuwoudt, P.J.R. Cronje, W. Saeys, and B. Nicolaï. 2012. "NIR spectroscopy applications for internal and external quality analysis of citrus fruit—A review." *Food and Bioprocess Technology* no. 5 (2):425–444. doi: https://doi.org/10.1007/s11947-011-0697-1.

Magwaza, L.S., S.I.M. Naidoo, S.M. Laurie, M.D. Laing, and H. Shimelis. 2016. "Development of NIRS models for rapid quantification of protein content in sweetpotato [Ipomoea batatas (L.) LAM.]." *LWT - Food Science and Technology* no. 72:63–70. doi: https://doi.org/10.1016/j.lwt.2016.04.032.

Magwaza, L.S., U.L. Opara, L.A. Terry, S. Landahl, P.J.R. Cronje, H.H. Nieuwoudt, A. Hanssens, W. Saeys, and B.M. Nicolaï. 2013. "Evaluation of Fourier transform-NIR spectroscopy for integrated external and internal quality assessment of Valencia oranges." *Journal of Food Composition and Analysis* no. 31 (1):144–154. doi: https://doi.org/10.1016/j.jfca.2013.05.007.

Marcelis, L.F.M., D.R. Verkerke, W.J. van der Burg, J.W. Aartse, R. Frankhuizen, and A. Boekestein. 1995. Non-destructive evaluation of internal quality of radish tubers using specific gravity, x-ray analysis and near infrared spectroscopy.

McGlone, V.A., D.G. Fraser, R.B. Jordan, and R. Künnemeyer. 2003. "Internal quality assessment of mandarin fruit by vis/NIR spectroscopy." *Journal of Near Infrared Spectroscopy* no. 11 (5):323–332. doi: https://doi.org/10.1255/jnirs.383.

McGlone, V.A., R.B. Jordan, and P.J. Martinsen. 2002. "Vis/NIR estimation at harvest of pre- and post-storage quality indices for 'Royal Gala' apple." *Postharvest Biology and Technology* no. 25 (2):135–144. doi: https://doi.org/10.1016/S0925-5214(01)00180-6.

McGlone, V.A., R.B. Jordan, R. Seelye, and C.J. Clark. 2003. "Dry-matter—a better predictor of the post-storage soluble solids in apples?" *Postharvest Biology and Technology* no. 28 (3):431–435. doi: https://doi.org/10.1016/S0925-5214(02)00207-7.

McGlone, V.A., and S. Kawano. 1998. "Firmness, dry-matter and soluble-solids assessment of postharvest kiwifruit by NIR spectroscopy." *Postharvest Biology and Technology* no. 13 (2):131–141. doi: https://doi.org/10.1016/S0925-5214(98)00007-6.

McGlone, V.A., and P.J. Martinsen. 2004. "Transmission measurements on intact apples moving at high speed." *Journal of Near Infrared Spectroscopy* no. 12 (1):37–43. doi: https://doi.org/10.1255/jnirs.406.

McGlone, V.A., P.J. Martinsen, C.J. Clark, and R.B. Jordan. 2005. "On-line detection of Brownheart in Braeburn apples using near infrared transmission measurements." *Postharvest Biology and Technology* no. 37 (2):142–151. doi: https://doi.org/10.1016/j.postharvbio.2005.04.011.

Mehinagic, E., G. Royer, R. Symoneaux, D. Bertrand, and F. Jourjon. 2004. "Prediction of the sensory quality of apples by physical measurements." *Postharvest Biology and Technology* no. 34 (3):257–269. doi: https://doi.org/10.1016/j.postharvbio.2004.05.017.

Mehl, P.M., Y.-R. Chen, M.S. Kim, and D.E. Chan. 2004. "Development of hyperspectral imaging technique for the detection of apple surface defects and contaminations." *Journal of Food Engineering* no. 61 (1):67–81. doi: https://doi.org/10.1016/S0260-8774(03)00188-2.

Mendoza, F., R. Lu, D. Ariana, H. Cen, and B. Bailey. 2011. "Integrated spectral and image analysis of hyperspectral scattering data for prediction of apple fruit firmness and soluble solids content." *Postharvest Biology and Technology* no. 62 (2):149–160. doi: https://doi.org/10.1016/j.postharvbio.2011.05.009.

Min, M., W.S. Lee, Y.H. Kim, and R.A. Bucklin. 2006. "Nondestructive detection of nitrogen in Chinese cabbage leaves using VIS–NIR spectroscopy." *HortScience* no. 41 (1):162–166.

Møller, S.M., S. Travers, H.C. Bertram, and M.G. Bertelsen. 2013. "Prediction of postharvest dry matter, soluble solids content, firmness and acidity in apples (cv. Elshof) using NMR and NIR spectroscopy--a comparative study." *European Food Research and Technology* no. 237 (6):1021–1024. doi: 10.1007/s00217-013-2087-6.

Nguyen-Do-Trong, N., J.C. Keresztes, B. De Ketelaere, and W. Saeys. 2016. "Cross-polarised VNIR hyperspectral reflectance imaging system for agrifood products." *Biosystems Engineering* no. 151:152–157 doi: http://dx.doi.org/10.1016/j.biosystemseng.2016.08.027.

Nguyen Do Trong, N., C. Erkinbaev, M. Tsuta, J. De Baerdemaeker, B. Nicolaï, and W. Saeys. 2014. "Spatially resolved diffuse reflectance in the visible and near-infrared wavelength range for non-destructive quality assessment of 'Braeburn' apples." *Postharvest Biology and Technology* no. 91:39–48. doi: https://doi.org/10.1016/j.postharvbio.2013.12.004.

Nicolaï, B.M., K. Beullens, E. Bobelyn, A. Peirs, W. Saeys, K.I. Theron, and J. Lammertyn. 2007. "Nondestructive measurement of fruit and vegetable quality by means of NIR spectroscopy---A review." *Postharvest Biology and Technology* no. 46 (2):99–118. doi: http://dx.doi.org/10.1016/j.postharvbio.2007.06.024.

Nicolaï, B.M., K.I. Theron, and J. Lammertyn. 2007. "Kernel PLS regression on wavelet transformed NIR spectra for prediction of sugar content of apple." *Chemometrics and Intelligent Laboratory Systems* no. 85 (2):243–252. doi: https://doi.org/10.1016/j.chemolab.2006.07.001.

Nicolaï, B.M., B.E. Verlinden, M. Desmet, S. Saevels, W. Saeys, K. Theron, R. Cubeddu, A. Pifferi, and A. Torricelli. 2008. "Time-resolved and continuous wave NIR reflectance spectroscopy to predict soluble solids content and firmness of pear." *Postharvest Biology and Technology* no. 47 (1):68–74. doi: https://doi.org/10.1016/j.postharvbio.2007.06.001.

Pan, L., R. Lu, Q. Zhu, J.M. McGrath, and K. Tu. 2015. "Measurement of moisture, soluble solids, sucrose content and mechanical properties in sugar beet using portable visible and near-infrared spectroscopy." *Postharvest Biology and Technology* no. 102:42–50. doi: http://dx.doi.org/10.1016/j.postharvbio.2015.02.005.

Pan, L., Q. Zhu, R. Lu, and J.M. McGrath. 2015. "Determination of sucrose content in sugar beet by portable visible and near-infrared spectroscopy." *Food Chemistry* no. 167:264–271 doi: http://dx.doi.org/10.1016/j.foodchem.2014.06.117.

Park, B., J.A. Abbott, K.J. Lee, C.H. Choi, and K.H. Choi. 2003. "Near-infrared diffuse reflectance for quantitative and qualitative measurement of soluble solids and firmness of delicious and gala apples." *Transactions of the ASAE* no. 46 (6):1721. doi: https://doi.org/10.13031/2013.15628.

Paz, P., M.-T. Sánchez, D. Pérez-Marín, J.-E. Guerrero, and A. Garrido-Varo. 2009. "Evaluating NIR instruments for quantitative and qualitative assessment of intact apple quality." *Journal of the Science of Food and Agriculture* no. 89 (5):781–790. doi: https://doi.org/10.1002/jsfa.3512.

Peirs, A., J. Lammertyn, K. Ooms, and B.M. Nicolaï. 2001. "Prediction of the optimal picking date of different apple cultivars by means of VIS/NIR-spectroscopy." *Postharvest Biology and Technology* no. 21 (2):189–199. doi: https://doi.org/10.1016/S0925-5214(00)00145-9.

Peirs, A., N. Scheerlinck, and B.M. Nicolaï. 2003. "Temperature compensation for near infrared reflectance measurement of apple fruit soluble solids contents." *Postharvest Biology and Technology* no. 30 (3):233–248. doi: https://doi.org/10.1016/S0925-5214(03)00118-2.

Peirs, A., N. Scheerlinck, K. Touchant, and B.M. Nicolaï. 2002. "PH—Postharvest technology—comparison of fourier transform and dispersive near-infrared reflectance spectroscopy for apple quality measurements." *Biosystems Engineering* no. 81 (3):305–311. doi: https://doi.org/10.1006/bioe.2001.0040.

Peirs, A., A. Schenk, and B.M. Nicolaï. 2005. "Effect of natural variability among apples on the accuracy of VIS-NIR calibration models for optimal harvest date predictions." *Postharvest Biology and Technology* no. 35 (1):1–13. doi: https://doi.org/10.1016/j.postharvbio.2004.05.010.

Peirs, A., J. Tirry, B. Verlinden, P. Darius, and B.M. Nicolaï. 2003. "Effect of biological variability on the robustness of NIR models for soluble solids content of apples." *Postharvest Biology and Technology* no. 28 (2):269–280. doi: https://doi.org/10.1016/S0925-5214(02)00196-5.

Pérez-Marín, D., P. Paz, J.-E. Guerrero, A. Garrido-Varo, and M.-T. Sánchez. 2010. "Miniature handheld NIR sensor for the on-site non-destructive assessment of post-harvest quality and refrigerated storage behavior in plums." *Journal of Food Engineering* no. 99 (3):294–302. doi: https://doi.org/10.1016/j.jfoodeng.2010.03.002.

Plans, M., J. Simó, F. Casañas, J. Sabaté, and L. Rodriguez-Saona. 2013. "Characterization of common beans (Phaseolus vulgaris L.) by infrared spectroscopy—Comparison of MIR, FT-NIR and dispersive NIR using portable and benchtop instruments." *Food Research International* no. 54 (2):1643–1651. doi: https://doi.org/10.1016/j.foodres.2013.09.003.

Pu, Y.-Y., Y.-Z. Feng, and D.-W. Sun. 2015. "Recent progress of hyperspectral imaging on quality and safety inspection of fruits and vegetables—A review." *Comprehensive Reviews in Food Science and Food Safety* no. 14 (2):176–188. doi: https://doi.org/10.1111/1541-4337.12123.

Qin, J., R. Lu, and Y. Peng. 2009. "Prediction of apple internal quality using spectral absorption and scattering properties." *Transactions of the ASABE* no. 52 (2):499. doi: https://doi.org/10.13031/2013.26807.

Qin, J., K. Chao, M.S. Kim, R. Lu, and T.F. Burks. 2013. "Hyperspectral and multispectral imaging for evaluating food safety and quality." *Journal of Food Engineering* no. 118 (2):157–171. doi: http://dx.doi.org/10.1016/j.jfoodeng.2013.04.001.

Qin, J., and R. Lu. 2008. "Measurement of the optical properties of fruits and vegetables using spatially resolved hyperspectral diffuse reflectance imaging technique." *Postharvest Biology and Technology* no. 49 (3):355–365. doi: https://doi.org/10.1016/j.postharvbio.2008.03.010.

Rady, A., and D. Guyer. 2015. "Utilization of visible/near-infrared spectroscopic and wavelength selection methods in sugar prediction and potatoes classification." *Journal of Food Measurement and Characterization* no. 9 (1):20–34. doi: https://doi.org/10.1007/s11694-014-9207-2.

Rady, A., D. Guyer, and R. Lu. 2015. "Evaluation of sugar content of potatoes using hyperspectral imaging." *Food and Bioprocess Technology* no. 8 (5):995–1010. doi: https://doi.org/10.1007/s11947-014-1461-0.

Rady, A.M., D.E. Guyer, W. Kirk, and I.R. Donis-González. 2014. "The potential use of visible/near infrared spectroscopy and hyperspectral imaging to predict processing-related constituents of potatoes." *Journal of Food Engineering* no. 135:11–25. doi: https://doi.org/10.1016/j.jfoodeng.2014.02.021.

Rajkumar, P., N. Wang, G. Eimasry, G.S.V. Raghavan, and Y. Gariepy. 2012. "Studies on banana fruit quality and maturity stages using hyperspectral imaging." *Journal of Food Engineering* no. 108 (1):194–200. doi: https://doi.org/10.1016/j.jfoodeng.2011.05.002.

Rinaldo, C., D. Cosimo, P. Antonio, T. Paola, T. Alessandro, V. Gianluca, R.-A. Margarita, V. Constantino, O. Coral, D. Colin, and J. David. 2001. "Time-resolved reflectance spectroscopy applied to the nondestructive monitoring of the internal optical properties in apples." *Applied Spectroscopy* no. 55 (10):1368–1374. doi: https://doi.org/10.1366/0003702011953496.

Rizzolo, A., M. Vanoli, P. Eccher Zerbini, S. Jacob, A. Torricelli, L. Spinelli, R.E. Schouten, and L.M.M. Tijskens. 2009. "Prediction ability of firmness decay models of nectarines based on the biological shift factor measured by time-resolved reflectance spectroscopy." *Postharvest Biology and Technology* no. 54 (3):131–140. doi: https://doi.org/10.1016/j.postharvbio.2009.05.010.

Rizzolo, A., M. Vanoli, L. Spinelli, and A. Torricelli. 2010. "Sensory characteristics, quality and optical properties measured by time-resolved reflectance spectroscopy in stored apples." *Postharvest Biology and Technology* no. 58 (1):1–12. doi: https://doi.org/10.1016/j.postharvbio.2010.05.003.

Saeys, W., M.A. Velazco-Roa, S.N. Thennadil, H. Ramon, and B.M. Nicolaï. 2008. "Optical properties of apple skin and flesh in the wavelength range from 350 to 2200 nm." *Applied Optics* no. 47 (7):908–919. doi: https://doi.org/10.1364/ao.47.000908.

Sanaeifar, A., A. Bakhshipour, and M. de la Guardia. 2016. "Prediction of banana quality indices from color features using support vector regression." *Talanta* no. 148:54–61. doi: https://doi.org/10.1016/j.talanta.2015.10.073.

Sánchez, M.-T., M.-J. De la Haba, J.-E. Guerrero, A. Garrido-Varo, and D. Pérez-Marín. 2011. "Testing of a local approach for the prediction of quality parameters in intact nectarines using a portable NIRS instrument." *Postharvest Biology and Technology* no. 60 (2):130–135. doi: https://doi.org/10.1016/j.postharvbio.2010.12.006.

Sánchez, T., H. Ceballos, D. Dufour, D. Ortiz, N. Morante, F. Calle, T. Zum Felde, M. Domínguez, and F. Davrieux. 2014. "Prediction of carotenoids, cyanide and dry matter contents in fresh cassava root using NIRS and Hunter color techniques." *Food Chemistry* no. 151:444–451 doi: https://doi.org/10.1016/j.foodchem.2013.11.081.

Scanlon, M.G., M.K. Pritchard, and L.R. Adam. 1999. "Quality evaluation of processing potatoes by near infrared reflectance." *Journal of the Science of Food and Agriculture* no. 79 (5):763–771. doi: https://doi.org/10.1002/(sici)1097-0010(199904)79:5<763::aid-jsfa250>3.0.co;2-o.

Schulz, H., H.-H. Drews, R. Quilitzsch, and H. Krüger. 1998. "Application of near infrared spectroscopy for the quantification of quality parameters in selected vegetables and essential oil plants." *Journal of Near Infrared Spectroscopy* no. 6 (A):A125–A130. doi: https://doi.org/10.1255/jnirs.179.

Siregar, S.T.W., W. Handayani, and A.H. Saputro. 2017. Bananas moisture content prediction system using Visual-NIR imaging. Paper read at *2017 5th International Conference on Instrumentation, Control, and Automation (ICA)*, 9–11 Aug. 2017.

Siripatrawan, U., Y. Makino, Y. Kawagoe, and S. Oshita. 2011. "Rapid detection of Escherichia coli contamination in packaged fresh spinach using hyperspectral imaging." *Talanta* no. 85 (1):276–281. doi: https://doi.org/10.1016/j.talanta.2011.03.061.

Steidle Neto, A.J., L. de Oliveira Moura, D. de Carvalho Lopes, L. de Almeida Carlos, L.M. Martins, and L. de Castro Louback Ferraz. 2017. "Non-destructive prediction of pigment content in lettuce based on visible–NIR spectroscopy." *Journal of the Science of Food and Agriculture* no. 97 (7):2015–2022. doi: https://doi.org/10.1002/jsfa.8002.

Su, W.-H., H.-J. He, and D.-W. Sun. 2017. "Non-Destructive and rapid evaluation of staple foods quality by using spectroscopic techniques—A review." *Critical Reviews in Food Science and Nutrition* no. 57 (5):1039–1051. doi: https://doi.org/10.1080/10408398.2015.1082966.

Su, W.-H., and D.-W. Sun. 2017. "Chemical imaging for measuring the time series variations of tuber dry matter and starch concentration." *Computers and Electronics in Agriculture* no. 140:361–373. doi: https://doi.org/10.1016/j.compag.2017.06.013.

Sun, X., H. Zhang, and Y. Liu. 2009. "Nondestructive assessment of quality of Nanfeng mandarin fruit by a portable near infrared spectroscopy." *International of Agricultural and Biological Engineering* no. 2 (1):65–70. doi: https://doi.org/10.3965/j.issn.1934-6344.2009.01.065-071.

Tarkosova, J., and J. Copikova. 2000. "Determination of carbohydrate content in bananas during ripening and storage by near infrared spectroscopy." *Journal of Near Infrared Spectroscopy* no. 8 (1):21–26. doi: https://doi.org/10.1255/jnirs.260.

Temma, T., K. Hanamatsu, and F. Shinoki. 2002a. "Development of a portable near infrared sugar-measuring instrument." *Journal of Near Infrared Spectroscopy* no. 10 (1):77–83. doi: https://doi.org/10.1255/jnirs.324.

Temma, T., K. Hanamatsu, and F. Shinoki. 2002b. "Measuring the sugar content of apples and apple juice by near infrared spectroscopy." *Optical Review* no. 9 (2):40–44. doi: https://doi.org/10.1007/s10043-002-0040-1.

Torres, I., D. Pérez-Marín, M.-J. De la Haba, and M.-T. Sánchez. 2017. "Developing universal models for the prediction of physical quality in citrus fruits analysed on-tree using portable NIRS sensors." *Biosystems Engineering* no. 153:140–148. doi: https://doi.org/10.1016/j.biosystemseng.2016.11.007.

Travers, S., M.G. Bertelsen, and S.V. Kucheryavskiy. 2014. "Predicting apple (cv. Elshof) postharvest dry matter and soluble solids content with near infrared spectroscopy." *Journal of the Science of Food and Agriculture* no. 94 (5):955–962. doi: https://doi.org/10.1002/jsfa.6343.

Tsuchikawa, S., E. Sakai, K. Inoue, and K. Miyamoto. 2003. "Application of time-of-flight near-infrared spectroscopy to detect sugar and acid content in satsuma mandarin." *Journal of the American Society for Horticultural Science* no. 128 (3):391–396.

Urraca, R., A. Sanz-Garcia, J. Tardaguila, and M.P. Diago. 2016. "Estimation of total soluble solids in grape berries using a hand-held NIR spectrometer under field conditions." *Journal of the Science of Food and Agriculture* no. 96 (9):3007–3016. doi: https://doi.org/10.1002/jsfa.7470.

Valero, C., P. Barreiro, M. Ruiz-Altisent, R. Cubeddu, A. Pifferi, P. Taroni, A. Torricelli, G. Valentini, D. Johnson, and C. Dover. 2005. "Mealiness detection in apples using time resolved reflectance spectroscopy." *Journal of Texture Studies* no. 36 (4):439–458. doi: https://doi.org/10.1111/j.1745-4603.2005.00026.x.

Van Beers, R., B. Aernouts, L.L. Gutiérrez, C. Erkinbaev, K. Rutten, A. Schenk, B. Nicolaï, and W. Saeys. 2015. "Optimal illumination-detection distance and detector size for predicting braeburn apple maturity from vis/NIR laser reflectance measurements." *Food and Bioprocess Technology* no. 8 (10):2123–2136. doi: https://doi.org/10.1007/s11947-015-1562-4.

Vanoli, M., P. Eccher Zerbini, L. Spinelli, A. Torricelli, and A. Rizzolo. 2009. "Polyuronide content and correlation to optical properties measured by time-resolved reflectance spectroscopy in 'Jonagored' apples stored in normal and controlled atmosphere." *Food Chemistry* no. 115 (4):1450–1457. doi: https://doi.org/10.1016/j.foodchem.2009.01.081.

Vaudelle, F., and J.-P. L'Huillier. 2015. "Influence of the size and skin thickness of apple varieties on the retrieval of internal optical properties using Vis/NIR spectroscopy—A Monte Carlo-based study." *Computers and Electronics in Agriculture* no. 116:137–149. doi: https://doi.org/10.1016/j.compag.2015.06.009.

Velasco, L., and W.J. Grüneberg. 1999. "Analysis of dry matter and protein contents in fresh yam bean tubers by near-infrared reflectance spectroscopy." *Communications in Soil Science and Plant Analysis* no. 30 (13–14):1797–1805. doi: https://doi.org/10.1080/00103629909370332.

Ventura, M., A. de Jager, H. de Putter, and F.P.M.M. Roelofs. 1998. "Non-destructive determination of soluble solids in apple fruit by near infrared spectroscopy (NIRS)." *Postharvest Biology and Technology* no. 14 (1):21–27. doi: https://doi.org/10.1016/S0925-5214(98)00030-1.

Villatoro-Pulido, M., R.M. Rojas, A. Muñoz-Serrano, V. Cardeñosa, M.Á.A. López, R. Font, and M. Del Río-Celestino. 2012. "Characterization and prediction by near-infrared reflectance of mineral composition of rocket (Eruca vesicaria subsp. sativa and Eruca vesicaria subsp. vesicaria)." *Journal of the Science of Food and Agriculture* no. 92 (7):1331–1340. doi: https://doi.org/10.1002/jsfa.4694.

Walsh, K.B., M. Golic, and C.V. Greensill. 2004. "Sorting of fruit using near infrared spectroscopy—Application to a range of fruit and vegetables for soluble solids and dry matter content." *Journal of Near Infrared Spectroscopy* no. 12 (3):141–148. doi: https://doi.org/10.1255/jnirs.419.

Wang, A., and L. Xie. 2014. "Technology using near infrared spectroscopic and multivariate analysis to determine the soluble solids content of citrus fruit." *Journal of Food Engineering* no. 143:17–24. doi: https://doi.org/10.1016/j.jfoodeng.2014.06.023.

Wang, J., J. Wang, Z. Chen, and D. Han. 2017. "Development of multi-cultivar models for predicting the soluble solid content and firmness of European pear (Pyrus communis L.) using portable vis–NIR spectroscopy." *Postharvest Biology and Technology* no. 129:143–151. doi: https://doi.org/10.1016/j.postharvbio.2017.03.012.

Wang, N.-N., D.-W. Sun, Y.-C. Yang, H. Pu, and Z. Zhu. 2016. "Recent advances in the application of hyperspectral imaging for evaluating fruit quality." *Food Analytical Methods* no. 9 (1):178–191. doi: https://doi.org/10.1007/s12161-015-0153-3.

Wang, Z., R. Van Beers, B. Aernouts, R. Watté, P. Verboven, B. Nicolai, B., and W. Saeys. 2019. "Microstructure affects light scattering in apples." *Postharvest Biologiy and Technology* no.159:110996. doi: 10.1016/j.postharvbio.2019.110996.

Wedding, B.B., C. Wright, S. Grauf, R.D. White, B. Tilse, and P. Gadek. 2013. "Effects of seasonal variability on FT-NIR prediction of dry matter content for whole Hass avocado fruit." *Postharvest Biology and Technology* no. 75:9–16. doi: https://doi.org/10.1016/j.postharvbio.2012.04.016.

Williams, P. and K. Norris. 2001. *Near-Infrared Technology in the Agricultural and Food Industries*, 2nd ed.. American Association of Cereal Chemists, Saint-Paul.

Wu, D., and D.-W. Sun. 2013. "Advanced applications of hyperspectral imaging technology for food quality and safety analysis and assessment—A review — Part II—Applications." *Innovative Food Science & Emerging Technologies* no. 19:15–28 doi: http://dx.doi.org/10.1016/j.ifset.2013.04.016.

Xia, J., X. Li, P. Li, Q. Ma, and X. Ding. 2007. "Application of wavelet transform in the prediction of navel orange vitamin C content by near-infrared spectroscopy." *Agricultural Sciences in China* no. 6 (9):1067–1073. doi: https://doi.org/10.1016/S1671-2927(07)60148-5.

Xie, C., B. Chu, and Y. He. 2018. "Prediction of banana color and firmness using a novel wavelengths selection method of hyperspectral imaging." *Food Chemistry* no. 245:132–140. doi: https://doi.org/10.1016/j.foodchem.2017.10.079.

Xing, J., C. Bravo, D. Moshou, H. Ramon, and J. De Baerdemaeker. 2006. "Bruise detection on 'Golden Delicious' apples by vis/NIR spectroscopy." *Computers and Electronics in Agriculture* no. 52 (1):11–20. doi: https://doi.org/10.1016/j.compag.2006.01.006.

Xue, L., and L. Yang. 2009. "Deriving leaf chlorophyll content of green-leafy vegetables from hyperspectral reflectance." *ISPRS Journal of Photogrammetry and Remote Sensing* no. 64 (1):97–106. doi: https://doi.org/10.1016/j.isprsjprs.2008.06.002.

Yang, X., J. Zhang, D. Guo, X. Xiong, L. Chang, Q. Niu, and D. Huang. 2016. "Measuring and evaluating anthocyanin in lettuce leaf based on color information." *IFAC-PapersOnLine* no. 49 (16):96–99. doi: https://doi.org/10.1016/j.ifacol.2016.10.018.

Ye, X., S. Oshita, X. Jin, Y. Makino, Y. Kawagoe, and H. Ando. 2009. "Non-destructive sensing of atp content as a potential indicator of freshness of spinach by Vis/Nir spectroscopy." *International Journal of Optomechatronics* no. 3 (1):30–40. doi: https://doi.org/10.1080/15599610902717843.

Young, M.W., D.K.L. Mackerron, and H.V. Davies. 1997. "Calibration of near infrared reflectance spectroscopy to estimate nitrogen concentration in potato tissues." *Potato Research* no. 40 (2):215–220. doi: https://doi.org/10.1007/bf02358246.

Zerbini, P.E., M. Grassi, R. Cubeddu, A. Pifferi, and A. Torricelli. 2002. "Nondestructive detection of brown heart in pears by time-resolved reflectance spectroscopy." *Postharvest Biology and Technology* no. 25 (1):87–97. doi: https://doi.org/10.1016/S09255214(01)00150-8.

Zhang, C., Q. Wang, F. Liu, Y. He, and Y. Xiao. 2017. "Rapid and non-destructive measurement of spinach pigments content during storage using hyperspectral imaging with chemometrics." *Measurement* no. 97:149–155. doi: https://doi.org/10.1016/j.measurement.2016.10.058.

Zhu, Q., J. Guan, M. Huang, R. Lu, and F. Mendoza. 2016. "Predicting bruise susceptibility of 'Golden Delicious' apples using hyperspectral scattering technique." *Postharvest Biology and Technology* no. 114:86–94 doi: https://doi.org/10.1016/j.postharvbio.2015.12.007.

Zhu, Q., C. He, R. Lu, F. Mendoza, and H. Cen. 2015. "Ripeness evaluation of 'Sun Bright' tomato using optical absorption and scattering properties." *Postharvest Biology and Technology* no. 103:27–34. doi: https://doi.org/10.1016/j.postharvbio.2015.02.007.

Zude, M., I. Birlouez-Aragon, P.-J. Paschold, and D.N. Rutledge. 2007. "Non-invasive spectrophotometric sensing of carrot quality from harvest to consumption." *Postharvest Biology and Technology* no. 45 (1):30–37. doi: https://doi.org/10.1016/j.postharvbio.2007.01.010.

Zude, M., M. Pflanz, L. Spinelli, C. Dosche, and A. Torricelli. 2011. "Non-destructive analysis of anthocyanins in cherries by means of Lambert–Beer and multivariate regression based on spectroscopy and scatter correction using time-resolved analysis." *Journal of Food Engineering* no. 103 (1):68–75. doi: https://doi.org/10.1016/j.jfoodeng.2010.09.021.

29 Grains

*Stephen R. Delwiche**
USDA Agricultural Research Service

CONTENTS

29.1 INTRODUCTION

This chapter is not meant to be an exhaustive compendium of near-infrared technology of grain as this topic has been covered in numerous reviews and the published studies over the past 50 plus years now number in the thousands. For recent reviews, the reader is referred to several works (Sendin et al. 2018, Fox and Manley 2014, Esteve Agelet 2014). Although this chapter does not altogether avoid reviewing the subject, an effort has been made to introduce these works in the context of either their recentness (last 10 years emphasized) or their novelty either in the analyte being measured or the technique in which measurement is made. That being said, however, it is impossible to avoid a small bit of history on the subject of NIR of grain, considering that this is arguably the origin of modern NIR spectroscopy, defining "modern" as beginning when digital computers became the means to control the capturing and processing of the spectra.

29.2 ORIGIN OF MODERN NIR PRACTICES

Several years ago, the author had a conversation with Karl Norris, then retired from the USDA, about the start of work on NIR of grain. From Norris's recollection, it began in the late 1950s at the request of the Weights and Measures Division of the U.S. National Institute of Standards and Technology, then known as the National Bureau of Standards (NBS). The concern at the time was to find a way to authenticate the weight of 5-pound flour packages, which were subject to weight loss caused by water evaporation during storage on the grocery shelf. Norris had recalled seeing reports on then recent work in Germany on moisture detection in paper by NIR. He suggested that NIR could be used to measure the moisture content of wheat flour. Because NBS researchers were using the Karl

* stephen.delwiche@usda.gov

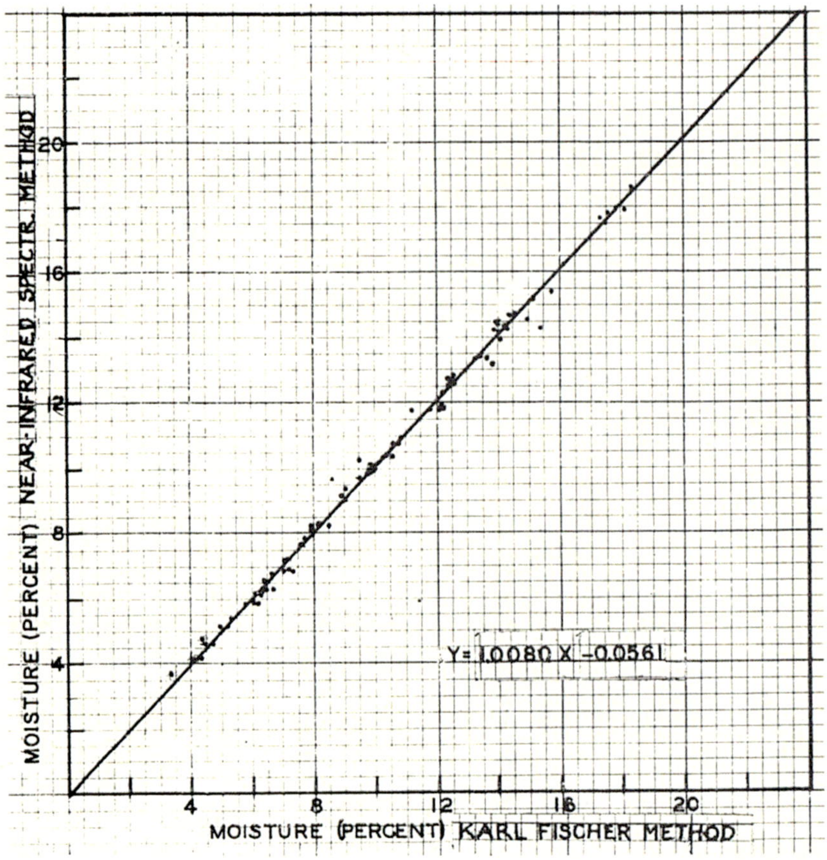

FIGURE 29.1 Very early application of NIR spectrometry on seeds. Two wavelength moisture calibration was based on the transmittance of moisture containing methanol extract of ground seed (oil seeds, grasses, and vegetables) as produced during Karl Fischer extraction. (Reprinted from Hart et al. 1962.)

Fischer methanol extract titration procedure for moisture measurement, a logical step was to examine whether NIR transmission of the extract could circumvent the remaining titration step in the Karl Fisher method. As seen by what may be the first published figure of a NIR calibration of moisture in seeds (Figure 29.1), Norris and coworkers (Hart et al. 1962) successfully demonstrated with just two wavelengths (a 1930 nm moisture band and a baseline response at 1800 nm) the possibilities of this technology that hitherto fore had been mostly a curiosity for spectroscopists. Based on the Norris team's success in measuring water in the methanol extract, work began on a more direct measurement of moisture in intact seeds that would not require methanol extraction. Carbon tetrachloride was chosen as a noninterfering immersion medium with a sufficiently close index of refraction to allow for low scattering of the energy and hence more direct transmission and absorption of light within the seeds. Although successful, the work was short lived because of awareness of the health dangers of working with organic solvents. By the late 1960s, work was well underway on the diffuse reflection approach at the USDA Beltsville laboratory, an approach that is widely practiced today.

29.3 ROADMAP OF GRAIN NIR APPLICATIONS

A general summary of the application of NIR spectroscopy to grain is shown in Figure 29.2. This figure is by no means a representation of all NIR grain applications; rather, it is intended to provide the reader with a summary of the more common developments. By far, moisture content and

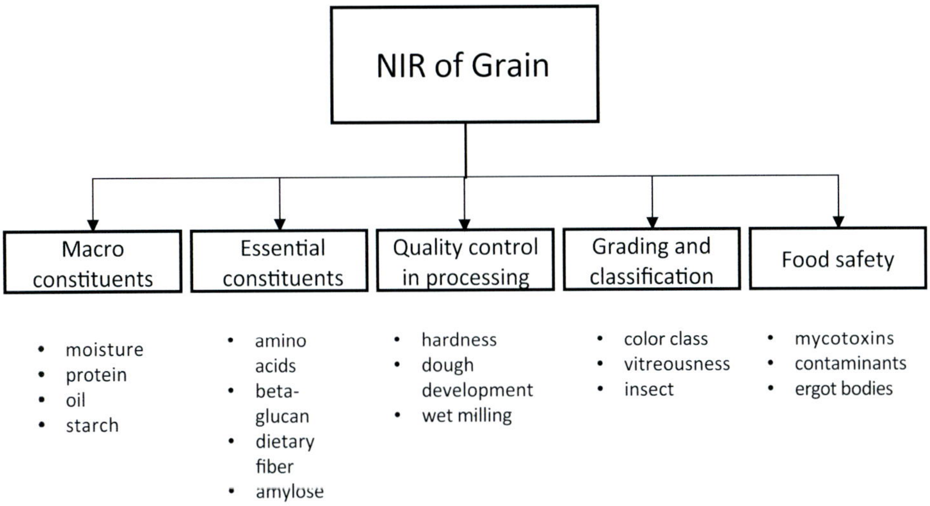

FIGURE 29.2 NIR of grain categories.

protein content are the most widely used of all applications and have achieved worldwide adoption because of the economic importance of these two constituents and the fact that the levels of accuracies of the respective NIR models approach those of their reference procedures. Other constituents, such as amino acids and fatty acids, have been shown to be amenable to NIR modeling although more at the level of plant genotype screening. Still other constituents, such as aflatoxin in corn or deoxynivalenol in wheat, have received the attention of NIR practitioners despite the concentrations of their natural occurrence in bulk lots, at sub-ppm levels, being below the detection limits of NIR reflection or transmission spectroscopy. Consequently, analyses must be performed on a single-seed basis, with the assumption that the mycotoxin contamination occurs on a very small fraction of the seeds within a lot, and second, that the NIR models may be sensitive to the fungi that produce the mycotoxin rather than the mycotoxin itself. New challenges arise with this route as (1) this means that hundreds of seeds per test sample must be evaluated and (2) the presence of the fungi is not a guarantee that the mycotoxin, a secondary metabolite whose synthesis is regulated by various environmental factors, will itself be present.

Regarding the grading and classification branch in Figure 29.2, NIR spectroscopy has a substantial input in a country's official grain grading operations, especially for contents of protein and moisture (wheat), oil (soybean), and starch (corn). In some countries, such as those in western Europe, Canada, and Australia, the protein content by NIR is a grade-determining component. In the United States, however, most, if not all, of the NIR calibration equations for raw grain are not used in determining official grade but rather as "official criteria" to which buyer and seller may specify in contracts of sale. Whole-grain NIR analysis in inspection operations typically consists of passing aliquots of raw grain through an optical chamber whereupon transmission or reflection measurements are collected, subjected to the calibration, and then averaged to report a single reading for a ~1-kg sample. Each country's system is unique, with details readily available from the governmental branch that covers food and agriculture. By way of example, official NIR usage operations of grain in the United States are detailed in the USDA Federal Grain Inspection Service's Near-Infrared Transmittance Handbook (USDA-FGIS 2006).

29.4 SCREENING TOOL

For plant geneticists of the cereals, especially wheat, NIR spectroscopy has long been desired as the salvation to the dilemma of developing varieties with good end-use property traits. These traits can describe the mixing properties of dough (elasticity, resistance to deformation, extensibility, mixing tolerance), the texture of noodles, the volume and crumb characteristics of baked bread, endosperm hardness, as well as the macro constituents (protein, fiber, lipids). To the breeder, the challenge is that these characteristics are typically unmeasurable without a sufficient quantity (i.e., hundreds of grams) available. Such quantities do not occur until the F5 or F6 generation, thus limiting the number of lines that can be realistically evaluated. At this point, many of the initial crosses have been culled without knowledge of their end-use quality potential. Attempts at using NIR spectroscopy as a breeder's tool go back decades (Williams 1979, Delwiche et al. 1996), with mixed success. The newest approach is to use the predictions from conventional NIR calibrations of these quality traits collectively as input in genomic models that select candidates at early stages (i.e., F2) of breeding (Hayes et al. 2017) and claiming that the accuracy of the NIR-enhanced (as well as NMR-enhanced) genomic models, defined as the correlation between the modeled prediction and the phenotype expression, improves from a range of 0–0.47 (genomic input alone) to 0–0.69 (genomic data augmented with spectroscopic data). Although such improvement may seem modest, these authors contend that the enhanced genomic models overcome a "major barrier" in wheat breeding programs that select for end-use quality (Hayes et al. 2017).

29.5 HARVESTING OPERATIONS

Advances in precision agriculture have fostered controlled planting, tillage, irrigation, and pesticide application operations to optimize field inputs. Concurrent with this advancement has been the gradual development of monitoring devices during harvest operations, with yield (tonnage) and moisture being the most important first goals. Ruggedization of NIR diode array-based instruments permitted the extension to nutritional constituents such as protein and oil contents. For example, the early work of Montes et al. (2006) demonstrated the potential of in-field on-combine NIR measurement of maize during harvest for estimation of dry matter content, protein content, and starch content, using a Zeiss Corona 45 diode array device (960–1690 nm) and partial least squares (PLS) regression modeling.

Newer directions on in-field measurement harvest operations have sought to establish cause-and-effect relationships between preharvest conditions and internal properties measured during harvest operations. For example, Long and McCallum (2015) conducted a field study to examine whether the grain yield of wheat was limited to nitrogen stress or water stress. The stress components were measured by a mass flow sensor on a combine for kernel yield, a LIDAR (i.e., an acronym for light detection and ranging) sensor for straw height and hence biomass yield, and a NIR spectrometer (600–1100 nm) for protein, all in real time and synchronized, thus allowing the field to be mapped for the degree of environmental stress. These same authors had previously demonstrated the ability to segregate wheat during combine harvest operations into high- and low-protein categories based on a PLS regression of spectra collected through a port of the bin-filling auger (Long et al. 2013).

29.6 NUTRITION

On one hand, concern about the possible ill health effects of gluten proteins from the standpoint of human enteropathies has kindled an interest in the nongluten-producing cereals. Traditionally, extensive NIR research has been conducted on the measurement of quality aspects of wheat and barley for breeding programs; much less has been devoted to sorghum. The popularity of sorghum has grown in recent years as the desire for gluten-free cereals has risen. The nutritional benefits of sorghum include antioxidant and anti-inflammatory properties from the presence of phenolic compounds and anticarcinogenic properties from tannins (Burdette et al. 2010). Dykes et al. (2014) successfully developed NIR calibrations on spectra (400–2500 nm) of bulk sorghum for total phenols, condensed tannins, and 3-deoxyanthocyanidins. Fox et al. (2012) reported on the use of conventional NIR reflectance spectroscopy of ground forage sorghum for the analysis of the antinutritional compound, hydrogen cyanide (HCN), a breakdown product of the cyanogenic glycoside dhurrin, a substance postulated to be part of the plant's defense mechanism against herbivores. The one unconventional aspect of the study was that the spectral measurements were made on the plant material, which then were regressed using PLS or multiple linear regression (MLR) to predict the potential level of HCN that could be produced under in situ conditions that would promote its release. Successful modeling, either by PLS of the visible through NIR region (400–2500 nm) or by MLR using wavelengths from the NIR region (1100–2500 nm), was possible for plant breeding screening.

On the other hand, a complete adoption of a gluten-free diet may lead to a reduction in diversity of the gut microbiota (De Palma et al. 2009). Although we have known for years that the use of NIR spectroscopy for wheat dough and end-use properties (largely a function of the developed protein-starch matrix generically known as gluten) has had limited success, the more direct approach of sensing the various endosperm storage proteins may be more fruitful. Potential beneficiaries of this approach include individuals afflicted with an immunological intolerance (celiac disease, CD) to certain cereal products such as wheat, rye, and barley. The benefit comes from the ability to consume wheat derived from wheat stock possessing low levels of gliadin, the prolamin fraction largely responsible for the immunological response. Using a large (hundreds) set of wheat samples spanning 5 years, with more than half the samples being of low-gliadin content transgenic lines accomplished through RNA interference, García-Molina et al. (2016) have demonstrated that low-gliadin and wild-type wheats are distinguishable using NIR reflectance of the grain in either bulk or ground meal formats. Provided the eventual acceptance of transgenic wheat by the public occurs, such a NIR method could be readily adopted by wheat handlers who will be confronted with the need to segregate low-gliadin lots.

Traditionally, research on amino acid composition in rice has focused on milled white rice flour (Wu et al. 2002), but newer work has looked at this topic applied to brown rice. Brown rice is produced during the first step in processing in which paddy rice is dehusked, leaving intact seeds that have the outer bran layer and germ. Zhang et al. (2011) applied conventional NIR reflectance spectroscopy on a diverse set of commercial varieties and breeders' advanced lines of *indica* rice and demonstrated reasonably accurate PLS calibration models (validation set SD/SEP ranging from 2.4 to 4.3) for 13 of 17 amino acids (excluding cysteine, methionine, and tyrosine), which typically have concentrations in the 0.2%–1.5% (w/w) range.

29.7 CONTAMINATION

Detection of grain contamination using NIR methodologies is usually straightforward when contamination levels are greater than 0.1% (w/w). Perhaps the most benign form of contamination in cereals arises from the harvesting process in which grain that is inadequately threshed becomes comingled with straw and chaff. Wallays et al. (2009) applied hyperspectral imaging (400–900 nm) on kernels of wheat as well as straw and chaff to identify a selection of wavelengths through a genetic algorithm that could be used to calculate the level of contamination. In its simplest form, the contamination level became the ratio of the number of pixels identified as nongrain to the total number of pixels. The contrast between nonwheat (straw and chaff) and kernel was sufficiently high such that five regions spread across 465–945 nm could be used to readily identify the two classes.

By far, the more concerning occurrences of grain contamination deal with the secondary metabolites (mycotoxins) of fungi that attack the seed during development as well as in storage after harvest. Some of these, such as aflatoxin, are carcinogenic and teratogenic to humans and livestock. Others, such as fumonisin and zearalenone, while less harmful, can cause sickness in humans and hinder weight gain in livestock. Conventional NIR spectroscopy studies on grain mycotoxins date back 20 years, while spectral imaging work is more recent. Firrao et al. (2010) constructed a multispectral imaging system consisting of 10 LEDs in the range of 720–940 nm and a CCD camera to capture images of ground maize samples of 30 g as representative of field lots brought to the first point of sale. Concentrations of fumonisin, aflatoxin, zearalenone, and deoxynivalenol were estimated through the development of artificial neural network (ANN) models. The significance of this work is its recognition of the need for deployment of rapid inspection systems at locations far removed from ideal laboratory conditions.

29.8 SINGLE-SEED ANALYSIS

Single-seed analysis by conventional, nonimaging technologies may be performed in transmission or reflection; however, signal to noise for transmission mode diminishes as wavelengths increase, particularly above 1100 nm, to the extent that conventional spectrometers operating above this wavelength are limited to diffuse reflection. Regardless of the mode of illumination, the ability to interrogate the object from multiple angles as accomplished by seed repositioning appears to be critical in quantitative models (Esteve Agelet et al. 2012b). Models of physical properties, such as seed weight, do not benefit from spectral preprocessing, whereas models of chemical constituents, such as oil content or protein content, are improved with routine spectral normalization (Esteve Agelet et al. 2012b).

Esteve Agelet and Hurburgh (2014) reviewed the state of the art of single-seed NIR analysis, noting that most applications have focused on wheat, corn, and soybeans. Conventional, spot analysis and hyperspectral imaging are addressed, whereby they emphasize the relevance of single-seed analysis for use in phenotyping in plant breeding programs, and the still unmet needs of the industry in sorting operations in which speed, automation, and robustness (hardware and software) will be necessary.

With respect to commercial production of the cereals and oilseeds, soybean (*Glycine max* L.) is the species that has most widely adopted transgenic breeding practices, primarily through the introduction of a gene from the bacteria *Agrobacterium tumefaciens* that enables the plant to be resistant to the herbicide glyphosate, thus imparting the condition known as "roundup ready" (RR). Although NIR technologies have been considered as possible candidates for screening, the basis for such screening will not lay in an ability to measure the expressed protein responsible for herbicide tolerance. Rather, secondary effects such as changes to the carbohydrate structure must be consistent to permit the spectral discernment of conventional and RR beans. Esteve Agelet et al. (2012a) emphasized this challenge by noting that NIR models using principal component analysis (PCA), ANN, or locally weighted principal component regression (PCR), when applied to fully independent samples, could achieve 79% accuracy at best. In a follow-up study (Esteve Agelet et al., 2013), the same researchers proposed that the differential manner between conventional and RR beans of moisture imbibed by the seed hull affects the overlapping spectral absorption bands of water and carbohydrates (cellulose, hemicellulose, and lignin) and thus permits spectrally based classification but only under highly controlled laboratory conditions.

Apart from the use of single-seed technology for the identification of diseased or nonbelonging objects, this technology may also be used to estimate the concentration of a macro constituent in the bulk lot. In such case, the issue of what constitutes an adequate sample to represent the lot becomes a more critical concern than in bulk grain or ground meal analysis because of the assumed vastly smaller number of single seeds that are being used to represent the lot as well as the way they are collected. General matters of sampling theory are described in detail in Gy (1998). Using wheat protein content as an example, Tønning et al. (2006) advanced Gy's principles on grain by explaining the sources of variance to be expected when NIR transmittance of single wheat seeds is used to estimate a lot's protein content. The ability to measure nitrogen content by a reference method such as Kjeldahl, as well as the known ability to develop highly accurate NIR models for protein content in conventionally prepared grain samples, made this combination particularly well suited for studying sampling error. They described overall, or global estimation error (*GEE*), as the sum of total sampling error (*TSE*) and the total analytical error (*TAE*). The *TSE* is defined as the sum of three error terms: $FSE+GSE+ISE$, in which the first of the components, *FSE*, represents the inherent error caused by nature (e.g., constituent content, particle mass, particle shape). The user has little control over this first component, whereas the remaining two components are under the user's control. Of the remaining two, *GSE* (grouping and segregation error) represents the imperfect physical distribution of particles within the lot, as caused during the filling or transport of a vessel. Tønning et al. (2006) remark that *GSE* may be improved (lowered) through the use of more material or by the crushing of the lot's material to improve homogeneity through mixing. The last component, *ISE* (incorrect sampling error), arises from poor sampling practices, such as what may occur when the material is gathered from the top layer of a vessel. By constructing an experiment in which a fixed number of single seeds used to represent a wheat lot was gathered using either good sampling procedures (i.e., using a riffle device) or poor ones (grab sampling), and forming multiple sets per lot for each procedure, these researchers were able to verify that the variances associated with the sampling components (*FSE*, *GSE*, and *ISE*) were 10 times higher than the variance associated with the analytical (NIR) method, thus emphasizing the importance and added complexity of sampling when working on a single-seed basis.

A way of pursuing transmission through single seeds at wavelength regions attributed to first overtones and combination bands of organic molecules, now possible, is by increasing the intensity of the light source. The design of a laser light source predispersive monochromator system is described in a recent publication by Ringsted et al. (2016). In this system, a supercontinuum laser, whose full operational range is 1500–5200 nm, is operated in the subrange of higher efficiency, 2260–2380 nm, to examine genetically related barley genotypes for spectral differences in starch, beta-glucan, and lipids. Signal through 1-mm-thick cross-sectional slices of barley seed endosperm is sufficiently strong to identify starch absorption bands at 2287 nm [O–H+C–C combination and O–H+C–O combination, as influenced by the crystallinity differences between amylopectin (more crystalline) and amylose (less crystalline)] and 2323 nm (C–H stretch, C–H deformation overtones) and a beta-glucan absorption band at 2349 nm.

Starch structure may be observed by NIR reflection spectroscopy in greater detail through phosphorylation (Hódsági et al. 2012) to produce resistant starch. Regions of carbohydrate absorption at 2080–2130 and 2275–2290 nm, which Hódsági and coworkers attributed to an O–H bending+C–O stretch combination band and O–H stretch+C–C stretch combination band, respectively, demonstrated a sensitivity to the relative amounts of the starch molecules amylose and amylopectin. The importance of such findings may take on added significance should the cereals industry find benefit in the use of waxy (very low amylose) wheat for food and industrial purposes (Graybosch 1998). One immediate application of NIR spectroscopy of waxy wheat is in identity preservation of lots purported to be pure in the waxy condition yet difficult to authenticate by visual examination alone. The challenge to discerning the spectral difference between conventional and waxy wheat can be seen in Figure 29.3, which shows a family of blend mixtures ranging from fully conventional to fully waxy at various stages of spectral preprocessing. Even with the spectral similarities, PLS regression is capable of predicting the mixture level, with a standard error of typically 5% units (Delwiche and Graybosch 2014). Beyond the above-mentioned observations of Hódsági's team, Delwiche and Graybosch's suggested reasons for the spectral differentiation of these very closely related starch molecules (with amylose and amylopectin structurally differing by the presence of α-(1–6) glycosidic linkages in amylopectin in addition to the α-(1–4) glycosidic linkages that are present in both molecules) include conformational effects as well as an affinity that amylose has toward lipids.

29.9 HYPERSPECTRAL IMAGING

Numerous recent reviews of hyperspectral imaging (HSI) research on agricultural and food products are found in the literature (Elmasry et al. 2012, Feng and Sun 2012, Dale et al. 2013, Sendin et al. 2017, Manley 2014). Use of NIR spectroscopy of single kernels and NIR hyperspectral imaging of whole grain for fungal infection has been recently reviewed by Orina et al. (2017). Cereal grains pose a challenge for the commercial application of hyperspectral image analysis because

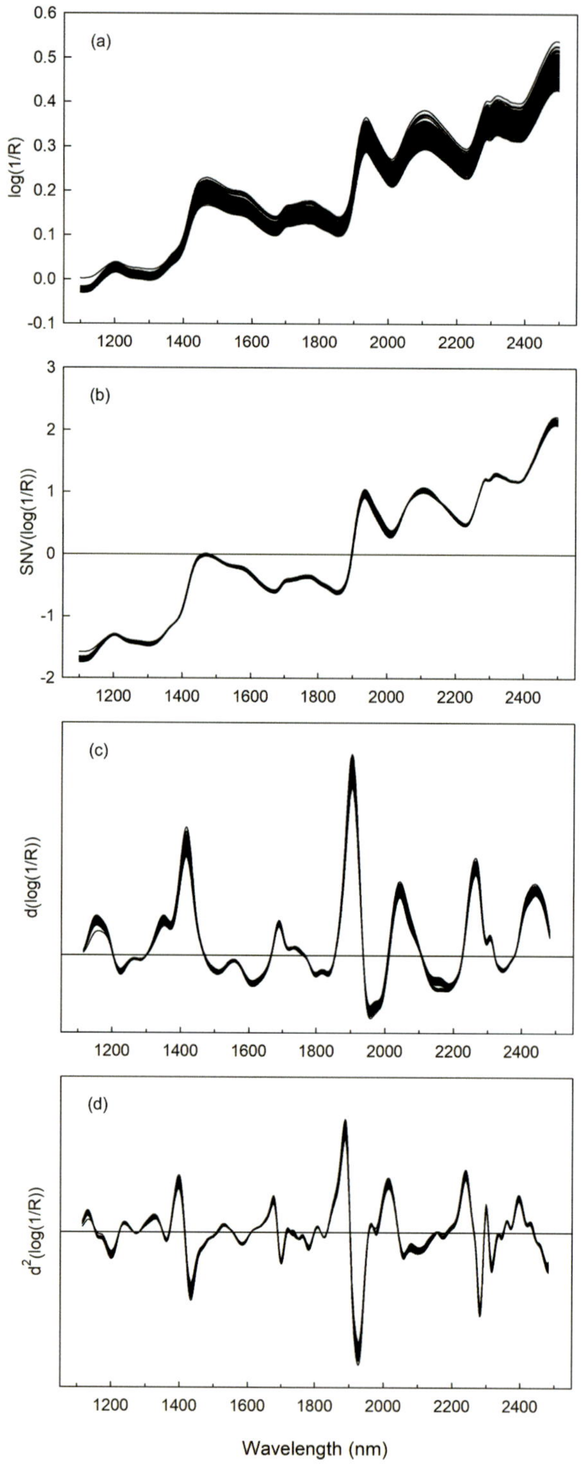

FIGURE 29.3 NIR spectra of wheat ground meal of binary mixtures of waxy and conventional wheat. Nine pairings, with each pair consisting of 29 mixture levels ranging from pure conventional to pure waxy. Typical spectral processing of the raw spectra (graph A) being standard normal variate (SNV, graph B), SNV+Savitzky-Golay first derivative (graph C), and SNV+Savitzky-Golay second derivative (graph D). (Reprinted from Delwiche and Graybosch, 2014).

of their small size and comparative low economic value. To date, HSI of grain has been limited to research investigations.

As in all imaging research, the first item for consideration is what defines the region of interest. Unlike larger agricultural commodities such as fresh produce, seeds of grain are small enough so that a tradeoff must be made between the number that can be imaged and the spatial resolution, thus defining the number of pixels representing a seed. Common imaging scenarios are posed as follows (Figure 29.4):

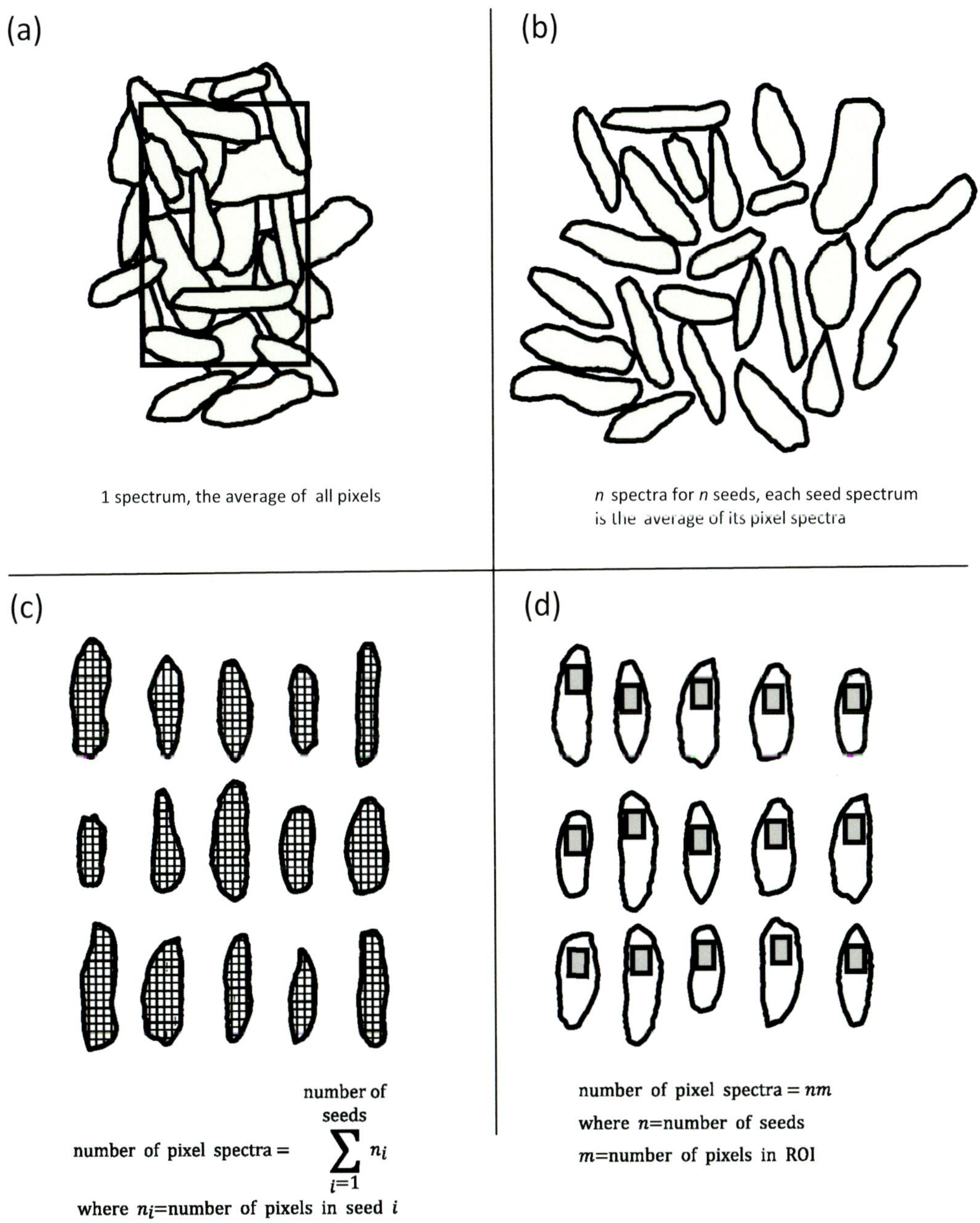

FIGURE 29.4 Schematic of the various approaches to objects in the field of view in hyperspectral imaging of grain applications. A = bulk lot, B = bulk sample single layer, C = individual seed, and D = subregion of a seed.

29.9.1 Bulk Lot

This is closest to the conventional whole-grain spectrometry. The camera lens is positioned above the surface of a lot of grain or a sample drawn from the lot. Ambient or controlled lighting may be used. The HSI system, through the averaging of hundreds to thousands of individual pixel spectra, reports the mean spectrum of the scanned region. This arrangement is akin to what a conventional spectrometer accomplishes by spot illumination. Because only the top surface of the lot is sensed by the camera, the interrogation of the sample is limited to the exposed surface, which may be disadvantageous compared to a whole-grain transmission analyzer that typically interrogates 400–500 g of seed. A variant of this is to maintain the individual pixel spectra, whereupon textural features of the image can be calculated to characterize the heterogeneity of the surface layer. However, the application of this analysis on grain is sparse in the literature.

29.9.2 Bulk Sample Single Layer

Seeds are spread out on a flat surface using a vibratory device to assist in their dispersal and ensure a single layer. In image processing, a universal preliminary step is needed to remove the background (conveyer belt in moving product applications, pan in static applications). Depending on the surface density, a second image processing step is applied to digitally separate touching seeds. Once individual seeds are identified, the entire seed, less any boundary pixels removed during an object erosion step, becomes the region of interest. The individual pixel spectra of a seed are averaged to produce one mean spectrum that represents the seed. Quantitative analysis of a constituent may be performed at the single-seed level or at the bulk sample level by averaging of each seed, with or without weighting by its size (as determined by pixel count and possibly a shape factor for the seed to convert an area measurement into a seed volume estimate).

29.9.3 Individual Seed

This configuration is like the previous configuration, with the difference being that pixel spectra within a seed are not averaged. This offers the potential advantage of visualizing spectral variation across the seed surface. However, this comes at the cost of processing time and computer memory as conceivably hundreds of pixel spectra are analyzed in lieu of a single-seed mean spectrum. Also, signal to noise is much lower with the pixel spectra, which may make it impossible to identify the desired feature differences within the seed. A compromise to the processing time and memory dilemma is to apply the calibration developed from seed average spectra to the individual pixel spectra, thus enabling a snapshot of the spatial distribution of the analyte across the seed. A recent example of this approach is found in Caporaso et al. (2018), in which protein in single wheat seeds is shown to concentrate near the embryo.

29.9.4 Subregion of a Seed

Seeds such as wheat are divided into various structural regions, such as the embryo or germ, the endosperm, endosperm halves as divided by the "crease" that bisects the seed along its length. Other seeds such as maize have regions within the endosperm that are translucent to light (typically termed vitreous) and other regions of endosperm called floury that are less dense and more opaque to light. With either type of seed, the region of interest (ROI) is smaller than the seed itself, a situation that is typical for most other food and agricultural applications of HSI. The region of interest may be defined by the user to be of precise geometry, as in the wheat kernel in Figure 29.5, or may be defined during image processing in which the general location of the region is known ahead of time (e.g., embryo region, region of blacktip fungus in wheat kernel, insect damage region in endosperm), with its precise location determined by image processing.

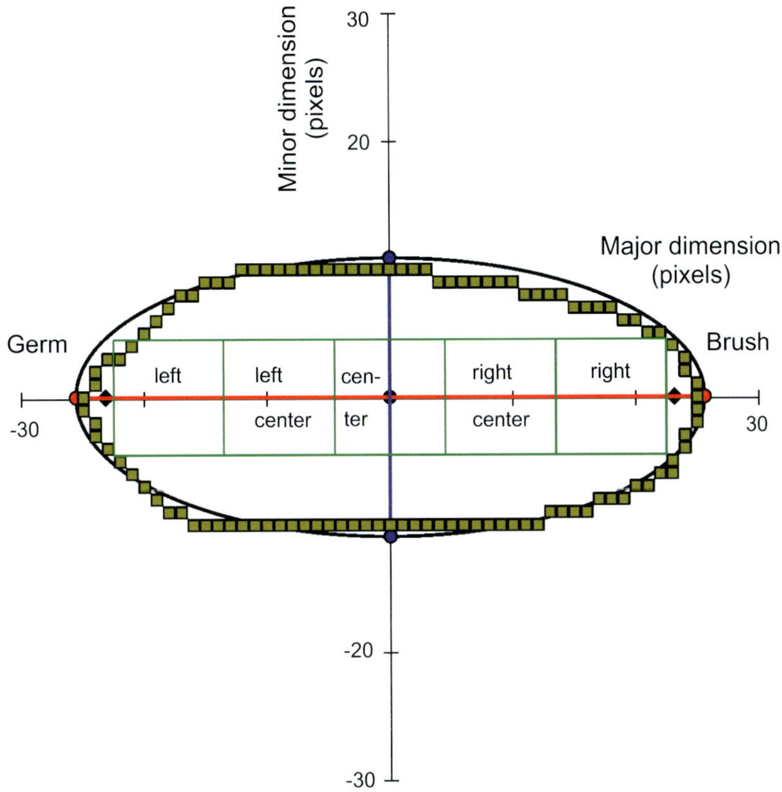

FIGURE 29.5 Anatomy of a seed for spectral and morphological image processing. Small filled boxes represent the pixels along the boundary, and the black curve is the ellipse of best fit with foci indicated as diamond symbols. Major and minor axis lengths are indicated by horizontal (red) and vertical (blue) line segments terminating in circle symbols. Five 9×9 pixel internal regions are shown as large interior boxes (left, left center, center, right center, and right), with the kernel centroid marked as a circle in the center box (Adapted from Delwiche et al. 2013.)

By a technique the authors call brushing, Manley et al. (2009) used the principal component (PC) score plots (factor *x* vs. factor *y*) of HSI spectra to delineate regions within the plots, whose pixels are then mapped back onto the original image in pseudocolor. An example of the application of this technique is shown in Figure 29.6. In an application involving HSI analysis of maize seeds, Manley and coworkers identified the floury (soft) and vitreous (harder) regions as well as a third type that they characterized as having different chemical and physical properties from floury and vitreous endosperms.

McGoverin et al. (2011) used HSI to explore viability in seeds of wheat, barley, and sorghum. Germination begins with the uptake of water in the embryonic region of the seed, followed by the activation of amylolytic and proteolytic enzymes that are responsible for hydrolyzing starches and proteins for plant development. Viability decreases as pregermination comes into play when seeds are exposed to the humid conditions of moisture incubation. Using a wavelength region of 1000–2498 nm, with PLS discriminant analysis (PLS-DA) and PLS regression, McGovern's team attributed the ability to distinguish the degree of viability from the 1920–1940 OH stretch deformation combination band of water as affected by the matrix, with differences becoming apparent in the fifth PC for barley and sorghum and the sixth PC for wheat on mean-centered, standard normal variate (SNV)-corrected log(1/*R*) spectra of freeze-dried seeds after soaking treatments of 0–24 h,

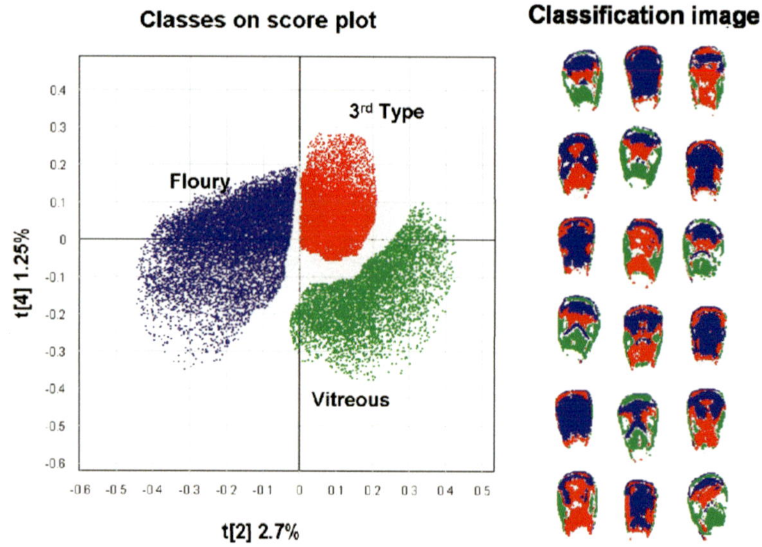

FIGURE 29.6 Example of "brushing" technique in hyperspectral imaging, in which the first two principal component scores (t[1] and t[2]) are plotted in a scatter plot, regions in the plot are manually delineated and assigned a color, and the pixels corresponding to each point are mapped back onto the image mask, in this case, kernels of maize. (Reprinted from Manley et al. 2009 with permission.)

with the lower components attributed to differences in macro chemical composition (endosperm vs. embryo) or physical (kernel curvature).

For grain, food safety inspection is probably the best application for HSI, whereby the benefit incurred by having a wholesome and contaminant-free food product will outweigh the cost and complexity of the imaging system. Food safety concerns in grain arise on both the macro and intra seed levels. At the former level, Vermeulen et al. (2012) describe the use of HSI to detect ergot bodies in raw grain. These bodies, which are the sclerotia of the fungus *Claviceps purpurea*, contain alkaloid compounds (ergocristine, ergotamine, ergocryptine, ergometrine, ergosine, and ergocornine) that have poisonous effects in humans and livestock, such that the European Union (EU) limits the contamination level of ergot bodies in unprocessed cereals (excluding rice and corn) at 0.5 g kg^{-1} (EU 2015). The responsible fungus particularly infects rye but is also found on wheat, triticale, barley, millet, and oats. Using PLS-DA and support vector machine (SVM) classification schemes, these authors were successful at identifying pixels of ergot at contamination levels as low as 0.1 g kg^{-1} and envisioned a throughput rate of 250 g/min using a push-broom imaging system. Even at a single wavelength, in this case, 1721 nm, there is sufficient contrast between wheat kernel and ergot body to allow for the identification of such bodies (Figure 29.7). Bulk grain analysis is advantageous in this situation because the ergot bodies are about the same size as the kernels. The same team of researchers demonstrated the transferability of the HSI system for ergot body detection from a laboratory level to a commercial system, with the ability to identify ergot bodies within various grains (wheat, barley, rye, and rapeseed) and then using morphological analysis and general knowledge of relative densities to estimate ergot weight concentrations as low as 0.2 g kg^{-1} (Vermeulen et al. 2013).

An aspect of HSI that is often overlooked or assumed to be of minimal concern is the surface shape of the object. Unlike conventional reflection spectroscopy of ground material in which the surface is flat, imaging of agricultural projects, especially grain kernels, invariably involves the issue of the camera recording a three-dimensional object in two dimensions. Curvature of the kernels may produce shadows due to imperfect lighting. Further, the closer the interior region of the object is to its boundary, the more oblique is the angle of returned radiation to the camera lens, with

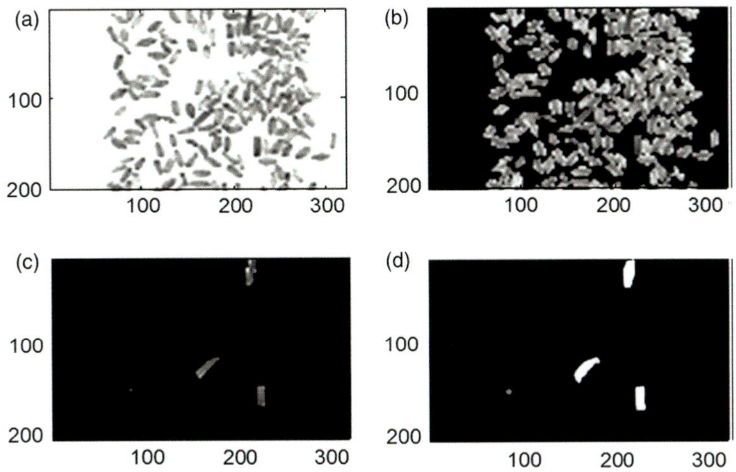

FIGURE 29.7 Example of hyperspectral imaging analysis steps taken to identify ergot bodies in a bulk wheat sample. In order, image (a) is a selected slice, (b) is the resulting image after removing the background (conveyer belt), (c) is the resulting image after removing wheat kernels, (d) is the final image in which ergot bodies are identified and assigned a white color. (Reprinted from Vermeulen et al. 2012 with permission.)

the possibility of a larger actual surface area being captured in a pixel. Attempts to reduce the edge effect are conducted by the erosion of the edge during image processing before spectral analysis, defining regions of interest that are removed from the edge, or through spectral preprocessing. Even when SNV or multiplicative signal correction (MSC) preprocessing was applied to hyperspectral images of wheat, barley, and sorghum kernels, the first PC typically is dominated by the physical effect of kernel shape (Manley et al. 2012).

An example of food safety HSI at the intra-seed level is that of Senthilkumar et al. (2016a), who studied the degree of fungal infection of barley kernels caused by *Aspergillus glaucus*, *Penicillium* spp., and *Ochratoxin* A contamination. Having selected significant wavelengths from PCA (1260, 1310, and 1360 for *A. glaucus* and *Penicillium* spp. and 1310, 1360, and 1480 nm for *Ochratoxin* A contamination), the distribution characteristics (mean, median, standard deviation, variance, minimum, maximum, and 10 histogram levels of reflectance) of the reflectance values of the pixels at each of these significant wavelengths were used in linear, quadratic, and Mahalanobis discriminant classifiers. They achieved very high accuracies in two class models (storage date a vs. storage date b) as well as in models involving all storage dates. Similar research was performed by these authors on canola (Senthilkumar et al. 2015) and wheat (Senthilkumar et al. 2016b). An even smaller amount of food safety HSI research has sought to use the technique as a microbiologist's plating and colony counting assistant. Williams et al. (2012) demonstrated that HSI score images could be used to visualize differences among Fusarium species colonies, which had been plated on potato dextrose sugar media, then identified by PLS-DA at the pixel level, thus permitting spatial detail that is not possible with conventional NIR spectrometry.

Whereas almost all hyperspectral imaging studies of grain have focused on the material after harvest, a recent study by Bauriegel et al. (2011) examined the incidence of Fusarium head blight in wheat during plant development. The rationale for performing such analysis before harvest is that decisions on harvest and load segregation can be made before the material becomes mixed in with the sound grain. Over the 400–1000 nm wavelength region examined, the most useful subregions for showing the contrast between healthy and diseased wheat spikes were 500–533, 560–575, 682–733, and 927–931 nm; however, the disease could not be detected immediately after the onset of flowering when infection typically occurs.

29.10 PARTING THOUGHTS

NIR spectroscopy of grain is one of the oldest applications of this technology and one of its greatest success stories. Grain is especially well suited for exploring new techniques owing to its abundance, modest cost, and adaptability to many formats (ground, whole, single seed, transmission, reflection). However, these features can also pose as vexing problems. For example, how does one determine the value of a cargo of grain based on the scan of a composite sample? In comments made more than 20 years ago, Geladi and Dåbakk 1995 stated that advances in chemometrics, as well as improvements in hardware design, will be tempered by issues of sampling and homogeneity of the samples. This remark still holds true today.

Macroconstituent analysis for economic value assessment (protein, moisture, oil, etc.) will most likely continue to drive the commercial market for NIR instruments in the grain industry for the near future. However, recent trends will contribute to the evolving path of NIR grain analysis, which are as follows:

1. Food safety—Naturally occurring mycotoxins in grain, such as deoxynivalenol, are regulated by most countries in the world. NIR methodologies applied to single seeds that are sensitive to either the mycotoxin itself (ideally) or the fungus producing the mycotoxin (alternatively, when the mycotoxin concentration is lower than the detectable limit) offer a possible alternative to expensive and time-consuming biochemical assays.
2. Phenotyping—Plant breeders, especially those working in the public breeding programs (wheat, oats, rice), are working with shrinking budgets to develop new varieties that are hearty, disease resistant, high yielding, and, at the same time, high in end-use quality. Any ability that a NIR method possesses to measure quality in the early generations of a plant breeding program will be immediately capitalized on by the breeding community.
3. Spectral imaging—Research on hyperspectral analysis has exploded in the past decade, with grain being no exception. Internal, chemical characterization of grain, as well as the morphological information of the seed, will continue to be the driving force for advancing spectral imaging grain research. The eventual reduction in the price of hyperspectral hardware and an increase in seed throughput capacity will lead to its commercial adoption.

REFERENCES

Bauriegel, E., Giebel, A., Geyer, M., Schmidt, U. and Herppich, W.B. 2011. Early detection of Fusarium infection in wheat using hyper-spectral imaging, computers electronic. *Agric.* 75:304–312.

Burdette, A., Garner, P.L., Mayer, E.P., Hargrove, J.L., Hartle, D.K. and Greenspan, P., 2010. Anti-inflammatory activity of select sorghum (Sorghum bicolor) brans. *J. Med. Food* 13:879–887.

Caporaso, N., Whitworth, M.B. and Fisk, I.D. 2018. Protein content prediction in single wheat kernels using hyperspectral imaging. *Food Chem.* 240:32–42.

Dale, L.M., Thewis, A., Boudry, C., Rotar, I., Dardenne, P., Beaten, V. and Pierna, J.A.F. 2013. Hyperspectral imaging applications in agriculture and agro-food product quality and safety control: A review. *Appl. Spectrosc. Rev.* 48:142–159.

Delwiche, S.R. and Graybosch, R.A. 2014. Measurement of blend concentrations of conventional and waxy hard wheats using NIR spectroscopy. *Cereal Chem.* 91:358–365.

Delwiche, S.R., McKenzie, K.S. and Webb, B.D. 1996. Quality characteristics in rice by near infrared reflectance analysis of whole grain milled samples. *Cereal Chem.* 73:257–263.

Delwiche, S.R., Souza, E.J. and Kim, M.S. 2013. Limitations of single kernel near-infrared hyperspectral imaging of soft wheat for milling quality. *Biosyst. Eng.* 115:260–273.

De Palma, G., Nadal, I., Collado, M.C. and Sanz, Y. 2009. Effects of a gluten-free diet on gut microbiota and immune function in healthy adult human subjects. *Br. J. Nutrit.* 102:1154–1160.

Dykes, L., Hoffmann, L., Portillo-Rodriguez, O., Rooney, W.L. and Rooney, L.W. 2014. Prediction of total phenols, condensed tannins, and 3-deoxyanthocyanidins in sorghum grain using near-infrared (NIR) spectroscopy. *J. Cereal Sci.* 60:138–142.

Elmasry, G., Kamruzzaman, M., Sun, D.W. and Allen, P. 2012. Principles and applications of hyperspectral imaging in quality evaluation of agro-food products: A review. *Crit. Rev. Food Sci. Nutrit.* 52:999–1023.

Esteve Agelet, L., Gowen, A.A., Hurburgh, C.R. and O'Donell, C.P. 2012a. Feasibility of conventional and Roundup Ready® soybeans discrimination by different near infrared reflectance technologies. *Food Chem.* 134:1165–1172.

Esteve Agelet, L., Armstrong, P.R., Romagosa Clariana, I. and Hurburgh, C.R. 2012b. Measurement of single soybean seed attributes by near-infrared technologies. A comparative study. *J. Agric. Food Chem.* 60:8314–8322.

Esteve Agelet, L., Armstrong, P.R., Tallada, J.G. and Hurburgh, C.R. 2013. Differences between conventional and glyphosate tolerant soybeans and moisture effect in their discrimination by near infrared spectroscopy. *Food Chem.* 141:1895–1901.

Esteve Agelet, L. and Hurburgh, C.R. 2014. Limitations and current applications of near infrared spectroscopy for single seed analysis. *Talanta*, 121:288–299.

European Union. 2015. Commission Regulation (EU) No 2015/1940 of 28 October 2015 amending Regulation (EC) 1881/2006 as regards maximum levels of ergots sclerotia in certain unprocessed cereals and the provisions on monitoring and reporting. OJ L 283, 29.10.2015, pp. 3–5.

Feng, Y.Z. and Sun, D.W. 2012. Application of hyperspectral imaging in food safety inspection and control: A review. *Crit. Rev. Food Sci. Nutr.* 52:1039–1058.

Firrao, G., Torelli, E., Gobbi, E., Raranciuc, S., Bianchi, G. and Locci, R. 2010. Prediction of milled maize fumonisin contamination by multispectral image analysis. *J. Cereal Sci.* 52:327–330.

Fox, G. and Manley, M. 2014. Applications of single kernel conventional and hyperspectral imaging near infrared spectroscopy in cereals. *J. Sci. Food Agric.* 94:174–179.

Fox, G.P., O'Donnell, N.H., Stewart, P.N. and Gleadow, R.M. 2012. Estimating hydrogen cyanide in forage sorghum (Sorghum bicolor) by near-infrared spectroscopy. *J. Agric. Food Chem.* 60:6183–6187.

García-Molina, M.D., García-Olmo, J. and Barro, F. 2016. Effective identification of low-gliadin wheat lines by near infrared spectroscopy (NIRS): Implications for the development and analysis of foodstuffs suitable for celiac patients. *PloS one* 11(3): e0152292.

Geladi, P. and Dåbakk, E. 1995. An overview of chemometrics applications in near infrared spectrometry. *J. Near Infrared Spectrosc.*, 3:119–132.

Graybosch, R.A. 1998. Waxy wheats: Origin, properties, and prospects. *Trends Food Sci. Technol.* 9:135–142.

Gy, P., 1998. *Sampling for Analytical Purposes.* John Wiley & Sons, New York.

Hart, J.R., Norris, K.H. and Golumbic, C. 1962. Determination of the moisture content of seeds by near-infrared spectrophotometry of their methanol extracts. *Cereal Chem.* 39:94–99.

Hayes, B.J., Panozzo, J., Walker, C.K., Choy, A.L., Kant, S., Wong, D., Tibbits, J., Daetwyler, H.D., Rochfort, S., Hayden, M.J. and Spangenberg, G.C. 2017. Accelerating wheat breeding for end-use quality with multi-trait genomic predictions incorporating near infrared and nuclear magnetic resonance-derived phenotypes. *Theo. Appl. Gen.* 130:2505–2519.

Hódsági, M., Gergely, S., Gelencsér, T. and Salgó, A. 2012. Investigations of native and resistant starches and their mixtures using near-infrared spectroscopy. *Food Bioprocess Technol.* 5:401–407.

Long, D.S., McCallum, J.D. and Scharf, P.A. 2013. Optical-mechanical system for on-combine segregation of wheat by grain protein concentration. *Agron. J.* 105:1529–1535.

Long, D.S. and McCallum, J.D., 2015. On-combine, multi-sensor data collection for post-harvest assessment of environmental stress in wheat. *Precis. Agric.* 16:492–504.

Manley, M. 2014. Near-infrared spectroscopy and hyperspectral imaging: Non-destructive analysis of biological materials. *Chem. Soc. Rev.* 43:8200–8214.

Manley, M., Williams, P., Nilsson, D. and Geladi, P., 2009. Near infrared hyperspectral imaging for the evaluation of endosperm texture in whole yellow maize (Zea maize L.) kernels. *J. Agric. Food Chem.* 57:8761–8769.

Manley, M., McGoverin, C.M., Engelbrecht, P. and Geladi, P., 2012. Influence of grain topography on near infrared hyperspectral images. *Talanta* 89:223–230.

McGoverin, C.M., Engelbrecht, P., Geladi, P. and Manley, M., 2011. Characterisation of non-viable whole barley, wheat and sorghum grains using near-infrared hyperspectral data and chemometrics. *Anal. Bioanalytic. Chem.* 401:2283.

Montes, J.M., Utz, H.F., Schipprack, W., Kusterer, B., Muminovic, J., Paul, C. and Melchinger, A.E., 2006. Near-infrared spectroscopy on combine harvesters to measure maize grain dry matter content and quality parameters. *Plant Breed.* 125:591–595.

Orina, I., Manley, M. and Williams, P.J., 2017. Non-destructive techniques for the detection of fungal infection in cereal grains. *Food Res. Int.* 100:74–86.

Ringsted, T., Dupont, S., Ramsay, J., Jespersen, B.M., Sørensen, K.M., Keiding, S.R. and Engelsen, S.B., 2016. Near-infrared spectroscopy using a supercontinuum laser: Application to long wavelength transmission spectra of barley endosperm and oil. *Appl. Spectrosc.* 70:1176–1185.
Sendin, K., Williams, P.J. and Manley, M. 2018. Near infrared hyperspectral imaging in quality and safety evaluation of cereals. *Crit. Rev. Food Sci. Nutr.* 58:575–590.
Senthilkumar, T., Jayas, D.S. and White, N.D.G. 2015. Detection of different stages of fungal infection in stored canola using near-infrared hyperspectral imaging. *J. Stored Prod. Res.* 63:80–88.
Senthilkumar, T., Jayas, D.S., White, N.D., Fields, P.G. and Gräfenhan, T. 2016a. Detection of fungal infection and Ochratoxin A contamination in stored barley using near-infrared hyperspectral imaging. *Biosys. Eng.* 147:162–173.
Senthilkumar, T., Jayas, D.S., White, N.D.G., Fields, P.G. and Gräfenhan, T. 2016b. Detection of fungal infection and Ochratoxin A contamination in stored wheat using near-infrared hyperspectral imaging. *J. Stored Prod. Res.* 65:30–39.
Tønning, E., Nørgaard, L., Engelsen, S.B., Pedersen, L. and Esbensen, K.H. 2006. Protein heterogeneity in wheat lots using single-seed NIT—A theory of sampling (TOS) breakdown of all sampling and analytical errors. *Chemom. Intel. Lab. Sys.* 84:142–152.
USDA-GIPSA 2006. Near-infrared transmittance handbook (NIRT). USDA-FGIS, Washington, D.C. (available from: http://www.gipsa.usda.gov/fgis/public_handbooks.aspx).
Vermeulen, P., Pierna, J.F., Egmond, H.V., Dardenne, P. and Baeten, V. 2012. Online detection and quantification of ergot bodies in cereals using near infrared hyperspectral imaging. *Food Addit. Contaminan. Part A* 29:232–240.
Vermeulen, P., Pierna, J.F., Van Egmond, H.P., Zegers, J., Dardenne, P. and Baeten, V. 2013. Validation and transferability study of a method based on near-infrared hyperspectral imaging for the detection and quantification of ergot bodies in cereals. *Anal. Bioanal. Chem.* 405:7765–7772.
Wallays, C., Missotten, B., De Baerdemaeker, J. and Saeys, W. 2009. Hyperspectral waveband selection for on-line measurement of grain cleanness. *Biosys. Eng.* 104:1–7.
Williams, P.C. 1979. Screening wheat for protein and hardness by near infrared reflectance spectroscopy. *Cereal Chem.* 56:169–172.
Williams, P.J., Geladi, P., Britz, T.J. and Manley, M. 2012. Near-infrared (NIR) hyperspectral imaging and multivariate image analysis to study growth characteristics and differences between species and strains of members of the genus Fusarium. *Anal. Bioanal. Chem.* 404:1759–1769.
Wu, J.G., Shi, C.H. and Zhang, X.M. 2002. Estimating the amino acid composition in the milled rice powder by near-infrared reflectance spectroscopy. *Field Crops Res.* 75:1–7.
Zhang, B., Rong, Z.Q., Shi, Y., Wu, J.G. and Shi, C.H., 2011. Prediction of the amino acid composition in brown rice using different sample status by near-infrared reflectance spectroscopy. *Food Chem.* 127:275–281.

30 Innovation in the Meat Industry Using Novel Near-Infrared Spectroscopy Sensors

Ana Garrido-Varo[*]*, Dolores Pérez-Marín, and Emiliano de Pedro*
ETSIAM, University of Córdoba

CONTENTS

30.1 INTRODUCTION

There are numerous papers in the literature discussing the use of NIRS for the analysis of meat products. There are also several recent reviews of the topic [1–4]. Rather than trying to set out an exhaustive review of the literature on the subject, this chapter therefore focuses on presenting information taken from the authors' research experience, acquired during the long process of developing robust models for meat analysis and for moving from at-line analysis to on-site analysis using novel handheld instrumentation.

The chapter is structured into two parts: the first sets out the methodology followed by the authors' group in order to obtain robust equations for analyzing the meat pieces produced under pig breeding programs and the second sets out how to transfer databases from at-line instruments to portable devices. Although all the findings presented herein refer to research carried out on Iberian pig muscles, the methodologies developed are all applicable to other animal breeds and species.

30.2 MONITORING NIRS CALIBRATIONS FOR USE IN ROUTINE MEAT ANALYSIS AS PART OF IBERIAN PIG-BREEDING PROGRAMS

One of the most painstaking aspects of using NIRS in laboratory and industrial contexts relates to maintaining the equations developed at a certain date such that they remain useful and applicable for future samples as time unfolds. This is not always easy, because various factors may change over the course of time, with a significant impact on the precision and accuracy of the NIRS models.

[*] palgavaa@uco.es

A breeding program for Iberian pigs, based on farm records, was designed by the Spanish Association of Iberian Pig Breeders (AECERIBER) in 1993, with one objective among others to select animals based on their fat quality [5–7]. The breeding program involves analyzing the meat of the large number of animals that join the program every year [8]. The economic and time investment required to obtain measurements of fat and protein by traditional wet chemistry were clearly inappropriate and beyond the resources of the Iberian pig-breeding program. This accounts for the fact that in 1999 research started into developing NIRS calibrations for the prediction of the main quality traits (fat, moisture and protein contents). Then, along several years, spectra of 325 meat samples were collected grounded using a FOSS NIR Systems 6500 (FNS) instrument (equipped with a spinning module) and analyzed by the reference method. Later on, calibrations were developed. It was decided to set a methodology that would enable a relatively simple annual recalibration involving minimal analytical cost. The samples, analysis and treatment of NIRS data were set out in greater detail in Zamora et al. [9]. Table 30.1 shows the number of samples and spectra accumulated from 1999 to 2009.

After spectral outliers had been removed, a set consisting of $N=310$ samples of different muscles (*gluteus medius*, *masseter* and *longissimus dorsi*) was used to develop the calibration models with samples from five consecutive years: 1999–2004. The set was divided into two: a training set containing about 70% of the samples ($N=213$) and a validation/test set containing the remaining 30% ($N=97$) (Table 30.2).

Tables 30.3 and 30.4 show the statistics for the calibration and test sets, respectively. Fat showed SEP (standard error of prediction) values ranging from 0.57% to 0.58%, which were higher than the SECV (standard error of cross-validation). Moisture showed an increment of 18.64% for the SEP with respect to the SECV. It was observed that the regression line between the reference and predicted values had a slope of around 1; however, the difference between the reference and predicted values for this parameter could be due to the storage conditions and freezing/thawing of the samples between the NIRS analysis and the reference analysis. Fresh meat is a highly moisture-sensitive product, and all the operational and storage factors affect this parameter [10]. Protein had

TABLE 30.1

Distribution of the Samples Used by Muscle Type and Campaign of Analysis

Muscle Type	1999	2000	2001	2003	2004	2009	Total
Gluteus medius	29	–	–	–	–		29
Masseter	–	–	–	96	–		96
Longissimus dorsi	36	25	87	28	24	514	714
Spinalis dorsi						464	464
Total	65	25	87	124	24	978	1303

TABLE 30.2

Descriptive Statistics for Training and Test Sets (Values Given in % of Wet Weight)

	Training Set (213 samples)				Test Set (97 samples)			
Parameter	Minimum	Maximum	Average	Standard Deviation	Minimum	Maximum	Average	Standard Deviation
Fat (%)	2.10	21.90	6.98	3.41	2.30	18.30	6.68	3.02
Moisture (%)	56.00	77.00	70.15	3.19	63.00	75.30	70.64	2.88
Protein (%)	16.80	26.20	21.52	1.74	16.90	24.70	21.39	1.56

TABLE 30.3
Calibration Statistics ($N=213$) for the Prediction Models Developed for Fat, Moisture and Protein

Parameter	Data Preprocessing	N° Factors	SEC	SECV	R^2_{CV}	RPD_{CV}
Fat (%)	SNV+DT (1,10,5,1)	7	0.32	0.34	0.98	10.03
	SNV+DT (2,5,5,1)	5	0.31	0.35	0.98	10.03
Moisture (%)	SNV+DT (1,10,5,1)	7	0.43	0.48	0.96	6.64
	SNV+DT (2,5,5,1)	5	0.44	0.49	0.96	6.51
Protein (%)	SNV+DT (1,10,5,1)	6	0.49	0.55	0.89	3.16
	SNV+DT (2,5,5,1)	6	0.46	0.55	0.90	3.16

SNV +DT, standard normal variate + detrending.

TABLE 30.4
External Validation Statistics of the Chemometric Models ($N=213$) Developed for Fat, Moisture and Protein Using the Training Set ($N=97$).

Parameter	Data Preprocessing	Bias	Slope	SEP	SEP(c)	RPD_V	GH	NH	$Mean_{NIRS}$	SD_{NIRS}
Fat (%)	SNV+DT (1,10,5,1)	0.046	0.97	0.58	0.58	5.88	0.92	0.56	6.65	3.13
	SNV+DT (2,5,5,1)	0.039	0.96	0.57	0.57	5.98	0.92	0.56	6.66	3.16
Moisture (%)	SNV+DT (1,10,5,1)	−0.105	1.04	0.59	0.58	5.40	0.92	0.56	70.88	2.63
	SNV+DT (2,5,5,1)	−0.109	1.05	0.60	0.59	5.32	0.92	0.56	70.88	2.58
Protein (%)	SNV+DT (1,10,5,1)	−0.098	0.93	0.59	0.58	2.95	0.92	0.56	21.44	1.57
	SNV+DT (2,5,5,1)	−0.093	0.93	0.60	0.59	2.90	0.92	0.56	21.43	1.57

a low RPD (ratio of performance to the deviation), which could be explained by the small SD (standard deviation) and range of the reference set (Table 30.2). The slope of the regression line between the reference and NIRS prediction values was also lower compared to the other parameters. The SEP(c) (standard error of prediction bias corrected) values were very similar to the SEP in all the parameters, showing that no systematic error was present in the measurements.

As expected, there was an increment of the SEP compared to the SECV values, which shows the need to obtain more robust models, although the RPD_v (for the prediction set) indicated that the calibrations were acceptable for fat and moisture. The importance of developing models that encompass the maximum variability of the product being analyzed, especially for a very heterogeneous product such as meat, and the importance of developing more accurate models to predict the chemical composition led to a decision to join both sets (calibration and test) together and develop new models. Table 30.5 shows the statistics related to the calibration models developed with the full set ($N=310$ samples).

The findings were very similar to those shown in Table 30.3, obtained using a relatively reduced number of samples. For each parameter, the number of PLS factors, R^2_{CV} and SECV were in the same range. Now the models have a greater number of samples and should be valid for samples within the ranges of approximately 2.10%–21.90% for fat, 56.00%–77.70% for water and 16.80%–26.20% for protein. The models can be considered useful according to the RPD_{CV} although protein showed the worst predictive capability.

Once the final calibrations were obtained, they were applied to a set of 514 *longissimus dorsi* samples derived from 1 year of the breeding program (2009). The 514 samples were all completely unknown (no lab data). Therefore, spectral outlier detection methods needed to be used in order to

TABLE 30.5
Statistics of the Chemometric Models Developed for Fat, Moisture and Protein with the Full Set (*N*=310 samples)

Parameter	Data Preprocessing	Average	Standard Deviation	Nº Factors	SEC	SECV	R^2_{CV}	RPD_{CV}
Fat (%)	SNV+DT (1,10,5,1)	6.89	3.31	7	0.33	0.36	0.98	9.19
Moisture (%)	SNV+DT (2,5,5,1)	70.35	3.08	7	0.37	0.42	0.97	7.33
Protein (%)	SNV+DT (1,10,5,1)	21.46	1.68	6	0.47	0.50	0.91	3.36

ascertain which of the unknown samples had to be sent to the wet chemistry lab and could be used for a future expansion of the existing training set. To evaluate the model's performance over time and recalibrate the models, a critical step is to choose which and how many samples have to be sent to the laboratory for wet chemistry analysis. A robust model is usually developed to analyze all reasonable samples of a given product, even from different years. These models require sample libraries that represent, spectrally and chemically, the variability of the constituents studied, and moreover, the set must cover an entire constituent range uniformly [11]. We therefore proposed the following criteria: use of the GH (global Mahalanobis distance) statistic to detect new spectral variability and the NH (neighbor Mahalanobis distance) statistic together with the initial NIR prediction for each constituent to detect new chemically interesting samples to be analyzed by wet chemistry and ensure a uniform range of each parameter in the libraries.

Furthermore, the number and type of samples to send for reference analysis are a compromise between the time and money available for that analysis. To reflect this, and on the basis of our previous experience in developing robust calibrations, we selected a set of 32 samples from the 2009 samples to be sent to the laboratory in accordance with the following criteria: two samples comprising spectral outliers with high GH and samples with high NH, values >1.0, and with a prediction value for one or more parameters not well represented in the models (30 samples).

Table 30.6 shows the recalibrated models including the 32 *longissimus dorsi* samples from 2009. The statistics indicated similar accuracy of the models in terms of cross-validation errors when compared to the previous models; however, more samples were included in an effort to achieve a balanced distribution of training samples throughout the entire range for a parameter. The RPD_{CV} for the recalibrated models (Table 30.6) showed an improvement in the accuracy of the fat models and similar values for moisture and protein contents compared to the nonrecalibrated models (Table 30.3).

TABLE 30.6
Statistics of the Chemometric Models Developed for Fat, Moisture and Protein with the Full Set Plus the 32 Samples Selected from 2009 (342 Samples)

Parameter	Data Preprocessing	Nº Factors	SEC	SECV	R^2_{CV}	RPD_{CV}
Fat (%)	SNV+DT (1,10,5,1)	6	0.36	0.37	0.98	9.37
	SNV+DT (2,5,5,1)	*4*	*0.37*	*0.40*	*0.98*	*8.675*
Moisture (%)	*SNV+DT (1,10,5,1)*	*5*	*0.48*	*0.50*	*0.96*	*6.72*
	SNV+DT (2,5,5,1)	5	0.47	0.51	0.96	6.58
Protein (%)	*SNV+DT (1,10,5,1)*	*7*	*0.48*	*0.49*	*0.91*	*3.38*
	SNV+DT (2,5,5,1)	4	0.57	0.58	0.87	2.86

Selected models are marked in italic.

A set comprising 464 *spinalis dorsi* samples from 2009 was used to compare the performance of the best initial models developed with the set of $N=310$ samples and the best recalibrated models ($N=310+32$ samples) for each parameter with samples of a muscle type (*spinalis dorsi*) not previously used in the training set (Figure 30.1). The SED (standard error of the differences) within both strategies were 0.12%, 0.07% and 0.09% for fat, moisture and protein contents, respectively. So there was no notable difference between the recalibrated models and the initial ones in terms of prediction values. However, it is worth making some comments about the GH and NH statistics when comparing the recalibrated and nonrecalibrated models. It was found that only two samples had GH values higher than 3, but the 78.59% of the 514 unknown samples had NH values above the limit (1.0), which indicates a low degree of similarity to the training samples. This means that although these *longissimus dorsi* samples were spectrally not too far from the center of the calibration population ($N=310$), they did not have enough neighboring samples. It should be remembered

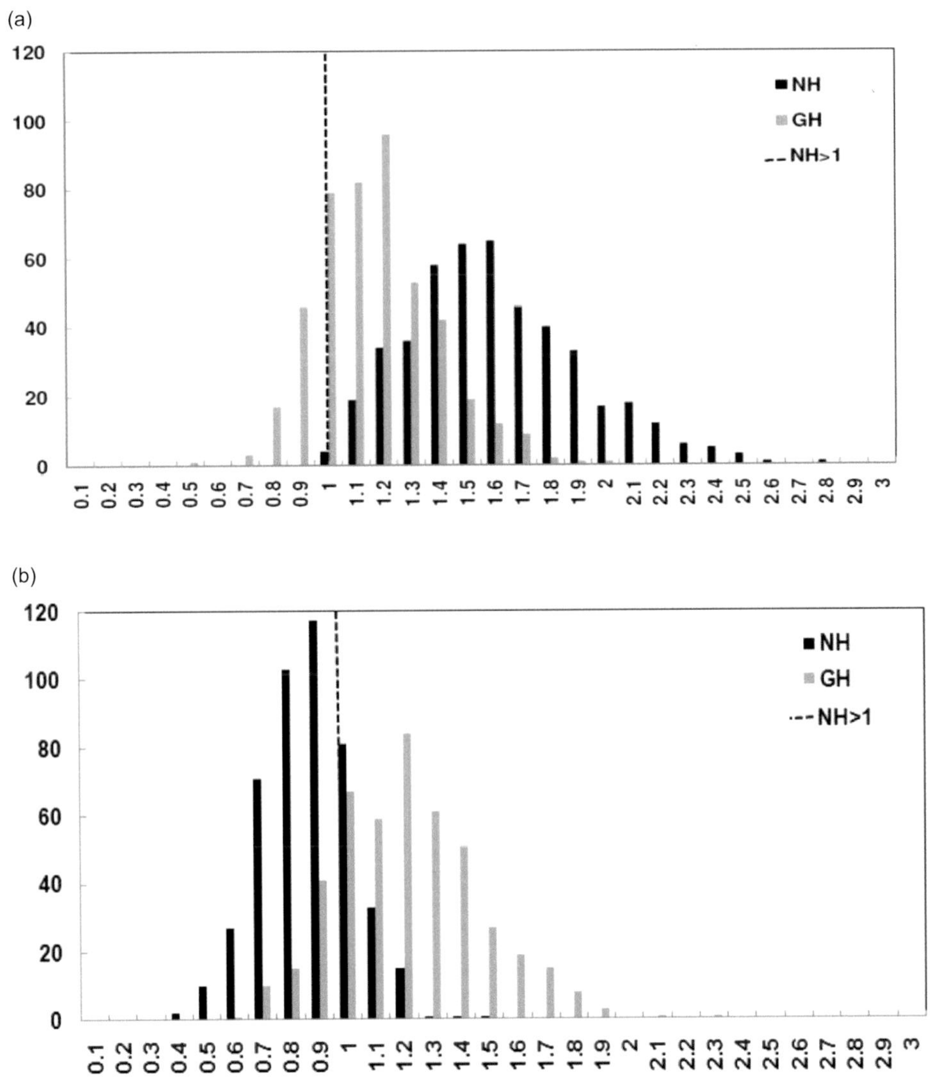

FIGURE 30.1 GH and NH distributions of the test set comprising 464 spinalis dorsi samples predicted with initial models (a) and with the recalibrated models (b).

that these samples were analyzed 4 years later, and some changes may have occurred in the instrumentation, sample preparation, environmental conditions in the NIRS laboratory, etc. In this case, recalibration was needed.

Comparing the initial and recalibrated models for the prediction of *spinalis dorsi* samples (Figures 30.1a and b), a reduction of the individual GH values for the 464 *spinalis dorsi* samples from 2009 was observed. It seemed that the conditions of when the analyses were performed (instrument status, operator, etc.) had an influence on the NIR spectra and the inclusion of some new and current samples in the routine analysis could help to detect outliers and explain the poor performance of the models previously developed. As far as the NH values were concerned, the *spinalis dorsi* samples showed an average value above the limit for this statistic, probably due to the spectral differences within muscle types because of not being included in the training set. When the predictions were performed with the recalibrated models, the NH values for the 464 *spinalis dorsi* samples were lower with 88.8% of the samples below the limit 1.0 (Figure 30.1b). So it can be said that the inclusion of new samples in routine models can help adjust them to the new performance of the instrument, operator, room conditions, etc. (changes that can occur even following the same sample preparation protocols) and make an efficient use of the information provided by the GH and NH statistics.

The majority of papers in this field used between 30 and 150 samples for calibration purposes and limited test sets. Here, a large data set showing a wide range in reference values was used, and more than 90 test samples for accuracy evaluation were used in the first step, reflected in the realistic robustness of the calibration models. Furthermore, we have shown how the GH and NH statistics can be used to interpret the prediction results of new samples in routine analysis and how a recalibration process including new samples to increase the variability present in the training set can improve the reliability of the predictions.

30.3 HANDHELD NIRS ANALYSIS FOR ROUTINE MEAT QUALITY CONTROL: DATABASE TRANSFER FROM AT-LINE INSTRUMENTS

Innovative advances in near-infrared spectroscopic instrumentation have enabled the development of new miniaturized spectrometers that combine NIRS technology with micro-electro-mechanical platforms, thus opening up new horizons for industrial NIRS applications.

Miniature Phazir (MPs) handheld near-infrared spectrometers are powerful instruments that are based on a combination of micro-electro-mechanical (MEMS) technology and digital transform spectroscopy (DTS), although there are other technologies that may be incorporated into this kind of device. They have several advantages over traditional instrument designs, especially for nondestructive on-line or in situ analysis. The MEMS silicon chip is a fixed, rapidly tunable, high-contrast pixelated optical filter or diffraction grating for wavelength selection. Used in conjunction with a single detector, it enables equipment costs to be lowered while ensuring the elimination of detector noise and providing a robust design with nonmoving parts [12]. A number of recent papers have evaluated this technology in the agro-food area [13–16].

The many databases constructed to date for at-line analysis of meat using nonportable instruments have required considerable investments in terms of time (years), resources, labor and analytical costs. Moreover, they yield specific samples, which are critical for the construction of robust models since they represent the variability encountered over a long period. It is clearly crucial to preserve these data sets in order to avoid having to research and develop NIRS applications from scratch every time a new instrument appears on the market. However, calibration models obtained using one instrument cannot yet be readily used on a different instrument, even where the optical and electronic designs are the same, since they do not produce exactly the same response for the same sample being analyzed [17], due to a number of constraints such as wavelength accuracy, dynamic range, distance of samples to the optics, detector photometric response and functionality changes with temperature [18].

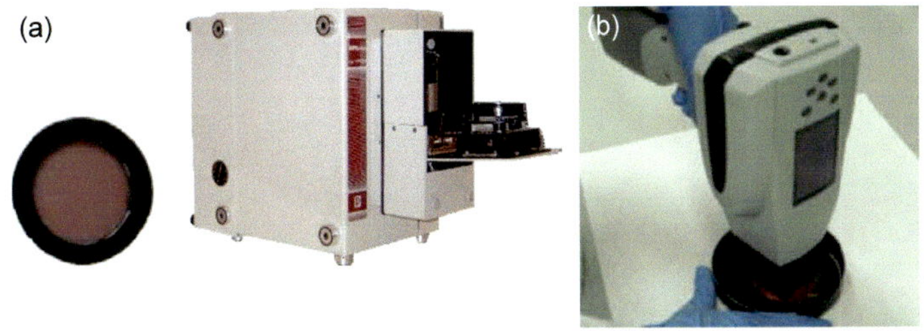

FIGURE 30.2 (a) Foss NIRSystems 6500 and spinning cup, 400–2500 nm, 2 nm (b) PhazirTM, 160–2400 nm, 8 nm.

Instead of transferring prediction models generated on the reference instrument—the procedure habitually used in earlier studies [18]—spectral databases generated on the FNS were transferred to the MP, with an objective to subsequently develop various prediction models on the portable instrument (Figure 30.2). This simplifies the routine work with the portable equipment, because it is not necessary to apply any cloning matrix when spectra are collected with the MP instrument.

A dataset of 342 Iberian pig muscle samples (the so-called recalibrated set) recorded using the FNS was transferred to the spectrometer. A separate set of 235 samples of ground Iberian pork muscle (*longissimus dorsi*) collected in 2010 was used for standardization and evaluation. This dataset was split—using the SELECT algorithm of Shenk and Westerhaus [19] to choose the most representative samples based on a Mahalanobis distance analysis—into a standardization set (30 samples, set S) and a validation test set (205 samples, set V_{spec}) to evaluate the database transfer spectrally. Sixty samples from the 205-sample validation set (setV_{spec}), also selected as the most representative using the SELECT procedure, were analyzed by wet chemistry for protein, fat and moisture contents. These 60 samples, henceforth referred to as setV_{pre}, were used for validation of the transferred calibration models. Further information about samples and NIR analysis can be found in Zamora-Rojas et al. [15].

The aim of our work was to obtain calibration models for the prediction of meat chemical composition (fat, moisture and protein contents) on the handheld MP that would be more affordable for use in R&D breeding programs, without significant loss of performance compared to NIRS monochromators.

The first point to highlight is the differences in spectral range and resolution between the two instruments (Figure 30.3). The FNS and the MP spectrometers differ in wavelength and spectral range; therefore, an interpolation based on a shape-preserving piecewise cubic function, which restores convexity to scattered convex data, was used to trim the NIRS scanning monochromator range to the 100 wavelengths of the device [13].

PLS models were developed with the FNS using the set of 342 samples collected from 1999 to 2004, considering the common wavelength range with the MP instrument (1600–2400 nm) and also the same wavelength interval (a nonconstant step of 8 nm). Table 30.7 shows the different models developed.

Similar accuracy and performance for all the chemical parameters were obtained when the wavelength range was trimmed to only 1600–2400 nm compared to the model (1100–2500 nm). Then, when the 400 wavelengths of the FNS in the 1600–2400 nm range were interpolated to the 100 wavelengths in the MEMS-NIRS device, the models developed on the NIRS scanning monochromator showed an increment in the number of PLS factors, except in the case of protein, and a reduction in the accuracy. The SEP values were 0.5% for fat, 0.63% for moisture and 0.55% for protein using measurements performed on the FNS instrument with the trimmed and interpolated PLS models. This represents an increment of 0.06% and 0.25% in the SEP values of the V_{pre} set compared to the reference models (1100–2500 and 2 nm wavelength interval) for fat and moisture, respectively. This indicates that the spectral range and wavelength step can have an effect on the model's

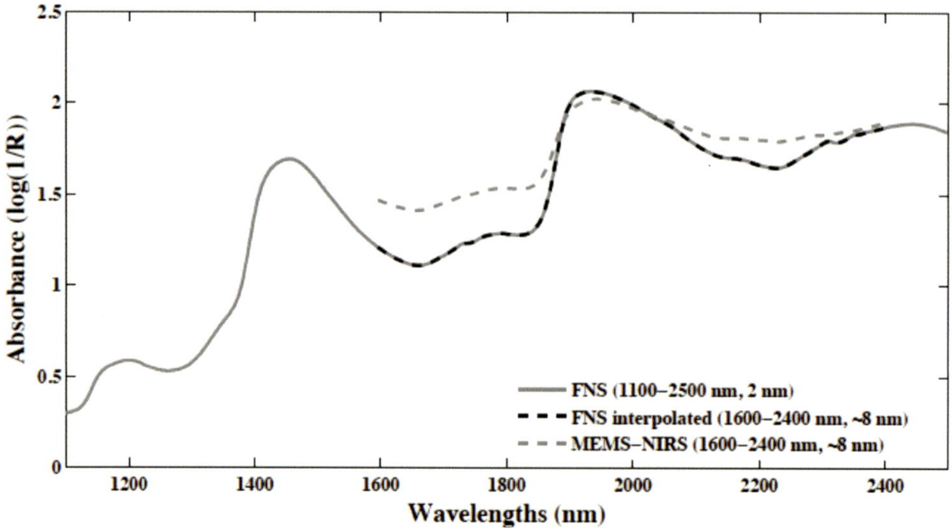

FIGURE 30.3 Mean spectra of the FNS (2 and 8 nm) and MP instruments (8 nm).

TABLE 30.7

PLS Model Statistics Developed for Fat, Moisture and Protein Using FNS Instrument

Parameter	Range (nm)/ Wavelength Interval (nm)	Mean (%)	SD (%)	No. PLS factors	SEC (%)	SECV (%)	R^2_{CV}	RPD_{CV}	SEP (%)	GH	NH
Fat	[a]1100–2500/2	7.03	3.51	6	0.34	0.35	0.99	10.02	0.44[b]	0.80[b]	0.46[b]
	1600–2400/2	6.82	3.30	6	0.32	0.33	0.99	10	0.38[b]	1.07[b]	0.31[b]
	1600–2400/~8	6.71	3.21	12	0.37	0.38	0.99	8.44	0.50[b]	1.14[b]	0.27[b]
									4.34[c]	30.91[c]	23.00[c]
Moisture	[a]1100–2500/2	70.07	3.35	7	0.44	0.46	0.97	7.28	0.38[b]	0.80[b]	0.46[b]
	1600–2400/2	70.51	2.71	5	0.44	0.45	0.97	6.02	0.48[b]	1.07[b]	0.31[b]
	1600–2400/~8	70.51	2.72	12	0.43	0.45	0.97	6.04	0.63[b]	1.14[b]	0.27[b]
									0.63[c]	30.91[c]	23.00[c]
Protein	[a]1100–2500/2	21.52	1.68	7	0.49	0.52	0.90	3.23	0.76[b]	0.80[b]	0.46[b]
	1600–2400/2	21.49	1.67	7	0.49	0.52	0.90	3.21	0.68[b]	1.07[b]	0.31[b]
	1600–2400/~8	21.51	1.67	5	0.61	0.62	0.86	2.69	0.55[b]	1.14[b]	0.27[b]
									2.57[c]	30.91[c]	23.00[c]

All models were pretreated with SNV+DT (1,10,5,1).

SD, standard deviation; SEC, standard error of calibration; SECV, standard error of cross-validation; R^2_{CV}, determination coefficient of cross-validation; RPD, residual predictive deviation; SEP, standard error of prediction, GH, global Mahalanobis distance; NH, neighbor Mahalanobis distance (SEP, GH and NH values were calculated with validation set V of 60 samples).

[a] Published by Zamora-Rojas et al. (2011).

[b] Spectra measured on FNS instrument.

[c] Spectra measured on MEMS-NIRS device.

predictive capacity. However, these trimmed and interpolated models developed in the FNS spectrometer were not suitable for spectra measured on the MEMS-NIRS instrument. The SEP values obtained were 4.34% for fat, 0.63% for moisture and 2.57% for protein, indicating that the moisture parameter may not require standardization, while this was not the case for the other parameters.

Nevertheless, the global Mahalanobis distance (GH) and neighbor Mahalanobis distance (NH) indicated very pronounced spectral differences within the FNS database and the MEMS-NIRS spectra measurements, with values of GH=30.91 and NH=23.00, respectively. This supported the need for standardization between instruments.

When investigating highly moisture-sensitive products such as meat, most authors recommend the use of real samples rather than generic standards to perform the standardization, since the latter cannot represent the entire experimental domain or the range of spectral patterns encountered in these products. Fearn and Feudale [20] have reviewed various mathematical algorithms and methodologies, known as calibration transfer or standardization strategies, to transform models or data measured on a given device so that they can be used by a different device.

In order to determine the best spectral adjustment method, the following factors were assessed:

- Three standardization algorithms—Direct standardization (DS), piecewise direct standardization (PDS) and spectral difference by wavelengths (SDWs)
- The use of different numbers of standardization samples (1, 2, 3, 4, 5, 6, 7, 8, 9, 10, 20 and 30)

The results indicated large differences in the spectra between FNS and MEMS-NIRS (Figure 30.4); however, these differences were reduced using one of the standardization algorithms with the optimized number of samples (SDW with five standardization samples, DS with four standardization samples or PDS with ten standardization samples). Although these results indicated a suitable performance of the standardization, reducing spectral differences between instruments for the validation-test set V_{spe} (Figure 30.4), when the different algorithms were applied to the large database (N=342) measured on the FNS instrument over the course of several years for the calibration development, each strategy had a different performance. Figure 30.5 shows the average spectra of the database (N=342) before (trimmed and interpolated FNS) and after the application of the different standardization strategies. While the SDW strategy yielded a good performance, the DS method showed a noisy tendency, especially at large wavelengths, and the PDS algorithm showed jumps at certain wavelengths, modifying the shape of the spectra and therefore yielding an unsatisfactory performance of the calibration transfer.

These visual results were corroborated by the fact that the calibration models developed using the PDS transfer matrices worked well to transfer the 205 samples of the V_{spe} set measured at the

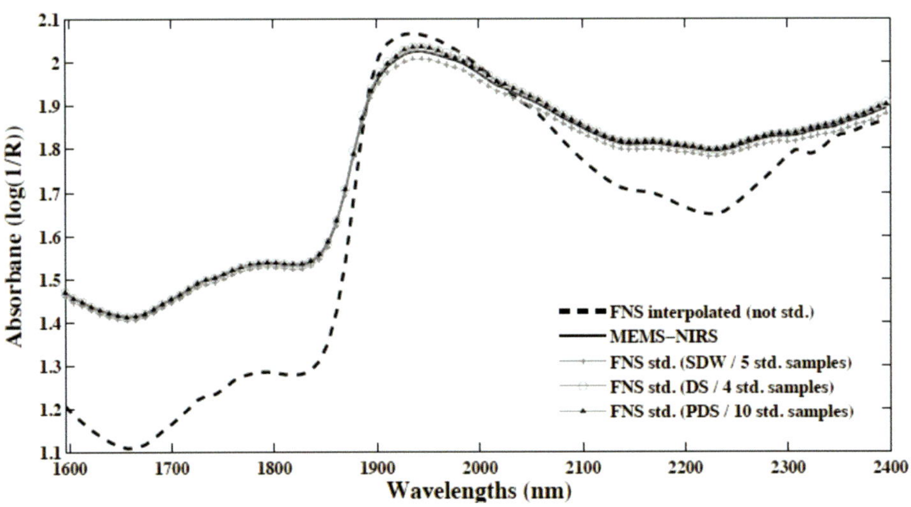

FIGURE 30.4 Average spectra of the N=*205 samples collected in 2010* (trimmed and interpolated FNS) and after the application of the different standardization strategies.

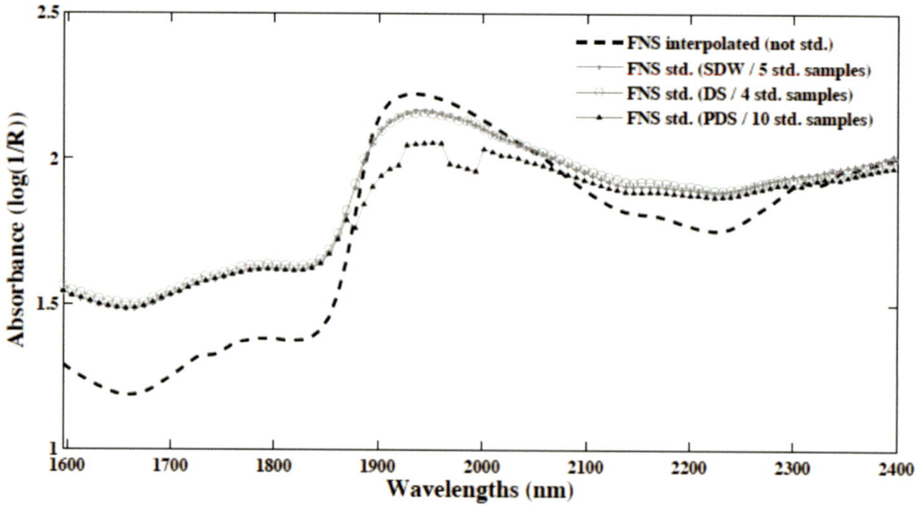

FIGURE 30.5 Average spectra of the $N=342$ samples collected from 1999, 2000 and 2001 (trimmed and interpolated FNS) and after the application of the different standardization strategies.

same time in both instruments (Figure 30.4), but the method did not work when samples measured in the NIRS monochromator in different years from the standardization moment were transferred (Figure 30.5). This could be attributable to the fact that samples measured in the FNS at different periods (years) underwent spectral changes due to instrumentation or highly moisture-sensitive sample conditions, etc. [9], which can affect the performance of the meat database standardization. This was observed in the standardization of the database ($N=342$ samples) from the FNS instrument to the MEMS-NIRS spectrometer, because samples collected in 2009—that is, a year not available to the collection of the standardization set—were transferred correctly, while this was not the case for samples collected in 1999, 2000 or 2001 (Figure 30.5).

The direct standardization (DS) algorithm showed a noisy tendency when it was used on the calibration set ($N=342$ samples from various years). The noisy tendency was found to be more severe with samples collected in years further from the standardization date. The reason for this may be the same as for the PDS results. Table 30.7 shows the transferred models when the DS method was used with the matrix calculated using four standardization samples, which exhibited the best spectral matching for both instruments. The results reveal that the models developed are not suitable for screening purposes in meat analysis, especially for moisture or protein prediction with large SEP values (9.33% and 1.16%, respectively), compared to those shown in Table 30.6. However, it was found that for the fat parameter, the transferred models had a very acceptable predictive capability, with a SEP of 0.59%.

The spectral differences by wavelength (SDW) algorithm showed a different pattern. The transferred spectra of both the validation and calibration databases preserved the shape in all samples with good signal-to-noise ratio, as the mean spectrum of the calibration set shows (Figures 30.5 and 30.6). This means that this strategy is more robust for its application to samples measured at different moments.

Therefore, the SDW method was applied to every standardization subgroup of $N=(1, 2, 3, 4, 5, 6, 7, 8, 9, 10, 20$ and $30)$, and the original database of $N=342$ samples was evaluated in each case. This means that a transferred database was developed individually with the added differences calculated from each standardization subgroup. To evaluate the performance of the transferred models, the validation set V_{pre} (60 samples) measured on the MEMS-NIRS device was predicted for each one and the SEP values for each analytical parameter were calculated. Figure 30.6 shows the SEP values

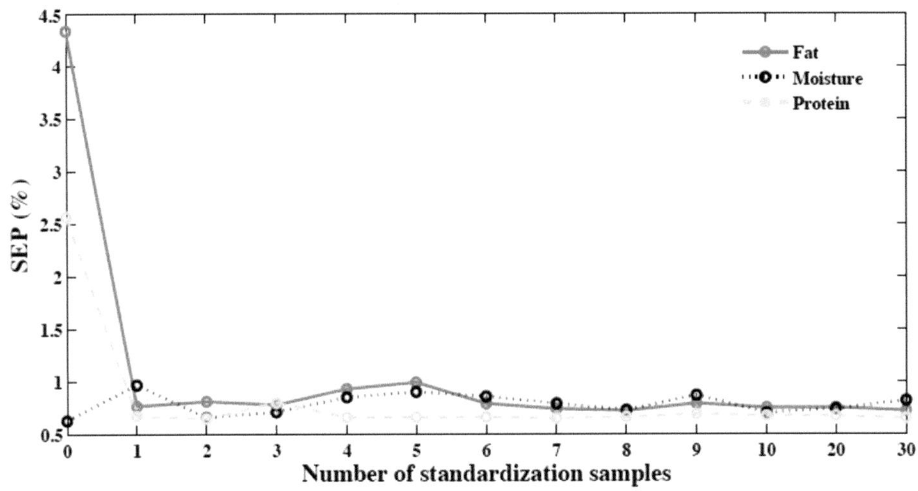

FIGURE 30.6 Standard error of prediction (SEP) of the validation set V_{pre} (60 samples) measured in the MEMS-NIRS device for the chemical composition prediction using the SDW transferred models.

calculated for each transferred model depending on the number of standardization samples used for the SDW algorithm. While the SEPs were 4.34% for fat, 0.63% for moisture and 2.57% for protein when the models were not standardized, the GH and NH were extremely large (Table 30.7); these SEP, GH and NH values were reduced using the SDW method with an optimum for eight standardization samples (Table 30.8).

Although five standardization samples showed the best spectral adjustment (results not shown) and a standardization subgroup of eight samples showed the best RMSEP values (0.72%, 0.73% and 0.66% for fat, moisture and protein, respectively) and a suitable spectral fit, as demonstrated by the values of GH=0.96 and NH=0.28 (Table 30.8). A similar study that sought to transfer a large feed (heterogeneous product) database from an FNS instrument to a MEMS-NIRS device was performed with 14 standardization samples although the authors did not attempt to optimize that number [13].

As far as the model statistics for the best transferred model (SDW and eight standardization samples) are concerned, the accuracy obtained comparing the FNS models developed with the same wavelength range and interval to the MEMS-NIRS instrument was very similar (Table 30.8). Moreover, the GH and NH statistics indicated good spectral matching with the library when the standardization was performed by SDW, reducing their values from 30.91 (GH) and 23.00 (NH) to 0.78 (GH) and 0.26 (NH) for the MEMS-NIRS spectra of the validation set V_{pre}.

Although the performance of the transferred models (Table 30.8) was slightly less accurate than the at-line models developed for the NIRS scanning monochromator (Table 30.7), both are suitable for meat analysis. Moreover, the loss in accuracy is offset when one considers the practical and technical advantages of the new device. The handheld MEMS-NIRS spectrometer enables in situ analysis with readily-available results for each individual sample featuring, easier sample presentation—that is, it is not necessary to fill a cup as required by the monochromator—and faster analysis—since more than a minute is required per subsample analysis in the FNS, while two seconds per replicate is sufficient in the MEMS-NIRS device. The difference in price between the two instruments should also be considered: the MEMS-NIRS instrument is cheaper and also requires minimal training of personnel for routine analysis. This makes the handheld MEMS-NIRS instrument more suitable and advantageous for industrial applications. Furthermore, it is expected that the model accuracy will be improved when new samples, analyzed directly by the MEMS-NIRS instrument, are included in the database.

TABLE 30.8
PLS Model Statistics Developed for Fat, Moisture and Protein Transferred to the MEMS-NIRS Device (1600–2400 nm with a Nonconstant step of 8 nm)

Parameter	Standardization	Mean (%)	SD (%)	No. PLS Factors	SEC (%)	SECV (%)	R^2_{CV}	RPD_{CV}	SEP (%)	GH	NH
Fat	DS method/four std. samples	6.53	2.88	3	0.61	0.61	0.96	4.72	0.59[a]	2.43[a]	0.09[a]
	SDW method/eight std. samples	6.66	3.15	13	0.37	0.39	**0.98**	8.08	**0.72**[a]	**0.96**[a]	**0.28**[a]
Moisture	DS method/four std. samples	70.40	2.75	4	1.62	1.64	0.64	1.67	9.33[a]	2.43[a]	0.09[a]
	SDW method/eight std. samples	70.48	2.73	12	0.44	0.47	**0.97**	5.81	**0.73**[a]	**0.96**[a]	**0.28**[a]
Protein	DS method/four std. samples	21.56	1.62	3	1.38	1.38	0.28	1.17	1.16[a]	2.43[a]	0.09[a]
	SDW method / 8 std. samples	21.50	1.66	5	0.62	0.66	**0.85**	2.51	**0.66**[a]	**0.96**[a]	**0.28**[a]

All models were pretreated with SNV+DT (1,10,5,1).
SD, standard deviation; SEC, standard error of calibration; SECV, standard error of cross-validation; R^2_{CV}*, determination coefficient of cross-validation; RPD, residual predictive deviation; SEP, standard error of prediction; GH, global Mahalanobis distance; NH, neighbor Mahalanobis distance (SEP, GH and NH values were calculated with validation set V of 60 samples).*
Bold values denote best model for each analytical parameter
^[a] Spectra measured on MEMS-NIRS device.

30.4 CONCLUSIONS AND PROSPECTS FOR THE FUTURE

The findings support the general statement that NIRS is suitable for the evaluation of meat chemical composition and offering a promising method for classifying individual animals in breeding programs on the basis of intramuscular fat content.

Most papers published in this field have used between 30 and 150 samples for calibration purposes and a limited number of samples for external validation. Here, the use of a large data set with a wide range of reference values, as well as over 90 test samples for initial accurate evaluation, ensures the realistic robustness of the calibration models.

A suitable calibration model for routine analysis requires good spectral libraries, a reliable validation process and regular testing of real performance. The GH and NH statistics can be used to interpret the prediction results for new samples in routine analysis. The method outlined here can be used to fine tune or recalibrate a model, using a small number of new unknown samples, and ensures reliable evaluation as well as the detection of highly extreme samples.

It has been scientifically demonstrated that large databases containing relevant samples collected over several years can be transferred successfully to new devices, which may be more suitable for in situ analysis in the industry.

ACKNOWLEDGMENTS

The work described in this chapter would not have been possible without the collaboration of many of our undergraduate and postgraduate students, laboratory technicians and representatives of several agro-food businesses. Special mention goes to Dr Eduardo Zamora-Rojas, author of a PhD thesis on the subject, and to our technicians, Mr. Antonio López, Mr. Manuel Sánchez and Ms. Maria del Carmen Fernández. Finally, the author wish to thank the editors for their invitation to contribute to this book and for their suggestions in preparing the manuscript.

REFERENCES

1. Downey, G. and Hildrum, K.I., 2004. Analysis of meats. In *Near-infrared Spectroscopy in Agriculture, Agronomy Monograph no. 44*, ed. Roberts, C.A., Workman Jr., J. and Reeves III, J.B., pp. 599–634. ASA, CSSA, SSSA, Madison, WI.
2. Prevolnik, M., Candek-Potokar M. and Skorjanc, D. 2004. Ability of NIR spectroscopy to predict meat chemical composition and quality - a review. *Czech J. Anim. Sci.*, 49 (11): 500–510.
3. Prieto, N., Roehe, R., Lavín, P., Batten, G. and Andrés, S. 2009. Application of near infrared reflectance spectroscopy to predict meat and meat products quality: A review. *Meat Sci.*, 83: 175–186.
4. Weeranantanaphan, J., Downey, G., Allen, P. and Sun, D.W. 2011. A review of near infrared spectroscopy in muscle food analysis: 2005–2010. *J. Near Infrar. Spectrosc.* 19: 61–104.
5. Silió, L. 2000. Iberian pig breeding programme. In *Developing Breeding Strategies for Lower Input Animal Production Environments*, ed. Galal, S., Boyazoghu, J. and Hammond, K., pp. 511–519. ICAR, Rome.
6. De Pedro, E.J., Rodríguez, C., Núñez, N., Garrido, A., García-Olmo, J., Dieguez, E. and García-Casco, J. 2002. Obtención de medidas de calidad de carne y grasa en el esquema de evaluación genética de cerdos Ibéricos mediante NIRS. *Solo Cerdo Ibérico*, 8: 7–16.
7. Fernández, A., De Pedro, E., Núñez, N., Silió, L., García-Casco, J. and Rodríguez, C. 2003. Genetic parameters for meat and fat quality and carcass composition traits in Iberian pigs. *Meat Sci.*, 64: 405–410.
8. Garrido, A., García-Olmo, J. and De Pedro, E.J. 2000. Espectroscopía de Infrarrojo Cercano (NIRS). Una metodología para implementar en sistemas de aseguramiento de la calidad y trazabilidad de productos derivados del cerdo ibérico. *Solo Cerdo Ibérico*, 4: 93–102.
9. Zamora-Rojas, E., Garrido-Varo, A., De Pedro-Sanz, E., Guerrero-Ginel, J.E. and Pérez-Marín, M.D. 2011. Monitoring NIRS calibrations for use in routine meat analysis as part of Iberian pig-breeding programs. *Food Chem.* 129(4): 1889–1897.
10. Cozzolino, D., De Mattos, V. and Martins, V. 2002. Visible/near infrared spectroscopy for predicting composition and tracing system of production of beef muscle. *Anim. Sci.*, 74: 477–484.
11. Shenk, J.S., Workman, J.J. and Westerhaus, M.O. 2001. Application of NIR spectroscopy to agricultural products. In *Handbook of Near-Infrared Analysis*, ed. Burns, D.A. and Ciurczak, E.W., pp. 419–474. Marcel Dekker, New York; Basel.
12. Pérez-Marín, D., Paz, P., Guerrero, J.E., Garrido-Varo, A. and Sánchez, M.T. 2010. Miniature handheld NIR sensor for the on-site non-destructive assessment of post-harvest quality and refrigerated storage behaviour in plums. *J. Food Eng.* 99(3): 294–302.
13. Fernández-Pierna, J.A., Vermeulen, P., Lecler, B., Baeten, V. and Dardenne, P. 2010. Calibration transfer from dispersive instruments to handheld spectrometers, *Appl. Spectrosc.* 64(6): 644–648.
14. Prado, N., Fernández-Ibáñez, V., González, P. and Soldado, A. 2011. On-site NIR spectroscopy to control the shelf life of pork meat, *Food Anal. Meth.* 4(4): 582–589.
15. Zamora-Rojas, E., Pérez-Marín, D., De Pedro-Sanz, E., Guerrero-Ginel, J.E. and Garrido-Varo, A. 2012. Handheld NIRS analysis for routine meat quality control: Database transfer from at-line instruments. *Chemom. Intell. Lab. Syst.* 114: 30–35.
16. Pérez-Marín, D., Garrido-Varo, A. and Guerrero, J.E. 2006. Remote near infrared instrument cloning and transfer of calibrations to predict ingredient percentages in intact compound feedstuffs. *J. Near Infrar. Spectrosc.* 14: 81–91.
17. Dardenne, P. 2002. Calibration transfer in near infrared spectroscopy. *NIRS News* 13(4): 3–7.
18. Fearn, T. (2001). Standardisation and calibration transfer for near infrared instruments: A review. *J. Near Infrar. Spectrosc.* 9: 229–244.
19. Shenk, J.S. and Westerhaus, M.O. 1991. Population structuring of near infrared spectra and modified partial least squares regression. *Crop Sci.* 31: 1548–1555.
20. Feudale, R.N., Woody, N.A., Tan, H.W., Myles, A.J., Brown, S.D. and Ferre, J. 2002. Transfer of multivariate calibration models: A review. *Chemom. Intell. Lab. Syst.* 64: 181–192.

31 NIR Analysis of Dairy Products

Rob Frankhuizen[*]
State Institute for Quality Control of Agricultural Products (RIKILT-DLO)

CONTENTS

31.1 INTRODUCTION

Economic and legal constraints require strict control of the composition of dairy products. For example, the level of butterfat, which is the most valuable ingredient in raw milk, is controlled by law or by customers' specifications in a whole range of dairy products. In the Netherlands, the quality payment scheme to farmers, which applies differential payments depending among others on fat and protein levels, forces the industry to test many thousands of raw milk samples. On the other hand, the worldwide market is becoming highly competitive, and there often is a great surplus of

* jworkman2002@yahoo.com

powder, butter, and cheese, particularly in Europe and New Zealand. The introduction of milk quotas means farmers will need rations even more finely tuned to produce the right levels of butterfat, protein, and casein. The industry also supplies milk-based baby foods and other dietary products. It is essential that, for nutritive reasons, these products comply with the claimed composition.

The official reference analytical procedures for moisture, fat, protein, and lactose all take too long to be satisfactory for controlling a manufacturing process. Consequently, there is an urgent need for instruments that can rapidly determine quality or other important economic parameters (Roberts 1977, Weaver 1978, 1984).

Near-infrared (NIR) reflectance spectroscopy (NIRS), in particular, potentially provides these benefits and has been practiced now for some years (Davies and Grant 1987).

Rapid analysis of milk for fat, protein, and total solids content by infrared (IR) absorption spectroscopy already has had a significant impact on the dairy industry (McGann 1978) and is an approved AOAC method (Biggs 1972). However, IR absorption spectroscopy is only suitable for liquid samples; consequently, milk powder, cheese, and other solid products have to be blended and homogenized by dissolving before analysis (Biggs 1979). Reflectance spectroscopy measures the intensity of light reflected from the surface of a sample, thereby eliminating the need for this sample preparation. In particular, NIRS makes use of this principle. NIR spectra for casein, butter, and dried milk have been determined by Goulden (1957), resulting in the use of several wavelengths to determine the sample composition.

The application of NIR to determine the chemical composition of milk powders has been thoroughly investigated (Bear et al. 1983, Ben-Gera and Norris 1968, Casado et al. 1978, Bear and Frank 1982, White et al. 1978). Birth and Washam improved the reproducibility of reflectance measurement on blue cheese (Birth and Washam 1977). Giangiacomo et al. reported the significant correlations between reflectance measurements and the composition of powdered, freeze-dried blue cheese (Giangiacomo et al. 1979). Data of Frank and Birth indicate extensive overlapping between NIR absorption bands of major milk constituents (Frank and Birth 1982).

This is typical for NIR spectra of agricultural products. Absorbance measurements at three to eight wavelengths are usually necessary to determine adequately the concentration of a particular constituent. Over the past few years, the State Institute for Quality Control of Agricultural Products (RIKILT) in the Netherlands investigated the suitability of NIRS for measuring the composition of some dairy products such as raw milk, butter, casein and caseinate, milk powders, and cheese. Correlations between major and some minor constituents analyzed by classical chemical methods and NIRS reflectance measurements were determined. For this, wavelengths were computer selected and unsuitable wavelengths were eliminated, based on statistical parameters and product or constituent information (Frankhuizen et al. 1983a, b, Frankhuizen 1985, 1986, Frankhuizen and van der Veen 1985).

Melamine as 2,4,6-triamino-1,3,5-triazine is a nitrogen-containing chemical substitute added as a counterfeit ingredient for falsifying protein analysis results in human and pet food products. This chemical is toxic to animals and humans and so has caused serious health risks when added to food products. Near-infrared (NIR) and mid-infrared (MIR) spectroscopic methods have been proposed for the rapid identification of melamine in complex dairy products. To date, neither NIR nor MIR has been reported to achieve a limit of detection (LOD) below 1 ppm on independent test sample sets. For various milk-based products, such as milk powder, liquid milk, and infant formula, an LOD below 1 ppm (0.76 ± 0.11 ppm) was reported using mathematical spectral preprocessing with partial least squares regression (PLS), polynomial PLS (poly-PLS), artificial neural network (ANN), support vector regression (SVR), and least squares support vector machine (LS-SVM). Published research has reported that near-infrared spectroscopy is useful as it is a regarded as a rapid, sensitive, and low-cost method for the analysis of milk-based products (Balabin and Smirnov 2011).

A published review describes the use of NIR spectroscopy for on-line and in-line analysis of dairy products, meat, fruit, grains, and beverages covering the literature from 1998 to 2008. The topics covered emphasize the methods designed for on/in-line measurement of data, chemometric treatment, and interpretation of the experimental results (Huang et al. 2008).

A review outlines the functional principles and applications of NIR hyperspectral imaging as applied to biological samples. Hyperspectral imaging using the NIR region provides a set of four-dimensional data cubes representing wavelength, amplitude (absorbance), and the *x* and *y* spatial dimensions. The advantage of this third-order hyperspectral imaging data is large when compared to basic first-order spectral data, using standard NIR spectroscopy. In the hyperspectral image, the additional spatial dimensions reveal chemical correlation when compared to structural location. This allows the identification of the position of chemical compounds in nonhomogeneous samples (Manley 2014).

Near-infrared spectroscopy is discussed in a review paper for its application to the analysis of powdered dairy products, liquid milk, cheese, butter, and fermented milk products. The review primarily discusses the determination of major components and the detection of adulteration (Rodriguez-Otero et al. 1997).

A comparison of near-infrared spectroscopy (NIRS), Raman spectroscopy (RS), and nuclear magnetic resonance (NMR) have been demonstrated for analysis of fat content and fatty acids (FAs) in dairy foods. Dairy products compared included milk, milk powder, and cheese. The NIRS technique was considered to be the most practical, and a summary of the analysis data using different NIRS methods is given in the tabular form (Tao and Ngadi. 2017).

31.1.1 Use of NIRS in the Dairy Industry

Near-infrared spectroscopy can be used in different areas of the dairy industry. It is possible to survey applications in three ways:

1. Analysis of the incoming milk in terms of fat, protein, and lactose for payments to farmers and standardization of the milk in terms of fat, protein, and solids.
2. Analysis of the spray-dried products on fat, moisture, and protein to get the right composition mixture for special products and analysis of the raw materials for the production of cream, cheese, casein, and whey.
3. Analysis of the finished products to control the products in terms of legal regulations and customers' or producers' specifications.

Governmental or other official laboratories for quality control of dairy products can use NIRS as a screening method for rapid analysis of the composition of unknown samples. Besides with NIRS, it might be possible to determine quality parameters where in the past no direct correlation with chemical or physical methods was apparent.

31.1.2 Standard Methods

The standard methods of laboratory analysis by wet chemistry vary from country to country. However, the following methods are generally accepted as standard methods in the dairy industry (Tuinstra-Lauwaars et al. 1985):

Component	Method
Moisture	Toluene
	Oven
	Karl Fisher
	Vacuum

(Continued)

Component	Method
Fat	Gerber
	Soxhlet extraction
	Gravimetric
	Röse–Gottlieb
	Schmid–Bondzynski–Ratzlaff extraction
	Mojonnier
	Babcock
Protein	Kjeldahl
	Dye binding
Lactose	Photometric
	Reductometric
	Polarimetric
	Titrimetric
	Enzymatic
Lactic acid and lactates	Photometric
	Enzymatic
Ash	Furnace
Salt	Potentiometric
	Titrimetric
	Gravimetric
pH	Electrometric

31.2 CALIBRATION PROCEDURE

Near-infrared spectroscopy measurements are as good as the calibration. That is why great accuracy and expense have to be made by calibrating a NIRS instrument. Calibration is achieved by taking a set of samples that have been analyzed by reference methods for parameters of interest. A minimum of 40 calibration samples normally is required in order to adequately calibrate the instrument. It is very important that a set of calibration samples represents the full range of the product being analyzed. For achieving a good "robust" calibration, it is therefore necessary that samples are collected over a period depending on composition, production process, production period, and production area. Definite minimum and maximum ranges are difficult to set; however, as a general guideline, the range should not go smaller than 2% and not exceed 20% for major constituents. It is important to know that most dairy products will change in composition over a period. Milk powder absorbs and dissipates moisture very readily. Products containing high levels of fat are liable to fat oxidation over a period of a few months, and products containing high levels of lactose or other carbohydrates may undergo some form of crystallization. The oxidation and crystallization processes can change the absorbance characteristics of fat and carbohydrates. Adequate storage will prevent significant changes in absorbance characteristics; however, it is better to use fresh samples and analyze the samples with the reference methods and NIRS at the same time. Accurate and consistent laboratory analysis are important in order to obtain the best results with the NIR technique. When sufficient samples with the required range are collected, a regression analysis between the analytical data and the optical density values of the samples can be carried out and will result in a set of constants (K values). Each product and constituent has its own set of constants. Statistical tests, incorporated in the software of the instrument, enables the user to choose the appropriate filter or wavelength combination for the calibration. Absorbance measurements at 3–8 wavelengths are usually necessary to determine adequately the concentration of a particular constituent. Standard deviation, correlation coefficient, and the t-test are the most important parameters for achieving a good calibration by using a multiple linear regression program. The ultimate test of any calibration is to check the validity of results obtained on samples that did not belong to the calibration set. Assuming a good conformity with the reference methods, the calibration can be used for routine measurements. It is important that the validity of the calibration is checked frequently. (Sample selection/handling and data analysis will be discussed in more detail in Chapter 8.)

A chemometrics-based paper describes the application of partial least square regression with laser correlation spectrometry of scattered light and for the study of the correlation between near-infrared spectra of dispersions and a particle density and size variations. Milk composition enriched by two different protein fractions, globulin and casein, was studied along with solutions of hydrated surfactant micelles. The well-known structure model of surfactant micelles and laser correlation spectrometry of scattered light was used for acquiring the reference sample set of dispersions. These values were standardized using the particle density and nanometer size value. This work should provide data for the development of a simple device to discriminate counterfeit milk protein-containing non-natural serum concentrates (Kalinin et al. 2009).

Near-infrared (NIR) analysis of liquid milk samples is complicated by the following two main factors: water absorption is large compared to fat, protein, and lactose absorption, and the naturally occurring fat globules in milk exist as light-scattering particles. The light scattering of fat globules creates spectral anomalies and inconsistencies. In one paper, different mathematical pretreatments were tested on spectra to reduce the effects of fat globules and to identify primary band locations for analyzing fat, protein, and lactose. Mean centering was reported to provide some reduction in spectral variation due to light scattering. For band wavelength locations, 1724, 1752, 2308, and 2344 nm were found to be optimum for the determination of fat content. The bands at 2050 and 2180 nm were assigned to protein composition. The absorption band at 2094 nm was assigned lactose content (Robert et al. 1987).

31.3 MILK

Milk has a NIR absorption spectrum very similar to that of water (Goulden 1964) (Figure 31.1). The difference between milk and water is that milk contains scattering particles in the form of fat globules and protein micelles. A beam of electromagnetic irradiation traveling through such a light-scattering medium takes a path longer than the sample thickness (Mestre 1935), with a subsequent increase in absorption, or intensification of the absorption bands (Ben-Gera and Norris 1968, Butler and Norris 1960). The introduction of light particles can result in erroneous information due to the wavelength dependence of the extinction (Latimer 1959) and an untrue, partial collection of the transmitted light (Norris 1965). These problems can be overcome using an integrating sphere to collect the transmitted light (Butler and Norris 1960).

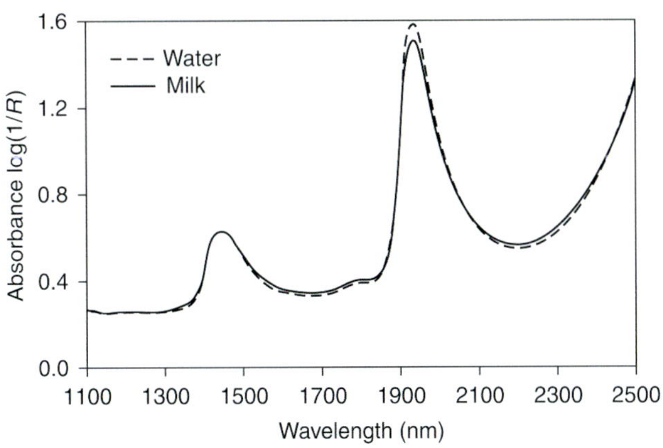

FIGURE 31.1 NIR reflectance spectra of milk and water.

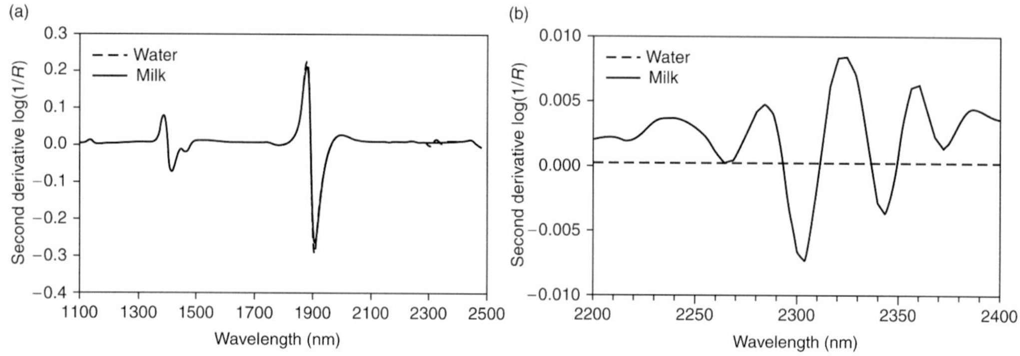

FIGURE 31.2 (a) Second-derivative NIR spectra of milk and water. (b) Second-derivative NIR spectra of milk and water with specific absorption bands of fat at 2304 and 2343 nm, respectively.

Another problem is the very broad and intensive absorption bands of water that overlap the absorption bands of protein and fat in the NIR milk spectrum for a great part (Figure 31.1). This overlap makes it difficult to correlate specific bands to the constituents from which they arise. Furthermore, the small variation of protein and fat content in standardized milk may cause only very slight changes in the NIR spectrum. Therefore, it is necessary to transform the raw data to first and second derivatives to interpret and correlate the overlapping absorption bands of protein and fat in the NIR milk spectrum (Figure 31.2a and b). In this way, reliable and accurate results can be obtained.

Milk and its derivatives can be analyzed by NIRS for fat, protein, lactose, and total solids content using a special drawer unit equipped with a liquid cell. It uses the principle of transflectance, which combines the advantages of transmission and reflectance. The reflected light measured in this case originates both from the surface of the liquid and from the rear of the cell. In the latter case, it has passed through the thin film of liquid contained in the cell. A high-pressure pump and a two-stage homogenization will homogenize the milk emulsion to particle size well below a micrometer. The unit also provides for temperature control and automatic wash from the NIR instrument. For achieving an accurate measurement, it is absolutely necessary that the samples be freshly analyzed after warming to about 45° in a waterbath and mixing with a mixer at the inlet of the homogenizer. Study of the suitability of NIRS for the determination of fat, protein, and lactose of milk (mixed samples of several individual cows) for payments and standardization purposes shows that the accuracy of NIRS is acceptable (Table 31.1).

Study of the usefulness of NIRS for the determination of fat, protein, and lactose of goat's milk shows that the standard errors of calibrations (SECs) of a test set originating from 30 individual goats (Table 31.2) is about five to ten times higher than the SECs of the separate calibration based

TABLE 31.1
NIRS Data (g/100 g) for Raw Cow Milk (Mixed Samples), Measured with a InfraAlyzer 500 Equipped with a Liquid Sample Cell and Homogenizer, Compared to Reference Methods

Constituent	*n*	SEP	Range	Accuracy Required
Fat	50	0.034	4.1–7.5	0.04
Protein	50	0.038	2.1–2.8	0.04
Lactose	50	0.043	3.5–4.1	0.05

n, number of samples; SEP, standard error of prediction; reference method fat: Gerber; protein: Kjeldahl; lactose: enzymatic.

TABLE 31.2
NIRS Calibration Data (g/100 g) for Raw Goat's Milk (Individual Goat Samples) Compared to Reference Methods

Constituent	*n*	SEC	Range
Fat	30	0.226	2.8–7.0
Protein	30	0.170	2.8–4.1
Lactose	30	0.281	3.3–5.1

TABLE 31.3
NIRS Calibration Data (g/100 g) for Raw Goat's Milk (Mixed Samples of Several Individual Goats) Compared to Reference Methods

Constituent	*n*	SEC	Range
Fat	20	0.022	3.0–6.3
Protein	20	0.028	3.0–3.9
Lactose	20	0.058	3.5–4.9

TABLE 31.4
NIRS Calibration Data (g/100 g) for Raw Goat's Milk (Mixed and Individual Samples) Compared to Reference Methods

Constituent	*n*	SEC	Range
Fat	50	0.096	2.8–7.0
Protein	50	0.079	2.8–4.1
Lactose	50	0.133	3.3–5.1

on mixed samples (Table 31.3). This is caused not only by an increased variation in the sample composition but especially by deviation of the fat and protein composition caused by udder disease (mastitis) of some goats. However, by the addition of the individual samples, including the deviating samples, to the calibration set and by an increase of the number of wavelengths, it was possible to calculate a calibration with more acceptable accuracy (Table 31.4).

It must be concluded that the measurement of fat, protein, and lactose of milk by NIRS can be successful. Our experience with cow and goat milk shows that the main problem is to find a representative sample set, a method of presenting the sample to the instrument that is exactly reproducible, and a method to identify unknown samples with an abnormal spectra. Study of Bertrand et al. (1987) illustrates that principal component analysis (PCA) has the potential to select the most relevant calibration samples (independent of the chemical composition) and to identify samples with abnormal spectra as an outlier.

Near-infrared spectroscopy (NIRS) has been tested for monitoring variation over time in milk composition for dairy cows, indicating potential dynamic changes in health and detection of nutritional imbalances. NIRS when combined with the diffuse reflectance sampling method has been used to monitor milk fat, protein, lactose, and urea in nonhomogenized milk. For one referenced paper, a total of 1610 of raw milk samples collected from 38 different cows were assessed. NIR spectra from 851 to 1649 nm were measured. Calibration and validation were performed, and results were reported as follows. The coefficients of determination as $R2 = 0.998$, 0.94, and 0.73, with standard error of prediction as SEP = 0.03, 0.07, and 0.09 for fat (%), protein (%), and lactose (%), respectively. It was noted that poor results were achieved for predicting urea in (mg l^{-1}) with a reported $R2$ = 0.31 and SEP = 40.23. The accuracy of predicted fat, protein, lactose, and urea content in milk was similar in results using either transmission or transflection modes. The NIRS analysis of raw milk constituents did not achieve laboratory accuracy but had the advantage that it can be done with higher frequency using the spectroscopic method (Melfsen et al. 2012).

Visible (Vis) and near-infrared (NIR) (400–995 nm) reflectance spectroscopy has been used for the analysis of total protein and milk fat in liquid milk. A fiber optic cable consisting of eight 200-μm diameter fibers in a linear array configuration was tested. An end fiber was used for illumination and other seven fibers were used for reflected light collection. Milk samples were measured with three variants of homogenization. Diverse sample sets were used for model testing, and the algorithm used was a joint variable selection and preprocessing optimization method based on the genetic algorithm approach. The root mean-square errors (RMSE) for several validation methods were <0.10% for fat and <0.08% for total protein content. For comparison, an LED-based unit was developed using nine LEDs for the analysis of fat and seven LEDs for protein analysis. Properly selected LEDs approximated the results for a full-scanning system (Bogomolov et al. 2017).

Visible (Vis) and short-wave near infrared (SW-NIR) (400–1100 nm) spectra were used for developing an analytical method for total protein and total fat content in raw milk. A sample set was used with varying degree of homogenization and an expanded range of nutrient composition. Using raw (uncorrected) spectral data combined with partial least squares (PLS) regression produced lower prediction errors than scatter-corrected spectra. The paper concluded that total protein and fat in raw milk are accurately determined using the combined Vis and SW-NIR spectra. The method was also reported to be robust against large variability of fat globule size (Bogomolov and Melenteva 2013).

Near-infrared (NIR) portable sensors were evaluated as a rapid monitoring or advisory analytical tool for real-time milk composition measurements. A portable NIR device was used for the determination of fat, protein, and solids-nonfat (SNF) for raw milk and for calibration transfer tests between two units. For the calibration and validation work, 542 fresh milk samples were measured over a spectral range of 1600–2400 nm using portable MicroPhazir™ (MP) NIR spectrometer. The optimum results were obtained using a 17 mm pathlength cuvette. For optimum results, the standard error of prediction (SEP)/standard error of cross-validation (SECV) were 0.96, 0.72, and 0.83 for fat, protein, and SNF contents, respectively. For testing calibration transfer, a transfer or standardization set of 10 samples was measured on both instruments. After applying a standardization matrix, the standard error of differences between master and transfer instruments was lowered using a standardization process and was reported to be useful for moving the calibrations across the instruments. The conclusion was the portable unit was useful for advisory levels of the measured parameters for farm-level use (de la Roza-Delgado et al. 2017).

Handheld near-infrared (NIR) spectroscopy was evaluated for its potential as a farm-level, on-site monitor for fatty acid (FA) profiles in raw milk. For the calibration test of this instrument, 108 liquid milk samples were measured at room temperature with reference analysis obtained using GC–MS. Calibration techniques were evaluated using various math pretreatments combined with partial least squares (PLS). Calibrations were externally validated and demonstrated predictive ability for total saturated, monounsaturated (MUFA), and polyunsaturated FA (PUFA) with coefficients of determination of cross-validation $\left(R_{cv}^2 > 0.8\right)$. Prediction of individual FA was demonstrated for caproic, capric, lauric, myristic, palmitic, and arachidic acids. Unsaturated FAs were predicted for oleic, conjugated linoleic acid, and omega-6 acids with R_{cv}^2 values ranging between 0.91 and 0.73. Validation statistics confirmed that the highest R_v^2 (coefficient of determination of external validation) values to quantify FA were for PUFA, linolenic acid ($R_v^2 = 0.92$), caproic acid, and MUFA ($R_v^2 = 0.87$). The authors reported successful application of farm-level analysis for fatty acid composition using the handheld device (Suárez et al. 2018).

Near-infrared (NIR) and mid-infrared (MIR) reflectance spectroscopy techniques were compared for prediction analysis of fatty acid (FA) composition of liquid and oven-dried forms.

For calibration and comparison studies, 250 bulk cow milk samples were collected from 70 farms in northwest Italy. The samples were measured using MIR for liquid milk and using NIR for liquid and oven-dried milk. For comparison in prediction accuracy, NIR and MIR FA predictions as g/100 g FA on oven-dried milk were similar. These parameters measured were for the sum of even chain-saturated FA (ECSFA), odd chain-FA (OCFA), unsaturated FA (UFA), conjugated linoleic acid (CLA), n-3 FA, and C18:1cis9 to C16 ratio.

NIR exhibited superior prediction results on oven-dried milk for monounsaturated FA (MUFA), *n*-6 to *n*-3 ratio, polyunsaturated FA (PUFA), and *n*-6 FA. However, MIR demonstrated superior predictions for almost all FA parameters when expressed as g/kg of milk. The NIR and MIR predictions for fresh liquid and oven-dried milk were similar, but the reliability decreased for NIR on thawed liquid milk. The researchers reported high-performance predictions using either NIR or MIR for routine milk FA analysis (Coppa et al. 2014).

In an attempt to demonstrate the feasibility of global models for fat and total protein content in milk, fresh milk samples were measured over the Vis-NIR range of 400–1100 nm from a large geographical area of Russia. It is well understood that changes in the chemistry of milk due to genetic, seasonal, regional, and other variations will cause potential predictive changes in NIR calibrations over time. It was reported by one research team that NIR predictions for total protein and fat content over a broad range of samples exhibited quite useful results to include root mean-square errors of prediction (RMSEP) of 0.09% and 0.10% for fat and total protein content, respectively. The demonstrated prediction results over such a diverse set point to the potential of creating global models for this application (Melenteva et al. 2016).

NIR spectroscopy was evaluated for the analysis of fatty acids (FAs), fat-soluble vitamin composition, and seasonal variation of ewe's milk. For this study, 219 ewe's milk samples from different breeds and diets were sampled monthly over 1 year. For reference data, FAs were analyzed using gas chromatography (GC), while the vitamins retinol and α- and γ-tocopherol were analyzed using liquid chromatography (HPLC). It was quickly determined that quantitative results were much improved by measuring samples dried on paper as compared to measuring the liquid milk samples. However, the vitamin analysis results were not improved using the samples dried on paper. The milk dried on paper prediction statistics was reported as good for capric, lauric, myristic, palmitoleic, stearic, and oleic acids and acceptable for caprilic, undecanoic, 9c, 11tCLA, ΣCLA, PUFA, ω3, ω6, retinol, and α-tocopherol. Discriminate analysis for seasonality used the partial least squares discriminant analysis (PLSDA) algorithm with 93% of winter samples and 89% of summer samples correctly classified. For classification, the milk dried on paper samples was used (Revilla et al. 2017).

A near-infrared (NIR) light backscatter sensor was combined with rheology measurements to evaluate the effects of protein (3%, 4%, and 5%), calcium (0.90 and 1.80 mm), inulin (2%, 5%, and 8%) concentrations on the hydrolysis, aggregation, and firming phases during enzymatic coagulation of low fat milk. The optical sensor was useful for the inline detection of the milk components as added during each of the coagulation phases (Arango et al. 2018).

Short-wave near-infrared (SW-NIR) spectroscopy was used for the analysis of biological fluids. Measurements were made in the 800–1100 nm spectral region for 100 raw milk samples. Calibrations using partial least squares (PLS) regression were used for the prediction of fat, protein, and lactose. Constituent band assignments were estimated using PLS-loading plots along with the regression *b*-vector coefficients. The correlation coefficient (*r*) and root-mean-square error of prediction (RMSEP) for fat were 0.996 and 0.087 wt %, respectively. Analysis of proteins and lactose was not as successful but was reported to be promising for this wavelength region. The work also described proposed band assignments for each of fat, protein, and lactose (Šašić and Ozaki 2001).

Near-infrared transmission (NIT) spectroscopy was tested for the analysis of fat, crude protein, true protein, and casein in cow milk. For calibration, partial least squares (PLS) were used for predictions on two sets of samples. The initial calibration included 76 nonhomogenized samples, whereas the second overall calibration used 96 homogenized and nonhomogenized samples. Standard errors of calibration were 0.12% for fat, 0.06% for crude protein, 0.04% for true protein, and 0.05% for casein in the overall calibration. Validation of the overall calibration with an independent set of samples gave standard errors of prediction of 0.07% for fat, 0.06% for crude protein and casein, and 0.05% for true protein.

The authors report that homogenization has an effect on fat determination using NIR and that nonhomogenized and homogenized samples are different spectrally with respect to fat determination (Laporte and Paquin1999).

31.4 MILK POWDER

31.4.1 General

Milk powder has been found to be one of the most suitable products to be analyzed using NIRS because of its uniform particle size and shape as well as its consistent formulation. Because milk powder has a small particle size, it does not require grinding. It is important, however, that the samples be well mixed and blended. Powder is simply inserted into a sample cup, leveled off, and sealed with a backing plate. For achieving good "robust" calibrations, we collected milk powders from the most important milk powder-producing EEC countries during a period of 2 years. Depending on the composition, production process, and production period, samples were selected for the calibrations. The samples were analyzed in duplicate according to the following methods: moisture was determined by oven drying (2 h at 102°C), protein (total nitrogen × 6.38) by the Kjeldahl method, and fat by the Rose–Gottlieb extraction method. Ash content was determined by dry ashing (1 h at 500°C), and enzymatic methods were used to measure lactose and lactate contents. The mean value for the skim milk powder content in calf-feed was calculated by using results from an enzymatic coagulation of the paracasein determined according to EEC method(s) (EEC regulation nr. 1725/79). Reflectance measurements were obtained by using an InfraAlyzer 400 after the powder had reached room temperature.

All combinations were tested by selected samples representative of the calibration. During the test period, it was not always possible to get a test set with the same range as the calibration set. Therefore, the *R* values of the test sets are sometimes smaller than those of the calibration sets.

To ensure that the correct wavelength was chosen, a spectral scan of a few different milk powders and of its isolated constituents protein, fat, and lactose was carried out using an InfraAlyzer 500 (Figures 31.3 and 31.4). The most useful interference filters for analyzing fat, protein, moisture, and lactose in milk powders are shown in Figure 31.4 and given in Table 31.5.

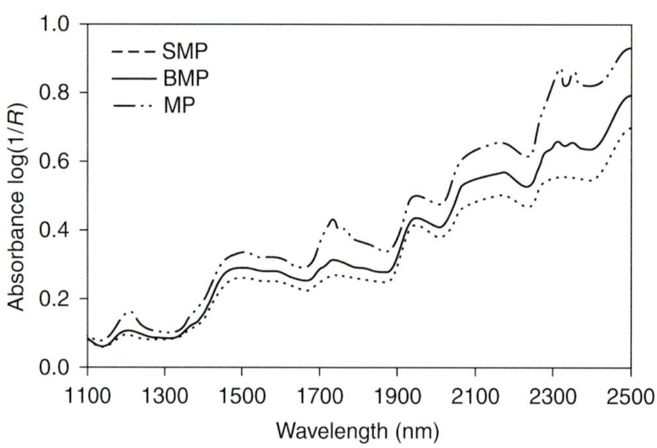

FIGURE 31.3 Typical NIR reflectance spectra of skim milk powder, buttermilk powder, and full cream milk powder.

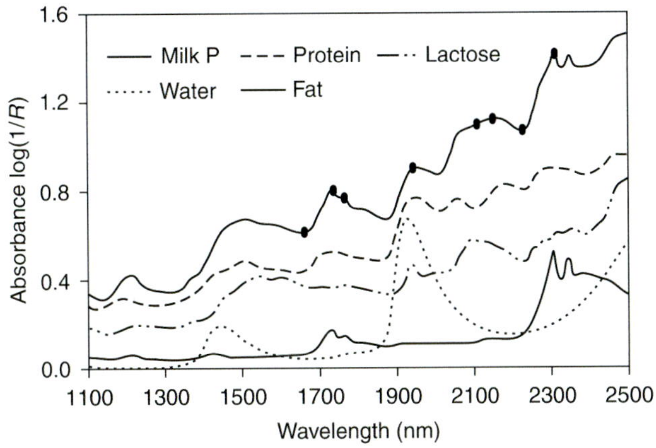

FIGURE 31.4 NIR reflectance spectra of skim milk powder and various isolated constituents (=selected wavelength).

TABLE 31.5
Absorption Bands of Milk Powder Constituents

Filter No.[a]	Absorption Band (nm)[b]	Constituents Contributing to Absorption
20	1650	Water, protein, lactose reference
16	1940	Water
17	1734	Protein, fat
15	1759	Fat, protein
14	2100	Protein, lactose
9	2139	Protein, lactose, water reference
6	2230	Fat, protein, lactose reference
4	2310	Fat, protein

[a] Filters of the Technicon InfraAlyzer 400 that correspond to the given wavelength bands.
[b] Approximate wavelength.

31.4.2 Results and Discussion

31.4.2.1 Protein

High correlations were observed between the protein percentages determined by the Kjeldahl method and those estimated by the InfraAlyzer 400 (Table 31.6). The square multiple correlation coefficients of 0.97–0.999 with SEC of 0.18%–0.34% are acceptable compared with the standard deviation of the reference method of about 0.15% for milk powders.

Furthermore, it is important to realize that the Kjeldahl method determines the crude instead of the true protein percentage. The crude protein percentage, however, can differ more or less from the true protein percentage, depending on the season and resulting in fluctuations in the percentage of nonprotein nitrogen (NPN). Probably NPN differs in the NIR absorption from the true protein, likely resulting in an increased SEC compared with a calibration against the true protein percentages.

Concerning the prediction sets (test sets), the results for the SEP and *R* values are comparable with those of the calibration sets. It can be concluded that the calibration is robust and reliable. Calibration procedures resulted in four to eight filters giving significant information, including filters for protein absorption at 2139 and 1734 nm.

TABLE 31.6
NIR Reflectance Data for Protein (g/100 g) in Some Milk Powders

	Calibration				Prediction			
Product	*n*	*R*	**SEC**	**Range**	*n*	*R*	**SEP**	**Range**
Skim milk powder	159	0.97	0.27	32.9–41.2	100	0.97	0.25	33.5–40.6
Buttermilk powder	116	0.97	0.21	28.9–34.6	50	0.97	0.21	29.2–34.6
Skim milk powder + buttermilk powder	274	0.98	0.34	28.9–41.2	150	0.97	0.36	29.2–40.6
Denatured milk powder	92	0.98	0.34	27.1–40.1	50	0.96	0.42	28.5–39.7
Milk powder with nonmilk fat	61	1.00	0.18	17.6–26.0	50	0.98	0.25	18.4–25.4

31.4.2.2 Moisture

Correlation coefficients were calculated between moisture determined by oven drying and moisture estimated by the InfraAlyzer 400 (R =0.97, tot. 0.99). Averaged SECs of 0.12% are acceptable (Table 31.7). The results obtained for the prediction sets are comparable with those of the calibration sets. The calibration procedure resulted in three to six filters giving significant information including the most significant filter for moisture absorption at 1940 nm.

Moisture determined by oven drying, often used for milk powder, is known to be affected by the relative air humidity, which has a negative influence on reproducibility. The InfraAlyzer 400 is calibrated with the results of oven drying; in other words, the accuracy of the NIR is highly limited by the accuracy of the reference method. The NIR linear relationship is, however, calculated with a large number of samples spread over a wide range of time. Under these circumstances, the influences of the relative air humidity are averaged and canceled out. To demonstrate this, 24 tins were filled with a homogeneous sample of spray-dried skim milk powder and air-tight sealed. Moisture in these samples was determined in duplicate on several days over a period of 2 months, according to the oven method and by the InfraAlyzer 400.

Figure 31.5 gives the relationship of moisture content vs. time for the two methods. The reproducibility of the NIR is obviously better than that of the oven method, whereas both methods give the same mean values as expected. Thus the moisture content determined by NIR gives more accurate results than the determination by the oven method.

31.4.2.3 Fat

For fat calibrations of skim milk powder, it was impossible to get samples with enough variation in fat content. Owing to the standardization of this product, the calibration set is not ideal. All calculated correlation coefficients are above 0.98 with an SEC of 0.09%–0.31% (Table 31.8). The latter 0.31% is found for a combined calibration of skim milk powder and buttermilk powder. The standard error is about twice as high as the separate calibrations, which is caused by an increased variation in the sample composition. On the other hand, samples of milk powder with nonmilk fat

TABLE 31.7
NIR Reflectance Data for Moisture (g/100 g) in Some Milk Powders

	Calibration				Prediction			
Product	*n*	*R*	**SEC**	**Range**	*n*	*R*	**SEP**	**Range**
Skim milk powder	159	0.97	0.12	2.9–5.8	100	0.97	0.11	3.1–5.7
Buttermilk powder	116	0.99	0.13	2.7–6.3	50	0.99	0.10	2.7–6.3
Skim milk powder + buttermilk powder	274	0.97	0.13	2.9–6.2	150	0.99	0.14	2.7–6.3
Denatured milk powder	100	0.97	0.11	2.8–6.3	50	0.96	0.16	3.1–6.2
Milk powder with nonmilk fat	64	0.98	0.13	1.3–4.4	50	0.96	0.14	1.5–4.3

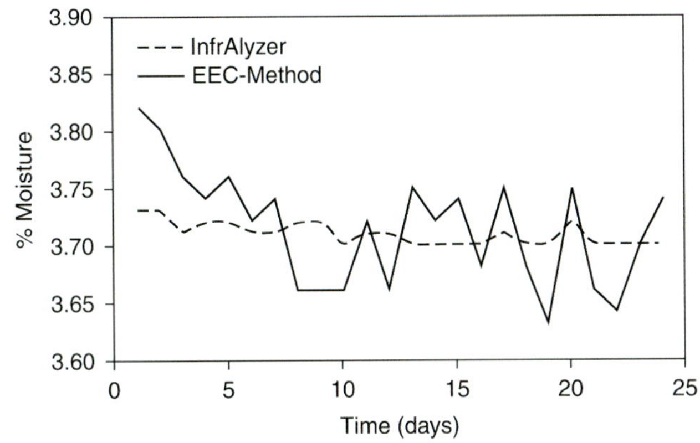

FIGURE 31.5 The relationship of moisture content vs. time for two methods in skim milk powder.

TABLE 31.8
NIR Reflectance Data for Fat (g/100 g) in Some Milk Powders

	Calibration				Prediction			
Product	*n*	*R*	**SEC**	**Range**	*n*	*R*	**SEP**	**Range**
Skim milk powder	159	0.98	0.09	0.8–2.8	100	0.98	0.10	0.5–1.3
Buttermilk powder	116	1.00	0.15	3.3–11.2	50	0.99	0.14	5.3–10.7
Skim milk powder + buttermilk powder	275	0.99	0.31	0.6–11.2	150	0.97	0.41	0.5–10.7
Denatured milk powder	92	0.99	0.19	0.6–7.5	50	0.96	0.24	0.8–6.5
Milk powder with nonmilk fat (overall)	62	1.00	0.30	26.2–50.4	50	0.99	0.34	27.2–49.9
Milk powder with nonmilk fat (lower range)	33	1.00	0.15	32.1–41.5	25	0.99	0.14	32.3–40.9
Milk powder with nonmilk fat (higher range)	29	0.99	0.13	47.4–50.4	25	0.99	0.14	47.9–49.9

give a wide variation of fat content. The SEC value is 0.30. By making two calibrations, the SEC value decreases to 0.14 on average. The values of SEP and *R* obtained for the prediction sets are comparable with those of the calibration sets.

The calibration procedure resulted in five to seven filters giving significant information, especially the filters for fat absorption at 2310 and 1759 nm.

31.4.2.4 Lactose

SECs of 0.38%, 0.35%, and 0.34% were calculated for the determination of lactose in skim milk powder, buttermilk powder, and milk powder with nonmilk fat (Table 31.9). These are acceptable in relation to the standard deviation of the reference method of about 0.25%. The SEC for the combined calibration of skim and buttermilk powder and for denatured milk powder is too high for accurate analysis. Similar results were observed for the test set. This is caused by the variation in the physical and chemical composition of the samples and by the few specific NIR bands of lactose in relation to the overall NIR spectrum. Therefore, many filters are needed for the calibration. About 10 filters gave significant information. Although each of the filters gave a low correlation coefficient, the combination of 10 filters resulted in an acceptable *R* value.

31.4.2.5 Lactate

Although there are no special claims for the lactate content in buttermilk powder, a calibration for this component was still made (Table 31.10) because it may give information about (1) the original lactose content, (2) the ratio between lactose and protein, or (3) the amount of addition of neutralizing compounds for the lactic acid. The latter influences the ash content.

TABLE 31.9
NIR Reflectance Data for Lactose (g/100 g) in Some Milk Powders

	Calibration				Prediction			
Product	*n*	*R*	SEC	Range	*n*	*R*	SEP	Range
Skim milk powder	159	0.95	0.38	42.6–51.6	100	0.94	0.44	43.6–50.2
Buttermilk powder	116	0.99	0.35	37.0–47.6	50	0.99	0.37	37.1–47.3
Skim milk powder + buttermilk powder	275	0.96	0.62	36.6–51.6	150	0.95	0.83	37.1–50.2
Denatured milk powder	92	0.96	0.62	36.8–49.0	50	0.94	0.75	37.5–48.2
Milk powder with nonmilk fat	64	1.00	0.34	23.4–37.6	50	0.98	0.38	24.5–36.1

TABLE 31.10
NIR Reflectance Data for Lactate (g/100 g) in Some Milk Powders

	Calibration				Prediction			
Product	*n*	*R*	SEC	Range	*n*	*R*	SEP	Range
Buttermilk powder	116	1.00	0.16	0.0–9.0	50	0.99	0.21	0.1–7.2
Skim milk powder + buttermilk powder	275	0.99	0.23	0.0–9.0	150	0.96	0.54	0.0–7.2
Denatured milk powder	92	0.99	0.10	0.1–4.2	50	0.97	0.19	0.3–4.1

The lactate content in skim milk powder of about 30–100 mg/100 g is too small to be determined by NIR. The combination of skim and buttermilk powder samples in one calibration resulted in an increased accuracy.

For the determination of lactate in buttermilk and denatured milk powder, the use of a lot of filters (12 in total) gives significant information. The need for so many filters is caused by the variation in physical and chemical composition of the samples and by the few specific NIR bands of lactate in relation to the overall NIR spectrum.

31.4.2.6 Ash

In contrast to most organic compounds, most inorganic compounds give no characteristic reflection signals in the NIR region. The correlation between the ash content determined by incineration at 500°C and the ash content predicted by NIRS (Table 31.11) can only be based on the presence of organic and bound milk salts. Probably the ash content predicted by the InfraAlyzer is correlated to the total amount of organic compounds and water because of the large number of filters (Biggs 1979, Goulden 1957, Bear et al. 1983, Ben-Gera and Norris 1968, Casado et al. 1978) giving significant information.

TABLE 31.11
NIR Reflectance Data for Ash (g/100 g) in Some Milk Powders

	Calibration				Prediction			
Product	*n*	*R*	SEC	Range	*n*	*R*	SEP	Range
Skim milk powder	159	0.76	0.08	7.7–8.3	100	0.73	0.11	7.6–8.2
Buttermilk powder	116	0.97	0.19	6.9–10.0	50	0.92	0.23	6.9–9.5
Skim milk powder + buttermilk powder	275	0.85	0.23	6.9–10.0	150	0.85	0.37	6.9–9.5
Milk powder with nonmilk fat	63	0.99	0.09	3.9–6.6	50	0.96	0.15	4.0–6.4
Denatured milk powder	100	0.86	0.18	6.6–9.5	50	0.79	0.27	6.7–9.4

For skim milk powder, the range in ash content is too small for determination by NIRS as can be seen from the relatively high SEC and SEP values and the small correlation coefficients, especially for the prediction set. For the other products, quantitative analysis is possible, however, with a relative low accuracy.

31.4.2.7 Skim Milk Powder and Powdered Milk Content

Skim milk powder can be converted to calf feed by adding nonmilk fat. Because this product is EEC aided, the determination of the percentage of skim milk powder in calf feed is prescribed by the EEC (EEC regulation nr. 1725/79). Determination of the skim milk powder content is done by enzymatic coagulation of paracasein, a time-consuming method. Moreover, the enzymatic method lacks accuracy. Therefore work has been done to determine the milk powder content by NIRS. The results improve if the skim milk powder content is based on the above-mentioned method together with the results of the protein, lactose, fat, and ash content. For these reasons, the calibration of the InfraAlyzer 400 is based on these parameters. The results show that the percentage of skim milk powder in a mixture of milk powder and nonmilk fat can be determined by NIR quantitatively (Table 31.12).

Calcium content in powdered milk was measured using both near-infrared (NIR) and mid-infrared (MIR) spectroscopic techniques. Variable selection was made using a hybrid spectral algorithm combined with a technique referred to as uninformation variable elimination (UVE). Successive projection algorithm (SPA) was combined with the UVE algorithm to select 11 NIR and 15 MIR variables from 2,756 NIR and 3,727 MIR data variables, respectively. The MIR method demonstrated predictions as 0.930 for the coefficient of determination (R 2), 3.703 for residual predictive deviation (RPD), 30.16 for root mean square error of prediction set (RMSEP), and 5.22% for relative errors of prediction (RSEP). For the NIR method, prediction results indicated 0.636 for R2, 1.587 for RPD, 78.82 for RMSEP, and 13.4% for RSEP. MIR spectroscopy was preferred as compared to NIR for the determination of calcium content in powdered milk (Wu et al. 2012).

Iron and zinc content in powdered milk was measured using both near-infrared (NIR) and mid-infrared (MIR) spectroscopy. Variable selection was completed using uninformation variable elimination (UVE) combined with the successive projections algorithm (SPA). For calibration, 18 NIR and 18 MIR variables were selected for iron content prediction and 17 NIR and 12 MIR variables for zinc content prediction. MIR provided improved prediction results as compared to NIR for both iron and zinc contents for both PLS and LS-SVM models. The iron content prediction results based on LS-SVM reported a coefficient of determination (R2) of 0.920, residual predictive deviation (RPD) was 3.32, and root-mean-square error of prediction (RMSEP) was 1.44. The zinc content prediction results based on LS-SVM yielded an R2 of 0.946, RPD of 4.36, and RMSEP equal to 0.32. MIR resulted in better prediction results as compared to NIR for iron and zinc contents in powdered milk (Wu et al. 2009).

TABLE 31.12
NIR Reflectance Data for Skim Milk Powder Content (g/100 g) in Milk Powder with Nonmilk Fat

	Calibration				Prediction			
Product	*n*	*R*	**SEC**	**Range**	*n*	*R*	**SEP**	**Range**
Milk powder with nonmilk fat	63	1.00	0.43	50.3–76.8	50	0.98	0.55	50.2–75.8

TABLE 31.13
Repeatability and Reproducibility (g/100 g) of InfraAlyzer 400 Method for Skim Milk Powder

	Mean	r	R_w	CV_{Rw}
Moisture	3.710	0.034	0.034	0.3
Fat	0.712	0.037	0.042	2.1
Protein	35.705	0.18	0.18	0.2
Ash	7.964	0.085	0.085	0.4
Lactose	48.779	0.39	0.49	0.4

r = repeatability = $2.83s_r$; R_w = reproducibility over the test period within the laboratory = $2.83s_{Rw}$; CV_{Rw} = coefficient of variation over the test period = s_{Rw}/mean × 100%.

TABLE 31.14
Repeatability and Reproducibility (g/100 g) of Reference Method for Skim Milk Powder

	Mean	r	R_w	CV_{Rw}
Moisture	3.712	0.068	0.15	1.4
Fat	0.718	0.037	0.042	2.1
Protein	35.802	0.207	0.327	0.3
Ash	7.952	0.029	0.075	0.4
Lactose	49.021	0.180	0.478	0.4

31.4.3 Repeatability and Reproducibility

Repeatability and reproducibility of the determination of moisture, fat, protein, ash, and lactose in skim milk powder with the InfraAlyzer 400 were calculated. For this purpose, 24 subsamples of skim milk powder were analyzed in duplicate on 24 successive working days with the InfraAlyzer 400 (Table 31.13) and with the reference methods (Table 31.14). From Tables 31.13 and 31.14, it can be concluded that the repeatability of the InfraAlyzer 400 for the determination of moisture, fat, and protein in milk powder is the same or better than that of the reference methods, whereas the reproducibility is the same or better for all components determined by NIR.

31.5 CASEIN AND CASEINATES

Caseins are produced by precipitating the protein from milk and drying the precipitate to approximately 8% moisture. Casein is a white crystalline substance and is manufactured in two forms depending on the process used for precipitation: acid casein and rennet casein. Caseins come in various mesh sizes, as well as unground casein. The most common sizes are 30, 60, 80, and 100 mesh. For the preparation of sodium or calcium caseinate, the casein is mixed with water, sodium hydroxide, or calcium hydroxide, and the mixture is spray dried to approximately 4% moisture. The powder produced is a very fine white powder that is very soluble. Because caseinates have small particle sizes, its grinding is not required. Also 60-, 80-, and 100-mesh casein does not require grinding; however, unground casein and 30-mesh casein require grinding for a short time (about 20 s) in a laboratory grinder. The different methods to produce casein and caseinates not only give products with differences in particle size, shape, and composition but also with differences in reflectance characteristics (Figure 31.6). Therefore it is better to make different calibrations for each category

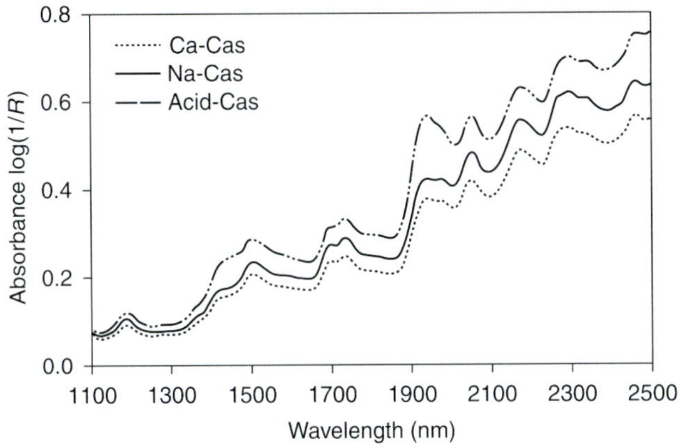

FIGURE 31.6 NIR reflectance spectra of sodium caseinate spray, calcium caseinate spray, and acid casein.

of product. However, obtaining a relevant number of samples of one category with a range of constituents can be a problem because of the tight specifications demanded. Some analysts therefore manipulate the composition of the calibration samples. A range for moisture can be obtained by drying out some samples and exposing other samples to the atmosphere to absorb moisture. Fat is another parameter that remains constant in casein and caseinates. However, manipulating the range by blending high- and low-range samples tends to become unrepresentative of the manufacturing process and therefore will not give accurate results. Another way to obtain samples with a sufficient range for calibration is to manipulate the process of producing casein and caseinates—a rather wasteful and expensive operation.

In order to calibrate our NIRS instruments, we have collected during a period of more than a year samples of casein and caseinate originating from the most important casein and caseinate producer in the Netherlands. However, especially for the fat and ash contents, it was not possible to get a calibration set with sufficient variation in range that is necessary to make a reliable and robust calibration. In order to get more variation in sample composition, we tried to make one "combined" calibration for both casein and caseinates. In spite of the earlier-mentioned differences in particle size and reflectance characteristics between casein and caseinate samples, the results for the calculated parameters were surprisingly good (Table 31.15). Emanating from the required accuracy of 0.1% for moisture, 0.05% for fat, 0.3% for protein, and 0.1% for ash, the accuracy of the determination of the mentioned parameters by NIRS is relative low. However, the NIRS linear relationship is calculated from the results of a large number of samples spread over a wide range of composition. By that, the calibration should be reliable and robust for screening all kinds of casein and caseinates on composition, which is confirmed by testing the calibration with 100 samples of divergent composition (Table 31.15).

TABLE 31.15
NIR Reflectance Data (g/100g) Measured with a InfraAlyzer 500 of a Combined Calibration of Casein and Caseinates

	Calibration				Predication			
Constituent	*n*	*R*	**SEC**	**Range**	*n*	*R*	**SEP**	**Range**
Moisture	100	0.99	0.14	3.2–8.9	100	0.99	0.15	3.5–8.7
Protein	100	0.95	0.n41	85.9–92.9	100	0.93	0.50	86.2–92.8
Fat	100	0.75	0.10	0.52–1.22	100	0.71	0.11	0.6–1.2
Ash	100	0.97	0.25	0.3–4.9	100	0.96	0.27	0.5–4.7

31.6 BUTTER

The composition of butter is regulated by law for parameters such as moisture and the solid nonfat content. The economics of manufacture demand that the final products are as near to the target values as possible. Traditionally, therefore, within the industry, a wide range of empirical tests have been built up to measure these parameters rapidly in order to be able to correct the production process immediately. Advantage of the NIRS technique is that the very speed of analysis often enables more tests to be carried out, thus providing tighter control than otherwise might be possible. To make a reliable butter calibration for measuring moisture and solid nonfat content, it is important to know that there are a few factors that can influence the accuracy negatively. The first one is the variation in composition. It is not easy to obtain a relevant number of samples with a sufficient range of moisture and solid nonfat content because of the tight specifications demanded. The second factor is the sample temperature. We have observed that a difference in temperature of about 5°C affects the accuracy, negatively caused by differences in absorbance level even after transforming the spectra in the second derivative spectra (Figure 31.7a and b). Therefore, it is absolutely necessary to standardize the sample temperature for NIR analysis. The third important factor is the sample preparation and the loading of the sample cup. Butter that has a very high milk fat content may undergo structural changes by kneading the sample resulting in differences in reflectance. Therefore, sample preparation and loading of the sample cup must be standardized. Because hardened butter will not spread easily into a cup with a quartz window without kneading of the sample, and because closed cups require more time to clean before measuring of the next sample, we used an open cup. With a little practice, it is possible to pack butter samples into an open cup using a spatula and to prepare a surface suitable for determinations of quality or composition.

We have collected 50 samples of hardened butter (sour uncolored) originating from four different factories and analyzed the samples in duplicate with both the reference methods and NIRs (InfraAlyzer 500). The calibration is tested by selected samples representative of the calibration set as far as possible. The calculated SEC of 0.07% for moisture and solid nonfat content (Table 31.16)

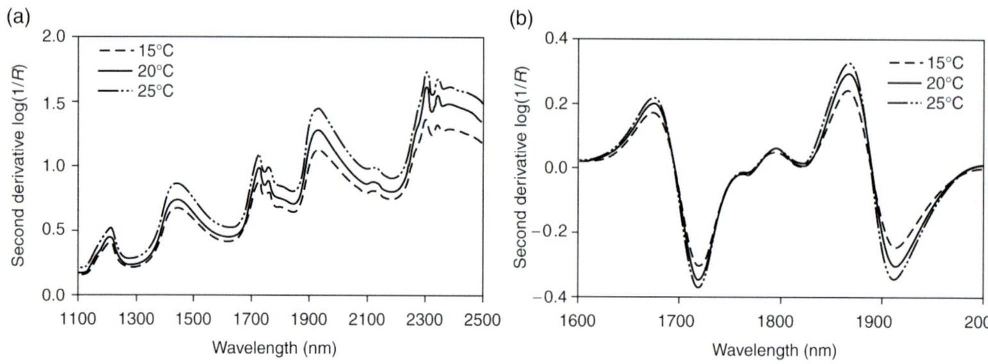

FIGURE 31.7 (a) NIR reflectance spectra for butter obtained by 15°C, 20°C, and 25°C. (b) Second-derivative NIR spectra of butter obtained by 15°C, 20°C, and 25°C.

TABLE 31.16
NIR Reflectance Data for Moisture and Solid Nonfat Content (g/100 g) of Butter

	Calibration				Prediction			
Constituent	*n*	*R*	**SEC**	**Range**	*n*	*R*	**SEP**	**Range**
Moisture	50	0.96	0.07	5.30–16.36	50	0.94	0.08	15.33–16.33
Solid nonfat content	50	0.94	0.07	1.59–2.41	50	0.88	0.09	1.63–2.37

is compared with the standard deviation of the reference method of about 0.05%. Concerning the prediction set (test set), the results for the SEP values are comparable with those of the calibration set. It can be concluded that the calibration is useful for screening butter samples on moisture and solid nonfat content. The calibration procedure for moisture resulted in three filters giving significant information, including the filter for water at 1940 nm. The procedure for solid nonfat content selected four filters, including the filter for fat at 1759 nm, for water at 1940 nm, and for protein at 2100 nm.

31.7 CHEESE

Cheese is one of the most difficult products to analyzed using NIRS because of the differences in process, physical and chemical composition, and the high level of moisture and milk fat. Therefore, it is very important to standardize the sampling and sample preparation in order to get an accurate and representative measurement. The purpose of our study was to investigate the feasibility of applying NIRS to measure some major and minor constituents, as well as the ripening time of the two most important Dutch cheese types: Edam and Gouda. To this purpose, we collected about 200 samples of cheese of both types from six producers in the Netherlands. The samples were prepared for analysis by removing the rind and ground, using a Hobart mill provided with a cheese head according to the Dutch standard. The samples were analyzed in duplicate according to the following methods. Moisture content, salt content, and pH were determined according to the Dutch standard. Fat content was measured by the Schmid–Bondzynski–Ratzlaff extraction method. Total nitrogen content was measured according to the Dutch standard making use of a Kjeltec automatic system. The protein content was calculated by multiplying the total nitrogen content with the factor 6.38. The soluble N contents were determined according to the method of Noomen (1977).

31.7.1 Sample Preparation

As a general rule, temperature differences of the samples will adversely affect the precision of the results. Products having high milk fat contents may undergo structural changes resulting in differences in reflectance. For similar reasons, ground cheese should not be exposed to room temperatures any longer that is strictly necessary and "spreading" movements when filling the sample cup should be avoided.

The RIKILT procedure for preparing cheese samples for NIR is as follows:

1. Take samples according to the Dutch standard.
2. Prepare the sample according to the Dutch standard. (*Note:* If analysis is postponed, the prepared sample should be stored in a cool place, that is, about 8°C. It should be removed from cold storage a few hours before commencing the analysis and be allowed to attain room temperature, that is, 20±2°C, without any active heating. In the sample container, the sample should be mixed with a spatula.)
3. Using a spatula, transfer sample to an open sample cup until it is amply filled.
4. Using the spatula, compress the (grained) sample to a level just below the rim of the sample cup. Avoid "spreading" manipulations.
5. Place the sample cup in the sample drawer of the NIR apparatus and close the sample drawer.
6. Carry out the measurement.
7. After measuring, remove the sample material and clean the sample cup.
8. For prediction purposes (composition of unknown samples), make measurements in triplicate (by repeating steps 3–6 inclusive).
9. Average the results obtained and calculate the percentages of the components by means of the calibration graph. (*Note:* For drawing the calibration graph, it is usually sufficient to perform single measurements.)

31.7.1.1 Effect on Precision

By taking average absorbance/reflectance values of three measurements, a reduction of SEC or SEP of about 30% is obtained. Temperature deviations of 5°C–10°C between the samples produce an increase of the SEC or SEP of 10%–50%, depending on the component to be determined.

31.7.1.2 Alternative Sample Preparation

In order to reduce the influence of the sample preparation on the accuracy of the NIRS measurements, we investigated an alternative sample preparation (described in the following text). This alternative procedure offers advantages over the previously described one because of the complete elimination of the variability in grain size, the uniform sample thickness (and thus the uniform height of filling the sample cup), whereas at the same time the (sub)samples will less readily "perspire." Samples prepared in this manner will produce a reduction in SEC and SEP of 20%–40% as compared to the SEC and SEP calculated from a similar set of samples prepared by the other method. Nonetheless, we have chosen the sample preparation first described for the following two reasons:

1. The sample material is similar to that used for the classical (reference) method.
2. The (sub)samples obtained are considered representative of the entire cheese.

The alternative sample preparation method is as follows:

- Follow the instructions of NEN 3752 up to 4.2.
- With the aid of a slicing apparatus for cheese or meat, cut a 7-mm-thick slice from the sample.
- With the aid of a cork borer, squeeze a small circular piece (of diameter 27 mm) out of the cheese.
- Place this piece in the center of an open sample cup and make the measurement.

31.7.2 Results: Major Constituents

As expected, high correlation coefficients between the reference methods, and NIRS were found for moisture, fat, and protein (Table 31.17). The SEC between the reference methods and NIRS is relatively low compared with the required accuracy of 0.2% for moisture, 0.2% for fat, and 0.25% for protein. Better results can be obtained by splitting up the calibration set into two, one for each type of cheese. In that case, the SEC values decrease by about 30%. Concerning the results obtained for the prediction sets, it can be concluded that the calibrations are robust and reliable for screening samples on major constituents. To give optimal results, four wavelengths per constituent were selected.

TABLE 31.17
NIR Reflectance Data for Moisture, Fat, and Protein (g/100 g) of a Combined Calibration of Gouda (48+) and Edam Cheese (40+) Obtained with an InfraAlyzer 500

	Calibration				Predication			
Product	***n***	***R***	**SEC**	**Range**	***n***	***R***	**SEP**	**Range**
Moisture	125	0.993	0.32	34.4–46.04	100	0.991	0.35	43.11–45.98
Fat	125	0.995	0.26	22.89–43.06	100	0.992	0.28	23.01–34.04
Protein	125	0.995	0.37	23.23–28.45	100	0.995	0.36	23.25–28.40

31.7.3 Results: Minor Constituents

Although NIRS is a new and exciting technique, the main potential of the technique is in the analysis of conventional major constituents such as moisture, fat, protein, and sugars in all kinds of products. Nevertheless, we also investigated the possibilities of NIRS to determine some minor constituents in cheese. This was done in order to evaluate the accuracy of NIRS for the analysis of salt and pH for control purposes and to evaluate the possibility of determining parameters such as the amount of nitrogen soluble in water or in trichloroacetic acid (TCA). Preliminary chemical investigations at our institute show that the last-mentioned parameters give a rather high correlation with the ripening age of Dutch cheese (Venema 1987). From these, it seems possible to control cheese on its declared degree of ripeness in terms of young ripened cheese by measuring the mentioned parameters.

Results for the calculated parameters salt, pH, sol. N/tot. N, TCA sol. N/tot. N, and water-soluble primary amines are surprisingly good (Table 31.18). For salt, the second-derivative method gives the best results. This could be expected because the determination of salt by NIRS is based on the change in the behavior of the water component in the spectrum.

Bernal and Fowler (1933) showed that the addition of electrolytes changes the spectrum of water in the IR overtone region. Luck (1974) showed that temperature variations caused similar spectral changes and attributed these changes to variations in the amount of hydrogen bonding. Begley et al. (1984) showed that the best calibration of NIR data to the salt content in meat occurs at a point in the spectrum where the salt-induced changes in the water spectrum can be mathematically isolated from other spectral variations by using a scanning NIRS instrument.

The low correlation coefficient of 0.747 for the pH is caused by the small pH range of about 0.6 pH unit. On the other hand, the determination of pH in cheese by NIRS is the result of many overlapping overtones that are also very weak. Consequently, it seems impossible to make a high precision calibration for the measurement of pH by NIRS. Mathematical manipulation of the raw NIR data of the cheese samples shows a high correlation or intercorrelation between the parameters—water sol. N/tot. N, TCA sol. N/tot. N, and water-soluble primary amines—and the total protein content. This is to be expected because during cheese ripening, proteins are broken down to peptides and amino acids, mainly by enzymatic processes. Because the measurement of protein by NIRS is based on the absorption of casein molecules as well as a variety of peptides and amino acids, it is very difficult to resolve the protein absorption bands in a lot of smaller bands and to correlate these bands to the constituents from which they arise. The only possibility probably is to use higher-order mathematical data transformations.

TABLE 31.18
NIR Reflectance Data for Some Minor Constituents of Gouda (48+) and Edam Cheese (40+)[a]

	Calibration				Predication			
Constituent	***n***	***R***	**SEC**	**Range**	***n***	***R***	**SEP**	**Range**
Salt (NaCl)	125	0.937	0.15	1.2–3.3	100	0.932	0.16	1.2–3.3
pH	125	0.747	0.08	5.1–5.8	100	0.713	0.08	5.1–5.7
Water sol. N/tot. N	125	0.916	2.07	11.1–34.7	100	0.892	2.50	11.1–36.0
TCA sol. N/tot. N	125	0.905	1.34	8.4–22.8	100	0.891	1.51	8.4–24.3
Water-soluble primary amines (μmol/g protein)	70	0.976	43.1	300–160	—	—	—	—

[a] Obtained with an InfraAlyzer 500.

31.7.4 Determination of Ripening Time

31.7.4.1 General

Dutch regulation for Gouda and Edam cheese requires a minimum ripening time of 28 days at a minimum temperature of 12°C before they may be sold. This is the only legal constraint set for the degree of ripening of these two types of cheese. For more aged cheese, there are no legal rules. Cheese being sold in retail is classified on the basis of its ripening time as follows: young (min. 28 days), young-matured (min. 2 months), matured (min. 4 months), extra matured (min. 7 months), old (min. 10 months), and extra old (min. 1 year). Because ripening conditions for cheese of the same type show little variation, a relationship between maturity (degree of ripening) and ripening time or category can be expected. Because the cost price is dependent on the category (the younger the cheaper), it is of interest to be able to verify the classification. Preliminary chemical investigations at our institute (Venema 1987) show that, from an analytical and economical point of view, an acceptable estimation of the ripening time of Edam and Gouda cheese can be made with a maximum of three independent variables. The three variables chosen by stepwise linear regression analysis were the TCA sol. N/tot. N ratio, which in all cases gave the highest correlation, the salt content, and the moisture content. The proteolytic parameter TCA sol. N/tot. N will increase with time, but the rate of the proteolytic degradation process will diminish with decreasing water content and increasing salt content. Therefore the variance in ripening temperature, humidity, enzyme activity, total composition, etc., influences the relationship between ripening stage and age negatively. The multiple correlation coefficients of the series were situated between 0.94 and 0.95. If discriminating between the young and the young-matured class of both Edam and Gouda cheese, the 95% confidence level is 12–13 days and 18 days if discriminating between the young-matured and matured class.

Next to the above-mentioned chemical investigations, we also studied the suitability of NIRA for the determination of the ripening stage of cheese. The same samples as for the chemical investigations were used. About 100 cheese samples with different ripening stages were collected from different producers. The known age of the cheese samples ranged from 25 to 412 days.

31.7.4.2 Results

For the total sample set, we found a correlation coefficient of 0.92 with an SEC of 28 days (Table 31.19). The reason for this high SEC is the very large range of 25–412 days for the cheese samples. It seems that in this large range, the relationship between ripening stage and age of cheese is not a linear one (Figure 31.8). The SEC is much lower when a calibration with a subset of samples with a range up to 160 days is made. Now a correlation coefficient of 0.96 is found with an SEC of about 11 days, somewhat higher than the value of about 8 days found with the chemical method.

TABLE 31.19
NIR Reflectance Data for Determination of the Ripening Time (Days) of Edam and Gouda Cheese Obtained with an InfraAlyzer 500

	Calibration			
Type of Cheese	*n*	*R*	SEC	Range
40% + 48% fat	100	0.92	28	25–412
40% + 48% fat	80	0.96	11	25–160
40% fat[a]	40	0.97	7	25–160
48% fat[b]	40	0.97	10	25–160

[a] Gouda cheese with 40% fat in dry matter.
[b] Edam cheese with 48% fat in dry matter.

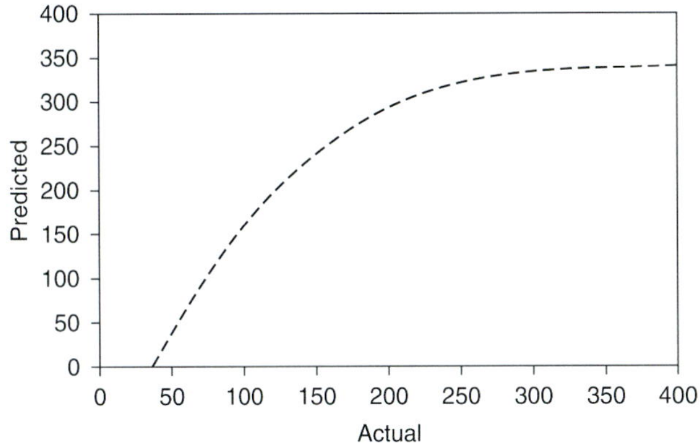

FIGURE 31.8 Relationship between the actual age and the predicted age of two types of Dutch cheese samples (range 25–412 days).

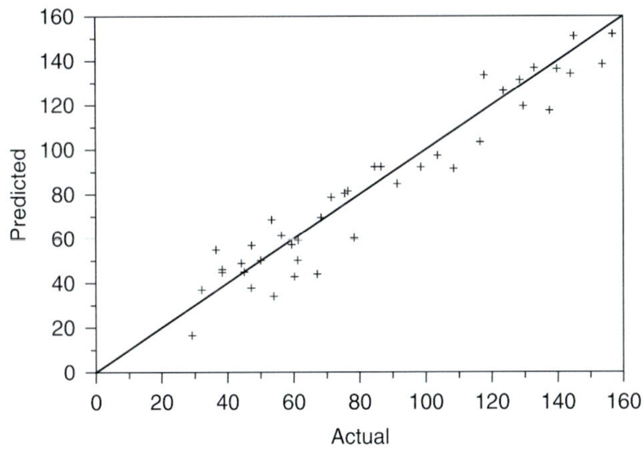

FIGURE 31.9 Relationship between the actual age and the predicted age of cheese samples with 40% fat (range 25–160 days).

Furthermore, we have divided the reduced set into two different sample sets because we used two different cheese types: 40% fat in dry matter and 48% fat in dry matter. A multiple correlation coefficient of 0.97 is calculated for both types of cheese with SEC values of 7.2 days (Figure 31.9) and 10 days, respectively (Table 31.19).

It was surprising that the selected wavelengths for the prediction of age are the same as used for predicting the ratio between TCA sol. N/tot. N, the percentage of water, and the percentage of salt. These are exactly the same parameters as are found in the chemical method. Because young-matured cheese is the main product, we tested the combined (40% + 48%) calibration for samples in the range of 25–160 days with 20 unknown samples. The results indicated that this calibration is insufficient for the quantitative determination of the age of young-matured or matured cheese. Because only the minimum ripening time of cheese with a significance of 5% (= 0.05) is of interest, a one-tail test is required. The 95% reliability interval of the lower tolerance limit, or rejection limit, for the cheese categories is 1.64 × residual standard error. For our test set, we calculated a SEP of about 25 days (1.64 × 15 = 25) for a 95% reliability interval. As was mentioned previously, the required analytical and economical accuracy in this region must be about 14 days. Therefore, we must conclude that the determination of the ripening time of cheese with NIRS is insufficient.

31.8 CONCLUSION

The increase in demands of quality control in the dairy industry requires the development of instruments for the rapid determination of quality-defining parameters. NIR reflectance instruments have the potential to provide these benefits. The main problem of NIRA is to find a representative sample set to calibrate the NIRA instrument. Sample preparation plays an important role. Particle size, homogeneity, temperature, and presentation of the sample have to be standardized.

The precision of NIRA is limited to a great extent by the precision of the reference methods used for calibration. Therefore, accurate and consistent laboratory analyses are very important.

Useful regression equations can be generated to determine the major constituents in milk, milk powder, casein, butter, and cheese. The accuracy of NIR analysis of major constituents in dairy products is similar to the accuracy obtained with the well-known reference methods. Repeatability and reproducibility of NIRA are the same or better than that of the reference methods.

By using a scanning instrument, the quantitative determination of parameters such as salt, pH, water sol. N/tot: N. TCA sol. N/tot. N, and water-soluble primary amines is also possible. The accuracy of the determination of these parameters with NIRS is relative low.

The quantitative determination of the age of cheese with NIRS is only possible with low accuracy.

REFERENCES

Arango, O., A.J. Trujillo, and M. Castillo. 2018. Monitoring the effect of inulin, protein, and calcium on milk coagulation phases using a fibre optic sensor. *Int. Dairy J.*, *81*: 80–86.

Balabin, R.M., and S.V. Smirnov. 2011. Melamine detection by mid-and near-infrared (MIR/NIR) spectroscopy: A quick and sensitive method for dairy products analysis including liquid milk, infant formula, and milk powder. *Talanta*, *85*(1): 562–568.

Bear, R.J., and J.F. Frank. 1982. Analysis of nonfat dry milk using near infrared reflectance spectroscopy. *Proceedings of the 2nd Annual NIRA Symposium*, Technicon, Tarrytown, NY.

Bear, R.J., J.F. Frank, and M. Loewenstein. 1983. *J. Assoc. Off. Anal. Chem.*, *4*: 66.

Begley, T.H., E. Lanza, K.H. Norris, and W.R. Hruschka. 1984. Determination of sodium chloride in meat by near-infrared diffuse reflectance spectroscopy. *J. Agric. Food Chem.*, *32*(5): 984–987.

Ben-Gera, I. and K.H. Norris. 1968. Influence of fat concentration on absorption spectrum of milk in near-infrared region. *Isr. J. Agric. Res.*, *18*(3): 117.

Bernal, J.D. and R.H. Fowler. 1933. A theory of water and ionic solution, with particular reference to hydrogen and hydroxyl ions. *J. Chem. Phys.*, *1*(8): 515–548.

Bertrand, D., P. Robert, D. Launay, and M.F. Devaux. 1987. Application of principal component analysis to the determination of fat and protein content of milk by near infrared spectroscopy, *Euro Food Chem. IV*, Leon, Norway, 519–523.

Biggs, D.A. 1972. Precision and accuracy of infrared milk analysis. *J. Assoc. Off. Anal. Chem.*, 55: 488.

Biggs, D.A. 1979. Applications of infrared instrumentation in cheese analysis. *Proc. 1st Biennial Marschall Int. Cheese Conference,* Madison, Wisconsin, 409–414.

Birth, G.S., and C.J. Washam. 1977. Optical properties of Blue cheese as affected by aging. *J. Dairy Sci.*, *60*: 57.

Bogomolov, A. and A. Melenteva. 2013. Scatter-based quantitative spectroscopic analysis of milk fat and total protein in the region 400–1100 nm in the presence of fat globule size variability. *Chemometrics and Intelligent Laboratory Systems*, *126*: 129–139.

Bogomolov, A., V. Belikova, V. Galyanin, A. Melenteva, and H. Meyer. 2017. Reference-free spectroscopic determination of fat and protein in milk in the visible and near infrared region below 1000 nm using spatially resolved diffuse reflectance fiber probe. *Talanta*, *167*: 563–572.

Butler, W.L., and K.H. Norris. 1960. The spectrophotometry of dense light-scattering material. *Arch. Biochem. Biophys.*, *87*: 31–40.

Casado, P., C. Blanco, A. Pozas, and L. Matorras. 1978. Determination of Fat, Protein and Moisture in milk-powders by means of reflectance spectroscopy in the NIR range. *Proc. Intl Dairy Congress*, 376.

Coppa, M., A. Revello-Chion, D. Giaccone, A. Ferlay, E. Tabacco, and G. Borreani. 2014. Comparison of near and medium infrared spectroscopy to predict fatty acid composition on fresh and thawed milk. *Food Chemistry*, *150*:49–57.

Davies, A.M.C. and A. Grant. 1987. Near infra-red analysis of food. *International Journal of Food Science & Technology*, *22*(3): 191–207.
de la Roza-Delgado, B., A. Garrido-Varo, A. Soldado, A.G. Arrojo, M.C. Valdés, F. Maroto, and D. Pérez-Marín. 2017. Matching portable NIRS instruments for in situ monitoring indicators of milk composition. *Food Control*, *76*: 74–81.
EEC regulation nr. 1725/79. Annex III, method L296/10. Quantitative analysis of skimmed milk powder in feedingstuffs by enzymatic coagulation of casein. Search subject at Link: https://eur-lex.europa.eu/legal-content/EN/TXT/?uri=CELEX%3A31980R2851.
Frank, J.F. and G.S. Birth. 1982. Application of near infrared reflectance spectroscopy to cheese analysis. *Journal of Dairy Science*, *65*(7): 1110–1116.
Frankhuizen, R. 1985. The use of NIRA for quality control of dairy products. *Proc. 8th Int. Symposium on Near Infrared Reflectance Analysis*. Technicon, Tarrytown, New York.
Frankhuizen, R. 1986. The Use of NIRS for Quality Control of Dairy Products. *Proc. Intl Near Infrared Diffuse Reflectance/Transmittance Spectroscopy Conference*, Budapest.
Frankhuizen, R., E.A.M. Boers, and H. Oortwijn. 1983a. Determination of the composition of dried skim milk by near infra-red reflectance spectroscopy (NIR). *Zuivelzicht*, *75*(9): 210–213.
Frankhuizen, R., E.A.M. Boers, and H. Oortwijn. 1983b. Determination of the composition of dried buttermilk by near infrared reflectance spectroscopy (NIR). *Zuivelzicht*, *75*: 547–549.
Frankhuizen, R., and N.G. van der Veen. 1985. Determination of major and minor constituents in milk powders and cheese by near infrared reflectance spectroscopy. *Neth. Milk Dairy J.*, *39*: 191–207.
Giangiacomo, R., D. Torreggiani, J.F. Frank, M. Loewenstein, and G.S. Birth. Analysis of Blue cheese using infrared reflectance spectroscopy. 1979. *J. Dairy Sci.*, *62*: 39.
Goulden, J.D.S. 1957. Diffuse reflexion spectra of dairy products in the near infra-red region. *J. Dairy Res.*, *24*(2): 242–251.
Goulden, J.D.S., 1964. Analysis of milk by infra-red absorption. *J. Dairy Res.*, *31*(3): 273–284.
Huang, H., H. Yu, H. Xu, and Y. Ying. 2008. Near infrared spectroscopy for on/in-line monitoring of quality in foods and beverages: A review. *J. Food Eng.*, *87*(3): 303–313.
Kalinin, A.V., V.N. Krasheninnikov, and A.V. Potapov. 2009. Effect of dispersion structure variation on chemometrical calibration of near-infrared spectrometer: Protein fractions in milk and reversed micelles solutions. *Chemom. Intell. Lab. Syst.*, *97*(1):33–38.
Laporte, M.F. and P. Paquin. 1999. Near-infrared analysis of fat, protein, and casein in cow's milk. *J. Agric. Food Chem.*, *47*(7): 2600–2605.
Latimer, P. 1959. *Influence of Selective Light Scattering on Measurements of Absorption Spectra of Chlorella. Plant Physiol.*, 34(3): 193–199.
Luck, W.A.P. 1974. *Structure of Water and Aqueous Solutions*. Hans Richaz, St. Augustine, West Germany, Ch. 3.
Manley, M., 2014. Near-infrared spectroscopy and hyperspectral imaging: Non-destructive analysis of biological materials. *Chem. Soc. Rev.*, *43*(24): 8200–8214.
McGann, T.C.A., 1978. Automated physico-chemical methods for the analysis of milk: A review of major advances (1960–1978). *Ir. J. Food Sci. Technol.*, *2*: 141–155.
Mestre, H., 1935. The absorption of radiation by leaves and algae. *Proc. Cold Spring Harbor Symp. Quant. Biol.*, *3*: 191–209.
Melenteva, A., V. Galyanin, E. Savenkova, and A. Bogomolov. 2016. Building global models for fat and total protein content in raw milk based on historical spectroscopic data in the visible and short-wave near infrared range. *Food Chem.*, *203*:190–198.
Melfsen, A., E. Hartung, and A. Haeussermann. 2012. Accuracy of milk composition analysis with near infrared spectroscopy in diffuse reflection mode. *Biosyst. Eng.*, *112*(3): 210–217.
Noomen, A., 1977. A rapid method for the estimation of the dissolved and undissolved nitrogen compounds in cheese. *Neth. Milk Dairy J.*, *31*: 163–176.
Norris, K.H. 1965. Measuring and using transmittance properties of plant materials. *Proc. Electromagnetic Radiation in Agriculture Conference*, Illuminating Engineering Society, New York, and American Society of Agricultural Engineers, St. Joseph, Michigan, 64–66.
Revilla, I., O. Escuredo, M.I. González-Martín, and C. Palacios. 2017. Fatty acids and fat-soluble vitamins in ewe's milk predicted by near infrared reflectance spectroscopy. Determination of seasonality. *Food Chem.*, *214*: 468–477.
Robert, P., D. Bertrand, M.F. Devaux, and R. Grappin. 1987. Multivariate analysis applied to near-infrared spectra of milk. *Anal. Chem.*, *59*(17): 2187–2191.

Roberts, G.M. 1977. Infrared analysis of milk powders and casein products. Technical Paper Pacific Scientific No. 4007.
Rodriguez-Otero, J.L., M. Hermida, and J. Centeno. 1997. Analysis of dairy products by near-infrared spectroscopy: A review. *J. Agri. Food Chem.*, *45*(8): 2815–2819.
Šašić, S. and Y. Ozaki. 2001. Short-wave near-infrared spectroscopy of biological fluids. 1. Quantitative analysis of fat, protein, and lactose in raw milk by partial least-squares regression and band assignment. *Anal. Chem.*, *73*(1): 64–71.
Suárez, P.L., A. Soldado, A. González-Arrojo, F. Vicente, and B. de la Roza-Delgado. 2018. Rapid on-site monitoring of fatty acid profile in raw milk using a handheld near infrared sensor. *J. Food Compos. Anal.*, *70*: 1–8.
Tao, F. and M. Ngadi. 2017. Applications of spectroscopic techniques for fat and fatty acids analysis of dairy foods. *Curr. Opin. Food Sci.*, *17*: 100–112
Tuinstra-Lauwaars, M., E. Hopkin, and S. Boelsma. 1985. Inventory of IDF/ISO/AOAC Adopted Methods of Analysis for Milk and Milk Products. *J. Assoc. Off. Anal. Chem.*, *68*: 1235.
Venema, D.P., H. Herstel, and H. L. Elenbaas. 1987. Determination of the ripening time of Edam and Gouda cheese by chemical analysis. *Neth. Milk Dairy J.*, 41: 215.
Weaver, R.W.V. 1978. Infrared reflectance analysis applied to the dairy industry. *Proc. 8th Technicon Intl NIRA Congress,* London.
Weaver, R.W.V. 1984. Near infrared reflectance analysis applied to dairy products. *Proc. of a Seminar,* University of Reading, London. Royal Society of Chemistry, 91–102.
White, C. H., G. A. Muck, M. Bulthaus, and P. Rotolo. 1978. The use of IR reflectance analyzer for measurement of fat and total solids of dairy products. *Proc. 73rd Annual Meeting of Am. Dairy Science Assoc.,* East Lansing, MI.
Wu, D., P. Nie, Y. He, and Y. Bao. 2012. Determination of calcium content in powdered milk using near and mid-infrared spectroscopy with variable selection and chemometrics. *Food Bioprocess Technol.*, *5*(4): 1402–1410.
Wu, D., Y. He, J. Shi, and S. Feng. 2009. Exploring near and midinfrared spectroscopy to predict trace iron and zinc contents in powdered milk. *J. Agric. Food Chem.*, *57*(5): 1697–1704.

SUGGESTED READINGS

Ahmed, W. and M. Goldstein. Analytical applications of reflectance spectroscopy in the near infrared region: A method for quality control of fat in dried milk, *Lab. Pract.*, *25*: 385 (1976).
Anon., Applications note on measurement of fat and total solids in milk samples using a GQA-31 EL, Pacific Scientific Technical Paper NIR 4000, May 18, 1978.
Anon., Infrared milk analysis, Interim official first action. *Official Methods of Analysis of the Association of Official Analytical Chemists*, Ed. Horwits, W., AOAC, Washington, DC, 13th ed., 1980, pp. 248–249.
Atherton, H.V. and J.A. Newlander. *Chemistry and Testing of Dairy Products.* AVI Publ. Co. Inc., Westport, CT, 4th ed., 1980, pp. 105–111.
Biggs, D.A., Instrumental infrared estimation of fat, protein and lactose in milk: Collaborative study. *J. Assoc. Off. Anal. Chem.*, *61*: 1015 (1978).
Biggs, D.A., Performance specifications for infrared milk analysis. *J. Assoc. Off. Anal. Chem.*, *62*: 1211 (1979).
Biggs, D.A., Infrared estimation of fat, protein and lactose in milk: Evaluation of Multispec instrument. *J. Assoc. Off. Anal. Chem.*, *62*: 1202 (1979).
Black, R.G., Analysis of raw milk by NIR, *Proc. NIR 84: An International Symposium on Near Infrared Reflectance Spectroscopy,* Melbourne, Australia, 1984, pp. 105–114.
Blanco, C., P. Casado, and A. Poza, Application of diffuse reflectance spectroscopy. It's application to the determination of powder milk composition. *Quim. Ind.* (Madrid), *26*: 361–363 (1980) (in Spanish).
Brown, R.J., Modern methods of analysis of milk and milk products: Rapid determination of main components, milk: The vital force, 507 (1987).
Collyer, M.J., An evaluation of the Neotec model 102 microcomputer controlled scanning spectrophotometer, for the measurement of major component in full cream milk powder. Dissertation submitted to the University of Reading, Department of Food Science, in partial fulfillment of the degree of Master in Science in Dairy Science, September 1982.
Diller-Zulauf, A. and H.R. Egli, Quantitative rapid determination of fat, nitrogen and dried substances in whey products with the InfraAlyzer. *Dtsch. Molk.-Ztg.*, *103*: 820–822 (1982) (in German).
Diller-Zulauf, A. and H.R. Egli, Quantitative Schnellbestimmung von Lactose, Fett-Stickstoff- und Trockensubstanze an Milch sowie des Milchfettgehaltes von Rahm mit dem InfraAlyzer 400 Dairy (DR). *Alimenta*, *5*: 107; *6*: 148 (1985).

De Vilder, J. and R. Bassuyt, Practical experiences with an InfraAlyzer 400 in determining the water, protein and fat content of milk powder, *Milchwissenschaft*, 38: 65 (1983).
Egli, H.R., Standardisation, quality insurance and control of milk products with the Technicon InfraAlyzer 400 Dairy, *NIRA International Symposium*, Technicon, Paris, November 1983.
Egli, H.R. and U. Meyhack, Measurements of the principal constituents of solid and liquid milk products by means of near infrared analysis. *Proceedings of a Seminar*, University of Reading, England, March 28–30, 1984. London, Royal Society of Chemistry, 1984.
Frank, J.F., Möglichkeiten zur Schnellbestimmung der Käsezusammensetzung durch die Anwendung der Reflektionsspektroskopie. *DMZ: Dtsch. Molk.-Ztg.*, *15*: 438 (1983).
Gilkison, I., An evaluation of the Technicon InfraAlyzer 400 Dairy for the analysis of milk and dairy products. *NIRA International Symposium,* Technicon, Paris, November 1983.
Goulden, J.D.S. and D.J. Manning, Determination of moisture in dairy products by near-infrared absorption of methanol extracts. *J. Dairy Res.*, *37*: 107 (1970).
Guiader, M., Cheese and meat analysis by NIR. *International Near Infrared Diffuse Reflectance/Transmittance Spectroscopy Conference*, Budapest, Hungary, May 1986.
Hammingh, A., Experience with the Technicon InfraAlyzer 400 R in dried dairy products, as well as on-line analysis of milkpowders, *NIRA International Symposium*, Technicon, Paris, November 1983.
Haugaard, C., Photometric determination of fat in milk. *J. Dairy Sci.*, *49*: 1185 (1966).
Hirschfeld, T., Sample area optimization in a diffuse reflectance near-infrared spectrophotometer. *Appl. Spectrosc.*, *39*: 1085 (1985).
Honigs, D.E., Near infrared analysis. *Anal. Instr.*, *14*: 1–62 (1985).
Honigs, D.E., G.M. Hieftje, and T. Hirschfeld, Number of samples and wavelengths required for the training set in near-infrared reflectance spectroscopy. *Appl. Spectrosc.*, *38*(6): 844 (1984).
Kaffka, K.J. and F. Kulscár, Attempts to determine egg content in pastry products using the NIR-technique. *Acta Alimentaria*, *11*: 47 (1982).
Lawrence, R. C. and J. Gilles, The assessment of the potential quality of young Cheddar cheese. *N. Zeal. J. Dairy Sci. Technol.*, *15*: 1–12 (1980).
Luchter, K., Interlaboratory milk study with the Technicon InfraAlyzer 400 Dairy, *NIRA International Symposium*, Technicon, Paris, November 1983.
McGann, T.C.A., Control of moisture in skim milk powders by IR reflectance using the Anocan model 106, Anim. Prod., The Agriculture Institute, 1975.
Mills, B.L. and F.R. Van de Voort, Evaluation of CH stretch measurement for estimation of fat in aqueous fat emulsions using infrared spectroscopy. *J. Assoc. Off. Anal. Chem.*, *65*: 1357 (1982).
Mitchell, D.J., R.W. Weik, and W. Hortritz, Collaborative studies of methods for butterfat in homogenized and chocolate milk. *J. Assoc. Off. Agric. Chem.*, *47*: 573 (1964).
Roy, R.B. and J. Bahl, Theoretical Basis for the Analysis of Fats and Oils by NIRA. *Presented at the 2nd Annual NIRA Symposium,* Technicon Corporation, Tarrytown, New York, June 1982.
Rudzik, L. and R. Wobbecke, Need for homogenization in infrared measurements on milk. *DZM: Dtsch. Molk.-Ztg.*, *36*: 298 (1982) (in German).
Sjaunja, L.-O., Studies on milk analysis of individual cow milk samples, thesis, Swedish University of Agriculture Sciences, 1982.
Toohey, F.S., Testing of dried milk by NIR, *Proc. NIR 84: An International Symposium on Near-Infrared Reflectance Spectroscopy,* Melbourne, Australia, 1984, pp. 114–119.
Whetsel, K.B., Near-infrared spectrophotometry, *Appl. Spectrosc. Rev.*, 2: 1–67 (1968).
Wollard, D.C., NIR technologie within the New Zealand dairy Industry: Its dramatic impact, *Proc. NIR 84: An International Symposium on Near-Infrared Reflectance Spectroscopy,* Melbourne, Australia, 1984, pp. 95–105.

32 Aquaphotomics

Roumiana Tsenkova[*] *and Jelena Munćan*
Kobe University

Zoltan Kovacs
Szent István University

CONTENTS

32.1 INTRODUCTION

Aquaphotomics (Tsenkova 2009) (*aqua-photo-omics* – all about water–light interaction) is a new scientific discipline, established by Prof. Roumiana Tsenkova and her group at Kobe University, Japan, which utilizes water–light interaction for qualitative and quantitative analyses of water molecular conformations, thus revealing their structure, function, and dynamics in aqueous and biological systems. The main aims in the establishment of aquaphotomics are the synthesis and systematization of all the knowledge about the structure–function relationship of water through water–light interaction revealed as a spectral pattern, an integrative marker.

Aquaphotomics was built on the foundation of near-infrared spectroscopy (NIRS) as a fast, non-destructive method allowing monitoring. Almost two decades ago, the NIRS laboratory at Kobe University, Japan, was interested in the prediction of milk composition and somatic cell count in healthy and mastitic dairy animals (Tsenkova et al. 2001, Tsenkova, Atanassova, and Toyoda 2001). Surprisingly, research results showed that the best accuracy of prediction of milk composition was achieved when models were built on separate datasets of milk spectra from healthy and mastitic animals. Apparently, the prediction accuracy was dependent on the physiological status of the animals. Loading vectors for milk protein measurements of healthy (low somatic cell count) and mastitic group (high somatic cell count) showed prominent spectral differences in the spectral regions around 1440 and 1912 nm (Figure 32.1). When regression vectors of the models for milk protein prediction built on separate datasets were compared, a distinct difference was observed again around the same absorbance bands (Figure 32.2).

These results clearly demonstrated that the regions of the water first overtone and combination region of OH stretching and bending vibrations (Ozaki 2002) carried information about the presence of disease (Tsenkova et al. 2001, Tsenkova, Atanassova, and Toyoda 2001). Subsequent research using other body fluids, urine and blood, and even tissues consistently showed the importance of

* rtsenkova@gmail.com.

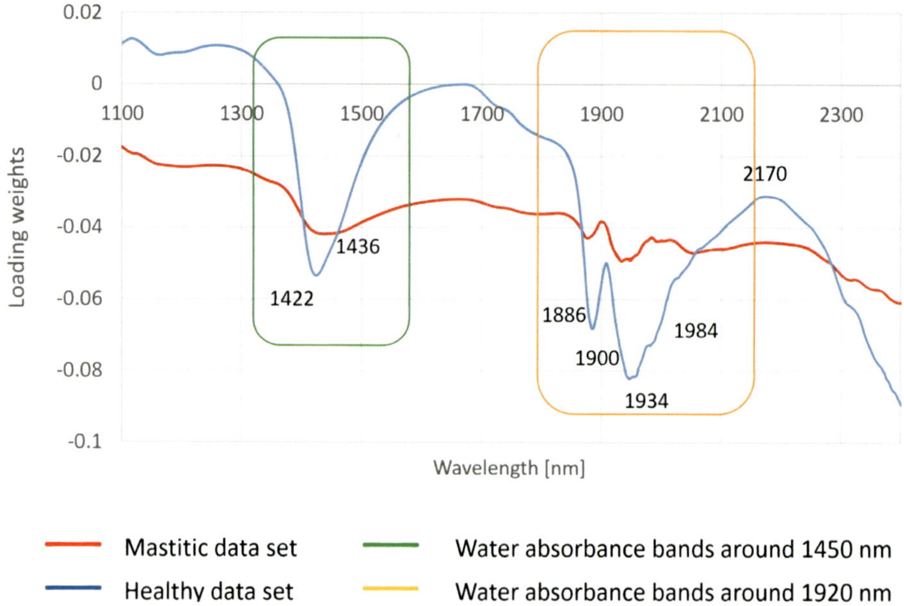

FIGURE 32.1 Factors' loadings in the partial least squares models, highly correlated with the regression coefficients and the protein content of milk of healthy and mastitic animals, reveal different functionality of water species. (Adapted from Tsenkova, R., Atanassova, S., Kawano, S., and Toyoda, K. 2001. Somatic cell count determination in cow's milk by near-infrared spectroscopy: A new diagnostic tool. *Journal of Animal Science* 79 (10):2550–2557.)

water bands for disease diagnosis and prediction of milk quality parameters (Tsenkova 2008a, Tsenkova and Atanassova 2002, Tsenkova et al. 2005).

With water being the common denominator in all the examined samples, there was only one possible explanation for the consistent importance of water absorbance bands (WABs) in the respective prediction models. The water molecular network is altered when composition of the aqueous system is changed, and this variation has been reflected in its near-infrared spectral pattern (Figures 32.1 and 32.2) (Tsenkova 2004). These studies on mastitis diagnostics using NIRS mark the beginning of the realization of the immense informational potential of water spectral patterns.

Numerous methods have been employed so far to study water: X-ray spectrometry (Huse et al. 2009, Nilsson et al. 2010, Shields et al. 2010), infrared spectroscopy (Fecko et al. 2003, Maréchal 2011, Robertson et al. 2003, Weber et al. 2000), Raman spectroscopy (Huang et al. 2009, Kitano et al. 2000, Mizoguchi, Hori, and Tominaga 1992), nuclear magnetic spectroscopy (Matubayasi, Wakai, and Nakahara 1997), dielectric relaxation spectroscopy (Okada et al. 1999), THz spectroscopy (Rønne, Åstrand, and Keiding 1999, Rønne et al. 1997), neutron diffraction (Ichikawa et al. 1991, Soper 2000) etc. Each method contributed to the understanding of water; however, its structure, let alone its functionality, is not yet understood. What early research on mastitis and many subsequent studies showed is that NIRS of bio- and aqueous systems can significantly contribute to the field of water science and better understanding of water molecular system (Tsenkova 2009).

After years of experience in Vis–NIRS of water and aqueous systems, Prof. Dr Roumiana Tsenkova from Biomeasurement Technology Laboratory, Kobe University, Japan, finally proposed the establishment of aquaphotomics (Tsenkova 2006a, 2006b, 2006c, 2006d, 2009) as a new "omics" discipline complementary to other "omics" disciplines, but solely devoted to the study of water through bio- and aqueous systems, to the systematization of already-abundant knowledge on water–light interaction, and to the discovery of new phenomena. What started as only a specialized approach to the analysis of near-infrared spectra, evolved into a new science.

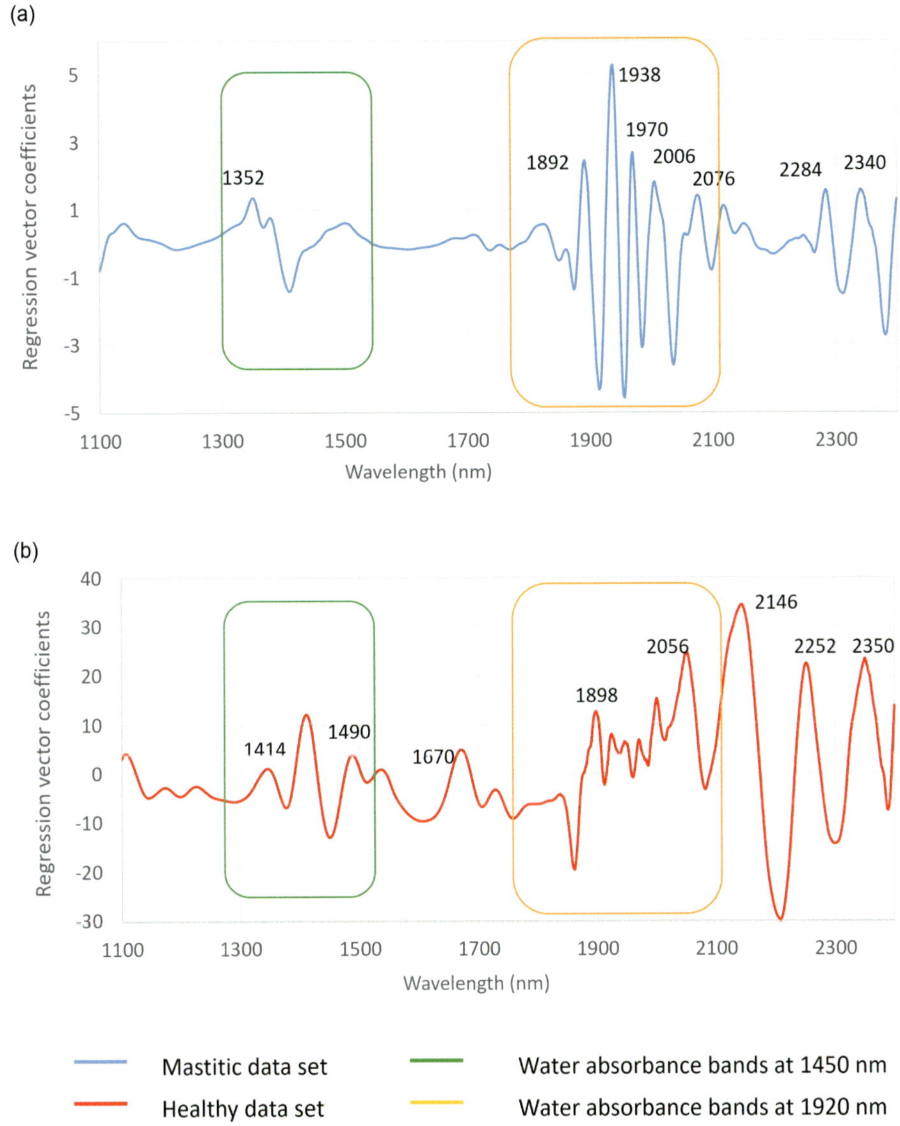

FIGURE 32.2 Water absorbance pattern shows changes of water matrix with the disease. Regression coefficients plots of the models for milk protein content prediction built on separate dataset of (a) spectra of milk samples with low somatic cell count obtained from healthy animals and (b) spectra of milk samples with high somatic cell count obtained from mastitic animals.

The intention of this chapter is to present the basic paradigm shift which the aquaphotomics is built upon, its terms and definitions, methods and techniques used to extract the information from the near-infrared spectra of aqueous systems, and to present numerous examples where aquaphotomics opened up new areas of thought, research, and applications.

32.2 INTRODUCING A NEW PARADIGM: NOVEL CONCEPTS AND TERMS

The analysis of water was present in the field of the NIRS ever since its inception (Hirschfeld 1985, Iwamoto, Uozumi, and Nishinari 1987, Grant, Davies, and Bilverstone 1989, Maeda et al. 1995). These early works analyzed and utilized the properties of water spectra, but the idea of water as a molecular system and biological matrix was not yet considered. Water has been often described as the "greatest enemy" of infrared and NIRS on account of its dominant absorption. There are two reasons for this attitude. For a long time, in biological sciences, the water was regarded as an inert medium. The living processes were described in the terms of genes, DNA, proteins, metabolites, or other single biomolecules. Understandably, the research methods were focused on extracting the signals related to the concentration of these biomolecules, and in the near-infrared region, dominant absorbance of water was seen as an obstacle to observe their absorbance bands. However, the field of water science has seen a significant progress in the last decades, which significantly changed the general opinion of water and its role in the living systems (Ball 2008a, 2008b, Chaplin 2006, Del Giudice, Spinetti, and Tedeschi 2010, Pollack, Cameron, and Wheatley 2006). Water is now recognized as an active solvent, adapting its structure to the solutes that it accommodates, and in biological systems, a biomolecule in its own right with an active role in the dynamics of biomolecular and cellular processes (Figure 32.3) (Ball 2008a).

Aquaphotomics is based on the so-called water collective mirror approach, where all the components of the aqueous system and surrounding energies influence the water structure. Any perturbation would result in changes of water molecular conformation, which will, in turn, be reflected in the water NIR spectrum. The shift in attitude towards water hence, moves focus of research methods to the analysis of signals from water. Analysis of only the changes of the water molecular matrix caused by changes in system functionalities as opposed to the analysis of individual components creates a possibility for a novel analytical approach and contributes complementary information to the "omics" concept (Figure 32.4) and to NIR-IR spectroscopy.

In order to accurately describe the subject of its exploration, aquaphotomics introduced some new terms and definitions, as well as abbreviations of these terms which are commonly used in publications, and it is worth to give a detailed explanation here.

Water matrix. Matrix is a word often used in aquaphotomics to emphasize the fact that water is a molecular network. This molecular network is comprised of different water molecular conformations, which due to the different strength of covalent bond as a result of hydrogen bonding, and the location, give rise to simultaneous existence of many structures and therefore various functionalities (Tsenkova 2009).

Water collective mirror approach or extended water mirror approach. Years of experience in NIRS of aqueous systems led to conclusion that the Vis–NIR spectral pattern captures all intricacies of the water matrix in great details (Tsenkova 2009). Given the extreme sensitivity of hydrogen bonds, even minute change in the environment and solute composition results in multitude of changes of the water matrix, which will be reflected in a mirror-like fashion in its absorbance spectral pattern. The information about the present solutes can thus be acquired indirectly by analyzing the changes at WABs. Aquaphotomics utilizes this informational and amplifying property of water, and that is why it is often described as a "water-mirror approach".

WAMACs – water matrix absorbance coordinates. The WAMACs are spectral ranges where specific WABs related to the respective functions are found with the highest probability (Tsenkova 2009). The word "coordinates" is used because these are the places in the electromagnetic (EM) spectrum, where the measured absorbance of water carries information about the various "building blocks" in its particular structure and allows the observation of the dynamics of aqueous systems. The WAMACs can be thought of as hubs of information and have been assigned to specific water molecular structures (Tsenkova 2009, Tsenkova, Kovacs, and Kubota 2015). For the water first overtone, 12 WAMACs (labeled C_i, $i = 1, 12$) have been experimentally discovered (each 6–20 nm width), and confirmed by overtone calculations of the already-reported water bands in the IR range (Tsenkova 2009). (Figure 32.5, Table 32.1).

WABs – water absorbance bands. One of the main objectives of aquaphotomics is the systematization of the already-existing information and discovery of new WABs related to specific structures. Various studies in the infrared range have identified the absorbance bands of numerous water species

(a)

(b)

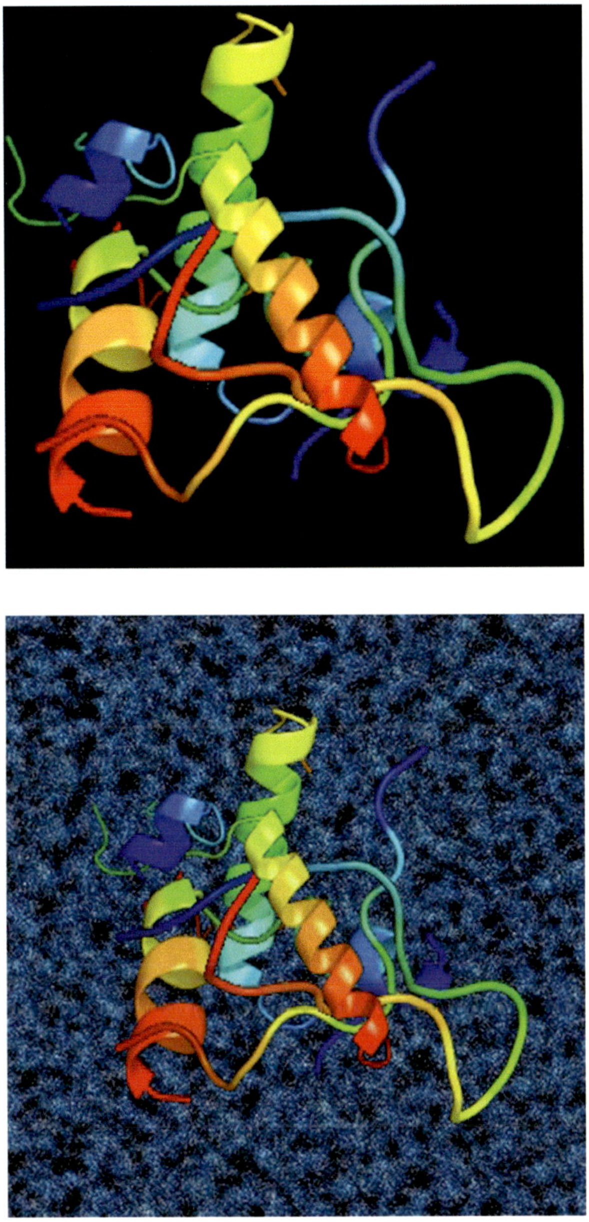

FIGURE 32.3 Shifting the paradigm: (a) From the concept of an inert, passive medium towards (b) water as an active solvent, adapting its structure and dynamics to the solutes that it accommodates and in biological systems a biomolecule in its own right (Ball 2008a).

(Buijs and Choppin 1963, Chandler 2002, Cupane, Levantino, and Santangelo 2002, Czarnik-Matusewicz et al. 1999, Doster et al. 1986, Fornés and Chaussidon 1978, Heiman and Licht 1999, Luck 1998, Maeda et al. 1995, Murayama et al. 2000, Robertson et al. 2003, Sartor, Hallbrucker, and Mayer 1995, Šašić, Segtnan, and Ozaki 2002, Segtnan et al. 2001). When their overtones were calculated, it was confirmed that, together with experimentally confirmed bands, these occur within the whole Vis–NIR range (Tsenkova 2005). So far, the spectral database of WABs has more than 500 bands in the area of the first, second, and third overtone of water (Tsenkova 2009, Tsenkova, Kovacs, and Kubota 2015).

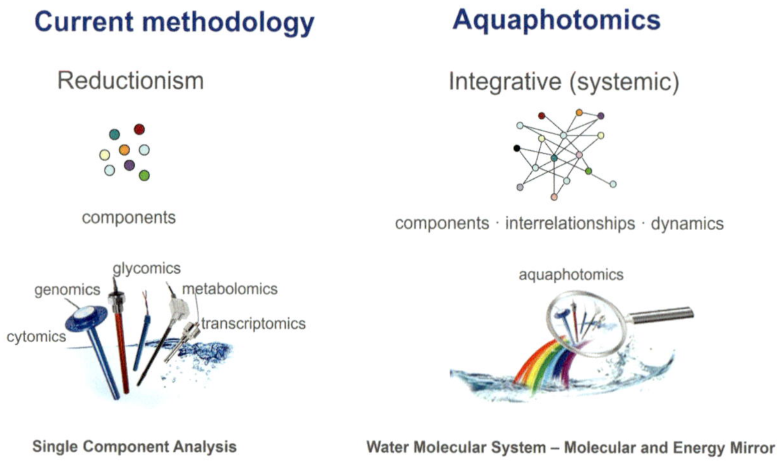

FIGURE 32.4 Aquaphotomics in relation to other "omics" concepts. While other "omics" disciplines provide single component analysis, aquaphotomics considers water a molecular matrix where all the rest of the biomolecules and energies insert their influence and leave their prints. Newly proposed concept uses water as a molecular and energy mirror, which in a holistic manner describes functionality of the whole system.

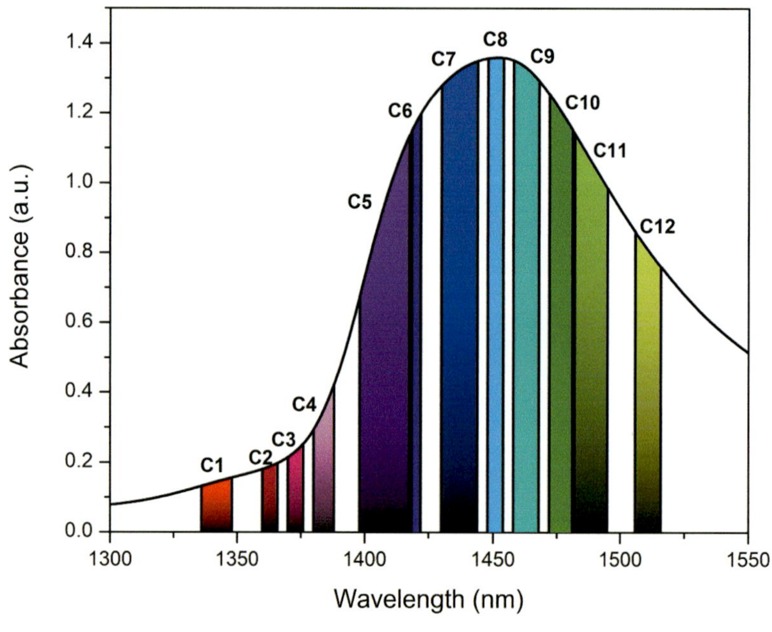

FIGURE 32.5 Water matrix coordinates – WAMACs: 12 characteristic water regions (C1–C12) found under the first overtone of water. (Created based on data from Tsenkova, R. Aquaphotomics: Dynamic spectroscopy of aqueous and biological systems describes peculiarities of water. *Journal of Near Infrared Spectroscopy* 17 (6):303–313.)

Activated water bands. If the respective perturbation of the aqueous system produces changes in the absorbance at specific WABs in the corresponding spectrum, these water bands are considered "activated". The exact location of activated WABs within WAMACs differs per bioaqueous system and perturbation.

WASP – water absorbance spectral pattern. The combination of the *activated water bands* depicts a characteristic spectral pattern, which reflects the condition of the whole water molecular

TABLE 32.1
Assignments of Water Molecular Structures Corresponding to the WAMACs in the First Overtone of Water (Tsenkova 2009)

WAMACs	Range (nm)	Assignment	Reference
C1	1336–1348	$\nu 3$, H_2O asymmetric stretching vibration	Ozaki 2002
C2	1360–1366	Water solvation shell, OH-$(H_2O)_{1,2,4}$	Robertson et al. 2003, Xantheas 1995
C3	1370–1376	$\nu 1 + \nu 3$, symmetrical stretching fundamental vibration and H_2O asymmetric stretching vibration	Ozaki 2002
C4	1380–1388	Water solvation shell, OH-$(H_2O)_{1,4}$ and superoxide, O_2-$(H_2O)_4$	Weber et al. 2000, Xantheas 1995
C5	1398–1418	Free water and free OH-(S0) or trapped water	Iwamoto, Uozumi, and Nishinari 1987, Kaffka et al. 1990, Kojić et al. 2014, Maeda et al. 1995
C6	1421–1430	Water hydration, H-OH bend and O…O	Esquerre et al. 2009, Tsenkova 2009, Tsenkova et al. 2004, Williams 2009
C7	1432–1444	Water molecules with one hydrogen bond (S1)	Cattaneo et al. 2009, Ozaki 2002, Segtnan et al. 2001
C8	1448–1454	$\nu 2 + \nu 3$, water solvation shell, OH-$(H_2O)_{4,5}$	Cattaneo et al. 2009, Ozaki 2002
C9	1458–1468	Water molecules with two hydrogen bonds (S2)	Franks 1973, Tsenkova et al. 2004
C10	1472–1482	Water molecules with three hydrogen bonds (S3)	Franks 1973, Ozaki 2002
C11	1482–1495	Water molecules with four hydrogen bonds (S4)	Gowen et al. 2009, Ozaki 2002, Segtnan et al. 2001
C12	1506–1516	$\nu 1$, $\nu 2$, symmetrical stretching fundamental vibration, and doubly degenerate bending fundamental or strongly bound water	Gowen et al. 2009, Headrick et al. 2005, Segtnan et al. 2001

Source: Adapted from Tsenkova, R. Aquaphotomics: Dynamic spectroscopy of aqueous and biological systems describes peculiarities of water. *Journal of Near Infrared Spectroscopy* 17 (6):303–313.

system. Within the water-mirror concept, WASP represents an integrative marker, which can contain a large amount of chemical and physical information about the respective aqueous system and is directly related to its functionality.

Aquagram. An aquagram (Tsenkova 2010) is a graphical representation of WASP (Figure 32.6). The graph displays normalized absorbance at the selected WABs. The radial axes of aquagram are defined by the selection of WABs. The normalized absorbance is calculated as follows:

$$A'_\lambda = \frac{A_\lambda - \mu_\lambda}{\sigma_\lambda}, \qquad (32.1)$$

where A'_λ –the normalized absorbance displayed on the aquagram, A_λ – absorbance after multiplicative scatter correction (MSC) (Martens, Jensen, and Geladi 1983) using the mean of the dataset as a reference spectrum, μ_λ – mean of all spectra, σ_λ – standard deviation of all spectra, λ – selected wavelengths from WAMAC regions corresponding to activated WABs (Figure 32.6). The exact number of axes, as well as which WABs will be chosen for display, depends on the specific system and perturbation. For the first overtone of water, the axes of the aquagram are usually based on previously discovered 12 WAMACs (Tsenkova 2009). The aquagrams are visually very convenient

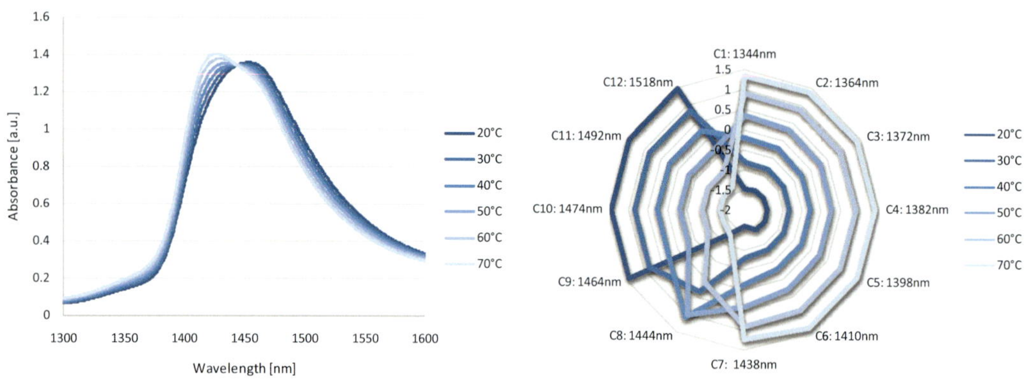

FIGURE 32.6 Spectra of water at different temperatures and corresponding aquagrams displaying the normalized absorbance at the selected water absorbance bands from the WAMAC regions in the area of first overtone of water at 1300–1600 nm.

to allow fast and comprehensive comparison of different systems or conditions under perturbation of the same system by comparison of their WASPs.

Aquaphotome. Aquaphotome can be defined as a comprehensive database of (1) all existing water bands with their assignments related to water molecular structures and (2) water spectral patterns with the interpretation of their functionality per system and perturbation (Tsenkova 2009). Each particular system has its own aquaphotome, relating the water structure as presented by WASP to the functionality it has in the system.

With the main terms and concepts introduced, the goals of aquaphotomics can be defined as:

- Achieving a complete understanding of the interaction between water and EM radiation through discovery and systematization of identified water bands and spectral regions (water matrix absorbance coordinates – WAMACs) where water interacts with light and thus can be monitored
- To use the WABs and water spectral patterns to explain different functionalities in various systems in terms of its water structure striving towards a complete understanding of water as a matrix of life (Tsenkova 2009).

32.3 DEVELOPING NEW TOOLBOX OF METHODS AND TECHNIQUES

The objective of each aquaphotomics study, regardless of the system analyzed, is to extract specific water absorbance spectral pattern (WASP) related to a particular perturbation from the respective spectrum (UV, Vis–NIR, IR etc.). In order to do this, special attention is given to the experimental design and acquisition of high-quality spectra, followed by the enhancement of subtle water signals and resolution of overlapping bands by employing various spectra pretreatment methods such as standard normal variate transformation, MSC, extended MSC, spectral averaging, spectral subtraction or derivatives, as well as an array of exploratory, classification, and regression analyses which can reveal the activated absorbance bands within highly influential variables in the loading vectors, regression vectors, or discriminating powers of the respective models.

Experimental design may vary depending on the type of aqueous systems involved, but should ensure that each measurement is taken using several consecutive spectra (consecutive illumination of the same sample), that each sample is presented with independent replicates, and that the order of replicates measurements is random. During the experiment, it is advisable to monitor major external influences like temperature and humidity as well as the temperature of the sample holder or the cuvette. Measurements of pure water as a control sample should always be repeated after a previously defined number of a group of samples. Measuring external environmental parameters can be very useful for the identification of major sources of perturbation, as well as for exploring the dynamics of different systems under the same environmental perturbations.

The NIR spectra of aqueous systems are very complex, and changes in its absorbance spectra caused by some perturbation will usually be very subtle, but persistent and consistent. All efforts in aquaphotomics data analysis are oriented towards the enhancement of small changes, removing unwanted influences on the spectra and resolving overlapped WABs in order to extract the consistent water spectral pattern caused by the perturbation of interest.

Perturbation is often used in aquaphotomics as opposed to traditional NIRS, which places emphasis on the control of the environment during the measurements. It can be defined as the "evaluation of a physical system by applying random changes to selected parameters and re-estimating the resulting system" (Tsenkova 2007). In practice, it means that the spectral acquisition is performed under various perturbations such as consecutive illuminations, changing temperature, or dilution in order to induce spectral changes of the respective systems. The use of intentional, artificially created perturbations can provide the change of entropy and result in revealing hidden spectral information (Tsenkova 2006c).

One of the first steps in aquaphotomics data analysis is usually spectral subtraction of raw or averaged spectra. The simplest approach is to subtract the average spectrum of pure water whose measurements were taken as a control, but other ways of performing spectral subtraction may reveal more information hidden in the spectral data such as subtracting the average spectrum along the perturbation of interest. These perturbations could include, for example, environmental temperature, humidity, and spectral consecutives (i.e., multiple consecutive illumination of the same sample). There are many other variations possible, for example, to group samples according to criteria (like high pH, low pH, or one solute type vs. another solute type, healthy vs. sick) and then perform the subtraction of average spectra of different groups. Thus, the calculated difference spectra reveal changes which are mainly caused by the criteria used for the selection of groups and can provide information on how a particular perturbation affects the water structure, i.e., the water spectral pattern, WASP. Subtraction of the first consecutive spectra from all other consecutive measurements, and subsequent analysis of thus acquired difference spectra, can reveal different response to illumination of different samples caused by the spectral measurement. This subtraction technique was used in a study of different prion protein isoforms in water solutions under the consecutive illumination perturbation (Tsenkova 2005, Tsenkova et al. 2004), where it was for the first time shown that illumination changes the water system and that each consecutive spectrum is different. The perturbation in the form of the absorbed photons creates an increased number of free water molecules which are available to interact with the present solutes in the aqueous system, thus performing "scanning" of the solutes and the rest of the water molecular system. This resulted in additional information being extracted, and in the case of the mentioned study, this approach revealed drastic differences in the free O-H absorbance bands and superoxides for different prion protein isoforms.

Another subtraction method, recently developed, proposes a "closest spectrum" subtraction (Kojić, Tsenkova, and Yasui 2017). This subtraction method involves creating all the possible pairs of differences (solution – pure solvent) and locating the closest pair (minimal difference) by selecting difference spectrum with the smallest area under the spectral curve. Pure solvent spectra can be acquired during the experiment, or found in a library which must be previously created of solvent spectra taken under various, mainly temperature, perturbations. The method provides on average a fourfold increase in precision compared to traditionally used average spectrum subtraction (Kojić, Tsenkova, and Yasui 2017).

Many of the well-known multivariate analysis methods such as principal component analysis (PCA), partial least squares regression (PLSR), and partial least squares discrimination analysis (PLS-DA), soft modeling of class analogies (SIMCA), multivariate curve resolution-alternative least squares (MCR-ALS), and two-dimensional correlation spectroscopy (2DCOS) are regularly used for the extraction of information of interest. Subtracted spectra, spectral derivatives, PCA loadings, PLSR regression vectors, SIMCA discriminating powers etc. extract the bands where spectral variations under controlled and uncontrolled perturbations could be observed. The obtained variables represent the activated WABs and as such, define the water spectral pattern – WASP, which in a holistic manner describes the analyzed aqueous system. Once the WASP is identified, discovery of new bands and the relation between specific water molecular structures and the defined perturbation are elucidated.

This field is still evolving, and with the growing knowledge, new developments in the experimental design and data analysis can be expected. Recently, a package "aquap2" (Pollner and Kovacs 2016) was developed for the R software environment (free download and instructions available at www.aquaphotomics.com), which enables automatic generation of experimental protocols for the use in aquaphotomics experiments and offers various multivariate analysis tools especially suited for aquaphotomics studies.

32.4 APPLICATIONS

Being a rapid and nondestructive, NIRS is a powerful method with a vast range of applications, and aquaphotomics expanded its horizons. The water-mirror approach provides measurements of concentrations previously thought impossible at ppm (Bázár et al. 2015, Bázár et al. 2014, Gowen, Amigo, and Tsenkova 2013, Sakudo et al. 2006, Tsenkova 2008b), even ppb levels under certain experimental conditions (Sakudo et al. 2005, 2006, Tsenkova 2008a, 2008b, Tsenkova, Iso, et al. 2007).

One of the proof-of-concept studies applied aquaphotomics spectral analysis for the purpose of measuring concentrations of different salts ranging from 0.002 to 0.1 mol L^{-1} (Gowen et al. 2015). Salts are used as model systems as they do not exhibit the NIR spectrum; therefore, the obtained results would be entirely based on the changes in the WABs. This was the first work examining the effect of salts in such a low concentration. Using three different spectrometer systems at three different locations on Earth (Japan, Italy, and Ireland), it was demonstrated that water-mirror approach enables predictions of concentration with the limit of detection at the 1000 ppm level. Other works reported successful measurements of concentration of metal ions in water in a range of 0–10 mg L^{-1} (Putra et al. 2010, Tsenkova, Fockenberg, et al. 2007) and polystyrene particles in aqueous suspensions (0.0001%–1%) (Tsenkova, Iso, et al. 2007). Water molecular mirror approach was further applied for the quantification of mono- and disaccharide solutes at 0.02–100 mM concentration (Bázár et al. 2015). This is the first research confirming the applicability of NIRS for qualitative and quantitative analyses of mono- and disaccharides at millimolar concentrations with the detection limit of 0.1–1 mM.

Recent aquaphotomics studies include the exploration of water state in different materials such as hydrogel materials of soft contact lenses (Munćan, Mileusnić, Šakota Rosić, et al. 2016, Šakota Rosić et al. 2016), and even nanomaterials, such as fullerene-based nanomaterials (Matija et al. 2012).

Aquaphotomics was also applied for the analysis of various biomolecule–water interactions providing insight into their dynamics and significant role the water plays in it. One of the first applications in this field was concerned with the identification of prion protein isoforms (Tsenkova et al. 2004). The binding of manganese (Mn) instead of copper (Cu) is thought to be one of the suggested mechanisms of converting the prion protein into a dysfunctional form. The aquaphotomics analysis of Mn and Cu isoforms revealed that Cu binding increases the protein stability in water, while Mn binding leads to instability and changes, which promote fibril formation. Another study investigated the formation of amyloid fibrils (Chatani et al. 2014). The study was done in parallel: The changes during fibrillation of insulin were monitored in the 2050 to 2350 nm and 1300 to 1600 nm spectral regions. The results showed that the entire process of the conformational changes of protein is reflected in the changes of water molecular matrix (Chatani et al. 2014). Aquaphotomics research on UV-induced changes in DNA (in DNA solutions with 5–20 μM concentration) reported the successful detection of UVC (UV light in the 240 to 260 nm range)-induced photolesions in single-stranded DNA based on the changes in water spectral pattern (Goto et al. 2015).

As a technique mostly suited for water systems analysis, one of the most natural applications of aquaphotomics is water quality control. Since measurements of very small concentrations of solutes were proved achievable, efforts were put into the detection of contaminants such as pesticides (alachlor and atrazine, concentration from 1.25 to 100 ppm, detection limit estimated to be 12.6 ppm for alachlor and 46.4 ppm for atrazine) (Gowen et al. 2011) and also filtration monitoring (Cattaneo et al. 2011). Moving on from the detection and quantification of individual solutes, aquaphotomics introduced a concept of water spectral pattern as an integrative marker (or biomarker) (Kovacs et al. 2016). The applicability of the proposed concept was evaluated by monitoring the water structural changes in different aqueous solutions such as acid, sugar, and salt solutions at millimolar concentration level, as well as in a real-life ground water system (Figure 32.7) (Kovacs et al. 2016). The success of this approach opened a new venue in water quality monitoring by offering a rapid and cost-effective method for continuous screening of water, where even small disturbances will be reflected in the WASP (Figure 32.7) and may serve as an indicator for the necessity of detailed conventional solute analysis.

Aquaphotomics also made a significant contribution to the field of microbiology by providing fast and nondestructive analysis (Nakakimura et al. 2012, Remagni et al. 2013, Slavchev et al. 2015, 2017). The example of such application was in growth monitoring of three different groups of bacterial strains: probiotic, nonprobiotic, and moderate (Slavchev et al. 2015). The results showed high accuracy of bacterial strains' classification according to their probiotic strength, where the most informative absorbance bands were found in the first overtone of water, suggesting the importance of water molecular system to the overall cell functionality. Comparing the groups based on their water spectral patterns, probiotic bacterial strains were found to be characterized by higher number of small protonated water clusters, free water molecules, and water clusters with weak hydrogen bonds

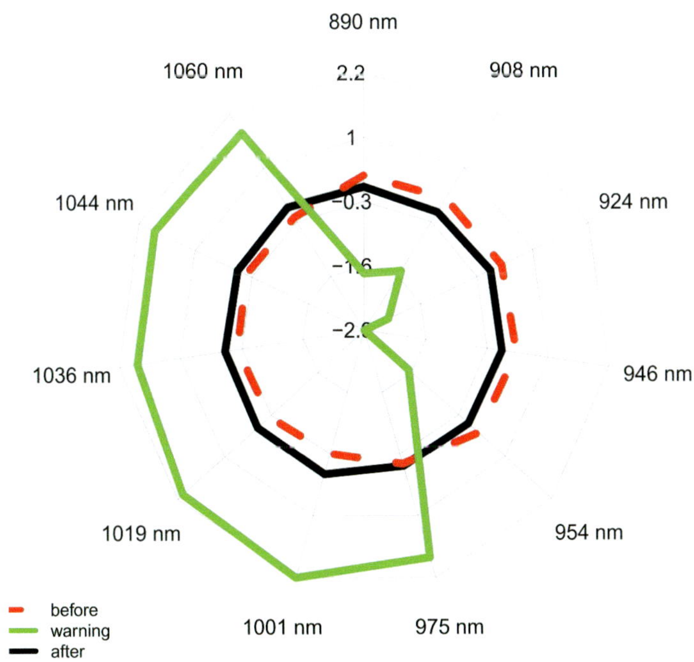

FIGURE 32.7 Aquagrams of ground water samples calculated at wavelengths found important in PCA-QCC in the third overtone range of water – data of 22 days of monitoring, where the average spectrum of the warning days is compared to those of the previous 10 days and the following 10 days.

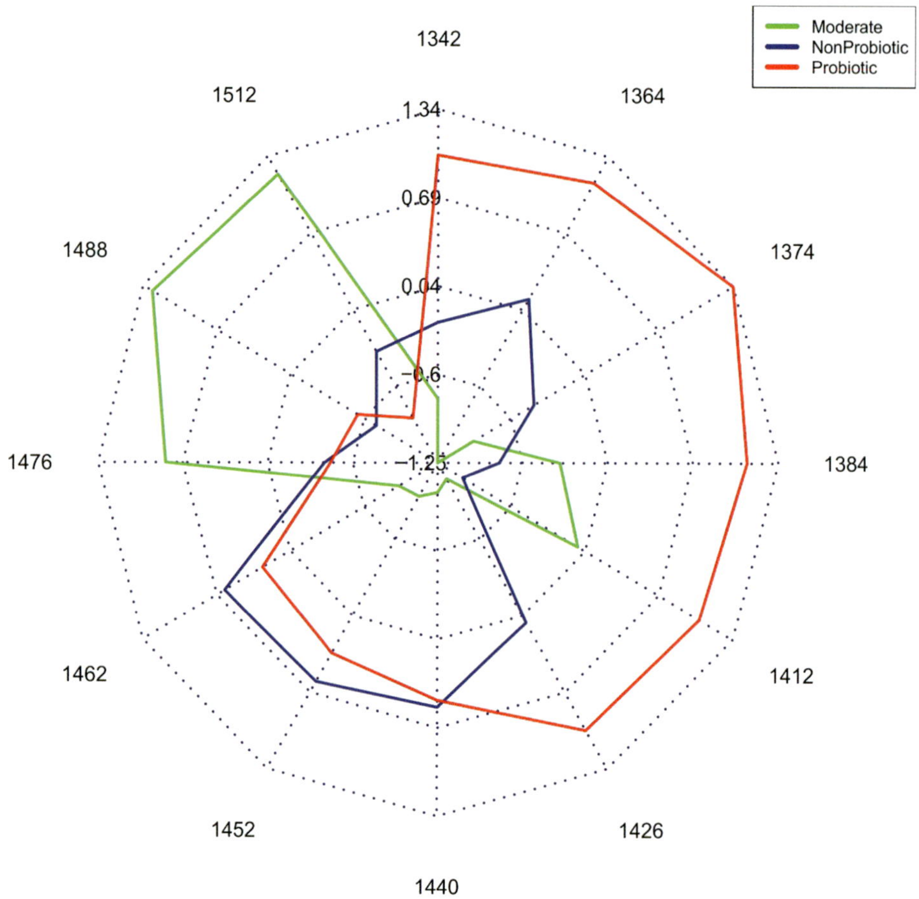

FIGURE 32.8 Water spectral pattern presented on aquagrams shows different water structure in different bacterial strains. (Reprinted with permission from Slavchev, A., Kovacs, Z., Koshiba, H., Nagai, A., Bázár, G., Krastanov, A., Kubota, Y., and Tsenkova, R. 2015. Monitoring of water spectral pattern reveals differences in probiotics growth when used for rapid bacteria selection. *PLoS One* 10 (7):e0130698. https://doi.org/10.1371/journal.pone.0130698.)

(Figure 32.8) (Slavchev et al. 2015). This work improved the understanding of the roles of specific water molecular species for living processes – it has been discovered that strong probiotic bacteria have a high number of less hydrogen-bonded water species, which may help in better understanding of their beneficial health effects. Similar study determined that aquaphotomics approach allows a successful, rapid selection of probiotic bacterial strains even in the water first overtone of combination vibrations range (1100–1300 nm) (Slavchev et al. 2017). The differences in water spectral pattern of different bacterial strains were shown to be a result of differences in extracellular metabolites which leave different imprint on water structure; in fact, it was found that 80% of information was based on the influence of metabolites on the water structure, not the cells (Nakakimura et al. 2012).

Aquaphotomics studies also contributed to better understanding of the development of diseases in biological systems such as plants and animals (Tsenkova 2009). For example, aquaphotomics discovered that the mosaic virus infection of soybean plants results in a decrease of less hydrogen-bonded water (Jinendra et al. 2010), a change observable even at symptomless stages of the disease. Also, different cold resistance abilities of genetically modified soybean plants were found to be associated with the specific water molecular structures (Jinendra 2011). Aquaphotomics-based disease diagnosis (Tsenkova 2007) in animals was first reported for mastitis in cows where disease was detected based on the specific water absorbance pattern of milk (Meilina et al. 2009, Tsenkova et al. 2001, Tsenkova, Atanassova, and Toyoda 2001), blood, and urine (Tsenkova and Atanassova 2002). Comparison of water spectral patterns of blood, milk, and urine of mastitic animals revealed that they all had common WABs activated as the consequence of a disease (Tsenkova 2008a). Aquaphotomics also provided the detection of ovulation period in giant panda using urine (Kinoshita et al. 2010, 2012), even when the hormonal changes during ovulation were in the range of 0.80–127.88 ng mL^{-1} (Kinoshita et al. 2010). Further study revealed that ovulation in giant panda resulted in an increased number of hydrogen-bonded species (Figure 32.9) (Kinoshita et al. 2012). Successful determination of fertile window in Bornean orangutan based on NIR spectra of urine (Kinoshita et al. 2016), and in dairy cows using NIR spectra of milk (Takemura et al. 2015), further confirmed that changes in water structure can be used as a marker of ovulation, instead of traditionally used, reagent-based methods. In the human medicine field, one of the earliest works utilized NIR spectra for the detection of HIV-1 virus in plasma (Sakudo et al. 2005). The results yielded

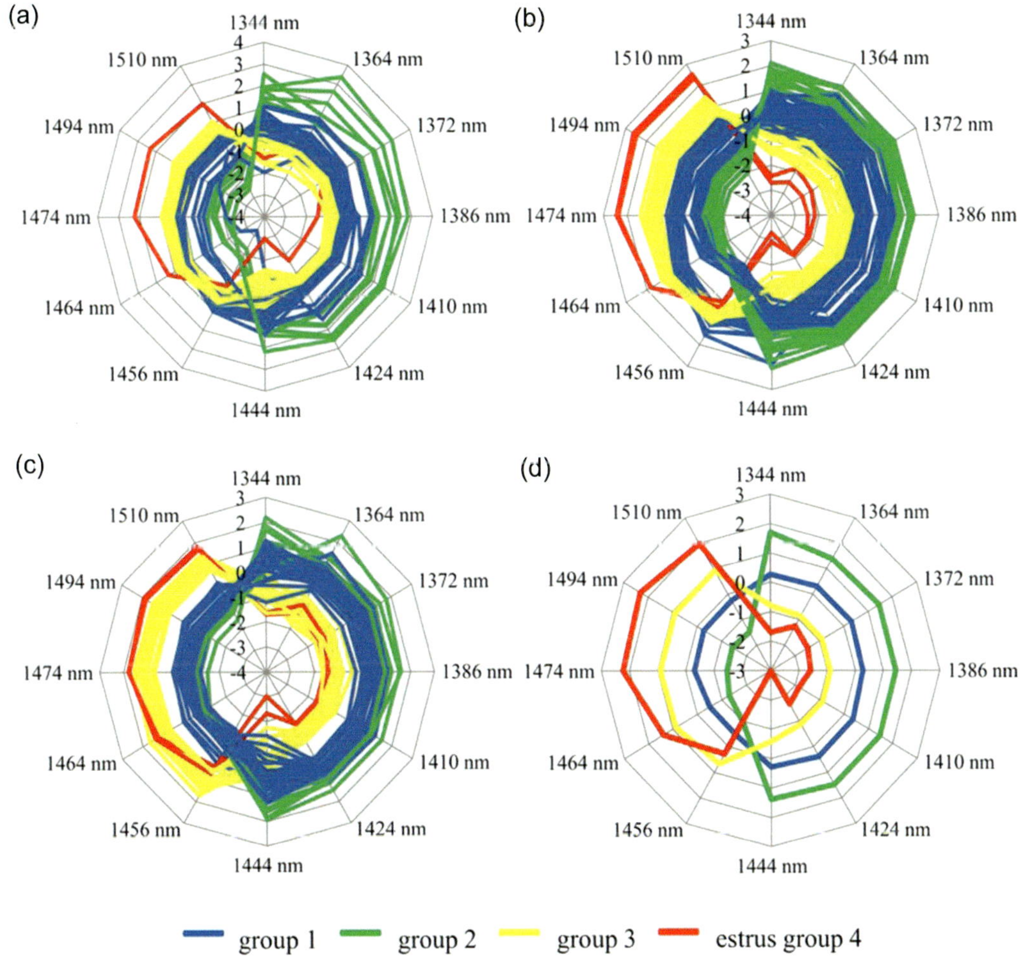

FIGURE 32.9 Aquagrams based on urine spectra from a female giant panda in each year (A: 2007, B: 2008, and C: 2009) and the aquagram of median values (D) of the groups classified according to the level of estrus from low level (group 1) to highest level (group 4), which corresponds to the time of the ovulation. The colors in the aquagrams correspond to groups 1, 2, 3 and estrus group 4. (Reprinted with permission from Kinoshita, K., Miyazaki, M., Morita, H., Vassileva, M., Tang, C., Li, D., Ishikawa, O., Kusunoki, H., and Tsenkova, R. 2012. Spectral pattern of urinary water as a biomarker of estrus in the giant panda. *Scientific Reports* 2.)

a good correlation with those obtained by the reference ELISA method (R^2 = 0.7319, SD/SECV = 1.856, SECV = 23.33 pg mL^{-1}), suggesting that NIRS could provide a rapid and very accurate screening method for HIV-1 infection, but for other viral diseases, too.

In other areas of biomedical science, applications of aquaphotomics could be expected. Studies on water–material interaction hold great promise in understanding some of the very elusive properties like wettability or biocompatibility. The structure of water has been recognized as fundamental for the biocompatibility of polymers as a number of recent studies point out (Morita, Tanaka, and Ozaki 2007, Morita et al. 2010, Tanaka, Hayashi, and Morita 2013). These works are excellent examples of other windows in EM spectra, which reveal structure–function relationship of water. Aquaphotomics was also proposed for in vivo monitoring of topical cream effects (Matija et al. 2013, 2017) in medicine and cosmetics. The same holistic approach, shown to be successful in water quality monitoring was proposed for the monitoring of dialysis efficacy (Munćan, Mileusnić, Matović, et al. 2016), which opens up new areas of aquaphotomics applications in the field of therapy monitoring.

32.5 CONCLUSION

The NIR range with its abundance of combination and overtone bands is very informationally rich and can provide an immense amount of information about water incomparable to other areas of the EM spectrum (Workman Jr 2000). In addition, NIRS provides a noninvasive, rapid method to study biological and aqueous systems. Introduction of aquaphotomics to this field pushed the limits of its previously envisioned possibilities.

Even though the NIR spectra of water and aqueous systems are characterized by broad and overlapped bands resulting in challenging and not-straightforward interpretation, aquaphotomics developed new approaches and techniques, tremendously helped spectral data analysis and understanding of the underlying phenomena. Utilization of aquaphotomics water-mirror approach resulted in the measurements of concentrations of solutes in aqueous systems at levels previously thought impossible. Systematization of the already-abundant information relating WABs to corresponding water conformations opened up new ways of interpretation of structure–function relationship of water and added up one more dimension in the description of aqueous and especially biological systems, in which water is seen as a significant, active biomolecule.

What started as a modest but ambitious project of one laboratory, today, includes research groups all across the globe which together work on achieving the common goal. Established website, www.aquaphotomics.com, offers a multitude of information about aquaphotomics methodology and applications, provides regular updates on publications within the field, downloadable resources such as software and data files, as well as upcoming events. Scientists from different backgrounds are applying aquaphotomics in a range of fields from agriculture, food engineering, chemical engineering, material science, biomedical science, and many more. Each scientific field contributes to the aquaphotome database with their specific pieces of the puzzle, slowly shaping up a more complete picture of water. Being a central life molecule, understanding of water will certainly lead to better understanding of the life itself.

ACKNOWLEDGMENTS

The author J.M. gratefully acknowledges the financial support of JSPS Postdoctoral Fellowship for Foreign Researchers (P17406). The authors would like to thank Lawrence K.F. Wong and Alexander Stoilov for their help with the artwork.

REFERENCES

Ball, P. 2008a. Water as a biomolecule. *ChemPhysChem* 9 (18):2677–2685.

Ball, P. 2008b. Water as an active constituent in cell biology. *Chemical Reviews* 108 (1):74–108.

Bázár, G., Kovacs, Z., Tanaka, M., Furukawa, A., Nagai, A., Osawa, M., Itakura, Y., Sugiyama, H., and Tsenkova, R. 2015. Water revealed as molecular mirror when measuring low concentrations of sugar with near infrared light. *Analytica Chimica Acta* 896:52–62.

Bázár, G., Kovács, Z., Tanaka, M., and Tsenkova, R. 2014. Aquaphotomics and its extended water mirror concept explain why NIRS can measure low concentration aqueous solutions. *Aquaphotomics, "Understanding Water in Biological World", The 5th Kobe University Brussels European Centre Symposium "Innovation, Environment, and Globalisation"*, Brussels, Belgium.

Buijs, K., and Choppin, G. R. 1963. Near-infrared studies of the structure of water. I. Pure Water. *The Journal of Chemical Physics* 39 (8):2035–2041.

Cattaneo, T. M. P., Cabassi, G., Profaizer, M., and Giangiacomo, R. 2009. Contribution of light scattering to near infrared absorption in milk. *Journal of Near Infrared Spectroscopy* 17 (6):337–343.

Cattaneo, T. M. P., Stefania, V., Elena, N., and Vittorio, E. 2011. Influence of filtration processes on aqueous nanostructures by NIR spectroscopy. *Journal of Chemistry and Chemical Engineering* 5 (11):1046–1052.

Chandler, D. 2002. Hydrophobicity: Two faces of water. *Nature* 417:491.

Chaplin, M. 2006. Do we underestimate the importance of water in cell biology? *Nature Reviews Molecular Cell Biology* 7 (11):861–866.

Chatani, E., Tsuchisaka, Y., Masuda, Y., and Tsenkova, R. 2014. Water molecular system dynamics associated with amyloidogenic nucleation as revealed by real time near infrared spectroscopy and aquaphotomics. *PLoS One* 9 (7):e101997.

Cupane, A., Levantino, M., and Santangelo, M. G. 2002. Near-infrared spectra of water confined in silica hydrogels in the temperature interval 365–5 K. *The Journal of Physical Chemistry B* 106 (43):11323–11328.

Czarnik-Matusewicz, B., Murayama, K., Tsenkova, R., and Ozaki, Y. 1999. Analysis of near-infrared spectra of complicated biological fluids by two-dimensional correlation spectroscopy: Protein and fat concentration-dependent spectral changes of milk. *Applied Spectroscopy* 53 (12):1582–1594.

Del Giudice, E., Spinetti, P. R., and Tedeschi, A. 2010. Water dynamics at the root of metamorphosis in living organisms. *Water* 2 (3):566–586.

Doster, W., Bachleitner, A., Dunau, R., Hiebl, M., and Lüscher, E. 1986. Thermal properties of water in myoglobin crystals and solutions at subzero temperatures. *Biophysical Journal* 50 (2):213–219.

Esquerre, C., Gowen, A., Tsenkova, R., O'Donnell, C., and Downey, G. 2009. Identification of water matrix coefficients in mushrooms (*Agaricus bisporus*) using robust ensemble of Monte Carlo uninformative variable elimination. *14th International Conference on Near Infrared Spectroscopy*, Bangkok, Thailand.

Fecko, C. J., Eaves, J. D., Loparo, J. J., Tokmakoff, A., and Geissler, P. L. 2003. Ultrafast hydrogen-bond dynamics in the infrared spectroscopy of water. *Science* 301 (5640):1698–1702.

Fornés, V., and Chaussidon, J. 1978. An interpretation of the evolution with temperature of the $\nu 2+\nu 3$ combination band in water. *The Journal of Chemical Physics* 68 (10):4667–4671.

Franks, F., ed. 1973. *Water: A Comprehensive Treatise* Vol. 3. New York: Plenum Press.

Goto, N., Bazar, G., Kovacs, Z., Kunisada, M., Morita, H., Kizaki, S., Sugiyama, H., Tsenkova, R., and Nishigori, C. 2015. Detection of UV-induced cyclobutane pyrimidine dimers by near-infrared spectroscopy and aquaphotomics. *Scientific Reports* 5.

Gowen, A., Amigo, J., and Tsenkova, R. 2013. Characterisation of hydrogen bond perturbations in aqueous systems using aquaphotomics and multivariate curve resolution-alternating least squares. *Analytica Chimica Acta* 759:8–20.

Gowen, A., Marini, F., Tsuchisaka, Y., De Luca, S., Bevilacqua, M., O'Donnell, C., Downey, G., and Tsenkova, R. 2015. On the feasibility of near infrared spectroscopy to detect contaminants in water using single salt solutions as model systems. *Talanta* 131:609–618.

Gowen, A., Tsenkova, R., Esquerre, C., Downey, G., and O'Donnell, C. 2009. Use of near infrared hyperspectral imaging to identify water matrix coordinates in mushrooms (*Agaricus bisporus*) subjected to mechanical vibration. *Journal of Near Infrared Spectroscopy* 17 (6):363–371.

Gowen, A., Tsuchisaka, Y., O'Donnell, C., and Tsenkova, R. 2011. Investigation of the potential of near infrared spectroscopy for the detection and quantification of pesticides in aqueous solution. *American Journal of Analytical Chemistry* 2 (8):53.

Grant, A., Davies, A. M. C., and Bilverstone, T. 1989. Simultaneous determination of sodium hydroxide, sodium carbonate and sodium chloride concentrations in aqueous solutions by near-infrared spectrometry. *Analyst* 114 (7):819–822.

Headrick, J. M., Diken, E. G., Walters, R. S., Hammer, N. I., Christie, R. A., Cui, J., Myshakin, E. M., Duncan, M. A., Johnson, M. A., and Jordan, K. D. 2005. Spectral signatures of hydrated proton vibrations in water clusters. *Science* 308 (5729):1765–1769.

Heiman, A., and Licht, S. 1999. Fundamental baseline variations in aqueous near-infrared analysis. *Analytica Chimica Acta* 394 (2):135–147.

Hirschfeld, T. 1985. Salinity determination using NIRA. *Applied Spectroscopy* 39 (4):740–741.

Huang, C., Wikfeldt, K. T., Tokushima, T., Nordlund, D., Harada, Y., Bergmann, U., Niebuhr, M., Weiss, T.M., Horikawa, Y., and Leetmaa, M. 2009. The inhomogeneous structure of water at ambient conditions. *Proceedings of the National Academy of Sciences* 106 (36):15214–15218.

Huse, N., Wen, H., Nordlund, D., Szilagyi, E., Daranciang, D., Miller, T.A., Nilsson, A., Schoenlein, R.W., and Lindenberg, A. M. 2009. Probing the hydrogen-bond network of water via time-resolved soft X-ray spectroscopy. *Physical Chemistry Chemical Physics* 11 (20):3951–3957.

Ichikawa, K., Kameda, Y., Yamaguchi, T., Wakita, H., and Misawa, M. 1991. Neutron-diffraction investigation of the intramolecular structure of a water molecule in the liquid phase at high temperatures. *Molecular Physics* 73 (1):79–86.

Iwamoto, M., Uozumi, J., and Nishinari, K. 1987. Preliminary investigation of the state of water in foods by near infrared spectroscopy. *Proceedings of the International NIR/NIT Conference*, Budapest, Hungary.

Jinendra, B. 2011. Near infrared spectroscopy and aquaphotomics: Novel tool for biotic and abiotic stress diagnosis of soybean. PhD Thesis, Kobe University.

Jinendra, B., Tamaki, K., Kuroki, S., Vassileva, M., Yoshida, S., and Tsenkova, R. 2010. Near infrared spectroscopy and aquaphotomics: Novel approach for rapid in vivo diagnosis of virus infected soybean. *Biochemical and Biophysical Research Communications* 397 (4):685–690.

Kaffka, K. J., Horváth, L., Kulcsár, F., and Váradi, M. 1990. Investigation of the state of water in fibrous foodstuffs by near infrared spectroscopy. *Acta Alimentaria* 11 (2):125–137.

Kinoshita, K., Kuze, N., Kobayashi, T., Miyakawa, E., Narita, H., Inoue-Murayama, M., Idani, G., and Tsenkova, R. 2016. Detection of urinary estrogen conjugates and creatinine using near infrared spectroscopy in *Bornean orangutans* (*Pongo pygmaeus*). *Primates* 57 (1):51–59.

Kinoshita, K., Miyazaki, M., Morita, H., Vassileva, M., Tang, C., Li, D., Ishikawa, O., Kusunoki, H., and Tsenkova, R. 2012. Spectral pattern of urinary water as a biomarker of estrus in the giant panda. *Scientific Reports* 2: 1–9.

Kinoshita, K., Morita, H., Miyazaki, M., Hama, N., Kanemitsu, H., Kawakami, H., Wang, P., Ishikawa, O., Kusunoki, H., and Tsenkova, R. 2010. Near infrared spectroscopy of urine proves useful for estimating ovulation in giant panda (*Ailuropoda melanoleuca*). *Analytical Methods* 2 (11):1671–1675.

Kitano, H., Sudo, K., Ichikawa, K., Ide, M., and Ishihara, K. 2000. Raman spectroscopic study on the structure of water in aqueous polyelectrolyte solutions. *The Journal of Physical Chemistry B* 104 (47):11425–11429.

Kojić, D., Tsenkova, R., Tomobe, K., Yasuoka, K., and Yasui, M. 2014. Water confined in the local field of ions. *ChemPhysChem* 15 (18):4077–4086.

Kojić, D., Tsenkova, R., and Yasui, M. 2017. Improving accuracy and reproducibility of vibrational spectra for diluted solutions. *Analytica Chimica Acta* 955:86–97.

Kovacs, Z., Bázár, G., Oshima, M., Shigeoka, S., Tanaka, M., Furukawa, A., Nagai, A., Osawa, M., Itakura, Y., and Tsenkova, R. 2016. Water spectral pattern as holistic marker for water quality monitoring. *Talanta* 147:598–608.

Luck, W. A. P. 1998. The importance of cooperativity for the properties of liquid water. *Journal of Molecular Structure* 448 (2):131–142.

Maeda, H., Ozaki, Y., Tanaka, M., Hayashi, N., and Kojima, T. 1995. Near infrared spectroscopy and chemometrics studies of temperature-dependent spectral variations of water: Relationship between spectral changes and hydrogen bonds. *Journal of Near Infrared Spectroscopy* 3 (4):191–201.

Maréchal, Y. 2011. The molecular structure of liquid water delivered by absorption spectroscopy in the whole IR region completed with thermodynamics data. *Journal of Molecular Structure* 1004 (1):146–155.

Martens, H., Jensen, S. A., and Geladi, P. 1983. Multivariate linearity transformation for near-infrared reflectance spectrometry. *Proceedings of the Nordic Symposium on Applied Statistics*, Norway.

Matija, L., Muncan, J., Mileusnic, I., and Koruga, D. 2017. Fibonacci nanostructures for novel nanotherapeutical approach. In: *Nano-and Microscale Drug Delivery Systems*, edited by A. M. Grumezescu; 49–74. Elsevier.

Matija, L., Tsenkova, R., Miyazaki, M., Banba, K., and Muncan, J. 2012. Aquagrams: Water spectral pattern as characterization of hydrogenated nanomaterial. *FME Transactions* 40 (2):51–56.

Matija, L., Tsenkova, R., Munćan, J., Miyazaki, M., Banba, K., Tomić, M., and Jeftić, B. 2013. Fullerene based nanomaterials for biomedical applications: Engineering, functionalization and characterization. *Advanced Materials Research* 633: 224–238.

Matubayasi, N., Wakai, C., and Nakahara, M. 1997. Structural study of supercritical water. I. Nuclear magnetic resonance spectroscopy. *The Journal of Chemical Physics* 107 (21):9133–9140.
Meilina, H., Kuroki, S., Jinendra, B. M., Ikuta, K., and Tsenkova, R. 2009. Double threshold method for mastitis diagnosis based on NIR spectra of raw milk and chemometrics. *Biosystems Engineering* 104 (2):243–249.
Mizoguchi, K., Hori, Y., and Tominaga, Y. 1992. Study on dynamical structure in water and heavy water by low-frequency Raman spectroscopy. *The Journal of Chemical Physics* 97 (3):1961–1968.
Morita, S., Tanaka, M., Kitagawa, K., and Ozaki, Y. 2010. Hydration structure of poly (2-methoxyethyl acrylate): Comparison with a 2-methoxyethyl acetate model monomer. *Journal of Biomaterials Science, Polymer Edition* 21 (14):1925–1935.
Morita, S., Tanaka, M., and Ozaki, Y. 2007. Time-resolved in situ ATR-IR observations of the process of sorption of water into a poly (2-methoxyethyl acrylate) film. *Langmuir* 23 (7):3750–3761.
Munćan, J., Mileusnić, I., Matović, V., Šakota Rosić, J., and Matija, L. 2016. The prospects of aquaphotomics in biomedical science and engineering. *Aquaphotomics: Understanding Water in Biology – 2nd International Symposium*, Kobe University, Kobe, Japan, 26–29 December.
Munćan, J., Mileusnić, I., Šakota Rosić, J., Vasić-Milovanović, A., and Matija, L. 2016. Water properties of soft contact lenses: A comparative near-infrared study of two hydrogel materials. *International Journal of Polymer Science* 2016.
Murayama, K., Czarnik-Matusewicz, B., Wu, Y., Tsenkova, R., and Ozaki, Y. 2000. Comparison between conventional spectral analysis methods, chemometrics, and two-dimensional correlation spectroscopy in the analysis of near-infrared spectra of protein. *Applied Spectroscopy* 54 (7):978–985.
Nakakimura, Y., Vassileva, M., Stoyanchev, T., Nakai, K., Osawa, R., Kawano, J., and Tsenkova, R. 2012. Extracellular metabolites play a dominant role in near-infrared spectroscopic quantification of bacteria at food-safety level concentrations. *Analytical Methods* 4 (5):1389–1394.
Nilsson, A., Nordlund, D., Waluyo, I., Huang, N., Ogasawara, H., Kaya, S., Bergmann, U., Näslund, L-Å., Öström, H., and Wernet, P. 2010. X-ray absorption spectroscopy and X-ray Raman scattering of water and ice; an experimental view. *Journal of Electron Spectroscopy and Related Phenomena* 177 (2):99–129.
Okada, K., Yao, M., Hiejima, Y., Kohno, H., and Kajihara, Y. 1999. Dielectric relaxation of water and heavy water in the whole fluid phase. *The Journal of Chemical Physics* 110 (6):3026–3036.
Ozaki, Y. 2002. Applications in chemistry. In: *Near-Infrared Spectroscopy: Principles, Instruments and Applications*, edited by H. W. Siesler, Y. Ozaki, S. Kawata and H. M. Heise. Weinheim, Germany: Wiley-VCH Verlag GmbH.
Pollack, G. H., Cameron, I. L., and Wheatley, D. N., eds. 2006. *Water and the Cell*. Dordrecht, Netherlands: Springer.
Pollner, B., and Kovacs, Z. 2016. Multivariate data analysis tools for R including aquaphotomics methods - aquap2. Available at: http://www.aquaphotomics.com/resources/aquap2/.
Putra, A., Faridah, F., Inokuma, E., and Santo, R. 2010. Robust spectral model for low metal concentration measurement in aqueous solution reveals the importance of water absorbance bands. *Jurnal Sains dan Teknologi Reaksi* 8 (1).
Remagni, M. C., Morita, H., Koshiba, H., Cattaneo, T. M. P., and Tsenkova, R. 2013. Near infrared spectroscopy and aquaphotomics as tools for bacteria classification. *NIR2013 Proceedings: Picking Up Good Vibrations*: 602.
Robertson, W. H., Diken, E. G., Price, E. A., Shin, J-W., and Johnson, M. A. 2003. Spectroscopic determination of the OH^- solvation shell in the $OH^-\cdot(H_2O)$ n clusters. *Science* 299 (5611):1367–1372.
Rønne, C., Åstrand, P-O., and Keiding, S.R. 1999. THz spectroscopy of liquid H_2O and D_2O. *Physical Review Letters* 82 (14):2888.
Rønne, C., Thrane, L., Åstrand, P-O., Wallqvist, A., Mikkelsen, K. V., and Keiding, S. R. 1997. Investigation of the temperature dependence of dielectric relaxation in liquid water by THz reflection spectroscopy and molecular dynamics simulation. *The Journal of chemical physics* 107 (14):5319–5331.
Šakota Rosić, J., Munćan, J., Mileusnić, I., Kosić, B., and Matija, L. 2016. Detection of protein deposits using NIR spectroscopy. *Soft Materials* 14 (4):264–271.
Sakudo, A., Tsenkova, R., Onozuka, T., Morita, K., Li, S., Warachit, J., Iwabu, Y., Li, G., Onodera, T., and Ikuta, K. 2005. A novel diagnostic method for human immunodeficiency virus Type-1 in plasma by near-infrared spectroscopy. *Microbiology and Immunology* 49 (7):695–701.
Sakudo, A., Tsenkova, R., Tei, K., Onozuka, T., Ikuta, K., Yoshimura, E., and Onodera, T. 2006. Comparison of the vibration mode of metals in HNO_3 by a partial least-squares regression analysis of near-infrared spectra. *Bioscience, Biotechnology, and Biochemistry* 70 (7):1578–1583.

Sartor, G., Hallbrucker, A., and Mayer, E. 1995. Characterizing the secondary hydration shell on hydrated myoglobin, hemoglobin, and lysozyme powders by its vitrification behavior on cooling and its calorimetric glass-->liquid transition and crystallization behavior on reheating. *Biophysical Journal* 69 (6):2679–2694.

Šašić, S., Segtnan, V. H., and Ozaki, Y. 2002. Self-modeling curve resolution study of temperature-dependent near-infrared spectra of water and the investigation of water structure. *The Journal of Physical Chemistry A* 106 (5):760–766.

Segtnan, V. H., Šašić, Š., Isaksson, T., and Ozaki, Y. 2001. Studies on the structure of water using two-dimensional near-infrared correlation spectroscopy and principal component analysis. *Analytical Chemistry* 73 (13):3153–3161.

Shields, R. M., Temelso, B., Archer, K. A., Morrell, T. E., and Shields, G. C. 2010. Accurate predictions of water cluster formation, (H_2O) n = 2 – 10. *The Journal of Physical Chemistry A* 114 (43):11725–11737.

Slavchev, A., Kovacs, Z., Koshiba, H., Bazar, G., Pollner, B., Krastanov, A., and Tsenkova, R. 2017. Monitoring of water spectral patterns of lactobacilli development as a tool for rapid selection of probiotic candidates. *Journal of Near Infrared Spectroscopy* 25(6): 423–431. 0967033517741133.

Slavchev, A., Kovacs, Z., Koshiba, H., Nagai, A., Bázár, G., Krastanov, A., Kubota, Y., and Tsenkova, R. 2015. Monitoring of water spectral pattern reveals differences in probiotics growth when used for rapid bacteria selection. *PLoS One* 10 (7):e0130698.

Soper, A. K. 2000. The radial distribution functions of water and ice from 220 to 673 K and at pressures up to 400 MPa. *Chemical Physics* 258 (2):121–137.

Takemura, G., Bázár, G., Ikuta, K., Yamaguchi, E., Ishikawa, S., Furukawa, A., Kubota, Y., Kovács, Z., and Tsenkova, R. 2015. Aquagrams of raw milk for oestrus detection in dairy cows. *Reproduction in Domestic Animals* 50 (3):522–525.

Tanaka, M., Hayashi, T., and Morita, S. 2013. The roles of water molecules at the biointerface of medical polymers. *Polymer Journal* 45 (7):701–710.

Tsenkova, R. 2004. NIRS-Nondestructive method for analysis of biological materials and diagnosis. *Advanced Biomaterials for Medical Applications* 180:143.

Tsenkova, R. 2005. Visible-near infrared perturbation spectroscopy: Water in action seen as a source of information. *12th International Conference on Near-infrared Spectroscopy*, Auckland, New Zealand.

Tsenkova, R. 2006a. Aquaphotomics. Aquaphotomics and chambersburg. *NIR News* 17 (6):12–14.

Tsenkova, R. 2006b. Aquaphotomics: Exploring water-light interactions for a better understanding of the biological world. Part 2: Japanese food, language and why NIR for diagnosis? *NIR News* 17 (5):8–14.

Tsenkova, R. 2006c. AquaPhotomics: Water absorbance pattern as a biological marker. *NIR News* 17 (7):13–23.

Tsenkova, R. 2006d. Tomas Hirschfeld Award Lecture. Orlando, FL.

Tsenkova, R. 2007. AquaPhotomics: Water absorbance pattern as a biological marker for disease diagnosis and disease understanding. *NIR News* 18 (2):14–16.

Tsenkova, R. 2008a. Aquaphotomics: Acquiring spectra of various biological fluids of the same organism reveals the importance of water matrix absorbance coordinates and the aquaphotome for understanding biological phenomena. *NIR News* 19 (1):13–15.

Tsenkova, R. 2008b. Aquaphotomics: The extended water mirror effect explains why small concentrations of protein in solution can be measured with near infrared light. *NIR News* 19 (4):13–14.

Tsenkova, R. 2009. Aquaphotomics: Dynamic spectroscopy of aqueous and biological systems describes peculiarities of water. *Journal of Near Infrared Spectroscopy* 17 (6):303–313.

Tsenkova, R. 2010. Aquaphotomics: Water in the biological and aqueous world scrutinised with invisible light. *Spectroscopy Europe* 22 (6):6–10.

Tsenkova, R., and Atanassova, S. 2002. Mastitis diagnotics by near infrared spectra of cow's milk, blood and urine using soft independent modelling of class analogy classification. *Near Infrared Spectroscopy: Proceedings of the 10th International Conference.*

Tsenkova, R., Atanassova, S., Kawano, S., and Toyoda, K. 2001. Somatic cell count determination in cow's milk by near-infrared spectroscopy: A new diagnostic tool. *Journal of Animal Science* 79 (10):2550–2557.

Tsenkova, R., Atanassova, S., and Toyoda, K. 2001. Near infrared spectroscopy for diagnosis: Influence of mammary gland inflammation on cow´ s milk composition measurement. *Near Infrared Analysis* 2 (1):59–66.

Tsenkova, R., Fockenberg, C., Koseva, N., Sakudo, A., and Parker, M. 2007. Aquaphotomics: Water absorbance patterns in NIR range used for detection of metal ions reveal the importance of sample preparation. *13th International Conference on Near Infrared Spectroscopy*, Umea-Vasa, Sweden & Finland.

Tsenkova, R., Iso, E., Parker, M., Fockenberg, C., and Okubo, M. 2007. Aquaphotomics: A NIRS investigation into the perturbation of water spectrum in an aqueous suspension of mesoscopic scale polystyrene spheres. *13th International Conference on Near Infrared Spectroscopy*, Umea-Vasa, Sweden & Finland.

Tsenkova, R., Kovacs, Z., and Kubota, Y. 2015. Aquaphotomics: Near infrared spectroscopy and water states in biological systems. In: *Membrane Hydration*, 189–211. Springer.

Tsenkova, R., Morita, H., Shinzawa, H., Hogeveen, H., Hillerton, J. E., and Ikuta, K. 2005. Near infrared spectroscopy for cow identification and in-vivo mastitis diagnosis. In: *Mastitis in Dairy Production. Current Knowledge and Future Solutions, 4th IDF International Mastitis Conference*, Maastricht, Netherlands.

Tsenkova, R. N., Iordanova, I. K., Toyoda, K., and Brown, D. R. 2004. Prion protein fate governed by metal binding. *Biochemical and Biophysical Research Communications* 325 (3):1005–1012.

Weber, J. M., Kelley, J. A., Nielsen, S. B., Ayotte, P., and Johnson, M. A. 2000. Isolating the spectroscopic signature of a hydration shell with the use of clusters: Superoxide tetrahydrate. *Science* 287 (5462):2461–2463.

Williams, P. 2009. Influence of water on prediction of composition and quality factors: The aquaphotomics of low moisture agricultural materials. *Journal of Near Infrared Spectroscopy* 17 (6):315–328.

Workman, J., Jr. 2000. *The Handbook of Organic Compounds: NIR, IR, Raman, and UV-Vis Spectra Featuring Polymers and Surfactants*. Elsevier.

Xantheas, S. S. 1995. Ab initio studies of cyclic water clusters $(H_2O)n$, n = 1–6. III. Comparison of density functional with MP2 results. *The Journal of Chemical Physics* 102 (11):4505–4517.

33 Bioprocesses

Benoît Igne[*]
GlaxoSmithKline

CONTENTS

33.1 INTRODUCTION

Bioprocesses use living organisms to transform raw materials into products of interest. They occur everywhere in the environment and are used to make products for human consumption such as food, wine and beer, fuels, active pharmaceutical ingredients, and vaccines. While the use of near-infrared (NIR) spectroscopy for beverages was covered in Chapters 24 and 31 of this Handbook, this chapter will focus on biofuels and biopharmaceutical products.

33.2 BIOFUELS

There are three types of biofuels. First-generation fuels use sugars and oils traditionally cultivated for human consumption as raw materials. The second generation is based on biomass that originates from by-products of human activities (plant parts and waste vegetable oils) or non-eatable plants. Third-generation fuels do not use raw materials from arable lands for production but rather algae. A comprehensive review of the use of NIR spectroscopy to the biomass energy conversion field was provided by Skvaril, Kyprianidis and Dahlquist (2017).

33.2.1 Raw Materials

Knowing the composition and energy potential of raw materials before processing has been a major focus of the NIR community. For first-generation biofuels, the analysis of grains, oils and flours is well documented in other chapters of this Handbook (Chapters 26 and 29). Readers can also refer

* igneben@gmail.com.

to Liebmann, Friedl, and Varmuza (2010), who demonstrated the characterization of starch, protein and moisture of wheat, rye, corn, barley, and triticale flours. Oil quality for biodiesels has also been investigated. Water content, acid number and iodine value are of importance for the quality of the fuel after transesterification. As an example, Baptista et al. (2007) reported the analysis of oil mixtures (soybean, rapeseed and palm) by NIR spectroscopy.

Second-generation biofuels require the characterization of heterogeneous samples from multiple sources. The most developed type of lignocellulose analysis by NIR spectroscopy was developed for forage analysis. Readers may refer to Chapter 27 in this Handbook on the topic. Sanderson et al. (1996) leveraged the knowledge developed on forages to evaluate the usability of NIR for the composition determination of woody and herbaceous feedstocks. They analyzed 121 samples for ethanol extractives, ash, lignin, uronic acids, arabinose, xylose, mannose, galactose, glucose, C, H, N and O. Partial least squares (PLS) models were then built, and using NIR spectroscopy, the authors concluded that based on their calibration and test set, ethanol extractives, lignin, and arabinose showed potential, with r^2 values of 0.96, 0.99 and 0.99 and prediction errors of 1.13, 0.53 and 0.12 wt%. Further work by Vogel et al. (2011), Sluiter and Wolfrum (2013), and Payne and Wolfrum (2015) showed the suitability of NIR spectroscopy for the analysis of products as varied as corn stover, sorghum, switchgrass, perennial cool-season grasses, switch grasses, rice straw, and miscanthus for parameters as varied as glucose, glucan, xylan, lignin and ash content and actual and theoretical ethanol yield. Using PLS2, Payne and Wolfrum (2015) reported prediction errors of 1.8, 1.0, 1.5 and 1.3 wt% for glucan, xylan, lignin, and ash, corresponding to an r^2 between 0.89 and 0.95.

Similarly, third-generation biofuels have used NIR spectroscopy to reduce the analytical costs. For example, Laurens and Wolfrum (2011) and Liu et al. (2015) used NIR spectroscopy to characterize algal lipids for triglycerides and phospholipids for several species. Figure 33.1 shows example spectra of different oleaginous *Chlorella* species containing different contents of lipids.

33.2.2 Fermentation and Distillation

The monitoring of fermentation processes is rather similar to other batch reactions with respect to the use of NIR spectroscopy. For a general discussion of process monitoring, readers may refer to Chapter 21 in this Handbook. The fermentation process has been characterized by NIR spectroscopy

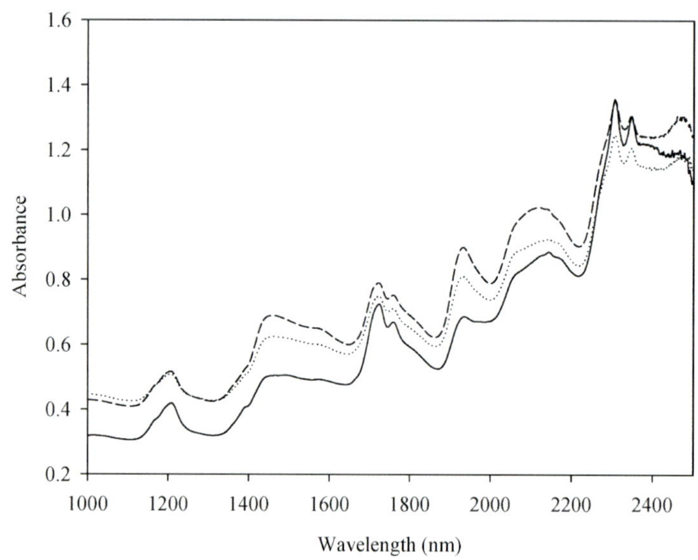

FIGURE 33.1 Spectra of *Chlorella vulgaris* (long dashed line), *Chlorella protothecoides* (dotted line) and *Chlorella zofingiensis* (solid line) samples. (Reproduced from Liu et al. (2015), open access copyright.)

through off-line/at-line analyses of samples. Newer applications include the introduction of probes directly into reactors for in-line/on-line measurements. While the main components of interest have often been glucose and ethanol, many other components have been monitored. A review article by Cervera et al. (2009) provides a comprehensive overview of the work carried out in this field.

Blanco, Peinado, and Mas (2004), Liebmann, Friedl, and Varmuza (2009), Liebmann, Friedl, and Varmuza (2010) and Pinto et al. (2016) demonstrated the use of NIR spectroscopy for the monitoring of glucose and ethanol from alcoholic mashes sampled from various time points during several fermentation processes. Liebmann showed the possibility to monitor the distillate and the stillage to evaluate fermentation yield and distillation efficiency.

In-line monitoring of the process was demonstrated by Blanco, Peinado, and Mas (2006), Corro-Herrera et al. (2016) and do Nascimento et al. (2017) for the prediction of biomass, ethanol, glucose and glycerol. Figure 33.2 shows the predicted vs. reference plots and kinetics plots for the measurement of biomass and glucose for a *Saccharomyces cerevisiae* fermentation. The regression model for biomass and glucose respectively yielded r^2 and RMSEP values of 0.99 and 0.276 g L^{-1} and 0.95 and 7.14 g L^{-1}. Commercial systems are available and are in use for fermentation monitoring of the corn-to-ethanol process.

33.2.3 Transesterification

Used in the manufacture of biodiesel, transesterification is a catalyzed reaction producing methyl ester and glycerol from oil and alcohol. Because methyl ester can be used directly as fuel, the amount of carryover glycerol, glycerides, alcohol, moisture and free fatty acids must be controlled in the final product. Near-infrared spectroscopy has been used in the monitoring and control of the transesterification.

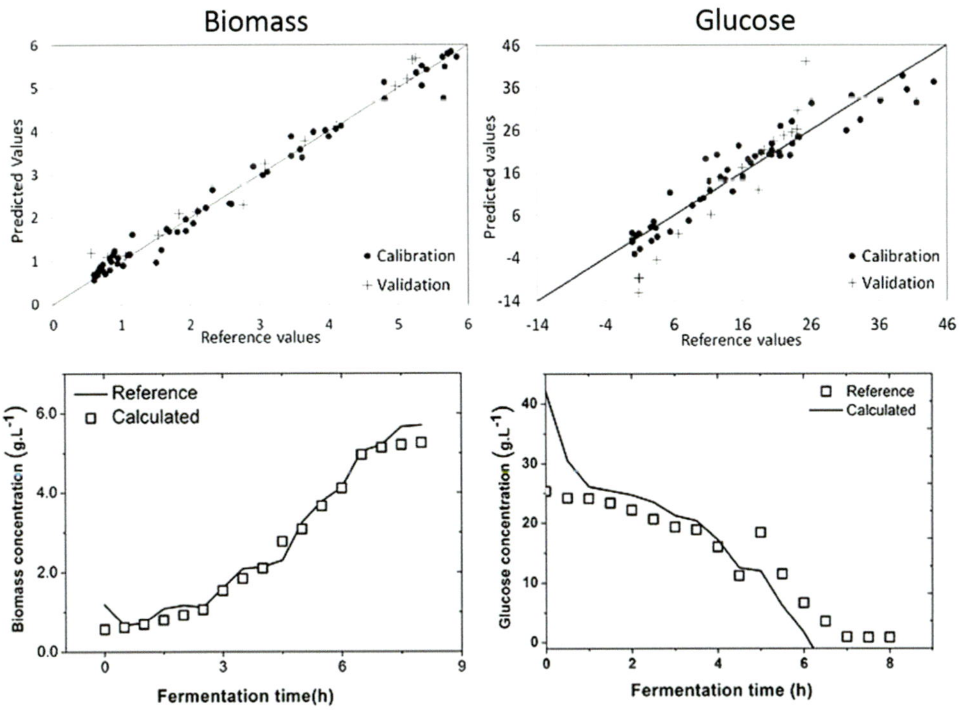

FIGURE 33.2 Prediction (top) and kinetics (bottom) plots for the monitoring of biomass and glucose. (Reproduced from do Nascimento et al. (2017) under CC BY 4.0, with minor modifications.)

A spectroscopic monitoring of the process based on spectral band tracking was published by de Oliveira et al. (2017), who used Jatropha oil to follow the evolution of characteristic bands as process monitoring. Multivariate curve resolution was used to perform a qualitative evaluation of an esterification process (Blanco, Castillo, and Beneyto 2007). Quantitative evaluations of products and contaminants (glycerol and alcohol) were performed by Knothe (1999) using partial least squares regression.

The in-line evaluation of the transesterification reaction was documented by many authors. Richard et al. (2011) and Killner, Rohwedder, and Pasquini (2011) looked at the performance of PLS models across various reaction temperatures. de Lima et al. (2014) used PLS and multiple linear regression to predict methyl ester content and mono-, di- and triglycerides over reaction times and parameters.

33.2.3 By-Product Analysis

Valuation of by-products from the production of biofuels is an economic need for the bioprocessing industry. Glycerol is produced in abundance from biodiesel production and can be used by the chemical industry. The bioethanol production outputs distillers' dried grains with solubles (DDGS) that can be sold as feedstock. DDGS are produced from corn, wheat, sorghum and barley and have a high protein content. Their quality can be determined by NIR spectroscopy (Tahir et al. 2012, Pekel et al. 2013, Harding et al. 2013). The possibility to identify the origin of DDGS has also been studied by Tena, Boix and von Holst (2015).

33.3 BIOPHARMACEUTICS

A biopharmaceutical process produces an active pharmaceutical ingredient using living organisms as the manufacturing engine. While the use of chicken eggs for the development of vaccines and recombinant DNA for the biosynthesis of insulin are early examples, most of the current work is focused on the development of complex therapeutic proteins produced by microorganisms or mammalian cells in large bioreactors. A biomanufacturing process typically consists of two steps: the growth of the microorganism and the harvest and conditioning of the protein of interest.

While bioprocesses, particularly in bioreactors, rely on in-line measurements to ensure cell health (continuous measurements of pH, dissolved oxygen, etc.), the monitoring and control of nutrients (i.e., glucose), byproducts (i.e., lactate, ammonia, ethanol) and products have always been challenging. For that reason, the industry still heavily relies on manual sampling and at-line measurements using bioassays and chromatographic tools. However, a significant body of literature exists on the use of in-line spectroscopy-based process analytical tools to monitor and control these processes.

As with other pharmaceutical processes, regulatory agencies have encouraged the use of process analytical tools. An article by Read et al. (2009) presents a top-level summary of process analytical technology (PAT) concepts and applications of NIR for upstream and downstream processing. The paper presents the principles and case studies used to illustrate PAT implementation for drug product manufacturing. The future of how PAT will be applied to biotech processes is addressed along with the integration of PAT with ICH Q8, Q9, and Q10 guidance documents.

Near-infrared spectroscopy has mainly been used in three areas of bioprocesses: raw material analysis, cell culture monitoring, and freeze-drying.

33.3.1 Raw Material Analysis

Difference in raw materials is a well-known source of variability for biological systems. It can cause unacceptable variability in cell culture performance and final product quality. The use of NIR spectroscopy allows the discrimination between good- and poor-performing media lots. Kirdar et al.

(2009) looked at the ability of NIR spectroscopy to identify the difference between soy hydrolysates lots and vendors. A comparison of NIR and Raman spectroscopy for raw material characterization was published by Trunfio et al. (2017).

33.3.2 Reactor Monitoring and Control

In their review of on-line monitoring and control tools for mammalian cell cultures, Teixeira et al. (2009) challenged the applicability of NIR spectroscopy for reactor monitoring:

> [...], the molar absorptivity in the NIR range is typically quite small; thus, the method is not the ideal one for diluted small components. This can be a limiting factor for the application to mammalian cell cultures, since a key process control objective during fed-batch operation is to keep glucose and glutamine at very low concentrations to prevent the accumulation of the toxic by-products ammonia and lactate. Furthermore, the overlapping of signals seen in the NIR range results in very broad peaks, leading to complex spectra and hindering the assignment of specific features to individual compounds. These characteristics require a chemometric data mining step to relate spectral information with the target compounds.

However, the use of NIR spectroscopy has been well documented in the literature (Hoehse et al. 2015).

33.3.2.1 Fermentations

Off-line use of infrared spectroscopy was first reported by Silman, Black and Norris (1983) for *A. awamori* and *P. oxalicum*. The authors sampled many 1 L fermentation flasks to collect samples for off-line analysis. One of the first applications of in situ monitoring of a fermentation process was reported by Cavinato et al. (1990). The authors monitored ethanol in mixtures of water, ethanol and yeasts and in bioreactors for an error of prediction from bioreactor to bioreactor of 0.4% in the range 0%–15%. Many additional elements were investigated such as cell density for an error of 1.6 g L^{-1} (Ge, Cavinato, and Callis 1994), lactic acid, and glucose (Vaccari et al. 1994), and acetate, biomass, ammonium, and glycerol with respective errors of 0.7 g L^{-1}, 1.4 g L^{-1}, 0.7 g L^{-1}, and 7 mmol L^{-1} (Hall et al. 1996).

Pichia pastoris is a microorganism species that has been particularly employed to produce monoclonal antibodies. As a consequence, there exists a large body of literature on the monitoring of the fermentation process for that organism by NIRS. Crowley et al. (2005) reported the use of NIRS at-line for the prediction of glycerol, methanol, biomass and the product in reflectance and transmittance modes. The authors used reflectance mode when transmittance was not possible due to very high absorption. Methanol was the only component that could be analyzed in both modes, with an error or 0.24 and 0.22 g L^{-1} for transmittance and reflectance, respectively. On-line monitoring for the same species was demonstrated by Goldfeld et al. (2014) and Kim et al. (2015). Goldfeld et al. (2014) used the ability to monitor methanol on-line (through a recirculation loop) and in real time to develop a feedback control loop to maintain the methanol at constant set points at 10, 5 and 1 g L^{-1}. Figure 33.3 shows the schematic of the measurement and control system.

High-throughput, in situ monitoring using mid-infrared (MIR) and NIR spectral analysis has been studied for biopharmaceutical analysis. MIR analysis of dehydrated cell pellets and in situ NIR analysis of whole culture broth were used to compare parameters related to plasmid production in recombinant *Escherichia coli* cultures. The MIR and NIR calibration models were generated using PLS regression to estimate plasmid production, host cell growth, glucose and glycerol consumption, and acetate by-product production and consumption. The predictive errors for biomass, plasmid, glucose, glycerol, and acetate concentrations based on MIR data were 0.7 g L^{-1}, 9 mg L^{-1}, 0.3 g L^{-1}, 0.4 g L^{-1}, and 0.4 g L^{-1}, respectively, whereas for NIR data the predictive errors obtained were 0.4 g L^{-1}, 8 mg L^{-1}, 0.3 g L^{-1}, 0.2 g L^{-1}, and 0.4 g L^{-1}, respectively. Both MIR and NIR methods were reported to be effective in monitoring these parameters (Sales et al. 2015).

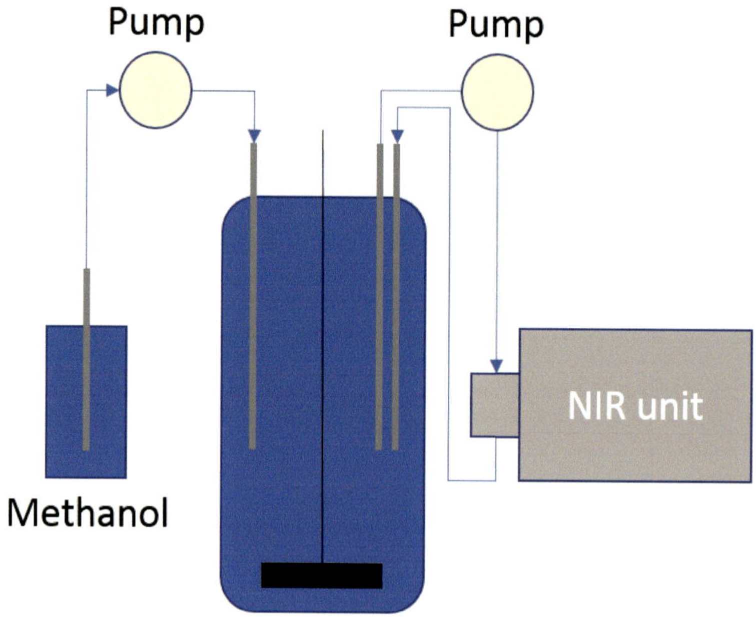

FIGURE 33.3 Schematic of the setup used by Goldfeld et al. (2014) to perform methanol feedback during *P. pastoris* fermentations.

33.3.2.2 Mammalian Cell Cultures

Chinese hamster ovary (CHO) cells are being increasingly used instead of yeasts to manufacture monoclonal antibodies. Unlike microbial organisms such as *P. pastoris*, CHO cell growth is a lot slower and the bioreactors runs two to three times longer. Many articles have been published on the monitoring of CHO cell bioreactors using NIR spectroscopy (McShane and Coté 1998, Lewis et al. 2000, Li et al. 2018).

A Fourier transform near-infrared (FT-NIR) spectrometer with multiplexed, transflectance fiber-optic measurement cells was used to monitor relevant culture parameters over two channels from ten independent batches of 12,500 L (Clavaud et al. 2013). Both principal components analysis and PLS regression methods were used. Seven different cell culture parameters were monitored, such as glucose concentration, osmolality, packed cell volume, product titer, viable cell density, integrated viable cell count, and integrated viable packed cell volume. Control charting and statistical process control methods were employ to monitor and control the culture media components.

Hakemeyer et al. (2012) and Li et al. (2018) looked into the scalability of NIR and Raman models (for the latter) from shake flasks to large-scale reactors. While results vary between the studies, Li et al. (2018) found that NIR models were not as scalable as Raman models for the larger scale investigated.

33.3.3 Freeze-Drying

To ensure that monoclonal antibodies remain stable and can be stored and shipped without damaging their therapeutic effects, it is often necessary to remove the water from the formulation. This can be achieved by freeze-drying or lyophilization. Water, having a strong NIR signal, is a tool of choice for in situ monitoring, through the container, during the water removal process (Kamat, Lodder, and DeLuca 1989, Brülls et al. 2007, De Beer et al. 2009, Kauppinen et al. 2013). The technique was applied to protein formulations and shown to have a similar error to the reference method Karl Fisher (Lin and Hsu 2002). A recent example of the use of chemical imaging to study the intra-vial moisture distribution during lyophilization was published by Brouckaert et al. (2018).

As the water is removed, however, it can generate stress on the protein structures and create aggregates or an irreversible modification of the protein structure. This is an undesirable quality attribute and the freeze-drying process must be monitored to understand the conditions that avoid aggregation. The suitability of NIR and Raman spectroscopy was investigated to monitor protein conformation. Pieters, De Beer, and Vander Heyden (2012) and Pieters et al. (2012) showed that NIR was an ideal tool for the monitoring of the sublimation process and the study of interactions between the protein and the lyoprotectant compounds. An example of study of the interaction between the formulation components and the proteins was discussed by Grohganz et al. (2011). Both near-infrared (NIR) and Raman spectroscopic methods were used for the solid-state analysis of freeze-dried protein formulations. These analytical techniques are useful for the rapid analysis of freeze-dried formulations of mannitol solid forms. Analysis of the spectroscopic data demonstrated the formation of δ-mannitol when protein was present and of β-mannitol in the absence of protein. The data analysis indicated that mannitol hemihydrate formation occurred when protein was absent. It was reported that the use of NIR and Raman spectroscopy were useful for monitoring excipient solid-state properties in pharmaceutical formulations.

33.4 SUMMARY

While a large body of literature is available on the use of NIR spectroscopy for pharmaceutical bioprocesses, implementation in industry still remains sparse. In a recent review, Jenzsch et al. (2017) proposed a potential explication:

> One possible reason for this apparent success of NIR in academia, but relatively poor acceptance in biopharma manufacturing, may be the lack of raw material variation seen in the limited number of fermentation runs performed by the academic institutions. This gives the impression that quantitative NIR measurement models in a fermenter are more robust and stable than is observed when these techniques are applied at an actual production facility, where variation in raw materials over an extended period of time includes spectral variability that may have not been present in the original calibration set.

This remark shows that despite significant progress made by not only NIR spectroscopy but also other spectroscopic techniques used in situ, much remains to be done to make these tools more robust and less complex for nonexpert users.

33.5 CONCLUSION

NIR spectroscopic methods have proven to be successful in monitoring critical parameters relative to bioprocessing regimes. The technique proves itself when in situ, real-time monitoring is required for process control, conformational analysis, and improvement of processes and product quality.

ACKNOWLEDGEMENT

The author would like to thank Charles R. Hurburgh Jr. and Jerome Workman, Jr. for their help in writing and reviewing the chapter.

REFERENCES

Baptista, P., P. Felizardo, J. C. Menezes, and M. Joana Neiva Correia. 2007. "Monitoring the quality of oils for biodiesel production using multivariate near infrared spectroscopy models." *Journal of Near Infrared Spectroscopy* 16 (5):445–454.

Blanco, M., M. Castillo, and R. Beneyto. 2007. "Study of reaction processes by in-line near-infrared spectroscopy in combination with multivariate curve resolution. Esterification of myristic acid with isopropanol." *Talanta* 72 (2):519–525. doi: 10.1016/j.talanta.2006.11.046.

Blanco, M., A. C. Peinado, and J. Mas. 2004. "Analytical monitoring of alcoholic fermentation using NIR spectroscopy." *Biotechnology and Bioengineering* 88 (4):536–542. doi: 10.1002/bit.20214.

Blanco, M., A. C. Peinado, and J. Mas. 2006. "Monitoring alcoholic fermentation by joint use of soft and hard modelling methods." *Analytica Chimica Acta* 556 (2):364–373. doi: 10.1016/j.aca.2005.09.066.

Brouckaert, D., L. De Meyer, B. Vanbillemont, P. J. Van Bockstal, J. Lammens, S. Mortier, J. Corver, C. Vervaet, I. Nopens, and T. De Beer. 2018. "Potential of near-infrared chemical imaging as process analytical technology tool for continuous freeze-drying." *Analytical Chemistry* 90 (7):4354–4362. doi: 10.1021/acs.analchem.7b03647.

Brülls, M., S. Folestad, A. Sparén, A. Rasmuson, and J. Salomonsson. 2007. "Applying spectral peak area analysis in near-infrared spectroscopy moisture assays." *Journal of Pharmaceutical and Biomedical Analysis* 44 (1):127–136. doi: 10.1016/j.jpba.2007.02.013.

Cavinato, A. G., D. M. Mayes, Z. Ge, and J. B. Callis. 1990. "Noninvasive method for monitoring ethanol in fermentation processes using fiber-optic near-infrared spectroscopy." *Analytical Chemistry* 62 (18):1977–1982. doi: 10.1021/ac00217a015.

Cervera, A. E., N. Petersen, A. E. Lantz, A. Larsen, and K. V. Gernaey. 2009. "Application of near-infrared spectroscopy for monitoring and control of cell culture and fermentation." *Biotechnology Progress* 25 (6):1561–1581. doi: 10.1002/btpr.280.

Clavaud, M., Y. Roggo, R. Von Daeniken, A. Liebler, and J.-O. Schwabe. 2013. "Chemometrics and in-line near infrared spectroscopic monitoring of a biopharmaceutical Chinese hamster ovary cell culture: Prediction of multiple cultivation variables." *Talanta* 111:28–38. doi: 10.1016/j.talanta.2013.03.044.

Corro-Herrera, V. A., J. Gómez-Rodríguez, P. M. Hayward-Jones, D. M. Barradas-Dermitz, M. G. Aguilar-Uscanga, and A. C. Gschaedler-Mathis. 2016. "In-situ monitoring of *Saccharomyces cerevisiae* ITV01 bioethanol process using near-infrared spectroscopy NIRS and chemometrics." *Biotechnology Progress* 32 (2):510–517. doi: 10.1002/btpr.2222.

Crowley, J. S., A. Arnold, N. Wood, L. M. Harvey, and B. McNeil. 2005. "Monitoring a high cell density recombinant *Pichia pastoris* fed-batch bioprocess using transmission and reflectance near infrared spectroscopy." *Enzyme and Microbial Technology* 36 (5):621–628. doi: 10.1016/j.enzmictec.2003.12.016.

De Beer, T. R. M., P. Vercruysse, A. Burggraeve, T. Quinten, J. Ouyang, X. Zhang, C. Vervaet, J. P. Remon, and W. R. G. Baeyens. 2009. "In-line and real-time process monitoring of a freeze drying process using Raman and NIR spectroscopy as complementary process analytical technology (PAT) tools." *Journal of Pharmaceutical Sciences* 98 (9):3430–3446. doi: 10.1002/jps.21633.

de Lima, S. M., B. F. A. Silva, D. Vieira Pontes, C. F. Pereira, L. Stragevitch, and M. F. Pimentel. 2014. "In-line monitoring of the transesterification reactions for biodiesel production using NIR spectroscopy." *Fuel* 115:46–53 doi: 10.1016/j.fuel.2013.06.057.

de Oliveira, A. D., A. F. de Sá, M. F. Pimentel, J. G. A. Pacheco, C. F. Pereira, and M. S. Larrechi. 2017. "Comprehensive near infrared study of Jatropha oil esterification with ethanol for biodiesel production." *Spectrochimica Acta Part A: Molecular and Biomolecular Spectroscopy* 170:56–64. doi: 10.1016/j.saa.2016.07.006.

do Nascimento, R. J. A., G. R. de Macedo, E. S. dos Santos, and J. A. de Oliveira. 2017. "Real time and in situ near-infrared spectroscopy (NIRS) for quantitative monitoring of biomass, glucose, ethanol and glycerine concentrations in an alcoholic fermentation." *Brazilian Journal of Chemical Engineering* 34:459–468.

Ge, Z., A. G. Cavinato, and J. B. Callis. 1994. "Noninvasive spectroscopy for monitoring cell density in a fermentation process." *Analytical Chemistry* 66 (8):1354–1362. doi: 10.1021/ac00080a023.

Goldfeld, M., J. Christensen, D. Pollard, E. R. Gibson, J. T. Olesberg, E. J. Koerperick, K. Lanz, G. W. Small, M. A. Arnold, and C. E. Evans. 2014. "Advanced near-infrared monitor for stable real-time measurement and control of *Pichia pastoris* bioprocesses." *Biotechnology Progress* 30 (3):749–759. doi: 10.1002/btpr.1890.

Grohganz, H., D. Gildemyn, E. Skibsted, J. M. Flink, and J. Rantanen. 2011. "Rapid solid-state analysis of freeze-dried protein formulations using NIR and Raman spectroscopies." *Journal of Pharmaceutical Sciences* 100 (7):2871–2875. doi: 10.1002/jps.22490.

Hakemeyer, C., U. Strauss, S. Werz, G. E. Jose, F. Folque, and J. C. Menezes. 2012. "At-line NIR spectroscopy as effective PAT monitoring technique in Mab cultivations during process development and manufacturing." *Talanta* 90:12–21. doi: 10.1016/j.talanta.2011.12.042.

Hall, J. W., B. McNeil, M. J. Rollins, I. Draper, B. G. Thompson, and G. MacAloney. 1996. "Near-infrared spectroscopic determination of acetate, ammonium, biomass, and glycerol in an industrial escherichia coli fermentation." *Applied Spectroscopy* 50 (1):102–108. doi: 10.1366/0003702963906726.

Harding, A. R., C. F. O'Neill, M. L. May, L. O. Burciaga-Robles, and C. R. Krehbiel. 2013. Evaluation of commercially available near-infrared reflectance spectroscopy prediction models for nutrient prediction of DDGS. Oklahoma State University.

Hoehse, M., J. Alves-Rausch, A. Prediger, P. Roch, and C. Grimm. 2015. "Near-infrared spectroscopy in upstream bioprocesses." *Pharmaceutical Bioprocessing* 3:153–172.

Jenzsch, M., C. Bell, S. Buziol, F. Kepert, H. Wegele, and C. Hakemeyer. 2017. "Trends in process analytical technology: Present state in bioprocessing." *Adv Biochem Eng Biotechnol* 165:211–252.

Kamat, M. S., R. A. Lodder, and P. P. DeLuca. 1989. "Near-infrared spectroscopic determination of residual moisture in lyophilized sucrose through intact glass vials." *Pharmaceutical Research* 6 (11):961–965. doi: 10.1023/A:1015997530367.

Kauppinen, A., M. Toiviainen, O. Korhonen, J. Aaltonen, K. Järvinen, J. Paaso, M. Juuti, and J. Ketolainen. 2013. "In-line multipoint near-infrared spectroscopy for moisture content quantification during freeze-drying." *Analytical Chemistry* 85 (4):2377–2384. doi: 10.1021/ac303403p.

Killner, M. H. M., J. J. R. Rohwedder, and C. Pasquini. 2011. "A PLS regression model using NIR spectroscopy for on-line monitoring of the biodiesel production reaction." *Fuel* 90 (11):3268–3273. doi: 10.1016/j.fuel.2011.06.025.

Kim, S., M. d'Anjou, K. J. Lanz, C. E. Evans, E. R. Gibson, J. T. Olesberg, M. Mallem, I. Shandil, A. Nylen, E. J. Koerperick, D. W. Cooley, G. A. Brower, G. W. Small, and M. A. Arnold. 2015. "Real-time monitoring of glycerol and methanol to enhance antibody production in industrial *Pichia pastoris* bioprocesses." *Biochemical Engineering Journal* 94:115–124. doi: 10.1016/j.bej.2014.12.002.

Kirdar, A. O., G. Chen, J. Weidner, and A. S. Rathore. 2009. "Application of near-infrared (NIR) spectroscopy for screening of raw materials used in the cell culture medium for the production of a recombinant therapeutic protein." *Biotechnology Progress* 26 (2):527–531. doi: 10.1002/btpr.329.

Knothe, G. 1999. "Rapid monitoring of transesterification and assessing biodiesel fuel quality by near-infrared spectroscopy using a fiber-optic probe." *Journal of the American Oil Chemists' Society* 76 (7):795–800. doi: 10.1007/s11746-999-0068-5.

Laurens, L. M. L., and E. J. Wolfrum. 2011. "Feasibility of spectroscopic characterization of algal lipids: Chemometric correlation of NIR and FTIR Spectra with exogenous lipids in algal biomass." *Bioenergy Research* 4 (1):22–35. doi: 10.1007/s12155-010-9098-y.

Lee, H. W., A. Christie, J. J. Liu, and S. Yoon. 2012. "Estimation of raw material performance in mammalian cell culture using near infrared spectra combined with chemometrics approaches." *Biotechnology Progress* 28 (3):824–832. doi: 10.1002/btpr.1536.

Lewis, C. B., R. J. McNichols, A. Gowda, and G. L. Coté. 2000. "Investigation of near-infrared spectroscopy for periodic determination of glucose in cell culture media in situ." *Applied Spectroscopy* 54 (10):1453–1457. doi: 10.1366/0003702001948592.

Li, M., B. Ebel, F. Chauchard, E. Guédon, and A. Marc. 2018. "Parallel comparison of in situ Raman and NIR spectroscopies to simultaneously measure multiple variables toward real-time monitoring of CHO cell bioreactor cultures." *Biochemical Engineering Journal* 137:205–213 doi: 10.1016/j.bej.2018.06.005.

Liebmann, B., A. Friedl, and K. Varmuza. 2009. "Determination of glucose and ethanol in bioethanol production by near infrared spectroscopy and chemometrics." *Analytica Chimica Acta* 642 (1):171–178. doi: 10.1016/j.aca.2008.10.069.

Liebmann, B., A. Friedl, and K. Varmuza. 2010. "Applicability of near-infrared spectroscopy for process monitoring in bioethanol production." *Biochemical Engineering Journal* 52 (2):187–193. doi: 10.1016/j.bej.2010.08.006.

Lin, T. P., and C. C. Hsu. 2002. "Determination of residual moisture in lyophilized protein pharmaceuticals using a rapid and non-invasive method: near infrared spectroscopy." *PDA Journal of Pharmaceutical Science and Technology* 56 (4):196–205.

Liu, B., J. Liu, T. Chen, B. Yang, Y. Jiang, D. Wei, and F. Chen. 2015. "Rapid characterization of fatty acids in oleaginous microalgae by near-infrared spectroscopy." *International Journal of Molecular Sciences* 16 (4):7045–7056. doi: 10.3390/ijms16047045.

McShane, M. J., and G. L. Coté. 1998. "Near-infrared spectroscopy for determination of glucose, lactate, and ammonia in cell culture media." *Applied Spectroscopy* 52 (8):1073–1078. doi: 10.1366/0003702981944968.

Payne, C. E., and E. J. Wolfrum. 2015. "Rapid analysis of composition and reactivity in cellulosic biomass feedstocks with near-infrared spectroscopy." *Biotechnology for Biofuels* 8 (1):43. doi: 10.1186/s13068-015-0222-2.

Pekel, A.Y., E. O. Cakir, M. Polat, G. Inan, and N. Kocabagh. 2013. "Correlations between chemical assays and near-infrared reflectance spectroscopy for nutrient components and correlations between nutrients and color scores of distillers dried grains with solubles." *Journal of Applied Poultry Research* 22:814–824.

Pieters, S., T. De Beer, J. C. Kasper, D. Boulpaep, O. Waszkiewicz, M. Goodarzi, C. Tistaert, W. Friess, J. P. Remon, C. Vervaet, and Y. Vander Heyden. 2012. "Near-infrared spectroscopy for in-line monitoring of protein unfolding and its interactions with lyoprotectants during freeze-drying." *Analytical Chemistry* 84 (2):947–955. doi: 10.1021/ac2022184.

Pieters, S., T. De Beer, and Y. Vander Heyden. 2012. "Near-infrared and Raman spectroscopy: Potential tools for monitoring of protein conformational instability during freeze-drying processes." *American Pharmaceutical Review* 15 (1).

Pinto, A. S. S., Sandra, C. Pereira, Marcelo P. A. Ribeiro, and C. S. Farinas. 2016. "Monitoring of the cellulosic ethanol fermentation process by near-infrared spectroscopy." *Bioresource Technology* 203:334–340. doi: 10.1016/j.biortech.2015.12.069.

Read, E. K., R. B. Shah, B. S. Riley, J. T. Park, K. A. Brorson, and A. S. Rathore. 2009. "Process analytical technology (PAT) for biopharmaceutical products: Part II. Concepts and applications." *Biotechnology and Bioengineering* 105 (2):285–295. doi: 10.1002/bit.22529.

Richard, R., Y. Li, B. Dubreuil, S. Thiebaud-Roux, and L. Prat. 2011. "On-line monitoring of the transesterification reaction between triglycerides and ethanol using near infrared spectroscopy combined with gas chromatography." *Bioresource Technology* 102 (12):6702–6709. doi: 10.1016/j.biortech.2011.03.111.

Sales, K. C., F. Rosa, P. N. Sampaio, L. P. Fonseca, M. B. Lopes, and C. R. Calado. 2015. "In situ near-infrared (NIR) versus high-throughput mid-infrared (MIR) spectroscopy to monitor biopharmaceutical production." *Applied Spectroscopy* 69 (6):760–772. doi: 10.1366/14-07588.

Sanderson, M. A., F. Agblevor, M. Collins, and D. K. Johnson. 1996. "Compositional analysis of biomass feedstocks by near infrared reflectance spectroscopy." *Biomass and Bioenergy* 11 (5):365–370. doi: 10.1016/S0961-9534(96)00039-6.

Silman, R. W., L. T. Black, and K. Norris. 1983. "Assay of solid–substrate fermentation by means of reflectance infrared analysis." *Biotechnology and Bioengineering* 25 (2):603–607. doi: 10.1002/bit.260250224.

Skvaril, J., K. G. Kyprianidis, and E. Dahlquist. 2017. "Applications of near-infrared spectroscopy (NIRS) in biomass energy conversion processes: A review." *Applied Spectroscopy Reviews* 52 (8):675–728. doi: 10.1080/05704928.2017.1289471.

Sluiter, A., and E. Wolfrum. 2013. "Near infrared calibration models for pretreated corn stover slurry solids, isolated and in situ." *Journal of Near Infrared Spectroscopy* 21 (4):249–257.

Tahir, M., M. Y. Shim, N. E. Ward, M. O. Westerhaus, and G. M. Pesti. 2012. "Evaluation of near-infrared reflectance spectroscopy (NIRS) techniques for total and phytate phosphorus of common poultry feed ingredients." *Poultry Science* 91 (10):2540–2547. doi: 10.3382/ps.2012-02211.

Teixeira, A. P., R. Oliveira, P. M. Alves, and M. J. T. Carrondo. 2009. "Advances in on-line monitoring and control of mammalian cell cultures: Supporting the PAT initiative." *Biotechnology Advances* 27 (6):726–732. doi: 10.1016/j.biotechadv.2009.05.003.

Tena, N., A. Boix, and C. von Holst. 2015. "Identification of botanical and geographical origin of distillers dried grains with solubles by near infrared microscopy." *Food Control* 54:103–110. doi: 10.1016/j.foodcont.2015.01.033.

Trunfio, N., H. Lee, J. Starkey, C. Agarabi, J. Liu, and S. Yoon. 2017. "Characterization of mammalian cell culture raw materials by combining spectroscopy and chemometrics." *Biotechnology Progress* 33 (4):1127–1138. doi: 10.1002/btpr.2480.

Vaccari, G., E. Dosi, A. L. Campi, G. Mantovani, A. González-Vara y R, and D. Matteuzzi. 1994. "A near-infrarod spectroscopy technique for the control of fermentation processes: An application to lactic acid fermentation." *Biotechnology and Bioengineering* 43 (10):913–917. doi: 10.1002/bit.260431003.

Vogel, K. P., B. S. Dien, H. G. Jung, M. D. Casler, S. D. Masterson, and R. B. Mitchell. 2011. "Quantifying actual and theoretical ethanol yields for switchgrass strains using NIRS analyses." *BioEnergy Research* 4 (2):96–110. doi: 10.1007/s12155-010-9104-4.

34 Near-Infrared Spectroscopy Imaging of Biological Materials and Systems

Mika Ishigaki
Shimane University

Daitaro Ishikawa
Fukushima University

Christian Huck
Leopold-Franzens University

*Yukihiro Ozaki**
Kwansei Gakuin University

CONTENTS

34.1 INTRODUCTION

In the last two decades or so, the developments of new NIR instruments such as Fourier transform (FT)-NIR spectroscopy and NIR spectroscopy imaging have extended the application of NIR spectroscopy dramatically (Ciurczak and Drennen 2002, Siesler et al. 2002, Roberts et al. 2004, Ozaki et al. 2006, Ozaki 2012, Workman and Weyer 2012, Jue and Matsuda 2016). Chemical imaging data such as NIR imaging data can be represented as a three-dimensional cube spanning one

* yukiz89016@gmail.com.

wavelength and two special dimensions (Lewis et al. 2004, Gowen et al. 2007, Grahn and Geladi 2007, Šašić and Ozaki 2010, Adu-Baffour 2012, Ishikawa et al. 2014, Salzer and Siesler 2014, Strlic 2018). This data construct is called a hypercube. To handle these hypercubes, one uses a variety of chemometric methods such as principal component analysis (PCA). NIR imaging collects data in a spatially resolved manner, so that using NIR imaging, one can learn from where on a sample a NIR spectrum is obtained. Therefore, information about the special distribution of sample components can be derived. These spatially resolved data provide qualitative and quantitative insights into the functionality of heterogeneous samples such as pharmaceuticals tablets, polymers, and agricultural and biological materials.

NIR imaging has recently been used in the following applications (Lewis et al. 2004, Gowen et al. 2007, Grahn and Geladi 2007, Šašić and Ozaki 2010, Adu-Baffour 2012, Ishikawa et al. 2014, Salzer and Siesler 2014, Strlic 2018):

1. Studies of polymer crystallization, blending, and degradation.
2. Diffusion process of a solvent into a polymer.
3. Water adsorption in organic materials' surface.
4. Imaging analysis of cells and tissues.
5. Studies of distributions and functions of biological molecules in biological systems.
6. Quality evaluation of foods and agricultural products
7. Quantitative and qualitative analysis and quality control of pharmaceutical tablets. This is concerned with PAT (process analytical technology) and QbD (quality by design).
8. Monitoring of water diffusion process into a pharmaceutical tablet.

The strong points of NIR imaging compared with IR and Raman imaging may be as follows:

i. NIR imaging is useful for probing thick samples or bulk materials with little or no sample preparation required
ii. It is easier to study and analyze an aqueous dispersion with NIR imaging than IR imaging.
iii. One can use an optical fiber for NIR measurement, while it is not easy to use it for IR measurement.

There is another category of NIR imaging, that is, functional NIR spectroscopy (fNIR or fNIRS). It has been used for functional neuroimaging. Using fNIR, brain activity can be monitored through hemodynamic responses associated with neuronal behavior (Surhone 2009, Haas 2010, Cui et al. 2011). NIR is a noninvasive imaging method involving the quantification of chromophore concentration resolved from the measurement of NIR light attenuation or temporal or phasic changes. The region from 700 to 900 nm is the region of the optical window in which skin, tissue, and bone are mostly transparent to NIR light while hemoglobin (Hb) and deoxygenated hemoglobin (deoxy-Hb) are stronger absorbers of light. Differences in the absorption spectra of deoxy-Hb and oxy-Hb allow the measurement of relative changes in hemoglobin concentration through the use of light attenuation at multiple wavelengths (Surhone 2009, Haas 2010, Cui et al. 2011).

The purpose of this chapter is to outline state-of-the-art of NIR imaging. Recent progress in NIR imaging devices and the applications of NIR imaging in biological sciences are described. We do not treat fNIR in this chapter.

34.2 NEWLY DEVELOPED NIR IMAGING SYSTEMS

Recently, several kinds of novel NIR imaging systems have been developed for a variety of purposes from industrial online monitoring to biomedical applications (Ishigaki et al. 2016a, 2016b, 2018a, 2018b, Ishikawa et al. 2012, 2013a, 2013b, 2013c, Murayama et al. 2013, 2015, 2018, Puangchit et al. 2017). Although positive features such as high sensitivity, high wavelength, and spatial resolutions are essential qualifications, issues concerning high-speed measurements and portability are also primary concerns in the development of the instruments (Ishikawa et al. 2012, 2013a, 2014). In this session, we introduce three kinds of new NIR imaging devices: a highly sensitive portable NIR imaging device (Ishikawa et al. 2012, 2013c, Murayama et al. 2013, 2015, 2018), a device with high-speed, wide-area monitoring over a broad NIR wavelength region (Ishikawa et al. 2013b, 2013c, 2016b, Puangchit et al. 2017, Ishigaki et al. 2018a), and an imaging-type two-dimensional Fourier spectroscopy (ITFS) system (Ishigaki et al. 2018b, Qi et al. 2015).

34.2.1 Portable NIR Imaging Device (D-NIRs)

D-NIRs is a compact ($151 \times 93 \times 120$ mm^3) portable NIR imaging instrument and enables one to develop images with high speed, high wavelength resolution (approximately 1 nm), and maximum spatial resolution (0.025 mm per pixel) (Ishikawa et al. 2012, 2013a). This system is suitable not only for PAT but also for other various industrial applications. The optical system of D-NIRs depends on a polychromator type of spectrometer, P-NIRs, which is equipped with a developed 640-element high-density photodiode array detector and a charge amplifier (Ishikawa et al. 2012, 2013a). It is noted that P-NIRs can obtain spectra in the 1000–1600 nm region with a 1 nm spectral resolution. Figure 34.1 depicts the imaging unit of D-NIRs consisting of light sources and a pair of Galvano mirrors. D-NIRs uses three halogen lamps of 5 W. The Galvano mirrors move to the direction of *x*- and *y*-axes, respectively, by two motors, and then, two-dimensional spectra data in the whole spectral region between 1000 and 1600 nm are obtained by a line scanning system. The design based on the combination of lamps and mirrors not only contributes to the portability of D-NIRs but also enables high-speed data acquisition at approximately 50 s per 100 pixels (in the case of 1 mm spatial resolution). The maximum spatial resolution of D-NIRs achieves 0.025 mm per pixel due to the special mechanical design for focus. Its maximum measurement speed is 5 ms per pixel.

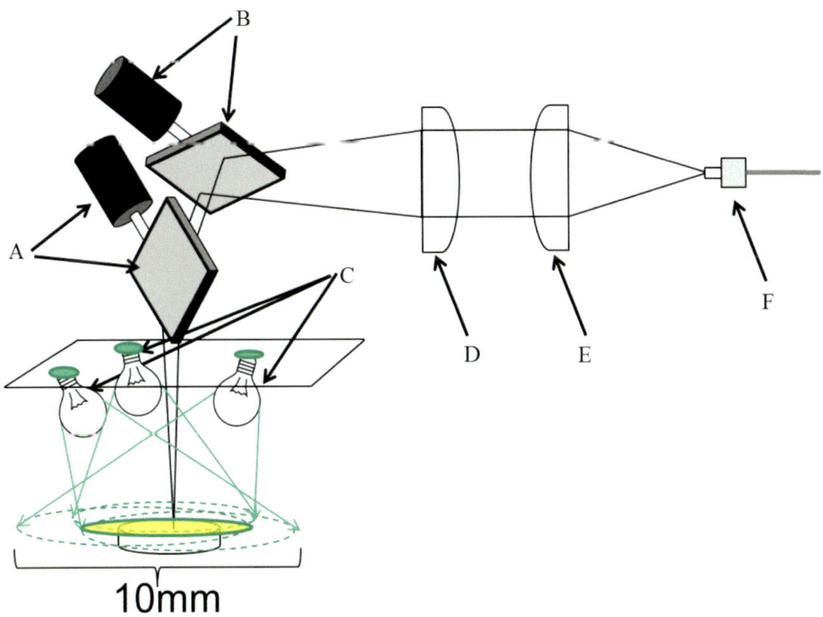

FIGURE 34.1 Schematic diagram of the imaging unit of the developed NIR imaging device (D-NIRs). The red lines indicate the diffuse-reflected light from the object. A: Galvano mirror and motor (*x*-axis); B: Galvano mirror and motor (*y*-axis); C: halogen lamps; D: focus lens 1; E: focus lens 2; and F: optical fiber. (Copyright: Ishikawa et al. 2012.)

34.2.2 A High-Speed and Wide-Area Monitoring NIR Imaging System with a Novel NIR Camera (Compovision)

Figure 34.2 shows a novel NIR imaging device with a newly developed hyperspectral camera (Compovision) by Sumitomo Electric Industries, Ltd (Ishikawa et al. 2013b, 2013c). This imaging system consists of a linear moveable stage, a halogen light source, and the NIR camera. The characteristic features of the camera include high-speed and wide-area monitoring over a broad NIR wavelength region thanks to a newly developed indium gallium arsenic (InGaAs) detector (Ishikawa et al. 2013b, 2013c). The detector is equipped with InGaAs and gallium arsenic antimony (GaAsSb) type II quantum wells (QWs), laminated on an indium phosphide (InP) substrate. Thus, one can measure an NIR spectrum in the 1000–2350 nm region of a 150×200 mm^2 area (approximately 100,000 pixels) within 2–5 s. Commercial NIR imaging devices measure into approximately 10×10 mm^2 area with several minutes. Thus, the performance of high-speed monitoring of the NIR imaging system is potentially powerful for various purposes from industrial process monitoring to biomedical applications. A wide-area NIR image that can be rapidly obtained is very attractive for a variety of practical applications.

34.2.3 Imaging-Type Two-Dimensional Fourier Spectroscopy (ITFS) System

ITFS uses the so-called phase-shift interferometry (Ishigaki et al. 2018b, Qi et al. 2015). The system adopts a characteristic mechanism of optical interference that gives spatial phase differences to the objected light from samples by a partial movable mirror as shown in Figure 34.3. The interference occurs only in the case where each light ray gathers to the same point in ITFS, and the interference between lights coming from out of the in-focus plane is not observed in alternate current component in the interferogram, detected as a direct current element. Therefore, by changing the focal plane, three-dimensional spectral data can be acquired using this system (Ishigaki et al. 2018b, Qi et al. 2015).

NIR measurements can be performed in both the reflectance and transmission modes by changing the positional relationship of the optical components such as light source and detection parts. In the former, we used the system consisting of a hyperspectral imaging unit (AOI Electronics Co., Ltd., NT00-T011), an InGaAs camera (Hamamatsu Photonics K.K., C10633-13), a halogen lamp, and an objective lens at 8× magnification. In the transmission mode, it was composed of a hyperspectral imaging unit (AOI Electronics Co., Ltd., NT00-T012), the same halogen lamp, an NIR camera (Sumitomo Electric Industries, Ltd., CVN800), and an objective lens at 4× magnification.

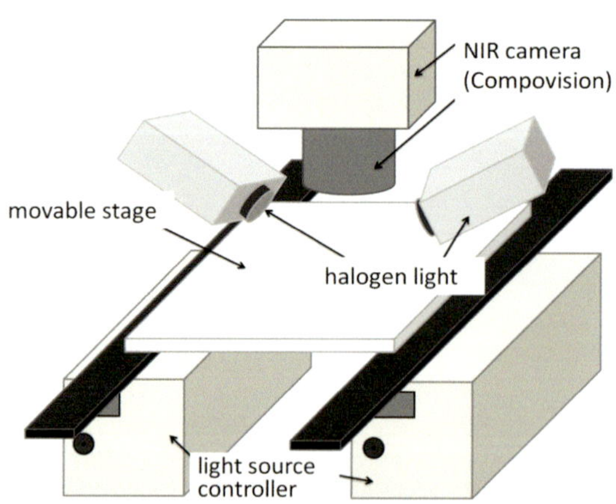

FIGURE 34.2 Outline view of the imaging instrument including the developed NIR camera. (Copyright: Ishikawa et al. 2013b.)

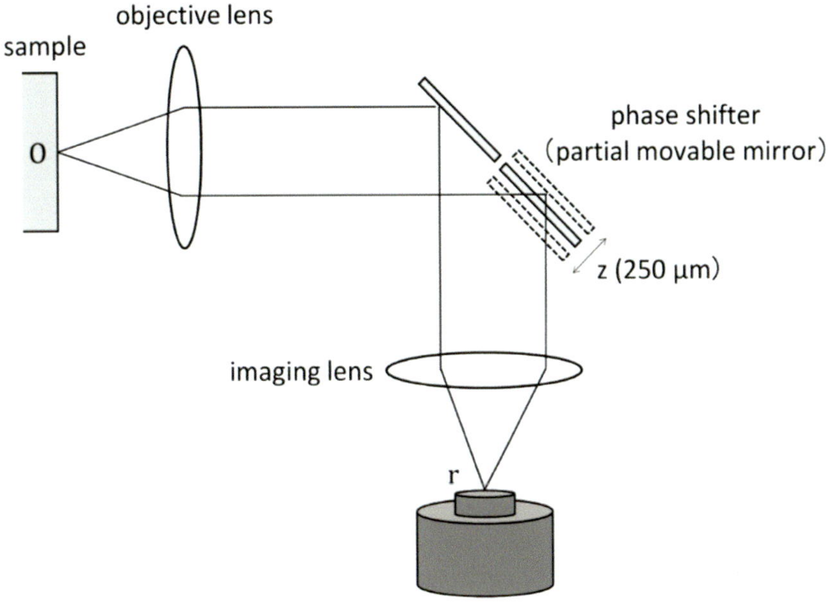

FIGURE 34.3 Optical conformation of the imaging-type two-dimensional Fourier spectroscopy system (ITFS; AOI Electronics Co., Ltd., NT00-T011). (Copyright: Ishigaki et al. 2018b.)

The spatial resolution with an objective lens at 8× magnification is about 8 μm. The system covers the region of 1000–2500 nm, and the wavelength resolution is 10 nm. The number of pixels in the two-dimensional image was approximately 81,000 points (319(X)×255(Y)). Of note is that the instrument adopts the wave-front split-type interferometer (Ishigaki et al. 2018b, Qi et al. 2015).

34.3 EXAMPLES OF NIR IMAGING STUDIES OF BIOLOGICAL MATERIALS AND SYSTEMS

NIR imaging has recently been used in pharmaceutical industry, agriculture, food research and technology, biology, and biomedical research (Lewis et al. 2004, Gowen et al. 2007, Grahn and Geladi 2007, Šašić and Ozaki 2010, Adu-Baffour 2012, Ishikawa et al. 2014, Salzer and Siesler 2014, Strlic 2018). For the applications in the pharmaceutical industry, for example, each pharmaceutical process such as blending, graduation, drying, and coating, the possibility of predicting quantity of components, evaluating the particle size, and determining the distribution of mixing components have been investigated (Lewis et al. 2004, Šašić and Ozaki 2010, Ishikawa et al. 2012, 2013b, 2014, Salzer and Siesler 2014, Murayama et al. 2013, 2015, 2018). In this chapter, two examples of applications to pharmaceutical industries utilizing novel types of NIR imaging devices are presented, and moreover, new types of biological applications and applications to studies of embryo development are introduced (Ishigaki et al. 2016b, 2018a, 2018b, Puangchit et al. 2017).

34.3.1 Pharmaceutical Tablet Monitoring by D-NIRs

To demonstrate the performance of D-NIRs as a PAT and QbD tool, the visualization of the distribution of tablet components was examined (Ishikawa et al. 2012, 2013b, Murayama et al. 2013, 2015, 2018). Figure 34.4 depicts a second-derivative image (a) and a standardized image (b) for the distribution of talc in the tablet developed by a peak at 1391 nm (Ishikawa et al. 2012). The standardized image was prepared by the mean and standard deviation of second derivative at 1391 nm to enhance the pattern observed in the image. Of note is the homogeneity of the component in the

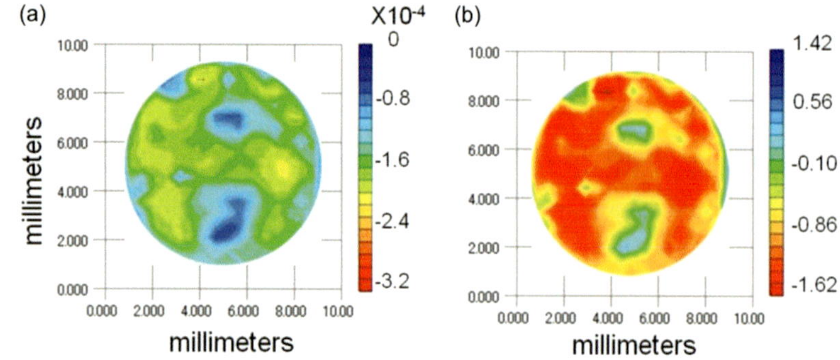

FIGURE 34.4 NIR images for the distribution of talc in a tablet developed by second-derivative spectra (a) and their standardization value (b). (Copyright: Ishikawa et al. 2012.)

tablet. Moreover, one should be recognized that D-NIRs has high portability, which enhances the potential as a PAT tool.

NIR images obtained by D-NIRs have also potential for the monitoring dissolution process of a table. Figure 34.5a and b shows NIR spectra and their second-derivative spectra of the tablet at a point of x (4.5 mm) and y (1.5 mm) during dissolution process, respectively (Ishikawa et al. 2013a). The intensity of a broad feature in the 1400–1600 nm region increases with the increase in bandwidth (Figure 34.5). Note in the second-derivative spectra that three bands appear at 1451, 1432, and 1420 nm in this region. However, a major reason for the intensity increase is water absorption because it increases with elapsed time. Figure 34.5c exhibits NIR images of the point mapped at 0, 120, and 240 min by the absorbance of 1361 nm due to ascorbic acid. One can clearly observe the dispersion of water into the tablet.

Figure 34.6 displays NIR images built by the intensity ration of two bands at 1361 and 1354 nm of the second-derivative spectra obtained by D-NIRs during the dissolution process (Ishikawa et al. 2013a). The images demonstrated that the concentration of ascorbic acid in the tablet changes gradually due to water penetration. These studies successfully presented that D-NIRs is an attractive tool for pharmaceutical applications.

34.3.2 Quantitative Analysis of a Model Tablet Containing Three Different Components Using the High-Speed Wide-Area Monitoring System with a Newly Developed NIR Camera

In many cases of high-speed NIR imaging devices, the improvement of quantitative accuracy of sample spectra obtained at line or in line is important to use them for various industrial situations. We demonstrated the practical potential of high-speed NIR imaging devices through the quantitative analysis for tablets using NIR spectra combined with chemometrics (Ishikawa et al. 2013b, 2013c). The diffuse reflectance (DR) NIR spectra of tablets containing ascorbic acid, talc, and cellulose were measured for the 1000–2300 nm region, and their images were developed. To evaluate the quantitative accuracy, moving-window PLS (MWPLS) was employed to determine the regions of interest. A root mean square error of prediction (RMSEP) of less than 4.1 wt% and an R^2 of more than 0.91 using one PLS factor were obtained as quantitative results (Ishikawa et al. 2013b). Although the quantitative accuracy of cellulose is relatively low (R^2=0.83, RMSE=4.71 wt%) comparing with others, it was found that the standard normal deviate (SNV) model is useful in such cases. Finally, the SNV model provided a better prediction of R^2 (0.86), and thus, one can say that its quantitative accuracy is almost as good as that obtained by a conventional NIR spectrometer.

For NIR imaging, sample tablets including talc, ascorbic acid, and cellulose were prepared. The sample contained a 3% food coloring reagent rather than the same proportion of cellulose.

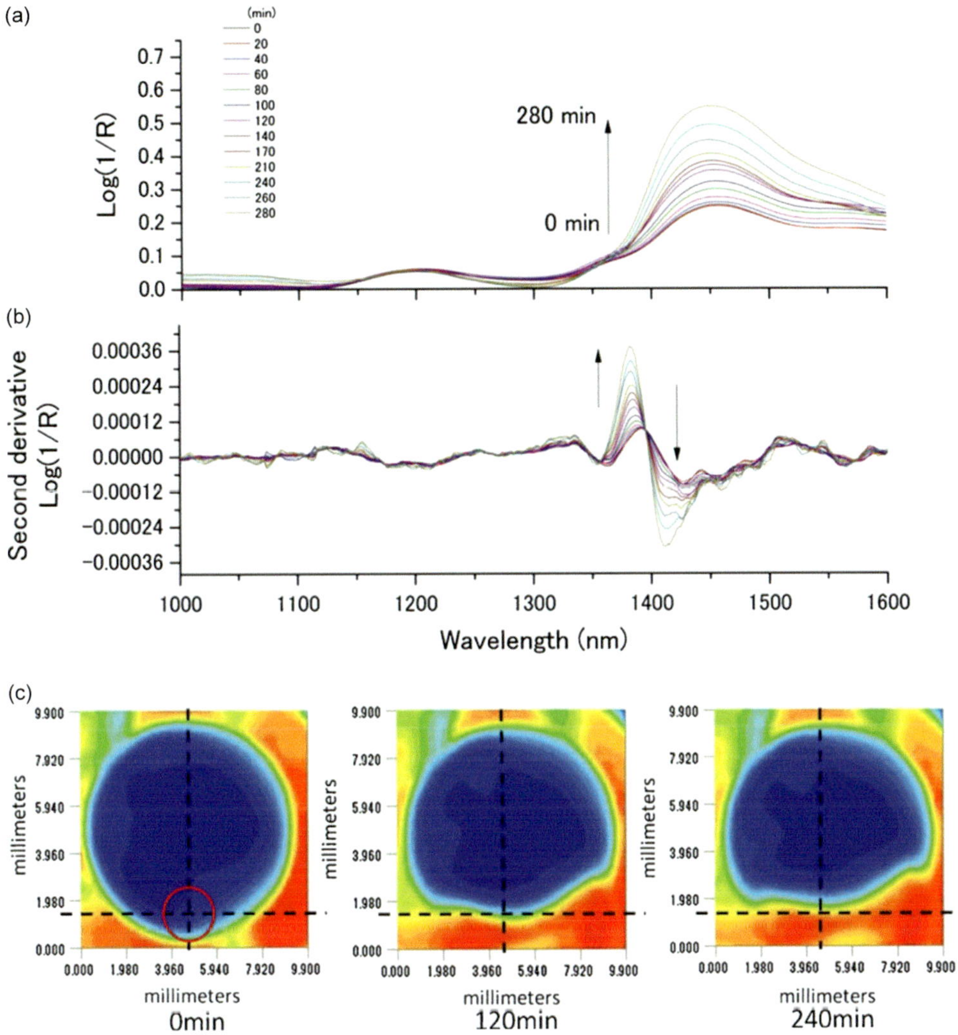

FIGURE 34.5 Time-dependent changes of (a) NIR spectra and (b) their second-derivative spectra of the tablet dissolution process at a point of x (4.5 mm) and y (1.5 mm). (c) NIR images of the point mapped at 0, 120, and 240 min by the absorbance of 1361 nm due to AsA. (Copyright: Ishikawa et al. 2013a.)

As shown in Figure 34.7a, each sample consisted of three separate parts that differed in component concentration. Figure 34.7b depicts images developed by the prediction results over each component concentration in the tablets. The contrast of each NIR image corresponds to the concentration of each component, and the homogeneity within the separated area of each tablet is demonstrated.

34.3.3 Noninvasive, High-Speed, NIR Imaging of the Biomolecular Distribution in Fertilized Fish Eggs

NIR imaging allows one to analyze the distribution of biomaterials and their molecular mechanisms in embryonic development of Japanese medaka fish nondestructively and noninvasively without staining using NIR imaging (Ishigaki et al. 2016a, 2016b, 2018a, 2018b, Puangchit et al. 2017). The high-speed and wide-area monitoring system was used in this research, enabling ultra-high-speed imaging; using this system, it is possible to acquire microscopic imaging data only in a few seconds.

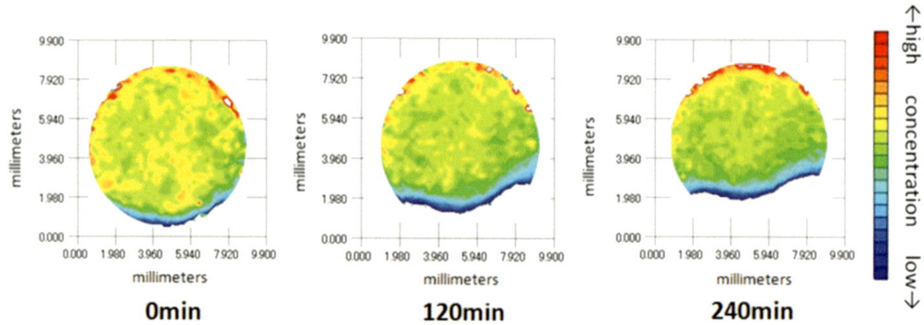

FIGURE 34.6 Changes in the peak-height ratio-based image of tablet dissolution developed by using the second-derivative intensities at 1361 and 1354 nm due to AsA and water, respectively. These ratio images were modified by arbitrary threshold to delete the color except the tablet part. (Copyright: Ishikawa et al. 2013a.)

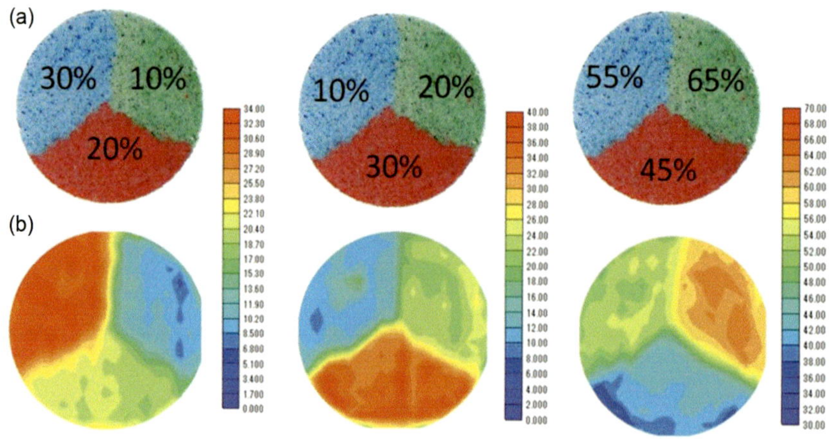

FIGURE 34.7 (a) Photos of the model tablet and (b) NIR images developed by the PLS regression. (Copyright: Ishikawa et al. 2013b.)

Figure 34.8 demonstrates the schematic view of the microscopic NIR imaging system (Ishigaki et al. 2018a). NIR measurements were performed in transmission mode. In order to regulate the optical path length, eggs were sandwiched between two glass slides with pinchcocks and spacers at 5 mm thickness. The objective lens at 5× magnification was placed in front of the camera and the spatial resolution was about 6.8 μm. The distributions of biomolecules were examined by mapping the intensities of NIR bands resulting from lipids, proteins, and water in two dimensions. The NIR imaging study succeeded in visualizing the structures of eyes, lipid bilayer membranes, micelles, and water-structure differences at the interface of different substances constituting different structures on the egg (Ishigaki et al. 2018a). Moreover, this investigation provided new insights on the metabolic mechanisms of lipids and membrane functions from the biased distribution of lipoproteins and the presence of unsaturated fatty acids in the egg membrane (Ishigaki et al. 2018a).

Figure 34.9a depicts an optical image of a fertilized medaka egg on the first day after fertilization (Ishigaki et al. 2018a). Soon after fertilization, the egg was structurally inhomogeneous. Many small oil droplets distributed throughout the egg began to coalesce and fuse into larger droplets. The cytoplasm at the same time began to form a blastodisc that would later transform into an embryonic disk, as shown in Figure 34.9b.

Figure 34.10a shows averaged NIR spectra in the 1000–2200 nm region of the three major parts (yolk, oil droplets, and embryo) of a fertilized medaka egg measured on the first day and the day

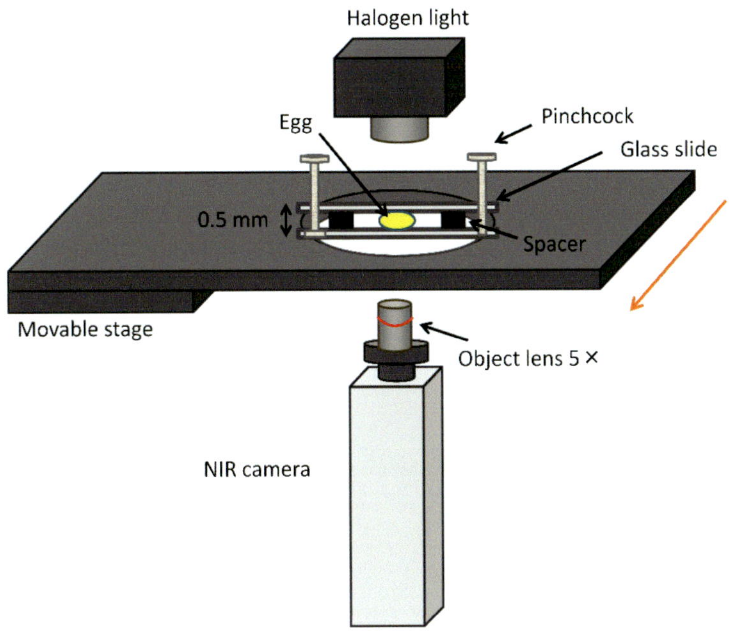

FIGURE 34.8 The microscopic NIR imaging system with a prepared slide of a medaka fish egg. (Copyright: Ishigaki et al. 2018a.)

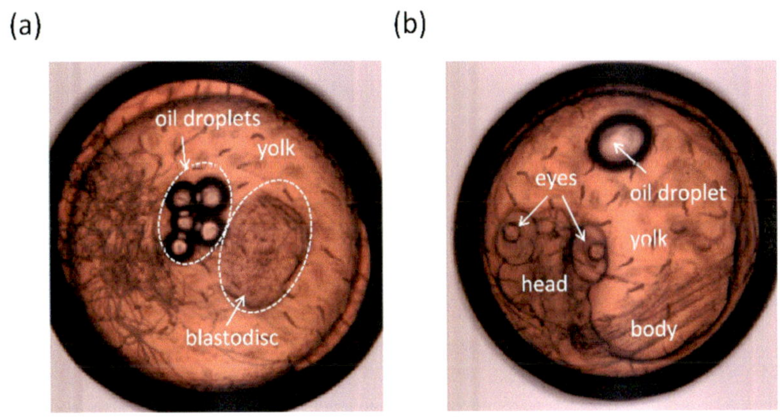

FIGURE 34.9 An optical image of a fertilized medaka egg on (a) the first day and (b) the fifth day after fertilization. (Copyright: Ishigaki et al. 2018a.)

just before hatching (JBH) (Ishigaki et al. 2018a). Figure 34.10b and c depicts second-derivative spectra of (a) in the 1100–1820 and 1800–2150 nm regions, respectively. The spectra from yolks and embryos have similar spectral patterns; however, the spectra from oil droplets have several characteristic peaks due to lipids. A band at 1217 nm arises from a second overtone of the C-H stretching mode of CH_2 groups, and two bands at 1716 and 1773 nm are attributed to the first overtone of the C-H stretching modes of CH_2 groups in hydrocarbons and aliphatic compounds (Ishigaki et al. 2018a). A broad feature centered at approximately 2050 nm in the yolk and embryo spectra (Figure 34.10c) originates from a combination of an N-H stretching mode and amide II. It is noted that the spectral shapes in the 1400–1500 and 1880–1980 nm regions due to water differ significantly between oil droplets and other parts of the eggs as shown in Figure 34.10b and c.

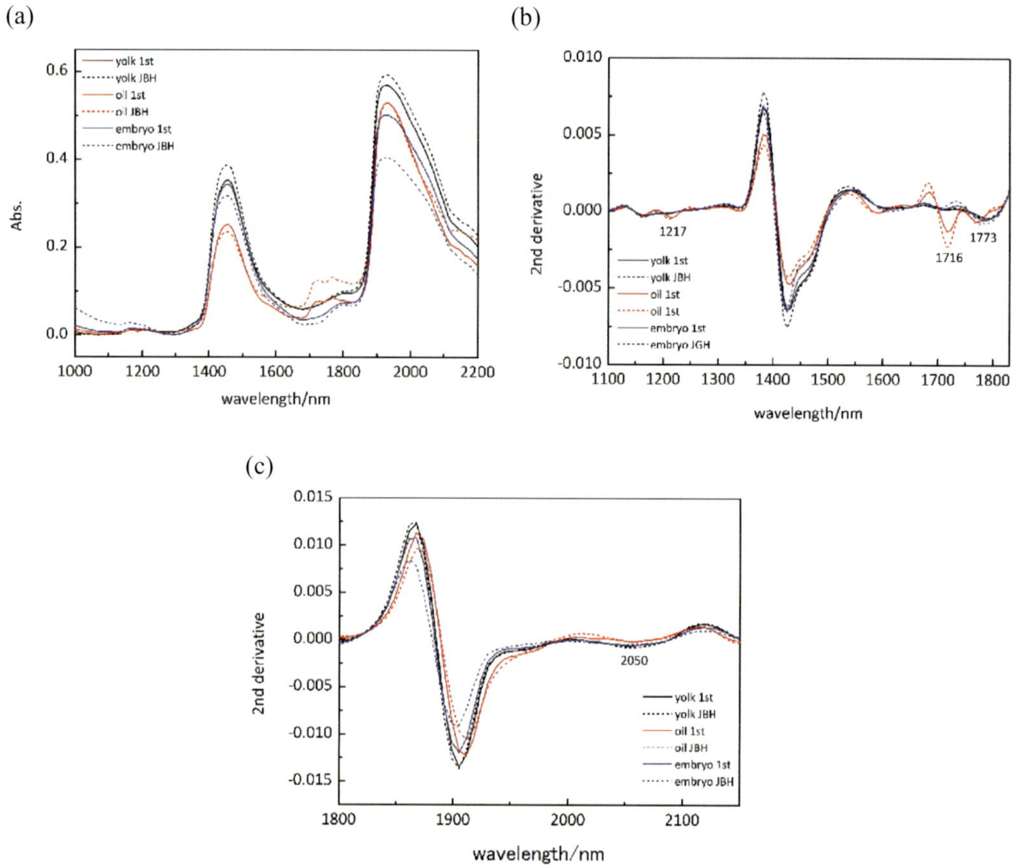

FIGURE 34.10 (a) NIR spectra in the 1000–2200 nm region of the yolk, oil droplets, and embryo of medaka eggs on the first day after fertilization (solid line) and the day just before hatching (dashed line) with baseline corrections. Second-derivative spectra in (b) the 1100–1820 nm region and (c) the 1800–2200 nm region. (Copyright: Ishigaki et al. 2018a.)

Figure 34.11a shows visible images of a medaka fish egg from the first day to the day JBH, and Figure 34.11(b)–(f) exhibits NIR images developed by plotting the second-derivative intensities of some notable bands and scores extracted from PCA. In Figure 34.11b, the shape of the egg membrane, the contour of oil droplets, and the outline of the embryo appear by plotting the intensity at 1767 nm. The selected wavelength is located in the middle of the two bands between egg membrane (1730 nm) and oil droplets (1773 nm), which are both attributed to the first overtone of the C-H stretching modes of aliphatic compounds, such as fatty acids and hydrocarbons (Ishigaki et al. 2018a). The fact that these images obtained with the information of C-H stretching in CH_2 groups showed clear membrane structures is consistent with the reported biomaterial constituents of cell membranes. Moreover, on the third day after fertilization, eye structures become clear. Eyes include a large amount of sphingolipids and glycerolipids with long carbon chains, which helps to clarify their structures in NIR images without staining.

Figure 34.11c displays the images developed by the C-H stretching band at 1716 nm of CH_2 groups included in aliphatic compounds. Of note is that this band has large contributions from the sp_2 CH_2 group in unsaturated fatty acids. Therefore, Figure 34.11c shows the distribution of particularly unsaturated fatty acids in various kinds of fatty acids, and the structures of oil droplets and egg membrane are highlighted. Using the characteristic peaks due to saturated or unsaturated

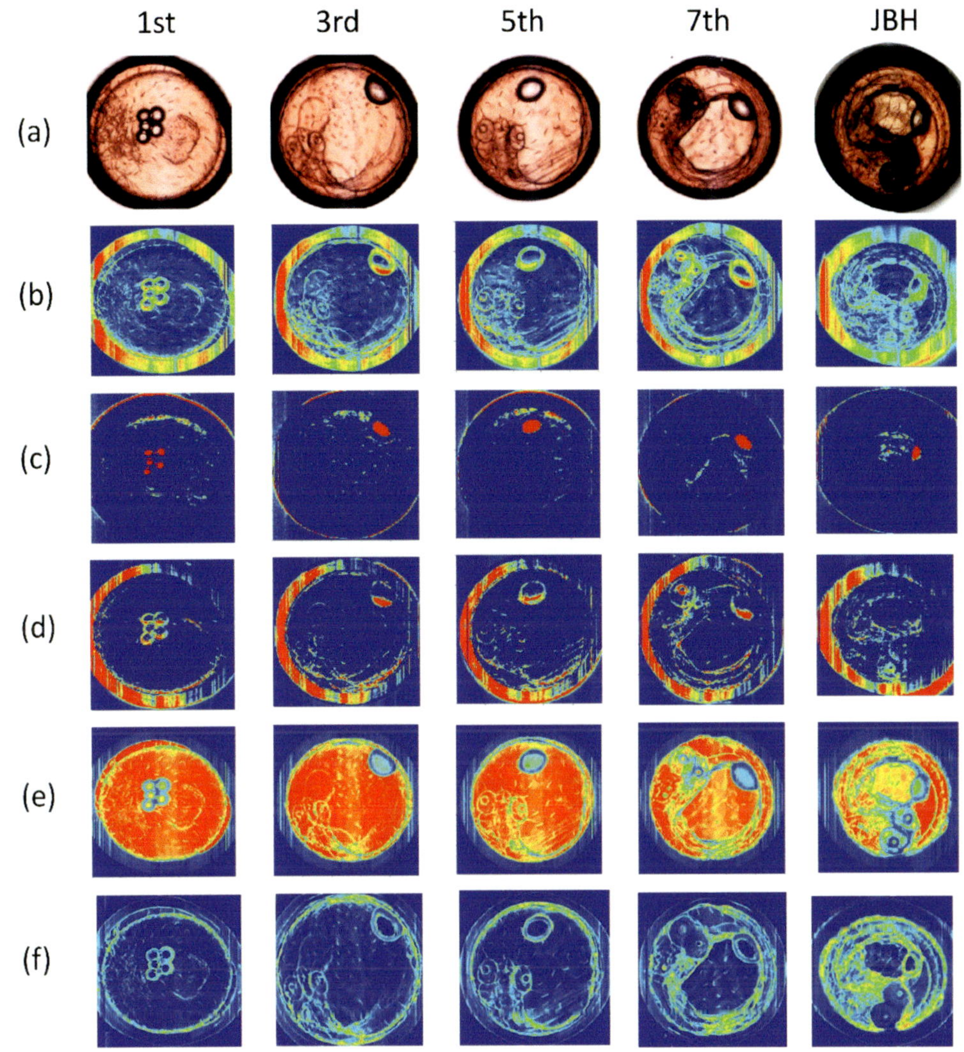

FIGURE 34.11 (a) Visible images of medaka eggs from the first day after fertilization to the day just before hatching, and near-infrared images of eggs based on band intensities in second-derivative spectra at (b) 1767 nm (aliphatic compounds), (c) 1716 nm (unsaturated fatty acids), (d) 1564 nm (proteins), and (e) 1460 nm (water), using (f) scores calculated by projecting PC1 loadings onto imaging data to identify the contribution of weakly hydrogen-bonded water species. (Copyright: Ishigaki et al. 2018a.)

structures, the distributions of different fatty acids can be selectively visualized. Of interest in these images is that the high oil contents within oil droplets and the thin-layer structure of the egg membrane are clearly visualized using this band. A terminal sp_2 CH_2 group exists at the ends of the molecules, and there are fewer sp_2 CH_2 than sp_3 CH_2 in fatty acids, which draw the membrane structure as shown in Figure 34.11b.

Images in Figure 34.11d were prepared based on the intensity of the band at 1564 nm arising from the first overtone of N-H stretching mode of amide groups (Ishigaki et al. 2018a). The structure of the egg membrane is clearly depicted because the second-derivative spectra obtained from the egg membrane show the significant peak. In cell membranes, proteins are embedded in the membrane structure and bound to lipids. These proteins are called membrane proteins, and these components are thought to be present in the image. Figure 34.11d appears to show that lipoproteins are heterogeneous in oil droplets. Heterogeneous components can also be seen in Figure 34.11b obtained by using a lipid band. The positions of these heterogeneous components overlap between Figure 34.11b and d, and these very likely are lipoproteins. Furthermore, the structures of eyes are shown in Figure 34.11d. For example, the vitreous body of the eye contains dense collagen, and collagen fibers have many amino groups. Therefore, eyes seem to be highlighted.

Figure 34.11e displays the images developed by plotting band intensities at 1460 nm due to strongly hydrogen-bonded water species. The images do reflect the slight differences in hydrogen-bonding states between the egg yolk and embryo, especially in the early stages of development. The concentration of strongly hydrogen-bonded water species, for example, is low in embryonic parts. It is very likely that this is caused by differences in the concentrations of proteins and lipids between the yolk and blastodisc. The yolk is a dense packet of energy and materials such as proteins. Water strongly interacts with dense materials in the yolk, and the environments surrounding water may be different between the yolk and embryo (Ishigaki et al. 2018a).

These results indicated the potential for NIR imaging in exploring the biological functions and metabolic systems of cells and embryos.

34.3.4 Nonstaining Blood Flow Imaging Using Optical Interference due to Doppler Shift

Ishigaki et al. further expanded the nonstaining NIR imaging of developing fish egg embryos. They not only obtained NIR images, but also detected interference signals caused by the Doppler effect with the ITFS system described above (Ishigaki et al. 2018b).

Using this system, absorption spectra of the egg substances were observed in the NIR region, and the information about the molecular composition of the embryo can be acquired noninvasively in situ. Of interest is that light reflected by a moving structure, which is associated with heart beat and blood flow, is also observed to show a slightly shifted frequency due to the Doppler shift. A beat signal is generated by heterodyne interference between the shifted and non-shifted frequency components. Figure 34.12a and b shows the interferograms obtained from the yolk part (static part) and from the heart part (moving part) of a medaka fish egg on the fifth day after fertilization, respectively (Ishigaki et al. 2018b). The interferogram from the static part exhibited the normal center burst, but the one obtained from the moving part, on the other hand, showed a noise-like pattern in addition to the center burst. The spectrum calculated by Fourier transformation of the interferogram from the heart part (Figure 34.12b) revealed three peaks at 3768, 1884, and 1256 nm (Figure 34.13). These new peaks, characteristic of the motion such as heart beat and blood flow,

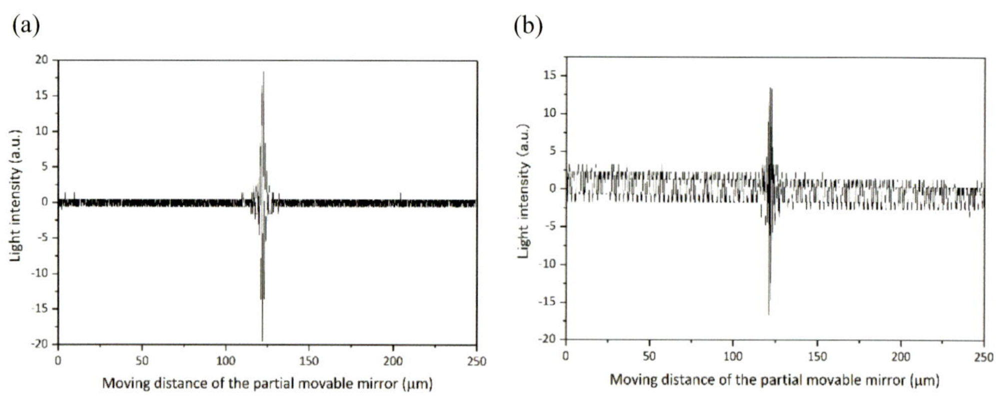

FIGURE 34.12 Interferogram (a) obtained from the yolk part (static part) and (b) from the heart part (moving part) of a medaka fish egg on the fifth day after fertilization. (Copyright: Ishigaki et al. 2018b.)

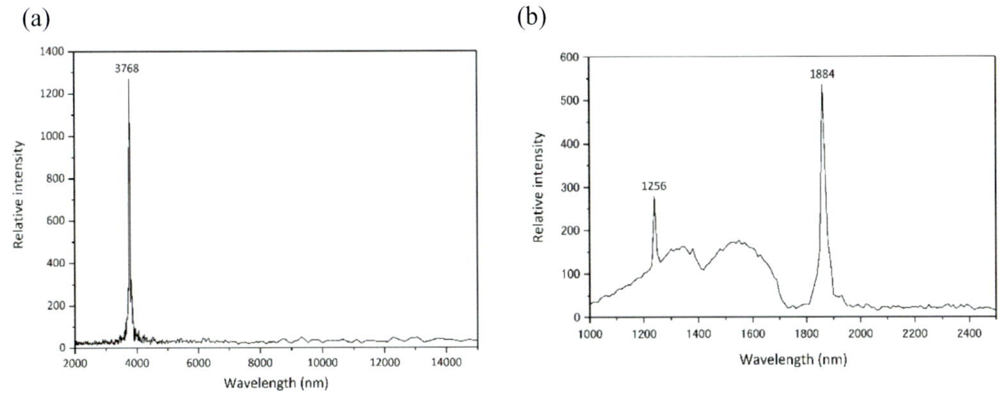

FIGURE 34.13 Spectral information calculated by Fourier transformation of the data in Figure 34.12 (b) in the wave number region of (a) 2000–15,000 nm and (b) 1000–2500 nm. (Copyright: Ishigaki et al. 2018b.)

seemed to be appearing in addition to those arising from molecular vibrations. In order to confirm their origin, the wavelength of the peaks λ [μm] after Fourier transformation of the interferogram (Figure 34.12b) was calculated. The number of periodic noise-like waveforms were counted as 19 during 6-s measurement, corresponding to the frequency as 3.2 Hz. The optical path difference given by the partial movable mirror was $250\sqrt{2}$ μm in 30 s. Therefore, λ was calculated as $\lambda[\mu\text{m}] = 250\sqrt{2}\,[\mu\text{m}]/3.2\,[\text{Hz}] \times 30$ s, resulting in $\lambda = 3.68$ μm. It shows good agreement with the new peak as shown in Figure 34.13a. Other two peaks (1.88 and 1.26 μm) in Figure 34.13b, on the other hand, correspond to the first and second overtones of the fundamental mode (3.68 μm), respectively.

Figure 34.14(A) and (B) shows the images developed by plotting the intensities in two dimensions at the wavelength of the second overtone (1256 nm) and the first overtone (1884 nm), respectively, by

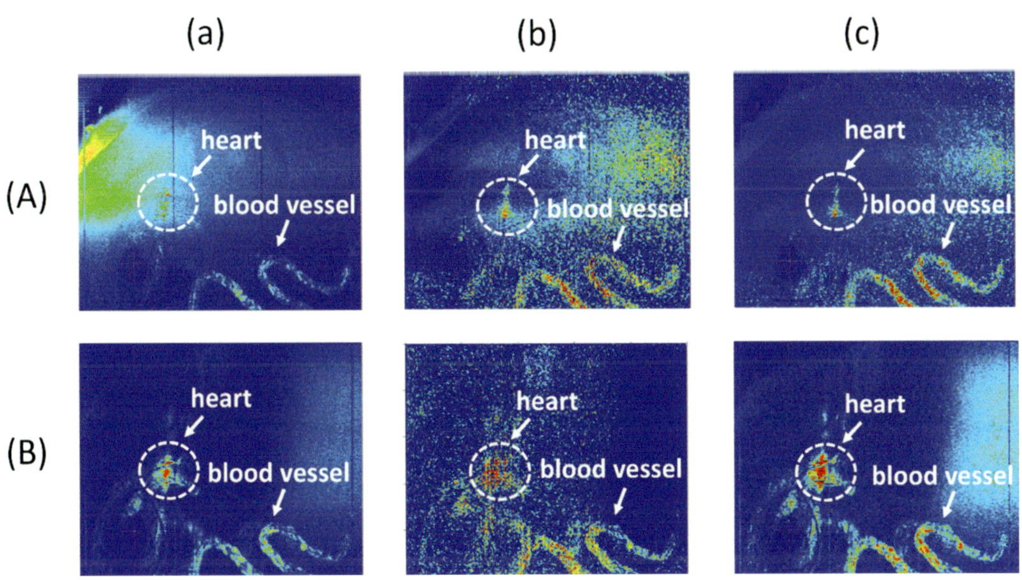

FIGURE 34.14 Blood flow images of a medaka fish egg on the fifth day after fertilization obtained by plotting (a) detected light intensity from the sample, (b) absorbance, and (c) relative intensity calculated as the light intensity divided by the white light reference at (A) 1260 and (B) 1860 nm. (Copyright: Ishigaki et al. 2018b.)

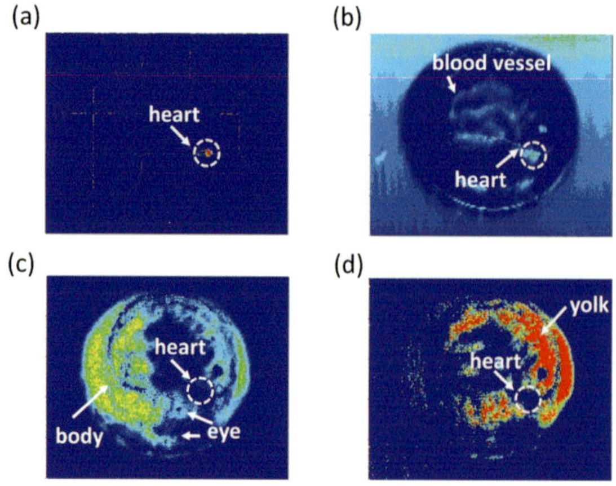

FIGURE 34.15 NIR images obtained by plotting (a) light intensity at 2260 nm and absorbance intensity at (b) 1880, (c) 1940, and (d) 2360 nm. (Copyright: Ishigaki et al. 2018b.)

using the detected light intensity (a), absorbance (b), and relative intensity (c) (Ishigaki et al. 2018b). It is noted that the location where the flow of red blood cells linked with the motion of the heart beat occurs is successfully visualized, and blood flow images can be created without staining.

Absorbance spectra obtained in a transmission mode show a large contribution due to water. Of note is that two dips originated from the heart beat are observed at ~1880 and 2260 nm. Figure 34.15 depicts NIR images developed by plotting the light intensity at 2260 nm (a) and absorbance at 1880 (b), 1940 (c), and 2360 nm (d). Figure 34.15a and b displays the exact part of the heart and the blood vessel stretched over the yolk in addition to the heart part (Ishigaki et al. 2018b). Figure 34.15c is produced using the water signals and scattering effect. Figure 34.15d is developed by the combination of the C–H stretching and bending modes of hydrocarbons and aliphatic compounds. The yolk has relatively high absorption due to the vibrational mode of C–H groups included in hydrocarbons and aliphatic compounds, and the yolk structure is highlighted in Figure 34.15d.

The technique described here has a variety of applications in biology and biomedical sciences such as cardiogenesis and differentiation of induced pluripotent stem (iPS) cells into cardiomyocytes. One may be able to catch the slight beating between them through optical interference in the very early stages of cardiogenesis and differentiation of iPS cells. It enables identification of the beat position and evaluation of the potential for proper differentiation of cells into cardiomyocytes.

34.4 FUTURE PROSPECTS

NIR imaging has become a powerful analytical tool; it has been utilized in various application fields. One of the possible new directions may be the use of NIR imaging for basic science. The investigations of embryo development described in this chapter are one good example. The NIR imaging studies of various kinds of cells including iPS cells hold considerable promise. NIR imaging may become a basic analytical tool in biology. As for practical applications, new attractive applications can be expected for almost all fields including pharmaceutical industry, food and agriculture, and medical sciences. Higher performance with more rapid measurement is expected to be developed. Although NIR imaging has been used for a wide range of art, historic, and archaeological materials, it is also starting to be used for studies of cultural heritage—the identification and quantitative characterization of heritage materials.

Recent advances in NIR instruments are of particular note. Potable NIR imaging systems should become more popular and popular. One can use it from characterizations, quality assessment, and growing monitoring in a field. Miniaturization of portable imaging systems is now in progress, so that the working places of portable NIR imaging devices should become wider and wider. Last but not at least, three-dimensional NIR imaging systems are now realistically expected.

REFERENCES

Adu-Baffour, B. *Near Infrared Imaging System*, **2012**, Lambert.

Ciurczak, E. W., Drennen, J. K., III. *Pharmaceutical and Medical Applications of Near-Infrared Spectroscopy*, **2002**, Boca Raton, FL: CRC Press.

Cui, X., Bray, S., Bryant, D. M., Glover, G. H., Reiss, A. L. *Neuroimage*, **2011**, *54(4)*, 2808–2821.

Gowen, A. A., O'Donnell, C., Cullen, P. J., Downey, G., Frias, J. M. *Trends Food Sci. Technol.*, **2007**, *18(12)*, 590–598.

Grahn, H., Geladi, P. (Eds.). *Techniques and Applications of Hyperspectral Image Analysis*, **2007**, Hoboken, NJ: John Wiley & Sons.

Haas, G. *Auditiv halluzinatorisches Paradigma fuer Magical Ideation und fNIRS*, **2010**, Riga, Latvia: VDM Verlag.

Ishigaki, M., Kawasaki, S., Ishikawa, D., Ozaki, Y. *Sci. Rep.*, **2016a**, *6*, 20066.

Ishigaki, M., Yasui, Y., Puangchit, P., Kawasaki, S., Ozaki, Y. *Molecules*, **2016b**, *21(8)*, 1003.

Ishigaki, M., Nishii, T., Puangchit, P., Yasui, Y., Huck, C. W., Ozaki, Y. *J. Biophotonics*, **2018a**, *11(4)*, e201700115.

Ishigaki, M. et al. *Anal. Chem.*, **2018b**, *90(8)*, 5217–5223.

Ishikawa, D., Murayama, K., Genkawa, T., Awa, K., Komiyama, M., Ozaki, Y. NIR News, **2012**, *23(8)*, 14–17.

Ishikawa, D., Murayama, K., Awa, K., Genkawa, T., Komiyama, M., Kazarian, S. G., Ozaki, Y. *Anal. Bioanal. Chem.*, **2013a**, *405(29)*, 9401–9409.

Ishikawa, D., Nishii, T., Mizuno, F., Kazarian, S. G., Ozaki, Y. *NIR News*, **2013b**, *24(5)*, 6–11.

Ishikawa, D., Nishii, T., Mizuno, F., Sato, H., Kazarian, S. G., Ozaki, Y. *Appl. Spectrosc.*, **2013c**, *67(12)*, 1441–1446.

Ishikawa, D., Shinzawa, H., Genkawa, T., Kazarian, S. G., Ozaki, Y. *Anal. Sci.*, **2014**, *30(1)*, 143–150.

Juc, T., Masuda, K. (Eds.). *Application of Near Infrared Spectroscopy in Biomedicine*, **2016**, Springer Science & Business Media.

Lewis, E. N., Schopperlrei, J., Lee, E., *Spectrosc*, **2004**, *19*, 28.

Murayama, K., Genkawa, T., Ishikawa, D., Komiyama, M., Ozaki, Y. *Rev. Sci. Instruments*, **2013**, *84*, 023104-1–8.

Murayama, K., Ishikawa, D., Genkawa, T., Sugino, H., Komiyama, M., Ozaki, Y. *Molecules*, **2015**, *20(3)*, 4007–4019.

Murayama, K., Ishikawa, D., Genkawa, T., Ozaki, Y. *Appl. Spectrosc.*, **2018**, *72(4)*, 551–561.

Ozaki, Y., McClure, W. F., Christy, A. A. (Eds.). *Near-Infrared Spectroscopy in Food Science and Technology*, **2006**, Hoboken, NJ: John Wiley & Sons.

Ozaki, Y. *Anal. Sci.*, **2012**, *28(6)*, 545–563.

Puangchit, P., Ishigaki, M., Yasui, Y., Kajita, M., Ritthiruangdej, P., Ozaki, Y. *Analyst*, **2017**, *142(24)*, 4765–4772.

Qi, W. Suzuki Y., Sato S., Fujiwara M., Kawashima N., Suzuki S., Abeygunawardhana P., Wada K., Nishiyama K., and Ishimaru I. *Appl. Opt.*, **2015**, *54(20)*, 6254–6259.

Roberts, C. A., Workman, J., and Reeeves J. *Near-Infrared Spectroscopy in Agriculture*, **2004**, Madison, WI: American Society of Agronomy.

Šašić, S. and Ozaki, Y. (Eds.). *Raman, Infrared, and Near-Infrared Chemical Imaging*, **2010**, Hoboken, NJ: John Wiley & Sons.

Salzer, R. and Siesler, H. W. (Eds.). *Infrared and Raman Spectroscopic Imaging*, 2nd Edition, **2014**, Wiley-VCH, Weinheim.

Siesler, H. W., Ozaki, Y., Kawata, S., Heise, H. M. (Eds.). *Near-Infrared Spectroscopy: Principles, Instruments, Applications*, **2002**, Hoboken, NJ: John Wiley & Sons.

Strlic, M. *Near Infrared Spectroscopy and Imaging for Cultural Heritage*, **2018**, London, UK: Royal Society of Chemistry.

Surhone, L. M. *Neuroimaging: History of Neuroimaging, Neuroscience, Psychology, Brain Mapping, Functional Neuroimaging, FNIR, Human Cognome Project, Magnetic Resonance Imaging, Magnetoencephalography*, **2009**, Beau Bassin, Mauritius: Miriam T. Timpledon Betascript Publishers.

Workman, J., Weyer, L. *Practical Guide and Spectral Atlas for Interpretive Near-Infrared Spectroscopy*, 2nd Edition, **2012**, Boca Raton, FL: CRC Press.

35 Blood Analysis by NIR Spectroscopy

Benoît Igne[*]
GlaxoSmithKline

Emil Ciurczak
Doramaxx Consulting

CONTENTS

35.1 INTRODUCTION

The physics of near-infrared (NIR) spectroscopy are favorable for biological applications, especially for in situ measurements. The low absorptivity inherent in NIR allows the radiation to penetrate deeper, resulting in longer pathlengths. The sources are intense, giving more radiation to work with; detectors are sensitive and nearly noise-free, giving a more precise and accurate spectrum. These physical realities allow for measurements through tissue, muscle, fat, and body fluids with great precision. Fiber-optic probes and, more recently, wearable systems complete the picture and make NIR spectroscopy a technique adaptable to any lab or clinic. This chapter will focus on the analysis of blood by NIR spectroscopy, specifically glucose and oxygenation.

35.2 BLOOD GLUCOSE

One of the most publicized and pursued uses of near-infrared in the life sciences is for in situ glucose measurements. Approximately 30,300,000 patients are believed to suffer from diabetes in the United States alone in 2015, with approximately 1.5 million new cases diagnosed every year. With the diagnostic market estimated in the multibillion-dollar range, the number of scientists seeking a spectroscopic solution to this analytical problem is significant. A large number of patents for blood-glucose-measuring devices relying on near-infrared spectroscopy have been issued. Examples of the patented technology are as follows:

- The patented device developed by Ham and Cohen [1]. In this transmission device, the light is passed through the finger. A neural network (NN) algorithm recognizes the spectral areas of greatest correlation with glucose concentration and builds the calibration equation automatically. It is hypothesized that because of the complex nature of blood chemistry, complex algorithms will likely be needed for determination of any component of the blood.

* igneben@gmail.com.

- The patented device developed by Acosta et al. [2]. In this reflectance/transflectance device, the glucose content is correlated with redistribution of fluids between vascular and extravascular and intra- and extracellular compartments.
- The patented device developed by Acosta et al. [3]. In this reflectance device, the glucose content is directly modeled from the collected spectra. Reference values come from glucose testers available to diabetic patients.
- The patented methodology developed by Chung et al. [4]. The proposed method uses a ratio of absorbance measurements at select wavelengths to develop a model that correlates directly the glucose content with the spectral information.

These few examples of patented technologies and methodologies point to the various approaches that have been selected by researchers to predict blood-glucose content: direct approaches that try to directly relate the glucose content to the NIR spectral information and indirect approaches that use a biological response to various glucose levels and attempt to relate these responses (more readily measurable by NIR) with glucose content. Along with the two types of approaches, significant work on chemometrics has been performed to enhance models' performance and robustness.

Modeling of the blood-glucose system has been an ongoing project for numerous researchers for several decades. In 1999, Khalil [5] stated in a review document that to date, no experiment proved that the signal measured by near-infrared spectroscopy was the actual blood-glucose concentration. Gary Small and Mark Arnold (University of Iowa), in particular, have published widely on the subject. In a 2005 review article [6], they surveyed the literature and discussed the merit of direct and indirect approaches to glucose modeling. Indirect glucose measurement is commonly based on the effect of glucose on the scattering properties of the tissue [7,8]. Arnold and Small note that in indirect measurements, the major challenge is to ensure the selectivity of the measurement. Selectivity, along with sensitivity, is also an issue with direct measurements. Glucose is present at millimolar concentration levels, and while it does possess a unique vibrational spectrum compared to the other major blood components, existing devices may not have the sufficient sensitivity to the analyte to develop robust tools.

Nevertheless, many authors have published on methods to predict blood glucose. In 1993 [9], Arnold and Small published a paper modeling the NIR measurement of glucose in a protein-containing matrix. The region from 2000 to 2500 nm was used for analysis of a series of glucose solutions ranging from 1.2 to 20.0 mM in a phosphate buffer closely resembling blood. The buffers contained such materials as bovine serum albumin (BSA). They found that a Gaussian-shaped Fourier filter combined with a partial least squares (PLS) regression delivered a reasonable standard error of calibration. In all cases, the glucose absorbance at 2270 nm gave the optimal correlation, with a standard error of 0.24 mM. In 1994, the same authors reported work on a temperature-insensitive glucose-measuring method [10]. In an approach similar to that just detailed [9], using a Fourier filter/PLS combination, a temperature range between 32°C and 41°C was investigated. The temperature fluctuations caused relatively large variations in the spectra due to the water band shifts. Fourier filtering effectively eliminated the differences, providing a standard error that was even better than for their previous work (0.14 versus 0.24 mM). The necessity of spectral pretreatment is clearly seen in this study. Such temperature correction is necessary should measurements be taken from someone with a fever or who is in shock. This particular point is emphasized in many of the patents for analytical devices. A system to ensure consistency of the temperature of the skin at the sampling site is often included so that interferences due to temperature are limited.

Small, Arnold, et al. performed further work in 1996, this time using physiological levels of glucose in the presence of protein and triglycerides [11]. The solutions used in the study contained between 1 and 20 mM of glucose. The interferences were varied within each level of glucose. It was seen that multivariate algorithms compensated for the chemical variations in the blood where the glucose level remained unchanged. The same researchers also performed some published work on the use of quadratic PLS and digital filtering techniques to account for non-glucose-related changes in the spectra [12]. They concluded that pretreatment helped reduce interferences and resulted in more robust equations. The instrumentation (either single-beam or double-beam) is made less important as interferences are accounted for in the NIR equation.

Two interesting papers by Arnold et al. were published in 1998 [13,14]. They were devoted to the calculations used in noninvasive blood-glucose measurements. The authors investigated neural networks and partial least squares as calibration approaches and examined such things as signal-to-noise enhancements through an understanding of how light is attenuated as it passes through tissue. The companion articles go into detail about the scattering effects of blood and tissue. Compensation schema are identified and proposed to alleviate some of these interferences. It is strongly recommended that fat, water, and tissue be compensated for in any model considered. This was further discussed by Uwadaira et al. [15] who investigated the effects of the interfering factors on the accuracy of noninvasive blood-glucose measurements.

A device developed in Germany by Schrader, where a laser illuminates the humor of the eye and measures the resulting absorbance spectrum was used to measure the amount of blood glucose in the patient [16]. The device is based on a patent developed by Backhaus et al. [17]. He found that the glucose levels in the anterior chamber of the eye closely follow changes in blood glucose with a latency of approximately 20 min. The equations developed by this instrumentation allow for noninvasive monitoring of physiological glucose levels with an error of $\pm 30\,mg\ dL^{-1}$. The basis for using NIR for determination of blood glucose by scanning through skin and muscle is that the blood-glucose level in the blood is similar, if not identical, to the glucose level in tissues. This was claimed by Fischer et al. in 1994 and later demonstrated through a series of measurements [18].

However, contradicting this work was a paper by Sternberg et al. [19]. This group claimed that tissue contained only 75% of the glucose level found in blood. Fortunately, NIR measurements are inclusive of blood and tissue. The calibration was based on a point of contact for each individual patient; thus, the ratio of tissue to venous/arterial blood will be a constant. Correlation of the spectra with blood-glucose readings is then acceptable.

In order to model more correctly the in vivo realities of human body chemistry, "phantoms" (simulated biological conditions, containing fat, blood, skin, etc.) were built for simulated in vivo testing. Arnold et al. built phantoms of water, fat, and muscle tissue to mimic the skin of a patient [20]. They found that in vivo overtone spectra collected across human webbing tissue with a thickness of 6.7 mm could be simulated with water layer thickness from 5.0 to 6.4 mm combined with fat layer thickness from 1.4 to 4.2 mm. For the purposes of this study, animal tissue and fat were used; there is little difference in composition between human and animal materials. They concluded that these phantom studies would help researchers develop patient-applicable methods.

This phantom work was continued by Arnold et al. in a later publication [21]. This is a negatively designed study, used as an object lesson to "warn" the inexperienced user about the pitfalls of chemometrics. They used an in vitro model of blood-simulated samples to build a model for blood-glucose determination. In this case, however, they carefully omitted any glucose from the samples. The samples were randomly assigned glucose values and a PLS regression developed. As with any PLS model, an equation was created to provide reasonable standard errors, regression coefficients, etc. Since there was no glucose present, this equation could not predict glucose when samples containing the sugar were tested. This paper discussed potential errors to avoid during calibration development and analysis.

An interesting observation was made by Maier et al. where it was observed that there exists a correlation between blood-glucose concentration and the reduced NIR scattering coefficient of tissue [22]. Using a frequency-domain NIR spectrometer, the scattering coefficient of tissue was measured with adequate precision to detect changes in glucose. The work was based on the theory that as the glucose concentration increased, the refractive index of the blood also increased in a predictable manner. This increased refractive index would then decrease the scattering coefficient of the blood and give an indication of the concentration. Some questions of the applicability of this work to in vivo measurements must be asked, but it does demonstrate one of the novel approaches being investigated in the field.

One other technique [23] employed a fiber-optic light pipe to measure blood glucose through the skin of a finger. The device uses a portion of the fiber, stripped of its cladding, as a virtual attenuated total reflectance (ATR) device positioned against the skin of the thumb. Because so much radiation is lost into the skin, white light was used with a monochromator situated postsample. This configuration gave better sensitivity than when first resolving the wavelengths. Postsampling indicates that "white" light is impinged on the skin and that the resultant emerging light is collected and submitted to a monochromator. A study by the team of Dr. Ozaki [24] investigated the use of a particular illumination and detection geometry with fiber optics that allowed them to successfully interrogate dermis tissue and reduce other interferences. Using partial least squares, they were able to obtain prediction error for all the subjects involved in the study of 23.7 mg dL^{-1}, thus showing similar results to other studies and showing the possibility to use the same model for different patients.

One "nonskin" application was published in a paper by Heise et al. [25], where a procedure for measuring blood glucose through the lip was described. They used the 1100–1800 nm wavelength range, and partial least squares was the algorithm of choice. The mean-square prediction error was estimated between 45 and 55 mg dL^{-1}. A lag time of approximately 10 min exists between the drawn blood values and the values derived from the lip tissue. The authors recommend using fiber optics for further developments in this field. An update to the study was presented in 2009 (Heise, Lampen, and Marbach [26]).

An example of indirect measurement of blood glucose was provided by Small and Arnold in a study investigating the change in blood scattering with respect to the change in the concentration of glucose [27]. The authors showed prediction errors in the 1–2.5 mM range depending on the part of the NIR signal considered.

A lot of work has been put into the development of chemometric methods that would extract the most relevant signal to allow an improved prediction performance and robustness of the glucose models. Besides the work done with artificial neural networks, literature exists on the use of the net analyte signal (NAS). Initially developed by Lorber as a means to calculate multivariate figures of merit [28], the NAS corresponds to the net signal of the analyte of interest. It can be used as a preprocessing step to focus the analysis on only the relevant signal. An example of the use of NAS for blood glucose can be found in Ren and Arnold [29]. On the same topic of removing interfering signal, techniques based on eliminating the signal that is not related to the signal of glucose have been published. Thus, Ozaki proposed to use a local orthogonal signal correction method to remove interferences from the main blood components in addition to searching for the best suited wavelength regions with a moving-window partial least squares [30,31]. Another technique utilizing the uninformative variable elimination combined with successive projections was intended by Li et al. [32]. A non-linear regression version of the uninformative variable elimination was later proposed by the same authors [33]. Approaches using Beer's law have also been investigated through the use of classical least squares. In an example by Maruo and Yamada [34], the pure component matrix was augmented by a constructed spectrum to correct for baseline drifts as a function of time. Researchers working at the University of Krakow have published several papers and given numerous talks and posters devoted to the math treatments of the complex spectra produced from the NIR examination of blood through the skin and muscle [35–37]. They have been working with neural networks in particular and have made some interesting observations.

To improve model robustness and performance, several authors have worked at optimizing the samples included in calibration. So et al. [38] used a Monte Carlo approach to determine the best set of samples that should be used. The use of simulated spectra has also been tested with results in in vivo experiments in the 12.3 mg dL^{-1} range [39].

Finally, with the development of wearable technologies, much work has been done on the development of new sensors. While not NIR-based, Verily, a Google subsidiary, first announced in 2014 the development of contact lenses before providing an update at the end of 2018 on their halt of the program (https://blog.verily.com/2018/11/update-on-our-smart-lens-program-with.html). Another recent article showed promising results prior to clinical trials (Park et al. [40]).

Other indirect measurements of the effect of glucose in the body have been used to measure glucose. Rachim and Chung [41] used the reflected optical signal of the change in arterial blood volume pulsation on a wristband as an indirect indication of glucose content. A more direct measurement of glucose using the first overtone band region (1500–1800 nm) was also recently proposed by Bae et al. [42].

35.3 BLOOD OXYGENATION

Venial and cranial blood oxygenation is simultaneously a (relatively) simple and yet nontrivial measurement. Early reports of NIR for diagnostic applications came from researchers such as Jobsis in 1977 [43]. He used NIR to monitor the degree of oxygenation of certain metabolites. Later, Ozaki et al. [44] examined venal blood to determine the level of deoxyhemoglobin. The back of the hand was illuminated, and the diffusely reflected light was captured by a miniature integrating sphere equipped with a lead sulfide (PbS) detector. The spectra were correlated with results from a CIBA Corning 278 blood-gas analyzer. The 760-nm band in the spectrum was seen to correlate quite well with deoxyhemoglobin, and a negative correlation exists with oxygenated hemoglobin.

Michael Sowa and his group [45] used NIR imaging as a noninvasive technique to monitor regional and temporal variations in tissue oxygenation. The purpose was to ascertain the effects of restricted blood outflow (venous outflow restriction) and interrupted blood inflow (ischemia). In this work, the software was the heart of the paper. Multivariate analyses of image and spectral data time courses were used to identify correlated spectral and regional domains. Fuzzy C-means clustering of image time courses revealed finer regional heterogeneity in the response of stressed tissues. The wavelength region from 400 to 1100 nm was monitored from 0 to 30 min, and a plot of these data was developed to produce a "topographical" representation of the phenomenon. Peaks and valleys were apparent where blood became oxygenated and deoxygenated. These standard wavelength-based values correlated well with the images developed by the 512×512 back-illuminated charge-coupled device (CCD) element.

Clustering results clearly showed areas of both low and high oxygenation. These results have important implications in the assessment of transplanted tissue viability. Mancini et al. [46] estimated skeletal muscle oxygenation by using the differential absorption properties of hemoglobin. Oxygenated and deoxygenated hemoglobin have identical absorptivities at 800 nm, while deoxygenated hemoglobin predominates at 760 nm. The effects of myoglobin on the readings were also investigated, and it was found that the readings' correlations were due to hemoglobin oxygenation. Venous oxygen saturation and absorption between 760 and 800 nm were correlated. Mancini and colleagues reached several conclusions: (1) Hypoxia in KCl-arrested hearts results in only moderate activation of anaerobic glycolysis; (2) oxygenation of the epicardial and midmural LV layers is similar; and (3) a large pO_2 gradient exists between vascular and intercellular space in beating and arrested crystalloid-perfused hearts.

Lin et al. examined the influence of fat layers on the determination of blood oxygenation in 1998 [47]. The phantom experiments showed fat influences patient-to-patient measurements. This is more easily compensated for in any individual patient. Yamamoto addressed the issue of fat interference with an oximeter that corrected for the influence of subcutaneous fat [48]. The wavelengths, again, were the key, as was the algorithm.

The effect of water on NIR determination of hemoglobin concentration in a tissue-like phantom was studied by Franceschini et al. in 1996 [49]. Their in vitro studies consisted of aqueous suspensions containing Liposyn, bovine blood, and yeast, buffered at pH 7.2. The optical coefficients of the mixture matched those of biological tissue in the NIR, and the hemoglobin concentration (23 μM) was also similar to that found in tissues. They oxygenated and deoxygenated the hemoglobin by sparging the mixture with either oxygen or nitrogen. They determined that water concentration must be taken into account to obtain accurate results of hemoglobin concentrations.

Jiang et al. presented a device that allowed noninvasive measurement of cerebral tissue oxygenation to be performed [50]. Again based on fiber optics, shorter NIR wavelengths were used. Another device allowing diffuse reflectance measurements of the skin was developed by Marbach and Heise [51]. The device presented has an on-axis ellipsoidal collecting mirror with efficient illumination for small sampling areas of bulky body specimens. The actual schematic is too complex to describe in this chapter. The researchers supported the optical design with a Monte Carlo simulation study of the reflective characteristics of skin tissue. While their work was centered on the 1600-nm peak associated with glucose (using the lip as the point of entry), the work is applicable to other tissue research.

Keiko Miyasaka presented some of his work [52] at a meeting in Toronto. As a worker in the field of critical care for children, he introduced what he calls a "Niroscope" for near-infrared spectroscopy. His work was performed during pediatric anesthesia and intensive care. Dr. Miyasaka found the Beer's law relationship not followed rigorously when the signal was passed through the cranium. Considering the massive scattering absorbed light, this was understandable. It did give, however, a semiquantitative or indicating equation. What Miyasaka was measuring was the intercranial levels of oxygenated hemoglobin (HbO_2), deoxygenated or reduced hemoglobin (Hb), and cytochrome redox status. Two methods were used: photon counting and a micro-type pulse laser. The photon-counting method is necessary because of the extreme attenuation of the incident radiation when traversing the cranium. The pulse laser was used to enhance the amount of light introduced into the brain.

Three conclusions may be reached from the NIR data: (1) Changes in HbO_2 levels reflect changes in arterial blood, (2) Hb changes are due to venous blood, and (3) total hemoglobin reflects changes in cerebral blood volume or intercranial pressure. This tool will be invaluable for emergency and operating room situations both for children and, someday, for adults. Van Huffel et al. used NIR to monitor brain oxygenation. They used the information to correlate with behavioral states of preterm infants and to understand the development of brain hemodynamic autoregulation [53]. The concentrations of HbO_2, Hb, and cytochrome aa3 (Cytaa3) are used to monitor the oxygenation level in infant brain blood. Some novel chemometrics were involved as well; windowed fast Fourier transform (WFFT) and wavelet analyses were employed. The purpose of the work was to see relationships between the computed chromophore concentrations and heart rate, breathing, and peripheral oxygen saturation. They presented similar work in 1998 as well [54].

Chris Cooper et al. [55] performed another study; this one aimed at the adult brain. In this work, NIR was used to determine the effects of changes in the rate of oxygen delivery on adult rat brain chemistry. Absolute levels of oxyhemoglobin, deoxyhemoglobin, and the redox state of the CuA center in mitochondrial cytochrome oxidase were determined. An interesting finding was that as the mean arterial blood pressure reached 100 mmHg, hemoglobin oxygenation began to fall, but the oxidized CuA levels only fell when cerebral blood volume autoregulation mechanisms failed at 50 mmHg. Hemoglobin oxygenation fell linearly with decreases in the rate of oxygen delivery to the brain, but the oxidized CuA concentration did not start to fall until this rate was 50% of normal. The results suggested that the brain maintained more-than-adequate oxygen delivery to mitochondria. Their conclusion was that NIR is a good measure of oxygen insufficiency in vivo.

A related study on human infants was performed by Wyatt et al. [56]. They used NIR to quantify the cerebral blood volume in human infants. Similar difficulties were encountered with the amount of light actually penetrating the cranial cavity, but useful equations were generated.

Kupriyanov et al. determined intracellular pO_2 in cardiac muscle by the balance between its diffusion from vascular to intercellular space and its uptake by mitochondria [57]. They reasoned that cessation of mechanical work decreased O_2 demand and should have reduced the O_2 gradient between vascular and intercellular spaces. For their research, they compared the effects of arterial pO_2 on myoglobin (Mb) oxygenation, O_2 uptake, and lactate formation rates in beating and KCl-arrested pig hearts.

Ischemia in the forearm was studied by Mansfield et al. in 1997 [58]. In this study, the workers used fuzzy C-means clustering and principal component analysis (PCA) of time series from the NIR imaging of volunteers' forearms. They attempted predictions of blood depletion and increase without a priori values for calibration. For those with a mathematical bent, this paper does a very nice job describing the theory behind the PCA and fuzzy C-means algorithms.

Another interesting paper was published by Wolf et al. in 1996 [59] where they used NIR and laser Doppler flowmetry (LDF) to study the effect of systemic nitric oxide synthase (NOS) inhibition on brain oxygenation. The study, performed on rats, demonstrated no effects on brain oxygenation during cortical spreading depression (CSD).

Doppler ultrasound was combined with NIR in another study [60]. Liem et al. used NIR and ultrasound to follow the cerebral oxygenation and hemodynamics in preterm infants treated with repeated doses of indomethacin. In addition to the normal concentrations of oxyhemoglobin, deoxyhemoglobin, and oxidized cytochrome aa3 measured by NIR, transcutaneous pO_2 and pCO_2, arterial O_2 saturation, and blood pressure were measured. Along with the cerebral blood volume, they were all used for diagnosis and research. Low oxygenation was then thought to be a possible contraindication for indomethacin treatment for preterm infants.

The physical placement of detectors on the scalp for brain blood oxygenation was studied by Germon et al. in a 1998 study [61]. Detectors placed 2.7 and 5.5 cm from a NIR emitter were compared for the determination of HHb, O_2Hb, oxidized cytochrome C oxidase, and total hemoglobin. The biological portion of the experiment was to measure the chemical changes with an induced reduction of the mean decrease in middle cerebral artery blood flow. The signal change per unit photon pathlength detected at 5.5 cm, for HHb, was significantly greater than at 2.7 cm. On the other hand, the increases in all chromophores detected at 5.5 cm during scalp hyperemia were significantly less than those detected at 2.7 cm. More work is indicated before meaningful applications can be designed from this work.

Using similar instrumentation, Henson et al. determined the accuracy of their cerebral oximeter under conditions of isocapnic hypoxia [62]. In healthy volunteers, dynamic end-tidal forcing was used to produce step changes in PET O_2, resulting in arterial saturation ranging from ~70% to 100% under conditions of controlled normocapnia (resting PTET O_2) or hypercapnia (resting plus 7–10 mmHg). Using standard methods, the O_2 concentrations for each patient under each condition were determined. Excellent correlation resulted in the rSO_2 and $[S_{circ}]_{j[vbar]}O_2$ for each individual patient. However, wide variability between patients was discovered. They concluded that under the current limitations of the equipment, the device was good for tracking trends in O_2 but could not be used as an absolute measure for different patients.

Numerous and disparate studies have been published or presented regarding the effect of various normal and pathologic conditions on blood oxygen:

1. Hoshi et al. investigated the neuronal activity, oxidative metabolism, and blood supply during mental tasks [63].
2. Okada presented work on impaired interhemispheric integration in brain oxygenation and hemodynamics in schizophrenia [64].
3. Hoshi looked into the features of hemodynamic and metabolic changes in the human brain during all-night sleep [65].
4. Akına et al. studied the clinical application of NIR in migraine patients [66]. They assessed the transient changes in brain tissue oxygenation during the aura and headache phases of a migraine attack.

Surgeons are concerned with brain blood flow to patients undergoing cardiopulmonary bypass surgery. An intensive study by Chow et al. was conducted where blood flows were restricted to patients from age 2 weeks to over 20 years [67]. Near-infrared was used to correlate blood flow rate with NIR spectra of the brain. Flows of 0.6, 1.2, and 2.4 L $m^{-2}min^{-1}$ were used. Their results showed that flow was related to mean arterial pressure, but did not correspond to pulsatility. This was interesting in that pulse rate is often used as a diagnostic to assure sufficient blood flow to the brain during surgery.

Totaro et al. published a detailed paper on the factors affecting measurement of cerebrovascular reactivity when measured by NIR [68]. Some of the points covered were the relative transparency of the skin, skull, and brain in the 700–1100 nm region and the oxygen-dependent tissue absorption changes of hemoglobin. Their study covered all relevant factors, such as age, sex, reproducibility, and venous return. The test was based on a 3-min baseline, a 3-min hypercapnia (5% CO_2 in air), and a 2-min recovery period.

Changes in NIR spectra and transcranial Doppler sonography parameters were significantly correlated with variations of end-tidal CO_2 ($P<0.005$). In addition, a significant correlation between the reactivity indexes (approximately, absorptivities) of NIR spectrometry parameters and flow velocity was found ($P<0.01$). High reproducibility was also found for deoxyhemoglobin ($r=0.76$), oxyhemoglobin ($r=0.68$), and flow velocity ($r=0.60$) reactivity indexes. No significant differences between the reactivity indexes of different body positions were found ($P<0.05$). The reactivity indexes of oxyhemoglobin decreased ($P>0.05$) and of deoxyhemoglobin increased ($P<0.01$) with age. Their overall conclusion was that NIR is a viable technique for evaluation of cerebrovascular reactivity for patients with cerebrovascular disease.

Some exciting work was reported by Hitachi [69]. The research, conducted at the Tokyo Metropolitan Police Hospital, used NIR to detect blood flow changes in the brain to determine sites of epileptic activity. The location of blood flow increases corresponded well with conventional methods such as intercranial electroencephalogram (EEG) or single-photon emission computed topography (SPECT). The technique was able to determine the side of the brain where the episode was taking place in all the patients on whom it was tried. This technique could replace the intrusive electrodes currently in use.

Near-infrared spectroscopy has been used to determine the activity of the brain using oxygenation levels. More particularly, a series of articles by Boas et al. [70,71] discusses the limitations that single-point NIR has in the analysis of focal brain activation compared to diffuse activation. Researchers found that it is possible using NIR spectroscopy to determine regions of the brain responsible for simple motor tasks such as moving the finger. However, the local nature of the focal activation is a challenge to NIR measurements. The main source of error was determined to be the cross talk between the pathlength of the entire area illuminated and the partial pathlength of the activated region, thus making it difficult to perform accurate measurements of the focal change. The authors determined that some wavelengths were more prone to cross talk (780- and 830-nm pairing compared to 690- or 760-nm pairing with 830 nm).

On the same topic, a significant amount of work has been done to study the level of oxygenation of the brain during exercise at sub-maximal [72] and maximal [73,74] levels. In a review by Cherie et al. [75], some very interesting results provided by NIR showed that oxygenation of the brain increases during exercise to hit a plateau and decline toward a baseline level at very hard exercise levels. The response was modified as a function of the training level with lower oxygenation levels attained by less fit people. Interestingly, the review ends by stating that improvements of NIR should allow a better understanding of local focal changes, a task undertaken a decade earlier by Boas et al. [70].

Cerebral oxygenation has been used in a multitude of other applications and helped develop the field of functional near-infrared spectroscopy. In addition to the topics already mentioned, fNIRS has found application in speech analysis and disorder [76], psychiatry and disease state [77], cerebral ischemia and hypoxia [78,79], brain injuries [80,81], and cardiac surgery [82,83], to name only a few. However, in a review, Highton et al. [84] noted that fNIRS is limited by the contamination of the signal by extracranial tissues. In addition, the large variability between individuals makes it difficult to set thresholds over or under which a problem is detected, thus making fNIRS ideal for trending and monitoring, but limits decision making to the individual level.

Improvement in the specificity of the measurements is now the focus of the research. Examples of measurements of nerve oxygenation have been presented [85].

Finally, a significant amount of work has been done with optoacoustic systems to image the vascular system. Using short pulses of near-infrared light, strong absorbers such as hemoglobin can be used to provide detailed images of blood vessels and are well suited for diagnosing and monitoring tissues pathologies such as those induced by tumors [86–89]. By modulating the wavelengths used, it is then possible to model the blood oxygen saturation and provide cancer researcher with the possibility to monitor tumor oxygenation [90].

In summary, oxygenation in general has been a major focal point for NIR medical research [91–117]. It continues to be a successful application for NIR.

REFERENCES

Blood Glucose

1. F. M. Ham and G. M. Cohen, "Non-invasive blood glucose monitoring", patent no. 5,553,616 (1997).
2. G. Acosta, J. R. Henderson, A. N. Abul-Haj, T. L. Ruchti, S. L. Monfre, T. B. Blank, and K. H. Hazen, "Compact apparatus for noninvasive measurement of glucose through near-infrared spectroscopy", patent no. US778787924 (2005).
3. G. Acosta, J. R. Henderson, A. N. Abul-Haj, T. L. Ruchti, S. L. Monfre, T. B. Blank, and K. H. Hazen, "Method and apparatus for noninvasive glucose concentration estimation through near-infrared spectroscopy", patent no. US20050107676 (2005).
4. J. W. Y. Chung, K. L. Fan, T. K. S. Wong, S. C. H. Lam, C. C. Cheung, C. M. Chan, and Y. K. Lau, "Method for predicting the blood glucose level of a person", patent no. US7409239 (2008).
5. O. S. Khalil, "Spectroscopic and Clinical Aspects of Noninvasive Glucose Measurements", *Clin. Chem.*, 45, 165 (1999).
6. M. A. Arnold, and G. W. Small, "Noninvasive Glucose Sensing", *Anal. Chem.*, 77, 5429 (2005).
7. M. Kohl, M. Cope, M. Essenpreis, and D. Boecker, "Influence of glucose concentration on light scattering in tissue-simulating phantoms", *Opt. Lett.*, 19, 2170–2172 (1994).
8. M. Kohl, M. Essenpreis, and M. Cope, "The influence of glucose concentration upon the transport of light in tissue-simulating phantoms", *Phys. Med. Biol.*, 40, 1267–1287 (1995).
9. M. A. Arnold, and L. A. Marquardt, "Near-Infrared Spectroscopic Measurement of Glucose in a Protein Matrix", *Anal. Chem.*, 65, 3271 (1993).
10. M. A. Arnold, G. Small, and L. A. Marquart, "Temperature-Insensitive Near-Infrared Spectroscopic Measurement of Glucose in Aqueous Solutions", *Appl. Spectrosc.*, 48(4), 477 (1994).
11. M. A. Arnold, S. Pan, H. Chung, and G. Small, "Near-Infrared Spectroscopic Measurement of Physiological Glucose Levels in Variable Matrices of Protein and Triglycerides", *Anal. Chem.*, 68(7), 1124 (1996).
12. G. W. Small et al., "Evaluation of Data Pretreatment and Model Building Methods for the Determination of Glucose from Near-Infrared Single-Beam Spectra", *Appl. Spectrosc.*, 53(4), 402 (1999).
13. J. J. Burmeister, M. A. Arnold, and G. W. Small, "Spectroscopic Considerations for Noninvasive Blood Glucose Measurements with Near-Infrared Spectroscopy", *IEEE Lasers Electro-Optics Soc*, 12, 6 (1998).
14. M. R. Riley, M. A. Arnold, and D. W. Murhammer, "Matrix-Enhanced Calibration Procedure for Multivariate Calibration Models with Near-Infrared Spectra", *Appl. Spectrosc.*, 52(10), 1339 (1998).
15. Uwadaira, Y., N. Adachi, A. Ikehata, and S. Kawano. 2010. "Factors affecting the accuracy of non-invasive blood glucose measurement by short-wavelength near infrared spectroscopy in the determination of the glycaemic index of foods." *Journal of Near Infrared Spectroscopy* 18 (5):291–300.
16. W. F. Schrader, "Non-Invasive Anterior Chamber Glucose Monitoring by Near-Infrared Absorption Spectroscopy, An Alternative to Blood-Glucose Monitoring in Diabetic Patients?" *Proc. 96th DOG Annual Meeting*, 1998.
17. J. Backhaus D. Boecker, B. Schrader, W. Schrader, H.U. Menzebach, E. Schmidt, "Device for the in vivo Determination of an Optical Property of the Aqueous Humor of the Eye", U.S. patent no. 5,535,743, July 16, 1996.

18. U. Fischer, K. Rebrin, T. Woedtke, and P. Able, "Clinical Usefulness of the Glucose Concentration in the Subcutaneous Tissue—Properties and Pitfalls of Electrochemical Biosensors", *Horm. Metab. Res.*, 26, 515 (1994).
19. F. Sternberg, C. Meyerhoff, F. J. Memel, H. Meyer, F. Bischoff, and E. F. Pfeiffer, "Subcutaneous Glucose Concentration: Its Real Estimation and Continuous Monitoring", *Diabetes Care*, 18, 1266 (1995).
20. M. A. Arnold, J. J. Burmeister, and H. Chung, "Phantoms for Noninvasive Blood Glucose Sensing with Near-Infrared Transmission Spectroscopy", *Photochem. Photobiol.*, 67(1), 50 (1998).
21. M. A. Arnold, J. J. Burmeister, and G. Small, "Phantom Glucose Calibration Models from Simulated Noninvasive Human Near-Infrared Spectra", *Anal. Chem.*, 70, 1773 (1998).
22. J. S. Maier, S. A. Walker, S. Fantini, M. A. Franceschini, and E. Gratton, "Possible Correlation between Blood Glucose Concentration and the Reduced Scattering Coefficient of Tissues in the Near-Infrared", *Opt. Lett.*, 19(24), 2026 (1994).
23. H. Shamoon, I. Gabriely, R. Wozniak, M. Mevorach, J. Kaplan, and Y. Aharm, "Transcutaneous Glucose Monitor During Hypoglycemia", paper 426, *5th Scientific Session, American Diabetes Association Meeting*, San Diego, June, 1999.
24. K. Maruo, M. Tsurugi, M. Tamura, and Y. Ozaki, "In Vivo Noninvasive Measurement of Blood Glucose by Near-Infrared Diffuse-Reflectance Spectroscopy", *Appl. Spectrosc.*, 57, 1236 (2003).
25. H. M. Heise, R. Marboch, G. Janatsch, and J. D. Kruse-Jarres, "Noninvasive Blood Glucose Assay by Near-Infrared Diffuse Reflectance Spectroscopy of the Human Inner Lip", *Appl. Spectrosc.*, 47(7), 875 (1993).
26. Heise, H. M., P. Lampen, and R. Marbach. 2009. Near-infrared reflection spectroscopy for non-invasive monitoring of glucose–established and novel strategies for multivariate calibration: Taylor & Francis Group LLC, CRC Press, FL.
27. A. K. Amerov, J. Chen, G. W. Small, and M. A. Arnold, "Scattering and Absorption Effects in the Determination of Glucose in Whole Blood by Near-Infrared Spectroscopy", *Anal. Chem.*, 77, 4587 (2005).
28. A. Lorber, "Error Propagation and Figures of Merit for Quantification by Solving Matrix Equations", *Anal. Chem.*, 58, 1167–1172 (1986).
29. M. Ren, and M. A. Arnold, "Comparison of Multivariate Calibration Models for Glucose, Urea, and Lactate from Near-Infrared and Raman Spectra", *Anal. Bioanal. Chem.*, 387, 879 (2007).
30. S. Kasemsumran, Y. P. Du, K. Maruo, and Y. Ozaki, "Selective Removal of Interference Signals for Near-Infrared Spectra of Biomedical Samples by Using Region Orthogonal Signal Correction", *Anal. Chim. Acta*, 526, 193 (2004).
31. S. Kasemsumran, Y. P. Du, K. Maruo, and Y. Ozaki, "Improvement of Partial Least Squares Models for In Vitro and In Vivo Glucose Quantifications by using Near-Infrared Spectroscopy and Searching Combination Moving Window Partial Least Squares", *Chemom. Intell. Lab. Syst.*, 82, 97 (2006).
32. L. N. Li, Q. B. Li, and G. J. Zhang, "A Weak Signal Extraction Method for Human Blood Glucose Noninvasive Measurement using Near Infrared Spectroscopy", *J Infrared Milli Terahz Waves*, 30, 1191 (2009).
33. L. N. Li, Q. B. Li, and G. J. Zhang, "A Nonlinear Model for Calibration of Blood Glucose Noninvasive Measurement using Near Infrared Spectroscopy", *Infrared Phys. Technol.*, 53, 410 (2010).
34. Maruo, K., and Y. Yamada. 2015. Near-infrared noninvasive blood glucose prediction without using multivariate analyses: Introduction of imaginary spectra due to scattering change in the skin. Vol. 20: SPIE.
35. K. Jagemann, C. Fischbacher, K. Danzer, U. A. Müller, and B. Mertes, "Applications of Near-Infrared Spectroscopy for Non-Invasive Determination of Blood/Tissue Glucose using Neural Networks", *Z. Phys. Chem.*, 191, 179 (1995).
36. C. Fischbacher, U. A. Muller, B. Mentes, K. U. Jagermann, and K. Danzerk, "Enhancing Calibration Models for Non-Invasive Near-Infrared Spectroscopic Blood Glucose Determination", *Fresenius J. Anal. Chem.*, 359, 78 (1997).
37. K. Danzer, U. A. Muller, B. Mertes, C. Fischbacker, and K. U. Jungemann, "Near-Infrared Diffuse Reflection Spectroscopy for Non-Invasive Blood-Glucose Monitoring", *IEEE LEOS Newslett.*, 4, 18 (1998).
38. C. F. So, J. W. Y. Chung, M. S. M. Siu, and T. K. S. Wong, "Improved Stability of Blood Glucose Measurement in Humans using Near Infrared Spectroscopy", *Spectroscopy*, 25, 137 (2011).
39. K. Maruo, T. Oota, M. Tsurugi, T. Nakagawa, H. Arimoto, M. Tamura, Y. Ozaki, and Y. Yamada, "New Methodology to Obtain a Calibration Model for Noninvasive Near-Infrared Blood Glucose Monitoring", *Appl. Spectrosc.*, 60, 441 (2006).

40. Park, J., J. Kim, S.-Y. Kim, W. H. Cheong, J. Jang, Y.-G. Park, K. Na, Y.-T. Kim, J. H. Heo, C. Y. Lee, J. H. Lee, F. Bien, and J.-U. Park. 2018. "Soft, smart contact lenses with integrations of wireless circuits, glucose sensors, and displays." *Science Advances* 4 (1):eaap9841. doi: 10.1126/sciadv.aap9841.
41. Rachim, V. P., and W. Y. Chung. 2019. "Wearable-band type visible-near infrared optical biosensor for noninvasive blood glucose monitoring." *Sensors and Actuators, B: Chemical* 173–180. doi: 10.1016/j.snb.2019.01.121.
42. Bae, J., V. V. Druzhin, A. G. Anikanov, S. V. Afanasyev, A. Shchekin, A. S. Medvedev, A. V. Morozov, D. Kim, S. K. Kim, H. Moon, H. Jang, J. Shim, and J. Park. 2017. A miniaturized near infrared spectrometer for non-invasive sensing of bio-markers as a wearable healthcare solution. Vol. 10116, *SPIE OPTO*: SPIE.

Blood Oxygenation

43. F. F. Jobsis, "Noninvasive, infrared monitoring of cerebral and myocardial oxygen sufficiency and circulatory parameters", *Science*, 198, 1264 (1977).
44. Y. Ozaki, A. Mizuno, T. Hayashi, K. Tashibu, S. Maraishi, and K. Kawauchi, "Nondestructive and Noninvasive Monitoring of Deoxyhemoglobin in the Vein by Use of a Near-Infrared Reflectance Spectrometer with a Fiber-Optic Probe", *Appl. Spectrosc.*, 46(1), 180 (1992).
45. M. G. Sowa J. R. Mansfield, G. B. Scarth, and H. H. Mantsch, "Noninvasive Assessment of Regional and Temporal Variations in Tissue Oxygenation by Near-Infrared Spectroscopy and Imaging", *Appl. Spectrosc.*, 51(2), 143 (1997).
46. D. M. Mancini, L. Bolinger, H. Li, K. Kendrick, B. Chance, and J. R. Wilson, "Validation of Near-Infrared Spectroscopy in Humans", *J. Appl. Physiol.*, 77(6), 2740 (1994).
47. L. Lin, and D. A. York, "Two-Layered Phantom Experiments for Characterizing the Influence of a Fat Layer on Measurement of Muscle Oxygenation using NIRS", paper 41, *Proc. SPIE, 3257 (Photonics West)*, San Francisco, January, 1998.
48. K. Yamamoto, M. Niwayama, L. Lin, T. Shiga, N. Kudo, M. Takahashi, "Near-Infrared Muscle Oximeter that Can Correct the Influence of a Subcutaneous Fat Layer" paper 17, *Proc. SPIE 3257 (Photonics West)*, San Francisco, January, 1998.
49. M. A. Franceschini, S. Fantini, A. Cerussi, B. Barbieri, B. Chance, and E. Gratton, *The Effect of Water in the Quantitation of Hemoglobin Concentration in a Tissue-Like Phantom by Near-Infrared Spectroscopy*, Optical Society of America, Washington, DC, 1996.
50. Z. X. Jiang, H. S. Lee, D. F. Wilson, D. F. Hanley, "Novel NIR Instrument for Non-Invasive Monitoring and Quantification of Cerebral Tissue Oxygenation," paper 44, *Proc. SPIE, 3257 (Photonics West)*, San Francisco, January, 1998.
51. R. Marbach, and H. M. Heise, "Optical Diffuse Reflectance Accessory for Measurements of Skin Tissue by Near-Infrared Spectroscopy", *Appl. Optics*, 34(4), 610 (1995).
52. K. Miyasaka, "NIRS Use in Pediatric Anesthesia and ICU", *96 PICU Conference*, Toronto.
53. S. Van Huffel, and P. Casaer, "Changes in Oxygenation and Hemodynamics in Neonatal Brain by Means of Near-Infrared Spectroscopy: A Signal Analysis Study", http://www.esat.kuleuven.be/sista/yearreport96/node6.html, 01/18/2014 (1997).
54. S. Van Huffel, P. Casaer, and M. Kessler, "Modeling and Quantification of Chromophore Concentrations, Based on Optical Measurements in Living Tissues", http://www.esat.kuleuven.be/sista/yearreport97/node33.html, 01/18/2014 (1998).
55. C. E. Cooper, J. Torres, M. Sharpe, and M. T. Wilson, "The Relationship of Oxygen Delivery to Absolute Hemoglobin Oxygenation and Mitochondrial Cytochrome Oxidase Redox State in the Adult Brain: A Near-Infrared Spectroscopy Study", *Biochem. J.*, 332, 627 (1998).
56. J. S. Wyatt, M. Cope, D. T. Delp, C. E. Richardson, A. D. Edwards, S. Wray, and E. O. Reynolds, "Quantitation of Cerebral Blood Volume in Human Infants by Near-Infrared Spectroscopy", *J. Appl. Physiol.*, 68, 1086 (1990).
57. V. V. Kupriyanov, R. A. Shaw, B. Xiang, H. Mantsch, and R. Deslauriers, "Oxygen Regulation of Energy Metabolism in Isolated Pig Hearts: A Near-IR Spectroscopy Study", *J. Mol. Cell. Cardiol.*, 29, 2431 (1997).
58. J. R. Mansfield, G. B. Scarth, R. L. Somorjai, H. H. Mantsch, "Fuzzy C-Means Clustering and Principal Component Analysis of Time Series from Near-Infrared Imaging of Forearm Ischemia" *Comput Med. Imag. Graph.*, 21(5), 299 (1997).
59. T. Wolf, U. Lindauer, H. Obrig, J. Drier, J. Back, A. Villringer, and O. Dirnagl, *J. Cereb. Blood Flow Metab.*, 16, 1100–1107 (1996).

60. K. D. Liem, J. C. Hopman, L. A. Kollee, and R. Oeseburg, "Effects of Repeated Indomethacin Administration on Cerebral Oxygenation and Hemodynamics in Pre-Term Infants: Combined Near-Infrared Spectrophotometry and Doppler Ultrasound Study", *Eur. J. Pediat.*, 153(7), 504 (1994).
61. T. J. Germon, P. D. Evans, A. R. Manara, N. J. Barnett, P. Wall, and R. J. Nelson, "Sensitivity of Near Infrared Spectroscopy to Cerebral and Extra-Cerebral Oxygen Changes Is Determined by Emitter-Detector Separation", *J. Clin. Monit.*, 10, 1 (1998).
62. L. C. Henson, C. Calalang, J. A. Temp, and D. S. Ward, "Accuracy of a Cerebral Oximeter in Healthy Volunteers under Conditions of Isocapnic Hypoxia", *Anesthesiology*, 88(1), 58 (1998).
63. Y. Hoshi, H. Onoe, Y. Watanabe, J. Andersson, M. Bergstram, A. Lilja, B. Langstrom, and M. Tamura, "Non-Synchronous Behavior of Neuronal Activity, Oxidative Metabolism, and Blood Supply During Mental Tasks in Man", *Neurosci. Lett.*, 172, 129 (1994).
64. F. Okada, Y. Tokumitsu, Y. Hoshi, and M. Tamura, "Impaired Interhemispheric Integration in Brain Oxygenation and Hemodynamics in Schizophrenia", *Euro. Arch. Psychiatry Clin. Neurosci.*, 244, 17 (1994).
65. Y. Hoshi, S. Mizukami, and M. Tamura, "Dynamic Features of Hemodynamic and Metabolic Changes in the Human Brain during All-Night Sleep as Revealed by Near-Infrared Spectroscopy", *Brain Res.*, 652, 257 (1994).
66. A. Akına, D. Bilensoya, U. E. Emira, M. Gülsoya, S. Candansayarc, and H. Bolay, "Cerebrovascular Dynamics in Patients with Migraine: Near-Infrared Spectroscopy Study", *Neurosci. Lett.*, 400, 86 (2006).
67. Chow et al., "The Relation between Pump Flow Rate and Pulsatility on Cerebral Hemodynamics during Pediatric Cardiopulmonary Bypass", *J. Thoracic Cardiovasc. Surg.*, 114(4), 1123 (1997).
68. R. Totaro, G. Barattelli, V. Quaresima, A. Carolei, and M. Ferrari, "Evaluation of Potential Factors Affecting the Measurement of Cerebrovascular Reactivity by Near-Infrared Spectroscopy", *Clin. Sci.*, 95, 497 (1998).
69. "Team Develops World's First Light-Based Procedure for Examining Epileptic Brain Sites," http://www.hitachi.com/New/cnews/E/1997/970710B.html, 01/18/2014.
70. D. A. Boas, T. Gaudette, G. Strangman, X. Cheng, J. J. A. Marota, and J. B. Mandeville, "The Accuracy of Near Infrared Spectroscopy and Imaging during Focal Changes in Cerebral Hemodynamics", *NeuroImage*, 13, 76–90 (2001).
71. G. Strangman, M. A. Franceschini, and D. A. Boas, "Factors Affecting the Accuracy of Near-Infrared Spectroscopy Concentration Calculations for Focal Changes in Oxygenation Parameters", *NeuroImage*, 18, 865–879 (2003).
72. K. Ide, A. Horn, and N. H. Secher, "Cerebral Metabolic Response to Submaximal Exercise", *J. Appl. Physiol.*, 87, 1604–1608 (1999).
73. Y. Bhambhani, R. Malik, and S. Mookerjee. "Cerebral Oxygenation Declines at Exercise Intensities above the Respiratory Compensation Threshold", *Respir. Physiol. Neurobiol.*, 156, 196–202 (2007).
74. A. Timinkul, M. Kato, T. Omori, C. C. Deocaris, A. Ito, T. Kizuka, Y. Sakairi, T. Nishijima, T. Asada, and H. Soya. "Enhancing Effect of Cerebral Blood Volume by Mild Exercise in Healthy Young Men: A Near-Infrared Spectroscopy Study", *Neurosci. Res.*, 61, 242–248 (2008).
75. C. R. Rooks, N. J. Thom, K. K. McCully, and R. K. Dishman, "Effects of Incremental Exercise on Cerebral Oxygenation Measured by Near-Infrared Spectroscopy: A Systematic Review", *Prog. Neurobiol.*, 92, 134–150 (2010).
76. M. J. Herrmann, A. C. Ehlis, and A. J. Fallgatter, "Frontal Activation during a Verbal-Fluency Task as Measured by Near-Infrared Spectroscopy", *Brain Res. Bull.*, 61, 51–56 (2003).
77. P.-H. Chou, and T.-H. Lan, "The Role of Near-Infrared Spectroscopy in Alzheimer's Disease", *J. Clin. Gerontol. Geriatr.*, 4(2), 33–36, (June 2013).
78. C. Pennekamp, M.L. Bots, L. J. Kappelle, G J de Borst, "The Value of Near-Infrared Spectroscopy Measured Cerebral Oximetry during Carotid Endarterectomy in Perioperative Stroke Prevention: A Review", *Eur. J. Vasc. Endovasc. Surg.*, 38, 539–545 (2009).
79. John K-J. Li, T. Wang, and H. Zhang, "Rapid Noninvasive Continuous Monitoring of Oxygenation in Cerebral Ischemia and Hypoxia", *Cardiovasc. Eng.*, 10, 213–217 (2010).
80. C. S. Robertson, S. P. Gopinath, and B. Chance, "Chance New Application for Near-Infrared Spectroscopy: Detection of Delayed Intracranial Hematomas after Head Injury", *J. Neurotrauma*, 1995; 12, 591–600.
81. A. Gill, K. Rajneesh, C. Owen, J. Yeh, M. Hsu, and D. K. Binder, "Early Optical Detection of Cerebral Edema In Vivo", *J Neurosurg* 2010; 114(2):470–447.

82. H. Vohra, A. Modi, and S. Ohri, "Does Use of Intra-Operative Cerebral Regional Oxygen Saturation Monitoring During Cardiac Surgery Lead to Improved Clinical Outcomes?", *Interact Cardiovasc Thorac Surg.*, 9, 318–322 (2009).
83. J. Murkin, "NIRS: A Standard of Care for CPB vs. an Evolving Standard for Selective Cerebral Perfusion?", *J. Extra Corpor. Technol.*, 41, 11–14 (2009). NIRS is being more widely used to optimize cerebral oxygenation during cardiac surgery, with the highest adoption rates in centers performing high-risk procedures.
84. D. Hightona, C. Elwellb, and M. Smitha, "Noninvasive Cerebral Oximetry: Is There Light at the End of the Tunnel?", *Curr. Opin. Anaesthesiol.*, 23(5), 576–581 (2010).
85. J. F. Jabre, G. M. Squintani, and K. K. H. Chui, "Oxyneurography: A New Technique for the Measurement of Nerve Oxygenation", *Muscle Nerve*, 45, 75–80 (2012).
86. A. A. Oraevsky, E. V. Savateeva, S. V. Solomatin, A. Karabutov, V. G. Andreev, Z. Gatalica, T. Khamapirad, and P. M. Henrichs, "Optoacoustic Imaging of Blood for Visualization and Diagnostics of Breast Cancer", *Proc. SPIE*, 4618, 81–94 (2012).
87. R. G. M. Kolkman, E. Hondebrink, W. Steenbergen, and F. F. M. de Mul, "In Vivo Photoacoustic Imaging of Blood Vessels using an Extreme-Narrow Aperture Sensor", *IEEE J. Sel. Top. Quantum Electron.*, 9, 343–346 (2003).
88. J. J. Niederhauser, M. Jaeger, R. Lemor, P. Weber, and M. Frenz, "Combined Ultrasound and Optoacoustic System for Real Time High Contrast Vascular Imaging In Vivo", *IEEE Trans. Med. Imaging*, 24, 436–440 (2005).
89. E. Z. Zhang, and P. C. Beard, "2D Backward-Mode Photoacoustic Imaging System for NIR (650–1200 nm) Spectroscopic Biomedical Applications", *Proc. SPIE*, 6086, 60860H-1–7 (2006).
90. J. Laufer, D. Delpy, C. Elwell, and P. Beard, "Quantitative Spatially Resolved Measurement of Tissue Chromophore Concentrations using Photoacoustic Spectroscopy: Application to the Measurement of Blood Oxygenation and Haemoglobin Concentration", *Phys. Med. Biol.*, 52, 141–168 (2007).
91. R. Boushel, F. Pott, P. Madsen, G. Radegran, M. Nowak, B. Quistroff, and N. Secher, "Muscle Metabolism from Near-Infrared Spectroscopy During Rhythmic Handgrip in Humans", *Eur. J. Appl. Physiol. Occup. Physiol.*, 79(1), 41–48 (1998).
92. Lassnigg, M. Hiemayr, P. Keznicki, T. Mullner, M Ehrlich, and G. Grubhofer, "Cerebral Oxygenation during Cardiopulmonary Bypass Measure by Near-Infrared Spectroscopy: Effects of Hemodilution, Temperature, and Flow", *J. Cardiothorac. Vasc. Anesth.*, 13(5), 544–548 (1999).
93. A. T. Lovell, H. Owen-Reece, C. E. Elwell, M. Smith, and J. C. Goldstone, "Continuous Measurement of Cerebral Oxygenation by Near-Infrared Spectroscopy During Induction of Anesthesia", *Anesth. Analg.*, 88(3), 554–558 (1999).
94. T. Higami, S. Kozawa, T. Asada, H. Obo, K. Gan, K. Iwahashi, and H. Nohara, "Retrograde Cerebral Perfusion Versus Selective Cerebral Perfusion as Evaluated by Cerebral Oxygen Saturation During Aortic Arch Reconstruction", *Ann. Thoracic Surg.*, 67(4), 1091–1096 (1999).
95. G. Liu, I. Burcev, F. Pott, K. Ide, I. Horn, and N. H. Sander, "Middle Cerebral Artery Flow Velocity and Cerebral Oxygenation During Abdominal Aortic Surgery", *Anaesth. Intensiv. Care*, 27(2), 148–153 (1999).
96. M. L. Blas, E. B. Lobato, and T. Martin, "Non-Invasive Infrared Spectroscopy as a Monitor of Retrograde Cerebral Perfusion during Deep Hypothermia", *J. Cardiothorac. Vasc. Anesthesia*, 13(2), 244–245 (1999).
97. M. J. Van de Ven, W. N. Colier, B. T. Kersten, B. Oeseburg, and H. Folgering, "Cerebral Blood Volume Responses to Acute $PaCO_2$ Changes in Humans, Assessed with Near-Infrared Spectroscopy", *Adv. Exp. Med. Biol.*, 471, 199–207 (1999).
98. M. Wolf, O. Weber, M. Keel, X. Golay, M. Scheidegger, H. U. Bucher, S. Kollias, P. Boesiger, and O. Banziger, "Comparison of Cerebral Blood Volume Measured by Near-Infrared Spectroscopy and Contrast Enhanced Magnetic Resonance Imaging", *Adv. Exp. Med. Biol.*, 471, 767–773 (1999).
99. J. Kytta, J. Ohman, P. Tanskanen, and T. Randell, "Extracranial Contribution to Cerebral Oximetry in Brain Dead Patients: A Report of Six Cases", *J. Neurosurg. Anesthesiol.*, 11(4), 252–254 (1999).
100. C. Casavola, L. A. Paunescu, S. Fantini, M. A. Franceshini, P. M. Lugara, and E. Gratton, "Application of Near-Infrared Tissue Oximetry to the Diagnosis of Peripheral Vascular Disease", *Clin. Hemorrheol. Microcirc.*, 21(3–4), 389–393 (1999).
101. W. N. Colier, V. Quaresima, B. Oeseburg, and M. Ferrari, "Human Motor-Cortex Oxygenation Changes Induced by Cyclic Coupled Movements of Hand and Foot", *Exper. Brain Res.*, 129(3), 457–461 (1999).
102. K. Krakow, S. Ries, M. Daffertshofer, and M. Hennerici, "Simultaneous Assessment of Brain Tissue Oxygenation and Cerebral Perfusion during Orthostatic Stress", *Eur. Neurol.*, 43(1), 39–46 (2000).

103. G. Fuchs, G. Schwartz, A. Kulier, and G. Litscher, "The Influence of Positioning on Spectroscopic Measurements of Brain Oxygenation", *J. Neurosurg. Anesthesiol.*, 12(2), 75–80 (2000).
104. A. J. Spielman, G. Zhang, C. M. Yang, S. Serizawa, M. Nagata, H. von Gizycki, and R. R. Alfano, "Intercerebral Hemodynamics Probed by Near-Infrared Spectroscopy in the Transition between Wakefulness and Sleep", *Brain Res.*, 866(1–2), 313–325 (2000).
105. Y. Nagashima, Y. Yada, M. Hattori, and A. Sakai, "Development of a New Instrument to Measure Oxygen Saturation and Total Hemoglobin Volume in Local Skin by Near-Infrared Spectroscopy and Its Clinical Application" *Int. J. Biometrology*, 44(1), 9–11 (2000).
106. B. A. McKinley, R. G. Marvin, Concanour, and F. A. Moore, "Tissue Hemoglobin O2 Saturation during Resuscitation of Traumatic Shock Monitored Using Near-Infrared Spectrometry", *J. Trauma-Injury Infection Crit. Care*, 48(4), 637–642 (2000).
107. R. Kragelj, T. Jarm, and D. Miklavcic, "Reproducibility of Parameters of Postocclusive Reactive Hyperemia Measured by Near-Infrared Spectroscopy and Transcutaneous Oximetry", *Ann. Biomed. Eng.*, 28(2), 168–173 (2000).
108. R. Boushel, H. Langberg, J. Olesen, J. Gonzales-Alonzo, J. Bülow, and M. Kjær. "Monitoring Tissue Oxygen Availability with Near Infrared Spectroscopy (NIRS) in Health and Disease", *Scand. J. Med. Sci. Sports*, 11(4), 213–222 (2001).
109. S. Goldman, F. Sutter, F. Ferdinand, and C. Trace, "Optimizing Intraoperative Cerebral Oxygen Delivery using Noninvasive Cerebral Oximetry Decreases the Incidence of Stroke for Cardiac Surgical Patients", *Heart Surg. Forum*, 7(5), 392–397 (2004).
110. G. M. Hoffman, "Pro: Near-Infrared Spectroscopy Should Be Used for All Cardiopulmonary Bypass", *J. Cardiothorac. Vasc. Anesth.*, 20(4), 606–612 (2006).
111. M. Wolf, M. Ferrari, and V. Quaresima, "Progress of Near-Infrared Spectroscopy and Topography for Brain and Muscle Clinical Applications", *J. Biomed. Opt.*, 12(6) (2007).
112. M. Wolf, G. Morren, D. Haensse, T. Karen, U. Wolf, J. C. Fauchère, and H. U. Bucher, "Near Infrared Spectroscopy to Study the Brain: An Overview", *Opto-Electronics Rev.*, 16(4), 413–419 (2008).
113. M. Schecklmann, A. C. Ehlis, M. M. Plichta, and A. J. "Fallgatter Influence of Muscle Activity on Brain Oxygenation during Verbal Fluency Assessed with Functional Near-Infrared Spectroscopy", *Neuroscience*, 171, 434–442 (2010).
114. J.-F. Georger, O. Hamzaoui, A. Chaari, J. Maizel, C. Richard, and J.-L. Teboul, "Restoring Arterial Pressure with Norepinephrine Improves Muscle Tissue Oxygenation Assessed by Near-Infrared Spectroscopy in Severely Hypotensive Septic Patients", *Intensive Care Med*, 36, 1882–1889 (2010).
115. Y. Shang, R. Cheng, L. Dong, S. J. Ryan, S. P. Saha, and G. Yu, "Cerebral Monitoring during Carotid Endarterectomy using Near-Infrared Diffuse Optical Spectroscopies and Electroencephalogram", *Phys. Med. Biol.*, 56, 3015–3032 (2011).
116. M. F. Oliveira, M. K. Rodrigues, E. Treptow, T. M. Cunha, E. M. V. Ferreira, and J. Alberto Neder, "Effects of Oxygen Supplementation on Cerebral Oxygenation during Exercise in Chronic Obstructive Pulmonary Disease Patients Not Entitled to Long-Term Oxygen Therapy", *Clin. Physiol. Funct. Imaging*, 32, 52–58 (2012).
117. S. M. Bailey, K. D. Hendricks-Muñoz, and P. Mally, "Splanchnic-Cerebral Oxygenation Ratio as a Marker of Preterm Infant Blood Transfusion Needs", *Transfusion*, 52, 252 (2012).

36 Analysis of Vascular Tissue Using NIR Spectrometry

*Robert A. Lodder**
University of Kentucky

CONTENTS

36.1 INTRODUCTION

Assessment of tissue oxygen saturation by near-infrared spectrometry was one of the earliest medical uses of NIRS, and studies employing the method continue to be published in the literature. Measurement of tissue oxygen saturation can be accomplished easily using only two short-wavelength NIR peaks, which helped accelerate the widespread acceptance of the method. On the other hand, papers describing NIR imaging in vascular tissue for atherosclerosis were published decades before medical instrumentation became available for that purpose. Hyperspectral imaging requires instrumentation that simply was not widely available 30 years ago. In the past few years, however, such instrumentation has become commercially available from many sources, and medical instruments using NIR imaging have been cleared for marketing by the US Food and Drug Administration. As a result, there are now many publications on NIR imaging of vascular tissue for atherosclerosis.

Atherosclerosis is not the only application of NIRS in the vascular area. NIRS is still used in agricultural assays of meat. NIR illumination is also being investigated for phototherapy of wounds, but such phototherapy is not yet accepted by conventional medicine. In a novel approach to NIR imaging, researchers continue to pursue X-ray-activated near-infrared persistent luminescent probes for deep-tissue in vivo bioimaging. These new optical probes are free of autofluorescence and provide self-sustained emission after X-ray excitation with a high signal-to-noise ratio.

36.2 NEAR-INFRARED IMAGING IN ATHEROSCLEROSIS

Imaging and autopsy studies have described atherosclerotic plaque in different vascular beds including varying degrees of lipid, fibrosis, and calcification. Recently, near-infrared spectroscopy has been validated as an accurate method for detecting lipid-core plaque in the coronary circulation. Invasive evaluation of plaque composition using near-infrared spectroscopy and intravascular ultrasound has not been reported in different peripheral arterial circulations. Invasive angiography and near-infrared spectroscopy in combination with intravascular ultrasound have been assessed in peripheral artery disease subjects prior to percutaneous revascularization (Abbas, 2017).

* lodder@g.uky.edu.

Imaging evaluation included measurements from angiography, intravascular ultrasound, and near-infrared spectroscopy. Results were compared among various vascular beds with regard to the appearance and degree of calcification and lipid-core plaque. One hundred twenty-six peripheral artery disease subjects were enrolled in the trial with a total of 149 lesions in their blood vessels, including the internal carotid, subclavian/axillary, renal, iliac, femoropopliteal, and infrapopliteal arteries. Plaque morphology was calcified in 89% of lesions and fibrous in 11%. Calcification ranged from 100% of renal artery stenoses (blockages) to 55% of subclavian/axillary artery stenoses. Lipid-core plaque was present in 32% of lesions, and prevalence ranged from 60% in carotid artery stenoses to 0% in renal artery stenoses. Lipid-core plaque was only noted in fibrocalcific plaque and was longitudinally and circumferentially surrounded by calcification. Near-infrared spectroscopy with intravascular ultrasound in stable peripheral artery disease subjects showed a high occurrence of calcified plaque and statistically significant differences in the occurrence of lipid-core plaque in various arterial beds. In the peripheral circulation, lipid-core plaque, when present, was always attendant with calcified plaque. The strong association of calcified plaque with lipid-core plaque in severe peripheral artery disease may help stabilize these plaques. Additional studies may help confirm this mechanism.

Near-infrared spectroscopy has been shown to be capable of identifying the constituents of lipid-rich plaques, and near-infrared imaging of lipid plaques has been used with computational modeling of low-density lipoprotein (LDL) transport in coronary arteries to good effect (Bampali, 2017). In this pilot study, five subjects received coronary angiography, intravascular ultrasound, and NIRS imaging. The borders of the lumen and adventitia were detected in the Intravascular Ultrasound (IVUS) images, and the 3D centerline was identified from the angiographic images. 3D reconstruction of the coronary arteries was accomplished by the integration of the detected borders with the 3D arterial centerline. Blood flow and LDL transport were modeled in the reconstructed sections in order to estimate the endothelial shear stress (ESS) and LDL accumulation in the arterial wall. ESS and LDL concentration maps were compared against the corresponding chemogram sections using NIRS imaging that marked the lipid-rich plaques. The mean ESS was 3.36 ± 3.03 Pa, and the normalized LDL concentration was 0.093 ± 0.005. The Pearson correlation coefficient of ESS with lumen area was -0.538 ($p < 0.0001$) and with plaque area was 0.111.

An elevated coronary artery calcium score (CACS) suggests a heightened risk for subsequent coronary events. However, it is not known whether subjects having a high CACS actually have large lipid-rich plaques characteristic of lesions giving rise to myocardial infarction (Bundy, 2017). Bundy et al. completed multi-vessel intracoronary near-infrared spectroscopy and intravascular ultrasound in 20 asymptomatic subjects with no previous history of coronary artery disease who had a screening CACS ≥300 Agatston units. (The Agatston score is named after its developer Arthur Agatston and is a measure of calcium on a coronary CT calcium scan. The score is calculated using a weighted value assigned to the highest density of calcification in a given coronary artery. The density is measured in Hounsfield units, and scores of 1 for 130–199 HU, 2 for 200–299 HU, 3 for 300–399 HU, and 4 for 400 HU and greater. This weighted score is then multiplied by the area (in square millimeters) of the coronary calcification.) A lipid-rich plaque was defined as a lesion having ≥1 bright yellow block on the near-IR chemogram. To measure the lipid burden of each lesion, the maximum lipid-core burden index in 4 mm ($maxLCBI_{4mm}$) was recorded. The average CACS of the study population was 764 ± 465 Agatston units. Lipid-rich plaque was not ubiquitous, and 40.0% subjects had no lipid-rich plaque discovered by near-infrared spectroscopy in any of their arteries. Large lipid-rich plaques having a $maxLCBI_{4mm}$ ≥400 (lipid-core burden index) were infrequent in general, found in only 25.0% of individuals and in only 1.9% of 268 10-mm coronary segments examined. This pilot study was the first to report near-infrared results in asymptomatic subjects with a high CACS, and showed that a high CACS is not necessarily suggestive of underlying lipid-rich plaque. The differences in coronary lipids suggest that near-infrared spectroscopic imaging might be able to improve risk stratification among asymptomatic individuals with a high CACS.

Serum lipids play an important part in the development of atherosclerosis. It is widely known that low-density lipoprotein cholesterol (LDL-C) level is an independent predictor of cardiovascular events, and reducing LDL-C markedly reduces the risk of major cardiovascular events (Honda, 2017). Other cholesterol fractions can have different effects. There is an inverse relationship between high-density lipoprotein cholesterol (HDL-C) levels and prospective risk of cardiovascular disease, in spite of LDL-C levels. But little is known about the connection between serum lipid measurements and change over time in plaque composition using in vivo coronary imaging. The aim of the Honda et al. study was to examine the association between serum lipids and change in coronary plaque lipid burden assessed by near-infrared spectroscopy (NIRS).

Serial near-infrared spectrometry–intravascular ultrasound studies were conducted in 49 subjects who received coronary angiography for an acute coronary syndrome (ACS) or stable ischemic symptoms. Autopsy studies have demonstrated that plaque rupture is commonly observed in subjects with acute coronary syndromes. These vulnerable plaques are distinguished by the existence of a large lipid pool. Sequential IVUS research has shown that LDL-C lowering with statins and proprotein convertase subtilisin/kexin type 9 (PCSK9) inhibitors slows the rate of atherosclerotic plaque progression, and provided evidence that the LDL-C levels were associated with the change in plaque volume.

Near-infrared/intravascular ultrasound imaging was carried out within the same coronary artery at baseline and at follow-up in Honda et al. After intracoronary administration of nitroglycerine (100–200 μg), a 3.2-Fr IVUS/NIRS (InfraReDx, Burlington, Massachusetts) catheter was moved forward into the target coronary artery and then withdrawn by automated pullback at 0.5 mm s^{-1} until arriving at the aorta. Intravascular ultrasound grayscale images and near-IR spectra were coregistered and stored together. LCBI and $maxLCBI_{4mm}$ were calculated automatically by the LipiScan analyzer software (LipiScan, InfraReDx, Burlington, MA, USA). The primary endpoint for the study was the change in $maxLCBI_{4mm}$. Univariate and multivariate linear regression analyses were used to explore the association between the serum lipid parameters and the change in lipid-core burden index at the 4-mm maximal segment ($maxLCBI_{4mm}$).

The average age of the subjects in the Honda study was 61 ± 9 years, 29% were female, and 35% presented with an acute coronary syndrome. The median low-density lipoprotein cholesterol was 101 mg dL^{-1}, the median high-density lipoprotein cholesterol (HDL-C) was 43 mg dL^{-1}, the median total cholesterol was 174 mg dL^{-1}, and the median triglyceride level was 133 mg dL^{-1}. Through the median follow-up period of 13 months, $maxLCBI_{4mm}$ declined significantly from 277 to 194. The percent change in HDL-C was negatively associated with the change in $maxLCBI_{4mm}$ in univariate analysis. There were no significant associations among the other lipid parameters and change in $maxLCBI_{4mm}$. Percent change in HDL-C remained significantly associated with the change in $maxLCBI_{4mm}$ in multivariate analysis.

The Honda study suggested that change in HDL-C is inversely correlated with the change in lipid-core plaque measured by NIRS. This correlation was independent of the use of statins by the subjects. Furthermore, none of the subjects whose HDL-C increased experienced growth of lipid-core plaque, while almost one-third of the subjects without an increase in HDL-C displayed plaque lipid growth even when using statin therapy. No significant correlations were found between other lipid parameters, including LDL-C level and change in lipid-core plaque. This finding emphasizes the influence of HDL on change in plaque composition. LDL-C lowering is a key therapeutic approach to reducing the risk of coronary artery disease, and the amount of LDL-C lowering is key. In addition, a larger drop in LDL-C by statin therapy and new PCSK9 inhibitors is correlated with a larger drop in plaque burden. However, the correlation between LDL-C levels and change in coronary plaque lipid burden over time is disputed. Some intracoronary imaging studies have communicated a positive correlation between LDL-C levels and lipid content of plaque. However, the Honda experiment and some other studies using radiofrequency intravascular ultrasound or near-IR spectroscopy demonstrated that serum LDL-C levels during statin therapy were not correlated with change in lipid burden of plaque. While the explanation for this difference is uncertain, a possible reason is that the majority of the study subjects (78%) were already on treatment with statins and the median baseline LDL-C levels were relatively low (101 mg dL^{-1}). Prior statin use may have weakened the correlation between LDL-C levels and change in coronary lipid burden over time. For this reason, the results suggest there are restrictions on the ability of LDL-C targeting therapy to alter the lipid composition of plaque in statin-treated subjects. Other therapeutic targets may be necessary to attain regression of lipid constituents within plaque.

IVUS and near-infrared spectroscopy were used to investigate characteristics of lesions that were determined to be histopathologically thin-cap fibroatheromas (TCFAs) (Inaba, 2017). Pathological studies have demonstrated that a TCFA is the crucial predecessor lesion of plaque rupture in subjects dying from acute myocardial infarction. A TCFA has been defined as a big necrotic core with an overlying thin fibrous cap less than 65 μm in thickness. Clinical analysis of characteristics of TCFA in vivo has encompassed several imaging modalities in the last 10 years: intravascular ultrasound, angioscopy, virtual histology or integrated backscatter-IVUS, and optical coherence tomography. But no single intravascular imaging method can detect all the features that histopathologically describe a TCFA. Near-infrared spectroscopy was developed and validated in human coronary artery autopsy studies for the in vivo identification of lipid-rich plaques (LRPs). Near-infrared spectroscopy has been combined with grayscale intravascular ultrasound in an integrated intravascular imaging device, with the capacity to increase the number of properties of a histopathologically defined TCFA that can be analyzed in vivo. The goal of the Inaba in vitro study was to investigate the recognition of pathologically defined TCFAs using grayscale intravascular ultrasound combined with near-IR spectrometry.

An Atlantis SR Pro 40-MHz catheter connected to an iLab system (Boston Scientific, Fremont, California) was moved forward along a 0.014-in. guidewire through the coronary artery mounted in a fixture. Intravascular ultrasound imaging was accomplished using motorized pullback at 0.5 mm s^{-1} to include proximal and distal luer connectors.

Employing the identical protocol as for intravascular ultrasound imaging, a 3.2-Fr InfraReDx (Burlington, Massachusetts) near-infrared spectrometric catheter was advanced into the distal coronary vessel. Automated mechanical pullback was performed at a speed of 0.5 mm s^{-1} and the light rotated at 240 rpm. The custom fixture's guideposts were marked during pullback to ensure registration with histology and intravascular ultrasound. Raw spectra were obtained at a rate of 1 spectrum every 25 ms.

The probability of lipid-rich plaque was depicted as a chemogram, which is a digital color-coded map of the location and probability of lipid as if observed from the luminal surface, with the X-axis designating the pullback position (1 pixel every 0.1 mm) and the Y-axis designating the circumferential position (1 pixel every 1°) after spatial filtering and image processing of the raw data as though the coronary vessel had been cut open along its longitudinal axis and positioned flat. Spectroscopic data at each pixel were constructed into a probability of lipid-rich plaque that was then mapped to a 128 (7-bit) red-to-yellow color scale with the lowest probability of lipid depicted as red and the highest probability of lipid depicted as yellow. Near-IR image analysis was accomplished offline by employing in-house, MATLAB-based software.

Intravascular ultrasound, near-infrared spectrometry, and histopathology evaluation included 39 left anterior descending arteries, 36 right coronary arteries, and 32 left circumflex arteries from 54 autopsied hearts. The median age of the subjects was 65 years, 80% had a cardiovascular cause of death, and 70% were male. A set of 271 IVUS-defined lesions were assessed. Of the 271 intravascular ultrasound-identified lesions, 73% of the lesions were mild to moderate with plaque burden less than 70%. Two hundred seventy-one lesions were histopathologically categorized and classified as 26 adaptive intimal thickenings, 65 noncalcified thick-cap fibroatheromas, 62 prefibroatheromas, 66 calcified thick-cap fibroatheromas, 42 calcified fibrous plaques, and 10 TCFAs.

Lesions identified as TCFAs showed the largest plaque burden, the highest remodeling index, and the greatest $maxLCBI_{4mm}$. Plaque burden equal to or greater than 69% (90% sensitivity,

75% specificity, and area under the curve 0.87), remodeling index equal to or greater than 1.07 (80% sensitivity, 79% specificity, and area under the curve 0.84), and $maxLCBI_{4mm}$ equal to or greater than 323 (80% sensitivity, 85% specificity, and area under the curve 0.84) predicted a histopathologic TCFA. Simply put, a large plaque burden and a high remodeling index assessed by intravascular ultrasound and lipid-rich plaque ascertained by near-infrared spectroscopy $maxLCBI_{4mm}$ are practical predictive markers of TCFA. Larger LCBI ($maxLCBI_{4mm} \geq 323$) is correlated with the TCFA in a nonruptured plaque. Real-time quantitative analysis of lipids within coronary artery plaques has become possible using NIRS.

Coronary artery disease is expected to remain the major cause of mortality and morbidity worldwide (Schuurman et al., 2018). Subjects with a history of coronary artery disease are at increased risk of later dangerous cardiovascular events, such as acute coronary syndrome. In about three-fourth of all cases, an acute coronary syndrome is produced by rupture of the fibrous cap covering vulnerable, lipid-rich core plaque in the coronary arteries. Coronary angiography is incapable of distinguishing lipid-rich core-containing plaques in the coronary artery wall, but lipid-rich plaques can be pinpointed by near-infrared spectroscopy. Using a catheter for intracoronary imaging via diffuse reflectance spectroscopy is becoming useful in recognizing subjects at increased risk of unfavorable cardiovascular outcome.

The European Collaborative Project on Inflammation and Vascular Wall Remodeling in Atherosclerosis (ATHEROREMO) and the Integrated Biomarker and Imaging Study 3 (IBIS-3) were created to probe phenotypes and vulnerability of plaque as determined by intravascular ultrasound and near-infrared spectroscopy. In the Schuurman study, long-term follow-up of both studies was accomplished with the intent to scrutinize the long-term predictive value of lipid-rich plaques as measured by near-infrared spectroscopy in subjects with coronary artery disease undergoing coronary angiography.

The most recent study (Schuurman et al., 2018) combined subjects from the ATHEROREMO-NIRS and IBIS-3-NIRS substudies. Both of these studies were executed in the Netherlands and had similar enrollment criteria and baseline study procedures. Subjects receiving diagnostic coronary angiography or percutaneous coronary intervention for acute coronary syndrome or stable angina pectoris (SAP) received baseline invasive imaging by near-infrared spectrometry and intravascular ultrasound and were followed up later on adverse cardiovascular events. The images collected were reviewed offline and were not utilized for subject care. ATHEROREMO-NIRS enrolled 203 subjects between April 2009 and January 2011, and IBIS-3-NIRS enrolled 131 subjects between January 2010 and June 2013. A total of 286 subjects were available because 48 subjects participated in both studies. Of these subjects, 275 subjects had baseline data available for both near-infrared spectroscopy and intravascular ultrasound and were therefore included in the Schuurman analysis.

Near-infrared spectroscopy was completed in a nonculprit artery of 275 subjects undergoing coronary angiography for acute coronary syndrome or stable angina. Lipid-core burden index was measured by an independent lab for the area of interest ($LCBI_{ROI}$) and the 4- and 10-mm-long sections with the maximum LCBI ($maxLCBI_{4mm}$ and $maxLCBI_{10mm}$). The primary endpoint was major adverse cardiac events (MACE), which was defined as the composite of all-cause death, nonfatal acute coronary syndrome, or unscheduled revascularization. Hazard ratios were adjusted for age, clinical risk factors, gender, and section plaque burden calculated from intravascular ultrasound. During a median follow-up of 4.1 years, 28.7% of subjects had a MACE. There was a statistically significant and independent continuous relationship between higher $maxLCBI_{4mm}$ values and a higher risk of MACE. Each 100-unit increase of $maxLCBI_{4mm}$ was associated with a 19% increase in MACE (hazard ratio 1.19, 95% confidence intervals: 1.07–1.32, P=0.001). Continuous $maxLCBI_{4mm}$ continued to be independently associated with MACE after exclusion of target lesion-related events as well as after exclusion of adverse events connected to the near-infrared spectrometry-imaged coronary segment.

Three study limitations are worthy of special discussion. First, the study population also contained subjects from IBIS-3, who were given large doses of rosuvastatin following the index procedure.

This cholesterol-lowering drug may have influenced the effect estimates. Nevertheless, a post hoc analysis did not reveal significant effect changes according to the study (ATHEROREMO or IBIS-3). Second, the follow-up questionnaire was finished by 90% of the subjects. Follow-up data were retrieved for most of the remaining subjects from accessible hospital records. However, the chance that loss to follow-up was at least in part biased cannot be completely excluded. On the other hand, a post hoc analysis of clinical and near-infrared spectroscopy attributes of the nonresponders as contrasted with those of the responders did not show any distinctions that suggested selective loss to follow-up. Third, the sample size of the single-center study was comparatively low. The authors examined LCBI as quartiles and as a continuous variable to study the association with adverse cardiac outcome after exclusion of target lesion-related and image section-related events. The Schuurman study concluded that LCBI, as measured by near-infrared spectroscopy in one non culprit coronary artery segment, predicts adverse cardiac outcome independent of clinical characteristics and intravascular ultrasound, during extended follow-up over 4 years in subjects referred for coronary angiography because of acute coronary syndrome or stable angina pectoris.

The subcutaneous vein network serves an important role in maintaining human skin tissue (Seker, 2017). In spite of advancements in optical imaging technologies, problems still remain with imaging of veins deep in the skin. Differing technologies have been applied to solve subcutaneous vein imaging issues. Venographic guidance supplies adequate contrast by employing special contrast agents for radiographic imaging. On the other hand, ultrasound-based imaging systems avoid exposing the subject to ionizing radiation, but much more training is required to use and interpret the data. For these reasons, diffuse optical methods are now being investigated as tools to replace the typical methods in some specific diagnostic studies.

Intravascular near-infrared fluorescence imaging (NIRF) can offer high-resolution observation of the human coronary artery. Animal studies have demonstrated feasibility of using NIRF for identifying inflammation and fibrin deposition in coronary arteries and other blood vessels of similar size. Near-infrared fluorescence (NIRF) molecular imaging offers an interesting new medical imaging implement that has more sensitivity than diffuse reflectance, and lacks the ionizing radiation of positron emission tomography and X-rays. Clinical applications of NIRF include using indocyanine green (ICG) as a nonspecific NIRF contrast agent to demarcate the blood and lymphatic vasculature. The literature contains an increasing number of publications on functional near-infrared imaging (fNIRI). Progress should continue in this nascent field. Soon FNIR should offer increased spatial and temporal resolution. fNIRI can be potentially integrated with additional imaging or sensing methods, for instance, fMRI, EEG, PET, and MEG. fNIRI can be reduced in size and made wearable to increase its scope of applications to such things as monitoring effects of drugs for depression, schizophrenia, or stroke patients. Real-time visualization of peripheral veins is also a potential use. Venipuncture is a medical procedure for gaining intravenous access. While a peripheral vein can be accessed in a single attempt, usually between 2 and 10 attempts are necessary. Excessive venipunctures are time-consuming, are uncomfortable for patients, and can increase cost through the use of extra supplies. Currently, there are a number of commercially available medical devices that utilize NIR to ease vascular access procedures.

Seker et al. (2017) assembled a near-infrared light source to gain the advantage of lower absorption of hemoglobin in this optical region. A reflectance geometry was employed to obtain vascular images. Vascular network analysis utilized recorded images for calculation of the width of vessels of interest and connectivities of selected regions. Seker et al. proposed that their imaging system had great advantages over other commercially available products, including the use of nonionizing radiation, an acceptable penetration depth of 0.5–3 mm, cost-effectiveness, and straightforward procedures for analyses.

The Seker et al. general optical system comprised an NIR LED-based light source, polarizers, infrared (IR) filter, and a camera (Xenics Infrared Camera: Xeva-1.7-320). The LED light source was implemented in both a disk and ring layout. Each LED module (disk/ring type) had 17 LEDs, some emitting 870 nm and some 940 nm.

Seker et al. devised two kinds of phantoms: one liquid-based phantom to imitate human blood and one pork gelatin-based solid phantom to imitate skin tissue. The absorbance of tissue and blood was imitated with 0.2% India ink, and 0.1% intralipid solution was used for scattering properties. Optical absorption and scattering coefficients were determined from the phantom media, using spectrometer and integrating sphere system-based measurements. Then, NIR optical imaging was performed on the solid (gelatin) phantoms in which artificial veins (plastic tubes) were inserted into the phantom base material. Different tube diameters were employed to imitate veins at a certain depth level (3.5 mm). The liquid phantom devised to imitate human blood was injected into the tubes to obtain vascular images. The Seker et al. method provided good results compared to the ultrasound-based methods employed in clinical practice.

Lipid-core plaque has been analyzed by near-infrared spectroscopy and compared to microcirculatory resistance (IMR) (Yang, 2017). Microvascular damage arising from distal embolization in the course of percutaneous coronary intervention is a significant source of periprocedural myocardial infarction, which is connected with poor clinical outcomes. Yang et al. evaluated the relationship between lipid-core plaque analyzed by NIRS and microvascular dysfunction analyzed invasively by the index of microcirculatory resistance (IMR). Yang et al. enrolled 36 subjects with left anterior descending artery plaques. NIRS was carried out before PCI, and fractional flow reserve (FFR), thermodilution coronary flow reserve (CFR), and IMR were carried out following PCI. The maximum value of LCBI for any of the 4-mm sections was computed, and the big-lipid-core plaque group was designated as those with $\text{maxLCBI}_{4\text{mm}}$ over 500. Microvascular dysfunction was designated as post-stent IMR over 25.

The mean age of the enrolled subjects was 60 years. Seven subjects (19.4%) had LCP that fell into the big group. The diameter of stenosis was similar in all subjects. The recorded intravascular ultrasound parameters of minimal lumen area (2.56 ± 0.58 vs. $2.34 \pm 0.37\,\text{mm}^2$) and lesion length ($24.6 \pm 10.2$ vs. 29.7 11.8 mm) were not significantly different. Pre-PCI FFR (0.73 ± 0.09 vs. 0.74 ± 0.08), post-PCI FFR (0.87 ± 0.05 vs. 0.86 ± 0.04), and post-PCI CFR (3.07 ± 1.83 vs. 2.06 ± 1.00) were not significantly different between the two groups. Post-PCI IMR was larger in the big-LCP group (19.6 ± 15.5 vs. 42.6 ± 17.6). Microvascular dysfunction was more frequent in the big-LCP group than in the small-LCP group (85.7% vs. 17.2%). In the course of elective PCI, large LCPs at the site of the target lesion were connected with post-procedural microvascular dysfunction.

36.3 TISSUE OXYGEN SATURATION

Intraoperative cerebral ($SctO_2$) and muscle tissue oxygen saturation ($SmtO_2$) established by NIRS are associated with postoperative complications and length of hospital stay (LOS) in subjects undergoing major spine surgery (Heringlake, 2017). They signed up 102 subjects undergoing lumbar or thoracolumbar spinal surgery and tracked tissue oxygen saturation bilaterally on the forehead and on a designated spot on both legs for ascertaining average SpO_2 and SaO_2, respectively. A ForeSight EliteVR (CAS Medical Systems, Branford, CT, USA) oximeter was used for the measurements. All measurements began when the subjects were anesthetized and moved to the horizontal position, and ended after a varying period of surgery.

Subjects were organized into classes with many or few complications and longer or shorter LOS using the number of complications and LOS registered in the hospital computer system. In spite of antecedent research in the surgical milieu, univariate and multivariate statistics uncovered no relationship between alterations in $SctO_2$ and complications or LOS. On the other hand, a number of $SmtO_2$ indices revealed feeble but statistically significant relationships with poor outcomes. Despite what might be anticipated on the basis of pathophysiology, subjects exhibiting adverse outcomes had higher $SmtO_2$ measurements than those with a less convoluted postoperative course. This surprising discovery is in clear contrast to previous research and was not explained in the published study. The authors did address a critical matter: the relationship between cerebral desaturation and clinical outcomes in subjects undergoing non-cardiac surgery. While there are still few data on perioperative cerebral oxygenation beyond the field of cardiac surgery, the study seemed to be acceptably powered.

From the perioperative time course of $SmtO_2$ classified by complications and LOS and the respective SDs, it was apparent that there was enormous variability in all observations (with SDs up to 20%) and a massive overlap among subject classes. Furthermore, the more lengthy the surgery, the higher were the variability of the results and the excursions of $SmtO_2$ values relative to baseline. Interestingly, the minimum SpO_2 values in subjects with a complicated course were measured when surgery in uncomplicated subjects had already completed. What it all means is not totally clear yet. Does somatic oxygenation really differentiate between subjects more than brain oxygenation between subjects with or without complications or longer LOS? Given the limitations of this study, it is hard to firmly draw such a conclusion. The data analyzed depict only the intraoperative time course and not excursions of cerebral and somatic oxygenation with regard to the non-anesthetized baseline. The data on the fractional inspired oxygen (FIO_2) administered during horizontal positioning of the subjects (when the measurements started) are not present; therefore, one cannot determine whether some subjects (i.e., those with cardiovascular risk factors or a high BMI) were preoxygenated with a greater FIO_2 before horizontal positioning, while others (i.e., with fewer risk factors or simply lean) were not, which is a facile way to inject confounders in saturation measurements. The relationship between bleeding and blood product use with poor outcomes was much more powerful than the association between alterations in $SmtO_2$ and complications or LOS. The publication does not state whether the anesthetists in charge used a greater FIO_2 in bleeding and unstable subjects than in stable subjects who were not bleeding. The lack of information on potentially important confounders means one can only speculate whether subject- or anesthesia-specific factors led to greater intraoperative $SmtO_2$, or the absence of outcome-relevant variations in $SctO_2$, or even both. Furthermore, the prominent changeability of $SmtO_2$, particularly in the class without complications, and the protracted assessments in the class with complications or prolonged LOS (with reduced $SmtO_2$ with reference to baseline) convey that it is arguable whether this experiment is frankly adequate to make consequential conjectures on the importance of somatic NIRS in this subject population. Meng and coworkers have communicated very weak, but statistically significant, correlations with outcomes. An enhanced multivariate prediction model might be extracted from such a data cloud, but whether anyone will be able to direct therapy in the operating room or at the bedside using the information contained within these noisy data is unclear. Nevertheless, even more can be learned from this experiment. The variability of the $SmtO_2$ measurements presents the question of whether it is suitable to use a technology that has been specially developed and calibrated for evaluating brain oxygen saturation somewhere else on the body, and then to require the assumption that these uncalibrated, off-label measurements provide more worthwhile information than those acquired from the measurement location for which the instrument was principally created and approved. The authors acknowledged that in spite of the results of many single-center experiments, a recent multicenter investigation in cardiac surgical subjects did not find any relationship between lowered perioperative cerebral oxygen saturation and poor outcomes. Even though NIRS sensors have been used upon different parts of the human body in observational studies with encouraging results, most of these investigations were conducted with instruments that had been specially designed for application at peripheral sites and not with cerebral oximeter probes. Prior to venturing further into somatic monitoring using cerebral sensors, it may be judicious to pay more thorough attention to the use of this technology for the indication for which it was primarily created and approved, applying the technology judiciously, and taking conscientious care of possible confounders.

Ultrasound-guided frequency-domain near-infrared spectroscopy (FD-NIRS) has been used to quantify placental oxygenation (Ko, 2017). Vigorous transport of oxygen across the placenta is vital for healthy fetal development. Noninvasive, transabdominal oximetry using continuous-wave near-infrared spectroscopy (CW-NIRS) has been tested in quantification of in vivo placental oxygenation in a small case study. Notwithstanding, the accuracy of NIRS techniques is highly dependent on tissue depth and variability in the optical properties of superficial tissue. CW-NIRS studies so far have been incapable of dealing with uncertainties related to the intervening abdominal and uterine tissue. Eventually, frequency-domain near-infrared spectroscopy (FD-NIRS) employing radio frequency (RF)-modulated light permitted subject-specific characterization of optical properties, and as a result, upgrading physiological accuracy.

In Ko et al., ultrasound-guided FD-NIRS was employed to quantify placental oxygenation in a singleton gestation with an anterior placenta. Imaging was conducted at two positions with the probe rotated 0°, 45°, and 90° relative to the sagittal plane at each position. Optical source–detector separations varied from 3.5 to 7 cm in the instrumentation to characterize superficial vs. deep-tissue contributions to the total observed signal. Depth sensitivity to placental oxygenation was validated using finite element modeling of collected optical data from tissues.

Physiologic assessments of total hemoglobin concentration (THC, mM) and tissue oxygen saturation (StO_2, %) were obtained at 20.4 weeks' gestation in a healthy mother with a BMI of 22.3. Placental depth, calculated from the ultrasound probe to the surface of the placenta, varied from 1.8 to 3.1 cm in this research. Simulations using collected superficial and placental optical properties validated sensitivity to placental tissue at depths up to 3.5 cm. Ko et al. concluded that noninvasive quantification of placental oxygenation might assist in pinpointing abnormal placental evolution. Trial measurements showed the usefulness of point-of-care, ultrasound-guided FD-NIRS oximetry. Including the most modern RF generation and detection hardware and analytical algorithms should further enhance the specificity and quantification accuracy of FD-NIRS placental oximetry.

Near-infrared spectroscopy was used to predict organ failure and outcome in sepsis in the Assessing Risk in Sepsis using a Tissue Oxygen Saturation (ARISTOS) study (Macdonald, 2018). Sepsis, described as a dysregulated host response to infection resulting in organ dysfunction, is connected to significant morbidity and mortality. In the emergency department, it is crucial to recognize the subject with early sepsis to make certain that the subject receives expedient treatment and appropriate positioning in the treatment order. Recognizing sepsis is difficult, considering the dynamic and heterogeneous nature of sepsis and the absence of dependable objective diagnostic criteria. Near-infrared spectroscopy is already used as a noninvasive technique for determining tissue oxygen saturation (StO_2), the percentage of oxygenated hemoglobin in skeletal muscle. NIRS can recognize subjects with reduced tissue perfusion and assist clinical identification of early sepsis. Sepsis leads to changes in the function of the microcirculation, and an abnormal StO_2 (normal range 75%–90%) was thought to be a possible indicator of microcirculation dysfunction. Even though reduced StO_2 is consistent with other indexes of oxygenation, the clinical value in the emergency department is still undetermined. Experiments have evaluated StO_2 for the identification of sepsis and severity/mortality, and to forecast the requirement for intensive care admission. New-onset organ dysfunction is the hallmark of sepsis and is key to the newly revised consensus definition. As a practical matter, sepsis is established by a rise in the sequential organ failure assessment (SOFA) score of 2 or more points from baseline. Furthermore, there is a bedside tool, the "quick sequential organ failure assessment" (qSOFA) score, that allocates one point for each of systolic blood pressure (SBP) less than 100 mmHg, respiratory rate (RR) of at least 22/min, and altered mental activity, which has been advanced to distinguish a greater risk of poor outcome.

The objective of the Macdonald experiment was to study whether StO_2 measurement is practical for diagnosis of sepsis, and to forecast clinical outcomes (mortality and prolonged ICU admission) among a group of emergency department subjects admitted to the hospital with infection. The inclusion criteria for study participants were constructed on the 2001 international consensus definition for sepsis that was valid at the time recruitment was initiated. The consensus definition necessitated the existence of at least two systemic inflammatory response criteria (temperature > 38°C or <36°C; pulse rate 90 beats/min; respirations > 20 breaths/min; white blood cell count $>12\times10^9$/L or $<4\times10^9$/L) in the environment of suspected or authenticated infection. Subjects to be treated by administration of intravenous antibiotics and admission to hospital were evaluated for eligibility throughout the hours of the research personnel. Exclusion criteria included age less than 18 years, taking intravenous antibiotics before presentation to research personnel (i.e., interhospital transfers), anticipated mortality by another cause (e.g., malignancy) within the next 90 days, preexisting limitation of care order, or other comorbidity prohibiting invasive organ support.

The primary endpoint was sepsis, as established by a SOFA score rise of 2 or more points within 72 h of admission, and integrated in-hospital mortality/ICU admission of at least 3 days. A group of 323 participants, with a median age of 64 (interquartile range: 47–77) years, was enlisted at three Australian hospitals. Forty-four percent met the criteria for sepsis, and 7% perished within 30 days. The mean ± SD StO_2 was 74% ± 8% in sepsis and 78% ± 7% without sepsis. StO_2 correlated with the peak sequential organ failure assessment score (Spearman's ρ −0.27, $P < 0.0001$). The area under the receiver operating characteristic curve was 0.66 (95% confidence interval: 0.60–0.72) for sepsis and 0.66 (0.58–0.75) for the composite outcome. $StO_2 < 75\%$ had an odds ratio of 2.67 (1.45–4.94) for the composite outcome compared with StO_2 at least 75%. NIRS-derived StO_2 corresponds to organ failure and is connected with the result in sepsis. Still, NIRS's ability to distinguish sepsis among emergency department subjects with infection is meager. NIRS cannot be endorsed for emergency department use in sepsis diagnosis. In logistic regression evaluation, a depressed StO_2 lower than 75% is an independent risk factor for mortality and/or protracted ICU admission. The result obtained that StO_2 combined with an increased lactate and the qSOFA score forecasts outcome is a potentially viable hypothesis and mandates independent prospective validation. The key is to identify the subject at risk of developing organ failure among those subjects appearing with characteristics of infection. Changes in microcirculation are a major constituent of the pathogenesis of sepsis. Having said that, the evaluation of microcirculatory dysfunction in the clinical setting is problematic. NIRS suggests a possible solution because NIRS can be implemented in a bedside test that is noninvasive and readily reproduced. Both an abnormally low StO_2 (<75%) and a high StO_2 (>89%) have been proposed to be connected with sepsis. The Macdonald data are predominantly consistent with preceding studies, which support a connection between abnormal StO_2 and organ failure/outcome in sepsis. For all that, the ability of StO_2 to distinguish between clinical outcome classes of interest is deficient, and NIRS cannot be endorsed in isolation for classification purposes on the basis of the Macdonald results.

36.4 RELATED TISSUES AND THERAPIES

Historically, near-infrared spectroscopy has been used in agriculture, and such uses persist to the present day. A recent paper (Roberts et al., 2017) described a feasibility study on the potential use of near-infrared reflectance spectroscopy to analyze muscles in living animals. Vascular tissues such as the aorta contains significant amounts of muscle. The reliability with which near-infrared spectroscopy penetrates living tissues was cited in the paper as a major reason for employing it to discriminate between fat and muscle tissue through the skin. In beef cattle, where postmortem carcass assessment or other meat composition analysis is conducted to assess the nutritional value of feeds on animal performance, a rapid, noninvasive, and nondestructive assay offers significant advantages to producers.

Short-wavelength NIR (SWNIR; 700–1100 nm; 13,900–9400 cm^{-1}) range can penetrate deeply into the skin, providing a possible spectral window for the analysis of animal and human tissues. A Fourier transform (FT) NIR instrument was used to collect data, and The Unscrambler software (version X; CAMO ASA, Oslo, Norway) was used for chemometric analysis (Roberts et al., 2017). The authors were able to differentiate between skin, muscle, fat, fat under skin, and muscle under skin. In the SWNIR region, spectral differences between these five groups were observed around the 10,750 and 10,288 cm^{-1} regions associated with water (O-H bonds). Absorbance bands around 11,900, 11,655, 10,929, and 9911 cm^{-1} were attributed to differences in water content (O-H bonds) and also associated with $-CH_2-$ third overtones. Absorbance bands between 10,929 and 10,800 cm^{-1} were correlated with lipid content of the tissue analyzed, while further bands associated with the

CH stretch harmonics were observed between 11,655 and 10,929 cm^{-1}. Absorbance bands between 10,400 and 9911 cm^{-1} were attributed mainly to water (O-H bonds). The most applicable changes were noticed in the long NIR wavelength region, where differences between the five types of samples were observed at 7100 cm^{-1} (O-H and C-H), 5263 cm^{-1} (O-H, mostly related to water), and between 4762 and 4350 cm^{-1} (CH combination tones). The results from the Roberts study indicated that the NIR spectrum has potential use to monitor some beef meat properties connected with fat or muscle noninvasively through the skin. Such a method would provide beef producers or industry with a useful tool to record the effects of health and nutrition on product characteristics with the goal of producing a consistent product and maximizing profits.

Spectral attenuation of brain and retina tissues has been determined in the near-infrared range using a fiber-based supercontinuum device (Saldaña-Díaz et al., 2017). The optical setup integrated a fiber-based supercontinuum light source with a straightforward fiber-optic collimator. This arrangement simultaneously permitted a broad spectral range of measurement and efficiently filtered the transmitted light while reducing the scattered light. The optical measurement of rat brain and retina tissues was provided to illustrate the performance. The attenuation coefficient for the brain and retina tissues was procured in the near-infrared region. The technique could be used in clinical research as a noninvasive method.

The many problems associated with chronic wounds have focused attention on the development of new painless, noninvasive, biophysical therapeutic interventions (Yadav, 2017). Near-infrared light-induced photobiomodulation (PBM) therapy is under investigation as a drug-free approach to promotion of wound healing, reducing inflammation and pain, and restoring function. The penetration power of near-IR light combined with its ability to beneficially modulate biochemical responses is potentially useful. Near-infrared light (800–830 nm) has been found to be the most effective and widely studied wavelength range, followed by red (630–680 nm) and near-infrared 904-nm superpulsed light.

PBM is a form of light therapy that can use coherent light sources (lasers), noncoherent light sources (comprising filtered lamps or light-emitting diodes), or sometimes a combination of both, generally in the visible and NIR spectral regions. It is a nonthermal process involving endogenous chromophores evoking photophysical and photochemical events (linear and nonlinear) at a mixture of biological scales. PBM is different from other light-based treatments in that it does not ablate tissue and it does not rely on heating. PBM is also different than photodynamic therapy (PDT), in which light excites endogenous chromophores or exogenously delivered nontoxic photosensitizers that then react with ambient oxygen to produce reactive oxygen species (ROS) that can kill infectious microorganisms and cancer cells or destroy undesirable tissues (e.g., atherosclerotic plaques in the arteries or choroidal neovascularization in the eyes).

An "optical window" exists in tissue between 700 and 1350 nm where the useful tissue penetration of light is maximized due to low absorption and scattering by the primary tissue components. The fact that NIR light can penetrate into a deep dermal tissue injury allows noninvasive treatment to be conducted without pharmaceuticals to improve healing processes. The effect of NIR light on tissue may depend mostly upon two factors: the amount of energy absorbed and the rate at which the energy is applied. The energy absorbed per unit volume of the tissue should be proportional to the product of the light intensity, the pulse duration, and the absorption coefficient of the tissue at the wavelength of the light. Successful therapeutic outcomes in PBM need ideal optical treatment protocols, including illumination parameters (e.g., wavelength, flux, power density, pulse intensity, and duty cycle) and the perfect treatment timetable.

NIR light is only slightly absorbed in the short-wavelength regions and has a substantial depth of tissue penetration (up to 30–40 mm), and so it provides increased accumulation of photons in the wound bed and more therapeutic healing efficacy. Either continuous-wave (CW) or pulsed-wave (PW) therapy can be used. Some report that PW is more effective than CW, because in PW there are pulse off-times of longer duration than the on-times, which lessens tissue heating. Superpulsed low-level light therapy emits extremely short pulses on the order of nanoseconds (10^{-9} s) or less, so very high-peak powers can be achieved without heating. These very short pulses may permit quick absorption at the cellular level, and Yadav et al. hypothesize that the period between pulses encourages better cell communication, leading to improved wound healing. The episodic nanosecond pulsing model in superpulsed low-level light therapy could build up multiple photodissociation events of NO from CCO, which in turn might boost mitochondrial respiration.

PBM is not yet accepted by conventional medicine. The major barriers to PBM utilization are as follows: (1) few well-controlled, randomized, double-blind clinical trials showing efficacy, (2) uncertainty regarding the cellular and molecular mechanisms transducing signals from the photons incident on the cells that exert the biological effects in the illuminated tissues, and (3) a large number of variables in dosing optically. There are different ways to dose for the same type of injury. To overcome these issues, researchers must focus on understanding of mechanisms, providing better guidelines and standardized protocols, and developing consistent illumination exposure parameters and well-controlled, randomized, double-blind clinical trials showing efficacy.

Researchers continue to pursue Xray-activated near-infrared persistent luminescent probes for deep-tissue in vivo bioimaging (Xue, 2017). Near-infrared persistent luminescence nanoparticles (PLNPs) are new optical probes that are free of autofluorescence and provide self-sustained emission after excitation with a high signal-to-noise ratio. Unfortunately, most NIR-emitting PLNPs have a short decay time and demand excitation by ultraviolet or visible light that itself poorly penetrates tissue to only a small depth. This, in turn, greatly reduces their applications for in vivo long-term imaging and biomarking. For these reasons, improved NIR-emitting PLNPs with in vivo activation features using new excitation sources with deeper penetration depths present a significant improvement to the PLNP technology. Xue et al. are developing a new type of X-ray-activated $ZnGa_2O_4$:Cr PLNPs (X-PLNPs) with efficient NIR persistent emission and rechargeable activation features, in which both the excitation wavelength and emission wavelength are able to penetrate deeply through tissue in vivo. X-PLNPs display enduring (up to 6 h) NIR emission at 700 nm after the discontinuation of the X-ray excitation source. Moreover, the devised X-PLNPs can be easily reactivated by a soft X-ray excitation source with little excitation power (45 kVp, 0.5 mA) to reinstate in vivo bioimaging signals at up to 20 mm depth. Renewable in vivo whole-body bioimaging was also demonstrated through intravenous injection/oral administration of X-PLNPs after in situ X-ray activation. This achievement is the first time that NIR-emitted PLNPs have been shown to be recharged by X-ray light for deep-tissue in vivo bioimaging. This finding clears the way for in vivo renewable bioimaging employing PLNPs and makes the PLNPs highly valuable in the bioimaging field.

36.5 CONCLUSION

Measurement of tissue oxygen saturation using near-infrared spectrometry continues to be a popular application. Simple, robust devices for determining tissue oxygen saturation are widely available. More complicated applications of NIR spectral imaging in vascular tissue are finally beginning to appear, decades after the technique was first described. Medical instruments using NIR imaging have been cleared for marketing by the US Food and Drug Administration. As a result, many publications on NIR imaging of vascular tissue for atherosclerosis appear annually.

NIRS is still employed in measurement of meat for agricultural applications. NIR light is also being studied for phototherapy of wounds, but such phototherapy is not yet accepted by mainstream medicine. In a recently developed approach to NIR imaging, researchers continue to pursue X-ray-activated near-infrared persistent luminescent probes for deep-tissue in vivo bioimaging. These new optical probes are free of autofluorescence and provide self-sustained emission after X-ray excitation with a high signal-to-noise ratio. NIRS will continue to be a growing method in medical applications and diagnostics.

REFERENCES

Abbas, Amr E., et al. "Invasive characterization of atherosclerotic plaque in patients with peripheral arterial disease using near-infrared spectroscopy intravascular ultrasound." *Catheterization and Cardiovascular Interventions* 90.3 (2017): 461–470.

Bampali, Theodora, et al. "The correlation of near-infrared-spectroscopy lipid pools with computationally measured accumulation of low density lipoprotein in coronary arteries." *Atherosclerosis* 263 (2017): e16–e17.

Bundy, Jacob, et al. "Variation in intracoronary lipid-rich plaque detected by near-infrared spectroscopy in asymptomatic individuals with high coronary artery calcium scores." *Journal of the American College of Cardiology* 69.11 Supplement (2017): 1178.

Heringlake, M., and H. Maurer. "Taking two steps at a time does not necessarily bring you forward! Monitoring peripheral tissue perfusion with near-infrared spectroscopy." *British Journal of Anaesthesia* 118:4 (2017): 485–486.

Honda, Satoshi, et al. "High-density lipoprotein cholesterol as a determinant of change in coronary plaque lipid burden assessed by near infrared spectroscopy." *Atherosclerosis* 263 (2017): e94–e95.

Inaba, Shinji, et al. "Intravascular ultrasound and near-infrared spectroscopic characterization of thin-cap fibroatheroma." *American Journal of Cardiology* 119.3 (2017): 372–378.

Ko, Tiffany, et al. "283: Quantifying placental oxygenation using ultrasound-guided frequency-domain near-infrared spectroscopy (FD-NIRS)." *American Journal of Obstetrics & Gynecology* 216.1 (2017): S172.

Macdonald, SPJ, Kinnear FB, Arendts G, Ho KM, Fatovich DM. "Near-infrared spectroscopy to predict organ failure and outcome in sepsis: the Assessing Risk in Sepsis using a Tissue Oxygen Saturation (ARISTOS) study." *European Journal of Emergency Medicine* 26(3) (2019): 174–179.

Roberts, J. J., et al. "A feasibility study on the potential use of near infrared reflectance spectroscopy to analyze meat in live animals: Discrimination of muscles." *Journal of Spectroscopy* 2017 (2017).

Saldaña-Díaz, José Eduardo, et al. "Spectral attenuation of brain and retina tissues in the near-infrared range measured using a fiber-based supercontinuum device." *Journal of Biophotonics* 10.9 (2017): 1105–1109.

Schuurman, Anne-Sophie, et al. "Near-infrared spectroscopy-derived lipid core burden index predicts adverse cardiovascular outcome in patients with coronary artery disease during long-term follow-up." *European Heart Journal* 39.4 (2018): 295–302.

Seker, Kübra, and Mehmet Engin. "Deep tissue near-infrared imaging for vascular network analysis." *Journal of Innovative Optical Health Sciences* 10.03 (2017): 1650051.

Xue, Zhenluan, et al. "X-ray-activated near-infrared persistent luminescent probe for deep-tissue and renewable in vivo bioimaging." *ACS Applied Materials & Interfaces* 9.27 (2017): 22132–22142.

Yadav, Anju, and Asheesh Gupta. "Noninvasive red and near-infrared wavelength-induced photobiomodulation: Promoting impaired cutaneous wound healing." *Photodermatology, Photoimmunology & Photomedicine* 33.1 (2017): 4–13.

Yang, Hyoung-Mo, et al. "Lipid-core plaque assessed by near-infrared spectroscopy and procedure related microvascular injury." *Atherosclerosis* 263 (2017): e56–e57.

37 Application of NIR Spectroscopy to Wood and Wood-Derived Products

L. R. Schimleck[*]
Oregon State University

S. Tsuchikawa
Nagoya University

CONTENTS

37.1 INTRODUCTION

In the late 1980s, efforts to utilize near-infrared (NIR) spectroscopy to estimate properties directly related to economics of the South African pulp and paper industry were reported in the scientific literature (Birkett and Gambino 1988, 1989). The motivation for exploring the technology was the high cost and time-consuming nature of traditional analytical methods. There was certainly interest in the sector in alternate measurement methods, but it is unknown why it took researchers so long to utilize NIR technology considering that it had been used in other industries utilizing plant-derived material for over 20 years. Birkett and Gambino (1988) expressed their surprise at the lack of research stating that "until recently very few, if any studies have been reported on applications of NIRS to the timber or pulping industries." They reported several preliminary investigations exploring potential applications of NIR spectroscopy including brightness, kappa number, lignin content,

* Laurence.Schimleck@oregonstate.edu

and pulp yield of ground wood samples and the estimation of burst and tear of handsheets. The interest generated by these early studies saw rapid growth in groups globally utilizing NIR spectroscopy for wood-related research, and in the number of NIR–wood-related publications as tracked by Tsuchikawa and Kobori (2015) in their recent review.

Initial studies led to various research directions pursued by multiple scientists over the last 30 years, and in this review, we summarize the various paths. Our review isn't all-encompassing; rather, we aim to identify some of the earliest work in each wood-related area, briefly report the research that followed, and conclude with the most recent studies that exist and studies that demonstrate how NIR technology has been utilized. The research areas we will examine are as follows:

- Pulp yield and wood chemistry
    - Estimation of genetic parameters related to pulp production
    - Nondestructive evaluation of standing trees
    - Wood property variation within trees
- Pulp and paper properties
- Physical/mechanical properties
    - Lumber
    - Reaction wood and wood defects
    - Natural durability
    - Wood decay/preservation
    - Moisture content
- Wood composites
- Bark
- Species discrimination and sorting wood products
- Resource assessment

37.2 PULP YIELD AND WOOD CHEMISTRY

Sections 37.2 and 37.3 relate directly to the pulp and paper industry, and over a span of 30 years, many attempts have been made to assess multiple properties important to this sector using NIR spectroscopy. This work has occurred at different points in the process of converting wood to pulp and ultimately to paper. Figure 37.1 summarizes the main steps in the production process and the properties evaluated by NIR spectroscopy at each point.

Pulp yield (the yield of chemically derived pulp from a given volume of wood) and cellulose and lignin, two wood components that largely determine pulp yield (Birkett and Gambino 1988, Schultz and Burns 1990, Wright et al. 1990, Michell 1995, Olsson et al. 1995), were a consistent focus of early NIR–wood-related research. For additional details, see Schimleck and Workman (2004), and Table 22.1 provides a summary of early NIR–wood chemistry studies.

An important development in this area was multi-site, multi-species calibrations. Schimleck et al. (2000) reported preliminary examples of multi-site pulp yield calibrations for plantation-grown Tasmanian bluegum (*Eucalyptus globulus* Labill.) and shining gum (*Eucalyptus nitens* (Deane & Maiden) Maiden), both internationally important eucalypt species for pulp production, from the states of Tasmania and Victoria in Australia. In a series of studies that aimed to produce calibrations applicable across stands, regions, and eucalypt species for temperate Australia, the scope of the calibrations was greatly expanded (Downes et al. 2009, 2010a, 2011). The work produced basic density, cellulose, and pulp yield calibrations based on 1138, 1800, and 1370 samples, respectively, and represents the greatest effort to date to develop "global" calibrations for these critical pulp industry parameters. Similar efforts in softwoods were reported by Hodge and Woodbridge (2004, 2010), with seven tropical and temperate pine species grown in Brazil, Chile, Colombia, South Africa, and the USA being used in the 2010 publication for cellulose and lignin calibrations.

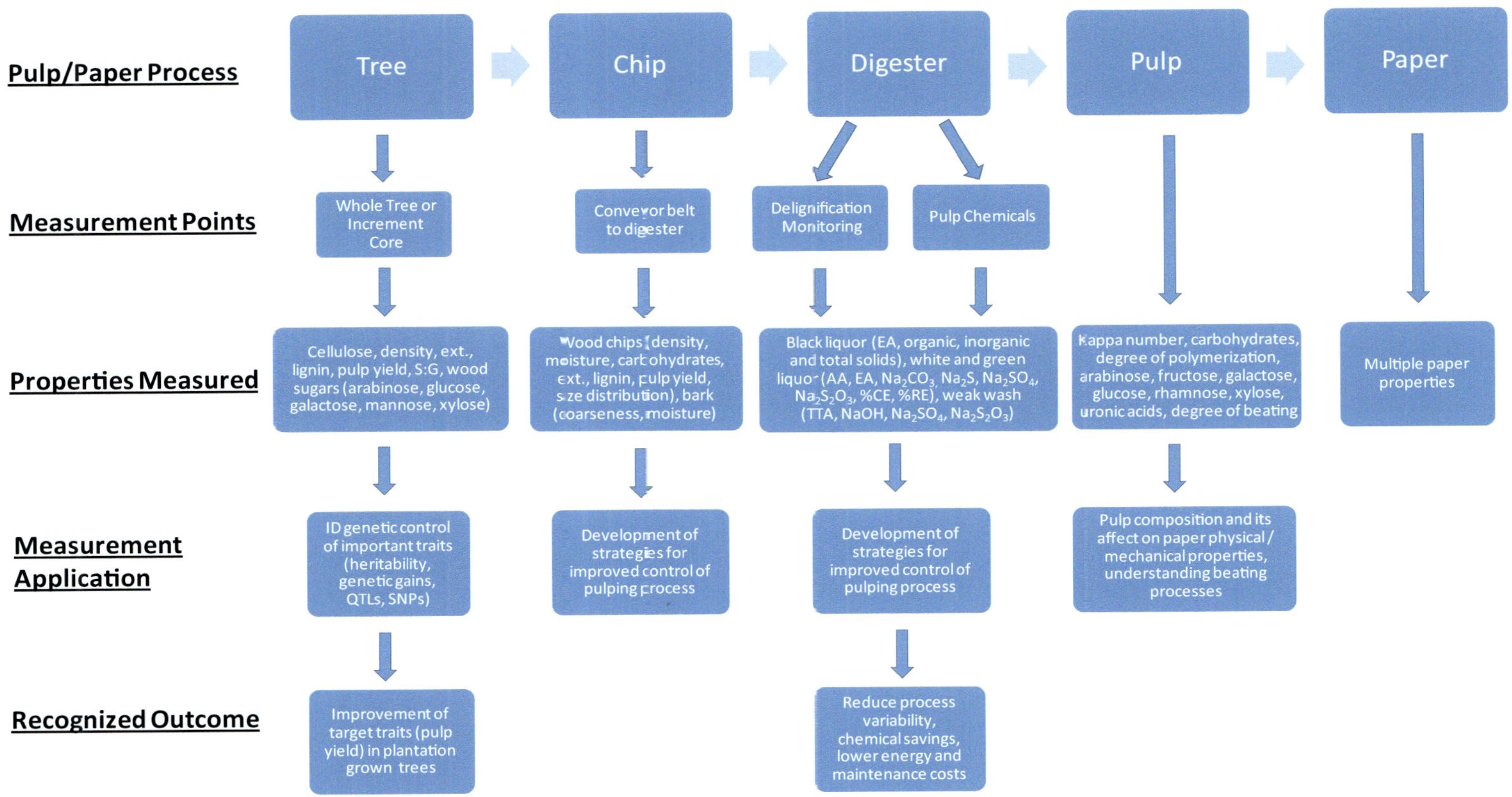

FIGURE 37.1 Properties evaluated by NIR spectroscopy at different stages in the production of pulp and paper from wood. How the estimated NIR data was utilized and final outcome (in terms of pulp and production) are also summarized.

The general approach has been to increase the size of the calibration set to incorporate all reasonably expected variation; however, in a large dataset many samples add no new variation, and algorithms exist for the selection of representative samples. Mora and Schimleck (2008) explored different approaches for this purpose and found that calibrations based on subsets of selected samples performed similarly to a model fitted using all samples, and importantly, "The identification of suitable samples for calibration based on spectral characteristics prior to laboratory analysis would provide substantial savings in analytical costs as well as the model." In a related study, de Sousa et al. (2011) used the Kennard–Stone algorithm to select samples. A total of 3369 samples were tested for density, lignin, and alcohol–toluene extractives and calibrations developed for these properties using all samples, 1000, 500, 200, and 100 samples. Aside from the 100-sample calibration, the others had similar statistics demonstrating the value of selecting representative samples.

37.2.1 Estimation of Genetic Parameters Related to Pulp Production

Pulp yield is very expensive to measure with a single estimate costing over USD800 (Stackpole et al. 2010) and, with density (mass divided by volume at a given moisture content), has the greatest impact on pulp mill economics (Borralho et al. 1993, Greaves et al. 1997). Owing to its importance, companies in the pulp and paper sector use tree breeding to improve pulp yield (in addition to other traits such as growth, form, and density). Gains in any trait via breeding rely on heritability of a given property (the ratio of genetic variance to the total variance), the accuracy with which a property is measured, and sampling intensity, hence the great interest in NIR spectroscopy for the estimation of pulp yield. Improving pulp yield has multiple benefits (Dean 1995), including the following:

- a reduction in wood requirement and hence wood costs;
- increased pulp production capacity for a given level of wood intake;
- lower chemical demand;
- improved pulp strength; and
- a reduction in fixed costs per ton.

Utilization of global calibrations in studies aiming to understand the genetic control of wood properties (Table 37.1) collectively represents one of the most valuable applications of NIR in wood. The importance of NIR in such studies was noted by Stackpole et al. (2011), who stated "we used NIR spectroscopy to estimate wood chemical components as this is the only practical method for measuring the large number of samples required for detailed quantitative genetic analyses."

Raymond et al. (2001) were the first to utilize NIR-generated data for this purpose. Their study used NIR-predicted pulp yield for the estimation of genetic parameters and genotype-by-environment interactions for pulp yield and pulpwood productivity (the product of basic density and pulp yield). Genetic parameters for cellulose content (used as a surrogate for pulp yield owing to the strong relationship between the two properties), extractives, lignin content, and syringyl-to-guaiacyl (S:G) ratio in Tasmanian bluegum were reported in subsequent studies (Raymond and Schimleck 2002, Poke and Raymond 2006, Stackpole et al. 2011). Genetic gains in cellulose content for shining gum determined with wet chemistry data and NIR-predicted data were compared by Schimleck et al. (2004). Estimated gains based on NIR-generated data and wet chemistry data were comparable, but by using NIR, only a fraction of the samples required wet chemical analysis, greatly reducing costs.

Studies that followed utilized calibrations reported by Downes et al. (2009, 2010a, 2011), to estimate genetic parameters, quantitative trait loci (QTL, a section of DNA, the locus, which correlates with variation in a phenotype, i.e., the quantitative trait), and in association mapping studies. Progressively, these studies have led to a greater understanding of the underlying genetic control

TABLE 37.1
Studies that Have Utilized NIR-Predicted Data to Understand the Genetic Control of Wood Properties

Publication	Species	Study Type	NIR-Estimated Wood Properties
Raymond et al. (2001)	*E. globulus*	Genetic parameters	PY
Raymond and Schimleck (2002)	*E. globulus*	Genetic parameters	Cellulose
Schimleck et al. (2004)	*E. nitens*	Genetic parameters	Cellulose
Thamarus et al. (2004)	*E. globulus*	QTL	PY
Poke and Raymond (2006)	*E. globulus*	Genetic parameters	Lignin, ext.
Freeman et al. (2009)	*E. globulus*	QTL	Cellulose, lignin, PY
Thumma et al. (2009)	*E. nitens*	Assoc. mapping	Cellulose
Stackpole et al. (2010)	*E. globulus*	Genetic parameters	Cellulose, PY
Southerton et al. (2010)	*E. nitens*	Assoc. mapping	Cellulose, lignin, PY
Thumma et al. (2010)	*E. nitens*	QTL	Cellulose, lignin, PY, ext.
Stackpole et al. (2011)	*E. globulus*	Genetic parameters	Lignin, S:G, ext.
Brawner et al. (2012)	*C. citriodora*	Genetic parameters	BD, PY
Dillon et al. (2012)	*C. citriodora*	Assoc. mapping	BD, PY
Freeman et al. (2013)	*E. globulus*	QTL	Cell, lignin, PY, ext, S:G
Thavamanikumar et al. (2014)	*E. globulus*	Assoc. mapping	Cellulose, lignin, PY, ext.
Hung et al. (2015)	*E. pellita*	Genetic parameters	BD, PY (MOE, MFA)
Hardner et al. (2016)	*E. dunnii*	Genetic parameters	BD, cellulose, PY
Varghese et al. (2017)	*E. camaldulensis*	Genetic parameters	Lignin, PY
Durán et al. (2017)	*E. globulus*	Genomic selection	BD

QTL, quantitative trait loci; BD, basic density; ext., extractives; PY, pulp yield; S:G, syringyl-to-guaiacyl ratio. Lignin can refer to Klason lignin (acid-insoluble lignin) or total lignin (Klason lignin+acid-soluble lignin).

of properties such as pulp yield, and as noted by Thamarus et al. (2004), "identification of QTLs affecting wood and fiber properties will lead to greater understanding of these traits and help in selection and manipulation of these traits in breeding programs."

In many of these studies, genetic and environmental variation was limited, and in an effort to expand on this research, Freeman et al. (2013) examined stability of growth and NIR-estimated wood property (cellulose, extractives, lignin, pulp yield, S:G ratio) QTLs for Tasmanian bluegum utilizing samples from multiple pedigrees and environments. A key finding of the study was the following: "Despite markedly different environments and pedigrees, many QTLs were stable, providing promising targets for the application of marker-assisted selection." Australian geneticists have led this work, but recently, Durán et al. (2017) reported the use of genomic selection for the prediction of basic density (as estimated by NIR) for Tasmanian bluegum grown in Chile.

Comparable work in softwoods has not been reported; however, Perez et al. (2007), in a study of maritime pine (*Pinus pinaster* Ait), estimated genetic parameters for several properties, including extractive content, Klason lignin, cellulose, hemicelluloses, individual monosaccharides (galactose, mannose, xylose), and S:G ratio.

37.2.2 Nondestructive Evaluation of Standing Trees

Early pulp yield–NIR research utilized whole-tree samples, where a felled tree was subsampled providing wood chips for pulping and NIR analysis. Destructive sampling is undesirable in tree improvement, and another important area of development has been nondestructive evaluation of standing trees. Nondestructive sampling typically involves removing an increment core and using it to estimate wood properties. A series of studies (Schimleck et al. 2005a, b, 2006) demonstrated that increment cores could be utilized to provide NIR-based whole-tree calibrations for pulp yield, and important related properties, allowing the studies that estimated genetic parameters, QTLs, etc., to be conducted with confidence.

Ideally, measurements from standing trees would be possible, negating the need to remove an increment core, and the emergence of field-portable instruments has facilitated the potential for on-site measurement. Meder et al. (2011) showed that pulp yield and cellulose calibrations obtained using spectra collected on a lab-based NIR instrument and two portable NIR spectrometers were similar; however, infield performance of the portable instrument provided moderate success. The seasonality of wood production is a key issue; i.e., at different times during the growing season, wood having different chemistries is produced increasing sampling variability. The authors concluded the following: "while there are still a number of in-forest sampling issues to be resolved around seasonal sampling, the implication is that low-cost, handheld NIR systems are a feasible alternative to laboratory-based NIR systems for routine assessment of Kraft pulp yield." Muneri et al. (2011) in a similar study of Tasmanian bluegum grown in Western Australia reported that estimates of pulp yield using milled increment cores and spectra obtained in the lab clearly outperformed the field-portable instrument. Sandak et al. (2016b) recently explored NIR spectroscopy (among several sensors) to detect log defects with the aim of determining quality indexes for log grading. While not directly related to studies of standing trees, it is a recent attempt of using NIR in the forest.

37.2.3 Wood Property Variation within Trees

There has been long-term interest in within-tree variation of wood properties; however, most properties are too expensive or difficult to measure to allow detailed examination. This is particularly true for pulp yield, and Schimleck and Michell (1998) utilized NIR spectroscopy to examine within-tree variation of pulp yield in shining gum trees at the ring level for multiple heights. Maps for low-, medium-, and high-yield families showed clear differences in patterns of within-tree variation at a level of detail impossible to achieve using lab-based pulping methods.

In studies that followed on radial strips, i.e., pith to bark, NIR-based calibrations for several wood properties were reported (Schimleck et al. 2002, Schimleck and Evans 2004). Many of these (density, microfibril angle or MFA, stiffness, wall thickness, coarseness) utilized data provided by the SilviScan instruments (Evans 1994), while others such as lignin and cellulose used wet chemistry data (Jones et al. 2006). Later studies reported the development of multi-site, multi-species calibrations for these properties that paralleled the pulp yield "global" model efforts (Jones et al. 2005a, b, Schimleck et al. 2010). These studies used breast height (approx. 1.4 m) samples, but in related work, trees were destructively sampled and multi-height calibrations developed that provided data used to map wood property (density, MFA, stiffness) variation within loblolly pine (*Pinus taeda* L.) trees (Mora and Schimleck 2009). In new research (Nabavi et al. 2018), models for loblolly pine tracheid length based on samples from across the SE USA were reported. These models are important as they allow wide-scale assessment and development of maps of within-tree variation of this critical pulp parameter.

NIR spectroscopy has also been utilized to explore radial variation in acetone-soluble and hot-water extractives, total phenolics, and lignin in European larch (*Larix decidua* Mill.) (Gierlinger and Wimmer 2004) and extractive content in white oak (*Quercus alba* L.) (Taylor et al. 2011). Taylor et al. (2011), in their study that examined the quality of *Q. alba* wood for wine and spirit barrel manufacture, also explored radial variation in compounds related to flavor profiles of aged wine and spirits, including vanillin, syringic acid, and syringaldehyde.

Low spatial resolution was an issue with early studies of radial variation by NIR spectroscopy (properties measured in 10 or 5 mm increments). Jones et al. (2007) explored different instrument options for improving resolution including FOSS (5000, XDS), Bruker, and ASD spectrometers and a hyperspectral imaging system (Matrix NIR). Density, MFA, and stiffness calibrations at 2 mm resolution were all noticeably weaker compared to results obtained using data collected in 5 mm increments with the calibrations for density being the most affected. Downes et al. (2010b, 2012) improved resolution to 1 mm in their studies of pulp yield variation in Tasmanian bluegum and shining gum radial strips, while Meder et al. (2010) also developed calibrations at 1 mm resolution for Tasmanian bluegum MFA and various wood sugars in radiata pine (*Pinus radiata* D. Don). Another issue has been nonlinearity in MFA and MOE models, and nonlinear calibration approaches such as least squares support vector machine (LS-SVM) regression (Cogdill et al. 2004, Mora and Schimleck 2010) and radial basis functions–partial least squares (RBF-PLS) (Mora and Schimleck 2010) have been explored. Generally, the nonlinear methods provided improved calibration and prediction statistics.

In studies of radial variation based on spatially consecutive NIR measurements of wooden strips, it is possible that the autocorrelation that exists among measurements could inflate calibration statistics. Mora et al. (2008) explored this possibility by incorporating serial correlation into the PLS algorithm. They were able to remove serial dependence by using an autoregressive correlation structure (second order) and found predictions of density were similar to those obtained using normal PLS as the initial dependence among measurements was not strong. A stronger autocorrelation was expected (owing to the nature of wood formation), and the authors suggested that dependence was reduced owing to the use of 10 mm increments rather than growth rings.

Hyperspectral imaging has recently been used to examine the spatial variation of lignin, cellulose, hemicellulose, and various wood sugars in radiata pine disks (rather than radial strips) at a resolution of 4 mm^2/pixel (Thumm et al. 2010, 2016) and intra-ring density in stone pine (*Pinus pinea* L.) (Fernandes et al. 2013a, b). In a related study, Ma et al. (2017) utilized a hyperspectral imaging system to examine density and MFA radial variation in sugi (*Cryptomeria japonica* (L.f.) D. Don).

37.3 PULP AND PAPER PROPERTIES

The aim of pulping, which utilizes several methods ranging from mechanical abrasion of wood to delignification using chemicals, is to separate wood into its constituent fibers (the pulp) to make paper. Birkett and Gambino (1988, 1989) identified many potential applications of NIR spectroscopy related to the sector, but it was in Sweden where much of the earliest research in this area occurred.

Wallbäcks et al. (1991a, b) successfully used NIR, in addition to ^{13}C nuclear magnetic resonance (^{13}C CP/MAS NMR) and MIR, to monitor changes in glucose, xylose, galactose, and Klason lignin contents in birch (*Betula* spp.) kraft pulp samples and to characterize seven bleached pine pulp samples obtained from different pulping processes. Wallbäcks et al. (1995) also used NMR and NIR on unbeaten and beaten pulps and found that both techniques could describe beating-related variation between samples. In related work, delignification (primarily the loss of lignin from wood) during laboratory-based kraft cooks was examined (Norman et al. 1994, Lindgren and Edlund 1998).

At a pulp mill, a conveyor belt transports wood chips into a digester where heated and pressurized cooking chemicals dissolve lignin present in the wood to produce pulp. Knowledge of chip properties can assist with optimizing the pulping process, and Axrup et al. (2000) investigated the potential of NIR spectroscopy to estimate properties of chips on a belt moving at 1 m s^{-1}. Calibrations were developed for wood chip moisture, extractives, Klason lignin, carbohydrates, and size distribution, in addition to bark coarseness and moisture.

Once the digestion, or cook, of chips is complete in an industrial setting, the pulp is washed, the cooking chemicals are recycled, and studies have investigated the utilization of NIR spectroscopy on both pulp and chemicals. Kappa number provides a standard measure of lignin remaining in pulp after digestion, and several authors (Birkett and Gambino 1988, Yuzak and Lohrke 1993, Olsson et al. 1995, Antti et al. 2000, Fardim et al. 2002) have reported its measurement using NIR spectroscopy. In addition to lignin, researchers have utilized NIR spectroscopy to measure the degree of polymerization (related to cellulose chain length) and carbohydrate composition, including various wood sugars and uronic acids of kraft pulps (Fardim et al. 2002, Monrroy et al. 2009, Tyson et al. 2010). Similarly, others (Vanchinathan et al. 1996, Hodges and Krishnagopalan 1999, Saucedo and Krishnagopalan 2000) have measured a range of components related to kraft pulp mill liquors. Continuous online measurement of liquor composition based on acousto-optic tunable filter (AOTF) technology has been reported (Hodges et al. 1999) and investigated in mill trials (Patrick 1998). In terms of commercialization, this sector has seen considerable activity with a number of instruments available (Trung et al. 2015).

Interest has also existed in estimating paper properties, typically measured on small paper sheets (handsheets) made from pulp, and tested under standardized conditions. Meder et al. (1994) explored the prediction of handsheet properties based on spectra collected from milled wood, while Antti et al. (1996) utilized NIR spectra obtained from handsheets. Antti et al. (1996) developed PLS calibrations for 17 properties, and of those, the level of beating, sheet thickness, porosity, density, tearing resistance, and short compression test (SCT) gave an R^2 greater than 0.85. Marklund et al. (1999) utilized NIR spectra collected from both milled wood and handsheets and reported that spectra of milled wood chips had nearly the same predictive ability as those of bleached pulps. More recently, Tyson et al. (2012) used eucalypt pulps obtained from a pulp mill to estimate handsheet physical and mechanical properties. All properties were poorly predicted, and the authors attributed this to the low variability of their dataset.

37.4 PHYSICAL/MECHANICAL PROPERTIES OF WOOD

The use of NIR spectroscopy to estimate many different properties related to the performance of lumber is discussed in Section 37.4, while Section 37.5 describes efforts to apply NIR to various composite materials. Figure 37.2 provides an overview of this research in terms of properties assessed and product type.

Two Danish papers, published in 1995, were the first to examine the utilization of NIR spectroscopy for estimating physical–mechanical properties of wood (Hoffmeyer and Pedersen 1995, Thygesen 1995). Like Birkett and Gambino, Hoffmeyer and Pedersen (1995), in their study of Norway spruce (*Picea abies* (L.) H. Karst.), were concerned with exploring a range of potential applications, while Thygesen (1995) utilized NIR to estimate basic density and dry matter content. The studies that followed focused on estimating properties important for understanding the performance of lumber when used in construction (for a summary, see Schimleck et al. 2018; Table 37.1). Properties examined included density, modulus of elasticity (MOE, a measure of the elastic deformation that wood undergoes when subject to an applied load), and modulus of rupture (MOR, a measure of the breaking strength of wood when subject to an applied load) and were mainly based on short, defect-free samples. Unlike the studies of pulp yield, research was largely based on softwood species that are widely used for construction.

37.4.1 LUMBER

NIR estimation of important properties using spectra collected directly from the surface of lumber has paralleled research on static bending samples. In the first of these studies (Meder et al. 2003), a Matrix-E FT-NIR spectrometer was used to collect spectra from green radiata pine cants that were then cut into lumber (multiple pieces of lumber were obtained from one cant). Lumber MOE was measured, and using these data, a calibration for stiffness was developed using NIR spectra from the appropriate cant. Relationships were moderate but considered strong enough for an economic segregation of cants into different sawing patterns.

Japanese researchers (Fujimoto et al. 2008, 2010a, b, Kobori et al. 2015) further investigated the application of NIR for estimating lumber properties and for grading. In the most recent study (Kobori et al. 2015), a robust spectrometer with the ability to acquire spectra rapidly along the

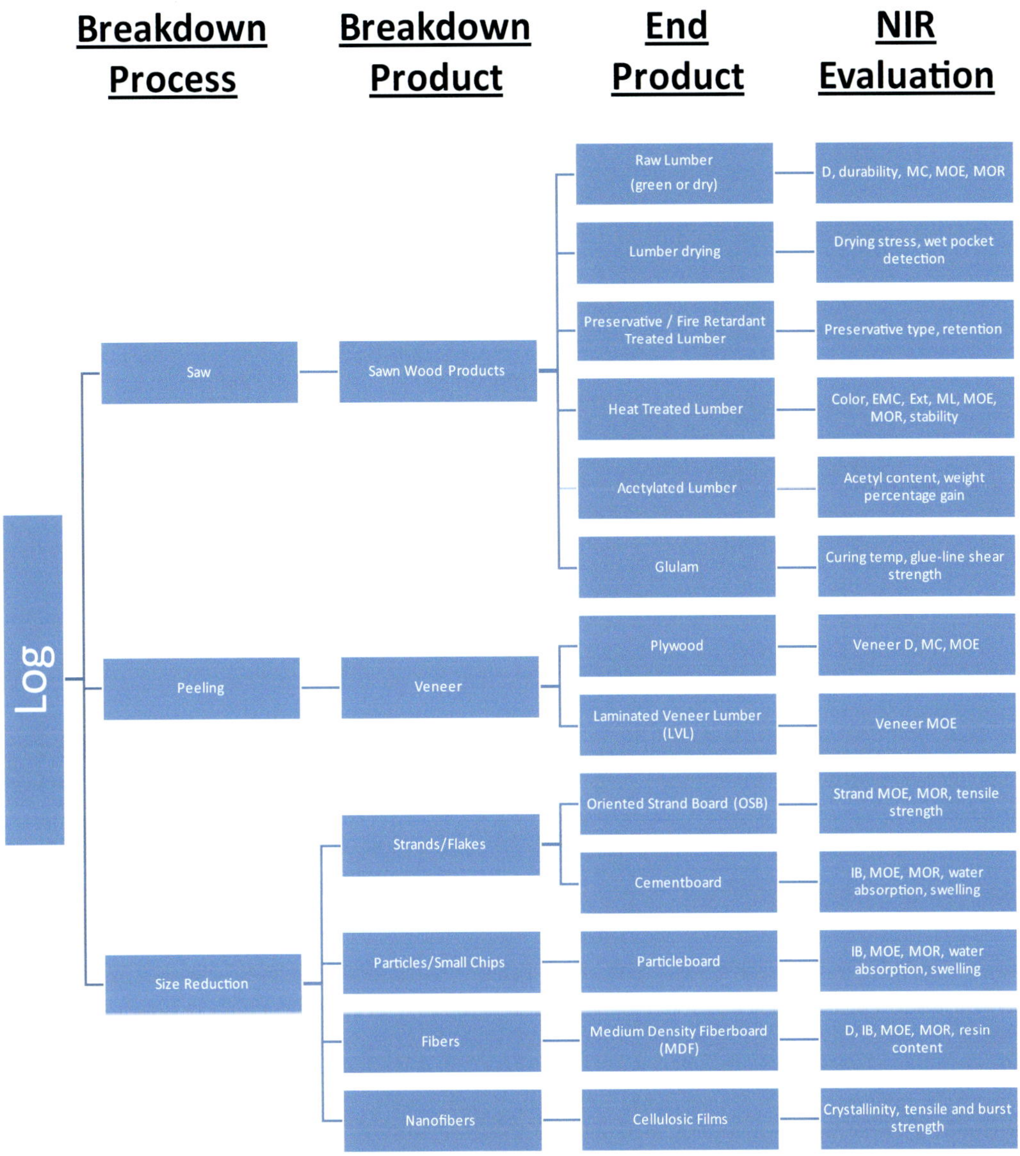

FIGURE 37.2 Properties evaluated by NIR spectroscopy in research on various wood products. Methods of log breakdown and breakdown products are included as they are critical intermediate steps in converting a log to an end product.

length of a piece of lumber (to collect representative spectra from moving samples) was used to scan the surface of 20 hinoki cypress (*Chamaecyparis obtusa* (Siebold & Zucc.) Endl.) boards. The calibration reported for moisture content was excellent, while that for MOE was very good. The authors concluded the following: "the developed device could be applied for the online quality certification of sawn lumber in commercial sawmills." A picture of the online grading system is shown in Figure 37.3.

In this study, multiple NIR spectra were collected from two areas of the longitudinal-tangential surface, i.e., the long flat face, of a moving board. Wood presents three different surfaces (radial, tangential, and transverse), and the configuration of equipment in a mill may determine the surface examined. In an alternative approach, Dahlen et al. (2017) utilized NIR collected from the

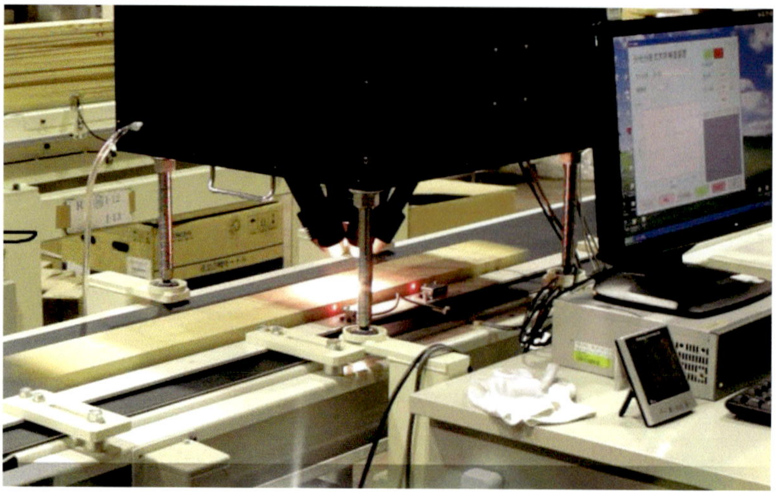

FIGURE 37.3 On-line fast grading NIR system.

end (or transverse surface) of loblolly pine boards obtained from sawmills in the southeast USA. In an earlier study of fresh-sawn red oak (*Quercus rubra* L.), spectra collected from transverse and radial surfaces provided better predictions than those collected from tangential surfaces (Defo et al. 2007).

Research has focused on measuring lumber prior to use, but in a recent study, Sandak et al. (2015b) explored the assessment of MOE and load for wood in service opening up a new area of research.

37.4.2 Reaction Wood and Wood Defects

Research has also utilized NIR spectroscopy for the identification of reaction wood and wood defects in lumber. Reaction wood is a feature common to all trees. It has a critical role in reinforcing and reorienting woody stems (Groover 2016) but, if severe, is undesirable in lumber as its properties are quite different to those of normal wood. The types of reaction wood produced differ between hardwoods (tension wood, present on the tension side of a leaning stem) and softwoods (compression wood, present on the compression side), with tension wood being quite difficult to detect visually. Researchers have used NIR spectroscopy to detect both compression wood (Meder and Meglen 2012, McLean et al. 2014) and non-recoverable collapse, a feature closely associated with tension wood in eucalypts (Wentzel-Vietheer et al. 2013, Downes et al. 2014). More recently, Thumm and Riddell (2017) utilized hyperspectral imaging to detect resin defects in appearance-grade radiata pine lumber with an accuracy rate of 80%—very promising results considering the presence of other wood features (sapstain, knots) that could be confused with resinous regions.

37.4.3 Natural Durability

Another important feature of wood is its natural durability, or resistance to degradation by fungi, various insects, and marine borers. Owing to the presence of extractives (various secondary metabolites), the heartwood of many species is highly durable (Scheffer and Morrell 1998) and valued for outdoor applications. However, with a shift toward short-rotation plantations to meet the fiber needs of society, the availability of naturally durable woods has become limited. NIR spectroscopy has been utilized to predict the natural durability (via weight loss) and extractive content in various species of larch (Gierlinger et al. 2002, 2003, Sykacek et al. 2006), Scots pine (*Pinus sylvestris* L.)

(Flaete and Haartveit 2004), and goncalo alves (*Astronium graveolens* Jacq.), a tropical hardwood recognized for its high durability (Taylor et al. 2008). Taylor et al. (2008) examined the susceptibility of goncalo alves to both a white rot (*Trametes versicolor* (L.) Lloyd) and a brown-rot fungus (*Gloeophyllum trabeum* (Pers.) Murrill) and found for *T. versicolor* NIR accurately predicted extractive content and susceptibility to decay, while *A. graveolens* was very resistant to attack by *G. trabeum*, but NIR spectra failed to accurately predict weight loss.

Recent efforts have focused on utilizing NIR-generated data to estimate the genetic parameters necessary to improve wood durability via breeding programs (e.g., Bush et al. (2011) for *Eucalyptus cladocalyx* F. Muell. and Pâques and Charpentier (2015) for hybrid larch) and to assess the durability of plantation-grown softwoods (Jones et al. 2011, 2013).

37.4.4 Wood Decay/Preservation

Wood that is nondurable will decay if used outdoors, and the process of wood decay in the presence of fungi has been the subject of many spectroscopic studies (Fackler and Schwanninger 2012). Specifically, NIR spectroscopy has been utilized to examine changes in wood during degradation by brown (Kelley et al. 2002, Fackler et al. 2007)- and white (Barton et al. 1995, Schwanninger et al. 2004b, Fackler et al. 2006, 2007)-rot fungi. In addition, NIR has been used to follow changes to the chemistry of loblolly pine wood chips treated with the white-rot fungi *Ceriporiopsis subvermispora* (Pilát) Gilb. & Ryvarden prior to pulping (Ferraz et al. 2004), while Via et al. (2008) examined spectral response to blue stain fungi. More recently, hyperspectral imaging has been used to examine blue stain fungi on coated and uncoated Norway spruce surfaces (Burud et al. 2014), and to assess weathering of thin wood samples (Smeland et al. 2016). FT-NIR spectroscopy has also been used to examine the weathering kinetics of wood veneers (Sandak et al. 2016c).

Nondurable wood can be treated with toxic preservatives to improve its durability. While this prolongs service life, it can be problematic if the wood is recycled or repurposed, and NIR spectroscopy has been utilized to identify wood treated with different preservative types (Feldhoff et al. 1998, So et al. 2004). It has also been used to estimate the concentration of boron-based preservatives in southern pine wood (Taylor and Lloyd 2007) and permethrin and bifenthrin in treated radiata pine (Meder and Edbon 2013). Mauruschat et al. (2016) explored the utilization of NIR spectroscopy for the detection and sorting of waste wood treated with preservatives and concluded the following: "NIR spectroscopy is capable of distinguishing between untreated wood and wood treated with both inorganic and organic preservatives. Further trials showed for the first time that also the kind of treatment (inorganic, organic or untreated) could be distinguished, even if the treated wood samples were of different moisture content." A hyperspectral imaging camera was recommended as the detection unit of choice for installation in an operational sorting facility. A related application is the detection of phosphorus-based flame retardants in wood by hyperspectral imaging (Stefansson et al. 2018).

Alternative approaches to improving the performance of wood in outdoor applications exist. A common method is heating lumber at temperatures ranging from 160°C to 260°C (Esteves and Pereira 2009). Temperature-dependent changes in wood chemistry occur, lowering the equilibrium moisture content of the wood and improving dimensional stability and durability (Esteves and Pereira 2009). Schwanninger et al. (2004a) explored the application of NIR spectroscopy to thermally modified European beech (*Fagus sylvatica* L.) and concluded that the technique could be used in quality control, while Esteves and Pereira (2008), in a study of heat-treated Tasmanian bluegum and maritime pine, also identified "potential for quality control and characterization of treated woods." More recently, Sandak et al. (2015a, 2016a) reported the application of NIR spectroscopy for the quality control of vacuum thermally modified wood. Another approach, acetylation, involves replacing OH functional groups in wood (H-bonding sites for water) with acetyl groups, and Stefke et al. (2008) and Schwanninger et al. (2011) utilized NIR spectroscopy (and FT-IR) to follow chemical changes in wood owing to acetylation.

37.4.5 Moisture Content

Wood moisture content (MC) is another critical feature in terms of its performance and durability. Wood is a hygroscopic material, and provided moisture content is below the fiber saturation point (FSP, approximately 30% MC for most species), it exhibits shrink/swell behavior with changes in mechanical or physical properties, for example, MOE as MC varies. In addition, if wood MC is too high, it becomes vulnerable to fungal degradation; hence, monitoring MC is very important for the forest products industry. As the NIR spectral region is strongly influenced by moisture, many researchers have used NIR to investigate its interaction with wood.

Initial efforts focused on the development of NIR-based calibrations for the estimation of MC in lab samples (Tsuchikawa et al. 1996). Thygesen and Lundqvist (2000a, b) investigated infield applicability wood moisture measurement by NIR under unstable temperature conditions. They systematically examined the thermal effects in NIR spectra of solid wood samples of Norway spruce between −20°C and +25°C and showed that temperature variation had to be considered when using NIR spectroscopy to determine wood MC, a critically important observation if NIR spectrometers are used in field settings exposed to temperature extremes. Leblon et al. (2013), in a review, provide a comprehensive summary of published MC calibrations.

MC calibrations for lumber have been reported (e.g., Kobori et al. 2015), while Mora et al. (2011) used hyperspectral images collected from the ends of loblolly pine logs to estimate whole-log MC. In lumber drying, NIR spectroscopy has been used to estimate drying stresses (Watanabe et al. 2012a) and to identify the presence of wet-pockets (isolated areas of very high moisture that do not dry sufficiently using normal kiln schedules) in both *Tsuga heterophylla* (Raf.) Sarg. (Watanabe et al. 2010) and *Abies lasiocarpa* (Hooker) Nuttall (Watanabe et al. 2012b). Watanabe et al. (2011) also utilized NIR to sort green hem-fir lumber based on estimated MC. They showed that NIR can indirectly estimate average moisture of green lumber; however, a limitation is that only surface MC is provided owing to limited depth penetration. Both developments are important and potentially can improve drying efficiency by ensuring that lumber of similar MC is dried at the same time. Recently, Tham et al. (2018) used a capacitive sensing method and NIR spectroscopy to predict wood density and MC simultaneously. Their findings suggested that combining the two methods makes it possible to predict density and MC more accurately.

Basic research exploring the nature of water in wood has also been conducted. Moisture diffusion processes in both softwoods (Tsuchikawa and Siesler 2003a) and hardwoods (Tsuchikawa and Siesler 2003b) have been monitored using NIR spectroscopy, while Tsuchikawa and Tsutsumi (1998) utilized NIR spectroscopy to detect the presence of capillary condensed water and adsorbed water. Vis–NIR hyperspectral imaging was used to visualize MC distribution in samples of European beech and Scots pine as they dried naturally (Kobori et al. 2013), while in new research, Sugii et al. (2017) utilized two-dimensional correlations to investigate the dynamic behavior of the chemical components of wood during drying. An interesting finding was the response of cell wall components and bound water (water in the cell wall) to drying even though the wood exceeded the fiber saturation point (it is accepted that bound water is removed only below the FSP).

37.5 WOOD COMPOSITES

Composite wood products encompass a wide range of wood products (particleboard, oriented strand board (OSB), plywood, laminated veneer lumber (LVL), etc.) manufactured using resin and wood in various forms (particles, strands, veneers). Some of the earliest NIR–wood-related work focused on the characterization and evaluation of resin-chip mixtures (Kniest 1992, Niemz et al. 1992, 1994), while later research focused on determination of stiffness, strength, and internal bond properties of panels (Rials et al. 2002, Hein et al. 2009, 2011) and process optimization (Sjoblom et al. 2004). Research has also sought to assess stiffness of veneers (Meder et al. 2002, Adedipe and Dawson-Andoh 2008a, b, Carneiro et al. 2010) used in the manufacture of plywood or LVL, to examine glulam gluelines (Gaspar et al. 2009) and to predict properties of strands used in the manufacture of oriented strand lumber (Kohan et al. 2012). In a recent study, the chemical structure of an adhesive post-cure was explored using FT-NIR spectroscopy (Ling et al. 2018). In other new research (Viana et al. 2016), NIR spectroscopy was used to estimate the crystallinity and strength of nanocellulosic films made from cellulose fibers and nanofibers obtained from *Pinus* kraft pulp (bleached and unbleached).

37.6 BARK

The bark of trees is an important source of energy in the forest products industry and as a raw material for the production of various compounds used in a range of applications that include leather tanning and the production of adhesives. Extractives from the bark of black wattle (*Acacia mearnsii* De Wild.) are widely used for this purpose, and estimation of quality parameters (tannin content, nontannins, extractives, polyflavonoid content, and Stiasny value) by NIR spectroscopy has been explored (Donkin and Pearce 1995, Schimleck and Yazaki 2003a). Similar properties (hot-water extractives, NaOH extractives, and Stiasny value) were determined for radiata pine bark (Schimleck and Yazaki 2003b), while So and Eberhardt (2006) used NIR to determine southern pine bark quality in terms of inner and outer bark proportions.

More recent bark-related studies have investigated the geographic origins of cork (Prades et al. 2012), the determination of fuel properties of solid biofuel mixtures (which include bark) (Skvaril et al. 2017), and the early identification of white ash (*Fraxinus americana* L.) trees infested by emerald ash borer (Finley et al. 2016).

37.7 SPECIES DISCRIMINATION AND SORTING WOOD PRODUCTS

Many authors have explored the utilization of NIR spectra for the discrimination of wood and wood products with considerable success owing to the differences that exist among species and genera in terms of taxonomy and chemical and physical properties. Many of the earliest studies focused on the ability to discriminate among species for reasons related to processing, including identification between species that were suitable or not for calcium bisulfite pulping (Eugster and Kucera 1995), the composition of wood chip mixtures for kraft pulping (Antti et al. 1996), and species identification prior to lumber production (Brunner et al. 1996). Other studies sought a rapid method to allow the identification of wood of different species (Nair and Lodder 1993, Brunner et al. 1996) or to discriminate among woods of different species (Tsuchikawa et al. 2003a, b) and different species within the same genus (Gierlinger et al. 2004). It has also been shown that provenance within species can be discriminated (Schimleck et al. 1996, Sandak et al. 2011).

In later studies, the identification of eucalypt hybrids using NIR spectra of leaves (Humphreys et al. 2008) and *Pinus* species hybrids using NIR spectra collected from ground foliage was reported (Espinoza et al. 2012, Dvorak et al. 2013). Related research has used portable NIR spectrometers to collect spectra that have facilitated field identification of hybrids of several *Corymbia* species (Abasolo et al. 2013) and *Pinus* species (Meder et al. 2014). This work is particularly important owing to the difficulty of confidently identifying pure species and hybrids of closely related species. However, great care is required with the field-based measurements to achieve good discrimination, and as noted by Abasolo et al. (2013), "The practical application of this technique for hybrid diagnosis of seedlings in the nursery will require careful attention to control environmental factors because seedling age and storage effects influenced the ability of NIR to identify hybrids."

A particularly important area of research that has emerged is the potential of NIR spectroscopy as a forensic tool in the fight against illegal logging (Dormontt et al. 2015). Studies have explored NIR spectroscopy to identify highly valued hardwoods such as mahogany (*Swietenia macrophylla* King) and to discriminate it from similar species (Braga et al. 2011, Pastore et al. 2011, Soares et al. 2017). Country of origin for mahogany has also been determined with confidence (Bergo et al. 2016) using discriminate analysis. NIR is considered to have excellent potential as a screening tool; however, more research is required to develop databases of samples to serve as chemical profiles for different locations, and statistical methods for classifying taxonomic differences (Dormontt et al. 2015).

In research on another high-value timber species, Casale et al. (2010) classified pernambuco (*Caesalpinia echinata* Lam.) wood quality using NIR spectroscopy. Pernambuco is used to make high-quality string instrument bows, but owing to concerns about the long-term survival of the species, it is now a Convention on International Trade in Endangered Species (CITES) appendix II listed species, and trade is limited. Not all pernambuco is suitable for bow manufacture, and linear discriminant analysis was used to successfully distinguish between samples recognized as having excellent and good-to-very good potential for making high-quality bows.

Another area of research has been automated online sorting systems based on NIR spectra. Examples include the classification and separation of red oak and white oak using NIR spectrometry and SIMCA (Adedipe et al. 2008) and an automatic system to separate wood wastes (plywood coated with resin, preservative-treated wood, hardwood and softwood), using Vis–NIR imaging analysis and chemometrics (Kobori et al. 2017).

37.8 RESOURCE ASSESSMENT

The utilization of cores/radial strips for the assessment of wood properties also allows NIR spectroscopy to be utilized for large-scale resource assessment. An excellent example of this application type was published by Giroud et al. (2017) who examined regional variation in wood density and modulus of elasticity for several boreal species (balsam fir, black spruce, jack pine, paper birch, and trembling aspen) in the Canadian province of Quebec. The study utilized increment cores collected during the inventory of Quebec's forests. In total, 30,159 cores representing 10,573 inventory plots were scanned using an automated NIR system (spot size 2 mm). SilviScan determined density and MOE variation for a subsample (1636), and NIR calibrations for both properties were developed. The calibrations were subsequently used to predict density and MOE for all cores, and the data were used to generate spatial maps for the study area (approximately 71°W–79°W and 45°N–51°N).

REFERENCES

Abasolo, M., D.J. Lee, C. Raymond, R. Meder and M. Shepherd. 2013. Deviant near-infrared spectra identifies *Corymbia* hybrids, *For. Ecol. Manage.*, 304, 121–131.

Adedipe, O.E. and B. Dawson-Andoh. 2008a. Predicting moisture content of yellow-poplar (*Liriodendron tulipifera* L.) veneer using near infrared spectroscopy, *For. Prod. J.*, 58(4), 28–33.

Adedipe, O.E. and B. Dawson-Andoh. 2008b. Prediction of yellow-poplar (*Liriodendron tulipifera*) veneer stiffness and bulk density using near infrared spectroscopy and multivariate calibration, *J. Near Infrared Spectrosc.*, 16(5), 487–496.

Adedipe, O.E., B. Dawson-Andoh, J. Slahor and L. Osborn. 2008. Classification of red oak (*Quercus rubra*) and white oak (*Quercus alba*) wood using a near infrared spectrometer and soft independent modelling of class analogies, *J. Near Infrared Spectrosc.*, 16(1), 49–57.

Antti, H., D. Alexandersson, M. Sjostrom and L. Wallbäcks. 2000. Detection of kappa number distributions in kraft pulps using NIR spectroscopy and multivariate calibration, *Tappi J.*, 83(3), 102–108.

Antti, H., M. Sjostrom and L. Wallbäcks. 1996. Multivariate calibration models using NIR spectroscopy on pulp and paper industrial applications, *J. Chemom.*, 10(5–6), 591–603.

Axrup, L., K. Markides and T. Nilsson. 2000. Using miniature diode array NIR spectrometers for analysing wood chips and bark samples in motion, *J. Chemom.*, 14(5–6), 561–572.

Barton, F.E., D.S. Himmelsbach, D.E. Akin, A. Sethuraman and K.-E.L. Erikson. 1995. Two-dimensional vibration spectroscopy III: Interpretation of the degradation of plant cell walls by white rot fungi, *J. Near Infrared Spectrosc.*, 3(1), 10.

Bergo, M.C.J., T.C.M. Pastore, V.T.R. Coradin, A.C. Wiedenhoeft and J.W.B. Braga. 2016. NIRS identification of *Swietenia macrophylla* is robust across specimens from 27 countries, *IAWA J.*, 37(3), 420–430.

Birkett, M.D. and M.J.T. Gambino. 1988. Potential applications for near infrared spectroscopy in the pulping industry, Pap. S. Afr., Nov/Dec, 5.

Birkett, M.D. and M.J.T. Gambino. 1989. Estimation of pulp kappa number with near-infrared spectroscopy, *Tappi J.*, 72(9), 193–197.

Borralho, N.M.G., P.P. Cotterill and P.J. Kanowski. 1993. Breeding objectives for pulp production of *Eucalyptus globulus* under different industrial cost structures, *Can. J. For. Res.*, 23(4), 648–656.

Braga, J.W.B., T.C.M. Pastore, V.T.R. Coradin, J.A.A. Camargos and A.R. da Silva. 2011. The use of near infrared spectroscopy to identify solid wood specimens of *Swietenia macrophylla* (CITES Appendix II), *IAWA J.*, 32(2), 285–296.

Brawner, J.T., R. Meder, M. Dieters and D.J. Lee. 2012. Selection of *Corymbia citriodora* for pulp productivity, *South. For.*, 74(2), 121–131.

Brunner, C.C., A.G. Maristany and D.A. Butler. 1996. Wood species identification using spectral reflectance, *For. Prod. J.*, 46(2), 82–85.

Burud, I., L.R. Gobakken, A. Flo, K. Kvaal and T.K. Thiis. 2014. Hyperspectral imaging of blue stain fungi on coated and uncoated wooden surfaces, *Int. Biodeterior. Biodegradation*, 88, 37–43.

Bush, D., K. McCarthy and R. Meder. 2011. Genetic variation of natural durability traits in *Eucalyptus cladocalyx* (sugar gum), *Ann. For. Sci.*, 68(6), 1057–1066.

Carneiro, M.E., W.L.E. Magalhaes, G.I.B. de Muniz and L.R. Schimleck. 2010. Near infrared spectroscopy and chemometrics for predicting specific gravity and flexural modulus of elasticity of *Pinus* spp. veneers, *J. Near Infrared Spectrosc.*, 18(6), 481–489.

Casale, M., L.R. Schimleck and C. Espey. 2010. Classification of pernambuco (*Caesalpinia echinata* Lam.) wood quality by near infrared spectroscopy and linear discriminant analysis, *J. Near Infrared Spectrosc.*, 18(6), 435–442.

Cogdill, R.P., L.R. Schimleck, P.D. Jones, G.F. Peter, R.F. Daniels and A. Clark. 2004. Estimation of the physical wood properties of *Pinus taeda* L. radial strips using least squares support vector machines, *J. Near Infrared Spectrosc.*, 12(4), 263–269.

Dahlen, J., I. Diaz, L. Schimleck and P.D. Jones. 2017. Near-infrared spectroscopy prediction of southern pine No. 2 lumber physical and mechanical properties, *Wood Sci. Technol.*, 51(2), 309–322.

de Sousa, L.C., J.L. Gomide, F.R. Milagres and D.P. de Almeida. 2011. Development of NIRS calibration models for minimization of Eucalyptus spp wood analysis, *Cienc. Florest.*, 21(3), 591–599.

Dean, G. 1995. Objectives for wood fibre quality and uniformity. In Eucalyptus plantations: improving fibre yield and quality. *CRC THF-IUFRO Conf.*, Potts, B.M., N.M.G. Borralho, J.B. Reid, R.N. Cromer, W.N. Tibbits and C.A. Raymond (eds.), (pp. 5–9), Cooperative Research Centre for Temperate Hardwood Forestry, Hobart, Tasmania.

Defo, M., A.M. Taylor and B. Bond. 2007. Determination of moisture content and density of fresh-sawn red oak lumber by near infrared spectroscopy, *For. Prod. J.*, 57(5), 68–72.

Dillon, S.K., J.T. Brawner, R. Meder, D.J. Lee and S.G. Southerton. 2012. Association genetics in *Corymbia citriodora* subsp. *variegata* identifies single nucleotide polymorphisms affecting wood growth and cellulosic pulp yield, *New Phytol.*, 195(3), 596–608.

Donkin, M.J. and J. Pearce. 1995. Tannin analysis by near-infrared spectroscopy, *J. Soc. Leather Technol. Chem.*, 79(1), 8–11.

Dormontt, E.E., M. Boner, B. Braun, G. Breulmann, B. Degen, E. Espinoza, S. Gardner, P. Guillery, J.C. Hermanson, G. Koch, S.L. Lee, M. Kanashiro, A. Rimbawanto, D. Thomas, A.C. Wiedenhoeft, Y.F. Yin, J. Zahnen and A.J. Lowe. 2015. Forensic timber identification: It's time to integrate disciplines to combat illegal logging, *Biol. Conserv.*, 191, 790–798.

Downes, G.M., R. Meder, C. Hicks and N. Ebdon. 2009. Developing and evaluating a multisite and multispecies NIR calibration for the prediction of Kraft pulp yield in eucalypts, *South. For.*, 71(2), 155–164.

Downes, G.M., R. Meder and C. Harwood. 2010a. A multi-site, multi-species near infrared calibration for the prediction of cellulose content in eucalypt woodmeal, *J. Near Infrared Spectrosc.*, 18(6), 381–387.

Downes, G.M., R. Meder, N. Ebdon, H. Bond, R. Evans, K. Joyce and S. Southerton. 2010b. Radial variation in cellulose content and Kraft pulp yield in *Eucalyptus nitens* using near-infrared spectral analysis of air-dry wood surfaces, *J. Near Infrared Spectrosc.*, 18(2), 147–155.

Downes, G.M., R. Meder, H. Bond, N. Ebdon, C. Hicks and C. Harwood. 2011. Measurement of cellulose content, Kraft pulp yield and basic density in eucalypt woodmeal using multisite and multispecies near infra-red spectroscopic calibrations, *South. For.*, 73(3–4), 181–186.

Downes, G.M., C.E. Harwood, J. Wiedemann, N. Ebdon, H. Bond and R. Meder. 2012. Radial variation in Kraft pulp yield and cellulose content in *Eucalyptus globulus* wood across three contrasting sites predicted by near infrared spectroscopy, *Can. J. For. Res.*, 42(8), 1577–1586.

Downes, G.M., M. Touza, C. Harwood and M. Wentzel-Vietheer. 2014. NIR detection of non-recoverable collapse in sawn boards of *Eucalyptus globulus*, *Eur. J. Wood Wood Prod.*, 72(5), 563–570.

Durán, R., F. Isik, J. Zapata-Valenzuela, C. Balocchi and S. Valenzuela. 2017. Genomic predictions of breeding values in a cloned *Eucalyptus globulus* population in Chile, *Tree Genet. Genomes*, 13(4), 74.

Dvorak, W.S., A. Nel and J.A. Espinoza. 2013. Evidence of low-levels of natural introgression between *Pinus jaliscana* and *Pinus oocarpa* in an open-pollinated progeny trial using near-infrared spectroscopy, *For. Ecol. Manage.*, 295, 20–27.

Espinoza, J.A., G.R. Hodge and W.S. Dvorak. 2012. The potential use of near infrared spectroscopy to discriminate between different pine species and their hybrids, *J. Near Infrared Spectrosc.*, 20(4), 437–447.

Esteves, B. and H. Pereira. 2008. Quality assessment of heat-treated wood by NIR spectroscopy, *Holz Roh Werkst.*, 66(5), 323–332.

Esteves, B.M. and H.M. Pereira. 2009. Wood modification by heat treatment: A review, *Bioresources*, 4(1), 370–404.

Eugster, R. and L.J. Kucera. 1995. Rapid and nondestructive determination of wood types using FT-NIR in cellulose production, *Holz Roh Werkst.*, 53(6), 392–392.

Evans, R. 1994. Rapid measurement of the transverse dimensions of tracheids in radial wood sections from *Pinus radiata*, *Holzforschung*, 48(2), 168–172.

Fackler, K., C. Gradinger, B. Hinterstoisser, K. Messner and M. Schwanninger. 2006. Lignin degradation by white rot fungi on spruce wood shavings during short-time solid-state fermentations monitored by near infrared spectroscopy, *Enzyme Microb. Technol.*, 39(7), 1476–1483.

Fackler, K. and M. Schwanninger. 2012. How spectroscopy and microspectroscopy of degraded wood contribute to understand fungal wood decay, *Appl. Microbiol. Biotechnol.*, 96(3), 587–599.

Fackler, K., M. Schwanninger, C. Gradinger, E. Srebotnikl, B. Hinterstoisser and K. Messner. 2007. Fungal decay of spruce and beech wood assessed by near-infrared spectroscopy in combination with uni- and multivariate data analysis, *Holzforschung*, 61(6), 680–687.

Fardim, P., M.M.C. Ferreira and N. Duran. 2002. Multivariate calibration for quantitative analysis of eucalypt kraft pulp by NIR spectrometry. *J. Wood Chem. Technol.*, 22(1), 67–81.

Feldhoff, R., T. Huth-Fehre and K. Cammann. 1998. Detection of inorganic wood preservatives on timber by near infrared spectroscopy, *J. Near Infrared Spectrosc.*, 6(A), A171–A173.

Fernandes, A., J. Lousada, J. Morais, J. Xavier, J. Pereira and P. Melo-Pinto. 2013a. Comparison between neural networks and partial least squares for intra-growth ring wood density measurement with hyperspectral imaging, *Comput. Electron. Agric.*, 94, 71–81.

Fernandes, A., J. Lousada, J. Morais, J. Xavier, J. Pereira and P. Melo-Pinto. 2013b. Measurement of intra-ring wood density by means of imaging VIS/NIR spectroscopy (hyperspectral imaging), *Holzforschung*, 67(1), 59–65.

Ferraz, A., R. Mendonca, A. Guerra, J. Ruiz, J. Rodriguez, J. Baeza, and J. Freer. 2004. Near-infrared spectra and chemical characteristics of *Pinus taeda* (Loblolly pine) wood chips biotreated by the white-rot fungus *Ceriporiopsis subvermispora*, *J. Wood Chem. Technol.*, 24(2), 99–113.

Finley, K., S. Chhin, P. Nzokou and J. O'Brien. 2016. Use of near-infrared spectroscopy as an indicator of emerald ash borer infestation in white ash stem tissue, *For. Ecol. Manage.*, 366, 41–52.

Flaete, P.O. and E.Y. Haartveit. 2004. Non-destructive prediction of decay resistance of *Pinus sylvestris* heartwood by near infrared spectroscopy, *Scand. J. For. Res.*, 19, 55–63.

Freeman, J.S., B.M. Potts, G.M. Downes, D. Pilbeam, S. Thavamanikumar and R.E. Vaillancourt. 2013. Stability of quantitative trait loci for growth and wood properties across multiple pedigrees and environments in *Eucalyptus globulus*, *New Phytol.*, 198(4), 1121–1134.

Freeman, J.S., S.P. Whittock, B.M. Potts, and R.E. Vaillancourt. 2009. QTL influencing growth and wood properties in *Eucalyptus globulus*, *Tree Genet. Genomes*, 5(4), 713–722.

Fujimoto, T., Y. Kurata, K. Matsumoto and S. Tsuchikawa. 2008. Application of near infrared spectroscopy for estimating wood mechanical properties of small clear and full length lumber specimens, *J. Near Infrared Spectrosc.*, 16(6), 529–537.

Fujimoto, T., Y. Kurata, K. Matsumoto and S. Tsuchikawa. 2010a. Feasibility of near-infrared spectroscopy for on-line grading of sawn lumber, *Appl. Spectrosc.*, 64(1), 92–99.

Fujimoto, T., Y. Kurata, K. Matsumoto and S. Tsuchikawa. 2010b. Feasibility of near-infrared spectroscopy for online multiple trait assessment of sawn lumber, *J. Wood Sci.*, 56(6), 452–459.

Gaspar, F., J. Lopes, H. Cruz, M. Schwanninger and J. Rodrigues. 2009. Application of near infrared spectroscopy and multivariate data analysis for the evaluation of glue lines of untreated and copper azole treated laminated timber before and after ageing, *Polym. Degrad. Stab.*, 94(7), 1061–1071.

Gierlinger, N., M. Schwanninger, B. Hinterstoisser and R. Wimmer. 2002. Rapid determination of heartwood extractives in Larix sp by means of Fourier transform near infrared spectroscopy, *J. Near Infrared Spectrosc.*, 10(3), 203–214.

Gierlinger, N., D. Jacques, M. Schwanninger, R. Wimmer, B. Hinterstoisser and L.E. Pâques. 2003. Rapid prediction of natural durability of larch heartwood using Fourier transform near-infrared spectroscopy, *Can. J. For. Res.*, 33(9), 1727–1736.

Gierlinger, N. and R. Wimmer. 2004. Radial distribution of heartwood extractives and lignin in mature European larch, *Wood Fiber Sci.*, 36(3), 387–394.

Gierlinger, N., M. Schwanninger, and R. Wimmer. 2004. Characteristics and classification of Fourier-transform near infrared spectra of the heartwood of different larch species (*Larix* sp.), *J. Near Infrared Spectrosc.*, 12(2), 113–119.

Giroud, G., J. Begin, M. Defo and C.H. Ung. 2017. Regional variation in wood density and modulus of elasticity of Quebec's main boreal tree species, *For. Ecol. Manage.*, 400, 289–299.

Greaves, B.L., N.M.G. Borralho and C.A. Raymond. 1997. Breeding objective for plantation eucalypts grown for production of kraft pulp, *For. Sci.*, 43(4), 465–472.

Groover, A. 2016. Gravitropisms and reaction woods of forest trees - evolution, functions and mechanisms, *New Phytol.*, 211(3), 790–802.

Hardner, C.M., A.L. Healey, G. Downes, M. Herberling and P.L. Gore. 2016. Improving prediction accuracy and selection of open-pollinated seed-lots in *Eucalyptus dunnii* Maiden using a multivariate mixed model approach, *Ann. For. Sci.*, 73(4), 1035–1046.

Hein, P.R.G., A.C.M. Campos, R.F. Mendes, L.M. Mendes and G. Chaix. 2011. Estimation of physical and mechanical properties of agro-based particleboards by near infrared spectroscopy, *Eur. J. Wood Wood Prod.*, 69(3), 431–442.

Hein, P.R.G., V.A. de Sa, L. Bufalino and L.M. Mendes. 2009. Calibrations based on near infrared spectroscopic data to estimate wood-cement panel properties, *Bioresources*, 4(4), 1620–1634.

Hodge, G.R. and W.C. Woodbridge. 2004. Use of near infrared spectroscopy to predict lignin content in tropical and sub-tropical pines, *J. Near Infrared Spectrosc.*, 12(6), 381–390.

Hodge, G.R. and W.C. Woodbridge. 2010. Global near infrared models to predict lignin and cellulose content of pine wood, *J. Near Infrared Spectrosc.*, 18(6), 367–380.

Hodges, R., G.A. Krishnagopalan and M. Pemberton. 1999. Implementation of AOTF-near-infrared spectroscopy for the on-line measurement of liquor composition in a continuous kraft digester, *Process Control News (for the Pulp and Paper Industries)*, 19(5), 2.

Hodges, R.E. and G.A. Krishnagopalan. 1999. Near-infrared spectroscopy for on-line analysis of white and green liquors, *Tappi J.*, 82(9), 101–106.

Hoffmeyer, P. and J.G. Pedersen. 1995. Evaluation of density and strength of Norway spruce wood by near-infrared reflectance spectroscopy, *Holz Roh Werkst.*, 53(3), 165–170.

Humphreys, J.R., J.M. O'Reilly-Wapstra, J.L. Harbard, N.W. Davies, A.R. Griffin, G.J. Jordan and B.M. Potts. 2008. Discrimination between seedlings of *Eucalyptus globulus*, E. nitens and their F(1) hybrid using near-infrared reflectance spectroscopy and foliar oil content, *Silvae Genet.*, 57(4–5), 262–269.

Hung, T.D., J.T. Brawner, R. Meder, D.J. Lee, S. Southerton, H.H. Thinh and M.J. Dieters. 2015. Estimates of genetic parameters for growth and wood properties in *Eucalyptus pellita* F. Muell. to support tree breeding in Vietnam, *Ann. For. Sci.*, 72(2), 205–217.

Jones, P.D., L.R. Schimleck, G.F. Peter, R.F. Daniels and A. Clark. 2005a. Nondestructive estimation of *Pinus taeda* L. tracheid morphological characteristics for samples from a wide range of sites in Georgia, *Wood Sci. Technol.*, 39(7), 529–545.

Jones, P.D., L.R. Schimleck, G.F. Peter, R.F. Daniels and A. Clark. 2005b. Nondestructive estimation of *Pinus taeda* L. wood properties for samples from a wide range of sites in Georgia, *Can. J. For. Res.*, 35(1), 85–92.

Jones, P.D., L.R. Schimleck, G.F. Peter, R.F. Daniels and A. Clark. 2006. Nondestructive estimation of wood chemical composition of sections of radial wood strips by diffuse reflectance near infrared spectroscopy, *Wood Sci. Technol.*, 40(8), 709–720.

Jones, P.D., L.R. Schimleck, C.L. So, A. Clark and R.F. Daniels. 2007. High resolution scanning of radial strips cut from increment cores by near infrared spectroscopy, *IAWA J.*, 28(4), 473–484.

Jones, T., R. Meder, C. Low, D. O'Callahan, C. Chittenden, N. Ebdon, A. Thumm and M. Riddell. 2011. Natural durability of the heartwood of coast redwood *Sequoia sempervirens* (D. Don) Endl. and its prediction using near infrared spectroscopy, *J. Near Infrared Spectrosc.*, 19(5), 381–389.

Jones, T.G., C.B. Low, R. Meder, D.R. O'Callahan, P.G. Milne, C.M. Chittenden, N. Ebdon and H.S. Dungey. 2013. Heartwood of *Cupressus lusitanica, C. macrocarpa*, Leyland and Ovens cypress and prediction of its durability using near-infrared spectroscopy, *Eur. J. Wood Wood Prod.*, 71(2), 183–192.

Kelley, S.S., J. Jellison and B. Goodell. 2002. Use of NIR and pyrolysis-MBMS coupled with multivariate analysis for detecting the chemical changes associated with brown-rot biodegradation of spruce wood, *FEMS Microbiol. Lett.*, 209(1), 107–111.

Kniest, C. 1992. Characteristic of urea resined wood particles by NIR spectroscopy, *Holz Roh Werkst.*, 50(2), 73–78.

Kobori, H., N. Gorretta, G. Rabatel, V. Bellon-Maurel, G. Chaix, J.M. Roger and S. Tsuchikawa. 2013. Applicability of Vis-NIR hyperspectral imaging for monitoring wood moisture content (MC), *Holzforschung*, 67(3), 307–314.

Kobori, H., T. Inagaki, T. Fujimoto, T. Okura and S. Tsuchikawa. 2015. Fast online NIR technique to predict MOE and moisture content of sawn lumber, *Holzforschung*, 69(3), 329–335.

Kobori, H., S. Higa, S. Tsuchikawa, Y. Kojima and S. Suzuki. 2017. Segregating wood wastes by repetitive principal component analysis of near infrared spectra, *J. Near Infrared Spectrosc.*, 25(3), 180–187.

Kohan, N.J., B.K. Via and S.E. Taylor. 2012. Prediction of strand feedstock mechanical properties with near infrared spectroscopy, *Bioresources*, 7(3), 2996–3007.

Leblon, B., O. Adedipe, G. Hans, A. Haddadi, S. Tsuchikawa, J. Burger, R. Stirling, Z. Pirouz, K. Groves, J. Nader and A. LaRocque. 2013. A review of near-infrared spectroscopy for monitoring moisture content and density of solid wood, *Forest. Chron.*, 89(5), 595–606.

Lindgren, T. and U. Edlund. 1998. Prediction of lignin content and pulp yield from black liquor composition using near-infrared spectroscopy and partial least squares regression, *Nord. Pulp Pap. Res. J.*, 13(1), 76–80.

Ling, Z., Y. Omura, N. Hori, T. Iwata and A. Takemura. 2018. In-situ chemical structure analysis of aqueous vinyl polymer solution-isocyanate adhesive in post-cure process by using Fourier transform near infrared spectroscopy, *Int. J. Adhes. Adhes.*, 81, 56–64.

Ma, T., T. Inagaki and S. Tsuchikawa. 2017. Calibration of SilviScan data of *Cryptomeria japonica* wood concerning density and microfibril angles with NIR hyperspectral imaging with high spatial resolution, *Holzforschung*, 71(4), 341–347.

Marklund, A., M. Paper, J.B. Hauksson and M. Sjostrom. 1999. Prediction of strength parameters for softwood kraft pulps. Multivariate data analysis based on orthogonal signal correction and near infrared spectroscopy, *Nord. Pulp Pap. Res. J.*, 14(2), 140–148.

Mauruschat, D., B. Plinke, J. Aderhold, J. Gunschera, P. Meinlschmidt and T. Salthammer. 2016. Application of near-infrared spectroscopy for the fast detection and sorting of wood-plastic composites and waste wood treated with wood preservatives, *Wood Sci. Technol.*, 50(2), 313–331.

McLean, J.P., G. Jin, M. Brennan, M.K. Nieuwoudt and P.J. Harris. 2014. Using NIR and ATR-FTIR spectroscopy to rapidly detect compression wood in *Pinus radiata*, *Can. J. For. Res.*, 44(7), 820–830.

Meder, R., S. Gallagher, M.O. Kimberley, J.M. Uprichard and K.L. Mackie. 1994. Prediction of wood chip and pulp properties via multivariate analysis of spectral data, *Proc. 48th Appita Gen. Conf.*, Melbourne, Australia, 479–484.

Meder, R., A. Thumm and H. Bier. 2002. Veneer stiffness predicted by NIR spectroscopy calibrated using mini-LVL test panels, *Holz Roh Werkst.*, 60(3), 159–164.

Meder, R., A. Thumm and D. Marston. 2003. Sawmill trial of at-line prediction of recovered lumber stiffness by NIR spectroscopy of *Pinus radiata* cants, *J. Near Infrared Spectrosc.*, 11(2), 137–143.

Meder, R., D. Marston, N. Ebdon and R. Evans. 2010. Spatially-resolved radial scanning of tree increment cores for near infrared prediction of microfibril angle and chemical composition, *J. Near Infrared Spectrosc.*, 18(6), 499–505.

Meder, R., J.T. Brawner, G.M. Downes and N. Ebdon. 2011. Towards the in-forest assessment of Kraft pulp yield: Comparing the performance of laboratory and hand-held instruments and their value in screening breeding trials, *J. Near Infrared Spectrosc.*, 19(5), 421–429.

Meder, R. and R.R. Meglen. 2012. Near infrared spectroscopic and hyperspectral imaging of compression wood in *Pinus radiata* D. Don, *J. Near Infrared Spectrosc.*, 20(5), 583–589.

Meder, R. and N. Edbon. 2013. *Quality Control of H2F Treated Timber via NIR Spectroscopy*, p. 45, Forest & Wood Products Australia, Melbourne.

Meder, R., D. Kain, N. Ebdon, P. Macdonell and J.T. Brawner. 2014. Identifying hybridisation in *Pinus* species using near infrared spectroscopy of foliage, *J. Near Infrared Spectrosc.*, 22(5), 337–345.

Michell, A.J. 1995. Pulpwood quality estimation by near-infrared spectroscopic measurements on eucalypt woods, *Appita J.*, 48(6), 425–428.

Monrroy, M., R.T. Mendonca, J. Ruiz, J. Baeza and J. Freer. 2009. Estimating glucan, xylan, and methylglucuronic acids in Kraft pulps of *Eucalyptus globulus* using FT-NIR spectroscopy and multivariate analysis, *J. Wood Chem. Technol.*, 29(2), 150–163.

Mora, C.R. and L.R. Schimleck. 2008. On the selection of samples for multivariate regression analysis: Application to near-infrared (NIR) calibration models for the prediction of pulp yield in *Eucalyptus nitens*, *Can. J. For. Res.*, 38(10), 2626–2634.

Mora, C.R. and L.R. Schimleck. 2009. Determination of within-tree variation of *Pinus taeda* wood properties by near infrared spectroscopy. Part 2: Whole-tree wood property maps, *Appita J.*, 62(3), 232–238.

Mora, C.R. and L.R. Schimleck. 2010. Kernel regression methods for the prediction of wood properties of *Pinus taeda* using near infrared spectroscopy, *Wood Sci. Technol.*, 44(4), 561–578.

Mora, C.R., L.R. Schimleck and F. Isik. 2008. Near infrared calibration models for the estimation of wood density in *Pinus taeda* using repeated sample measurements, *J. Near Infrared Spectrosc.*, 16(6), 517–528.

Mora, C.R., L.R. Schimleck, S.C. Yoon and C.N. Thai. 2011. Determination of basic density and moisture content of loblolly pine wood disks using a near infrared hyperspectral imaging system, *J. Near Infrared Spectrosc.*, 19(5), 401–409.

Muneri, A., T. Asada, K. Tomita, K. Kusunoki and C. Szota 2011. Between-tree variation in stem volume, wood density, fibre length and Kraft pulping properties of *Eucalyptus globulus* and the utility of field-portable NIR Spectroscopy and wood cores in evaluating pulpwood quality properties of standing trees, *Appita J.*, 64(4), 356–361.

Nabavi, M., J. Dahlen, L. Schimleck, T.L. Eberhardt and C. Montes. 2018. Regional calibration models for predicting loblolly pine tracheid properties using near infrared spectroscopy, *Wood Sci. Technol.*, 52(2), 445–463.

Nair, P. and R.A. Lodder. 1993. Near-IR identification of woods for restoration of historic buildings and furniture, *Appl. Spectrosc.*, 47(3), 287–291.

Niemz, P., F. Dutschmann and B. Stolken. 1994. Using NIR-spectroscopy for evaluating resin content in chip mixtures, *Holz Roh Werkst.*, 52(1), 6–8.

Niemz, P., S. Korner, O. Wienhaus, W. Flamme and M. Balmer. 1992. Applying NIR spectroscopy for evaluation of the hardwood softwood ratio and resin content in chip mixtures, *Holz Roh Werkst.*, 50(1), 25–28.

Norman, E., T. Iversen, T. Lindgren and U. Edlund 1994. Addition of methanol to a kraft cook. A study of its effect on delignification, *Nord. Pulp Pap. Res. J.*, 9(1), 49–52.

Olsson, R.J.O., P. Tomani, M. Karlsson, T. Joseffson, K. Sjoberg and C. Bjorklund. 1995. Multivariate characterization of chemical and physical descriptors in pulp using NIRR, *Tappi J.*, 78(10), 158–166.

Pâques, L.E. and J.P. Charpentier. 2015. Perspectives for genetic improvement in heartwood size and extractive content in relation to natural durability and aesthetics in interspecific hybrid larch (*Larix* x *eurolepis*), *Eur. J. For. Res.*, 134(5), 857–868.

Pastore, T.C.M., J.W.B. Braga, V.T.R. Coradin, W.L.E. Magalhaes, E.Y.A. Okino, J.A.A. Camargos, G.I.B. de Muniz, O.A. Bressan and F. Davrieux. 2011. Near infrared spectroscopy (NIRS) as a potential tool for monitoring trade of similar woods: Discrimination of true mahogany, cedar, andiroba, and curupixa, *Holzforschung*, 65(1), 73–80.

Patrick, K.L. 1998. New sensor technology holds promise for better control of chemical fiberline, Part 1, *Pulp Pap.*, 72(3), 39–42, 47.

Perez, D.D., A. Guillemain, P. Alazard, C. Plomion, P. Rozenberg, J.C. Rodrigues, A. Alves and G. Chantre. 2007. Improvement of *Pinus pinaster* Ait elite trees selection by combining near infrared spectroscopy and genetic tools, *Holzforschung*, 61(6), 611–622.

Poke, F.S. and C.A. Raymond. 2006. Predicting extractives, lignin, and cellulose contents using near infrared spectroscopy on solid wood in *Eucalyptus globulus*, *J. Wood Chem. Technol.*, 26(2), 187–199.

Prades, C., I. Gomez-Sanchez, J. Garcia-Olmo and J.R. Gonzalez-Adrados. 2012. Discriminant analysis of geographical origin of cork planks and stoppers by near infrared spectroscopy, *J. Wood Chem. Technol.*, 32(1), 54–70.

Raymond, C., L. Schimleck, A. Muneri and A.J. Michell. 2001. Genetic parameters and genotype-by-environment interactions for pulp yield predicted using near infrared reflectance analysis and pulp productivity in *Eucalyptus globulus*, *Forest Genetics*, 8(3), 213–224.

Raymond, C.A. and L.R. Schimleck. 2002. Development of near infrared reflectance analysis calibrations for estimating genetic parameters for cellulose content in *Eucalyptus globulus*, *Can. J. For. Res.*, 32(1), 170–176.

Rials, T.G., S.S. Kelley and C.L. So. 2002. Use of advanced spectroscopic techniques for predicting the mechanical properties of wood composites, *Wood Fiber Sci.*, 34(3), 398–407.

Sandak, A., J. Sandak and M. Negri. 2011. Relationship between near-infrared (NIR) spectra and the geographical provenance of timber, *Wood Sci. Technol.*, 45(1), 35–48.

Sandak, A., J. Sandak and O. Allegretti. 2015a. Quality control of vacuum thermally modified wood with near infrared spectroscopy, *Vacuum*, 114, 44–48.

Sandak, A., J. Sandak and M. Riggio. 2015b. Estimation of physical and mechanical properties of timber members in service by means of infrared spectroscopy, *Constr. Build. Mater.*, 101, 1197–1205.

Sandak, A., O. Allegretti, I. Cuccui, J. Sandak, L. Rosso, G. Castro, F. Negro, C. Cremonini and R. Zanuttini. 2016a. Thermo-vacuum modification of poplar veneers and its quality control, *Bioresources*, 11(4), 10122–10139.

Sandak, A., J. Sandak, K. Bohm, A. Zitek and B. Hinterstoisser. 2016b. Near infrared spectroscopy as a tool for in-field determination of log/biomass quality index in mountain forests, *J. Near Infrared Spectrosc.*, 24(6), 587–594.

Sandak, A., J. Sandak, I. Burud and L.R. Gobakken. 2016c. Weathering kinetics of thin wood veneers assessed with near infrared spectroscopy, *J. Near Infrared Spectrosc.*, 24(6), 549–553.

Saucedo, V.M. and G.A. Krishnagopalan. 2000. Application of in-situ near infrared analysis for the measurement of cooking liquor components during kraft pulping, *J. Pulp Pap. Sci.*, 26(1), 25–30.

Scheffer, T.C. and J.J. Morrell. 1998. *Natural Durability of Wood: A Worldwide Checklist of Species*, p. 58, Oregon State University, Corvallis, OR.

Schimleck, L.R., A.J. Michell and P. Vinden. 1996. Eucalypt wood classification by NIR spectroscopy and principal components analysis, *Appita J.*, 49(5), 319–324.

Schimleck, L.R. and A.J. Michell. 1998. Determination of within-tree variation of kraft pulp yield using near-infrared spectroscopy, *Tappi J.*, 81(5), 229–236.

Schimleck, L.R., C.A. Raymond, C.L. Beadle, G.M. Downes, P.D. Kube and J. French. 2000. Applications of NIR spectroscopy to forest research, *Appita J.*, 53(6), 458–464.

Schimleck, L.R., R. Evans, J. Ilic and A.C. Matheson. 2002. Estimation of wood stiffness of increment cores by near-infrared spectroscopy, *Can. J. For. Res.*, 32(1), 129–135.

Schimleck, L.R. and Y. Yazaki. 2003a. Analysis of black wattle (*Acacia mearnsii* de wild) bark by near infrared spectroscopy, *Holzforschung*, 57(5), 527–532.

Schimleck, L.R. and Y. Yazaki. 2003b. Analysis of *Pinus radiata* D. Don bark by near infrared spectroscopy, *Holzforschung*, 57(5), 520–526.

Schimleck, L.R. and R. Evans. 2004. Estimation of *Pinus radiata* D. Don tracheid morphological characteristics by near infrared spectroscopy, *Holzforschung*, 58(1), 66–73.

Schimleck, L.R. and J.J. Workman. 2004. In *Near-Infrared Spectroscopy in Agriculture, A Tri-Societies Monograph*, C. Roberts, J.J. Workman and J. Reeves (eds.), pp. 635–646, American Society of Agronomy, Crop Science Society of America and Soil Science Society of America, Madison, WI.

Schimleck, L.R., P.D. Kube and C.A. Raymond. 2004. Genetic improvement of kraft pulp yield in *Eucalyptus nitens* using cellulose content determined by near infrared spectroscopy, *Can. J. For. Res.*, 34(11), 2363–2370.

Schimleck, L.R., P. Payne and R.H. Wearne. 2005a. Determination of important pulp properties of hybrid poplar by near infrared spectroscopy. *Wood Fiber Sci.*, 37(3), 462–471.

Schimleck, L.R., P.D. Kube, C.A. Raymond, A.J. Michell and J. French. 2005b. Estimation of whole-tree kraft pulp yield of *Eucalyptus nitens* using near infrared spectra collected from increment cores. *Can. J. For. Res.*, 35(12), 2797–2805.

Schimleck, L.R., G.D.S.P. Rezende, B.J. Demuner and G.M. Downes. 2006. Estimation of whole-tree wood quality traits using near infrared spectra collected from increment cores. *Appita J.*, 59(3), 231–236.

Schimleck, L.R., G.R. Hodge and W. Woodbridge. 2010. Toward global calibrations for estimating the wood properties of tropical, sub-tropical and temperate pine species, *J. Near Infrared Spectrosc.*, 18(6), 355–365.

Schimleck, L.R., J.L.M. Matos, R. Trianoski and J.G. Prata. 2018. Comparison of methods for estimating mechanical properties of wood by NIR spectroscopy, *J. Spectrosc.* Article ID 4823285, 10 pages.

Schultz, T.P. and D.A. Burns. 1990. Rapid secondary analysis of lignocellulose - Comparison of near-infrared (NIR) and Fourier-transform infrared (FTIR), *Tappi J.*, 73(5), 209–212.

Schwanninger, M., B. Hinterstoisser, N. Gierlinger, R. Wimmer and J. Hanger. 2004a. Application of Fourier transform near infrared Spectroscopy (FT-NIR) to thermally modified wood, *Holz Roh Werkst.*, 62(6), 483–485.

Schwanninger, M., B. Hinterstoisser, C. Gradinger, K. Messner and K. Fackler. 2004b. Examination of spruce wood biodegraded by *Ceriporiopsis subvermispora* using near and mid infrared spectroscopy, *J. Near Infrared Spectrosc.*, 12(6), 397–409.

Schwanninger, M., B. Stefke and B. Hinterstoisser. 2011. Qualitative assessment of acetylated wood with infrared spectroscopic methods, *J. Near Infrared Spectrosc.*, 19(5), 349–357.

Sjoblom, E., B. Johnsson and H. Sundstrom. 2004. Optimization of particleboard production using NIR spectroscopy and multivariate techniques, *For. Prod. J.*, 54(6), 71–75.

Skvaril, J., K. Kyprianidis, A. Avelin, M. Odlare and E. Dahlquist. 2017. Fast determination of fuel properties in solid biofuel mixtures by near infrared spectroscopy, *Energy Procedia*, 105, 1309–1317.

Smeland, K.A., K.H. Liland, J. Sandak, A. Sandak, L.R. Gobakken, T.K. Thiis and I. Burud. 2016. Near infrared hyperspectral imaging in transmission mode: Assessing the weathering of thin wood samples, *J. Near Infrared Spectrosc.*, 24(6), 595–604.

So, C.L. and T.L. Eberhardt. 2006. Rapid analysis of inner and outer bark composition of southern yellow pine bark from industrial sources, *Holz Roh Werkst.*, 64(6), 463.

So, C.L., S.T. Lebow, L.H. Groom and T.G. Rials. 2004. The application of near infrared (NIR) spectroscopy to inorganic preservative-treated wood, *Wood Fiber Sci.*, 36(3), 329–336.

Soares, L.F., D.C. da Silva, M.C.J. Bergo, V.T.R. Coradin, J.W.B. Braga and T.C.M. Pastore. 2017. Evaluation of a NIR handheld device and PLS-DA for discrimination of six similar Amazonian wood species, *Quim. Nova*, 40(4), 418–426.

Southerton, S.G., C.P. MacMillan, J.C. Bell, N. Bhuiyan, G. Downes, I.C. Ravenwood, K.R. Joyce, D. Williams and B.R. Thumma. 2010. Association of allelic variation in xylem genes with wood properties in *Eucalyptus nitens*, *Aust. Forestry*, 73(4), 259–264.

Stackpole, D.J., R.E. Vaillancourt, A. Alves, J. Rodrigues and B.M. Potts. 2011. Genetic variation in the chemical components of *Eucalyptus globulus* Wood, *G3*, 1(2), 151–159.

Stackpole, D.J., R.E. Vaillancourt, G.M. Downes, C.E. Harwood and B.M. Potts. 2010. Genetic control of kraft pulp yield in *Eucalyptus globulus*, *Can. J. For. Res.*, 40(5), 917–927.

Stefansson, P., I. Burud, T. Thiis, L. Gobakken and E. Larnøy. 2018. Estimation of phosphorus-based flame retardant in wood by hyperspectral imaging—a new method, *J. Spectral Imaging*, 7(1), a3.

Stefke, B., E. Windeisen, M. Schwanninger and B. Hinterstoisser. 2008. Determination of the weight percentage gain and of the acetyl group content of acetylated wood by means of different infrared spectroscopic methods, *Anal. Chem.*, 80(4), 1272–1279.

Sugii, S., T. Fujimoto, H. Tsutsumi, T. Inagaki and S. Tsuchikawa. 2017. Dynamic behavior of wood chemical components under drying process measured by near infrared spectroscopy, *J. Near Infrared Spectrosc.*, 25(6), 400–406.

Sykacek, E., N. Gierlinger, R. Wimmer and M. Schwanninger. 2006. Prediction of natural durability of commercial available European and Siberian larch by near-infrared spectroscopy, *Holzforschung*, 60(6), 643–647.

Taylor, A. and J. Lloyd. 2007. Potential of near infrared spectroscopy to quantify boron concentration in treated wood, *For. Prod. J.*, 57(1–2), 116–117.

Taylor, A.M., C. Freitag, E. Cadot and J.J. Morrell. 2008. Potential of near infrared spectroscopy to assess hot-water-soluble extractive content and decay resistance of a tropical hardwood, *Holz Roh Werkst.*, 66(2), 107–111.

Taylor, A.M., N. Labbe and A. Noehmer. 2011. NIR-based prediction of extractives in American white oak heartwood, *Holzforschung*, 65(2), 185–190.

Tham, V.T.H., T. Inagaki and S. Tsuchikawa. 2018. A novel combined application of capacitive method and near-infrared spectroscopy for predicting the density and moisture content of solid wood, *Wood Sci. Technol.*, 52(1), 115–129.

Thamarus, K., K. Groom, A. Bradley, C.A. Raymond, L.R. Schimleck, E.R. Williams and G.F. Moran. 2004. Identification of quantitative trait loci for wood and fibre properties in two full-sib pedigrees of *Eucalyptus globulus*, *Theor. Appl. Genet.*, 109(4), 856–864.

Thavamanikumar, S., L.J. McManus, P.K. Ades, G. Bossinger, D.J. Stackpole, R. Kerr, S. Hadjigol, J.S. Freeman, R.E. Vaillancourt, P. Zhu and J.F.G. Tibbits. 2014. Association mapping for wood quality and growth traits in *Eucalyptus globulus* ssp. *globulus* Labill identifies nine stable marker-trait associations for seven traits, *Tree Genet. Genomes*, 10(6), 1661–1678.

Thumm, A., M. Riddell, B. Nanayakkara, J. Harrington and R. Meder. 2010. Near infrared hyperspectral imaging applied to mapping chemical composition in wood samples, *J. Near Infrared Spectrosc.*, 18(6), 507–515.

Thumm, A., M. Riddell, B. Nanayakkara, J. Harrington and R. Meder. 2016. Mapping within-stem variation of chemical composition by near infrared hyperspectral imaging, *J. Near Infrared Spectrosc.*, 24(6), 605–616.

Thumm, A. and M. Riddell. 2017. Resin defect detection in appearance lumber using 2D NIR spectroscopy, *Eur. J. Wood Wood Prod.*, 75(6), 995–1002.

Thumma, B.R., B.A. Matheson, D.Q. Zhang, C. Meeske, R. Meder, G.M. Downes, and S.G. Southerton. 2009. Identification of a *cis*-acting regulatory polymorphism in a eucalypt *COBRA*-like gene affecting cellulose content, *Genetics*, 183(3), 1153–1164.

Thumma, B.R., S.G. Southerton, J.C. Bell, J.V. Owen, M.L. Henery and G. Moran. 2010. Quantitative trait locus (QTL) analysis of wood quality traits in *Eucalyptus nitens*, *Tree Genet. Genomes*, 6(2), 305–317.

Thygesen, L.G. 1995. Determination of dry matter content and basic density of Norway spruce by near infrared reflectance and transmittance spectroscopy, *J. Near Infrared Spectrosc.*, 2(3), 127–135.

Thygesen, L.G. and S.-O. Lundqvist. 2000a. NIR measurement of moisture content in wood under unstable temperature conditions. Part 1. Thermal effects in near infrared spectra of wood, *J. Near Infrared Spectrosc.*, 8(3), 183–189.

Thygesen, L.G. and S.-O. Lundqvist. 2000b. NIR measurement of moisture content in wood under unstable temperature conditions. Part 2. Handling temperature fluctuations, *J. Near Infrared Spectrosc.*, 8(3), 191–199.

Trung, T., B. Conroy and D. Wilson. 2015. NIR spectroscopy: A paradigm shift in pulp properties measurements, *Pulp Pap. Canada.*, 116(6), 24–25.

Tsuchikawa, S., M. Torii and S. Tsutsumi. 1996. Application of near infrared spectrophotometry to wood. 4. Calibration equations for moisture content, *Mokuzai Gakkaishi* 42(8), 743–754.

Tsuchikawa, S. and S. Tsutsumi. 1998. Adsorptive and capillary condensed water in biological material, *J. Mater. Sci. Lett.*, 17(8), 661–663.

Tsuchikawa, S. and H.W. Siesler 2003a. Near-infrared spectroscopic monitoring of the diffusion process of deuterium-labeled molecules in wood. Part I: Softwood, *Appl. Spectrosc.*, 57(6), 667–674.

Tsuchikawa, S. and H.W. Siesler. 2003b. Near-infrared spectroscopic monitoring of the diffusion process of deuterium-labeled molecules in wood. Part II: Hardwood, *Appl. Spectrosc.*, 57(6), 675–681.

Tsuchikawa, S., K. Inoue, J. Noma and K. Hayashi. 2003a. Application of near-infrared spectroscopy to wood discrimination, *J. Wood Sci.*, 49(1), 29–35.

Tsuchikawa, S., K. Yamato and K. Inoue. 2003b. Discriminant analysis of wood-based materials using near-infrared spectroscopy, *J. Wood Sci.*, 49(3), 275–280.

Tsuchikawa, S. and H. Kobori. 2015. A review of recent application of near infrared spectroscopy to wood science and technology, *J. Wood Sci.*, 61(3), 213–220.

Tyson, J.A., L.R. Schimleck, A.M. Aguiar, J.I.M. Abad, G. Rezende and O.M. Filho. 2012. Development of near infrared calibrations for physical and mechanical properties of eucalypt pulps of mill-line origin, *J. Near Infrared Spectrosc.*, 20(2), 287–294.

Tyson, J.A., L.R. Schimleck, A.M. Aguiar, J.I.M. Abad, G. Rezende, O.M. Filho and C.R. Mora. 2010. Comparison of sample preparation methods for NIR analysis of carbohydrate content of unbleached eucalyptus pulps, *J. Wood Chem. Technol.*, 30(3), 283–298.

Vanchinathan, S., S. Ananth, J.M. Jarvis and G.A. Krishnagopalan. 1996. Kraft-liquor alkali analysis using near-infrared spectroscopy, *Tappi J.*, 79(10), 187–191.

Varghese, M., C.E. Harwood, D.J. Bush, B. Baltunis, R. Kamalakannan, P.G. Suraj, D. Hegde and R. Meder. 2017. Growth and wood properties of natural provenances, local seed sources and clones of *Eucalyptus camaldulensis* in southern India: Implications for breeding and deployment, *New Forest.*, 48(1), 67–82.

Via, B.K., C.L. So, L.G. Eckhard, T.F. Shupe, L.H. Groom and M. Stine. 2008. Response of near infrared diffuse reflectance spectra to blue stain and wood age, *J. Near Infrared Spectrosc.*, 16(1), 71–74.

Viana, L.C., G.I.B. de Muniz, P.R.G. Hein, W.L.E. Magalhaes and M.E. Carneiro. 2016. NIR spectroscopy can evaluate the crystallinity and the tensile and burst strengths of nanocellulosic films, *Maderas-Cienc. Tecnol.*, 18(3), 493–504.

Wallbäcks, L., U. Edlund, B. Norden and I. Berglund. 1991a. Multivariate characterization of pulp using solid-state 13CNMR, FTIR, and NIR, *Tappi J.*, 74(10), 201–206.

Wallbäcks, L., U. Edlund, B. Nordén, T. Iversen and U. Mohlin. 1991b. Multivariate characterization of pulp, Part 2. Interpretation and prediction of beating processes, *Nord. Pulp Paper Res. J.*, 6(3), 104–109.

Wallbäcks, L., U. Edlund, T. Lindgren and R. Agnemo. 1995. Multivariate characterization of pulp Part 3. Evaluation of physical and spectroscopic data on unbeaten and beaten kraft pulp samples, *Nord. Pulp Paper Res. J.*, 10, 88–93.

Watanabe, K., F. Hart, S.D. Mansfield and S. Avramidis. 2010. Detection of wet-pockets on the surface of *Tsuga heterophylla* (Raf.) Sarg. by near infrared (NIR) spectroscopy, *Holzforschung*, 64(1), 55–60.

Watanabe, K., S.D. Mansfield and S. Avramidis. 2011. Application of near-infrared spectroscopy for moisture-based sorting of green hem-fir timber, *J. Wood Sci.*, 57(4), 288–294.

Watanabe, K., I. Kobayashi, S. Saito, N. Kuroda and S. Noshiro. 2012a. Nondestructive evaluation of drying stress level on wood surface using near-infrared spectroscopy, *Wood Sci. Technol.*, 47(2), 299–315.

Watanabe, K., S.D. Mansfield and S. Avramidis. 2012b. Wet-pocket classification in *Abies lasiocarpa* using spectroscopy in the visible and near infrared range, *Eur. J. Wood Wood Prod.*, 70(1–3), 61–67.

Wentzel-Vietheer, M., R. Washusen, G.M. Downes, C. Harwood, N. Ebdon, B. Ozarska and T. Baker. 2013. Prediction of non-recoverable collapse in *Eucalyptus globulus* from near infrared scanning of radial wood samples, *Eur. J. Wood Wood Prod.*, 71(6), 755–768.

Wright, J.A., M.D. Birkett and M.J.T. Gambino. 1990. Prediction of pulp yield and cellulose content from wood samples using near infrared reflectance spectroscopy, *Tappi J.*, 73(8), 164–166.

Yuzak, E. and C. Lohrke. 1993. *At-Line Kappa Number Measurement by Near-Infrared Spectroscopy*, pp. 663–671, Tappi Press, Atlanta, GA.

38 NIR Spectroscopy Applied to the Oil and Gas Industry

Sébastien Gourvénec and Nicolas Bernard*
Total Refining & Chemicals

CONTENTS

38.1 INTRODUCTION

Applications of NIR spectroscopy in the oil and gas industry are usually mature and used in the field for quite a long period already. The use of NIR spectroscopy in petroleum industries has indeed dramatically increased in the past 25 years. This can be explained by the fact that the NIR technology has the capability for fast, nondestructive, online, and simultaneous multi-compositional/properties analyses. Refinery or petrochemical products are highly complex. Their chemical compositions usually depend on many different factors, and samples can be gases, liquids such as gasoline, or even solids such as polymers. This chapter will focus on aspects that are extremely important for successful applications of the technology in the oil and gas sector. Some keys steps, ranging from sampling and hardware to chemometrics, are crucial to avoid unsatisfactory experiences with NIR applications. It is also very useful to be fully aware and clear about some important considerations to prevent from unsuccessful deployment of the technology. Efforts are especially to be made on communication (not to oversell the capabilities of the NIR technology, not to see it as a "black box") and training (to address expertise requirements). NIR technology will then be accepted by the business and will deliver the expected benefits. This chapter is particularly focused on liquids and solids, and not on gases which were deeply described in the previous edition of this Handbook.

* sebastien.gourvenec@total.com

38.2 GENERALITIES ON THE USE OF NIR SPECTROSCOPY IN THE REFINING INDUSTRY

38.2.1 NIR Spectra and Regions of Interest

In the refining industry, as in almost every other application, NIR spectra cannot be used directly to obtain the information of interest. Whereas with mid-infrared (MIR) or Raman spectroscopy, a specialist can identify the molecular bonds present in the sample just by inspecting the position of bands in the spectrum, it is almost impossible to do so in the NIR region. This is due to the fact that a NIR spectrum (from 780 nm/12,800 cm^{-1} to 2500 nm/4000 cm^{-1}) shows the harmonics and combinations of the stretching and bending vibrations seen in MIR, resulting in broad and undefined absorption bands. A single band can be attributable to several combinations and overtone vibrations.

The most dominant harmonic and combination bands in NIR come from X-H vibrations (C-H, O-H, N-H bonds). Since refining products are composed of a wide range of hydrocarbon molecules (aromatics, paraffins, olefins, etc.), NIR spectroscopy is well suited to the analysis of most of these products. It can be interesting to have a look at spectra of some pure hydrocarbons that can be found in refining products. Figure 38.1 presents spectra of five hydrocarbon molecules representative of the different families of molecules present in refining products (decane, which is representative of the paraffin family; iso-octane of iso-paraffins; octene of olefins; methylcyclohexane of naphthenes; and toluene of aromatics).

Four major areas with absorption bands can be noticed: combination bands: 4000–5000 cm^{-1}; first overtone: 5200–6400 cm^{-1}; second overtone: 6400–7600 cm^{-1} and 8000–9000 cm^{-1}. It is important to keep in mind that it is not possible to successfully exploit all of them at the same time because of the important differences in absorbance intensities. In the industry, different cell pathlengths are used for each of these areas in order to optimize the absorbance intensity to be around one. As an example, typical pathlengths used for clear liquid products are 10 mm for the second overtones, 1 or 2 mm for the first overtones, and 0.5 mm for combinations. For dark and black products (crude oils, greases, etc.), the pathlength would be shorter. The spectra shown in Figure 38.1 have been acquired

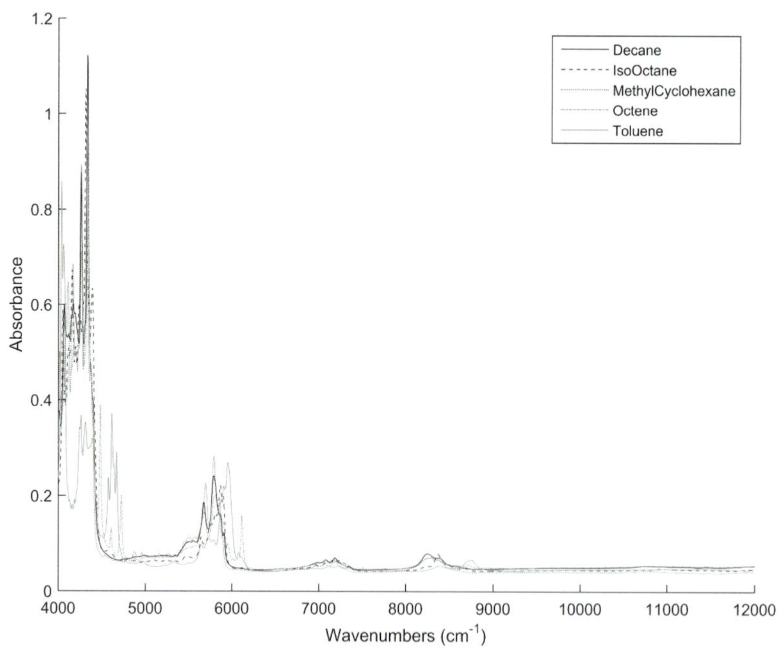

FIGURE 38.1 NIR spectra of different hydrocarbons.

with a 0.5 mm pathlength. The combination region is the most selective one, allowing resolution of minor spectral features [1].

38.2.2 Acquisition Systems

When it comes to analyzing liquids or solids by NIR spectroscopy, spectral acquisition quality must be taken into account and optimized. This is extremely important to obtain NIR spectra of good quality that can later on reveal a lot of useful information using chemometrics.

Several types of instruments, as well as sampling systems, exist (detailed in the instrumentation chapters). FT-NIR instruments are the most commonly used, as they provide accurate and precise spectra and are fast and quite robust. Most of the considerations of this chapter apply to these types of instruments. NIR instruments also allow remote sampling capabilities. Using optical fibers is indeed a possibility to analyze remotely, away from hazardous, radioactive, or explosive environments. As previously described in Section 38.2.1 of this chapter, as important information is present in combination band regions, samples are often sent to the spectrometer, avoiding the use of optical fibers. Even if there are different qualities of optical fibers, there is often a cutoff region that does not provide any satisfactory quality spectra in the combination band regions. Instead, (liquid) samples, such as crude oils and clear products, are usually measured in a transmission cell, with an adjustable pathlength. The pathlength is a key parameter of the measurement and has to be optimized depending on the spectral region of interest, the product to be analyzed, and the application (Section 38.2.1 of this chapter). For solids, such as polymers, greases, and bitumen, the situation is different as measurement is usually performed using reflection of the light on the sample. NIR spectra could be obtained using optical fibers or with specific accessories depending on vendors (rotating cup, for instance). Some general principles remain, though, such as averaging several spectra to improve the quality of the information. A "best" compromise is to be found between the number of scans, some acquisition steps, and the total acquisition time. Also, if possible, especially on heterogeneous samples, spectra must be acquired on different parts of the sample. This will lead to more representative spectra of the sample. What is important when acquiring spectra is the adequacy between best spectral quality that can be obtained on a given sample and the way it is measured with appropriate accessories (probes, cells, etc.).

38.2.3 Deployment and Installation

Generally speaking, NIR spectroscopy is deployed in refineries in strategic points to measure continuously and rapidly, in real time, and in a nondestructive way, some key properties. They are used for monitoring, controlling, or even optimizing refining processes. This technology can then be applied on various products, ranging from crudes oils to finished products, not forgetting intermediate products. All these samples contain a highly complex mixture of hydrocarbons and heteroatomic organic compounds of varying molecular weights and polarities. NIR technology, through the use of correlative chemometric models, can be utilized for the measurement of diverse properties in various processes. It has the advantage of being "multi-properties"; i.e., for a given set of spectra, if several models are developed, a single analyzer can provide many properties simultaneously and bring advantages in terms of costs, footprint, etc., over many other analyzers.

Deploying the NIR technology requires particular attention, though, as it deals with transferring the whole NIR method. The NIR method is composed of, to make it simple, equipment with an appropriate measurement accessory, a measurement method with optimized parameters, and a model. Traditional approach consists in developing the measurement method (which comprises the spectral acquisition and the modeling through chemometrics) in a lab and transferring it to a pilot or a plant. Preparing and developing the application offline can allow the optimization the acquisition of spectra and preparing the calibration database (described in Section 38.3.3 of this chapter). When the application is mature enough, it can be transferred at-line (close to the process), on-line, or in-line. Transfer of the method can be performed directly in some particular cases (standardization of the equipment and/or of the measurement method) or with appropriate methodology. This methodology involves usually pretreatment of spectra as well as chemometric model transfer techniques that minimize the impact of the difference between the lab and the analyzer methods [2–4]. It is to be noted that if the calibration model is altered during the transfer, it needs to be validated again. In the oil and gas industry, in-line applications (measuring censor is introduced in the process) are quite rare, mainly for security reasons. Refining environments are indeed extremely dangerous and several regulations are in place when dealing with analyzers (such as NIR spectrometers) systems. Risks and toxicological analysis must be performed, and ATEX (explosive atmosphere) constraints must be taken into account. NIR spectrometers can be installed in analyzer shelters to prevent explosions from occurring. Most of the equipments are for on-line use, which is, as described in Section 38.2.2 of this chapter, when the samples are sent to the instrument, in real time through a sample loop or re-circulating loop. This is the most deployed solution for liquids in refining environments. It has to be mentioned that for a single instrument, multiple cells can be used for different lines in a process or even different products. This minimizes the cost by sharing an instrument for different measurements that are made alternatively and sequentially. For solids, both on-line (when feasible) and at-line (measurement is made very close to the process with sample taken from it) exist and are used.

38.3 SPECIFICITIES OF NIR SPECTROSCOPY ON PETROLEUM PRODUCTS

Main specificities of the use of NIR spectroscopy on petroleum products (compared to other fields of application) are linked to the nature of the product itself. Some constraints linked to the nature of the product and to the refining process create difficulties and limitations in the use of NIR spectroscopy. This part of the chapter summarizes these main specificities and gives some tips to handle these difficulties.

38.3.1 Constraints Inherent to Petroleum Products

Petroleum products, from crude oil to finished products, are quite specific products to handle and to analyze. The main reason is the complexity of the products: a crude oil is a mixture of about hundreds of thousands hydrocarbon molecules, which are split by boiling ranges, and then transformed into several products (gas, naphtha, gasoline, kerosene, bitumen, etc.). This means that the precise composition of the products to be analyzed is not known in detail nor entirely controlled (in opposition, e.g., to pharmaceutical or cosmetic products that are a mixture of well-known, synthesized molecules in defined yields).

Still linked to the nature of refining and petrochemical products, care must be taken when handling the samples. Lots of petroleum products contain dangerous and volatile molecules, and they need to be analyzed and stored in specific conditions to avoid any risks (see Section 38.2.3 of this chapter).

Another specificity, explaining also somehow why the detailed composition of the products is not precisely known, is the properties of interest of the products. It is mainly physical properties, such as density, distillation points, flash point, freeze point, and octane number, that will define the quality of the final products. Chemical content is measured mostly to obtain the overall yields of molecular families (see Section 38.2.1 of this chapter) but not detailed compositions of specific molecules.

In refining and petrochemicals, products to be analyzed are not known in detail and highly depend on the crude oil refined, the process scheme of the plant, and the unit parameters used. This also means that the sources of variability in the product are numerous (especially for final products) and not under the control of the analyst developing the NIR models. The main constraint in developing the NIR analysis method in refining is the fact that it entirely depends on the production and the raw material (crude oil). This is similar to some other industries (food industry, for instance), but not all of them. It is almost always impossible to create representative standards at a defined level for a property to develop or enhance a calibration database because it is either too complex or too costly, or simply not even possible sometimes.

38.3.2 Difficulties and Limitations

These constraints result in some difficulties and limitations for developing and using NIR spectroscopy to analyze petroleum products. As a result of the chemical complexity of products and the properties of interest, several samples with different compositions, and hence different NIR spectra, can have the same values for several properties to be measured. Figure 38.2 is an example of two gasoline NIR spectra in the combination bands. Both have a RON (research octane number) measured at 94, but still their chemical composition is different, and so are their spectral features. The spectral information linked to properties of interest can be quite complex to extract, and chemometric methods are mandatory to create calibrations.

The main, and maybe hardest, difficulty in developing NIR spectroscopy in the refining industry is the calculation of the chemometric model that will interpret samples' spectra and give the analysis result. Indeed, to achieve good model performances, all sources of compositional variability related to the process and the crude oil need to be included in the data. The samples used to develop a calibration need to be representative of the process, e.g., to span all the range of possible composition (spectral information) and of property values that will be encountered during the life of the model. As a result of the numerous variability sources in production, it can be hard and can take a lot of time to gather sufficiently representative data. In fact, it is usually impossible to do it from the start and it is necessary to update the models regularly in order to take into account sources of variability that could not be included before, because they are either new or not sufficiently represented in the data.

It is therefore very important to monitor NIR analysis performances using a well-defined control plan (see Section 38.5.3 of this chapter). Some of these variability sources that can have an impact

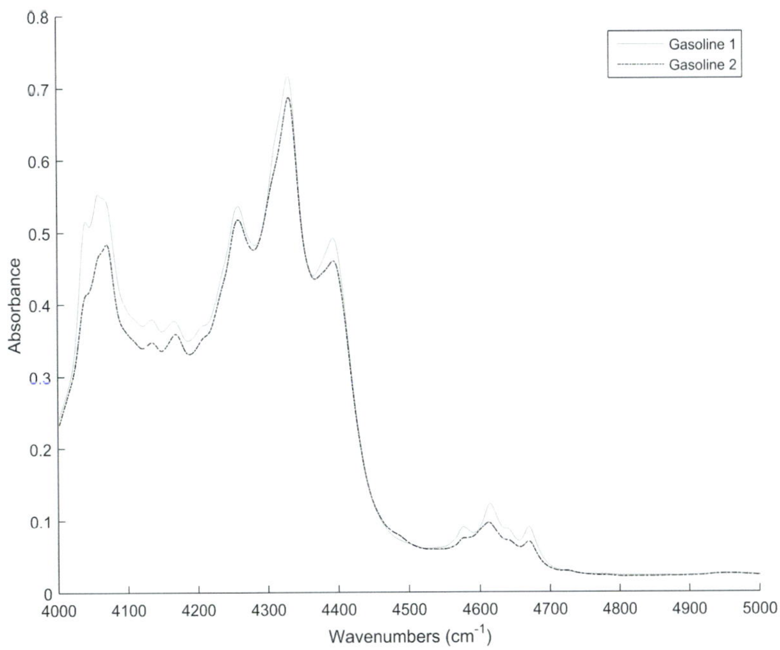

FIGURE 38.2 NIR spectra of two gasoline samples with the same RON measurement (RON=93).

on the accuracy and precision of the NIR analysis results are quite well known. It can be a change in the process (from a new process scheme to a simple change in an upstream unit parameter), the crude oil being refined, a new input in the unit, changes in the production specifications, etc. The seasonality of the product can also be a variability source that needs to be anticipated and representatively included in the calibration database. Once samples have been analyzed (NIR spectra and conventional or reference analysis), algorithms can be used to help select a representative database. They are described in Section 38.3.3 of this chapter. Still, a difficulty in developing and controlling NIR analysis models is the need to be aware of this information from the process that can impact the samples' composition and to take them into account. Users of NIR spectroscopy also need to be aware that a model will not last forever and an update will eventually be necessary necessary to avoid analysis error.

Other limitations arise from the NIR spectroscopy technology itself. As stated before in this chapter, the dark products are harder to handle than the clear ones because most of the signal is absorbed. For example, good NIR spectra of crude oil, atmospheric residues, bitumen, or grease are hard (sometimes impossible) to obtain and require specific sample preparation (filtration, dilution, etc.) and/or specific hardware setup. Because water absorption is quite high in NIR region and water does not mix well with petroleum products, its presence in a sample will most probably result in an unusable spectrum.

Finally, when using NIR spectroscopy, the sample degradation is an important aspect to keep in mind. Some refining products are sensitive to heat, evaporation, or light exposition. If not handled with the appropriate care and analyzed quickly with NIR and the conventional method, its properties can degrade, creating a systematic error in the NIR method development or its control.

38.3.3 Dealing with Representativity and Databases

Due to the constraints and limitations described in Sections 38.3.1 and 38.3.2 of this chapter, building up models for extracting useful and interpretable information from NIR spectra must be performed with caution. One particular key aspect of the methodology is to get all the necessary data available prior to modeling. This is a crucial prerequisite in the sense that the model calculated later on will be applied to many different products and molecules. So a large enough database is to be built before modeling. Large enough means as diverse as possible, big enough in terms of the number of samples, and also representative of the usage of the model. In other words, a properly organized set of data, often called a database, must be put in place before jumping into modeling. This database should contain, as a minimum, names and time stamps of available samples, spectra, and properties to be modeled. Again, all available data must cover a representative period of time and enough fluctuations of the process and of the produced products. This will make models more robust and more performing in the future. Complementary data, which are the repeatability and the reproducibility of the reference method with which the property to be modeled was measured, should also be in this database. This will allow comparing the performance of the model with the metrics of the reference method and will help in deciding whether predictions of models based on NIR spectra are equivalent or not to the reference method. All other additional information such as batch references and grades of products could also be listed in this database as it will be of great help for interpreting results of some modeling steps. As in the oil and gas industry it is not really possible to create or design samples (see Section 38.3.1 of this chapter), it is recommended to collect as many samples as possible over time, i.e., acquiring a NIR spectrum each time a sample is analyzed in the lab with a reference method. It is very frequently asked about the number of required samples before starting building a model. The answer is not simple nor obvious. All is a matter of representativity, and it is probably better to have 100 samples of very diverse compositions and properties, rather than 500 samples of similar origin, close in time, etc. However, that said, it is classically admitted that about 100 samples could be an acceptable minimal number of samples to start investigating a model for prediction. It does not mean that models cannot be built with less than

100 samples, but this provides a rough guidance for being able to calibrate and validate a model. Validating a model is probably one of the most important parts of the methodology and will be detailed in Section 38.4.1 of this chapter. Nevertheless, it must be said that all the available data for modeling must be separated in at least two subsets of data prior to modeling. The first one will be used for calibrating the model, while the second one will be used for (external) validation. Again, it is extremely important to split the initial dataset into two subsets, as the calibration dataset should be representative of the population for which future predictions are to be made [5]. Validation set will be used for estimating the performance of a model on unknown samples. Selection can be either manual or random or with specific algorithms. Several algorithms exist, and the most popular ones are probably the Duplex algorithm [6] and the Kennard–Stone algorithm [7], also known sometimes as CADEX. The Kennard–Stone algorithm is based on an iterative selection of spectra which maximizes distances with spectra previously already selected. It creates a single dataset (typically the calibration set) with a predefined number of samples, the rest being the second subset (the two subsets are built sequentially, one after the other). Since only one subset is composed of selected samples, there is a risk that the other one lacks representativity, especially if lots of samples are selected by the algorithm. The Duplex algorithm shares similar principles but builds up the two subsets in parallel, by putting spectra alternatively in calibration subset and validation subset until a stopping criterion (number of points in the smallest subset) is reached. The rest of the samples, not selected, are then put either in calibration or in validation according to preferences and needs. This method allows avoiding the issues described with the Kennard–Stone algorithm. Other algorithms exist [8–10], but it is extremely important to keep in mind that the quality of a predictive NIR model greatly depends on the quality of the calibration sample selection (hence on the validation set as well). Overall, organized databases, good representativity, and proper selection of spectra are necessary for optimal modeling methodology. They are sometimes somehow forgotten but may constitute significant hurdles to successful modeling of NIR spectra in oil and gas applications.

38.4 CHEMOMETRICS ON NIR SPECTRA

The main objective when analyzing petroleum products with NIR spectroscopy is to obtain, quickly and with the minimum of sample to handle, diverse physical and/or chemical properties of interest. To do so, and as stated before in this chapter, chemometric tools are necessary. The most commonly used technique to obtain these properties from NIR spectra is the PLS (partial least squares) regression [11,12]. It is a multivariate explanatory method used to predict variables, such as the aromatic contents of gasoline, using multiple other correlated variables, here the spectral signal at different wavelengths. A regression coefficient is attributed to each explanatory variable to obtain the variable to predict. What is usually called "model" is the resulting equation. PLS regression is often associated with PCA (principal component analysis) [12], an exploratory method that is used to extract and visualize structure and trends by decomposing the data into new variables (linear combinations of the original ones) describing the main sources of variability. It can also be used to identify clusters and classify data.

This section details some methodology and provides recommendations on how to use these two techniques, how to develop PLS models on petroleum products, and how to be confident in their performances. The focus is on PLS but can be applicable to other quantitative and qualitative methods. Theory of PLS and PCA is not in the scope of this chapter.

38.4.1 METHODOLOGY

Since NIR analysis is now widely used in the refining and petrochemical industry, two American Society for Testing and Materials (ASTM) standards have been written to help developing and validating chemometric models on these data. The first one, the ASTM E1655 "Standard Practices for Infrared Multivariate Quantitative Analysis" [13], deals with the steps of multivariate model development from the spectra acquisition to the validation. The second one, ASTM D6122 "Standard Practice for Validation of the Performance of Multivariate Online, At-Line, Field and Laboratory Infrared Spectrophotometer, and Raman Spectrometer Based Analyzer Systems" [14], is focused on the validation of the analysis and models once calibrated. Methodology to develop good PLS regression models on petroleum products can be summarized in the eight steps presented on the flowchart (Figure 38.3). Each one is important to obtain a valid and relevant prediction. Some advices and tips are also given as a result of experience in managing NIR models on petroleum products.

The first step is often forgotten but is really important. Before starting, the needs should be defined as clearly as possible. This step should answer, at least, the following questions:

- What are the properties to be modeled? On which products? (This obvious question is necessary and should define the range of the property to be analyzed and category of products to use in the models. For example, a model for the RON on unleaded gasoline 95, 98, with or without ethanol?)
- What is the reference analysis method associated with each property to be modeled? What is its performance? (This question can lead to the conclusion that the reference method is not accurate or precise enough to calibrate a model and that another one should be used.)
- What performances are expected from each model? (It is necessary to have, at this step, at least an approximation of the required performances in order to objectively judge on the model at the validation step.)

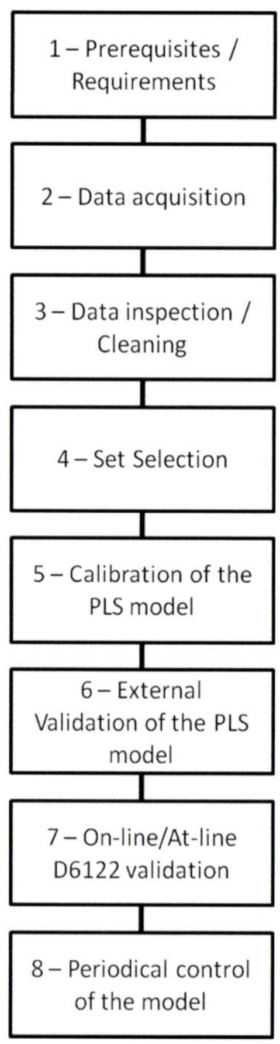

FIGURE 38.3 PLS model development methodology flowchart.

The second step is the data acquisition, i.e., spectra of samples and associated reference value for each property to be modeled. The samples acquired during this step should represent the widest sources of variability that will be encountered by the model during routine analysis: seasonality of products, change of recipe, different crude oils, etc.

The third step consists in a preliminary inspection and cleaning of the data, before modeling. This step aims at gaining time for the next ones, and optimizing the selection of samples in step 4. Inspection of data means identifying possible errors (of acquisition or of reference analysis) or atypical samples. It can be done by performing some general statistics on the properties and PCA on spectra and properties (if several). It is also possible to estimate the feasibility of a model at this step, by comparing the variability contained in the samples' property values and the reproducibility of the reference analysis method. If the first one is not high enough compared to the second one, no regression will be possible and more samples will be required.

The fourth step is the selection of the calibration and validation sets of samples. Both need to be representative of the range of application of the model. This aspect is more detailed in Section 38.3.3 of this chapter.

The fifth step is the calibration of the model itself. It should contain at least these three parts: (1) the choice of the optimal number of latent variables to explain the property of interest. This can be done, for instance, by looking at the property explained variance for each latent variable added. Performing a cross-validation and using the root mean square error of calibration (RMSEC) and of cross-validation (RMSECV), trying to obtain the simplest model with the best possible performances (Figure 38.4) is another possibility. Caution must be taken at this step (more details are given in Section 38.4.2 of this chapter). (2) Once defined, atypical samples in the model should be identified and handled with care. If some samples have too high influence on the model, it is necessary to investigate them and determine if they are representative of the samples to predict.

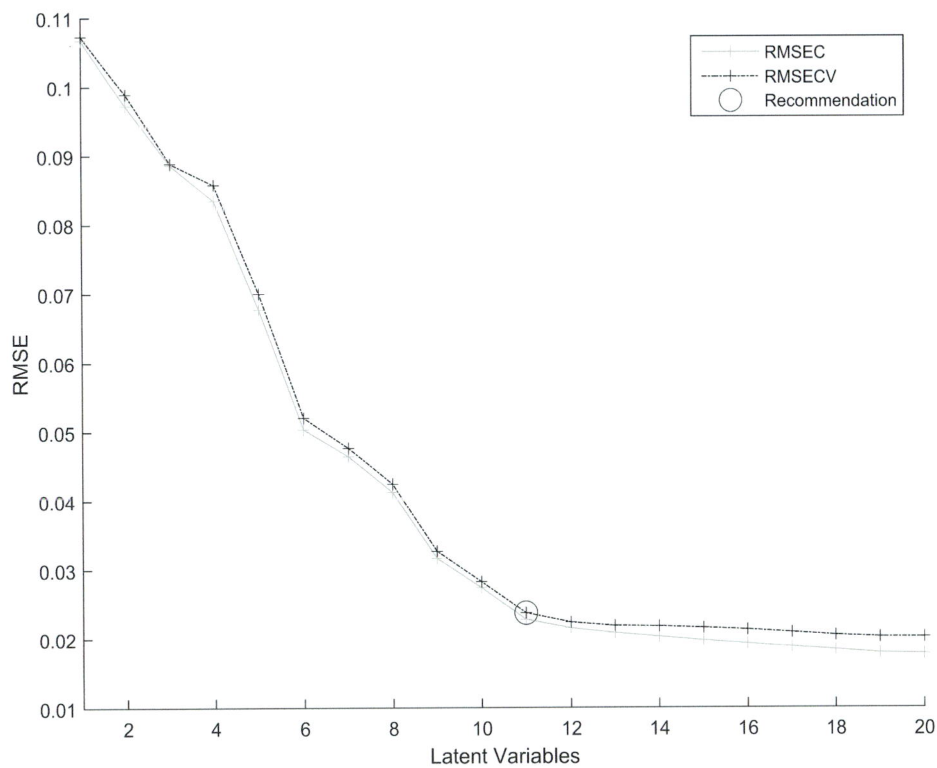

FIGURE 38.4 Selection of optimal number of latent variables.

If not, it might be necessary to remove them from the calibration and go back to (1). (3) Then, if the performances obtained in terms of predictions are satisfactory compared to the requirements, step 6 is to be performed. This can be achieved by looking at RMSEC, at RMSECV, and at the regression plot between reference measured values and predicted values (sometimes denoted as $Y_{observed}=f(Y_{predicted})$ plot). It is to be noted that the way this regression between observed and predicted values is done is important in the evaluation of the model performances [15,16].

The validation of the model is the sixth step. The aim is to validate if the number of latent variable chosen previously is really the optimal one, if the model is able to understand the information contained in unknown spectra, and its predictive performances on unknown validation samples. Root mean square error of prediction (RMSEP) is the best estimate of the prediction error made by the regression model. The prediction residual distribution should also be taken into account, together with the prediction bias, the slope of the regression line between reference measured values and predicted values, and its coefficient of determination.

The seventh step is also a validation step, but following the ASTM D6122 standard tests, and it should be performed on-line or at-line. During this step, 30 samples, new ones from the production or a subset of the validation set used in step 6 (recommended), are analyzed using the model. The results, if valid, are used to compute statistics (bias, normality of prediction residuals, etc.) to validate the NIR analysis results. Control charts are also computed.

The last step is in fact the routine periodical validation of the models. Like the first step, it is often forgotten. But once a model is calibrated and validated, it does not mean it will always work on the products. In fact, it is not rare that process parameters change and impact the analyzed product spectra and properties. If one or more parameter changes have never been included in the model (similar production condition not included in calibration samples), the model might not work as expected anymore. Control samples should be analyzed on regular schedule with the NIR spectroscopy and with the reference method and results included in the control charts created in step 7.

38.4.2 Main Pitfalls to Avoid

Even when following scrupulously the methodology described in Section 38.4.1 of this chapter, there are some points of attention that need to be carefully looked at when developing chemometric models. Some mistakes can indeed be made, leading to some prediction problems and sometimes even disappointment or disillusion about NIR technology and associated chemometrics. A few pitfalls are to be avoided to get to the successful point where models will deliver what they are supposed to in many different situations with the expected performance.

The performance of a model (also sometimes denoted as the accuracy or precision in the common language) is indeed crucial when deploying a model based on NIR spectra. As already mentioned several times, NIR spectroscopy could replace one or many other (more lengthy, more expensive) analyses, and it is very important not to lose result performance in this process. That means that models will provide, in the end, predictions that must be as good as the original reference method. Unfortunately, the performance of the original method is somehow forgotten or not properly taken into account. Repeatability (r) and reproducibility (R) are classically not precisely known. Prediction results are consequently improperly challenged against this method as soon as results are different. This leads to a prerequisite in the first step of the methodology: The reference method that will be replaced by a model from NIR data has to be well assessed (r, R, range validity, etc.). As a consequence, predictions from models will then be challenged against these metrics in a fair and objective way.

Another classical mistake that is currently made is overfitting. Overfitting models means that non-desired noise (or even small artifacts in spectra) has been modeled, as well as the main information. This is typically revealed by a high number of latent variables in PCA or PLS. This number can be denoted as the complexity of the model, and one pitfall to avoid is to have models with too high complexity. These models will indeed be less robust and also less interpretable and will not provide, over time, good-quality predictions. There are a lot of criteria and diagnostics that could help detecting overfitted models, and also to choose the right complexity. Figures 38.5 and 38.6 represent typical situations of overfitting. While on Figure 38.5, it is clear that predicted and measured reference values are almost identical on the training set, Figure 38.6 shows noisy regression coefficients. Both figures can be interpreted as signs of overfitting.

This could seem obvious, but if too many atypical points (also sometimes denoted as outliers) are present in the training set, models will not be as optimal as they should be. They will indeed try to take into account the atypicity and will consequently be mathematically different than if no outliers were present. Outliers could have a significant impact on the models and on the predictions. A right compromise between "keep all points" and "eliminate too many points" has to be found when developing a model. Again, different criteria exist to detect outliers, but it has to be reminded that the final decision about eliminating points belongs to the person in charge of building the model (according to the root cause of the atypicity, if available, and the knowledge of the data). Among the different criteria, residuals or leverage could be considered, but many others are available [14].

Last but not least, validating a model will be a key step for success. Validating can be seen as the ultimate step that will ensure that a model will predict well unknown samples, i.e., samples that are not part of the training set. This is called external validation. It is sometimes confused with cross-validation (see Section 38.4.1 of this chapter, step 5) that also reflects partly the ability of the model to predict. Cross-validation, however, is usually more useful to decide on the complexity of the model and should only be used during the calibration step. Validation of models must be performed on unknown samples as this is really representative of the future usage of the model. Additionally, validating with unknown samples is useful to evaluate the future anticipated performance of the model.

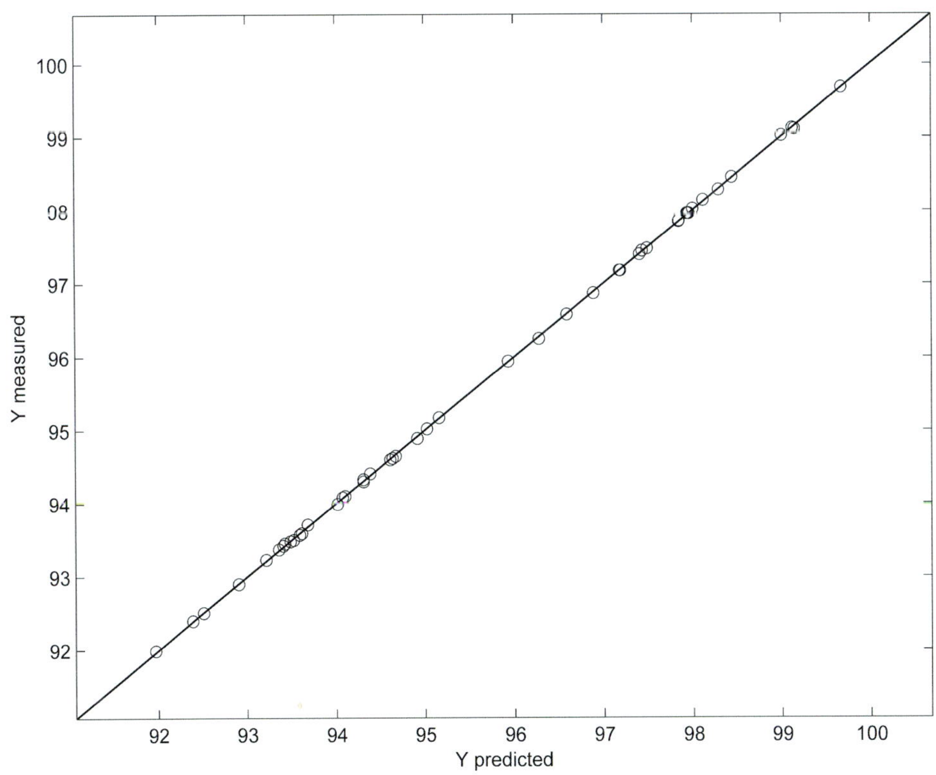

FIGURE 38.5 $Y_{observed}$ vs $Y_{predicted}$ for an overfitted PLS model.

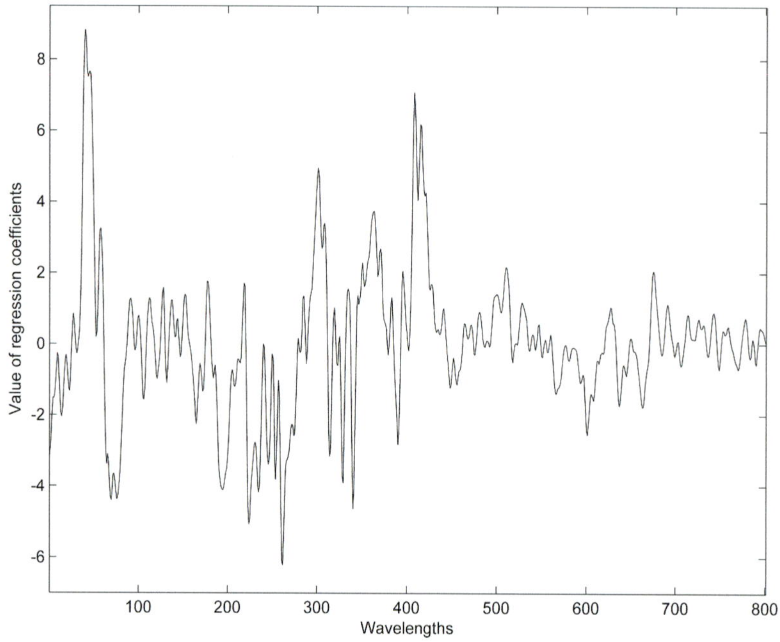

FIGURE 38.6 Regression coefficients for an overfitted PLS model.

As described previously, there are a few elements that can compromise the chance of building a successful model. Monitoring a deployed model is also important but not detailed here. Other misunderstandings that are described in Ref. [17] should also be taken into account when dealing with chemometric models.

38.5 EXAMPLE OF ONE TYPICAL APPLICATION

Several applications of NIR spectroscopy exist in the oil and gas industry as described in Figure 1 of Ref. [18]. Some of them are quite rare, while some are really common and deployed in several companies. This chapter details one typical application that is about blending and more especially gasoline blending.

38.5.1 Gasoline Blending

Gasoline is made up of different components that come from different units in a refinery. Economically, selecting the optimal combination of components to produce final gasoline product is one of the most critical and most important steps for the refiner. Gasoline blending is indeed much more complicated than a simple mixing of components. A typical refinery has different streams (up to 15 sometimes) that can feed up the blender, and those have to be blended together to meet simultaneously 10–15 different quality specifications. Finished gasoline product properties, which are commonly denoted as qualities, can range from color to more physical or chemical properties such as vapor pressure, boiling point, octane number, or aromatics contents, for instance. As any individual component contributes uniquely to each of these qualities and has a different cost of production, the proper allocation of each component into the blend is crucial and extremely important economically. The objective is to get the final product in the most optimized way. While several solutions are available and proposed offline (*a priori*, before blending), NIR spectroscopy is commonly used for real-time blending. It offers the possibility to give access to the different qualities of the finished product throughout the whole blending process. These qualities can then be fed up to a multivariable

TABLE 38.1
Typical Expected Performances for NIR Gasoline Models

Properties	Units	Property Range	Model Complexity	RMSEC	RMSEP	R^2
RON	N/A	90–100	4–6	0.2–0.4	0.2–0.4	≥ 0.90
MON	N/A	80–90	4–6	0.2–0.5	0.2–0.5	≥ 0.90
Benzene	%v/v	0.3–1.0	9–10	0.02–0.03	0.02–0.04	≥ 0.90
Aromatics	%v/v	15–40	5–6	0.2–0.4	0.3–0.5	≥ 0.98
Olefines	%v/v	0–35	5–6	0.3–0.8	0.3–0.9	≥ 0.90
Oxygen	%m/m	0–4	5–7	0.03–0.08	0.03–0.08	≥ 0.95
Ethanol	%v/v	0.0–0.6	5–7	0.02–0.04	0.02–0.04	≥ 0.95

optimizer and controller to reach the final step as targeted. Table 38.1 provides an overview of some key qualities available with the NIR technology with typical expected performance (supposing that the key aspect of chemometric modeling has been well managed).

38.5.2 Valorization and Added Value

As different possibilities exist to produce finished gasoline product from various components, a real-time optimizer is extremely useful for gasoline blending. The added value of the NIR technology for this application lies in the fact that it can give access to several qualities, through several models most of the time, in real time. These predicted qualities can be fed up to the optimizer that can adjust properly the rate of addition of each component. Valorization of the use of the NIR technology is then demonstrated through the capability of providing very rapidly several qualities (from 10 to 20 typically) at a time, but also through the optimization of the blend. NIR spectroscopy can even be sometimes used for on-line certification of blends, which is also very beneficial to the refiner: less lab quality analyses, blends released more rapidly, etc. One of the other advantages is that one NIR analyzer can replace several individual conventional analyzers (up to 10), such as knock engine, distillation, and chromatographs, decreasing the high cost of investment and maintenance. NIR analyzers are known as being mature and proven, reliable, highly accurate (but model-dependent), and reasonable in terms of maintenance cost.

Typically, and as an example, it is admitted that blending systems controlled by NIR analyzers can be operated at octane set points closer to the specification than systems based on knock engines, for instance. The value per octane unit was estimated to be \$5/m^3 (\$0.79/bbl) in Ref. [19], which provides a cost saving of around \$150,000/month for a production of 100,000 m^3/month and a gain of 0.3 octane unit. The entire cost of an FT-NIR system, including sample conditioning and analyzer shelter, can be paid out in 3–4 months.

Barsamian [20] also quoted the benefit range from \$4 to 8 million/year, depending on the number and size of blenders.

38.5.3 Life Cycle and Performance Monitoring

Once a model is developed and implemented on a spectrometer, the work is not over. To ensure full NIR analysis valorization and that a model is still performing according to the requirements, it is necessary to control it at regular time intervals. A model is a static representation of the product and property analyzed, built with the calibration data. If the product evolves because of a new source of variability, not accounted in the calibration data, its predictions might become erroneous. A model is an equation, so it will always give a result, even if absurd. That is why it is mandatory to periodically control the validity of a model on new production samples. This control can and should be performed according to statistical standards, as described in ASTM D6122 [14], for example.

One important point to understand here is that periodical control charts are not based on standard samples, like it is usually done to control more conventional analysis methods. The aim is to control not only the spectrometer itself, but also model performance. For that reason, it needs to be performed on new and actual production samples. These samples are analyzed with both the conventional method and the NIR method to calculate the differences between the two. The frequency of these control sample analyses should be appropriately chosen according to the production rate and importance of the analysis for the business (once a week, every two weeks, etc.). It should also be kept in mind that the conventional analysis method is necessary to perform these control tests or when spectra are outliers. For routine analysis, the NIR method is sufficient.

Performances of the spectrometer have to be checked according to manufacturers' procedures (spectroscopic noise, photometric linearity, etc.) [14]. The validity of a model during its lifetime is controlled by performing statistical tests on prediction residuals (NIR analysis - Reference analysis). Applying ASTM D6122 norm implies performing tests on the prediction bias and residual variance and monitoring three control charts: the individual values (on prediction residuals), the MR2 (moving range of 2), and the EWMA (exponentially weighted moving average), to detect tendencies. Only samples that are not outliers are used in this step (see Section 38.4 of this chapter). The number of outlier samples (during routine analysis) is also a good indicator to follow when monitoring NIR models.

The most common sign of a need for update is when a prediction bias appears, often linked to a change in process parameters or issues with a unit. For some products, it can also be quite common to see on these charts some "waves" that are linked to product seasonality. Caution must be taken when abnormal prediction errors or bias appears, because it can also be due to the reference method analysis. To clarify this point, the reference analysis method control charts should always be available. If erroneous reference analysis is used to update the model and correct for a reference method bias, the error will be included in the model. That should never happen. The number of samples resulting in an outlying NIR analysis can also increase dramatically when the model needs to be updated.

To update models, control samples are to be included in the database and model rebuilt. This can be done either directly by splitting the new samples between calibration and validation sets or by performing all the sample selection processes with the merged old and new samples. To optimize the update, it is often useful to understand why the current model is no longer performing as expected.

The average lifetime of models in the refining industry depends of the product (and stability of the process), the representativity of calibration samples, and the robustness of the model (number of latent variables used, complexity of regression coefficients, etc.). For some stable applications, such as reformate, models can last several years; for others that are more prone to process changes, it can be around 6–12 months.

38.6 CONCLUSIONS

NIR spectroscopy is very much used in the oil and gas industry. As detailed in the different sections of this chapter, some important and key aspects have to be well managed so that the technology can be successfully deployed and expected benefits obtained. About chemometrics, a lot of other techniques than PLS exist, and are probably newer in the field but also less common [21]. It can also be mentioned that not only NIR can be used as a single spectroscopy, but also NIR and MIR combined together can be used [22].

One of the current trends that can be seen is also about miniaturization of the technology so that more measurement NIR points could be accessible in a refinery with the possibility to install mini- or micro-spectrometers closer to the process.

REFERENCES

1. Cho, S., Kwon, K. and Chung, H. 2006. Varied performance of PLS calibration using different overtone and combination bands in near-infrared region. *Chemometrics and Intelligent Laboratory Systems* 82: 104–106.
2. Bouveresse, E. and Massart, D.L. 1996. Improvement of the piecewise direct standardisation procedure for the transfer of NIR spectra for multivariate calibration. *Chemometrics and Intelligent Laboratory Systems* 32: 201–213.
3. Bouveresse, E. and Massart, D.L. 1996. Standardisation of near-infrared spectrometric instruments: A review. *Vibrational Spectroscopy* 11: 3–15.
4. Pereira, C.F., Pimentel, M.F., Galvão, R.K., Honorato, F.A., Stragevitch, L. and Martins, M.N., 2008. A comparative study of calibration transfer methods for determination of gasoline quality parameters in three different near infrared spectrometers. *Analytica Chimica Acta* 611: 41–47.
5. Vandeginste, B.G.M., Massart, D.L., Buydens, L.M.C., De Jong, S., Lewi, P.J. and Smeyers-Verbeke, J. 1998. *Handbook of Chemometrics and Qualimetrics: Part B*. Elsevier Science B.V: Amsterdam, The Netherlands.
6. Snee, R.D. 1977. Validation of regression models: Methods and examples. *Technometrics* 19(7): 415–428.
7. Kennard, R.W. and Stone, L.A. 1969. Computer aided design of experiments. *Technometrics* 11(1): 137–148.
8. Galvão, R.K.H., Araújo, M.C.U., José, G., Pontes, M., Silva, E. and Saldanha, T. 2005. A method for calibration and validation subset partitioning. *Talanta* 67(4): 736–740.
9. Walid, G. and Mohamed, L. 2016. A kernel distance-based representative subset selection method. *Journal of Statistical Computation and Simulation* 86(11): 135–148.
10. Clark, R.D. 1997. OptiSim: An extended dissimilarity selection method for finding diverse representative subsets. *Journal of Chemical Information and Computer Sciences* 37: 1181–1188.
11. Wold, S., Sjostrom, M., and Eriksson, L. 2001. PLS-regression: A basic tool of chemometrics. *Chemometrics and Intelligent Laboratory Systems* 58: 109–130.
12. Martens, H. and Naes, T. 1991. *Multivariate Calibration*. John Wiley & Sons: Hoboken, NJ.
13. ASTM E1655-05. American Society for Testing & Materials. Standard practice for Infrared Multivariate Quantitative Analysis. West Conshohocken, 2012.
14. ASTM D6122-15. American Society for Testing & Materials. Standard practice for validation of the Performance of Multivariate Online, At-Line, and Laboratory Infrared Spectrometer Based Analyzer Systems. West Conshohocken, 2015.
15. Fearn, T. 1998. Which way round? A tale of two regressions. *NIR News* 9(3): 6–7.
16. Piñeiro, G., Perelman, S., Guerschman, J.P. and Paruelo, J.M. 2008. How to evaluate models: Observed vs. predicted or predicted vs. observed? *Ecological Modelling* 216: 16–322.
17. Kjeldahl, K. and Bro, R. 2010. Some common misunderstandings in chemometrics. *Journal of Chemometrics* 24: 558–564.
18. Chung, H. 2007. Applications of near infrared spectroscopy in refineries and important issues to address. *Applied Spectroscopy Reviews* 42: 251–285.
19. Espinosa, A., Sanchez, M., Osta, S., Boniface, C., Gil, J., Martens, A., Descales, B., Lambert, D. and Valleur, M. 1994. On-line NIR analysis and advanced control improve gasoline blending. *Oil and Gas Journal* 92(42). ISSN 0030-1388.
20. Barsamian, A. 2001. Get the most out of your NIR analyzers. *Hydrocarbon Processing* 80: 69–72.
21. Lee, S., Choi, H., Cha, K. and Chung, H. 2013. Random forest as a potential multivariate method for near-infrared (NIR) spectroscopic analysis of complex mixture samples: Gasoline and naphtha. *Microchemical Journal* 110: 739–748.
22. Felıcio, C.C., Bras, L.P., Lopes, J.A., Cabrita, L. and Menezes, J.C. 2005. Comparison of PLS algorithms in gasoline and gas oil parameter monitoring with MIR and NIR. *Chemometrics and Intelligent Laboratory Systems* 78: 74–80.

39 NIR Applied to Polymer Analysis

*Jean Guilment**
Laboratoire d'Etude des Matériaux (LEM)

CONTENTS

39.1 INTRODUCTION

Near-infrared (NIR) spectroscopy is a part of vibrational spectroscopy in the same way as the mid-infrared (MIR) spectroscopy and Raman spectroscopy. It is a physicochemical method of analysis based on the interaction between light radiation in the NIR range (12,500–4000 cm^{-1} or 800–2500 nm) and the molecules in the sample.

However, unlike MIR spectroscopy and Raman spectroscopy, which are interested in the fundamental vibrations of the molecule, the NIR will probe the harmonics and combinations of these modes of vibration. The spectra are weak and generally poorly resolved but can be obtained very easily on unprepared samples. It shows a certain complexity that can explain why this technique is rarely taught in the academic field while it has a great attractiveness in the industrial field.

NIR spectroscopy is very commonly used by a large number of industries and gives rise to a wide variety of analytical applications. Many products can be studied without preparation allowing their identification, their classification, or even their qualitative discrimination (quality control of raw materials, for example (Cinier and Guilment, 1998)). Complex mixtures content can be determined with great precision thanks to the chemometric methods available. NIR also makes it possible to work through optical fibers and thus opens the possibility of online control of processes in industry or in vivo analysis in medicine (Bertrand, 2002; Fontange and Chauchard, 2009).

This chapter presents the specificities of the NIR with respect to the polymer analysis. Today polymers scientists rely on many different analytical methods to characterize macromolecules (infrared spectroscopy, nuclear magnetic resonance (NMR), mass spectrometry (MS), gel permeation chromatography (GPC), rheology, …). Their goals vary from the assessment of the quality of

* jean.guilment@arkema.com

incoming monomers, the control of processes, and the control of the quality of intermediates and final products. It is also sometimes necessary to troubleshoot processing problems and to analyze competitors' products.

39.2 HISTORY

The discovery of the NIR region is attributed to William Herschel in 1800. But it was not until 1881 that Abbott and Festing recorded the spectra of several organic liquids and 1905 that Coblentz recorded the spectra of 19 substances and postulated that these bands can be part of a series of harmonics. Initially for theoretical and instrumental reasons (quartz prism), Kaye had assigned the zone 3.5–0.7 μm (2800–14,285 cm^{-1}) to NIR. But this region contains fundamental vibrations CH, OH, and NH of strong absorptivity which mask the harmonics even quite intense like $2\times\nu$ C=O towards 3400 cm^{-1}. Since the last fundamental vibration ν H-F lies around 3960 cm^{-1}, we will generally attribute the area 4000–12,500 cm^{-1} to the NIR (2500–800 nm). The bands of combinations and harmonics are 10–1000 times less intense than the fundamental bands.

The first analytical application of NIR was developed in 1962 by Karl Norris (Hart et al., 1962). In this first study, the authors describe a method for measuring the moisture content of cereal samples. The water was first extracted from the grains by dissolution in methanol, and the spectral measurements were taken in transmission on the methanolic extract. In 1968, Ben-Gera and Norris (1968) showed that it was possible to obtain diffuse reflection spectra that can be exploited directly without extraction. They also came up with the idea of using multiple linear regression to establish predictive equations linking spectral values to laboratory analyses.

In 1972, under the direction of P. Williams, the NIR technique was adopted as a routine method for the analysis of protein in wheat as a replacement for the Kjeldahl assay to avoid the use of highly aggressive chemical reagents. Shortly thereafter, J.S. Shenk et al. (1981) also developed the concept of a network of spectrometers in which the devices are interconnected and calibrated to give the same spectral response. A lot of work has been done on verifying the correct operation of the devices and the transfer of the calibration equations from one spectrometer to another.

From this period, manufacturers have put on the market many types of spectrometers (filters, monochromator, Fourier transform, etc.) dedicated to the most diverse analytical applications. More recently, advances in chemometrics provide powerful tools for extracting useful information from spectral data.

Since the 1990s, under the impetus of major industrial groups, the technique has been developing continuously, particularly in the fields of chemistry, pharmaceuticals, petroleum, and polymers. However, the teaching of NIR remains very limited in the chemistry departments of universities where generally only the MIR spectroscopy and Raman spectroscopy are presented.

In 2004, the US Food and Drug Administration (FDA) decided to give a new dynamic to the pharmaceutical sector by encouraging industries and laboratories to implement a new concept dedicated to improving production processes and R & D: PAT (Process Analytical Technology). The use of real-time analysis tools is put forward and should lead to a better knowledge of the production chain to allow control of processes and optimization of productivity. In this context, NIR and Raman spectroscopies have appeared as particularly appropriate analytical tools because they allow the realization of fast, nondestructive, and noncontact analyses.

39.3 A COMPLEMENTARY TECHNIQUE OF MIR AND RAMAN SPECTROSCOPIES

When an absorption spectrum is drawn over the entire range of the infrared (0–12,500 cm^{-1}), a larger number of bands are generally obtained than expected based on the formula "3 N-6"; with N the number of atoms. This is due in the first place to the so-called harmonic bands, which are transitions between the ground state and the upper energetic states of the oscillator provided that it is anharmonic (Dalibart and Servant, 2000). In addition to overtones, the combination bands, resulting from the interaction of two or more modes of vibration for the same functional group, give rise to NIR absorption. In these transitions, a photon excites two or more vibrations simultaneously by increasing the quantum number of each energy level. The photon that has the corresponding energy ($E=h\nu$) will induce a transition at a frequency approximately equal to the sum or product of the frequencies of the transitions that exist independently. This results from the shape of the potential anharmonic energy curve (Morse potential).

The chemical groups having an absorption in the region of the NIR are mainly of the form X-H, where X corresponds to carbon, oxygen, or nitrogen atoms, and H denotes the hydrogen atom (combination and overtones from at least bands shown in the 2500–4000 cm^{-1} of the MIR region and bands from the fingerprint region). In general, it is more difficult to interpret the NIR spectra than their equivalent in the MIR region. However, the NIR region is well suited to the development of analytical applications.

39.3.1 Ability to Analyze Samples without Preparation

Unlike MIR spectroscopy, the recording of NIR spectra does not generally require preparation because of the low intensity of the spectra. The fundamental bands on the MIR spectrum (right part) and the harmonic and combination bands on the NIR spectrum (left part) can be seen in Figure 39.1.

Polyethylene (PE) is characterized in the MIR spectrum by elongation vibrations of CH_2: $\upsilon_a CH_2=2920$ cm^{-1} and $\upsilon_s CH_2=2850$ cm^{-1} and a deformation vibration at 1459 cm^{-1}.

These bands are saturated on the 300-µm film MIR spectrum. They are only accessible on a very thin film (<30 µm) or by attenuated total reflectance (ATR) analysis.

In the NIR spectrum of the 300-µm PE film, all bands are observable without saturation. The combination bands at 4324 and 4251 cm^{-1} are at slightly lower values than the simple sum of the wave numbers ($\upsilon+\delta$), which is, respectively, 4379 and 4309 cm^{-1} because of the anharmonicity. Similarly, we can observe the bands of first overtones at 5665 and 5777 cm^{-1} (instead of $2\times2850=5700$ cm^{-1} and $2\times2920=5840$ cm^{-1}).

Absorption in the NIR region is much smaller than in the MIR region. This partly explains the lack of consideration of this area of the spectrum for a very long time. Today, the low intensity of harmonic bands and combination is rather considered as an advantage to study directly in transmission or diffuse reflection samples up to several millimeters thick.

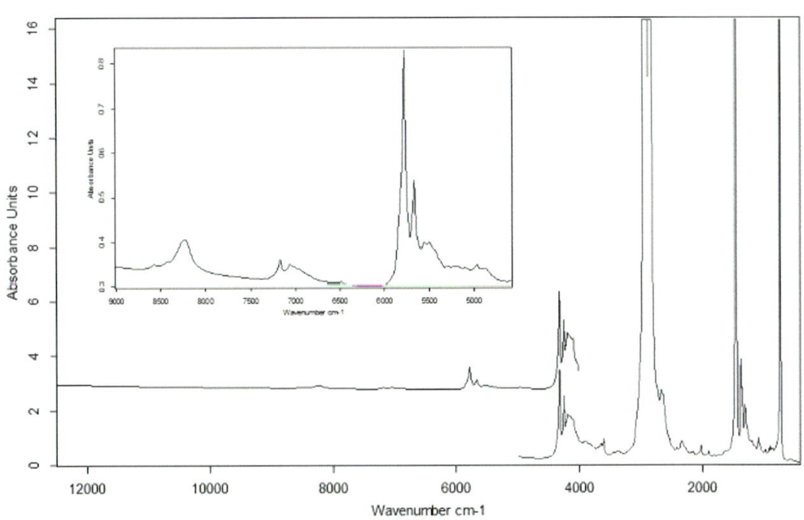

FIGURE 39.1 MIR and NIR absorption spectra of a 300-µm film of polyethylene. The NIR spectrum is shifted upward for better observation.

The intensity of the vibrations varies from one to several orders of magnitude when we go from fundamental bands to higher harmonics. For example, for a cell with a one-millimeter pathlength, the absorbance of the CH bond of the liquid chloroform will be about 17 for the fundamental vibration: 1.5 for the first overtone, 0.05 for the second, and only 0.002 for the third, which makes it possible to use fairly long optical paths adapted to the industrial problem to be treated.

In Figure 39.2, it can be seen that transmission can be carried out on polyamide samples up to 20 mm without saturation depending on the spectral range used, whereas in Figure 39.3, the diffuse reflectance seems limited to samples with smaller thickness or size (a few hundred microns in the area of the first harmonics).

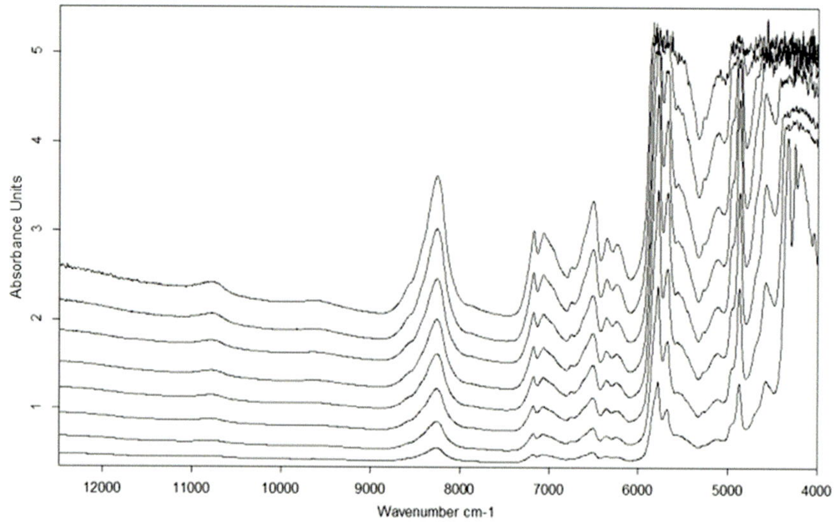

FIGURE 39.2 Absorption spectra of polyamide plates: from 2 to 20 mm recorded in transmission mode (from bottom to top).

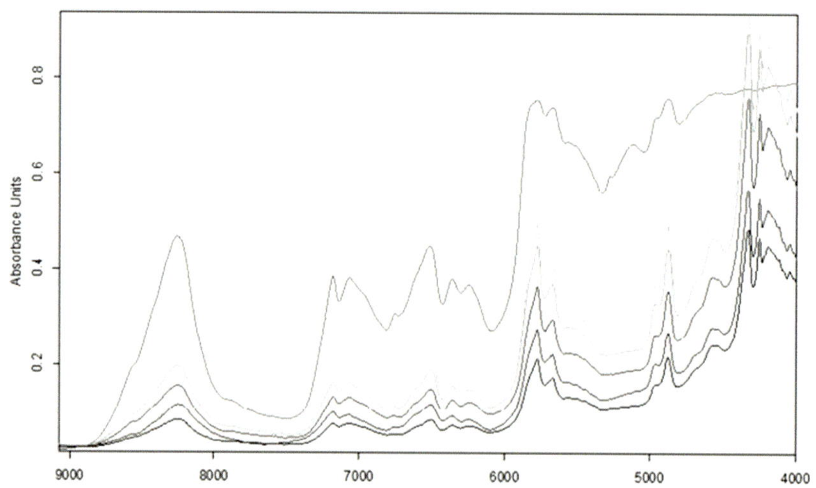

FIGURE 39.3 Absorption spectra of polyamide powders: from 20 to 120 μm (spectra from the bottom in ascending order) and granules of a few mm (top spectrum) recorded in diffuse reflection mode.

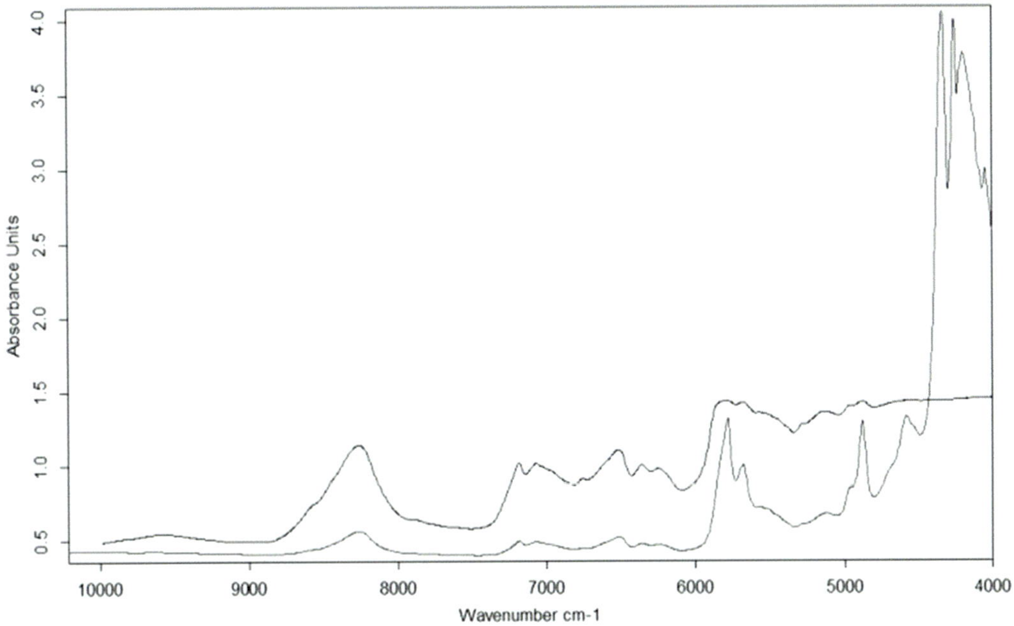

FIGURE 39.4 Absorption spectrum of polyamide. High: pellet of a few mm recorded in diffuse reflection mode; Low: 2-mm plate recorded in transmission mode.

In the case of diffuse reflectance, we observe, in addition to an effect due to the size of the particles, a baseline effect and a saturation effect of the spectrum due to the loss of light signal by diffraction and some reflected light on the surface of the grains (Figure 39.4).

39.3.2 Qualitative Analysis

NIR spectroscopy having been first used in the field of agrochemistry, it is often remembered that spectra are composed of wide bands (large presence of water) and are very difficult to interpret from a spectroscopic point of view. We therefore conclude quite quickly that it is necessary to use chemometrics and that NIR spectroscopy resembles a black box.

Without minimizing the contribution of chemometrics to extract the information from the spectra, it is, however, interesting to look at these more closely. For example, in Figure 39.5, it is easy, without mathematical intervention, to identify the aromatic bands of xylene (4605, 5950, 8720 cm^{-1}) with respect to the pentane spectrum or the contributions of OH vibrations in the spectrum of ethanol (4800, 5150, 6330, 6756 cm^{-1}). Compared to the MIR or Raman spectra, the differences are often even enhanced by the multiplicative factor inherent in the transition from fundamental vibrations to overtones.

In the case of polymers, the classification by families is also relatively easy. On Figure 39.6a, the difference between PE and polypropylene (PP) is clearly seen in the CH harmonics region around 5800 cm^{-1} (CH_2 vibrations in the PE and CH/CH_3 vibrations in the PP). On the other end, the PE/PP copolymer looks very similar to PP at a broad-level approach.

A more refined study makes it possible to identify (and thus quantify) the presence of a PE copolymer in the PP (Figure 39.6b). This possibility is subject to obtaining spectra of good quality and with a spectral resolution sufficient (4 or 8 cm^{-1}) to separate the contributions.

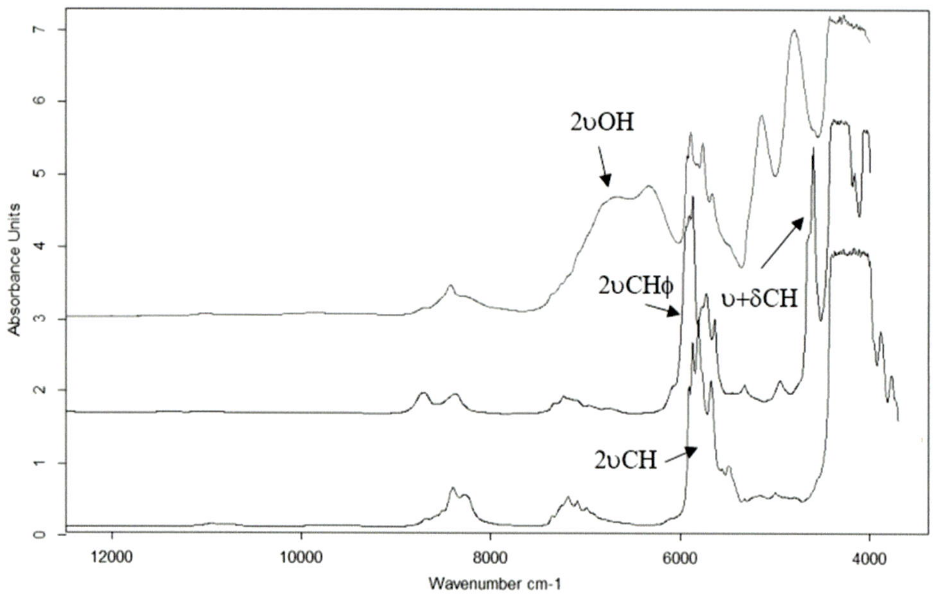

FIGURE 39.5 NIR absorption spectra of liquids (Bottom: pentane, Middle: xylene, Top: ethanol).

39.3.3 Quantitative Analysis

The ease of obtaining NIR spectra with a very high signal-to-noise ratio, without sample preparation and using silica-based optical fibers, makes it a technique of choice for quantitative analysis. It is still necessary to temper this advantage by some important observations to take into account:

- First of all, the Beer–Lambert law is systematically false in the case of the NIR since it applies only to highly diluted samples and without interactions between compounds. Nevertheless, there is always a relation of proportionality between the NIR signal and the concentrations. The precision and accuracy of the method may depend on the choice of a given concentration range.
- The NIR signal is strongly influenced by the physical state of the samples (granularity, crystallinity, shape of the objects, color, etc...). This may be an advantage if one wishes to predict these properties but requires a very important collection of product spectra if one wants to make an independent quantitative analysis of these physical properties.
- Finally, spectral information very rich in NIR is often convoluted, and it is necessary to use chemometrics tools to establish robust and reliable quantitative models.

Once these points are taken into consideration, NIR spectroscopy is by far the best spectroscopic method for performing quantitative analyses.

39.4 METHODOLOGY AND EXAMPLES OF APPLICATIONS IN THE POLYMER FIELD

In contrast to the spectroscopy in the MIR, which requires a certain ingenuity to obtain usable spectra, we have seen that the acquisition of the NIR spectra is easy either in transmission up to thicknesses of several millimeters (or even centimeters depending on the spectral range) or in diffuse reflection with objects of shape and size that can be very variable (granules or polymer powders, grains of wheat or corn, fodder, soil samples, etc.). This ease of acquisition of the spectra is found

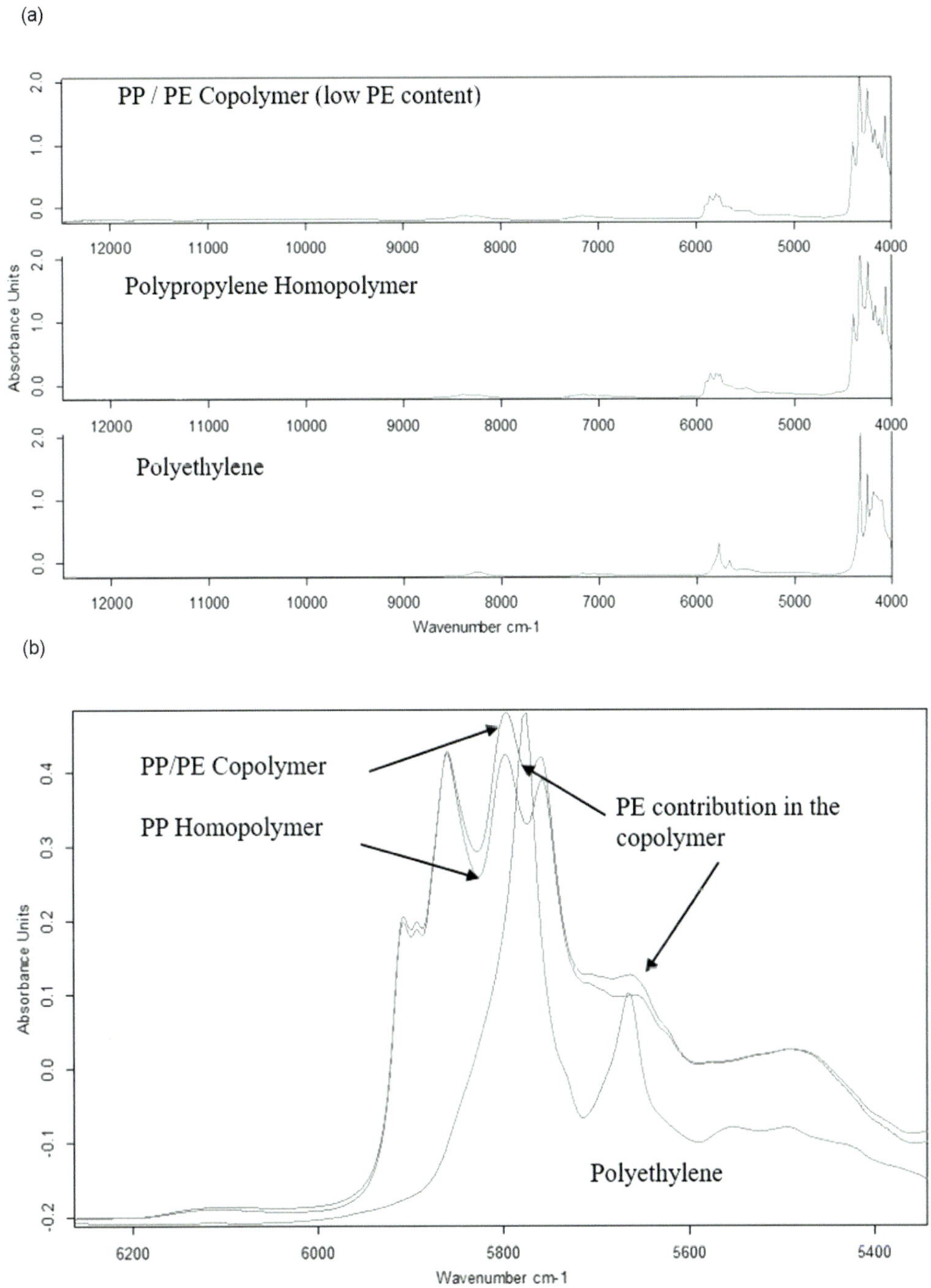

FIGURE 39.6 (a) NIR absorption spectra of polyolefins. (b) NIR absorption spectra of polyolefins (magnification).

at the same level with Raman spectroscopy but with the disadvantage of a relatively low signal-to-noise ratio and the presence of fluorescence in a large number of cases.

In addition, as in Raman spectroscopy, the measurement can be easily taken remotely through the use of optical fibers. More particularly in the context of the analysis of industrial processes, it is common to define the types of analysis as in Figure 39.7.

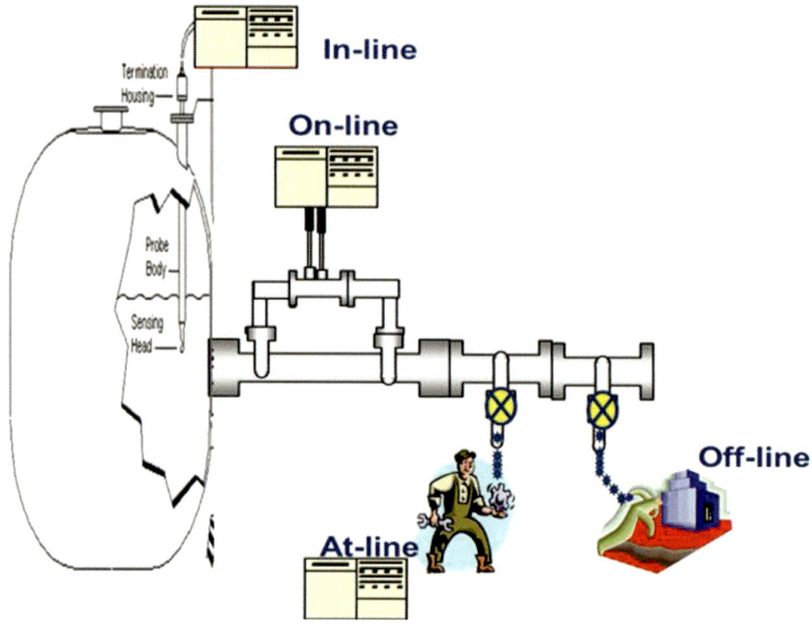

FIGURE 39.7 Different ways to control an industrial process.

- Off-line corresponds to the analysis that can be done at the control laboratory. It often uses a reference technique that can require a rather long and tedious implementation. This type of analysis is preferred to release batches of finished products. The use of NIR off-line has become quite common for the control of raw materials.
- At-line corresponds to the analysis carried out directly at the foot of the production line by the operators. The method should be simple and provide statistical tools to indicate to the operator that the result is likely to be valid or alert them if in doubt.
- On-line and in-line are two types of in situ measurements in the process to follow. This type of measurement eliminates human intervention but requires preparation and implementation rather tedious. The term "on-line" is used instead when the analysis takes place on a recirculation loop, whereas the term "in-line" implies that the measurement is done directly in the process reactor.

39.4.1 Polymer-Specific Questions

Industrial plastics are usually a complex mixture of macromolecules (themselves structurally heterogeneous), with adjuvants and possibly fillers. It is easy to predict that, in practice, analysis issues may arise at different levels, ranging from the basic identification to the study of a particular details of the microstructure of the polymer, or the determination of impurities.

In this field, NIR has to find its own place. Its ease of use and rapidity orient the applications of NIR in the process control domain where it can help in improving the yield and the quality of the products. On the other end, one should not expect NIR spectroscopy to be a very explicit technique allowing to solve problems from scratch. However, it is interesting to start collecting NIR information as soon as possible during the research phase of a product development to have a range of variations wide enough when moving to the industrialization step.

The chapter on polymer analysis published in the previous edition of the NIR handbook (Kradjel and Lee, 2007) gives a very extensive list of applications of NIR in the field of polymers. No major development has happened since.

We would like to give more emphasis to spectroscopic considerations of the NIR related to polymers. First of all, NIR being a spectroscopic technique, the information contained in the spectra is directly related to the chemistry of the products studied. It is then quite obvious to understand the capability of NIR to distinguish between various polymers and allow to selectively sort them. The question becomes how fast you can go and how selective you can be. In the waste business, it is complicated by the variety of multilayer packaging with various amounts of fillers. Acousto-optic tunable filter (AOTF) systems or diode-based systems can be very fast in this application.

To follow on these lines, products where the properties are directly linked with the copolymer content (ethylene/acetate, ethylene/acrylate, polyether block amide, etc.) are rather straightforward to tackle. Usually, the bands are well defined (see Figures 39.13 and 39.14 in the case of ethylene/vinyl acetate (EVA) copolymer) and the proportions of co-monomers are rather large with a specification range at a few % level. It is just necessary to have a precise reference method available (NIR can never be better than the reference method as it is calibrated against it even if it is very often more repeatable). Usually, NMR is the method of choice, but its cost limits the development of calibrations. The NIR calibration is very often based on a PLS (partial least squares) regression, and its precision largely depends on the number of samples used. To have a good statistical representation of the working field, several hundreds of samples are usually required.

It becomes more perilous when one wants to predict indirect properties, which are widely used in the polymer field. A lot of polymerization processes are controlled with simple methods available at-line or in a control laboratory. It can be an acid index or a isocyanate group (–N=C=O–) (NCO) index, a solution viscosity, or a melt fluidity index. These indexes give an indication on the progress of the polymerization reaction with quite a number of unsaid limitations. When one wants to use NIR to follow the reaction and express the results keeping the index representation, it is very important to determine what is laying below.

As an example, in the case of polyamides, melt flow index (MFI) and viscosity are widely used as they give an indication of how the product can be used in the transformation process later on. The viscosity is, of course, linked to the amount of polymerization of the product. The Mark–Houwink equation gives a relation between intrinsic viscosity $[\eta]$ and molecular weight M:

$$[\eta] = K M^a \text{ or } \log[\eta] = \log K + a \log M. \qquad (39.1)$$

From this equation, the molecular weight of a polymer can be determined from data on the intrinsic viscosity, and vice versa. The values of the Mark–Houwink parameters, a and K, depend on the particular polymer–solvent system. It can be seen readily that this equation is not linear, and one will have to either work in logarithms or reduce the range to make it linear. The NIR is directly correlated with M through the ratio of the end groups to the rest of the chain. Moreover, in the case of a PA X.Y polyamide (PA6.6 as an example), it is possible to have the same viscosity with an excess of amine or acid end groups or an even situation, but the properties of the final product will be extremely different. The viscosity does not give indications on the stoichiometry of the starting materials when the NIR properly calibrated will be able to give both the acid and amine end groups. Reference methods such as NMR or potentiometric titration are required. The same kind of considerations can be expected with the other indexes.

Finally, most of the polymers are stabilized or modified with various products, and it is of interest to know the amount of additives used. The main problem in this case is about the detection limits and the precision of the NIR compared to a chromatographic method. Most of the times, the NIR bands corresponding to the additives arise on the top of the polymer spectrum and there will be a compromise between the detection limit and the precision (similar to what is described in the agro-food field (Abbas et al., 2013)). Generally speaking, one can expect a detection limit of 50–100 ppm or a quantification limit 3 times higher with a rather poor precision (of course, there can be good cases sometimes).

39.4.2 Analysis in the Liquid State

The analysis of liquids will be made directly in glass or quartz cells in the laboratory. They may also be taken using optical fibers equipped with probes in transmission or in transflexion (see Figure 39.8). The use of monofibers with large diameters allows us to monitor processes at distances up to 100 m from the spectrometer.

For liquids, the important criterion is to optimize the optical path for the spectral range containing the information on the product to be determined.

In the case of a qualitative analysis, a glass tube of 5–6 mm in diameter generally makes it possible to discriminate most of the products by obtaining spectra with an unsaturated absorbance. As seen in Figure 39.5, the spectra of liquids generally have fairly fine and well-separated bands.

In the case of quantitative analysis, it is relatively simple to carry out a prestudy in the laboratory on samples (preferably industrial) representative of the ranges to be studied. The case of water and to a lesser extent the alcohols and products having strong hydrogen bonds is to be treated with caution. Indeed, the water has a very high absorption coefficient in the NIR, which allowed the development of the technique for humidity measurements in different environments. The water spectrum is characterized by strong absorption bands at about 5160, 6950, 8403, and 10,310 cm^{-1}, i.e., 1940, 1440, 1190, and 970 nm in the pure state. The very intense band at 1940 nm (5160 cm^{-1}) is a combination band of shear and elongation vibrations. The 1440 nm (6950 cm^{-1}) and 970 nm (10,310 cm^{-1}) bands are, respectively, the first and second harmonics of the elongation vibrations. The position and shape of the water absorption bands are affected by many factors such as temperature, solvent/solute

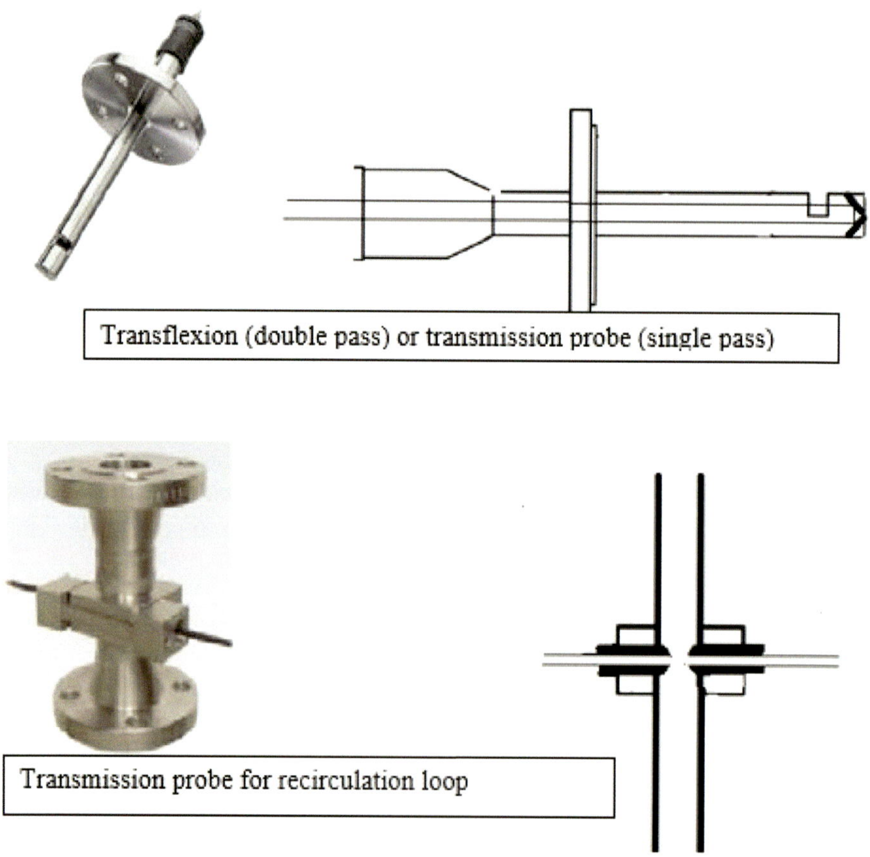

FIGURE 39.8 Example of probes used for the analysis of liquids in an industrial environment.

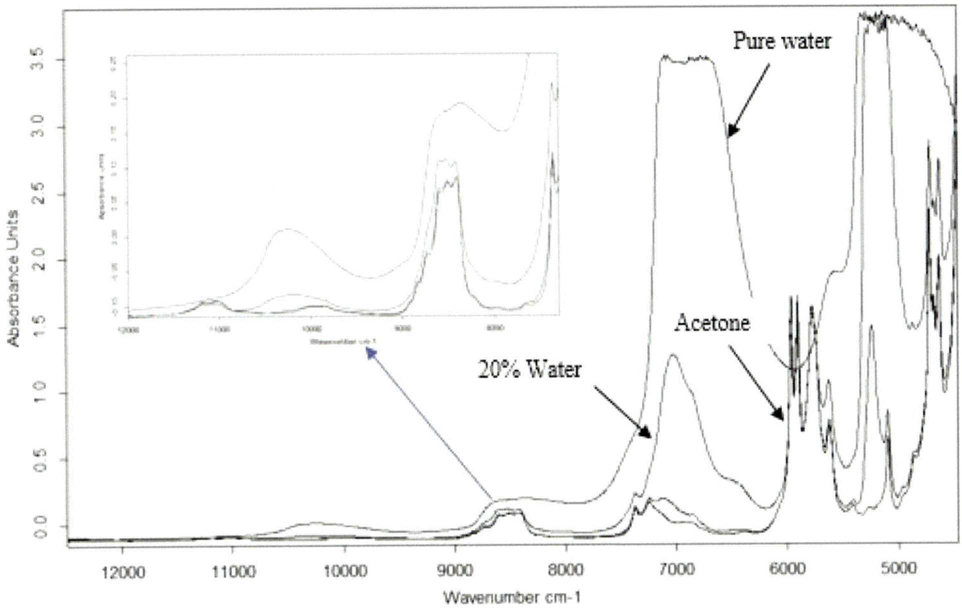

FIGURE 39.9 Spectra of pure water and water in acetone at 20%, 2%, and 0%.

concentration, and sample size. Thus, it can be seen in Figure 39.9 that the water strips in acetone stand out at about 5240, 7040, 8500, and 10,250 cm^{-1}. These values are fairly characteristic of water as a solute at room temperature.

When the temperature is varied, the hydrogen bonds decrease in intensity and the vibration frequencies of the molecules vary.

In the case of water (Figure 39.10), the vibrations $\nu+\delta$ H_2O and 3ν H_2O have their positions go from, respectively, 8350–8670 cm^{-1} and 10,130–10,400 cm^{-1}, while their width at half height decreases sharply from about 1000 to 200 cm^{-1}. The effect of the temperature is always one of the points to check when one is interested in the NIR spectroscopy of liquids.

The great advantage of NIR spectroscopy over other vibrational spectroscopies is that the information is present several times in the spectrum. It will therefore be easy depending on the target concentrations to choose optimal spectral ranges to perform a measurement. This point is illustrated in Figure 39.11 with the determination of water in acetone. If we are interested in high water concentrations (e.g., between 0% and 20%, Figure 39.11a), we will be able to use the spectral range between 5500 and 8000 cm^{-1} in which the band of first harmonic (2ν H_2O) has an intensity less than or equal to 2 OD (optical density) with an S/N ratio of the order of 10^5 or 10^6, which will allow a quantitative analysis with great precision.

In case we want to measure smaller amounts of water (0%–2.0% Figure 39.11b), we can look at the spectral zone from 4500 to 6200 cm^{-1} for which the vibration $\nu+\delta$ H_2O at 5240 cm^{-1} has ODs between 0.5 and 2.0 with a S/N similar to the previous example. Overall, this means that we gain a factor of 10 in sensitivity just by changing the spectral range. Given the intensity observed in Figure 39.11b for the lowest water concentration (5000 ppm), it can be extrapolated that it will be quite possible to gain a factor of 10–100 on the limit of detection. This limit of quantification is of the order of 10 ppm for water in acetone using a chemometric model of PLS type and a suitable range.

The examples above apply in the polymer domain for raw materials and solvents purification. It is also important to keep in mind these considerations when it comes to on-line studies of polymerization reactions in the melted state (Watari, 2014; Alig et al., 2010; Watari et al., 2004).

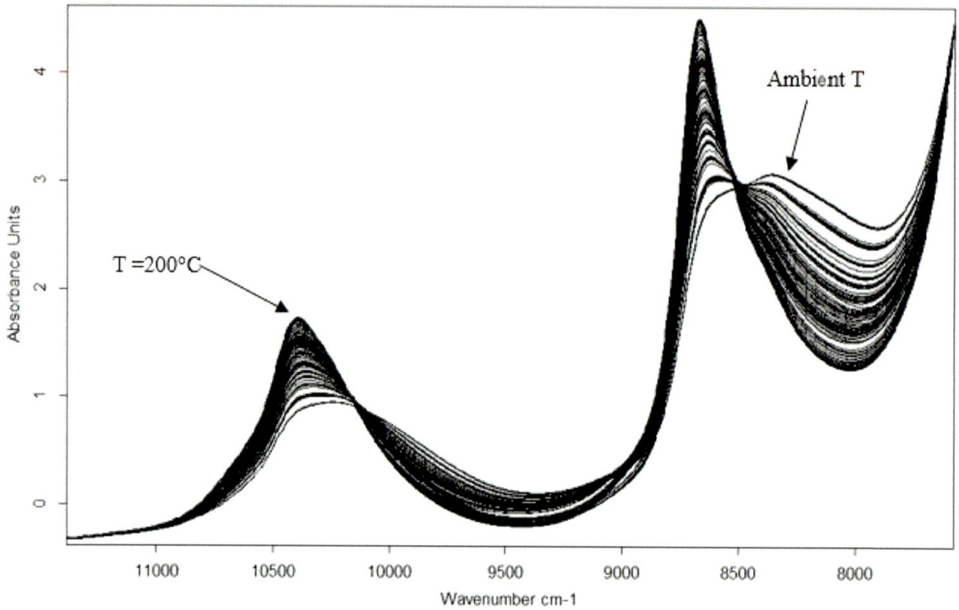

FIGURE 39.10 Spectra of pure water as a function of the ambient temperature up to $T = 200°C$.

The development of quantitative and qualitative models is being discussed at length in other chapters of this handbook; it will not be discussed here.

39.4.3 Analysis in the Solid State

The analysis of solids is slightly more complex than that of liquids to the extent that we will add a physical factor (shape, size, crystallinity, etc...) to the chemical signature of the product. This factor can be mitigated by making thin films or pressed pellets if the manufacturing methodology of these samples is sufficiently repeatable and there is no wish for analysis on-line. The most popular solution for doing at-line or on-line measurements is diffuse reflection (Figure 39.12).

The integrating sphere (Figure 39.12a) makes it possible to measure all types of solid samples that can be put in a cup or a glass or quartz flask. The principle is that an operator takes a sample and brings it to the window of the sphere to measure the spectrum. The analysis only takes a few seconds and does not require a trained technician. Some examples of spectra are given in Figures 39.3 and 39.4 for powders or granules of polymers.

The systems shown in Figure 39.12b–d make it possible to measure the solid product spectra directly online. The multifiber probe system (Figure 39.12b and d) allows analysis directly in a reactor (e.g., solid-phase polymerization of polyamide powders). The system comprises a bundle of about 50 fibers over a surface of a few millimeters. The light is brought to the sample by ten fibers, and the scattered light is collected by the remaining fibers. The system of Figure 39.12c allows a noncontact measurement of objects whose size can vary from a millimeter to a few centimeters. This system lends itself well to the analysis of polymer granules in a conveying circuit.

NIR analysis of solids is used in a large number of agro-food applications (protein content in cereals, nutritional value of feeds, etc.), in pharmaceutics (qualification of raw materials, quality of homogeneity of mixtures, quantitative analysis of the active ingredients, crystalline phases, etc.), or in chemistry (contents of copolymers, residual monomers, rate of polymerization, etc.)

Starting from the example proposed in Figure 39.13, one can try to comprehend some of the notions mentioned before as to the impact of the observable physical and chemical properties on

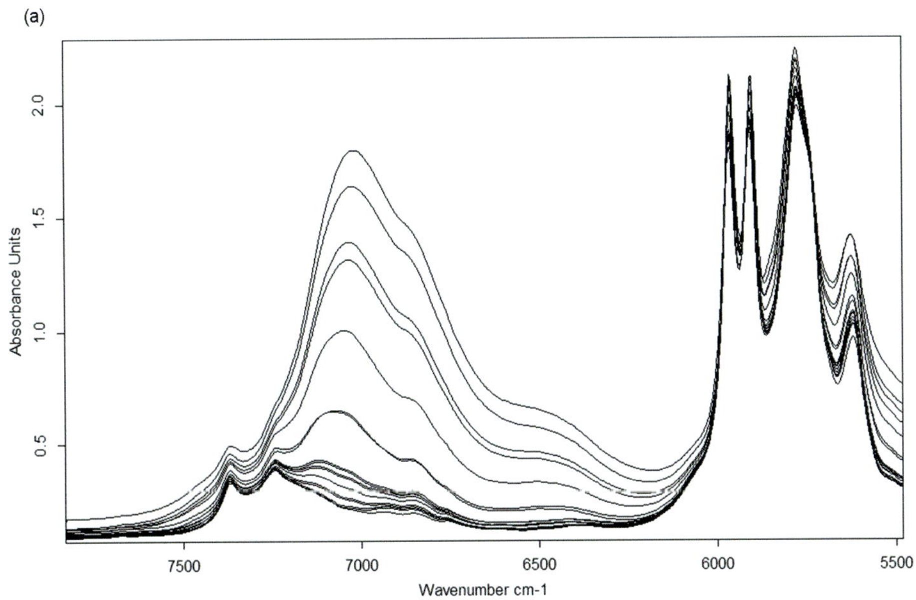

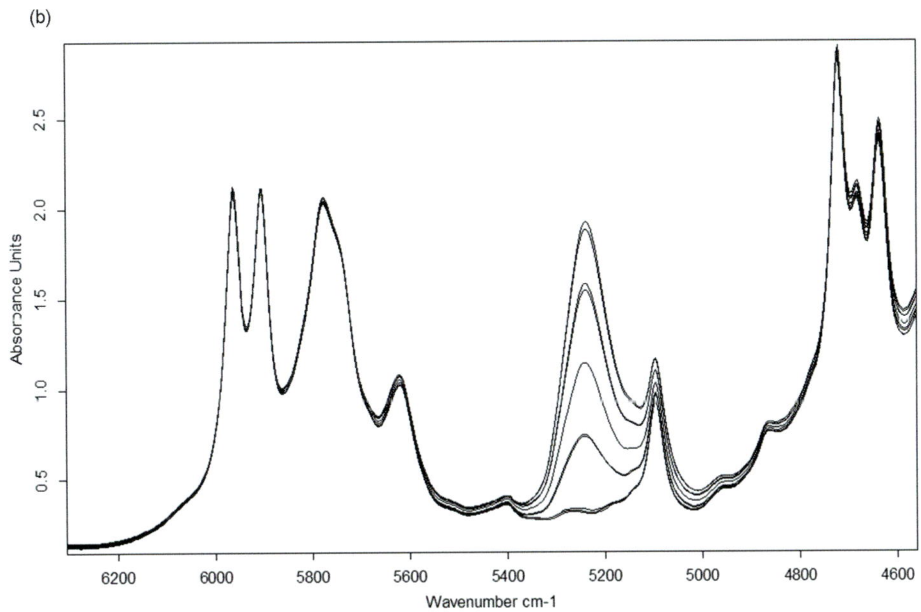

FIGURE 39.11 (a) Water spectra in acetone at 0%, 0.5%, 1.0%, 1.5%, 2.0%, 5%, 10%, 15%, and 20% (bottom to top). (b) Spectra of water in acetone at 0%, 0.5%, 1.0%, 1.5%, and 2.0% (bottom to top).

the spectra. At first reading, one can see the effect of the size of the granules which brings global intensity variations on the spectra and the saturation of the harmonic bands of CH_2 and CH_3 in the region between 5700 and 6000 cm^{-1}. Unlike the saturation observed on the spectra of the liquids in transmission, one observes a crushing effect of the bands with a maximum intensity, which rarely exceeds 2.0 of OD. This effect is to be taken into account in normalization procedures that may be distorted by this phenomenon.

By normalizing the spectra over an unsaturated zone (Figure 39.14), it is possible to highlight the spectral contribution of the vinyl acetate copolymer (e.g., $\delta CH_2 + \nu C{=}O$ band at 4760 cm^{-1}, $2\nu CH_3$

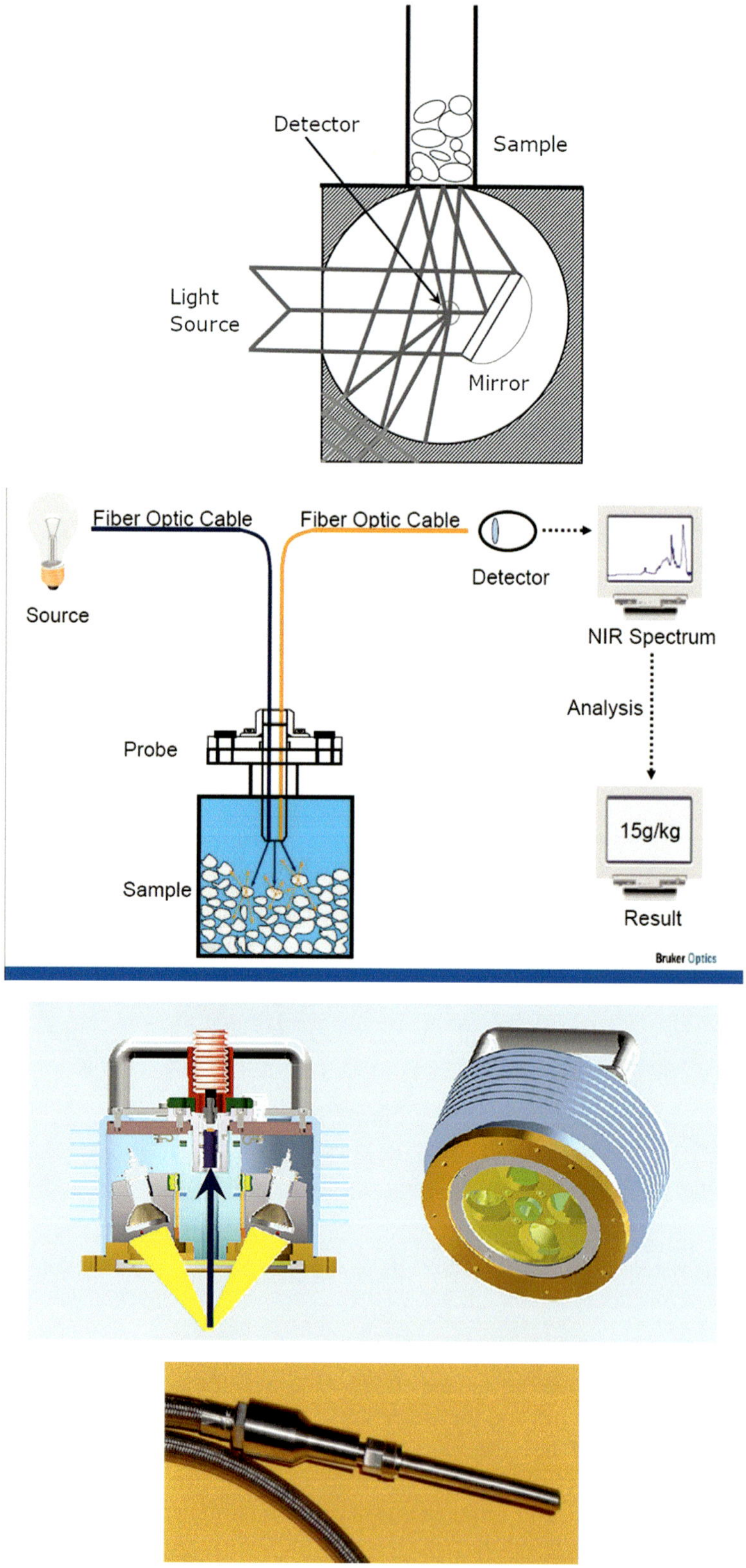

FIGURE 39.12 Different techniques for measuring the spectra of solids by NIR.

at 5940 cm^{-1}, $2\nu CH_2+\nu C=0$ at 7400 cm^{-1}). Although the development of quantitative methods in NIR will be based on a PLS approach or similar, observation of the spectra ensures that the spectral zone used contains relevant information.

From the spectra recorded at-line by the operators and from a reference method (IR analysis according to ISO standard 8985 – revision 1998), a PLS calibration line can be established as shown in Figure 39.15. The root mean square error of prediction (RMSEP) of the method is very close to the variability of the reference method ($1\sigma=0.3\%$).

However, it is interesting to note that in addition to the level of vinyl acetate present in these copolymers, the NIR analysis does not involve any sample preparation, which makes it possible to measure the level of residual monomers present in these copolymers' products. Indeed, the relatively volatile vinyl acetate monomer disappears during the manufacture of hot films for the measurement of MIR spectra. This monomer is characterized by a sharp band at approximately 6100 cm^{-1} corresponding to the 2 $\nu C=C$ elongation easily visible in Figure 39.13.

In addition, it is possible in the case of polymers to use the NIR spectra to estimate the degree of polymerization. This degree of polymerization measured in spectrometry by the ratio between the amount of chain ends and the total chain is generally measured by indirect methods such as size exclusion chromatography (molecular weight distribution) or melt viscosity measurements (the simplest is the MFI) or in solution (inherent viscosity). The ideal reference measurement for this physical property important to the application is NMR whose cost is prohibitive for a routine measurement.

The fact of being able to analyze samples in the form of powders, granules, plates, chips, etc. without preparation makes it possible to have qualitative information on the physical organization of these samples.

In Figure 39.16, we can see the NIR spectra of a photographic coupler (114EMZ whose structure is given in the Figure), which can have two distinct crystalline forms involving the NH bond in free or bound form (around 6600 cm^{-1}). This chemical feature of little interest for this product can be critical in the case of the active ingredients of drugs. In this case, NIR spectroscopy can be a very effective means of verifying that the right product is present even directly in the finalized drugs.

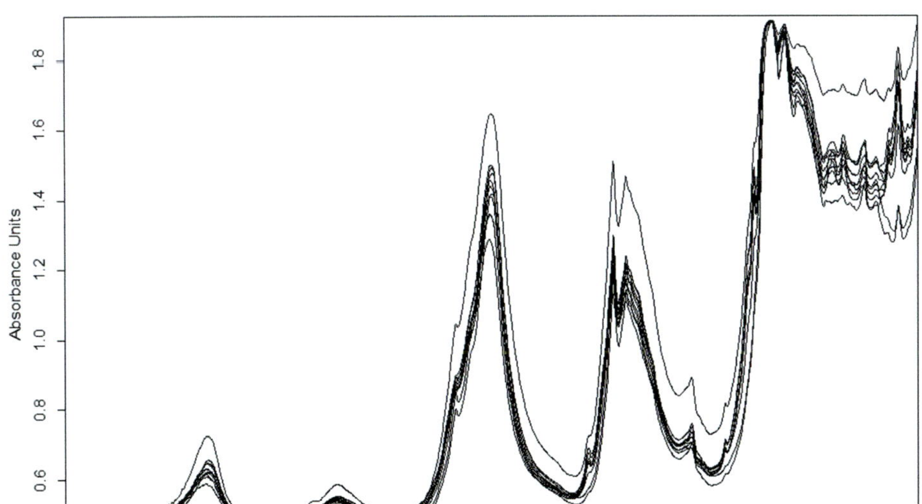

FIGURE 39.13 NIR diffuse reflectance spectra of ethylene/vinyl acetate copolymer granules ranging from 5% to 42% of acetate copolymer.

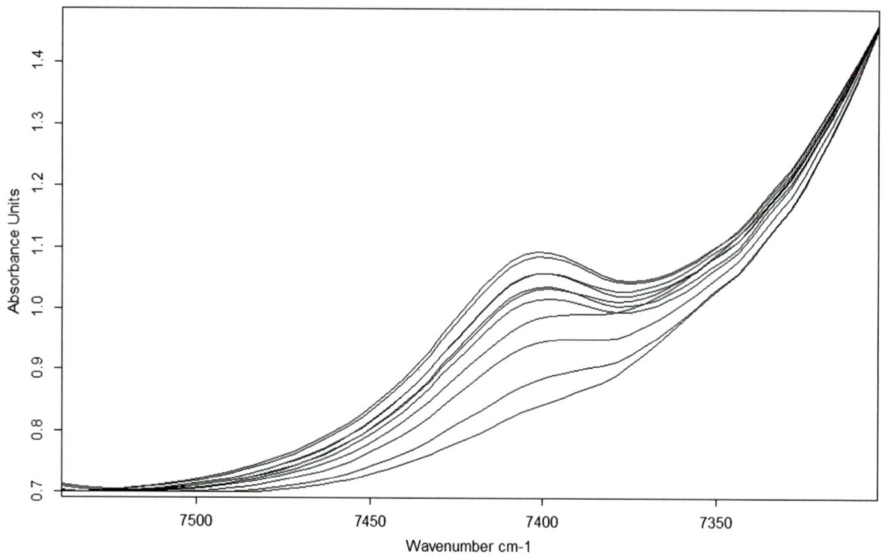

FIGURE 39.14 NIR diffuse reflectance spectra of ethylene/vinyl acetate copolymer granules ranging from 5% to 42% of acetate copolymer (from bottom to top, respectively).

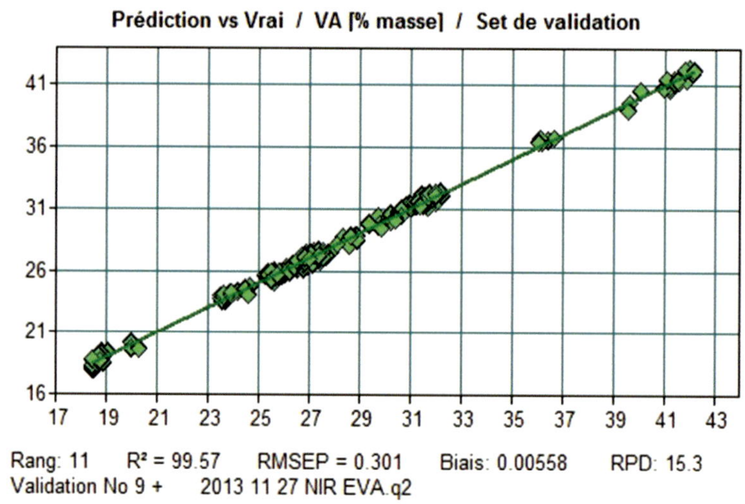

FIGURE 39.15 Quantitative analysis of ethylene/vinyl acetate copolymers (PLS method).

The use of discriminant chemometric methods, such as the SIMCA (soft independent modeling of class analogy) approach, will make it possible to identify and classify products according to their chemical nature and possibly their crystalline form (Figure 39.17). Note in Figure 39.17, the two polymorphic forms of 114EMZ.

39.4.4 Gas Analysis

The analysis of the gases is not really favorable in the field of NIR since the signal measured for the harmonics and the combinations of the modes of vibration is one to several orders of magnitude smaller than the signal corresponding to the vibrations fundamental. However, the sensitivity of the

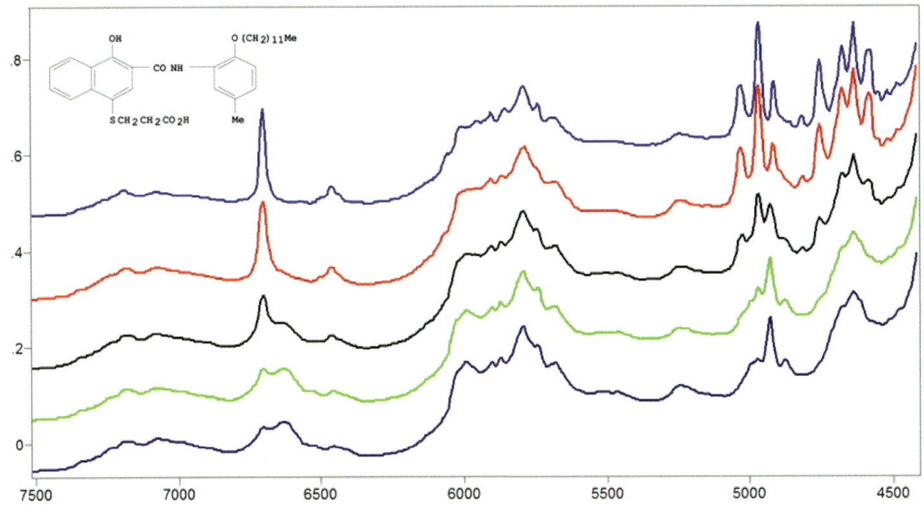

FIGURE 39.16 NIR spectra of powder samples with different crystallization form.

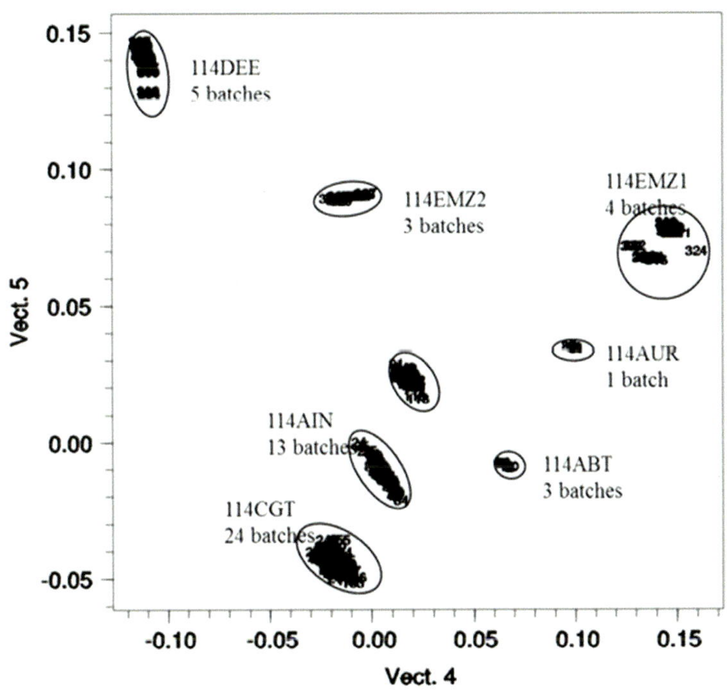

FIGURE 39.17 Classification of photographic couplers from their NIR spectra.

detectors and the power of the lamps used in the NIR field makes it possible to partially compensate for this handicap and to consider using it in a few particular cases.

Figures 39.18 and 39.19 show the gas-phase NIR spectra in an acrylic acid stream. The optical path used is 2 m, which gives ODs of the order of 0.2–0.3 approximately. As in the case of MIR spectroscopy, the elongation of the optical path using multireflection systems could make it possible to gain an order of magnitude in intensity.

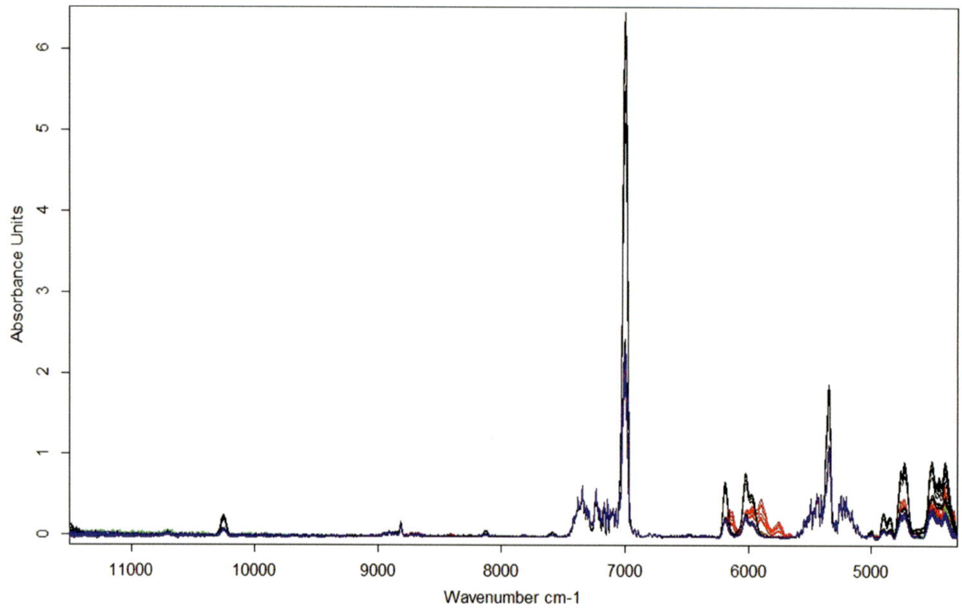

FIGURE 39.18 NIR spectra of an acrylic acid gas stream at about 10% in a nitrogen gas stream.

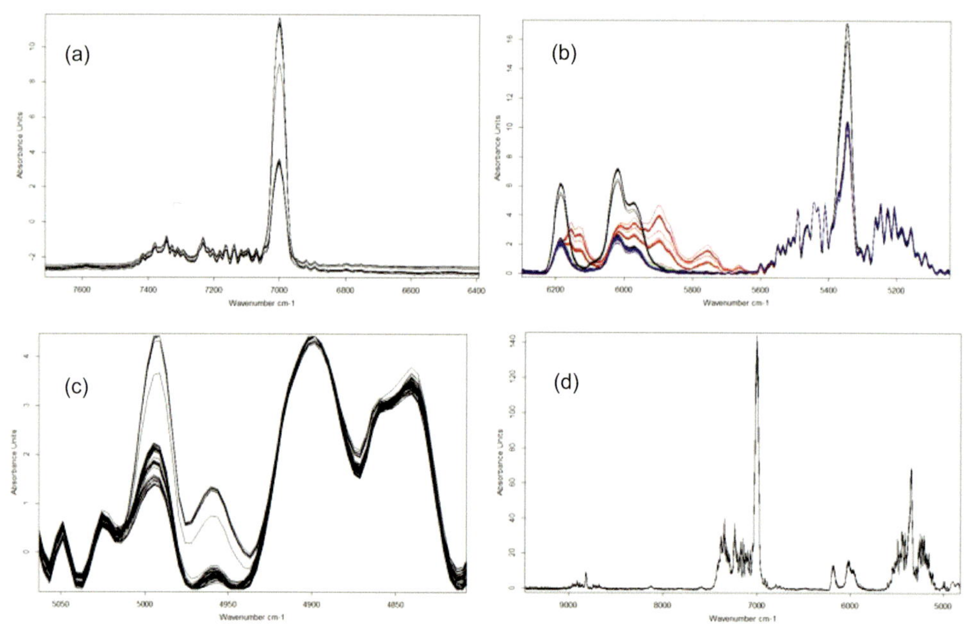

FIGURE 39.19 NIR spectra of a flux of acrylic acid gas at about 10% in a flow of nitrogen gas: (a) acrylic acid zone, (b) variations around propylene, (c) variations around CO_2 and (d) variations around water vapor.

The variations shown in Figure 39.19 indicate that it is possible to visualize relative changes of about 10% on initial concentrations. The initial concentrations are ranging from about 1% to 10% depending on the compounds.

However, the use of NIR spectroscopy in the gas field is underdeveloped. For Raman, some new probes are emerging.

39.5 CONCLUSION

NIR spectroscopy is a spectroscopy technique in its own right that allows qualitative and quantitative information to be obtained quickly and without sample preparation.

The interpretation of the spectra is complicated by the overlapping of bands due to the multiplication of the possible modes (harmonics and combinations of active modes in MIR and/or in Raman). It is, nevertheless, important to take the time to observe the spectra to define the zones containing relevant information, especially in the idea of applying the technique to industrial processes.

The lack of sample preparation is a significant advantage in using chemometric tools to process the data as the variations that will be observed are likely to be related to the nature of the product rather than the way in which the spectra were obtained (which is a problem in the MIR and even more so for the NMR, the MS, and the chromatographic techniques).

Even if this chapter has only dealt with the spectroscopic aspect of NIR, it is generally recognized that the combination of NIR and chemometrics is a tool of first choice for quantitative and qualitative analyses in an industrial environment. In particular, it is much easier to use than chromatography techniques (liquid or gaseous) online (simplified maintenance, cost reduction). The sensitivity of the NIR is probably smaller than those of chromatographic-based methods, but its speed of response can really make it possible to monitor the processes while keeping the separation techniques for the final quality control of the products.

REFERENCES

Abbas, O., Lecler, B., Dardenne, P., Baeten, V. 2013. "Detection of melamine and cyanuric acid in feed ingredients by near infrared spectroscopy and chemometrics." *Journal of Near Infrared Spectroscopy*, 21(3):183–194.

Alig, I., Steinhoff, B., Lellinger, D. 2010. "Monitoring of polymer melt processing." *Measurement Science and Technology*, 21(6):062001.

Ben-Gera, I., Norris, K. H., 1968. "Determination of the moisture content in soybeans by direct spectrophotometry." *The Israel Journal of Agricultural Research*, 18(3): 125–132.

Bertrand, D. 2002. "La spectroscopie proche infrarouge et ses applications dans les industries de l'alimentation animale." *Production Animale*, 15(3):209–219.

Cinier, R., Guilment, J. 1998. "Qualification of solid raw materials using near infrared spectroscopy." *Journal of Near Infrared Spectroscopy*, 6:A231–A237.

Dalibart, M., Servant, L. 2000. "Spectroscopie dans l'infrarouge", Techniques de l'ingénieur, P2845 v3, https://www.techniques-ingenieur.fr/base-documentaire/archives-th12/archives-techniques-d-analyse-tiata/archive-3/spectroscopie-dans-l-infrarouge-p2845, last accessed 9-Dec 2018.

Fontange, C., Chauchard, F. 2009. "Introduction à l'apport des techniques proche infrarouge pour le contrôle des procédés de l'industrie pharmaceutique et démarche PAT" SPECTRA ANALYSE, 269. http://www.spectraanalyse.fr/wp-content/uploads/2013/07/SA269_27-35.pdf, last accessed 9-Dec 2018.

Hart, J. R., Norris, K. H., Golumbic, C. 1962. "Determination of the moisture content of seeds by near-infrared spectrophotometry of their methanol extracts." *Cereal Chemistry* 39:94–99.

Kradjel, C., Lee, K. A. 2007. NIR analysis of polymers. In E. Ciurczak, D.A. Burns (eds.) *Handbook of Near-Infrared Analysis*, 3rd Edition, pp. 529–568. CRC Press, Boca Raton, FL.

Shenk, J. S., Landa, I., Hoover, M. R., Westerhaus, M. O. 1981. "Description and evaluation of a near infrared reflectance spectrocomputer for forage and grain analysis." *Crop Science*, 21(3): 355–358.

Watari, M. 2014. "A review of online real-time process analyses of melt-state polymer using the near-infrared spectroscopy and chemometrics." *Applied Spectroscopy Reviews* 49(6):462–491.

Watari, M., Higashiyama, H., Mitsui, N., Tomo, M., Ozaki, Y. 2004. "On-line monitoring of the density of linear low-density polyethylene in a real plant by near-infrared spectroscopy and chemometrics." *Applied Spectroscopy*, 58(2): 248–255.

40 Pharmaceutical Analysis

Benoît Igne[*]
GlaxoSmithKline

Emil Ciurczak
Doramaxx Consulting

CONTENTS

40.1 INTRODUCTION

Near-infrared spectroscopy (NIRS) has been used in the field of pharmaceutical analysis since the early 1980s for raw material analysis (Ciurczak, Torlini, and Demkowicz 1986), and qualitative (Ciurczak and Maldacker 1986) and quantitative analyses of powders and solids (Lodder, Selby, and Hieftje 1987). A review by Ciurczak provides an overview of the early applications (Ciurczak 1987). In 1988, Lodder discussed the situation that the sampling cells of equipment available at the time were too large, required grinding of the sample or with a spot size too small to accommodate heterogeneous samples, and proposed a new sample holder and reflector to accommodate for intact tablet analysis (Lodder and Hieftje 1988a). Since these first developments in the pharmaceutical industry with instruments mainly designed for agricultural product analysis, there has been a significant adoption of the technology by pharmaceutical companies and academic groups. Today, instruments located in technical rooms adjacent to the manufacturing floors, or self-contained systems (wireless units),

[*] igneben@gmail.com

are used to monitor and control manufacturing processes for raw material identification, active ingredients, and final dosage form manufacturing. In this chapter, a review of the use of NIRS in the pharmaceutical industry will be provided. Section 40.2 will focus on use of the technology for raw material identification. Section 40.3 will discuss the monitoring and control of chemical reactions used in the manufacture of active pharmaceutical ingredients. Section 40.4 will describe how processes involved in the drug product manufacturing are evaluated by NIRS.

40.2 RAW MATERIAL IDENTIFICATION

Before being used to manufacture products, the identity and quality of the raw materials should be confirmed upon receipt to ensure that only the correct materials are used. Historically, the identity test has been performed by infrared (IR) spectroscopy in a laboratory adjacent to the receiving area, with further physical tests performed in a QC laboratory. However, it is more and more desirable to perform the test directly on the incoming materials, and thus, hand-held devices, based on NIRS and Raman spectroscopy, are being used for that purpose. While Raman and IR spectroscopies have a specificity advantage over NIR, a large body of literature exists on its use for raw material identification. Moreover, NIRS is more sensitive to physical parameters, especially particle size.

Much has been written on the topic, and there are many approaches to performing identification by NIR. A comprehensive review was written by Luypaert, Massart, and Vander Heyden (2007). However, there are three main recurring themes that outline the work done in this area:

- The creation of a relevant spectral library
- The chemometrics used to accurately determine the membership of a sample
- The transfer of these methods across a network of instruments.

Mainly for a qualitative application, the usual first step is to collect a relevant body of data representing the variability of the material to identify, a.k.a. "a library". Unlike a quantitative model whose goal is to represent potential sources of variations that a process or product could encounter, a qualitative model for identification will try to be sensitive to differences in polymorphic forms, particle size, degree of crystallinity, or purity. Thus, care must be taken to include enough sources of variations to avoid false negatives as the model would be unreliable when new lots of raw material, with differences not observed before, are measured; however, a spectral space too large decreases the ability of the model to correctly identify outliers.

Once a library has been created, it may prove difficult to rely solely on band location to confirm the identify of a material. This is mostly due to the limited specificity of NIR as compared with IR and Raman spectroscopies where peak location is often sufficient. Multivariate techniques can be used to develop reliable methods. Blanco and Romero (2001) provided an overview of the methods available for classification. Methods based on discrimination of the raw spectra (Dreassi, Ceramelli, Corti, et al. 1995), discriminant analysis with principal component analysis (PCA) and partial least-squares (PLS) (Andre 2003, Dreassi, Ceramelli, Savini, et al. 1995), correlation (Plugge and van der Vliest 1992), conformity index and distance (Ritchie, Mark, and Ciurczak 2003), and soft independent modeling of class analogy (SIMCA) (Krämer and Ebel 2000), have mostly been used to develop identification methods. With progress in machine learning methods, artificial neural networks and support vector machine (SVM)-based algorithms are also being investigated (Yang et al. 2011, Sun et al. 2016). Sun et al. (2016) looked at 253 different pharmaceutical compounds and compared algorithm performance across lots and spectrometers. Their results showed that when using a single instrument and similar lots, all algorithms performed very well. However, when including lot-to-lot variability and instrument variability, the performance of classical methods was significantly affected, but the machine learning-based methods were more robust.

This example touches on a very critical part of the use of any identification method. Once built on an instrument or a set of instruments, it is critical that its transfer onto other instruments be planned.

The method should be readily available and proved equivalency between units in case of a failure of the original instrument or transfer of a product to another manufacturing site. Readers should refer to the calibration transfer chapter for more information on this topic.

Counterfeit detection should also be considered as an identification problem that uses many of the same tools mentioned above. A comprehensive review was published by Krakowska et al. (2016).

40.3 DRUG SUBSTANCE

The manufacturing of pharmaceutical dosage forms usually involves two steps: the synthesis and purification of the active ingredient followed by the large-scale manufacturing of oral dosage forms (tablets, capsules). They are normally seen as two individual steps, often taking place at different manufacturing sites, but there are examples of end-to-end manufacturing (Lakerveld et al. 2015).

Because the NIR light penetrates relatively deep in the sample and is not as easily absorbed as the fundamental IR vibrations, it can be used to monitor synthetic reactions (monophasic and biphasic) and isolation methods such as crystallization and spray drying. In this section, examples of reaction monitoring and control will be provided.

40.3.1 Reaction and Process Monitoring

By simply introducing a NIR probe in a reactor or a process, a large amount of information can be gathered. Chemists and engineers are usually interested in the kinetics of reaction, by-product formation, and/or reaction end-point determination. While these goals are traditionally achieved by collecting off-line samples (after ensuring the quenching does not affect the chemicals of interest), followed by high performance liquid chromatography (HPLC) analysis, it can take several hours to obtain a simple result. In many situations, NIR can be used to track a reaction and gain insights into what is taking place in the vessel. In the following sections, a few examples will be covered, ordered by increasing complexity of the modeling approach.

The simplest approach to reaction monitoring is band or peak tracking. If a particular species is well resolved from the other chemicals, it is possible to follow a process by simply looking at the absorption value(s) at a particular or set of wavelengths as a function of time. An example of such process is the monitoring of lyophilization. Brülls et al. (2007) used NIRS and the area under the curve at peak maximum for the first OH overtone (5200 cm^{-1}, 1940 nm) to follow a lyophilization process. Figure 40.1 shows the change of the OH combination band as a function of the moisture level in the samples. After the identification of the peak's minima and maximum, to ensure appropriate baseline correction, the peak area was regressed against the known moisture values measured by Karl-Fisher to build a simple univariate model that was used to predict the water content as a function of drying time.

Qualitative determination of the ratio of chemicals using multivariate curve resolution (MCR) was demonstrated by Blanco, Castillo, and Beneyto (2007) using the acid-catalyzed esterification of myristic acid by isopropanol. The monitoring of the conversion rates and yield as a function of time were provided by MCR. Figure 40.2 presents the obtained trends that were later confirmed by the off-line measurements. MCR is based on the principle of Beer's law and uses whole or partial spectrum to reconstitute the spectrum of the chemical species at various time points and their relative concentration during the reaction (Tauler, Kowalski, and Fleming 1993). While PCA, another qualitative method, can certainly be used for reaction monitoring, MCR ensures specificity and depending on the constraints employed, can yield quantitative results (Antunes et al. 2002). PCA will decompose the spectra as a function of variance. If there are intermediate species that form, the shape of the principal components may be distorted and not allow the appropriate monitoring of the reaction. MCR will identify that new species are formed using the equality constraint that ensure specificity for known species. For liquid systems with a limited number of reagents, MCR is a very useful method to extract the relevant information from NIR spectra.

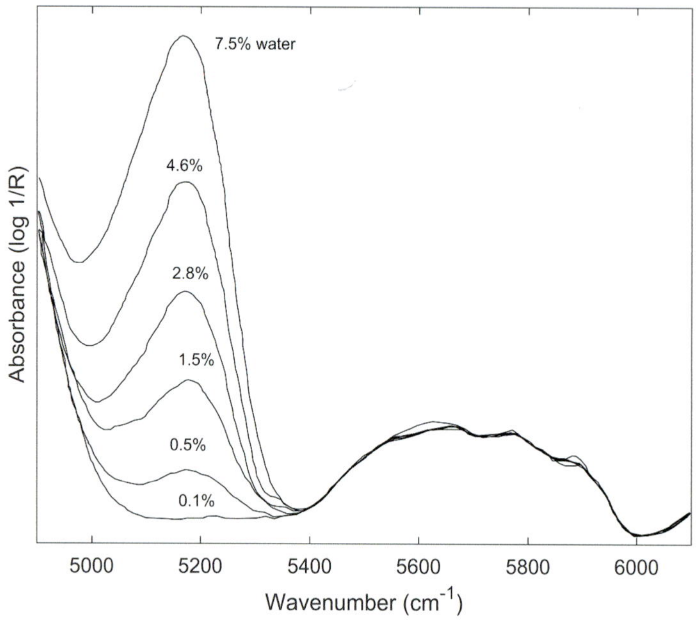

FIGURE 40.1 Standard normal variate pretreated spectra of samples of hydroxypropyl-β-cyclodextrin at various moisture levels. (Reproduced with permission from Brülls et al. (2007).)

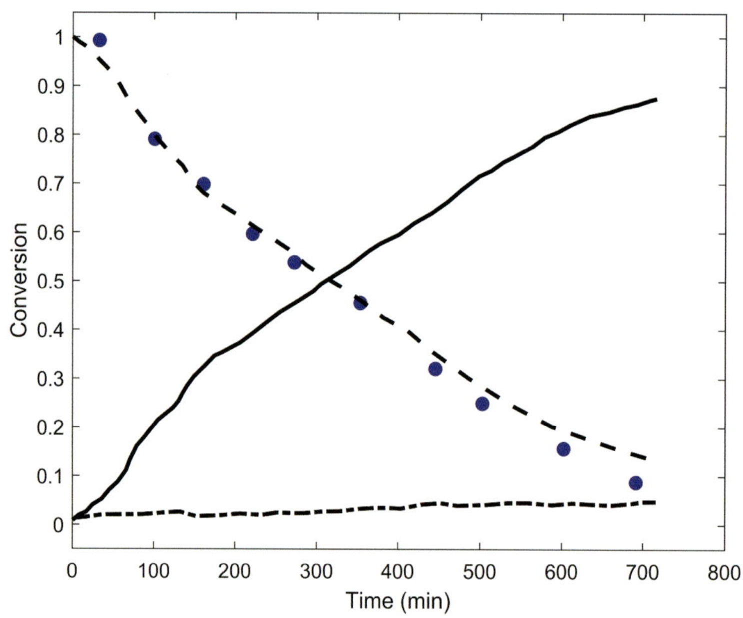

FIGURE 40.2 Predicted concentration profiles. (—) myristic acid; (--) isopropyl myristate; (-.-) isopropanol; (•): acid concentrations obtained by titration. (Reproduced with permission from Blanco, Castillo, and Beneyto (2007).)

In distillation processes, knowing the solvent composition is often a critical information to product quality. Taking off-line samples for gas chromatography (GC) is the traditional method of analysis but in many circumstances, it can be difficult and can risk operators safety. In an example provided by Chanda et al. (2015), a distillation process involving the removal of ethyl acetate and its replacement by isopropyl alcohol (IPA) was monitored by NIR. The authors developed a quantitative model for the distillate in a process that involved stepwise additions of IPA. A design of experiment was conducted to build the calibration model, with GC used to determine the solvent composition off-line. Figure 40.3 shows the solvent swap profile as predicted by NIR, which allowed the reaction to be controlled within lower and upper limits linked to the operating space of the next unit operation.

In a similar example, PLS was used to determine the solvent composition of a crystallization system to determine the seeding point (Hamilton et al. 2015). Using NIRS, the authors built a chemometrics model with spectra collected at various scale reactors (1 L and 20 L). Nuclear magnetic resonance spectroscopy was used to determine the off-line sample composition. Figure 40.4 presents the spectral variation taking place during a 1-L reaction (Figure 40.4a) and the cross-validation results of the model built from the 1-L and 20-L reactor data (Figure 40.4b).

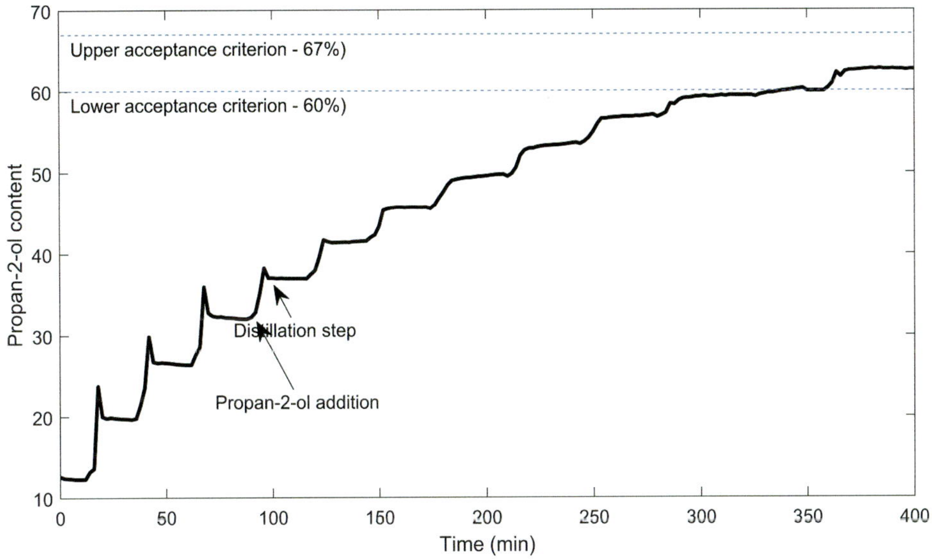

FIGURE 40.3 Distillation and solvent swap profile over time. (Reproduced from Chanda et al. (2015); permissions related to the material excerpted should be directed to the American Chemical Society.)

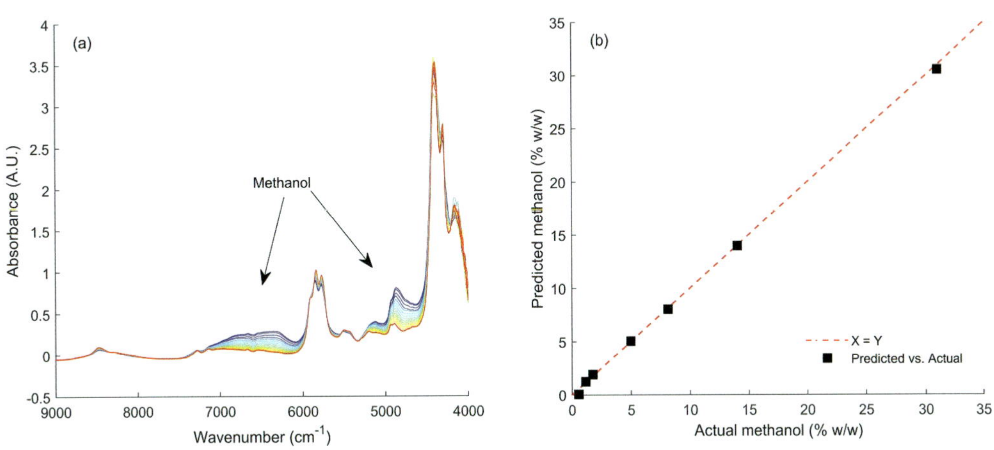

FIGURE 40.4 Example NIR spectra recorded during a 1-L scale process (a). Observed-vs-predicted plot using data from 1-L and 20-L scales (b). (Reproduced with permission from Hamilton et al. 2015.)

40.3.2 Purification and Crystal Form

The last step of a synthesis is usually an isolation and purification process. The most common processes used in the pharmaceutical industry are crystallization and spray drying. In this step, the molecules of interest are separated from their solvent either through the formation of crystals by using temperature and solubility as the main driver for precipitation of the molecules out of solution or through evaporation of the solvents. In crystallization, the form of the crystal and the yield are two of the most important parameters for which NIR can be used. The form is important because the stability and solubility of an active are based on the energy level of its crystal structure. Thus, during crystal formation or transition reactions, it can be necessary to monitor the habit of the crystal being formed. As the crystallization progresses, ensuring that a maximum of active ingredient has left, the liquid state can also be of interest as extended crystallization times may affect the amount of residual solvents.

The first analyses of polymorphism by NIR were reported by Ciurczak (1987). A large number of studies have since shown the power of NIR for off-line analysis for the quantification of crystalline and/or amorphous forms in powder mixtures or tablets (Aldridge et al. 1996, Blanco et al. 2000, Aaltonen et al. 2003, Heinz et al. 2007, Otsuka, Kato, and Matsuda 2001). A method to determine polymorphic purity of azithromycin was proposed by Blanco et al. (2004). Since the amorphous phase is anhydrous but the crystalline form is a hydrate, the authors used the water bands to build PLS models to quantitatively determine the purity of the samples.

As the instrumentation allowed analysis *in situ*, NIR was used to monitor polymorphs in the reactors. A study by Févotte et al. (2004) looked at using NIR to explore the effect of crystallization parameters onto the crystallization and form transition process of an active ingredient. Using a transflectance probe, processing parameters were varied to understand their effect on reaction kinetics. The effects of temperature, habit and size of seed crystals, and quality of solvents were studied with NIR predicting for each condition the concentration of Form II of the drug in the vessel. Barnes et al. (2010) provided an example of application of the monitoring of a polymorphic transformation process from pilot to industrial scale. The authors built a model for the prediction of the form of interest with PLS. The reaction monitoring used a fiber optic probe equipped with a purge system that allowed nitrogen to flow in from of the sapphire window to prevent probe fouling.

40.3.3 Summary

These examples are a high-level overview of what has been done by industry and academia to monitor drug manufacturing processes using NIR. A large body of literature exists on how NIR can be used to control pharmaceutical processes but this is beyond the scope of this chapter. An example of control, including feedback, was published by Cervera-Padrell et al. (2012) using NIR to feedback control a reaction intermediate in a continuous active ingredient manufacturing process.

40.4 DRUG PRODUCT

Active ingredient manufacture typically aims to achieve a particular bioavailability profile and stability. Crystals are usually not designed with flowability in mind. Additionally, depending on the potency of the molecule, the drug substance may need to be formulated with other ingredients to deliver a reasonable size product to patients (some medicines are mostly made of active ingredients, but others have less than a percent drug load). When considering drug product (capsules, tablets), one of the most important manufacturing requirements is powder flowability. If powders cannot be delivered to a tablet press or an encapsulator for high-speed and large-scale use, the properties of the powder must be modified using inactive excipients and processing unit operations. Three main processes are usually employed:

- Direct compression for the well-flowing active ingredients
- Granulation (dry or wet) for poorly flowing active ingredients
- Extrusion for when bioavailability of the drug may need to be increased, when blend homogeneity cannot be assured or when tablet compressibility is difficult.

The final product may be coated to protect the active ingredient from the environment for stability or mask an unpalatable taste. In this chapter, the monitoring and control of unit operations involved in direct compression, granulation, extrusion, tableting, and coating will be presented in the context of batch and continuous manufacturing. The final dosage form analysis will also be discussed.

40.4.1 Blending

Blending, also called mixing, is used to combine active and inactive pharmaceutical ingredients to obtain a powder that can further be processed into tablets/capsules or be granulated/extruded. The main quality parameters of a blend are its assay and uniformity. For batch blenders, an assay consists of measuring the content of a representative amount of powder, and the uniformity represents the standard deviation of samples collected at various locations. For continuous blenders, the uniformity of the powder coming out is evaluated for mean and standard deviation as a function of time. Powder blending can be challenging in reason of the interaction of ingredient physical and chemical properties (particle size, shape, density, moisture, electrostatic charges) that can cause agglomeration and segregation. In addition, thief sampling, the traditional method for evaluating blend quality in batch processes, is known for inducing errors in the evaluation of quality parameters (Poux et al. 1991).

For these reasons, a large body of literature exists on the use of NIR for monitoring and controlling powder blends. Unlike traditional sampling that usually measures powder samples at the end of the blending process at ten or more different locations, NIR provides the possibility, using wireless units, to look at the spectrum of one location at every blender rotation. The nature of the measurements is significantly different but provides information not obtainable using traditional sampling methods: understanding of blend kinetics, comparison of blend trends between batches, measurements of not only the active ingredient but also the excipients (most excipients are spectrally active in the NIR range), and a possibility to call the blend end-point, which may vary from batch to batch based on differences in powder properties and loading. To bridge the traditional sampling method and NIR sampling, El-Hagrasy and Drennen (2006) measured the spectrum of several locations of a V-blender at various blending time points to assess the blending kinetics within and across shells. The authors noted that "intrashell mixing occurred at a fraction of the time needed to achieve intershell" mixing. Osorio et al. (2014) used NIR imaging to look at not only the macromixing (that single point NIR provides) but also micromixing events taking place during blending. A NIR camera was used to image 1 cm^2 of the blend at each rotation. The evolution of aggregates (formation and breakages) was observable, and the authors proposed that this method could be used to optimize formulations and blending times. Similar work was performed by Ma and Anderson (2008) after compression of entire blends and imaging of their cross sections.

Blends have mostly been monitored and controlled using either a comparison of spectra within a batch to identify when the various components stop changing (qualitative evaluation) or the prediction of ingredient concentration as a function of time (quantitative evaluation). These will now be discussed.

40.4.1.1 Qualitative Analysis for Blend Monitoring and Control

The spectra have directly been used to monitor and stop blends. In 1988, Lodder et al. used a bootstrap method on each variable to group spectra with similar features and applied a confidence interval around the cluster to determine the membership of new spectra to that space (Lodder and Hieftje 1988b). This method requires to have representative spectra of homogeneous material. The self-modeling SIMPLISMA approach was tested in 1991 (Windig and Guilment 1991). SIMPLISMA aims at selecting wavelengths with the highest selectivity with respect to the parameter of interest. The authors noted that "assuming that Beer's law is followed, each pure wavelength is proportional to the concentration of the compound absorbing". The evolution of the variance of the pure wavelengths was then used to estimate the evolution of homogeneity. This approach, however, only works when there is at least a unique peak for the active, unfaceted by excipients, which is far from common with many pharmaceutical formulations.

The previous analyses were performed on sampled powder from the blend and off-line analysis by NIR. Hailey et al. and Sekulic et al. reported the on-line monitoring of powder blend homogeneity by NIRS in 1996 (Hailey et al. 1996, Sekulic et al. 1996). Sekulic et al. (1996) used a moving block standard deviation calculation to identify the time at which mixing was complete. Figure 40.5 shows an example of NIR data for a blending process and the application of the moving block standard deviation approach (Figure 40.5b).

PCA has extensively been used to characterize blends. From the first publications in 1987 (Ciurczak 1987), it has been used in a multitude of ways. Wargo and Drennen (1996) used PCA prior to bootstrap analysis to identify small changes in the spectra that bootstrap on the variables could not properly monitor. SIMCA was tested to determine the blend end-point (Sekulic et al. 1998). The authors showed that by developing a PCA model by only selecting spectra corresponding to homogeneous powder, the prediction of similar end-point for additional runs of the same composition was possible. An F-test on the residual variance was employed to estimate if a new sample belonged to the target class. PCA model diagnostics, such as Hoteling's T^2, have also been used to determine the membership of a spectrum to the desired homogeneity state (De Maesschalck et al. 1998, Igne et al. 2014). Figure 40.5c and d illustrates PCA scores and Hotelling's T^2 trends for a batch.

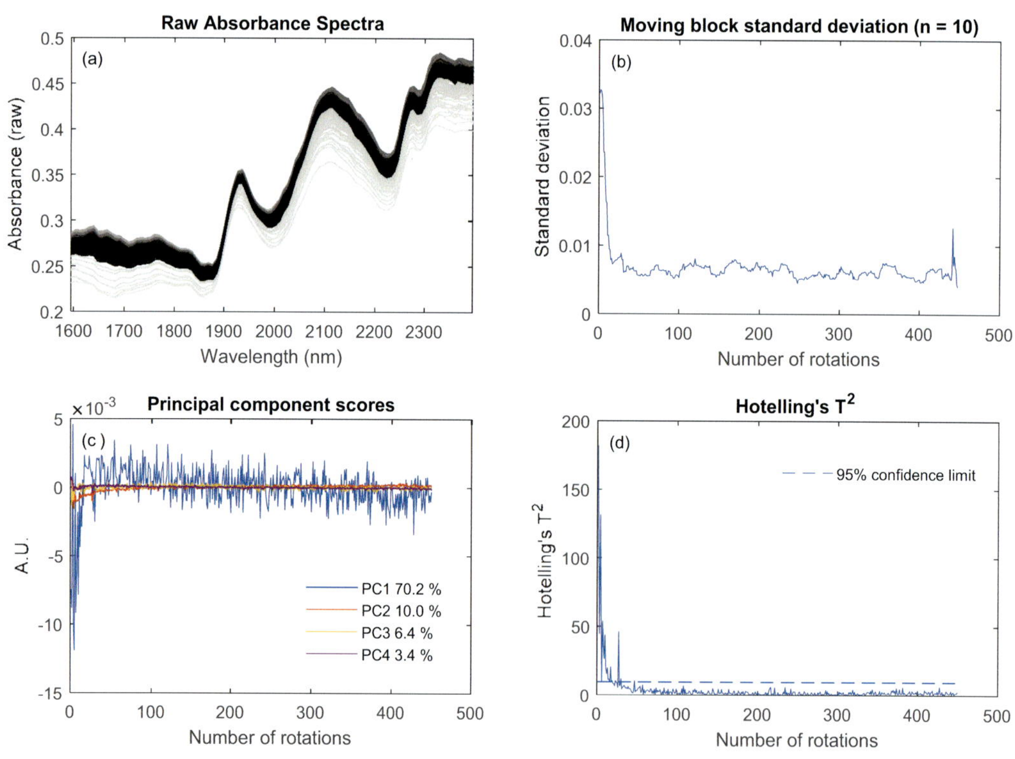

FIGURE 40.5 Examples of blend monitoring algorithms. (a) Raw NIR data (gray to black as function of blending time, (b) moving block standard deviation, (c) principal component analysis scores, and (d) Hoteling's T^2 trend from the PCA.

Puchert et al. (2011) proposed an extension of the use of PCA with the principal component score distance analysis approach. In their study, they calculated the Euclidian distance between two successive spectra considering the first three factors. A moving block standard deviation was then applied onto the distance terms. In addition, the authors used a SIMCA-like approach by developing successive PCA models with only the most stable spectra in calibration. It allowed them to obtain a better resolution when projecting new samples.

Originally proposed for the determination of multivariate figures of merit (Lorber 1986) and later employed as a preprocessing method for quantitative models (Goicoechea and Olivieri 2002), net analyte signal (NAS) was used to characterize the blend mixing processes. The NAS represents the part of the signal related to the component of interest that is orthogonal to the subspace formed by the other components of a mixture. A NAS value can be calculated for each sample. Skibsted et al. (2006) calculated the NAS of a "golden batch", considered homogeneous, and used it during subsequent blending experiments to compare the variance of the NAS between the golden batch and the batch to characterize.

Most of the methods described above have provided either an evaluation of a particular component (peak trending with SIMPLISMA) or the whole batch (moving block standard deviation, distance methods, and PCA). However, it can often be useful to be able to resolve the trends of the individual components, especially when they have functional impacts on the final tablet. MCR was used to that aim. Jaumot et al. (2013) used MCR to track the evolution of all the components in a formulation. Figure 40.6a shows an example of MCR monitoring of a blend.

40.4.1.2 Quantitative Analysis for Blend Monitoring and Control

Shi et al. (2008) noted the following: "While it is frequently argued that the concentration variation is implicit in qualitative methods, the expression of variation is explicit in quantitative methods due to the specificity of a calibration model for a particular concentration variation". In recent years, blend characterization by quantitative means has become particularly popular. Quantitative approaches require a calibration model to capture the concentration variation during the blending process. A quantitative model is able to express blending processes in terms of concentration variations, which is comparable to the standard criteria of current regulatory requirements.

Compared to qualitative methods that aim at detecting changes or rather the lack of change in spectral variation to determine end-point, the evolution of the predicted values is used when

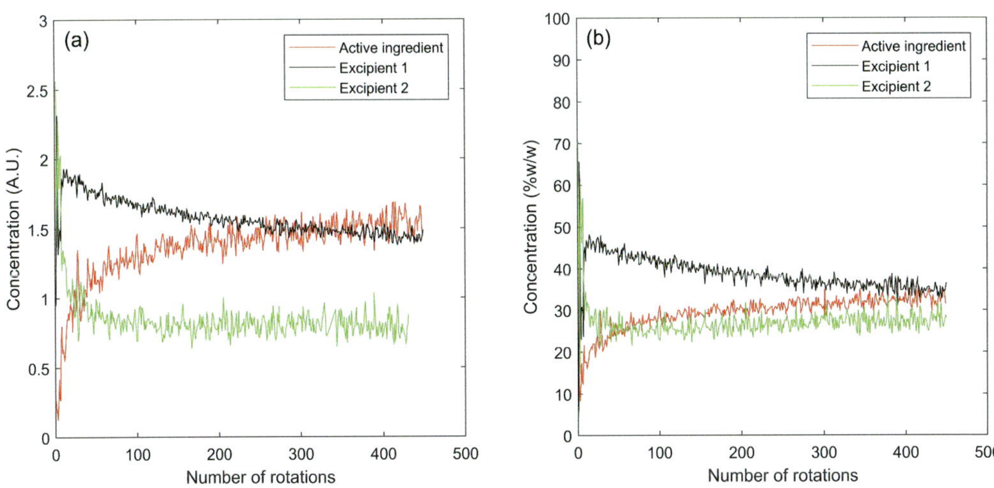

FIGURE 40.6 Examples of blend monitoring by MCR. (a) MCR-resolved concentration profiles. (b) MCR-resolved blending concentration profiles when correlation constraint is imposed. (Reproduced with permission from Jaumot et al. 2013.)

employing quantitative approaches. PLS regression has been the primary method used to develop quantitative models (Berntsson et al. 2002, Popo et al. 2002, Shi et al. 2008). However, other less popular modeling methods have been employed and shown to perform adequately (Wu et al. 2009, Zacour et al. 2011). Zacour et al. (2011) compared PLS to various implementations of classical least squares regression (involving augmentation procedures of the K matrix with derived shapes of interferences). MCR was also tested while using the correlation constraint that allows quantitative results to be obtained (Jaumot et al. 2013) (Figure 40.6b).

In most situations, reported results focused on the monitoring of the active ingredient alone as it is often the only parameter that is being measured by reference analyses. However, it is not the only ingredient to have been monitored. The prediction of the distribution of lubricant (magnesium stearate) was found to be possible at levels as low as 0.5% w/w (Duong et al. 2003). Recognizing the importance of excipient homogeneity for the performance of the delivery form, Shi et al. (2008) introduced a pooled statistics to not only consider the distribution of the active, but also of the excipients.

The root mean square error from the nominal value (RMSNV) is a weighted, cumulative, pooled standard deviation indicator that takes into account the deviation of the predicted concentration of the major components of a mixture to their target concentration, over a given number of rotations. For an n component system, it is defined as

$$\text{RMSNV}_{t-i}^{t} = \sqrt{\frac{\sum_{j=t-i}^{t} w_1\left(\hat{Y}_{1j}-Y_{1j}\right)^2+\sum_{j=t-i}^{t} w_2\left(\hat{Y}_{2j}-Y_{2j}\right)^2+\cdots+\sum_{j=t-i}^{t} w_n\left(\hat{Y}_{nj}-Y_{nj}\right)^2}{i\left(\sum_{j=t-i}^{t} w\right)}},$$

where w is the weight given to a component, $\hat{Y}_{nj}$ is the predicted and Y_{nj} is the target value for the nth component, i is the number of data points used in a single window, t is the latest collected data point, and j is an individual prediction time. The RMSNV allows the evaluation of the trends and distance from target concentrations for not only the active pharmaceutical ingredient (API), but also of the major excipients of the formulation. In the original paper, the end-point was determined based on the distribution of RMSNV values. In Zacour et al. (2011), the root mean square error of calibration for the components of interests was used as threshold under which the blend was determined homogeneous. Example trends of RMSNV for a blend monitoring by a sensor on each arm of a V-blender are presented in Figure 40.7, showing different mixing kinetics.

The major drawback of quantitative models based on PLS or similar inverse methods is the need for calibration samples. Blends made of different compositions need to be acquired so a model can be built. This is often feasible at the small scale but rarely at pilot and manufacturing scales. However, this brings a major issue regarding calibration transfer. Models developed on small-scale blenders may not be directly transferable to the larger-scale blender because of the significant difference in density between scales. To address this problem, Mohan et al. (2018) used a testing system equipped with a load cell to apply various amounts of force to small amounts of powders to mimic different manufacturing scales.

40.4.1.3 Considerations for Other Types of Blenders

The techniques described above have mostly been applied onto bin or V-shell processes as they have been and still are the most employed blending methods. However, other types of mixers exist and should be considered. For instance, the monitoring of high shear blenders was discussed by Berntsson et al. (2002). Continuous blenders are another type of mixer. They are increasingly utilized in support to the development of continuous manufacturing lines. Colón et al. (2014) presented a comprehensive study on the quantitative model development for a continuous blender.

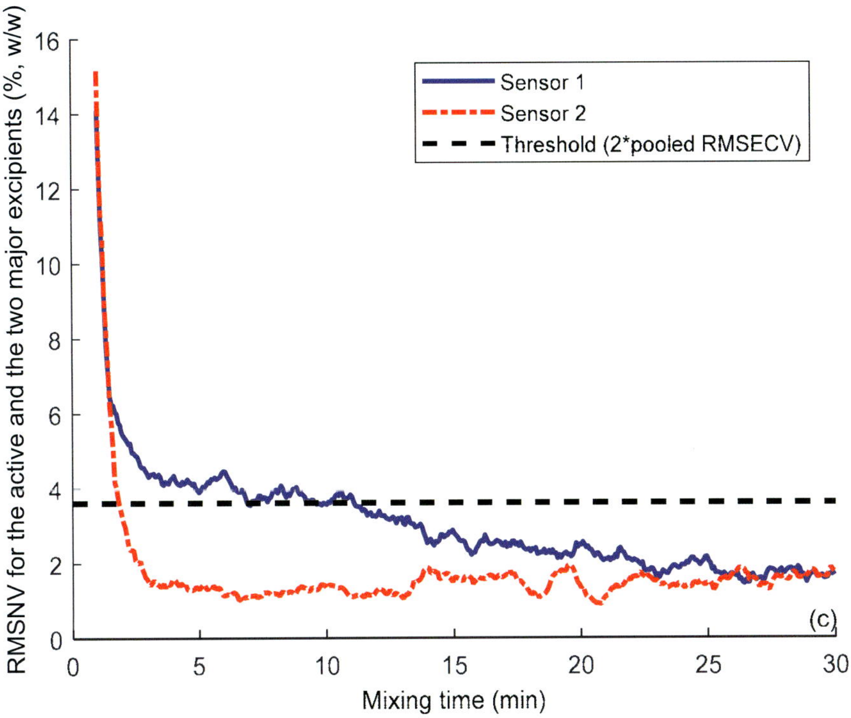

FIGURE 40.7 RMSNV example combining the API and two excipient predictions to evaluate the pooled blending trend from two sensors on each arm of a V-blender. (Reproduced with permission from Igne et al. 2013.)

Unlike batch blenders into which all the ingredients are dispensed at once and a large quantity of powder is mixed, continuous mixers (low or high shear) only mix small quantities of powder in a continuous fashion. While for batch blending, mixing kinetics are usually tracked for an end-point, continuous mixers rely on delivering homogeneous powder at a steady state of homogeneity. Finally, while in most batch blenders, spectra can only be collected every few seconds (when powder is against the sampling window), continuous mixers allow continuous monitoring. These differences require an understanding of sampling size and frequency to ensure that the relevant measurements are taken.

Effective sampling volume is important for both batch and continuous blenders. In batch blending, it allows us to determine the number of rotations that should be considered when trying to assess blend homogeneity. If a spectrum at rotation *n* only sampled 20 mg of powder, it is necessary to take into account several rotations to ensure that decisions are made on a relevant mass of powder (often related to the size of the final dosage form). In continuous blending, this is not different, except for the fact that the sampling frequency can be adjusted. Thus, it is important to understand how much powder is being analyzed by each spectrum (mass of powder passing in front of the probe during the collection of a spectrum) and the number of spectra to average to obtain a relevant measure. This was studied by Alam et al. (2017). The authors related powder flow rate to powder thickness for various angles for measurements.

40.4.2 Granulation and Drying

Granulation is usually employed when a formulation cannot be directly used to make final dosage forms. By creating larger particles, granulation improves flow characteristics, and compaction tendencies, and helps achieve and maintain content uniformity. There are several granulation processes: wet granulation, dry granulation, and extrusion.

40.4.2.1 Wet Granulation

During wet granulation, the dry solid products are blended before the addition of a solution. A binder is often used to help granule formation and can be either dispensed as part of the formulation or in the granulation solution. Wet granulation can be achieved by high shear granulators, fluid-bed granulators, or twin-screw granulators. The granulation must eventually be dried. Drying is completed by fluid-bed drying, tray drying, or microwave drying.

With the possibility to monitor moisture content and particle size, NIR has shown to be a valuable tool for the monitoring and control of wet granulations. Although granulation is not usually considered to be a mixing procedure, content uniformity is certainly a concern. NIR may be used to monitor the uniformity of granules.

Frake et al. (1997) used an on-line NIR system to monitor a fluid-bed granulation process. By tracking the water band, the authors were able to determine the granulation end-point. The relationship between the NIR signal and water content during granulation was confirmed by Rantanen et al. (1998), who used Karl Fisher to determine the water content corresponding to different granulation time points. Miwa and Makado (2009) used NIRS to not only track a granulation process but also determine the amount of water needed to achieve granulation. The authors used moisture levels predicted by NIR to estimate a lower and upper limit for the amount of water needed to granulate a multicomponent system. Li et al. (2007) used NIRS to troubleshoot and optimize a wet granulation process. During the development phases, the authors realized that out-of-trend content uniformity values were obtained for some sieve cuts due to the interaction of the excipients and the binding solution during the coalescence process. A qualitative method was put in place to determine within different sieve fractions the change in content uniformity.

The application of NIRS to high shear wet granulation was investigated by Jorgensen et al. (2004), who used the change in spectral slope of the NIR spectra to correlate with the various phases of the granulation process as determined by the torque of the impeller. The various phases of granulation were also studied by Kuriyama et al. (2018). The authors related the spectral baseline to the formation of API polymer granules.

Unlike high shear and fluid-bed granulation, twin-screw granulation is a continuous process. An example used chemical imaging to understand the effect of the process parameters onto the quality of the granules, particularly its water content (Vercruysse et al. 2014).

After granulation, the solvent added to enable binding needs to be removed. Rantanen et al. (1998) showed, in addition to the suitability of NIR to monitor fluid-bed granulation on-line, the possibility to use the technique to monitor drying operations. Zhou et al. (2003) discussed the possibility of using PCA to differentiate between the removal of surface water and bond water and the precise determination of end-points in situations where several hydrates coexisted. Considerations of model validation for the prediction of water loss during drying were studied by Peinado, Hammond, and Scott (2011). The authors describe the approach taken to determine the figures of merit required by external guidelines and discuss some of the interpretations of the published methods to accommodate for the accuracy of the end-point method. Rantanen and Yliruusi (1998) confirmed earlier findings by Ciurczak, Torlini, and Demkowicz (1986) about the possibility to predict the granule particle size by NIR during granulation. Similar results were obtained by Nieuwmeyer et al. (2007), who were able to develop a prediction model using PLS for the prediction of median granule size. The prediction of bulk density was also found possible (Alcalà et al. 2010). Figure 40.8 presents a qualitative NIR monitoring of a fluid-bed granulation and drying process by PCA.

40.4.2.2 Dry Granulation

Usually achieved by roller compaction and milling, dry granulation uses compression to consolidate pharmaceutical powders. Density of the ribbons and particle size of the resulting granules are usually the parameters of interest. This was studied by Gupta et al., who monitored on-line the properties of the ribbons as they formed and the particle size after milling (Gupta et al. 2005). Miller,

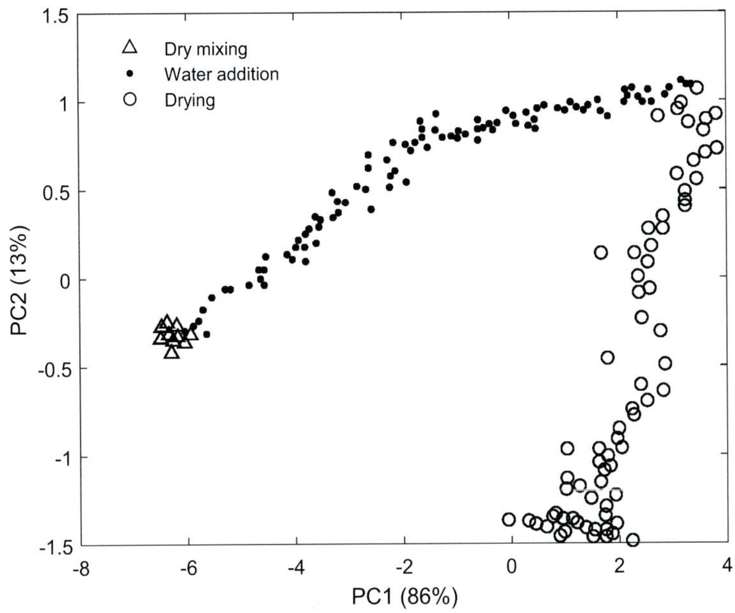

FIGURE 40.8 PCA score trends for a fluid-bed granulation and drying process. (Reproduced with permission from Alcalà et al. 2010.)

Morris, and Gupta (2006) used NIRS to facilitate scale-up of the process to larger equipment. Talwar et al. (2017) used compacts to develop a quantitative model using PLS to predict the solid fraction of ribbons. The authors studied the relationship between the chemical information and the physical information present in the NIR spectra and how it impacted the modeling efforts.

40.4.2.3 Hot-Melt Extrusion

Hot-melt extrusion is a form of granulation in which the active is embedded with polymers to achieve a higher bioavailability for poorly soluble drugs. It can also be used to develop a sustained release. Drug and polymer are heated and exposed to intense shear to form an extrudate. A NIR probe can be mounted on the exit die of the extruder to monitor the process. Drug and polymer concentrations and solid state are usually the parameters of interest (Saerens et al. 2012, Smith-Goettler, Gendron, and Meyer 2014, Hitzer et al. 2017).

40.4.3 Compression

In the context of batch manufacturing, NIR analysis around compression was usually performed on the final dosage forms. However, with the development of continuous lines, a need arose to monitor the final blend uniformity that was usually measured by thief sampling or NIR on blenders. Two types of analysis have been reported: analyses in the feed-frame where a port was open in the ceiling of the feed-frame to position a probe right above the paddles, and analyses in the feed-chute, as powder is dispensed into the feed-frame.

The initial report of monitoring powder uniformity into a tablet feed-frame was published by Liu and Blackwood (2012). The authors related the trend of first-derivative spectra using PCA to the content of tablet cores. The authors also reported significant disturbance patterns caused by the paddles that could be reduced by adjusting the height of the probe into the feed-frame. The same group of authors performed a follow-up study where other factors were analyzed: tablet press throughput, feed-frame speed, and the powder residence time into the feed-frame (Ward et al. 2013, Šašić et al. 2015).

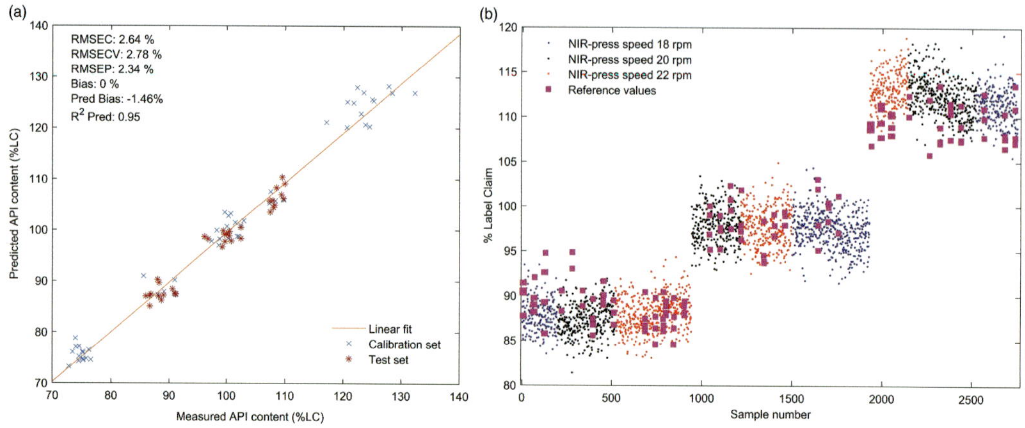

FIGURE 40.9 Predicted vs. reference plot for feed-frame predictions from individual tablet assay for calibration and test sets (a). Validation predictions for different press speeds and concentration levels (b).

The first quantitative analysis of powder homogeneity was reported in 2014 (Wahl et al. 2014). Using a mixture design to create component variability and PLS to build a model, the authors reported an error of 2%w/w for a target active concentration of 34% of the formulation. The validation of a feed-frame method was published by De Leersnyder et al. (2018). Using an 8% target active formulation, Li et al. (2019) calibrated and validated a quantitative method with an error of 2.65% w/w when exposing the model to variability in paddle speed. Figure 40.9 shows the calibration, test, and validation results for that low-dose application. As developing models onto a press takes a large amount of powder, benchtop systems were developed to allow model development with a fraction of the quantity needed (Mateo-Ortiz et al. 2014, Hetrick et al. 2017).

The use of NIRS in the feed-chute was published for the monitoring of content and density (Singh et al. 2015, Van Snick et al. 2017). In the former, the authors used the predicted bulk density to improve the operation of the press through a feed-forward/feedback controls of the tablet hardness and weight.

40.4.4 Coating

Tablets are usually coated for cosmetic or functional purposes or to mask unpalatable tastes. Andersson et al. (1999) collected the off-line spectra of tablets coated for various amounts of time and predicted the thickness of coating with a limit of quantification around 0.2 mm. Intertablet coating variation was studied off-line by Moes et al. (2008). In both studies, NIRS was found to be a suitable method to monitor the coating process and was a tool of choice to determine if process variations existed between manufacturing scales. Gendre et al. (2011) developed an extensive study in which the mass of coating material and film thickness was monitored in-line by PCA and PLS. The authors found that NIR results were comparable to Terahertz pulse imaging measurements. Möltgen et al. (2012) compared the usefulness of hyperspectral imaging along with off-line and in-line NIR measurements to control a coating process for inter- and intratablet analysis. The images allowed a better understanding of the coating variability within each tablet. The prediction in line and in real time of tablet weight gain and moisture was demonstrated by Igne et al. (2016).

The coating in pan and fluid bed is a chaotic process. Tablets are in constant movement, and significant sampling problems exist due to the difficulty to ensure the "correct" positioning of samples in front of the probe during on-line measurements. To bypass this issue, Lee et al. (2011) describe a method based on averaging and clustering to isolate the most meaningful spectra. The proposed method was able to yield coating thicknesses as good as 3% deviated from actual measurements.

Sampling issues related to coating monitoring was studied by Igne et al. (2016) who looked at the relationship between time and tablet coating for quantitative model development.

Unlike coated tablets, pellets are an intermediary material that may need to be coated to obtain functional properties such as delayed release. Pelletization technology has provided pharmaceutical formulation scientists with tremendous flexibility during solid oral dosage form development. Unfortunately, pharmaceutical pelletization may often involve lengthy and expensive manufacturing processes. Andersson et al. reported the possibility to predict pellet coating thickness with an error as low as 2.2 μm (Andersson et al. 2000). Lee et al. (2011) reported very good agreement between actual measurements and NIR predicted pellet film thickness for the real-time monitoring and end-point determination of the coating end-point. Additional examples of successful pellet coating monitoring by NIRS were published by Marković et al. (2014). The authors used a large number of batches to build a hierarchical PLS model.

40.4.5 Capsule and Tablet Analysis

In the earliest NIR assays, tablets and capsules were not analyzed intact. Before NIR spectral collection, drugs were extracted from the matrix into solution. One of the first reported uses of NIRS for tablets was by Zappala and Post (1977) for meprobamate tablets. Very good and exhaustive reviews by Forina, Casolino, and De La Pezuela Martinez (1998), Reich (2005), Roggo Roggo et al. (2007), and Luypaert, Massart, and Vander Heyden (2007) summarize the recent development in pharmaceutical assays and more particularly API quantification. It is, however, necessary to briefly emphasize several challenging applications that have been successfully completed. The prediction of lubricant (Magnesium Stearate) was found to be possible at levels as low as 0.5% w/w (Duong et al. 2003). Ito et al. (2010) showed the feasibility to use NIRS to predict the content of API in intact bilayer tablets and the research team showed that it was unnecessary to control the side that was irradiated when working in transmittance configuration. The prediction of API through coating was found possible by Boiret, Meunier, and Ginot (2011). An example of method validation for tablet assay was provided by Blanco and Alcalá (2006).

Capsule work is sparser in the literature, but an example of use of NIR for quantitative analysis was provided by Ritchie et al. (2002).

40.4.6 Considerations for Intact Dosage form Analysis

Sample positioning is the single largest source of error in NIR analysis of tablets in diffuse reflectance or transmittance measurements. Hardware, methodology, and mathematics may be used to reduce this error. Tablet-specific sample cells allow a consistent positioning of tablets and reduce this error. Repeating measurements and using the mean (or median) spectrum for each tablet significantly reduces the spectral variability. The median calculation results in less weighting by odd spectra than does the mean.

Tablet spectra must usually be corrected for baseline shifting prior to analysis. Many techniques have been attempted, but derivatives and scattering correction calculations are most common. This baseline correction is critical even if an average tablet spectrum is used.

Curved surfaces and embossing affect the spectrum of a tablet as positioning is varied, but the effect of such factors can be reduced by the methods just discussed. Natural variations in tablet mass and hardness also affect a tablet's spectrum through baseline shifting.

40.5 CONCLUSION

This chapter tried to provide an overview of the many applications for which NIRS has been used for in the pharmaceutical industry. However, it is important to recognize that NIR has limitations and that the best tool should be used for each intended application. NIR is part of a toolbox that has proven to be highly reliable and effective. But other tools are available (Simon et al. 2015). Practitioners should evaluate the conditions of a particular measurement and determine if NIRS is best suited. Nevertheless, it is an extremely sensitive technology, which provides chemical and physical information about the sample of interest. It is readily deployable and is one of the most installed spectroscopic technologies in manufacturing plants.

REFERENCES

Aaltonen, J., J. Rantanen, S. Siiriä, M. Karjalainen, A. Jørgensen, N. Laitinen, M. Savolainen, P. Seitavuopio, M. Louhi-Kultanen, and J. Yliruusi. 2003. "Polymorph screening using near-infrared spectroscopy." *Analytical Chemistry* 75 (19):5267–5273. doi: 10.1021/ac034205c.

Alam, M. A., Z. Shi, J. K. Drennen, and C. A. Anderson. 2017. "In-line monitoring and optimization of powder flow in a simulated continuous process using transmission near infrared spectroscopy." *International Journal of Pharmaceutics* 526 (1–2):199–208. doi: 10.1016/j.ijpharm.2017.04.054.

Alcalà, M., M. Blanco, M. Bautista, and J. M. González. 2010. "On-line monitoring of a granulation process by NIR spectroscopy." *Journal of Pharmaceutical Sciences* 99 (1):336–345. doi: 10.1002/jps.21818.

Aldridge, P. K., C. L. Evans, H. W. Ward Ii, S. T. Colgan, N. Boyer, and P. J. Gemperline. 1996. "Near-IR detection of polymorphism and process-related substances." *Analytical Chemistry* 68 (6):997–1002. doi: 10.1021/ac950993x.

Andersson, M., S. Folestad, J. Gottfries, M. O. Johansson, M. Joseffson, and K. G. Wahlund. 2000. "Quantitative analysis of film coating in a fluidized bed process by in-line NIR spectrometry and multivariate batch calibration." *Analytical Chemistry* 72 (9):2099–2108. doi: 10.1021/ac990256r.

Andersson, M., M. Josefson, F. W. Langkilde, and K. G. Wahlund. 1999. "Monitoring of a film coating process for tablets using near infrared reflectance spectrometry." *Journal of Pharmaceutical and Biomedical Analysis* 20 (1–2):27–37. doi: 10.1016/S0731-7085(98)00237-4.

Andre, M. 2003. "Multivariate analysis and classification of the chemical quality of 7-aminocephalosporanic acid using near-infrared reflectance spectroscopy." *Analytical Chemistry* 75 (14):3460–3467. doi: 10.1021/ac026393x.

Antunes, M. C., J. E. J. Simão, A. C. Duarte, and R. Tauler. 2002. "Multivariate curve resolution of overlapping voltammetric peaks: Quantitative analysis of binary and quaternary metal mixtures." *Analyst* 127 (6):809–817. doi: 10.1039/b200243b.

Barnes, S. E., T. Thurston, J. A. Coleman, A. Diederich, D. Ertl, J. Rydzak, P. Ng, K. Bakeev, and D. Bhanushali. 2010. "NIR diffuse reflectance for on-scale monitoring of the polymorphic form transformation of pazopanib hydrochloride (GW786034); Model development and method transfer." *Analytical Methods* 2 (12):1890–1899. doi: 10.1039/c0ay00376j.

Berntsson, O., L. G. Danielsson, B. Lagerholm, and S. Folestad. 2002. "Quantitative in-line monitoring of powder blending by near infrared reflection spectroscopy." *Powder Technology* 123 (2–3):185–193. doi: 10.1016/S0032-5910(01)00456-9.

Blanco, M., and M. Alcalá. 2006. "Content uniformity and tablet hardness testing of intact pharmaceutical tablets by near infrared spectroscopy: A contribution to process analytical technologies." *Analytica Chimica Acta* 557 (1–2):353–359. doi: 10.1016/j.aca.2005.09.070.

Blanco, M., M. Castillo, and R. Beneyto. 2007. "Study of reaction processes by in-line near-infrared spectroscopy in combination with multivariate curve resolution. Esterification of myristic acid with isopropanol." *Talanta* 72 (2):519–525. doi: 10.1016/j.talanta.2006.11.046.

Blanco, M., J. Coello, H. Iturriaga, S. Maspoch, and C. Pérez-Maseda. 2000. "Determination of polymorphic purity by near infrared spectrometry." *Analytica Chimica Acta* 407 (1–2):247–254. doi: 10.1016/S0003-2670(99)00828-4.

Blanco, M., and M. A. Romero. 2001. "Near-infrared libraries in the pharmaceutical industry: A solution for identity confirmation." *Analyst* 126 (12):2212–2217. doi: 10.1039/b105012p.

Blanco, M., D. Valdés, M. S. Bayod, F. Fernández-Marí, and I. Llorente. 2004. "Characterization and analysis of polymorphs by near-infrared spectrometry." *Analytica Chimica Acta* 502 (2):221–227. doi: 10.1016/j.aca.2003.10.016.

Boiret, M., L. Meunier, and Y. M. Ginot. 2011. "Tablet potency of Tianeptine in coated tablets by near infrared spectroscopy: Model optimisation, calibration transfer and confidence intervals." *Journal of Pharmaceutical and Biomedical Analysis* 54 (3):510–516. doi: 10.1016/j.jpba.2010.09.029.

Brülls, M., S. Folestad, A. Sparén, A. Rasmuson, and J. Salomonsson. 2007. "Applying spectral peak area analysis in near-infrared spectroscopy moisture assays." *Journal of Pharmaceutical and Biomedical Analysis* 44 (1):127–136. doi: 10.1016/j.jpba.2007.02.013.

Cervera-Padrell, A. E., J. P. Nielsen, M. J. Pedersen, K. M. Christensen, A. R. Mortensen, T. Skovby, K. Dam-Johansen, S. Kiil, and K. V. Gernaey. 2012. "Monitoring and control of a continuous Grignard reaction for the synthesis of an active pharmaceutical ingredient intermediate using inline NIR spectroscopy." *Organic Process Research & Development* 16 (5):901–914. doi: 10.1021/op2002563.

Chanda, A., A. M. Daly, D. A. Foley, M. A. Lapack, S. Mukherjee, J. D. Orr, G. L. Reid, D. R. Thompson, and H. W. Ward. 2015. "Industry perspectives on process analytical technology: Tools and applications in API development." *Organic Process Research and Development* 19 (1):63–83. doi: 10.1021/op400358b.

Ciurczak, E. W. 1987. "Uses of near-infrared spectroscopy in pharmaceutical analysis." *Applied Spectroscopy Reviews* 23 (1–2):147–163. doi: 10.1080/05704928708060445.

Ciurczak, E. W., R. P. Torlini, and M. P. Demkowicz. 1986. "Determination of particle size of pharmaceutical raw materials using near-infrared reflectance spectroscopy." *Spectroscopy* 1 (7):36–40.

Ciurczak, E. W., and T. A. Maldacker. 1986. "Determination of enantiomeric purity of valine by near infrared analysis." *Spectroscopy* 1 (1):36–39.

Colón, Y. M., M. A. Florian, D. Acevedo, R. Méndez, and R. J. Romañach. 2014. "Near infrared method development for a continuous manufacturing blending process." *Journal of Pharmaceutical Innovation* 9 (4):291–301. doi: 10.1007/s12247-014-9194-1.

De Leersnyder, F., E. Peeters, H. Djalabi, V. Vanhoorne, B. Van Snick, K. Hong, S. Hammond, A. Y. Liu, E. Ziemons, C. Vervaet, and T. De Beer. 2018. "Development and validation of an in-line NIR spectroscopic method for continuous blend potency determination in the feed frame of a tablet press." *Journal of Pharmaceutical and Biomedical Analysis* 151:274–283. doi: 10.1016/j.jpba.2018.01.032.

De Maesschalck, R., F. Cuesta Sánchez, D. L. Massart, P. Doherty, and P. Hailey. 1998. "On-line monitoring of powder blending with near-infrared spectroscopy." *Applied Spectroscopy* 52 (5):725–731. doi: 10.1366/0003702981944148.

Dreassi, E., G. Ceramelli, P. Corti, S. Lonardi, and P. L. Perruccio. 1995. "Near-infrared reflectance spectrometry in the determination of the physical state of primary materials in pharmaceutical production." *The Analyst* 120 (4):1005–1008. doi: 10.1039/AN9952001005.

Dreassi, E., G. Ceramelli, L. Savini, P. Corti, P. L. Perruccio, and S. Lonardi. 1995. "Application of near-infrared reflectance analysis to the integrated control of antibiotic tablet production." *The Analyst* 120 (2):319–323. doi: 10.1039/AN9952000319.

Duong, N. H., P. Arratia, F. Muzzio, A. Lange, J. Timmermans, and S. Reynolds. 2003. "A homogeneity study using NIR spectroscopy: Tracking magnesium stearate in Bohle bin-blender." *Drug Development and Industrial Pharmacy* 29 (6):679–687. doi: 10.1081/DDC-120021317.

El-Hagrasy, A. S., and J. K. Drennen. 2006. "A process analytical technology approach to near-infrared process control of pharmaceutical powder blending. Part III: Quantitative near-infrared calibration for prediction of blend homogeneity and characterization of powder mixing kinetics." *Journal of Pharmaceutical Sciences* 95 (2):422–434. doi: 10.1002/jps.20465.

Févotte, G., J. Calas, F. Puel, and C. Hoff. 2004. "Applications of NIR spectroscopy to monitoring and analyzing the solid state during industrial crystallization processes." *International Journal of Pharmaceutics* 273 (1–2):159–169. doi: 10.1016/j.ijpharm.2004.01.003.

Forina, M., M. C. Casolino, and C. De La Pezuela Martinez. 1998. "Multivariate calibration: Applications to pharmaceutical analysis." *Journal of Pharmaceutical and Biomedical Analysis* 18 (1–2):21–33. doi: 10.1016/S0731-7085(98)00153-8.

Frake, P., D. Greenhalgh, S. M. Grierson, J. M. Hempenstall, and D. R. Rudd. 1997. "Process control and end-point determination of a fluid bed granulation by application of near infra-red spectroscopy." *International Journal of Pharmaceutics* 151 (1):75–80. doi: 10.1016/S0378-5173(97)04894-1.

Gendre, C., M. Genty, M. Boiret, M. Julien, L. Meunier, O. Lecoq, M. Baron, P. Chaminade, and J. M. Péan. 2011. "Development of a Process Analytical Technology (PAT) for in-line monitoring of film thickness and mass of coating materials during a pan coating operation." *European Journal of Pharmaceutical Sciences* 43 (4):244–250. doi: 10.1016/j.ejps.2011.04.017.

Goicoechea, H. C., and A. C. Olivieri. 2002. "Chemometric assisted simultaneous spectrophotometric determination of four-component nasal solutions with a reduced number of calibration samples." *Analytica Chimica Acta* 453 (2):289–300. doi: 10.1016/S0003-2670(01)01232-6.

Gupta, A., G. E. Peck, R. W. Miller, and K. R. Morris. 2005. "Real-time near-infrared monitoring of content uniformity, moisture content, compact density/tensile strength, and young's modulus of roller compacted powder blends." *Journal of Pharmaceutical Sciences* 94 (7):1589–1597. doi: 10.1002/jps.20375.

Hailey, P. A., P. Doherty, P. Tapsell, T. Oliver, and P. K. Aldridge. 1996. "Automated system for the online monitoring of powder blending processes using near-infrared spectroscopy Part I. System development and control." *Journal of Pharmaceutical and Biomedical Analysis* 14 (5):551–559. doi: 10.1016/0731-7085(95)01674-0.

Hamilton, P., M. J. Sanganee, J. P. Graham, T. Hartwig, A. Ironmonger, C. Priestley, L. A. Senior, D. R. Thompson, and M. R. Webb. 2015. "Using PAT to understand, control, and rapidly scale up the production of a hydrogenation reaction and isolation of pharmaceutical intermediate." *Organic Process Research and Development* 19 (1):236–243. doi: 10.1021/op500285x.

Heinz, A., M. Savolainen, T. Rades, and C. J. Strachan. 2007. "Quantifying ternary mixtures of different solid-state forms of indomethacin by Raman and near-infrared spectroscopy." *European Journal of Pharmaceutical Sciences* 32 (3):182–192. doi: 10.1016/j.ejps.2007.07.003.

Hetrick, E. M., Z. Shi, L. E. Barnes, A. W. Garrett, R. G. Rupard, T. T. Kramer, T. M. Cooper, D. P. Myers, and B. C. Castle. 2017. "Development of near infrared spectroscopy-based process monitoring methodology for pharmaceutical continuous manufacturing using an offline calibration approach." *Analytical Chemistry* 89 (17):9175–9183. doi: 10.1021/acs.analchem.7b01907.

Hitzer, P., T. Bäuerle, T. Drieschner, E. Ostertag, K. Paulsen, H. van Lishaut, G. Lorenz, and K. Rebner. 2017. "Process analytical techniques for hot-melt extrusion and their application to amorphous solid dispersions." *Analytical and Bioanalytical Chemistry* 409 (18):4321–4333. doi: 10.1007/s00216-017-0292-z.

Igne, B., H. Arai, J. K. Drennen, and C. A. Anderson. 2016. "Effect of sampling frequency for real-time tablet coating monitoring using near infrared spectroscopy." *Applied Spectroscopy* 70 (9):1476–1488. doi: 10.1177/0003702816662622.

Igne, B., A. D. Juan, J. Jaumot, J. Lallemand, S. Preys, J. K. Drennen, and C. A. Anderson. 2014. "Modeling strategies for pharmaceutical blend monitoring and end-point determination by near-infrared spectroscopy." *International Journal of Pharmaceutics* 473 (1–2):219–231. doi: 10.1016/j.ijpharm.2014.06.061.

Igne, B., S. Talwar, J. K. Drennen, and C. A. Anderson. 2013. "Online monitoring of pharmaceutical materials using multiple NIR sensors - Part II: Blend end-point determination." *Journal of Pharmaceutical Innovation* 8 (1):45–55. doi: 10.1007/s12247-012-9146-6.

Ito, M., T. Suzuki, S. Yada, H. Nakagami, H. Teramoto, E. Yonemochi, and K. Terada. 2010. "Development of a method for nondestructive NIR transmittance spectroscopic analysis of acetaminophen and caffeine anhydrate in intact bilayer tablets." *Journal of Pharmaceutical and Biomedical Analysis* 53 (3):396–402. doi: 10.1016/j.jpba.2010.04.029.

Jaumot, J., B. Igne, C. A. Anderson, J. K. Drennen, and A. De Juan. 2013. "Blending process modeling and control by multivariate curve resolution." *Talanta* 117:492–504. doi: 10.1016/j.talanta.2013.09.037.

Jorgensen, A. C., P. Luukkonen, J. Rantanen, T. Schaefer, A. M. Juppo, and J. Yliruusi. 2004. "Comparison of torque measurements and near-infrared spectroscopy in characterization of a wet granulation process." *Journal of Pharmaceutical Sciences* 93 (9):2232–2243. doi: 10.1002/jps.20132.

Krakowska, B., D. Custers, E. Deconinck, and M. Daszykowski. 2016. "Chemometrics and the identification of counterfeit medicines—A review." *Journal of Pharmaceutical and Biomedical Analysis* 127:112–122. doi: 10.1016/j.jpba.2016.04.016.

Krämer, K., and S. Ebel. 2000. "Application of NIR reflectance spectroscopy for the identification of pharmaceutical excipients." *Analytica Chimica Acta* 420 (2):155–161. doi: 10.1016/S0003-2670(00)00877-1.

Kuriyama, A., J. Osuga, Y. Hattori, and M. Otsuka. 2018. "In-line monitoring of a high-shear granulation process using the baseline shift of near infrared spectra." *AAPS PharmSciTech* 19 (2):710–718. doi: 10.1208/s12249-017-0882-2.

Lakerveld, R., B. Benyahia, P. L. Heider, H. Zhang, A. Wolfe, C. J. Testa, S. Ogden, D. R. Hersey, S. Mascia, J. M. B. Evans, R. D. Braatz, and P. I. Barton. 2015. "The application of an automated control strategy for an integrated continuous pharmaceutical pilot plant." *Organic Process Research and Development* 19 (9):1088–1100. doi: 10.1021/op500104d.

Lee, M. J., D. Y. Seo, H. E. Lee, I. C. Wang, W. S. Kim, M. Y. Jeong, and G. J. Choi. 2011. "In line NIR quantification of film thickness on pharmaceutical pellets during a fluid bed coating process." *International Journal of Pharmaceutics* 403 (1–2):66–72. doi: 10.1016/j.ijpharm.2010.10.022.

Li, W., J. Cunningham, H. Rasmussen, and D. Winstead. 2007. "A qualitative method for monitoring of nucleation and granule growth in fluid bed wet granulation by reflectance near-infrared spectroscopy." *Journal of Pharmaceutical Sciences* 96 (12):3470–3477. doi: 10.1002/jps.20974.

Li, Y., C.A. Anderson, J.K. Drennen, C. Airiau, and B. Igne. 2019. "Development of an in-line near-infrared method for blend content uniformity assessment in a tablet feed frame." *Applied Spectroscopy*. doi: 10.1177/0003702819842189.

Liu, Y., and D. Blackwood. 2012. "Sample presentation in rotary tablet press feed frame monitoring by near infrared spectroscopy." *American Pharmaceutical Review*.

Lodder, R. A., M. Selby, and G. M. Hieftje. 1987. "Detection of capsule tampering by near-infrared reflectance analysis." *Analytical Chemistry* 59 (15):1921–1930. doi: 10.1021/ac00142a008.

Lodder, R. A., and G. M. Hieftje. 1988a. "Analysis of intact tablets by near-infrared reflectance spectrometry." *Applied Spectroscopy* 42 (4):556–558. doi: 10.1366/0003702884429201.

Lodder, R. A., and G. M. Hieftje. 1988b. "Detection of subpopulations in near-infrared reflectance analysis." *Applied Spectroscopy* 42 (8):1500–1512.

Lorber, A. 1986. "Error propagation and figures of merit for quantification by solving matrix equations." *Analytical Chemistry* 58 (6):1167–1172. doi: 10.1021/ac00297a042.

Luypaert, J., D. L. Massart, and Y. Vander Heyden. 2007. "Near-infrared spectroscopy applications in pharmaceutical analysis." *Talanta* 72 (3):865–883. doi: 10.1016/j.talanta.2006.12.023.

Ma, H., and C. A. Anderson. 2008. "Characterization of pharmaceutical powder blends by NIR chemical imaging." *Journal of Pharmaceutical Sciences* 97 (8):3305–3320. doi: 10.1002/jps.21230.

Marković, S., K. Poljanec, J. Kerč, and M. Horvat. 2014. "In-line NIR monitoring of key characteristics of enteric coated pellets." *European Journal of Pharmaceutics and Biopharmaceutics* 88 (3):847–855. doi: 10.1016/j.ejpb.2014.10.003.

Mateo-Ortiz, D., Y. Colon, R. J. Romañach, and R. Méndez. 2014. "Analysis of powder phenomena inside a Fette 3090 feed frame using in-line NIR spectroscopy." *Journal of Pharmaceutical and Biomedical Analysis* 100:40–49. doi: 10.1016/j.jpba.2014.07.014.

Miller, R. W., K. R. Morris, and A. Gupta 2006. "Roller compaction scale-up." In *Pharmaceutical Process Scale-Up,* 2nd Ed., edited by M. Levin, pp. 237–266. Boca Raton, FL: Taylor & Francis.

Miwa, A., and K. Makado. 2009. "A method for predicting the amount of water required for wet granulation using NIR." *International Journal of Pharmaceutics* 376 (1–2):41–45. doi: 10.1016/j.ijpharm.2009.04.013.

Moes, J. J., M. M. Ruijken, E. Gout, H. W. Frijlink, and M. I. Ugwoke. 2008. "Application of process analytical technology in tablet process development using NIR spectroscopy: Blend uniformity, content uniformity and coating thickness measurements." *International Journal of Pharmaceutics* 357 (1–2):108–118. doi: 10.1016/j.ijpharm.2008.01.062.

Mohan, S., W. Momose, J. M. Katz, M. N. Hossain, N. Velez, J. K. Drennen, and C. A. Anderson. 2018. "A robust quantitative near infrared modeling approach for blend monitoring." *Journal of Pharmaceutical and Biomedical Analysis* 148:51–57. doi: 10.1016/j.jpba.2017.09.011.

Möltgen, C. V., T. Puchert, J. C. Menezes, D. Lochmann, and G. Reich. 2012. "A novel in-line NIR spectroscopy application for the monitoring of tablet film coating in an industrial scale process." *Talanta* 92:26–37. doi: 10.1016/j.talanta.2011.12.034.

Nieuwmeyer, F. J. S., M. Damen, A. Gerich, F. Rusmini, K. Van Der Voort Maarschalk, and H. Vromans. 2007. "Granule characterization during fluid bed drying by development of a near infrared method to determine water content and median granule size." *Pharmaceutical Research* 24 (10):1854–1861. doi: 10.1007/s11095-007-9305-5.

Osorio, J. G., G. Stuessy, G. J. Kemeny, and F. J. Muzzio. 2014. "Characterization of pharmaceutical powder blends using in situ near-infrared chemical imaging." *Chemical Engineering Science* 108:244–257. doi: 10.1016/j.ces.2013.12.027.

Otsuka, M., F. Kato, and Y. Matsuda. 2001. "Determination of indomethacin polymorphic contents by chemometric near-infrared spectroscopy and conventional powder X-ray diffractometry." *Analyst* 126 (9):1578–1582. doi: 10.1039/b103498g.

Peinado, A., J. Hammond, and A. Scott. 2011. "Development, validation and transfer of a Near Infrared method to determine in-line the end point of a fluidised drying process for commercial production batches of an approved oral solid dose pharmaceutical product." *Journal of Pharmaceutical and Biomedical Analysis* 54 (1):13–20. doi: 10.1016/j.jpba.2010.07.036.

Plugge, W., and C. van der Vliest. 1992. "The use of near infrared spectroscopy in the quality control laboratory of the pharmaceutical industry." *Journal of Pharmaceutical and Biomedical Analysis* 10 (10–12):797–803. doi: 10.1016/0731-7085(91)80083-L.

Popo, M., S. Romero-Torres, C. Conde, and R. J. Romañach. 2002. "Blend uniformity analysis using stream sampling and near infrared spectroscopy." *AAPS PharmSciTech* 3 (3). doi: 10.1007/BF02830622.

Poux, M., P. Fayolle, J. Bertrand, D. Bridoux, and J. Bousquet. 1991. "Powder mixing: Some practical rules applied to agitated systems." *Powder Technology* 68 (3):213–234. doi: 10.1016/0032-5910(91)80047-M.

Puchert, T., C. V. Holzhauer, J. C. Menezes, D. Lochmann, and G. Reich. 2011. "A new PAT/QbD approach for the determination of blend homogeneity: Combination of on-line NIRS analysis with PC scores distance analysis (PC-SDA)." *European Journal of Pharmaceutics and Biopharmaceutics* 78 (1):173–182. doi: 10.1016/j.ejpb.2010.12.015.

Rantanen, J., S. Lehtola, P. Rämet, J. P. Mannermaa, and J. Yliruusi. 1998. "On-line monitoring of moisture content in an instrumented fluidized bed granulator with a multi-channel NIR moisture sensor." *Powder Technology* 99 (2):163–170. doi: 10.1016/S0032-5910(98)00100-4.

Rantanen, J., and J. Yliruusi. 1998. "Determination of particle size in a fluidized bed granulator with a near infrared set-up." *Pharmacy and Pharmacology Communications* 4 (2):73–75.

Reich, G. 2005. "Near-infrared spectroscopy and imaging: Basic principles and pharmaceutical applications." *Advanced Drug Delivery Reviews* 57 (8):1109–1143. doi: 10.1016/j.addr.2005.01.020.

Ritchie, G. E., H. Mark, and E. W. Ciurczak. 2003. "Evaluation of the conformity index and the Mahalanobis distance as a tool for process analysis: A technical note." *AAPS PharmSciTech* 4 (2). doi: 10.1208/pt040224.

Ritchie, G. E., R. W. Roller, E. W. Ciurczak, H. Mark, C. Tso, and S. A. MacDonald. 2002. "Validation of a near-infrared transmission spectroscopic procedure: Part B: Application to alternate content uniformity and release assay methods for pharmaceutical solid dosage forms." *Journal of Pharmaceutical and Biomedical Analysis* 29 (1–2):159–171. doi: 10.1016/S0731-7085(02)00010-9.

Roggo, Y., P. Chalus, L. Maurer, C. Lema-Martinez, A. Edmond, and N. Jent. 2007. "A review of near infrared spectroscopy and chemometrics in pharmaceutical technologies." *Journal of Pharmaceutical and Biomedical Analysis* 44 (3 SPEC. ISS.):683–700. doi: 10.1016/j.jpba.2007.03.023.

Saerens, L., L. Dierickx, T. Quinten, P. Adriaensens, R. Carleer, C. Vervaet, J. P. Remon, and T. De Beer. 2012. "In-line NIR spectroscopy for the understanding of polymer-drug interaction during pharmaceutical hot-melt extrusion." *European Journal of Pharmaceutics and Biopharmaceutics* 81 (1):230–237. doi: 10.1016/j.ejpb.2012.01.001.

Šašić, S., D. Blackwood, A. Liu, H. W. Ward, and H. Clarke. 2015. "Detailed analysis of the online near-infrared spectra of pharmaceutical blend in a rotary tablet press feed frame." *Journal of Pharmaceutical and Biomedical Analysis* 103:73–79. doi: 10.1016/j.jpba.2014.11.008.

Sekulic, S. S., J. Wakeman, P. Doherty, and P. A. Hailey. 1998. "Automated system for the on-line monitoring of powder blending processes using near-infrared spectroscopy. Part II. Qualitative approaches to blend evaluation." *Journal of Pharmaceutical and Biomedical Analysis* 17 (8):1285–1309. doi: 10.1016/S0731-7085(98)00025-9.

Sekulic, S. S., H. W. Ward Ii, D. R. Brannegan, E. D. Stanley, C. L. Evans, S. T. Sciavolino, P. A. Hailey, and P. K. Aldridge. 1996. "On-line monitoring of powder blend homogeneity by near-infrared spectroscopy." *Analytical Chemistry* 68 (3):509–513. doi: 10.1021/ac950964m.

Shi, Z., R. P. Cogdill, S. M. Short, and C. A. Anderson. 2008. "Process characterization of powder blending by near-infrared spectroscopy: Blend end-points and beyond." *Journal of Pharmaceutical and Biomedical Analysis* 47 (4–5):738–745. doi: 10.1016/j.jpba.2008.03.013.

Simon, L. L., H. Pataki, G. Marosi, F. Meemken, K. Hungerbühler, A. Baiker, S. Tummala, B. Glennon, M. Kuentz, G. Steele, H. J. M. Kramer, J. W. Rydzak, Z. Chen, J. Morris, F. Kjell, R. Singh, R. Gani, K. V. Gernaey, M. Louhi-Kultanen, J. Oreilly, N. Sandler, O. Antikainen, J. Yliruusi, P. Frohberg, J. Ulrich, R. D. Braatz, T. Leyssens, M. Von Stosch, R. Oliveira, R. B. H. Tan, H. Wu, M. Khan, D. Ogrady, A. Pandey, R. Westra, E. Delle-Case, D. Pape, D. Angelosante, Y. Maret, O. Steiger, M. Lenner, K. Abbou-Oucherif, Z. K. Nagy, J. D. Litster, V. K. Kamaraju, and M. S. Chiu. 2015. "Assessment of recent process analytical technology (PAT) trends: A multiauthor review." *Organic Process Research and Development* 19 (1):3–62. doi: 10.1021/op500261y.

Singh, R., A. D. Román-Ospino, R. J. Romañach, M. Ierapetritou, and R. Ramachandran. 2015. "Real time monitoring of powder blend bulk density for coupled feed-forward/feed-back control of a continuous direct compaction tablet manufacturing process." *International Journal of Pharmaceutics* 495 (1):612–625. doi: 10.1016/j.ijpharm.2015.09.029.

Skibsted, E. T. S., H. F. M. Boelens, J. A. Westerhuis, D. T. Witte, and A. K. Smilde. 2006. "Simple assessment of homogeneity in pharmaceutical mixing processes using a near-infrared reflectance probe and control charts." *Journal of Pharmaceutical and Biomedical Analysis* 41 (1):26–35. doi: 10.1016/j.jpba.2005.10.009.

Smith-Goettler, B., C. M. Gendron, and R. F. Meyer. 2014. "Understanding hot melt extrusion via near infrared spectroscopy." *NIR News* 25 (7):10–12. doi: 10.1255/nirn.1481.

Sun, L., C. Hsiung, C. G. Pederson, P. Zou, V. Smith, M. Von Gunten, and N. A. O'Brien. 2016. "Pharmaceutical raw material identification using miniature near-infrared (MicroNIR) spectroscopy and supervised pattern recognition using support vector machine." *Applied Spectroscopy* 70 (5):816–825. doi: 10.1177/0003702816638281.

Talwar, S., C. Nunes, T. Stevens, V. Nesarikar, P. Timmins, C. A. Anderson, and J. K. Drennen. 2017. "Understanding the impact of chemical variability and calibration algorithms on prediction of solid fraction of roller compacted ribbons using near-infrared (NIR) spectroscopy." *Applied Spectroscopy* 71 (6):1209–1221. doi: 10.1177/0003702816671960.

Tauler, R., B. Kowalski, and S. Fleming. 1993. "Multivariate curve resolution applied to spectral data from multiple runs of an industrial process." *Analytical Chemistry* 65 (15):2040–2047. doi: 10.1021/ac00063a019.

Van Snick, B., J. Holman, C. Cunningham, A. Kumar, J. Vercruysse, T. De Beer, J. P. Remon, and C. Vervaet. 2017. "Continuous direct compression as manufacturing platform for sustained release tablets." *International Journal of Pharmaceutics* 519 (1–2):390–407. doi: 10.1016/j.ijpharm.2017.01.010.

Vercruysse, J., M. Toiviainen, M. Fonteyne, N. Helkimo, J. Ketolainen, M. Juuti, U. Delaet, I. Van Assche, J. P. Remon, C. Vervaet, and T. De Beer. 2014. "Visualization and understanding of the granulation liquid mixing and distribution during continuous twin screw granulation using NIR chemical imaging." *European Journal of Pharmaceutics and Biopharmaceutics* 86 (3):383–392. doi: 10.1016/j.ejpb.2013.10.012.

Wahl, P. R., G. Fruhmann, S. Sacher, G. Straka, S. Sowinski, and J. G. Khinast. 2014. "PAT for tableting: Inline monitoring of API and excipients via NIR spectroscopy." *European Journal of Pharmaceutics and Biopharmaceutics* 87 (2):271–278. doi: 10.1016/j.ejpb.2014.03.021.

Ward, H. W., D. O. Blackwood, M. Polizzi, and H. Clarke. 2013. "Monitoring blend potency in a tablet press feed frame using near infrared spectroscopy." *Journal of Pharmaceutical and Biomedical Analysis* 80:18–23. doi: 10.1016/j.jpba.2013.02.008.

Wargo, D. J., and J. K. Drennen. 1996. "Near-infrared spectroscopic characterization of pharmaceutical powder blends." *Journal of Pharmaceutical and Biomedical Analysis* 14 (11):1415–1423. doi: 10.1016/0731-7085(96)01739-6.

Windig, W., and J. Guilment. 1991. "Interactive self-modeling mixture analysis." *Analytical Chemistry* 63 (14):1425–1432. doi: 10.1021/ac00014a016.

Wu, H., M. Tawakkul, M. White, and M. A. Khan. 2009. "Quality-by-Design (QbD): An integrated multivariate approach for the component quantification in powder blends." *International Journal of Pharmaceutics* 372 (1–2):39–48. doi: 10.1016/j.ijpharm.2009.01.002.

Yang, C. I. W., S. Chen, F. Ouyang, I. C. Yang, and C. Y. Tsai. 2011. "A robust identification model for herbal medicine using near infrared spectroscopy and artificial neural network." *Journal of Food and Drug Analysis* 19 (1):9–17.

Zacour, B. M., B. Igne, J. K. Drennen Iii, and C. A. Anderson. 2011. "Efficient near-infrared spectroscopic calibration methods for pharmaceutical blend monitoring." *Journal of Pharmaceutical Innovation* 6 (1):10–23. doi: 10.1007/s12247-011-9096-4.

Zappala, A. F., and A. Post. 1977. "Rapid near IR spectrophotometric determination of meprobamate in pharmaceutical preparations." *Journal of Pharmaceutical Sciences* 66 (2):292–293. doi: 10.1002/jps.2600660246.

Zhou, G. X., Z. Ge, J. Dorwart, B. Izzo, J. Kukura, G. Bicker, and J. Wyvratt. 2003. "Determination and differentiation of surface and bound water in drug substances by near infrared spectroscopy." *Journal of Pharmaceutical Sciences* 92 (5):1058–1065. doi: 10.1002/jps.10375.

41 Soil Sensing by Visible and Infrared Radiation

Fabrício da Silva Terra
Federal University of Pelotas

Rodnei Rizzo
University of São Paulo

Eyal Ben Dor
Tel Aviv University

*José A. M. Demattê**
University of São Paulo

CONTENTS

* jamdemat@usp.br

41.1 INTRODUCTION

Soil is one of the most important natural resources in the environment, and efforts for its preservation are needed. The world population is expected to exceed 8 billion people by 2050 (FAO, 2013). Today, we have about 867 million chronically undernourished people in the world. At present, more than 1.5 billion ha (about 11%) of the globe's land surface (13.4 billion ha) is used for crop production, 28% (3.7 billion ha) is under forest, and 35% (4.6 billion ha) comprises grasslands and woodland ecosystems (FAO, 2011, 2013). To accomplish the task of feeding the entire world population, one of the most important strategies is closing the yield gap (Godfray et al., 2010), which requires appropriate land use planning of existing production areas and upcoming ones too. If, on the one hand, we need soil to produce food and fiber, on the other hand, we are degrading it. As determined by Lal (2003), soils have been eroded at a rate of 201.1×10^{12} kg year^{-1}.

Due to the heterogeneity of soil types, it is impossible to understand, evaluate, or map them using manpower only. Technological solutions are available but nothing like spectral soil sensing, which has emerged as one of the most influential technique. Spectral soil sensing is based on information acquisition of an object without contact and can be divided into remote (sensors at airborne vehicles or satellites) and proximal (sensors at field or laboratory) sensing (Figure 41.1). Reflectance data from visible (Vis: 350–750 nm), near-infrared (NIR: 750–1100 nm), shortwave-infrared (SWIR: 1100–2500 nm), and mid-infrared (mid-IR or MIR: 2500–25,000 nm or 4000–400 cm^{-1}) spectral ranges have been widely used for soil reflectance spectroscopy (Stenberg et al., 2010; Soriano-Disla et al., 2014). In summary, in the 1960s, Obukhov and Orlov (1964) and Bowers and Hanks (1965) pioneered publications on the influence of soil particle size, moisture, and organic matter on soil reflectance. Recently, hundreds of studies corroborate the importance of soil sensing in all domains from the laboratory (Viscarra Rossel et al., 2016) to orbital levels (Mulder et al., 2011). Several soil properties can be analyzed by reflectance spectroscopy, such as water content, particle size distribution, mineralogy (oxides and phyllosilicates), organic matter content, organic and inorganic carbon contents, and cation exchange capacity (Stenberg et al., 2010; Soriano-Disla et al., 2014). Applications of spectral sensing in soil science cover soil physical and chemical properties modeling, pedogenesis and weathering, soil classification, mapping, conservation, mineralogy, pollution, fertility, and monitoring (Terra et al., 2018).

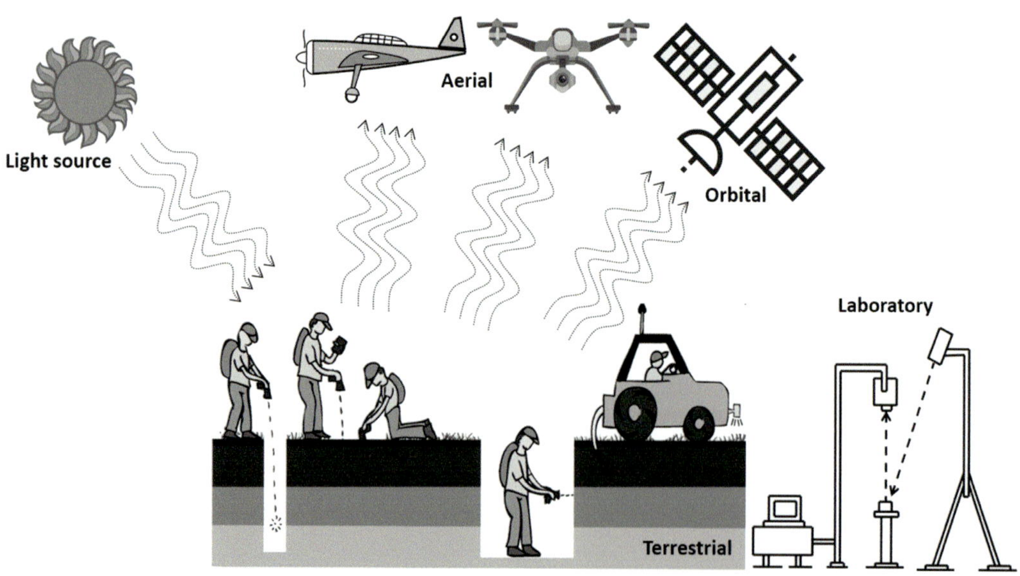

FIGURE 41.1 Remote and proximal soil spectral sensing.

In general, spectral sensing has provided a general and flexible set of instruments and tools for users who need soil information. In fact, it also constitutes a part of the soil science domain under a rapid development. Relevant soil spectral sensing applications can contribute to understanding soil-forming processes and horizontal and vertical distribution patterns (Terra et al., 2018). Spectral sensing technologies are in constant evolution together with powerful computing and statistical tools that will support soil researchers to reach the "next level" in soil assessment and mapping. It is a matter of time until soil spectral sensing helps the soil science achieve its primary goal: to understand and preserve soils for future generations through sustainable land use and management. Therefore, this chapter discusses the application of reflectance spectroscopy in soil science.

41.2 SOIL SPECTRAL BEHAVIOR

Generally, soils are composed of solid fractions and pores in a half-to-half proportion (Weil and Brady, 2017). These solid fractions are made of about 45% mineral particles in different sizes (sand, silt, and clay) and of about 5% organic matters (humus, roots, and organisms). The mineral particles are mainly iron and aluminum oxides (hematite, goethite, and gibbsite), phyllosilicates (montmorillonite, illite, hydroxy-interlayered vermiculite, and kaolinite), carbonates (calcite and dolomite), and quartz. Water (dissolved salts in soil solution) and gases (CO_2, O_2, and N_2) fill the pores in the same proportion.

After interactions with electromagnetic radiation (EMR), the behavior of the reflected radiance (reflectance) from soils along the studied spectral range is affected by the diversity, concentration, and size of their inorganic and organic components (Figure 41.2). So, the soil spectrum is a cumulative property resulting from different combinations of the spectral behavior of each soil component (Bowers and Hanks, 1965), where changes in soil composition have affected the final spectrum of soil (Figure 41.3). The paths of the incoming EMR in a soil sample, mainly, reflectance and absorptance, will be dependent on phenomena of macro- and microscopic interactions, respectively.

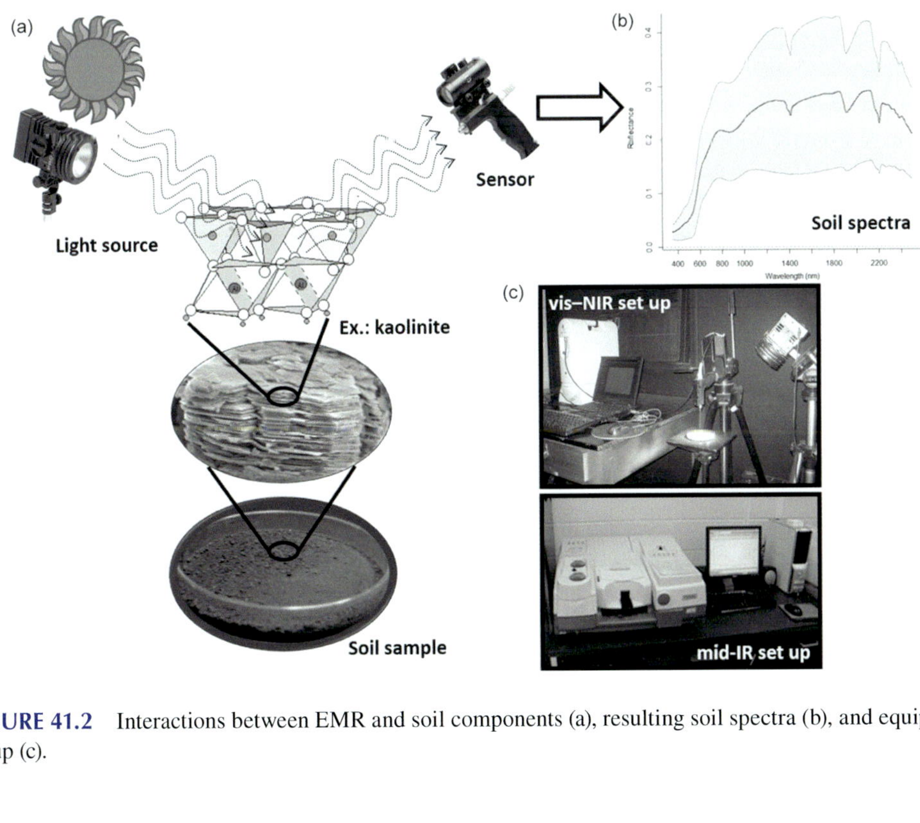

FIGURE 41.2 Interactions between EMR and soil components (a), resulting soil spectra (b), and equipment setup (c).

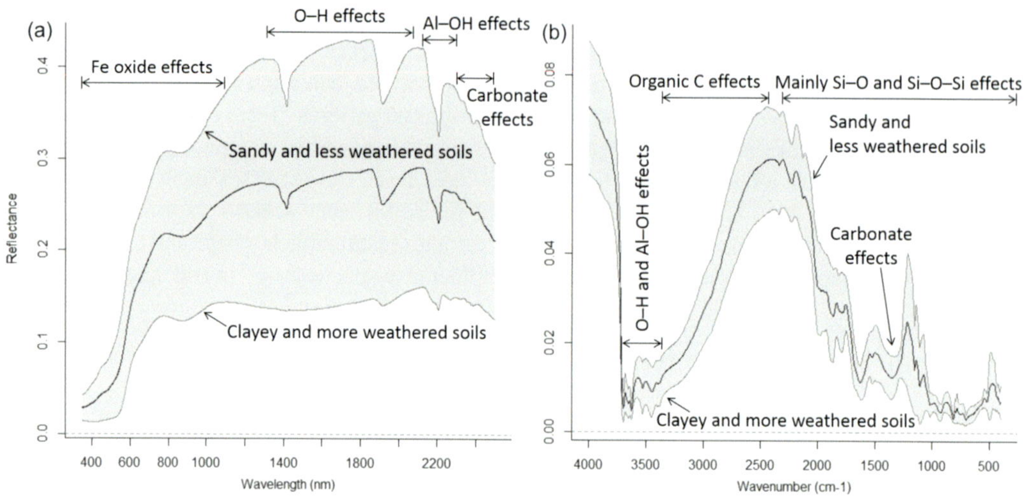

FIGURE 41.3 Vis-NIR (a) and mid-IR (b) spectral behaviors of dry mineral soils highlighting changes in reflectance intensities and absorption features with their causes due to different mineral and organic compositions.

41.2.1 Macroscopic Interactions

Interactions at the macroscopic level are directly related to EMR diffraction and scattering phenomena and consequently depend on soil physical and textural properties. These interactions are observed in the overall reflectance intensity (albedo) and depend on changes in soil particle distribution, structure (arrangement of particles), and compaction that affects the distribution of soil porosity and, mainly, the surface roughness (microrelief) (Figure 41.3). The rougher the soil surface, the lower the reflected radiance, since diffuse (scattered) reflectance and shading occur, attenuating its intensity (Adams and Filice, 1967; Townsend, 1987). With the reduction of roughness, the directional (specular) reflectance increases as well as its intensity.

Another important factor contributing to increasing soil reflectance intensity is the presence of transparent minerals that do not absorb energy, such as quartz present in sand fraction. In general, soil minerals are trans-opaques, which simultaneously reflect and absorb radiance in different spectral ranges. However, increases in contents of soil organic matter and moisture produce an overall attenuation of the soil reflectance intensity due to the absorbance of these components (Demattê and Terra, 2014; Terra et al., 2015, 2018).

41.2.2 Microscopic Interactions

Microscopic interactions are responsible for absorption features (negative peaks) in soil spectra and are dependent on soil mineral and organic composition (Figure 41.3). These interactions are limited to the first 10–50 μm of soil thickness and happen due to EMR absorption by atoms and molecules (functional groups) of soil mineral and organic compounds (Baumgardner et al., 1985). These functional groups resulting from weathering, pedogenesis, and organic matter decomposition are the same ones responsible for chemical (adsorption, desorption, cations and anion exchange, etc.), physical (water retention, aggregation, etc.), and biological (mineralization, enzymatic activity, etc.) interactions in soils. The main soil organic functional groups are aromatics (C-H), amines (N-H), alkyls (C-H), carboxyls (C=O), amides (C=O), aliphatics (C-H), methyls (C-H), phenolics (C-OH), polysaccharides (C-O), and carbohydrates (C-O). The main inorganic functional group is hydroxyl (-OH) exposed on the outer surface of phyllosilicate and oxidic clay minerals (silanol: Si]-OH, aluminol: Al]-OH, and ferrol: Fe]-OH) (Sparks, 2002) and present in water molecules.

Interactions at the microscopic level are dependent on specific phenomena related to the amounts of energy at characteristic wavelengths (or wavenumbers) and are conditioned by the atomic (quantum model, electronic distribution, and ionic radius size) and molecular structures of soil components. The strength of their chemical bonds and ionic impurities contained in functional groups, such as ions adsorbed on them, play also an important part (Hollas, 2004; Bernath, 2005). These phenomena refer to specific absorptions produced by atomic electronic transitions and molecular vibrations if there is enough energy. Although soils are complex systems and their mineral and organic compounds occur mixed in different proportions, each component will be represented by specific absorption features at the selected wavelength (or wavenumber) positions. However, these absorptions may also be slightly changed in their positions since organic and mineral molecules do not have a total harmonic behavior in soils (see, e.g., Ben-Dor and Banin, 1990a). Some absorption overlaps can also be observed mainly between soil water content and phyllosilicate clay minerals due to their hydroxyl groups.

41.2.2.1 Soil Minerals

41.2.2.1.1 Iron Oxides

In the visible (350–750 nm) and part of NIR (750–1100 nm) spectral ranges, microscopic interactions take place at the electronic level when there is enough radiation to produce changes in energy levels of electrons in ions that make the crystalline structure of soil minerals (Sherman and Waite, 1985). Charge-transfer and crystal-field effects are the primary mechanisms that cause a rearrangement of these valence electrons and produce different absorption features in soil spectra with very intense bands from 350 to 550 nm and, after that, broad bands up to 1100 nm, respectively (Hunt et al., 1971). In soils, these electronic transitions occur mostly in iron oxides, since Fe is a transition metal and the ferric ion (Fe^{3+}) is present in hematite (Fe_2O_3) and goethite (FeOOH), which are the most common (Figure 41.4). Absorption features of hematite are centered at 445 nm [$^6A_1 \rightarrow {}^4E,{}^4A_1$], 530 nm [$2(^6A_1) \rightarrow 2(^4T_1)$], 650 nm [$^6A_1 \rightarrow {}^4T_2$], and 885 nm [$^6A_1 \rightarrow {}^4T_1$], while absorption features of goethite are centered at 435 nm [$^6A_1 \rightarrow {}^4E,{}^4A_1$], 480 nm [$2(^6A_1) \rightarrow 2(^4T_1)$], 650 nm [$^6A_1 \rightarrow {}^4T_2$], and 917 nm [$^6A_1 \rightarrow {}^4T_1$] (Figure 41.3) (Sherman and Waite, 1985). In the visible range, these absorptions with different intensities explain the reddish and yellowish colors that hematite and goethite give to soils, respectively (Figure 41.4)

Magnetite ($FeO.Fe_2O_3$) and ilmenite ($FeTiO_3$) are iron oxides common in tropical soils derived from mafic rocks. Their reflectance spectra present very low albedo (<5%) and no clear absorption features between 350 and 1100 nm, since they are opaque minerals, and the absorbed radiance is due to ionic ($Fe^{2+} \rightarrow Fe^{3+}$, $Fe^{2+} \rightarrow O$, $Fe^{3+} \rightarrow O$, $Fe^{2+} \rightarrow Ti^{4+}$, and $Ti^{4+} \rightarrow O$) and intraionic (Fe^{2+} and Fe^{3+}) charge transfer (Hunt et al., 1971). Other transition metals (Ti, Cu, Mn, Cr, Zn, Ni, etc.) may be present in soils as free ions or oxides but in lesser quantities than Fe.

41.2.2.1.2 Phyllosilicates

In the other part of the NIR (1100–2500 nm), also known as shortwave-infrared (SWIR), microscopic interactions occur due to nonfundamental molecular vibrations (overtones and combinations of fundamentals), since there is enough energy level to change the stationary state of the molecules of the soil functional groups (Hunt, 1977). Although these interactions need less energy than those in the previous range, they produce stronger absorption features. In soils, nonfundamental vibrations (overtone and combination modes) are basically related to stretching overtone of the OH bond at 1400 nm (fundamental O-H at 3575 cm^{-1}), bending–stretching combination of molecular H_2O at 1900 nm (fundamentals H-O-H at 1645 cm^{-1} and O-H at 3484 cm^{-1}), and bending–stretching combination of aluminum–hydroxyl bond (Al-OH) at 2200 nm (fundamentals Al-OH at 915 cm^{-1} and O-H at 3620 cm^{-1}) (Hunt and Salisbury, 1971a) (Figure 41.3). This last wavelength can be slightly shifted toward mid-IR based on substitution of Al by Fe or Mg (Ben-Dor and Banin, 1990b). Water and mainly hydroxyl molecules are found in the structures of phyllosilicate clay minerals, which are formed by a combination (a layer) of one silicon tetrahedral sheet (O^{2-} groups coordinated around Si^{4+}) and one alumina octahedral sheet (OH^- groups coordinated around Al^{3+}). Soil phyllosilicates

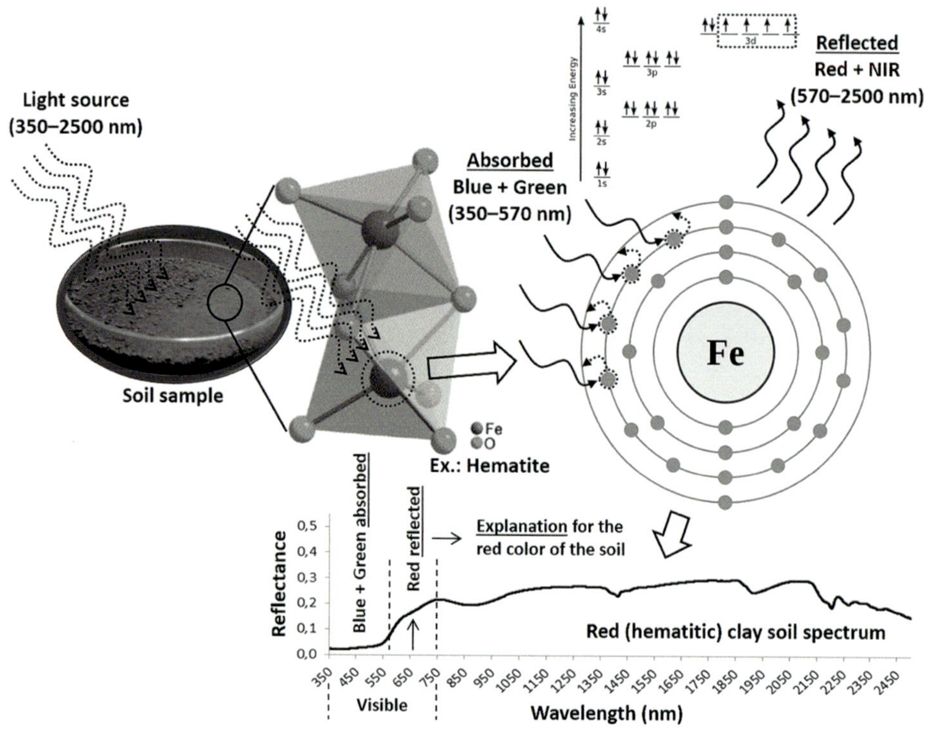

FIGURE 41.4 Representation of electronic transitions of iron ions in the hematite (iron oxide), and the visible spectral behavior of the hematitic soil explaining its red color.

(secondary or pedogenic minerals) are divided into 1:1 (Si-tetrahedron + Al-octahedron) and 2:1 (Si tetrahedron + Al-octahedron + Si-tetrahedron) clay minerals represented, respectively, by kaolinite and by mica (illite and hydroxyl-interlayered vermiculite) and smectite (montmorillonite) groups (Hunt and Salisbury, 1971a).

Although all soil phyllosilicates share absorptions at 1400, 1900, and 2200 nm, their distinction is possible based on shape, width, and intensity of their features (Figure 41.3). Generally, water molecules are strongly adsorbed on the interlayer surfaces of 2:1 clay minerals, and therefore, they present strong absorption features at 1400 and 1900 nm. Kaolinite presents a dissymmetry at 2205 nm towards higher wavelengths but weaker absorptions at these three positions compared to 2:1 minerals. Soil weathering leads to mineral transformation due to hydrolysis where water and hydroxyls are removed from phyllosilicate structures, where the 2:1 mineral can be transformed into 1:1 mineral. The latter can also be transformed into aluminum oxide reducing the intensity of those three absorption features in soil spectra (Figure 41.3) (Demattê and Terra, 2014; Terra et al., 2015, 2018).

41.2.2.1.3 Aluminum Oxide

Gibbsite [$Al(OH)_3$] is the main aluminum oxide present in heavily weathered soils as a result of total hydrolysis in tropical environments where it may be the predominant mineral in the clay fraction. In soil spectra, the absorption feature identifying this mineral is positioned at 2260 nm (Figure 41.3) due to the bending–stretching combination of aluminum–hydroxyl bond (Al-OH). Nonetheless, the O-H fundamental stretching vibration is shifted from 3620 to 3520 cm^{-1} because of the molecular structure of this oxide and strength of its hydrogen bonds (Hunt, 1977).

41.2.2.1.4 Carbonates

Calcite ($CaCO_3$) and dolomite [$CaMg(CO_3)_2$] are carbonate minerals common in incipient soils from regions where chemical weathering process is low, or they may be present in weathered soils due to fertility management practices, such as liming. Calcite has its absorption feature positioned at 2350 nm in soil spectra (Figure 41.3) produced by a third stretching overtone of C-O bonds (fundamental absorption of CO_3^{-2} at 1415 cm^{-1}) (Hunt and Salisbury, 1971b).

41.2.2.1.5 Silica

In soils, silica (SiO_2) is present in sand fraction as quartz and in the silicon tetrahedral sheet of phyllosilicate clay minerals. However, the related absorption bands are only present in the fingerprint region of mid-IR spectra due to Si-O stretching modes from 2000 to 1000 cm^{-1} and due to Si-O-Si bending modes from 1000 to 400 cm^{-1} (Figure 41.3). Quartz has characteristic bands at 2000, 1870, 1790, 1080, 800, 700, and 400 cm^{-1} (Nguyen et al., 1991). Other absorptions in the fingerprint can be assigned to other silicate structures masked by overlaps of quartz.

41.2.2.2 Soil Organic Matter

Soil organic matter is represented by organic carbon compounds with variable physicochemical characteristics from plant decomposition. Chlorophyll, tannins, and humic substances are examples of organic chromophores present in soils (Del Vecchio and Blough, 2004). Even in spectra of "mineral soils" where organic carbon contents are generally lower than 12% (IUSS Working Group WRB, 2015), it is possible in some cases to detect absorptions across NIR-SWIR ranges assigned to stretching overtones as following compounds: (1) aromatics at 825, 1100, and 1650 nm (fundamental C-H at 3030 cm^{-1}); (2) amines at 751, 1000, 1500, and 2060 nm (fundamental N-H at 3330 cm^{-1}); (3) alkyls at 853, 1138, and 1706 nm (fundamental C-H at 2930 cm^{-1}) and at 877, 1170, and 1754 nm (fundamental C-H at 2850 cm^{-1}); (4) carboxylic acids at 1449 and 1930 nm (fundamental C=O at 1725 cm^{-1}); (5) amides at 1524 and 2033 nm (fundamental C=O at 1640 cm^{-1}); (6) aliphatics at 1706 and 2275 nm (fundamental C-H at 1465 cm^{-1}); (7) methyls at 1730 and 2307 nm (fundamental C-H at 1445 cm^{-1}) and at 1852 and 2469 nm (fundamental C-H at 1350 cm^{-1}); (8) phenolics at 1961 nm (fundamental C-OH at 1275 cm^{-1}); (9) polysaccharides at 2137 nm (fundamental C-O at 1170 cm^{-1}); and (10) carbohydrates at 2381 nm (fundamental C-O at 1050 cm^{-1}) (Viscarra Rossel and Behrens, 2010; Cécillon et al., 2012; Viscarra Rossel and Hicks, 2015). However, soil organic carbon peaks may be masked by stronger absorptions of oxide and phyllosilicate clay minerals, quartz, and water.

41.2.2.3 Water Content (Moisture)

In general, the soil has been spectrally analyzed under dry conditions, since its moisture affects the characteristic bands of other soil components. However, residual water remains strongly adsorbed on organic and mineral particle surfaces even after drying, and it may also be contained in quartz and 2:1 phyllosilicate structures. In these situations, the presence of water is identified by the two deep and narrow bands at 1400 and 1900 nm, as already stated. As the water content increases, these two bands become deeper and broader with their maximum absorptions shifted to 1450 and 1940 nm, followed by a general decrease of the soil albedo as effect of the macroscopic interaction (Bowers and Hanks, 1965; Bowers and Smith, 1972). At soil saturation conditions, typical absorptions of liquid water at 760, 970, and 1190 nm (stretching combination of fundamentals O-H at 3484 and 3220 cm^{-1}) can be observed mainly in spectra of quartz sand soil with low organic matter content (Tian and Philpot, 2015), since these water absorptions may be masked by organic matter and/or iron oxides in darker soils.

41.3 SPECTROSCOPIC MODELING OF SOIL PROPERTIES

Predictions of soil physical and chemical properties based on reflectance spectroscopy are possible since the soil spectral behavior is sensitive to changes in concentrations of its organic and mineral components and of elements adsorbed on their surfaces. Comprehensive reviews about Vis–NIR and mid-IR reflectance spectroscopy for soil property predictions are presented by Stenberg et al.

(2010) and Soriano-Disla et al., (2014). The performance of spectroscopic models is, generally, assessed by R^2 (coefficient of determination), RMSE (root mean square error), RPD (ratio of performance to deviation), and RPIQ (ratio of performance to interquartile) (Bellon-Maurel et al., 2010), and depends on the size and spectral variability of soil datasets (local, regional, country, or global) (Ramirez-Lopez et al., 2013a, 2013b). Choice of spectral ranges is also crucial for determining the quality of models, where mid-IR data have presented better results compared to Vis–NIR because of their fundamental interactions with soil properties (Viscarra Rossel et al., 2006; Knox et al., 2015; Terra et al., 2015).

Due to the multivariate nature of the soil spectra, chemometrics modeling is required through the use of linear or nonlinear regression models: partial least squares regression, multiple linear regression, principal component regression, support vector machines, artificial neural networks, boosted trees, random forest, etc. Viscarra and Behrens (2010) present a comparison between multivariate calibration models to quantify the important soil quality attributes.

Generally, soil sample preparation for reflectance spectroscopy analysis includes only drying, grinding, and sieving without the addition of chemicals. To that end, soils have been oven-dried at 45°C for 48 h and sieved to <2 mm for Vis–NIR spectra, and to <200 μm for mid-IR spectra, in order to reduce the particle size effect (Ben-Dor and Banin, 1995a, 1995b). Protocols establishing rules for spectral acquisition geometries and soil preparation are available in order to standardize the soil spectroscopy, with the intent to enable sharing spectral data, and compare prediction modeling (Ben-Dor et al., 2015; Viscarra Rossel et al., 2016; Romero et al., 2018). For field applications, factors such as surface roughness, sample heterogeneity, and mainly moisture content still affect any soil spectral modeling. However, some authors have used spectral processing methods to achieve comparable performance to laboratory analyses (Stevens et al., 2008; Viscarra Rossel et al., 2009; Stenberg, 2010; Minasny et al., 2011; Ji et al., 2016a).

Some mathematical functions are commonly used on soil reflectance data before modeling to ensure a direct relation between absorbed energy and properties of interest and to correct for nonlinear light scattering and noise effects. The most common preprocessing is the log-transform from reflectance (R) into absorbance ($A = \log_{10} 1/R$) preserving a linear correlation between radiance and concentrations. The Kubelka–Munk function is also used to deal with scattering effects.

After the spectra have been linearized to the parameters of interest, other preprocessing methods have been used to remove spectral noise (averaging spectra, Savitzky–Golay smoothing (Savitzky and Golay, 1964), and moving average filters) and enhance spectral features (continuum removal (Clark and Roush, 1984), baseline correction, the first- and second-derivative processing, multiplicative scatter correction (Geladi et al., 1985), standard normal variate correction, and wavelet transform (Ricker, 1953)).

41.3.1 First-Order Modeling

Soil properties with the first-order spectral relations directly affect soil spectra modifying their overall reflectance intensities and absorption features due to macroscopic and microscopic interactions, respectively. Generally, these properties are strongly correlated with reflectance or absorbance data, and their concentrations are better quantified by spectroscopic modeling (Terra et al., 2015). Examples of the first-order soil properties are particle size distribution (sand, silt, and clay), mineralogy, organic content, and moisture. A review with more recent soil property predictions is presented in Table 41.1.

41.3.1.1 Particle Size Distribution

Soil particle size distribution refers to the proportions of sand (2–0.05 mm), silt (0.05–0.002 mm), and clay (<0.002 mm) in soil samples. This distribution has been conventionally analyzed by a specific methodology based on Stokes' law and chemical dispersing agents (calcium hexametaphosphate and sodium hydroxide) from which the proportions of sand and clay are determined

TABLE 41.1
R^2 Values for Some Soil Properties from Vis–NIR and Mid-IR Spectroscopic Modeling, where *N* is the Number of Predictions

Attribute	Vis–NIR			Mid-IR			References
	Range	*N*	Mean	Range	*N*	Mean	
							Particle Size Distribution
Clay	0.05–0.94	25	0.79	0.20–0.98	16	0.84	Ben-Dor and Banin (1995a); Janik and Skjemstad (1995); Janik and Skjemstad (1998); Chang et al. (2001); Shepherd and Walsh (2002); Cozzolino and Morón (2003); Islam et al. (2003); Brown et al. (2006); Madari et al. (2006); Vašát et al. (2014); Viscarra Rossel et al. (2006); Waiser et al. (2007); Janik et al. (2009); Minasny et al. (2009); Viscarra Rossel et al. (2009); Bricklemyer and Brown (2010); Viscarra Rossel and Behrens (2010); Summers et al. (2011); Xie et al. (2012); Sarathjith et al. (2014); Vohland et al. (2014); Wenjun et al. (2014); Hicks et al. (2015); Ludwig et al. (2015); Terra et al. (2015); Veum et al. (2015); Wang et al. (2015b); Demattê et al. (2016a); Ji et al. (2016b); O'Rourke et al. (2016b); Viscarra Rossel et al. (2016); Conforti et al. (2017); Hu et al. (2017); Pinheiro et al. (2017); Yang et al. (2012); Franceschini et al. (2018); Riedel et al. (2018); Xu et al. (2018)
Sand	0.09–0.92	22	0.77	0.13–0.97	12	0.81	
							Mineralogy
Clay mineralogy	0.40–0.94	22	0.53				Fernandes et al. (2004); Brown et al. (2006); Viscarra Rossel et al. (2009, 2010a); Viscarra Rossel (2011)
Si/Al/Fe/Ti/Mn oxides	0.46–0.71	7	0.61	0.50–0.61	5	0.57	Janik and Skjemstad (1995); Brown et al. (2006); Ben-Dor et al. (2008a); Reeves and Smith (2009); Summers et al. (2011); Terra et al. (2015)
$CaCO_3$	0.69–0.90	4	0.75	0.83–0.98	4	0.95	Ben-Dor and Banin (1990a,b); Janik and Skjemstad (1995); Ben-Dor et al. (1995a,b); Brown et al. (2006); Ben-Dor et al. (2008a); Reeves and Smith (2009); Summers et al. (2011)
OC_{total}	0.41–0.91	24	0.78	0.71–0.97	12	0.89	McCarty et al. (2002); Madari et al. (2005); Van Waes et al. (2005); Reeves III et al. (2006); Viscarra Rossel et al. (2006); Janik et al. (2007a); Morgan et al. (2009); Vasques et al. (2009); Stenberg (2010); Vasques et al. (2010); Vohland et al. (2011); Yang et al. (2012); Stevens et al. (2013); Terra et al. (2015); Viscarra and Hicks (2015); Ludwig et al. (2016); Viscarra Rossel et al. (2016), Zhang et al. (2017)
Moisture	0.40–0.98	14	0.84	0.55–0.95	7	0.83	Dalal and Henry (1986); Ben-Dor and Banin (1995a,b); Janik et al. (1998); Hummel et al. (2001); Slaughter et al. (2001); Chang et al. (2005); Mouazen et al. (2005); Janik et al. (2007b); Ben-Dor et al. (2008a); Christy (2008); Zornoza et al. (2008); Janik et al. (2009); Kodaira and Shibusawa (2013); Ackerson et al. (2015); Franceschini et al. (2018)

(Continued)

TABLE 41.1 (*Continued*)

R^2 Values for Some Soil Properties from Vis–NIR and Mid-IR Spectroscopic Modeling, where *N* is the Number of Predictions

	Vis–NIR			Mid-IR			
Attribute	**Range**	***N***	**Mean**	**Range**	***N***	**Mean**	**References**
pH	0.02–0.92	35	0.57	0.01–0.92	12	0.68	Janik and Skjemstad (1995); Janik et al. (1998); Reeves et al. (1999); Chang et al. (2001); Reeves and McCarty (2001); Shepherd and Walsh (2002); Islam et al. (2003); He et al. (2007); Mouazen et al. (2007); Zornoza et al. (2008); Janik et al. (2009); Kuang and Mouazen (2011); Xie et al. (2012); Sarathjith et al. (2014); Vohland et al. (2014); Wenjun et al. (2014); Hicks et al. (2015); Ludwig et al. (2015); Terra et al. (2015); Veum et al. (2015); Wang et al. (2015b); Demattê et al. (2016a); Ji et al. (2016b); O'Rourke et al. (2016b); Conforti et al. (2017); Hu et al. (2017); Pinheiro et al. (2017); Yang et al. (2017); Franceschini et al. (2018); Riedel et al. (2018); Xu et al. (2018)
Exch. Al	0.30–0.78	5	0.52	0.64–0.90	7	0.81	Cohen et al. (2005); Minasny et al. (2009); Reeves and Smith (2009); Soriano-Disla et al. (2013); Sarathjith et al. (2014); Terra et al. (2015); Zhang et al. (2017); Riedel et al. (2018)
							Macronutrients
Ca	0.01–0.90	18	0.67	0.29–0.96	14	0.78	Janik et al. (1998); Chang et al. (2001); Shepherd and Walsh (2002); Cozzolino and Morón (2003); Islam et al. (2003); Janik et al. (2009); Soriano-Disla et al. (2013); Vašát et al. (2014); Araújo et al. (2015); Hicks et al. (2015); Terra et al. (2015); Veum et al. (2015); Demattê et al. (2016a); Ji et al. (2016a); O'Rourke et al. (2016a, 2016b); Peng et al. (2016); Sila et al. (2016); Pinheiro et al. (2017); Soriano-Disla et al. (2017); Zhang et al. (2017); Franceschini et al. (2018); Riedel et al. (2018)
Mg	0.12 – 0.90	20	0.67	0.55 – 0.97	7	0.76	Janik et al. (1998); Janik et al. (2009); Reeves and Smith (2009); Soriano-Disla et al. (2013); Vašát et al. (2014); Askari et al. (2015); Terra et al. (2015); Veum et al. (2015); Cozzolino (2016); Demattê et al. (2016a); Ji et al. (2016a); O'Rourke et al. (2016a, 2016b); Pinheiro et al. (2017); Zhang et al. (2017); Franceschini et al. (2018)
K	0.00–0.90	27	0.52	0.13–0.83	10	0.53	Janik et al. (1998); Chang et al. (2001); Cozzolino and Morón (2003); Daniel et al. (2003); Islam et al. (2003); He et al. (2007); Zornoza et al. (2008); Reeves and Smith (2009); Mouazen et al. (2010); Soriano-Disla et al. (2013); Chacón Iznaga et al. (2014); Gras et al. (2014); Sarathjith et al. (2014); Wenjun et al. (2014); Terra et al. (2015); Veum et al. (2015); Wang et al. (2015b); Hicks et al. (2015); Demattê et al. (2016a); Ji et al. (2016b); La et al. (2016); O'Rourke et al. (2016a, 2016b); Zhang et al. (2017); Riedel et al. (2018); Xu et al. (2018)
P	0.11 – 0.81	27	0.54	0.07 – 0.91	9	0.51	Janik et al. (1998); Daniel et al. (2003); Bogrekci and Lee (2007); He et al. (2007); Mouazen et al. (2007); Zornoza et al. (2008); Janik et al. (2009); Mouazen et al. (2010); Kuang and Mouazen (2011); Soriano-Disla et al. (2013); Chacón Iznaga et al. (2014); Gras et al. (2014); Sarathjith et al. (2014); Wenjun et al. (2014); Camargo et al. (2015); Hicks et al. (2015); Ludwig et al. (2015); Terra et al. (2015); Veum et al. (2015); Wang et al. (2015b); Demattê et al. (2016a); Ji et al. (2016a, 2016b); La et al. (2016); Mouazen and Kuang (2016); O'Rourke et al. (2016b); Pinheiro et al. (2017); Zhang et al. (2017); Xu et al. (2018)

(*Continued*)

TABLE 41.1 (*Continued*)

R^2 Values for Some Soil Properties from Vis–NIR and Mid-IR Spectroscopic Modeling, where *N* is the Number of Predictions

Attribute	Vis–NIR Range	Vis–NIR *N*	Vis–NIR Mean	Mid-IR Range	Mid-IR *N*	Mid-IR Mean	References
N	0.09–0.99	32	0.79	0.22–0.98	15	0.84	Dalal and Henry (1986); Janik and Skjemstad (1995); Janik et al. (1998); Reeves et al. (1999); Chang et al. (2001); Reeves and Mccarty (2001); Viscarra Rossel et al. (2006); Zornoza et al. (2008); Kuang and Mouazen (2011); Kodaira and Shibusawa (2013); Gras et al. (2014); Vohland et al. (2014); Wenjun et al. (2014); Hicks et al. (2015); Li et al. (2015); Ludwig et al. (2015); Veum et al. (2015); Wang et al. (2015a); Ji et al. (2016a); Ludwig et al. (2016); Conforti et al. (2017); de Souza Bahia et al. (2017); Linsler et al. (2017); Nie et al. (2017); Zhang et al. (2017); Xu et al. (2018)
S		1	0.67		1	0.85	Ludwig et al. (2015); O'Rourke et al. (2016)
							Micronutrients
B	0.40–0.64	3	0.36		1	0.22	Sarathjith et al. (2014); Terra et al. (2015)
Co		1	0.39	0.65–0.90	2	0.77	Reeves and Smith (2009); Soriano-Disla et al. (2013); O'Rourke et al. (2016)
Cu	0.01–0.71	8	0.32	0.09–0.53	6	0.44	Janik et al. (1998); Janik et al. (2009); Minasny et al. (2009); Reeves and Smith (2009); Soriano-Disla et al. (2013); Vašát et al. (2014); Terra et al. (2015); Veum et al. (2015); Ji et al. (2016a); O'Rourke et al. (2016); Riedel et al. (2018)
Fe	0.19–0.90	18	0.62	0.08–0.88	10	0.67	Janik et al. (1998); Chang et al. (2001); Cozzolino and Morón (2003); Islam et al. (2003); Cohen et al. (2005); Janik et al. (2009); Minasny et al. (2009); Reeves and Smith (2009); Xie et al. (2012); Soriano-Disla et al. (2013); Sarathjith et al. (2014); Camargo et al. (2015); Hicks et al. (2015); Terra et al. (2015); Veum et al. (2015); Ji et al. (2016a); Mohanty et al. (2016); de Souza Bahia et al. (2017); Zhang et al. (2017); Riedel et al. (2018)
Mn	0.01–0.70	11	0.42	0.15–0.66	6	0.53	Janik et al. (1998); Chang et al. (2001); Christy (2008); Minasny et al. (2009); Soriano-Disla et al. (2013); Vašát et al. (2014); Terra et al. (2015); Veum et al. (2015); O'Rourke et al. (2016); Zhang et al. (2017); Riedel et al. (2018)
Mo		1	0.09				O'Rourke et al. (2016)
Na	0.08–0.68	6	0.39	0.02–0.85	8	0.64	Janik et al. (1998); Chang et al. (2001); Islam et al. (2003); Zornoza et al. (2008); Minasny et al. (2009); Reeves and Smith (2009); Soriano-Disla et al. (2013); Hicks et al. (2015); Waruru et al. (2015); Demattê et al. (2016b); Ji et al. (2016a); Mouazen and Kuang (2016); Riedel et al. (2018)
Ni		1	0.48	0.70–0.98	3	0.73	Reeves and Smith (2009); Soriano-Disla et al. (2013); Riedel et al. (2018)

(*Continued*)

TABLE 41.1 *(Continued)*

R^2 Values for Some Soil Properties from Vis–NIR and Mid-IR Spectroscopic Modeling, where N is the Number of Predictions

	Vis–NIR			Mid-IR			
Attribute	**Range**	***N***	**Mean**	**Range**	***N***	**Mean**	**References**
Zn	0.26–0.57	8	0.37	0.15–0.66	4	0.18	Chang et al. (2001); Christy (2008); Minasny et al. (2009); Xie et al. (2012); Soriano-Disla et al. (2013); Sarathjith et al. (2014); Terra et al. (2015); Veum et al. (2015); O'Rourke et al. (2016); Zhang et al. (2017); Riedel et al. (2018)
							Heavy Metals
Cu	0.25–0.95	13	0.79				Janik et al. (1998); Wang et al. (2000); Minasny et al. (2009); Janik et al. (2009); Reeves and Smith (2009); Soriano-Disla et al. (2013); Vašát et al. (2014); Veum et al. (2015); Ji et al. (2016a); O'Rourke et al. (2016); Riedel et al. (2018)
Fe	0.23–0.71	4	0.41				Gholizadeh et al. (2015); Demattê et al. (2016b); Tayebi et al. (2017)
Mn	0.50–0.80	3	0.58				Pandit et al. (2010); Gholizadeh et al. (2015); Demattê et al. (2016b)
Mo	0.64–0.97	4	0.70				Luce et al. (2017)
Ni	0.50–0.92	7	0.86				Samiei Fard and Matinfar (2016); St. Luce et al. (2017); Sun et al. (2017); Wang et al. (2017)
Zn	0.09–0.93	14	0.62				Wu et al. (2005); Pandit et al. (2010); Gholizadeh et al. (2015); Demattê et al. (2016b); St. Luce et al. (2017); Sun et al. (2017); Wang et al. (2017)
Hg	0.48–0.93	3	0.59				Wu et al. (2005); Song et al. (2012); Wang et al. (2017)
Pb	0.01–0.74	8	0.67	0.00–0.42	2	0.21	Wu et al. (2005); Pandit et al. (2010); Song et al. (2012); Xie et al. (2012); Soriano-Disla et al. (2013); Pietrzykowski and Chodak (2014); Gholizadeh et al. (2015); Wang et al. (2017); Samiei Fard and Matinfar (2016); Riedel et al. (2018)
As	0.18–0.89	6	0.43	0.05–0.54	2	0.29	Wu et al. (2005); Soriano-Disla et al. (2013); Gholizadeh et al. (2015); Samiei Fard and Matinfar (2016); Shi et al. (2017); Riedel et al. (2018)
Cd	0.30–0.97	10	0.70				Pandit et al. (2010); Song et al. (2012); Xie et al. (2012); Gholizadeh et al. (2015); Samiei Fard and Matinfar (2016); St. Luce et al. (2017); Wang et al. (2017); Riedel et al. (2018)
Cr	0.44–0.98	5	0.75	0.86–0.90	2	0.92	Reeves and Smith (2009); Araújo et al. (2014); O'Rourke et al. (2016); Wang et al. (2017)

(EMBRAPA, 2017). Silt content is obtained from the difference between 100% and the sum of sand and clay. Contents of sand and especially clay have considerable importance in agricultural and environmental soil functions, since they influence soil structure, water and aeration dynamics, chemical reactions, and management practices.

Sand and clay fractions are, respectively, represented by quartz, phyllosilicates, and oxides, which are pedogenetic minerals and produce direct effects on soil spectral behaviors. Their spectroscopic predictions started in the mid-1990s (Ben-Dor and Banin, 1995a; Janik and Skjemstad, 1995; Janik et al., 1998) with adequate performance: $R^2 \geq 0.75$ in Vis–NIR and $R^2 \geq 0.80$ in mid-IR (Table 41.1). Sand and clay spectral models are on average more accurate compared to other soil properties (Table 41.1). Spectral variables important for sand and clay modeling include mineral absorption features (Viscarra Rossel et al., 2016), since they are the first-order properties. However, macroscopic interactions that affect the overall reflectance intensity must also be considered to explain the selected spectral variables.

41.3.1.2 Mineralogy

Generally, soil clay minerals represented by phyllosilicates (2:1 and 1:1) and oxides (Al and Fe) are directly responsible for soil physical and chemical functions, such as aggregation (structure), water retention, cation exchange capacity, chemical adsorption, and calcium and potassium availability (Weil and Brady, 2017). Since minerals are about 45% of soil matrix and quartz is chemically inert, types and concentrations of clay minerals determine the reactivity and specific surface area of soil particles, although soil organic matter is also involved. Conventionally, types of clay minerals have been identified by X-ray diffraction (XRD) (Harris and White, 2008) and their concentrations quantified by sulfuric acid digestion in soil samples (EMBRAPA, 2017).

Phyllosilicates and oxides have been successfully quantified by reflectance spectroscopy based on different techniques such as continuum removal (Viscarra Rossel et al., 2009; Viscarra Rossel, 2011), normalized indices (Viscarra Rossel et al., 2010a), ordinal unit of reference XRD values (Brown et al., 2006), simple linear regression between mineral concentration and absorption feature width (Fernandes et al., 2004), and band shifting position of the metal-OH bending around 2200 nm (Ben-Dor and Banin, 1990b) (Table 41.1). Oxides extracted by acid digestion, such as SiO_2, Fe_2O_3, Al_2O_3, TiO_2, and MnO, have also been accurately predicted by multivariate regression models and soil spectra (Ben-Dor and Banin, 1995b; Ben-Dor et al., 2008a; Summers et al., 2011; Terra et al., 2015). Concentrations of calcium carbonates ranging from 10% to 75% were satisfactorily estimated by Ben-Dor and Banin (1990a) with standard errors of 5.9% and 7.9% for calibration and prediction models, respectively. Other performances of mineralogical modeling based on soil spectra are found in Table 41.1.

41.3.1.3 Total Organic Carbon

Soil organic matter is represented by total organic carbon compounds and plays an important role affecting soil physical and chemical properties, such as soil aggregation, reactivity, acidity, and fertility, as well as clay. Besides that, soil organic matter is an important indicator of climate changes too, where soil organic carbon sequestration is associated with mitigation of greenhouse gas emissions (Lal and Follett, 2009). For total organic carbon analyses, some methods have conventionally been used for semiquantification and quantification, such as loss-on-ignition (Schulte and Hopkins, 1996), hydrogen peroxide digestion (Robinson, 1927), dry combustion (Nelson and Sommers, 1996), and mainly the Walkley–Black method based on potassium dichromate and sulfuric acid (Walkley and Black, 1934).

Spectroscopic modeling has been successfully applied for total organic carbon quantifications due to its direct effects on soil spectra (Madari et al., 2005; Reeves et al., 2006, Janik et al., 2007a; Stevens et al., 2013); however, their performances are variables (Viscarra Rossel et al., 2016). As observed by Terra et al. (2015), better results for the total organic carbon modeling have been achieved with mid-IR spectral data compared to Vis–NIR (Table 41.1). Fractions of soil organic carbon (such as particulate, humic, and resistant fractions) and soil organic matter composition (such as carbohydrate, protein, lignin, lipid and black carbon) have been accurately modeled by reflectance spectroscopy (Cécillon et al., 2012; Viscarra and Hicks, 2015).

41.3.1.4 Moisture

In soils, water in its free liquid state flows and is stored in structural macropores and is adsorbed on mineral and organic functional groups of soil particles in micropores (Weil and Brady, 2017). Soil water content is essential for plant development and hydrological processes, and its effects on reduction of the overall soil reflectance intensity have been previously discussed. Conventionally, the moisture of a soil sample can be determined by gravimetric analysis where the proportion of soil water is calculated by the ratio between water and dry soil masses. Due to a significant correlation between soil moisture and albedo, the modeling of water content at soil field capacity has been well modeled by laboratory Vis–NIR and mid-IR reflectance spectroscopy with similar performances (mean $R^2 > 0.80$) in both ranges for different mineral soils (Table 41.1), according to Demattê et al. (2006), Ben-Dor et al. (2008a), and Janik et al. (2007b). Additional studies are needed on the spectroscopic quantification of soil water content for field applications to reduce prediction errors compared to validation of other soil properties, such as clay and organic carbon (Stenberg, 2010).

41.3.2 Second-Order Modeling

There is a great interest in employing reflectance spectroscopic analysis to predict plant nutrients in soils, in particular macro- and micronutrients. However, if one is interested in quantifying the concentration of elements in the soil solution, corresponding to the fraction available for plant absorption, then spectroscopy will probably not provide robust models (Janik et al., 1998). This same issue is observed for predictions of heavy metal concentrations in soils because there are no spectral features directly related to these in the Vis–NIR and mid-IR soil spectral behaviors.

Second-order soil properties do not directly affect soil spectra, where they do not produce absorption features or change reflectance intensities. However, these properties are generally adsorbed or depend on functional groups of minerals and organic compounds. Since the second-order properties are strongly correlated with spectrally active first-order properties, their concentrations may be indirectly modeled by reflectance spectroscopy. Some researchers have described good correlations between conventional analyses and IR quantifications, where the performance of these models is related to a local covariation between the predicted element and the first-order property (Sarathjith et al., 2014).

41.3.2.1 pH

The acidity of soils is represented by pH, which is the negative logarithm of H^+ activity in the soil solution. Water, calcium ($CaCl_2$), and potassium chlorides (KCl) are the solutions used for soil pH analysis. The spatiotemporal variability of soil pH is directly related to many factors, including mineral composition (proton-rich clays, Al oxyhydroxide minerals and sulfides), weathering reactions (hydrolysis), organic matter content, and soil management (e.g., ammonium fertilization/nitrification and liming) (Janik et al., 1998). The soil reflectance spectroscopy is not able to directly predict the H^+ activity in the soil solution, even though satisfactory models have been described using laboratory measurements of the spectra (Table 41.1). Clay fraction content and mineralogy, organic matter content, and carbonates are some of the most relevant first-order attributes for pH predictions (Stenberg et al., 2010). *In situ* spectroscopic modeling of pH is also becoming increasingly common (Franceschini et al., 2018; Ji et al., 2016a; Wenjun et al., 2014; Zhang et al., 2017a). The variability in model performances is extremely high, with R^2 values ranging from 0.02 to 0.92. In general, mean values of R^2 are reasonably satisfactory for both Vis–NIR (0.57) and mid-IR models (0.68).

41.3.2.2 Macronutrients

Soil macronutrients are essential for plant growth. There have been many attempts to predict soil nutrients based on IR spectroscopy, especially calcium (Ca), magnesium (Mg), potassium (K), nitrogen (N), phosphorus (P), and sulfur (S) (Table 41.1). If suitable, the technique can provide great support in everyday and precision agriculture (PA) activities, including variable rate applications of fertilizers. Results described in the literature have indicated highly variable R^2 values in both Vis–NIR and mid-IR ranges for Ca (0.01–0.9 and 0.29–0.96), Mg (0.12–0.89 and 0.55–0.97), K (0–0.9 and 0.13–0.83), P (0.02–0.92 and 0.07–0.91), and N (0.09–0.9 and 0.22–0.98) (Table 41.1). Among these elements, nitrogen has presented better results in both Vis–NIR and mid-IR ranges with a mean R^2 of 0.79 and 0.84, respectively. In most cases, the wide R^2 range of macronutrient's models is related to the inability of spectral information to describe the variability of these elements in soils. In addition, the different laboratory methods used to define the standard values, as well as the different fractions assessed (e.g., extractable, exchangeable, and total), probably contribute to the large variation described here (Chang et al., 2001). Furthermore, fractions most related to the concentration of the elements in solution, rather than to the chemistry of the soil matrix, will probably be poorly predicted (Soriano-Disla et al., 2014).

The good predictions of soil macronutrients are related to covariations between element assessed and the first-order soil attributes. Ca, Mg, and K are usually predicted based on the covariation with mineralogical properties (Chang et al., 2001). N is highly correlated with soil organic matter and carbon concentrations, mostly because of the amide groups. Accurate spectral predictions of phosphorous concentration in soils are related to its covariation with clay mineralogy and organic matter content (Chacón Iznaga et al., 2014; La et al., 2016; Sarathjith et al., 2014). Furthermore, there are strong absorption bands in the mid-IR region related to phosphate minerals (He et al., 2007), which may aid in the quantification.

41.3.2.3 Micronutrients

Soil micronutrients are as crucial as macronutrients for plant nutrition and production. The appropriate management of crops requires not only adequate concentrations of soil macronutrients but also others such as iron (Fe), manganese (Mn), zinc (Zn), copper (Cu), boron (B), molybdenum (Mo), and chlorine (Cl). These elements are denominated micronutrients and are usually required by the plants in concentrations relatively lower than macronutrients. Cobalt (Co), sodium (Na), and nickel (Ni) are not considered micronutrients, but aid in the optimum development of plants. Due to the lowest natural concentration of these micronutrients in soils, they would probably not be detected by the technique even if they had specific spectral absorptions in the Vis–NIR and mid-IR ranges (Soriano-Disla et al., 2014).

The most common micronutrients evaluated in soil spectroscopic studies are Cu, Mn, Fe, and Zn (Table 41.1). However, the performances of their models are highly variable, and their mean R^2 values are usually low but satisfactory for B (R^2 ranging between 0.4 and 0.64 and mean of 0.36), Cu (0.01–0.71 and 0.32), Fe (0.19–0.9 and 0.62), Mn (0.01–0.7 and 0.42), Na (0.08–0.68 and 0.39), and Zn (0.26–0.57 and 0.37). The models calibrated with mid-IR data also presented a limited performance for Cu (0.09–0.53 and 0.44), Fe (0.08–0.88 and 0.55), Mn (0.15–0.66 and 0.53), and Na (0.02–0.85 and 0.64) (Table 41.1). The good performance of these prediction models is directly related to the covariation of micronutrients with mainly clay content and mineralogy (oxides and phyllosilicates) and soil organic matter, as observed by Vašát et al. (2014), Terra et al. (2015), and Riedel et al. (2018).

41.3.2.4 Heavy Metals

Heavy metal contamination in soils has been considered a worldwide ecological issue due to its potential and persistent threats to the environment (Wu et al., 2015). Heavy metals accumulate in soils and are hardly removed by natural conditions (Wang et al., 2017). Due to their persistent nature and long biological half-lives, they can accumulate in the food chain and eventually influence human health (Xie et al., 2012; Gholizadeh et al., 2015). Heavy metal monitoring in soils has become increasingly important, especially in regions under industrial activities and intensive agriculture (Shi et al., 2014).

Many efforts have been made to evaluate reflectance spectroscopy as a potential technique for heavy metal contamination assessment in soils. Researchers have mostly focused on arsenic (As), cadmium (Cd), copper (Cu), nickel (Ni), lead (Pb), and zinc (Zn), although predictions of mercury (Hg) and chrome (Cr) are also reported (Table 41.1). Significant correlations have been reported between heavy metals and absorption features in Vis–NIR and mid-IR attributed to iron oxides, clays, and organic matter content, suggesting a strong bond with these soil constituents (Kemper and Sommer, 2002; Song et al., 2012; Luce et al., 2017). Although heavy metals are not spectrally active elements, some good results have been observed. As previously mentioned for soil nutrients, the R^2 for Vis–NIR models are also highly variable and their mean values are usually satisfactory for Cd (0.3–0.97 and 0.7), Cr (0.44–0.98 and 0.75), Cu (0.25–0.95 and 0.79), Mo (0.64–0.97 and 0.7), and Ni (0.5–0.92 and 0.86) (Table 41.1). In general, spectroscopic models for predictions of other heavy metals (As, Fe, Mn, Pb, and Zn) have presented unsatisfactory mean R^2 (Table 41.1). Among the conventional chemical analyses employed to characterize soil metal concentrations, Soriano-Disla et al. (2014) describe the aqua regia digestion method with the best results for mid-IR modeling.

41.4 APPLICATIONS

41.4.1 Soil Spectral Libraries

One of the significant gaps in effective soil monitoring is the availability of calibration models capable of assessing soil attributes from a wide range of soil types (Cécillon et al., 2009). Researchers have suggested that the development of large soil spectral databases may facilitate the use of reflectance spectroscopy. This is mainly by reducing the number of calibration samples required for local or regional applications, as well as facilitating soil monitoring from remote or proximal spectral sensing techniques (McCarty et al., 2002; Shepherd and Walsh, 2002; Brown, 2007). These large databases described herein were defined by Brown et al. (2006) as Soil Spectral Libraries (SSL), which provide laboratory measurements of spectral and physicochemical information about a given soil at a given location. SSLs can be local, regional, or global based on their geographical extents and their represented soil variability. Local SSLs refer to field-scale datasets, while regional ones include larger geographic extents, such as physioclimatic zones. Global SSLs are developed from samples of the world's major soil taxa from multiple continents (Brown et al., 2006; Viscarra Rossel et al., 2016). SSLs are intended for two primary purposes: (1) for spectral soil sensing procedures as *in situ* measurements for calibration and/or end-members selection in data processing, or (2) for a rapid assessment of soil properties by reflectance spectroscopy (Brown, 2007; Selige et al., 2006; Viscarra Rossel et al., 2006).

One of the first SSLs is described by Condit (1970), who measured and studied the Vis–NIR spectra of 160 samples derived from 36 states in the United States of America. Later, Stoner and Baumgardner (1981) created a dataset with spectral information from 485 soil samples of the United States and Brazil. Nowadays, there is a considerable number of Vis–NIR and mid-IR SSLs spread around the world for national and continental uses with significant amounts of soil samples (>1000) per library, such as in Africa (Shepherd and Walsh, 2002; Stevens et al., 2013), the United States of America (Brown et al., 2006), Europe (Stevens et al., 2013), Czech Republic (Brodský et al., 2011), Belgium (Genot et al., 2011), Denmark (Knadel et al., 2012), China (Shi et al., 2015), Brazil (Bellinaso et al., 2010; Terra et al., 2015), and Australia (Viscarra Rossel et al., 2008; Viscarra Rossel and Webster, 2012). At the world scale, Brown et al. (2006) measured and analyzed the spectra of 4184 soil samples with great compositional diversity from across the Americas, Africa, and Asia for global soil characterization. The combined effort of many researchers around the globe allowed the creation of a global spectral library with more than 23,600 soil spectra from 12,509 sites (Viscarra Rossel et al., 2016).

It is important to assure that SSLs are measured under the same data acquisition geometry and protocol to enable merging datasets. Whereas one SSL is measured under a well-defined protocol, another SSL may be using a different setup and prevent data sharing among SSLs. To solve this problem, Ben-Dor et al. (2015) have suggested using the internal soil standard (ISS) concept allowing all SSLs worldwide to be shared. Romero et al. (2018) developed an ISS method to compile reflectance data collected from different sources to build the Brazilian Soil Spectral Library. Another important question refers to the development of methods for transferring calibration models across equipment (sensors) to compensate for instrument variation to make spectral data sharing among SSLs possible.

41.4.2 A Cost-Effective Alternative for Soil Wet Chemistry

Appropriate management and provision of soil ecosystem services require an efficient soil monitoring system, which demands low-cost and fast analytical methods. Such techniques must be replicable as well as provide accessible results for the end users. So far, soil assessments have been mostly done by traditional laboratory analyses (i.e., wet chemistry). Vis–NIR and mid-IR reflectance spectroscopies have become a viable alternative to the traditional methods since it is fast, environmentally friendly, nondestructive, reproducible, and repeatable (Viscarra Rossel et al., 2006; Nocita et al., 2015; Soriano-Disla et al., 2014). Due to the increased performance of computers, development of data mining techniques, availability of cost-effective radiometric sensors, reliable performances of spectroscopic soil modeling, and development of SSLs, reflectance spectroscopy can be considered a surrogate for soil wet chemistry analyses.

The Vis–NIR spectroscopy has been routinely used by the Africa Soil Information Service (AfSIS), which has provided services and developed national soil information systems in many African countries (e.g., Ethiopia, Ghana, Nigeria, and Tanzania). Such initiative aims to provide affordable soil testing and advisory services to farmers (AfSIS, 2018). The Joint Research Centre of the European Commission contracted a laboratory to analyze about 20,000 samples for the Land Use/Cover Area frame Survey (LUCAS) soil monitoring network (Tóth et al., 2013). The cost for SOC spectral analysis was reduced by one-third compared to using the reference analyses (Nocita et al., 2015). According to the World Agroforestry Center (ICRAFT), soil spectroscopy might reduce carbon analysis costs by 56%, when compared to the thermal oxidation method. Spectral analysis becomes even more attractive when the daily measurement capacity is considered, where the technique is capable of processing 350 samples per day, while a C/N Thermal analyzer provides 30 acidified or 60 unacidified samples (ICRAFT, 2018).

Recently, technological packages such as SpecSolo (2018) and SoilCares (2018) focusing on the quantification of soil properties by reflectance spectroscopy have been sold to farmers and other stakeholders. Customers can choose between sending their soil samples for accredited laboratories and acquiring their analytical infrastructure. Based on spectroscopic models, it is possible to predict soil macro- and micronutrients, pH, and organic matter (SoilCares, 2018).

Although reflectance spectroscopy has been mostly employed for soil carbon inventories and fertility diagnosis, studies have suggested it as a useful tool in soil contamination monitoring. Schwartz et al. (2012) indicated that sampling and analytical analysis represent a large share of the cost of soil contamination assessment. These authors also describe the potential of spectral analysis to monitor petroleum contamination in soils, which provided better results at a lower cost compared to traditional analytical methods.

41.4.3 Precision Agriculture

Agricultural fields are heterogeneous regarding soil properties, varying in space and time (Ge et al., 2011). Conventional soil fertility recommendations are based on elaborate sampling procedures, which provide an "average" value from a specific area. Such procedure usually results in either too much or too little fertilizer locally applied (Stoorvogel et al., 2015). The advent of global positioning systems (GPS) enabled the development of the PA concept, which aims to improve the efficiency of agricultural inputs using locally variable rate applications (Stenberg et al., 2010). PA requires highly efficient and reliable techniques for gathering and processing site-specific data (Ge et al., 2011). Information needed by PA includes high-resolution (<10 m) maps of soil properties, such as plant available water and nutrient status in the rooting zone.

Soil Vis-NIR and mid-IR spectroscopies can provide a detailed spatial sampling over a site, generating an excellent continuous description of soil properties variability (Demattê et al., 2015). Acquiring data at higher density and measuring soil surface at adjustable depth are the main advantages of on-the-go reflectance measurements (Franceschini et al., 2018). Vågen et al. (2006) tested the potential of Vis–NIR SSLs for predicting and mapping soil fertility in the highlands of Madagascar. Awiti et al. (2008) evaluated the use of NIR spectroscopy as a tool for diagnosing soil condition for agriculture and environmental management. Viscarra Rossel et al. (2010b) developed a soil fertility index for sugar cane production and modeled it spatially using Vis–NIR spectra and terrain attributes derived from a digital elevation model. All authors described herein indicated laboratory spectroscopy as a useful tool to describe soil fertility and quality, as well as indicate areas for different management.

Using an AgroSpec (Tec5) mobile and Vis–NIR spectrometer for *in situ* measurements of soil pH, Tekin et al. (2013) produced variable rate lime recommendation maps with $R^2=0.81$, RMSEP=0.20, and RPD=2.14. Mouazen et al. (2007) used a spectrometer attached to a subsoiler to satisfactorily measure soil organic carbon ($R^2=0.73$), moisture content ($R^2=0.89$), pH ($R^2=0.71$), and P ($R^2=0.69$). Kusumo et al. (2008) and Roudier et al. (2015) developed a portable Vis–NIR spectrometer with a prototype probe to assess organic carbon, nitrogen, and root density of soil cores in the field. Kodaira and Shibusawa (2013) developed a real-time soil sensor mounted on a tractor and equipped with one micro-CCD camera and two spectrometers to investigate soil moisture content, organic matter, pH, electrical conductivity, cation exchange capacity, total carbon, nitrogen, available phosphorus, and phosphorus absorptive coefficient. The accuracy of these spectroscopic models ranged from 0.45 to 0.93 and from 1.0 to 3.6 for R^2 and RPD, respectively.

41.4.4 Field Applications

Currently, there is a great demand for the use of field reflectance spectroscopy where *in situ* measurements of soil spectra promptly provide information about field conditions. Although potentially less accurate, field measurements can facilitate the acquisition of larger amounts of spectral data resulting in a better description of soil variation. The users are capable of retrieving essential soil data (e.g., clay or organic matter contents) in real time by combining readings from portable or on-the-go sensors to SSLs. Such field applications are also cost-effective and less time-consuming, having fewer associated method steps (e.g., storing, drying, and sieving) (Viscarra Rossel et al., 2009).

Recently, Ben-Dor et al. (2017) developed an assembly that can be mounted to any field spectrometer to obtain soil spectral data in the field that equal to those obtained in the laboratory. The assembly named SoilPRO has measured the soil spectra without disturbing the soil surface and with quality under any environmental and operational conditions. The SoilPRO device was developed to overcome problems such as atmospheric attenuation, sun radiation, and operator experience that significantly deteriorate the quality of field reflectance measurements and hence might affect the final proxy model and reduce its accuracy.

Using soil reflectance spectroscopy in field applications is still a relatively recent topic in the literature (Roudier et al., 2017). A limited number of studies have focused on the prediction of some soil attributes under field domain (e.g., organic carbon, pH, P, N, moisture, mineralogy, particle size distribution) (Mouazen et al., 2007; Waiser et al., 2007; Ben-Dor et al., 2008a; Christy, 2008; Viscarra Rossel et al., 2009; Igne et al., 2010; Kodaira and Shibusawa, 2013; Ji et al., 2016a; Tekin et al., 2016; Roudier et al., 2017; Franceschini et al., 2018). Results from these studies presented an extremely variable accuracy, with R^2 ranging from 0.2 to 0.9. Furthermore, field-based approaches are usually described as substantially less accurate (Kuang et al., 2012).

The lowest performance of field modeling is attributed to several external parameters, or "*in situ*" factors, such as soil-to-sensor distance and inclination, illumination conditions, contamination or smearing of other targets, and field natural heterogeneity due to soil moisture and roughness (Mouazen et al., 2009; Bricklemyer and Brown, 2010; Kodaira and Shibusawa, 2013; Ackerson et al., 2015; Poggio et al., 2015; Nawar and Mouazen, 2017; Franceschini et al., 2018). The most discussed "*in situ*" factor is the soil moisture, since varying levels of water content mask out most of the soil spectral behavior, as previously discussed, and consequently reduce the accuracy of soil property predictions (Morgan et al., 2009; Minasny et al., 2011; Nocita et al., 2013; Rienzi et al., 2014; Roudier et al., 2017). Moreover, SSLs based on processed (air-dried, grinded, and sieved) samples would not be suitable for predictions of soils in field conditions (Roudier et al., 2017). This problem has also arisen from the surface appearance in the field, which is disturbed by the sampling process required for laboratory analysis.

Recently, different approaches have been developed and employed to reduce the effects from environmental factors on field soil spectra. Minasny et al. (2011) implemented the external parameter orthogonalization (EPO) (Roger et al., 2003) to remove the effects of moisture in spectra of rewetted soils, improving predictions of organic carbon from $R^2=0.57$ to $R^2=0.87$. Ackerson et al. (2015) used the EPO to remove effects of soil water content, roughness, aggregation, and temperature variations and to model clay content from field spectra ($R^2=0.59$). Another methodology that has been used to remove the "*in situ*" factors is the direct standardization (DS). Ji et al. (2016a,b) used DS to reduce the effects of environmental factors in field spectra, aiming to improve the performance of organic carbon models from the Chinese SSL. Ji et al. (2016b) could find good results for predictions based on DS-transformed spectra ($R^2=0.70$, RPD = 1.79). Besides these methods, Franceschini et al. (2018) also tested the orthogonal signal correction (OSC) to improve data acquired from a Vis–NIR spectrometer, which had the best overall performance for the evaluated soil properties compared to the other ones.

41.4.5 Soil Imaging Spectroscopy

Imaging spectroscopy (IS) brings a new dimension to the field of remote sensing in general and for soil remote sensing in particular (Ben-Dor, 2002). It enlarges the envelope of point spectrometry into a spatial domain and provides a tangible perspective by adding a spatial detail to spectral information (Ben-Dor et al., 2008b). Accordingly, the IS technology enhances the thematic application of spectral recognition algorithms, which are established and innovative approaches. This capability can work from either long or short distances such as those acquired by satellites, airborne, drones, field, and laboratory domains without touching the samples. It can cover the optical (400–2500 nm) and the thermal spectral regions (3000–14,000 nm) (Sayão et al., 2018). After years of accumulating evidence from point spectroscopy domain of organic and inorganic materials, the IS technology is now entering many fields while soil analysis is still considered novel. This is mainly due to the substantial evidence that the spectral information can predict soil attributes from a-far and from experience gained in the IS technology to exploit the spectral information for mapping the earth surface (Ben-Dor, 2002; Demattê et al., 2009, 2017). IS can generate deteriorated datasets as a result of nonideal conditions such as atmosphere attenuation, clouds, relative low signal-to-noise ratio (due to the movement of the platform), and differential sun elevation.

A pioneering study of Ben-Dor et al. (2002) showed the capability to map several soil attributes from airborne IS sensor (DAIS7915) over selected fields in Israel. Prediction models were generated after soil spectral sampling over a 5 × 5-m pixel area per sample associated with precisely locating the sampling point geographically, and wet chemical analysis. The raw airborne spectral data were atmospherically filtered and used as a source to run the previous models. The authors demonstrated that organic content, hygroscopic moisture, and electric conductivity of the soil "paste" extraction could be modeled. Applying their models on a pixel-by-pixel basis revealed the spatial distribution for each property within the field in question. It was concluded that IS has a significant potential to map soil surface. In another study, Selige et al. (2006) were able to obtain reasonable top soil maps of organic matter and texture generated from the HyMap sensor in a rapid and nondestructive manner using a multivariate approach. Both studies were successful based on a precise sampling of soil within fields under questions and applying multivariate spectral-based models. To that end, they used field data (spectroscopy and chemistry), and after extracting the spectral-based model for each attribute, they implemented the model on a pixel-by-pixel basis.

Recently, Castaldi et al. (2018) showed that using LUCAS SSL enabled a direct mapping of soil organic matter in a ploughed field in Belgium. That was a pioneer study which showed a direct utilization of SSL for spatial mapping of agriculture field with airborne IS data. The IS technology with the SSL databases combined with data mining machinery (e.g., PARACUDA-II®, Carmon and Ben-Dor, 2017) provides a new dimension in which quantitative analysis of the soil spectra can be performed on a spatial domain.

As the IS technology enables the collection of spectral information from every pixel in the image while covering a large area with the high temporal cycle, it brings a new dimension to the soil remote sensing arena (Ben-Dor et al., 2004). If the IS spectral information is of high quality and a proper pretreatment is used, then the chemometric approach can be successfully applied, providing unique spectral spatially based soil information (Demattê et al., 2007). We can thus say that if done correctly and all difficulties are resolved, the IS technology can serve as a novel tool for the soil science discipline. Although there is no doubt that IS carries a high potential for soil applications, not many researchers have used this technology so far on a routine basis. This is mainly because of the objective problem of delivering high-quality IS data to analysts as well as the very few professional personnel available who know how to process raw IS into true reflectance values. This leads to the conclusion that with more IS activity in soil science, then progress in solving the above restrictions will be achieved.

41.5 CONCLUSION AND FINAL REMARKS

Without a doubt, soil sensing by Vis–NIR and mid-IR reflectance spectroscopy has high potential and applicability in qualitative and mainly quantitative soil analyses, since it is an environmentally friendly and cost-effective method. It may be an excellent alternative to replace some conventional soil analyses, such as particle size distribution or soil organic carbon, which require reagents and produce wastes. Only by the soil spectral behavior, much information about soil mineral and organic composition can be obtained at once. When soil spectra are combined with multivariate models, several essential soil properties can be efficiently modeled due to direct (first-order) and indirect (second-order) relations. In general, mid-IR spectral data have produced the most suitable models to quantify soil properties when compared to the Vis–NIR spectral range, which is also a reliable option.

When different soil spectra are collected in an SSL, their applications are improved by reducing calibration samples to quantify soil properties and by facilitating soil monitoring from different areas. In fact, the higher the soil sample variability of an SSL, the more significant the range of its applications. In addition to laboratory applications, soil reflectance spectroscopy is useful for soil assessment in the field and real time by proximal or remote spectral sensing techniques. However, it is essential to continue developing and testing different methods capable of removing external parameters ("*in situ*" factors) affecting soil spectra, in unique soil moisture, to improve their modeling performance for field applications. By using IS, spatial details are added to spectral assessments of soils enlarging the utilization of point spectroscopy into a spatial domain.

As discussed, soil reflectance spectroscopy interacts with different areas of soil science and can be helpful for soil survey, mapping, classification, pedogenesis, management practices, etc. Soil sensing by reflectance spectroscopy in the Vis–NIR and mid-IR regions is a reliable and advantageous analysis method and will help achieve sustainable agricultural practices.

REFERENCES

Ackerson, J.P., Demattê, J.A.M., Morgan, C.L.S. 2015. Predicting clay content on field-moist intact tropical soils using a dried, ground VisNIR library with external parameter orthogonalization. *Geoderma*, 259–260, 196–204.

Adams, J.B., Filice, A.L. 1967. Spectral reflectance 0.4 to 2.0 microns of silicate rock powders. *Journal of Geophysical Research*, 72, 5705–5715.

AfSIS. 2018. Rapid and low-cost technologies generate high values soils information for planners and farmers. Retrieved from http://africasoils.net/2016/10/07/rapid-low-cost-technologies-generate-high-value-soils-information-for-planners-and-farmers-2/.

Araújo, S.R., Söderström, M., Eriksson, J., Isendahl, C., Stenborg, P., Demattê, J.A.M. 2015. Determining soil properties in Amazonian Dark Earths by reflectance spectroscopy. *Geoderma*, 237, 308–317.

Araújo, S.R., Wetterlind, J., Demattê, J.A.M., Stenberg, B. 2014. Improving the prediction performance of a large tropical vis-NIR spectroscopic soil library from Brazil by clustering into smaller subsets or use of data mining calibration techniques. *European Journal of Soil Science*, 65, 718–729.

Askari, M.S., O'Rourke, S.M., Holden, N.M. 2015. Evaluation of soil quality for agricultural production using visible-near-infrared spectroscopy. *Geoderma*, 243–244, 80–91.

Awiti, A.O., Walsh, M.G., Shepherd, K.D., Kinyamario, J. 2008. Soil condition classification using infrared spectroscopy: A proposition for assessment of soil condition along a tropical forest-cropland chronosequence. *Geoderma*, 143, 73–84.

Baumgardner, M.F., Silva, L.F., Biehl, L.L., Stoner, E.R. 1985. Reflectance properties of soils. *Advances in Agronomy*, 38, 1–43.

Bellinaso, H., Demattê, J.A.M., Romeiro, S.A. 2010. Soil spectral library and its use in soil classification. *Brazilian Journal of Soil Science*, 34, 861–870.

Bellon-Maurel, V., Fernandez-Ahumada, E., Palagos, B., Roger, J.M., McBratney A. 2010. Critical review of chemometric indicators commonly used for assessing the quality of the prediction of soil attributes by NIR spectroscopy. *Trends in Analytical Chemistry*, 29, 1073–1081.

Ben-Dor, E. 2002. Quantitative remote sensing of soil properties. *Advances in Agronomy*, 75, 173–243.

Ben-Dor, E., Banin, A. 1990a. Near-infrared reflectance analysis of carbonate concentration in soils. *Applied Spectroscopy*, 44, 1064–1069.

Ben-Dor, E., Banin, A. 1990b. Diffuse reflectance spectra of smectite minerals in the near infrared and their relationship to chemical composition/Spectres de réflectance diffuse de smectites dans le proche infrarouge, en relation avec la composition chimique. *Sciences Géologiques, bulletins et mémoires*, 43, 117–128.

Ben-Dor, E., Banin, A. 1995a. Near-infrared analysis as a rapid method to simultaneously evaluate several soil properties. *Soil Science Society of America Journal*, 59, 364–372.

Ben-Dor, E., Banin, A. 1995b. Near infrared analysis (NIRA) as a method to simultaneously evaluate spectral featureless constituents in soils. *Soil Science*, 159, 259–270.

Ben-Dor, E., Goldshalager, N., Braun, O., Kindel, B., Goetz, A.F.H., Bonfil, D., Agassi, M., Margalit, N., Binayminy, Y., Karnieli, A. 2004. Monitoring of infiltration rate in semiarid soils using airborne hyperspectral. *Technology International Journal of Remote Sensing*, 25, 1–18.

Ben-Dor, E., Granot, A., Notesco, G. 2017. A simple apparatus to measure soil spectral information in the field under stable conditions. *Geoderma*, 306, 73–80.

Ben-Dor, E., Heller, D., Chudnovsky, A. 2008a. A novel method of classifying soil profiles in the field using optical means. *Soil Science Society of America Journal*, 72, 1113–1123.

Ben-Dor, E., Ong, C., Lau, I.C. 2015. Reflectance measurements of soils in the laboratory: Standards and protocols. *Geoderma*, 245–246, 112–124.

Ben-Dor, E., Patkin, K., Banin, A., Karnieli, A. 2002. Mapping of several soil properties using DAIS-7915 hyperspectral scanner data: A case study over clayey soils in Israel. *International Journal of Remote Sensing*, 23, 1043–1062.

Ben-Dor, E., Taylor, G., Hill, J., Demattê, J.A.M., Chabrillat, S., Sommer, S., Aloisi, R.R. 2008b. Imaging spectroscopy for soil applications. *Advances in Agronomy*, 97, 321–392.

Bernath, P.E. 2005. *Spectra of atoms and molecules*, 2nd Edition. Oxford: Oxford University Press. 454 p.

Bogrekci, I., Lee, W.S. 2007. Comparison of ultraviolet, visible, and near infrared sensing for soil phosphorus. *Biosystems Engineering*, 96, 293–299.

Bowers, S.A., Hanks, J.R. 1965. Reflection of radiant energy from soil. *Soil Science*, 100, 130–138.

Bowers, S.A., Smith, S.J. 1972. Spectrophotometric determination of soil water content. *Soil Science Society of America Proceedings*, 36, 978–980.

Brodský, L., Klement, A., Penížek, V., Kodešová, R., Boruvka, L. 2011. Building soil spectral library of the Czech soils for quantitative digital soil mapping. *Soil and Water Research*, 6, 165–172.
Brown, D.J. 2007. Using a global VNIR soil-spectral library for local soil characterization and landscape modeling in a 2nd-order Uganda watershed. *Geoderma*, 140, 444–453.
Brown, D.J., Shepherd, K.D., Walsh, M.G., Mays, M.D., Reinsch, T.G. 2006. Global soil characterization with VNIR diffuse reflectance spectroscopy. *Geoderma*, 132, 273–290.
Bricklemyer, R.S., Brown, D.J. 2010. On-the-go VisNIR: Potential and limitations for mapping soil clay and organic carbon. *Computers and Electronics in Agriculture*, 70, 209–216.
Camargo, L.A., Marques, J., Barrón, V., Alleoni, L.R.F., Barbosa, R.S., Pereira, G.T. 2015. Mapping of clay, iron oxide and adsorbed phosphate in Oxisols using diffuse reflectance spectroscopy. *Geoderma*, 251–252, 124–132.
Carmon N., Ben-Dor, E. 2017. An advanced analytical approach for spectral-based modelling of soil properties. *International Journal of Emerging Technology and Advanced Engineering*, 7, 90–97.
Castaldi, F., Chabrillat, S., Jones, A., Vreys, K., Bomans, B., van Wesemael, B. 2018. Soil organic carbon estimation in croplands by hyperspectral remote APEX data using the LUCAS topsoil database. *Remote Sensing*, 10, 1–19.
Cécillon, L., Barthès, B.G., Gomez, C., Ertlen, D., Genot, V., Hedde, M., Stevens, A., Brun, J.J. 2009. Assessment and monitoring of soil quality using near-infrared reflectance spectroscopy (NIRS). *European Journal of Soil Science*, 60, 770–784.
Cécillon, L., Certini, G., Lange, H., Forte, C., Strand, L.T. 2012. Spectral finger print of soil organic matter composition. *Organic Geochemistry*, 46, 127–136.
Chacón Iznaga, A., Rodríguez Orozco, M., Aguila Alcantara, E., Carral Pairol, M., Díaz Sicilia, Y.E., de Baerdemaeker, J., Saeys, W. 2014. Vis/NIR spectroscopic measurement of selected soil fertility parameters of Cuban agricultural Cambisols. *Biosystems Engineering*, 125, 105–121.
Chang, G. W., Laird, D. A., Hurburgh, G. R. 2005. Influence of soil moisture on near-infrared reflectance spectroscopic measurement of soil properties. *Soil Science*, 170, 244–255.
Chang, C., Laird, D., Mausbach, M.J. 2001. Near-infrared reflectance spectroscopy – principal components regression analyses of soil properties. *Soil Science Society of America Journal*, 65, 480–490.
Christy, C.D. 2008. Real-time measurement of soil attributes using on-the-go near infrared reflectance spectroscopy. *Computers and Electronics in Agriculture*, 61, 10–19.
Clark, R.N., Roush, T.L. 1984. Reflectance spectroscopy: Quantitative analysis techniques for remote sensing applications. *Journal of Geophysical Research*, 89, 6329–6340.
Cohen, M.J., Prenger, J.P., DeBusk, W.F. 2005. Visible-near infrared reflectance spectroscopy for rapid, nondestructive assessment of wetland soil quality. *Journal of Environmental Quality*, 34, 1422–1434.
Condit, H. 1970. The spectral reflectance of American soils. *Photogrammetric Engineering*, 36(9), 955–966.
Conforti, M., Matteucci, G., Buttafuoco, G. 2017. Using laboratory Vis-NIR spectroscopy for monitoring some forest soil properties. *Journal of Soils and Sediments*, 18, 1009–1019.
Cozzolino, D. 2016. Near infrared spectroscopy as a tool to monitor contaminants in soil, sediments and water - State of the art, advantages and pitfalls. *Trends in Environmental Analytical Chemistry*, 9, 1–7.
Cozzolino, D., Morón, A. 2003. The potential of near-infrared reflectance spectroscopy to analyse soil chemical and physical characteristics. *Journal of Agricultural Sciences*, 140, 65–71.
Dalal, R.C., Henry, R.J. 1986. Simultaneous determination of moisture, organic carbon, and total nitrogen by near infrared reflectance spectrophotometry. *Soil Science Society of America Journal*, 50, 120–123.
Daniel, K.W., Tripathi, N.K., Honda, K. 2003. Artificial neural network analysis of laboratory and in situ spectra for the estimation of macronutrients in soils of Lop Buri (Thailand). *Australian Journal of Soil Research*, 41, 47–59.
de Souza Bahia, A.S.R., Marques, J., La Scala, N., Pellegrino Cerri, C.E., Camargo, L.A. 2017. Prediction and mapping of soil attributes using diffuse reflectance spectroscopy and magnetic susceptibility. *Soil Science Society of America Journal*, 81, 1450–1462.
Del Vecchio, R., Blough, N.V.B. 2004. On the origin of the optical properties of humic substances. *Environmental Science & Technology*, 38, 3885–3891.
Demattê, J.A.M., Bellinaso, H., Araújo, S.R., Rizzo, R., Souza, A.B. 2016a. Spectral regionalization of tropical soils in the estimation of soil attributes. *Revista Ciência Agronômica*, 47, 589–598.
Demattê, J.A.M., Fiorio, P.R., Ben-Dor, E. 2009. Estimation of soil properties by orbital and laboratory reflectance means and its relation with soil classification. *The Open Remote Sensing Journal*, 2, 12–23.
Demattê, J.A.M., Galdos, M.V., Guimarães, R.V., Genú, A.M., Nanni, M.R., Zulu, J. 2007. Quantification of tropical soil attributes from ETM+/LANDSAT-7 data. *International Journal of Remote Sensing*, 28, 3813–3829.

Dematté, J.A.M., Morgan, C.L.S., Chabrillat, S., Rizzo, R., Franceschini, M.H.D., Terra, F. d., Vasques, G.M., Wetterlind, J. 2015. Spectral sensing from ground to space in soil science: State of the art, applications, potential and perspectives. In: Prasad S. Thenkabail. (Org.). *Land Resources Monitoring, Modeling, and Mapping with Remote Sensing*. 1 ed. Boca Raton, FL: CRC Press Taylor & Francis Group, 2015, vol. 3, pp. 647–708.

Dematté, J.A.M., Oliveira, J.D.C., Tavares, T.R., Lopez, L.R., Terra, F.D.S., Araújo, S.R., Fongaro, C.T., Maia, S.M.F., Mello, F.F.D.C., Rizzo, R., Vicente, S., de Melo Bortolleto, M.A., Cerqueira, P.H.R. 2016b. Soil chemical alteration due to slaughterhouse waste application as identified by spectral reflectance in São Paulo State, Brazil: An environmental monitoring useful tool. *Environmental Earth Sciences*, 75, 1277–1287.

Dematté, J.A.M., Sousa, A.A., Alves, M.C., Nanni, M.R., Fiorio, P.R., Campos, R.C. 2006. Determining soil water status and other soil characteristics by spectral proximal sensing. *Geoderma*, 35, 179–195.

Dematté, J.A.M., Sayão, V.M., Rizzo, R., Fongaro, C.T. 2017. Soil class and attribute dynamics and their relationship with natural vegetation based on satellite remote sensing. *Geoderma*, 302, 39–51.

Dematté, J.A.M., Terra, F.S. 2014. Spectral pedology: A new perspective on evaluation of soils along pedogenetic alterations. *Geoderma*, 217–218, 190–200.

EMBRAPA. 2017. *Manual de métodos de análise de solo/Centro Nacional de Pesquisa de Solos*, 3rd Edition. Rio de Janeiro: EMBRAPA CNPS, p. 212.

FAO. 2011. *The State of the World's Land and Water Resources for Food and Agriculture - Managing Systems at Risk*. Rome: FAO, 47 p.

FAO. 2013. *Statistical Yearbook 2013 World food and agriculture*. Rome: FAO, 289 p.

Fernandes, R.B.A., Barrón, V., Torrent, J., Fontes, M.P.F. 2004. Quantification of iron oxides in Brazilian latosols by diffuse reflectance spectroscopy. *Brazilian Journal of Soil Science*, 28, 245–257.

Franceschini, M.H.D., Dematté, J.A.M., Kooistra, L., Bartholomeus, H., Rizzo, R., Fongaro, C.T., Molin, J.P. 2018. Effects of external factors on soil reflectance measured on-the-go and assessment of potential spectral correction through orthogonalisation and standardisation procedures. *Soil & Tillage Research*, 177, 19–36.

Ge, Y., Thomasson, J.A., Sui, R. 2011. Remote sensing of soil properties in precision agriculture: A review. *Frontiers of Earth Science*, 5, 229–238.

Geladi, P., Macdougall, D., Martens, H. 1985. Linearization and scatter-correction for near-infrared reflectance spectra of meat. *Applied Spectroscopy*, 39, 491–500.

Genot, V., Colinet, G., Bock, L., Vanvyve, D., Reusen, Y., Dardenne, P. 2011. Near infrared reflectance spectroscopy for estimating soil characteristics valuable in the diagnosis of soil fertility. *Journal of Near Infrared Spectroscopy*, 19, 117–138.

Gholizadeh, A., Borůvka, L., Vašát, R., Saberioon, M., Klement, A., Kratina, J., Tejnecký, V., Drábek, O. 2015. Estimation of potentially toxic elements contamination in anthropogenic soils on a brown coal mining dumpsite by reflectance spectroscopy: A case study. *PLoS One*, 10, 1–14.

Godfray, H.C.J., Beddington, J.R., Crute, I.R., Haddad, L., Lawrence, D., Muir, J.F., Pretty, J., Robinson, S., Thomas, S. M., Toulmin, C. 2010. Food security: The challenge of feeding 9 billion people. *Science*, 327, 812–818.

Gras, J.P., Barthès, B.G., Mahaut, B., Trupin, S. 2014. Best practices for obtaining and processing field visible and near infrared (VNIR) spectra of topsoils. *Geoderma*, 214–215, 126–134.

Harris, W., White, G.N. 2008. X-ray Diffraction Techniques for Soil Mineral Identification. In: Ulery, A.L., Drees, L.R. (eds.) *Methods of Soil Analysis, Part 5. Mineralogical Analysis*. Wisconsin: American Society of Agronomy. pp. 677.

He, Y., Huang, M., García, A., Hernández, A., Song, H. 2007. Prediction of soil macronutrients content using near-infrared spectroscopy. *Computers and Electronics in Agriculture*, 58, 144–153.

Hicks, W., Viscarra Rossel, R.A., Tuomi, S. 2015. Developing the Australian mid-infrared spectroscopic database using data from the Australian Soil Resource Information System. *Soil Research*, 53, 922–931.

Hollas, J.M. 2004. *Modern Spectroscopy*, 4th Edition. Chichester: John Willey and Sons. p. 482.

Hu, B., Chen, S., Hu, J., Xia, F., Xu, J., Li, Y., Shi, Z. 2017. Application of portable XRF and VNIR sensors for rapid assessment of soil heavy metal pollution. *PLoS One*, 12, 1–13.

Hummel, J.W., Sudduth, K.A., Hollinger, S.E. 2001. Soil moisture and organic matter prediction of surface and subsurface soils using an NIR soil sensor. *Computers and Electronics in Agriculture*, 32, 149–165.

Hunt, G.R. 1977. Spectral signatures of particulate minerals in the visible and near-infrared. *Geophysics*, 42, 501–513.

Hunt, G.R., Salisbury, J.W. 1971a. Visible and near-infrared spectra of mineral and rocks: I. Silicate minerals. *Moderns Geology*, 1, 283–300.

Hunt, G.R., Salisbury, J.W. 1971b. Visible and near-infrared spectra of mineral and rocks: II. Carbonates. *Moderns Geology*, 2, 23–30.
Hunt, G.R., Salisbury, J.W., Lenhoff, C.J. 1971. Visible and near-infrared spectra of mineral and rocks: III. Oxides and hydroxides. *Moderns Geology*, 2, 195–205.
ICRAFT. 2018. Cost Error Analysis. World Agroforestry Center. Retrieved from http://intranet.worldagroforestry.org/soc/costerror.html.
Igne, B., Reeves, J.B., McCarty, G., Hively, W.D., Lundc, E., Hurburgh, C.R. 2010. Evaluation of spectral pretreatments, partial least squares, least squares support vector machines and locally weighted regression for quantitative spectroscopic analysis of soils. *Journal of Near Infrared Spectroscopy*, 18, 167–176.
Islam, K., Singh, B., McBratney, A.B. 2003. Simultaneous estimation of several soil properties by ultra-violet, visible, and near-infrared reflectance spectroscopy. *Soil Research*, 41, 1101–1114.
IUSS Working Group WRB. 2015. World Reference Base for Soil Resources 2014, Update 2015. International Soil Classification System for Naming Soils and Creating Legends for Soil Maps. World Soil Resources Reports No. 106. FAO, Rome.
Janik, L.J., Forrester, S.T., Rawson, A. 2009. The prediction of soil chemical and physical properties from mid-infrared spectroscopy and combined partial least-squares regression and neural networks (PLS-NN) analysis. *Chemometric and Intelligent Laboratory Systems*, 97, 179–188.
Janik, L.J., Merry, R.H., Forrester, S.T., Lanyon, D.M., Rawson, A. 2007b. Rapid prediction of soil water retention using mid infrared spectroscopy. *Soil Science Society of America Journal*, 71, 507–514.
Janik, L.J., Merry, R.H., Skjemstad, J.O. 1998. Can mid infrared diffuse reflectance analysis replace soil extractions? *Australian Journal of Experimental Agriculture*, 38, 681–696.
Janik, L.J., Skjemstad, J.O. 1995. Characterization and analysis of soils using mid-infrared partial least-squares. II. Correlations with some laboratory data. *Australian Journal of Soil Research*, 33, 637–650.
Janik, L.J., Skjemstad, J.O., Shepherd, K.D., Spouncer, L.R. 2007a. The prediction of soil carbon fractions using mid-infrared-partial least square analysis. *Australian Journal of Soil Research*, 45, 73–81.
Ji, W., Adamchuk, V.I., Biswas, A., Dhawale, N.M., Sudarsan, B., Zhang, Y., Viscarra Rossel, R.A., Shi, Z. 2016a. Assessment of soil properties in situ using a prototype portable MIR spectrometer in two agricultural fields. *Biosystems Engineering*, 152, 14–27.
Ji, W., Li, S., Chen, S., Shi, Z., Viscarra Rossel, R.A., Mouazen, A.M. 2016b. Prediction of soil attributes using the Chinese soil spectral library and standardized spectra recorded at field conditions. *Soil & Tillage Research*, 155, 492–400.
Kemper, T., Sommer, S. 2002. Estimate of heavy metal contamination in soils after a mining accident using reflectance spectroscopy. *Environmental Science & Technology*, 36, 2742–2747.
Knadel, M., Deng, F., Thomsen, A., Greve, M.H. 2012. Development of a Danish national Vis-NIR soil spectral library for soil organic carbon determination. *Digital Soil Assessments and Beyond*. doi:10.1201/b12728-79.
Knox, N.M., Grunwald, S., McDowell, M.L., Bruland, G.L., Myers, D.B., Harris, W.G. 2015. Modelling soil carbon fractions with visible near-infrared (VNIR) and mid-infrared (MIR) spectroscopy. *Geoderma*, 239–240, 229–239.
Kodaira, M., Shibusawa, S. 2013. Using a mobile real-time soil visible-near infrared sensor for high resolution soil property mapping. *Geoderma*, 199, 64–79.
Kuang, B., Mahmood, H.S., Quraishi, M.Z., Hoogmoed, W.B., Mouazen, A.M., van Henten, E.J. 2012. Sensing soil properties in the laboratory, in situ, and on-line. A review. *Advance in Agronomy*, 114, 155–223.
Kuang, B., Mouazen, A.M. 2011. Calibration of visible and near infrared spectroscopy for soil analysis at the field scale on three European farms. *European Journal of Soil Science*, 62, 629–636.
Kusumo, B.H., Hedley, C.B., Hedley, M.J., Hueni, A., Tuohy, M.P., Arnold, G.C. 2008. The use of diffuse reflectance spectroscopy for in situ carbon and nitrogen analysis of pastoral soils. *Australian Journal of Soil Research*, 46, 623–635.
La, W.J., Sudduth, K.A., Kim, H.J., Chung, S.O. 2016. Fusion of spectral and electrochemical sensor data for estimating soil macronutrients. *Transactions of the ASABE*, 59, 787–794.
Lal, R. 2003. Soil erosion and the global carbon budget. *Environment International*, 29, 437–450.
Lal, R., Follett, R. 2009. *Soil Carbon Sequestration and the Greenhouse Effect*, 2nd Edition. Madison: Soil Science Society of America Special Publication. p. 410.
Li, S., Ji, W., Chen, S., Peng, J., Zhou, Y., Shi, Z. 2015. Potential of VIS-NIR-SWIR spectroscopy from the Chinese Soil Spectral Library for assessment of nitrogen fertilization rates in the paddy-rice region, China. *Remote Sensing*, 7, 7029–7043.
Linsler, D., Sawallisch, A., Höper, H., Schmidt, H., Vohland, M., Ludwig, B. 2017. Near-infrared spectroscopy for determination of soil organic C, microbial biomass C and C and N fractions in a heterogeneous sample of German arable surface soils. *Archives in Agronomy and Soil Science*, 63, 1499–1509.

Ludwig, B., Linsler, D., Höper, H., Schmidt, H., Piepho, H.P., Vohland, M. 2016. Pitfalls in the use of middle-infrared spectroscopy: Representativeness and ranking criteria for the estimation of soil properties. *Geoderma*, 268, 165–175.

Ludwig, B., Sawallisch, A., Heinze, S., Joergensen, R.G., Vohland, M. 2015. Usefulness of middle infrared spectroscopy for an estimation of chemical and biological soil properties - Underlying principles and comparison of different software packages. *Soil Biology and Biochemistry*, 86, 116–125.

Madari, B.E., Reeves, J.B., III, Coelho, M.R., Machado, P.L.O.A., De-Polli, H., Coelho, R.M., Benites, V.M., Souza, L.F., McCarthy, G.W. 2005. Mid- and near-infrared spectroscopic determination of carbon in a diverse set of soils from the Brazilian national Soil collection. *Spectroscopic Letters*, 38, 721–740.

Madari, B.E., Reeves, J.B., III, Machado, P.L.O.A., Guimarães, C.M., Torres, E., McCarty, G.W. 2006. Mid- and near-infrared spectroscopic assessment of soil compositional parameters and structural indices in two Ferralsols. *Geoderma*, 136, 245–259.

McCarty, G.W., Reeves, J.B., III, Reeves, V.B., Follett, R.F., Kimblem, J.M. 2002. Mid- infrared and near-infrared diffuse reflectance spectroscopy for soil carbon measurement. *Soil Science Society of America Journal*, 66, 640–646.

Minasny, B., McBratney, A.B., Bellon-Maurel, V., Roger, J.-M., Gobrecht, A., Ferrand, L., Joalland, S. 2011. Removing the effect of soil moisture from NIR diffuse reflectance spectra for the prediction of soil organic carbon. *Geoderma*, 167–168, 118 124.

Minasny, B., Tranter, G., McBratney, A.B., Brough, D.M., Murphy, B.W. 2009. Regional transferability of mid-infrared diffuse reflectance spectroscopic prediction for soil chemical properties. *Geoderma*, 153, 155–162.

Mohanty, B., Gupta, A., Das, B.S. 2016. Estimation of weathering indices using spectral reflectance over visible to mid-infrared region. *Geoderma*, 266, 111–119.

Morgan, C.L.S., Waiser, T.H., Brown, D.J., Hallmark, C.T. 2009. Simulated in situ characterization of soil organic and inorganic carbon with visible near-infrared diffuse reflectance spectroscopy. *Geoderma*, 151, 249–256.

Mouazen, A. M., De Baerdemaeker, J., and Ramon, H. 2005. Towards development of on-line soil moisture content sensor using a fibre-type NIR spectrophotometer. *Soil & Tillage Research*, 80, 171–183.

Mouazen, A.M., Kuang, B. 2016. On-line visible and near infrared spectroscopy for in-field phosphorous management. *Soil & Tillage Research*, 155, 471–477.

Mouazen, A.M., Kuang, B., De Baerdemaeker, J., Ramon, H. 2010. Comparison among principal component, partial least squares and back propagation neural network analyses for accuracy of measurement of selected soil properties with visible and near infrared spectroscopy. *Geoderma*, 158, 23–31.

Mouazen, A.M., Maleki, M.R., Cockx, L., Van Meirvenne, M., Van Holm, L.H.J., Merckx, R., De Baerdemaeker, J., Ramon, H. 2009. Optimum three-point linkage set up for improving the quality of soil spectra and the accuracy of soil phosphorus measured using an on-line visible and near infrared sensor. *Soil & Tillage Research*, 103, 144–152.

Mouazen, A.M., Maleki, M.R., De Baerdemaeker, J., Ramon, H. 2007. On-line measurement of some selected soil properties using a VIS-NIR sensor. *Soil & Tillage Research*, 93, 13–27.

Mulder, V. L., De Bruin, S., Schaepman, M. E., Mayr, T.R. 2011. The use of remote sensing in soil and terrain mapping - A review. *Geoderma*, 162, 1–19.

Nawar, S., Mouazen, A.M. 2017. Predictive performance of mobile vis-near infrared spectroscopy for key soil properties at different geographical scales by using spiking and data mining techniques. *Catena*, 151, 118–129.

Nelson, D.W., Sommers, L.E. 1996. Total Carbon, Organic Carbon, and Organic Matter. In: Black, C.A., Ed., *Methods of Soil Analysis. Part 3. Chemical Methods*. Madison: Soil Science of America and American Society of Agronomy. pp. 961–1010.

Nie, P., Dong, T., He, Y., Qu, F. 2017. Detection of soil nitrogen using near infrared sensors based on soil pre-treatment and algorithms. *Sensors*, 17, 1102–1115.

Nocita, M., Stevens, A., Noon, C., Van Wesemael, B. 2013. Prediction of soil organic carbon for different levels of soil moisture using Vis-NIR spectroscopy. *Geoderma*, 199, 37–42.

Nocita, M., Stevens, A., van Wesemael, B., Aitkenhead, M., Bachmann, M., Barthés, B., Ben-Dor, E., Brown, D.J., Clairotte, M., Csorba, A., Dardenne, P., Demattê, J.A.M., Genot, V., Guerrero, C., Knadel, M., Montanarella, L., Noon, C., Ramirez-Lopez, L., Robertson, J., Sakai, H., Soriano-Disla, J.M., Shepherd, K.D., Stenberg, B., Towett, E.K., Vargas, R., Wetterlind, J. 2015. Soil spectroscopy: An alternative to wet chemistry for soil monitoring. *Advances in Agronomy*, 132, 139–159.

Nguyen, T.T., Janik, L.J., Raupach, M. 1991. Diffuse reflectance infrared Fourier transformed (DRIFT) spectroscopy in soil studies. *Australian Journal of Soil Research*, 29, 49–67.

Obukhov, A.I., Orlov, D.S. 1964. Spectral reflectivity of the major soils group and possibility of using diffuse reflection in soil investigation. *Society Soil Science*, 1, 174–184.

O'Rourke, S.M., Minasny, B., Holden, N.M., McBratney, A.B. 2016a. Synergistic use of Vis-NIR, MIR, and XRF spectroscopy for the determination of soil geochemistry. *Soil Science Society of America Journal*, 80, 888–899.

O'Rourke, S.M., Stockmann, U., Holden, N.M., McBratney, A.B., Minasny, B. 2016b. An assessment of model averaging to improve predictive power of portable vis-NIR and XRF for the determination of agronomic soil properties. *Geoderma*, 279, 31–44.

Pandit, C.M., Filippelli, G.M., Li, L. 2010. Estimation of heavy-metal contamination in soil using reflectance spectroscopy and partial least-squares regression. *International Journal of Remote Sensing*, 31, 4111–4123.

Peng, J., Ji, W., Ma, Z., Li, S., Chen, S., Zhou, L., Shi, Z. 2016. Predicting total dissolved salts and soluble ion concentrations in agricultural soils using portable visible near-infrared and mid-infrared spectrometers. *Biosystems Engineering*, 152, 94–103.

Pietrzykowski, M., Chodak, M. 2014. Near infrared spectroscopy - A tool for chemical properties and organic matter assessment of afforested mine soils. *Ecological Engineering*, 62, 115–122.

Pinheiro, É., Ceddia, M., Clingensmith, C., Grunwald, S., Vasques, G. 2017. Prediction of soil physical and chemical properties by visible and near-infrared diffuse reflectance spectroscopy in the central amazon. *Remote Sensing*, 9, 293–315.

Poggio, M., Brown, D.J., Bricklemyer, R.S. 2015. Laboratory-based evaluation of optical performance for a new soil penetrometer visible and near-infrared (VisNIR) foreoptic. *Computers and Electronics in Agriculture*, 115, 12–20.

Ramirez-Lopez, L., Behrens, T., Schmidt, K., Stevens, A., Dematte, J.A.M., Scholten, T. 2013a. The spectrum-based learner: A new local approach for modeling soil vis-NIR spectra of complex datasets. *Geoderma*, 195–196, 268–279.

Ramirez-Lopez, L., Behrens, T., Schmidt, K., Viscarra Rossel, R., Dematte, J.A.M., Scholten, T. 2013b. Distance and similarity-search metrics for use with soil vis-NIR spectra. *Geoderma*, 199, 43–53.

Reeves, J.B., Follett, R.F., McCarty, G.W., Kimble, J.M. 2006. Can near or mid-infrared diffuse reflectance spectroscopy be used to determine soil carbon pools? *Communications in Soil Science and Plant Analysis*, 37, 2307–2325.

Reeves, J.B., Mccarty, G.W. 2001. Quantitative analysis of agricultural soils using near infrared reflectance spectroscopy and a fibre-optic probe. *Journal of Near Infrared Spectroscopy*, 34, 25–34.

Reeves, J.B., McCarty, G.W., Meisinger, J.J. 1999. Near infrared reflectance spectroscopy for the analysis of agricultural soils. *Journal of Near Infrared Spectroscopy*, 7, 179–193.

Reeves, J.B., Smith, D.B. 2009. The potential of mid- and near-infrared diffuse reflectance spectroscopy for determining major- and trace-element concentrations in soils from a geochemical survey of North America. *Applied Geochemistry*, 24, 1472–1481.

Ricker, N. 1953. The form and laws of propagation of seismic wavelets. *Geophysics*, 18, 10–40.

Riedel, F., Denk, M., Müller, I., Barth, N., Gläßer, C. 2018. Prediction of soil parameters using the spectral range between 350 and 15,000 nm: A case study based on the Permanent Soil Monitoring Program in Saxony, Germany. *Geoderma*, 315, 188–198.

Rienzi, E.A., Mijatovic, B., Mueller, T.G., Matocha, C.J., Sikora, F.J., Castrignanò, A. 2014. Prediction of soil organic carbon under varying moisture levels using reflectance spectroscopy. *Soil Science Society of America Journal*, 78, 958–967.

Robinson, W.O. 1927. The determination of organic matter in soils by means of hydrogen peroxide. *Journal of Agricultural Research*, 34, 339–356.

Roger, J.M., Chauchard, F., Bellon-Maurel, V. 2003. EPO-PLS external parameter orthogonalisation of PLS application to temperature-independent measurement of sugar content of intact fruits. *Chemometrics and Intelligent Laboratory Systems*, 66, 191–204.

Romero, D.J., Ben-Dor, E., Demattê, J.A.M., Souza, A.B., Vicente, L.E., Tavares, T.R., Martello, M., Strabeli, T.F., Barros, P.P.S., Fiorio, P.R., Gallo, B.C., Sato, M.V., Eitelweind, M.T. 2018. Internal soil standard method for the Brazilian soil spectral library: Performance and proximate analysis. *Geoderma*, 312, 95–103.

Roudier, P., Hedley, C.B., Lobsey, C.R., Viscarra Rossel, R.A., Leroux, C. 2017. Evaluation of two methods to eliminate the effect of water from soil vis–NIR spectra for predictions of organic carbon. *Geoderma*, 296, 98–107.

Roudier, P., Hedley, C.B., Ross, C.W. 2015. Prediction of volumetric soil organic carbon from field-moist intact soil cores. *European Journal of Soil Science*, 66, 651–660.

Samiei Fard, R., Matinfar, H.R. 2016. Capability of vis-NIR spectroscopy and Landsat 8 spectral data to predict soil heavy metals in polluted agricultural land (Iran). *Arabian Journal of Geosciences*, 9, 745–755.

Sarathjith, M.C., Das, B.S., Wani, S.P., Sahrawat, K.L. 2014. Dependency measures for assessing the covariation of spectrally active and an active soil properties in diffuse reflectance spectroscopy. *Soil Science Society of America Journal*, 78, 1522

Savitzky, A., Golay, M.J.E. 1964. Smoothing and differentiation of data by simplified least squares procedures. *Analytical Chemistry*, 36, 1627–1639.

Sayão, V.M., Demattê, J.A.M., Bedin, L.G., Nanni, M.R., Rizzo, R. 2018. Satellite land surface temperature and reflectance related with soil attributes. *Geoderma*, 325, 125–140.

Schulte, E.E., Hopkins, B.G. 1996. Estimation of Organic Matter by Weight Loss-on-Ignition. In *Soil Organic Matter: Analysis and Interpretation (SSSA Spec. Publ. 46)*, ed. F. R. Magdoff, et al., 21–31. Madison: Soil Science Society of America Special Publication.

Schwartz, G., Ben-Dor, E., Eshel, G. 2012. Quantitative analysis of total petroleum hydrocarbons in soils: Comparison between reflectance spectroscopy and solvent extraction by 3 certified laboratories. *Applied and Environmental Soil Science*, 2012, 751956, 11 pages, doi:10.1155/2012/751956.

Selige, T., Bohner, J., Schmidhalter, U. 2006. High resolution topsoil mapping using hyperspectral image and field data in multivariate regression modeling procedures. *Geoderma*, 136, 235–244.

Shepherd, K.D., Walsh, M.G., 2002. Development of reflectance spectral libraries for characterization of soil properties. *Soil Science Society of America Journal*, 66, 988–998.

Sherman, D.M., Waite, T.D. 1985. Electronic spectra of Fe^{+3} oxides and oxide hydroxides in the near IR to near UV. *American Mineralogist*, 70, 1262–1269.

Shi, T., Chen, Y., Liu, Y., Wu, G. 2014. Visible and near-infrared reflectance spectroscopy – An alternative for monitoring soil contamination by heavy metals. *Journal of Hazardous Materials*, 256, 166–176.

Shi, T., Liu, H., Chen, Y., Fei, T., Wang, J., Wu, G. 2017. Spectroscopic diagnosis of arsenic contamination in agricultural soils. *Sensors*, 17, 1036–1051.

Shi, Z., Ji, W., Viscarra Rossel, R.A., Chen, S., Zhou, Y. 2015. Prediction of soil organic matter using a spatially constrained local partial least squares regression and the Chinese vis-NIR spectral library. *European Journal of Soil Science*, 66, 679–687.

Sila, A.M., Shepherd, K.D., Pokhariyal, G.P. 2016. Evaluating the utility of mid-infrared spectral subspaces for predicting soil properties. *Chemometric and Intelligent Laboratory Systems*, 153, 92–105.

Slaughter, D. C., Pelletier, M. G., Upadhyaya, S. K. 2001. Sensing soil moisture using NIR spectroscopy. *Applied Engineering in Agriculture*, 17, 241–247.

SoilCares. 2018. SoilCares: Soil Analyses for farmers. Retrieved from https://www.agrocares.com/.

Song, Y., Li, F., Yang, Z., Ayoko, G.A., Frost, R.L., Ji, J. 2012. Diffuse reflectance spectroscopy for monitoring potentially toxic elements in the agricultural soils of Changjiang River Delta, China. *Applied Clay Science*, 64, 75–83.

Soriano-Disla, J.M., Janik, L.J., Allen, D.J., McLaughlin, M.J. 2017. Evaluation of the performance of portable visible-infrared instruments for the prediction of soil properties. *Biosystems Engineering*, 161, 24–36.

Soriano-Disla, J.M., Janik, L., McLaughlin, M.J., Forrester, S., Kirby, J., Reimann, C., Albanese, S., Andersson, M., Arnoldussen, A., Baritz, R., Batista, M.J., Bel-Lan, A., Birke, M., Cicchella, D., Demetriades, A., Dinelli, E., De Vivo, B., De Vos, W., Dohrmann, R., Duris, M., Dusza-Dobek, A., Eggen, O.A., Eklund, M., Ernstsen, V., Filzmoser, P., Finne, T.E., Flight, D., Fuchs, M., Fugedi, U., Gilucis, A., Gosar, M., Gregorauskiene, V., Gulan, A., Halamid, J., Haslinger, E., Hayoz, P., Hobiger, G., Hoffmann, R., Hoogewerff, J., Hrvatovic, H., Husnjak, S., Johnson, C.C., Jordan, G., Kivisilla, J., Klos, V., Krone, F., Kwecko, P., Kuti, L., Ladenberger, A., Lima, A., Locutura, J., Lucivjansky, P., Mackovych, D., Malyuk, B.I., Maquil, R., Meuli, R.G., Miosic, N., Mol, G., Négrel, P., O'Connor, P., Oorts, K., Ottesen, R.T., Pasieczna, A., Petersell, V., Pfleiderer, S., Poňavič, M., Prazeres, C., Rauch, U., Salpeteur, I., Schedl, A., Scheib, A., Schoeters, I., Sefcik, P., Sellersjö, E., Skopljak, F., Slaninka, I., Šorša, A., Srvkota, R., Stafilov, T., Tarvainen, T., Trendavilov, V., Valera, P., Verougstraete, V., Vidojevid, D., Zissimos, A.M., Zomeni, Z. 2013. The use of diffuse reflectance mid-infrared spectroscopy for the prediction of the concentration of chemical elements estimated by X-ray fluorescence in agricultural and grazing European soils. *Applied Geochemistry*, 29, 135–143.

Soriano-Disla, J.M., Janik, L.J., Viscarra Rossel, R.A., MacDonald, L.M., McLaughlin, M.J. 2014. The performance of visible, near-, and mid-infrared reflectance spectroscopy for prediction of soil physical, chemical and biological properties. *Applied Spectroscopy Reviews*, 49, 139–186.

Sparks, D.L. 2002. *Environmental Soil Chemistry*, 2nd Edition. San Diego: Academic Press. p. 352.

SpecSolo. 2018. SpecSolo: Tecnologia Inovadora em solos [SpecSolo: Soil inovative technology. Retrieved from http://www.specsolo.com.br/.

St. Luce, M., Ziadi, N., Gagnon, B., Karam, A. 2017. Visible near infrared reflectance spectroscopy prediction of soil heavy metal concentrations in paper mill biosolid- and liming by-product-amended agricultural soils. *Geoderma*, 288, 23–36.

Stenberg, B. 2010. Effects of soil sample pretreatments and standardised rewetting as interacted with sand classes on Vis-NIR predictions of clay and soil organic carbon. *Geoderma*, 158, 15–22.

Stenberg, B., Viscarra Rossel, R. A., Mouazen, A. M., Wetterlind, J. 2010. Visible and near infrared spectroscopy in soil science. *Advances in Agronomy*, 107, 164–206.

Stevens, A., Nocita, M., Tóth, G., Montanarella, L., van Wesemael, B. 2013. Prediction of soil organic carbon at the European scale by visible and near infrared reflectance spectroscopy. *PLoS One*, 8, e66409.

Stevens, A., van Wesemael, B., Bartholomeus, H., Rosillon, D., Tychon, B., Ben-Dor, E. 2008. Laboratory, field and airborne spectroscopy for monitoring organic carbon content in agricultural soils. *Geoderma*, 144, 395–404.

Stoner, E.R., Baumgardner, M.F. 1981. Characteristic variations in reflectance of surface soils. *Soil Science Society of America Journal*, 45, 1161–1165.

Stoorvogel, J.J., Kooistra, L., Bouma, J. 2015. Managing soil variability at different spatial scales as a basis for precision agriculture. In *Soil - Specific Farming: Precision Agriculture*, eds. R. Lal and B. A. Stewart. Boca Raton: CRC Press Taylor & Francis Group pp. 37–72.

Summers, D., Lewis, M., Ostendorf, B., Chittleborough, D. 2011. Visible near-infrared reflectance spectroscopy as a predictive indicator of soil properties. *Ecological Indicators*, 11, 123–131.

Sun, W., Zhang, X., Zou, B., Wu, T. 2017. Exploring the potential of spectral classification in estimation of soil contaminant elements. *Remote Sensing*, 9, 632–651.

Tayebi, M., Naderi, M., Mohammadi, J., Tayebi, M.H. 2017. Comparing different statistical models for assessing Fe-contaminated soils based on VNIR/SWIR spectral data. *Environmental Earth Sciences*, 76, 734–744.

Tekin, Y., Kuang, B., Mouazen, A.M. 2013. Potential of on-line visible and near infrared spectroscopy for measurement of pH for deriving variable rate lime recommendations. Sensors (Switzerland). *Sensors*, 13, 10177–10190.

Tekin, Y., Tumsavaş, Z., Ulusoy, Y., Mouazen, A.M. 2016. On-line Vis-NIR sensor determination of soil variations of sodium, potassium and magnesium. *IOP Conference Series: Earth and Environmental Sciences*, 4, 1–10.

Terra, F.S., Demattê, J.A.M., Viscarra Rossel, R.A. 2015. Spectral libraries for quantitative analyses of tropical Brazilian soils: Comparing vis-NIR and mid-IR reflectance data. *Geoderma*, 255–256, 81–93.

Terra, F.S., Demattê, J.A.M., Viscarra Rossel, R.A. 2018. Proximal spectral sensing in pedological assessments: Vis-NIR spectra for soil classification based on weathering and pedogenesis. *Geoderma*, 318, 123–136.

Tian, J., Philpot, W.D. 2015. Relationship between surface soil water content, evaporation rate, and water absorption band depths in SWIR reflectance spectra. *Remote Sensing of Environment*, 169, 280–289.

Tóth, G., Jones, A., Montanarella, L. 2013. The LUCAS topsoil database and derived information on the regional variability of cropland topsoil properties in the European Union. *Environmental Monitoring and Assessment*, 185, 7409–7425.

Townsend, T.E. 1987. Discrimination of iron alteration minerals in visible and near-infrared reflectance data. *Geophysical Research*, 92, 1441–1454.

Vågen, T.G., Shepherd, K.D., Walsh, M.G. 2006. Sensing landscape level change in soil fertility following deforestation and conversion in the highlands of Madagascar using Vis-NIR spectroscopy. *Geoderma*, 133, 281–294.

Van Waes, C., Mestdagh, I., Lootens, P., Carlier, L. 2005. Possibilities of near infrared reflectance spectroscopy for the prediction of organic carbon concentrations in grassland soils. *Journal of Agricultural Sciences*, 143, 487–492.

Vašát, R., Kodešová, R., Borůvka, L., Klement, A., Jakšík, O., Gholizadeh, A. 2014. Consideration of peak parameters derived from continuum-removed spectra to predict extractable nutrients in soils with visible and near-infrared diffuse reflectance spectroscopy (VNIR-DRS). *Geoderma*, 232–234, 208–218.

Vasques, G.M., Grunwald, S., Harris, W.G. 2010. Spectroscopic models of soil organic carbon in Florida, USA. *Journal of Environmental Quality*, 39, 923–934.

Vasques, G.M., Grunwald, S., Sickman, J.O. 2009. Modeling of soil organic carbon fractions using visible–near-infrared spectroscopy. *Soil Science Society of America Journal*, 73, 176–184.

Veum, K.S., Sudduth, K.A., Kremer, R.J., Kitchen, N.R. 2015. Estimating a soil quality index with VNIR reflectance spectroscopy. *Soil Science Society of America Journal*, 79, 637–649.

Viscarra Rossel, R.A. 2011. Fine-resolution multiscale mapping of clay minerals in Australian soils measured with near infrared spectra. *Journal of Geophysical Research*, 116, F04023. doi:10.1029/2011JF001977.

Viscarra Rossel, R.A., Behrens, T. 2010. Using data mining to model and interpret soil diffuse reflectance spectra. *Geoderma*, 158, 46–54.

Viscarra Rossel, R.A., Behrens, T., Ben-Dor, E., Brown, D.J., Demattê, J.A.M., Shepherd, K.D., Shi, Z., Stenberg, B., Stevens, A., Adamchuk, V., Aïchi, H., Barthès, B.G., Bartholomeus, H.M., Bayer, A.D., Bernoux, M., Böttcher, K., Brodský, L., Du, C.W., Chappell, A., Fouad, Y., Genot, V., Gomez, C., Grunwald, S., Gubler, A., Guerrero, C., Hedley, C.B., Knadel, M., Morrás, H.J.M., Nocita, M., Ramirez-Lopez, L., Roudier, P., RufastoCampos, E.M., Sanborn, P., Sellitto, V.M., Sudduth, K.A., Rawlins, B.G., Walter, C., Winowiecki, L.A., Hong, S.Y., Ji, W. 2016. A global spectral library to characterize the world's soil. *Earth-Science Reviews*, 155, 198–230.

Viscarra Rossel, R.A., Bui, E.N., De Caritat, P., McKenzie, N.J. 2010a. Mapping iron oxides and the color of Australian soil using visible–near-infrared reflectance spectra. *Journal of Geophysical Research*, 115, F04031. doi:10.1029/2009JF001645.

Viscarra Rossel, R. A., Cattle, S. R., Ortega, A., Fouad, Y. 2009. In situ measurements of soil colour, mineral composition and clay content by vis-NIR spectroscopy. *Geoderma*, 150, 253–266.

Viscarra Rossel, R.A., Hicks, W.S. 2015. Soil organic carbon and its fractions estimated by visible–near infrared transfer functions. *European Journal of Soil Science*, 66, 438–415.

Viscarra Rossel, R.A., Jeon, Y.S., Odeh, I.O.A., and McBratney, A.B. 2008. Using a legacy soil sample to develop a mid-IR spectral library. *Australian Journal of Soil Research*, 46, 1 16.

Viscarra Rossel, R.A., Rizzo, R., Demattê, J.A.M., Behrens, T. 2010b. Spatial modeling of a soil fertility index using visible–near-infrared spectra and terrain attributes. *Soil Science Society of America Journal*, 74, 1293–1300.

Viscarra Rossel, R.A., Walvoort, D.J.J., McBratney, A.B., Janik, L.J., Skjemstad, J.O. 2006. Visible, near infrared, mid infrared or combined diffuse reflectance spectroscopy for simultaneous assessment of various soil properties. *Geoderma*, 131, 59–75.

Viscarra Rossel, R.A., Webster, R. 2012. Predicting soil properties from the Australian soil visible–near infrared spectroscopic database. *European Journal of Soil Science*, 63, 848–860.

Vohland, M., Besold, J., Hill, J., Fründ, H-C. 2011. Comparing different multivariate calibration methods for the determination of soil organic carbon pools with visible to near infrared spectroscopy. *Geoderma*, 166, 198–205.

Vohland, M., Ludwig, M., Thiele-Bruhn, S., Ludwig, B. 2014. Determination of soil properties with visible to near- and mid-infrared spectroscopy: Effects of spectral variable selection. *Geoderma*, 223–225, 88–96.

Walkley, A., Black, I.A. 1934. An examination of Degtjareff method for determining soil organic matter and a proposed modification of the chromic acid titration method. *Soil Science*, 37, 29–37.

Wang, D., Chakraborty, S., Weindorf, D.C., Li, B., Sharma, A., Paul, S., Ali, M.N. 2015a. Synthesized use of VisNIR DRS and PXRF for soil characterization: Total carbon and total nitrogen. *Geoderma*, 243–244, 157–167.

Wang, Y., Huang, T., Liu, J., Lin, Z., Li, S., Wang, R., Ge, Y. 2015b. Soil pH value, organic matter and macronutrients contents prediction using optical diffuse reflectance spectroscopy. *Computers and Electronics in Agriculture*, 111, 69–77.

Wang, F., Li, C., Wang, J., Cao, W., Wu, Q. 2017. Concentration estimation of heavy metal in soils from typical sewage irrigation area of Shandong Province, China using reflectance spectroscopy. *Environmental Science and Pollution Research*, 24, 16883–16892.

Waiser, T.H., Morgan, C.L.S., Brown, D.J., Hallmark, C.T. 2007. In situ characterization of soil clay content with visible near-infrared diffuse reflectance spectroscopy. *Soil Science Society of America Journal*, 71, 389–396.

Waruru, B.K., Shepherd, K.D., Ndegwa, G.M., Sila, A., Kamoni, P.T., Shepherd, K.D., Ndegwa, G.M., Sila, A., Kamoni, P.T. 2015. Estimation of stability-related soil properties using mid-infrared spectroscopy for non-agricultural land use in Lake Victoria basin, Kenya. *East African Agricultural and Forestry Journal*, 81, 79–94.

Wenjun, J., Zhou, S., Jingyi, H., Shuo, L. 2014. In situ measurement of some soil properties in paddy soil using visible and near-infrared spectroscopy. *PLoS One*, 9, e105708. doi:10.1371/journal.pone.0105708.

Weil, R.R., Brady, N. 2017. *Nature and properties of soils*, 15th Edition. Harlow: Pearson. p. 1104.

Wu, Q., Leung, J.Y.S., Geng, X., Chen, S., Huang, X., Li, H., Huang, Z., Zhu, L., Chen, J., Lu, Y. 2015. Heavy metal contamination of soil and water in the vicinity of an abandoned e-waste recycling site: Implications for dissemination of heavy metals. *Science of Total Environment*, 506–507, 217–225.

Wu, Y., Chen, J., Wu, X., Tian, Q., Ji, J., Qin, Z. 2005. Possibilities of reflectance spectroscopy for the assessment of contaminant elements in suburban soils. *Applied Geochemistry*, 20, 1051–1059.

Xie, X.L., Pan, X.Z., Sun, B. 2012. Visible and near-infrared diffuse reflectance spectroscopy for prediction of soil properties near a copper smelter. *Pedosphere*, 22, 351–366.

Xu, D., Ma, W., Chen, S., Jiang, Q., He, K., Shi, Z. 2018. Assessment of important soil properties related to Chinese Soil Taxonomy based on vis–NIR reflectance spectroscopy. *Computers and Electronics in Agriculture*, 144, 1–8.

Yang, X.M., Xieb, H.T., Drurya, C.F., Reynoldsa, W.D., Yanga, J.Y., Zhang, X.D. 2012. Determination of organic carbon and nitrogen in particulate organic matter and particle size fractions of Brookston clay loam soil using infrared spectroscopy. *European Journal of Soil Science*, 63, 177–188.

Zhang, Y., Biswas, A., Ji, W., Adamchuk, V.I. 2017a. Depth-specific prediction of soil properties in situ using vis-nir spectroscopy. *Soil Science Society of America Journal*, 81, 993–1004.

Zhang, L., Yang, X., Drury, C., Chantigny, M., Gregorich, E., Miller, J., Bittman, S., Reynolds, D., Yang, J. 2017b. Infrared spectroscopy prediction of organic carbon and total nitrogen in soil and particulate organic matter from diverse Canadian agricultural regions. *Canadian Journal of Soil Science*, 98, 77–90.

Zornoza, R., Guerrero, C., Mataix-Solera, J., Scow, K.M., Arcenegui, V., Mataix-Beneyto, J. 2008. Near infrared spectroscopy for determination of various physical, chemical and biochemical properties in Mediterranean soils. *Soil Biology ad Biochemistry*, 40, 1923–1930.

42 NIR Analysis of Textiles

*Subhas Ghosh**
Eastern Michigan University

James Rodgers
Solutia Inc.

CONTENTS

42.1 INTRODUCTION

Near-infrared (NIR) spectroscopy has been found to be a useful technique to characterize raw materials and finished textile products, and NIR methods and techniques continue to find increasingly diverse and wide-ranging quantitative and qualitative applications in the textile industry. Quantitative analyses determine the amount (or quantity) of the property/species of interest in a substance or material. Qualitative analyses can be used to identify either a specific species or a substance present in a material (i.e., coating on a fiber), the type of material itself (i.e., cotton, nylon, or polyester), or the quality of the material. NIR quantitative and qualitative methods allow the user to rapidly, accurately, and precisely monitor key chemical, physical, and morphological properties of textile

* sghosh@emich.edu

fibers, yarns, fabrics, and chemical textile auxiliaries. Chemical properties are specific chemical species or groups present in the material (i.e., CH, OH, NH) that result in NIR spectral absorbencies at distinctive wavelength(s). Physical properties are nonchemical molecular or physical properties (i.e., turbidity, light scattering) that result in changes or movement in NIR spectral absorbencies. Morphological changes are changes in the material's molecular structure or fiber's morphology (i.e., hydrogen bonding, crystalline state, and structure) that result in changes in NIR spectral absorbencies.

NIR analyses are useful because a sample may be rapidly, accurately, and precisely tested without destroying its integrity or sample matrix. Because textile test specimens are normally in solid form, diffuse reflectance spectroscopy provides a special advantage by eliminating extensive sample preparation for testing textile materials. Additional productivity advantages are simplicity and ease of operation, minimum training requirements, and multiple property/species analyses. A high level of accuracy and precision are achieved through the use of advanced chemometric software/modeling, stable instrumental platforms, and flexible instrumental settings (scan frequency, scan averaging, etc.).

This chapter will be presented in two sections: one for natural fibers (cotton, wool, blends, etc.) and one for synthetic fibers (nylon, polyester, blends, etc.). Each section will review many of the more common NIR analyses, methods, and techniques for textile products. Both laboratory and at-line/on-line NIR analyses will be reviewed. In addition, a more detailed description will be provided in each section for a few of the more common NIR textile techniques. Even though most textile readers may have a prior knowledge of some of these detailed applications, it cannot be assumed that all readers are familiar with all of them. Thus, background and a brief preparatory description of the textile process and materials prior to each detailed application will be given, as well as specific references for those who wish to study the subject in greater detail.

42.2 NATURAL FIBERS

Spectroscopy has been found to be a useful technique to characterize raw materials and finished textile products. Because textile test specimens are in solid form, diffuse reflectance spectroscopy provides a special advantage by eliminating extensive sample preparation for testing textile materials such as fiber, yarn, and fabric. NIR analysis is useful because a sample may be rapidly tested without destroying its integrity.

Perhaps, some preparatory description of the textile process and materials prior to each application will be useful. Even though most textile readers are expected to have a prior knowledge of some of these applications, it cannot be assumed that they are familiar with all of them. Therefore, essential background is provided for each application as well as some references for those who wish to study the subject in greater detail. In this part of this chapter, a discussion of the determination of cotton textile and synthetic textile properties is presented.

NIR was used to develop a discrimination method for classifying wool, cashmere, yak, and angora rabbit fibers in both raw and combed sliver states. NIR was also tested for quantitative determination of percent cashmere in cashmere and wool blend mixtures. NIR spectra of hair vary highly due to the pigmentation and average fiber diameter than to bulk chemical composition. The method used relied on the soft independent modeling by class analogy (SIMCA) algorithm approach. The SIMCA approach using NIR spectra allowed the classification of various fiber groups into separate types, such as within white fiber types (e.g., white cashmere and wool) to between white and pigmented fibers, with discriminant distances from 12.64 for the nearest classes to above 1000 for the most distinct types of fibers. The classification method was found to be accurate with the exception of a single yak sample. The approach was reported both as accurate for classification and as a quick quantitative estimate of the amount of cashmere in wool and cashmere mixtures (Zoccola et al. 2013).

42.2.1 Amount of Reducing Sugar from Cotton Surface

Native cotton contains approximately 95% cellulose and some other materials such as waxes, pectins, protein, organic acids, and sugars. Stickiness of cotton during processing is often attributed to the high sugar content level that is in excess of 0.5%. The problem becomes evident when deposits are built up on rolls and other parts of the machine. This situation is further aggravated by unfavorable climatic conditions. Most common observations show deposits on the cards, draw frames, combers, and spinning frames. In addition, more frequent roller lapping has been observed. The well-known causes of stickiness are the white fly (*Benesia tabi*) and aphid (*Aphis gossypii*) secretions commonly referred to as "honeydew." Stickiness of cotton is also attributed to other plant sugars and seed oils (Roberts et al. 1976). Sugars mainly found in cotton include glucose, mannose, fructose, and pentoses; these are collectively called "reducing sugars." The amount of reducing sugars and other related substances present in cotton is usually determined by comparing the reducing ability of the hot water extraction of the cotton to that of a reducing substance like glucose.

Perkins' (Roberts et al. 1976) procedure can be used as a reference method to calibrate the NIR instrument. In the Perkins' method, the aqueous sample extract of cotton, which mostly contains reducing substances, is reacted with an excess of potassium ferricyanide in the presence of sodium carbonate. The potassium ferricyanide oxidizes the reducing materials and in the process is reduced to potassium ferrocyanide. The amount of potassium ferrocyanide produced is determined by titrating with standard ceric sulfate in an acid solution using *o*-phenanthroline/ferrous sulfate complex (ferroin) as an indicator. The amount of ceric sulfate consumed is proportional to the amount of potassium ferrocyanide, which in turn corresponds to the amount of reducing materials present in original samples. Perkins' method is calibrated by using known amounts of glucose in standard solution.

Raw cotton contains both simple and complex carbohydrates. Simple carbohydrates include mono- and disaccharides. Collectively, water-soluble simple carbohydrates are classified as reducing sugars. Disaccharides are easily hydrolyzed to monosaccharides. In general, reducing sugars (monosaccharides) are often unbranched and regarded as straight-chain polyhydroxyaldehydes (aldoses) or ketones (ketoses); the former have the carbonyl group on C-1 (carbon atom 1), while the latter on C-2. The remaining carbon atoms are hydroxylated. The mid-infrared examinations of reducing sugars show a broad band with shoulder in the region of 3.2–2.6 μm, which is due to OH groups that are hydrogen-bonded in various degrees. Because all reducing sugars have similar chemical structures, their NIR scans are expected to be quite similar, as shown in Figure 42.1.

Figure 42.1 shows the scanning of cotton fiber and various sugars in powder form on a Technicon InfraAlyzer 500. Samples were scanned from 1100 to 2500 nm at 4-nm steps. It can be seen by the comparison of the spectra in Figure 42.1 that slight differences do exist, especially in the C–H combination region. Table 42.1 shows the principal bands in NIR spectra of raw cotton, sucrose, glucose,

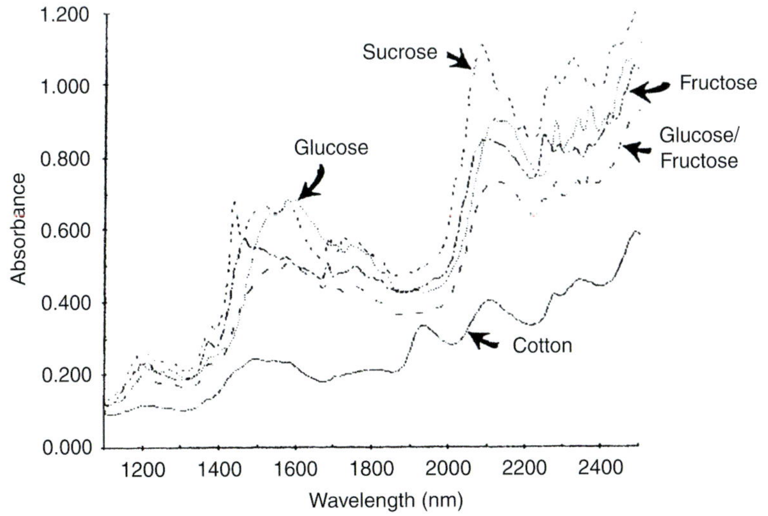

FIGURE 42.1 NIR spectra of cotton, glucose, fructose, sucrose, and a dry mixture of glucose and fructose.

TABLE 42.1
Principal Bands in NIR Spectra of Raw Cotton, Glucose, Fructose, and a Mixture of Glucose and Fructose[a]

Cotton	Glucose	Fructose	Glucose/Fructose
2340	2340	2370	2370
2270	2255	2250	2260
	2150		
2100			2100
	2085	2080	
1920	1950	1945	1940
	1902		
1820			
	1730	1740	1740
1585	1590	1580	1580
1550	1520		1520
1490		1450	
1375	1385	1380	1380
1220	1195		1195

[a] Wavelength expressed as λ_{max} (nm). Molecular ratio 1:1.

fructose, and a mixture (dry powder) of glucose and fructose (molar ratio 1:1). Band positions were obtained from an individual scan of cotton, glucose, fructose, and a mixture of glucose and fructose.

In Table 42.1, there are slight differences in the C–H combination and the C–H first overtone region of sucrose, glucose, and fructose. The assignment of absorption bands that are used for the generation of calibration equations can be made by finding the proper combination of the fundamental vibrations of functional groups associated with reducing sugars.

The coefficient of determination (R^2) was 0.98 with a standard error of prediction (SEP) of 0.02%. A 2-g cotton fiber in a solid sample cup was used for scanning. A four-wavelength regression model including 1445, 1722, 1905, and 2230 nm was generated from the correlation of reflected log(1/*R*) data to the sugar values obtained from the titration method (Figure 42.2).

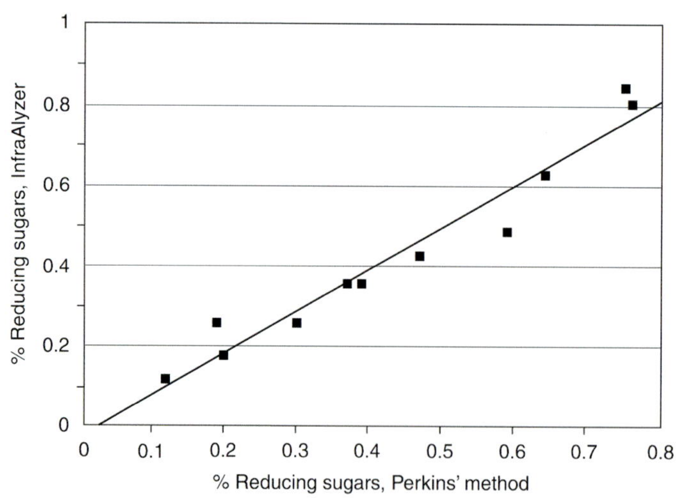

FIGURE 42.2 Relationship between the amount of reducing sugars using Perkins' method and as determined on the InfraAlyzer 500; R^2=0.98.

The absorbance bands occurring at 1445 and 1905 nm are assigned, respectively, to overcome the combination band of OH group stretching vibrations (Ghosh and Roy 1988). Usually, the free OH groups show greater absorbance due to the combination band around 1905 nm, while the hydrogen-bonded OH group exhibits the maximum absorbance at longer wavelengths. In addition, the combination band is more often utilized for the determination of OH groups because its absorptivity is approximately twice as great as that of the first overtone band.

Giangiacoma et al. (1981a, 1981b) used the NIR method successfully to analyze glucose, fructose, and sucrose from dry powdered samples. Reducing sugars present in aqueous solution were concentrated onto glass fiber by evaporation prior to analysis by the NIR instrument (Meurens 1984).

Because in aqueous solution there is more water–sugar than sugar–sugar interaction, Lanza and Li (1984) were unable to use the NIR method to determine individual reducing sugars from fruit juices. However, by using slightly modified NIR methods that included separate calibration equations for each class of samples, these authors were able to analyze total sugars from a wide variety of fruit juices and wines.

Because cotton composition and physical characteristics vary widely from one growth area to another, it is necessary to collect the calibration sample set from a wide variety of cottons representing different growth regions. Cotton samples taken directly from the bale can be used for calibration and analysis. However, samples with a very high level of trash should be cleaned through a trash analyzer. A comparative measurement of reducing sugars in cotton by near-infrared reflectance analysis (NIRA) and Perkins' manual method is shown in Figure 42.2.

The combination band at 2230 nm may be attributed to the CH_2 stretching and bending vibrations. This band position also corresponds to the combination band due to CO and OH groups' stretching vibrations. Because the most reducing sugars show the absence of a CO group absorption in the UV region, the combination band at 2230 nm may be entirely due to $-CH_2$ bending and stretching vibrations. The absorbance band at 1722 nm results from the first overtone of CH stretching vibrations. Apparently more than one assignment is possible to give the observed band because of the complex sugar scans in the NIR region.

42.2.2 Quantitative Analysis of Polyester/Cotton Blends

It is a common practice in textile manufacturing to blend fibers of two or more generic types to achieve the desired fabric characteristics. For example, cotton is often blended with polyester terephthalate (PET) fibers to enhance durability and make the fabric easy to care for. The proportion of each type of fiber in the total blend must remain constant in the manufacturing process. Normally, polyester/cotton blend analysis is performed by dissolving cotton in 70% H_2SO_4 solution. This procedure requires 8 h before the final result is obtained. NIRA is used to perform a blend analysis within 2 min in textile laboratories.

Both cotton and polyester (polyethylene terephthalate) fibers show the characteristic bands in the NIR region. Principal bands in NIR spectra of cotton and polyester are listed in Table 42.2. Band positions are calculated from the individual spectra of cotton and PET, as shown in Figure 42.3. Usually, a regression model consisting of three or four wavelengths can accurately determine the composition of a blend.

Two methods of blending are employed in the textile industry: intimate and drawing blending. Intimate blending is performed at the very beginning of the process in the opening line, and drawing blending is done at an intermediate processing stage in the drawing operation. The sample presentation differs slightly, depending on the type of blended stock, because the homogeneity of the blend depends on the method of blending. Intimate blending samples require two readings with different sample orientations with respect to the source (0° and 90° rotation); an average of the two readings should be used. Drawing blending samples, however, require four determinations: After two determinations (at 0° and 90°), the specimen must be turned over and two more readings taken at different orientations.

TABLE 42.2
Principal Bands in NIR Spectra of Cotton and Polyester Terephthalate

Cotton, λ_{max} (nm)		Polyester Terephthalate, λ_{max} (nm)		
1216	1776	1128	1828	2256
1270	1824	1172	1868	2336
1372	1930	1368	1908	2396
1444	2104	1412	1952	
1490	2276	1616	2088	
1550	2340	1660	2132	
1590		1716	2156	
1708		1800	2184	

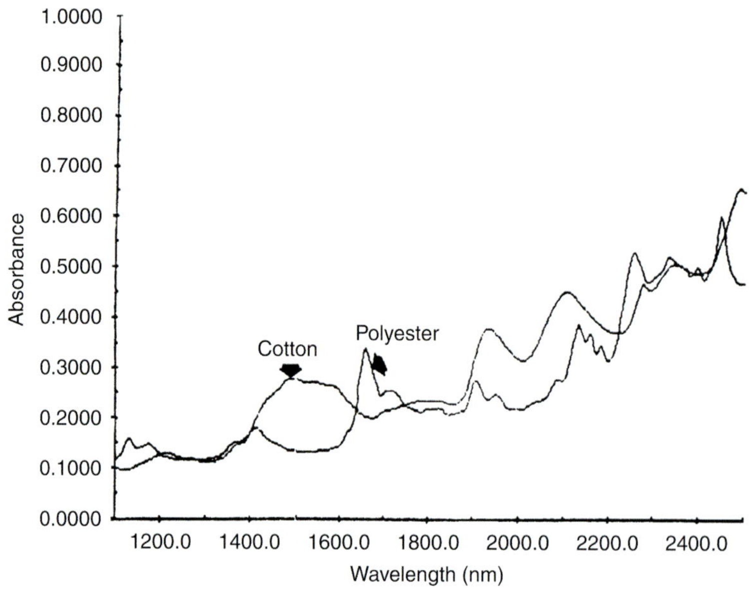

FIGURE 42.3 NIR spectra of polyester (PET) and cotton blend sample. An example of the spectral reconstruction is shown in Figure 42.4 that can be used to interpret the method.

A set of samples was analyzed using the NIR and a gravimetric method to determine the percent polyester in the blend, as shown in Table 42.3. Because the complex NIR spectra of the polyester/cotton are difficult to analyze, a *spectra reconstruction* (Honigs et al. 1984) method can be used to generate separate spectra for polyester and a spectra of cotton for a spectrum of the polyester/cotton.

Note that NIR spectroscopy has been tested to monitor the esterification reaction of polyester resins. The reaction is monitored by determining the acid value (AV) and hydroxyl value (OHV) during the reaction process. The standard reference laboratory methods used for the NIR calibrations are based on titrimetry. Calibration models once developed were also transferred to a second instrument using piecewise direct standardization (PDS). Following calibration transfer, the second instrument was reported to produce similarly good results for AV and OHV parameters (Blanco et al. 2008).

42.2.3 Degree of Mercerization

In the mercerization process, cotton fabrics are usually treated with 20% NaOH solution under tension. Its purpose is to enhance the fabric's characteristics such as dye affinity, dimensional

TABLE 42.3
Blend Analysis

Percent PET H_2SO_4	Dissolution Percent PET by NIRA
54.15	54.11
57.31	57.10
58.48	58.27
64.48	65.25
68.85	69.05
67.64	66.18
73.09	73.53
74.06	74.87

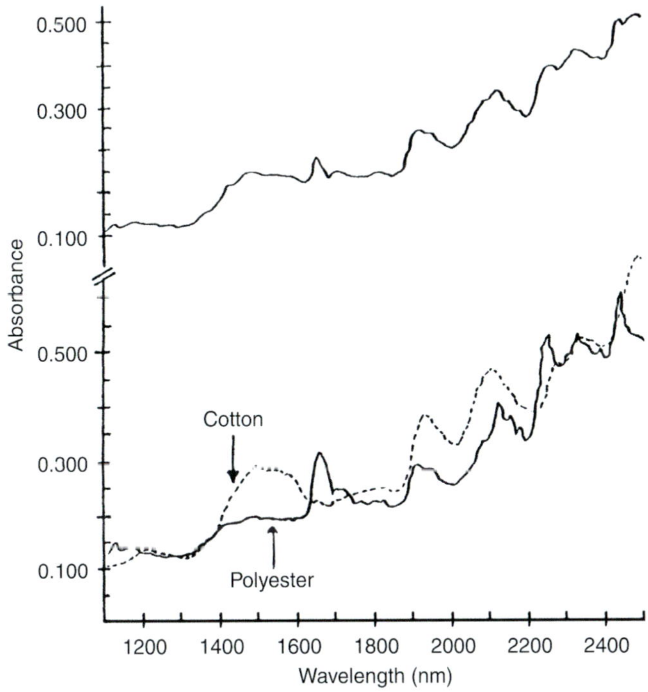

FIGURE 42.4 Spectral scan of 50/50 polyester/cotton blend (a) and spectral reconstruction of polyester/cotton blend (b).

stability, tensile strength, and luster (Edelstein 1936, Olson 1983). The process must be controlled to ensure fabric quality. Dye shade variation is the most common quality problem related to mercerized fabrics. The conventional method for determining the degree of mercerization, barium activity number (BAN), is laborious and requires 6 h before a result is obtained, which makes the test unsuitable for process control. Determining the BAN has been the only accepted method for measuring the degree of mercerization. The BAN of mercerized fabric is determined by boiling the samples of mercerized and unmercerized fabrics in a soap and soda ash solution. Both samples are washed until a neutral pH is obtained. Fabrics are treated with a barium hydroxide solution for a definite time period. The ratio of the amount of barium hydroxide absorbed by the mercerized specimen to that absorbed by the unmercerized specimen is determined by titrating against 0.1 HCl.

NIR spectra of unmercerized and mercerized cotton have been compared by Nikitin (1950). It has been stated that mercerization caused band shifts at specific wavelengths and the weakening of the intermolecular interaction of the cellulose structure along with an increase in the OH–O spacing. The band shifts, however, could also be attributed to a structural rearrangement of the cellulose macromolecules. Untreated native cotton contains mainly cellulose I structure; however, conversion of cellulose I to cellulose II occurs when the native cellulose is treated with NaOH. The conversion is associated with crystal lattice transformation and a decrease in order due to the imperfection in the cellulose crystallites. Equatorial X-ray diffraction trace of the mercerized and unmercerized cotton, which was prepared using an Ni-filtered Cuk α radiation, is shown in Figure 42.5. It is interesting to observe that a shoulder begins to appear at $2\theta = 20°$, consisting of the (Nikitin 1950) line of cellulose II in the mercerized sample. The degree of conversion of cellulose I to II, owing to the NaOH treatment, is illustrated in Figure 42.6. The conversion ratio was calculated from the diffractogram after correcting for the change in order using a procedure reported by Gjonnes and Norman (1960).

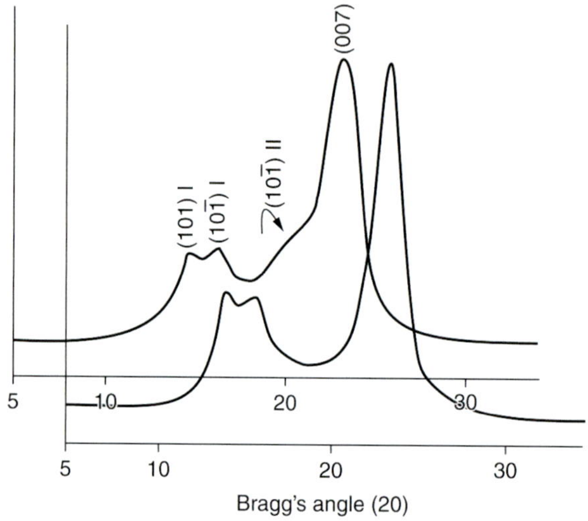

FIGURE 42.5 Comparison of equatorial scans of cellulose I and forming cellulose II.

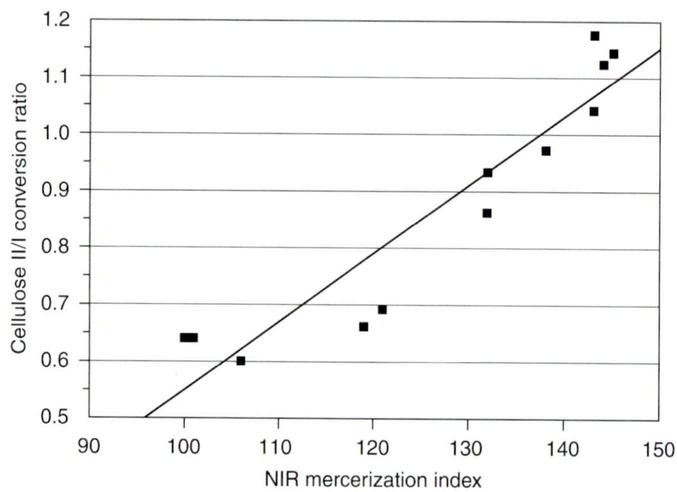

FIGURE 42.6 Relationship between the cellulose II/I conversion ratio and mercerization index.

A three-wavelength regression model can be used to determine the degree of mercerization of cotton fabric. There is an excellent agreement between the BAN and NIR mercerization index as illustrated in Figure 42.7.

NIR spectra of native and mercerized cotton are shown in Figure 42.8. The hydroxyl group overtone region near 1445 nm shows a more pronounced band in mercerized cotton, and some of the bands have disappeared because of the NaOH treatment by removing various noncellulosic materials from the cotton surface. The second-derivative spectra of the mercerized cotton (Figure 42.9) show a change in peak ratio in the hydroxyl overtone region, indicating a conformational change in the OH group. Moreover, there is a change in absorption at the 1820-nm band as the degree of mercerization increases.

It is assumed (Nikitin 1950) that an alcoholate-type derivative is formed:

$$\text{Cellulose–OH} \rightarrow \text{Cellulose–ONa}$$

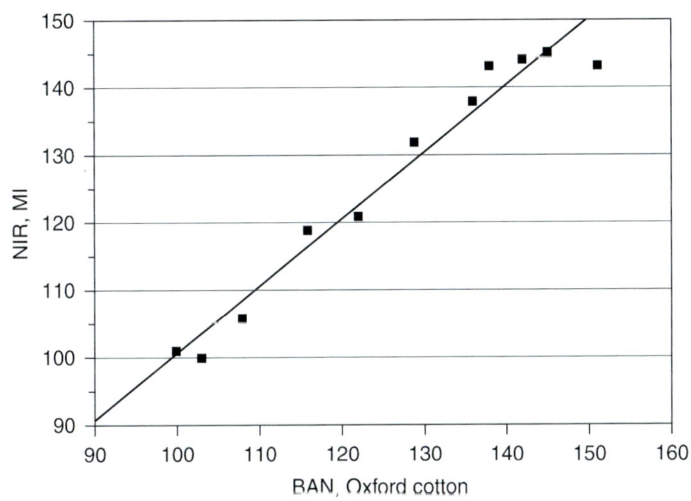

FIGURE 42.7 Relationship between BAN and mercerization index measured by NIRS.

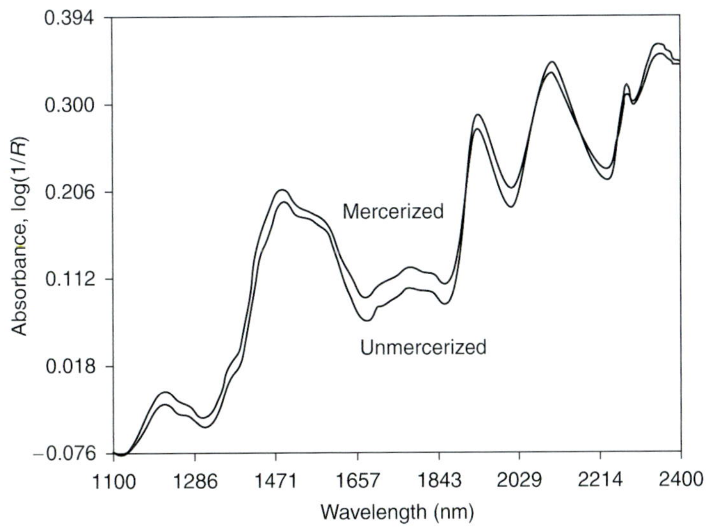

FIGURE 42.8 NIR spectra of native (low)- and high-mercerized cotton.

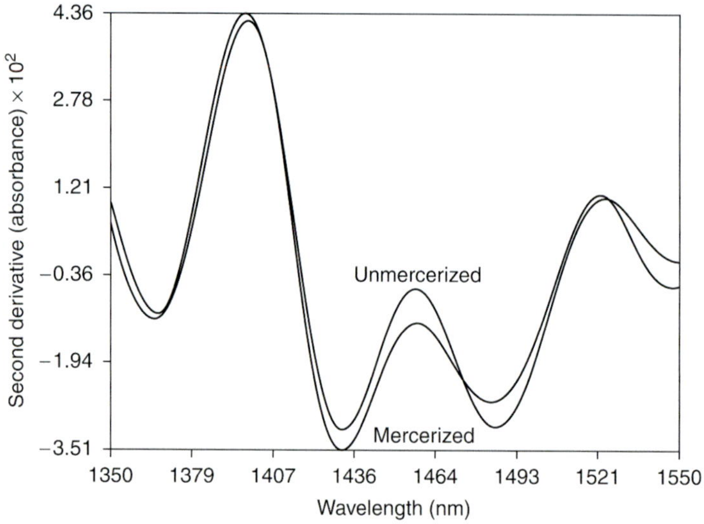

FIGURE 42.9 Second-derivative spectra of high- and low-mercerized cotton.

when the cotton fabric is treated with NaOH. After rinsing with weak acid and neutralizing the original structure, cellulose–OH is retained; however, the conformation of the macromolecule changes, and thus, the hydroxyl groups involved in the hydrogen bonding are affected.

42.2.4 Cotton Fiber Maturity

Cotton is by far the most important fiber in the apparel textile industry. Cotton fiber maturity is an important fiber property that influences end-use properties of raw cotton. Fiber maturity does not indicate the length of the period of fiber growth. It means that the fiber wall has developed to an acceptable level of thickening. The cotton fiber initially grows out from the seed wall with just the primary (outside) wall being formed. Only after the fiber has grown to almost its full length does the secondary wall begin to form, starting on the outside wall and growing inward, causing the fiber to thicken. Immature cotton results when the normal wall thickening processes are interrupted because of insects, plant disease, and bad weather. The degree of maturity of cotton fiber must be determined before it is processed in the textile plant because immature cotton produces an unacceptable product.

A typical chemical composition of native cotton is given in Table 42.4 (Guthrie et al. 1949). The main constituent is cellulose. Some other materials, such as waxes, pectin, sugars, organic acid, and protein, are also present. The relative amount of pectins and other noncellulosic substances reduces as the cotton matures (Guthrie et al. 1949).

Ward (1955) pointed out that crystal growth occurs during the maturity of cotton fibers. Therefore, better order and greater crystallinity of the cellulose molecule are expected for the matured fibers.

Usually, two methods are used to determine cotton fiber maturity: Causticaire and the Shirley Maturity Tester. These time-consuming methods also confuse fiber fineness with maturity. Therefore, they provide biased maturity values. A more fundamental method has been developed by Thibodeaux, USDA, where maturity of cotton is determined by taking the ratio of the fiber wall area to the area of a circle of the same perimeter. This is accomplished by using a microscope and an image analyzer. NIR is a better alternative in determining fiber maturity because it can be calibrated to measure maturity at different levels of fiber fineness and is fast in testing as a routine procedure. Two important criteria must be considered while selecting samples (a "trial set") for calibration. The characteristics of cotton vary considerably from one growth area to another and from one generic variety to another. Therefore, a trial set should contain samples from different growth areas and

TABLE 42.4
Composition of Typical Mature Cotton

	Percent
Cellulose	94.0
Protein ($N \times 6.25$)	1.3
Pectic substances	0.9
Ash	1.2
Wax	0.6
Malic, citric, and other organic acids	0.8
Total sugars	0.5
Pigment	Trace
Other	0.9

Source: H. Howell and J. Davis, *Textile Chem. Color*, *23*: 69 (1991).

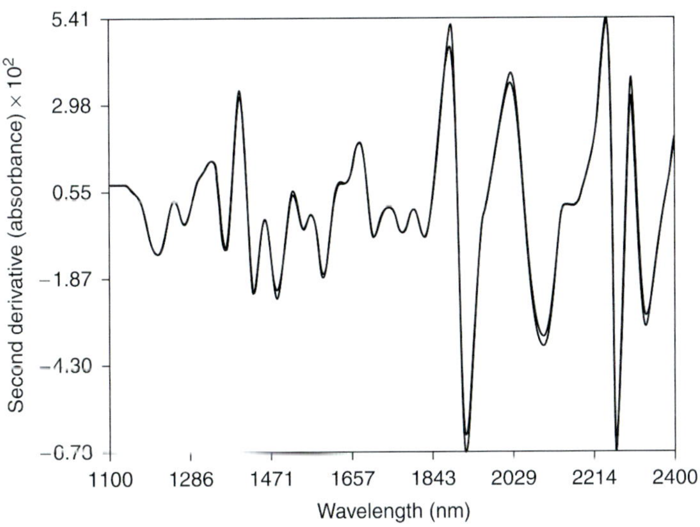

FIGURE 42.10 Derivative spectra of matured and immature cotton.

varieties to obtain a robust calibration. Moreover, it is necessary to categorize samples in different groups of fiber fineness to use an appropriate bias for each group. A sample presentation does not require any special technique. A 2-g sample can be packed in a standard solid sample cup for presentation.

The second-derivative spectra of immature and matured cotton are illustrated in Figure 42.10. Absorbance change due to a change in maturity can be observed at various wavelengths. More than one model can be used to predict the fiber maturity. However, a four-wavelength regression model was developed by correlating absorbance log(1/*R*) values with microscopic maturity values.

Fiber samples collected from the various textile plants were tested by the NIR and microscopic methods. The relationship between the two sets of data is shown in Figure 42.11, where agreement was excellent with $r^2=0.97$. NIR cotton maturity tests were performed on four varieties of cotton during their various stages of growth as illustrated in Table 42.5. This method of determining cotton maturity provides a technique for textile manufacturing to quickly detect cotton samples that could create quality problems in textile products.

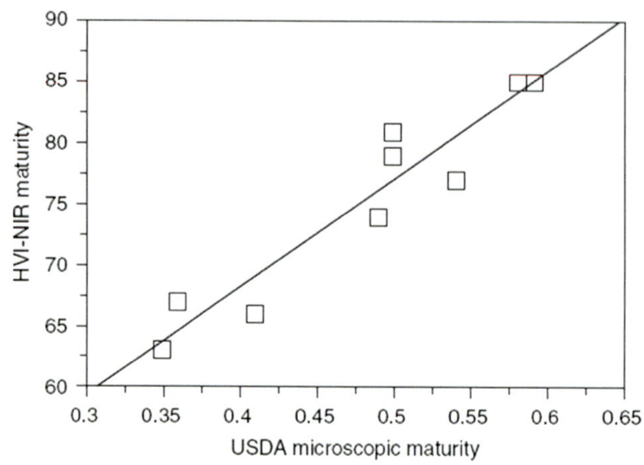

FIGURE 42.11 Relationship between the ITT NIR maturity and microscopic maturity by USDA.

TABLE 42.5
Cotton Fiber Growth Period and Maturity

	NIR Maturity Index			
Days after Pink Flower	**STVL 213**	**STVL 825**	**DPL 50**	**DBS 119**
25	50.2	42.7	47.4	49.5
30	63.1	59.7	57.7	63.7
35	64.0	60.9	61.6	66.0
40	68.9	68.8	69.1	69.4
45	69.9	75.9	72.0	68.6
50	75.5	79.2	76.7	75.0
55	75.7	78.7	77.5	77.5
r	0.97	0.98	0.99	0.95

42.2.5 Summary and Other Applications of Cotton Analysis

NIRA is a useful technique for characterizing textile raw materials, fiber, yarns, and fabrics. It is a nondestructive quantitative analysis that is simple to use and allows a rapid testing of the sample. Its ability to determine multiple components of the sample simultaneously and eliminate extensive sample preparation are the major advantages of NIRA in the characterization of textile materials. Many innovative mathematical treatments, for example, discriminant analysis and spectral reconstruction, have been developed by instrument manufactures and software companies. These instruments not only aid in the quantitative analysis of the data but also allow morphological investigations of fibers and yarns and a rapid, qualitative identification of specific sample sets.

Several NIRA applications were presented on the quantitative characterization of textile raw materials, fibers, yarns, and fabrics. NIRA has been used to determine the percent moisture and finish-on-fiber (FOF) of fibers and yarns, the amount of reducing sugars and maturity of cotton fibers, the degree of mercerization of cotton fabrics, the percent polyester in polyester/cotton blends, and the heatset temperature of nylon carpet yarns. In each analysis, excellent agreement has been observed between the standard laboratory method and NIRA results. Various mathematical treatments of the data, for example, derivative math, enable the investigator to predict the probable type and extent of morphological changes occurring within the textile samples.

Discriminant analysis has been used to identify and classify similar but different fibers, polymers, and yarns based on the differences in their properties and morphology. It has been used to identify and classify different nylon polymer types (nylon66 and 6) and heatsetting methods (Suessen and Superba) by polymer type, by heatsetting method, and simultaneously by polymer type and heatsetting method. Further, discriminant analysis has been used to identify and classify staple polyester fibers by fiber producer, by tenacity level, and simultaneously by producer and tenacity level.

As the advances in NIRA methodology and instrumentation continue, increased use of the technique for textile applications will also continue to grow. One area of high potential is that of on-line real-time analysis of textile samples in a production line. NIRA exemplified the use of sophisticated late 20th-century instruments in an established, relatively conservative industry. It does so by characterizing textile materials accurately, quickly, with simple and easy-to-use procedures.

Micronaire, a property of cotton fiber quality, is considered an indicator of cotton fiber fineness and maturity. Over the years, NIR has been reported to be useful in determining cotton properties. However, past studies have not provided a definitive conclusion regarding the capabilities of NIR spectroscopy as an accurate general technique for cotton quality analysis. In a 2010 paper, bench top and portable NIR systems were evaluated for predictions of cotton fiber micronaire, maturity, and fineness. Part of the experiment was to determine if a "universal" or robust calibration model is possible for NIR measurements of these specific cotton fiber properties. For this experiment, cotton sample spectra were determined using six different instruments: five commercial and one custom-built spectrophotometer. For all instruments, the NIR results were reported as "very acceptable" for determining the tested cotton fiber properties (Rodgers et al. 2010b).

A portable NIR instrument was evaluated for monitoring the most important fiber properties of cotton for laboratory, at-line, and field use. The initial testing was to determine the suitability of the portable NIR to compare to the reference laboratory measurement of cotton fiber micronaire. Micronaire is the primary indicator of cotton fiber fineness and maturity. The portable NIR instrument was capable of reporting spectral differences between samples with varying micronaire. A comparative evaluation was performed to determine the optimum instrumental conditions for laboratory cotton micronaire measurements. The optimum measurements were taken by using a glass window on the sample observation port, while increasing detector gain. Predictive NIR models indicated the satisfactory R^2 values, with low residuals and with corresponding small numbers of outlier samples (i.e., less than 12%). Portable NIR instrumentation was demonstrated to be useful and fast for the measurement of cotton fiber micronaire (Rogers et al. 2010a).

NIR was tested for the measurement of dyed or printed cotton–polyester (PET) blends. Previous publications have focused on the measurements of yarn slivers and greige fabrics rather than dyed fabrics. A rapid measurement method has been needed for the fiber content in dyed cotton–PET fabrics. To develop this method, the samples were measured in two locations using two NIR instruments; the sample spectra were measured in reflectance. The 100% cotton and 100% PET samples were easily discriminated using NIR spectra. Spectral variation was observed for the dyed cotton–PET samples with changing cotton content. Baseline adjustments in the spectra were completed using a spectral normalization algorithm. The 5-min NIR analysis was reported to be in agreement with the reference method within $\pm$3.0% cotton for nearly 90% of the validation samples (Rogers and Beck 2009).

High-volume instrumentation (HVI) is most readily applied for measuring instrument-based cotton quality but is unable to identify the specifics types of "trash" content in field cotton samples. In a published study, NIR spectroscopy was reported to identify 100% homogeneous individual samples of botanical trash with a 98% accuracy for the identification of cotton lint, botanical, and field trash in botanical and field trash mixtures (Fortier et al. 2012).

There is interest in a rapid and reliable method to discriminate cotton source of origin, whether from USA or from outside USA. This classification method would be valuable for levying the appropriate tariffs for non-American cotton. The discriminating method would also prevent the problem of cotton identity theft. One paper uses NIR spectroscopy to identify domestic and foreign cotton fibers. The technique described uses NIR spectra in combination with principal component analysis (PCA). The paper reports that the method developed is able to accurately discriminate cotton groups from USA and foreign sources (Fortier and Rodgers 2014).

42.3 PROPERTIES OF SYNTHETIC FIBERS, YARNS, AND OTHER TEXTILE PRODUCTS

42.3.1 Overview

Numerous quantitative and qualitative NIR techniques have been developed and reported for synthetic fibers, yarns, and other textile products. Many of these applications can be grouped into five major categories: heat/thermal history, "coatings," fiber identification, moisture content, and at-line/on-line measurements. Several authors have used NIR techniques and methodology to monitor the heat history of synthetic fibers and yarns. Ghosh, Rodgers, Tincher, and coworkers (Ghosh and Rodgers 1985, Rogers and Lee 1989, 1991, Rodgers 1984, Tincher et al. 1986) reported on NIR techniques to measure the heat history and heatset temperature of nylon carpet yarns. Ghosh and Tincher also studied the heat history and setting temperature of polyester (PET)-textured yarns and carpet yarns (Ghosh 1986b, (Dorrity et al. 1995).

NIR investigations and monitoring of the level of coating applied to the textile substrate (FOF, binder, size, lube, etc.) were performed by Drews, Lemere, Rodgers, Hill, and Harper and coworkers. Drews and Harris (1990) examined by NIR the latex binders on PET-nonwoven materials. Lemere (1989) applied NIR techniques to monitor warp size add-on and composition for PET blends. Rodgers (1984) reported on the NIR measurement of FOF applied to nylon yarns. Hill (1993) measured the lube/coating level applied to sewing thread. Harper (1993) studied the application of FOF to rayon yarns.

Fiber identification not only includes the specific type of polymer (nylon66, PET, etc.) and processing (i.e., type heatsetting) for a textile substrate, but also can include the quality and consistency of the material relative to standard production and previous historical samples. Howell and Davis (1991) reported on the identification of dyed and greige fibers using NIR spectroscopy. Ghosh and Rodgers (1988) used NIR discriminant analysis to differentiate between nylon66 and nylon6 carpet yarns that had been either Suessen or Superba heatset – joint identification of polymer type and heatset processing type. Ghosh (1986a) identified and classified staple PET by producer and tenacity level with the use of NIR discriminant analysis. Considerable research has been performed in the identification of the polymer type in textile materials – especially carpets – for recycling purposes. Costello and Knepper (1996) discussed a new, customized NIR analyzer that quickly and accurately determined the carpet fiber polymer type in a manufacturing environment. Jasper and Kovacs (1994) used neural networks and NIR spectroscopy to identify various synthetic and natural fibers from 17 fiber groups. Beck and coworkers (1996) also developed a fiber identification library that differentiated between 18 fiber groups of synthetic and natural fibers (390 samples), using PCA. Rodgers (2002) investigated the influences of carpet and instrumental parameters on carpet face fiber identification by NIR, using both static/laboratory and fiber-optic probe sampling systems.

One of the most thoroughly researched applications of NIR to textile substrates is the measurement of substrate moisture content. Most of the textile moisture applications are for natural fiber products – wool, cotton, and so forth. For synthetic textile products, Elliott et al. (1954) studied the spectral locations and the spectral influence of water on synthetic fiber spectra. Rodgers (1984) and Rodgers et al. (2002) determined the moisture content of nylon fibers by NIR spectroscopy both in the laboratory and at-line/"pseudo" on-line.

One of the most recent applications of NIR methodology for textile substrates is in the area of at-line and on-line analyses. At-line analyses are those in which the NIR analyzer is located in the manufacturing area and the samples are brought to the analyzer for analysis by area operators rather than sending the sample to the laboratory. On-line analyses are those in which the NIR analyzer is placed in the manufacturing process, and the NIR measurement, sample handling, and sample preparation are automatically controlled by a computer. Rodgers and Ghosh (1994, 1995, 2000) and Rodgers et al. 2001) have reported on the transference of the NIR measurement of nylon carpet yarn heatset temperature and heat history from the laboratory to at-line in the manufacturing area. Ghosh and coworkers (Ghosh 1993, Ghosh and Brodmann 1993) measured on-line the size content on PET blend warp yarns and the durable press content on PET-blend fabrics. Jerry (1996) studied the pigment concentration during colored fiber extrusion using the on-line NIR spectroscopy. Rodgers et al. (2002) monitored the moisture content and surrogate FOF of textile bobbins during spinning using a portable NIR analyzer.

To explore how NIR methodology can be used for textile products and to demonstrate the versatility and robustness of the technology, a more detailed description is provided in Sections 42.3.2–42.3.6 for a few of the more common NIR textile techniques. Examples of textile processes and measurements are discussed and reviewed for the areas of moisture content, FOF, heatset temperature, textile product identification, and at-line/on-line analyses.

42.3.2 Moisture Content in Nylon Fibers

The moisture content in synthetic fibers, yarns, and fabrics is a critical variable that can have a significant impact on physical properties, processing behavior, and manufacturing productivity. For hydrophilic materials, like nylon, the moisture content in the fiber, yarn, and fabric serves as a complementary variable to temperature and heating time (Saunders 1984, Starkweather 1973). It is known that temperature and heating time significantly influence the morphology of the synthetic fiber, which in turn impacts its dyeing and physical properties. For nylon fibers, it has been shown that an increase in the water content (i.e., increased relative humidity or increased moisture add-on in processing) of the fiber yields the similar results as with an increase in temperature or heating time. Hence, it is of critical importance in the textile manufacturing process to monitor and control the moisture content of the nylon material.

There are several methods that can be used to determine the moisture content of nylon materials. These methods include such well-known techniques as titration, dielectric capacitance, resistance, heated balance, and microwave devices to name a few. One of the better-known standard methods for determining the total moisture content of textile fibers and yarns is the Karl Fischer reagent (KFR) titration method. The KFR method is the quantitative reaction of water with the methanolic KFR solution of I_2, O_2, and pyridine (or pyridine substitute) (Scholz 2012). This method consists of placing the yarn sample in a flask containing the KFR solution and then back-titrating the system with KFR solution to match a blank's (no yarn) color. The end point of the titration can be visually observed in the back titration by the color change from yellow to rust brown, or it can be determined colorimetrically. Although the KFR method is very accurate, it is time-consuming, requires careful laboratory and sampling techniques, and normally uses odorous chemicals. NIRA can be performed on nylon yarns in less than 5 min in a textile laboratory and avoids the aforementioned difficulties.

Water has very strong absorbance bands in the NIR spectral regions (Mitchell 1976, Rotolo 1979, Mark 1984). A laboratory investigation comparing the percent moisture on nylon66 spun fiber as measured by NIR to that measured by the KFR method is shown in Figure 42.12 (Rodgers 1984). The sample size was 3.5 ± 0.5 g, and two packings were made for each sample. The calibration samples ranged from 0.60% to 10.87% moisture. From the calibration data, the regression models used in the investigation were a two-wavelength (1722 and 2083 nm) and a three-wavelength (1445, 1905, and 1940 nm) model. The SEPs were excellent, being 0.5% H_2O for the three-wavelength model and 0.3% for the two-filter model. The three-wavelength model was expected, as the 1445-, 1905-, and 1940-nm wavelengths have been previously assigned as OH group stretching vibrations. The precision and sensitivity of the two-wavelength model were unexpected. The 1722-nm

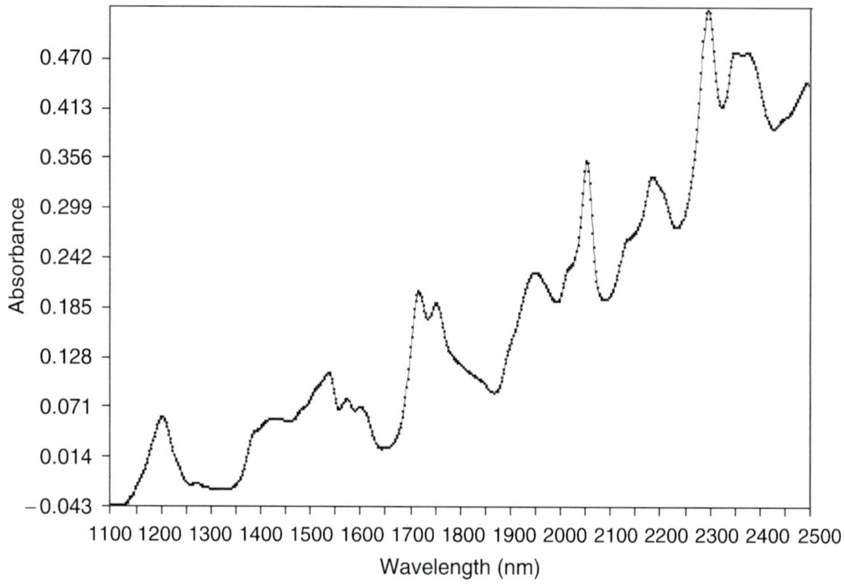

FIGURE 42.12 Percent moisture on nylon66 spun yarn by the NIR and KFR methods, three-wavelength model. (Redrawn from information in Rodgers, *Proc. 7th International NIRA Symposium*, Technicon Technical Center, Tarrytown, NY, 1984.)

wavelength has been assigned as the first overtone of CH stretching vibrations, and the 2083-nm wavelength has been assigned as the C=O/N–H stretch combination. Since neither wavelength corresponds to an expected OH band assignment, the two-wavelength model must be responding to a particular change in the nylon66 yarn rather than the moisture level alone. Since nylon66 is a hydrophilic material, it has a strong attraction for water and is capable of having hydrogen bonding between the water and nylon66 molecules. Thus, the hydrogen bonding forces and molecular structure of nylon66 are somewhat dependent on the percent moisture content. One possible explanation for the two-wavelength model is that changes in the percent moisture content of the fiber result in changes in the crystalline environment of the CH and amide species, and it is this change that the NIR method is detecting. An at-line/on-line measurement of fiber moisture will be presented in Section 42.3.6.

42.3.3 Finish-on-Fiber Measurements

It is a common practice in the manufacture of synthetic fibers to add a small amount of lubricating material to the fiber to assist its performance and runnability in downstream textile manufacturing processes (Saunders 1984, Billica 1984, Crossfield 1984, King 1980). The lubricating material is often referred to as "finish." The three primary functions of the finish are lubrication of the fiber surface, static protection, and filament cohesion. The thin layer of finish is an interface between the fiber and other surfaces. The finish acts as a lubricant to reduce frictional wear and damage when the fiber passes over other contact surfaces, such as guides, rolls, and needles. The finish provides a static protection and filament adhesion to the fiber bundle so as to maintain the fiber bundle integrity and uniformity, properties that are critical to the success of the downstream textile manufacturing of the fiber bundle. The two basic methods for finish application to the fiber surface are the roll and metered finish methods. The more traditional roll application consists of a rotating roll applicator, partially submerged in the finish solution, taking the finish solution from the roll to the fiber surface. This application method is very simple and inexpensive but normally not uniform in its application of finish. The metered finish method consists of an accurately regulated flow, or "spray," of finish through a metering pump directly onto the fiber surface, yielding a more uniform application of finish. It is also desirable to "spread" the fiber bundle prior to finish application, so individual filaments are coated as uniform as possible.

The amount of finish on the fiber surface, called "finish-on-fiber" or FOF, is of critical importance in the textile manufacturing process. Too little finish on the fiber will not give the fiber bundle acceptable lubrication, static protection, and filament cohesion for downstream processing, resulting in "flaring" and excessive mechanical wear on the fiber bundle that will cause interruptions and lose productivity, in downstream processing. The application of excessive finish to the fiber may cause the fiber bundle to adhere to contact surfaces, causing interruptions in downstream processing, and often results in deposits on contact surfaces that not only affect the downstream process efficiency but also lead to increased costly cleaning of the processing machinery. An environmental concern of excessive finish application is increased chemical oxygen demand (COD) on water supplies. Hence, the control and monitoring of the finish level on the synthetic fiber is critical to the maximum productivity and processability in textile manufacturing processes.

There are several methods that can be used to determine the FOF of synthetic fibers, but the traditional methods for nylon fibers consist of the extraction of the finish oils from the fiber and the subsequent quantitative determination of the amount of finish present on a known weight of fiber (FOF). The finish oils are extracted with either hot or cold organic solvents (i.e., CC14) from the fiber. Two methods of determining the FOF are a gravimetric and an infrared (IR) method. In the gravimetric methods, the solvent is vaporized by either heat or steam, and the finish residue is weighed. The FOF is the weight of the finish residue compared to the initial sample weight. The gravimetric method is time-consuming and may be influenced by the evaporation of volatile oils present in the finish during the vaporizing procedure. A more common FOF method consists of measuring the IR absorption of the various CH species in the extracted finish oils at 3.3- to 3.5-μm wavelength, and this absorption is compared to a known standard to obtain the FOF. Although both methods are accurate, they do require solvent extraction of the finish from the fiber and careful, extensive laboratory, and calibration procedures. NIR methodology can be used to perform the analysis in less than 5 min without laborious extraction, laboratory, and calibration procedures.

A set of nylon66 spun yarn samples was analyzed using the NIR method and the IR solvent extraction method, using a filter-based NIR analyzer (Rodgers 1984). This example demonstrates a unique capability of the NIR technique – the ability to perform nondestructive multiple measurements on a single sample simultaneously. Here, both percent moisture and FOF could be determined on the same 3.5-g sample. A four-wavelength NIR multiple linear regression (MLR) model consisting of 546, 1818, 2230, and 2270 nm was used in the FOF evaluation. As shown in Table 42.6, there was good agreement between the IR extraction and NIR methods, with an SEP of 0.04% FOF. The 546-nm wavelength can be attributed to the finish base, the 2230- and 2270-nm wavelengths are due to overtone and combination CH stretches from the various CH groups in the finish oils, and the 1818-nm wavelength is due to a C=O/O–H stretch combination.

TABLE 42.6
Finish-on-Fiber Comparison (%)

Item	IR	NIRA
X	1.06	1.12
Sx	0.032	0.048
SEP	—	0.04
95% CI	0.09	0.03
	10	

Source: Adapted from Dorrity and Tincher et al., *Textile Res. J.*, *65*: 583 (1995).

Similar results for FOF on nylon fibers were obtained with a scanning NIR analyzer, using a fiber-optic remote sensing head (Application Note #26, FOSS). The scanning unit offers the ability to utilize advanced chemometric modeling techniques – specifically, derivative math, partial least-squares (PLS), and principal component regression (PCR) analyses. The samples were scanned over the 1100- to 2500-nm spectral range, and distinct second-derivative spectral differences with varying FOF levels were observed. A derivative PLS NIR calibration (1100–2300 nm spectral range) was developed, yielding excellent predictive results on a small verification sample set.

42.3.4 Heatset Temperature Measurements

Heatsetting, or twist setting, of synthetic yarns is one of the most critical processes in textile manufacturing. Heatsetting is a heat-treating process where heat is uniformly applied to the yarn under specific conditions of temperature, yarn tension, moisture, and heating time in order to obtain satisfactory fabric aesthetics, properties, and quality (Arghyros and Backer 1982, Denton 1983, Ghosh and Rodgers 1986). Heat treating the yarn is normally performed for stabilization against shrinkage, relaxation of internal stresses, bulk development, dye fixation, and twist setting or permanence. It gives the yarn a thermal and morphological "memory." Heat has a significant impact on the molecular structure of synthetic yarns, which in turn influences the yarns' chemical and physical properties and the morphological response of the yarns to further manufacturing steps. Type and quality of heatsetting can significantly affect the dye level, dyeing strike rate, dye bath exhaustion, dyeing variability, bulk appearance, and dye fastness of the yarn and resulting fabric.

Heatsetting consistency and uniformity are of critical importance in carpet yarn manufacturing. Nonuniform and inconsistent heatsetting can result in streaks in the finished carpet. A streak is a visual pattern parallel to the tufting direction that is caused by physical, optical, or dye difference, either within the yarn tuft-line or between adjacent yarn tuft-lines. One of the major components of heatsetting is the temperature at which the yarn is heatset – the heatset temperature. Several studies in carpet yarn manufacturing have shown that a small, but significant, variation in the heatset temperature (normally, $\pm 5^{\circ}$ from target) of the synthetic yarn during heatsetting can result in streaks in the finished carpet (Baeder 1981, Gosline 1984).

Two forms of heatsetting are primarily used in carpet yarn manufacturing – the hank process (autoclave) and the continuous processes (Billica 1984). The autoclave process is labor-intensive, requires multiple processing steps, and is subject to heatsetting variations and nonuniformity. The continuous processes are single-step, require little labor, and are more attractive from both a technical and economic standpoint. Since the early 1970s, the major continuous heatsetting processes are the Suessen and Superba heatsetting systems. The Suessen is a dry method, while the Superba is a wet method. In the Superba process, the yarn is heatset with saturated steam; in the Suessen process, the yarn is heatset with dry heat, in which the small amount of steam present serves as a conditioning medium and to reduce oxidation and yarn discoloration. Superba and Suessen heatsetting systems result in yarns with vastly different morphological and dye properties (Gosline 1984, Hotfeld and Shepand 1977, Miller and Gibson 1983). Moisture lowers the glass transition temperature of the fiber and favors the growth of large crystallites. Dry heat favors an initial nucleation of the fiber structure, forming small crystallites up to a given temperature, where crystal growth is favored. The result is that wet heatset yarns have more open structure and increased crystallinity than dry heatset yarns, which leads to an increased dyeing rate for the Superba yarns.

For hydrophilic yarns, such as nylon66 and nylon6, the differences in the dyeability of the Superba and Suessen heatset yarns can be dramatic. Carpet yarns heatset by continuous processes are susceptible to streakiness due, in part, to (1) the large number of settings and controls that can deviate from standard conditions by operator choice or by normal system fluctuations and (2) the more uniform yarn produced by these systems can present a more perfect background for yarn defects to appear in the carpets. Thus, proper control and monitoring of the major parameters of the heatset process, especially the heatset temperature, is critical to the quality of the finished carpet.

Various techniques, such as X-ray diffraction, fiber density, dyeing or dye rate methods, and thermal methods of analysis, used to estimate the heat history and heatset temperature of synthetic yarns are not suitable for routine quality control/process control because of the time and extensive sample preparation required to perform the test. Therefore, NIR methodology offers a simple, rapid, and accurate technique for measuring the heatset temperature of carpet yarns.

Several authors have demonstrated the NIR method's superior capabilities in the measurement of nylon carpet yarn heatset temperature (Ghosh and Rodgers 1985, Rogers and Lee 1989, 1991, Rodgers 1984, Tincher et al. 1986). A typical NIR spectrum for nylon is shown in Figure 42.13. Examples, using both filter-based and scanning NIR analyzers, will be presented for Superba and Suessen heatset systems, for undyed (greige) yarns, and for dyed yarns. The calibration samples ranged over a very wide (50°) range for both the Superba and Suessen models. Although several NIR prediction models have been proposed and utilized for the Superba and Suessen heatset nylon carpet yarns, they contain two consistent features. First, the 2130-nm wavelength is common to many of the published models. Second, the wavelengths used in the prediction model normally occur in the same NIR spectral region from 1960 to 2310 nm. The prediction models are not necessarily "universal" models; rather, they are explicit models that will predict the heatset temperature differences within a given set or type of samples. Sample preparation and presentation, environmental, instrumental, and manufacturing process differences between different locations or mills may result in slight differences in the absolute prediction models for each location or mill. Many of these differences can be easily corrected with a slight bias change for each location or mill.

Many published filter-based NIR prediction models contain the 2100-, 2130-, and 2270-nm wavelengths. Studies of the various NIR spectra obtained from various Suessen heatset nylon66 carpet yarns and comparison of the NIR spectra to other polymers with similar functional groups have yielded much information on the probable wavelength assignments and the probable rationale for the 2130-nm wavelength commonality in the published NIR prediction equations (Rogers and Lee 1989). The 2130-nm wavelength is due to both the C–H stretch/C=O stretch combination and N–H deformation/N–H stretch combination; the 2100-nm wavelength is due to C=O stretch combinations; and the 2270-nm wavelength is due to various CH and CH_2 stretches, bends, and deformations. It has been proposed that a consistent "slope" change is occurring at the 2100- and

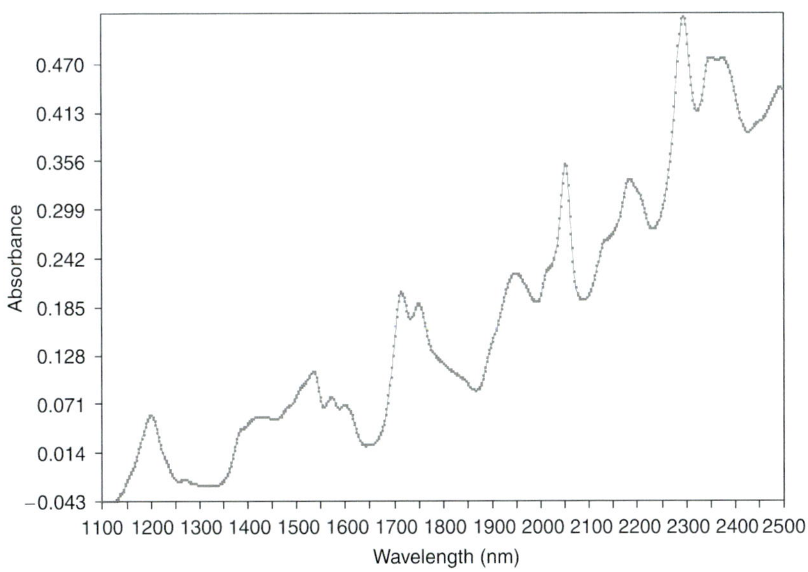

FIGURE 42.13 NIR spectra of heatset nylon yarn, 1100–2500 nm. (Redrawn from information in Rodgers and Ghosh, *Textile Res. J.*, *70*: 519 (2000).)

2130-nm wavelengths, with increasing absorption at 2100 nm and decreasing absorption at 2130 nm with increasing heatset temperature. This slope change is most likely the driving force enabling the NIR method to measure the heatset temperature. The 2270-nm wavelength is most likely a reference or baseline band for the NIR spectra, as nonsignificant absorption differences were observed with increasing heatset temperature.

In addition to the above spectroscopic approach, the effects of changes in heat history on the NIR spectra, and on the prediction models for heatset temperature, of nylon carpet yarns have been studied from a crystallographic approach (Rogers and Lee 1991). Various crystalline, thermal, and laser light scattering parameters were determined. It has been shown that the NIR spectrum and prediction models were responding to changes in the nylon fiber's crystalline perfection with changing heatset temperature. The two components of the crystalline perfection index for nylon66 are the hydrogen bonding plane (d100) and the Van der Waals plane (d010). The d010 spacing, after 190°C, correlated well with increasing heatset temperature, while the d100 spacing exhibited a random pattern. Thus, the NIR spectrum and prediction models for heatset nylon66 carpet yarns were responding to changes in the C–H stretches in the Van der Waal's plane.

42.3.4.1 Heatset Temperature Prediction

42.3.4.1.1 Greige Nylon Carpet Yarns

For Superba heatset carpet yarns, there was an excellent agreement between the Superba actual measured process heatset temperature and the NIR-predicted heatset temperatures for both nylon66 and nylon6 yarns (Rodgers 1984). Figure 42.14 shows a comparison of a nylon66 Superba heatset yarn lot that was heatset over a wide temperature range, and it demonstrates the excellent agreement between the NIR method and the process heatset temperature. The multiple coefficients of determination (R^2) and SEP were very good, being 0.99 and 1.8°F, respectively.

A majority of the work to date has been on Suessen heatset nylon66 or nylon6 greige carpet yarns, where the NIR prediction model measured the Suessen heatset temperature of samples that had been Suessen heatset at different temperatures. Table 42.7 shows the NIR method's capabilities for nine nylon66 sample sets. The NIR method did accurately measure the effective heatset temperature of these Suessen samples (Rodgers 1984). The NIR method and measurement variability were very good and comparable to the thermal methods of analysis, with an R^2 of 0.97 and a pooled

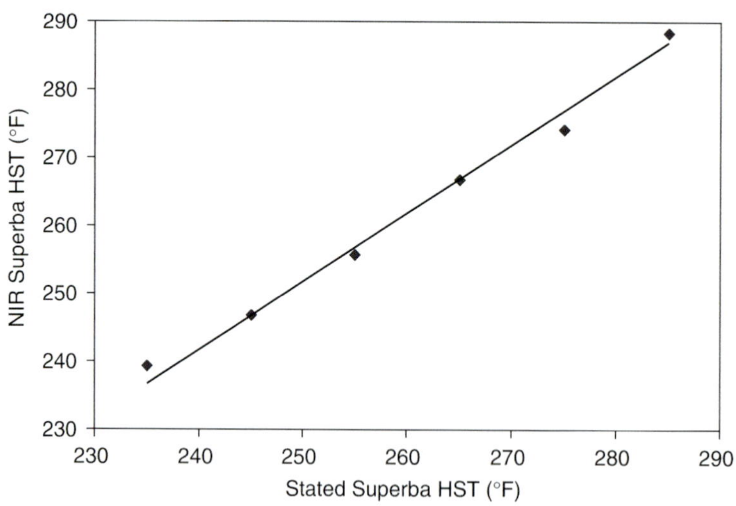

FIGURE 42.14 The relationship between Superba heatset temperature and the NIR method for nylon66 Greige yarn. (Redrawn from information in Rodgers, *Proc. 7th International NIRA Symposium*, Technicon Technical Center, Tarrytown, NY, 1984.)

TABLE 42.7
Suessen Heatset Temperature of Greige Nylon66 Yarns

	SEP (°C)			
	All		190°C–210°C	
HST range (°C)[a]	NIRA	Thermal	NIRA	Thermal
196–211	1.2	1.4	1.2	1.4
205–215	0.6	2.3	0.6	1.3
190–210	0.9	1.4	0.9	1.4
190–210	1.9	1.4	1.9	1.4
178–210	2.6	7.9	0.6	3.4
176–210	2.4	4.8	1.6	2.1
174–210	3.2	—	1.3	—
184–224	2.4	1.9	1.3	1.7
180–210	4.2	3.8	1.1	3.4
Pooled SEP (°C)	2.3	3.7	1.2	2.2

[a] HST, heatset temperature.

SEP of 2.3°C over the entire measurement range. These results were well within the temperature limits that are normally required to prevent dye streaks in the finished carpets. In many cases, it was observed that the NIR model's and thermal analysis method's prediction capability decreased below 190°C and above 220°C. This decrease resulted in higher-than-normal SEPs for the sample sets that contain the very low and very high heatset temperatures. However, the Suessen heatset temperature of primary interest to the carpet manufacturer normally lies in the region between 190°C and 210°C. Within this region, the NIR method's agreement with the Suessen heatset temperature was excellent, with an R^2 of 0.97 and a pooled SEP of 1.2°C.

The deviation from linearity in the NIR model's prediction capability for nylon66 yarns below 190°C was investigated with various crystalline, thermal, and laser light scattering parameters (Rogers and Lee 1991). These deviations were normally small to moderate (<5°C) down to 185°C, but they consistently appeared. Strong similarities were observed in the plots, vs. Suessen heatset temperature, of the NIR measured heatset temperature, thermal parameter phr (peak height ratio, the ratio of the peak heights from the two peaks obtained in a differential scanning calorimeter [DSC] scan of heatset nylon66 yarns), laser light scattering intensity, and X-ray diffraction d010 spacing. For all of these methods, a definite change in the morphological response as a function of heatset temperature was exhibited in the plots by a marked change in slope at about 190°C. The distinct crystalline response transition occurring in nylon66 at about 190°C is the Brill transition (Brill 1956, Itoh 1976, Starkweather and Jones 1981). Thus, the slight deviation in linearity in the NIR model's response below heatset temperatures of 190°C was primarily due to the Brill transition.

42.3.4.1.2 Dyed Nylon Carpet Yarns

The difference in dyed vs. greige nylon carpet yarns is that the dyed yarns have undergone the additional textile manufacturing processes such as tufting, dyeing, and finishing. These additional processes can impact the yarns' morphological properties, so it was by no means certain that the greige yarn calibration equations would perform satisfactorily for the dyed yarns. A comparative study is presented in which several samples of nylon66 yarns that had been Suessen heatset at different temperatures were dyed by several different laboratory and manufacturing processes. The NIR method, with only minor modifications to the three-wavelength MLR prediction equation, did accurately measure the effective heatset temperature of the dyed samples (Rodgers 1984). The method

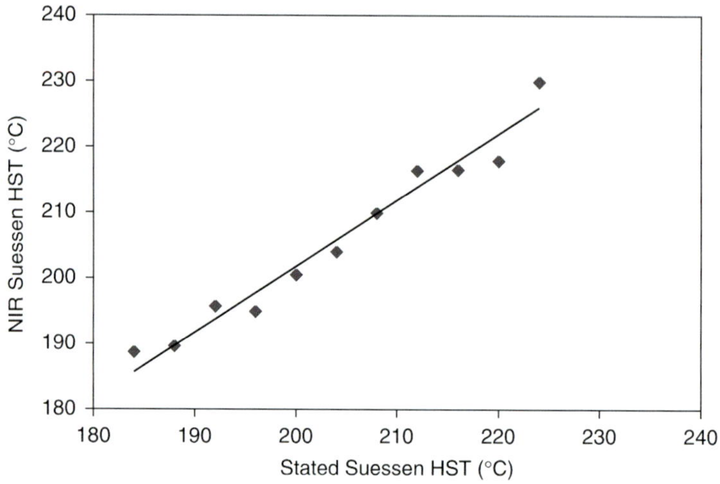

FIGURE 42.15 The relationship between Suessen heatset temperature and the NIR method for dyed nylon66 yarn. (Redrawn from information in Rodgers, *Proc. 7th International NIRA Symposium*, Technicon Technical Center, Tarrytown, NY, 1984.)

and measurement variability were good. Figure 42.15 shows this good agreement, on a nylon66 lot, with an R^2 of 0.96 and an SEP of 2.7°C. As observed with the greige yarn nylon66 samples, the model's prediction capability decreased below 190°C and above 220°C. The SEP within the 190°C–210°C range, the Suessen temperature region of primary interest to the most nylon carpet yarn manufacturers, was 2.3°C.

42.3.4.2 Process Control Measurements, Nylon Carpet Yarns

In the previously mentioned heatset temperature examples, the work centered on "model prediction" evaluations, where the NIR method measured the heatset temperature of nylon66 and nylon6 carpet yarns that had been heatset at different temperatures. The second broad area of heatset temperature measurements is "process control" evaluations, in which the NIR methodology monitors, on a routine basis, the heatset temperature differences on routine heatset samples. The process control measurement of heatset temperature will be illustrated using Suessen heatset nylon66 carpet yarns. Many carpet yarn manufacturers attempt to control the Suessen heatset temperature to within ±3°C of target (normally, 200°C) using the Suessen system controllers. As mentioned previously, heatset temperature values outside of ±5°C from target are of primary concern for streak propensity (Gosline 1984). Therefore, NIR methodology can be used as a diagnostic tool to (1) monitor and identify off-temperature Suessen tunnels that have significantly low (or high) heatset temperatures (outlier/exception mode) and (2) monitor the effective Suessen heatset temperature over time of the Suessen unit (time-monitoring mode). In both cases, the NIR methodology is used to *monitor* the Suessen unit for off-standard conditions.

A three-wavelength NIR prediction equation was used to monitor two large sets of routine process nylon66 Suessen heatset samples (Rodgers 1986). A total of 528 samples representing 88 work-shifts over a 30-day period were evaluated. The NIR method successfully measured the Suessen heatset temperature of both lots of routine production, as shown for the second lot in Table 42.8. No NIR values, shift or daily average, were outside the ±5°C "streak limit," indicating excellent process control monitoring. The NIR method predicted that no Suessen heatsetting streaks should occur in the finished carpets. The NIR method also detected the known slight heatset temperature differences between the six tunnels in a Suessen unit. The yarns from both lots were tufted, dyed, and finished into carpets. As the NIR method had predicted, no streaks were observed in the carpets of both lots.

TABLE 42.8
NIRA Measurement of Suessen Heatset Temperature, Routine Production[a]

	NIRA, by Shift (°C)			NIRA, Daily (°C)
Day	S1	S2	S3	S
1	201	202	202	202
2	200	199	201	200
3	197	198	201	199
4	199	201	201	200
5	197	201	199	199
6	201	200	202	201
7	199	202	201	201
8	200	201	202	201
9	202	200	197	200
10	199	200	199	199
11	200	201	199	200
12	200	200	201	200
13	198	199	200	199
14	201	200	201	201
15	203	200	201	201
16	202	200	201	201
17	201	198	201	200
X	200	200	201	
Sx	1.7	1.1	1.3	
X		200		
Sxp[b]	1.4			

Source: Adapted from Rodgers and Ghosh, *Eastern Analytical Symposium*, Somerset, NJ, 1995.

[a] Target = 200°C for Suessen unit.

[b] Sxp = pooled Sx, all shifts.

42.3.5 Textile Product Identification

Identification of different polymers, their properties, and their morphological differences normally requires extensive testing and complicated time-consuming analysis. Many statistical analysis methods have been combined with NIR to identify dissimilar textile products, including PCA, linear or multiple regression, PLS, derivative math, Mahalanobis distance (i.e., discriminate analysis), nearest neighbors techniques, and neural networks. Most textile fibers, yarns, and fabrics have chemical structures that yield complex NIR spectra, and as such, these species normally require three or more wavelengths to classify the material. The following examples are given to demonstrate the potential and high flexibility of the various identification software programs.

42.3.5.1 Nylon and Heatsetting-Type Identification

It is well known that different nylon polymer types (nylon66 and nylon6) and different heatsetting methods (Suessen and Superba) will have different fiber morphology and molecular structures, and these differences can often result in dye streaks if yarns of different heatsetting methods or nylon polymer types are mixed. The NIR method, using both filter-based and scanning NIR analyzers with the discriminate analysis technique, has been shown to quickly and accurately identify nylon66 and nylon6 yarns that have been either Suessen or Superba heatset (Ghosh and Rodgers 1988). The discriminate analysis technique and theory have been previously discussed

(Mark and Tunnell 1985). The discriminate analysis technique develops calibration identification equations that can be used to identify samples that are similar to the calibration samples. The evaluation consisted of three separate identification sets: nylon polymer type, heatsetting type, and specific identity. For the nylon polymer-type and heatsetting-type identification, the discriminate analysis software was being asked to identify the standard one variable (polymer type or heatsetting type) based on the detection of the known morphological differences between nylon66 and nylon6 yarns and between Suessen and Superba heatset yarns. The specific-identity classification presented a much more complex problem to the qualitative software, for it was being asked to identify a sample based on two variables: polymer type and heatsetting type. Three- and four-wavelength calibration equations were obtained for the identification sets. The Mahalanobis distances between the identification sets were large. The larger the Mahalanobis distances between the groups, the higher the probability for the discriminate analysis technique to correctly classify the sample sets. The size of the Mahalanobis distances between groups indicated that the molecular structure differences determined by the discriminate analysis technique were greater for the polymer types than for the heatsetting types. The calibration equations were then evaluated for predictability using four sample sets of 34 yarns. The 34 unknowns were classified according to polymer type, heatsetting type, and specific identity (Table 42.9). The classification of the unknowns was excellent, with 100% correct identification for all three calibrations.

42.3.5.2 Polyester Fiber by Different Producers

It has been observed that polyester staple fibers that are of different tenacity (response to tensile stress placed on the fiber) levels and that are produced by different fiber producers will have different fabric dyeing and manufacturing production criteria. Both regular and high-tenacity staple polyester fibers were obtained from four fiber producers (Ghosh 1986a). The NIR method, using both filter-based and scanning analyzers with discriminate analysis software, successfully identified and classified the staple polyester samples by producer, by tenacity level, and simultaneously by producer and tenacity (Table 42.10). The fiber identification was successful even when the polyester fiber was blended with cotton in a 50/50 blend. The eight groups of fibers were dyed. Variations

TABLE 42.9
Discriminant Analysis Calibration Equations

		Mahalanobis Distances (MD)		
Calibration Set	**Wavelengths (nm)**	**From**	**To**	**MD**
Nylon polymer type	2472	Nylon66	Nylon6	65.60
	2486			
	2500			
Heatsetting type	1884	Suessen	Superba	6.96
	2094			
	2122			
	2500			
Specific identity	1506	Suessen/nylon66	Suessen/nylon6	45.44
	1534		Superba/nylon66	11.11
	2472		Superba/nylon6	37.15
	2500	Suessen/nylon6	Superba/nylon66	56.28
			Superba/nylon6	8.38
		Superba/ nylon66	Superba/nylon6	47.97

Source: Adapted from Rodgers and Ghosh, *Textile Res. J.*, *71*: 135 (2001).

TABLE 42.10
Identification of Nylon Heatset Unknowns

		Unknowns		Correct Classification		
Sample Set	Number	Polymer Type	Heatsetting Type	Polymer Type	Heatsetting Type	Specific Identity
A	10	Nylon66	Suessen	10	10	10
B	10	Nylon6	Suessen	10	10	10
C	4	Nylon66	Superba	4	4	4
D	10	Nylon6	Superba	10	10	10

Source: Adapted from Rodgers and Ghosh, *Textile Res. J.*, *71*: 135 (2001).

in dye uptake, which would have resulted in quality problems, were present between the various tenacity levels and producers. This NIR method provided a quick and accurate technique for the identification of the polyester fiber before quality problems could occur.

42.3.5.3 Fiber/Polymer-Type Identification

Qualitative analyses can be used to identify and classify fibers of dissimilar or different polymeric type. Often, the spectral differences between the different fiber types can be quite small and minor (i.e., nylon66 vs. nylon6). Several qualitative measurements for fiber identification have been reported, but most of these use the more traditional spectral analysis and spectral pattern recognition techniques (Howell and Davis 1991, Ghosh and Rodgers 1988, Ghosh 1986a, Costello and Knepper 1996, Beck 1996, Rodgers 2002). A novel approach to fiber identification was the application of neural networks to the spectral analysis and spectral pattern recognition programs (Jasper and Kovacs 1994). A scanning NIR unit measured the spectra of 390 synthetic and natural fiber samples from 17 different fiber types/groups. Using backpropagation for the NIR network learning rule and mean subtraction, a neural network was developed that included 700 input absorbance wavelengths, 21 hidden nodes, and 17 output nodes. The neural network not only successfully identified fibers from the 17 fiber groups, but also correctly identified the fiber components in blends (i.e., linen–rayon blend) from fabrics that had not been included in the training sets.

42.3.5.4 Carpet Fiber Identification for Recycling

Recycling programs are growing in popularity and importance, especially for the carpet industry. A key component of a potential carpet recycling program is the rapid and accurate identification of the fiber/polymer type used in the carpet face fiber in the returned carpet. Several fiber producers have developed their own in-house NIR carpet fiber identification programs, and several NIR analyzers specifically designed to identify the carpet face fiber are commercially available. An example of the flexibility of the NIR technique for carpet face fiber identification can be shown by evaluations performed with one of the first commercially available units – the Carpet Analyzer (Costello and Knepper 1996). The Carpet Analyzer is a mobile, fiber-optic probe, scanning NIR unit that scans the sample quickly (5 s or less analysis time) from 1200 to 2400 nm. Derivative PCA calibrations and an internal fiber identification library were developed to accurately identify the polymer type of most carpet face fibers, including nylon66, nylon6, polyester, polypropylene, wool, and acrylic. The hand-held probe is placed against the carpet surface, the NIR spectrum is collected, and the carpet spectrum is compared to the fiber types present in the precalibrated internal library. Using similar procedures established and demonstrated by the Carpet Analyzer, many other commercial NIR/FT-NIR analyzers are available today for the identification of carpet face fiber.

Investigations were also conducted to obtain a better understanding of the impact of key carpet component and instrumental variables on the identification of carpet face fiber by NIR (Rodgers

2002). The improved understanding of these influences on the NIR calibrations could explain why some NIR identification models appear to be more robust and "capable" than others. Static (load sample into a sample cell or place the same against the NIR unit's port) and fiber-optic probe analyses were performed on an extensive series of carpets and yarns of nylon66, nylon6, polypropylene, polyester, and acrylic polymer. The carpet components evaluated were fiber type, color, carpet and yarn construction, and dyeing/coloration method. The instrumental variables evaluated were sampling system (static vs. probe) and measurement speed (scans per sample). As expected, distinct spectral differences were observed between the nylons, polyester, polypropylene, and acrylic fibers (Figure 42.16). These differences were accentuated with the use of second-derivative spectra, and the second-derivative spectra were able to differentiate between nylon66 and nylon6 (Figure 42.17). Color and construction differences did result in changes in the intensity of the NIR absorbance and,

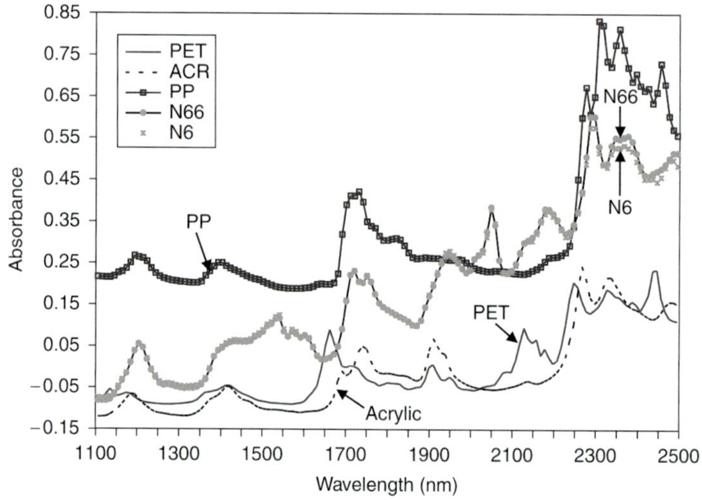

FIGURE 42.16 NIR absorbance spectra for polyester (PET), acrylic (ACR), polypropylene (PP), nylon66 (N66), and nylon6 (N6) fiber types, 1000–2500 nm. (Redrawn from information in Rodgers, *AATCC Review*, 2: 27 (2002).)

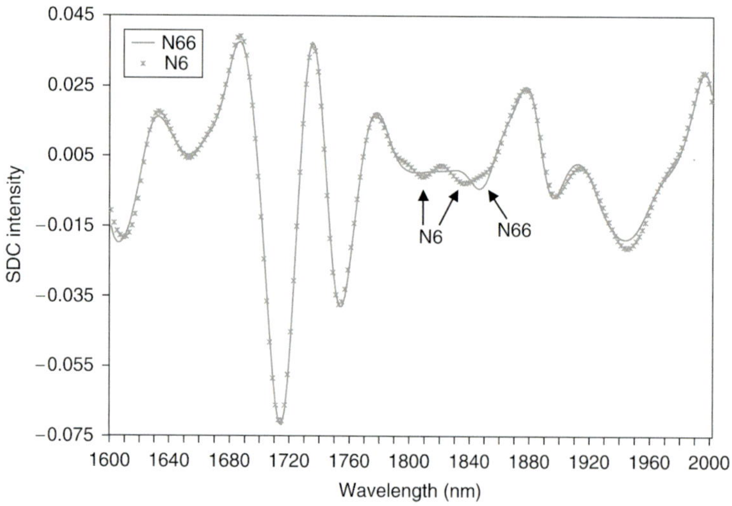

FIGURE 42.17 NIR second-derivative spectra for nylon66 (N66) and nylon6 (N6) fiber types, 1600–2000 nm. (Redrawn from information in Rodgers, *AATCC Review*, 2: 27 (2002).)

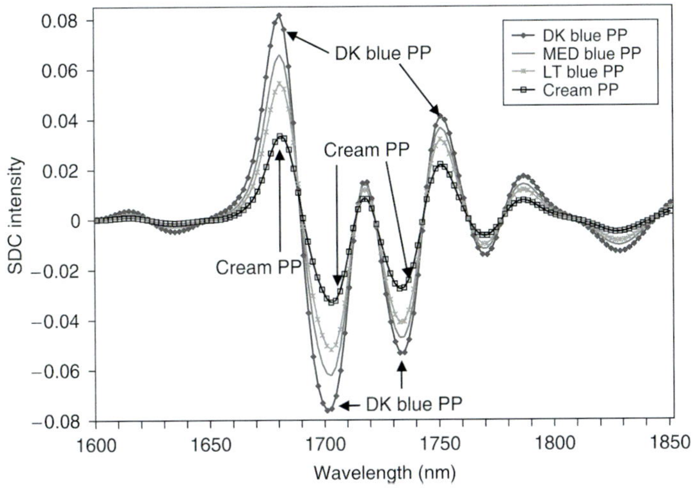

FIGURE 42.18 NIR second-derivative spectra for different pigment color polypropylene (PP) fiber, 1600–1850 nm. (Redrawn from information in Rodgers, *AATCC Review*, 2: 27 (2002).)

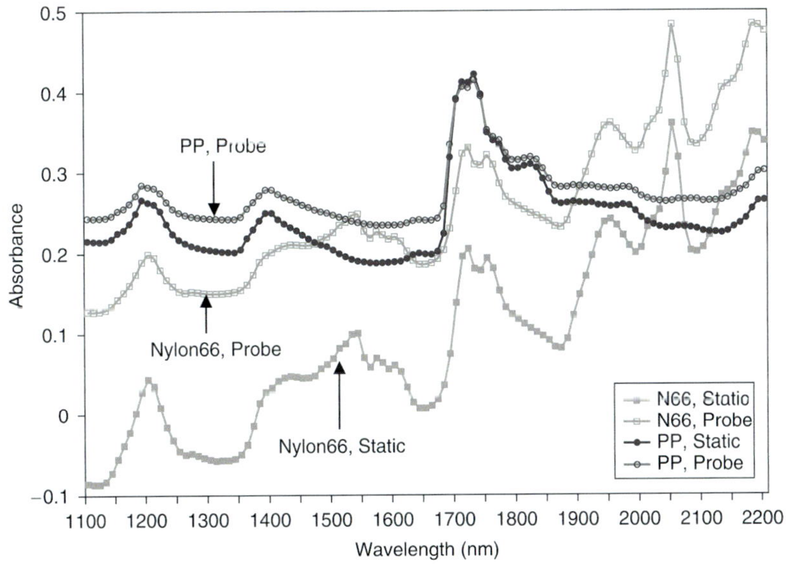

FIGURE 42.19 NIR absorbance spectra for polypropylene (PP) and nylon66 (N66) fiber types, static/bench-top vs. fiber-optic probe sampling systems, 1000–2200 nm. (Redrawn from information in Rodgers, *AATCC Review*, 2: 27 (2002).)

in some instances, noticeably and distinct shifts in the spectral baseline (i.e., pigments and type carpet surface). However, while spectral changes were observed with moderate to large changes in the carpet components, the second-derivative spectra demonstrated that shifts in the polymer type's characteristic wavelengths did not occur (Figure 42.18 for an example). The major instrumental parameter to influence the spectral curve was the sampling system – static vs. probe. Only minor spectral baseline movements were observed with changes in analysis time (number of scans per sample). Significant spectral differences were observed in the spectral intensities of the polymers between the static and probe systems, with the probe yielding lower absorbance differences (peak-to-baseline amplitude) (Figure 42.19). However, once again, the second-derivative spectra demonstrated that shifts in the polymer type's characteristic wavelengths did not occur as a result of these

TABLE 42.11
Identification of Staple Polyester Tenacity Unknowns

	Unknowns			Correct Classification		
Sample Set	**Number**	**Fiber Producer**	**Tenacity Level**	**Fiber Producer**	**Tenacity Level**	**Simultaneous**
1	5	DuPont	High	5	5	5
2	5	DuPont	Regular	5	5	5
3	5	Celanese	High	5	5	5
4	5	Celanese	Regular	5	5	5
5	5	Eastman	High	5	5	5
6	5	Eastman	Regular	5	5	5
7	5	Hoeschst	High	5	5	5
8	5	Hoeschst	Regular	5	5	5

Source: Adapted from Ghosh, *J. Textile Institute*, *84*: 85 (1993).

instrumental parameters. Thus, for a robustly calibrated system (many carpet components), carpet component and instrumental parameters resulted in minimal effects on fiber-type identification. Optimal polymer-type identification was made with the second-derivative NIR calibrations. The NIR calibrations were validated on almost 200 carpet and fiber samples, with 100% correct identification rates using the second-derivative calibration (Table 42.11).

Carpet is reprocessed into recycled yarn for reuse as synthetic fiber. The material taken from the used carpet is shaved face yarn. The chemical composition of the face yarn is determined using NIR to classify it into one of several category types to include: polypropylene, polyester, polylactic acid, polyamide, or natural fibers like wool. A process for reusing the carpet fibers and identifying them with NIR has been patented (Hoover et al. 2011).

42.3.6 At-Line/On-Line Analyses

The advent of stringent financial and cost controls, manufacturing efficiencies and productivity, and "re-engineering" have placed a strong (and increasing) pressure on laboratory analyses and laboratory resources. Although these financial and productivity pressures tend to restrict the number and usage of laboratory analyses, the need for "just-in-time," accurate, and precise analytical results on key physical and chemical properties of a product is of an even greater importance today. One answer to this need for fast, precise, and accurate analytical results in today's business environment is the use of at-line and on-line analyses. At-line analyses are those in which the NIR analyzer is located in the manufacturing area, and the samples are brought to the analyzer for analysis by area operators rather than sending the sample to the laboratory. On-line analyses are those in which the NIR analyzer is placed in the manufacturing process, and the NIR measurement, sample handling, and sample preparation are automatically controlled by a computer. Recent instrumental advances by manufacturers of NIR instrumentation have resulted in an increasing use of NIR reflectance measurements on textile substrates, both at-line and on-line.

There are several procedures that can be utilized to evaluate and implement at-line and on-line measurements. One of the most common NIR analysis processes consists of three phases: bench-top unit evaluation, pilot plant or laboratory evaluation of the at-line/on-line instrumentation, and manufacturing area evaluation. The bench-top phase establishes the method definition and analysis potential. The pilot plant phase verifies the method and instrumentation for the analysis under consideration, determines the optimum sampling system, and identifies the operational parameters and concerns. The manufacturing area phase optimizes data treatment, optimizes operational and instrumental parameters, and proceeds with the method implementation.

At-line and on-line NIR measurements present unique complexities and operational concerns (Rodgers and Ghosh 1994, 1995, 2000, Rodgers et al. 2001). Several major operational concerns for at-line and on-line measurements of textiles must be considered and planned for before NIR evaluations. The major operational concerns are as follows:

- Sample matrix effects
- Type of data treatment
- Temperature and environmental impacts
- Instrumental parameters
- Details of the experimental process.

Sample matrix effects can include the number of scans required to obtain the desired accuracy, precision, and analysis time; the number of readings or packings of the sample; sample presentation and conformity (bobbin wind, fabric weave, surface hardness, etc.); fabric surface effects (additives, denier, fiber cross section, etc.); and where on the textile surface to measure the sample. Data treatment parameters include the type of mathematical or chemometric model (MLR, PLS, PCR, derivative, etc.) and the type and strategic emphasis of statistical parameters (within-sample variability, between-sample variability, R^2, residual analyses, etc.). Temperature and environmental parameters consist of the effect of process/measurement temperature on the NIR spectra, type of material (fiber type, hazards, etc.), chemical groups/environment present, and potential absorption deviations due to chemical or physical parameters present in the sample (water, additives, etc.). Instrumental parameters involve the spectral range, the size and weight of the sampling system, the expected use of the measurement (ergonometric effects, at-line or on-line, etc.), simplicity and ease of operation, and data manipulation and collation (averaging, automatic saving of spectra, background subtraction, etc.). The experimental process can include the quantity of process detail required to obtain an acceptable at-line/on-line NIR method, the quality and composition of the calibration and prediction sample sets (natural or prepared samples, accuracy and precision of reference method, sample selection techniques, etc.), the number of calibration and prediction samples, and a defined end-state (conditions and requirements to be completed for a successful implementation).

Three examples, namely, one at-line, one on-line, and one at-line/"pseudo" on-line, are presented to demonstrate the flexibility and versatility of the NIR spectroscopy and instrumentation for process measurements.

42.3.6.1 At-Line Measurement of Carpet Yarn Heatset Temperature

The importance of the uniform and consistent application of heat/heat history during the heatsetting of carpet yarns has been described previously. The heatset temperature is one of the most important operating variables in the heatsetting process. Previous investigations described earlier emphasized bench-top and laboratory NIR measurements of heatset temperature. This project's objective was to demonstrate the transference of the measurement of Suessen and Superba heatset temperature for nylon66 carpet yarns from bench-top to at-line, and to provide manufacturing with a simple and rapid at-line NIR method to monitor the heatset temperature of Suessen heatset nylon carpet yarn *directly on the bobbin* (Rodgers and Ghosh 1994, 1995, 2000, Rodgers et al. 2001). The defined end-state was the development of a method with good at-line accuracy and precision (residual error/SEP of <4°C for Suessen yarns and <4°F for Superba yarns). The initial measurements were taken in the laboratory with a scanning NIR analyzer and two sampling systems: a fiber-optic probe ("probe") and a fiber-optic remote sensing head ("head") (Rodgers and Ghosh 2000). Second-derivative NIR spectra of the heatset yarns were compared, as a function of heatset temperature, for major areas

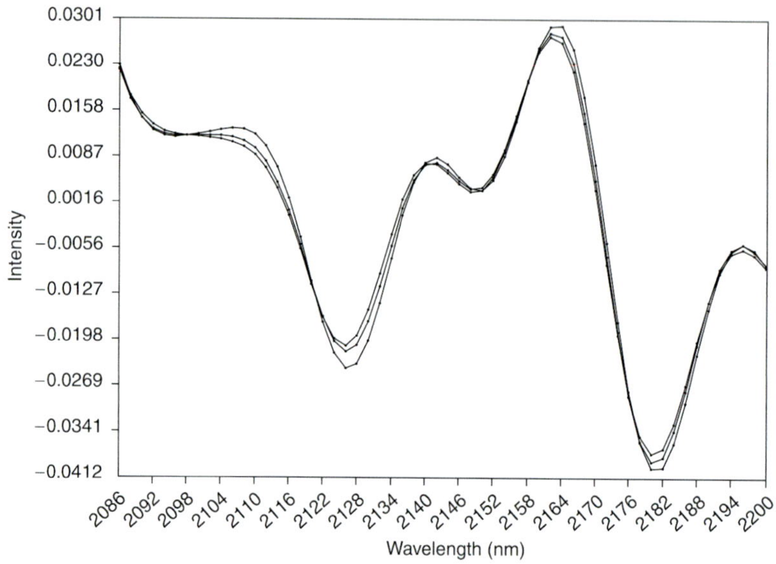

FIGURE 42.20 At-line second-derivative NIR spectra of Suessen heatset nylon66 yarn, 2086–2200 nm. (Redrawn from information in Rodgers and Ghosh, *Textile Res. J.*, *70*: 519 (2000).)

TABLE 42.12
At-Line NIR Measurements of Heatset Temperature

Sampling System	Parameter	Validation Set Results	
Superba Heatset		**260°F–285°F**	**All**
Probe[a]	Bias (°F)	−0.6	0.3
	R^2	0.98	0.99
	SEP (°F)	1.08	1.87
Head[a]	Bias (°F)	0.5	1.7
	R^2	0.99	0.99
	SEP (°F)	0.98	1.80
Suessen Heatset		**190°C–210°C**	**All**
Probe[b]	Bias (°C)	−0.6	0.2
	R^2	0.98	0.99
	SEP (°C)	0.72	2.20
Head[a]	Bias (°C)	0.1	2.0
	R^2	0.98	0.98
	SEP (°C)	0.78	2.30

Source: Rodgers and Ghosh, *Textile Res. J.*, *70*: 519 (2000).
[a] Second-derivative MLR calibration equation.
[b] Second-derivative PLS calibration equation.

of spectral difference (Figure 42.20). Spectral differences were primarily observed in the 1960- to 2200-nm region, and this spectral region was selected for calibration development. Excellent agreement was obtained between the stated heatset temperatures and the NIR-determined heatset temperatures for both sampling systems, with high R^2s and low SEPs (Table 42.12). The probe system was the most promising overall system for *at-line* heatset temperature measurements on the bobbin. The probe system yielded a simpler operation, lower analyzer weight, faster analysis time, increased

TABLE 42.13
At-Line Determination of Production Lot Quality by NIR

Sample Set	Production Lot Quality	SEP (°F)
Superba 1	Good	0.78
Superba 2	Problem	2.31
Superba 3	Unknown	2.03

Source: Rodgers and Ghosh, *Textile Res. J.*, *71*: 135 (2001).

flexibility, superior measurements on small/low weight bobbins, and acceptable accuracy. The head system was best suited for *on-line* heatset temperature measurements (larger sampling area and greater precision).

The NIR analyzer, with probe sampling system, was evaluated in the manufacturing environment at several carpet manufacturers (Rodgers et al. 2001). Bobbin/at-line NIR calibrations were developed in the process area laboratory and transferred to manufacturing locations. Very good heatset temperature agreement was obtained for both wide-range and routine production samples, with low SEPs and high R^2s. Quality differences in routine production bobbins from different lots of Superba heatset yarns were observed. On the basis of the SEP of each lot (consistency of heatset temperature for bobbins in a lot), the at-line NIR technique correctly identified and segregated the "good-quality" lot from a "problem-quality" lot at one manufacturer (Table 42.13). In addition, the NIR method determined that an "unknown-quality" lot would be marginally acceptable for shipment – an assessment subsequently confirmed by the manufacturer. At another manufacturer, yarn product/heatset-type differences (regular Suessen vs. Hi-twist/stufferbox Suessen) and environmental influences were observed. Hi-twist heatset yarns yielded different diffuse reflectance than regular Suessen heatset yarns. The expected slight difference in NIR-determined heatset temperature was observed, but the precision for each yarn type was very similar, thus implying comparable product quality (Figure 42.21).

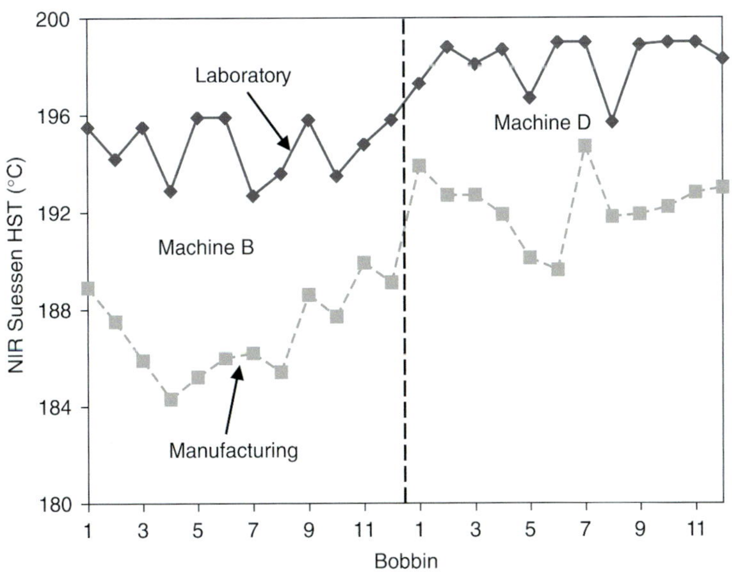

FIGURE 42.21 Effect of environmental measurement conditions on the NIR-determined Suessen heatset temperature, regular (Machine D) vs. Hi-twist (Machine B) heatsetting machines. (Redrawn from information in Rodgers and Ghosh, *Textile Res. J.*, *71*: 135 (2001).)

The major environmental influence was area temperature – a 15°F–20°F temperature difference existed between the laboratory and manufacturing location measurements. As expected, significant shifts in NIR-determined heatset temperature were observed, but the precision for each yarn (SEP) remained excellent and comparable. Thus, only a slight bias adjustment to target would be required to use the laboratory at-line NIR calibration in the manufacturing area. The at-line NIR method provided a rapid, accurate, and precise process control/quality control measurement of nylon66 carpet yarn heatset temperature.

42.3.6.2 On-Line Measurement of PVA Size on Warp Yarns

Size is added to warp yarns to minimize warp end breaks during weaving. The amount of polyvinyl alcohol (PVA) and uniform application of size to the warp (size add-on) can significantly influence weaving efficiency. A common sizing agent used for PET blend warp yarns is PVA. An on-line NIR method was developed that accurately and precisely monitored the PVA add-on on the warp sizing (slasher) machine (Ghosh 1993). NIR spectra were collected from 1100 to 2500 nm using a scanning NIR analyzer with a remote fiber-optic sensing head. Distinct second-derivative spectral differences were observed between 1700 and 1800 nm with increasing size add-on (Figure 42.22). Both derivative MLR and PLS NIR calibrations were developed and evaluated. The best NIR-calculated PVA add-on agreement was obtained with a derivative PLS model, with a R^2 of 0.99 and an SEP of 0.23% size for laboratory slasher samples (Figure 42.23). The MLR models also performed well on the laboratory slasher, but a long-term drift was observed for samples prepared and measured on a commercial slasher (manufacturing environment). This long-term drift resulted in much greater SEPs for the MLR equations than observed for the PLS equation. Thus, the derivative PLS equation was the best and most stable NIR calibration for PVA add-on for commercial slashers. The PLS model was also evaluated in a process control mode. The PLS calibration correctly monitored the slasher as it was operating normally (only one outlier) and as it was operating during periods of process upsets and machine changes. The changes in warping efficiency and size add-on during these process upsets were observed.

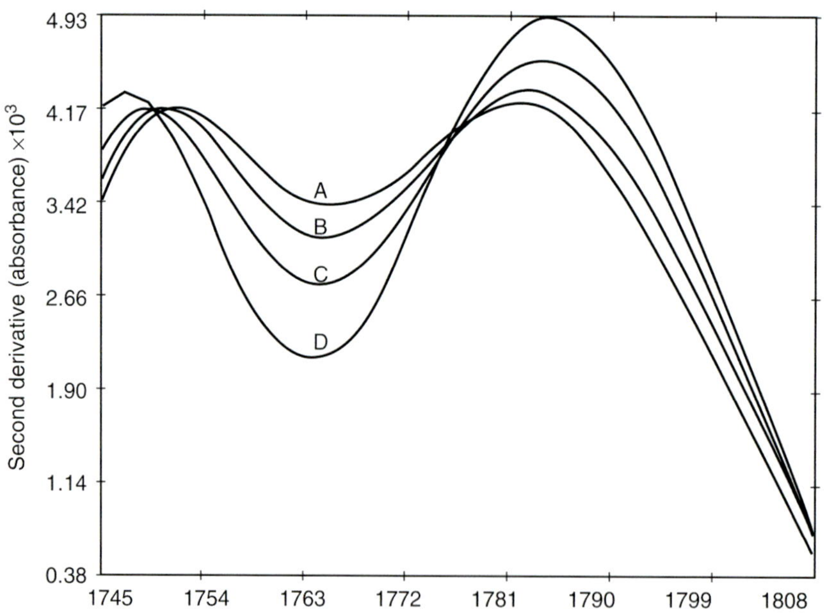

FIGURE 42.22 On-line second-derivative NIR spectra of PVA-sized PET yarn, 1700–1800 nm. A=9.0, B=11.7, C=14.6, D=20.1 %Size. (Redrawn from information in Ghosh, *J. Textile Institute*, *84*: 85 (1993).)

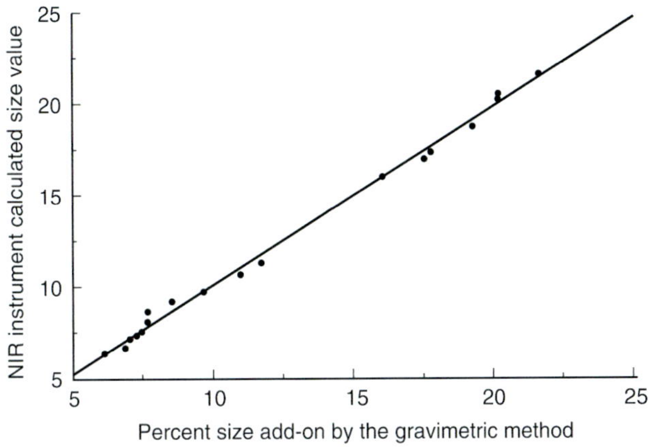

FIGURE 42.23 Relationship between laboratory-analyzed percent size on warp and on-line NIR-calculated percent size using a PLS calibration. (Redrawn from information in Ghosh, *J. Textile Institute*, *84*: 85 (1993).)

42.3.6.3 At-Line/"Pseudo" On-Line Measurement of Tire and Carpet Yarn Moisture and Finish-on-Fiber

The importance of accurately and precisely determining the moisture content and FOF applied to textile products has been described previously. The moisture content and FOF applied to the surface of textile fibers can significantly influence the fiber's physical properties, manufacturing processes and control, quality, and productivity. Many investigations have emphasized the specific moisture or FOF analyses using sophisticated bench-top or on-line NIR analyzers. One investigation studied the feasibility of using a portable, low-cost NIR analyzer to determine the moisture of textile bobbins in the manufacturing area, during preparation of the bobbins in spinning (Rodgers et al. 2002). The noncontact measurement was taken in the spinning manufacturing area, and the NIR unit was moved from position to position down the manufacturing unit. Thus, this form of process analysis can be considered as "pseudo" on-line – an at-line analyzer measuring bobbins on each position sequentially.

The moisture content and the FOF applied to the fiber are often critical control variables, and they can significantly influence and affect the fiber's physical properties, manufacturing processes and quality, and productivity. A recurring difficulty and concern in the production of nylon yarn bobbins is the rapid detection of outlier or out-of-specification moisture and FOF bobbins and machine positions at or near the application of the finish in manufacturing. As the identification of moderate process deviations that would lead to outlier bobbins was the primary concern of the manufacturing area, the high degree of accuracy normally associated with a full-scanning NIR analyzer was not required for this application. A portable, low-cost, noncontact, semiquantitative NIR analyzer was used to determine moisture, directly on tire and carpet yarn bobbins, in various manufacturing areas (at-line). Moisture content calibrations were developed with bobbins of wide moisture ranges for tire yarn and carpet yarn products. The NIR unit successfully monitored moisture differences between bobbins in several manufacturing areas (spinning and lag areas, spun and drawn yarns, etc.) and in the laboratory after the bobbins had acclimated. Using the same NIR calibration for the manufacturing area and laboratory moisture analyses, very good method agreement was observed between the NIR and thermal/weight loss (reference) moisture results for the manufacturing area and laboratory analyses, well within the desired moisture agreement between the two techniques of ±0.5% moisture for over 90% of the samples analyzed (Table 42.14 for an example). Outlier bobbins and off-standard manufacturing positions were readily and quickly identified.

In spinning, it may be feasible to use the spun fiber moisture measurement as a surrogate measurement for bobbin FOF. For spun fiber production, FOF normally increases/decreases in the

TABLE 42.14
At-Line Moisture, Portable NIR Unit, Laboratory, and Manufacturing Areas

	Manufacturing Area Moisture (%)			Laboratory Moisture (%)[a]		
Samples	NIR	Brabender	Delta	NIR	Brabender	Delta
1	2.74	3.12	0.38	4.76	4.79	0.03
2	2.62	2.90	0.28	4.86	4.97	0.11
3	2.62	2.86	0.24	5.00	5.03	0.03
4	2.59	2.88	0.29	4.91	4.87	–0.04
5	2.14	2.27	0.13	4.81	4.75	–0.06
6	2.61	2.80	0.19	4.79	4.82	0.03
7	2.52	2.80	0.28	5.16	4.97	–0.19
8	2.62	2.68	0.06	5.01	5.02	0.01
AVG	2.56	2.79	0.23	4.91	4.90	–0.01
SD	0.31	0.31	0.30	0.21	0.17	0.13

Source: Rodgers, *AATCC Review*, *2*: 27 (2002).

[a] Laboratory moisture determined after a 24-h lag.

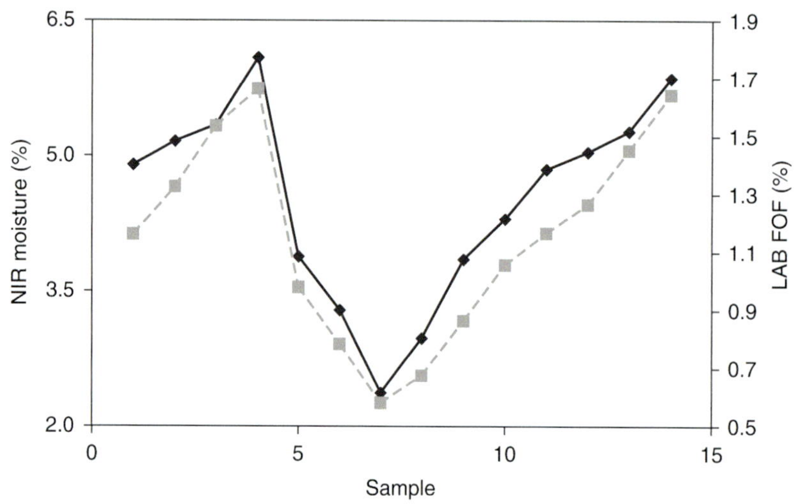

FIGURE 42.24 The relationship between laboratory and at-line NIR finish-on-fiber for nylon66 bobbins. (Redrawn from information in Rodgers, et al., *Pittsburgh Conference 2002*, New Orleans, LA, 2002.)

same direction as the moisture. A strong correlation was observed between on-the-bobbin moisture obtained by the NIR unit during spinning and FOF (Figure 42.24). Surrogate, direct-read FOF calibrations were developed for tire and carpet yarn products. The robustness of the NIR surrogate FOF calibrations was evaluated on bobbins from different products (different denier/size, tints/"color," end uses, manufacturing conditions, plant locations, etc.). Very good method agreement was observed between the NIR and laboratory FOF results (Figures 42.25 and 42.26). Of critical importance to the manufacturing area was the rapid at-line identification of several "outlier" bobbins of low FOF due to manufacturing upsets, preventing their contamination of downstream processes. The influences and effects of all yarn differences, manufacturing locations, and operational and environmental conditions were slight and manageable.

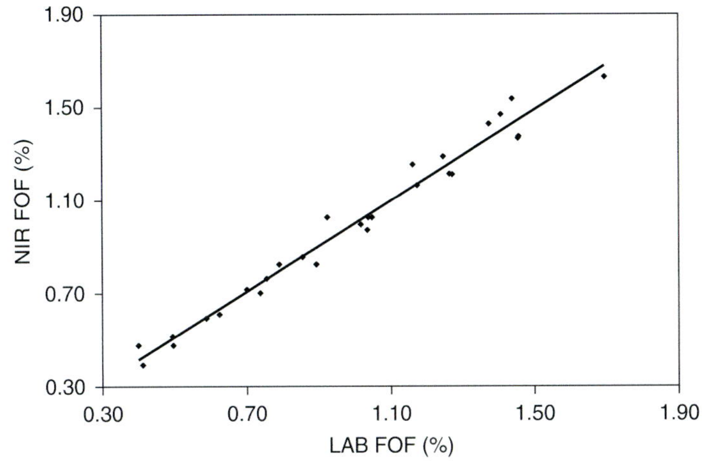

FIGURE 42.25 The relationship between laboratory and at-line NIR finish-on-fiber for nylon66 BCF carpet bobbins (a wide range of finish-on-fiber). (Redrawn from information in Rodgers, et al., *Pittsburgh Conference 2002*, New Orleans, LA, 2002.)

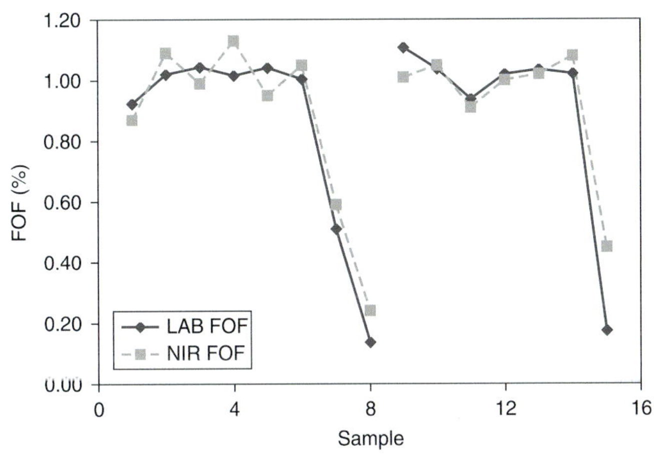

FIGURE 42.26 The relationship between laboratory and at-line NIR finish-on-fiber for nylon66 BCF carpet routine production bobbins. (Redrawn from information in Rodgers, et al., *Pittsburgh Conference 2002*, New Orleans, LA, 2002.)

REFERENCES

Arghyros, S. and Backer, S. 1982. Mechanics of texturing thermoplastic yarns: Part VIII: An experimental study of heat setting. *Textile Research Journal*, *52*(5), 295–312.

Application Note #26, FOSS NIRSystems, Silver Springs, MD.

Baeder, J. 1981. Operation Techniques on Continuous Heatsetting Machines, Part II: Early Detection of Process Upsets, *Proc. 2nd Symposium on Solving Tufted Textile Problems*, Textile Eng. Dept, Auburn University, Auburn, AL.

Beck, K. 1996. An Overview of Textile Applications of NIR Technology: Theory/Instrumentation and Qualitative Applications. *In AATCC Textile Applications of NIR Technology Symposium, Asheville, NC.*

Billica, H. 1984. "Practical Consideration in Finish Use" by Harry Billica. *Fiber Producer*, 21.

Blanco, M., J. Cruz, and M. Armengol. 2008. Control production of polyester resins by NIR spectroscopy. *Microchemical Journal*, *90*(2), 118–123.

Brill, V.R. 1956. Beziehungen zwischen Wasserstoffbindung und einigen Eigenschaften von Polyamiden. *Die Makromolekulare Chemie: Macromolecular Chemistry and Physics*, *18*(1), 294–309.

Costello, M. and S. Knepper. 1996. Identification of Post-Consumer Carpets by NIRA, *AATCC Textile Applications of NIR Technology Symposium*, Asheville, NC.

Crossfield, R. 1984. *Fiber World*, 21.

Denton, M. 1983. *Textile Month*, 57.

Dorrity, J. L., A. Kumar, and W.C. Tincher. 1995. Determining the heat-set temperature for poly (ethylene terephthalate) carpet yarn with near-infrared reflectance spectroscopy. *Textile Research Journal*, *65*(10), 583–586.

Drews, M. and S. Harris. 1990. Using NIR to Check Latex Binders in Nonwovens, *AATCC NIR Technology in Textiles Symposium*, Charlotte, NC.

Edelstein, S. M. 1936. *Am. Dyes Stuff Ref.*, *25*, 458.

Elliott, A., et al. 1954. *British Journal of Applied Physics*, *5*, 337.

Fortier, C., and J.E. Rodgers. 2014. Preliminary examinations for the identification of US domestic and international cotton fibers by near-infrared spectroscopy. *Fibers*, *2*(4), 264–274.

Fortier, C., J.E. Rodgers, J. Foulk, and D. Whitelock. 2012. Main content area Near-infrared classification of cotton lint, botanical and field trash. *Journal of Cotton Science*, *16*(1), 72–79.

Ghosh, S. 1986a. Applicable Textile Process Control Analysis for Monitoring by NIRA, *NIRA Seminar*, Charlotte, NC.

Ghosh, S. 1986b. Applications of Heat History Measurements of Polyester Yarn, *Proc. 10th International NIRA Symposium*, Technicon Technical Center, Tarrytown, NY.

Ghosh, S. 1993. On-line measurement of polyvinyl alcohol size on warp yarns using a near-IR diffuse reflectance spectroscopy method. *Journal of the Textile Institute*, *84*(1), 85–98.

Ghosh, S., and G. Brodmann. 1993. Online Measurement of Durable Press Resin on Fabrics using the NIR Spectroscopy Method. *Textile Chem. Color*, *25*(4), 11–14.

Ghosh, S., and J.E. Rodgers. 1985. Determining heatset temperature by near-infrared reflectance spectroscopy. *Textile Research Journal*, *55*(9), 556–560.

Ghosh, S., and J.E. Rodgers. 1986. *Melliand Textilberichte*, *67*, 26.

Ghosh, S., and J.E. Rodgers. 1988. *Meilland Textilberichte*, *69*: 361.

Ghosh, S. and R.B. Roy. 1988. Quantitative near-infra-red analysis of reducing sugar from the surface of cotton. *Journal of the Textile Institute*, *79*(3), 504–510.

Giangiacomo, R. et al. 1981a. *Encyclopedia of Analytical Chemistry*, Ed. Robert A. Meyers, John Wiley & Sons Ltd., Chester.

Giangiacomo, R., J.B. Magee, G.S. Birth, and G.G. Dull. 1981b. Predicting concentrations of individual sugars in dry mixtures by near-infrared reflectance spectroscopy. *Journal of Food Science*, *46*(2), 531–534.

Gjonnes, J. and N. Norman. 1960. X-ray investigations on cellulose II and mixtures of cellulose I and II. *Acta Chemica Scandinavica*, *14*(3), 683–688.

Gosline, K. 1984. An Overview of the Carpet Yarn Twist Setting Process, *Proc. Carpet Manufacturing Technology Conference*, Clemson University, Clemson, SC.

Guthrie, J. D., C. L. Hoftpanir, M. F. Stansbury and W.A. Reevers. 1949. *Survey of the Chemical Composition of Cotton Fibers, Cotton Seed, Peanuts, and Sweet Potatoes, A Literature Review*, United States Department of Agriculture, Agricultural Research Administration, Bureau of Agriculture and Industrial Chemistry, AIC-61 (revised March).

Harper, J. 1993. Applications of Near Infrared Reflectance Spectroscopy to Rayon Yarns, *AATCC Textile Applications of NIR Technology Symposium*, Asheville, NC.

Hill, B. 1993. Evaluation of a NIR Measuring Process for Sewing Thread, *AATCC Textile Applications of NIR Technology Symposium*, Asheville, NC.

Honigs, D.E., G.M. Hieftje, and T. Hirschfeld. 1984. A new method for obtaining individual component spectra from those of complex mixtures. *Applied Spectroscopy*, *38*(3), 317–322.

Hoover, G.T., Jr, M.K. Roark, B.F. McSheehy Jr, and M.B. Studholme. 2011. *Method of Separation and Cleaning of Post Consumer Carpet Face Yarn from Carpet Backing and Yarn Product Produced Therefrom*. U.S. Patent 8,017,662.

Hotfeld, W., and M. Shepand. 1977. *Can. Textile J.*, 72.

Howell, H.E. and J. R. Davis. 1991. Qualitative identification of fibers using NIR spectroscopy. *Textile Chemist & Colorist*, *23*(9), 69.

Itoh, T. 1976. Change with temperature in crystal structures of nylons 6, 66 and 610. *Japanese Journal of Applied Physics*, *15*(12), 2295.

Jasper, W.J. and E.T. Kovacs. 1994. Using neural networks and NIR spectrophotometry to identify fibers. *Textile Research Journal*, *64*(8), 444–448.
Jerry, R. 1996. On-line NIR Measurement of Pigment Concentration during Color Fiber Extrusion, *AATCC Textile Applications of NIR Technology Symposium*, Asheville, NC.
King, C. 1980. *Fiber Producer*, 17.
Lanza, E. and B.W. Li. 1984. Application for near infrared spectroscopy for predicting the sugar content of fruit juices. *Journal of Food Science, 49*(4), 995–998.
Lemere, L. 1989. *Textile Chem. Color*, *21*: 19.
Mark, H. 1984. Chemical Assignments of Near Infrared Absorption Bands. *Customer Application Group*, Technicon, Tarrytown, NY.
Mark, H.L. and D. Tunnell. 1985. Qualitative near-infrared reflectance analysis using Mahalanobis distances. *Analytical Chemistry*, *57*(7), 1449–1456.
Meurens, M. 1984. *Proceedings of 3rd Annual Users Conference for NIR Research*, Pacific Scientific Inc., Silver Springs, MD.
Miller, R., and M. Gibson. 1983. Effects of Process Induced Morphology Changes on Dyeing of Nylon6,6 Carpets, AATCC RA91, *International Dyeing Symposium*, Atlanta, GA.
Mitchell, J., Jr. 1976. Methods for the determination of water in polymers. *Analytica Chimica Acta*, *81*(2), 231–263.
Nikitin, V. I. 1950. *Vestn. Leningr*, 3, 33.
Olson, E. S. 1983. *Wet Processing, Volume I, Preparation of Fibers and Fabrics*, Noyes, Park Ridge, NJ, 164.
Roberts, C.W., H.S. Koenig, R.G. Merrill, P.S.R. Cheung, and H.H. Perkins Jr. 1976. Implications of monosaccharides in sticky cotton processing. *Textile Research Journal*, *46*(5), 374–380.
Rodgers, J.E., 1984. Specific Applications of NIRA in the Textile Industry, *Proceedings of 7th International NIRA Symposium*, Technicon Technical Center, Tarrytown, NY.
Rodgers, J.E., 1986. NIRA Potential for the Process Monitoring of Heatset Temperature, *Proceedings of 10th International NIRA Symposium*, Technicon Technical Center, Tarrytown, NY.
Rodgers, J.E., 2002. Influences of carpet and instrumental parameters on the identification of carpet face fiber by NIR. *AATCC Review*, *2*(6), 27.
Rodgers, J.E., and K. Beck. 2009. NIR characterization and measurement of the cotton content of dyed blend fabrics. *Textile Research Journal*, *79*(8), 675–686.
Rodgers, J.E., and S. Ghosh. 1994. NIR Analysis of Textile Fibers and Yarns — from Bench-top to On-Line, *7th International Diffuse Reflectance Conference*, Chambersburg, PA.
Rodgers, J.E., and S. Ghosh. 1995. At-Line Application of NIR Spectroscopy for Textiles, *Eastern Analytical Symposium*, Somerset, NJ.
Rodgers, J.E., and S. Ghosh. 2000. Measuring nylon carpet yarn heat history by remote NIR spectroscopy: Part I: Development of laboratory remote fiber-optic measurements. *Textile Research Journal*, *70*(6), 519–524.
Rodgers, J.E., and S. Lee. 1989. NIRA primary wavelength investigations of nylon yams. *Textile Research Journal*, *59*(9), 513–518.
Rodgers, J.E., and S. Lee. 1991. Structural properties of Nylon 66 carpet yarns observed with near infrared analysis. *Textile Research Journal*, *61*(9), 531–536.
Rodgers, J.E., C. Fortier, J. Montalvo, X. Cui, S.Y. Kang, and V. Martin. 2010a. Near infrared measurement of cotton fiber micronaire by portable near infrared instrumentation. *Textile Research Journal*, *80*(15), 1503–1515.
Rodgers, J.E., J. Montalvo, G. Davidonis, and T. VonHoven. 2010b. Near infrared measurement of cotton fiber micronaire, maturity and fineness-A comparative investigation. *Textile Research Journal*, *80*(9), 780–793.
Rodgers, J.E., S. Ghosh, and W.D. Cardwell. 2001. Measuring nylon carpet yarn heat history by remote NIR spectroscopy. Part II: Applying remote fiber optic NIR techniques to the manufacturing environment. *Textile Research Journal*, *71*(2), 135–144.
Rodgers, J.E. et al. 2002. Process Monitoring of the Moisture and Finish-on-Fiber of Textile Products Using a Portable Near InfraRed Analyzer, *Pittsburgh Conference 2002*, New Orleans, LA.
Rotolo, P. 1979. Near infrared reflectance instrumentation [new research tool for the cereal chemist]. *Cereal Foods World*, *24*, 94.
Saunders, J.H. 1984. Polyamides, Fibers, in *Encyclopedia of Textiles, Fibers, and Nonwoven Fabrics* (M. Grayson, ed.), John Wiley & Sons, New York, 347.
Scholz, E. 2012. *Karl Fischer Titration: Determination of Water*, Springer Science & Business Media.

Starkweather, H.W. 1973. Transitions and Relaxations in Nylons, *Nylon Plastics* (M. Kohan, ed.), John Wiley & Sons, New York, 307.

Starkweather, H.W., G.A. Jones. 1981. Crystalline transitions in powders of nylon 66 crystallized from solution. *Journal of Polymer Science Part B: Polymer Physics*, *19*(3), 467–477.

Tincher, W. et al. 1986. *Textile Chem. Color*, *18*, 25.

Ward, K., Jr. 1955. *Chemistry and Chemical Technology of Cotton*, Interscience, New York, 15.

Zoccola, M., N. Lu, R. Mossotti, R. Innocenti, and A. Montarsolo. 2013. Identification of wool, cashmere, yak, and angora rabbit fibers and quantitative determination of wool and cashmere in blend: A near infrared spectroscopy study. *Fibers and Polymers*, *14*(8), 1283–1289.

Index

Note: **Bold** and *Italic* page numbers refer to **tables** and *figures* and page numbers followed by 'n' refer to notes.

For Product Safety Concerns and Information please contact our EU representative GPSR@taylorandfrancis.com Taylor & Francis Verlag GmbH, Kaufingerstraße 24, 80331 München, Germany

Batch number: 10753435

Printed by Printforce, the Netherlands